PATTY'S TOXICOLOGY

Fifth Edition

Volume 3

**Metals and Metal Compounds
Compounds of Inorganic Nitrogen,
Carbon, Oxygen, and Halogens**

PATTY'S TOXICOLOGY

Fifth Edition
Volume 3

EULA BINGHAM
BARBARA COHRSSEN
CHARLES H. POWELL
Editors

CONTRIBUTORS

Eula Bingham
B. Dwight Culver
Slawomir Czerczak
J. Michael Davis
L. Faye Grimsley
Jan Gromiec

Sverre Langård
George D. Leikauf
Melissa McDiarmid
Jay F. Murray
Stevan W. Pierce
Daniel R. Prows

Konrad Rydzynski
Katherine S. Squibb
Philip Strong
Daniel Thau Teitelbaum
Vickie L. Wells
W. H. (Bill) Wells Jr.

A Wiley-Interscience Publication
JOHN WILEY & SONS, INC.
New York / Chichester / Weinheim / Brisbane / Singapore / Toronto

SOUTH UNIVERSITY
709 MALL BLVD.
SAVANNAH, GA 31406

DISCLAIMER: Extreme care has been taken in preparation of this work. However, neither the publisher nor the authors shall be held responsible or liable for any damages resulting in connection with or arising from the use of any of the information in this book.

This book is printed on acid-free paper. ∞

Copyright © 2001 by John Wiley & Sons, Inc. All rights reserved.

Published simultaneously in Canada.

No part of this publication may be reproduced, stored in a retrieval system or transmitted in any form or by any means, electronic, mechanical, photocopying, recording, scanning or otherwise, except as permitted under Sections 107 or 108 of the 1976 United States Copyright Act, without either the prior written permission of the Publisher, or authorization through payment of the appropriate per-copy fee to the Copyright Clearance Center, 222 Rosewood Drive, Danvers, MA 01923, (978) 750-8400, fax (978) 750-4744. Requests to the Publisher for permission should be addressed to the Permissions Department, John Wiley & Sons, Inc., 605 Third Avenue, New York, NY 10158-0012, (212) 850-6011, fax (212) 850-6008, E-Mail: PERMREQ @ WILEY.COM.

For ordering and customer service, call 1-800-CALL-WILEY.

Library of Congress Cataloging in Publication Data:

Patty's toxicology / [edited by] Eula Bingham, Barbara Cohrssen, Charles H. Powell.— 5th ed.
    p. ; cm.
  "A Wiley-Interscience publication."
  Includes bibliographical references and index.
  ISBN 0-471-31934-1 (cloth: v. 3 : alk.paper); 0-471-31943-0 (set)
    1. Industrial toxicology—Encyclopedias. I. Bingham, Eula. II. Cohrssen, Barbara.
III. Powell, Charles H. IV. Patty's industrial hygiene and toxicology
  [DNLM: 1. Occupational Medicine. 2. Occupational Diseases. 3. Poisons. 4. Toxicology. WA 400 P3222 2000]
RA1229 .P38 2000
613.6′2—dc21   99-053898

Printed in the United States of America.

10 9 8 7 6 5 4 3 2 1

Contributors

Eula Bingham, Ph.D., Kettering Laboratory, University of Cincinnati, Cincinnati, Ohio

B. Dwight Culver, MD, Department of Medicine, University of California, Irvine, California

Slawomir Czerczak, Ph.D., Nofer Institute of Occupational Medicine, Lodz, Poland

J. Michael Davis, Ph.D., U.S. EPA National Center for Environmental Assessment, Research Triangle Park North Carolina

L. Faye Grimsley, MSPH, CIH, University of Cincinnati, Cincinnati, Ohio

Jan Gromiec, Ph.D., Nofer Institute of Occupational Medicine, Lodz, Poland

Sverre Langård, MD, Center for Occupational and Environmental Medicine, The National Hospital, Oslo, Norway

George D. Leikauf, Ph.D., University of Cincinnati, Cincinnati, Ohio

Melissa McDiarmid, MD, MPH, University of Maryland, School of Medicine, Baltimore, Maryland

Jay F. Murray, Ph.D., Murray Associates, San Jose, California

Stevan W. Pierce, CIH, SCP, PE(CA), EOHS, Arlington, Texas

Daniel R. Prows, Ph.D., University of Cincinnati, Cincinnati, Ohio

Konrad Rydzynski, Ph.D., Nofer Institute of Occupational Medicine, Lodz, Poland

Katherine S. Squibb, Ph.D., University of Maryland, School of Medicine, Baltimore, Maryland

Philip L. Strong, Ph.D., U.S. Borax, Valencia, California

Daniel Thau Teitelbaum, MD, Medical Toxicology, Denver, Colorado

Vickie L. Wells, MS, CIH Belmont, California

W. H. (Bill) Wells, Jr., Ph.D., CIH Belmont, California

Preface

In this Preface to the Fifth Edition, we acknowledge and note that it has been built on the work of previous editors. We especially need to note that Frank Patty's words in the preface of the second edition are cogent:

> This book was planned as a ready, practical reference for persons interested in or responsible for safeguarding the health of others working with the chemical elements and compounds used in industry today. Although guidelines for selecting those chemical compounds of sufficient industrial importance for inclusion are not clearly drawn, those chemicals found in carload price lists seem to warrant first consideration.
>
> Where available information is bountiful, an attempt has been made to limit the material presented to that of a practical nature, useful in recognizing, evaluating, and controlling possible harmful exposures. Where the information is scanty, every fragment of significance, whether negative or positive, is offered the reader. The manufacturing chemist, who assumes responsibility for the safe use of his product in industry and who employs a competent staff to this end, as well as the large industry having competent industrial hygiene and medical staffs, are in strategic positions to recognize early and possibly harmful exposures in time to avoid any harmful effects by appropriate and timely action. Plant studies of individuals and their exposures regardless of whether or not the conditions caused recognized ill effects offer valuable experience. Information gleaned in this manner, though it may be fragmentary, is highly important when interpreted in terms of the practical health problem.

While we have not insisted that chemical selection be based on carload quantities we have been most concerned about agents (chemical and physical) in the workplace that are of toxicological concern for workers. We have attempted to follow the guide as expressed by Frank Patty in 1962 regarding practical information.

The expansion of this edition to include biological agents, e.g., wood dust, Histoplasma, not previously covered, reflects our concern with their toxicology and potential for adverse health effects in workers. In the workplace of the new century, physical agents and human factors appear to be of more concern. Traditionally, these agents or factors, ergonomics, biorhythms, vibration, and heat and cold stress were centered on how one measures them. Today, understanding the toxicology of these agents (factors) is of great importance because it can assist in the anticipation, recognition, evaluation, and control of the physical agent. Their mechanisms of actions and the assessment of adverse health effects are as much a part of toxicology as dusts and the heavy metals.

Chapters on certain topics such as reproduction and development, and neurotoxicology reflect the importance of having at hand for practical use such information to help those persons who are responsible for helping to safeguard health to better understand toxicological information and tests reported for the various chemicals. As noted in Chapter One, the trend in toxicology is increasingly focused on molecular biology and, for this "decade of the genome," molecular genetics. Therefore, it seemed crucial to have a chapter that would help to explain the dogma of our teachers in industrial toxicology that, frequently, there are two workers side by side, and one develops an occupational disease and the other does not. Hence the chapter on genetics was authored by an expert in environmental genetics.

The thinking and planning of this edition was a team effort by us: Charlie, Barbara, and Eula. Over many months we worked on the new framework and selected the contributors. When Charlie died in September, 1998, we (Barbara and Eula) knew that we had a road map and, with the help of our expert contributors, many of whom the three of us have known for 10, 20, or even 30 years, would complete this edition. The team effort was fostered among the current editors by many of the first contributors to Patty's such as Robert A. Kehoe, Francis F. Heyroth, William B. Deichmann, and Joseph Treon, all of whom were at Kettering Laboratory at the University of Cincinnati sometime during their professional lives. The three of us have a long professional association with the Kettering Laboratory: Charles H. Powell received a ScD., Barbara Cohrssen received a MS, and Eula Bingham has been a lifetime faculty member. Many of the authors were introduced to us through this relationship and association.

The authors have performed a difficult task in a short period of time for a publication that is as comprehensive as this one is. We want to express our deep appreciation and thanks to all of them.

Kettering Laboratory, Cincinnati, Ohio EULA BINGHAM, Ph.D.

San Francisco, California BARBARA COHRSSEN, MS

CHARLES H. POWELL, ScD.

Contents

37	**Vanadium, Niobium, and Tantalum** Konrad Rydznski, Ph.D.	1
38	**Chromium, Molybdenum, and Tungsten** Sverre Langård, MD	75
39	**Manganese and Rhenium** J. Michael Davis, Ph.D.	129
40	**Iron and Cobalt** L. Faye Grimsley, MSPH, CIH	169
41	**Nickel, Ruthenium, Rhodium, Palladium, Osmium, and Platinum** Slawomir Czerczak, Ph.D., and Jan Gromiec, Ph.D.	195
42	**Uranium and Thorium** Melissa McDiarmid, MD, MPH, and Katherine S. Squibb, Ph.D.	381
43	**The Lanthanides, Rare Earth Metals** W. H. (Bill) Wells Jr., Ph.D., CIH and Vickie L. Wells, MS, CIH	423
44	**Phosphorus, Selenium, Tellurium, and Sulfur** Eula Bingham, Ph.D.	459

45	**Boron**	519
	B. Dwight Culver, MD, Philip Strong, Ph.D., and Jay F. Murray, Ph.D.	
46	**Alkaline Materials, Sodium, Potassium, Cesium, Rubidium, Francium, and Lithium**	583
	Stevan W. Pierce, CIH, SCP, PE(CA)	
47	**Inorganic Compounds of Carbon, Nitrogen, and Oxygen**	607
	George D. Leikauf, Ph.D. and Daniel R. Prows, Ph.D.	
48	**The Halogens**	731
	Daniel Thau Teitelbaum, MD	
Subject Index		827
Chemical Index		855

USEFUL EQUIVALENTS AND CONVERSION FACTORS

1 kilometer = 0.6214 mile
1 meter = 3.281 feet
1 centimeter = 0.3937 inch
1 micrometer = 1/25,4000 inch = 40 microinches
  = 10,000 Angstrom units
1 foot = 30.48 centimeters
1 inch = 25.40 millimeters
1 square kilometer = 0.3861 square mile (U.S.)
1 square foot = 0.0929 square meter
1 square inch = 6.452 square centimeters
1 square mile (U.S.) = 2,589,998 square meters
  = 640 acres
1 acre = 43,560 square feet = 4047 square meters
1 cubic meter = 35.315 cubic feet
1 cubic centimeter = 0.0610 cubic inch
1 cubic foot = 28.32 liters = 0.0283 cubic meter
  = 7.481 gallons (U.S.)
1 cubic inch = 16.39 cubic centimeters
1 U.S. gallon = 3.7853 liters = 231 cubic inches
  = 0.13368 cubic foot
1 liter = 0.9081 quart (dry), 1.057 quarts
  (U.S., liquid)
1 cubic foot of water = 62.43 pounds (4°C)
1 U.S. gallon of water = 8.345 pounds (4°C)
1 kilogram = 2.205 pounds

1 gram = 15.43 grains
1 pound = 453.59 grams
1 ounce (avoir.) = 28.35 grams
1 gram mole of a perfect gas ≎ 24.45 liters
  (at 25°C and 760 mm Hg barometric pressure)
1 atmosphere = 14.7 pounds per square inch
1 foot of water pressure = 0.4335 pound per
  square inch
1 inch of mercury pressure = 0.4912 pound per
  square inch
1 dyne per square centimeter = 0.0021 pound per
  square foot
1 gram-calorie = 0.00397 Btu
1 Btu = 778 foot-pounds
1 Btu per minute = 12.96 foot-pounds per second
1 hp = 0.707 Btu per second = 550 foot-pounds
  per second
1 centimeter per second = 1.97 feet per minute
  = 0.0224 mile per hour
1 footcandle = 1 lumen incident per square foot
  = 10.764 lumens incident per square meter
1 grain per cubic foot = 2.29 grams per cubic meter
1 milligram per cubic meter = 0.000437 grain per
  cubic foot

To convert degrees Celsius to degrees Fahrenheit: °C (9/5) + 32 = °F
To convert degrees Fahrenheit to degrees Celsius: (5/9) (°F − 32) = °C
For solutes in water: 1 mg/liter ≎ 1 ppm (by weight)
Atmospheric contamination: 1 mg/liter ≎ 1 oz/1000 cu ft (approx)
For gases or vapors in air at 25°C and 760 mm Hg pressure:
  To convert mg/liter to ppm (by volume): mg/liter (24,450/mol. wt.) = ppm
  To convert ppm to mg/liter: ppm (mol. wt./24,450) = mg/liter

CONVERSION TABLE FOR GASES AND VAPORS[a]
(Milligrams per liter to parts per million, and vice versa; 25°C and 760 mm Hg barometric pressure)

Molecular Weight	$\frac{1}{\text{mg/liter}}$ ppm	1 ppm mg/liter	Molecular Weight	$\frac{1}{\text{mg/liter}}$ ppm	1 ppm mg/liter	Molecular Weight	$\frac{1}{\text{mg/liter}}$ ppm	1 ppm mg/liter
1	24,450	0.0000409	39	627	0.001595	77	318	0.00315
2	12,230	0.0000818	40	611	0.001636	78	313	0.00319
3	8,150	0.0001227	41	596	0.001677	79	309	0.00323
4	6,113	0.0001636	42	582	0.001718	80	306	0.00327
5	4,890	0.0002045	43	569	0.001759	81	302	0.00331
6	4,075	0.0002454	44	556	0.001800	82	298	0.00335
7	3,493	0.0002863	45	543	0.001840	83	295	0.00339
8	3,056	0.000327	46	532	0.001881	84	291	0.00344
9	2,717	0.000368	47	520	0.001922	85	288	0.00348
10	2,445	0.000409	48	509	0.001963	86	284	0.00352
11	2,223	0.000450	49	499	0.002004	87	281	0.00356
12	2,038	0.000491	50	489	0.002045	88	278	0.00360
13	1,881	0.000532	51	479	0.002086	89	275	0.00364
14	1,746	0.000573	52	470	0.002127	90	272	0.00368
15	1,630	0.000614	53	461	0.002168	91	269	0.00372
16	1,528	0.000654	54	453	0.002209	92	266	0.00376
17	1,438	0.000695	55	445	0.002250	93	263	0.00380
18	1,358	0.000736	56	437	0.002290	94	260	0.00384
19	1,287	0.000777	57	429	0.002331	95	257	0.00389
20	1,223	0.000818	58	422	0.002372	96	255	0.00393
21	1,164	0.000859	59	414	0.002413	97	252	0.00397
22	1,111	0.000900	60	408	0.002554	98	249.5	0.00401
23	1,063	0.000941	61	401	0.002495	99	247.0	0.00405
24	1,019	0.000982	62	394	0.00254	100	244.5	0.00409
25	978	0.001022	63	388	0.00258	101	242.1	0.00413
26	940	0.001063	64	382	0.00262	102	239.7	0.00417
27	906	0.001104	65	376	0.00266	103	237.4	0.00421
28	873	0.001145	66	370	0.00270	104	235.1	0.00425
29	843	0.001186	67	365	0.00274	105	232.9	0.00429
30	815	0.001227	68	360	0.00278	106	230.7	0.00434
31	789	0.001268	69	354	0.00282	107	228.5	0.00438
32	764	0.001309	70	349	0.00286	108	226.4	0.00442
33	741	0.001350	71	344	0.00290	109	224.3	0.00446
34	719	0.001391	72	340	0.00294	110	222.3	0.00450
35	699	0.001432	73	335	0.00299	111	220.3	0.00454
36	679	0.001472	74	330	0.00303	112	218.3	0.00458
37	661	0.001513	75	326	0.00307	113	216.4	0.00462
38	643	0.001554	76	322	0.00311	114	214.5	0.00466

CONVERSION TABLE FOR GASES AND VAPORS (*Continued*)
(*Milligrams per liter to parts per million, and vice versa;*
25°C and 760 mm Hg barometric pressure)

Molecular Weight	1 mg/liter ppm	1 ppm mg/liter	Molecular Weight	1 mg/liter ppm	1 ppm mg/liter	Molecular Weight	1 mg/liter ppm	1 ppm mg/liter
115	212.6	0.00470	153	159.8	0.00626	191	128.0	0.00781
116	210.8	0.00474	154	158.8	0.00630	192	127.3	0.00785
117	209.0	0.00479	155	157.7	0.00634	193	126.7	0.00789
118	207.2	0.00483	156	156.7	0.00638	194	126.0	0.00793
119	205.5	0.00487	157	155.7	0.00642	195	125.4	0.00798
120	203.8	0.00491	158	154.7	0.00646	196	124.7	0.00802
121	202.1	0.00495	159	153.7	0.00650	197	124.1	0.00806
122	200.4	0.00499	160	152.8	0.00654	198	123.5	0.00810
123	198.8	0.00503	161	151.9	0.00658	199	122.9	0.00814
124	197.2	0.00507	162	150.9	0.00663	200	122.3	0.00818
125	195.6	0.00511	163	150.0	0.00667	201	121.6	0.00822
126	194.0	0.00515	164	149.1	0.00671	202	121.0	0.00826
127	192.5	0.00519	165	148.2	0.00675	203	120.4	0.00830
128	191.0	0.00524	166	147.3	0.00679	204	119.9	0.00834
129	189.5	0.00528	167	146.4	0.00683	205	119.3	0.00838
130	188.1	0.00532	168	145.5	0.00687	206	118.7	0.00843
131	186.6	0.00536	169	144.7	0.00691	207	118.1	0.00847
132	185.2	0.00540	170	143.8	0.00695	208	117.5	0.00851
133	183.8	0.00544	171	143.0	0.00699	209	117.0	0.00855
134	182.5	0.00548	172	142.2	0.00703	210	116.4	0.00859
135	181.1	0.00552	173	141.3	0.00708	211	115.9	0.00863
136	179.8	0.00556	174	140.5	0.00712	212	115.3	0.00867
137	178.5	0.00560	175	139.7	0.00716	213	114.8	0.00871
138	177.2	0.00564	176	138.9	0.00720	214	114.3	0.00875
139	175.9	0.00569	177	138.1	0.00724	215	113.7	0.00879
140	174.6	0.00573	178	137.4	0.00728	216	113.2	0.00883
141	173.4	0.00577	179	136.6	0.00732	217	112.7	0.00888
142	172.2	0.00581	180	135.8	0.00736	218	112.2	0.00892
143	171.0	0.00585	181	135.1	0.00740	219	111.6	0.00896
144	169.8	0.00589	182	134.3	0.00744	220	111.1	0.00900
145	168.6	0.00593	183	133.6	0.00748	221	110.6	0.00904
146	167.5	0.00597	184	132.9	0.00753	222	110.1	0.00908
147	166.3	0.00601	185	132.2	0.00757	223	109.6	0.00912
148	165.2	0.00605	186	131.5	0.00761	224	109.2	0.00916
149	164.1	0.00609	187	130.7	0.00765	225	108.7	0.00920
150	163.0	0.00613	188	130.1	0.00769	226	108.2	0.00924
151	161.9	0.00618	189	129.4	0.00773	227	107.7	0.00928
152	160.9	0.00622	190	128.7	0.00777	228	107.2	0.00933

CONVERSION TABLE FOR GASES AND VAPORS (*Continued*)
(Milligrams per liter to parts per million, and vice versa;
25°C and 760 mm Hg barometric pressure)

Molecular Weight	1 mg/liter ppm	1 ppm mg/liter	Molecular Weight	1 mg/liter ppm	1 ppm mg/liter	Molecular Weight	1 mg/liter ppm	1 ppm mg/liter
229	106.8	0.00937	253	96.6	0.01035	277	88.3	0.01133
230	106.3	0.00941	254	96.3	0.01039	278	87.9	0.01137
231	105.8	0.00945	255	95.9	0.01043	279	87.6	0.01141
232	105.4	0.00949	256	95.5	0.01047	280	87.3	0.01145
233	104.9	0.00953	257	95.1	0.01051	281	87.0	0.01149
234	104.5	0.00957	258	94.8	0.01055	282	86.7	0.01153
235	104.0	0.00961	259	94.4	0.01059	283	86.4	0.01157
236	103.6	0.00965	260	94.0	0.01063	284	86.1	0.01162
237	103.2	0.00969	261	93.7	0.01067	285	85.8	0.01166
238	102.7	0.00973	262	93.3	0.01072	286	85.5	0.01170
239	102.3	0.00978	263	93.0	0.01076	287	85.2	0.01174
240	101.9	0.00982	264	92.6	0.01080	288	84.9	0.01178
241	101.5	0.00986	265	92.3	0.01084	289	84.6	0.01182
242	101.0	0.00990	266	91.9	0.01088	290	84.3	0.01186
243	100.6	0.00994	267	91.6	0.01092	291	84.0	0.01190
244	100.2	0.00998	268	91.2	0.01096	292	83.7	0.01194
245	99.8	0.01002	269	90.9	0.01100	293	83.4	0.01198
246	99.4	0.01006	270	90.6	0.01104	294	83.2	0.01202
247	99.0	0.01010	271	90.2	0.01108	295	82.9	0.01207
248	98.6	0.01014	272	89.9	0.01112	296	82.6	0.01211
249	98.2	0.01018	273	89.6	0.01117	297	82.3	0.01215
250	97.8	0.01022	274	89.2	0.01121	298	82.0	0.01219
251	97.4	0.01027	275	88.9	0.01125	299	81.8	0.01223
252	97.0	0.01031	276	88.6	0.01129	300	81.5	0.01227

[a] A. C. Fieldner, S. H. Katz, and S. P. Kinney, "Gas Masks for Gases Met in Fighting Fires," *U.S. Bureau of Mines, Technical Paper No. 248*, 1921.

PATTY'S TOXICOLOGY

Fifth Edition

Volume 3

Metals and Metal Compounds
Compounds of Inorganic Nitrogen,
Carbon, Oxygen, and Halogens

CHAPTER THIRTY-SEVEN

Vanadium, Niobium, and Tantalum

Konrad Rydzynski, Ph.D.

Vanadium (V), niobium (Nb), and tantalum (Ta) are transition metals from group V. They have partly filled *d* shells, so they are defined as transition elements. Vanadium and niobium are widely distributed in earth's crust, but there are few concentrated deposits of these elements. Tantalum is less abundant in the earth's crust; it occurs in the same minerals as niobium, and their separation is complex. The main commercial sources of both are the *columbite–tantalite* series of minerals [(Fe/Mn)(Nb/Ta)$_2$O$_6$], with various Fe/Mn and Nb/Ta ratios (1).

Pure or almost pure elements in massive form are gray-colored, ductile metals with high (V, Ta) or moderate (Nb) hardness and very high melting points. Vanadium group elements are resistant to chemicals, and this resistance increases with the atomic number. At room temperature they are not affected by air, water, or alkalies. Vanadium dissolves in oxidizing acids (e.g., nitric acid, concentrated sulfuric acid, aqua regia) and in hydrofluoric acid. Niobium and tantalum can be dissolved by HNO$_3$/HF mixture, and they are slowly attacked by hydrofluoric acid. All these elements dissolve very slowly in fused alkalies, producing salts: vanadates, niobates or tantalates, and hydrogen. Vanadium, niobium, and tantalum pentaoxides are the main products of air oxidation at high temperatures; vanadium can also form trioxide and tetraoxide under these conditions. At elevated temperatures, metals combine with some other nonmetals, for example, with hydrogen, nitrogen, carbon, and silica, giving compounds, many of which are interstitial and non-stoichiometric (1, 2). All of these elements have five valence electrons; however, electronic configuration of valence orbitals is different — vanadium and tantalum have two *s* and

---

Patty's Toxicology, Fifth Edition, Volume 3, Edited by Eula Bingham, Barbara Cohrssen, and Charles H. Powell.
ISBN 0-471-31934-1 © 2001 John Wiley & Sons, Inc.

three d valence electrons, but niobium has only one s and four d valence electrons. Vanadium can exist in oxidation states from -1 to $+5$, the most common oxidation states being the valences of $+3$, $+4$, and $+5$. The most prevalent oxidation state of niobium and tantalum is $+5$; compounds containing these elements in lower oxidation states ($+3$ or $+4$) are much more difficult to obtain (1).

Vanadium compounds with vanadium in the $+2$ oxidaton state are the ionic compounds, and they have properties similar to those of iron $+2$ compounds, but they are stronger reducing agents. Vanadium $+3$ compounds are ionic, too, and they are often analogous to iron $+3$ compounds. The covalence of vanadium compounds increases with the oxidation state. Vanadium oxide (VO) has alkaline character. Vanadium oxides with vanadium in the $+4$ or $+5$ oxidation states are amphoteric; vanadium pentaoxide V_2O_5 is more acidic than basic. It is an anhydride of many acids analogous to phosphoric acid.

Analogical compounds of tantalum and niobium in the $+4$ or $+5$ oxidation state are very similar to each other because the same atomic and ionic radius exists in both elements. Niobium and tantalum have chemistries in the $+5$ oxidation state that are very similar to those of typical nonmetals. They have virtually no cationic chemistry but form numerous anionic species. Their pentaoxides — Nb_2O_5 and Ta_2O_5 — are amphoteric, but they have a more basic character than V_2O_5.

Some physicochemical parameters of group 5 elements are shown in Table 37.1.

Vanadium compounds are the most toxic among all the three elements; tantalum compounds are practically nontoxic. Reported LC_{50} values for vanadium pentoxide are between 70 and 200 mg/m^3; instillation to trachea of 50 mg/rat of tantalum hydride caused only mild pathological changes and there is no data on niobium. Vanadium compounds are moderately toxic when given orally; reported LD_{50} values are in hundreds mg/kg body weight. Niobium and tantalum compounds given orally are practically nontoxic; reported LD_{50} values are in thousands mg/kg body weight. All elements and

Table 37.1. Some Physicochemical Parameters of Group V Elements

	Vanadium (V)	Niobium (Nb)	Tantalum (Ta)
Electronic configuration (valence orbitals)	$3d^34s^2$	$4d^45s^1$	$5d^36s^2$
Atomic number	23	41	73
Atomic mass	50.9	92.9	180.9
Melting point, (°C)	1700	2500	3000
Boiling point, (°C)	—	5000	>4000
Density, (g/cm^3)	6.1	8.6	16.7
Electronegativity	1.6	1.6	1.5
Atomic radius	1.34	1.46	1.46
Ionic radius Me^{5+}	0.59	0.69	0.68
Ionic radius Me^{4+}	0.63	0.74	—
Ionic radius Me^{2+}	0.88	—	—
Normal potential Me/Me^{5+} (V)		-0.6	-0.7
Normal potential Me/Me^{4+} (V)	-1.5	—	

their compounds are absorbed from the respiratory tract, and eliminated through the kidney. Their absorbtion from the gastrointestinal (GI) tract is poor. They are distributed to internal organs and there are data indicating that vanadium and tantalum might accumulate in bone. Vanadium and niobium have an irritant effect on mucous membranes and skin. Therefore, irritant effects on the upper respiratory tract and lungs are observed when animals are exposed by inhalation to vanadium and niobium compounds; however, vanadium compounds have stronger effects.

Many studies have documented the mitogenic potential of vanadium compounds. The results of mutagenicity studies of vanadium are conflicting, and no firm conclusions can be drawn. There is no such information on niobium and tantalum. However, there are no data indicating that any of these compounds may have carcinogenic potential. Indeed, there are studies suggesting that some vanadium and niobium compounds may have antitumor activity.

Vanadium crosses the blood–testis barrier and accumulates in the testis. Some studies indicate that this metal is a male reproductive toxicant and that the degree of its toxicity depends on the route of exposure, oxidation state, period of dosing, and dose. Vanadate (V^{5+}) and vanadyl (V^{4+}), as well as niobium, are capable of crossing the placenta and reach the fetus. In some studies it was documented that vanadium compounds reveal embryofetotoxic and teratogenic effects. There are no such indications for niobium and tantalum compounds.

Acute vanadium toxicity in humans is characterized by a latency period, which depends on the concentration of vanadium, the individual sensitivity of the subject, and the properties of the specific vanadium compound. Condensation aerosol of vanadium pentoxide is more toxic than a disintegration aerosol. Acute vanadium effects can be divided into "mild," "moderate," and "severe" forms.

The clinical features of mild toxicity include rhinitis with a profuse and often bloody discharge, sneezing, and an itching and burning sensation in the throat. The rhinitis may be followed by the development of a dry cough with expectoration of small amounts of viscid sputum, general weakness, and exhaustion. Conjunctivitis is frequently observed. Other symptoms may include diarrhea. In moderate toxicity, in addition to conjunctivitis and irritation of the upper respiratory tract, there is bronchitis with expiratory dyspnea and bronchospasm. There are frequent disturbances in GI tract activity, including vomiting and diarrhea. Some affected persons have cutaneous manifestations of toxicity in the form of rashes and eczema with itching papules and dry patches. Severe toxic effects include bronchitis and bronchopneumonia. Other symptoms may also be more prominent, such as headache, vomiting, diarrhea, palpitations, sweating, and general weakness. Disorders of the nervous system include severe neurotic states and tremor of the fingers and hands.

Chronic vanadium exposure results mainly in pathological changes in the lung. Workers who were exposed for a long time to vanadium complained of coughing; eye, nose, and throat irritation; breathing difficulties during physical exertion; and headache. Clinical findings often revealed intense hyperemia of the nasal, laryngeal, and pharyngeal mucosa and rhinitis (of a simple catarrhal form but also of a hypertrophic, subatrophic, or atrophic form). Sometimes epistaxis was present. The main respiratory diseases diagnosed were chronic bronchitis with or without emphysema and diffuse pneumosclerosis. The changes in the lung were often accompanied by cardiovascular disturbances (arrhythmia, bradycardia, unspecific changes in electrocardiogram). A statistically significant increase in the incidence of enlarged liver with a decrease in functional tests were seen in exposed workers. Systemic effects, such as tendency toward anaemia and leukopenia, and basophilic granulation of leukocytes, have been reported. There is no acute and chronic human toxicity data on niobium. From a very few reports on tantalum toxicity in workers and long-term experience from its use in surgery, it might be concluded that tantalum metal, metal dust, and powders are physiologically inert.

The current occupational exposure limits expressed as time-weighted averages (TWAs) for vanadium and tantalum are similar in most countries and established as 0.05 and 5 mg/m^3, respectively. Similar values for niobium compounds as for tantalum are recommended in some countries.

1.0 Vanadium

1.0.1 CAS Number: *[7440-62-2]*

1.0.2 Synonyms: NA

1.0.3 Trade Names: NA

1.0.4 Molecular Weight: *50.942*

1.0.5 Molecular Formula: *V*

1.1 Chemical and Physical Properties

1.1.1 General

Vanadium is a white-to-gray metal with compounds widely distributed at low concentrations in the earth's crust. The element was discovered in 1801 in Mexico by Andreas Manuel del Rio, who erroneously thought it was chromium. Thirty years later, the Swedish chemist Nils Gabriel Sefström rediscovered it and named the element "Vanadis," the Scandinavian name for Venus, as he was dazzled by its luster and brightness as well as by the variety of striking colors of its salts (3).

Vanadium is widely, but sparsely, distributed in the earth's crust (4, 5). It may be found at levels as great as 0.07% in the lithosphere and exists in the form of over 50 different mineral ores (6). Very pure vanadium is rare because it is quite reactive toward oxygen, nitrogen, and carbon at the elevated temperatures used in conventional thermometallurgy. It is used in alloy steels and cast iron, to which it transfers ductility and shock resistance. One of the most important commercial products containing vanadium is an iron alloy, *ferrovanadium* (2). Elemental vanadium does not occur in nature, but its compounds exist in different mineral ores and in association with fossil fuels. The principal ores are *patronite* (a complex sulfide), *vanadinite* [$Pb_5(VO_4)_3Cl$], *carnotite* [$K_2O \cdot 2UO_3 \cdot V_2O_5 \cdot 3H_2O$], and *roscoelite* (a composite mineral that contains silicate) (4–7). Vanadium is also found in phosphate rock, some iron ores, and crude petroleum deposits. Fuel deposits from oil-fired furnaces have been found to contain up to 50% vanadium pentoxide (6–9).

Vanadium is usually bound with other elements such as oxygen, sodium, sulfur, or chloride in a wide variety of complex ions and coordination complexes. Vanadium salts have characteristic colors in solution (blue, yelow, green, red, orange, black). The CAS number, molecular formula, and temperatures and other parameters of some vanadium compounds are presented in Table 37.2

Vanadium has an atomic number of 23, an atomic weight of 50.942, a melting point of 1890°C, a boiling point of 3380°C at 1 atm, and specific gravity of 6.11 at 18.7°C. Natural vanadium has two isotopes, ^{50}V and ^{51}V; the former is slightly radioactive with a half-life of 6×10^{15} years. Seven other isotopes ($^{46-49}V, ^{52-54}V$) of the element have been synthesized. The V_b group of elements, which includes vanadium, has electrons in the d orbitals. Vanadium has a low lying 3 d electron orbital configuration, which characterizes its unusual oxidation features. It has six oxidation states ($-1, 0, +2, +3, +4,$ and $+5$); $+3, +4,$ and $+5$ are the most common. The most stable oxidation state is $+4$. Under physiological conditions, at typical cellular concentrations of 10^{-6}–10^{-7}, vanadate anion ($H_2VO_4^{1-}$), and vanadyl cation (VO^{2+}) are the most frequent vanadium ions. In biological systems, vanadium is present in the extracellular fluid in the $+5$ redox state as vanadate ($H_2VO_4^{1-}$), which is intracellularly reduced to $+4$ redox state vanadyl (VO^{2+}). However, in one paper it was shown that vanadate ($+5$) is not reduced to vanadyl ($+4$) inside the cell, and that vanadyl sulfate ($+4$) is capable of a spontaneous oxidation to vanadate ($+5$) *in vivo* (10).

Vanadium can also have the oxidation states of $-1, 0, +2$. At higher oxidation states, the element forms negatively charged oxyanions (e.g., VO_4^{3-}, HVO_4^{2-}, $H_2VO_4^{1-}$) similar to phosphate compounds. In acidic solutions the element occurs as a monovalent cation (VO^{2+}), in basic solutions it becomes VO_4^{3-}. In neutral solutions, vanadium is present mostly as $H_2VO_4^{1-}$ (11). It can form oligomers in aqueous solutions, and the degree of polymerization is both concentration- and pH-dependent (3). Vanadium can form complexes with biologically essential compounds such as GDP, ADP, NAD$^+$, gluthatione, ribosides, catecholamines, hemoglobin, and transferrin (12). Vanadium forms an alloy with iron (ferrovanadium) and a very hard and stable carbide, V_4C_3, in carbon and most alloy steel.

1.1.2 Odor and Warning Properties

The scarce available literature data indicate that vanadium and most of its compounds are odorless. Vanadium tetrachloride (VCl$_4$) and vanadium trichloride (VCl$_3$) have pungent odors because they decompose slowly in room temperature with Cl$_2$ formation.

1.2 Production and Use

Vanadium occurs primarily as a by-product or a coproduct during the extraction of other compounds such as iron, titanium, phosphate, or petroleum. It is extracted from carnotite, phosphate rock deposits, titaniferous magnetites, and vanadiferous clays. A process called *salt roasting* during the initial stage of extraction produces the oxide concentrate. The ores, petroleum residues, iodide thermal decomposition products, and slags formed during ferrovanadate production are crushed, dried, finely ground, mixed with a sodium salt, and

Table 37.2. Chemical Identity of Vanadium and Its Compounds

Compound	CAS Number	Molecular Formula	Molecular Weight	Boiling Point (°C)	Melting Point (°C)	Specific Gravity	Solubility in Water at 68°F	Vapor Pressure (mm Hg)	Flammability	Refractive Index (at 20°C)
Vanadium	[7440-62-2]	V	50.942	3380	1890±10	6.11 g/cm^3 at 18.7°C	Insoluble	—	—	3.03
Vanadium dichloride	[10580-52-6]	VCl$_2$	121.84	—	1350	3.23 at 18°C	Reacts violently	—	—	—
Vanadium trichloride	[7718-98-1]	VCl$_3$	157.29	—	>4 decomposes	3.00 at 18°C	Insoluble	—	—	—
Vanadium oxydichloride	[10213-09-9]	VOCl$_2$	137.84	—	—	—	—	—	—	—
Vanadium oxytrichloride	[7727-18-6]	VOCl$_3$	173.29	126.7	−77	1.81 at 32°C	—	—	—	—
Vanadium tetrachloride	[7632-51-1]	VCl$_4$	192.74	140	−28	1.816 at 30°C	Soluble with decomposition	—	—	—
Vanadium trioxide	[1314-34-7]	V$_2$O$_3$	149.9	—	1940	4.87 at 18°C	Insoluble	—	—	—
Vanadium pentoxide	[1314-62-1]	V$_2$O$_5$	181.9	1750 (decomposes)	690	3.357 g/cm^3 at 18°C	1g/125 mL	7.8 mbar at 20°C	Not flammable	1.46; 1.52; 1.76
Vanadium tribromide	[13470-26-3]	VBr$_3$	290.67	—	Decomposes	—	—	—	—	—
Vanadium trioxybromide	[13520-90-6]	Br$_3$OV	306.65	170	−59	—	Reacts violently	—	—	—
Vanadyl sulfate	[27774-13-6]	VOSO$_4$	163.0	—	—	—	Soluble	0 at 20°C	Not flammable	—
Vanadyl sulfate pentahydrate	[12439-96-2]	SVO$_5$×5H$_2$O	253.1	—	—	—	Slightly soluble	—	—	—
Sodium metavanadate	[13718-26-8]	NaVO$_3$	121.93	—	630	—	211 g/L at 25°C	—	—	—
Sodium orthovanadate	[13721-39-6]	Na$_3$VO$_4$	183.91	—	850–866	—	Soluble	—	—	—
Sodium tetravanandate	[12058-74-1]	Na$_2$V$_4$O$_4$	425.74	—	—	—	—	—	—	—
Amonium metavanadate	[7803-55-6]	NH$_4$VO$_3$	116.98	—	200	2.326 g/cm^3 at 20°C	Slightly soluble	—	Not flammable	—
Ammonium hexafluorovanadate	[13815-31-1]	F$_6$V$_3$NH$_4$	219.09	—	—	—	—	—	—	—
Ferrovanadium	[12604-58-9]	Ferro V	—	1482–1521	—	—	Insoluble	—	—	—

roasted. The hot ore, containing sodium metavanadate, precipitates as a red cake, is then mixed with sulfuric acid, and the resultant precipitate is dried to form vanadium pentoxide (4, 6, 8, 13). The vanadium pentoxide can then be processed further to form the required vanadium compound. Pure vanadium is difficult to obtain as it tends to be readily contaminated by other elements. Methods to extract pure vanadium include iodide refining, electrolytic refining in a fused salt, and electrotransport. The highest purity vanadium has been purified by the electrotransport technique.

World production of vanadium from ores, petroleum concentrates, and slags has remained fairly constant since the late 1990s and is around 34,000–40,000 tons (14). The major producers of vanadium are South Africa, Russia, the United States, China, and Finland. The European countries, Japan and the United States, together use 85% of the total output.

Vanadium and its compounds are currently used for a wide variety of purposes. The annual consumption of vanadium within the United States increased from 3277 tons in 1983 to 4883 tons by 1985 and remained around this level. Approximately 83–87% of vanadium consumed in the United States is utilized as an alloying agent for ferrovanadium alloy in the steel industry (14). To produce various high resistance carbon steels, vanadium is combined with chromium, nickel, manganese, boron, titanium, tungsten, and other metals. Vanadium is added to steel either in the form of ferrovanadium (an iron/vanadium alloy containing 400–800 g·V/kg) or vanadium carbide. These steels are used in a variety of products, such as automobile parts, springs, and ball bearings. These are invaluable in the manufacture of jet aircraft engines. Likewise, the nonferrous titanium alloys are essential in the manufacture of supersonic aircrafts. Vanadium compounds also have an important role as industrial catalysts. Vanadium-containing catalysts are used in several oxidation reactions such as the manufacture of phthalic anhydride and sulfuric acid as well as in the production of pesticides and black dyes, inks, and pigments, which are used by the textile, printing, and ceramic industries. An important use of vanadium is as an oxidation catalyst in automobile catalytic converters. Other minor functions of vanadium compounds include their use as color modifiers in mercury-vapor lamps, driers in paints and varnishes, corrosion inhibitors in flue-gas scrubbers, and as components in photographic developers. Future applications of vanadium compounds may include an increased number of uses as a catalyst, a potential role in superconductors, thermal or light-activated resistor–conductors, vanadate glasses, electrooptical switches, and the induction of high magnetic fields (6–9,13–15).

1.3 Exposure Assessment

A number of analytical techniques have been used to determine ppm to ppt levels of vanadium in different media. These include neutron activation analysis (NAA), graphite furnace atomic absorption spectrometry (GFAAS), flame atomic absorption spectrophotometry (FAAS), isotope dilution thermal ionization–mass spectrometry (IDMS), direct current plasma-atomic emission spectrometry (DCP-AES), and inductively coupled plasma–atomic emission spectrometry (ICP-AES).

In general, biological and environmental samples may be prepared prior to vanadium quantification by acid digestion with nitric acid. Sample dilution with nitric acid or other agents to solubilize vanadium from the sample matrix can also be employed. If the concentration of vanadium in the dissolved sample is very low, preconcentration techniques such as chelation or extraction may be used. Chelation and extraction efficiency will vary with the technique used.

1.3.1 Air

Both AAS and AES methods are commonly used to detect vanadium in the air. Trace levels of vanadium (as vanadium pentoxide) have been detected in air samples by GFAAS (16). A detection limit of 0.25–0.40 µg of vanadium/m^3 for an air sample of 2000 m^3 was achieved. The DCP-AES method, originally employed for vanadium measurements in workplace air (see Section 1.3.3), is also available to measure vanadium in ambient air.

1.3.2 Background Levels

The levels of vanadium measured in ambient air vary widely between rural and urban locations. The concentrations measured over the South Pole ranged from 0.001 to 0.002 ng of vanadium/m^3 and are frequently two orders of magnitude lower than those over the ocean at middle latitudes (17). For example, measurements taken at nine rural sites located in the eastern Pacific area averaged 0.1 ng vanadium/m^3 (range 0.02–0.8 ng vanadium/m^3).

Background vanadium concentrations in unpolluted air taken at five different rural sites in northwest Canada were found to average 0.72 ng vanadium/m^3 (range 0.21–1.9 ng vanadium/m^3) (18). The average ambient concentrations in rural air in the United States ranged from < 1 to 40 ng vanadium/m^3 (4), although some rural areas may have levels as high as 64 ng vanadium/m^3, due to localized burning of fuel oils with a high vanadium content (17). Vanadium levels in ambient urban air vary extensively with the season and location (17, 18). It was found that some, primarily northeastern — U.S. cities, — have concentrations of 150–1400 ng vanadium/m^3 with an average of 620 ng vanadium/m^3. In the second group of cities, widely distributed throughout the United States, ambient-air concentrations ranged from 3 to 22 ng vanadium/m^3 with an average of 11 ng vanadium/m^3. The variation is attributed to the use of large quantities of residual fuel oil by the cities of the second group for the generation of heat and electricity, particularly during winter months (18).

The freshwater levels of vanadium illustrate geographic variations produced by differences in effluents and leachates, from both anthropogenic and natural sources. Measurements of natural freshwater vanadium in such U.S. rivers as the Animas, Colorado, Green, Sacramento, San Joaquin, and San Juan, as well as some freshwater supplies in Wyoming, range from 0.3 to 200 µg vanadium/L (4, 19). The presence of naturally occurring uranium ores in the rivers in the Colorado Plateau resulted in vanadium levels of ≤70 µg/L, and of 30–220 µg/L in Wyoming (4). Some municipal waters have been found to contain the levels of 1–6 µg vanadium/L (19), although levels of 19 µg vanadium/L have been reported in nine New Mexico municipalities (4). Seawater vanadium levels are considerably lower than freshwater levels because much of the vanadium is precipitated (4, 19). The concentrations measured usually average 1–3 µg vanadium/L

(19), although levels as high as 29 µg/L have been reported (4). The total content of seawater vanadium has been estimated to be 7.5×10^{12} kg (7.5×10^9 metric tons) (4).

The level of vanadium measured in soil is closely related to the parent rock type (19, 20). A range of 3–310 mg/kg has been observed, with tundra podsols and clays exhibiting the highest concentration, 100 mg/kg and 300 mg/kg, respectively (4). The average vanadium content of soils in the United States is 200 mg/kg (4) and seems to be most abundant in western regions, especially the Colorado Plateau (6, 21).

The majority of foods have naturally occurring low concentrations of vanadium, many 1 ng/g or less (22). Food items containing the highest levels of vanadium include ground parsley (1800 ng/g dry weight), freeze-dried spinach (533–840 ng/g), wild mushrooms (50–2000 ng/g dry weight), and oysters (455 ng/g wet weight) (22). In general, seafoods have been found to be richer in vanadium than terrestrial animal tissues (17).

1.3.3 Workplace Methods

NIOSH (method 7504) for detecting vanadium and 7300 for other elements in workplace air are recommended. A working range of 2.5–1000 µg of vanadium/sample (5.0–2000 µg/m^3) and estimated limit of detection of 3.2 µg/sample was obtained (23). However, this was a simultaneous elemental analysis, not a compound-specific one. A method for the determination of vanadium in workplace air using DCP-AES was reported by Pyy et al. (24).

A detection limit of 4 µg of vanadium/m^3 for a 25-L sample and a practical working range of 0.01–100 mg of vanadium/L sample were obtained. DCP-AES was shown to have the sensitivity and precision similar to that of GFAAS recommended by NIOSH (method 173 or 290) (25).

The ICP-AES method is useful for monitoring the blood of workers exposed to vanadium as well as concomitantly exposed to several other metals (NIOSH method 8005) (26). A working range of 10–10,000 µg of vanadium/100 g blood and 2–2000 µg/g tissue was obtained, and the detection limits were estimated to be 1 µg/100 g blood and 0.2 µg/tissue.

The NAA technique has also been widely used to measure vanadium in biological samples (27–30). Detection limits as low as ppb (µg/L) levels of vanadium in blood and urine samples were obtained (27, 28, 30).

GFAAS can also be used for measuring trace levels of vanadium in the serum and urine of humans (30, 31). Detection limits of 0.08 µg/L in serum and 0.06 µg/L in urine were achieved (31). The GFAAS technique is as sensitive as NAA, and it is also rapid, simple, relatively free from interference, and relatively inexpensive (31, 32).

1.3.4 Community Methods

Vanadium concentrations in air are measured with sufficient accuracy and precision with the GFAAS or DSP-AES methods (see Section 1.3.1). GFAAS and FAAS are the techniques (NIOSH methods 7911 and 7910) recommended by EPA's Office of Solid Waste and Emergency Response for measuring low levels of vanadium in water and waste water (33). Detection limits of 4 and 200 µg of vanadium/L of sample were obtained using GFAAS and FAAS techniques, respectively.

Spectrophotometry has also been employed to measure ppm levels of vanadium in aqueous media (34, 35). Spectrophotometry is commonly employed to analyze for the presence of vanadium in soil. Detection of low ppm concentrations in the soil has been reported (34, 36). IDMS and spectrophotometry have been used for measuring low ppm (g/g) levels of vanadium in plant and marine animal tissues (34, 36, 37).

1.3.5 Biomonitoring/Biomarkers

Several biomarkers of exposure have been identified for vanadium, but none of them can be used to quantitatively determine the exposure levels. There are no well-documented biomarkers of effect specific for vanadium. The primary effects of exposure to vanadium dusts are different respiratory effects (coughing, wheezing, breathing difficulties), which can follow inhalation of many types of dust.

1.3.5.1 Blood. Vanadium in the blood plasma can exist in a bound and unbound form as vanadyl or vanadate reversibly binds to human serum transferrin. There is a wide divergence in blood serum/plasma vanadium levels reported in the literature, from 0.42 mg/L (38) down to 0.67 ng/L (39). In general, these levels tend to decline as analytical techniques improve and become more sensitive. Studies on occupationally exposed populations have shown a poor, if any, correlation between individual levels and exposure. For other information, see sections 1.4.1.3, 1.4.2.2.3, and 1.4.2.3.3.

1.3.5.2 Urine. The principal route of excretion of absorbed vanadium is through the kidneys. Vanadium is found in the urine of exposed workers as well as patients taking vanadium salts. However, poor or no correlation between urinary vanadium levels and oral dose or vanadium concentrations in air were found. For other information, see Sections 1.4.1.3, 1.4.2.2.3 and 1.4.2.3.3.

1.3.5.3 Other. It is estimated that the total body content of vanadium in healthy, adult humans approximates 100 µg.

The concentration of vanadium in human milk was determined to be 0.1–0.2 µg/g. Some vanadium workers develop a characteristic green tongue, as a result of direct accumulation of vanadium dusts on the tongue (40). One report from the 1950s states that vanadium exposure was associated with decreased cystine content in the fingernails of vanadium workers (41). However, alterations in cystine levels can also be associated with dietary changes and with other diseases, so this effect does not seem to be specific for vanadium exposure. No other commonly measured cellular changes have been identified with vanadium exposure.

1.4 Toxic Effects

1.4.1 Experimental Studies

1.4.1.1 Acute Toxicity. In the study designed to determine the LD_{50}, rabbits were exposed to vanadium pentoxide dust at concentrations of 77, 109, 205, or 525 mg/m^3 for periods of 7, 4, 7, or 1 h, respectively (42). Two of four rabbits died following an acute exposure to 205 mg V_2O_5/m^3 (114 mg vanadium/m^3). There was marked tracheitis, pulmonary edema, and bronchopneumonia. Conjunctivitis, enteritis, and fatty infiltration of the liver were also observed.

In acute studies by Roshchin (43, 44), albino rats were exposed to vanadium pentoxide fume at 10–70 mg/m^3, or dust at 80–700 mg/m^3, and ferrovanadium as dust at 1000–10,000 mg/m^3. The minimum concentration of vanadium pentoxide (fume) that gave rise to mild signs of acute poisoning was 10 mg/m^3. The LC$_{50}$ concentration for fume was 70 mg/m^3. One-hour exposures to V$_2$O$_5$ pulverized dust at 80 mg/m^3 caused acute intoxication; the lethal concentration was 700 mg/m^3 in one hour. Acute inhalation toxicity in rats was characterized by an irritation of the respiratory mucosa and nasal discharge that sometimes contained blood. Animals breathed with difficulty, and there were crepitations. In cases of severe poisoning, diarrhoea, paralysis of the hind limbs, and respiratory failure were followed by death. Animals that died or were killed at various times after exposure showed severe congestion and small hemorrhages in all internal organs, signs of increased intracranial pressure, and fatty degeneration of the liver. In the lungs, there was capillary congestion, edema, bronchitis, and focal interstitial pneumonia. Inhalation of dispersion aerosols of ferrovanadium did not produce any toxic effects, probably because the particles were too large.

A gavage study has shown that 41 mg vanadium/kg as sodium metavanadate is the LD$_{50}$ (14 days) for rats, and the value for mice is 31.2 mg/kg (45). The lethal dose by single intravenous administration to the rabbit is about 1.5 mg V$_2$O$_5$/kg (46).

1.4.1.2 Chronic and Subchronic Toxicity

1.4.1.2.1 Respiratory Effect. Male cynomolgus monkeys exposed by inhalation to vanadium pentoxide dust at concentrations of 0.3 or 2.8 mg/m^3 for 6 h/day, 1 day/week, for 2 weeks, showed significant central and peripheral airflow restriction one day after each exposure (47). They also had a dramatic increase in polymorphonuclear leukocytes in bronchioalveolar lavage, thus increasing total cell counts. No such changes occurred at 0.3 mg/m^3.

Rats that breathed uncoated bismuth orthovanadate dust (0.11 or 1.2 mg/L) or silica-coated bismuth orthovanadate (0.15 or 1.3 mg/L) for 6 h/day for 2 weeks showed an increase in lung weight, and alveolar proteinosis manifested by an increased accumulation of alveolar macrophages, lung lipids, and type II pneumocytes. Six months after exposure, cholesterol granulomas and foamy macrophages were observed; these changes were reduced but still persisted after one year (48).

Rabbits exposed for one hour daily for 8 months to 20–40 mg V$_2$O$_5$/m^3 had developed difficulty in breathing (42). At autopsy, pathological changes included chronic rhinitis and tracheitis, emphysema, patches of lung atelectasis, bronchopneumonia, and sometimes, pyelonephritis and conjunctivitis.

Continuous inhalation exposure of rabbits and guinea pigs to 10–30 mg V$_2$O$_5$/m^3 for 10 months caused bronchitis, pneumonia, weight loss, and bloody diarrhea (17, 49). Pathological changes in the lungs of rats exposed to vanadium pentoxide fume at 3–5 mg/m^3 for 2 h on alternate days, for 3 months, or to V$_2$O$_5$ dust at 10–30 mg/m^3 for 4 months included enlarged blood vessels with swollen endothelium, congestion, oedema, lymphostasis, small haemorrhages, and connective-tissue proliferation. In some animals purulent bronchitis, pneumonia, or abscesses developed (43).

In a study by Pazhynich (50) where rats were exposed for 70 days to V_2O_5 fume at 0.027 or 0.002 mg/m,^{3} histopathological changes following high level inhalation exposure included marked lung congestion, focal lung hemorrhages, and extensive bronchitis. Roshchin found that animal exposure to ferrovanadium dust (1 h on alternate days, for 2 months) gave rise to serious pathologic changes only at very high concentrations (1000 to 2000 mg/m^3). The pulmonary effects consisted in chronic bronchitis and perialveolitis, namely, chronic inflammation of the lungs (51). In a report (52) of another study by Roshchin (53) it was stated that a 2-month exposure of rats at 40–80 mg/m^3 ferrovanadium dust caused bronchitis, interstitial sclerosis, and perivascular edema. No other details were given.

1.4.1.2.2 Gastrointestinal Effects. Rabbits exposed for 8 months to high levels (205 mg/m^3) of vanadium pentoxide dust showed little histopathological damage to the GI system (42).

1.4.1.2.3 Hepatic Effects. Rabbits exposed for 8 months to vanadium pentoxide dust revealed some fatty degeneration of the liver (42). However, the liver function was not tested, and the author stated, without explanation, that the liver changes were of no special significance.

1.4.1.2.4 Renal Effects. Minor renal effects (as indicated by increased plasma urea, and mild histological changes) were seen in rats after oral exposure to sodium metavanadate for 3 months at levels of $\leq$10% of the oral LD$_{50}$ (54). The author reported a dose-related trend, but quantitative histopathological data were not provided.

Albuminuria was reported after intravenous injection of sodium metavanadate at 2.5–5 mg/kg to male dogs (40). Fatty changes in the kidney of the rat and rabbit were observed after inhalation of 10–70 mg VCl$_3$/m^3 for 2 h/day for 9–12 months (44).

Pazhynich (50) reported granular degeneration of the epithelial cells of the convoluted tubules, with necrotic areas, in rats exposed by inhalation to V_2O_5 fume at concentrations of 0.002 and 0.027 mg/m^3 for 70 days.

Rabbits exposed acutely or chronically to vanadium pentoxide dust showed fatty degeneration of the kidney, but the author, without explaining, did not attribute this effect to vanadium (42).

Summary of Key studies on animal toxicity of vanadium and its compounds are summarized in Table 37.3.

1.4.1.3 Pharmacokinetics, Metabolism, and Mechanisms

1.4.1.3.1 Adsorption. The absorption and distribution of vanadium compounds depend on the route of entry and their solubility in body fluids. The higher the solubility in water and biological media, the more toxic the compound, presumably because the better the absorption.

Table 37.3. Summary of Key Studies on Toxicity of Vanadium and Vanadium Compounds in Animals

Compound	CAS Number	Species	Route of Exposure	Concentration or Dose	Exposure	Observed Effects	Ref.
Acute toxicity							
		Rabbit	Inhalation	77 mg/m^3 109 205 525	7 h 4 h 7 h 1 h	NOAEL 77 mg/m^3; at 205 mg/m^3 — death (2/4), marked tracheitis, pulmonary edema, bronchopneumonia, conjunctivitis, enteritis, fatty infiltration of the liver	42
Vanadium pentoxide	[1314-62-1]	Rat	Inhalation (fume)	10–70 mg/m^3	Acute	LC$_{50}$ 70 mg/m^3; minimum effective concentration 10 mg/m^3; hemorrhagic inflammation in lungs, hemorrhages in internal organs, paralysis, pneumonia, respiratory failure	43
		Rat	Inhalation (dust)	80–700 mg/m^3	Acute	1-h exposures at 80 mg/m^3 caused acute intoxication (hemorrhagic inflammation in lungs, hemorrhages in internal organs, paralysis, pneumonia, respiratory failure); the lethal concentration was 700 mg/m^3 in 1 h;	43
Ferrovanadium	[12604-58-9]	Rat	Inhalation	1000–10000 mg/m^3	Acute	No effects	43
Sodium metavanadate	[13718-26-8]	Rat	Gavage	41 mg · V/kg	Acute	LD$_{50}$	45
		Mouse	Gavage	31.2 mg · V/kg	Acute	LD$_{50}$	45

Table 37.3. (*Continued*)

Compound	CAS Number	Species	Route of Exposure	Concentration or Dose	Exposure	Observed Effects	Ref.
Vanadyl sulfate	[27774-13-6]	Guinea pig	Subcutaneous	800 mg/kg	Acute	LD$_{100}$	135
		Guinea pig	Subcutaneous	560 mg/kg	Acute	LD$_{50}$	135

Repeated-Dose Toxicity

		Monkey	Inhalation	0.3 mg/m^3 2.8 mg/m^3	6 h/day at weekly intervals	At 2.8 mg/m^3 central and peripheral airflow restriction one day after each exposure, increase in polymorphonuclear leukocytes in bronchoalveolar lavage at 0.3, no effect	47
		Rat	Inhalation	0.002 (fume) 0.027 mg/m^3	70 days	0.027 mg/m^3 — lung congestion, focal lung hemorrhages, bronchitis; decrease in motor chron-axie^a ratio of muscles of tibia; 0.002 or 0.027 mg/m^3 - degeneration of the epithelial cells of the convoluted tubules with areas of necrosis	50
V$_2$O$_5$	[1314-62-1]	Rat	Inhalation	3–5 mg/m^3 (fume)	2 h, on alternate days for 3 months	In lungs: engorged blood vessels with swollen endothelium, congestion, edema, lymphostasis, small hemorrhages	43
		Rat	Inhalation	10–30 mg/m^3 (dust)	2 h, on alternate days for 4 months	In lungs: engorged blood vessels with swollen endothelium, congestion, edema, lymphostasis, small hemorrhages	43

	Rabbit	Inhalation	10–30 mg/m^3	10 months	Bronchitis, pneumonia, loss of weight, bloody diarrhoea	49
	Rabbit	Inhalation	20–40 mg/m^3	1 h/day for 8 months	Conjunctivitis, chronic rhinitis, tracheitis, emphysema, patches of lung atelectasis, bronchopneumonia, and sometimes pyelonephritis	42
Bismuth orthovanadate (BiVO$_4$) [53801-77-7]	Rat	Inhalation	Uncoated BiVO$_4$: 0.11 or 1.2 mg/L; silica-coated BiVO4: 0.15 or 1.3 mg/L	6 h/day, 5 days/week for 2 weeks	Increases in lung weight, alveolar proteinosis	48
Sodium orthovanadate [13721-39-6]	Rat	In diet	100–200 mg/kg body weight	≤ 56 weeks	Dose-dependent increase in systolic pressure	131
Sodium metavanadate [13718-26-8]	Rat	In drinking water	1, 10, 140 µg/mL	6 or 7 months	Increased systolic and diastolic pressure	134
Ferrovanadium [12604-58-9]	Rat	Inhalation	1000–2000 mg/m^3	1 h on alternate days for 2 months	Chronic bronchitis and perialveolitis	51
	Rat	Inhalation	40–80 mg/m^3	2 months	Bronchitis, interstitial sclerosis, and perivascular edema	42

[a]The *chronaxie* ratio is the number expressing the sensitivity of a nerve to electrical stimulation.

Indirect evidence of absorption after inhalation of vanadium in animals is indicated in the studies in which vanadium was administered intratracheally. Soluble vanadium compounds that are inhaled and deposited are readily absorbed. A rapid absorption of vanadium was reported in rats receiving ^{48}V-labeled vanadyl chloride or vanadyl pentoxide (55). The highest rate of absorption of intratracheally administered ^{48}V-vanadium nitrate (0.4 or 20 mg/kg body weight) was found to occur 5 min after administration (56). After intratracheal instillation of 200 ng/kg body weight of ^{48}V-labeled pentavalent and tetravalent vanadium, pulmonary clearence was initially rapid: 80 and 85% of the tetravalent (V^{4+}) and pentavalent (V^{5+}) forms of vanadium, respectively, were removed from the lungs within 3 h (57). After 24 h, more than 50% of vanadyl oxychloride was cleared from the lungs of male rats (58), and at 3 days, 90% of vanadium pentoxide was eliminated from the lungs of female rats (55). However, at 12 days about 2% of the element was still present in the lung. In another study 50% of 40 µg vanadium pentoxide in 0.9% saline solution instilled intratracheally was cleared within 18 min, and the rest within a few days (59).

The absorption of vanadium through the GI tract of animals is low. Less than 0.1% of an intragastric dose of ^{48}V, equal to 0.2 LD_{50}, was detectable in the blood of rats at 15 min postexposure, and less than 0.2% at 1 h (56). It was reported that only 2.6% of an orally administered radiolabeled dose of vanadium pentoxide was absorbed 3 days after exposure in rats (55). Young rats that consumed vanadium in their drinking water and feed were found to have higher tissue vanadium levels 21 days after birth than they did 115 days after birth (60).

Since vanadium is a metal, absorption through the skin is thought to be minimal due to its low solubility. No penetration with ^{48}V of human skin samples *in vitro* was found.

1.4.1.3.2 Distribution. After intratracheal administration, vanadium is rapidly distributed in tissue. The pentavalent and tetravalent forms of vanadium compounds were found to have similar distribution patterns. Three hours after inhalation exposure to pentavalent or tetravalent vanadium, 15–17% of the absorbed dose was detected in the lung, 2.8% in the liver, and 2% in the kidney (57). Although the kidney levels are high after exposure, the bone had a higher retention of vanadium. Skeletal levels of vanadium peaked 1–3 days postexposure (55, 56, 59), accounted for 30% of administered vanadium dose by day 7, and were reported to persist after 63 days (58). Within 15 min after intratracheal administration of 0.36 mg/kg ^{48}V-vanadium oxychloride, radiolabeled vanadium was detectable in all organs investigated, except the brain. The highest concentration was in the lungs, followed by the heart and kidney. The other organs had low levels. Maximum concentrations were reached in most tissues between 4 and 24 h (58). Vanadium is found to have a biphasic lung clearance after a single acute exposure (58, 59). The initial phase is rapid, and a large percentage of the absorbed dose is distributed to most organs and blood 24 h postexposure, followed by a slower clearance phase.

Acute oral studies with rats showed the highest vanadium concentration to be located in the skeleton. Male rats had approximately 0.05% of the administered ^{48}V in bones, 0.01% in the liver, and less than 0.01% in the kidney, blood, testis, or spleen after 24 h (57). Similar findings were noted by other authors who reported that the bone had the highest concentration of radiolabeled vanadium, followed by the kidney (56). Conklin et al. (55)

reported that after 3 days, 25% of the absorbed vanadium pentoxide was detectable in the skeleton and blood of female rats.

Oral exposure of an intermediate duration produced the highest accumulation of vanadium in the kidney. In young male rats at 3 weeks of age, the kidney, heart, and lung had the highest levels immediately after exposure (60). However, vanadium levels in these organs decreased significantly at 115 days of age. It is speculated that the higher levels of vanadium in the young rat tissues may be due to a higher retention capacity of the undeveloped tissues, or to a greater nonselective permeability of the undeveloped intestinal barrier.

Adult rats exposed to 50 mg/L of vanadyl sulphate or sodium orthovanadate in drinking water for 3 months had the highest vanadium levels in the kidney, followed by bone, liver, and muscle (61). All tissue levels plateaued at the third week of exposure. A possible explanation for the initially higher levels in the kidney during an intermediate-duration exposure is the daily excretion of vanadium in urine. At the cessation of treatment, vanadium was mobilized rapidly from the liver and slowly from the bones. Other tissue levels decreased rapidly after exposure was discontinued; however, vanadium retention was much longer in the bone (60, 61). These findings are supported with recent data from the experiment in which male Wistar rats were given vanadyl sulfate in drinking water at concentrations of 0.5–1.5 mg/L for a period of one year, after which time there was a significant concentration-dependent increase in vanadium levels in organs. The bone had the highest concentration, followed by kidney, testis, liver, pancreas, and brain. Although only a negligible amount of vanadium could be detected in plasma samples 16 weeks after vanadyl withdrawal, there were still considerable amounts of vanadium accumulated in various organs; the highest concentration was found in bone, followed by testis, kidney, liver, pancreas, and brain. The concentrations of vanadium in these organs varied from about 10 to 60% (bone) of the corresponding values before vanadyl withdrawal (62). Comparative studies on the biodistribution of ^{48}V and ^{32}P in rats suggested the possibility that vanadium is taken up by bones via the same pathway as phosphorus, and thus the retention of vanadium in bone may have been due to phosphate displacement. This view is supported by the structural similarities between the pentavalent elements: vanadate and phosphates (63).

No studies were located regarding vanadium distribution in animals after dermal exposure. Low amounts of ^{48}V-vanadium nitrate were detected, mainly in kidney and bone, 16 days after subcutaneous injection to Wistar rats.

In intraperitoneal administration to rats, vanadium is distributed to all organs. After 24 h, the highest concentrations can be found in the bone and kidney, although the initial levels, observed 3 h after injection, are highest in the kidney (56, 64, 65). The bone uptake rates increase with time up to 72 h, and remain high after 168 h. This is similar to the distribution following inhalation and oral exposures.

1.4.1.3.3 Metabolism. Vanadium is an element, and as such, is not metabolized. However, in the body, there is an interconversion of two oxidation states of vanadium: the tetravalent form, vanadyl (V^{4+}); and the pentavalent form, vanadate (V^{5+}). In the plasma, vanadium can exist in a bound or unbound form (66). Vanadium as vanadyl (67) or vanadate (68) reversibly binds to human serum transferrin at two metal binding sites on the protein, and is then taken up into erythrocytes. These two factors may affect the biphasic clearance of vanadium that occurs in the blood. With intravenous administration of vanadate or vanadyl, there is a short lag time for vanadate binding to transferrin, but, at 30 h, the association is identical for the two vanadium forms (V^{4+} and V^{5+}) (69). Vanadium–transferrin binding is most likely to occur with the vanadyl form as this complex is more stable (69). The transferrin-bound vanadium is cleared from the blood at a slower rate than unbound vanadium, which explains the biphasic clearance pattern (70). Specific intracellular vanadyl–ferritin complexes might be also formed in rat liver, spleen, and kidney. The metabolic pathway appears to be independent of the route of exposure (57).

Oral administration of vanadium in rats interfered with copper metabolism, probably by inhibiting the intestinal absorption of copper (71).

Exogenous vanadyl and vanadate ions both mimic the biological action of insulin by stimulating glucose transport and oxidation in rat adipocytes and skeletal muscle, stimulating glycogen synthesis in rat liver and diaphragm, and inhibiting gluconeogenesis in hepatocytes. Furthermore, *in vitro* studies demonstrated that vanadate induced lipogenesis and inhibited lipolysis in rat adipocytes (72, 73). It is speculated that vanadate (+5) and vanadyl mimic the responses of insulin through alternative signaling pathways, not involving insulin receptor activation. It has been shown that vanadyl sulfate stimulates glycogen synthesis in vitro, which is associated with an activation of phosphatidylinositol 3-kinase. However, it has also been suggested that the inhibition of protein tyrosine phosphatase by vanadium salts may mimic insulin action by activating insulin receptor protein tyrosine kinase activity, thus preventing dephosphorylation of the insulin receptor (74). However, the question as to whether vanadate (+5), vanadyl (+4), or both are the active species is still open. It was found that the effect of $VOSO_4$ (+4) is only 25–35% of that of vanadate or insulin. The issue is further complicated since it was found that vanadate (+5) is not reduced to vanadyl (+4) inside the cell, and that vanadyl sulfate (+4) is capable of a spontaneous oxidation to vanadate (+5) *in vivo* (10).

Many authors have confirmed that vanadium compounds could influence the acivity of different enzymes. However, the mechanism for most of the effects observed remains obscure. Table 37.4 summarizes the effects of vanadium compounds on some enzymes.

1.4.1.3.4 Excretion. Vanadium administered intratracheally to rats was excreted predominantly in the urine (58) at levels twice as high as in the feces (59).

Since vanadium is poorly absorbed in the GI tract, a large percentage of vanadium is excreted unabsorbed in the feces in rats following oral exposure. More than 80% of the administered dose of ammonium metavanadate accumulated in the feces after 6 days (67). After 2 weeks of exposure, approximately 60% of sodium metavanadate was found in the feces (75). However, the principal route of excretion of the small absorbed portion of vanadium in animals is through the kidney. The half-life for vanadium elimination from the kidney of Wistar rats fed vanadyl sulfate trihydrate (0.75 mg/mL) in drinking water for 3 weeks was found to be 11.7 days (76). It is supported by another study in which the half-lives were determined of vanadium elimination from various organs of male Sprague–Dawley rats treated with sodium orthovanadate or vanadyl sulfate at concentrations ranging from 1.6 to 160 μmol·V/kg body weight per day for a week. The values varied greatly among organs; the liver and kidney have the shortest (∼3–4 days) and the testis the longest (~14–16 days) half-lives. Almost all the tissues showed a slightly longer vanadium half-life in the group of rats fed sodium orthovanadate compared to those receiving vanadyl sulfate (77).

No animal studies were located regarding vanadium excretion after dermal exposure.

Female Syrian hamsters were injected a single dose of 30 μg GFAAS vanadium standard solution (prepared together with GFAAS aluminium standard solution) intraperitoneally or intramuscularly. The recovery of vanadium excreted in urine was 53% after 24 h and 78% after 72 h. Only a small amount was found in the blood, and the level in the organs was below the detection limits. The total recovery of the injected vanadium was 98% (78).

In a study on rats that received 12 intraperitoneal injections of sodium metavanadate (1.84 mg/kg per injection) or vanadyl sulfate trihydrate (6.35 mg/kg per injection) within 4 weeks, the effects of different chelating agents administered daily for 5 days after vanadium salts treatment were examined. It was found that only Tiron (4,5-dihydroxy-1,3-benzenedisulfonic acid disodium salt) significantly decreased the tissue concentration and urine elimination of vanadium. Ascorbic acid and 2-mercaptosuccinic acid were not effective, and the effectiveness of deferoxamine was uncertain (79).

1.4.1.4 Reproductive and Developmental. Vanadium crosses the blood–testis barrier and accumulates in the testis (80). Some studies indicate that this metal is a male reproductive toxicant; and that its degree of toxicity depends on the route of exposure, oxidation state, period of dosing, and the dose. Subcutaneous injection of vanadium sulphate at the dose of 0.08 mM/kg in mice significantly reduced mean testes weights, but no necrotic changes were observed (81).

Table 37.4. Effects of Vanadium Compounds on Enzymes

Inhibitory Effects	Stimulatory Effects
Na^+-K^+ATPase	Phospholipase C
Ca^{+2}-ATPase	Adenyl cyclase
H^+-K^+ATPase	Thyrosine kinase
H^+-ATPase	Adenyl cyclase
K^+-ATPase	Mitogen-activated protein kinase
Ca^+-Mg^+ATPase	Phosphatidylinositol 3-kinase
Dynein ATPase	NADPH oxidase
Actomyosin ATPase	Glycogen synthase
Protein tyrosine phosphatase	Lipoprotein lipase
Glutamine dehydrogenase	
Alkaline phosphatase	
Acid phosphatase	
Glucose-6-phosphatase	
Phosphofructokinase	
Alanine aminotransferase	
Aspargine aminotransferase	

In the study of Roshchin (56) vanadium (the chemical form and the route of exposure not specified) at the dose of 0.05 LD_{50} for a period of 20 days induced changes in the morphology and function of spermatozoa (reduced motility and increased number of dead spermatozoa) and the morphology of seminiferous epithelium. After administration of vanadium to rats on day 10 of pregnancy, considerable increase in embryonal mortality due to an enhancement in preimplantation death of embryos was observed.

Altamirano (82) suggested that vanadium toxicity in prepubertal rats is higher in males than in females. Vanadium pentoxide (12.5 mg/kg) injected intraperitoneally to female prepubertal rats every 2 days from birth to day 21 of life induced a decrease in the ovulation rate without significant changes in the number of ova shed by ovulating animals as compared to the controls. The same treatment in male rats induced an increase in the weight of seminal vesicles, thymus, and submandibular glands.

Sodium metavanadate given to male mice in the drinking water at concentrations of 0, 0.1, 0.2, 0.3, or 0.4 mg/mL for 64 days before mating with untreated females to obtain dosages of 0, 20, 40, 60, or 80 mg/kg per day induced a significant decrease in sperm count at 40, 60, and 80 mg/kg per day without affecting its motility and appearance. The pregnancy rate was decreased at 60 and 80 mg/kg per day (83).

Vanadium pentoxide treatment (IP injections at the dose of 8.5 μg/g, every 3 days during 60 days) of CD-1 male mice resulted in a decrease in fertility rate, implantations, live fetuses, and fetal weight, and an increase in the number of resorptions per dam (84). Sperm count, motility, and morphology were impaired with the advancement of treatment, probably due to DNA damage in the testis cells depending on the dose, as assessed by COMET assay.

Vanadate (V^{5+}) and vanadyl (V^{4+}) are capable of crossing the placenta and reaching the fetus. It is considered that placenta represents only a partial barrier to vanadium exposure. Edel and Sabbioni (85) showed that the amounts of vanadium found in the placental unit were remarkably higher than those found in fetuses. The route of exposure plays a very important role in embryofetotoxicity of vanadium. Metavandate given orally to rats on days 6–14 of pregnancy at doses of 20 mg/kg per day or lower was neither embryolethal nor teratogenic (86). However, daily doses of 4 or 8 mg/kg given IP to mice on gestational days 6–15 induced an increased number of resorptions and dead fetuses, an increased percentage of postimplantational loss, and malformations (87).

Vanadyl sulfate administered by gavage to pregnant Swiss mice at doses of 0, 37.5, 75, or 150 mg/kg per day on days 6–15 of pregnancy induced a significant increase in the number of early resorptions per litter at all dose levels, signs of fetotoxicity as evidenced by lower fetal weights and fetal lengths, and the presence of developmental variations (88). Malformation incidence was also increased. Maternal toxicity was observed in the vanadium-treated animals, as evidenced by reduced weight gain, reduced body weight on gestation day 18, and decreased absolute liver and kidney weights at 75 and 150 mg/kg per day. The no-observable-effect level (NOEL) for maternal toxicity, embryofetotoxicity, and teratogenicity for vanadyl sulfate under these test conditions was estimated to be below 37.5 mg/kg per day for Swiss mice.

Adult male rats were given metavanadate (V^{5+}) by gavage sodium at doses of 5, 10, or 20 mg/kg per day for 60 days before mating with females that received the same doses during 14 days prior to mating and during the periods of gestation and lactation. It was found that body weight, body length, and tail length in the offspring were significantly decreased from birth and during the entire lactation period (89).

Pregnant Wistar rats were injected IP with vanadium pentoxide at 0.33, 1.0, or 3.0 mg/kg per day on days 6–15 of gestation, and maternal toxic symptoms, decrease in weight gain during treatment and in placenta weight, increase in incidence of embryofetus mortality and external or skeletal malformation, and fetal growth retardation were observed in the high dosed group (90). Increased incidence of embryofetus mortality and external or skeletal malformation, delayed ossification of bone, and decreased placenta weight were observed in the 1.0-mg/kg group. It was concluded that vanadium may exert a direct effect on the embryofetus or a "double effect" on both placenta function and the embryofetus.

The treatment of CD-1 pregnant mice with vanadium pentoxide (8.5 mg/kg per day) on days 6–15 of pregnancy resulted in the reduction of fetal weight and the number of ossification points in forelimbs and hind limbs, and an increase in the frequency of abnormal fetuses (91).

National Institutes of Health (NIH) mice were injected vanadium pentoxide (5 mg/kg body weight, IP) at different times of gestation (on days 1–5, 6–15, 7, 8, 9, 10, 11, or 14–17). Increased frequency of resorptions or fetal deaths were observed after injections on days 6–15, 7, or 14–17 of gestation (92). Delayed ossification of bones was noted after injections on days 6–15, 8, 10, or 14–17 of gestation.

The administration of sodium metavanadate to rats during pregnancy and lactation (75 µg·V/g diet through day 21 postpartum) resulted in a decreased survival of pups from the vanadium-treated group (93). The surviving pups gained less weight than did control pups, despite similar birth weights. They had higher relative liver, brain, and testis weights than did the controls.

It has been shown that oral vanadate treatment was ineffective during diabetic pregnancies in animals. Moreover, it produced a toxic effect, reduced reproductive capacity, and interfered with fetal growth and development in both normal and diabetic animals (94).

Lung collagen content in rats continuously exposed for two generations to 20 ppm vanadium as sodium metavanadate drinking solutions for both F_1 term fetuses and 21-day-old F_1 pups did not differ from that of control rats. However, total collagen and soluble collagen within lungs of adult F_1 rats and F_2 term fetuses were significantly lower than in nontreated controls (95).

1.4.1.5 Carcinogenesis. A chronic, 8-month inhalatory study, designed to test the effects other than cancer in rabbits, did not reveal any increases in tumors (42).

The results obtained by Levy et al. (96), who investigated vanadium solids using intrabronchial pellet implantation in the lower left bronchus of rats, have not shown the carcinogenic effect. The materials produced chronic inflammatory changes in 44/100 rats, bronchial inflammation in 50/100, squamous metaplasia in 10/100, and one bronchial carcinoma in a male rat after 645 days.

No studies were located that specifically studied cancer in animals after oral exposure to vanadium. However, some studies designed to test other endpoints indicated no increase in tumor frequency in rats and mice chronically exposed to 0.5–4.1 mg vanadium/kg as vanadyl sulfate in drinking water (97–99). Although the results of these oral studies were negative for carcinogenicity, they were inadequate for evaluating the carcinogenic effects because insufficient numbers of animals were used, it was not determined whether a maximum tolerated dose was achieved, a complete histological examination was not performed, and only one exposure dose per study was evaluated.

Ammonium vanadate (10 or 20 mg/L) in drinking water had no influence on tumor development of large-bowel neoplasms in mice treated with 1,2-dimethylhydrazine (DMH) given by subcutaneous injection for 20 weeks (100). Although thymidine incorporation was increased, ammonium vanadate did not have any effect on the incidence or type of tumor induced by DMH.

Studies carried out since the late 1990s suggest that vanadium could be considered representative of a new class of nonplatinum metal antitumor agents (101). It has been shown that metallocene dichlorides, $(C_2H_5)_2MCl_2$ (where M = titanium, vanadium, molybdenum, or niobium), exhibit cancerostatic activity against the Erlich ascites tumour system in mice, and that treatment with such substances cured the tumor (102). Dietary vanadium (25 mg V^{4+} sulfate/kg) was found to block the induction of rat mammary carcinogenesis by 1-methyl-1-nitrosourea during the postinitiation stages of the neoplastic process. Both the cancer incidence and the average number of cancers per rat were reduced by the vanadium diet without inhibiting the overall growth of animals (103). Peroxo heteroligand vanadates (+5) were shown to exert antitumor activity against L1210 murine leukemia (104). The special nature of electron transfer within the V^{5+}-peroxo moiety was proposed to be responsible for this phenomenon. Sardar et al. (105) documented the potential of vanadium to inhibit the growth of a transplantable murine lymphoma. Further, it was shown that vanadium can prevent cellular proliferation by stimulating certain haematological indices, including erythropoietin (106).

Bishayee and Chatterjee (107) and Bishayee et al. (108) demonstrated that vanadium at 0.5 ppm (as ammonium monovanadate) in drinking water was very effective in arresting the development of diethylnitrosamine (DENA)-induced hepatocarcinogenesis (DENA; 200 mg kg^{-1}, IP) in rats without any toxic manifestations. Vanadium during the entire experiment, before initiation and during promotion, greatly reduced the incidence, multiplicity, and size of visible persistent nodules (numerous observations support the concept that the nodules are the precursors of hepatic cancer) with a concurrent arrest in the number and spread of γ-glutamyltranspeptidase-positive hepatic foci in total liver parenchyma. The anticarcinogenic potential of vanadium was primarily observed at the initiation phase and only secondarily at the promotion stage.

1.4.1.6 Genetic and Related Cellular Effects Studies. Vanadium at micromolar concentrations stimulates DNA synthesis and cell proliferation in human fibroblasts (109), embryonic chicken bone cells (110), interleukin-3-dependent mast cells (111), and Swiss mouse 3T3 fibroblasts (112). Vanadate can induce transformation in Balb/3T3 cells (113), hamster embryo cells (114), and bovine papilloma virus DNA-transfected C3H 10T1/2 cells (115). Vanadium ions have been found to be mitogenic and stimulate DNA synthesis in quiescent human fibroblasts and Swiss mouse 3T6 cells (116). At concentrations lower than 10^{-10} M, vanadate stimulates colony formation of human breast and lung cancer cells, whereas at concentrations above 10^{-10} M the colony formation is inhibited (117). Vanadium, administered orally by gavage as ammonium monovanadate in concentrations of 0.01–0.8 μg vanadium/day, has been found to stimulate tumor cell proliferation in mice bearing a transplantable ascitic lymphoma (118), which indicates the mitogenic potential of vanadium.

The results of mutagenicity studies of vanadium with bacterial assays are conflicting, and no firm conclusions can be drawn (Table 37.5).

Vanadium demonstrated positive effects in tests with *Bacillus subtilis* (119) and WP2, WP2uvrA, and Cm-981 strains of *E. coli* (120). In other strains of *E. coli* (ND-160, MR102) however, vanadium was shown to be nonmutagenic (120). Results of assays employing *S. typhimurium* present conflicting data. Although some data indicate no mutagenic effects of vanadium in TA1535, TA1537, TA98, and TA100 strains (120), Arlauskas (17) found a positive effect of NH_4NO_3 in TA1535 strain of *S. typhimurium*.

Ammonium metavanadate increased the convertant and revertant frequencies in the D7 strain of *Saccharomyces cerevisiae*; the highest activity was observed without metabolic activation (121).

Significant levels of induced micronuclei were observed in two strains of mice (615 and Kunming albino) given vanadium pentoxide by IP injection at doses 6.4, 2.13, or 0.17 mg/kg body weight for 5 consecutive days (120). Both the subcutaneous injection of vanadium pentoxide solution (0.25, 1, or 4 mg/kg) and the inhalation of vanadium pentoxide dust (0.5, 2, or 8 mg/m^3) induced micronuclei in the 615 mouse strain. However, negative results were obtained following oral administration of vanadium pentoxide at doses of 1.4, 2.8, 5.7, or 11.3 mg/kg body weight, daily, for 6 weeks, to Kunming albino mice.

The micronucleus test was found to be positive for vanadyl sulfate, sodium orthovanadate, and ammonium metavanadate in bone marrow of mice following intragastric treatment (122). In contrast, except for vanadyl sulfate, no difference was found between controls and treated animals in the structural chromosome aberration test

Table 37.5. Results of Mutagenicity Studies of Vanadium with Bacterial Assays

Compound	Species	Endpoint	Result	Ref.
	Bacillus subtilis	Rec assay	Weak positive	118
	Escherichia coli	Gene mutation		119
	WP2		+	
	WP2uvrA		+	
	Cm-981		+	
V_2O_5	ND-160		−	
	MR102		−	
	Salmonella typhimurium	Gene mutation		119
	TA1535		−	
	TA1537		−	
	TA98		−	
	TA100		−	
$VOCl_2$	*Bacillus subtilis*	Gene mutation	Weak positive	118
NH_4VO_3	*Salmonella typhimurium*	Gene mutation		120
	TA1535		+	

performed 24 and 36 h after treatment. At the same sampling intervals, second metaphases were positively scored for the induction of numerical chromosome aberrations for all three vanadium salts. In addition, the frequency of hypoploid and hyperploid cells was shown to be statistically different from the control value. Polyploid cells were also induced by all compounds, but their frequency was not statistically significant.

In a dominant lethal mutation assay where vanadium pentoxide was administered daily by subcutaneous (SC) injections to male mice for 3 months at doses of 0.2, 1, or 4 mg/kg body weight, the results obtained were considered negative for the induction of dominant lethal mutations (120).

The activities of vanadium trioxide, vanadyl sulfate, and ammonium metavanadate in inducing sister chromatid exchange (SCE) and chromosomal aberrations (CAb) were assayed in Chinese hamster ovary cells (123). The toxic concentrations (TC_{50}) for these compounds were found to be 25, 23, and 16 µg elemental vanadium/mL, respectively. At doses 0.02–0.25 TC_{50}, vanadium compounds were able to induce significant increases ($p < .01$) in the SCE frequency with or without the presence of rat hepatic S9 mix. These compounds also induced CAb in the cells at doses closely equivalent to the TC_{50}. In studies with ammonium metavanadate at concentrations of 5–40 µM, weak mutagenesis was demonstrated at the *hprt* gene of Chinese hamster V79 cells, and at the *gpt* locus of $hprt^-/gpt^+$ transgenic cell line G12 (124).

Although the biochemical mechanism of vanadium carcinogenicity and toxicity is still not fully understood, studies have indicated that vanadium-mediated generation of hydroxyl radical (.OH) and related oxygen species may play an important role (125–127). Byczkowski and Kulkarni (128) have reported that vanadium (IV) initiated lipid peroxidation and catalyzed cooxygenation of benzo[*a*]-pyrene-7,8-dihydrodiol to its ultimate carcinogenic metabolite. They suggested that lipid peroxidation may be crucial for toxicity in organs with limited endogenous lipid peroxidation. Vanadium (IV) is able to cause molecular oxygen-dependent 2'-deoxyguanosine (dG) hydroxylation and DNA strand breaks. It also reacts with *t*-butyl-OOH to produce *t*-butyl-OOH free radicals, which cause dG hydroxylation and DNA strand breaks (129).

1.4.1.7 Other: Neurological, Pulmonary, Skin Sensitization

1.4.1.7.1. Cardiovascular Effects. In rats, massive exposure by inhalation of vanadium oxides and salts leads to arrythmias, extrasystoles, and ECG abnormalities (43). Some changes in ECG can also be observed in dogs after iv administration of sodium metavanadate at 2.5 mg/kg (40).

Vanadium sulfate (500 mg/kg diet, 6 weeks) mobilized excess arteical cholesterol in rabbits previously maintained on a cholesterol rich diet (130).

Rats fed sodium orthovanadate in diet [100 or 200 mg/kg body weight (bw)] for up to 56 weeks developed a gradual increase in systolic blood pressure. The elevated pressure was sustained in a dose-dependent manner and positively correlated with plasma levels of vanadium (0.04–0.27 mg/L) (131).

Rats fed 15 mg vanadium/kg as ammonium vanadate for 2 months showed increased right ventricular pressure and pulmonary hypertension, but no changes in systemic circulation (132). This laboratory also showed that the addition of sodium orthovanadate in their food for 6 months did not alter heart rate or blood pressure, but did induce vasoconstriction (133).

Increased systolic and diastolic blood pressure was noted at the end of the exposure period in rats given sodium metavanadate in drinking water at concentrations of 1, 10, or 40 µg/mL for 6 or 7 months (134). Pathological changes in the myocardium (perivascular swelling and fatty degeneration) were observed in rats and rabbits following long-term inhalation exposure to V_2O_3, V_2O_5, and VCl_3 (1–70 mg/m^3, 2 h/day, for 9–12 months) (44).

1.4.1.7.2 Hematological Effects. Some studies indicate a stimulative effect of vanadium on hemoglobin and erythrocyte levels. Kopylova (135) observed increases in the erythrocyte count and hemoglobin level in rabbits injected SC with vanadyl sulfate (1 mg/kg daily for 2 months). Similarly, Trummert and Boehm (136) reported an increase in erythrocyte count following iv injection of vanadium gluconate (0.3–1.5 mg/kg daily for 40 days); hemoglobin level was not significantly affected.

Adverse effects on hematological parameters were observed by Roshchin (44). Hemoglobin levels in rabbits exposed to V_2O_3 (40–75 mg/m^3, 2 h/day, for up to 12 months) decreased from 75 to 68% of normal levels. The number of leukocytes declined by the end of the test from 7000–8000 to 5000/mm^3; no change was noted in controls. Sokolov (137) noted an increase in the leukocyte count in rats intoxicated with V_2O_5 by inhalation.

In rats drinking 0.15 mg V/mL ammonium metavanadate solution for 4 weeks, a statistically significant increase in leukocyte count was found as a result of an increased number of both neutrophils and lymphocytes (138).

Rabbits exposed acutely or chronically to vanadium pentoxide dusts showed no bone marrow changes on histological examination (42).

1.4.1.7.3 Immunological Effects. Rabbits under acute or chronic exposure to vanadium pentoxide dusts did not show histopathological changes in the spleen (42). Mice exposed to vanadium in drinking water (1, 10, or 50 mg vanadium/L) for 1–3 months showed a dose-related but nonsignificant decrease in the antibody-forming cells in the spleen when challenged with sheep erythrocytes. Splenic lymphocytes showed increased DNA synthesis *in vitro* (139). Mild spleen hypertrophy and hyperplasia were seen in rats treated with vanadium in drinking water for 3 months (54), but further immunological tests were not performed. Female mice given ammonium metavanadate ip at doses of 2.5, 5, or 10 mg/kg, every 3 days for 3, 6, or 9 weeks, showed a dose-related increase in resistance to *E. coli* endotoxin lethality up to 6 weeks and a dose-related decrease in resistance to *Listeria* lethality (140). Enlargement of the liver and spleen with enhanced formation of splenic megakaryocytes and red blood cell precursors was also observed.

The selective immunotoxic effects (depression of phagocytosis, splenotoxicity, enlargement of spleen, elevation of peripheral blood leukocytes, and T- and B-cell activation) were observed in rats given V_2O_5 in drinking water (1 or 100 mg·V/L) for 6 months, and mice given V_2O_5 in 0.2 mL water (6 mg·V/kg) by gavage 5 days a week for 6 weeks (141).

1.4.1.7.4 Neurological Effects. It is suggested that vanadium has a selective effect on adrenergic pathways as it decreased noradrenaline levels in brain of rats (142) and in hypothalamus of mice (143). Roshchin (43) observed neurophysiological disturbances (impaired conditioned reflexes and neuromuscular excitability) in dogs and rabbits exposed orally or subcutaneously to V_2O_3, V_2O_5, VCl_3, or NH_4VO_3. In studies where vanadium pentoxide and ammonium vanadate were administered orally to rats and mice in doses of 0.005–1 mg vanadium/kg for 21 days (higher levels) to 6 months (lower levels), a dose of 0.05 mg vanadium/kg was found to be the threshold for functional disturbances in the conditioned reflex activity in both rats and mice (144). The motor chronaxy ratio of antagonistic muscles of the tibia in rats exposed by inhalation to V_2O_5 at 0.027 mg/m^3 decreased from 1.5 at the beginning of the study to 1.0 on day 20 and 0.5 on day 30. The decrease continued until a level of about 0.25 was reached. The chronaxie ratio returned to normal about 18 days after cessation of exposure on day 70 (50). Rabbits exposed to vanadium pentoxide for 8 months did not show pathological changes in the brain (42). At least some of these effects might be related to the inhibitory effects of vanadium salts on different ATP-ase enzymes (see Section 1.4.1.3.4).

1.4.1.7.5 Body Weight. Rats given sodium metavanadate in food for an intermediate period of time showed a slight decrease in body weight (145, 146). Mice chronically exposed to vanadyl sulfate in drinking water showed a slight, but not statistically significant, weight increase as compared to controls (99), but experiments from the same laboratory under comparable exposure conditions showed no weight changes in either mice or rats (98,147), respectively. Chronic exposures of ≤4.1 mg vanadium/kg as vanadyl sulfate in food or water did not affect mortality in rats or mice, respectively (98, 147).

1.4.2 Human Experience

1.4.2.1 General Information. Elemental vanadium does not occur in nature; however, vanadium compounds exist in over 50 different mineral ores and in association with fossil fuels. The toxicologically significant compounds are vanadium pentoxide (V_2O_5), sodium metavanadate ($NaVO_3$), sodium orthovanadate (Na_3VO_4), vanadyl sulfate ($VOSO_4$), and ammonium vanadate (NH_4VO_3). Vanadium pentoxide dust is usually encountered in occupational settings, and humans can be exposed via the inhalation route.

1.4.2.2 Clinical Cases

1.4.2.2.1 Acute Toxicity. No studies were located regarding death in humans after inhalation exposure to vanadium. Acute vanadium toxicity is characterized by a latency period, which depends on the concentration of vanadium, the individual sensitivity of the subject, and the properties of the specific vanadium compound. The more soluble salts of vanadium pentoxide have a more rapid action than do the vanadium oxides. Condensation aerosol of vanadium pentoxide is more toxic than is a disintegration aerosol.

The first report of acute occupational exposure to vanadium was made by Dutton in 1911 (148). He reported that men exposed to vanadium oxide as dust and fume complied of dry cough and eye irritation. Williams described vanadium poisoning in 12 workers cleaning oil-fired boilers (149). Air sampling showed most of the dust particles to be smaller than 1 µm. Vanadium concentration ranged from 17.2 mg/m^3 in a superheater chamber to 58.6 mg/m^3 in a combustion chamber. The primary symptoms usually occurred between half and one hour after the beginning of work, but in some cases were delayed for 12 hours. They consisted in rhinorhoea, sneezing, watering of the eyes, and soreness of the throat. The secondary symptoms, they were dry cough, wheezing, severe dyspnea, lassitude, and depression, appeared after a period of 6–24 h. A greenish-black coating developed on the tongue and faded 2–3 days after contact with petroleum soot ceased. These symptoms continued while at work and became less severe only 3 days after ceasing work. X-ray films of the chest, electrocardiograms, and routine examination of the urine for albumin, sugar, and blood revealed no abnormalities.

Brown described 12 cases of vanadium poisoning in men working in a gas turbine heat exchanger (150). The symptoms, which appeared between 1 and 14 days of exposure, consisted in acute irritation of the respiratory tract, with nosebleed and blackening of the tongue and teeth. Irritation of the skin and eyes was also reported. Vanadium was in the form of a silicate, vanadate, or pentoxide, and made up 11–20% of the deposit on some tubes that the men were cutting away.

In 1956 Gul'ko (49) reported symptoms of respiratory tract irritation (sneezing, minor epistaxis, and coughing) and eye irritation experienced by men occupationally exposed to dust containing a mixture of vanadium pentoxide, sodium, and ammonium metavanadate. These men were exposed to vanadium-bearing dust concentrations ranging from 3.6 to 25.1 mg/m^3, 95% of which consisted of particles below 5 µm. According to Roshchin (44), acute vanadium effects can be divided into "mild," "moderate," and "severe" forms.

The clinical features of mild toxicity include rhinitis with a profuse and often bloody discharge, sneezing, and an itching and burning sensation in the throat. The rhinitis may be followed by the development of a dry cough with expectoration of small amounts of viscid sputum, general weakness, and exhaustion. Conjunctivitis is frequently observed. Other symptoms may include diarrhea.

In moderate toxicity, in addition to conjunctivitis and irritation of the upper respiratory tract, there is bronchitis with expiratory dyspnea and bronchospasm. There are frequent disturbances in the activity of the GI tract, including vomiting and diarrhea. Some affected persons have cutaneous manifestations of toxicity in the form of rashes and eczema with itching papules and dry patches.

Severe toxic effects include bronchitis and bronchopneumonia. Other symptoms may also be more prominent, such as headache, vomiting, diarrhea, palpitations, sweating, and general weakness. Disorders of the nervous system include severe neurotic states and tremor of the fingers and hands.

Pronounced reductions in forced vital capacity (mean 0.5 L), forced expiratory volume (mean 0.5 L), and forced midexpiratory flow (mean 1.16 L/s) occurred in 17 men within 24 h after exposure to dust during cleaning of power station boilers, and had not returned to preexposure levels by the eight day (151). The men were exposed to a time-weighted average (TWA) respirable dust (2% of particles < 10 µm) of 523 µg/m^3, containing 15.3% vanadium (total dust concentration ~26 mg/m^3). Sixteen of the men wore respirators, which subsequently were found to have peak leakages of up to 9%, while one volunteer had a one-hour exposure wearing only a compressed paper oronasal mask. Four weeks after exposure no residual deficits were present. A urinary vanadium concentration of 280 μg/L was found in the volunteer, but none of the others had concentrations above the test threshold of 40 μg/L.

Zenz and Berg (152) studied the effects of vanadium exposure in nine healthy volunteers aged 27–44 years, for whom baseline lung function data were available. Two volunteers were inadvertently exposed to 1 mg/m^3 of vanadium pentoxide dust for an 8-h period. Some sporadic coughing developed after the fifth hour and was believed to be psychologic at the time. Near the end of the seventh hour, more frequent coughing developed in both subjects. By evening, persistent coughing began and remained for 8 days. There were no other signs of irritation. Lung function tests performed immediately following exposure and repeated once weekly for 3 weeks revealed no differences. There were no alterations of the white blood cell counts, differential cell patterns, or urinalyses. Three weeks after the original exposure, those same two volunteers were accidentally reexposed for a 5-min period to a heavy cloud of vanadium pentoxide dust. Within 16 h, marked coughing and sputum production developed. The following day, rales and expiratory wheezes were present throughout the entire lung field. However, pulmonary functions were normal. After administration of a therapeutic dose of isoproterenol (1 : 2000) under positive pressure, coughing was relieved for about one hour but then resumed and continued for one week. Five volunteers were exposed for an 8-h period to vanadium pentoxide at an average concentration of 0.2 mg/m^3 (> 98% of dust particles was < 5 μm). Even at this concentration, all men developed a loose cough the following morning. Physical examination were nonrevealing. Subjects stopped coughing after 7–10 days. Spirometry was repeated at the end of 2 weeks without detectable changes from preexposure studies. The greatest amount of vanadium was found in the urine 3 days after exposure (0.13 mg/L), and none was detectable one week following exposure. Maximum fecal vanadium was 0.003 mg/g. None was found after 2 weeks. Two new volunteers not previously exposed to vanadium pentoxide were exposed to this compound at a concentration of 0.1 mg/m^3 for 8 h. No symptoms occurred during or immediately after exposure. However, within 24 h production of mucus increased. This mucus was easily cleared by slight coughing, which increased after 48 h, subsided within 72 h, and disappeared completely after 4 days. Pulmonary function tests and differential white blood counts remained normal in all exposed volunteers.

Pazhynich (153) studied the irritant effects of vanadium pentoxide condensation aerosol on 11 volunteers. At a concentration of 0.4 mg/m^3, all reported a tickling or itching sensation and a feeling of dryness in the region of the root of the tongue, the posterior wall of the pharynx, and the fauces, as well as a slight prickling sensation in the nose and posterior pharyngeal wall. These symptoms were easily tolerated. A concentration of 0.16 mg/m^3 caused mild signs of irritation in only five volunteers, and a concentration of 0.08 mg/m^3 was not noticed by any volunteer. It was concluded that the mean perceptible concentration for human beings is 0.27 mg/m^3 and that 0.16 mg/m^3 is imperceptible.

Todaro et al. (154) described an episode of acute intoxication due to inhalation of vanadium-containing dust (concentrations and time of exposure not specified) in a group of 10 workers during maintainance work inside a boiler of an oil-fired electricity power station. Irritative symptoms of the upper respiratory tract, green tongue (in 6 out of 10 subjects), and the mean vanadium values in urine of 92 µg/L were observed. Two weeks after the episode there was a complete remission of the symptoms and urinary vanadium. Checks made after 6 months and 1 and 2 years later did not reveal any alterations in the general blood chemistry parameters and the urinary vanadium.

1.4.2.2.2 Chronic and Subchronic Toxicity. Chronic vanadium exposure results mainly in pathological changes in the lung. Workers who were exposed for a long time to vanadium complained of coughing; eye, nose, and throat irritation; breathing difficulties during physical exertion; and headache. Clinical findings often revealed intense hyperemia of the nasal, laryngeal, and pharyngeal mucosa, rhinitis (of a simple catarrhal form but also of a hypertrophic, subatrophic, or atrophic form). Sometimes epistaxis was present. The main respiratory diseases diagnosed were chronic bronchitis with or without emphysema and diffuse pneumosclerosis. The changes in the lung were often accompanied by cardiovascular disturbances (arrhythmia, bradycardia, unspecific ECG changes). Biochemical disturbances such as hypervitaminosis with dysproteinaemia and an increase in the serum concentration of sulfhydryl groups were observed. A statistically significant increase in the incidence of enlarged liver with a decrease in functional tests were seen in exposed workers. Systemic effects, such as a tendency toward anemia and leukopenia, and basophilic granulation of leukocytes, have been reported (155).

In another study 36 men of aged 20–50 years employed in a metallurgical plant producing vanadium pentoxide were exposed to vanadium dust particles that were relatively large in size (39% < 12 µm, 22% < 8 µm) (42). It was estimated that a concentration of 6.5 µg V_2O_5/m^3 represented the worst exposure conditions. Of these workers, 22 reported a dry cough, wheezing sounds could be detected in 31, and 27 were short of breath. One man developed acute pneumonitis, and 4 others developed bronchopneumonia. There was no evidence of systemic toxicity. A dry eczematous dermatitis developed in 9 men, but only 1 man showed a positive patch test.

Lewis (40) studied 24 male workers in an environment in which the maximum exposure was 0.925 mg vanadium (as $V_2O_5)/m^3$ of air. In most cases, the exposure was to 0.3 mg vanadium/m^3. More than 92% of the dust particles were smaller than 0.5 µg in every process area sampled. Symptoms of cough with sputum production, eye, nose, throat irritation, and wheezing were related to physical findings of wheezes, rales, or rhonchi, injected pharynx, and green tongue. All of these symptoms and physical findings were statistically significant compared to those in 45 referents.

Characteristic features of vanadium poisoning in 53 workers performing emergency repair work on oil-fired power station boilers exposed to vanadium pentoxide (1.2–11 mg/m^3) (but also to manganese, calcium, and nickel oxides, and sulfur compounds) for ~380–600 hours/year included upper respiratory tract irritation symptoms: rhinitis, pharyngeal catarrh, laryngitis, and changes in the paranasal sinuses in 45%, and increased lung markings in 24.5% by X-ray analysis. Bradycardia was observed in 22% of cases with no other changes in electrocardiograms (156).

It is suggested that exposure to vanadium increases bronchial responsiveness even without clinical appearance of bronchial symptoms (157). In a group of 11 workers exposed to vanadium pentoxide during periodical works in boilers of an oil-fired power station, none had symptoms of bronchial inflammation or significant airway obstruction. However, bronchial responsiveness, investigated using a metacholine challenge test, was significantly higher in the exposed group compared to the control group. It was shown that human volunteers (assumed body weight 70 kg) given 0.47–1.3 mg vanadium/kg as ammonium vanadyl capsules for 45–68 days had intestinal cramping and diarrhea. They had no hematological abnormalities (WBC, differential count, platelets, and reticulocytes), changes in urinanalysis, or changes in serum glutamic oxaloacetic transferase, cholesterol, triglyceride, or phospholipid levels (158).

1.4.2.2.3 Pharmacokinetics, Metabolism, and Mechanisms. Several occupational studies indicate that absorption can occur in humans following inhalation exposure. An increase in urinary vanadium levels was found in workers exposed to less than 1 ppm of vanadium (40, 159–161). The vanadium concentration in serum was also reported to be higher following exposure to vanadium pentoxide dust than in the nonoccupationally exposed controls (160). The rate and extent of vanadium absorption in humans is not known; however, it has been estimated that about 25% of soluble vanadium compounds may be absorbed via the respiratory tract. There is also a possibility that oral exposure (mucociliary clearance) contributed to vanadium levels in the serum. No systemic toxic effects were observed in volunteers who consumed vanadium as ammonium vanadyl tartrate in capsules, suggesting that it may be poorly absorbed (158).

No specific studies were located regarding absorption in humans after dermal exposure to vanadium, although absorption by this route is generally considered to be very low (17). Vanadium is a metal; therefore, absorption through the skin is thought to be quite minimal because of its low solubility.

No data have been located regarding the distribution of vanadium in humans immediately following exposure. At autopsy, vanadium has been detected in the lung (in 52% of the cases) and intestine (in 16% of the cases) of humans with no known occupational exposure (38). This is probably accumulation from chronic breathing of vanadium from naturally occurring dusts or air contaminated with fuel oil combustion waste products. The amount detected in the intestines probably resulted from swallowing the dusts. The heart, aorta, brain, kidney, muscle, ovary, and testis were found to have no detectable vanadium concentrations. Bone was not tested. The study was of limited value because exposure levels were not determined and insensitive detection methods were used. Serum vanadium levels in occupationally exposed workers were highest within a day after exposure followed by a rapid decline in levels on cessation of exposure (159, 160). Analytical studies have shown low levels of vanadium in human kidneys and liver, with even less in brain, heart, and milk. Higher levels were detected in hair, bone, and teeth (22).

No studies were located regarding distribution in humans after oral exposure to vanadium. It is known that vanadium in the plasma can exist in a bound or unbound form (66). Vanadium as vanadyl or vanadate (68) reversibly binds to human serum transferrin. No studies were located regarding excretion in humans after oral or dermal exposure to vanadium.

1.4.2.2.4 Reproductive and Developmental. Autopsy data have not provided detectable levels of vanadium in human reproductive organs. It is unlikely that the reproductive system is a sensitive indicator for vanadium toxicity in humans.

1.4.2.2.5 Carcinogenesis. No studies regarding the carcinogenicity of vanadium in humans were located. Workers who have been exposed to vanadium dust did not show an increased number of cancer deaths (42, 161, 162), although detailed studies were not performed.

1.4.2.2.6 Genetic and Related Cellular Effects Studies. The results of studies on the effects of vanadium compounds on human leukocytes present somewhat contradictory data. McLean et al. (163) did not report any genotoxic effects of vanadyl chloride at a concentration of 5×10^{-5} M on human peripheral white blood cells. Similarly, in an *in vitro* study on human peripheral lymphocyte cultures with vanadium pentoxide concentrations of 0.047, 0.47, or 4.7 M, no increases in the frequency of sister chromatid exchange were observed (120).

However, other data indicate that vanadium has the potential for genotoxicity in humans. Human leukocytes have been shown to have DNA strand breaks from exposure to vanadate (164). Sodium metavanadate, ammonium metavanadate, sodium ortovanadate, and vanadyl sulfate were not found to increase the frequency of structural chromosome aberrations, whereas a significant increase in numerical aberrations, micronuclei, and satellite associations was found. Fluorescence *in situ* hybridization (FISH) applied to the human lymphocyte micronucleus assay, by means of an alphoid centromere-specific DNA probe, confirmed the aneuploidogenic potentiality of vanadium (165). The induction of aneuploid cells by vanadium pentoxide (concentrations ranging from 0.001 to 0.1 μM) was investigated using FISH with chromosome-specific probes by Ramirez et al. (166). V_2O_5-treated cells from different donors exhibited substantial interindividual variability in hyperdiploid frequencies. The maximal induction of hyperdiploidy was approximately 2–4% for chromosomes 7 and 1 in lymphocytes from one donor, whereas the maximal induction was only approximately 1% for the same chromosomes in cells from a different donor. In spite of this variability, dose-dependent trends in hyperdiploid frequency were observed in the lymphocyte cultures from all donors for both chromosomes with increasing concentrations of V_2O_5. This effect may occur through a disruption of microtubule function.

Rojas et al. (167) evaluated the genotoxicity of vanadium pentoxide (3000, 30, and 0.3 μM) directly in whole-blood leukocytes and in human lymphocyte cultures using the single-cell gel electrophoresis assay (comet assay). This chemical produced a clear dose–response relationship in DNA migration in whole-blood leukocytes and a significant positive effect only with the highest tested concentration in human lymphocyte cultures. These results indicate that V_2O_5 is capable of inducing DNA single-strand breaks and/or alkali-labile damage.

1.4.2.3 Epidemiology Studies

1.4.2.3.1 Acute Toxicity. Rajner (168) has investigated 30 vanadium workers in a metallurgical plant. He observed that when a new production process was introduced, the symptoms of acute vanadium poisoning developed in three workers after 16 h of work, including severe respiratory difficulties, headache, dejection, and loss of appetite. Acute inflammatory changes of the upper respiratory tract with copious mucous production, edema of the vocal cords, and profuse epistaxes were reported. In these acutely poisoned workers, vanadium values were about 4000 µg/L urine.

1.4.2.3.2 Chronic and Subchronic Toxicity. A report by Rajner (168) on 30 vanadium workers in a metallurgical plant described particularly severe signs and symptoms. All workers who had been exposed for a long time (up to 22 years in 27 subjects, mostly in ferrovanadium and vanadium pentoxide smelting operations) complained of coughing and eye, nose, and throat irritation, breathing difficulties during physical exertion ("more than two-thirds of the workers"), and headache (12 cases). Clinical findings included intense hyperemia of the mucosa of the nasal septum in 20 workers; perforation of the nasal septum was seen in four workers exposed for an average of 18 years. Intense hyperemia of the mucosa of the throat and larynx with dilated fine capillaries was found in 50% of the workers. Bronchoscopy indicated the presence of chronic bronchitis, and bronchial smears revealed sloughed epithelium. The average value of vanadium in urine among permanent employees was 45 µg/L; vanadium pentoxide smelter workers had maximum values of about 400 µg/L.

Matantseva (169) studied 77 workers in contact with vanadium pentoxide as dust and fume at concentrations exceeding the MAC value (dust = 0.5 mg/m^3; fumes = 0.1 mg/m^3) for periods ranging from 1 to 12 years. Nearly all the subjects had various complaints relating to the upper respiratory tract, including unpleasant sensations in the nose; a liquid mucous discharge from the nose; obstructed nasal breathing; a sensation of burning and dryness in the nasopharynx; scratching, dryness, and tickling in the throat; hoarseness of the voice; and cough. Physical examination showed rhinitis, which was of a simple catarrhal form in workers exposed for less than 3 years, a hypertrophic and subatrophic form if the exposure was for more than 3 years, and an atrophic form if the exposure was between 7 and 12 years. Examination of the lungs revealed acute and chronic lesions in the form of bronchitis, peribronchitis, and pneumosclerosis. Hyperventilation and an elevated basal metabolic rate were noted.

In a study by Roshchin (170) on the effects of exposure to vanadium-containing Bessemer slag dust (concentrations in air 5–150 mg/m^3) on 45 workers, subatrophic rhinitis, bronchitis, and pneumosclerosis were seen in subjects with longterm occupational exposure (11 workers). Chronic bronchitis was found in every worker employed for 5 years or more. Clinical and X-ray examination of all 45 subjects showed radiological changes in 24 employed for 10 years or more; in 11 subjects, pneumoconiosis of stage I–II was diagnosed. Changes in the lung were accompanied by changes in the cardiovascular and nervous systems, biochemical disturbances (hypervitaminosis with dysproteinemia and an increase in the serum concentration of sulfhydryl groups), a tendency to anaemia and leukopenia, and biochemical markers of liver function.

In another study, Roshchin (171) described chronic effects of vanadium in 193 workers who had been exposed to aerosols of free vanadium pentoxide: 127 worked in vanadium metallurgy and 66 were power plant boiler cleaners. The length of occupational contact with vanadium was > 10 years for 60%, 5–10 years for 30%, and < 5 years for the remaining 10%. Practically all workers complained of irritation of the nasal and pharyngeal mucosa, including itching, a profusely running nose (especially during work), and unpleasant sensations in the throat and nose. Epistaxis was frequent in 20%. Physical examination revealed a high incidence of changes in the nasal mucosa: dryness (40%), erosion (23%), scars (8%), perforation (4%), hyperemia (10%), and hypertrophy (7%). Also noted were dryness of the pharynx (5%), hyperemia of the pharynx (5%), hyperemia of the larynx (4%), and tonsillitis (5%). The most common pathological changes in the upper respiratory tract were subatrophic rhinitis (40%) and destructive changes in nasal mucosa (35%); hypertrophic rhinitis was seen less frequently (7%). The overwhelming majority had a dry cough; cough with viscid sputum was less common. Workers with longer occupational exposure complained of shortness of breath, which appeared sometimes after 5 but mostly after 10 years of work in the industry. Almost all workers complained of aching or shooting pains in the chest and of lassitude and weakness. The main respiratory diseases diagnosed were chronic bronchitis (40%) and diffuse pneumosclerosis (13%). Hematological tests showed γ-globulins to be raised (19.4% compared with 12.2% in controls), and the albumin:globulin ratio to be 1:1–1:2 (1:9 in controls). Regular observations over a period of 14 years showed that the chronic bronchitis tended to get worse, with development of bronchospasm. After a long period of time, some subjects developed pneumosclerosis; in others, the disease progressed slowly from chronic bronchitis to diffuse pneumosclerosis and pulmonary emphysema.

Holzhauer and Schaller (172) examined 121 chimney sweepers with an average exposure duration of 19 years (± 5 years). Vanadium exposure as determined by personal sample measurements was between 0.73 and 13.7 mg vanadium pentoxide/day (4 µg in the general population). Urinary excretion was determined to be between 0.15 and 13 µg/L, which was significantly higher than the values in 31 referents. The main complaints of the subjects were wheezing, rhinitis, conjunctival irritation, cough with sputum, dyspnea, and hoarseness; there were no skin symptoms.

Kiviluoto et al. (160,173–175) and Kiviluoto (176) reported the results of a cross-sectional study on 63 males exposed to vanadium-containing dust in a vanadium factory; a reference group matched for age and smoking was selected from a magnetite ore mine. The workers had been exposed to vanadium dust for an average of 11 years at concentrations ranging from 0.1 to 3.9 mg/m^3 (estimated average exposure levels of 0.2–0.5 mg/m^3); the respirable fraction (< 5 µm) was 20%. Nasal biopsies and lung function tests were taken at the end of the summer holidays (2–4 weeks). Nasal smears and biopsies were repeated in 31 workers, 7–11 months later, after hygienic improvements had reduced the exposure levels to 0.01–0.04 mg/m^3. Microscopic examination of nasal smears revealed a significant increase in neutrophils, and the biopsy of nasal mucosa showed significantly elevated numbers of plasma and round cells in the exposed workers. There was no further change in the cell findings after 10 months of exposure to 0.01–0.04 mg/m^3 vanadium dust; eosinophils did not show any differences between the exposed and the referents. The authors attributed these findings to "an irritating effect of vanadium dust on the mucous membranes of the upper respiratory tract." Biopsies from workers with the longest exposures (170–241 months) showed "a zone-like sub-epithelial infiltration of mononuclear cells and frequent papillarity in the mucous membrane surface with its hyperaemic capillaries." A similar pattern was seen in vanadium-exposed rabbits (42). A random sample of 12 nasal biopsies was further investigated for the amount and classes of immunoglobulins (IgE, IgG, IgM, and IgD). IgG subclasses were not studied. There were no differences between the 12 workers and their referents (160). Pulmonary condition was assessed by means of questionnaires, X-ray, and pulmonary function testing. The only one significant difference between the workers exposed for an average of 11 years to 0.1–3.9 mg/m^3 (estimated average, 0.2–0.5 mg/m^3) and at the time of investigation to 0.01–0.04 mg/m^3, and their matched referents were more common complaints of wheezing in the exposed worker group (176). The importance of this finding remains doubtful. It may reflect the respiratory findings mentioned above, since upper respiratory irritation may be accompanied by transient reflex bronchospasm. A series of laboratory tests designed to evaluate various clinical parameters showed no differences between the worker and control groups (175).

1.4.2.3.3 Pharmacokinetics, Metabolism, and Mechanisms. Occupational studies showed significantly increased urianary vanadium levels in the exposed workers (40, 159–161, 177). Male and female workers exposed to 0.1–0.19 mg/m^3 vanadium in a manufacturing company had significantly higher urinary levels (20.6 g/L) than did the nonoccupationally exposed control subjects (2.7 g/L) (161). The correlation between ambient vanadium levels and urinary vanadium is difficult to determine from these epidemiological studies (158). In most instances, no other excretion routes were monitored. Analytical studies have shown very low levels in human milk (22). Evidence from animal studies supports the occupational findings (see Section 1.4.1.3). Epidemiological studies and animal studies suggest that elimination of vanadium following inhalation exposure is primarily in the urine.

1.4.2.3.4 Reproductive and Developmental. No studies were found in the literature.

1.4.2.3.5 Carcinogenesis. Epidemiological studies on the effects of vanadium on human health were published using a correlational approach, which has well-known limitations, although it imitates a population-based cohort study. Stock (178) reported the results of a study in which airborne concentrations of 13 trace elements were correlated with mortality from lung cancer, pneumonia, and bronchitis in 23 localities in the United Kingdom. At concentrations ranging from 1.1 to 42 µg/1000 m^3, vanadium showed a weak association with mortality from lung cancer (considering the population density, sex, and age), with a correlation coefficient of 0.347. Airborne vanadium levels were also correlated with mortality from pneumonia in males, with a correlation coefficient for mortality from pneumonia of 0.443. For mortality involving bronchitis, vanadium gave a correlation coefficient of 0.563. Vanadium also showed an association with mortality from cancers other than lung cancer in males, but not in females. However, in this study, as usually in the studies of this kind, it is not certain that cases of interest (lung cancer, pneumonia) had been exposed at all. There are also the uncertainties of mortality data and failure to consider confounding factors.

In another study, Hickey et al. (179) considered 10 metals in the air, including vanadium, in 25 communities in the United States. Various techniques, including canonical analysis, were used to correlate airborne metal concentrations with mortality indices for 1962 and 1963 involving eight disease categories. The mean atmospheric concentrations for vanadium at various locations ranged from 0.001 to 0.672 µg/m^3. The incidence of several diseases, including "diseases of the heart," nephritis, and "arteriosclerotic heart," could be correlated reasonably well with air levels of vanadium and other metals. A high correlation between vanadium and nickel was unexplained.

1.4.2.3.6 Genetic and Related Cellular Effects Studies. No studies were found in the literature.

1.4.2.3.7 Other: Neurological, Pulmonary, Skin Sensitization, etc. Most of statistical studies point to negative correlations between environmental levels of vanadium and certain other trace elements and the incidence of cardiovascular disease.

Schroeder (180) reported a significant negative correlation between the vanadium content of municipal waters and the death rates due to arteriosclerotic heart disease. In a study by Voors (181) on the correlation between seven metals (calcium, chromium, lithium, zinc, manganese, nickel, vanadium) and arteriosclerotic heart disease, a low vanadium intake was associated significantly with a higher incidence of arteriosclerotic heart disease in nonwhite populations, but no direct correlation was demonstrated for white populations. In a joint WHO and IAEA study on the role of trace elements in the etiology of cardiovascular diseases in 20 countries, a significant role was shown for environmental lack of vanadium as well as chromium, zinc, manganese, calcium, and magnesium (182). Conversely, Hickey et al. (179) noted a positive correlation between airborne vanadium levels and the incidence of cardiovascular disease.

1.5 Standards, Regulations, or Guidelines of Exposure

The current occupational exposure limits expressed as time-weighted averages (TWAs) for vanadium are similar in most countries and established as 0.05 mg/m^3. They are based on the determination of vanadium pentoxide [*1314-62-1*] concentrations as fumes and/or dusts. In some countries, as a result of irritative properties of vanadium and its compounds, short-term limits of exposure were also established. Table 37.6 summarizes the main current regulations and guidelines.

1.6 Studies on Environmental Impact

Natural sources of atmospheric vanadium include continental dust, marine aerosol, and volcanic emissions (4, 18, 19). The quantities entering the atmosphere from each of these sources are uncertain; however, continental dust is believed to account for the largest portion of naturally emitted atmospheric vanadium followed by marine aerosols. Contributions from volcanic emissions are believed to be negligible when compared with the other two sources (18).

Anthropogenic releases of vanadium to the air account for approximately two-thirds of all vanadium emissions (18).

The global biogeochemical cycling of vanadium is characterized by releases to the atmosphere, water, and land by natural and anthropogenic sources, long-range

Table 37.6. Regulations and Guidelines for Exposure to Vanadium and Its Compounds[a]

	OSHA PEL	ACGIH TLV	German Standard	Sweden Standard	British Standard	Polish Standard	WHO	EPA	NIOSH
Vanadium and Compounds (Measured as Vanadium Pentoxide)									
TWA (mg/m^3)	0.05 respirable; 0.1 as fume	0.05 (respirable dust and fume)	0.05	0.2 (total dust)	0.5 (total dust); 0.05 (fume and respirable dust)	0.5	1 µg/m^3 (air quality guideline for vanadium (24 h))		
STEL (mg/m^3)				0.05 (respirable dust)					3
REL (mg/m^3; 15 min ceiling limit for the total dust and fume for vanadium compounds except the metal and vanadium carbide)									0.05
IDLH (mg/m^3)									70 (dust and fume)
RFD								9×10^{-3} mg/kg per day (oral)	
Ferrovanadium Dust									
TWA (mg/m^3)	1	1	1						1
STEL (mg/m^3)	3								3

[a]*Abbreviations*: ACGIH — American Conference of Governmental Industrial Hygienists; OSHA- Occupational Safety and Health Administration; NIOSH — National Institute for Occupational Safety and Health; TWA — Time-weighted average; REL — recommended exposure limit; STEL — short term exposure limit; PEL — permissible exposure limit; TLV — threshold limit value; IDLH — immediately dangerous to life or health level; RfD — reference dose.

transportation of particles in both air and water, wet and dry deposition, adsorption, and complexing. Vanadium generally enters the atmosphere as an aerosol. From natural sources, vanadium is probably in the form of mineral particles; it has been suggested that these may frequently be in the less soluble trivalent form (4, 18). From synthetic (human-made) sources almost all the vanadium released to the atmosphere is in the form of simple or complex vanadium oxides (4). The transport and partitioning of vanadium in water and soil is influenced by pH, redox potential, and the presence of particulates. In freshwater, vanadium generally exists in solution as the vanadyl ion (V^{4+}) under reducing conditions; and the vanadate ion (V^{5+}), under oxidizing conditions, or as an integral part of, or adsorbed onto, particulate matter (183). Thus, vanadium is transported in water in one of two ways: solution or suspension. It has been estimated that only 13% is transported in solution, while the remaining 87% is in suspension (17).

Natural sources of vanadium release to water include wet and dry deposition, soil erosion, and leaching from rocks and soils. The largest amount of vanadium release occurs naturally through water erosion of land surfaces. It has been estimated that approximately 32,300 tons of vanadium are dissolved and transported to the oceans by water, and an additional 308,650 tons are thought to be transported in the form of particulate and suspended sediment (19).

Anthropogenic releases to water and sediments are far smaller than natural sources (19). Such sources of vanadium in water may include leaching from the residue of ores and clays, vanadium-enriched slags, urban sewage sludge, and certain fertilizers, all of which are subjected to rain and groundwater drainage, as well as leachate from ash ponds and coal preparation wastes (4, 19). Leaching may potentially occur from landfills and from the airborne particulate matter that is deposited in areas with high residual fuel-oil combustion, although neither of these release sources is documented.

On entering the ocean, vanadium in suspension or sorbed onto particulate is deposited on the seabed (17). The fate of the remaining dissolved vanadium is more complex. Only about 0.001% of vanadium entering the oceans is estimated to persist in a soluble form (4). Some marine organisms, in particular the ascidians (sea squirts), bioconcentrate vanadium very efficiently, attaining body concentrations approximately 10,000 times greater than the ambient seawater (4).

In general, marine plants and invertebrates contain higher levels of vanadium than terrestrial plants and animals. No data are available regarding biomagnification of vanadium within the food chain, but human studies suggest that it is unlikely; most of the 1–2% vanadium that appears to be absorbed by humans following ingestion is rapidly excreted in the urine with no evidence of long-term accumulation (184). Biotransformation is not considered to be an important environmental fate process.

Natural releases of vanadium to soil result from weathering of rock-bearing vanadium minerals, precipitation of vanadium particulate from the atmosphere, deposition of suspended particulate from water, and plant and animal wastes. The largest amount of vanadium released to soil occurs through the natural weathering of geologic formations (4, 18). Anthropogenic releases of vanadium to soil are less widespread than are natural releases and occur on a smaller scale. These include the use of certain fertilizers containing materials with a high vanadium content such as rock phosphate (10–1000 mg/kg vanadium), superphosphate (50–2000 mg/kg vanadium), and basic slag (1000–5000 mg/kg vanadium) (19) as well as disposal of industrial wastes such as slag heaps and mine tailings.

2.0 Niobium

2.0.1 CAS Number: [7440-03-1]

2.0.2 Synonyms: Columbium

2.0.3 Trade Names: VN1

2.0.4 Molecular Weight: 92.91

2.0.5 Molecular Formula: Nb

2.1 Chemical and Physical Properties

2.1.1 General

Niobium was discovered by Charles Hatchett in 1801 and isolated by Christian Blomstrand of Sweden in 1964. Its name was given after the Greek mythological figure Niobe, the daughter of Tantalos; tantalum always was associated with niobium. For many years the terms "niobium" and "columbium" were used interchangeably; however, the name "niobium" was officially adopted by International Union of Pure and Applied Chemistry (IUPAC) in 1950. Niobium is not a very rare element; its crustal abundance is 24 ppm, which is similar, or greater than, those of many common elements, such as lead or cobalt.

The physical and chemical properties of niobium and a few of its compounds are listed in Table 37.7.

Niobium is a shiny white, soft, and malleable metal. The element is inert to HCl, HN0$_3$, or aqua regia at room temperature, slightly soluble in HF, but is attacked by alkali hydroxides or oxidizing agents at all temperatures. In pure form niobium is ductile, unless it is allowed to associate at elevated temperatures with common gases: N_2, H_2, or O_2. Thus, when processed, it must be placed in a protective environment. Niobium, as can other metals from group V, can assume valences of +1, +2, +3, +4, but +5 is the most common and stable state. Numerous forms of niobium occur, including pentachloride, pentafluoride, pentaoxide, and potassium oxypentafluride salts. One natural isotope of niobium exists: ^{93}Nb; numerous artificial isotopes include $^{88-92}$Nb and $^{94-101}$Nb. One of the isotopes, namely, ^{95}Ni, exists in the transient equilibrium with ^{95}Zr, which is formed as a result of neutron activation of zirconium-based cladding and as a fission product in a nuclear power reactor. ^{95}Zr decays by beta emission to ^{95m}Nb, which in turn decays by internal transition to ^{95}Nb. It has a physical half-life of 35.15 days; it also decays by beta emission, and characteristic gamma emission of 765.8 keV (185). Pure niobium is oxidized rapidly in high temperatures; but its oxidation resistance can be increased by alloying. The principal use is that of ferroniobium used to alloy carbon and stainless steel, nonferrous metals, and metals used in arc-welding rods. The addition of niobium to extra-low carbon steels increases their strength, cryogenic ductility, and formability. Resistance to thermal shock is also improved. Elemental niobium does not occur in nature, but its

Table 37.7. Chemical Identity of Niobium and Its Compounds

Compound	CAS Number	Molecular Formula	Molecular Weight	Physical State	Melting point (°C)	Boiling Point (°C)	Specific Gravity	Solubility in Water	Vapor Pressure	Flammability
Niobium	[7440-03-7]	Nb	92.91	Gray lustrous metal	2468	4727	8.57	Insoluble in water	—	In the form of dust, moderately explosive
Niobium pentaoxide or pentoxide	[1313-96-8]	Nb_2O_5	265.81	White crystals	1520	—	4.47	Insoluble in water; soluble in H_2SO_4, HF, bases	—	—
Niobium dioxide	[12034-59-2]	NbO_2	124.9	Black crystals	1772	—	7.28	Insoluble in water; soluble in acids, bases	—	—
Niobium pentachloride	[10026-12-7]	$NbCl_5$	270.2	Yellow crystals	204.7	250	2.75	Decomposed by water; soluble in HNO_3, CCl_4, ethanol	—	—
Niobium pentafluoride	[7783-68-8]	NbF_5	187.9	Strongly refractive deliquescent crystals	80	234.9	2.7	Hydrolize in water	—	—
Niobium carbide	[12069-94-2]	NbC	104.9	Dark blue luminous needle crystals	3500	—	7.6	Insoluble in water	—	—
Niobium potassium fluoride	[17523-77-2]	F_5NbO_2K	282.1	Lustrous leaflets	—	—	—	Soluble in hot water, soluble in cold water 1/13	—	—
Potassium niobate	[12030-85-2]	$KNbO_3$	179.99	—	—	—	—	Soluble in water	—	—
Niobium hydride	[12655-93-5]	NbH	93.91	—	infusable	—	6.6	Insoluble in water	—	—

39

compounds exist in different mineral ores, associated with tantalum as the oxide. The most important commercial sources are columbite (niobite, or tantalite), $(Fe,Mn)(Nb,Ta)_2O_6$; tantalocolumbite (niobite–tantalite):$(Fe,Mn)(Ta,Nb)_2O_6$; and pyrochlore, $NaCaNb_2O_6F$. The ores are found primarily in Brazil, Canada, Zaire, Malaysia, Nigeria, and the former Soviet Union. Pyrochlore represents the major source of known world reserves of niobium (186).

2.1.2 Odor and Warning Properties

No data were found in the literature; however, it might be assumed that niobium and most of their compounds are odorless. Niobium pentachloride ($NbCl_5$) have pungent odor, because it decomposes slowly when heated, with Cl_2 formation. Niobium in the form of dust is moderately explosive, when exposed to flame, or by chemical reaction.

2.2 Production and Use

The extracting and refining process for niobium from ore is extremely complex and consist of a series of operations, starting from upgrading the ores by concentration. Disruption of the niobium-containing matrix is performed then by an ore-opening procedure with hot HF or fusion with alkali fluxes. The next steps include pure niobium compound preparation and reduction to metalic niobium, followed by refining, consolidation, and fabrication of the metal. Niobium is so closely associated with tantalum that they must be separated either by fractional crystallization or by solvent extraction before purification (187). For other details, see discussion of tantalum in Section 3.2.

Niobium physicochemical characteristics lead to numerous industrial applications, especially in alloys. Niobium is added to carbon and stainless steel to form ferroniobium. These alloys provide greater strength, cryogenic ductility, and formability; thus have been used in pipeline construction. The niobium–germanium and niobium–titanium alloys are superconductive up to 23K and are used in special wire for superconducting magnets to obtain high magnetic fields for use in communications and containment of thermonuclear fusion plasmas. In addition, niobium in alloys has numerous biomedical applications in different alloys for prostheses and implants, including titanium–aluminum–niobium and titanium–zirconium–niobium–tantalum–palladium. Radioactive niobium microspheres are used in experimental studies of blood flow.

2.3 Exposure Assessment

Very few reports deal with analysis of niobium in different media. Industrial exposures to niobium are rather limited.

2.3.1 Air

No data have been found in the literature concerning ambient air.

2.3.2 Background Levels

Niobium levels in terrestrial plants of less than 0.4 mg/kg dry weight were reported. However, plants located near niobium deposits demonstrated a marked capacity to accumulate and concentrate this element to levels higher than 1 mg/kg (188). Niobium occurs in seawater in the form of $Nb(OH)_6^-$ at approximately 0.005 ppb (189). Marine plants, such as seaweeds, appear to concentrate niobium from seawater contaminated with nuclear fallout (190).

2.3.3 Workplace Methods

A spectrochemical procedure following extraction in hexanone was employed for measuring workplace niobium levels. Niobium concentrations are estimated colorimetrically, using hydroquinone in concentrated sulfuric acid. It was shown that an average niobium recovery of 99.9% can be obtained from solutions of known amounts of niobium and various other metals. Of 31 metallic ions tested, only tantalum and large amounts of molybdenum interfered seriously (191).

The ICP-AES method is useful for monitoring the blood of workers exposed to niobium as well as several other metals simultaneously (NIOSH method 8005) (26), even if niobium is not listed in the original NIOSH manual.

2.3.4 Community Methods

Determination of niobium in water samples with limits of detection in the 0.1–0.2-ppb range has been achieved with high dispersion spectrography, inductively coupled plasma atomic emission spectroscopy, and X-ray emission spectroscopy (192).

2.3.5 Biomonitoring/Biomarkers

2.3.5.1 Blood. From a very limited number of analyses of human male tissues, blood serum values of 0.53–0.74 ppm Nb, compared with that in red blood cells of 4.19–6.4 ppm, were reported by Schroeder and Balassa (193). In another study mean blood concentrations of 4–4.7 pg/L were found in healthy humans from United Kingdom (194). However, these scarcely data do not allow for any conclusion on the use of blood niobium concentrations for biomonitoring.

2.3.5.2 Urine. The principal route of excretion of absorbed niobium is through the kidneys. Schroeder and Balassa (193) have reported urinary levels of 0.25 ppm in three young females. However, no data on correlation between urinary niobium levels and oral dose or the element concentrations in air were found.

2.3.5.3 Other. Niobium can be determined in biological tissue photometrically using 3,5,7,2′,4′-pentahydroxyflavone (morin) as a reagent. Elimination of the major interfering ions has been succesfully achieved by CCl_4 extraction with the use of diethyldithiocarbamate (195). Schroeder and Balassa (193) used a colorimetric method with 4-(2-pyridylazo) resorcinol (PAR) as a reagent. Their method had a sensitivity of < 0.04 µg/mL solution of ash. PAR effectively eliminated some 40 interfering ions and the addition of a small amount of zinc supressed the color from tantalum. They found the highest niobium content, 8.8 ppm (wet weight), in the one of investigated liver samples.

It was estimated that total body intake of niobium by healthy, adult humans approximates 600 µg, leading to an averge total body burden of 112 mg Nb (193), but this value may be overestimated (192).

2.4 Toxic Effects

2.4.1 Experimental Studies

2.4.1.1 Acute Toxicity. Acute toxicity of niobium is relatively high once niobium has gained access to the blood stream, but is low via the GI tract, from which it is poorly absorbed (see Section 2.4.1.3.1). Schubert (196) reported that 50 mg/kg of sodium niobate produced unspecific, severe intoxication in the rat. Cochran et al. (197) noted considerably greater toxicity for niobium pentachloride at 14 mg/kg by the intraperitoneal route in rats. However, potassium niobate given by gavage was pratically nontoxic for this species: oral LD_{50} was 3000 mg/kg.

Haley et al. (198) using niobium chloride of high purity, reported the intraperitoneal LD_{50} for male CF strain as 61 mg/kg (56–64 mg/kg); by comparison, the oral LD_{50} was 940 mg/kg (930–950 mg/kg). The observed symptoms of toxicity were urination, defecation, a milky exudate from the anus, abdominal stretching with cavation of the lower abdominal area, decreased respiration, and lethargy; the first deaths occurred at 48 h and then variably depending on dosage until day 7 postexposure. The acute toxicity of potassium niobate and niobium pentachloride was investigated in mice, rats, rabbits, and dogs by ip, iv and oral routes by Downs et al. (199). Single intravenous doses of ≤20 mg Nb/kg to all investigated species except mice were not fatal. The LD_{50} value for mice and rats given a single ip injection of sodium niobate was 13 and 92 mg Nb/kg, respectively; death occurred within 3–5 days postexposure. The oral LD_{50} value for the rat was 725 mg Nb/kg. Renal injury was observed in all species following parenteral injection at doses of 20–50 mg Nb/kg.

2.4.1.2 Chronic and Subchronic Toxicity. Toxicological effects consisting of body weight and histopathological changes have been studied on rats by Downs et al. (199). Niobium pentachloride was administered intraperitoneally at daily (5 days/week) doses of 10, 30, or 50 mg Nb/kg to female Wistar rats. A total of 21 doses was administered to rats given doses of 10 or 30 mg Nb/kg. There were no effects on body weight of rats injected with a dose of 10 mgNb/kg; a moderate loss of weight was noted at the higher two levels. Increasing mortality occured with increasing dosage; 3/5 (three of five) rats died after 15 injections of 30 mg Nb/kg, and all died or were sacrificed moribund after 12–14 injections of 50 mg/Nb/kg. Renal changes consisting of increased weight, slight epithelial proliferation, and intratubular brown pigmentation resulted from 21 injections of 10 mg Nb/kg, whereas the kidneys of those receiving of 30 or 50 mg Nb/kg showed, in addition, tubular necrosis and regenerative proliferation. The liver, spleen, adrenal glands, lungs, and bone marrow showed no histopathological changes.

Similar effects were seen in male Wistar rats given ip daily injections of potassium niobate at dose of 10, 29, 57, or 95 mg Nb/kg. No rat survived 4 injections of 95 mg Nb/kg, and 2/5 died from 5 doses of 57 mg Nb/kg. All survived the 11 injections of 10 or 29 mg Nb/kg. Renal effects resembling those from niobium pentachloride. Similar renal effects were also observed when niobium pentachloride was given in 3–5 doses by iv injections to rabbits and dogs (199). No effects on growth was noted when weanling Wistar rats were fed dietary levels of niobium pentachloride or potassium niobate as high as 1% for 7 weeks. No lesions were observed in the liver, kidney, or spleen of the rats fed potassium niobate; rats dosed by niobium pentachloride were not subjected to histopathological studies (199).

In another study, in which niobium pentachloride was fed to CFN rats at dietary levels of 0.01, 0.1, or 1% for 12 weeks, the animals did not appear to differ from controls, except for effects on the liver (198). These effects consisted of perinuclear vacuolation of the parenchymal cells and coarse granulation of the cytoplasm. However, it should be noted that the liver effects were less frequently seen at 1% level, than at 0.1 or 0.01%. Sodium niobate was given to Long–Evans rats and Charles River CD mice in drinking water at concentration of 5 ppm over their lifespans. Diets were carefully prepared with added vitamins and ferrous sulfate and assessed for essential components and metals. It was found that mean niobium content in diet was 1.62 ppm contributed mainly by corn oil; the content of zirconium was 2.66 ppm; of vanadium, 3.2 ppm; and of lead, 0.2 ppm (98). Enhanced growth rate in males, but not in females, and surprisingly, reduced longevity among males, expressed as the mean age of the last surviving 10%, was seen. A 22% decrease in serum cholesterol was observed in female rats, but not in males. In a similar study using mice, authors reported a decreased median lifespan and longevity associated with suppression of growth of older animals. A twofold increased incidence of fatty degeneration of the liver was also seen (200).

2.4.1.3 Pharmacokinetics, Metabolism, and Mechanisms

2.4.1.3.1 Absorption. Indirect evidence of absorption after inhalation of niobium in animals is indicated. Inhalation exposure of dogs to ^{95}Nb oxalate and oxide aerosols resulted in 60% absorption of the total dose (201). Two single-exposure inhalation studies on female rats were carried out for studying the absorption, distribution, and excretion with time, using one aerosol containing only ^{95}Nb as a radioactive tracer and another one to which stable ^{93}NbCl$_5$ at high concentrations had been added (202). Rats were nose-only exposed to aerosols produced from a solution of oxalic acid as a carrier for 30 min. This exposure resulted in absorption of 36 and 71% of the total dose of tracer ^{95}Nb and particulate ^{95}Nb, respectively. It was found that the initial deposition in the lung was about twice as great for the tracer as for the carrier (39.7 *vs.* 20.7%); but the rate of loss was faster. Conversely, whole-body deposition of the carrier aerosol was twice as much as tracer (71 *vs.* 36%). These data showed the greater early lung deposition of tracer aerosol as a fraction of total body burden, but also the large difference in the opposite direction at longer times.

Niobium is poorly absorbed to the blood. ^{95}Nb content, expressed as percentage of initial whole-body count, did not exceed 2% within 12 h, was 0.7% on day 1, and was removed by day 17 in rats exposed to tracer aerosol (202).

The absorption of niobium through the GI tract of animals is low. Only a small amount of niobium (~1%) was absorbed from the gut after oral dosage to mice, rats, monkeys, and beagles (203, 204). The behavior of neutron-irradiated, stimulated Chernobyl UO$_2$ particles containing ^{95}Nb as well as ^{141}Ce, ^{144}Ce, ^{95}Zr, and ^{103}Ru in the GI tract were studied (205). None of the radionuclides were detected in liver, kidney, muscle, bone, brain, blood, and urine, indicating that they were not absorbed. Several studies have demonstrated that niobium absorption might be greater in newborn animals than in adults, but variable results have been obtained in different species. Although ^{95}Nb was poorly absorbed following oral administration to adult rats, the nuclide was absorbed to a considerable extent by suckling rats resulting in 10–1000-fold higher whole-body retention than in adults (206). Mraz and Eisele (207) measured ^{95}Nb uptake of about 6 and 4% in newborn and 7-day-old rats, respectively, compared with about 0.1% in 21-day-old weaned rats. Higher values of 40 and 35% were obtained for 1-day-old pigs and sheep, respectively, compared with values of less than 0.5% in weaned animals (208). In the more recent study (209) the GI absorption of ^{95}Nb ingested in milk by adult guinea pigs on the milk-supplemented diet was estimated as 0.8%. A value of 1.4% was obtained for guinea pigs fasted for 24 h before and 2 h after the oral administration. The absorption in 2-day-old animals was estimated as 1.5%. These authors in studies of ligated rat GI tract segments in situ demonstrated a sixfold higher absorption of ^{95}Nb from the stomach compared to duodenal and jejunal segments (210). Because niobium is a metal, absorption through the skin is thought to be quite minimal due to its low solubility.

2.4.1.3.2 Distribution. It was shown in the experiment described above (202) that after inhalatory exposure niobium is distributed to internal organs and pelt. ^{95}Nb activity in five tissues (bone, kidney, spleen, testes and liver) showed increasing values during the first 7 days after intraperitoneal injection to Sprague–Dawley rats (204). Thereafter, activity gradually decreased until day 45 (when last measurements were made), although it was still high in the above-mentioned tissues. Following parenteral administration of ^{95}Nb, 50% of dose was localized in bone and resided in that tissue with an effective half-life of 30 days (211). Mice exposed to 5 ppm sodium niobate in drinking water from weaning until natural death (approximately 2 years) accumulated niobium primarily in spleen > heart > liver > lung > kidney (200). A similar study in rats resulted in tissue accumulation in spleen > kidney > heart > lung > liver (98). Pregnant rats and rabbits had different tissue distribution patterns of ^{95}Nb following iv injection (212). In rats, all maternal tissues exhibited higher concentrations compared to the corresponding fetal organs; the highest fetal/maternal ratio was 0.6 for bone. In rabbits, the fetal/maternal ratio was 3.5 for bone.

The biodistribution of ^{95}Nb was investigated in tumor-bearing rats and mice (213). It was found that ^{95}Nb was more dominant in the connective tissue (especially inflammatory tissue) than in the other categories of tumor tissue. It accumulated rapidly into mitochondrial fraction (containing lysosomes) of the liver, reaching about 50% at 48 h. The main binding substance for the nuclide was the acid mucopolysaccharide whose molecular mass exceeded 40000 daltons. This is in agreement with another study in which niobium pentachloride was injected intraperitoneally at a daily dose of 5 mg/kg twice weekly for 1 month (214). Results demonstrated that niobium was detected with phosphorus in cytoplasm lysosomes of cells of the proximal convoluted tubule after 1 week. In samples taken after 1 month, numerous lysosomes containing dense deposits as well as free deposits were seen in the tubular lumen.

2.4.1.3.3 Excretion. Detailed information on excretion of both stable ^{93}NbCl$_5$, and ^{95}Nb as a radioactive tracer with oxalic acid as a carrier has been reported by Thomas et al. (202), following single inhalation exposure to rats. The rates of clearance from lung were highly dependent on ^{95}Nb chemical form and particle size (MMD). The authors also found that blood niobium values of carrier and tracer differed in time; on day 1 the percent on initial whole-body content of carrier was 1.47 and of tracer, 0.71. By comparison, lung values on day 1 were 23 times greater for carrier and 47.3 times greater for tracer, showing the faster lung clearance rate for carrier particles. ^{95}Nb tissue contents — kidney, bone, liver — on day 1 of carrier were somewhat less than that of blood. Tracer values of liver were comparable, but in bone were double, and in kidney, half of the blood value. After 4 days blood, lung, and bone niobium values of carrier were about one-third those of day 1. The values of liver and kidney were approaching the limit of detection when tracer values of the lung were still 92% of day 1. During this time kidney and liver values had decreased only slightly and bone had fallen by half. At 69 days niobium levels in blood, liver, and kidney were below the limit of detection. However, it was until 141 days that all tissues with tracer, except lung, had values below detection limit. It was documented that fecal excretion of niobium following inhalatory exposure to the larger particles (0.43 um MMD) of the carrier is higher than urinary excretion throughout, and after one day remains quite constant at an average ratio 2.5:1–3:1. This ratio for the tracer was considerably lower and shows an approximately equal excretion by both routes. The greater fecal excretion of niobium in the carrier animals probably represents the removal and swallowing of a greater amount of material from the respiratory tract by ciliary action. In this experiment a half-life retention time of 27.5 days was established for the longer exponential component. This is equivalent to a biological half-life of 120 days. It was also found that when five succesive daily exposures were carried out, the whole-body burdens closely followed the pattern predicted from single exposure kinetics. However, in other studies (201, 215) of pulmonary retention of ^{95}Nb in beagle dogs, based on whole-body counting, it was found that the half-time of long-term pulmonary clearance was close to a year, a considerably longer time than demonstrated in rats after inhalatory exposure.

Whole-body activity after single oral and iv application of carrier-free ^{95}Nb oxalate to RF mice, Sprague–Dawley rats, rhesus monkeys, and beagle dogs and after ip injections in mice and rats were measured (204). It was found that effective half-lives of long components after iv injection were 32, 29, 26, and 28 days for mice, rats, monkeys, and dogs, respectively. Urinary excretion was predominant after parenteral injection; it was about the same for three of the species (14–17%), except for the mouse, in which excretion was more than double for both iv and ip routes. Fecal excretion was small, within the rather narrow limits of 3–8% for both parenteral routes in all four species. In general, urinary excretion of niobium exceeded fecal excretion by a factor of 3–9 in all species studied. For the oral administration, the fecal excretory route predominated with values of 97–99% for all four species. It is consistent with results from studies on Chernobyl UO$_2$ particles containing ^{95}Nb as well as ^{141}Ce, ^{144}Ce, ^{95}Zr, and ^{103}Ru, which revealed that approximately 98% of the total administered radioactivity was excreted with feces within 3 days (205). A twofold greater intestinal retention of about 6% of ^{95}Nb than for other radionuclides was observed 1 day after administration.

The retention of ^{95}Nb was studied in adult guinea pigs and in 2-day-old neonates (209). It was found that whole-body retention of ^{95}Nb after intraperitoneal administration was slightly lower in 2-day-old guinea pigs than in adults, with rapid secretion of about 50% over the first day compared with 40% in adults. A subsequent clearence was similar in both groups with a retention half-life of about 30 days.

Potassium niobate was found to inhibit mouse liver succinic dehydrogenase and adenosine triphosphatase in vitro (197).

2.4.1.4 Reproductive and Developmental. It was found that ^{95}Nb following iv injection crosses placenta in pregnant rats and rabbits and may accumulate in fetal internal organs (212). In rats, all maternal tissues exhibited higher concentrations compared to the corresponding fetal organs; the highest fetal/maternal ratio was 0.6 for bone. In rabbits, however, the fetal/maternal ratio was 3.5 for bone.

No other studies referring to the reproductive toxicity were found in the literature.

2.4.1.5 Carcinogenesis. No evidence was found that niobium is carcinogenic. Indeed, there are some studies suggesting its antitumor activity. In the mouse study of Schroeder et al. (200) occurence of 23.6% of tumors in the niobium-treated group (5-6.62 ppm niobium in drinking water and diet for a lifetime) versus 34.8% for the controls were documented. In another study, four air-stable niobocene complexes $[(C_2H_5)_2NbCl_2]^+ X^-$ with $X = BF_4$, AsF_6, SbF_6, SO_3CF_3] were investigated for antitumor properties against Ehrlich ascites tumor in female CF1 mice (216). A niobocene complexes containing tetrafluoroborate, or trifluoromethanesulfonate as anions induced a maximal cure rate of 50% and led to increases in lifespan of 182 and 178%, respectively, following application of optimal doses. The other two niobocene complexes with hexafluoroarsene or hexalfuoroantimonate as anions effected a maximal cure rate of 100% and increases in lifespan of 346 and 376, respectively, on day 90 of transplantation.

2.4.1.6 Genetic and Related Cellular Effects Studies. Niobium pentachloride was negative for mutagenic effects in the Kada recombination assay (217) performed in wild and recombination-deficient strains of *B. subtilis*.

2.4.1.7 Other: Neurological, Pulmonary, Skin Sensitization. The effect on the cardiovascular system in the cat was described by Haley et al. (198). A transient hypotension of 50 -60 mm Hg in both carotid and femoral blood pressure coupled with a decreased femoral blood flow after iv administration of 2 mg/kg niobium pentachloride was reported. Doses of 0.5-1 mg/kg produced no observable effects. At the 2-mg/kg dose, electrocardiographic changes occurred in the QRS complex and in the T wave, accompanied by transient skipped beats. Electrocardiogram changes included transient ventricular fibrillation, changes in the P and T waves, heart block, and cardiac standstill. Immediate complete cardiovascular collapse and respiratory paralysis resulted from a dose of 5 mg/kg. The effects of niobium pentachloride mentioned above could not be modified by atropinization, and cardiovascular collapse could be not counteracted by epinephrine.

It was documented that niobium compounds are irritant to skin and eye. Haley et al. (198) described irritation effects of niobium pentachloride on the skin and the eye. Direct application to unabraded rabbit skin produced an irritation index of 4 within 24 h for both edema and erythrema, which persisted for 72 h. Within 7 days an eschar was formed with a loss of skin in the area. On the abraded skin the reaction was even more severe. A maximal irritation index of 8 occured within 24 h; perforating ulcers with penetration to the muscle layers developed in 7 days. Healing had not occured by the time animals were killed at day 14.

Intradermal injection in guinea pigs of the niobium pentachloride in concentrations of $1:10$ to $1:10^6$ produced dose-dependent irritant changes in the skin. In the highest concentration of 1:10, the 24-h irritation index for erythrema plus edema was 8; eschar of 13 mm in diameter and scar started to form afterward.

Suprisingly, this compound produced very slight ocular irritation. Direct application of 1 mg of niobium pentacholride to the rabbit eye resulted only in an increase in blinking rate and redness of the conjunctiva. At 24 h there were no evidence of corneal or iris damage, conjunctivitis, chemosis, or increased lacrimation.

2.4.2 Human Experience

2.4.2.1 General Information: NA

2.4.2.2 Clinical Cases

2.4.2.2.1 Acute Toxicity. No cases of niobium and/or its compounds acute poisoning were reported in the literature.

2.4.2.2.2 Chronic and Subchronic Toxicity. No data were found in the literature.

2.4.2.2.3 Pharmacokinetics, Metabolism, and Mechanisms. Limited data exist in the literature concerning human lung retention, metabolism, and excretion of niobium. In 1963 Colfield (218) described a case history of female worker exposed by inhalation to ^{95}Zr–^{95}Ni (chemical compound not known). The lung burden was followed for more than 250 days. Calculations made, using the ICRP publication 30 lung model, showed that the best fit appears to be for a class Y compound (219).

In 1971 Waligora (215) reported an exposure to ZrO_2 (^{95}Nb) of a laboratory technician who was involved in the decontamination of aerosol generating equipment used in animal experimentation. The author used an exponential fitting procedure to derive a biological half-life in the lung of 224 days and on that basis suggested that the oxides of zirconium (Nb) be treated as inhalation class Y. It was confirmed by other calculations (220). Wrenn et al. (221) have reported lung burden measurements from autopsy samples of five humans exposed to atmospheric fallout from atomic weapon explosions. ^{95}Zr–^{95}Ni (chemical compound not known) were collected on air fliters and measured during 6 months. They concluded, using the ICRP publication 2 model, that the biological half-life in the lung was 67 days, which indicates a class W behavior. It was confirmed by other calculations (220). In one incident at the National Reactor Testing Station, Idaho (USA), an individual received a body burden of 13.5 µCi of ^{95}Zr–^{95}Ni as well as 36 µCi of ^{141}Ce and 27 µCi of ^{144}Ce, supposedly by ingestion. Over 99.7% of his total body burden was eliminated in 4 days. Analysis of feces and urine showed that elimination took place exclusively through

GI tract with no trace of these nuclides being detectable in any 24-h sample (222). A case of a contract worker exposed to $^{95}Zr-^{95}Ni$ (probably their oxides) during explosive plugging of boiler tubes was described (220). The body burden was followed over a 6 months period using whole-body counter, which showed that the total activity in the body burden best fits a class Y model. A limited number of urine and fecal samples were taken and analysed for these radionuclides at 2–3 days postexposure. They found the fecal to urinary excretion ratio in the early phase to be in the order of 104:1, which is in contradiction to data for Reference Man given in ICRP Publication 23 (1975) (223). For stable niobium the urinary and fecal losses quoted in ICRP are 0.36 and 0.26 mg/day, respectively. In the authors opinion, such a high ratio is partly due to mechanical clearence from the lung during the early phase after inhalation.

2.4.2.2.4 Reproductive and Developmental. No studies on reproductive toxicology were found in the literature.

2.4.2.2.5 Carcinogenesis. No studies on cancerogenic effects of niobium and its compounds were found.

2.4.2.2.6 Genetic and Related Cellular Effects Studies. No studies concerning genetic and related cellular effects were found.

2.4.2.2.7 Other: Neurological, Pulmonary, Skin Sensitization, etc. No studies were found.

2.4.2.3 Epidemiology Studies

2.4.2.3.1 Acute Toxicity. No studies concerning niobium and/or its compounds acute poisoning were reported in the literature.

2.4.2.3.2 Chronic and Subchronic Toxicity. No studies were reported in the literature.

2.4.2.3.3 Pharmacokinetics, Metabolism, and Mechanisms. No studies were found.

2.4.2.3.4 Reproductive and Developmental. No epidemiological studies on reproductive toxicology of niobium were found in the literature.

2.4.2.3.5 Carcinogenesis. In one study about 50 excess lung cancer cases per million person-years at risk per working-level months were observed at a niobium mine. However, it was connected to other hazards, mainly radon and thoron daughter exposure (224).

2.5 Standards, Regulations, or Guidelines of Exposure

Recommended maximum allowable concentrations of niobium hydride in air was suggested to be 6 mg/m^3 (225). The maximum allowable concentrations of niobium nitride in the former Soviet Union is 10 and 5 mg/m^3 for niobium carbide in Switzerland. A limit of 0.01 mgNb/L has been established in the former Soviet Union for drinking water (192). The USEPA published effluent limitations and wastewater pretreatment standards for primary Nb/Ta production (192).

2.6 Studies on Environmental Impact

Elemental niobium has an average value of 24 ppm in the earth's crust. Coal slags from 12 sources contained niobium with values of 10–24 ppm (226). Similar values were obtained from mineral slag samples from copper and nickel smelters. Analysis of terrestrial plants revealed niobium levels of < 0.4 mg/kg dry weight; however, plants located near niobium deposits demonstrated a marked capacity to accumulate and concentrate the metal to niobium level of > 1 mg/kg (188). Marine plants, such as seaweeds, appear to concentrate niobium from seawater contaminated with nuclear fallout (190). Niobium occurs in seawater in the form of $Nb(OH)_6^-$ at approximately 0.005 ppb (189).

3.0 Tantalum

3.0.1 CAS Number : [7440-25-7]

3.0.2 Synonyms: Tantalum-181

3.0.3 Trade Names: NA

3.0.4 Molecular Weight: 180.95

3.0.5 Molecular Formula: Ta

3.0.6 Molecular Structure:

Tantalum is a grey, heavy, and hard metallic element of a group Va element in the periodic table, the name of which is derived from the Greek mythological figure, Tantalos, father of Niobe, because of the almost invariable occurrence of tantalum with another group Va element, niobium. In 1802 Gust Ekeberg discovered tantalum. It was confused with niobium until Heinrich Rose showed in 1844 that niobic and tantalic acids were different. Tantalum is one of the rarer elements, ranking 54th in order of concentration in the earth's surface. It is normally associated with niobium (columbium), which is about 11 times more concentrated. The most important commercial source is a ferrous manganese tantalate niobate, $(Fe,Mn)(Ta,Nb)_2O_6$, called tantalite or niobate (columbite), depending on which metal predominates. Tantalum also may be found in the rare-earth mineral euxenite, $(Y,Ce,Cs,U,Th)(Nb,Ta,Ti)_2O_6$. Tantalum occurs uncombined with niobium in microlite, a fluoride-containing oxide of sodium, calcium, and tantalum, $(Na,Ca)_2Ta_2O_6(O,OH,F)$.

3.1 Chemical and Physical Properties

3.1.1 General

The physical and chemical properties of tantalum and some of its compounds used in industry are listed in Table 37.8. Tantalum is extremely resistant to chemical action. Its

Table 37.8. Chemical Identity of Tantalum and Its Compounds

Compound	CAS Number	Molecular Formula	Molecular Weight	Physical State	Melting Point (°C)	Boiling Point (°C)	Specific Gravity	Solubility in Water	Vapor Pressure	Flammability
Tantalum	[7440-25-7]	Ta	180.95	Gray solid or powder	2996	5429	16.69; powder: 14.491	Insoluble in water	—	Autoignition temperature—minimum ignition temperature 300°C (layer), 630°C (cloud); explosive limits—minimum explosion dust concentration. < 0.2 g/L
Tantalum oxide	[1314-61-0]	Ta_2O_5	441.9	White microcrystalline powder	1870–1880	—	8.735	Insoluble in water	—	—
Tantalum carbide	[12070-06-3]	TaC	192.95	Black crystalline	3880	5500	13.9	Insoluble in water	—	—
Tantalum chloride	[7721-01-9]	$TaCl_5$	358.2	White or light-yellow crystalline powder	216–220	239.3	3.68 (27°C)	Decomposed by water	—	—
Tantalum fluoride	[7783-71-3]	TaF_5	275.95	Colorless prisms	96.8	229.5	4.74 (20°C)	Soluble in water	100 mm at 130°C	—

melting point (2996°C) is among the highest values encountered and is exceeded only by that of tungsten and rhenium.

Tantalum is one of the most inert of all metals that reacts with chemicals below 150°C. Its inertness is caused by a tenacious, self-healing film of Ta_2O_5, which makes it serviceable in capacitors as an insulator. In the fifth group of the periodic table in the $5d$ transition series, its valence electron configuration is $5d^3 6s^2$, which accounts for its maximum valence of +5. Oxidation states of +4, +3, and +2 are known, but are unstable forms.

About 36 metal systems form alloys with tantalum. To enhance high temperature applications of tantalum, it is alloyed with high temperature melting-point metals, niobium, molybdenum, hafnium, titanium, tungsten, vanadium, and zirconium. Among the tantalum alloys receiving most attention is the one with tungsten; alloys of tantalum with 10, 20, and 30% W show unusually high temperature hardness up to 1000°C. Tantalum carbide, a hard, extremely heavy, refractory crystalline solid, has a hardness of 1800 kg/mm^2 and is extremely resistant to chemical action, except at elevated temperatures.

3.1.2 Odor and Warning Properties

From existing data, it might be concluded that tantalum and most of its compounds are odorless.

3.2 Production and Use

Tantalum ores are found in Australia, Brazil, Canada, Mozambique, Nigeria, Portugal, Republic of Zaire, and Thailand (48% of the global production). The ores are concentrated by hand separation, washing, tabling, and electrostatic and electromagnetic means. The concentrates ordinarily contain 60% or more of combined oxides (Ta_2O_5 and Nb_2O_5); and associated impurities are iron, tin, titanium, zirconium, silica, and manganese. For the concentrates imported into the United States, the ratio of Ta_2O_5 to Nb_2O_5 varies from 12:1 to 3:4, averaging about 1:1 for the bulk of the material. About half of the annual production of tantalum pentaoxide is from tin slags; the rest, from ores.

The separation of tantalum from niobium requires several complicated procedures. In the metallurgical extraction of tantalum from its ores, solvent extraction procedures have largely replaced the fractional crystallization methods of the past. In one such process the aqueous solution of metal ions containing 0.5 N HCl and 3.3 N HF is treated with an organic solvent such as methyl isobutyl ketone (MEK). The tantalum species are extracted preferentially into the organic phase, with the niobium species concentrating in the aqueous phase. The organic phase is back-extracted with water to recover tantalum. Tantalum and niobium are recovered as the oxides from the aqueous phases by complexing the fluoride with boric acid and precipitating the oxide with aqueous ammonia. The precipitate obtained from the initial aqueous layer is 98% niobium oxide; that from the organic layer is 99.5% tantalum oxide. In the absence of hydrofluoric acid and at very high hydrochloric acid concentrations, the niobium concentrates in the organic phase. The tantalum remains largely in the aqueous phase. Metallic tantalum is prepared by the fused electrolysis of K_2TaF_7 or by reduction of the oxide with active metals or carbon. A powder is obtained that is compacted into bars after being washed and then sintered in a vacuum furnace with the bar acting as the heating element. The bar is then cold-rolled to sheets or wire.

Most tantalum is used for making electrolytic capacitors and vacuum furnace parts; it also replaces platinum in chemical, surgical, and dental apparatus and instruments. The phenomenal corrosion resistance of tantalum over wide ranges of temperature and concentration has led to its application in chemical equipment. Because it is easily formed and welded, tantalum per se is used to form many components of vacuum tubes, as well as a 92.5Ta/7.5W alloy and a Ta-Ni alloy containing up to 30 percent Ta. The dielectric oxide film makes the metal useful as an electric current rectifier. A property of tantalum exploited for diagnostic radiology is its high value of the atomic number, 73, making it more visible on X-rays than iodine or barium. Thus, the element is used in angiography, laryngography, for airway visualization, particle deposition, and lung morphology. Several properties of tantalum, such as malleability, light weight, strength, and relative biological inertness, make the metal ideal for use in surgery. It is well-suited for use in cranioplasty because of an oxide coating which protects it from corosion, even after reshaping via cutting, bending, hammering, and drilling (227). Tantalum is also used to fabricate aircraft parts and missile parts, in spinnerets in rayon industry, in chemical industry for acid-proof equipment, in rubber industry as catalyst in synthesis of butadiene. Tantalum oxide is used to make special glass with high index of refraction for camera lenses.

Tantalum carbide chief use is in cutting tools; it is also a constituent of cobalt-cemented carbide, carballoy. Tantalum chloride because of its high reactivity, is used for the chlorination of organic substances and the production of tantalum metal. Tantalum pentafluoride and potassium fluorotantalate (K_2TaF_7) occur as intermediates in the preparation of pure tantalum.

3.3 Exposure Assessment

Industrial exposures to tantalum are rather limited. Tantalum metal exposures in the fabrication of metal ingots or metal parts constitute a certain hazard, as do the preparation and handling of $TaCl_5$. Fluoride exposure constitutes the only hazard anticipated during tantalum extraction.

3.3.1 Air

No information found.

3.3.2 Background Levels

The tanatalum metal is present at about 1 ppm in the earth's crust. There is very little geographic variability or environmental mobilization of tantalum, possibly because it is insoluble in water (228). Seawater levels seem to be lower than 0.004 ppb, and tantalum is found in the form of $Ta(OH)_5^0$ (189).

3.3.3 Workplace Methods

The method recommended in the United States by NIOSH for determining tantalum concentrations in air is plasma emission spectroscopy method S201 (229) (also NIOSH method 0500). A known volume of air is drawn through a mixed cellulose ester membrane filter to trap tantalum. A sample filter is digested with nitric and perchloric acids, and the tantalum is dissolved with nitric and hydrofuoric acids. The solution is then analyzed by plasma emission spectrometry. There is another spark source mass spectrometry method evaluated by NIOSH (23) even if tantalum is not listed there. This method has the capability of determining tantalum (and 77 other elements) in submilligram samples of ores, dusts on filters, and biological specimens. This is a determination that is essentially impossible by other methods. Its unique capabilities for determining tantalum in bulk dust and particulates on membrane filters have been demonstrated (230).

3.3.4 Community Methods

No information found.

3.3.5 Biomonitoring/Biomarkers

Analytical methods for tantalum in ores and biological tissues and fluids developed since the late 1960s have used photometric techniques. Spectrophotometric determination of tantalum in alloys using malachite green was reported (230), along with numerous other dyes. With improvement in sensitivity and selectivity over other commonly used dyes (pyrogallol, phenylfluorone, hydroquinone), the method has a range of ≤ 150 µg Ta. Another photometric method uses methyl green for determining tantalum as a fluorocomplex with a claimed sensitivity of ≤ 3 µg/mL (231). Spectrographic methods suffered from low spectral sensitivity and hence were replaced by the more sensitive photometric analysis and neutron activation analysis (NAA). NAA has been used to detect tantalum levels less than 1 ppm in marine invertebrates (232) and in grass (233). Tantalum concentrations of 33 ppm were detected in lung tissue of a worker occupationally exposed to tungsten carbide dusts (234). The method used in the United States for determining tantalum in biological samples is spark source mass spectrometry, worked out by NIOSH (26), even if tantalum is not listed there. Its unique capabilities for determining tantalum in bulk dust, particulates on membrane filters, acid-ashed and low-temperature ashed lung tissue, homogenized dried lung tissue, lymph nodes, blood, and urine have been demonstrated.

A German method for determining tantalum in 100–200-mg freeze-dried biologic samples is based on reactor irradiations of the tissue and γ spectrometry with a sensitivity down to 0.005 ppm (235). Tantalum has been similarly determined in ores by radioactivating 50–100-mg samples and measuring the organically extracted ^{182}Ta activity in a scintillation γ-ray spectrometer (236). Radiochemical determination of ^{182}Ta has a sensitivity of 20 ng/g (237).

3.3.5.1 Blood.
Tantalum concentration of 0.2–0.45 ng/mL in blood of three workers exposed to hard metal dusts, containing 4.45% of this element, were reported (238). In one case, Ta blood level was as high as 45 ng/mL.

In another study it was documented that a single normal blood sample contained 1.6 µg Ta/100 g, a content about one-tenth that of the lead content of the same individual (239).

3.3.5.2 Urine. Della Torre et al. (238) found Ta urine levels of 0.2–0.3 ng/mL. In the highly exposed case concentration of Ta in urine was 1 ng/mL.

Jacobs (239) found two normal urine specimens to be widely disparate for some unknown reason: 5 and 240 µg/L.

3.3.5.3 Other. Tantalum concentrations of ≤ 140 and 520 ng/g were found in pubic hair and toenails, respectively (238). According to these authors, hair and nails may be used as an index of continuous chronic tantalum exposure or to measure previous exposure. Among other biological samples analyzed, two adult male lungs contained 7.5 and 14 ppm Ta, respectively, of ash; the normal lung of a female resident in the same area contained 3.1 ppm. The tantalum content of the pulmonary lymph nodes showed no evidence of accumulation; two samples contained 1.7 and 1.8 ppm, of ash.

Of 11 samples of coal miners' lungs analyzed, tantalum values ranged from 0.4 to 16 ppm of lung ash, mean 5.0 ppm. By comparison, the range of lead values from the same lungs was 0.1–2 ppm, with a mean of 3.5 ppm of ash, showing that tantalum as a trace environmental element is more prevalent than realized thus far (239).

The concentration of tantalum in lung tissue was found to be far higher than in bronchoalveolar lavage (BAL) fluid, but the factor is so variable that BAL fluid cannot be taken as representative of the concentration of elements in lung tissue. High concentrations in tissues or body fluids are indicative for exposure, but not for intoxication. In the light of available data, there are no levels above which development of intoxication is inevitable (240).

3.4 Toxic Effects

3.4.1 Experimental Studies

3.4.1.1 Acute Toxicity. Tantalum oxide may be considered nontoxic, since doses as high as 8000 mg/kg given orally produced no effects in rats (197). This is probably due to poor absorption of tantalum oxide, since its solubility is extremly low. The rat LD_{50} values obtained after single oral administration of potassium tantalum fluoride, tantalum chloride, and tantalum oxide were 2500, 1900, and 8000 mg/kg, respectively, indicating a low toxicity. However, the LD_{50} values obtained after ip injection of tantalum chloride and potassium tantalum fluoride to rats were 75 and 375 mg/kg, respectively. The LD_{50} value obtained after iv injection of tantalum fluoride to mice was 110 mg/kg (241). Tantalum hydride, instilled intratracheally in doses of 50 mg/rat, produced mild pathological changes consisting in pneumoconiosis, as well as dystrophic changes in lungs and parenchymatous organs (242).

Acute toxicity studies are summarized in Table 37.9.

3.4.1.2 Chronic and Subchronic Toxicity. Within one month, the intratracheal introduction of the 10% suspension of tantalum oxide dust (100 mg) to guinea pigs produced focal reactions whose severity apparently was quantitatively related to the topographic disposition of the metallic oxide (243). Histopathological changes in lung included transient bronchitis, interstitial pneumonitis, and hyperemia (244). After one year

Table 37.9. Summary of Toxicity of Tantalum and Its Compounds in Animals

Compound	CAS Number	Species	Route of Exposure	Concentration or Dose	Exposure	Effects	Ref.
Tantalum oxide	[1314-61-0]	Rat	Gavage	>8000 (>6500 mg Ta/kg)	50% suspension	LD_{50}	197
		Guinea pig	Intratracheal	100 mg/animal	Single dose		243
Potassium tantalum fluoride K_2TaF_7	[16924-00-8]	Rat	Gavage	2500 (1150 mg Ta/kg)	50% suspension	LD_{50}	197
		Rat	Intraperitoneal	375 (173 mg Ta/kg)	50% suspension	LD_{50}	197
Tantalum chloride	[7721-01-9]	Rat	Gavage	1900 (958 mg Ta/kg)	50% solution	LD_{50}	197
		Rat	Intraperitoneal	75 (38 mg Ta/kg)	50% solution	LD_{50}	197
Tantalum fluoride	[7783-71-3]	Mouse	Intravenous	110 (56 mg Ta/kg)	Acute	LD_{50}	242

the bronchi and bronchioles appeared to have undergone almost complete healing, but there was slight hyperplasia of the epithelium. Perhaps in consequence of earlier sustained bronchial or bronchiolar damage, foci of hypertrophic emphysema were found. The dust was not fibrogenic. Although there was so little residual effect after a year, the author warned against classifying tantallum oxide as innocuous dust.

The passivity of tantalum metal for biological tissues has been amply demonstrated by its long-time use in surgical procedures in animals. Implants of tantalum plates and screws have produced no inflammation or oedema (245). Insertion of tantalum plates into the bones of dogs and rabbits has also been shown to cause no microscopic or X-ray evidence of bone or soft-tissue irritation after periods of 3 weeks to 3 months (246).

In spite of these observations, there are some reports indicative of a delayed inflammatory reaction to tantalum. In one study, tantalum stock was implanted in the subperiosteal region of the mandible, buccal mucosa, and the subcutaneous paravertebral region of the back of monkeys (247). Tissue responses assessed after 3 weeks were variable and appeared to be a function of the surface treatment of the material prior to implantation. Increased cellularity, including fibroblasts and multinucleated giant cells, was the predominant feature in the tissues. In two other studies, 18 months after tantalum bronchography in dogs (248) and 6 months after this procedure in cats and monkeys (249), granulomatous changes were evident in the lungs. In another study, however, no pathological changes were observed in lungs of dogs 2 years after insufflation with tantalum dust (250).

3.4.1.3 Pharmacokinetics, Metabolism, and Mechanisms

3.4.1.3.1 Absorption. Absorption of tantalum compounds depends largely on its solubility in water. Cochran et al. (197) concluded that orally administered tantalum oxide was not absorbed, since doses of $\leq$ 8000 mg/kg were nontoxic. The same workers found that tantalum chloride and potassium tantalum fluoride were at least 6–8 times more toxic than the oxide when given under the same conditions. They attributed this difference to the greater solubility of the chloride and fluoride, which resulted in increased absorption of the tantalum halide compounds. Fleshman et al. (251) found that radiotantalum (^{182}Ta) administered orally to rats as potassium tantalate is excreted rapidly through the GI tract and that only < 1% is retained in the body. ^{182}Ta$_2$O$_3$ was injected into rats iv or intramuscularly (im) either alone or in combination with sodium citrate as a complexing agent (252,253). Of all the citrate-complexed ^{182}Ta, 70% was absorbed from the site of the im injection at 30 days as opposed to 15% for the noncomplexed ^{182}Ta, although the relative distributions were essentially the same. A little information on absorption, distribution, and excretion of ^{182}Ta in young and adult rats from oral administration was reported (254). The amount of tantalum nuclide absorbed was several orders of magnitude greater in young suckling rats than in adults, with an initial rapid loss of the nuclide by weaning time (3 weeks). Distribution of ^{182}Ta was greatest in the ileum, kidney, and bone.

3.4.1.3.2 Distribution. Absorbed ^{182}Ta is widely distributed throughout the body. Fleshman et al. (251) observed in rats administered orally 25 µCi ^{182}Ta that, with exception of day 1, bone had accumulated more of the absorbed tantalum than any other tissue. Even though it makes up only approximately 6% of the total body weight, it is the deposition site for over 40% of the total tantalum present in the body from the 14th day onward. Pelt, muscle, and gastrointestinal tract also contained significant amounts of ^{182}Ta. Other tissues containing at least 1% of the body burden were liver, kidney, testes and epididymis, and spleen. On a per-gram basis, bone and kidneys showed the highest concentration of ^{182}Ta. This is in contrast with results of Scott and Crowley (255), who found liver to be the organ of highest concentration (%/g) of ^{182}Ta after iv administration to rats.

The biodistribution and binding substances of ^{182}Ta were investigated using tumor-bearing rats and mice (213). The concentrations of ^{182}Ta were more dominant in the connective tissue (especially inflammatory tissue) than in the other categories of tumor tissue. This nuclide accumulated rapidly into the mitochondrial fraction (containing lysosome) of the liver, reaching about 50% after 48 h, but existed relatively uniformly in the tumor cells. The main binding substance of ^{182}Ta in the above-mentioned tissues was the acid mucopolysaccharide whose molecular mass exceeded 40,000 daltons.

The distribution of ^{182}Ta was determined in freshwater clams (*Anodonta nuttaliana*) after a 10–12-day exposure in aquarium water (256). Transfer across the gut and/or exterior surfaces and subsequent circulation by body fluid was evident since ^{182}Ta was localized in the stomach/digestive gland (51% of absorbed dose), intestine/gonad (14%), body fluid (6%), and body wall (5%).

3.4.1.3.3 Metabolism. No studies were found in the literature.

3.4.1.3.4 Excretion. Doull and Dubois (257) found that ^{182}Ta, after oral administration of the tantalum oxide, was totally eliminated in the feces within 2 days. They detected no activity whatever in any urine samples or in the carcass after 10 days. When ^{182}Ta was given orally as tantalate to dogs, more than 97% was excreted within 8 days and all but 1% appeared in the feces (R. J. Chertok and S. Lake Lawrence, Radiation Laboratory, Livermore, CA, unpublished data, 1970). Fleshman et al. (251) showed that orally administered ^{182}Ta as tantalate is rapidly excreted by rats through the GI tract. After 7 days, more than 97% is accounted for in the feces and less than 1% in the urine. Elimination of tantalum is assumed to follow a curve of at least three components: an early phase, reflecting rapid excretion ($t_{1/2} = 0.25$ day) from the GI tract of unabsorbed compound; a second phase ($t_{1/2} = 2$ to 5 days), which may reflect a loss of ^{182}Ta that is loosely bound in tissues; and a third phase, accounting for elimination of ^{182}Ta that has been absorbed and accumulated within tissue ($t_{1/2} = 62$ days in males and 119 days in females). Tantalum is present in the body in detectable amounts for up to 106 days after exposure.

Bianco et al. (258) examined the clearance times of tantalum metal powder following its introduction into the respiratory tract. The tantalum was inhaled or insufflated in female beagle dogs as a partly neutron-activated ^{182}Ta powder with a specific activity of ~ 2 mCi/g for inhalation studies; for insufflation, it was blended with tantalum powder with a specific activity of ~ 5 µCi/g. Airborne concentrations approximated 1 g/m^3 at an aerodynamic mass of 4 µm. Exposures were usually less than 1 h, and were nasal only. ^{182}Ta retention measurements indicated a rapid, early tracheobronchial passage, later followed by a prolonged alveolar clearance phase. After inhalation there was, on average, 7 times more tantalum in the alveolar phase, compared to the amount insufflated. This "alveolarization" leads to prolonged retention with a mean biologic removal half-life greater than 2 years. The dominant clearance mechanism, mucociliary transport, was slower than that of most other insoluble dusts and appeared independent of the presence of tantalum in the respiratory tract.

Pulmonary clearance of tantalum dust following insufflation in dogs was dependent on particle size; a 1-μm powder was removed from the alveolar regions with a clearance half-life of 2.1 years and 5 or 10-μm powders were removed with a half-time of 333 days (250). Rapid postinsufflation uptake by the pulmonary lymph nodes was observed with up to 12% of the initial alveolar burden present in the lymph nodes at 240 days and 6% present at 816 days.

3.4.1.4 Reproductive and Developmental. No studies were found in the literature.

3.4.1.5 Carcinogenesis. Although Oppenheimer (259), using embedded metal foil technique, have elicited two malignant fibrosarcomas in 50 embeddings of tantalum metal in 25 Wistar rats after a latent period of 714 days, these results remain of controversial issue.

3.4.1.6 Genetic and Related Cellular Effects Studies. No studies were found in the literature.

3.4.1.7 Other: Neurological, Pulmonary, Skin Sensitization. On the basis of an *in vitro* study with tantalum oxide using rabbit alveolar macrophage cells, Matthay et al. (260) reported that tantalum may exert a toxic effect on these cells.

3.4.2 Human Experience

3.4.2.1 General Information. Both acutely and chronically, tantalum and its compounds are practically nontoxic by all routes and concentrations encountered under industrial conditions.

3.4.2.2 Clinical Cases

3.4.2.2.1 Acute Toxicity. No studies were found in the literature.

3.4.2.2.2 Chronic and Subchronic Toxicity. Apart from a study of adverse tantalum effects in Russian chemical workers and welders, no reports on the health effects in workers are available in the literature. Workers of tantalum/niobium smelter were exposed to air concentrations of tanatalum dust from few to 120 mg/m^3 for 3–24 years. At some locations smelter workers were exposed to mercury dusts. In 10 out of 12 smelters the urine mercury measurements were established to be 0.002–0.02 mg/l. Clinical studies made on 22 chemical workers and welders handling both tantalum and niobium showed little evidence of poisoning, apart from radiological signs of early pulmonary fibrosis and, in one or two cases, chronic atrophic rhinitis (261). Miller et al. (262) concluded from case reports of workers exposed to tantalum contained in cobalt-cemented tungsten carbide, a product of the hard-cutting tool industry, that TaC does not provoke a necrotizing or fibrosing response in that tissue and thus acts as a physiologically inert substance. This conclusion was supported by histological observations of rat lung.

From four workers employed in sharpening and grinding operations of hard metal tools, one, a 37-year-old female exposed for 7 years to hard metal dust (W = 71.4%, Co = 24%, Ta = 4.45%), developed hard metal pneumoconiosis, which progressed to death (238). Cytology of BAL showed a high number of eosinophils, representing more than 30% of the cell population. Biopsy of the lung revealed interstitial fibrosis with hyperplasia of the secondary-type pneumocytes and inflammatory cellular infiltration of the interstitium. High tungsten and tantalum concentrations were determined in the admission broncho-alveolar lavage (BAL) and in the lung biopsy four months later by neutron activation analysis.

The physiological inertness of tantalum metal, metal dust, and powders is demonstrated by its long use in surgery (263). A 10-year study of tantalum gauze use in the repair of hernias revealed no effects from the gauze itself (264).

There are some case studies that report allergic responses to tantalum. In one of these studies, the onset of chronic urticaria 10 months after surgical implantation of tantalum staples in humans was reported. Intradermal testing with tantalum produced an urticarial response. The urticaria did not improve with antihistamines treatment, but responded dramatically to disulfiram chelation therapy and resolved completely after surgical removal of the tantalum staples (265). Another case of urticaria from surgical clips of tantalum alloys has been described by Romaguera and Vilaplana (266). Two months after operation, urticarial symptoms appeared on the overlying skin, which did not improve with treatment. The lesions disappeared 6 weeks after the prosthesis was removed.

Patients with orbital implants covered with tantalum mesh developed pain, headache, mucopurulant discharge, diffuse conjunctival inflammation, and erosion of tissues surrounding the implant 10–15 years postimplantation (267).

3.4.2.2.3 Pharmacokinetics, Metabolism, and Mechanisms. Distribution of ^{182}Ta following contamination of a worker with an estimated 30 µCi of 115-day ^{182}Ta and 150 µCi of 5-day ^{183}Ta during a nuclear reactor accident was reported by Sill et al. (222). On the first day postexposure, the scan showed distinct maxima in the area of the nasopharynges and in the section from the lower end of the sternum to a point 10 in. lower. On the second day after exposure the activity had left the nasopharyngeal area and had centered in the midpoint of the sternum, with the larger percentage of activity in the vicinity of the lower GI tract in the descending colon. On the third day part of the activity was removed from the lower GI tract, but the remaining activity was still centered there. Of four additional scans made at 8, 28, 35, and 63 days, each showed identical distribution patterns, indicating that the activity did not move or decrease significantly during this time. Regarding elimination in the worker, 93% of the long-lived ^{182}Ta component disappeared from the body in 7 days. In these 7 days, the total ^{182}Ta activity decreased from 30 to 2 µCi, but only 5.5 µCi was excreted in the feces; the major part of the difference was attributed to external contamination. No tantalum activity could be detected in the urine even from ^{183}Ta. The remaining activity, amounting to about 0.05%/day in the feces, showed a decrease similar to the 115-day radiological half-life of ^{182}Ta. In view of the experimental uncertainties, the authors concluded that the biologic half-life was greater than 1000 days. Gamzu et al. (268) were the first to report on the use of tantalum metal powder as an insufflating agent in human patients for assessing the rates of clearance from differently sized pulmonary airways without adverse effects, thus demonstrating the inertness of tantalum metal powder. Pulmonary clearance rates were established for tantalum metal powder in 26 human subjects, using a radiographic method. The tantalum powder, with a mean mass diameter of 2.4 µm, was introduced into selected areas of the tracheobronchial tree in amounts varying from about 7 to 15 g via a nasotracheal catheter. In eight patients studied by tracheography, clearance was complete within 20 h. In 18 patients studied by bronchography, radiographic clearance showed an orderly progression from large to small airways, except for distal bronchioles, which showed no clearance for the first 24–48 h. In terminal airways, radiopacity usually increased during the first 10–48 h, but was followed by no significant amount of clearance for 15 months.

The clearance of activated tantalum oxide accidentally inhaled by three reactor workers was studied. The whole-body retention after 7 days was about 1% of the initial deposit. In one subject studied for a further 424 days, the residual activity in the thorax was cleared with a biological half-life of about 1400 days. The results support indications from other studies that tantalum powder administered by inhalation has prolonged alveolar retention (269).

3.4.2.2.4 Reproductive and Developmental. No studies were found in the literature.

3.4.2.2.5 Carcinogenesis. No studies were found in the literature.

3.4.2.2.6 Genetic and Related Cellular Effects Studies. No studies were found in the literature.

3.4.2.2.7. Other: Neurological, Pulmonary, Skin Sensitization, etc. Tantalum wire stents induce expression of activation-dependent epitopes: CD62p (GMP140) and CD63 (GP53) on platelets, what potentially may result in thrombus formation (270).

3.4.2.3 Epidemiology Studies

3.4.2.3.1 Acute Toxicity. No information was found.

3.4.2.3.2 Chronic and Subchronic Toxicity. No information was found.

3.4.2.3.3 Pharmacokinetics, Metabolism, and Mechanisms. No information was found.

3.4.2.3.4 Reproductive and Developmental. No studies were found in the literature.

3.4.2.3.5 Carcinogenesis. No studies were found in the literature.

VANADIUM, NIOBIUM, AND TANTALUM

Table 37.10. Regulations and Guidelines for Exposure to Tantalum and Compounds

Concentration (mg/m^3)	OSHA PEL	ACGIH TLV	German Standard	British Standard	Polish Standard	NIOSH
Tantalum						
TWA	5	5 (metal/oxide dust)	5 (total dust)		5	
STEL				10		10
REL						5
IDLH						2500 (as Ta)
Tantalum fluoride						
TWA	2.5 (as F)	2.5 (as F)	2.5 (as F)	2.5 (as F)	1 (as HF)	
REL						2.5 (as F)
STEL					3 (as HF)	

3.4.2.3.6 Genetic and Related Cellular Effects Studies. No studies were found in the literature.

3.4.2.3.7 Other: Neurological, Pulmonary, Skin Sensitization, etc. No studies were found in the literature.

3.5 Standards, Regulations, or Guidelines of Exposure

The current occupational exposure limits expressed as a time-weighted averages (TWAs) for tantalum metal [*7440-25-7*] and tantalum oxide [*1314-61-0*] dusts are similar in most countries and established as 5 mg/m^3. In some countries, due to irritative properties of tantalum and its compounds, short-term limits of exposure were also established. Table 37.10 summarizes the main current regulations and guidelines.

3.6 Studies on Environmental Impact

Application of the spark source mass spectrometry has revealed an entirely unrealized general distribution of tantalum as a trace metal in the environment and in humans (240). Environmental samples, chrysotile and cristobalite asbestos, contained 0.18 and 0.27 ppm Ta by weight, respectively, and a sample of cement dust (U.S.), 0.9 ppm. Coal dust samples varied considerably, not only by geographic area but also from pit to pit. Bituminous coal samples from Pennsylvania varied from < 0.03 to 0.31 ppm Ta to almost 1.0 ppm for a similar coal sample from Utah. Airborne foundry and welding dusts, however, contained quite similar amounts of tantalum, 2.2 and 1.75 ppm, respectively.

Seawater levels seem to be very low (lower than 0.004 ppb), and tantalum is found in the form of Ta(OH)$_5^0$ (189). Concentrations of tantalum ranging from < 1 to > 400 ppb have been found in a variety of marine invertebrates of the class Ascidians, or sea squirts (232).

BIBLIOGRAPHY

1. F. A. Cotton and G. Wilkinson, *Advanced Inorganic Chemistry: A Comprehensive Text*, 3rd ed., Wiley Interscience, New York, 1972, pp. 819–829, 934–944.
2. Registry of Toxic Effects of Chemical Substances (RTECS), *Data-base on CD-ROM*, National Institute for Occupational Safety and Health, Washington, DC, 1998.
3. B. R. Nechay, Mechanisms of action of vanadium. *Ann. Rev. Pharmacol. Toxicol.* **24**, 501–524 (1984).
4. R. U. Byerrum et al., *Vanadium*, National Academy of Sciences, Washington, DC, 1974.
5. M. Windholz, ed., *The Merck Index*, 10th ed., Merck & Co., Rahway, NJ; 1983, pp. 1417–1419.
6. M. Grayson, in *Kirk-Othmer Encyclopedia of Chemical Technology*, 3rd ed., Vol. 23, Wiley, New York; 1983, pp. 688–704.
7. R. C. Weast, *Chemical Rubber Company Handbook of Chemistry and Physics*, 50th ed., CRC Press, Cleveland, OH; 1969, pp. B-144–B-145, B-261.
8. S. M. Brooks, Pulmonary reactions to miscellaneous mineral dusts, man-made mineral fibers, and miscellaneous pneumoconioses. *Occupational Respiratory Diseases*, DHHS (NIOSH) Publ. No. 86(102), U.S. Department of Health and Human Services, National Institute of Occupational Safety and Health, Division of Respiratory Disease Studies, Washington, DC, 1986, pp. 401–458.
9. H. J. Symanski, Vanadium, alloys and compounds. *Encycl. Occup. Health Saf.* **2**, 2240–2241 (1983).
10. J. Li et al., Evidence for the distinct vanadyl (+4)–dependent activating system for manifesting insulin-like effects. *Biochemistry* **35**, 8314–8318 (1996).
11. E. Dafnis and S. Sabatini, Biochemistry and pathophysiology of Vanadium. *Nephron* **67**, 133–143 (1994).
12. Y. Shechter and A. J. Ron, Effect of depletion of phosphatate and bicarbonate ions on insulin action in rat adipocytes. *J. Biol. Chem.* **261**, 14951–14954 (1986).
13. E. Browning, *Toxicity of Industrial Metals*, 2nd ed., Appleton-Century-Crofts, New York; 1969, pp. 340–343.
14. H. E. Hilliard, *Vanadium. The Minerals Yearbook — Minerals and Metals*, 1987, pp. 917–927.
15. F. W. Mackinson, R. S. Stricoff, and L. J. Partridge, Jr., *NIOSH/OSHA — Occupational Health Guidelines for Chemical Hazards*, DHHS (NIOSH) Publ. U.S. Department of Health and Human Services; National Institute of Occupational Safety and Health, Washington, DC, 1978.
16. N. Quickert, A. Zdrojewski, and L. Dubois, The accurate measurement of vanadium in airborne particulates. *Int. J. of Environ. Anal. Chem.*, **3**, 229–238 (1974).
17. World Health Organization (WHO), Vanadium. *Environmental Health Criteria 81*, WHO, Geneva, 1988.
18. W. H. Zoller et al., The sources and distribution of vanadium in the atmosphere. *Adv. Chem. Ser.* **123**, 31–47 (1973).

19. B. Van Zinderen and J. F. Jaworski, *Effects of Vanadium in the Canadian Environment*, National Research Council of Canada, *Associate Committee of Scientific Criteria for Environmental Quality*, Ottawa, Canada; 1980.
20. M. D. Waters, Toxicology of vanadium. In R. A. Goyer and M. A. Melhman, eds., *Toxicology of Trace Elements. Advances in Modern Toxicology*, Vol. 2, Halsted Press, New York; 1977, pp. 147–189.
21. H. L. Cannon, The biogeochemistry of vanadium. *Soil Sci.* **98**, 196–204 (1963).
22. A. R. Byrne and L. Kosta, Vanadium in foods and in human body fluids and tissues. *Sci. Total Environ.* **10**, 17–30 (1978).
23. National Institute for Occupational Safety and Health (NIOSH), *NIOSH Manual of Analytical Methods*, U.S. Department of Health and Human Services, Washington, DC, 1984, pp. 7300.1–7300.5
24. L. Pyy, L. H. J. Lajunen, and E. Hakala, Determination of vanadium in workplace air by DCP emission spectrometry. *Am. Ind. Hyg. Assoc. J.* **44**, 609–614 (1983).
25. National Institute for Occupational Safety and Health, (NIOSH), *NIOSH Manual of Analytical Methods*, Nos. 173 and 290. U.S Department of Health and Human Services, Washington, DC, Nos. 173 and 290. 1977.
26. National Institute for Occupational Safety and Health (NIOSH), *NIOSH Manual of Analytical Methods*,No. 8005, U.S Department of Health and Human Services, Washington, DC, 1977.
27. R. O. Allen and E. Steinnes, Determination of vanadium in biological materials by radiochemical neutron activation analysis. *Anal. Chem.* **50**, 1553–1555 (1978).
28. N. Lavi and Z. B. Alfassi, Determination of trace amounts of titanium and vanadium in human blood serum by neutron activation analysis: Coprecipitation with Pb/PDC/2 or Be/PDC/3. *J. Radioanal. Nucl. Chem. Lett.* **126**, 361–374 (1988).
29. D. M. Martin and N. D. Chasteen, Vanadium. *Methods Enzymol.* **158**, 402–421 (1988).
30. F. Mousty et al., Atomic-absorption spectrometric, neutron-activation and radioanalytical techniques for the determination of trace metals in environmental, biochemical and toxicological research. Part I. Vanadium. *Analyst (London)* **109**, 1451–1454 (1984).
31. O. Ishida et al., Improved determination of vanadium in biological fluids by electrothermal atomic absorption spectrometry. *Clin. Chem. (Winston-Salem, N.C.)* **35**, 127–130 (1989).
32. S. S. Krishnan, S. Quittkat, and D. R. Crapper, Atomic absorption analysis for traces of aluminium and vanadium in biological tissue. A critical evaluation of the graphite furnace atomizer. *Can. J. Spectrosc.* **21**, 25–30 (1976).
33. U.S. Environmental Protection Agency (USEPA), *Test Methods for Evaluating Solid Waste*, Vol. lA; USEPA, Office of Solid Waste and Emergency Response, Washington, DC, 1986, pp. 7910.1–7911.3.
34. S. A. Abbasi, Pollution due to vanadium and a new spot test for detection of traces of vanadium in water, plants, soils, and rocks. *Int. J. Environ. Stud.* **18**, 51–52 (1981).
35. A. R. Jha et al., A sensitive and selective spot test method for the detection of vanadium(V) in a water sample. *Int. J. Environ. Stud.* **14**, 235–236 (1979).
36. K. R. Paul and V. K. Gupta, Toxicology, solvent extraction and spectrophotometric determination of vanadium in complex materials. *Am. Ind. Hyg. Assoc. J.* **43**, 529–532 (1982).
37. J. D. Fassett and H. M. Kingston, Determination of nanogram quantities of vanadium in biological material by isotope dilution thermal ionization mass spectrometry with ion counting detection. *Anal. Chem.* **57**, 2474–2478 (1985).

38. H. A. Schroeder, J. J. Balassa, and I. H. Tipton, Abnormal trace metals in man: Vanadium. *J. Chronic Dis.* **16**, 1047–1071 (1963).
39. M. Simonoff, C. Conri, and G. Simonoff, Vanadium in depressive states. *Acta Pharmacol. Toxicol.* **59**(7), 463–466 (1986).
40. C. E. Lewis, The biological effects of vanadium. II. The signs and symptoms of occupational vanadium exposure. *AMA Arch. Ind. Health* **19**, 497–503 (1959).
41. J. T. Mountain, F.R. Stockell, and H. E. Stokinger, Studies in vanadium toxicology: III. Fingernail cystine as an early indicator of metabolic changes in vanadium workers. *AMA Arch. Ind. Health* **12**, 494–502 (1955).
42. S. G. Sjöberg, Vanadium pentoxide dust — A clinical and experimental investigation on its effects after inhalation. *Acta Med. Scand. Suppl.* **238**, 1–188 (1950).
43. A. V. Roshchin, Vanadium. In Z. I. Izraelson, ed., *Toxicology of the Rare Metals*, Israel Program for Scientific Translations, Jerusalem, 1967, pp. 52–59.
44. A. V. Roshchin, *Vanadium and its Compounds*, Medicina Publishing House, Moscow 1968 (in Russian).
45. J. M. Llobet and J. L. Domingo, Acute toxicity of vanadium compounds in rats and mice. *Toxicol. Lett.* **23**, 227–231 (1984).
46. T. G. F. Hudson, Vanadium. *Toxicology and Biological Significance*, Monogr. No. 36, Elsevier, New York, 1964, pp. 74–77, 126–133.
47. E. A. Knecht et al., Pulmonary effects of acute vanadium pentoxide inhalation in monkeys. *Am. Rev. Respir. Dis.* **132**, 1181–1185 (1985).
48. K. P. Lee and P. J. Gillies, Pulmonary response and intrapulmonary lipids in rats exposed to bismuth orthovanadate dust by inhalation. *Environ. Res.* **40**, 115–135 (1986).
49. A. G. Gul'ko, On the characteristics of vanadium as an industrial poison. *Gig. Sanit.* **21**(11), 24–28 (1956).
50. V. M. Pazhynich, Maximum permissible concentration of vanadium pentoxide in the atmosphere. *Gig. Sanit.* **31**, 6–11 (1966).
51. A. V. Roshchin, Hygienic characteristics of industrial vanadium aerosol. *Gig. Sanit. Iss.* **11**, 49–53 (1952).
52. H. E. Stokinger, Vanadium. In G. D. Clayton, and F. E. Clayton, eds., *Patty's Industrial Hygiene and Toxicology*, 3rd ed., Vols. 2A, 2B, 2C, Wiley, New York; 1981, pp. 2016–2030.
53. A. V. Roshchin, L. V. Zhidkova, and A. Dushenkina, The pathogenic properties of industrial vanadium, ferrovanadium and vanadium carbide dusts (an experimental study). *Gig. Tr. Prof. Zabol.* **10**, 21–25 (1966).
54. J. L. Domingo et al. Short-term toxicity studies of vanadium in rats. *J. Appl. Toxicol.* **5**, 418–421 (1985).
55. A. W. Conklin et al., Clearance and distribution of intratracheally instilled vanadium-48 compounds in the rat. *Toxicol. Lett.* **11**, 199–203 (1982).
56. A. V. Roshchin, E. K. Ordzhonikidze, and I. V. Shalganova, Vanadium-toxicity, metabolism, carrier state. *J. Hyg. Epidemiol. Microbiol. Immunol.* **24**, 377–383 (1980) (in Russian).
57. J. Edel and E. Sabbioni, Retention of intratracheally instilled and ingested tetravalent and pentavalent vanadium in the rat. *J. Trace Elem. Electrolytes Health Dis.* **2**, 23–30 (1988).
58. S. G. Oberg, R. D. R. Parker, and R. P. Sharma, Distribution and elimination of an intratracheally administered vanadium compound in the rat. *Toxicology* **11**, 315–323 (1978).

59. K. Rhoads and C. L. Sanders, Lung clearance, translocation, and acute toxicity of arsenic, beryllium, cadmium, cobalt, lead, selenium, vanadium, and ytterbium oxides following deposition in rat lung. *Environ. Res.* **36**, 359–378 (1985).
60. J. Edel et al., Disposition of vanadium in rat tissues at different age. *Chemosphere* **13**, 87–93 (1984).
61. R. D. R. Parker and R. P. Sharma, Accumulation and depletion of vanadium in selected tissues of rats treated with vanadyl sulfate and sodium orthovanadate. *J. Environ. Pathol. Toxicol.* **2**, 235–245 (1978).
62. S. Dai et al., Toxicity studies on one-year treatment of non-diabetic and sreptozocin-diabetic rats with vanadyl sulfate. *Pharmacol. Toxicol.* **75**, 265–273 (1994).
63. A. Shaver et al., The chemistry of peroxvanadium compounds revelant to insulin mimesis. *Mol. Cell. Biochem.* **153**, 17–24 (1995).
64. R. P. Sharma, S. G. Oberg, and R. D. Parker, Vanadium retention in rat tissues following acute exposures to different dose levels. *J. Toxicol. Environ. Health* **6**, 45–54 (1980).
65. R. Amano, S. Emomoto, and M. Nobuta, Bone uptake of vanadium in mice: Simultaneous tracing of V, Se, Sr, Y, Zr, Ru and Rh using a radioactive multitracer. *J. Trace Med. Biol.* **10**(3), 145–148 (1996).
66. M. Bruech et al., Effects of vanadate on intracellular reduction equivalents in mouse liver and the fate of vanadium in plasma, erythrocytes, and liver. *Toxicology* **31**, 283–295 (1984).
67. B. W. Patterson et al., Kinetic model of whole-body vanadium metabolism: Studies in sheep. *Am. J. Physiol.* **251**, R325-R332 (1986).
68. W. R. Harris and C. J. Carrano, Binding of vanadate to human serum transferrin. *J. Inorg. Biochem.* **22**, 201–218 (1984).
69. W. R. Harris, S. B. Friedman, and D. Silberman, Behavior of vanadate and vanadyl ion in canine blood. *J. Inorg. Biochem.* **20**, 157–169 (1984).
70. E. Sabbioni and E. Marafante, Metabolic patterns of vanadium in the rat. *Bioinorg. Chem.* **9**, 389–408 (1978).
71. D. Witkowska, R. Oledzka, and B. Markowska, Effect of intoxication with vanadium compounds on copper metabolism in the rat. *Bull. Environ. Contam. Toxicol.* **40**, 309–316 (1988).
72. Y. Shechter, Insulin-mimetic effects of vanadate: Possible implication for future treatment of diabetes. *Diabetes*, **39**, 1–5 (1990).
73. Y. Shechter et al., Insulin-like actions of vanadate are mediated in an insulin-receptor-idependent manner via non-receptor protein tyrosine kinases and protein phosphotyrosine phosphatases. *Mol. Cell. Biochem.* **153**, 39–47 (1995).
74. S. K. Pandey, M. B. Anand-Sirivastava, and A. K. Srivastava, Vanadyl sulfate-stimulated glycogen synthesis is associated with activation of phosphatidylinositol 3-kinase and is independent of insulin receptor tyrosine phosphorylation. *Biochemistry* **37**, 7006–7014 (1998).
75. J. D. Bogden et al., Balance and tissue distribution of vanadium after short-term ingestion of vanadate. *J. Nutr.* **112**, 2279–2285 (1982).
76. S. Ramanadham et al., The distribution and half-life for retention of vanadium in the organs of normal and diabetic rats orally fed vanadium (IV) and vanadium (V). *Biol. Trace Elem. Res.* **30** (1991).
77. F. G. Hamel and W. C. Duckworth, The relationship between insulin and vanadium metabolism in insulin target tissues. *Mol. Cell. Biochem.* **153**, 95–102 (1995).

78. K. Merritt, R. W. Margevicius, and S. A. Brown, Storage and elimination of titanium, aluminium and vanadium salts, in vivo. *J. Biomed. Mater. Res.* **26**, 1503–1515 (1992).
79. M. Gomez et al., Evaluation of the efficacy of various chelating agents on urinary excreation and tissue distribution in rats. *Toxicol. Lett.* **57**, 227–234 (1991).
80. R. D. R. Parker, R. P. Sharma, and S. G. Oberg, Distribution and accumulation of vanadium in mice tissues. *Arch. Environ. Contam. Toxicol.* **9**, 393–403 (1980).
81. V. P. Kamboj and A. B. Kar, Antitesticular effect of metallic and rare earth salts. *J. Reprod. Fertil.* **7**, 21–28 (1964).
82. M. Altamirano et al., Sex differences in the effects of vanadium pentoxide administration to prepubertal rats. *Med. Sci. Res.* **19**, 825–826 (1991).
83. J. M. Llobet et al., Reproductive toxicity evaluation of vanadium in male mice. *Toxicology* **80**, 199–206 (1993).
84. L. M. Altamirano et al., Reprotoxic and genotoxic studies of vanadium pentoxide in male mice. *Teratog. Carcinog. Mutagen.* **16**, 7–17 (1996).
85. J. Edel and E. Sabbioni, Vanadium transport across placenta and milk of rats to the fetus and newborn. *Biol. Trace Elem. Res.* **22**, 265–267 (1989).
86. J. L. Paternain et al., Embryotoxic effects of sodium metavanadate administered to rats during organogenesis. *Rev. Esp. Fisio.* **43**, 223–228 (1987).
87. M. Gomez et al., Embryotoxic and teratogenic effects of intraperitoneally administered metavanadate in mice. *J. Toxicol. Environ. Health* **37**, 47–56 (1992).
88. J. L. Paternain et al., Developmental toxicity of vanadium in mice after oral administration. *J. Appl. Toxicol.* **10**, 181–186 (1990).
89. J. L. Domingo et al., Influence of chelating agents on the toxicity, distribution, and excretion of vanadium in mice. *J. Appl. Toxicol.* **6**, 337–341 (1986).
90. T. Zhang et al., A study on developmental toxicity of vanadium pentoxide in Wistar rats. *Hua Hsi I Ko Ta Hsueh Hsueh Pao* **24**, 92–96 (1993).
91. L. M. Altamirano, L. Alvarez Barrera, and E. Roldan Reyes, Cytogenetic and teratogenic effects of vanadium pentoxide on mice. *Med. Sci. Res.* **21**, 711–713 (1993).
92. T. Zhang, X. Gou, and Z. Yang, A study on developmental toxicity of vanadium pentoxide in NIH mice. *Hua Hsi I Ko Ta Hsueh Hsueh Pao* **22**, 192–195 (1991).
93. M. Elfant and C. L. Keen, Sodium vanadate toxicity in adult and developing rats. Role of peroxidative damage. *Biol. Trace Elem. Res.* **14**, 193–208 (1987).
94. S. Ganguli et al., Effects of maternal vanadate treatment on fetal development. *Life Sci.* **55**, 1267–1276 (1994).
95. M. Kowalska, The effect of vanadium on lung collagen content and composition in two successive generations of rats. *Toxicol. Lett.* **41**, 203–208 (1988).
96. L. S. Levy, P. A. Martin, and P. L. Bidstrup, Investigation of the potential carcinogenicity of a range of chromium containing materials on rat lung. *Br. J. Ind. Med.* **43**(4), 243–256 (1986).
97. M. Kanisawa and H. A. Schroeder, Life-term studies on the effects of arsenic, germanium, tin, and vanadium on spontaneous tumours in mice. *Cancer Res.* **27**, 1192–1195 (1967).
98. J. A. Schroeder, M. Mitchener, and A. P. Nason, Zirconium, niobium, antimony, vanadium, and lead in rats: Life term studies. *J. Nutr.* **100**, 59–68 (1970).
99. H. A. Schroeder and M. Mitchener, Life-time effects of mercury, methyl mercury, and nine other trace metals on mice. *J. Nutr.* **10**, 452–458 (1975).

100. A. N. Kingsnorth et al., Vanadate supplements and 1,2-dimethylhydrazine induced colon cancer in mice: Increased thymidine incorporation without enhanced carcinogenesis. *Br. J. Cancer* **53**, 683–686 (1986).

101. P. Kopf-Maier, Cytostatic non-platinum metal complexes: New perspective for the treatment of cancer? *Naturwissenschaften* **74**, 374–382 (1987).

102. P. Kopf-Maier, P. Hesse, and H. Kopf, Tumor inhibition by metallocenes: Effect of titanocene and hafnocene dichlorides on Ehrlich ascites tumor in mice. *J. Cancer Res. Clin. Oncol.* **96**(1) 43–51 (1980).

103. H. J. Thompson, N. D. Chasteen, and L. D. Meeker, Dietary vanadyl (IV) sulfate inhibits chemically-induced mammary carcinogenesis. *Carcinogenesis (London)* **5**, 849–851 (1984).

104. C. Djordjevic and G. L. Wampler, Antitumor activity and toxicity of peroxo heteroligand vanadates(V) in relation to biochemistry of vanadium. *J. Inorg. Biochem.* **25**, 51–55 (1985).

105. S. Sardar et al., Protective role of vanadium in the survival of hosts during the growth of a transplantable murine lymphoma and its profound effects on the rates and patterns of biotransformation. *Neoplasma* **40**, 27–30 (1993).

106. A. Chakraborty and M. Chatterjee, Enhanced erythropoietin and suppresion of gamma-glutamyl transpeptidase (GGT) activity in murine lymphoma following administation of vanadium. *Neoplasma* **41**, 291–296 (1994).

107. A. Bishayee and M. Chatterjee, Inhibitiory effect of vanadium on rat liver carcinogenesis initated with diethylnitrosamine and promoted by phenobarbital. *Br. J. Cancer* **71**, 1214–1220 (1975).

108. A. Bishayee et al., Vanadium-mediated chemiprotection against chemical hepatocarcinogenesis in rats: Haematological and histological characteristics. *Eur. J. Cancer Prev.* **6**, 58–70 (1997).

109. G. A. Jamieson et al., Effects of phorbol ester on mitogen and orthovanadate stimulated responses of cultured human fibroblasts. *J. Cell. Physiol.* **134**, 220–228 (1988).

110. K. H. W. Lau, H. Tanimoto, and D. J. Baylink, Vanadate stimulates bone cell proliferation and bone collagen synthesis in vitro. *Endocrinology (Baltimore)* **123**, 2858–2867 (1988).

111. A. Tojo et al., Vanadate can replace interlukin-3 for transient growth of factor-dependent cells. *Exp. Cell Res.* **171**, 16–23 (1987).

112. J. B. Smith, Vanadium ions stimulated DNA synthesis in Swiss mouse 3T3 and 3T6 cells. *Proc. Natl. Acad. Sci. U.S.A.* **80**, 6162–6166 (1983).

113. E. Sabbioni et al., Cellular retention, cytotoxicity and morphological transformation by vanadium (IV) and vanadium (V) in BALB/3T3 cell lines. *Carcinogenesis (London)* **12**, 47–52 (1991).

114. E. Rivedal, L. E. Roseng, and T. Sanner, Vanadium compounds promote the induction of morphological transformation of hamster embryo cells with no effect on gap junctional communication. *Cell Biol. Toxicol.* **6**, 303–314 (1990).

115. L. A. Kowalski, S. S. Tsang, and A. J. Davidson, Vanadate enhances tranformation of bovine papillomavirus DNA-transfected C3H/10T1/2 cells. *Cancer Lett.* **64**, 83–90 (1992).

116. G. Carpenter, Vanadate, epidermal growth factor and the stimulation of DNA synthesis. *Biochem. Biophys. Res. Commun.* **102**, 1115–1121 (1981).

117. U. Hanauske et al., Biphasic effect of vanadium salts on in vitro tumor colony growth. *Int. J. Cell Cloning* **5**, 170–178 (1987).

118. A. Chakraborty et al., Vanadium: A modifier of drug-metabolizing enzyme patterns and its critical role in cellular proliferation in transplantable murine lymphoma. *Oncology*, **52**, 310–314 (1995).
119. N. Kanematsu, M. Hara, and T. Kada, Rec assay and mutagenicity studies on metal compounds. *Mutat. Res.* **77**, 109–116 (1980).
120. Sun Mianling, *Toxicity of Vanadium and its Environmental Health Standard*, West China University of Medical Science, 1987, p. 20.
121. G. Bronzetti et al., Vanadium: Genetical and biochemical investigations. *Mutagenesis* **5**, 293–295 (1990).
122. R. Ciranni, M. Antonetti, and L. Migliore, Vanadium salts induce cytogenetic effects in in vivo treated mice. *Mutat. Res.* **343**, 53–60 (1995).
123. Y. J. Owusu et al., An assessment of the genotoxicity of vanadium. *Toxicol. Lett.* **50**, 327–336 (1990).
124. C. B. Klein et al., Metal mutagenesis in transgenic Chinese hamster cell lines. *Environ. Health Perspect.* **102**, 63–67 (1994).
125. A. J. Carmicheal, Vanadyl-induced Fenton-like reaction in RNA: An ESR and spin trapping study. *FEBS Lett.* **261**, 165–170 (1990).
126. A. J. Carmicheal, Reaction of vanadyl with hydroperoxide: An ESR spin trapping study. *Free Radical Res. Commun.* **10**, 37–45 (1990).
127. H. Sakurai, Vanadium distribution in rats and DNA cleavage by vanadyl complex: Implication for vanadium toxicity and biological effects. *Environ. Health Perspect.* **102**, 35–36 (1994).
128. J. Z. Byczkowski and A. P. Kulkarni, Lipid peroxidation and benzo[a]pyrene derivative co-oxygenation by environmental pollutants.
129. X. Shi et al., Vanadium (IV)-mediated free radical generation and related $2'$-deoxyguanosine hydroxylation and DNA damage. *Toxicology* **106**, 27–38 (1996).
130. G. L. Curran and R. L. Costello, Reduction of excess cholesterol in the rabbit aorta by inhalation of endogenous cholesterol synthesis. *J. Exp. Med.* **103**, 49–56 (1956).
131. R. P. Steffen et al., Effect of prolonged dietary administration of vanadate on blood pressure in the rat. *Hypertension* **3**; I-173–I-178 (1981).
132. D. Susic and D. Kentera, Effect of chronic vanadate administration on pulmonary circulation in the rat. *Respiration* **49**, 68–72 (1986).
133. D. Susic and D. Kentera, Dependence of the hypertensive effect of chronic vanadate administration on renal excretory function in the rat. *J. Hypertens.* **6**, 199–204 (1988).
134. P. Boscolo et al., Renal toxicity and arterial hypertension in rats chronically exposed to vanadate. *Occup. Environ. Med.* **51**, 500–503 (1994).
135. L. M. Kopylova, A study of some blood indicators following prolonged administration of vanadium sulfate. *Tr. Voronezh. Gos. Med. Inst.* **85**, 62–65 (1971).
136. W. Trummert and G. Boehm, The trace element vanadium and its haemopoietic action. *Blut* **3**, 210–216 (1957) (in German).
137. S. M. Sokolov, Gigienitcheskaja ocenka nepreryvnogo i preryvistogo vozdiejstvija pjatiokisi vanadija. *Gig. Sanit.* **9**, 77–79 (1983).
138. H. Zaporowska and W. Wasilewski, Haematological results of vanadium intoxication in Wistar rats. *Comp. Biochem. Physiol. C* **101C**, 57–61 (1992).
139. R. P. Sharma et al., Effects of vanadium on immunological functions. *Am. J. Ind. Med.* **2**, 91–99 (1981).

140. M. D. Cohen et al., Effect of ammonium metavanadate on the murine immune response. *J. Toxicol. Environ. Health* **19**, 279–298 (1986).
141. A. Mravcova et al., Effects of orally administered vanadium on the immune system and bone metabolism in experimental animals. *Sci. Total Environ. Suppl.*, pp. 663–669 (1993).
142. D. Witkowska and J. Brzezinski, Alteration of brain noradrenaline, dopamine and 5-hydroxytryptamine levels during vanadium poisoning. *Pol. J. Pharmacol. Pharm.* **31**, 393–398 (1979).
143. R. P. Sharma, R. A. Coulombe, Jr., and B. Srisuchart, Effects of dietary vanadium exposure on levels of regional brain neurotransmitters and their meatabolites. *Biochem. Pharmacol.* **35**(3), 461–465 (1986).
144. K. P. Seljankina, Data for determining the maximum permissible content of vanadium in water basins. *Gig. Sanit.* **26**(10), 6–12 (1961).
145. K. W. Franke and A. L. Moxon, The toxicity of orally ingested arsenic, selenium, tellurium, vanadium and molybdenum. *J. Pharmacol. Exp. Ther.* **61**, 89–102 (1937).
146. J. T. Mountain, L. L. Delker, and H. E. Stokinger, Studies in vanadium toxicology: Reduction in the cystine content of rat hair. *Arch. Ind. Hyg. Occup. Med.* **8**, 406–411 (1953).
147. H. A. Schroeder and J. J. Balassa, Arsenic, germanium, tin and vanadium in mice: Effects on growth, survival and tissue levels. *J. Nutr.* **92**, 245–252 (1967).
148. W. F. Dutton, Vanadiumism. *J. Am. Med. Assoc.* **56**(22), 1648 (1911).
149. N. Williams, Vanadium poisoning from cleaning oil-fired boilers. *Br. J. Ind. Med.* **9**, 50–55 (1952).
150. R. C. Brown, Vanadium poisoning from gas turbines. *Br. J. Ind. Med.* **12**, 57–59 (1955).
151. R. E. Lees, Changes in lung function after exposure to vanadium compounds in fuel oil ash. *Br. J. Ind. Med.* **37**, 253–256 (1980).
152. C. Zenz and B. A. Berg, Human responses to controlled vanadium pentoxide exposure. *Arch. Environ. Health* **14**, 709–712 (1967).
153. V. M. Pazhynich, Experimental basis for the determination of maximum allowable concentration of vanadium pentoxide in atmospheric air. In V. A. Rjazanov, ed., *The Biological Effects and Hygienic Importance of Atmospheric Pollutants*, Medicina Publishing House, Moscow, 1967, pp. 201–217.
154. A. Todaro et al., Acute exposure to vanadium–containing dusts: The health effects and biological monitoring in a group of workers emoloyed in boiler maintenance. *Med. Lav.* **82**(2), 142–147 (1991).
155. H. Watanabe, H. Murayama, and S. Yamaoka, Some clinical findings in vanadium workers. *Jpn. J. Ind. Health* **8**(7), 23–27 (1966) (in Japanese).
156. J. Izycki et al., Evaluation of professional exposure to vanadium compounds and other environmental factors of workers employed in cleaning of oil-fired boilers with special regards to the state of respiratory tract. *Med. Pr.* **22**(4), 421–431 (1971).
157. R. Pistelli et al., Increase in non-specific bronchial responsiveness following occupational exposure to vanadium. *Med. Lav.* **82**, 270–275 (1991).
158. E. G. Dimond, J. Caravaca, and A. Benchimol, Vanadium: Excretion, toxicity, lipid effect in man. *Am. J. Clin. Nutr.* **12**, 49–53 (1963).
159. B. Gylseth et al., Vanadium in the blood and urine of workers in a ferroalloy plant. *Scand. J. Work Environ. Health* **5**, 188–194 (1979).
160. M. Kiviluoto, L. Pyy, and A. Pakarinen, Serum and urinary vanadium of workers processing vanadium pentoxide. *Int. Arch. Occup. Environ. Health* **48**, 251–256 (1981).

161. P. Orris, J. Cone, and S. McQuilkin, *Health Hazard Evaluation Report HETA 80–096–1359*, NTIS-PBBS-163574, Eureka Company, Bloomington, IL, U.S. Department of Health and Human Services, National Institute of Occupational Safety and Health, Washington, DC, 1983.
162. F. J. Vintinner et al., Study of the health of workers employed in mining and processing of vanadium ore. *AMA Arch. Ind. Health* **12**, 635–642 (1955).
163. J. R. McLean et al., Rapid detection of DNA strand breaks in human peripheral blood cells and animal organs following treatment with physical and chemical agents. In K. C. Bora et al., eds., *Chemicals Mutagenesis, Human Population Monitoring and Genetic Risk Assessment*, Progr. Mutat. Res., Vol. 3, Elsevier, Amsterdam, 1982, p. 137.
164. H. C. Birnboim, A superoxide anion induced DNA strand-break metabolic pathway in human leukocytes: Effects of vanadate. *Biochem. Cell Biol.* **66**, 374–381 (1988).
165. L. Migliore et al., Cytogenetic damage induced in human lymphocytes by four vanadium compounds and micronucleus analysis by fluorescence in situ hybridization with a centromeric probe. *Mutat. Res.* **319**, 205–213 (1993).
166. P. Ramirez et al., Disruption of microtubule assembly and spindle formation as a mechanism for the induction of aneuploid cells by sodium arsenite and vanadium pentoxide. *Mutat. Res.* **386**, 291–298 (1997).
167. E. Rojas et al., Genotoxicity of vanadium pentoxide evaluate by the single cell gel electrophoresis assay in human lymphocytes. *Mutat. Res.* **359**, 77–84 (1996).
168. V. Rajner, Effect of vanadium in the respiratory tract. *Cesk. Otolaryngol.* **9**, 202–204 (1960).
169. E. I. Matantseva, The state of the respiratory organs in workers coming into contact with vanadium pentoxide. *Gig. Tr. Prof. Zabol.* **7**, 41–44 (1960).
170. A. V. Roshchin, Hygienic evaluation of dust from vanadium-containing slag. *Gig. Sanit.* **12**, 23–29 (1963).
171. A. V. Roshchin, Vanadium metallurgy in the light of industrial hygiene and questions with the prevention of occupational diseases and intoxication. *Gig. Tr. Prof. Zabol.* **8**, 3–10 (1964).
172. K. P. Holzhauer and K. H. Schaller, Hazards at the workplace and occupation-linked health damage. In *Occupational Medicine Studies in Chimney Sweeps*. Thieme, Stuttgart, 1977.
173. M. Kiviluoto et al., Effects of vanadium on the upper respiratory tract of workers in a vanadium factory: A macroscopic and microscopic study. *Scand. J. Work Environ. Health* **5**, 50–58 (1979).
174. M. Kiviluoto, L. Pyy, and A. Pakarinen, Serum and urinary vanadium of vanadium-exposed workers. *Scand. J. Work Environ. Health* **5**, 362–367 (1979).
175. M. Kiviluoto, L. Pyy, and A. Pakarinen, Clinical laboratory results of vanadium-exposed workers. *Arch. Environ. Health* **36**, 109–113 (1981).
176. M. Kiviluoto, Observations on the lungs of vanadium workers. *Br. J. Ind. Med.* **37**, 363–366 (1980).
177. C. Zenz, J. P. Bartlett, and W. H. Thiede, Acute vanadium pentoxide intoxication. *Arch. Environ. Health* **5**, 542–546 (1962).
178. P. Stock, On the relations between atmospheric pollution and rural localities and mortality from cancer, bronchitis, pneumonia, with particular reference to 3,4-benzopyrene, beryllium, molybdenum, vanadium and arsenic. *Br. J. Cancer* **14**, 397–418 (1960).
179. R. J. Hickey, E. P. Schoff, and R. C. Clelland, Relationship between air pollution and certain chronic disease death rates. *Arch. Environ. Health* **15**, 728–738 (1967).

180. H. A. Schroeder, Municipal drinking-water and cardiovascular death rates. *J. Am. Med. Assoc.* **195**, 81–85 (1966).
181. A. W. Voors, Minerals in the municipal water and atherosclerotic heart death. *Am. J. Epidemiol.* **93**, 259–266 (1971).
182. R. Masironi, Trace elements and cardiovascular diseases. *Bull. W. H. O.* **40**, 305–312 (1969).
183. B. Wehrli and W. Stumm, Vanadyl in natural waters: Adsorption and hydrolysis promote oxygenation. *Geochim. Cosmochim. Acta* **53**, 69–77 (1989).
184. S. Fox, Assessment of cadmium, lead and vanadium status of large animals as related to the human food chain. *J. Anim. Sci.* **65**, 1744–1752 (1987).
185. International Commission on Radiological Protection (ICRP), *Radionucleide Transformations: Energy and Intesity of Emissions*, Pergamon, Oxford; ICRP Pub. No. 38, Ann. ICRP 19(11–13), 1983, pp. 216–237.
186. E. S. Bartlett, Niobium and compounds. In C. A. Hampel and G. G. Hawley, eds., *The Encyclopedia of Chemistry,* 3rd ed., Van Nostrand Reinhold, New York, 1973, pp. 711–712.
187. A. Fichte, H. J. Reisisdorf, and H. Rothmans, Niobium and niobium compounds. In *Ullmanns Encyclopedie der Technischen Chemie*, 4th ed., Vol. 17, Verlag Chemie, Weinheim, 1979, pp. 303–314.
188. N. A. Tyutina, V. B. Aleskovskii, and P. L. Vasilev, *Geochemistry* **6**, 668–671 (1959).
189. K. W. Bruland, Trace elements in seawater. *Chem. Oceanogr.* **8**, 208 (1983).
190. A. Yamato, N. Miyagawa, and N. Miyanaga, Radioactive nuclides in the marine environment— distribution and behaviour of ^{95}Zr, ^{95}Nb originated from fallout. *Radioisotopes* **33**, 449–455 (1984).
191. G. R. Waterbury and C. E. Dick, *Anal. Chem.* **30**, 1007 (1958).
192. R. Wennig and N. Kirsch, Niobium. In H. G. Seiler, H. Sigel, and A. Sigel, eds., *Handbook on Toxicity of Inorganic Compounds*, Dekker, New York, 1988, pp. 469–473.
193. H. A. Schroeder and J. J. Balassa, Abnormal trace metals in man: Niobium. *J. Chronic Dis.* **18**, 229 (1965).
194. E. J. Hamilton, N. J. Minski, and J. J. Cleary, The concentration and distribution of some stable elements in healthy human tissues from United Kingdom. *Sci. Total Environ.* **1**, 341–347 (1972/1973).
195. K. Burger, *Organic Reagents in Metal Analysis,* Pergamon, New York, 1973, pp. 111–126, 196.
196. J. Schubert, An experimental study of the effect of zirconium and sodium citarte treatment on the metabolism of plutonium and radioyttrium. *J. Lab. Clin. Med.* **34**, 313 (1949).
197. K. W. Cochran et al., Acute toxicity of zirconium, columbium, strontium, lanthanum, cesium, tantalum and yttrium. *Arch. Ind. Hyg. Occup. Med.* **1**, 637–650 (1950).
198. T. J. Haley, N. Komesu, and K. Raymond, Pharmacology and toxicology of niobium chloride. *Toxicol. Appl. Pharmacol.* **4**, 385–392 (1962).
199. W. L. Downs et al., The toxicity of niobium salts. *Am. Ind. Hyg. Assoc. J.* 337–346 (1965).
200. H. A. Schroeder et al., Zirconium, niobium, antimony and fluorine in mice: Effects on growth, surival and tissue levels. *J. Nutr.* **95**, 95–101 (1968).
201. R. G. Cuddihy, Deposition and retention of inhaled niobium in beagle dogs.*Health Phys.* **34**, 167–176 (1978).
202. R. G. Thomas, R. L. Thomas, and J. K. Scott, Distribution and excretion of niobium following inhalation exposure of rats. *Am. Ind. Hyg. Assoc. J.* **28**, 1–7 (1967).

203. C. R. Flechter, The radiological hazards of zirconium-95 and niobium-95. *Health Phys.* **16**, 209–220 (1969).
204. J. E. Furchner and G. A. Drake, Comparative metabolism of radionuclides in mammals. VI. Retention of ^{95}Nb in the mouse, rat, monkey and dog. *Health Phys.* **21**, 173–180 (1971).
205. S. Lang and T. Raunemaa, Behavior of neutron-activated uranium dioxide dust particles in the gastrointestinal tract of the rat. *Radiat. Res.* **126**, 273–279 (1991).
206. Y. Shiraishi and R. Ichikawa, Absorption and retention of ^{144}Ce and ^{95}Zr, ^{95}Nb in newborn, juvenile and adult rats. *Health Phys.* **22**, 373–378 (1972).
207. F. R. Mraz and G. R. Eisele, Gastrointestinal absorption of ^{95}Nb by rats of different ages. *Radiat. Res.* **69**, 591–593 (1977).
208. F. R. Mraz and G. R. Eisele, Gastrointestinal absorption, tissue distribution and excretion of ^{95}Nb in newborn and weanling swine and sheep. *Radiat. Res.* **72**, 533–536 (1977).
209. J. D. Harrison, J. W. Haines, and D. S. Popplewell, The gastrointestinal absorption and retention of niobium in adult and newborn guinea pigs. *Int. J. Radiat. Biol.* **58**(1), 177–186 (1990).
210. G. R. Eisele and F. R. Mraz, Absorption of ^{95}Nb from ligated segments of the gastrointestinal tract of the rat. *Health Phys.* **40**, 235–238 (1981).
211. J. G. Hamilton, The metabolism of the fission products and the heaviest elements. *Radiology* **49**, 325–333 (1947).
212. M. Schneidereit and H. Kriegel, Comparative distribution of niobium-95 in maternal and fetal rats. *Experientia* **42**, 619–620 (1986).
213. A. Ando and I. Ando, Biodistribution of 95Nb and 182Ta in tumor-bearing animals and mechanisms for accumulation in tumor and liver. *J. Radiat. Res.* **31**(1), 97–109 (1990).
214. J. P. Berry, F. Bertrand, and P. Galle, Selective intralysosomal concentration of niobium in kidney and bone marrow cells: A microanalytical study. *Biometals* **6**, 17–23 (1993).
215. S. J. Waligora, Jr., Pulmonary retention of zirconium oxide (^{95}Nb) in man and beagle dogs. *Health Phys.* **20**, 89–91 (1971).
216. P. Kopf-Maier and T. Klapotke, Antitumor activity of ionic niobocene and molybdenocene complexes in high oxidation states. *J. Cancer Res. Clin. Oncol.* **118**(3), 216–221 (1992).
217. T. Kada, K. Tutikawa, and Y. Sadaie, In vitro and host-mediated rec-assay procedures for screening chemical mutagens; and phloxine, a mutagenic red dye detected. *Mutat. Res.* **16**(2), 165–174 (1972).
218. R. E. Cofield, In vivo gamma spectroscopy for inhalations of neptunium-237-protactinium-233, cobalt-60, and zirconium-95-niobium-95. *Health Phys.* **9**, 283–292 (1963).
219. International Commission on Radiological Protection (ICRP), *Limits for Intakes of Radionucleides by Workers,* ICRP Publ. No. 30, Part 1, Ann. ICRP 2 (3/4), Pergamon, Oxford, 1979, pp. 79–80.
220. K. S. Thind, Retention and excretion of ^{95}Zr- ^{95}Nb in humans. *Health Phys.* **69**(6), 957–960 (1995).
221. M. E. Wrenn, R. Mowafy, and G. R. Laurer, ^{95}Zr- ^{95}Nb in human lungs from fallout. *Health Phys.* **10**, 1051–1058 (1964).
222. C. W. Sill et al., Two studies of acute internal exposure to man involving gerium and tantalum radioisotopes. *Health Phys.* **16**, 325–332 (1969).
223. International Commission on Radiological Protection (SCRP), *Report of the Task Group on Reference Man,* ICRP Publ. 23, Pergamon, Oxford, 1975.

224. H. M. Solli et al., Cancer incidence among workers exposed to radon and thorn daughters at a niobium mine. *Scand. J. Work Environ. Health* **11**, 7–13 (1985).
225. P. H. Payton, Niobium and niobium compunds. In *Kirk-Othmer Concise Encyclopedia of Chemical Technology,* 3rd ed., Wiley, New York, 1985, pp. 783–785.
226. L. E. Stettler, H. M. Donaldson, and G. C. Grant, Chemical composition of coal and other mineral slags. *Am. Ind. Hyg. Assoc. J.* **43**, 225–238 (1982).
227. J. T. McFadden, Neurosurgical metallic implants. *J. Neurosurg. Nurs.* **3**, 123–130 (1971).
228. U. Forstner, Metal pollution of terrestrial waters. In J. O. Nriagu, ed., *Changing Metal Cycles and Human Health,* Springer-Verlag, New York, 1984, pp. 71–94.
229. National Institute of Occupational Safety and Health (NIOSH), *Manual of Analytical Methods,* 2nd ed., Vol. 5, Method No. S201, U.S. Department of Health Education and Welfare, NIOSH, Cincinnati, OH, 1977.
230. A. R. Eberle and M. W. Lerner, Spectrophotometric determination of tantalum in boron, uranium, zirconium, and uranium-zircaloy-2 alloy with malachite green. *Anal. Chem.* **39**, 662–664 (1967).
231. V. M. Tarayan et al., *Dokl. Akad. Nauk Arm. SSR* **48**, 52 (1969).
232. J. D. Burton and K. S. Massie, *J. Mar. Biol. Assoc. U.K.* **51**, 679–683 (1971).
233. De Meester, Tantalum. In H. G. Seiler, H. Sigel, and A. Sigel, eds., *Handbook on Toxicology of Inorganic Compounds,* Dekker, New York, 1988, pp. 661–663.
234. J. Edel et al., Trace metal lung disease: Hard metal pneumoconiosis. A case report. *Acta Pharmacol. Toxicol.* **59**(Suppl. 7), 52–55 (1986).
235. F. Lux and R. Zeister, *Anal. Chem.* **261**, 314 (1972).
236. Glukhov et al., *Radiokhimiya* **12**, 534 (1970).
237. H. F. Haas and V. Krivan, *Z. Fresenius Anal. Chem.* **314**, 532–537 (1983).
238. F. Della Torre et al., Trace metal lung diseases: A new fatal case of hard metal pneumoconiosis. *Respiration* **57**, 248–253 (1990).
239. M. L. Jacobs, Evaluation of spark source mass spectrometry in the analysis of biologic samples. *HEW Publ. (NIOSH) (U.S.),* pp.75–186 (1975).
240. G. Rizzato et al., Multi-element follow up in biological specimens of hard metal pneumoconiosis. *Sarcoidosis* **9**, 104–117 (1992).
241. B. Venugopal and T. I. Kuckey, *Metal Toxicity in Mammals,* Vol. 2, Plenum, New York, 1978, pp. 229–231.
242. G. A. Shurko and I. T. Brakhova, Studies on the influence on organism of transitional metals of group V hydrides with correlations to its electrochemical and crystalline structure. *Gig. Tr. Prof. Zabol.* **1**, 43–45 (1975).
243. A. B. Delahant, An experimental study of the effects of rare metals on animal lungs. *AMA Arch. Ind. Health* **1**, 116–120 (1955).
244. G. W. H. Schepers, The biological action of tantalum oxide. *AMA Arch. Ind. Health* **1**, 121–123 (1955).
245. H. M. Carney, An experimental study with tantalum. *Proc. Soc. Exp. Biol. Med.* **51**, 147–148 (1942).
246. G. L. Burke, The corrosion of metals in tissues; an introduction to tantalum. *Can. Med. Assoc. J.* **43**, 125–128 (1940).
247. M.A. Meenaghan et al., Tissue response to surface-treated tantalum implants: Preliminary observations in primates. *J. Biomed. Mater. Res.* **13**, 631–643 (1979).

248. W. E. Weller and E. Kammler, Long-term effect of tantalum dust in connection with inhalation bronchography. *Respiration* **30**, 430–442 (1973).
249. R. Masse, R. Ducousso, and D. Nolibe, Etude expérimentale de la pulmonaires des particles métalliques: Application à la bronchographie au tantale. *Rev. Fr. Mal. Respir.* **1**, 1063–1066 (1973).
250. P. E. Morrow et al., Pulmonary retention and translocation of insufflated tantalum. *Radiology* **121**, 415–421 (1976).
251. D. G. Fleshman, A. J. Silva, and B. Shore, The metabolism of tantalum in the rat. *Health Phys.* **21**, 385–392 (1971).
252. P. W. Durbin, K. G. Scott, and J. G. Hamilton, The distibution of radioisotopes of some heavy metals in the rat. *Lawrence Radiat. Lab., Berkeley, Rep.* **UCRL-3607**, 17–19 (1956).
253. J. G. Hamilton et al., The metabolic properties of various elements. *Lawrence Radiat. Lab., Berkeley, Rep.* **UCRL-1143** (AECD-3200), 18–22 (1950).
254. Y. Shiraishi and R. J. Ichikawa, *Radiat. Res. (Tokyo)* **13**, 14 (1972).
255. K. G. Scott and J. Crawley, Tracer studies. *Lawrence Radiat. Lab., Berkeley, Rep.* **UCRL-1143** (1951).
256. F. L. Harrison and D. J. Quinn, Tissue distribution of accumulated radionuclides in freshwater clams. *Health Phys.* **23**, 509–517 (1972).
257. J. Doull and K. Dubois, *Metabolism and Toxicity of Radioactive Tantalum*, Part 2, Q. Prog. Rep. No. 2, University of Chicago Toxicity Lab., Chicago, 1949, p. 12.
258. A. Bianco et al., *Radiology* **112**, 549 (1974).
259. B. S. Oppenheimer et al., Carcinogenic effect of metals in rodents. *Cancer Res.* **16**, 439–441 (1956).
260. R. A. Matthay et al., Tantalum oxide, silica and latex: Effects on alveolar macrophage viability and lysozyme relase. *Invest. Radiol.* **13**, 514–518 (1978).
261. J. L. Egorov, Materials for hygienic characteristics of rare elements dusts: Tantalum and niobium. *Gig. Tr. Prof. Zabol.* **1**, 16 (1957).
262. C. W. Miller et al., Pneumoconiosis in the tungsten-carbide tool industry. *AMA Arch. Ind. Health* **8**, 453–465 (1953).
263. A. R. Koontz and R. C. Kimberly *Ann. Surg.* **137**, 833 (1953).
264. F. Mitchell-Heggs, *Br. J. Surg.* **50**, 907 (1963).
265. B. S. Werman and R. L. Rietschel, Chronic urticaria from tantalum samples. *Arch. Dermatol.* **117**(7), 438 439 (1981).
266. C. Romaguera and J. Vilaplana, Contact dermatitis from tantalum. *Contact Dermatitis* **32**, 184 (1995).
267. V. A. Przybyla, Jr. and F. G. LaPiana, Complications associated with use of tantalum-mesh-covered implants. *Ophthalmology* **89**, 121–123 (1982).
268. G. Gamzu, R. M. Weintraub, and J. A. Nadel, Clearance of tantalum from airways of different caliber in man evaluated by a roentgenographic method. *Am. Rev. Respir. Dis.* **107**, 214–224 (1973).
269. D. Newton, Clearance of radioactive tantalum from the human lung after accidental inhalation. *Am. J. Roentgenol.* **32**, 327–328 (1977).
270. K. Gutensohn et al., Flow cytometric analysis of coronary stent-induced alterations of platelet antigens in an in vitro model. *Thromb. Res.* **86**, 49–56 (1997).

CHAPTER THIRTY-EIGHT

Chromium, Molybdenum, and Tungsten

Sverre Langård, MD

1.0 Chromium

1.0.1 CAS Number: [7440-47-3]

1.0.2 Synonyms: Chrome

1.0.3 Trade Names: NA

1.0.4 Molecular Weight: 52

1.0.5 Molecular Formula: Cr

1.1 Chemical and Physical Properties

1.1.1 General

The physical and chemical characteristics of chromium and some of its compounds are summarized in Table 38.1.

The term *chromium* is derived from the Greek word for color, because most chromium compounds are brightly pigmented. The element chromium was discovered in 1798 by N. L. Vauquelin, but it had already been used in swords by the Hittits about 1300 B.C. Chromium occurs in nature in bound-form chromite ore, which is the only chromium ore of any importance, and it makes up 0.1–0.3 ppm of the earth's crust. The red color of rubies and green color of emeralds, serpentine, and chrome mica are produced by chromium.

Patty's Toxicology, Fifth Edition, Volume 3, Edited by Eula Bingham, Barbara Cohrssen, and Charles H. Powell.
ISBN 0-471-31934-1   © 2001 John Wiley & Sons, Inc.

Table 38.1. Physical and Chemical Properties of Selected Chromium Compounds

Compound	Mol. Formula	At. or Mol. Wt.	Sp. Gr.	M.P. (°C)	B.P. (°C)	Reactivity
Chromium	Cr	51.996	7.2	1857	2672	Reacts with HCl, H_2SO_4, not HNO_3
Chromyl chloride	CrO_2Cl_3	154.92	1.91	116	—	Miscible with carbon tetrachloride, CS_2; reacts with H_2O to form CrO_3, HCl, $CrCl_3$, Cl_2
Chromic oxide (chromium sesquioxide)	Cr_2O_3	152.02	5.21	2435	4000	Insol. water; sl. sol. acids, alkalies
Chromic trioxide, "chromic acid"	CrO_3	100.1	2.70	196	Dec.	Sol water, H_2SO_4
Chromic chloride	$CrCl_3$	158.35	2.76	1150 subl.	1300	Sol. water, alcohol; insol. ether
Chromous chloride	$CrCl_2$	122.92	2.75	824	—	V.sol. water; insol. ether
Potassium dichromate	$K_2Cr_2O_7$	292.19	2.676	—	Dec. 500	Sl. sol. water; insol. alcohol

Chromium exhibits a valence of 2+, 3+, and 6+ in its compounds. It is a blue-white, hard metal that is not oxidized in moist air or by heating, and it oxidizes to only a slight extent. In an atmosphere of CO_2 it oxidizes to Cr_2O_3; in HCl, to $CrCl_2$. Chromium combines directly with nitrogen, carbon, silicon, and boron. A passive form of the metal is conferred by oxidizing acids, attributable to a film of Cr_2O_3. Bivalent chromium compounds are basic, the trivalent compounds amphoteric, and the hexavalent compounds acidic. The chromate ion in acidic solution is a powerful oxidizing agent. Divalent chromium compounds closely resemble Fe^+ compounds; Cr_3^+ compounds resemble those of Al_3^+ (1).

1.1.2 Odor and Warning Properties: None

1.2 Production and Use

Chromite was mined in the United States up to 1961. In 1989, 248.6 million lb of chromium was produced in the United States; 94.6 million lb was imported, and 35.2 million lb was exported. It is mined in African countries, Russia, Turkey, and other countries (2).

Chromium metal is prepared by reducing the ore in a blast furnace with carbon (coke) or silicon to form an alloy of chromium and iron called ferrochrome, which is used as the starting material for the many iron-containing alloys that employ chromium. Chromium to be used in iron-free allloys is obtained by reduction or electrolysis of chromium compounds. Chromium is difficult to work in the pure metal form; it is brittle at low temperatures, and its high melting point makes it difficult to cast.

The use of chromium in stainless steel (18%+) is a major use of the element. Chromium is alloyed with iron to enhance its resistance to corrosion, its hardness, and its workability. Other metals (e.g., vanadium, manganese, tungsten, and molybdenum) are added to these alloys to obtain special properties. Stainless steel always contains nickel and chromium, generally 8–11 or 18–21%, respectively. Super corrosion-resistant types of steel, such as those used for furnaces, heat exchangers, and burner heads, contain about 30% chromium. In the early 1990s stainless and heat-resistant steel accounted for about four-fifths of the United States consumption of chromium.

Significant nonferrous (iron-free) chromium alloys include stellite, which contains cobalt and tungsten and is used in cutting, lathing, and milling tools, and NiCr (nichrome), which is used in resistance wire in electrical heaters, irons, and toasters. Smaller amounts of chromium are used in drilling muds, in water treatment as rust and corrosive inhibitors, for toners in copying machines, and in magnetic tapes.

Chrome plating, which creates a hard, wear-resistant, attractive surface, is another major area of chromium use. Chrome plating can be performed by immersion or by electrolysis. The latter method allows very thin layers to be deposited, but uses a good deal of current; the cathode current efficiency is only 10–15%.

Most manmade compounds of chromium contain chromium in 3+ and 6+. Chromium compounds often have a green color; but yellow, blue, red, and violet compounds are also known. Chromium trioxide, Cr_2O_3, is one of significant importance and is used as a pigment (chromic oxide green). Chrome alum forms beautiful violet crystals and is used in the tanning of leather and in textile dyeing. A number of other chromium salts are utilized in the textile industry as mordants.

Compounds of chromium with a valance of 6+ are named chromates. Most of these have a yellow color, and all are toxic to cells, animals, and humans. Chromates are used as anticorrosives in water-cooling systems, sometimes giving rise to release of Cr(6+) to the environment. Chrome yellow is a pigment used to a large extent and is toxic because it contains Cr(6+) and lead. Because these elements are toxic, their runoff into rivers may exert severe effects on river flora. Trivalent chromium is the most abundant form of chromium brought into the environment by human activity.

Chromyl chloride [*14977-61-8*] is used in organic oxidations and chlorinations, as a solvent for CrO_3 and in the making of chromium complexes.

The U.S. National Occupational Exposure Survey estimated that a total of about 200,000 workers, including about 30,000 women, were potentially exposed to hexavalent chromium compounds (3). The typical airborne concentrations in various industrial operations are given in Table 38.2; however, the combustion of coal and oil is the largest single source of air pollution (4).

Chromium in the trivalent form is an essential trace element to humans. It is involved in the metabolism of glucose. Chromium deficiency may result in impaired glucose tolerance, peripheral neuropathy, and elevated serum insulin, cholesterol, and triglycerides, similar to those symptoms observed in diabetic patients (5). The U.S. National Research Council has recommended a daily chromium intake of 50 to 200 mg, based on an average intake of 50 mg and resulting in no signs of deficiency (6, 7).

Table 38.2. Chromium Concentration in Various Industrial Operations

Type of Industrial Process	Range of Airborne Concentrations ($\mu g\,Cr/m^3$)
Stainless steel welding	50–400
Chromate production	100–500
Chrome plating	5–25
Ferrochrome alloys	10–140
Chrome pigment	60–600
Tanning	10–50 (nearly all soluble trivalent Cr)

Chromium trioxide (Cr 6+) is still being used in some countries as an agent to stop nosebleeds. However, health authorities in most countries have recommended to stop that use due to the toxic effects of Cr(6+).

1.3 Exposure Assessment

1.3.1 Air

Chromium is almost ubiquitous in nature, and chromium in the air may originate from erosion of shales, clay, and many other kinds of soil. In countries with chromite mining, production processes may contribute to airborne chromium. Endpoint production of chromium compounds probably constitutes the most important source of chromium in air. There is only limited information on chromium concentrations in the environmental atmosphere.

1.3.2 Background Levels

Measurements carried out above the North Atlantic, north of latitude 30°N, thousands of kilometers from major land masses, showed concentrations from 0.07 to 1.1 ng Cr/m^3, and the concentrations above the South Pole were slightly lower (8). In the Shetland Islands and Norway 0.7 ng Cr/m^3 has been measured, 0.6 ng/m^3 in north-west Canada, 1–140 ng Cr/m^3 in continental Europe, 20–70 ng Cr/m^3 in Japan, and 45–67 ng Cr/m^3 in Hawaii (9). Monitoring of the ambient air in 1977–1980 in many urban and rural areas of the United States showed concentrations to be in the range of 5.2 to 156.8 ng Cr/m^3, rural and urban annual average, respectively. The maximum concentration determined in the United States in any one measurement was about 684 ng Cr/m^3 (24-h average) (10). Mass median diameters of chromium particles in the ambient air have been reported to be in the range 1.5–1.9 μm (10). Ranges of levels in the European Union member states were given as follows: remote areas: 0–3 ng Cr/m^3; urban areas: 4–70 ng Cr/m^3; and industrial areas: 5–200 ng Cr/m^3 (11). As the toxicity and carcinogenicity is completely different for Cr(6+) and Cr(3+) compounds, monitoring of air concentrations of chromium in air should be made separately for the two species.

1.3.3 Workplace Methods

Monitoring chromium in the workplace is performed either by area sampling or by using personal samplers. Personal samples are collected on a 37-or 25-mm, 0.8-mm-pore size, cellulose ester membrane filter. To avoid reduction from chromium (6+) to chromium (3+) on the filter, hexavalent chromium should be sampled on PVC filters. Sampling pumps are generally calibrated at 2.0 L/min. The samples are analyzed using atomic absorption spectrometry (AAS) or inductively coupled plasma atomic emission spectrometry. For AAS the wavelength is 357.9 nm. An air–acetylene flame is generally used, although in the presence of iron and nickel a nitrous oxide–acetylene flame is recommended. Care should be taken in this analysis because some of the filters may contain chromium. This description is for NIOSH Method 7024. For hexavalent chromium (+6), NIOSH Methods 7600 or 7604 are recommended for determining workplace exposures (11a).

1.3.4 Community Methods

1.3.5 Biomonitoring/Biomarkers

Hair samples and lung tissue samples have been taken for the purpose of monitoring chromium exposure and the body burden of chromium, respectively, but so far these attempts have been unsuccessful for monitoring exposure. However, such analyses may still prove to be useful for research purposes.

1.4 Toxic Effects

Trivalent (chromic) and hexavalent (chromate) are the two species of chromium of prime interest for toxicity. Chromates, the most biologically active species, are — particularly the water-soluble compounds — readily taken up by living cells and reduced within the cell via reactive intermediates to stable Cr(3+) species. Reduction of chromates also takes place in various body fluids. Hexavalent compounds are characterized by variability in water solubility, differences that seem to be of major significance to the detoxification and the bioavailability, hence the toxicity and carcinogenicity, of the different hexavalent chromium compounds.

1.4.1 Experimental Studies

1.4.1.1 Acute Toxicity. Short-term inhalation, oral, or dermal studies suggest that female animals are more susceptible to the lethal effects of Cr(6+). Whether this relationship exists in the human population is unclear (4).

1.4.1.2 Chronic and Subchronic Toxicity. Calcium chromate, chromium trioxide, lead chromate, strontium chromate, and zinc chromate are shown to be able to induce sarcomas after intrabronchial implantation, intramuscular, intrapleural, or subcutaneous injection. Few if any sarcomas have been produced by barium chromate, sodium chromate, sodium dichromate, or chromic acetate (13).

1.4.1.3 Pharmacokinetics, Metabolism, and Mechanisms. The uptake of chromium after Cr(6+) exposure in the airways depends on the concentration in the inhaled air, the chemical properties of the inhaled compound, and the different interactions between the Cr-containing aerosols and concomitant exposure factors, such as tobacco smoking. Such interactions are reflected in the excretion of chromium in the urine; current smoking apparently increases the excretion of chromium, probably resulting from an enhanced retention in the bronchial tree resulting in enhanced uptake of chromium. Such an enhancement was observed by Kalliomüki et al. (14) and by Sjögren et al. (15), both studies revealing an increased urinary total chromium excretion among smoking welders as compared with nonsmoking welders when the two groups were exposed to comparable air levels of fumes from stainless steel welding, which are known to contain Cr(6+) compounds. Stridsklev et al. (16) also found enhanced excretion of total chromium in the urine of smoking compared with nonsmoking MMA/SS (metal arc welding/stainless steel) welders.

Detoxification of Cr(6+) compounds by reduction to the Cr(3+) state takes place in the saliva, epithelial lining fluid, and pulmonary alveolar macrophages, red blood cells, and in the liver cells and it has been shown that reduction of Cr(6+) in body fluids greatly reduces its potential toxicity and genotoxicity. For example, the Cr(6+) reductive capacity of whole blood is 187--234 mg per individual and the reducing capacity of red blood cells is at least 93–128 mg, explaining why the element has no systemic toxicity, except at very high doses. It has also been shown that reduction takes place in the saliva (0.7–2.1 mg) and the gastric juice (at least 84–88 mg) (17, 18). The reduction of Cr(6+) also explains the absence of cancer occurring at a distance from the *port of entry* into the organism. The Cr(6+) that escapes reduction in the digestive tract will be detoxified in the blood and then in the liver, having an overall reducing capacity of 3300 mg. The reductive processes of the GI tract seem to explain the relatively low oral toxicity of Cr(6+) compounds and the absence of cancer resulting from ingestion of Cr(6+) compounds, that is, after swallowing reflux from the respiratory tract (17). Similarly in terminal airways Cr(6+) is reduced in the epithelial lining fluid (0.9–1.8 mg) and in pulmonary alveolar macrophages (136 mg). These reduction hurdles may only be overwhelmed under conditions of massive exposure by inhalation, as occurs in certain work environments prior to the implementation of suitable industrial hygiene measures (17). This explains why most of the toxic effects and the carcinogenic outcomes take place subsequent to inhalation of Cr(6+) compounds, and why Cr(6+) taken in by the oral route is likely to be effectively reduced to noncarcinogenic Cr(3+) compounds (19). Should chromate cross the cell membranes, reduction takes place in the mitochondria, in the mixed-function oxidase system in cell cytoplasm, and in the nuclear membrane (20, 21).

The types of reactive intermediates generated upon reduction of chromium (6+) by glutathione, ascorbic acid, or hydrogen peroxide and the possible resulting DNA damage have been studied and determined. Reaction of chromium (6+) with glutathione *in vitro* led to formation of two Cr(6+) complexes and the glutathione thioyl radical. When Cr(6+) was reacted with DNA in the presence of glutathione, chromium–DNA adducts with no DNA strand breakage were obtained. The level of chromium–DNA adduct formation correlated with chromium (6+) formation. Reaction of chromium (6+) with hydrogen peroxide led to formation of hydroxyl radical. No Cr(6+) was detectable at 24°C; however, low levels of the tetraperoxochromium (6+) complex were detected at much lower temperatures. Reaction of Cr(6+) with DNA in the presence of hydrogen peroxide produced significant DNA strand breakage and the 8-hydroxydeoxyguanosine adduct, whose formation correlated with hydroxyl radical production. No significant chromium–DNA adduct formation was found. Thus the nature of Cr(6+)-induced DNA damage appears to depend on the reactive intermediates, that is, Cr(6+) or hydroxyl radical produced during the reduction of Cr(6+) (22).

1.4.1.3.1 Absorption

1.4.1.3.1.1 AIRWAYS. Animal studies on uptake in the airways have been shown to be difficult to carry out; hence there are only few animal studies demonstrating uptake of chromium subsequent to inhalation of chromates (23).

1.4.1.3.1.2 THE GASTROINTESTINAL (GI) TRACT. There are significant differences between the rate of uptake of compounds of Cr(3+) and Cr(6+) in the GI tract (6). About 0.5% of ingested compounds of Cr(3+) seem to be taken up in the GI tract, while a much greater proportion of ingested Cr(6+) compounds is taken up. The rate of uptake of the Cr(6+) compounds seems to be comparable in humans and animals (23, 24). To permit uptake in the GI tract, Cr(3+) must be present as an organic compound (**d1**). There is also evidence that the water solubility influences the uptake of organic Cr(3+) from the GI tract (6).

1.4.1.3.1.3 SKIN. Hexavalent chromium compounds are taken up through the skin to a limited extent, while Cr(3+) compounds penetrate with great difficulty. Skin uptake of Cr(6+) compounds could be a prerequisite for the allergic skin reactions to Cr compounds to occur (25, 26).

1.4.1.3.2 Distribution. After IV delivery of ^{51}CrCl$_3$, a large proportion of the ^{51}Cr is accumulated in the lungs, the liver, and the kidneys, and only about one-third of the administered Cr remains in the bloodstream after 1 h (6, 12). After IV administration of Na$_2$^{51}CrO$_4$ to rats, the distribution to the organs of ^{51}Cr seemed to be dependent on the dose; high doses resulted in relatively more ^{51}Cr retention in lung tissue, while the retention is most pronounced in the liver and the kidneys during the first few days and even later in the spleen. The ^{51}Cr was quite evenly distributed between plasma and erythrocytes shortly after IV administration, but there was a quick shift to the erythrocytes; and after 24 h little ^{51}Cr was left in the plasma (20). Chromium from Cr(3+) compounds is taken up in the erythrocytes to a very limited extent (27). The tissue distribution of the two species of chromium does not differ significantly, except for the uptake in red blood cells, which is easily explained by the fact that Cr(6+) is readily reduced to Cr(3+) in various tissues (28, 29).

Hexavalent chromium deposited in the bronchial tree is reduced locally, except when the reductive capacity of the bronchial lining fluids, macrophages, and epithelial cells is overwhelmed by too large a dose (17, 18). Hence, the chromium deposited as Cr(6+) is taken up and transported to the circulating blood as Cr(3+), except that Cr(6+) is so abundant that it overwhelms the reductive capacity or for other reasons passes through the epithelial lining as Cr(6+) and into the bloodstream. Cr(6+) is subsequently reduced to

Cr(3+) by the red blood cells and other constituents of the blood (27). Although some chromium transportation may take place via the lymph, the blood is the main transportation medium for chromium.

Subsequent to IV administration of $^{51}CrCl_3$, a significant proportion of the ^{51}Cr is accumulated in the lungs, the liver, and the kidneys, and only about one-third of this Cr remains in the blood after 1 h. After IV administration of $Na_2^{51}CrO_4$ to rats, the distribution to the organs of ^{51}Cr appears to differ with the magnitude of the dose; the higher the dose, the more retention in the lung tissue, while the retention is most pronounced in the liver and the kidneys during the first few days and later in the spleen (20). Immediately after IV administration ^{51}Cr is quite evenly distributed between plasma and erythrocytes, but there is a quick shift to the erythrocytes; after 24 h only a little ^{51}Cr is left in the plasma (20). Trivalent chromium compounds are taken up in the erythrocytes only to a limited extent (27). Thus the tissue distribution of the two species of chromium is not significantly different, except for the uptake in red blood cells and the accumulation of ^{51}Cr in the lungs after administration of $Na_2^{51}CrO_4(r)$. A reason for this apparent equality may result from the fact that chromium does not bind to organic molecules in the hexavalent state and must be reduced to trivalent before binding (30).

1.4.1.3.3 Excretion. Most of the excretion takes place through the kidneys (6, 12), but also to some extent through the liver and bile (31). As cannulation of the bile duct for bile sampling is not versatile in humans, and chromium in the feces is derived from mixed chromium sources, urine and blood are the only body fluids that are versatile for monitoring of chromium exposure (6, 15, 16, 24).

1.4.1.4 Reproductive and Developmental. Long-term (12 weeks) ingestion of Cr(6+) (1000 mg/L) in drinking water was shown to reduce the number of females impregnated by males, and both the number of viable fetuses and the number of implantations were reduced. Similarly the number of dead fetuses was reduced in exposed females. Similar results were found when adding Cr(3+) compounds (5000 mg/L) to the drinking water (32). Exposure to Cr(6+) (1000 ppm) and Cr(3+) (1000 ppm) in the drinking water seemed to reduce the sexual activity of male rats (11).

1.4.1.5 Carcinogenesis. Studies in animals have given some evidence that Cr(6+) compounds characterized by low water insolubility may cause cancer in various tissues where these compounds have been administered. However, Cr(6+) compounds are among the few chemicals that seem to be more potent cancer inducers in humans than in animals. Studies where animals inhaled chromates, which is comparable to human exposure situations, have not been successful in generating cancer. Nettesheim et al. (33) succeeded in inducing bronchiogenic carcinomas after long-term inhalations of chromates, however, only after combined exposures to viruses.

Most other studies have reported on the application of different Cr(6+) and Cr(3+) compounds intramuscularly, by implantation, and intratracheally. Many of the studies have caused tumors at the site of injection, as reviewed by IARC (34). Studies in animals have not given as convincing results on the causality between exposure to Cr(6+) compounds as have epidemiological studies on exposed humans. Animal studies where chromates were

CHROMIUM, MOLYBDENUM, AND TUNGSTEN 83

administered by the IM or the IP route have led to significant excesses of malignant tumors.

1.4.1.6 Genetic and Related Cellular Effects Studies. Chromic compounds (Cr 3+) do not readily cross cell membranes and are relatively nontoxic *in vivo*. However, intracellular trivalent chromium may react slowly with both nucleic acids and proteins and may be genotoxic. The genotoxicity of trivalent chromium was investigated *in vitro* using a DNA replication assay and *in vivo* by CaC_2-mediated transfection of chromium-treated DNA into *Escherichia coli*. When DNA replication was measured on a trivalent chromium-treated template using purified DNA polymerases (either bacterial or mammalian), both the rate of DNA replication and the amount of incorporation per polymerase binding event were increased. When transfected into *E. coli*, trivalent chromium-treated bacteriophage DNA showed a dose-dependent increase in mutation frequency. These results suggest that Cr(3+) alters the interaction between the DNA template and the polymerase so that the binding strength of the DNA polymerase is increased and the fidelity of DNA replication is decreased. These interactions may contribute to the mutagenicity of chromium ions *in vivo* and suggest that trivalent chromium could contribute to chromium-mediated carcinogenesis (35).

1.4.1.7 Other: Neurological, Pulmonary, Skin Sensitization. This (also includes any toxic effects as a result of chemotherapeutic exposure during manufacture or use.) Attempts have been made to study skin effects in animals, but no appropriate models were found.

1.4.2 Human Experience

Chromium may enter the human body by three different routes: through the airways, through the gastrointestinal (GI) tract, or through the skin, in particular through injured skin. The airway route is the port of entry most significant for untoward effects of chromates, while the oral route is essential for chromium as an essential element. As chromium has at least three different half-lives in the body, the kinetics of the most prominent chromium species, Cr(6+) and Cr(3+), should be taken into account when monitoring chromium in blood and urine. The levels of total chromium in blood and urine are governed by all these half-lives; the water solubility of the chromium compounds(s) of concern, and the distribution of the particle size of the inhaled areosol influences the rate of uptake. Once taken up in cells lining the bronchial tree or the GI tract, the kinetics of these two significant species are different and should be accounted for (6, 12).

The above phenomenon makes it complicated to evaluate the significance of a given concentration of chromium found in blood in an exposed subject. By the same token, a given concentration of chromium in blood (i.e., whole blood or plasma) cannot be applied directly for evaluation of those health hazards that are characterized by an intermediate or long-term latency period. For the same reasons and because there is no direct relationship between the concentration of chromium in blood and the exposure-related hazards or disease risk of a given exposed subject or group of workers, "biological threshold values" for blood are virtually not applicable for chromium compounds. However, monitoring of chromium levels of whole blood, erythrocytes, or plasma may still be useful for research purposes and for the purpose of judging whether or not a subject has been heavily exposed to chromates shortly prior to monitoring. Chromium (6+) molecules after a heavy exposure to chromate compounds may have escaped reduction to Cr(3+) at the site of deposition or entry, and hence reach the bloodstream, are quickly taken up by red blood cells and are reduced to Cr(3+), which stays intracellular. Therefore, a high concentration of chromium in erythrocytes most likely indicates the magnitude of the acute exposure hours to days before, while the plasma chromium level is likely to be informative about long-term uptake of Cr(3+).

For the reasons indicated it is also difficult to evaluate the significance of a given concentration of chromium found in urine. A stable high level of chromium in the urine is likely to reflect the total body burden of chromium, while short-term fluctuations in the urinary chromium level more likely reflect short-term exposure to chromates.

1.4.2.1 General Information. Chromium (3+) is an essential nutrient that plays a role in glucose, fat, and protein metabolism by potentiating insulin action (5–7). The biologically active chromium compound for this action is composed of chromium, nicotinic acid, and amino acids. This compound is called a glucose tolerance factor. Hexavalent chromium compounds have a strong oxidative power.

1.4.2.2 Clinical Cases. Kidney necrosis was reported many times 80–90 years ago after oral intake of gram amounts of Cr(6+) compounds, starting with tubular necrosis and undamaged glomeruli, sometimes with concomitant diffuse necrosis of the liver and sebsequent loss of architecture (36, 37). Some reports also described bleeding from ulcerations of the intestinal mucosa, with cardiovascular shock as a possible complication (36). When the patient survived, chronic kidney or liver damage could occur.

1.4.2.2.1 Acute Toxicity. A number of reviews are published on such untoward effects (34, 38–40). As the exposure levels in work places currently are quite low in the Western world, except in some rare instances, the likelihood of occurrence of local skin ulcers, liver cell necrosis, and tubular necrosis in the kidneys, is quite small. However, such effects may possibly occur among heavily Cr(6+)-exposed workers in developing countries, whenever the reductive capacity in the port of entry is overwhelmed.

1.4.2.2.2 Chronic and Subchronic Toxicity. Deep, chronic skin ulcers were reported early this century after chromates had been deposited on skin, primarily resulting in minor damage. Perforation of the nasal septum was reported frequently after chromates were deposed on the nasal septum. In 1827 Cumin (41) observed ulceration of the skin among workers handling potassium dichromate. In 1884 (42) perforation of the nasal septum was reported among workers exposed to potassium bichromate. It is unlikely that perforation of the nasal septum is related to the air concentration of Cr(6+), but rather to deposition of large particles containing Cr(6+) on the mucosa.

It has been shown in both experimental and epidemiological studies that exposure to high levels of Cr(6+) may induce toxic effects on tissues and mucous membranes. The correlation between the degree of local toxic effects and the chemical state of chromium was demonstrated in both the macro- and the microscopic investigations and, in particular, in the cytological examinations. Cases of atypia were found only in workers exposed to Cr(6+) compounds. Evidence of atypia could indicate that Cr(6+) compounds may act as a carcinogenic agent on the rhinosinusal mucosa. For this reason, the introduction of cytological nasal examination in health surveillance for such workers has been considered. Sample collection from the nasal mucosa by brushing is the method of choice because it is simple, noninvasive, and gives appropriate diagnostic results (43).

1.4.2.2.3 Pharmacokinetics, Metabolism, and Mechanisms. In the general population the urinary concentration has been lowered over time, from 150 µg Cr/24 h in the 1960s to 0.3 µg Cr/24 h in the 1980s, in accored with improved analytical methods (44). The uptake of chromium in man after ingestion of Cr(6+) compounds varies with the dietary constituents (45). There are no reports published on the uptake of chromium compounds in human airways. However, there are reports on the urinary excretion of chromium among Cr(6+) exposed workers that provide indirect evidence on uptake of Cr(6+) by inhalation. Results have been presented on workers exposed to aerosols from Cr_2O_3 and welding fumes generated from stainless steel welding (14, 16, 46, 47).

Excretion of chromium in the urine of chrome platers was proposed for biologic monitoring of ongoing exposure. Urinary chromium declines after discontinued exposure in welders welding on chromium-alloyed stainless steel, but not in chrome platers. Lindberg and Vesterbeg (48) calculated half-times ($t_{1/2}$ from 10 chrome platers over a weekend, and from 23 chrome platers over a 31-d vacation. They considered that their results indicated that the excretion of chromium can be approximated by a two-compartment model. Estimated from the median values, an initial rapid phase with an assumed $t_{1/2}$ of 2–3 d is followed by a phase with a $t_{1/2}$ of approximately a month.

Chromate exposure of 103 German stainless steel welders who were using manual metal arc welding (MMA), metal inert gas welding (MIG), and both methods as measured by ambient and biologic monitoring. At the workplaces the maximum CrO_3 concentrations were 80 mg/m^3. The median values were 4 mg/m^3 for MMA and 10 mg/m^3 for MIG. The median chromium concentrations in erythrocytes, plasma, and urine of all welders were less than 0.60, 9.00, and 32.50 mg/L. These authors indicated that for biologic purposes of monitoring, chromium concentrations in erythrocytes and, simultaneously, in plasma seem to be suitable parameters. According to these results, chromium levels in plasma and urine on the order of 10 and 40 mg/L correspond to an external exposure of 100 mg CrO_3/m^3. Chromium concentrations in erythrocytes greater than 0.60 mg/L indicate an external chromate exposure greater than 100 mg CrO_3/m^3 (46). However, there is no general agreement on the conclusions from this group of researchers.

In another study hair and urine samples were collected from 34 males tannery workers and from 12 normal adults. Eighteen of the workers dealt directly with chromium, and the remaining 16 (controls) worked in the offices and kitchen of the same plant. When compared with normal adult values, urinary chromium concentration, chromium/creatinine ratio, daily chromium excretion, and hair chromium concentrations were significantly higher in the workers; and urinary b-2-microglobulin/creatinine ratios, significantly lower in both tannery workers and in controls. A negative correlation was found between urinary b-2-microglobulin/creatinine and Cr/creatinine ratios of tannery workers and controls. No correlations between the duration of exposure to chromium and hair and urinary chromium values were found. High values observed in workers with short exposures suggested that chromium is readily absorbed through the respiratory system (49).

Kishi et al. (50) scrutinized the chromium content in the organs of six workers who had worked in a chromate manufacturing plant, had been exposed to a considerable amount of chromium for more than 10 years, and had died of lung cancer. The total chromium in the lungs of the workers averaged 51.5 mg/g (range 24.8–210 mg/g), versus levels in the lungs of unexposed of 0.07–1.01 mg/g. Organs other than the lungs of the workers also had higher levels of chromium than those of the controls. It was clear that the metal remained in the lungs long after exposure to chromate had ceased (50).

1.4.2.2.4 Reproductive and Developmental. It was suggested by Bonde (51) that reduced spermatogenesis, spontaneous abortion, congenital malformation, and childhood malignancy following exposure to Cr(6+) among stainless steel welders, alternatively with that radiant heat during welding could reduce the semen quality. In initial studies exposure to welding was reported with a higher prevalence during periods of infertility than prior to conception in the case-referent study (OR 2.0), which was considered to be consistent with the findings in a cross-sectional study showing reduced semen quality in welders. The reduced semen quality and fertility were, however, attributable to the welding of mild steel; hence, no relationship was found between such measures of exposure to chromium and parameters of semen quality.

A group of Danish researchers studied the occurrence of spontaneous abortion from 1977 through 1987 among 2520 pregnancies of spouses of 1715 married metal workers. Work histories were compiled by the use of a postal questionnaire, and information on live born children, spontaneous abortions, and induced abortion was obtained from national medical registers. The proportion of spontaneous abortions was not increased for pregnancies considered to be at risk from stainless steel welding when compared with pregnancies not at risk (odds ratio 0.78). Hence, the authors concluded that the results were not in agreement with suggestions that spouses of stainless steel welders have increased risk of spontaneous abortion (52).

A Danish study aimed to estimate whether welding by males has an impact on couple fecundity (53). For this purpose a sample of Danish couples without previous reproductive experience was recruited. Among 430 couples that were studied, 201 males were metal workers and 130 were welders, and the couples were followed for a maximum of six menstrual cycles from termination of birth control until a clinical pregnancy. Compared with nonwelding metal workers, the fecundity odds ratio (OR) of male exposure in welding was 0.86. An interaction between male smoking and welding was found; within smokers the OR for welding was 0.40 and within nonsmokers it was 1.22. The authors concluded that the decreased fecundity among smoking welders could be attributable to both current and previous welding exposure.

1.4.2.2.5 Carcinogenesis. Long-term exposure to Cr(6+) compounds has been observed in many epidemiological studies to enhance the risk of cancer of the respiratory organs among the exposed. The relationship between employment in industries producing Cr(6+) compounds from chromite ore and enhanced risk of lung cancer is well established. There is agreement in several studies that long-term exposure to some chromium-based pigments enhance the risk of lung cancer. An association has also been observed between exposure to chromic acid in hard plating and lung cancer, but that association is not strong. Some studies have weakly indicated excesses of cancer of the GI tract, but the results are inconsistent and are not confirmed in well-designed studies. There is no indication that chromite ore does have an associated enhanced risk of cancer. Although it has not yet been identified which chromium compound (or compounds) is (are) responsible for enhanced risk of cancer in respiratory organs, there is general agreement that it is the Cr(6+) species that are responsible for the elevated cancer risks and that the Cr(6+) species are not.

1.4.2.2.6 Genetic and Related Cellular Effects Studies. Thirty-nine electric welders exposed to chromium and nickel were compared with 18 controls standardized for age, gender, and smoking with respect to the prevalence of sister chromatid exchange (SCE) and DNA strand breakage and cross-linking in their blood lymphocytes. A correlation was found between the prevalence of SCEs and of individual DNA strand breakage versus the concentration of chromium in the urine (54). Elevated levels of chromosomal aberrations was observed by Bigaliev et al. (55); Sarto et al. (56), and Knudsen et al. (57) in lymphocytes from workers exposed to different Cr(6+) compounds. Studies on untoward cytogenetic effects in MMA welders were, however, considered inconclusive by IARC (34).

1.4.2.2.7 Other: Neurological, Pulmonary, Skin Sensitization, etc. Work-related exposure to chromium may result in dermatitis. Perforation of the nasal septum was reported frequently after chromates were deposited on the nasal septum. In 1827 Cumin (41) observed dermatitis among workers handling potassium dichromate.

1.4.2.3 Epidemiology Studies. A number of cohort and case studies among workers exposed to chromates have been carried out and have shown clear evidence of enhanced risk of cancer of the respiratory organs among those heavily exposed for a long time period (34, 58). Among the workers at an Ohio chromate manufacturing plant there was an elevated risk of lung cancer, with a high proportion of deaths due to this type of cancer. The exposure was mixed $0.01-0.15\,\text{mg/m}^3$ water-soluble Cr(6+) and $0.1-0.58\,\text{mg/m}^3$ insoluble Cr(6+).

To scrutinize whether workers newly employed in sections of the factory, constructed to reduce hazardous exposures, had a reduced respiratory cancer risk as compared with workers employed at an old facility, Hayes et al. (59) studied a group Baetjer (60) had studied. The cohort comprised of 2101 workers initially employed between 1945 and 1974 with 90 days-employment. Vital statistics were ascertained by mid-1977 for 88% ($n = 1803$) of the cohort, and these subjects were used in the analyses. Age, race, and time-specific rates for mortality of Baltimore males were used as reference entity. The SMR for cancer deaths of the trachea, bronchus, and lung (ICD 162) was 202 [95% CI = 155–263], based on 59 deaths versus 29.2 expected. The SMR for lung cancer was

180 [95% CI = 110–270] (20 deaths), for employees hired between 1945 and 1949 with less than 3 years work. Those with three or more years work, hired 1945–1949, had an SMR of 300 [95% CI = 160–520] (13 deaths). Lung cancer SMR for workers hired 1950 to 1959, with less than 3 years work at a new facility, was 70 (2 deaths), while workers hired 1950 to 1959, employed less than 3 years at the older facility, had an SMR = 180 [95% CI = 90–310] (12 deaths). Workers with 3 years+ in the new facility, first employed 1950–1959, had SMR for lung cancer at 400 (three deaths) [95% CI = 80–1170], while their collegaues in the old facility had SMR of 340 (nine deaths) [95% CI = 260–650]. The same author followed the same cohort later (348) and found that the excess risk for lung cancer related to duration of exposure to chromate dusts for subjects followed for 30 years or more after initial employment. For this subgroup, the SMRs were 81, 139, 201, and 321 for the subjects with 0 years, less than 1 year, 1–9 years, and 10+ years of exposure to Cr(6+) dusts ($p < .01$, for trent), respectively. The risk for digestive cancer was only weakly associated with exposure. Tobacco smoking was not considered as a confounder.

Some of the studies among workers in the chromium plating industry (61, 62) have been inconclusive, and the findings of a Japanese study of chrome platers were nonpositive (63). A historical prospective cohort study was conducted in nine Italian chrome-plating plants to examine the mortality of workers employed for at least 1 year from January 1951 to December 1981. The study group included 178 subjects, 116 of whom were from "hard" and 62 from "bright" chrome-plating plants. The vital statistics were ascertained to be 97% of them. The total number of deaths was close to expected: 15 observed versus 15.2 expected, while deaths from cancers exceeded the expected number: 8 observed versus 4.2 expected. The cohort members were separated into two subcohorts depending on the intensity of exposure, which was higher in "hard" than in "bright" chrome plating. Most cancer deaths occurred among "hard" chromium platers: 7 observed versus 2.7 expected, $p = .02$. All deaths from lung cancer occurred in this subgroup; 3 observed, 0.7 expected, $p = .03$(64).

A study in the United Kingdom (65) on the mortality among 2689 chromium electroplaters recruited in the period 1946–1983, mainly thin plating of bumpers and overriders, indicated an excess incidence of deaths from stomach cancer: 25 deaths due to stomach cancer ($E = 16.2$; RR $= 25/16.2 = 1.54$). However, the excess was observed in males only and there was no mention of exposure–response relationship, and smoking habits could only partly be accounted for. The authors had not accounted for latent periods or analyzed the relation between time windows with heavy exposure in relation to outcome 20–35 years later. The whole cohort consisted of 2689 workers (males/females = 1288/1401) first employed between 1946-1975 and employed more than 6 months as chromium platers. Some exposure measurements had been taken before 1973, presenting gradually falling air concentrations of chromium trioxide (Cr_2O_3) as high as 8.0, 1.6, and 0.4 mg Cr/m^3, respectively. After 1973 air concentrations were generally below 50 µg Cr/m^3, but the observation period was too short after that to observe a possible reduction in occurrence of deaths due to stomach cancer after those improvements in the exposure situation.

A historical prospective cohort study was conducted among 415 small-scale chrome-plating plants in Japan to study the mortality among platers employed between 1970 and 1976. In all 1193 male metal platers were identified in 1976 and divided into a chromium-plater subgroup with 626 workers and a non-chromium-plater subgroup of 567. Both subgroups were followed from 1976 through December 1987. Only lung cancer was found to be significantly higher than expected for all platers: 16 observed versus 8.9 expected; SMR 179; 95% CI 102 to 290. This elevated SMR, however, was not statistically significant in either of the two subgroups. The SMR for lung cancer of the chromium-plater subgroup was highest among those exposed for the shortest duration and among those exposed earlier (66).

Hexavalent chromium fumes are generated during welding on stainless steel. A cohort consisting of 234 welders working on stainless steel and exposed to high levels of chromium was selected. According to an earlier survey, the Cr(6+) exposure of these welders may have exceeded 20 mg/m^3. A comparison cohort consisting of 208 railway-track welders exposed to low levels of chromium was also selected. The participants of both cohorts had welded for at least 5 years sometime between 1950 and 1965 and were followed for mortality until December 1984. Among the welders exposed to Cr(6+), five deaths occurred owing to pulmonary tumors. This number is significantly elevated compared to one death that occurred among the welders exposed to low levels of chromium, but not significantly in excess of the corresponding mortality of the general population. The authors concluded that exposure to stainless steel welding fumes might be associated with an increased incidence of pulmonary tumors (67).

The International Agency for Research on Cancer (IARC) carried out a multinational study comprising of mild steel (MS) as well as SS welders (68). The total number of 11,092 welders from nine European countries participated, recruited mainly during the 1960s and the 1970s. There were 116 lung cancer deaths versus 86.8 expected. There were also 15 deaths of bladder cancer versus 7.9 expected, and 12 versus 8.6 deaths of cancer of the kidneys, both cancer sites generally considered to be related to smoking. There were, however, only 42 deaths from illnesses in the respiratory system versus 61 expected, which could indicate that tobacco smoking did not play a major role as cause of death in this study group. By years of first employment among SS welders the SMR for lung cancer was 157, which could indicate that SS welders were more likely to die from this type of cancer than was the whole group of welders (59).

The incidence of cancer was studied in a historical prospective cohort study among 2957 boiler welders recruited from 1942 to 1981, including a subcohort of 606 stainless steel welders (69). There were 269 cancer cases observed versus 264 expected, of which 50 lung cancer versus 37.5 expected and 3 cases of pleural mesotheliomas versus 1.1 expected. In the subcohort of stainless steel welders there were six cases of lung cancer versus 5.8 expected, and 1 case of pleural mesothelioma versus 0.2 expected. Hence, the studied welders carried a small excess risk of lung cancer that did not seem to be associated with stainless steel welding. Smoking and asbestos exposure could be confounders.

The results of studies of ferrochromium workers (70, 71) were inconclusive as to lung cancer risk. More recently a cohort study on the incidence of cancers and crude death rates in ferrochromium and ferrosilicon workers provided some evidence for carcinogenicity in this industry segment. The whole cohort was observed from January 1953 to December 1985. Two sets of results were presented, one restricted to workers first employed before 1960 and one to workers first employed before 1965. The latter cohort consisted of 1235 workers. The total mortality in the whole cohort was low (SMR = 81), as was the overall incidence of cancers (SIR = 84). An excess of lung cancer (SIR = 154) and cancer of the prostate (SIR = 151) was observed in the ferrochromium workers employed before 1965. Cancer of the kidney was also in excess (SIR = 273) in the ferrochromium group, with a mean latency time of 39 years. Two cases of malignant melanomas had occurred, versus 0.19 expected, in a small subgroup of workers in electrical shops and an electric power station (72).

Studies in humans indicate that chromates of intermediate water solubility (e.g., zinc chromates) could be more potent enhancers of the risk of lung cancer than are chromates of low water solubility.

1.4.2.3.3 Pharmacokinetics, Metabolism, and Mechanisms. Chromium in urine, blood, and seminal fluid was determined among 60 welders and 45 controls. The concentration of chromium in urine and blood did not change across a work shift or after a 3-week break in exposure. However, stainless steel and mild steel welders who were exposed to low levels of chromium and steel welders who were mildly exposed had significantly increased concentrations of chromium in post-shift urine (mean 2.1 nmol/mmol creatinine), compared with the controls (mean 0.7 nmol/mmol creatinine). Preshift blood chromium concentrations showed a similar variation between exposed workers and referents. Subgroups of stainless steel welders had very high levels of chromium in seminal fluid. The investigators (73) suggested that attention should focus on the potential risk of delayed health effects among stainless steel and mild steel welders who heretofore had not been considered at risk from chromium exposure.

1.4.2.3.4 Carcinogenesis. Chromium (6+) compounds — calcium, strontium, zinc, and lead chromates — were studied for cytotoxicity and morphological transformation in Syrian hamster embryo (SHE) cells in relation to their solubilization in cell culture conditions and intracellular chromium concentration. Calcium, strontium, and zinc chromates were completely solubilized after 1 d of incubation in cell cultures; for lead chromate, 20–36% chromium was solubilized only after 7 d. In two parallel transformation assays, the SHE cells were treated with suspensions or with corresponding supernatants (containing only solubilized chromium) of these compounds. A (statistically significant) relationship was observed between the concentration of chromium and the amount of chromium per cell, irrespective of the compound (except for suspensions of lead chromate). The cytotoxicity resulted from extracellular solubilized chromium because treatments with either supernatants or suspensions of calcium, strontium, and zinc chromates gave the same LC_{50}. A clear dose–response relationship was observed for the induction of morphological transformation for each compound, either previously solubilized or in suspension. The expression of the transformation incidence as a function of the chromium concentration/cell revealed that (1) the transformation incidence relates to the chromium concentration/cell, irrespective of the chromium compound calcium, strontium, or zinc chromate; and (2) the transformation incidence induced by solubilized lead chromate is higher than that induced by the other compounds at the same concentration of chromium/cell. A double treatment with solutions of chromium and lead at corresponding concentrations induced the same transformation incidence as the solubilized lead chromate. The results show that the solubilization of particulate

CHROMIUM, MOLYBDENUM, AND TUNGSTEN

chromium (6+) compounds is a critical step for the transforming activities; the concentration of intracellular soluble chromium is strongly associated with the transforming activity of calcium, strontium, and zinc chromates, whereas lead appears to act synergistically with chromium in inducing the transformation by lead chromate (74).

1.4.2.3.5 Other: Neurological, Pulmonary, Skin Senstization, etc. Chromium, especially potassium dichromate, is a very common skin sensitizing agent, secondary only to that of nickel. The resulting hypersensitivity results from chromium itself as a hapten (75).

Five patients with asthma related to chromium salts exposure in their work area are presented. All of them were nonatopics and exhibited a history of contact dermatitis (possitive patch tests for potassium dichromate) previous to the onset of bronchial asthma. Solutions of $K_2Cr_2O_5$ were prepared in normal saline at 0.01, 0.1, and 1 mg/mL for skin prick tests (SPT) and bronchial provocation tests (BPT). Immediate cutaneous reaction by SPT was negative for controls and patients. BPT were performed by the tidal breathing method, with possitive results in all patients. A negative response was recorded in four control unexposed asthmatics. An attempt to inhibit BPT with sodium chromoglycate was unsuccessful. The diversity of reactions (immediate, dual, and late) registered in BPT support the theory that bronchial reactivity can be induced specifically by inhalation of chromium salts. The followup data indicate a good prognosis, provided that patients remain away from exposure (76). A worker who developed de novo asthma after plating with nickel and chromium, but not other metals, was subjected to inhalation challenge and immunoserologic tests to evaluate this association. He developed acute asthma when exposed to chromium sulfate and a biphasic asthmalike response to nickel sulfate. Radioimmunoassays incorporating the challenge materials revealed specific IgE antibodies to the provocative agents, but not to another metal (gold), which he could tolerate. The findings support the postulates that (1) bronchial reactivity can be specifically induced by fumes of metallic salts, even in a previously nonallergic individual, and (2) an IgE type I immunopathogenic mechanism is involved (77).

Kidney disease is often cited as one of the adverse effects of chromium, yet chronic renal disease due to occupational or environmental exposure to chromium has not yet been reported. Occasional cases of acute tubular necrosis (ATN) following massive absorption of chromate have been described. Chromate-induced ATN has been extensively studied in experimental animals following parenteral administration of 15 mg/kg body weight potassium chromate (6+). The chromate is selectively accumulated in the convoluted proximal tubule where necrosis occurs. An adverse long-term effect of low-dose chromium exposure on the kidneys is suggested by reports of low-molecular-weight (LMW) proteinuria in chromium workers. Excessive urinary excretion of b-2 microglobulin, a specific proximal tubule brush border protein, and retinol-binding protein has been reported among chrome platers and welders, However, LMW proteinuria occurs after a variety of physiological stresses, is usually reversible, and cannot alone be considered indicative of chronic renal disease. Chromate-induced ATN and LMW proteinuria among chromium workers nevertheless raise the possibility that low-level, long-term exposure may produce persistent renal injury. The absence of evidence for chromate-induced chronic renal disease cannot be interpreted as proof of the absence of such injury (78).

Table 38.3. Air Standards (mg/m^3) and Classification for Various Chromium Compounds

Substance	ACGIH	OSHA	NIOSH
Chromic acid and chromates (as CrO$_3$)		Ceiling 0.01	0.001 as Cr carcinogen
Chromite ore processing (chromate) as Cr	0.05, A1		
Chromium (II) compounds as Cr	0.5	0.5	0.5
Chromium (III) compounds as Cr	0.5	0.5	0.5
Chromium (VI) compounds as Cr water soluble	0.05		
Chromium (VI) compounds as Cr certain water insoluble	0.01, A1		
Lead chromate as Cr	0.012, A2		
Strontium chromate as Cr[a]	0.0005, A2		
Zinc chromates as Cr	0.01, A1	Ceiling 1 as CrO$_3$	0.001 carcinogen

[a]Proposed, 1992.

1.5 Standards, Regulations, or Guidelines of Exposure

For chromium (II) and chromium (III), the following are the exposure limits: ACGIH TLV TWA = 0.5 mg/m^3, OSHA PEL TWA = 1 mg/m^3, and NIOSH REL TWA = 0.5 mg/m^3, IDLH = 250 mgCr(II)/m^3. ACGIH, OSHA, and NIOSH treat chromium and its compounds in slightly different ways (79), as shown in Table 38.3. It should be noted that both IARC and NTP classify certain chromium compounds as carcinogens (80). The ACGIH proposed lowering the TLV for strontium chromate to 0.0005 mg/m^3 because it is a significantly more potent carcinogen than other chromates (81). The exposure limits for the chromates (+6) is 0.1 mg/m^3 for OSHA as a ceiling value, and 0.001 mg/m^3 as a NIOSH REL. The ACGIH has a number of recommended TLVs depending upon the chromium compound.

To protect worker health, fully protective industrial hygiene procedures should be initiated, despite Hathaway's (19) suggestion that the apparent threshold based on detoxification by reduction to trivalent chromium "precludes any risk" from occupational exposures that do not involve excessive short-term exposures and that are below 0.05 mg/m^3. Bolla et al. (43) have recommended the introduction of cytological nasal examination into health surveillance for workers. Sample collection from the nasal mucosa by brushing is simple, is noninvasive, and gives good diagnostic results. Medical surveillance should be directed to early detection of irritation of the mucous membranes and skin or hypersensitivity reactions. Tests that might detect the onset of respiratory tract tumors should be included as part of the annual physical examination.

The ACGIH biologic exposure determinants for water-soluble hexavalent chromium (the TLV is 0.05 mg/m^3) in the urine is 10 mg/g creatinine for an increase during the work shift and 30 mg/g of creatinine for the end of the shift at the end of the work week. The ACGIH noted that there is some background for those not occupationally exposed (82).

The ACGIH TLV for chromyl chloride is 0.025 mg/m^3. The NIOSH REL for chromyl chloride, which they treat as a carcinogen, is 0.001 mg/m^3.

CHROMIUM, MOLYBDENUM, AND TUNGSTEN

1.6 Studies on Environmental Impact

There are only few studies on possible chromium-related human diseases related to release of chromium into the environment. An early study was carried out at Tokushima University in Japan among residents living close to a newly started chromate manufacturing plant. The research concluded that no definite signs could be found indicating health injury related to chromate released from the new plant (84). The first one carried out in Western countries was performed among the residents in proximity to the ferrochromium smelter in Trollhättan on the Swedish west coast in 1980 (71). No difference was observed between cancer mortality among the population living close to the smelter, as compared to that in the population living farther away.

A study reported from the Peoples Republic of China was hampered by poor characterization of the exposure to Cr(6+) compounds, by a poorly defined study population, and a too short observation period to permit revealing a possible relationship that might have been present between exposure Cr(6+) and occurrence of cancers (85). Therefore, the nonpositive result from that study permits no conclusions on a possible relationship nor on the absence of relationship between Cr(6+) exposure and occurrence of cancer. Another study (86) on a possible association between alleged chromium exposure and childhood leukemia in Massachusetts has been retracted (87). There are no further studies on possible relationships between exposure to Cr(6+) in the environment and occurrence of cancer in the population.

2.0 Molybdenum

2.0.1 CAS Number: [7439-98-7]

2.0.2 Synonyms: NA

2.0.3 Trade Names: NA

2.0.4 Molecular Weight: 95.94

2.0.5 Molecular Formula: Mo

2.1 Chemical and Physical Properties

2.1.1 General

Mo is a dark-gray, or a black powder with a metallic luster and a chemical element of the second transition series. The name is derived from the Greek molybdos, meaning "lead." In 1778 Carl Scheele of Sweden recognized molybdenite as a distinct ore of a new element. Hjelm in 1782 prepared an impure form of the metal.

The pure metal is prepared by the reduction of purified molybdic trioxide or ammonium molybdate with hydrogen. Molybdenite concentrates are produced by flotation, resulting in a rough concentrate of 10 or 12 to 1; it is reground and refloated, yielding a final concentrate of 190:1. When Mo is a by-product of copper mining, a concentrate of copper and molybdenum is first produced, and the two ores are later separated by differential flotation. Government specifications of molybdenite concentrates require 80% MoS_2 (min), 1% Cu (max), 0.3% Pb (max), and 0.2% (max) As, P, and Sn combined.

Table 38.4. Chemical and Physical Characteristics of Molybdenum and Some of its Salts

Compound	Mol. Formula	At. or Mol. Wt.	Sp. Gr.	M.P. (°C)	B.P. (°C)	Solubility
Molybdenum	Mo	95.95	10.28	2629	4612	Insol. water; sol. conc. HNO_3, H_2SO_4; sl. sol. HCl, insol. HF, dil. H_2SO_4
Molybdic oxide	MoO_3	143.95	4.50 (19.5°C)	795	Subl. 1115	1.066 g/L (18°C); sol. acids, alkalies, NH_4OH
Molybdenum disulfide	MoS_2	160.08	4.8 (14°C)	1185	Dec. in air	Insol. water; sol. HNO_3, H_2SO_4, aqua regia, insol. dil. acids
Ammonium molybdate	$(NH_4)_2MoO_4$	196.03	2.276 (25°C)	Dec.	—	400 g/L (20°C); dec. hot water; sol. acid; insol. alkalies
Calcium molybdate	$CaMoO_4$	200.01	4.38–4.53	—	—	Insol. cold water; dec. hot water; sol. acid; insol. alcohol, ether
Sodium molybdate	Na_2MoO_4	205.92	3.28 (18°C)	687	—	Insol. water and dilute acids; sol. most organic acids

The physical and chemical properties of molybdenum and some of its compounds are listed in Table 38.4.

Molybdenum exhibits oxidation states of 0, 1+, 2+, 3+, 4+, 5+, and 6+. Of the valence forms, 6+ is the most stable, but several industrially important compounds exist, for example, MoS_2 and MoO, with intermediate valences.

The metal is very hard, but more ductile than the chemically similar element tungsten. Mo has a high elastic modulus, and of the more readily available metals only tungsten and tantalum have higher melting points.

The mechanical properties of molybdenum depend on its history of processing; as produced by power-metallurgy methods or by arc melting, Mo is brittle, but can be made ductile by heating at 1000–1300°C. Molybdenum can maintain reasonably high levels of hardness at extremely high temperatures and is superior to the best "superalloys" of nickel for heat-resistant applications. Mo and tungsten form a series of solid solutions with melting points higher than that of Mo. Mo also alloys with many lighter alloying substances, cobalt, iron, aluminum, chromium, and silicon, which effectively increases its strength in lesser amounts than tungsten. Molybdenum is competely miscible with niobium and tantalum and with the high-temperature form of titanium.

Molybdenum is resistant to acid attack of HF and HCl. Eight fluorides, 11 chlorides, 5 bromides, and 2 iodides of molybdenum are known.

Molybdenum forms a very complex series of compounds; with the exception of some of the halides, sulfides, and oxides, few simple salts of Mo are known. There are several reasons for the complexity. There is a strong tendency for Mo to form complex compounds. Shifts between different coordination numbers (88–90) result from relatively minor differences in conditions. Many metal oxides, from lithium to lead, react readily with MoO_3 to form normal molybdates or polymolybdates, depending on the stoichiometry of the system. Like the oxides, they vary in color from white to violet, depending on the color-forming characteristics of non-Mo metal.

Hexavalent molybdenum has the striking tendency to form isopoly and heteropoly acids and salts. Molybdenum also forms a series of oxy acids and salts.

2.1.2 Odor and Warning Properties

No data are available.

2.2 Production and Use

Molybdenite (MoS_2) is the only important mineral source at present, but small quantities of powellite ($CaMoO_4$) are mined from time to time; in the past, deposits of wulfenite ($PbMoO_4$) were worked. Molybdenite is commonly associated with copper ores; thus production of Mo is regulated to a considerable degree by demand for copper. Of the total output in 1974, about 34% was a by-product from ores of copper, tungsten, and uranium; the remaining 66% was recovered from ores processed for molybdenum.

Free molybdenum does not occur in nature, but it is extracted from molybdenite, wulfenite, and powellite and is recovered as a by-product of copper and tungsten mining operations. Molybdenum is found in many parts of the world, but relatively few deposits are rich enough to warrant recovery costs. By far the largest and richest deposits occur in the western hemisphere, with the United States contributing the major share. A large Mo deposit has recently been found in the Misty Fiords area in southeast Alaska's Tongass National Forest (91).

Molybdenite concentrates are roasted to produce technical-grade oxide, considerable amounts of which are used directly in steel; the rest is converted to other molybdenum products. MoO_3 of higher purity is made by sublimation of the technical-grade oxide or from $(NH_4)_2MoO_4$. FerroMo is made from the oxide by ignition with aluminum, iron ore, ferrosilicon, lime, and fluorspar.

The roasting process requires precise temperature control to prevent material losses from sublimation and volatilization, which occur at about 700°C or higher. Technical-grade oxide may be further purified by heating to about 1000°C, at which temperature Mo oxide volatilizes readily, thereby effecting a separation from impurities that form more stable oxides. The volatilized oxide (generally 99.97% MoO_3) is collected in bag filters. A second purifying step is sometimes inserted at this stage. It involves dissolving pure oxide in ammonium hydroxide, filtering, and then evaporating to crystallize out the pure compound ammonium molybdate. Pure oxide or ammonium molybdate may be converted to metal powder by a two-step hydrogen reduction at 680°C (forming MoO_2) and at 1090°C.

Molybdenum metal powder is consolidated into metallic bodies by one of two processes:

1. *Arc casting.* Prefabricated or continuously compacted and sintered electrodes made from metal powder are arc melted in a water-cooled cooper mold. The process is usually carried out in a vacuum, although in some instances inert atmosphere melting is employed. The process is particularly well suited for the preparation of Mo-base alloys by virtue of the flexibility available in altering the composition of the melting electrode.
2. *Power metallurgy.* Molybdenum powder charges are compacted in a mold of the desired size and shape, and the green compacts are sintered (usually in vacuum or hydrogen) to produce bodies generally of 95–98% of theoretical density. Further densification is accomplished during subsequent hot-working sequences. Technical molybdic oxide may also be converted to ferroMo, a product favored by many foundries and steel producers. The smelting reaction is exothermic and is completed in approximately 20 min. Some ferromolybdenum is prepared by direct reaction in electric furnaces.

By far the largest use of MoO_3 is for various types of steels, of which ferroMo is the most important. Most of the molybdenum concentrates, except for purified MoS_2 used in the production of the lubricant, is converted to MoO_3 for direct use in steels, chemicals, or ceramics.

Molybdenum steel is used as wire, rod, and sheet, as super-strength and ultra-high-strength steel, and particularly for Mo shapes in manufacturing parts for aircraft and missiles. Small amounts of Mo metal are used in electronic parts, in induction heating elements, as electrodes for glass melting, and in metal-spray applications for steel and other metals. Molybdenum film provides bonding for subsequently sprayed metal deposits when applied as a thin layer on ferritic steel. Mo compounds are used as lubricants (MoS_2), catalysts (Mo, MoO_3, MoO_2), and colors (Mo inorganic and organic complexes). Unlike graphite, MoS_2 does not require water film for lubrication and thus can be used in vacuum. As catalysts, molybdenum compounds are used in hydrogenation, cracking, alkylation, and reforming of petroleum fractions, in the Fischer–Tropsch synthesis, and in various oxidation–reduction reactions. Because traces of sulfur poison other catalysts, but not Mo, this property extends the use of Mo, particularly in reactions involving petroleum and natural gas.

A comprehensive bibliography on the use of Mo catalysts has been prepared (92). Mo is used in colors as (I) molybdate pigments, (II) Mo-colored complexes with animal fibers, (III) insoluble dye complexes with phosphomolybdic acid, and (IV) mordants. In ceramics, molybdates permit bonding of vitreous enamels to steel by virtue of their property of lowering the surface tension of silicate melts, increasing their fluidity, improving welding power, and increasing adherence to metal. Lead molybdate is used in labeling glass containers. Small tonnages of Mo are used in fertilizers as a trace element, in making electric contacts, and in Mo-bearing titanium alloys. $Mo(OH)_3$ is used in electroplating to give black (moly-black) protective coatings; trivalent Mo compounds may be used to tan skins.

2.3 Exposure Assessment

Work-related exposure during production and fabrication of Mo products are to dusts and fume of Mo, its oxides, and its sulfides, chiefly from electric furnace or other high-temperature treatment. MoS_2 as a lubricant may be applied to metal surfaces at 700°F. Spraying of Mo may provide a hazard, and loss of Mo catalysts to the air adds to the metal burden of contaminated atmospheres. The sublimation characteristics of MoO_3 (above 800°C) present a fume hazard.

In addition to its industrial hygiene significance, Mo is of considerable biological importance as an essential trace element in the Mo-flavoprotein enzyme xanthine oxidase, in which it functions as an electron transport agent (93). It is also necessary for the fixation of nitrogen in the soil by bacteria; cattle and sheep can be poisoned feeding on herbage that has taken up Mo in abnormal quantities.

2.3.1 Workplace Methods

NIOSH recommends (94) inductively coupled plasma atomic absorption spectrometry (AAS) for the analysis of samples of workplace airborne concentrations of Mo (NIOSH Method 7300). AAS has the tendency to replace classical colorimetric procedures. The dithiol method has been revamped as a simple, rapid, precise, and most sensitive procedure for determining Mo in biological, geochemical, and steel samples (95).

2.3.2 Community Methods

The AAS method has also been used to determine Mo and vanadium in tap water and mineral waters after anion-exchange separation. After absorption on the exchange resin as citrate complexes, Mo is eluted with perchloric–hydrochloric acid mixture, and the eluates are analyzed by AAS. Amounts on the order of 0.2 to 13 mg/L were found (97).

2.3.3 Biomonitoring/Biomarkers

Biological samples are ashed at 550°C and leached with HCl. The dithiol–Mo complex is extracted with isoamyl acetate. Ascorbic and citric acids eliminate interferences from iron and tungsten, and KI provides high tolerance to copper. Sensitivity for biological samples is 0.05 ppm, with a relative standard deviation better than 7% and with complete recovery of Mo. A similar dithiol colorimetric procedure was reported by Cardenas and Mortenson (96) for the determination of Mo in the presence of tungsten in biologic materials. Interference from constituents present in biological materials was negligible.

Determination of Mo in guinea pig tissues (lung, liver, kidney, spleen, bone) following 25 days inhalation of sub-LD_{50} doses of the sulfide, the dust and fume of the trioxide, and $CaMoO_4$ showed rather uniform distribution in the tissues, including bone, for all substances at the end of exposure (98). Values of 10–60 mg/10 g fresh tissue were common for all substances except MoO_3 and $CaMoO_4$ dusts, which were higher by a two- to fourfold factor in the kidney, lung, spleen, and bone. Other exceptions were high lung retention for the sulfide and molybdate (40 and 18 times, respectively) over that by other tissues, according to the thiocyanate method of analysis. Tissue content of Mo determined

2 weeks later showed decreases up to fivefold for all five tissues, including bone, indicating temporary storage in this tissue. Again, $CaMoO_4$ and the sulfide were the exceptions; lung retention was still 72% of that at the end of exposure. Control Mo values for these tissues were from 10 to 50 mg/10 g, the higher value occurring in the spleen.

Mutant Long–Evans rats with a cinnamon coat color (LEC rats) have been established as an animal model for Wilson disease, a genetic disorder of copper (Cu) metabolism (99). Systemic disposition of Mo and altered distributions of Cu were compared in eight organs between LEC rats and Wistar rats (normal) at different times after a single intraperitoneal injection of tetrathiomolybdate (TTM) for chelation therapy. Excretion through urine and feces was also examined. Hepatic disposition of Mo was significantly increased in LEC rats, indicating that the interaction of TTM with Cu results in enhanced uptake of Mo. Concentrations of Mo and Cu decreased in the liver of LEC rats over time, whereas those in the spleen increased. Although the concentration of Mo in the kidneys decreased over time after an initial increase in both rats, Cu concentration increased over time. Cu was not redistributed to the brain. The excretion of Mo into urine was decreased and that into feces was increased in LEC rats compared with those in Wistar rats. These results were interpreted to indicate that TTM is taken up by the liver depending on the Cu content, and the Cu and Mo removed from the liver are mostly excreted through feces. Redistribution of Cu was observed in the spleen and kidneys, but not in the brain.

2.4 Toxic Effects

2.4.1 Experimental Studies

Investigations into the causes of Mo disease in ruminants, particularly cattle and sheep, have provided a mechanism for its action, at least in these species. High Mo content of herbage in pastures, first found in 1938 (100), caused "scouring" disease (diarrhea) in cattle (which copper deficiency aggravated, but administration of copper salts alleviated) (101). These findings provided a link with other studies reporting a reciprocal relationship between Mo and copper.

Whenever the intake of Mo in cattle is enhanced, the storage of metabolically active copper in the liver declines proportionally, whereas copper in nonutilizable form tends to accumulate there (102). Symptoms of hypocuprosis occurred in cattle grazing on pastures with high Mo content. Evidence indicates that excess Mo promotes binding of copper to a serum protein, thus limiting tissue uptake of copper (103). The resistant rat does not display great changes in hepatic copper storage and shows greater capacity to eliminate absorbed Mo than sensitive cattle (104, 105).

2.4.1.1 Acute Toxicity. Signs of Mo poisoning are loss of appetite, listlessness, diarrhea, and reduced growth rate. Death from injected doses occurs in guinea pigs in 2 h to 4 d, depending on the dose. Anemia is characteristic of Mo toxicity, with low hemoglobin concentration and reduced red cell counts (106). Cattle, rabbits, and chicks on high dietary levels of Mo showed joint deformities of the extremities (107). Histopathologically, livers and kidneys of severely poisoned animals showed fatty degeneration. Bronchial and alveolar exudate, in moderate amounts, was present in animals exposed by inhalation.

2.4.1.2 Chronic and Subchronic Toxicity. The antagonism that is known to exist between Mo and copper is influenced in a complex way by dietary sulfate, depending on the species, as indicated by the following observations (102): (1) sulfate intake causes a redistribution of Mo in the blood characterized by red cell depletion; (2) sulfate increases elimination of Mo from the body; (3) depletion of hepatic copper with associated signs of copper deficiency from Mo occurs only at dietary sulfate levels above some critical value and is thus has an antagonistic effect of sulfate on Mo metabolism and toxicity in the rat. Paradoxically, sulfate intensifies Mo toxicity in the ruminant (108–111). This paradox was resolved by Daniel and Gray (108), who showed that Mo can act in two different ways. In the copper-depleted rat Mo produces a still greater copper deficiency that is exacerbated by sulfate, but prevented by copper.

In animals with normal copper stores, Mo does not induce copper deficiency; its toxicity is completely prevented by sulfate, which Dick (102) believes causes a decrease in both absorption of Mo from the gut and its reabsorption from the kidney tubules, lowering the level of tissue Mo. Thus the contradictory effects of sulfate on molybdenosis in ruminants and nonruminants can be explained on the basis of the copper status of the animals, the ruminant with its multiple stomachs having a greater capacity to retain copper than the nonruminant.

The toxicity of Mo is reduced by thiol compounds such as cysteine and methionine, which are oxidized to sulfate by sulfoxidase on which Mo has an inhibitory effect. However, Mo-induced depletion of sulfoxidase can be reversed by excess sulfate (in rats); copper, which acts as a cofactor for sulfoxidase, tends to maintain high levels of sulfoxidase activity (108–111). Another factor in the mechanism is xanthine oxidase (XO), for which Mo is a cofactor and which is further induced by exogenous Mo (112). Molybdenum catalyzes sulfite to sulfate and, upon increased intake, diminishes the availability of sulfite. This leads to excessive production of hydrogen sulfide, thus rendering copper inactive by forming insoluble CuS (113).

The only definite animal study of Mo toxicity that is of industrial health interest is that of Fairhall et al. (98). These authors added to the toxicologic information derived from laboratory animals. The authors found molybdenite, MoS_2, by inhalation or oral routes, to be "practically nontoxic." Guinea pigs exposed 1 h daily for 25 d at the very high level of 490 mg/m^3 of molybdenite ore (230 mg Mo/m^3) showed only increased respiration; rats ingesting as much as 500 mg daily for 44 d showed no toxic signs, and all gained weight. However, guinea pigs injected intraperitoneally with 800 mg/kg showed 17% mortality in 4 d and 25% in 4 mo; survivors remained well and gained weight, indicating a possible physical effect from the mass injection rather than a toxic reaction to the ore. In contrast, all hexavalent compounds [MoO_3, $(NH_4)_2MoO_4$, and $CaMoO_4$] tested were increasingly fatal over the same dose range; all oral daily doses in excess of 100 mg/d being fatal.

Daily repeated 1-h inhalation exposure schedule at 250 mg MoO_3/m^3 (164 mg Mo/m^3) was observed to be irritating to the guinea pigs. Loss of appetite and weight occurred; diarrhea and muscular incoordination were observed, as well as loss of hair. Of the 51 animals exposed, 26 died subsequent to the tenth exposure. On the other hand, exposure to freshly generated MoO_2 fume under the same conditions was shown to be less toxic with only 8.3% mortality compared to 51% mortality with the dust; and no mortality when the exposure level was reduced to about one-third (57 mg Mo/m^3). This finding could be explained by the more rapid solution and elimination of the large surface area fume particle (98). CaMoO$_4$ dust at 915 mg/m^3 (125 mg Mo/m^3) under the same exposure conditions was shown to be fatal to 20.8% of 24 guinea pigs, however, showing any toxic signs.

Eye and skin irritation tests on calcium and zinc molybdates showed no irritation to intact or abraded skin, and no significant eye irritation in the conventional rabbit test animal (1). Na$_2$MoO$_4$, on the other hand, caused primary irritation at 24 h but cleared at 72 h. A 20% solution in the eye caused redness of the conjunctiva with discharge, but no irritation to the cornea and iris. The compound was not a sensitizer. The dihydrate, Na$_2$MoO$_4 \cdot$ 2H$_2$O, upon continued feeding to rabbits at dietary levels of 0.1% and higher was uniformly fatal within a few weeks. Addition of copper to the diet prevented the development of toxicity and was therapeutically effective in combating Mo poisoning (114).

In a recent study the micronucleus (MN) assay in human lymphocytes and mouse bone marrow and the dominant lethal assay in mice were applied to assess possible genotoxic effects of Mo salts *in vitro* and *in vivo* (115). Two salts of Mo were tested in whole blood cultures. Ammonium molybdate appeared to be more potent than sodium molybdate in causing a dose-dependent decrease in viability and replicative index and an increase in MN formation in binucleated lymphocytes ($p < 0.001$). A dose–response relationship was observed in both kinetochore-positive MN (caused by chromosome lagging) and kinetochore-negative MN (associated with chromosome breakage). Based on the results of a toxicity study of sodium molybdate, two doses, 200 and 400 mg/kg, were assessed in the bone marrow MN assay in mice (two IP injections 24 and 48 h prior to euthanasia). A small but statistically significant increase in MN prevalence in polychromatic erythrocytes was observed ($p < 0.05$). The same protocol was used to analyze dominant lethality. A dose-related increase in postimplantation loss represented mostly by early resorptions was observed the first week after treatment ($p = 0.003$). These results could indicate that sodium molybdate induces dominant lethality at the postmeiotic stage of spermatogenesis.

2.4.1.3 Pharmacokinetics, Metabolism, and Mechanisms. Excretory patterns of Mo were determined after oral intake only (98). Following rapid absorption from the gastrointestinal tract (guinea pig), MoO$_3$ was deposited rather uniformly in the critical organs within 4 h; the blood and bile contained relatively high levels of Mo. Rabbits showed a similarly rapid absorption of Mo following ingestion, with quickly rising blood levels.

2.4.1.3.1 Distribution. Molybdenum is rapidly eliminated by the kidneys, returning essentially to normal values in 72 h after exposure. Rats stored relatively more Mo from ingested MoO$_3$ than from CaMoO$_4$. Storage was least from MoS$_2$. Significant storage above normal of Mo in bone was noted in all cases. The distribution of microgram quantities of injected radioactive ^{99}Mo in dogs was selectively concentrated in the liver and kidney, with high concentrations in the endocrine glands (pancreas, pituitary, and, especially, the thyroid and adrenal glands). Brain, white marrow, and fat contained negligible amounts (117).

2.4.1.3.2 Excretion. South African weathers ($n = 5$) were each dosed with 2.5 g copper oxide needles (118). Three weeks later these sheep were placed on metabolic crates to permit urine and fecal collection. Ammonium tetrathiomolybdate (TTM) (1.7 mg kg^{-1}) was given, and IV blood samples were collected at intervals. The concentration of serum Mo was best described by a two-compartment open model with first-order rate constants. An elimination half-life of 396.8 min, steady-state volume of distribution of 0.8 e kg^{-1} were observed using nonlinear compartmental analysis. A statistical significant ($p \leq 0.05$) increase in Mo excretion in the feces occurred at 24 and 48 h following TTM administration. No increase in fecal and urinary copper excretion was found, probably because the sheep were not copper loaded.

Fecal elimination, which is about half that of the urinary, similarly returned to normal in 72 h.

2.4.1.4 Reproductive and Developmental. The so-called camel sway disease in the Hexi Corridor of Gansu province was studied (119). The contents of eight minerals in soils, in forage, and in the blood and hair of bactrian camels from this region were determined. The related blood indices were also measured. The Mo concentration in soils and forage was 4.8 ± 0.02 and 4.8 ± 0.25 µg/g (dry matter), respectively, the copper to Mo ratio in the forage being only 1.3. The Cu concentration in blood and hair from the camels was 0.28 ± 0.17 µg/mL and 3.50 ± 1.00 µg/g, respectively. There was a hypochromic microcytic anaemia and a low level of ceruloplasmin in the blood. It is therefore suggested that sway disease of bactrian camels in this region is caused by secondary copper deficiency, mainly due to the high Mo content in soils and forage. The copper deficiency in the camels was aggravated during reproduction. Oral administration of copper sulfate can prevent and cure the disease.

Guinea pigs in groups of eight and their offspring were given drinking water containing Mo as ammonium molybdate (AM) or thiomolybdate (TM) throughout and subsequent to pregnancy (120). All adult females had estrous cycles, and conception rates were unaffected. Fetal death was common in groups given the high dose of TM. The Cu concentration in liver was reduced in all groups at all ages except for pups killed at birth from animals given AM. The concentration of Mo was elevated in liver and kidney of all groups. The concentration in plasma of Cu, Mo, and copper insoluble in trichloroacetic acid was elevated in all groups. Superoxide dismutase activity was significantly reduced in dams and 6-week-old pups in which TM administration commenced before mating. Histological damage occurred in the pancreas of animals given AM or TM. The effects on the fetus and pancreas were considered to result from Cu deficiency rather than toxicity of Mo.

2.4.1.5 Carcinogenesis. Maltoni has noted the production of subcutaneous sarcomas in rats from the local injection of Mo orange (lead and Mo chromates) (121); 36 of 40 rats (90%) had tumors, an incidence greater than that from either chrome yellow or chrome orange (lead chromates), which elicited 65% tumors in the same number of rats. Dosage and time of tumor appearance were not stated, and injection-site sacromas are poor indicators of carcinogenicity.

It has been indicated that Mo may prevent the carcinogenic effects of *N*-nitroso compounds (122). Male Wistar rats weighing 170–190 g were pretreated with Na$_2$MoO$_4$ (1.24 mmol/kg body weight, IP once a day) for 3 d and on day 4, they were exposed to NDEA (50 mg/kg body weight, once, IP). Na$_2$MoO$_4$ pretreatment prevented both *N*-nitrosodiethylamine (NDEA)-induced DNA strand breaks and disruption of the metabolism of those cations but rather enhanced lipid peroxidation. The results suggested that Mo prevented NDEA-induced DNA damage by preventing disruption of intracellular Ca metabolism while stimulating the metabolism of the nitroso compound via a nontoxic pathway.

2.4.1.6 Genetic and Related Cellular Effects Studies. No data found in the literature.

2.4.1.7 Other: Neurological, Pulmonary, Skin Sensitization. No data found in the literature.

2.4.2 Human Experience

2.4.2.1 General Information. In human metabolism, Mo is a trace element involved in protein synthesis and oxidation reactions. It is also an essential trace element for plants and is important for soil fertility. Mo is considered essential because it is part of a complex called Mo cofactor that is required for the three mammalian enzymes XO, aldehyde oxidase (AO), and sulfite oxidase (SO) (123). XO participates in the metabolism of purines, AO catalyzes the conversion of aldehydes to acids, and SO is involved in the metabolism of sulfur-containing amino acids. Mo deficiency is not found in free-living humans, but deficiency is reported in a patient receiving prolonged total parenteral nutrition with clinical signs characterized by tachycardia, headache, mental distrubances, and coma. XO deficiency is relatively benign, still patients with isolated deficiencies of SO or mMo cofactor exhibit mental retardation, neurologic problems, and ocular lens dislocation. XO and AO may also participate in the inactivation of some toxic substances, inasmuch as studies suggest that Mo deficiency is a factor in the higher incidence of esophageal cancer in populations consuming food grown in Mo-poor soil. In Mo stimulation of XO, elevated levels of uric acid, blood, and urine occur (124), leading to abnormalities in uric acid metabolism, and eventually acting as a predisposing factor for the development of gout.

A clear antagonism exists between isomorphic tungstate and molybdate, as shown in the chick and the rat (125). This antagonism, which depressed growth rates in the chick and resulted in 25% mortality, did not affect growth rates in rats or alter oxidation of xanthine to uric acid and allantoin, despite all tissues being depleted of Mo and XO. The tungstate diet, however, did not interfere with the absorption of Mo from the rat intestine. Neither chromate (126) nor vanadate (112) affected Mo metabolism or created Mo deficiency. From a study of biochemical changes in workers exposed to Mo dust, it would appear that humans stand midway between the chick and the rat in respect to our metabolism of Mo.

Glutaminases, which cause the release of ammonia, are significantly inactivated by the brain and liver using small doses of chronically administered molybdate salts (127). Similarly, small amounts of Mo salts like those of copper, vanadium, and selenium affect vitamin A status by impairing the efficiency of intestinal utilization of precursor carotenes (128).

Not all the actions of Mo with trace elements are antagonistic or adverse to bodily functions, however. Molybdenum, which aids retention of fluorine in bone and soft tissue of old rats (128), in a later study acted synergistically with fluorine to decrease caries incidence in this species by 52% (129). Also, small amounts of Mo salts, like those of copper, cobalt, tungsten, and mercury, increase antibody formation, specifically increasing the agglutinins against diphtheria and typhoid bacilli (130). If the mechanism of action of Mo just cited seems extremely complex, it is probably no more so than that of any other essential trace element where so many mechanistic facts have been developed.

Cu, Zn, Se, and Mo are involved in many biochemical processes supporting life. The most significant of these processes are cellular respiration, cellular utilization of oxygen, DNA and RNA reproduction, maintenance of cell membrane integrity, and sequestration of free radicals. Cu, Xn, and Se are involved in destruction of free radicals through cascading enzyme systems. Superoxide radicals are reduced to hydrogen peroxide by superoxide dismutases in the presence of Cu and Zn cofactors. Hydrogen peroxide is then reduced to water by the selenium–glutathione peroxidase couple. Conversely, excess intake of these trace elements may result in toxic effects; that is, a fine balance is essential for healthiness. Patients deficient in trace elements usually present with common symptoms such as malaise, loss of appetite, anemia, infection, skin lesions, and low-grade neuropathy, thus complicating the diagnosis. Symptoms of intoxication by trace elements may be general such as flulike and CNS symptoms, fever, coughing, nausea, vomiting, diarrhea, anemia, and neuropathy. Serum, plasma, and erythrocytes may be used for the evaluation of copper and zinc status, whereas only serum or plasma is recommended for selenium. Whole blood is preferred for Mo. Mo levels are best determined by neutron activation and highly sensitive inductively coupled plasma-mass spectrometry.

2.4.2.2 Clinical Cases

2.4.2.2.1 Acute Toxicity. A survey of a Mo-roasting plant in Colorado was made by Walravens et al. (131), wherein 25 males workers were examined for biochemical abnormalities from exposure to soluble Mo dusts. The 8-h TWA exposure to soluble Mo (mainly MoO_3 and other Mo oxides) was about 9.5 mg Mo/m^3, approximately double the recommended TLV. The mean age of the workers was 28.3 years (range 19–44 years), and the mean duration of employment, 4.0 years (range 0.5–20 years). Apart from some generalized medical complaints, the only adverse biochemical findings were (1) large elevations in serum ceruloplasmin (average 50.47 versus 30.50 mL for controls) and (2) smaller increments in serum uric acid concentrations of 5.90 ppm (0.24 mg/100 mL versus 5.01 mL for controls). Significant Mo absorption was shown by plasma concentrations of 0.9–36.5 mg/ 100 mL versus 0 to 3.4 mg/100 mL and 120–11,000 mg Mo/L urine versus 20–230 mg for controls. Urine copper levels of 4–347 mg/L versus 40 for controls were, with the exception of the one high value, within the range of control values. Despite high plasma and urine Mo levels and moderate elevations of serum uric acid levels, no evidence for Mo-induced gout was found in the medical questionnaire. The high employee turnover rate could explain this lack, for Kovalskii et al. (132) and Akopyan (133) found signs of gout in factory workers and among inhabitants of Mo-rich areas in Armenia. The main features were joint pains in the knees, hands, and feet. Articular deformities, erythema, and edema of the joint areas were noted. The syndrome occurred at a dietary consumption level that produced the daily Mo body burden from 10 mg of inhaled dust. The serum uric acid levels in the U.S. study, however, did not reach those in the USSR study. The high employee turnover rate did permit a judgment as to the suitability of the TLV of 5 mg Mo/m^3 for soluble compounds.

2.4.2.2.2 Chronic and Subchronic Toxicity. Work-related diseases of the skin have been detected in 19.6% of 352 Russian workers engaged in Mo production. These diseases are characterized by a relatively low incidence of the dermatitis transformation into eczema. There are weak skin reactions to Mo tests (20% aqueous solution of ammonium paramolybdate). Experimental and clinical immunologic studies have revealed that mostly humoral immune mechanisms with a relatively weak involvement of the T-lymphocytes contribute to the pathogenesis of the dermatoses due to Mo exposure. This fact is responsible for poor clinical manifestation of the delayed-type hypersensitivity reactions and, at the same time, a manifest IgE-dependent pattern of allergic reactions. A clear-cut correlation between the frequency of the dermatoses and the Mo level in environmental dust was detected (134).

These findings are in contrast to the statements repeated by Stokinger (1) from the medical officer at the plant in Climax, Colorado, indicating that in 50 years he had seen no dermatitis, no internal disorder, or any other pathological condition attributable to the Mo compounds used in Climax.

The favorable experience during more than 25 years operation at Langeloth, Pennsylvania, where Mo disulfide is converted to molybdic oxide, ferroMo, and other products was also recounted by Stokinger (1). Furthermore, there were no unusual incidences of any disease-including cancer—either among people working at Langeloth or among people who resided in the nearby environment. Still, work-related inhalation of Mo has caused pneumoconiosis and goutlike symptoms in Austria (135).

2.4.2.2.3 Pharmacokinetics, Metabolism, and Mechanisms. The fate of ^{99}Mo in humans (four cases) following intravenous injection showed 5-d cumulative urinary excretion ranging from 16.6 to 27.2% of the dose (50–100 mCi) (909). The main excretory pathway was via the kidney, the same as that from orally ingested Mo in animals, cited above. The 10 d fecal excretion was 6.8% in one patient and less than 1% in another. Molybdenum clears the blood rapidly (98); less than 0.5% remains at 24 h. The whole blood concentration is slightly greater than in plasma.

Normal Mo values in humans currently are known only for blood, lung, and lymph nodes. Allaway et al. (136) determined the Mo content in the blood of adult males at 19 U.S. collection sites in 15 states at different periods in the year. Of 229 samples, 48 had detectable amounts as determined by the thiocyanate method, which has a detection limit of 0.5 mg Mo/100 mL blood. These were confined to nine geographic sites with mean values from 0.50 to 15.73 mg/100 mL and a range of 0.5–41.0, the unusually high values coming from Missoula, Montana (41 mg/100 mL) and Rapid City, South Dakota

(14.2 mg). Values at all other sites were between 0.50 and 3.37 mg/100 mL. There was no evidence of a broad geographic pattern of distribution.

In normal blood, Mo is firmly bound to red blood cells and plasma proteins, with somewhat greater amounts associated with the red cells (137). The investigators found that in leukemia the Mo content of the red cells increases, without concomitant change in its distribution in the plasma constituents. In all types of anemia, especially those due to iron deficiency and cancer, there is a substantial decrease in the amount of Mo in both red cell and plasma.

The Mo content of adult lungs taken at autopsy from 100 urban U.S. residents showed a range of 0.04–3 mg Mo/g dry tissue; hilar lymph node values were from 0.05 to 320 mg/g, the latter being approximately the same as those for manganese, a similarity between Mo and manganese already noted in contents of other body tissues (1). Tissue levels depend in part on the content of sulfate in the diet relative to the amount of Mo intake.

A study of Mo absorption, excretion, and balance was conducted in four young men fed five amounts of dietary Mo, ranging from 22 to 1490 micrograms/d, for 24 d each (138). The study was conducted to obtain a data base for a recommendation on dietary Mo intake for healthy young men. Stable isotopes of Mo were used as tracers. ^{100}Mo was fed five times during the study and ^{97}Mo was infused three times. ^{94}Mo was used to quantify the Mo isotopes and total Mo in urine, fecal collections, and diets by isotope dilution. Adverse effects were not observed at any of the dietary intakes. Mo was very efficiently absorbed, 88–93%, at all dietary Mo intakes, and absorption was most efficient at the highest amounts of dietary Mo. The amount and percentage of Mo excreted in the urine increased as dietary Mo increased, suggesting that Mo turnover is slow when dietary Mo is low and increases as dietary Mo increases. We conclude from these results that dietary intakes between 22 and 1500/d by adult men are safe for $\mu g \geq$ 24 d and that Mo retention is regulated by urinary excretion. Mo is conserved at low intakes, and excess Mo is rapidly excreted in the urine when intake is high.

2.4.2.2.4 Absorption. Human red blood cells were incubated in the presence of Na$_2$MoO$_4$ and the initial rate of Mo uptake was measured (116). Mo uptake appeared to be inhibited by external chloride, bicarbonate, sulfate, and phosphate in the range of concentrations previously described for anion carrier fluxes. Trace elements, previously described to be translocated by the anion carrier (i.e., copper, zinc, and cadmium) appeared to slightly inhibit uptake of Mo. Mo uptake was stimulated by acidification, suggesting that the monovalent HMoO$_4$-anion species can be more rapidly translocated than the divalent anion complex MoO$_4$(2-), the predominant form at physiological pH. It was concluded that the anion carrier can catalyze rapid Mo movements across red cell membranes, supporting reports of an enterohepatic circulation of Mo, with red blood cells acting as Mo carrier between the intestine and the liver.

2.5 Standards, Regulations, or Guidelines of Exposure

The OSHA PEL and ACGIH TLV for the soluble compounds of Mo are both 5 mg Mo/m^3. The ACGIH TLV for the insoluble compounds is 10 mg Mo/m^3, whereas the OSHA PEL is 15 mg Mo/m^3 for total dust (79, 139).

The TLVs, originally based on the experimental animal studies of Fairhall et al. (98), have now been fairly well substantiated for the soluble compounds by the survey of Walravens et al. (131) and for the insoluble forms by the long experience of Amax, Inc., formerly Climax Mo Co (1).

2.6 Studies on Environmental Impact

Lovely (140) suggested in a review that microorganisms can enzymatically reduce molybdenum in the environment, but reduction of this metal has not been studied sufficiently.

3 TUNGSTEN

3.0 Tungsten

3.0.1 CAS Number: *[7440-33-7]*

3.0.2 Synonyms: Wolfram

3.0.3 Trade Names: NA

3.0.4 Molecular Weight: 183.9

3.0.5 Molecular Formula: W

3.1 Chemical and Physical Properties

The physical and chemical properties of tungsten and some of its compounds are listed in Table 38.5.

The chemistry of tungsten and its compounds is similar to that molybdenum. Tungsten exists in several states of oxidation, 0, 2+, 3+, 4+, 5+, and 6+. The most stable is 6+, the lower valence states being relatively unstable. Bivalent tungsten exists only as halogen compounds. Like molybdenum, tungsten has a strong tendency to form complexes exemplified by a large series of heteropoly acids formed with oxides of phosphorus, arsenic, vanadium, and silicon, among others (e.g., phosphotungstic acid, $H_3PW_{12}O_4 \cdot 14H_2O$). In addition, compounds of tungsten exist in which tungsten occurs in more than one valence state [$2WNùW(NH_2)_2$]. Tungsten forms a series of oxyhalides (e.g., $WOCl_4$, WO_2CL_2, and $WOBr_4$).

The properties of tungsten alloys offer more limited uses than those of most metals. One of the chief alloys, ferrotungsten (70–80% W with up to 0.6% C) has a melting range of 3000–5000°F and is added to steel to obtain strength, wear resistance, and special resistance to tempering, oxidation, and high temperatures. A tantalum alloy, 90:10 Ta–W, first produced in the 1930s, is still used in missile, nuclear, and rocket engineering because of its ability to withstand substantial structural loads at temperatures well above 3000°F while still lending itself to fabrication (141). A rhodium alloy (28% Rh) is considered the most suitable for providing increased ductility of tungsten at temperatures up to 300°C and for improving its oxidation resistance at high temperatures. Small alloying additions of

CHROMIUM, MOLYBDENUM, AND TUNGSTEN

Table 38.5. Physical and Chemical Properties of Tungsten and Some of its Important Compounds

Compound	Mol. Formula	At. or Mol. Wt.	Sp. Gr.	M.P. (°C)	B.P. (°C)	Solubility
Tungsten	W	183.85	19.35 (20°C)	3410	5660	Insol. water; sol. HF + HNO$_3$; insol. KOH, HF
Tungsten, trioxide	WO$_3$	231.85	7.16	1473	—	Insol. water, acids; sol. hot alkalies, HF
Tungstic acid (ortho)	H$_2$WO$_4$	249.86	5.5	−H$_2$O, 100	1473	Insol. cold H$_2$O; sl. sol. hot H$_2$O; sol. alkalies, HF, NH$_3$; insol. acids
Tungsten carbide	WC	195.86	15.83 (18°C)	2780	6000	Insol. water; sol. aqua regia, HF + HNO$_3$
Tungsten hexachloride	WCl$_6$	396.57	3.52 (25°C)	346.7	—	Dec. 60°C H$_2$O; sol. ether, alcohol, benzene, CCl$_4$; v.sol. CS$_2$
Tungsten oxytetrachloride	WOCl$_4$	341.66	—	211	227.5	Dec. water: sol. CS$_2$, SNCl$_2$, benzene
Tungsten disulfide	WS$_2$	247.98	7.5 (10°C)	Dec. 1250	—	Insol. cold water; sol. HF + HNO$_3$; insol. alcohol

titanium, vanadium, zirconium, and carbon improve the tensile ductility of tungsten at 3500°F, and if added jointly, more than double the tensile strength at that temperature.

The oxide, WO$_3$, and acid, H$_2$WO$_4$, both heavy, yellow powders, insoluble in water, are utilized as pigments in ceramics and as color-resistant mordants for textiles and fireproofing fabrics. The sodium salt, which solubilizes the acid to a colorless crystalline substance, is similarly used for fireproofing. The carbide, with the unusual Mohs hardness of 9+, is used in cemented carbide tools. Among the simple tungsten salts, WCl$_6$, a dark blue, relatively volatile substance, has a greater number of uses than most in coatings on base metals, in vapor deposition for bonding metals, for formation of single crystal tungsten wire, as an additive to SnO$_2$ for electrically conducting coating for glass, and as a catalyst for olefin polymerization. Its colorless, more highly volatile counterpart, the hexafluoride, has uses limited to vapor-phase deposition of tungsten and as a fluorinating agent. Tungsten disulfide is used as a solid lubricant. Phosphotungstic acid, a strong acid oxidizing agent, has a wide range of uses from an analytic reagent to imparting water resistance to plastics, adhesives, and cement, to catalyst in organic reactions, and antistatic agent in textiles. Ammonium tungstate is used in its preparation.

3.1.1 General

Tungsten occurs principally in the minerals wolframite (Fe,Mn)WO$_4$, scheelite (CaWO$_4$), ferberite (FeWO$_4$), and hubnerite (MnWO$_4$). They are found in China, the Buryat

Republic of Russia, Kazakhstan, South Korea, Bolivia, and Portugal. Wolframite is the most important ore worldwide; scheelite is the principal domestic U.S. ore. Scheelite, when pure, contains 80.6% WO_3, the most common impurity being MoO_3. The percentages of FeO and MnO in wolframite vary considerably; hubnerite is the term applied to ore containing more than 20% MnO, and ferberite, to ore containing more than 20% FeO. Intermediate samples are called wolframite.

Tungsten is a member of the third series of transition metals. The name is derived from the Swedish "tung sten," meaning "heavy stone." The symbol is W for wolframite, the mineral from which the element was first recognized in 1779 by the English chemist Peter Woulfe. The metal tungsten was first isolated in 1783 by Spanish scientists Jose and Fausto d'Elhuyar through the reduction, by means of charcoal, of the tungstic acid found in wolframite.

3.1.2 Odor and Warning Properties

No information in the literature.

3.2 Production and Use

Most domestic production in the United States comes from two mines, one in California and the other in Colorado. In addition, a number of mines in eight western states and in North Carolina had small production. Tungsten is mined in the United States by opencut operations similar to those for gold, but most tungsten in this country comes from underground workings. The United States imports tungsten ore and concentrates, mostly from Canada.

Modern tungsten mills are of the gravity or gravity-flotation type, which attain recoveries of WO_3 as high as 70–90%. Production of finished concentrates may involve acid-leaching (HCl) to remove phosphorus, roasting to remove arsenic or sulfur, or magnetic separation to improve grade. Calcite ($CaCO_3$) often complicates flotation procedures used with scheelite because of like densities. Copper, arsenic, antimony, bismuth, phosphorus, sulfur, lead, molybdenum, tin, zinc, and manganese are the most common impurities in concentrates used for direct charging for steel. Some scheelite concentrates require digestion and precipitation to remove combined copper and molybdenum. Scheelite is "nodulized" to minimize dust losses. Ferrotungsten is made in electric furnaces with carbon or silicon as the reducing agent, or by alumino- or silicothermic methods.

Manufacture of the tungsten metal (powder tungsten) chemical group uses more than half the total tungsten. A solution of tungsten from the concentrates is required from which tungsten may be precipitated as tungstic acid or left in solution as Na_2WO_4 or K_2WO_4. Both forms require additional purification. An important method from the work hygiene standpoint is the treatment with NH_4OH to form $(NH_4)_2WO_4$, followed by heating to form ammonium paratungstate. Because of the possible untoward effects of associated impurities noted previously, elaborate steps are taken to minimize their presence in the final products; for example, arsenic cannot be tolerated at a concentration higher than 0.02%, and phosphorus at more than 0.05%.

There are two divisions of the tungsten metal powder industry, hydrogen reduction and carbon reduction. Hydrogen reduction is used for metal powder for most WC for all filament wire; carbon reduction is resorted to when tungsten powder is used for welding rods, coating oil-well tools, and for some WC. In hydrogen reduction, tungstic acid or ammonium paratungstate in iron or nickel boats is placed in tubes through which hydrogen flows, usually in a two-stage reduction. Carbon-reduced tungsten is made similarly from tungstic acid, but lampblack or natural or manufactured gas is the reducer.

Vapor deposition of tungsten is possible by the pyrolysis of gaseous compounds such as tungsten carbonyl, $W(CO)_6$, which sublimes at 50°C and boils at 175°C, tungsten hexafluoride, WF_6, which boils at 19.5°C, or tungsten hexachloride, WCl_6, which boils at 346.7°C. Vapor deposition of tungsten on a substrate of graphite or copper by heating WF_6 with hydrogen yields a fine-grained deposit. The deposits, which are 100% dense, contain less than 50 wt. ppm of interstitial impurities and less than 100 wt. ppm of metallics. The fluorine content, which is at least 15 wt. ppm, increases the ductile-to-brittle transition temperature. For the production of tungsten wire, rod, and sheet, hydrogen-reduced powder is compressed, sintered, heated to incipient fusion by passage of electric current, and swaged, drawn, or rolled.

Tungsten carbide (WC) is produced from either hydrogen- or carbon-reduced powder by heating with lampblack, or WC crystals may be prepared directly from the ore. A mixed WTiC may be produced. WC tools are produced by sintering a mixture of WC and cobalt, or nickel, or both in dies of the desired shape. TiC and TaC are both often blended with WC before sintering with cobalt.

The prime use of tungsten is in cutting and wear-resistant materials (65%), mill products (12%), specialty steels, tools, stainless, and alloys (9%), hard-facing rods (8%), super alloys (3%), and chemicals (2%). It can reasonably be inferred from the small usage of ferro and super alloys relative to other uses, that tungsten, unlike most metals, forms relatively few alloys with properties superior to those of others.

3.3 Exposure Assessment

Exposure to tungsten-containing compounds may occur during production and uses of tungsten, its alloys, and compounds, rather than to tungsten itself. It is, however, still not clear precisely what role tungsten plays in the exposures. In the WC cutting-tool industry cobalt, exposure to fume and dust implies exposure to WC, TiC, TaC, and NiC. In the manufacture of tungsten metal, hazardous exposures to associated metals in the ore are chiefly to arsenic, antimony, manganese, lead, copper, bismuth, tin, and molybdenum.

3.3.1 Workplace Methods

The adoption of AAS for quantitative microanalysis of metals continues with tungsten, NIOSH analytical method 7074 (144). This is the most economical method for analysis of airborne tungsten. After collection of the sample on a cellulose membrane filter and washing with a mixture of HNO_3 and HF, the dissolved sample is aspirated into a nitrous oxide–acetylene flame of the AAS and read at the characteristic lines of 255 and 401 nm. Optimal working conditions are 500 mg/mL, and lowest detectable concentration is 3 mg/mL.

The sensitivity is 35 mg W/mL at 1% absorbance. Vanadium, molybdenum, manganese, chromium, and nickel interfere by enhancing tungsten absorption; but a 2% addition of Na_2SO_4 eliminates these interferences.

3.3.2 Community Methods

The preceding methods may also be applied for these purposes.

3.3.3 Biomonitoring/Biomarkers

Cardenas and Mortenson (96) proposed a spectrophotometric method for the quantitative determination of tungsten in biological materials. That method involves the selective formation and extraction of the tungsten complex with 4-methyl-1,2-dimercaptobenzene. Absorbance is read at 630 nm for 5 to 100 nmol. Interferences are negligible. Sensitivity, precision, and accuracy were not given. An X-ray fluorescence procedure was employed by Baudouin (142) to determine tungsten in biologic tissues.

The tungstate ion, WO_2, is isomorphic with molybdate ion, MoO_4; therefore, tungstate antagonizes the normal metabolic action of molybdate in its role as metal carrier in the enzymes xanthine dehydrogenase (125), sulfite oxidase, aldehyde oxidase (112), and nitrate reductase (145). Administration of tungstate in the drinking water of rats at levels from 1 to 100 ppm resulted in proportionate decreases of sulfite oxidase activities, as well as in those of xanthine oxidase (112). The pool of excess molybdenum could be replaced by tungsten at the 100-ppm level, but 1 ppm Mo almost completely reversed the effect of the 100 ppm W, indicating mutual antagonism of tungsten and molybdenum.

The electrical properties of membranes — both biological and artificial — were studied in the presence of a number of negatively charged tungsten carbonyl complexes, such as $[W(CO)_5(CN)]$-, $[W(CO)_5(NCS)]$-, $[W_2(CO)_{10}(CN)]$-, and $[W(CO)_5(SCH_2C_6H_5)]$-, using the single-cell electrorotation and the charge-pulse relaxation techniques (146). Most of these negatively charged tungsten complexes introduced mobile charges into the membranes, as judged from electrorotation spectra and relaxation experiments, meaning that the tungsten derivatives act as lipophilic anions. They greatly contributed to the polarizability of the membranes and led to a marked dielectric dispersion.

3.3.3.1 Blood. In a study designed to identify a safe threshold of sodium tungstate in drinking water (147), the threshold of tungsten accumulation in the tissues of animals fed sodium tungstate was determined from the accumulation of tungsten in a series of increasing doses. Deposition followed a dose–response pattern, rats and rabbits receiving 5 and 0.5 mg/kg had the highest tissue concentrations of tungsten. Less pronounced, but significant, accumulation of tungsten was found in tissues of animals given 0.05 mg/kg, while a dose of 0.005 mg/kg produced increases only in blood and intestinal tungsten. No accumulation of sodium tungstate appeared to occur below 0.005 mg/kg W as tungstate because elimination kept pace with intake.

3.3.3.2 Urine. An analytical method (143) was applied to analyze 14 urine samples from persons not exposed at work to metals to determine the concentration of antimony (Sb), bismuth (Bi), lead (Pb), cadmium (Cd), mercury (Hg), palladium (Pd), platinum (Pt), tellurium (Te), tin (Sn), thallium (Tl), and tungsten (W) in urine. The purpose was to develop a method that is versatile for the determination of environmentally as well as work-related element excretion. The detection limits for these elements appeared to be between 5 and 50 ng/L. For some of these elements ICP-MS appeared to be more sensitive than atomic absorption spectrometry (AAS).

3.3.3.3 Other. Exposure to tungstate activates brain glutaminase and inactivates liver glutaminase, whereas molybdate inactivates both sites. This shows that tungstate has the capacity to act at more than one enzyme site; and the biologic action of tungstate is independent of its isomorphic replacement of molybdate. It is also of interest that the tungstate effects on glutaminase were elicited by microamounts, 0.5 mg injected intraperitoneally daily for 100 d, with effects occurring as early as 40 d (at 100 d, 172% activation of brain glutaminase, 226% inactivation of liver glutaminase). Levels required to inactivate xanthine dehydrogenase (125) or to decrease effectively the availability of the serum SH group (147) are several orders of magnitude greater, showing that glutaminase is the most sensitive enzyme to tungstate thus far observed.

3.4 Toxic Effects

Many investigations on the physiologic effects of tungsten followed the marketing of cobalt-cemented WC just before 1940. Hence most of the investigations concern the toxicity and health effects of cemented WC and its constituents, particulary in humans, rather than tungsten and its compounds themselves, all of which may blur the true toxicity of tungsten. The most significant exposure-related disease is mostly referred to as hard metal pneumoconiosis. The few determinations of toxicity of tungsten and its compounds made before 1950 clearly showed a difference between soluble and insoluble forms. Soluble compounds were distinctly more toxic than insoluble forms, resulting in two separate permissible limits for industrial exposure.

3.4.1 Experimental Studies

3.4.1.1 Acute Toxicity. The toxicity of tungsten metal dust and dusts of WC, both highly insoluble in body fluids, was determined using intratracheal injection by several investigators, including Miller et al. (148), Lundgren and Swensson (149), and Dehalant and Schepers (150).

The relative sensitivity of enzymes and enzyme groupings was examined by Nadeenko in rabbits given daily oral doses of 5, 0.5, 0.05, and 0.005 mg/kg of sodium tungstate, and the activity of various biological parameters (147). Most sensitive during an 8-month period was blood alkaline phosphatase; 0.05 mg/kg doses produced minimal (ca. 18%) and sporadic decreases at 4–8 m. All other parameters measured, serum SH groups, blood cholinesterase, and liver glycogen function, showed minimal (16–23%) decreases only at levels of 0.5 and 5 mg/kg, with no changes at 0.05 and 0.005 mg/kg.

3.4.1.2 Chronic and Subchronic Toxicty. Dietary Na_2WO_4, equivalent to a W : Mo ratio of 100 : 1 completely inhibited the deposition of intestinal xanthine oxidase in the rat and markedly reduced xanthine dehydrogenase and molybdenum in the liver. Na_2WO_4

produced a molybdenum deficiency in the chick. Half the uric acid normally excreted by the chick was replaced by xanthine and hypoxanthine. All effects of tungstate were reversed by small amounts of molybdate, thus indicationg a true antagonism.

Tungstate is, likewise, a competitive inhibitor of molybdenum utilization by bacteria and plants (145). Molybdenum, specific and essential for the induction of nitrate reductase involving de novo protein synthesis, is antagonized by tungsten as a result of the formation of a tungstoprotein analogue. The tungsten–protein analogue is lacking in nitrate reductase activity, but is active as a diaphorase.

Sodium tungstate *in vivo* activates brain glutaminase, increasing ammonia release, in contrast to $(NH_4)_2MoO_4$, which inhibits it (151). The tungsten activation is due either to elevation of phosphorus ion concentration in the brain or to activation of glutaminase by direct action on its molecule, just as in liver aldehyde oxidase activation by tungsten ions (152). It has been shown that hard metal dust, which consists of a mixture of cobalt and WC, is more toxic toward mouse peritoneal and rat alveolar macrophages than pure cobalt (Co) or WC. A study was aimed at investigating the toxic effects of Co and hard metal dust on alveolar epithelial type II cells (AT-II), and to compare these with alveolar macrophages (153). Freshly isolated rat and human AT-II and rat alveolar macrophages were exposed for 18 h to particles of Co, WC, or Co/WC. Release of lactate dehydrogenase was measured as an indicator for cell toxicity. For rat AT-II, TD_{50} values per 10^5 cells were 672 μg (95% C.I. = 264–1706 μg) for pure Co and 101 μg (95% C.I. = 59–172 μg) for Co in Co/WC mixture. For rat alveolar macrophages, TD_{50} values per 10^5 cells were 18 μg (95% C.I. = 15–24 μg) for pure Co and 5 μg (95% C.I. = 5–6 μg) for Co in Co/WC mixture. WC only at high concentrations caused an increase in lactate dehydrogenase. No toxicity was found in human AT-II for either Co, WC, or Co/WC. These results could indicate (1) rat AT-II are less sensitive to Co than rat alveolar macrophages, (2) human AT are less sensitive to Co than rat AT-II, and (3) the toxicity of Co is enhanced by the presence of WC.

3.4.1.3 Pharmacokinetics, Metabolism, and Mechanism

3.4.1.3.1 Absorption. The kinetics of tungstic oxide following inhalation were determined in dogs by the use of radio-labeled $^{181}WO_3$ (154). After inhalation of $^{181}WO_3$ mist of 98 mCi/mL specific activity for 6 h, 1.9 to 8.0 mCi was deposited in the respiratory tract, 60% of which was in the lower part of the tract. Measurements made over the lung area showed that about 69% of the activity was lost, with a biological half-life of 4 h. Measurements taken on the lower half of the body showed longer retention — 94% had a half-life of 9h; 4.1 percent a half-life of 6.3 d, with the remaining 1.6% of the radioactivity having a half-life of 139 d. Radioactivity was lost rapidly from the blood.

3.4.1.3.2 Distribution. The first study (in 1924) made of the kinetics of sodium tungstate (155) detected tungsten in the blood and urine in addition to finding it in the liver, kidneys, lungs, stomach, and intestines of guinea pigs treated either orally or subcutaneously with 500 mg/animal. No elimination studies were made. Of the organs tested, the lungs and kidneys retained maximal radioactivity; other tissues contained only about 10% as much as these organs. In terms of total body burden of radioactivity, most was found in the skeleton (37%), lungs (31%), kidneys (15%), liver (9.7%), and skeletal muscle (5.7%).

A dietary metabolic study carried out in 1945 in rats (156) of tungstic oxide and sodium tungstate equivalent to 0.1% W, NH_4 paratungstate at 0.5% W, and W metal at 2 and 10% W revealed after 100 d on the diets that bone and spleen retained the most W. Concentrations averaged 115 and 75 ppm, respectively. Less than 10 ppm was present in liver, kidney, and skin. Blood, lung, testes, and muscle showed traces of tungsten in occasional samples. Except for a single instance in each organ, the heart, brain, and uterus were free of tungsten by the method of analysis employed by Aull and Kinard (157): 99.9% average recoveries at 0–0.1 mg. According to the investigators, the distribution patterns of the several compounds differed only quantitatively.

Whole-body autoradiography and impulse counting experiments were used to study the distribution of radioactivity in the pregnant mouse after administration of [^{185}W] tungstate. Rapid uptake was found in a number of tissues and skeleton, red pulp of the spleen, adrenal, liver, thyroid, pituitary, and ovary and in the intestine and kidneys, through which it was rapidly excreted. Tungsten was also readily transported from mother to fetus, although markedly more in late rather than in early gestation. The most significant retention of the element was found in the maternal skeleton, kidneys, and spleen and in the visceral yolk sac epithelium and the skeleton of the fetus. Also, *in vitro* cytotoxicity experiments showed inhibition by tungstate of cartilage production in limb bud mesenchymal cultures at concentrations similar to those found *in vivo* (158).

3.4.1.4 Reproductive and Developmental. There are no data available.

3.4.1.5 Carcinogenesis. In 1986 it was reported that tungsten added to the drinking water (150 ppm) of rats increased the incidence of *N*-nitroso-*N*-methylurea-induced mammary carcinoma among female Sprague–Dawley rats (160). According to Schepers (161), neither tungsten, WC, nor $MnWO_4$ induced tumors in guinea pigs in which they were intratracheally introduced.

3.4.1.6 Genetic and Related Cellular Effects Studies. A study was carried out using cobalt metal, a mixture of cobalt with tungsten carbide and cobalt chloride, to compare their DNA-damaging capacity (161). Concentrations from 0 to 6.0 μg Co-equivalent/mL were tested. These three compounds were all able to induce DNA damage in isolated human lymphocytes from three donors, in a dose- and time-dependent way. However, a significant interexperimental and interdonor variability in response was observed, which was ascribed to technical parameters and unidentified individual factors. The potential for DNA damaging of the Co–WC mixture was higher than that of cobalt metal and cobalt chloride. No significant increase of DNA migration was observed when the DNA of cells treated with Co–WC or WC were incubated with the oxidative lesion-specific enzyme formamidopyrimidine DNA glycosylase. That was interpreted to suggest that during the short treatment period no substantial oxidative damage to DNA was produced.

3.4.1.7 Other: Neurological, Pulmonary, Skin Sensitization. No experimental animal studies of skin effects have been reported. Ocular effects of tungsten in animals are covered by two reports. Lauring and Wergeland (162) inserted a particle of tungsten less than 1 mm diameter into the midvitreous region of the eye of two rabbits and, from weekly and monthly examinations and a final pathological examination at the end of 1 year, classified tungsten metal as "completely inert." Sodium tungstate, when applied directly to the corneal stroma in rabbit's eyes without protection of the epithelium, however, is toxic in the range of pH 7 to 9 (163). Practically, neither tungstic acid nor the soluble tungstates present much of a hazard to the eyes (152).

In the experiments of Miller et al. pure WC resulted in pigment deposition in the peribronchial connective tissue and lymph nodes, but no increase in reticulum or collagen. Lundgren and Swensson (149) also failed to show any fibrogenic activity resulting from tungsten. In Dehalant's experiments with tungsten metal and carbide, the dust was introduced into the trachea as a suspension in isotonic saline — a dosage of 150 mg in three equal quantities at weekly intervals. Both the metal and the carbide (unless mixed with cobalt) proved relatively inert, even at these massive doses. Schepers (164) examined the acute and the chronic response of the lungs to particulate tungsten metal and cobalt and found focal interstitial pneumonitis and bronchiolitis as immediate response. After 1 y there were some slight residual lesions, chiefly in the form of a minor degree of atrophic emphysema. A comparison of these resuls with those of cobalt metal administration would suggest that the toxic effects of tungsten carbide dust on the lungs of animals are due not so much to tungsten itself.

After these authors in previous studies in the rat had demonstrated the greater acute pulmonary toxicity of a WC–Co mixture compared to Co or WC alone, they undertook a study to compare in the same animal model the delayed lung response after intratracheal administration of Co or WC–Co particles (cobalt particle 6.3 wt%) (165). The outcome effects were also compared with those obtained after treatment with arsenic trioxide and crystalline silica producing an acute toxic insult and a progressive fibrogenic response, respectively. Cellular and biochemical parameters (LDH, N-acetyl-beta-D-glucosaminidase, total protein, albumin, fibronectin, and hyaluronic acid) were measured in bronchoalveolar lavage fluid following single and repeated intratracheal instillations. The study results indicated that the delayed lung response observed after WC–Co is different from that after cobalt metal alone. A single intratracheal dose of WC–Co induced an acute alveolitis that persisted for at least 1 mo. However, 4 mo after a single instillation of WC–Co, no clear histological lung fibrosis could be seen, indicating a reversibility. Following repeated intratracheal instillations, enhanced lung hydroxyproline content and histopathological evidence of interstitial fibrosis were observed after WC–Co, but not after administration of each component separately. The mechanism of the fibrotic reaction resulting from WC–Co seems different from the progressive inflammatory reaction induced by crystalline silica, and the authors suggested that it might result from a scarring reaction elicited by repeated acute insults.

Hard metal alloys are made of a mixture of tungsten carbide particles (WC, more than 80%) cemented in cobalt metal powder (Co, 5–10%). Inhalation of hard metal particles may result in an interstitial pulmonary disease (166). In a macrophage culture model, butylated hydroxytoluene (1 mM) protected form the cytotoxicity of hard metal particles, suggesting a possible involvement of lipid peroxidation in the toxicity of these powders. In a biochemical system, a mixture of Co and WC particles stimulated the production of thiobarbituric acid–reactive substances from arachidonic acid. Using a spin-trapping system applied to aqueous particulate suspensions and electrochemical techniques, the authors presented experimental evidence that the association of Co and carbide particles represents a specific toxic entity producing large amounts of activated oxygen species. The mechanism of this interaction takes place through the oxidation of cobalt metal catalyzed at the surface of carbide particles and resulting in the reduction of dissolved oxygen. This physicochemical property was suggested to provide a new basis for interpreting their inflammatory action and their possible carcinogenic effect on the lung.

The role of reactive radical formation in the toxicity of these particles has been assessed in a macrophage culture model (167). Catalase, superoxide dismutase, sodium azide, sodium benzoate, mannitol, taurine, and methionine were all unable to protect against the cytotoxic effects of cobalt ions and cobalt metal alone or mixed with tungsten carbide. The authors concluded that no evidence was found that production of reactive oxygen species contributes to the elective toxicity of carbide–cobalt mixtures.

3.4.2 Human Experience

Reports, surveys, and epidemiological studies of worker exposures to tungsten have centered on the cemented WC industry, as it is called in the United States, the "hard metal industry" in Great Britain and elsewhere. The cobalt-cemented WC hard metal is produced by powder metallurgy from tungsten and carbon, with cobalt as a binder. At times tantalum, titanium, vanadium, molybdenum, and chromium may be added, each ending as carbide in the finished product. A typical product contains 80–90% WC, 8–18% TiC, and 5–25% Co, but most grades contain an average of 7% Co.

Cases of pulmonary involvement have been reported from the United States and in other countries. Bech et al. (151) reviewed the older reports of pneumoconiosis beginning in 1940 in Germany and later reports from Hungary, Sweden, and the Soviet Union, in which several hundreds of hard metal workers showed a high prevalence of respiratory disorders and radiographic changes.

The first report of typical, hard metal disease occurring from grinding cemented WC came from Miller et al. in the United States (148), who described breathlessness and radiological changes in three workers employed in grinding hard metal for 3.5, 6, and 7 years, respectively, using Al_2O_3 grit SiC and diamond wheels. Breathing zone dust counts varied from 3.7 to 7.7 mppcf for workers doing fine grinding, 38 to 40 mppcf at rough-grinding machines, and general room air, 1.4 mppcf. Cobalt analyses were $0.1–0.2 \, mg/m^3$ in the breathing zones of operators. Sampling problems appear to have prevented analyses for tungsten. The plant manufacturing WC tools for 18 years employed 200–300 men, about 150 of whom over the years had been engaged in grinding operations.

From this report and others it is impossible to estimate the incidence of typical hard metal disease with pulmonary fibrosis because (1) only cases with severe respiratory complaints were counted, or (2) those with initial symptoms left employment. A prevalence estimate of 5% could be deduced from the study of Baudouin et al. (142), who reported five cases of diffuse interstitial pulmonary fibrosis among 100 workers in a hard metal factory where dusts of tungsten, titanium, tantalum, and nickel metals, of their carbides and oxides, and of cobalt metal were in the exposure atmosphere during the various operations. Analysis by X-ray fluorescence of the lung of one worker with clinical and radiological changes showed large amounts of tungsten, amounts of titanium greater than normal, and small amounts of tantalum and niobium. The carbides of tungsten and titanium were demonstrated in the lung by X-ray diffraction.

In his 1974 review (152) Bech found accounts of hard metal disease from the United States, Great Britain, Australia, France, and Switzerland to be remarkably similar in describing (1) a moderate incidence of cough, dyspnea, and wheezing; (2) typical hypersensitivity asthma in a few, sometimes developing after only short exposure; (3) the occurrence of a relatively high incidence of minor radiological abnormalities; and (4) marked radiological changes in a small number.

The symptoms of cough, dyspnea, and wheezing usually develop gradually after years of exposure and may remit following cessation of exposure. Those workers who develop asthma usually change their occupation, and their symptoms generally dissppear following removal from exposure. There is no correlation between the onset of symptoms, the length of exposure, and the development of interstitial fibrosis. There are reports of the disease developing after short periods of exposure (168–170).

The minor radiological abnormalities observed are slight increases of lung markings and very sparse linear opacities, which are best seen in the lower and middle zones. These radiological changes may represent "inert" dust shadows from accumulations of dust derived from hard metal, predominantly tungsten carbide, and from the grinding tools. Those with marked radiological changes show varying degrees of linear, micronodular, and nodular opacities concentrated particularly in the mid and lower zones, prominence of hilar shadows, and sometimes, honeycombing.

There appears not to be any correlation between the onset of symptoms, the length of exposure, and the development of interstitial fibrosis. There are reports of the disease developing after comparatively short periods of exposure (168–170). The disease may be arrested by early removal from exposure. Radiological clearing was reported by Miller et al. (148), Joseph (169), and Coates and Watson (168). This would appear to occur where the disease had developed recently. Early recognition of the disease is important, because it is progressive and potentially lethal.

Treatment with corticosteriods and removal from exposure arrest progression of the disease. To emphasize the seriousness of hard metal disease, Bech (152) stated that of 12 employees with the disease, 8 died of secondary complications. Four of the 8 died of corpulmonale and cardiac failure, and the other four died of emphysema. Except for differences in degree and duration of the symptoms, the general course of the disese was similar in all 12 cases, enforcing the belief that individual hypersensitivity plays a role in this disease. Baudouin and associates (142) reported that immunologic tests showed a cellular, rather than a humoral, immune deficiency in some workers who had reached the fibrotic stage.

A fatal case of hard metal pneumoconiosis was reported by Della Torre et al. (171) in 1990. Four subjects working in sharpening and grinding operations of hard metal tools were involved. Only one worker, a 37-year-old female exposed for 7 years to hard metal dusts, developed hard metal pneumoconiosis, which rapidly progressed to death. Cytology of the bronchoalveolar lavage (BAL) showed a high number of eosinophils, more than 30% of the cell population. Biopsy of the lung revealed interstitial fibrosis with hyperplasia of the pneumocytes of the second type and inflammatory cellular infiltration of the interstitium. High tungsten and tantalum concentrations were determined in the admission BAL and in the biopsy 4 mo later by neutron activation analysis; cobalt levels were near normal. The content of cobalt and tungsten in blood and urine and particularly in pubic hair and toenails of the patients was significantly higher than the normal values. This suggests that such biologic specimens could be used as indicators of chronic exposure to hard metal dusts.

Sprince et al. (172) carried out a cross-sectional survey of 1039 WC production workers and found work-related wheeze in 113 participants (10.9%). The occurred ratio of work-related wheeze were 2.1 times for present cobalt exposures exceeding 50 mg/m^3 compared with exposures less than or equal to 1/0 were 5.1 times for average lifetime cobalt exposures exceeding 100 mg/m^3, compared with exposures less than or equal to 100 mg/m^3 in those with latency exceeding 10 years. Interstitial lung disease was also present in three workers with very low average lifetime exposures (less than 8 mg/m^3). Grinders of hard carbide had lower mean DLCO than nongrinders, even though their cobalt exposures were lower. The same authors (173) had previously indicated that interstitial and obstructive lung disease occur in WC workers in correlation with elevated peak air concentrations of cobalt.

3.4.2.1 General Information. Reports, surveys, and epidemiological studies of worker exposures to tungsten have centered on the cemented WC industry, as it is called in the United States, the "hard metal industry" in Great Britain and elsewhere. The cobalt-cemented WC hard metal is produced by powder metallurgy from tungsten and carbon, with cobalt as a binder. At times tantalum, titanium, vanadium, molybdenum, and chromium may be added, each ending as carbide in the finished product. A typical product contains 80–90% WC, 8–18% TiC, and 5–25% Co, but most grades contain an average of 7% Co.

3.4.2.2 Clinical Cases

3.4.2.2.1 Acute Toxicity. A healthy 19-year-old recruit in a French artillery regiment was reported to have drunk 250 mL of a mixture of beer and wine that he had rinsed in a hot 155-mm gun barrel (174). He complained of nausea followed by seizures 15 min later and turned comatose for 24 h, presenting sings of encephalopathy. A moderate renal failure was seen initially and worsened to an extensive tubular necrosis with anuria on the second day. Inductively coupled plasma (ICP) emission spectrometry revealed very high concentrations of tungsten in the "beverage" as well as in gastric content, blood and urine (1540, 8, 5, and 101 mg/L, respectively). An ICP quantitative assay of tungsten in biological fluids, hair, and nails was then developed, showing high blood levels (>0.005 mgL) until day 13 in spite of six hemodialyses, and in urine until day 33. Tungsten was also found in hair and nails. The clinical evolution was satisfactory over weeks, and the patient was declared totally cured after 5 mo.

3.4.2.2.2 Chronic and Subchronic Toxicity. Some information, although incomplete, has been developed on normal tungsten values in inorganic forms and in human tissues and fluids. A limited testing of spark source mass spectrometric procedures (175) on inorganic substances and biologic samples showed that a sample of chrysotile asbestos and cement contained less than 0.03 ppm of tungsten. A sample of crocidolite asbestos contained about 10 times as much, while bituminous coal dust from Pennsylvania and Utah varied from insignificant traces to 0.32 ppm W. Tungsten concentration in coal miners' lungs varied from a low of about 0.5 ppm to a high of 10 ppm, and W concentration of the pulmonary lymph nodes from these lungs showed no evidence of tungsten accumulation in phagocytosis. The tungsten concentrations in the lungs of nonmining residents in the area were comparable to those in the lungs of coal miners with the lowest values. In comparing tungsten values in coal miners' lungs with those in coal dust, it would appear that tungsten from sources other than coal dust exposures contributed to tissue levels. These sources are currently unknown, but edible vegetation cannot tolerate appreciable levels of tungsten because it interferes with the essential role of molybdenum in nitrate reductase activity in food plants (145).

The possible genotoxic activity of hard metal particles as compared with Co and WC alone was studied in human peripheral lymphocytes incubated with Co or WC–Co. A dose- and time-dependent increased production of DNA single-strand breaks (ssb) was evidenced by alkaline single-cell gel electrophoresis (SCGE) and modified alkaline elution (AE) assays (176). WC alone did not produce DNA ssb detectable by the AE assay, but results obtained with the SCGE assay could suggest that it either allows some uncoiling of the chromatin loops or induces the formation of slowly migrating fragments. The results seemed consistent with the implication of an increased production of hydroxyl radicals when Co is mixed with WC particles. The SCGE results also seemed to suggest that WC may modify the structure of the chromatin, leading to an increased DNA sensitivity to clastogenic effects. Both mechanisms are not mutually exclusive and may concurrently contribute to the greater clastogenic activity of WC–Co dust. This property of WC–Co particles could account for the lung cancers observed in some hard metal workers.

3.4.2.2.3 Pharmacokinetics, Metabolism, and Mechanisms. In a fatal case with hard metal disease the content of cobalt and tungsten in blood and urine and particularly in pubic hair and toenails of the patients was significantly higher than the normal values (172). This suggests that such biological specimens could be used as indicators of chronic exposure to hard metal dusts. See also Ref. 5 (174).

3.4.2.2.4 Reproductive and Developmental. No data found.

3.4.2.2.5 Carcinogenesis. Except for an indicative result in a hypothesis-generating study by Moulin et al. (177), no human studies have indicated a carcinogenic potential for either soluble or insoluble tungsten compounds. Two cases (178) were reported separately in the hard metal industry [a bronchogenic carcinoma in a 63-year-old worker in Great Britain (152) and pulmonary fibroadenomatosis in a 57-year-old French worker], both without smoking histories, the incidence is so low as to place the association into question. Schepers (165) found bronchial hyperplasia and metaplasia in animals given high doses of WC and cobalt.

3.4.2.2.6 Genetic and Related Cellular Effects Studies. The genotoxic potential compared of pure cobalt powder (Co) and a cobalt-containing alloy, cobalt–tungsten carbide (WC-Co), involved in specific lung disorders, in parallel with the alkaline single-cell gel electrophoresis (SCGE) assay (comet assay) and the cytokinesis-blocked micronucleus (MN) test, both carried out *in vitro* on isolated human leukocytes (179). The comet assay indicated that the WC–Co mixture produced a higher level of DNA damage than Co alone; WC alone was not able to induce a dose-dependent DNA breakage effect as was seen for Co and WC–Co. Results from the MN test seemed to confirm these observations.

3.4.2.2.7 Other: Neurological, Pulmonary, Skin Sensitization, etc. Jordan et al. (180) reported memory deficits related to difficulties in attention and verbal memory among workers suffering from hard metal disease. This is one of the few suggestions of the effect of tungstate on the CNS among workers. They (180) reported memory deficits related to difficulties in attention and verbal memory among workers suffering from hard metal disease. This is one of the few confirmations of the metabolic effect of tungstate on the CNS among workers.

Cases of pulmonary involvement have been reported from the United States and abroad. Bech et al. (151) reviewed the older reports of pneumoconiosis beginning in 1940 in Germany and later reports from Hungary, Sweden, and the Soviet Union, in which several hundreds of hard metal workers showed high incidences of respiratory disorders and radiographic changes. Bech et al. contributed six additional cases from Great Britain.

The first report of hard metal disease occurring from grinding cemented WC came from Miller et al. in the United States (148), reporting on breathlessness and radiological changes in three workers employed in grinding hard metal for 3.5, 6, 7 years, respectively, using Al_2O_3 grit SiC and diamond wheels. Breathing zone dust counts varied from 3.7 to 7.7 mppcf for workers doing fine grinding, 38 to 40 mppcf at rough-grinding machines, and general room air, 1.4 mppcf. Cobalt analyses were 0.1 to 0.2 mg/m^3 in the breathing zones of operators, but sampling problems seemed to have prevented analyses for tungsten. The plant manufacturing WC tools for 18 years employed 200 to 300 men, of whom about 150 had been engaged in grinding operations.

From this report and others it is not possible to estimate the incidence of typical hard metal disease with pulmonary fibrosis because (1) only cases with severe respiratory complaints were counted, or (2) those with initial symptoms left employment. A possible incidence estimate of 5% may be deduced from the study of Baudouin et al. (142), who reported five cases of diffuse interstitial pulmonary fibrosis among 100 workers in a hard metal factory where dusts of tungsten, titanium, tantalum, and nickel metals, of their carbides and oxides, and of cobalt metal were in the exposure atmosphere during the various operations. Analysis by X-ray fluorescence of the lung of one worker with clinical and radiological changes showed large amounts of tungsten, amounts of titanium greater than normal, and small amounts of tantalum and niobium. The carbides of tungsten and titanium were demonstrated in the lung by X-ray diffraction.

The symptoms of cough, dyspnea, and wheezing usually develop gradually after years of exposure, tend to be less after a holiday or the weekend, and may remit following cessation of exposure. Workers who develop asthma usually change their occupation, and their symptoms generally disappear following removal from exposure.

In early-onset cases, development of the disease may be arrested by removal from exposure. Radiological clearing was reported by Miller et al. (148), Joseph (169), and Coates and Watson (168). This would appear to occur where the disease has developed recently. Early recognition of the disease is important, because it is progressive and potentially lethal.

As in chronic beryllium lung disease, treatment with corticosteroids and removal from exposure arrest progression of the disease. To emphasize the seriousness of hard metal disease, Bech (152) stated that of 12 employees with the disease, 8 died of secondary complications. Four of the 8 died of corpulmonale and cardiac failure, and the other 4 died of emphysema. Except for differences in degree and duration of the symptoms, the general course of the disease was similar in all 12 cases, enforcing the belief that individual hypersensitivity plays a role in this disease. Baudouin and associates (142) reported that immunologic tests showed a cellular, rather than a humoral, immune deficiency in some workers who had reached the fibrotic stage.

In addition to pulmonary effects in the hard metal industry, skin disorders occur. Of 1200 employees examined, 20 had an erythematous, papular type of dermatitis, limited primarily to the sides of the neck, flexor parts of the forearm, and backs of the hands. All the workers had exposures to the carballoy dust of at least 1 mo before the dermatitis appeared (181).

Patch tests on six of the workers with dermatitis revealed no reaction to the oxides of tungsten, tantalum, or titanium, or to carbon. All six, however, reacted to metallic cobalt, to the unfused WC, and to other materials used in the cemented product. The authors concluded that the sensitivity to cobalt caused the dermatitis, which was accentuated by the abrasiveness of the dust. Although no mention was made whether those with dermatitis had any pulmonary symptoms, the conclusion agrees with clinical findings in the pulmonary cases.

The incidence of skin disorders among workers in a Swedish hard metal industry was considerably higher than that found earlier in the United States (34 of 360 employees, 9.5 versus 1.7%), although the industry employed good general and local exhaust ventilation and cleaning facilities (182), Patch tests on 26 employees indicated contact eczema in 16; pruritis without skin lesions, 8; folliculitis, 6; and neurodermatitis, 5. Eczema and itching were localized on those body areas with high dust contact, that is, the face and extremities. Only eczema and itching were attributed to occupational exposure. Patch tests on 14 with contact eczema revealed that three were sensitive to cobalt, and two were sensitive to finished hard metal as well. No reaction was shown by seven of eight workers with pruritis and the five and neurodermatitis.

3.4.2.3 Epidemiology Studies. Monitoring of Co, Ta, and W, using mainly neutron activation analysis, was carried out on the urine, blood, pubic hair, and toenails of 251 subjects exposed at work to hard metal dusts (159). Cobalt and tungsten air levels at workplaces varied widely and frequently exceeded the TLV (0.05 mg Co/m^3). Tungsten and tantalum in pubic hair (WH and TaH) and toenails (WN and TaN) also seem to be useful indicators in proving hard metal exposure qualitatively. The determination of these two elements, rather than cobalt, in the bronchoalveolar lavage (BAL) seems useful in complementing the diagnosis of hard metal disease. The analysis of the BAL subfractions showed that W and Ta were firmly incorporated into the macrophage fraction. Multielement analysis of biological specimens from diseased subjects suggests that hard metal disease does not relate to Co, W, and Ta levels in the specimens considered. These findings may support the hypothesis on the possible haptenic properties of Co, which may induce hypersensitivity and immunorelated toxic effects.

3.5 Standards, Regulations, or Guidelines of Exposure

Soluble compounds as W, ACGIH TLV TWA = 1 mg/m^3, STEL/CEIL = 3mg/m^3, NIOSH REL TWA = 1mg/m^3, STEL/CEIL = 3 mg/m^3. Insoluble compounds as W. ACGIH TWA = 5mg/m^3, STEL/CEIL = 10 mg/m^3, NIOSH REL TWA = 5 mg/m^3, STEL/CEIL = 10 mg/m^3. The systemic toxicity of the soluble compounds is the basis for the difference between the two. Russia adopted a limit of 2 mg/m^3 (Jan 93).

BIBLIOGRAPHY

1. H. E. Stokinger, in G. Clayton and F. Clayton, eds., *Patty's Industrial Hygiene and Toxicology*, 3rd ed., Vol. 2A, Wiley, New York, 1981, pp. 1493–2060.
2. U.S. Bureau of Mines, *Mineral Commodity Summaries*, U.S. Bur. Mines, Washington, DC, 1990.
3. National Institute for Occupational Safety and Health (NIOSH), *National Occupational Exposure Survey*, NIOSH, Cincinnati, OH, 1990.
4. Agency for Toxic Substances and Disease Registry (ATSDR), *Toxicological Profile for Chromium*, ATSDR, Atlanta, GA, 1992.
5. R. A. Andersen, Chromium as an essential nutrient for humans. *Regul. Toxicol. Pharmacol.* **26**, 835–841 (1997).
6. W. Mertz, Chromium occurrence and function in biological systems. *Physiol. Rev.* **49**, 163–239 (1969).
7. R. A. Andersen and A. S. Koslovsky, Chromium intake, absorption, and excretion of subjects consuming self-generated diets. *Am. J. Clin. Nut.* **41**, 1177–1183 (1985).
8. R. A. Ducer et al., Atmospheric trace metals at remote northern and southern hemisphere sites: Pollution or natural? *Science* **187**, 59–61 (1975).
9. H. J. M. Bowen, *Environmental Chemistry of the Elements*, Academic Press, London, 1979.
10. U.S. Environmental Protection Agency (USEPA), *Health Assessment Document for Chromium*, Final Rep. No. EPA600/8-83-014F, USEPA, Research Triangle Park, NC, 1984.
11. E. Lahmann et al., *Heavy Metals: Identification of Air Quality and Environmental Problems in the European Community*, Rep. No. EUR 10678 EN/I and EUR 10678 EN/II), Vols. 1 and 2, Commission of the Europeans Communities Luxembourg, 1986.
11a. NIOSH, *Manual of Analytical Methods*, 4th ed., 1994.
12. C. Onkelinx, Compartment analysis of metabolism of chromium (III) in rats of various ages. *Am. J. Physiol.* **232**, E478–E484 (1977).

13. National Institute for Occupational Safety and Health (NIOSH) *Manual of Analytical Methods*, NIOSH, Cincinnati, OH, 1984.
14. P. L. Kalliomäki et al., Lung retained contaminants, urinary chromium and nickel among stainless steel welders. *Int. Arch. Occup. Environ. Health* **49**, 67–75 (1981).
15. B. Sjögren, L. Hedström, U. Ulfvarson, Urine chromium as an estimator of air exposure to welding fumes. *Int. Arch. Occup. Environ. Health* **51**, 347–354 (1983).
16. I. C. Stridsklev et al., Biological monitoring of chromium and nickel among stainless steel welders using the manual metal arc method. *Int. Arch. Occup. Environ. Health* **65**, 209–219 (1993).
17. S. A. De Flora et al., Estimates of the chromium(VI) reducing capacity in human body compartments as a mechanism for attenuating its potential toxicity and carcinogenicity. *Carcinogenesis (London)* **18**, 531–537 (1997).
18. S. A. De Flora, Threshold mechanisms and site specificity in chromium (VI) carcinogenesis. *Carcinogenesis (London)* (1999) (submitted for publication).
19. J. A. Hathaway, *Sci. Total Environ.* **86**, 169 (1989).
20. S. Langård, The time related subcellular distribution of chromium in the rat liver cell after intravenous administration of Na$_2$ ^{51}CrO$_4$. *Biol. Trace Elem. Res.* **1**, 45–54 (1979).
21. S. A. De Flora et al., *Cancer Res.* **45**, 3188 (1985).
22. J. Aiyar et al. *Environ. Health Perspect.* **92**, 53 (1991).
23. S. Langård and A. L. Nordhagen, Small animal inhalation chambers and the significance of dust ingestion when exposing rats to zinc chromates. *Acta Pharmacol. Toxicol.* **46**, 43–46 (1980).
24. L. L. Hopkins, Distribution in the rat of physiological amounts of injected ^{51}Cr^{III} with time. *Am. J. Physiol.* **209**, 731–735 (1965).
25. N. Bang-Pedersen, The effects of chromium on the skin. In S. Langård, ed., *Biological and Environmental Aspects of Chromium*, Elsevier/North-Holland Biomedical Press, Amsterdam, 1982, pp. 249–275.
26. J. Polak, J. L. Turk, J. R. Frey, Studies on contact hypersensitivity to chromium compounds. *Prog. Allerg.* **17**, 145–226 (1973).
27. S. J. Gray, and K. Sterling. The tagging of red cells and plasma proteins with radioactive chromium. *J. Clin. Invest.* **29**, 1604–1613 (1950).
28. S. A. De Flora et al., Metabolic reduction of chromium by alveolar macrophages and its relationship to cigarette smoke. *Clin. Invest.* **77**, 1917–1924 (1986).
29. S. A. De Flora et al., Metabolic reduction of chromium, as related to its carcinogenic properties. *Biol. Trace Elem. Res.* **21**, 179–187 (1989).
30. S. Langård, Absorption, transport, and excretion of chromium in man and animals. In S. Langård, ed., *Biological and Environmental Aspects of Chromium*, Elsevier/North-Holland Biomedical Press, Amsterdam, 1982, pp. 169–194.
31. M. Cikrt and V. Bencko, Biliary excretion and distribution of ^{51}Cr(III) and ^{51}Cr(VI) in rats. *J. Hyg. Epidemiol. Microbiol. Immunol.* **23**, 241–248 (1979).
32. A. Elbetieha and M. H. Al-Hamood, Long-term exposure of male and female mice to trivalent and hexavalent chromium compounds: Effect on fertility. *Toxicology* **116**, 34–47 (1997).
33. H. P. Nettesheim et al., Effect of calcium chromate dust, influenza virus, and 100 R whole-body X-radiation on lung tumor incidence in mice. *J. Natl. Cancer Inst. (U.S.)* **47**, 1129–1138 (1971).

34. International Agency on Research of Cancer (IARC) *Monograph on the Evaluation of Carcinogenic Risks to Humans*, Vol. 49, IARC, Lyon, France, 1990.
35. E. T. Snow, *Environ. Health Perspect.* **92**, 75 (1991).
36. H. Brieger, Zur Klinik der akuten Chromatvergiftung. *Z. Exp. Pathol. Ther.* **21**, 393–408 (1920).
37. D. B. Kaufman, W. DiNicola, and R. McIntosh, Acute potassium dichromate poisoning treated by peritoneal dialysis. *Am. J. Dis. Child.* **119**, 374–376 (1970).
38. S. Langård, S. ed., *Biological and Environmental Aspects of Chromium*, Elsevier Biomedical Press. Amsterdam; 1982.
39. D. Burrows, ed., *Chromium Metabolism and Toxicity*, CRC Press, Boca Raton, FL 1983.
40. E. Nieboer and S. L. Show, Mutagenic and other genotxic effects of chromium compounds. In J. O. Nriagu and E. Nieboer, eds., *Chromium in the Natural and Human Environment*, Wiley, New York, 1988, pp. 399–441.
41. W. Cumin, *Edinburgh Med. Surg. J.* **28**, 295 (1827).
42. N. J. Mackenzie, *J. Am. Med. Assoc.* **3**, 601 (1884).
43. I. Bolla et al., *Med. Lav.* **81**, 390 (1990).
44. B. E. Guthrie, The nutrional role of chromium. In S. Langård, ed., *Biological and Envronmental Aspects of Chromium*, Elsevier/North-Holland Biomedical Press, Amsterdam, 1982, pp. 117–148.
45. R. M. Donaldson and R. F. Barreras, Intestinal absorption of trace quantities of chromium. *J. Lab. Clin. Med.* **68**, 484–493 (1966).
46. J. Angerer et al., Occupational chronic exposure to metals I. Chromium exposure of stainless steel welders—biological monitoring. *Int. Arch. Occup. Environ. Health* **59**, 503–512 (1987).
47. C. Minoia and A. Cavalleri, Chromium in urine, serum and red blood cells in the biological monitoring of workers exposed to different chromium valency states. *Sci. Total Environ.* **71**, 323–327 (1988).
48. E. Lindberg and O. Vesterberg, *Am. J. Ind. Med.* **16**, 485 (1989).
49. G. Saner et al., *Br. J. Ind. Med.* **41**, 263 (1984).
50. R. Kishi et al., Chromium content of organs of chromate workers with lung cancer. *Am. J. Ind. Med.* **11**, 67–74 (1987).
51. J. P. Bonde, The risk of male subfecundity attributable to welding of metals. Studies of semen quality, infertility, fertility, adverse pregnancy outcome and childhood malignancy. *Int. J. Androl.* **1993**, 16 (Suppl 1) 1–29.
52. N. H. Hjollund, J. P. Bonde, and K. S. Hansen, Male-mediated risk of spontaneous abortion with reference to stainless steel welding. *Scand. J. Work Environ. Health* **21**(4), 272–276 (1995).
53. N. H. Hjøllund et al., A follow-up study of male exposure to welding and time to pregnancy. *Reprod. Toxicol.* **12**, 29–37 (1998).
54. W. Popp et al., *Int. Arch. Occup. Environ. Health* **63**, 115 (1991).
55. A. B. Bigaliev et al., Cytogenetic examination of workers engaged in chrome production. *Genetika* **13**, 545–547 (Engl. summ.) (1977).
56. F. Sarto et al., Increased incidence of chromosomal aberrations and sister chromatid exchanges in workers exposed to chromic acid (CrO_3) in electroplating factories. *Carcinogenesis (London)* **3**, (9), 1011–1016 (1982).

57. L. E. Knudsen, et al., Biomonitoring of genotoxic exposure among stainless steel welders. *Mut. Res.* **279**, 129–143 (1992).
58. S. Langård, One hundred years of chromium and cancer; A review of the epidemiological evidence and selected case reports. *Am. J. Ind. Med.* **17**, 189–215 (1990).
59. R. Hayes, A. M. Lilienfeld, and L. M. Snell, Mortality in chromium chemical production workers: A prospective study. *Int. J. Epidemiol.* **8**, 365–374 (1979).
60. A. M. Baetjer, Pulmonary carcinoma in chromate workers. II. Incidence on basis of hospital records. *Arch. Ind. Hyg. Occup. Med.* **2**, 505–516 (1950).
61. H. Royle, *Environ. Res.* **10**, 141 (1975).
62. M. Silverstein et al., *Scand. J. Work Environ. Health* **7**(Suppl. 4), 156 (1981).
63. T. Okubo and K. Tsuchiya, *Biol. Trace Elem. Res.* **1**, 35 (1979).
64. I. Franchini et al., *J. Scand. Work Environ. Health* **9**, 27 (1983).
65. T. Sorahan, D. C. L. Burges, and J. A. H. Waterhouse, A mortality study of nickel/chromium platers. *Br. J. Ind. Med.* **44**, 250–258 (1987).
66. K. Takahashi and T. Okubo, *Arch. Environ. Health* **45**, 107 (1990).
67. B. Sjögren et al., *Scand. J. Work Environ. Health* **13**, 247 (1987).
68. L. Simonato et al., A historical prospective study of European stainless steel, mild steel, and shipyard welders. *Br. J. Ind. Med.* **48**, 145–154 (1991).
69. T. E. Danielsen, S. Langård, and A. Andersen, Incidence of cancer among Norwegian boilerwelders. *Occup. Environ. Health* **53**, 321–324 (1996).
70. S. Langård et al., *Br. J. Ind. Med.* **37**, 114 (1980).
71. G. Axelsson, R. Rylander, and A. Schmidt, *Br. J. Ind. Med.* **37**, 121 (1980).
72. S. Langård et al., *Br. J. Ind. Med.* **47**, 14 (1990).
73. J. P. Bonde and J. M. Christensen, *Arch. Environ. Health* **46**, 225 (1991).
74. Z. Elias et al., *Carcinogenesis (London)* **12**, 1811 (1991).
75. J. Descotes, *Immunotoxicology of Drugs and Chemicals*, Elsevier, Amsterdam, 1989, pp. 297–336.
76. J. M. Olaguibel and A. Basomba, *Allergol. Immunopathol.* **17**, 133 (1989).
77. H. S. Novey et al., *J. Allergy Clin. Immunol.* **72**, 407 (1983).
78. R. P. Wedeen and L. F. Qian, *Environ. Health Perspect.* **92**, 71 (1991).
79. American Conference of Governmental Industrial Hygienists (ACGIH), *Guide to Occupational Exposure Values*, ACGIH, Cincinnati, OH, 1990.
80. National Toxicology Program (NTP), *Sixth Annual Report on Carcinogens*, NTP, Research Triangle Park, NC, 1991.
81. TLV Committee, *Appl. Occup. Environ. Hyg.* **7**, 135 (1992).
82. American Conference of Governmental Industrial Hygienists (AGCIH), *Threshold Limit Values and Biological Exposure Indices, 1992-1993*, ACGIH, Cincinnati, OH, 1992.
83. American Conference of Governmental Industrial Hygienists (ACGIH), *Documentation of Threshold Limit Values and Biological Exposure Indices*, 5th ed., ACGIH, Cincinnati, OH, 1986.
84. H. M. Kondo et al., On the health injury for inhabitants in the vicinity of chromate producing factory. *Ann. Rep. Tokushima Pref. Inst. Publ. Health* **10**, 45–65 (1971).
85. J. D. Ziang and S. K. Li, Cancer mortality in a Chinese population exposed to hexavalent chromium in water. *J. Occup. Environ. Med.* **39**, 315–319 (1997).

86. J. Durant et al., Elevated incidence of childhood leukemia in Woburn, Massachusetts: NIEHS Superfund Basis Research Program Searches for Causes. *Environ. Health Perspect.* **103**, 93 (1995).
87. C. E. Rogers et al., Hair analysis does not support hypothesized arsenic and chromium exposure from drinking water in Woburn, Massachusetts. *Environ. Health Perspect.* **105**, 1090–1997 (1998).
88. P. J. Shuler and P. J. Bierbaum, *Environmental Surveys of Al Reduction Plants*, HEW Publ. No. (NIOSH) 74–101, U.S. Department of Health, Education and Welfare, Washington, DC, 1974.
89. G. H. Farrar and M. L. Moss, in I. M. Kolthoff and P. J. Elving, eds., *Treatise on Analytic Chemistry*, Vol. 4, Part 2, Wiley-Interscience, New York, 1966, p. 367.
90. D. W. Lander et al., *Appl. Spectrosc.* **25**, 270 (1971).
91. Alaska's Tangass National Forest, *Science* **198**, 473 (1977).
92. S. H. Killeffer and A. Linz, *Molybdenum Compounds*, Interscience, New York, 1952.
93. H. R. Mahler and D. E. Green, *Science* **120**, 7 (1954).
94. National Institute for Occupational Safety and Health, (NIOSH) *Pocket Guide to Hazardous Chemicals*, NIOSH, Cincinnati, OH, 1990.
95. B. F. Quin and R. R. Brooks, *Anal. Chim. Acta* **74**, 75 (1975).
96. J. Cardenas and L. E. Mortenson, *Anal. Biochem.* **60**, 372 (1974).
97. J. Korkisch and H. Krivanec, *Anal. Chim. Acta* **83**, 111 (1976).
98. L. T. Fairhall et al., *U.S. Public Health Bull.* Date.
99. Y. Ogra, M. Ohmichi, and K. T. Suzuki, Systemic disposition of molybdenum and copper after *J. Trace Elem. Med. Biol.* **9**, 165–169 (1995).
100. W. S. Ferguson et al., *Nature (London)* **141**, 553 (1938).
101. J. B. Neilands et al., *J. Biol. Chem.* **172**, 431 (1948).
102. A. T. Dick, *Mo and Cu Relationships in Animal Nutrition, Symp., Inorg. Nitrogen Metab.*, Johns Hopkins University Press, Baltimore, MD, 1956.
103. B. S. W. Smith and H. Wright, *Clin. Chim. Acta* **62**, 55 (1975).
104. R. Compere and E. Francois, *Bull. Rech. Agron. Grembloux* **1**, 534 (1966).
105. M. A. Macilese et al., *J. Nutr.* **99**, 77 (1969).
106. J. D. Burke et al., *Blood* **8**, 1105 (1953).
107. W. S. Ferguson et al., *J. Agric. Sci.* **33** 40 (1943).
108. L. J. Daniel and L. F. Gray, *Proc. Soc. Exp. Biol. Med.* **83**, 487 (1953).
109. A. Ramaiah and E. R. B. Shanmugasundrum, *Biochem. Biophys. Acta* **60**, 373 (1962).
110. C. F. Mills et al., *J. Nutr.* **65**, 129 (1958).
111. R. F. Miller and N. O. Price, *Fed. Proc., Fed. Am. Soc. Exp. Biol.* **15**, 564 (1956).
112. J. L. Johnson and K. V. Rajagopalan, *J. Biol. Chem.* **249**, 859 (1974).
113. A. W. Halverson et al., *J. Nutr.* **71**, 95 (1960).
114. L. R. Arrington and G. K. Davis, *J. Nutr.* **51**, 295 (1953).
115. N. Titenko-Holland et al., *Environ. Mol. Mutagen.* **32**, 251–259 (1998).
116. I. Gimenez, R. Garay, and J. O. Alda, Molybdenum uptake through the anion exchanger in human erythrocytes. *Pfluegers Arch.* **424**, 245–249 (1993).
117. A. Bru et al., *C. R. Seaners Soc. Biol. Ses Fil.* **237**, 279 (1953).

118. C. J. Botha, G. E. Swan, and P. P. Minnaar, *J. S. Afr. Vet. Assoc.* **66**, 6–10 (1995).
119. Z. P. Liu, Z. Ma, and Y. J. Zhang, *Vet. Res. Commun.* **18**, 251–260 (1994).
120. J. M. Howell, Y. Shunxiang, and J. M. Gawthorne, *Res. Vet. Sci.* **55**(2), 224–230 (1993).
121. C. Maltoni, *Int. Congr. Ser — Excerpta Med.* **322**, 19 (1973).
122. T. Koizumi et al., *Biol. Pharm. Bull.* **18**, 460–462 (1995).
123. V. M. Sardesai, *Nutr. Clin. Pract.* **8**, 277–281 (1993).
124. E. V. Gusev, *Gig. Sanit.* **34**, 63 (1969).
125. E. S. Higgins et al., *J. Nutr.* **59**, 539 (1956).
126. S. Capilina and E. Ghizari, *Acad. Repub. Pop. Rom. Stud. Cercet. Fiziol.* **7**, 471 (1962).
127. G. E. Mitchell, Jr. et al., *Int. Z. Vitaminforsch.* **38**, 308 (1968).
128. G. K. Stookey and J. C. Muhler, *Proc. Soc. Exp. Biol. Med.* **109**, 268 (1962).
129. G. K. Stookey et al., *Proc. Soc. Exp. Biol. Med.* **109**, 702 (1962).
130. D. G. Devyatka et al., *Gig. Sanit.* **36**, 133 (1971).
131. P. A. Walravens et al., *Arch. Environ. Health* **34**, 302 (1979).
132. V. V. Kovalskii et al., *Zh. Obshch. Biol.* **33**, 179 (1961).
133. A. O. Akopyan, *Zh. Otd. Vyp. Farm. Khim. Svedstva Toksikol.* No. 154791 (1966).
134. L. A. Dueva and S. S. Stepanian, *Vestn. Dermatol. Vererol.* **10**, 47 (1989).
135. R. Lindner et al., *Pneumonologie* **44**, 898 (1990).
136. W. H. Allaway et al., *Arch. Environ. Health* **16**, 342 (1968).
137. Y. M. Bala and V. M. Lifshits, *Fed. Proc. Trans. Suppl.* **35**, 370 (1966).
138. J. R. Turnlund, W. R. Keyes, and G. L. Peiffer, *Am. J. Clin. Nutr.* **62**, 790–796 (1995).
139. American Conference of Governmental Industrial Hygienists (ACGIH), *Documentation of Threshold Limit Values — 1999*, ACGIH, Cincinnati OH, 1999.
140. D. R. Lovley, *Annu. Rev. Microbiol.* **47**, 263–290 (1993).
141. *Chem. Eng. News*, 52 (Oct. 19, 1959).
142. J. Baudouin et al., *Nouv. Presse Med.* **4**, 1353 (1975).
143. P. Schramel, I. Wendler, and J. Angerer, The determination of metals (antimony, bismuth, lead, cadmium, mercury, palladium, platinum, tellurium, thallium, tin and tungsten) in urine samples by inductively coupled plasma-mass spectrometry. *Int. Arch. Occup. Environ. Health* **69**, 219–223 (1997).
144. National Institute for Occupational Safety and Health (NIOSH), *Criteria for a Recommended Standard for Occupational Exposure to Tungsten and Cemented Carbide*, NIOSH Publ. No. 77–127, NIOSH, Cincinnati, OH, 1977.
145. B. A. Notton and E. J. Hewitt, *Biochem. Biophys. Res. Commun.* **44**, 702 (1971).
146. M. Kurschner et al., Interaction of lipophilic ions with the plasma membrane of mammalian cells studies by electrorotation. *Biophy. J.* **74**(6), 3031–3043 (1998).
147. V. G. Nadeenko, *Hyg. Sanit.* **31**, 197 (1966).
148. C. W. Miller et al., *Arch. Ind. Hyg.* **8**, 453 (1953).
149. K. D. Lundgren and A. Swensson, *Acta Med. Scand.* **146**, 20 (1953).
150. A. B. Dehalant and G. W. H. Schepers, *Arch. Ind. Health* **12**, 120 (1955).
151. A. O. Bech et al., *Br. J. Ind. Med.* **19**, 239 (1962).
152. A. O. Bech, *J. Soc. Occup. Med.* **24**, 11 (1974).

153. G. Roesems et al., *In vitro* toxicity of cobalt and hard metal dust in rat and human type II pneumocytes. *Pharmacol. Toxicol.* **81**, 74–80 (1997).
154. R. L. Aamodt, *Health Phys.* **28**, 733 (1975).
155. T. Karantassis, *Ann. Med. Leg.* **5**, 44 (1924) (in French).
156. F. W. Kinard and J. C. Aull, *J. Pharmacol. Exp. Ther.* **83**, 53 (1945).
157. J. C. Aull and F. W. Kinard, *J. Biol. Chem.* **135**, 119 (1940).
158. M. Wide et al., *Environ Res.* **40**, 487 (1986).
159. E. Sabbioni et al., Metal determinations in biological specimens of diseased and non-diseased hard metal workers. *Sci. Total Environ.* **150**(1-3), 41–54 (1994).
160. H. J. Wei et al., *J. Natl. Cancer Inst.* **74**, 469 (1985).
161. M. De Boeck, D. Lison, and M. Kirsch-Volders, Evaluation of the *in vitro* direct and indirect genotoxic effects of cobalt compounds using the alkaline comet assay. Influence of interdonor and interexperimental variability. *Carcinogenesis (London)* **19**, 2021–2029 (1998).
162. L. Lauring and F. L. Wegeland, Jr., *Mil. Med.* **135**, 171 (1970).
163. J. S. Fredenwald et al., *Arch. Ophthalmol. (Chicago)* **35**, 98 (1946).
164. G. W. H. Schepers, *Arch. Ind. Health* **12**, 140 (1955).
165. G. Lasfargues et al., The delayed lung responses to single and repeated intratracheal administration of pure cobalt and hard metal powder in the rat. *Environ. Res.* **69**, 108–121 (1995).
166. D. Lison et al., Physicochemical mechanism of the interaction between cobalt metal and carbide particles to generate toxic activated oxygen species. *Chem. Res. Toxicol.* **8**, 600–606 (1995).
167. D. Lison and R. Lauwerys, Evaluation of the role of reactive oxygen species in the interactive toxicity of carbide-cobalt mixtures on macrophages in culture. *Arch. Toxicol.* **67**, 347–351 (1993).
168. E. O. Coates and J. H. L. Watson, *Ann. Intern. Med.* **75**, 709 (1971).
169. M. Joseph, *Aust. Radiol.* **12**, 92 (1968).
170. J. Rochemaure et al., *J. Fr. Med. Chir. Thorac.* **26**, 305 (1972).
171. F. Della Torre et al., *Respiration* **57**, 248 (1990).
172. N. L. Sprince et al., *Am. Rev. Respir. Dis.* **138**, 122 (1988).
173. N. L. Sprince et al., *Chest* **86**, 549 (1984).
174. P. Marquet et al., Tungsten determination in biological fluids, hair and nails by plasma emission spectrometry in a case of severe acute intoxication in man. *J. Forensic Sci.* **42**, 527–530 (1997).
175. M. L. Jacobs, *Evaluation of Spark Source Mass Spectrometry in the Analysis of Biologic Samples*, NIOSH Res. Rep., HEW Pub. No. (NIOSH) 75–186, U.S. Department of Health, Education and Welfare, Washington, DC, 1975.
176. D. Anard et al., *In vitro* genotoxic effects of hard metal particles assessed by alkaline single cell gel and elution assays. *Carcinogenesis (London)* **18**, 177–184 (1997).
177. J. J. Moulin et al., Lung cancer risk in hard-metal workers. *Am. J. Epidemiol.* **148**, 241–248 (1998).
178. A. Collet et al., *Rev. Tuberc. Pneumol.* **27**, 357 (1963).
179. F. Van Goethem, D. Lison, and M. Kirsch-Volders, Comparative evaluation of the *in vitro* micronucleus test and the alkaline single cell gel electrophoresis assay for the detection of

DNA damaging agents: Genotoxic effects of cobalt powder, tungsten carbide and cobalt-tungsten carbide. *Mutat. Res.* **392**, 31–43 (1997).
180. C. Jordan et al., *Toxicol. Lett.* **54**, 241 (1990).
181. L. Schwartz et al., *J. Allergy* **16**, 51 (1945).
182. E. Skog, *Ind. Med. Surg.* **32**, 266 (1963).

CHAPTER THIRTY-NINE

Manganese and Rhenium

J. Michael Davis, Ph.D.

INTRODUCTION TO CLASS OF CHEMICALS

Manganese (Mn, atomic number 25) and rhenium (Re, atomic number 75) are group 7 (VIIB) transition elements. Before the discovery and confirmation of the existence of rhenium predicted by Mendeleev's periodic law, rhenium was provisionally termed dvi-manganese because of its expected resemblance to manganese (1). Manganese and rhenium share many of the general chemical characteristics of metals in the transition series, including multiple valency, the ability to form stable complex ions, paramagnetism, and catalytic properties. However, the second and third elements in the transition series generally have chemical properties more similar to each other than to the first member. Thus, in many respects, rhenium is chemically more similar to technetium than to manganese (2).

Inhalation of particulate Mn constitutes the dominant route through which toxicity is expressed under most occupational conditions. Manganese is notably toxic to the central nervous system (CNS) and also has effects on the respiratory system and on reproductive function. Numerous clinical cases of frank Mn toxicity denote a characteristic syndrome that may include psychiatric symptoms, dystonia and rigidity, impaired manual dexterity, and gait disturbances. Several epidemiological studies provide a coherent pattern of evidence of neurotoxicity from occupational exposure to Mn at average concentrations around 1 mg/m^3 or lower. The primary effects observed in such workers pertain to motor function, especially hand steadiness, eye–hand coordination, and rapid coordinated movements, which imply involvement of the CNS extrapyramidal system. Although a growing body of literature is devoted to medical applications of the radioactive isotopes

---

Patty's Toxicology, Fifth Edition, Volume 3, Edited by Eula Bingham, Barbara Cohrssen, and Charles H. Powell.
ISBN 0-471-31934-1 © 2001 John Wiley & Sons, Inc.

[186]Re and [188]Re, very limited information is available on the toxicity of rhenium itself, which makes it difficult to characterize its toxicity with confidence. The few studies conducted thus far suggest that acute administrations of Re may have relatively low toxicity, at least by noninhalation routes. It has been described as "relatively inert" in the body and produces transient changes in blood pressure (both hypo- and hypertensive), tachycardia, sedation, and ataxia. In one comparative study, the lethal oral dose of Re was about eight times higher than that of molybdenum. However, one report suggests that it could be more potent as an inhalation toxicant. If true, rhenium and manganese might share the feature of having much greater toxicity by inhalation than by ingestion.

1.0 Manganese (Mn)

1.0.1 CAS Number: [7439-96-5]

Manganese is one of the group 7 (VIIB) transition elements and has an atomic number of 25 and an atomic weight of 54.938049. Although its use apparently dates from antiquity, manganese was recognized as an element in the early 1770s by Swedish chemist Carl W. Scheele and others and was isolated by a co-worker, Johan G. Gahn. The term manganese is thought to derive from the Greek word mangania for magic and/or the Latin word for magnet, based on the supposed magnetic properties of the manganese ore, pyrolusite. Synonyms include elemental manganese, manganum, colloidal manganese, and cutaval.

1.1 Chemical and Physical Properties

Manganese is widely distributed in the earth's crust, and it is the twelfth most abundant element and the fifth most abundant metal. As a pure metal, manganese is white-gray or silver in color. Manganese does not occur naturally in elemental form but is widely distributed in numerous mineral forms. The most commonly occurring mineral forms of manganese are oxides (pyrolusite, manganite, psilomelane, and hausmannite), silicates (braunite and rhodonite), sulfides (manganese blende and hauserite), and carbonates (manganoan calcite and rhodocrosite).

Manganese has a melting point of 1233°C, and its density is 7.2 g/cm^3 at 20°C. It is chemically reactive and decomposes in cold water. Depending on the allotropic form of the pure metal, its physical characteristics vary from hard and brittle to soft and flexible. Manganese forms compounds in oxidation states ranging from -3 to $+7$, but only the $+2$ and $+3$ states are usually found in living organisms (3). Some of the more common compounds of manganese include salts, oxides, and organomanganese compounds, which differ in their solubility and other characteristics. The physical and chemical properties of a few of the more common forms of manganese are summarized in Table 39.1. Manganese isotopes including the natural isotope ^{55}Mn and eight radioisotopes, ^{49}Mn to ^{57}Mn, have half-lives that range from less than a second to 4×10^6 yr.

1.1.2 Order and Warning Properties

No information on its odor or other warning properties was found.

Table 39.1. Physical and Chemical Properties of Manganese and Selected Manganese Compounds

Compound	CAS	Molecular Formula	Mol. Wt.	Boiling Point (°C)	Melting Point (°C)	Density	Solubility in Water (at 68°F)	Refractive Index (20°C)	Vapor Pressure (mmHg)	UEL	LEL
Manganese	[7439-96-5]	Mn	54.94	1962	1233	7.2	Decomposes		1 mmHg at 1292°C		
Manganese dioxide	[1313-13-9]	MnO$_2$	86.94	No data	535	5.03	Insoluble; soluble in HCl		No data		
Manganese tetroxide	[1317-35-7]	Mn$_3$O$_4$	228.82	No data	1564	4.86	Insoluble; soluble in HCl		No data		
Manganese carbonate	[598-62-9]	MnCO$_3$	114.94	No data	Decomposes	3.12	Insoluble; soluble in dilute acid		No data		
Manganese sulfate	[7785-87-7]	MnSO$_4$	151.00	850	700	3.25	520 g/L (5°C); 700 g/L (70°C)		No data		
Manganese chloride	[7773-01-5]	MnCl$_2$	125.84	1190	650	2.98	723 g/L (25°C)		10 mm Hg at 778°C		
Manganese acetate	[638-38-0]	C$_4$H$_6$MnO$_4$	173.03	?	?	1.59	Soluble in water and alcohol; decomposes in cold water		?		
Methylcyclopentadienyl manganese tricarbonyl	[12108-13-3]	CaH$_7$MnO$_3$	218.10	232.8	2.2	1.39	29 mg/L at 25°C; hydrocarbon soluble		0.047 at 20°C		
Potassium permangante	[7722-64-7]	KMnO$_4$	158.03	No data	Decomposes at 240°C	No data	63.8 g/L (20°C)		No data		

1.2 Production and Use

The major use of manganese is in the production of ferrous and nonferrous alloys. Iron and steel production accounts for more than 90% of the worldwide consumption of manganese (4). As an alloying agent in finished steel, manganese contributes increased strength, hardness, and abrasion resistance. Manganese is also used as an alloying element in aluminum manufacturing. The major application for aluminum–Mn alloys is aluminum beverage containers, to which they provide strength, hardness, and stiffness.

The major nonmetallurgical use of manganese is as manganese dioxide (MnO_2) for producing dry-cell batteries. Manganese sulfate ($MnSO_4$) and manganese oxide (MnO) are incorporated into fertilizers and animal feeds. Compounds of manganese are also used in glass and ceramic materials. Organic compounds of manganese are used in various applications, including the pesticide manganese ethylene-bis-dithiocarbamate (maneb). Another widely used organomanganese compound is methylcyclopentadienyl manganese tricarbonyl (MMT), which is primarily an additive to gasoline for increasing the octane number. As a fuel additive, MMT has been used in the United States and Canada since the 1970s, but it was generally not permitted in unleaded U.S. gasoline until 1995 (5). The maximum allowable concentration of MMT is 0.03125 g M/gal gasoline in the United States and 0.018 g Mn/L in Canada.

1.3 Exposure Assessment

OSHA recognizes two analytical methods for manganese and its compounds (including manganese fume) in air. Each method determines total manganese in the sample and does not distinguish among different chemical forms. Both methods are applicable to air, wipe, or bulk samples. For air samples, an appropriate volume of air is drawn through a mixed-cellulose ester membrane filter using a calibrated sampling pump. Wipe and bulk samples are collected by standard grab sampling techniques.

Method ID-121 uses atomic absorption spectroscopy to determine total manganese in the sample. For air and wipe samples, the filter or wipe that contains the sample is digested by heating it in concentrated HNO_3. After the sample is digested, one or two drops of concentrated HCl are added to facilitate dissolution of any remaining particulate matter. Then, the solution is diluted with water to a final concentration of 4% HNO_3. The solution is then analyzed by atomic absorption spectroscopy. The analytical detection limit for this method is 0.01 mg/mL, and, because air samples are diluted to 25 mL, the detection limit for the method for an air sample of 10 L would be 0.025 mg/m^3.

Method ID-125G is performed by inductively coupled argon plasma-atomic emission spectroscopy to determine total elemental Mn in a digested sample. In this method, the samples are digested using 1:1 H_2SO_4, to which several drops of 30% H_2O_2 are added. The resulting solution is analyzed by inductively coupled argon plasma-atomic emission spectroscopy. The quantitative detection limit for manganese by this method is 0.04 mg/mL, and air samples are generally diluted to 50 mL. Thus, this method could detect a manganese air concentration of 0.020 mg/m^3 in a 10-L air sample.

NIOSH also has an analytical method that it recommends for determining workplace exposures, method #7300 (5a).

1.4 Toxic Effects

Reviews of Mn toxicity often begin by noting that Mn is considered a nutritionally essential element that is critical for many physiological processes to function properly. Several enzyme systems are either activated by or require Mn, and Mn plays an important role in metabolism, the nervous system, and other key systems and functions (6). Although certainly significant, this fact has little or no bearing on Mn toxicity related to occupational exposure. There is no reason to believe that occupational Mn exposure, particularly by inhalation, would serve to obviate a deficiency state, especially given that it has not been conclusively established that Mn deficiency has ever occurred in humans except possibly for some select, nonoccupational populations (7). Moreover, the efficiency of homeostatic mechanisms for regulating oral Mn intake and elimination makes it unlikely that toxicity would result from ingestion under typical occupational exposure conditions, assuming that dietary intake is otherwise within an accepted range of 2–5 mg/day (8). Although some uptake may routinely result from ingesting Mn cleared from the respiratory tract by the mucociliary escalator, the relative contribution of this ingested portion to a potentially toxic internal dose is most likely quite minor in view of the fact that only about 5% of gastrointestinal Mn is absorbed and homeostatic mechanisms normally regulate Mn levels very effectively. Although dermal contact with Mn is also undoubtedly frequent in many occupational contexts, the solubility of Mn oxides in water is relatively low, and even solutions of Mn salts pose little likelihood of dermal uptake. Thus, the major focus of this review is on the effects of inhalation exposure to Mn. The reader is referred to other sources for information on Mn toxicity from dietary intake (7, 9).

Given the varied physicochemical properties of Mn and its compounds, the toxic effects of Mn in different forms and by different routes of exposure would be expected to vary as well. These varied effects may be modulated in several ways. Factors such as particle size, solubility, and oxidative capacity affect both uptake and effects at the portal of entry (e.g., lung). When inhaled and absorbed into the blood, Mn is also transported to the central nervous system (CNS) before its first pass through the liver, whereas Mn absorbed via the gastrointestinal tract passes directly to the liver first. This distinction between the inhalation and oral routes of exposure to Mn is presumably of critical importance in accounting for the greater risk of toxicity by inhalation than by ingestion. Much remains to be learned about the details of the pharmacokinetics and pharmacodynamics of Mn by different routes of uptake and for different oxidation states of Mn (10, 11). Thus, it is not clear that studies of one form of Mn by one route are necessarily directly comparable to studies of other forms or by other routes, and apart from comparisons of the uptake and distribution of some selected inorganic compounds, few studies have attempted to characterize quantitatively the comparative toxicity of different forms of Mn and Mn compounds. Nevertheless, some basic features of the toxic effects of Mn are apparent, and these are the focus of this review.

Much more has been reported about Mn toxicity by inhalation in humans than in experimental animals. This may be due in part to the relatively long experience with Mn toxicity in workers and in part to continuing questions about the suitability of nonprimate animal models, at least for Mn neurotoxicity (12–14). Consequently, this review focuses first on human evidence, with animal studies primarily in a supporting, albeit very important, role, especially with regard to understanding the mechanisms of Mn toxicity and its effects on systems other than the CNS. In this connection, also note that the neurobehavioral effects of Mn have tended to overshadow other aspects of Mn toxicity in the medical and toxicological literature. Notwithstanding some evidence that the CNS may be more sensitive to Mn than other systems are (15, 16), care should be taken to avoid focusing on CNS function to the exclusion of other systems. The risk is that researchers may tend to "look for the keys where the light is," when in fact Mn has multiple effects that should be investigated as more sensitive methods for various end points are developed.

Couper (17) is generally credited with first documenting some of the features of Mn toxicity in workers in 1837. The men he examined had developed muscle weakness, staggering walk, faint speech, vacant expression, and increased salivation after a period of breathing dust from grinding Mn oxide to make bleaching powder. Since then, numerous reports from around the world have described these and other signs and symptoms of Mn toxicity or manganism. The most prominent features of Mn toxicity from occupational exposure have been reflected in the CNS and respiratory system, although reproductive and immune functions have also been implicated. The majority of reports of manganism have concerned occupational exposure to Mn dust and fumes. Such toxicity can be severe and permanently disabling but has not generally been considered directly lethal. Severe, possibly even lethal, cases of toxicity from noninhalation and nonoccupational Mn exposure have been noted, including poisonings from ingesting contaminated drinking water (18), excessive Mn through parenteral nutrition (19–22), and other accidental conditions.

1.4.1 Acute Toxicity

The toxicity of Mn in humans has been primarily observed in relation to subchronic or chronic exposure conditions. This may be due in part to the typically gradual onset of symptoms and signs of Mn toxicity. However, some types of Mn exposure are more likely to be associated with acute effects either because of the form of Mn encountered or the nature of the exposure, for example, the industrial production and occupational use of certain organomanganese compounds. Accidental or intentional ingestion of Mn compounds is not addressed as such here, but information from experimental animal studies using oral dosing is included as appropriate to provide basic toxicity information.

1.4.1.1 Inorganic Manganese. Although no reports of toxicity in humans from acute inhalation exposure to Mn were found, respiratory effects would be expected as the most immediate effects from acute inhalation exposure, especially fumes from welding operations. The term "metal fume fever" has sometimes been applied in this connection, but this condition, marked by a flu-like reaction within a few hours of exposure, is a nonspecific condition that results from a variety of metal oxide welding fumes, as well as other air contaminants (23). Of the several experimental animal studies that have examined the acute effects of Mn exposure by inhalation, a few are described here. In a range finding study, Shiotsuka (24) observed concentration-related effects on the respiratory system in rats that had been exposed to MnO_2 at concentrations starting at 68 mg/m^3 for 6 hours/day, 5 days/week for two weeks. The effects included increased lung weight and a focal pneumonitis in the alveolar ductal regions with interstitial hypercellularity. At higher concentrations, the pneumonitis became more diffuse, and the interstitial lesions progressed to granulomas. No hematologic effects or body weight changes were observed. Bergstrom (25) reviewed prior work on the acute respiratory toxicity of MnO_2 and conducted further experiments with guinea pigs exposed to 22 mg/m^3 for 24 hours. He concluded that all of the evidence taken together indicated that a primary inflammatory reaction can occur in the respiratory tract but that this reaction is of limited duration after an acute exposure (less than 7 days under the conditions of this experiment). Adding a bacterial challenge enhanced the reaction and also showed that MnO_2 exposure can diminish the effectiveness of bacterial clearance mechanisms in the lungs. However, biphasic changes occurred in certain measures. The number of macrophages was significantly decreased immediately after exposure but increased over the next 7 days; during the first day, bacterial clearance was also increased at 1 hour postexposure but declined at 3, 5, and 24 hours following exposure. Adkins et al. (26, 27) exposed mice for two hours to Mn_3O_4 at concentrations of 0.879 or 1.837 mg Mn/m^3 and found immediate reductions in alveolar macrophages, along with various other biochemical changes related to pulmonary defense mechanisms. Adkins et al. (28) also investigated the effects of bacterial challenge in conjunction with a single inhalation exposure of mice to Mn_3O_4 at 0.22–2.65 mg Mn/m^3 for 2 hours. Susceptibility to respiratory bacterial infection was significantly related to Mn concentration, and resultant mortality exceeded zero at concentrations above approximately 0.5 mg/m^3 (95% confidence limit). Maigetter et al. (29) similarly evaluated the effects in mice of exposures to MnO_2 for 1–3 hour at 109 mg/m^3 for up to 4 days in conjunction with bacterial and viral challenges. Susceptibility to the infectious agent, measured either as an inflammatory reaction with alveolar septal edema and congestion or as survival time and mortality rate, depended on exposure duration.

1.4.1.2 MMT. As an organic compound, MMT is likely to present physicochemical and toxic properties somewhat different from the inorganic Mn compounds to which workers are more commonly exposed. Acute occupational exposure to this fuel additive could occur by either the dermal or inhalation route, although the likelihood of exposure to MMT by inhalation may be reduced because of its low vapor pressure and rapid photodegradation (30). The health risks associated with the combustion products of MMT in gasoline have been evaluated in depth (5, 31–33) and relate primarily to low-level chronic exposure to particulate inorganic Mn rather than the organomanganese additive itself. Very little information exists on the toxic effects of MMT in humans. It has been described as moderately irritating to the eyes but is not known to cause irritation or sensitization of the skin (34). An incident involving contact of about 5–15 mL of MMT on the hand and wrist of a worker was said to have "allegedly caused 'thick tongue,' giddiness, nausea, and headache within 3 to 5 minutes" (34).

In a series of basic toxicity studies, Hinderer (30) exposed nine groups of Sprague-Dawley rats (10 per group) by inhalation for either 1 hour or 4 hours to an unspecified range of MMT concentrations to determine the LC_{50} values and make limited observations on the gross toxicity. The LC_{50} values were 247 mg/m^3 for 1 hour and 76 mg/m^3 for 4 hours, with 95% confidence limits of 229–271 mg/m^3 and 67–87 mg/m^3, respectively. Among the reactions noted were decreased general activity, conjunctivitis, dyspnea, weight loss for 7 days, and focal hemorrhages in the lungs. Dermal exposure studies were conducted with groups of rabbits, half of which were shaved so the material could be applied directly to the skin. Plastic sleeves were wrapped around the areas to which MMT was applied. After 24 hours of exposure, the animals were observed for 14 days. These dermal studies were conducted in different laboratories and yielded LC_{50} values ranging from 140 mg/kg to 795 mg/kg across laboratories. Several significant toxic effects were noted, including red spots on the lung; discoloration of the liver, kidney, and spleen; congested kidneys; swollen liver, kidney, and spleen; and bloody diarrhea. Erythema and edema of the skin, as well as body weight loss, occurred in most of the animals. Other, more variable reactions were described as polyapnea, vocalization, excitation, ataxia, tremors, cyanosis, and convulsions. Using the Draize scoring method (35), skin irritation was judged moderate, and low or no eye irritation occurred. The oral LC_{50} was 58 mg/kg in the rat (consistent with Ref. 36) and 230 mg/kg in the mouse. Necropsies performed 14 days after oral gavage revealed that the lungs were "dark red." Female mice and rats had lower LC_{50} values than males.

More extensive investigations of the acute toxic effects of MMT have demonstrated that the compound is selectively toxic to the lungs, regardless of the route of administration (37–42). These studies have shown that the injury is characterized by a severe interstitial inflammation and edema of the alveolar region and only mild damage to the bronchiolar region, and by large increases in bronchoalveoloar lavage fluid albumin and other protein content, as well as smaller increases in lactate dehydrogenase. Minimal or no hepatic or renal injury was observed (41). Other studies focused on the neurotoxic effects of MMT, particularly its seizure-inducing action. Fishman et al. (43) observed seizure activity in mice within 0.5–1.5 minutes after intraperitoneal injection of 50 mg/kg or more MMT. Comparable doses administered subcutaneously did not induce seizures. Administration of $MnCl_2$ at doses sufficient to yield equivalent brain Mn concentrations also produced no sign of seizure activity, suggesting that the organic parent compound or a metabolite was responsible for the seizures rather than inorganic Mn. The seizure activity was correlated with brain Mn levels rather than administered dose (cf. 44).

Hanzlik et al. (37, 38) identified two major metabolites of MMT, $(CO)_3MnC_5H_4CO_2H$ and $(CO)_3MnC_5H_4CO_2OH$, and investigated their role in the toxic effects of orally administered MMT on the lung. They concluded that phenobarbital-induced liver activation of MMT protected against lung injury and mortality because the LD_{50} was 2.5 times higher after phenobarbital treatment. Later work on cyclopentadienyl manganese tribcarbonyl (CMT), a closely related organomanganese compound, has suggested a different conclusion. In rats, CMT produces pneumotoxic and neurotoxic effects that are quite similar to those of MMT (45), albeit at levels about half those of MMT in its toxic effects on the lung (42). By selectively modulating hepatic and pulmonary metabolism of CMT *in vivo*, Blanchard et al. (46) compiled evidence suggesting that the toxic effects observed in the alveolar region were likely to be a function of a metabolite of CMT produced by cytochrome P-450 activation in the lung. Given that inhibition of pulmonary metabolism protected against CMT-induced pneumotoxicity *in vivo*, the lung protective effect associated with an induction of hepatic metabolism may have been due to increased clearance and biliary excretion of the parent compound, resulting in a smaller delivered dose of CMT to the lung. This line of reasoning is not inconsistent with the fact that phenobarbital induction of CYP-450 activation in the liver did not increase the neurotoxicity of CMT (45), but it remains to be seen whether *in situ* activation occurs in the CNS.

1.4.2 Chronic Toxicity

Numerous cases of chronic Mn toxicity in workers have been described in the literature during the past century (47–52). One of the characteristics of Mn toxicity that repeatedly emerges from these accounts is the high degree of individual differences in susceptibility and the course of the disorder. Two workers may labor side by side, yet one will manifest signs and symptoms of Mn toxicity, and the other will not. One worker may shown signs of toxicity after only 2–3 months of exposure, whereas another may not experience any indications of toxicity until after several years of work under the same conditions. Two workers may also show comparable early signs of Mn toxicity but follow rather different and variable courses of disease over time. Several factors may contribute to these individual differences. Differences in nutrition, especially iron status, have long been thought to play a role in Mn toxicity (53). Because iron and Mn share absorption pathways, gastrointestinal absorption of Mn can increase in a state of iron deficiency as homeostatic mechanisms attempt to enhance iron absorption and Mn substitutes for the missing iron (54). Other elements, including calcium (55), zinc (56), phosphorus (57), and copper (58), have also been noted for interactive effects with Mn, as well as other nutritional factors such as protein deficiency (59, 60). Co-exposure to other pollutants could also affect susceptibility to Mn toxicity (61, 62). Genetic factors are another likely basis for differences in susceptibility, as illustrated by an autosomal recessive trait in Belgrade rats that results in impaired iron transport and produces greater hepatic uptake of Mn but lesser uptake in brain and other organs (63). Several studies have also implicated a variant allele for cytochrome P-450 debrisoquine hydroxylase as a risk factor for Parkinson's disease (64–69), which raises the question of whether this or a similar genetic defect could figure into susceptibility to Mn neurotoxicity. Chronic liver failure has also been associated with excessive Mn uptake and neurotoxicity (70–74). Of course, variables such as age, gender, smoking, and other characteristics may also influence susceptibility to metal toxicity (75).

1.4.2.1 Neurotoxicity

1.4.2.1.1 Clinical Cases. As summarized in a recent review by Pal et al. (76), the clinical characteristics of Mn neurotoxicity have been generally divided into three stages: (1) behavioral changes, (2) motoric features, and (3) dystonia with severe gait disturbances. The course, severity, and duration of these features vary greatly among individuals, perhaps depending on differences in exposure conditions, as well as individual susceptibility factors. The first stage may last for a few weeks or months and include a psychiatric component (locura manganica or manganese madness) in which individuals experience mood swings (spontaneous, uncontrollable laughing then crying), nervousness, irritability, compulsive behavior, and aggressiveness. Such psychiatric disturbances may be more characteristic of miners than of industrial workers (77). Patients also often report fatigue, headache, apathy, insomnia, somnolence, memory loss, diminished libido, impotence, excessive salivation, loss of appetite, hallucinations, and slowed movements. With or without these symptoms, motor signs indicative of extrapyramidal dysfunction may appear insidiously. These include an expressionless face, low-volume monotone speech, impaired manual dexterity, and a notable difficulty in walking backwards (52). As these signs progress, a Parkinsonian syndrome may emerge with hypertonic muscular rigidity and dystonia from contraction of antagonist muscles, along with slowed movement (bradykinesia or hypokinesia). A "cog-wheel" rigidity, described as a passive resistance of a limb to movement giving way to accelerated movement, is often noted. Also frequently reported is a characteristic gait described as a "cock walk" because the extended trunk, flexed arms, and walking on the toes or balls of the feet resemble the appearance of a chicken walking. Spasmodic movements of various muscles, particularly the cranial musculature, may manifest as eye tics, grimaces, torticollis, and other involuntary contractions. Tremor is variable in occurrence, possibly involving resting as well as action types (78). A low-amplitude, low-frequency resting tremor usually affects only the upper limbs, if at all, and can be exacerbated by emotion, fatigue, or stress. An action tremor is more prevalent and often associated with slowed movement and rigidity and may manifest in shaky handwriting. Some case reports indicate that early removal of a worker from exposure soon after symptoms or signs of Mn toxicity first become evident can prevent progression or even produce regression of the disease (79, 80). However, Pal et al. (76), reinforcing the clinical observations of many other physicians, reported that in their experience "[o]nce these signs and symptoms have become established,... the disorder is progressive, even when patients are removed from the environment with high Mn content..."

Despite some shared features of manganism and idiopathic Parkinson's disease (PD), including involvement of the dopaminergic and extrapyramidal systems, the two diseases differ in several respects (76, 81–83). Among the distinguishing features, the "cock walk," the difficulty in walking backwards and tendency to fall backward when displaced, and, sometimes, a fine nonresting tremor characterize manganism clinically. In addition, response to dopamine therapy is short-lived or altogether absent in manganism patients. Neuropathologically, PD is characterized by degeneration of the nigrostriatal pathway, whereas the most prominent pathological change in manganism occurs in the pallidum. Lewy bodies in the substantia nigra are diagnostic of PD but virtually absent in manganism. Computerized tomography (CT) and magnetic resonance imaging (MRI) provide little diagnostic information for PD. However, given the paramagnetic properties of Mn, high intensity signals in the basal ganglia, hypothalamus, pituitary, and other areas may be evident by MRI—but not necessarily, especially if exposure has previously terminated (84, 85). Positron emission tomography (PET) also provides differentiation. Flourodopa PET scans have shown reduced uptake in the striatum of PD patients, consistent with the loss of dopaminergic cells in the nigrostriatal pathway, whereas in manganism, the fluorodopa PET scan is normal, reflecting the fact that the pallidum is the primary target (84).

1.4.2.1.2 Epidemiology Studies. Most of the epidemiological studies of Mn health effects have focused on workers who had long-term occupational exposure to Mn via inhalation and have especially investigated neurobehavioral performance because of the well-known neurotoxic properties of Mn. Only a few of these studies are reviewed here;

for a more detailed review of recent studies of Mn neurotoxicity in relation to occupational exposure, see Iregren (86).

Although they have various strengths and limitations, several studies in recent years have provided a rather consistent pattern of results pointing to perturbations in neurobehavioral performance associated with average occupational exposure levels around 1 mg/m^3 or lower. One of the most informative studies, particularly with respect to exposure characterization, was conducted by Roels et al. (16). In this study of 92 male workers from an alkaline battery manufacturing plant in Belgium and a matched group of 101 controls, the Mn-exposed group had been exposed to MnO_2 for an average of 5.3 years (range: 0.2–17.7 years). The geometric means of the workers' time-weighted average (TWA) airborne Mn concentrations, as determined by personal sampler monitoring at the breathing zone, were 0.215 mg Mn/m^3 for respirable dust and 0.948 mg Mn/m^3 for total dust. No data on particle size or purity were presented, but the median cut point for the respirable dust fraction was 5 µm according to information provided by Roels (16, 87). Total and respirable dust concentrations were highly correlated ($r = 0.90, p < .001$) with the Mn content of the respirable fraction representing, on average, 25% of the Mn content in the total dust. The authors noted that the personal monitoring data were representative of the usual exposure of the workers because work practices had not changed during the last 15 years of the operation of the plant.

Occupational-lifetime integrated exposure to Mn was estimated for workers by multiplying the current airborne Mn concentration for their respective job classification by the number of years in which they held that classification and adding the resulting (arithmetic) products for each job position a worker had held. The geometric mean occupational-lifetime integrated respirable dust (IRD) concentration was 0.793 mg Mn/m^3 × year (range: 0.040–4.433), and the geometric standard deviation was 2.9 mg Mn/m^3 × year, based on information provided by Roels (87). The geometric mean occupational-lifetime integrated total dust (ITD) concentration was 3.505 mg Mn/m^3 × year (range: 0.191–27.465). Geometric mean blood Mn (MnB: 0.81 µg/dL) and urinary Mn (MnU: 0.84 µg/g creatinine) concentrations were significantly higher in the Mn-exposed group than in the control group, but on an individual basis, no significant correlation was found between either MnB or MnU and various external exposure parameters. Current respirable and total Mn dust concentrations correlated significantly with geometric mean MnU on a group basis (Spearman $r = 0.83, p < .05$).

A self-administered questionnaire focused on occupational and medical history, neurological complaints, and other symptoms. Neurobehavioral function was evaluated by tests of audioverbal short term memory, visual simple reaction time, hand steadiness, and eye-hand coordination. Blood samples were assayed for several hematologic parameters (erythrocyte count, leukocyte count, hemoglobin, hematocrit, mean corpuscular volume, mean corpuscular hemoglobin, platelets, differential leukocyte count), Mn, lead, zinc protoporphyrin, and serum levels of calcium, iron, follicle-stimulating hormone (FSH), luteinizing hormone (LH), and prolactin. Urinary Mn, cadmium, and mercury concentrations were also determined.

Responses to the questionnaire indicated no significant differences between groups in neurological symptoms. However, Mn workers performed worse than controls on several measures of neurobehavioral function. Visual reaction time was consistently and significantly slower in the Mn-exposed workers measured in four 2-minute periods, and slowing was especially pronounced over the total 8-minute period. Also, significantly greater variability was evident in the reaction times of the Mn-exposed group. Abnormal values (defined as $\geq$ 95th percentile of the control group) for mean reaction times were significantly more prevalent in the Mn-exposed group during three of four 2-min intervals of the 8-min testing period.

Five measures of eye–hand coordination (precision, % precision, imprecision, % imprecision, uncertainty) reflected more erratic control of fine hand–forearm movement in the Mn-exposed group than in the controls, and mean scores on all five measures were significantly different for the two groups. There was also a significantly greater prevalence of abnormal values for these five measures in the Mn-exposed group. The hole tremormeter test of hand steadiness indicated a consistently greater amount of tremor in the Mn-exposed workers, and performance for two of the five hole sizes showed statistically significant impairment.

Roels et al. (16) performed an exposure–response analysis by classifying IRD and ITD values into three groups and comparing the prevalence of abnormal scores for visual reaction time, hand steadiness, and eye–hand coordination to controls. This analysis indicated that the prevalence of abnormal eye–hand coordination values was significantly greater in workers whose IRD levels were less than 0.600 mg Mn/m^3 × year or whose ITD values were less than 2.500 mg Mn/m^3 × years. However, the relationship between exposure and response was not linear across groups. Visual reaction time and hand steadiness showed linear exposure-related trends but did not achieve statistical significance except at levels of >1.200 (IRD) and >6.000 (ITD) mg Mn/m^3 × year. Overall, the authors' analysis of the data on a group basis did not identify a threshold concentration of airborne Mn for the effects they observed. [Subsequent analyses of these data by the U.S. Environmental Protection Agency (31), when evaluating potential health risks related to the gasoline additive MMT, provided a very wide range of possible "effect levels," depending on the choice of mathematical models and assumptions. From this wide range of effect levels, the EPA identified concentrations of 0.09–0.2 µg/m^3 as likely candidate values for an alternative to the existing Reference Concentration (RfC) of 0.05 µg/m^3. The RfC is defined as an estimate (whose uncertainty spans about an order of magnitude) of a continuous inhalation exposure level for the human population (including sensitive subpopulations) that is likely to be without appreciable risk of deleterious noncancer effects during a lifetime. The value was derived by adjusting the occupational exposure value for daily lifetime exposure of the general population and dividing by selected uncertainty factors (33).]

Roels and colleagues (15) also conducted an earlier cross-sectional study of 141 male workers exposed to manganese dioxide, tetroxide, and various salts (sulfate, carbonate, and nitrate). A matched group of 104 male workers was selected as a control group. The two groups were matched for socioeconomic status and background environmental factors; in addition, both groups had comparable workload and work-shift characteristics. The TWA of total airborne manganese dust ranged from 0.07 to 8.61 mg/m^3, and the overall arithmetic mean was 1.33 mg/m^3, the median was 0.97 mg/m^3, and the geometric mean was 0.94 mg/m^3. The duration of employment ranged from 1 to 19 years, and the mean was 7.1 years. The particle size and purity of the dust were not reported. Neurological examination, neurobehavioral function tests (simple reaction time, short-term memory, eye–hand coordination, and hand tremor), and a self-administered questionnaire were used to assess the possible neurotoxic effects of manganese exposure.

Significant differences in mean scores between Mn-exposed and reference subjects were found for objective measures of visual reaction time, eye–hand coordination, hand steadiness, and audioverbal short-term memory. The prevalence of abnormal scores on eye–hand coordination and hand steadiness tests showed a dose-response relationship with blood manganese levels; short-term memory scores were related to years of manganese exposure but not blood manganese levels. The prevalence of subjective symptoms was greater in the exposed group than in controls for 20 of 25 items on the questionnaire, and four items were statistically significant: fatigue, tinnitus, trembling of fingers, and irritability.

The findings of Roels et al. (15, 16) are supported by other recent reports. Mergler et al. (88) conducted a cross-sectional study of 115 male ferromanganese and silicomanganese alloy workers in southwest Quebec. A matched-pair design was employed because of presumptively high environmental pollutant levels; 74 pairs of workers and referents were matched for age, educational level, smoking status, number of children, and length of residency in the region. Air concentrations of respirable and total dust were sampled by stationary monitors during silicomanganese production. The geometric mean of a series of 8-hr time-weighted averages was 0.035 mg Mn/m^3 (range: 0.001–1.273 mg Mn/m^3) for respirable dust and 0.225 mg Mn/m^3 (range: 0.014–11.480 mg Mn/m^3) for total dust. The authors noted that past dust levels at certain job sites had been considerably higher. The mean duration of the workers' manganese exposure was 16.7 years and included Mn fumes, as well as mixed oxides and salts of Mn. MnB was significantly higher in the Mn alloy workers, but MnU did not differ significantly between exposed workers and controls.

The number of discordant pairs, in which workers reported undesirable symptoms on a self-administered questionnaire but their matched pairs did not, was statistically significant for 33 of 46 items, including fatigue; emotional state; memory, attention and concentration difficulties; nightmares; sweating in the absence of physical exertion; sexual dysfunction; lower back pain; joint pain; and tinnitus. Workers did not report symptoms typical of advanced manganism (e.g., hand tremor, changes in handwriting, loss of balance when turning, difficulty in reaching a fixed point) significantly more than referents, which suggests that the other reported symptoms were probably not due to bias on the part of the workers.

The greatest differences in neurobehavioral function were evident in tests of motor function, especially tests such as the diadochokinesimeter that require coordinated alternating and/or rapid movements. Diadochokinesis refers to the ability to perform rapidly alternating movements such as supination and pronation of the forearm and is an indicator of extrapyramidal function. Workers performed significantly worse on the motor scale of a neuropsychological test battery in both overall score and eight subscales of rapid sequential or alternating movements. Worker performance was also significantly worse in tests of hand steadiness, parallel line drawing performance, and ability to rapidly identify and mark specified alphabetic characters within strings of letters. Performance in a variety of other tests of psychomotor function, including simple reaction time, was worse in Mn-exposed workers but marginally significant ($.05 < p < .10$). In addition, Mn alloy workers differed significantly from referents in measures of cognitive flexibility and emotional state. Olfactory sensitivity was also significantly enhanced in the Mn-alloy workers (89).

The matched-pair design of Mergler et al. (88) helped reduce differences between exposed and referent subjects that might otherwise have confounded the study. However, to the extent that the referents were exposed to Mn in the ambient atmosphere, such exposure may have reduced the differences in neurobehavioral performance between workers and referents. This possibility is supported by the fact that the finger-tapping speed of both workers and referents on a computerized test was slower than that of Mn-exposed workers assessed in the same test by Iregren (90) in Sweden and by subsequent findings of neurobehavioral dysfunction in the general population surrounding the smelter in Quebec (91).

Workers exposed to Mn in two Swedish foundries (15 from each plant) were evaluated in a study first reported by Iregren (90). The exposure to manganese varied from 0.02 to 1.4 mg/m^3 (mean = 0.25 mg/m^3; median = 0.14 mg/m^3) for 1–35 years (mean = 9.9 years). Earlier monitoring measurements in both factories suggested that essentially no changes in exposure had occurred in either factory for the preceding 18 years. Each exposed worker was matched in age, geographical area, and type of work to two workers not exposed to manganese in other industries. Neurobehavioral function was assessed by eight computerized tests and two manual dexterity tests. There were significant differences between exposed and control groups for simple reaction time, the standard deviation of reaction time, and finger tapping speed of the dominant hand. In addition, digit span short-term memory, speed of mental addition, and verbal (vocabulary) understanding differed significantly between exposed and control groups. The difference in verbal understanding suggested that the two groups were not well matched in general cognitive abilities. When verbal performance was used as an additional matching criterion, the differences between the groups in simple reaction time, the standard deviation of reaction time, and finger tapping speed remained statistically significant, despite a decrease in statistical power due to the reduced size of the reference group to 30 workers. Further analyses using verbal test scores as a covariate also indicated that these same three measures of neurobehavioral function were statistically different in exposed and control workers. No significant correlation was found within the exposed group to establish a concentration–response relationship.

Additional reports of neurobehavioral and electrophysiological evaluations of these same Swedish workers were published by Wennberg et al. (92, 93). Although none of the latter results achieved statistical significance at $p = .05$, increased frequency of self-reported health symptoms, increased prevalence of abnormal electroencephalograms, slower brainstem auditory evoked potential latencies, and slower diadochokinesometric performance were found in the exposed workers. An increase in P-300 latency, as suggested by these results, was also observed in patients who had Parkinsonism, according to Wennberg et al. (92), who viewed these results in Mn-exposed workers as early (preclinical) signs of disturbances similar to Parkinsonism.

Among other recent studies, Lucchini and colleagues (89, 94, 95) conducted a series of investigations of workers at an Italian ferroalloy plant. In their most extensive study to date, they administered a battery of neuropsychological tests to a group of 61 male workers and 87 controls who were closely comparable in age, alcohol and caffeine consumption, smoking, and other variables, but differed in education, night shift assignment, and noisy work environment. Air concentrations of Mn (as MnO_2 and Mn_3O_4) changed by more than an order of magnitude in some work areas during the 16 years preceding the study. A cumulative exposure index was calculated for each alloy worker based on the worker's employment history and the annual surveys of different work areas in the plant. The geometric mean cumulative exposure to total Mn dust was 1.205 mg/m^3 × year, which when divided by the average length of exposure (15.2 years) yielded an average annual exposure level of 0.071 mg/m^3. Exposed workers had significantly higher MnB and MnU levels than controls. Although MnB showed a significant positive correlation with airborne Mn, neither MnB nor MnU was correlated with the cumulative exposure index. The exposed workers reported more symptoms of irritability, loss of equilibrium, and rigidity. After correction for age, shift work, blood lead, and other confounding variables, the alloy workers also performed more poorly than controls on tasks requiring coordinated, rapid, alternating movements, thus corroborating the results of Mergler et al. (88) and Iregren (90) for the same tests. Certain measures of tremor characteristic of manganese neurotoxicity were also altered in the alloy workers. Dividing the exposed workers into three groups representing low, mid, and high cumulative exposure index values provided indications of significant concentration–response relationships for two cognitive functions (symbol digit and digit span) and for finger tapping in both dominant and nondominant hands.

Numerous other studies have provided positive evidence of diminished neurobehavioral performance in Mn-exposed workers across a range of exposure conditions (48, 49, 51, 52, 80, 96–104). Every study of Mn neurotoxicity has limitations, and a few reports (105) even show no significant relationship between low-level Mn exposure and neurobehavioral measures. Nevertheless, all of these studies taken together provide a coherent pattern of evidence indicating that neurotoxicity is associated with low-level occupational manganese exposure at average concentrations around 1 mg/m^3 or lower. In particular, the most frequently demonstrated neurobehavioral effects pertain to motor function, especially hand steadiness, which is consistent with involvement of the extrapyramidal system. Tasks that require coordinated, sequential, alternating movements at maximum speed are the most sensitive to lower level Mn exposure (106). However, for group screening purposes, a battery of tests should include measures of motor function and response speed and also tests of memory and rating scales for mood and subjective symptoms because the latter also differentiate Mn-exposed workers with some regularity (86).

Although evidence supports an association between low-level Mn exposure and neurotoxicity in otherwise asymptomatic workers, the data are less clear for issues such as the influence of duration of exposure and/or age on such effects and the reversibility or irreversibility of preclinical neurobehavioral outcomes. Note that the mean period of exposure ranged from ~5 years in the study of Roels et al. (16) to ~17 years in Mergler et al. (88). Although these periods constitute chronic exposures by definition, they are not as long as many actual work tenures and therefore may not provide as great an opportunity for the expression of Mn neurotoxicity. Also, the average age of workers in these studies ranged from approximately 31 years (16) to 46 years (90). Although some statistical analyses have not demonstrated a trend toward poorer neurobehavioral performance in Mn-exposed workers as a function of work tenure or age (107), other findings do suggest that vulnerability to Mn toxicity increases with age (18, 91; see also Ref. 108 for animal evidence).

Data are limited and somewhat mixed in the reversibility of preclinical signs of impaired neurobehavioral performance. Roels et al. (109) reported an 8-year follow-up study of battery plant workers who had been first described in their 1992 report. The workers were tested annually on eye–hand coordination, hand steadiness, and simple visual reaction time by the same personnel who administered these tests for the earlier (16) report. Ongoing air monitoring provided the basis for estimating the time course of average Mn exposure levels, which fluctuated from year to year but declined overall during the 8-year period. Of the original 92 subjects, 34 remained in the cohort at the end of the follow-up, and full testing were results were available for 21 of these workers. In addition, 24 workers whose Mn exposure ceased between two and five years after the original testing were tested annually. Controls ($n = 39$) came from the original control group. Of the three tests, eye–hand coordination showed the strongest and most significant relationship to exposure changes over time, and performance improved or declined in conjunction with decreases or increases in annual average air Mn levels. These relationships were evident despite corrections for age and other potentially confounding variables. The deficits in performance were not fully reversible, however, in any of the workers but those with the lowest average exposure levels. Results for the workers whose exposure had terminated paralleled those who had ongoing exposure: eye–hand coordination performance improved significantly, but hand steadiness and simple visual reaction time performance showed no improvement after at least 3 years of no Mn exposure. Other reports of follow-up evaluations of Mn workers have indicated variously that performance may remain stable (neither improve nor worsen) with ongoing exposure (95, 107) or may worsen despite removal from exposure (110, 111). Table 39.2 provides a summary of recent selected epidemiology studies of Mn-exposed workers.

1.4.2.1.3 Experimental Animal Studies of Mn Neurotoxicity Mechanisms. Nonhuman primates exposed to high concentrations of Mn manifest various signs of Mn toxicity that are qualitatively similar in various respects to those observed in humans: dystonic posture, gait disorders, action tremor, hyperactivity, and emotional displays (112–115). Inhaled manganese also produces significant alterations in dopamine levels in the caudate and globus pallidus of Rhesus monkeys (112, 116). Rodent studies of Mn neurotoxicity have focused primarily on neurochemical end points rather than behavioral effects. Alterations of striatal dopamine and other biogenic amines were among the most commonly investigated effects. As Newland (13) noted, however, the findings from animal studies vary a great deal for several reasons, including differences in species and strain of experimental animals, and also in protocols, forms of Mn administered, routes of exposure, dosage (across several orders of magnitude), and rates of exposure. The results from experimental animal studies are nonetheless of value in confirming the qualitative features of manganism and also in helping to illuminate the mechanistic basis of such effects, particularly when coupled with human data.

The mechanisms underlying Mn neurotoxicity have in fact been the subject of much investigation and speculation (112, 117–122). It is well established that the basal ganglia are a primary site of neuronal degeneration associated with Mn exposure, particularly the putamen and globus pallidus, based on autopsy results for Mn-exposed humans (121) and nonhuman primates (112, 116), as well as brain imaging studies in humans (84, 85) and

Table 39.2. Selected Epidemiological Studies of Mn-exposed Workers

Study	Type of Industry	N	Age ($\bar{x}$ yrs.)	Exposure ($\bar{x}$ yrs.)	Mn Species	Conc[a] (mg/m^3)	Primary Neurobehavioral Decrements
Mergler et al. (88)	Alloy plant	115 exposed (74 matched pairs)	43	17	Mixed oxides, salts, fumes	0.035 PM$_5$ (g.m.)	Speed/coordination of rapid alternating movements of hands/fingers; hand steadiness; parallel line drawing; and others
Roels et al. (16)	Alkaline battery plant	92 exposed 102 control	31	5	MnO$_2$	0.150 (g.m.)	Hand steadiness; eye–hand coordination; visual reaction time
Iregren (90)	Foundry	30 exposed 30–60 control	46	10	Not stated	0.140 (mdn.)	Simple reaction time; standard deviation of reaction time; finger tapping speed
Roels et al. (15)	Production facility	141 exposed 104 control	34	7	Mixed oxides and salts	0.940 (g.m.)	Visual reaction time; eye–hand coordination; hand steadiness; audioverbal short-term memory
Lucchini et al. (95)	Alloy plant	61 exposed 87 control	42	15	MnO$_2$, Mn$_3$O$_4$	0.071 (g.m.)	Finger tapping speed; tremor; short-term memory and mental arithmetic

[a]g.m. = geometric mean; mdn. = median.

nonhuman primates (114, 115, 123, 124). However, these regions do not necessarily show an increase in Mn concentration (114, 121), so it is not clear that Mn necessarily acts directly in these areas to produce its characteristic neurotoxicity. These observations have led to some speculation that the effects of Mn may be mediated by other metals with which it interacts, namely iron, copper, or aluminum (114, 122, 125).

In terms of the neurochemistry of manganese toxicity, several studies have shown that dopamine levels are affected by manganese exposure in humans, monkeys, and rodents, often with an initial increase in dopamine followed by a longer term decrease (81, 116, 126, 127). Some theories of manganese neurotoxicity have focused on the role of excessive Mn in oxidizing dopamine that results in free radicals and cytotoxicity (81, 119). However, recent work brings into question the notion that Mn neurotoxicity is necessarily mediated by oxidative damage (128) and even suggests that Mn may protect against oxidative damage (125). Nevertheless, the loss of dopamine in the globus pallidus is well documented and would be expected to result in a disinhibitory increase in tonic activity that would in turn feed back to thalamic motor nuclei and produce the rigidity of manganism (129).

Another possible mechanism of Mn neurotoxicity involves the role of Mn in mitochondrial energy metabolism (129–131). Brouillet et al. (127) have suggested that the mitochondrial dysfunctional effects of Mn result in various oxidative stresses on cellular defense mechanisms (e.g., GSH) and, secondarily, free radical damage to mitochondrial DNA. In view of the slow release of Mn from mitochondria (131), such an indirect effect would help account for a progressive loss of function in the absence of ongoing Mn exposure, because manganese toxicity is reported to progress in humans despite the termination of exposure (49, 132). In addition, Ca levels within the mitochondria increase, in part because Mn inhibits cytosolic Ca efflux, resulting in destruction of the mitochondria (129). In this excitotoxic state, cells in the globus pallidus die, and the result is the dystonia characteristic of manganism. In summarizing the above evidence, Verity (122) has speculated on the possible use of iron chelation and increased antioxidant as therapeutic measures for excessive Mn exposure.

1.4.2.2 Respiratory, Reproductive, and Other Effects

Respiratory Toxicity. The respiratory system is another primary target for manganese toxicity; numerous reports of "manganese pneumonia" and "pneumonitis" and other effects on the respiratory system have appeared in the literature, dating back at least to 1921 (133). Lloyd-Davies (134) reported an increased incidence of pneumonia in men employed at a potassium permanganate manufacturing facility during an 8-year period in which the number of workers in the facility varied from 40 to 124. Dust characteristics were well described in terms of collection conditions and particle size and composition, but actual exposure levels were not measured. Calculated air concentrations ranged from 9.6 to 83.4 mg/m^3 as manganese dioxide, which constituted 41–66% of the dust. The incidence of pneumonia in the workers was 26 per 1000, compared to an average of 0.73 per 1000 in a reference group of over 5000 workers. Workers also complained of bronchitis and nasal irritation. In a continuation of this study, Lloyd-Davies and Harding (135) reported the results of sputum and nasopharyngeal cultures for four men diagnosed as having lobar or bronchopneumonia. Except for one of these cases, they concluded that manganese dust, without the presence of bacterial infection or other factors, caused the observed pneumonitis.

In their cross-sectional study of workers exposed to mixed manganese oxides and salts, described above, Roels et al. (15) included a lung function test (forced vital capacity, forced expiratory volume, peak expiratory flow rate and maximal expiratory flow rate at 50% and 75% of the FVC), blood and urine tests, and a self-administered questionnaire that was designed to detect respiratory symptoms, as well as CNS effects. The authors found that significantly greater prevalences of coughs during the cold season, dyspnea during exercise, and recent episodes of acute bronchitis were self-reported by the Mn-exposed group. However, objectively measured lung function parameters were only slightly ($< 10\%$) altered and only in Mn-exposed smokers (also see Ref. 136 regarding a possible synergism between manganese and smoking in producing respiratory symptoms). In their later study of alkaline battery plant workers, Roels et al. (16) again used a self-administered questionnaire that included respiratory symptoms, and they evaluated lung function by spirographic measures of forced vital capacity. Responses to the questionnaire indicated no significant differences between groups in respiratory symptoms. Nor were spirometric, hormonal, or calcium metabolism measurements significantly different for the two groups. However, differences in the forms of manganese (MnO_2 versus mixed Mn oxides and salts) to which the workers in these two studies were exposed make it difficult to compare the results of these two studies.

Experimental animal data qualitatively support human study findings in that manganese exposure results in an increased incidence of pneumonia in rats exposed to 68–219 mg/m^3 MnO_2 for two weeks (24), pulmonary congestion in monkeys exposed to 0.7 or 3.0 mg/m^3 MnO_2 for five months (137), pulmonary emphysema in monkeys exposed to 0.7–3.0 mg/m^3 MnO_2 for 10 months (138), and bronchiolar lesions in rats and hamsters exposed to 0.117 mg/m^3 Mn_3O_4 for 56 days (139). Lloyd-Davies and Harding (135) also induced bronchiolar epithelium inflammation, widespread pneumonia, and granulomatous reactions in rats administered 10 mg MnO_2 by intratracheal injection and pulmonary edema in rats administered 5–50 mg $MnCl_2$ in the same fashion. However, no significant pulmonary effects were detected in other studies of rats and monkeys exposed to as much as 1.15 mg Mn/m^3 as Mn_3O_4 for 9 months (140–142) and rabbits exposed to as much as 3.9 mg Mn/m^3 as $MnCl_2$ for 4–6 weeks (143).

Reproductive Toxicity. Effects on reproductive function, most notably impotence and loss of libido, were frequently self-reported by male workers chronically exposed to Mn (49, 51, 97, 144–146). Despite this historical experience, relatively few epidemiological studies have investigated the incidence of sexual dysfunction and reproductive outcomes in Mn workers, and these only in male workers; no known studies to date have investigated female worker reproductive behavior or function. Jiang et al. (147) compared responses (it is not stated how the responses were elicited) of 314 male Mn workers and 314 male controls from six factories and mines in China. The two groups were matched in age, work tenure, reproductive age, smoking, cultural background, and other factors. The geometric mean airborne MnO_2 concentration was 0.145 mg/m^3. Sexual dysfunction, including impotence and hyposexuality (but not abnormal ejaculation), was significantly greater in the Mn-exposed workers, and a significant trend toward greater dysfunction occurred with increasing work tenure. However, there was no significant difference in reproductive outcomes for the two groups.

Lauwerys et al. (148) reported the results of a fertility questionnaire administered to male factory workers ($n = 85$) exposed to manganese dust. This study involved the same population of workers in which Roels et al. (15) measured neurobehavioral performance. The range of manganese levels in the breathing zone was 0.07–8.61 mg/m^3, and the median concentration was 0.97 mg/m^3. The average length of exposure was 7.9 years (range of 1–19 years). A group of workers ($n = 81$) who had a similar workload was used as a control group. The number of births expected during different age intervals of the workers (16–25, 26–35, 36–45 years) was calculated on the basis of the reproductive experience of the control employees during the same periods. A statistically significant decrease was observed in the number of children born to workers exposed to manganese dust during the ages of 16–25 and 26–35. No difference in the sex ratio of the children was found. However, a more recent report from the same group of investigators (149), based on 70 of the alkaline battery plant workers evaluated by Roels et al. (16), indicated that the probability of fathering a live birth was not different between the Mn-exposed and control workers. Also, in the study by Roels et al. (16), serum levels of certain hormones related to reproductive function (FSH, LH, prolactin) were not significantly different for the full group of 92 Mn workers compared to 102 controls. The latter results are partially supported by a preliminary report by Alessio et al. (150), who found that serum FSH and LH levels were not significantly different in 14 workers generally exposed to < 1 mg Mn/m^3 compared to controls. However, prolactin and cortisol levels were significantly higher in these Mn-exposed workers. It is possible that differences in the forms of Mn to which workers were exposed in these studies may have contributed to the differences in the results, but insufficient information exists to substantiate this speculation. Average concentrations of airborne Mn also differed slightly in the reports of Gennart et al. (149) and Roels et al. (16), evidently because only a subset of Mn workers, presumably with different job functions, was used in the Gennart et al. (149) analysis. The median respirable dust concentration was 0.18 mg/m^3, and the median total dust concentration [comparable to Roels et al. (15) and Lauwerys et al. (148)] was 0.71 mg/m^3.

Few animal studies have investigated reproductive effects of Mn exposure, regardless of route. Intratracheal instillation administration of a single dose of MnO_2 in rabbits yielded degeneration of the seminiferous tubules and sterility over a period of 4–8 months (151, 152). Reduced testosterone levels were found in rats orally administered Mn_3O_4 (153), and slowed maturation of the testes was found in mice given Mn_3O_4 orally (154). However, adult rat reproductive function and sperm morphology were not affected (153, 155, 156). Female reproductive function was largely unaffected in terms of ovulation, resorptions, fetal weights, and litter size, although a slight decrease in pregnancy rate was observed (153). However, the reduction in pregnancy rate is difficult to interpret because both males and females were exposed to Mn before breeding. Female mice exposed by inhalation to MnO_2 before conception had significantly larger litters than controls (157). Kontur and Fechter (158) observed no effects on litter size or weight except for reductions at Mn concentrations in drinking water so high that water intake by the dams was reduced and likely accounted for the effects. Manganese readily crosses the placenta, and some developmental effects have been found in offspring of Mn-exposed dams. Reduced activity levels and growth were observed in mice whose mothers had inhaled MnO_2 during pregnancy (157). Neurochemical data for rats treated with Mn during early development were mixed (158–161).

Other Systemic Effects and Carcinogenicity. Two reports indicate that blood pressure, either systolic (162) or diastolic (163), was reduced in Mn-exposed workers. However, electrocardiograms were normal in the workers with lower diastolic pressure (163). A large amount of literature exists on the cardiovascular effects of divalent Mn in animal models, because of its use as an experimental method to manipulate calcium channel function in heart muscle. However, both positive and negative inotropic effects of Mn have been observed, and the various cardiac effects of Mn, which have been described as "complex and difficult to interpret" (164), are beyond the scope of this review. Roels et al. (16) found that erythropoietic parameters and serum iron concentrations were consistently and significantly lower in Mn-exposed workers, albeit within the normal range of values. Various other studies have noted elevated white blood cell counts in Mn workers (15, 89).

Evidence for the carcinogenicity of Mn is mixed and inconclusive. Stoner et al. (165) found some limited indications of tumors in a mouse lung adenoma screening assay using manganous sulfate by intraperitoneal injection, but only at the highest dose. Furst (166) found no increase in tumor responses to manganese powder or manganese dioxide by gavage or intramuscular injection in either rats or mice. Co-administration of Mn dust with known inducers of sarcoma appeared to be protective (167–169). Witschi et al. (170) found lung cell proliferation but no increase in tumor incidence in mice given intraperitoneal injections of MMT. EPA concluded that the available studies were inadequate to assess the carcinogenicity of Mn and assigned it a D classification (not classifiable as to human carcinogenicity) (171).

1.4.3 Pharmacokinetics and Metabolism

A great deal of investigation has been devoted to the absorption of dietary Mn through the gastrointestinal tract and the homeostatic mechanisms that regulate its uptake and elimination. It is generally accepted that Mn absorption via ingestion is typically of the order of 3–5% in normal adults (172). However, the inhalation pharmacokinetics of manganese and its compounds have not been described quantitatively. Qualitatively, it is clear that, after absorption via the respiratory tract, manganese is transported through the bloodstream directly to the CNS, bypassing the liver and the opportunity for first-pass hepatic clearance. This direct path from the respiratory tract to the CNS is thought to be the primary reason for the differential toxicity of inhaled and ingested manganese and helps account for the greater sensitivity of the CNS to Mn than that shown by other systems. The complex roles and interplay among factors such as oxidation states, endogenous redox reactions, transfer proteins, transport to brain regions via the blood brain barrier and/or the cerebrospinal fluid and choroid plexus, not to mention competitive interactions with iron and other elements, need to be better understood before the pharmacokinetics and pharmacodynamics of Mn in its different forms can be fully explicated (173). Depending on the form to which one is exposed, Mn may be found within the body in at least two or three oxidation states ($+2$, $+3$, and potentially $+4$ if, for example, the exposure was to MnO_2). Depending on its oxidation state, it may bind with greater or lesser affinity to transferrin, alpha2-macroglobulin, albumin, or possibly some other protein. It may also undergo oxidation or reduction to a higher or lower state. As it is carried in the blood, Mn distributes in varying degrees to specific organs, for example, bone, liver, kidney, brain, and other tissues. Bound or unbound to a carrier protein, Mn may transfer across the blood–brain barrier or via the cerebrospinal fluid and the choroid plexus to various brain regions (173). Despite the fact that the amount of Mn that ultimately reaches the brain is small in relation to organs such as bone and liver, even such relatively small amounts of Mn in one form or another apparently can disrupt the proper functioning of key areas of the CNS.

Using ^{54}Mn radiolabeled Mn chloride and Mn oxide administered to human subjects in nebulized solutions, Mena et al. (53) found that 60% of the radioactivity initially measured in the lung was recovered in the feces within 4 days. This finding does not differentiate between the amount absorbed through the respiratory tract and the amount that may have been removed from the respiratory tract via mucociliary clearance and ingested. Nor does it indicate how much of the Mn absorbed through the respiratory tract was retained in various compartments. Newland et al. (123) administered ^{54}Mn to rhesus macaques by inhalation in a single episode and observed retention in the lung for 500 days, which was about twice as long as fecal radioactivity could be detected. Although solubility along with particle size would be expected to influence the rate of absorption, Drown et al. (174) found little difference in the amounts of relatively insoluble and soluble forms of Mn ultimately distributed to the brain after intratracheal instillation in rats.

Blood levels of Mn in occupationally exposed workers are generally higher than the levels in nonoccupationally exposed persons, but on an individual basis, blood Mn levels have not proven useful as a bioindicator of exposure. Mergler and Baldwin (175) have suggested that blood Mn levels might more accurately reflect body burden due to occupational exposure if the influence of current exposure can be eliminated, for example, by removing workers from exposure for a few weeks. A source of error in blood Mn measurements may also be the contamination introduced by anticoagulants, including heparin, commonly used in collecting blood samples, which suggests that serum Mn might be more useful (9). As yet, however, no fully satisfactory biomarker of exposure to Mn has been identified, but attention is being directed at biochemical markers of effects (176). More than 90% of the Mn excreted from the body is eliminated in the feces, either by passing through the gastrointestinal tract without being absorbed or, if absorbed, by biliary excretion. The rate of clearance from different compartments varies considerably. The longest elimination half-times are generally observed in brain and bone, although how much of the radioactivity in the head region reflects brain versus cranial bone is uncertain (177). In monkeys, Newland et al. (123) measured half-times of 223–267 days for the head region versus 94–187 days for the slowest term for the chest. These long half-times, it was thought, reflect both slower clearance of brain stores and replenishment from other organs, particularly the respiratory tract. In rats, Drown et al. (174) also observed slower clearance of labeled Mn from the brain than from the respiratory tract.

The bioavailability of different forms of manganese is also relevant to the pharmacokinetics of Mn. Roels et al. (16) in their 1992 report, noted that geometric mean blood and urinary Mn levels of workers exposed only to manganese dioxide were lower (MnB: 0.81 µg/dL; MnU: 0.84 µg/g creatinine) than those of workers exposed to mixed oxides and salts in their 1987 (15) report (MnB: 1.22 µg/dL; MnU: 1.59 µg/g creatine), even though airborne total dust levels were approximately the same (geometric means of 0.94 and 0.95 mg/m^3, respectively). However, Mena et al. (53) observed no difference between the absorption of 1 µm particles of MnCl$_2$ and Mn$_2$O$_3$ in healthy adults. Experimental animal studies comparing the uptake of soluble salts and relatively insoluble oxides of Mn also provide mixed evidence on the distribution of different forms of Mn to the brain. For example, Roels et al. (16) found higher concentrations of Mn in the striatum following intratracheal instillation of MnCl$_2$ compared to MnO$_2$. However, in a similar study using intratracheal instillation of MnCl$_2$ and Mn$_3$O$_4$ in rats, Drown et al. (174) found that, although the soluble chloride initially cleared four times faster than the insoluble oxide from the respiratory tract and reached a peak brain level within 1 day verus 3 days for the oxide, after 2 weeks the amounts of labeled Mn in the respiratory tract and whole brain were similar for the two compounds. Gianutsos et al. (178) also compared the distribution of Mn to the brain in mice after single subcutaneous injections of MnCl$_2$ and Mn$_3$O$_4$ and found Mn levels consistently higher from the chloride over a 22-day period. Komura and Sakamoto (179), comparing different forms of Mn in mouse diet, also suggested that less soluble forms such as MnO$_2$ were taken up to a significantly greater degree in cerebral cortex than the more soluble forms of MnCl$_2$ and Mn(HC$_3$COO)$_2$; however, the corpus striatal binding characteristics of the +4 valence state of Mn in MnO$_2$ were not substantially different from those of the divalent forms in MnCl$_2$, Mn(CH$_3$COO)$_2$, and MnCO$_3$. Because different oxidation states of certain metals (e.g., chromium, nickel, mercury) are known to have different toxicities, some investigators have suggested that manganese can have quite different neurotoxic properties depending on its oxidation state (117, 119). There have been suggestions that the higher valence states of Mn (Mn^{+3}, Mn^{+4}) and the higher oxides in ores (Mn$_2$O$_3$ and Mn$_3$O$_4$) are more toxic than other forms (180), but further work is needed to determine the existence and nature of any such differential toxicity.

Also of interest here is the potential for direct CNS uptake of Mn from the nasal region via the olfactory tract. Several studies have demonstrated that Mn and other metals can be transported along the primary olfactory neurons to the olfactory bulbs, but thus far Mn is the only metal that passes to other regions of the brain by neuronal transport (181). In rats, Mn applied to the intranasal region travels over time to large areas of the telencephalon and into the diencephalon (182, 183). These findings raise the possibility that coarse particles or even particles larger than 10 µm MMAD could be pertinent to the health risks associated with occupational exposure to particulate Mn. At present, it is not clear whether such neuronal transport occurs under normal inhalation and how factors such as particle size and solubility may influence CNS uptake by this pathway.

1.5 Standards

Occupational exposure limits that have been established by various organizations for manganese in various forms are listed in Table 39.3. The current ACGIH Threshold Limit Value for elemental manganese [CAS Number: *7439-96-5*] and inorganic compounds (as Mn) is an 8-h TWA concentration of 0.2 mg/m^3 in workplace air. The OSHA PEL for manganese compounds and fume (as Mn) is 5 mg/m^3 as a ceiling value.

Table 39.3. Occupational Exposure Limits

	OSHA PEL mg/m^3	NIOSH REL mg/m^3	ACGIH TLV mg/m^3	AIHA WEEL	ANSI STD.	Germany MAK mg/m^3	Sweden NGV / TGV mg/m^3	U.K. TWA mg/m^3	Poland MAC mg/m^3
Mn and compounds [7439-95-5]									
TWA		1	0.2			5	2.5	1	0.3
STEL		3					2.5	3	
Ceiling	5						(respirable)		5
Mn fume [7439-95-5]									
TWA	5	1	0.2						
STEL	5	3	3						
Ceiling	5								
MMT [12108-13-3]									
TWA			0.2 (skin)						
STEL		0.2 (skin)							
Ceiling	5								
CMT [12079-65-1]									
TWA			0.1 (skin)						
STEL	5								
Ceiling	5								
Mn tetroxide [1317-35-7]									
TWA		1							
STEL	5								
Ceiling	5								

MANGANESE AND RHENIUM

NIOSH recommends limiting exposure to manganese compounds and fume (as Mn) to a TWA concentration of 1 mg/m^3, and a STEL of 3 mg/m^3.

The ACGIH TLV for manganese fume (elemental and inorganic) is 0.2 mg/m^3 as an 8-h TWA and a STEL (15 min) of 3 mg/m^3. From 1979 to the present, the ACGIH TLV-TWA for manganese fume is 1 mg/m^3, as Mn, and the STEL is 3 mg/m^3, as Mn. In 1992, the TLV-TWA for manganese, elemental and inorganic compounds, was proposed as 0.2 mg/m^3, as Mn. ACGIH no longer has a TLV specific for manganese tetroxide (Mn$_3$O$_4$) because exposure to this compound is limited by the TLV for manganese and its compounds, that is, the 8-hour TLV for Mn$_3$O$_4$ fume is 1 mg/m^3 and for Mn$_3$O$_4$ dust, it is 5 mg/m^3. The ACGIH also recommends an 8-h TWA of 0.2 mg Mn/m^3 for MMT [CAS Number: *12108-13-3*], and 0.1 mg Mn/m^3 for CMT [CAS Number: *12079-65-1*], and has indicated that special precautions should be taken to protect workers from dermal contact with both of these compounds.

2.0 Rhenium (Re)

2.0.1 CAS Number: *[7440-15-5]*

Rhenium, one of the group 7 (VIIB) transition elements, has an atomic number of 75 and an atomic weight of 186.207. Its discovery is generally credited to Walter Karl Noddack, Ida Eva Tacke-Noddack, and Otto Berg in 1925, although others independently identified the element around the same time (1). Rhenium, the last naturally occurring element discovered, was found by X-ray spectrographic analysis. The name derives from the Greek word "Rhenus," in reference to the Rhine River or Rhineland in Germany.

2.1 Chemical and Physical Properties

Depending on the process used to isolate and process it, rhenium may appear as a brown-black powder or a silvery white solid metal. Rhenium is among the least common of the natural elements; it generally occurs as a trace element in molybdenite, columbite, gadolinite, and platinum ores. A mineral form of rhenium apparently exists but is very rare (2, 184).

Rhenium has eight valence states that range from 0 to 7. There are two naturally occurring isotopes, 185 and 187, and several radioisotopes whose half-lives range from 2 min to 70 d (185). Rhenium is characterized by a very high melting temperature (3180°C) and density (21.02 g/cm^3 at 20°C). It has high electrical resistivity over a wide temperature range, low vapor pressure, good shock resistance, and low friction. Despite its density and hardness, it remains ductile after recrystallization.

2.2 Production and Use

Rhenium is extracted from copper and molybdenite ores, usually as a by-product of copper mining. Rhenium is often alloyed with tungsten or molybdenum to improve the properties of these refractory metals. Rhenium is used in aircraft turbine blades, electrical filaments (especially for mass spectrometers), welding rods, thermocouples, cryogenic magnets, photographic flashbulb filaments, and as a catalyst in the crack distillation of petroleum (186). In recent years, its material properties have been of interest for applications involving space rocket propulsion chambers (187). It is available from commercial sources in powder, rod, pellet, foil, wire, and other forms.

Among the compounds that can be formed with rhenium are sulfides, fluorides, chlorides, bromides, iodides, halides, and oxides. The most frequently encountered compound of rhenium is potassium perrhenate, which can be used to recover rhenium from either natural sources or laboratory residues because of its solubility properties and the ease of oxidizing most rhenium compounds to perrhenate (185). Rhenium can also form organometallic compounds with carbonyl or cyclopentadienyl ligands, and shares some similarities with manganese in this respect (188).

Radioisotopes of rhenium have been used as therapeutic agents to treat cancers. ^{186}Re has a half-life of 3.7 days, which is long enough for delivery of an adequate dose to a tumor site while minimizing systemic toxicity (189). Other advantageous characteristics of ^{186}Re as a radiopharmaceutical include the following: 90% of its energy (91% abundance beta particles) is delivered within 2 mm of the source; its gamma radiation is well suited for imaging to confirm tissue distribution; the low energy and abundance, as well as small fraction of higher energy gamma photons, reduces radiation exposure to nontarget organs and to medical personnel; and its rapid decay to stable daughters together with the forgoing features facilitates waste disposal. Various rhenium complexes have been developed and considered for radiopharmaceutical applications (190). For example, ^{186}Re, conjugated to hydroxyethylenediphosphonic acid (HEDP) to ensure selective skeletal localization, has been used to alleviate pain associated with metastatic bone cancer (191). ^{186}Re and ^{188}Re have also been covalently attached to monoclonal antibodies as antitumor agents at various sites (192). ^{188}Re is similar to ^{186}Re but has a shorter half-life of only about 17 h. Bone marrow toxicity is the dose-limiting factor in using ^{186}Re and ^{188}Re in such therapeutic applications (193, 194). Other possible uses of ^{186}Re include treatment for rheumatoid arthritis (195). The biomedical chemistry of rhenium has been reviewed by Dilworth and Parrott (196).

2.3 Exposure Assessment

Incidental information on concentrations of Re dust measured in a (presumably Russian) Re production facility was contained in an abstract by Suvorov (197). During treatment of KReO$_4$, dust concentrations were 0.2–3.5 mg/m^3; during treatment of Re-tungsten and Re-molybdenum alloys, the concentrations were 0.8–45 and 1–100 mg/m^3, respectively. Another report indicates that "In foundries, mechanical industries, and the manufacture of glow lamps, various operations can lead to aerosols of rhenium at concentrations of 0.2–3.5 mg/m^3" [Haguenoer and Furon (198), cited in Jacquet (199)]. Rhenium has not been detected, even at trace concentrations, in any organisms (200). Atomic absorption analytic test methods for Re have been published by the U.S. Environmental Protection Agency (201).

2.4 Toxic Effects

2.4.1 Experimental Studies

Relatively little information exists on the toxicity of rhenium. Hurd et al. (202) performed the first known experiments on the acute toxicity and kinetics of rhenium. Nine mice and six rats were given intraperitoneal injections of potassium perrhenate (KReO$_4$) corresponding to doses of 0.05–3 mg and 2.5–50 mg rhenium, respectively. The authors reported that all of the mice except one "recovered" within 12 h, but they did not identify or describe any effects from which the animals recovered other than water intoxication as an immediate effect of the injection due to the low water solubility of the salt and thus relatively large amounts of the solution that had to be injected. The one fatality occurred immediately upon injection and was not considered an effect of the agent.

In the rats, "no unusual effects were observed in the week following" the injections. Starting at one week after the initial treatment, the rats were given additional injections amounting to 5 mg rhenium, followed by injections of 2.5 mg rhenium daily for 9 days. A single rat was administered 50 mg rhenium, or approximately 200 mg/kg body weight. "No unusual effects were observed in any of the rats" over several weeks.

Hurd et al. (202) also investigated the distribution of rhenium in two rabbits given intravenous injections of the salt. One rabbit received 25 and 45 mg rhenium one-half hour apart and a second rabbit received two injections of 100 mg one-half hour apart. Spectrographic analysis of organs collected 1.5 h after the first injection indicated that rhenium was present in urine, testes, heart, kidney, liver, spleen, and possibly adrenals, but not brain.

Maresh et al. (203) subsequently investigated the toxicity of rhenium in various forms in rodents and dogs. Eleven "immature female rats" were given intraperitoneal injections of sodium perrhenate (NaReO$_4$) solution containing 42 mg rhenium. Six of the rats that received doses ranging from 310 to 600 mg Re/kg body weight showed no visible effects; three rats that received 830–890 mg/kg suffered from respiratory and neurological effects for a few hours; and two rats that received 1050 and 1380 mg/kg died 60 and 30 min later. Effects following the sublethal doses included increased respiration rate and amplitude, labored breathing, and rigid extension of the tail and hind limbs, sometimes alternating with convulsions. These effects lasted a few hours or overnight and disappeared completely without any residual paralysis. The erythropoietic effects of Re were also evaluated by subcutaneous injection of 40–230 mg Re/kg body weight as an aqueous solution of sodium perrhenate (NaReO$_4$) for eight weeks to six male albino rats. Body weight, hemoglobin, erythrocytes, and leucocytes were measured daily in two rats and weekly in the remainder. No effects were evident compared to a baseline period before treatment. A "slight progressive decrease in erythrocytes and hemoglobin" was attributed to a "secondary anemia due to frequent bleedings."

Maresh et al. (203) also administered solutions of rhenium salts (K$_2$ReCl$_6$ and ReCl$_3$) by intraperitoneal injection to mice. The solutions were described as "toxic," that is, presumably lethal, albeit at an unspecified dosage. The authors noted that the toxicity may have been mediated by the liberation of HCl during hydrolysis of the rhenium salts, because "[e]quivalent injections of HCl or FeCl$_3$ killed the mice in the same manner."

The authors also reported the effects of intravenous injections of sodium perrhenate (NaReO$_4$) solution on blood pressure in two dogs under anesthesia and premedicated with morphine. In a 5.5 kg dog given 86 mg Re/kg body weight, mean femoral pressure rose from 108 to 130 mm, and the heart rate rose from 150 to 198 beats per minute. An equal volume of normal saline yielded a 4-mm rise in blood pressure before the Re treatment and a 10-mm rise subsequent to the treatment. Mean pressure in a 13.7 kg dog given 62 mg

Re/kg body weight increased from 118 to 144 mm (compared to a 4 mm increase with saline), and heart rate decreased from 102 to 96 beats per minute (whereas saline yielded an increase from 96 to 102 beats per minute). The injections produced no changes in respiratory rate and amplitude.

After comparable treatments with a solution of sodium molybdenate in 20 rats, the authors concluded that the lethal dose of molybdenum was about one-eighth that of rhenium, 114–117 mg Mo/kg body weight compared to 900–1000 mg Re/kg body weight. They noted that "rhenium surprises observers by its relatively low toxicity and general inertness in the body" compared to some other rare elements (203). In rats, for example, "it lacks the hemopoietic stimulus of cobalt or germanium, develops visible symptoms only in large doses, and produces only small transitory changes in blood pressure."

The comparative toxicity of potassium perrhenate ($KReO_4$) and rhenium trichloride (Re_2Cl_6) was evaluated by Haley and Cartwright (204) in a series of experiments. The LC_{50} in male mice given intraperitoneal injections of $KReO_4$ was 1.8 g Re/kg body weight (range: 1.6–2 g Re/kg body weight) for acute toxicity manifested over a 7-day period as sedation and severe ataxia. The LD_{50} for Re_2Cl_6, was 178 mg Re/kg body weight (range: 168–189 mg Re/kg body weight), and the toxicity was manifested as sedation and abdominal irritation. The perrhenate caused no detectable eye irritation in rabbits by the method of Draize et al. (35); the trichloride was immediately irritating but no effects were evident after 24 h and no permanent damage resulted. Neither the perrhenate nor the trichloride had any effect on abraded or unabraded skin in rabbits, although the trichloride produced a permanent black stain. No effects of the perrhenate were found in isolated rabbit ileum or guinea pig enteric ganglia preparation. The trichloride had no effect at lower doses, but at a dose of 1.5 mg Re/mL the trichloride reportedly decomposed and liberated hydrochloric acid (HCl) that caused the ileum to go into spasm.

Haley and Cartwright (204) also found that $KReO_4$ produced transient hypertension and tachycardia but no change in respiration in cats administered 6.4–32 mg Re/kg body weight. Hypotension and bradycardia occurred at 38–45 mg Re/kg, leading to death by cardiovascular collapse and respiratory failure in four of five cats at 45 mg Re/kg. Atropine (2 mg/kg) increased the doses necessary for these effects by 6–13 mg Re/kg. Various electrocardiographic changes were also induced by $KReO_4$, although atropine had no effect on the cardiac responses. Control studies with KCl indicated that the K^+ ion contributed little to these effects. Contraction of the nictitating membrane in nonatropinized but not atropinized preparations indicated a direct effect of $KReO_4$ at the superior cervical ganglion. Although Re_2Cl_6 administered to cats produced a slight hypertension at 6 mg Re/kg and was lethal at 13 mg Re/kg, these effects, it was concluded, were a function of the HCl acid liberated by decomposition of Re_2Cl_6 because the effects were duplicated by administering equivalent concentrations of HCl to control animals.

Suvorov (197, 205) reported the results of a series of experiments in which Re was administered to rats by different routes. According to an English abstract (197), inhalation of Re dust for 3–6 months yielded a "thick fibrosis in the lungs." A "single inhalation" of Re condensates at 20 mg/m^3 resulted in an "acute interstitial process in the lungs." At 6 mg/m^3, limited toxicity was indicated by "indexes of protein exchange." The LD_{50} of finely dispersed Re by intraperitoneal injection was 10,000 mg/kg in mice; the LC_{50} for

MANGANESE AND RHENIUM

KReO$_4$ was 692 mg/kg. Although Re metal dust was judged comparable to tungsten in toxicity, the aerosol of the condensates had "noticeably high toxicity."

2.4.1.1 Pharmacokinetics, Metabolism, and Mechanisms. A few studies focused on the metabolism and pharmacokinetics of rhenium. Baumann et al. (206), Shellabarger (207), and Durbin et al. (200) administered radiolabeled Re to rats by intraperitoneal or intravenous injection. In all three studies, Re was rapidly and actively taken up by the thyroid to a significantly greater extent than by any other organ, the peak tissue concentration was generally measured at 1-2 hours after injection, and concentrations declined rapidly thereafter. In the study by Baumann et al. (206), the ratio of Re in thyroid to blood ranged from 25 to 100 in normal and thiouracil-treated rats after a single intraperitoneal injection, whereas the organ to blood ratio of Re in lung, liver, and muscle was consistently less than one. Shellabarger (207) observed even greater ratios and concluded that no significant amounts of Re were found in any organ or tissue other than the thyroid and gastrointestinal tract. Shellabarger also found, however, that the thyroid took up 10–500 times more coadministered radioactive iodine than Re, and concluded that, unlike iodine, Re is not organically bound by the thyroid. Most of the Re was excreted in urine. Baumann et al. (206) recovered 92–96% of injected Re from three rabbits over two days by expressing the bladders daily. Fecal elimination was negligible. Durbin et al. (200) found that 92% of intravenously injected Re was excreted in urine of rats within 24 hours. Over a 16-day period, 94% was recovered in urine and 5% in feces. Bardfeld and Shulman (208) focused on the transport and distribution of Re in the CNS of adult cats as a potential method for choroid plexus ablation in the control of hydrocephalus. Rhenium-188 perfused via the lateral ventricle concentrated in the choroid plexus; it was also found to a much lesser extent (80:1) in the cerebral cortex and cerbrospinal fluid, and a comparable amount was found extracranially in the thyroid. The pharmacokinetics of ^{186}Re were investigated in patients who had bone cancer (209, 210). Renal clearance was approximately 70% within 1-3 days, but distribution in these patients is likely to be different from that of normal individuals.

2.4.2 Human Experience

Nadler (211) asserted without documentation that "People who have worked over a long period of time in the extraction and production of rhenium metal and its most important compounds, perrhenic acid and ammonium perrhenate, have not shown any toxic effects." Jacquet (199) stated that "There are no reports of cases of human toxicity." The only other information on the effects of rhenium on humans concerns therapeutic applications of radioactive isotopes and their side effects. A Material Safety Data Sheet issued by the Rembar Company, Inc. (212) states that the dust or powder form of rhenium may cause eye and mucus membrane irritation. The MSDS indicates that no information is available regarding health hazards.

2.5 Standards, Regulations, or Guidelines of Exposure

No information on regulatory or guidance values for rhenium was located.

2.6 Studies on Environmental Impact

No information on environmental impacts of rhenium was located.

BIBLIOGRAPHY

1. J. G. F. Druce, (1948) *Rhenium: Dvi-manganese, the Element of Atomic Number 75*, Cambridge University Press, Cambridge, UK, 1948.
2. P. M., Treichel, Jr., Rhenium and rhenium compounds. In J. I. Kroschwitz, ed., *Kirk-Othmer Encyclopedia of Chemical Technology*, 4th ed., Vol. 21, Wiley, New York, 1997, pp. 335–346.
3. C. L. Keen, and S. Zidenberg-Cherr, Manganese toxicity in humans and experimental animals. In D. J. Klimis-Tavantzis, ed., *Manganese in Health and Disease*, CRC Press, Boca Raton, FL, 1994, pp. 193–205.
4. L. R. Matricardi and J. Downing, Manganese and manganese alloys. In J. I. Kroschwitz and M. Howe-Grant, eds., *Kirk-Othmer Encyclopedia of Chemical Technology*, 4th ed., Vol. 15, Wiley, New York, 1995, pp. 963–990.
5. J. M. Davis, Methylcyclopentadienyl manganese tricarbonyl: Health risk uncertainties and research directions. *Environ. Health Perspect.* **106**(Suppl. 1), 191–201 (1998).
5a. NIOSH *Manual of Analytical Methods*, 4th ed., 1994.
6. F. C. Wedler, Biochemical and nutritional role of manganese: An overview. In D. J. Klimistavantzis, ed., *Manganese in Health and Disease*. CRC Press, Boca Raton, FL, 1994, pp. 1–37.
7. C. L. Keen et al., Nutritional aspects of manganese from experimental studies. *NeuroToxicology* **20**, 213–224 (1999).
8. National Research Council, *Recommended Dietary Allowances*, 10th ed., National Academy of Sciences, Washington, DC, 1989, pp. 230–235.
9. J. L. Greger, Nutrition versus toxicology of manganese in humans: Evaluation of potential biomarkers. *Neurotoxicology* **20**, 205–212 (1999).
10. M. Aschner, Manganese homeostasis in the CNS. *Environ. Res.* **80**, 105–109 (1999).
11. E. A. Malecki et al., Existing and emerging mechanisms for transport of iron and manganese to the brain. *J. Neurosci. Res.* **56**, 113–122 (1999).
12. D. E. McMillan, A brief history of the neurobehavioral toxicity of manganese: Some unanswered questions. *NeuroToxicology* **20**, 499–508 (1999).
13. M. C. Newland, Animal models of manganese's neurotoxicity. *NeuroToxicology* **20**, 415–432 (1999).
14. W. K. Boyes and D. B. Miller, A review of rodent models of manganese neurotoxicity. In J. S. Cranmer, D. Mergler and M. Williams-Johnson, eds., *Manganese: Are There Effects from Long-term, Low-level Exposures?*, Proc. 15th Int. Neurotoxicol. Conf., Little Rock, AR, 1997.
15. H. Roels et al., Epidemiological survey among workers exposed to manganese: Effects on lung, central nervous system, and some biological indices. *Am. J. Ind. Med.* **11**, 307–327 (1987).
16. H. A. Roels et al., Assessment of the permissible exposure level to manganese in workers exposed to manganese dioxide dust. *Br. J. Ind. Med.* **49**, 25–34 (1992).
17. J. Couper, On the effects of black oxide of manganese when inhaled into the lungs. *Br. Ann. Med. Pharm.* **1**, 41–42 (1837).

18. R. Kawamura et al., Intoxication by manganese in well water. *Kitasato Arch. Exp. Med.* **18**, 145–169 (1941).
19. S. A. Mirowitz, T. J. Westrich, and J. D. Hirsch, Hyperintense basal ganglia on T1-weighted MR images in patients receiving parenteral nutrition. *Radiology* **181**, 117–120 (1991).
20. A. Ejima et al., Manganese intoxication during total parenteral nutrition. *Lancet* **8790**, 426 (1992).
21. H. Komaki et al., Tremor and seizures associated with chronic manganese intoxication. *Brain Dev.* **21**, 122–124 (1999).
22. S. Nagatomo et al., Manganese intoxication during total parenteral nutrition: Report of two cases and review of the literature. *J. Neurol. Sci.* **162**, 102–105 (1999).
23. P. D. Blanc, Lesson 1, Vol. 12 — Inhalation fever. Pulmonary and Critical Care Update Online (1999). Available: www.chestnet.org/education/pccu/
24. R. N. Shiotsuka, (1948) *Inhalation Toxicity of Manganese Dioxide and a Magnesium Oxide — Manganese Dioxide Mixture*, U.S. Army Project Order No. 3803, Brookhaven Nat. Lab. Rep. No. BNL 35334, U.S. Army Medical Research and Development Command, 1984. Available from: NTIS, Springfield, VA (AD-A148 868-3-XAD).
25. R. Bergstrm, Acute pulmonary toxicity of manganese dioxide. *Scand. J. Work Environ. Health* **3**(Suppl.), 1–41 (1977).
26. B. Adkins, Jr. et al., Biochemical changes in pulmonary cells following manganese oxide inhalation. *J. Toxicol. Environ. Health* **6**, 445–454 (1980).
27. B. Adkins, Jr., G. H. Luginbuhl, and D. E. Gardner, Acute exposure of laboratory mice to manganese oxide. *Am. Ind. Hyg. Assoc. J.* **41**, 494–500 (1980).
28. B. Adkins, Jr. et al., Increased pulmonary susceptibility to streptococcal infection following inhalation of manganese oxide. *Environ. Res.* **23**, 110–120 (1980).
29. R. Z. Maigetter et al., Potentiating effects of manganese dioxide on experimental respiratory infections. *Environ. Res.* **11**, 386–391 (1976).
30. R. K. Hinderer, Toxicity studies of methylcyclopentadienyl manganese tricarbonyl (MMT). *Am. Ind. Hyg. Assoc. J.* **40**, 164–167 (1979).
31. U.S. Environmental Protection Agency (USEPA), *Reevaluation of Inhalation Health Risks Associated with Methylcyclopentadienyl Manganese Tricarbonyl (MMT) in Gasoline* (final), EPA Rep. no. EPA/600/R-94/062, USEPA, Office of Research and Development, Washington, DC, 1994.
32. J. M. Davis, Inhalation health risks of manganese: An EPA perspective. *NeuroToxicology* **20**, 511–518 (1999).
33. J. M. Davis et al., The EPA health risk assessment of methylcyclopentadienyl manganese tricarbonyl (MMT). *Risk Anal.* **18**, 57–70 (1998).
34. American Conference of Governmental Industrial Hygienists (ACGIH), Magnesite; magnesium oxide fume. In *Documentation of the Threshold Limit Values and Biological Exposure Indicies*, 6th ed., ACGIH, Cincinnati, OH; 1991, pp. 867–870.
35. J. H. Draize, G. Woodard, and H. O. Calvery, *J. Pharmacol. Exp. Ther.* **82**, 377 (1944).
36. D. K. Hysell et al., Oral toxicity of methylcyclopentadienyl manganese tricarbonyl (MMT) in rats. *Environ. Res.* **7**, 158–168 (1974).
37. R. P. Hanzlik et al., Biotransformation and excretion of methylcyclopentadienyl manganese tricarbonyl in the rat. *Drug Metab. Dispos.* **8**, 428–433 (1980).

38. R. P. Hanzlik, R. Stitt, and G. J. Traiger, Toxic effects of methylcyclopentadienyl manganese tricarbonyl (MMT) in rats: Role of metabolism. *Toxicol. Appl. Pharmacol.* **56**, 353–360 (1980).
39. P. J. Hakkinen and W. M. Haschek, Pulmonary toxicity of methylcyclopentadienyl manganese tricarbonyl: Nonciliated bronchiolar epithilial (clara) cell necrosis and alveolar damage in the mouse, rat, and hamster. *Toxicol. Appl. Pharmacol.* **65**, 11–22 (1982).
40. P. J. Hakkinen et al., Potentiating effects of oxygen in lungs damaged by methylcyclopentadienyl manganese tricarbonyl, cadmium chloride, oleic acid, and antitumor drugs. *Toxicol. Appl. Pharmacol.* **67**, 55–69 (1983).
41. P. A. McGinley et al., Disposition and toxicity of methylcyclopentadienyl manganese tricarbonyl in the rat. *Toxicol. Lett.* **36**, 137–145, (1987).
42. R. J. Clay and J. B. Morris, Comparative pneumotoxicity of cyclopentadienyl manganese tricarbonyl and methylcyclopentadienyl manganese tricarbonyl. *Toxicol. Appl. Pharmacol.* **98**, 434–443 (1989).
43. B. E. Fishman, P. A. McGinley, and G. Gianutsos, Neurotoxic effects of methylcyclopentadienyl manganese tricarbonyl (MMT) in the mouse: Basis of MMT-induced seizure activity. *Toxicology* **45**, 193–201 (1987).
44. J. Komura and M. Sakamoto, Chronic oral administration of methylcyclopentadienyl manganese tricarbonyl altered brain biogenic amines in the mouse: Comparison with inorganic manganese. *Toxicol. Lett.* **73**, 65–73 (1994).
45. D. A. Penney et al., The acute toxicity of cyclopentadienyl manganese tricarbonyl in the rat. *Toxicology* **34**, 341–347 (1985).
46. K. T. Blanchard, R. J. Clay, and J. B. Morris, Pulmonary activation and toxicity of cyclopentadienyl manganese tricarbonyl. *Toxicol. Appl. Pharmacol.* **136**, 280–288 (1996).
47. L. Casamajor, An unusual form of mineral poisoning affecting the nervous system: Manganese? *J. Am. Med. Assoc.* **69**, 646–649 (1913).
48. R. H. Flinn, P. A. Neal, and W. B. Fulton, Industrial manganese poisoning. *J. Ind. Hyg. Toxicol.* **23**, 374–387 (1941).
49. J. Rodier, Manganese poisoning in Moroccan miners. *Br. J. Ind. Med.* **12**, 21–35 (1955).
50. I. Mena et al., Chronic manganese poisoning: Clinical picture and manganese turnover. *Neurology* **17**, 128–136 (1967).
51. D. G. Cook, S. Fahn, and K. A. Brait, Chronic manganese intoxication. *Arch. Neurol. (Chicago)* **30**, 59–64 (1974).
52. C.-C. Huang et al., Chronic manganese intoxication. *Arch. Neurol. (Chicago)* **46**, 1104–1106 (1989).
53. I. Mena et al., Chronic manganese poisoning: Individual susceptibility and absorption of iron. *Neurology* **19**, 1000–1006 (1969).
54. B. Lonnerdal, Dietary factors affecting trace element absorption in infants. *Acta Paediatr. Scand., Suppl.* **351**, 109–113 (1989).
55. V. A. Murphy et al., Elevation of brain manganese in calcium-deficient rats. *NeuroToxicology* **12**, 255–264 (1991).
56. A. M. Scheuhammer and M. G. Cherian, The distribution and excretion of manganese: The effects of manganese dose, L-dopa and pretreatment with zinc. *Toxicol. Appl. Pharmacol.* **65**, 203–213 (1982).

57. K. J. Wedekind et al., Phosphorus, but not calcium, affects manganese absorption and turnover in chicks. *J. Nutr.* **121**, 1776–1786 (1991).
58. W. J. Weiner, P. A. Nausieda, and H. L. Klawans, The effect of levodopa, lergotrile, and bromocriptine on brain iron, manganese, and copper. *Neurology* **28**, 734–737 (1978).
59. R. S. Srivastava, R. C. Murthy, and S. V. Chandra, Effect of manganese on some bioantioxidants in various organs of protein-deficient rats. *Biochem. Int.* **18**, 903–912 (1989).
60. R. K. Randhawa and B. L. Kawatra, Effect of dietary protein on the absorption and retention of Zn, Fe, Cu and Mn in pre-adolescent girls. *Nahrung* **37**, 399–407 (1993).
61. K. Kostial et al., Effect of a metal mixture in diet on the toxicokinetics and toxicity of cadmium, mercury and manganese in rats. *Toxicol. Ind. Health* **5**, 685–698 (1989).
62. G. S. Shukla and S. V. Chandra, Concurrent exposure to lead, manganese, and cadmium and their distribution to various brain regions, liver, kidney, and testis of growing rats. *Arch. Environ. Contam. Toxicol.* **16**, 303–310 (1987).
63. A. C. G. Chua and E. H. Morgan, Manganese metabolism is impaired in the Belgrade laboratory rat. *J. Comp. Physiol. B* **167**, 361–369 (1997).
64. A. Barbeau et al., Ecogenetics of Parkinson's disease: 4-hydroxylation of debrisoquine. *Lancet* **2**, 1213–1216 (1985).
65. I. Kondo and I. Kanazawa, Association of Xba I allele (Xba I 44 kb) of the human cytochrome P-450dbl (CYP2D6) gene in Japanese patients with idiopathic Parkinson's disease. In T. Nagatsu, H. Narabayashi, and M. Yoshida, eds., *Parkinson's Disease: From Clinical Aspects to Molecular Basis*, Springer-Verlag, New York, 1991 pp. 111–117.
66. I. Kondo and I. Kanazawa, Debrisoquine hydroxylase and Parkinson's disease. *Adv. Neurol.* **60**, 338–342 (1993).
67. M. Armstrong et al., Mutant debrisoquine hydroxylation genes in Parkinson's disease. *Lancet* **339**, 1017–1018 (1992).
68. C. A. D. Smith et al., Debrisoquine hydroxylase gene polymorphism and susceptibility to Parkinson's disease *Lancet* **339**, 1375–1377 (1992); (published erratum: *Ibid.* **340**, 64) (1992).
69. Y. Tsuneoka et al., A novel cytochrome P-450IID6 mutant gene associated with Parkinson's disease. *J. Biochem. (Tokyo)* **114**, 263–266 (1993).
70. R. A. Hauser et al., Blood manganese correlates with brain magnetic resonance imaging changes in patients with liver disease. *Can. J. Neurol. Sci.* **23**, 95–98 (1996).
71. D. Krieger et al., Manganese and chronic hepatic encephalopathy. *Lancet* **346**, 270–274 (1995).
72. W.-C. Sue, C.-Y. Chen, and C.-C. Chen, Dyskinesia from manganism in a hepatic dysfunction patient. *Zhonghua Minguo Xiaoerke Yixuehui Zazhi* **37**, 59–64 (1996).
73. T. F. Barron, A. G. Devenyi, and A. C. Mamourian, Symptomatic manganese neurotoxicity in a patient with chronic liver disease: Correlation of clinical symptoms with MRI findings. *Pediatr. Neurol.* **10**, 145–148 (1994).
74. E. A. Malecki et al., Iron and manganese homeostasis in chronic liver disease: Relationship to pallidal T1-weighted magnetic resonance signal hyperintensity. *NeuroToxicology* **20**, 647–652 (1999).
75. M. Gochfeld, Factors influencing susceptibility to metals. *Environ. Health Perspect., Suppl.* **105**(4), 817–822 (1997).
76. P. K. Pal, A. Samii, and D. B. Calne, Manganese neurotoxicity: A review of clinical features, imaging and pathology. *NeuroToxicology* **20**, 227–238 (1999).

77. J. R. Sanchez-Ramos, Toxin-induced parkinsonism. In M. B. Stern and W. C. Koller, eds., *Neurological Disease and Therapy*, Vol. 18, Dekker, New York, 1993, pp. 155–171.
78. F. Hochberg et al., Late motor deficits of Chilean manganese miners: A blinded control study. *Neurology* **47**, 788–795 (1996).
79. G. Jonderko, A. Kujawska, and H. Langauer-Lewowicka, Problems of chronic manganese poisoning on the basis of investigations of workers at a manganese alloy foundry. *Int. Arch. Arbeitsmed.* **28**, 250–264 (1971).
80. L. T. Smyth et al., Clinical manganism and exposure to manganese in the production and processing of ferromanganese alloy. *J. Occup. Med.* **15**, 101–109 (1973).
81. A. Barbeau, Manganese and extrapyramidal disorders (a critical review and tribute to Dr. George C. Cotzias). *NeuroToxicology* **5**, 13–36 (1984).
82. J. W. Langston, I. Irwin, and G. A. Ricaurte, Neurotoxins, Parkinsonism and Parkinson's disease. *Pharmacol. Ther.* **32**, 19–49 (1987).
83. D. B. Calne et al., Manganism and idiopathic parkinsonism: Similarities and differences. *Neurology* **44**, 1583–1586 (1994).
84. E. C. Wolters et al., Positron emission tomography in manganese intoxication. *Ann. Neurol.* **26**, 647–651 (1989).
85. K. Nelson et al., Manganese encephalopathy: Utility of early magnetic resonance imaging. *Br. J. Ind. Med.* **50**, 510–513 (1993).
86. A. Iregren, Manganese neurotoxicity in industrial exposures: Proof of effects, critical exposure level, and sensitive tests. *NeuroToxicology* **20**, 315–324 (1999).
87. H. Roels, [Letter to Dr. M. Davis, U.S. EPA, on definitions of "respirable," "total," and "inhalable" dusts]. Bruxelles, Belgium: Université Catholique de Louvain, Unité de Toxicologie Industrielle et Médecine du Travail, 1993.
88. D. Mergler et al., Nervous system dysfunction among workers with long-term exposure to manganese. *Environ. Res.* **64**, 151–180 (1994).
89. R. Lucchini et al., Motor function, olfactory threshold, and hematological indices in manganese-exposed ferroalloy workers. *Environ. Res.* **73**, 175–180 (1997).
90. A. Iregren, Psychological test performance in foundry workers exposed to low levels of manganese. *Neurotoxicol. Teratol.* **12**, 673–675 (1990).
91. D. Mergler et al., Manganese neurotoxicity, a continuum of dysfunction: Results from a community based study. *NeuroToxicology* **20**, 327–342 (1999).
92. A. Wennberg et al., Manganese exposure in steel smelters a health hazard to the nervous system. *Scand. J. Work Environ. Health* **17**, 255–262 (1991).
93. A. Wennberg, M. Hagman, and L. Johansson, Preclinical neurophysiological signs of parkinsonism in occupational manganese exposure. *NeuroToxicology* **13**, 271–274 (1992).
94. R. Lucchini et al., Neurobehavioral effects of manganese in workers from a ferroalloy plant after temporary cessation of exposure. *Scand. J. Work Environ. Health* **21**, 143–149 (1995).
95. R. Lucchini et al., Long term exposure to "low levels" of manganese oxides and neurofunctional changes in ferroalloy workers. *NeuroToxicology* **20**, 287–298 (1999).
96. B. Sjgren, P. Gustavsson, and C. Hogstedt, Neuropsychiatric symptoms among welders exposed to neurotoxic metals. *Br. J. Ind. Med.* **47**, 704–707 (1990).
97. J.-D. Wang et al., Manganese induced parkinsonism: An outbreak due to an unrepaired ventilation control system in a ferromanganese smelter. *Br. J. Ind. Med.* **46**, 856–859 (1989).

98. A. B. N. Badawy and A. A. Shakour, Chronic manganese intoxication (neurological manifestations). *Trace Elem. Man Anim. —TEMA 5, Proc. 5th Inter. Sympo., 1984* (1985), pp. 261–263.

99. P. Sigel and K. D. Bergert, Eine frühdiagnostische berwachungsmethode bei Manganexposition [An early diagnostic control method at exposition with manganese]. *Z. Gesamte Hyg. Ihre Grenzgeb.* **28**, 524–526 (1982).

100. S. V. Chandra et al., An exploratory study of manganese exposure to welders. *Clin. Toxicol.* **18**, 407–416 (1981).

101. M. Saric, A. Markicevic, and O. Hrustic, Occupational exposure to manganese. *Br. J. Ind. Med.* **34**, 114–118 (1977).

102. A. M. Emara et al., Chronic manganese poisoning in the dry battery industry. *Br. J. Ind. Med.* **28**, 78–82 (1971).

103. S. Tanaka and J. Lieben, Manganese poisoning and exposure in Pennsylvania. *Arch. Environ. Health* **19**, 674–684 (1969).

104. P. Schuler et al., Manganese poisoning: Environmental and medical study at a Chilean mine. *Ind. Med. Surg.* **26**, 167–173 (1957).

105. J. P. Gibbs et al., Focused medical surveillance: A search for subclinical movement disorders in a cohort of U.S. workers exposed to low levels of manganese dust. *NeuroToxicology* **20**, 299–314 (1999).

106. A. Iregren, Using psychological tests for the early detection of neurotoxic effects of low level manganese exposure. *NeuroToxicology* **15**, 671–677 (1994).

107. K. S. Crump and P. Rousseau, Results from eleven years of neurological health surveillance at a manganese oxide and salt producing plant. *NeuroToxicology* **20**, 273–286 (1999).

108. M. S. Desole et al., Cellular defence mechanisms in the striatum of young and aged rats subchronically exposed to manganese. *Neuropharmacology* **34**, 289–295 (1995).

109. H. A. Roels et al., Prospective study on the reversibility of neurobehavioral effects in workers exposed to manganese dioxide. *NeuroToxicology* **20**, 255–272 (1999).

110. C.-C. Huang et al., Progression after chronic manganese exposure. *Neurology* **43**, 1479–1483 (1993).

111. C.-C. Huang et al., Long-term progression in chronic manganism: Ten years of follow-up. *Neurology* **50**, 698–700 (1998).

112. H. Eriksson et al., Effects of manganese oxide on monkeys as revealed by a combined neurochemical, histological and neurophysiological evaluation. *Arch. Toxicol.* **61**, 46–52 (1987).

113. Y. Suzuki et al., Study of subacute toxicity of manganese dioxide in monkeys. *Tokushima J. Exp. Med.* **22**, 5–10 (1975).

114. C. W. Olanow et al., Manganese intoxication in the rhesus monkey: A clinical imaging, pathologic, and biochemical study. *Neurology* **46**, 492–498 (1996).

115. H. Shinotoh et al., MRI and PET studies of manganese-intoxicated monkeys. *Neurology* **45**, 1199–1204 (1995).

116. E. D. Bird, A. H. Anton, and B. Bullock, The effect of manganese inhalation on basal ganglia dopamine concentrations in rhesus monkey. *NeuroToxicology* **5**, 59–65 (1984).

117. F. S. Archibald and C. Tyree, Manganese poisoning and the attack of trivalent manganese upon catecholamines. *Arch. Biochem. Biophys.* **256**, 638–650 (1987).

118. J. Donaldson and A. Barbeau, (1985) Manganese neurotoxicity: Possible clues to the etiology of human brain disorders. In S. Gabay, J. Harris, and B. T. Ho, eds., *Metal Ions in Neurology and Psychiatry*, Neurol. Neurobiol. Ser., Vol. 15, Alan R. Liss, New York, pp. 259–285.
119. J. Donaldson, D. McGregor, and F. LaBella, Manganese neurotoxicity: A model for free radical mediated neurodegeneration? *Can. J. Physiol. Pharmacol.* **60**, 1398–1405 (1982).
120. H. Eriksson et al., Manganese induced brain lesions in *Macaca fascicularis* as revealed by positron emission tomagraphy and magnetic resonance imaging. *Arch. Toxicol.* **66**, 403–407 (1992).
121. M. Yamada et al., Chronic manganese poisoning: A neuropathological study with determination of manganese distribution in the brain. *Acta Neuropathol.* **70**, 273–278 (1986).
122. M. A. Verity, Manganese neurotoxicity: A mechanistic hypothesis. *NeuroToxicology* **20**, 489–498 (1999).
123. M. C. Newland et al., The clearance of manganese chloride in the primate. *Fundam. Appl. Toxicol.* **9**, 314–328 (1987).
124. D. K. Dastur, D. K. Manghani, and K. V. Raghavendran, Distribution and fate of ^{54}Mn in the monkey: Studies of different parts of the central nervous system and other organs. *J. Clin. Invest.* **50**, 9–20 (1971).
125. I. Sziraki et al., Implications for atypical antioxidative properties of manganese in iron-induced brain lipid peroxidation and copper-dependent low density lipoprotein conjugation. *NeuroToxicology* **20**, 455–466 (1999).
126. G. C. Cotzias et al., Interactions between manganese and brain dopamine. *Med. Clin. North Am.* **60**, 729–738 (1976).
127. E. P. Brouillet et al., Manganese injection into the rat striatum produces excitotoxic lesions by impairing energy metabolism. *Exp. Neurol.* **120**, 89–94 (1993).
128. K. A. Brenneman et al., (1999) Manganese-induced developmental neurotoxicity in the CD rat: Is oxidative damage a mechanism of action? *NeuroToxicology* **20**, 477–488 (1999).
129. C. E. Gavin, K. K. Gunter, and T. E. Gunter, Manganese and calcium transport in mitochondria: Implications for manganese toxicity. *NeuroToxicology* **20**, 445–454 (1999).
130. M. Aschner and J. L. Aschner, Manganese neurotoxicity: Cellular effects and blood-brain barrier transport. *Neurosci. Biobehav. Rev.* **15**, 333–340 (1991).
131. C. E. Gavin, K. K. Gunter, and T. E. Gunter, Manganese and calcium efflux kinetics in brain mitochondria: Relevance to manganese toxicity. *Biochem. J.* **266**, 329–334 (1990).
132. G. C. Cotzias et al., Chronic manganese poisoning: Clearance of tissue manganese concentrations with persistence of the neurological picture. *Neurology* **18**, 376–382 (1968).
133. National Academy of Sciences, *Manganese*, National Academy of Sciences, Washington, DC, (1973).
134. T. A. Lloyd Davies, Manganese pneumonitis. *Br. J. Ind. Med.* **3**, 111–135 (1946).
135. T. A. Lloyd Davies and H. E. Harding, Manganese pneumonitis: Further clinical and experimental observations. *Br. J. Ind. Med.* **6**, 82–90 (1949).
136. M. Saric and S. Lucic-Palaic, Possible synergism of exposure to airborne manganese and smoking habit in occurrence of respiratory symptoms. In W. H. Walton, ed., *Inhaled Particles IV* (in two parts), Proc. Int. Symp. Part 2, Pergamon, Oxford, UK, 1977, pp. 773–779.
137. K. Nishiyama et al., [Effect of long-term inhalation of manganese dusts. II. Continuous observation of the respiratory organs of monkeys and mice]. *Nippon Eiseigaku Zasshi* **30**, 117 (1975).

138. Y. Suzuki et al., Effects of the inhalation of manganese dioxide dust on monkey lungs. *Tokushima J. Exp. Med.* **25**, 119–125 (1978).
139. W. Moore et al., Exposure of laboratory animals to atmospheric manganese from automotive emissions. *Environ. Res.* **9**, 274–284 (1975).
140. C. E. Ulrich, W. Rinehart, and W. Busey, Evaluation of the chronic inhalation toxicity of a manganese oxide aerosol. I. introduction, experimental design, and aerosol generation methods. *Am. Ind. Hyg. Assoc. J.* **40**, 238–244 (1979).
141. C. E. Ulrich et al., Evaluation of the chronic inhalation toxicity of a manganese oxide aerosol. II. Clinical observations, hematology, clinical chemistry and histopathology. *Am. Ind. Hyg. Assoc. J.* **40**, 322–329 (1979).
142. C. E. Ulrich, W. Rinehart, and M. Brandt, Evaluation of the chronic inhalation toxicity of a manganese oxide aerosol. III. Pulmonary function, electromyograms, limb tremor, and tissue manganese data. *Am. Ind. Hyg. Assoc. J.* **40**, 349–353 (1979).
143. P. Camner et al., Rabbit lung after inhalation of manganese chloride: A comparison with the effects of chlorides of nickel, cadmium, cobalt, and copper. *Environ. Res.* **38**, 301–309 (1985).
144. Y. Suzuki et al., A labor hygiene survey of ferromanganese workers. Part I. *Shikoku Acta Med.* **29**, 412–424 (1973).
145. Y. Suzuki et al., A labor hygiene survey of ferromanganese workers. Part 2. *Shikoku Acta Med.* **29**, 433–438 (1973).
146. Y. Suzuki et al., Studies on the amounts of manganese in the blood and urine of ferromanganese workers. *Shikoku Acta Med.* **29**, 425–432 (1973).
147. Y. Jiang et al., Effects of manganese on the sexual function and reproductive outcome of male exposed workers. *Zhonghua Laodong Weisheng Zhiyebing Zazhi* **14**, 271–273 (1996).
148. R. Lauwerys et al., Fertility of male workers exposed to mercury vapor or to manganese dust: A questionnaire study. *Am. J. Ind. Med.* **7**, 171–176 (1985).
149. J. P. Gennart et al., Fertility of male workers exposed to cadmium, lead, or manganese. *Am. J. Epidemiol.* **135**, 1208–1219 (1992).
150. L. Alessio et al., Interference of manganese on neuroendocrinal system in exposed workers: Preliminary report. *Biol. Trace Elem. Res.* **21**, 249–253 (1989).
151. S. V. Chandra et al., Sterility in experimental manganese toxicity. *Acta Biol. Med. Ger.* **30**, 857–862 (1973).
152. P. K. Seth et al., Effects of manganese on rabbit testes. *Environ. Physiol. Biochem.* **3**, 263–267 (1973).
153. J. W. Laskey et al., Effects of chronic manganese (Mn_3O_4) exposure on selected reproductive parameters in rats. *J. Toxicol. Environ. Health* **9**, 677–687 (1982).
154. L. E. Gray, Jr. and J. W. Laskey, Multivariate analysis of the effects of manganese on the reproductive physiology and behavior of the male house mouse. *J. Toxicol. Environ. Health* **6**, 861–867 (1980).
155. M. Hejtmancik et al., *The Chronic Study of Manganese Sulfate Monohydrate (CAS No. 10034-96-5) in F322 Rats*, Report to National Toxicology Program, Research Triangle Park, NC, by Battelle Columbus Laboratories, Columbus, OH, 1987.
156. M. Hejtmancik et al., *The Chronic Study of Manganese Sulfate Monohydrate (CAS No. 10034-96-5) in B6C$_3$F$_1$ Mice*, Report to National Toxicology Program, Research Triangle Park, NC, by Battelle Columbus Laboratories, Columbus, OH, 1987.

157. B. A. Lown et al., Effects on the postnatal development of the mouse of preconception, postconception and/or suckling exposure to manganese via maternal inhalation exposure to MnO$_2$ dust. *NeuroToxicology* **5**, 119–131 (1984).
158. P. J. Kontur and L. D. Fechter, Brain regional manganese levels and monoamine metabolism in manganese-treated neonatal rats. *Neurotoxicol. Teratol.* **10**, 295–303 (1988).
159. R. Deskin, S. J. Bursian, and F. W. Edens, Neurochemical alterations by manganese chloride in neonatal rats. *NeuroToxicology* **2**, 65–73 (1980).
160. R. Deskin, S. J. Bursian, and F. W. Edens, The effect of chronic manganese administration on some neurochemical and physiological variables in neonatal rats. *Gen. Pharmacol.* **12**, 279–280 (1981).
161. J. C. K. Lai, T. K. C. Leung, and L. Lim, Differences in the neurotoxic effects of manganese during development and aging: Some observations on brain regional neurotransmitter and non-neurotransmitter metabolism in a developmental rat model of chronic manganese encephalopathy. *NeuroToxicology* **5**, 37–47 (1984).
162. M. Saric and O. Hrustic, Exposure to airborne manganese and arterial blood pressure. *Environ. Res.* **10**, 314–318 (1975).
163. Y. Jiang et al., Effects of managanese exposure on egg and blood pressure. *Zhonghua Laodong Weisheng Zhiyebing Zazhi* **22**, 341–343 (1996).
164. H. Brurok et al., Manganese and the heart: Acute cardiodepression and myocardial accumulation of manganese. *Acta Physiol. Scand.* **159**, 33–40 (1997).
165. G. D. Stoner et al., Test for carcinogenicity of metallic compounds by the pulmonary tumor response in strain A mice. *Cancer Res.* **36**, 1744–1747 (1976).
166. A. Furst, Tumorigenic effect of an organomanganese compound on F344 rats and Swiss albino mice: Brief communication. *J. Natl. Cancer Inst.* **60**, 1171–1173 (1978).
167. F. W. Sunderman, Jr., T. J. Lau, and L. J. Cralley, Inhibitory effect of manganese upon muscle tumorigenesis by nickel subsulfide. *Cancer Res.* **34**, 92–95 (1974).
168. F. W. Sunderman, Jr. et al., Effects of manganese on carcinogenicity and metabolism of nickel subsulfide. *Cancer Res.* **36**, 1790–1800 (1976).
169. F. W. Sunderman, Jr. et al., Manganese inhibition of sarcoma induction by benzo(a)pyrene in Fischer rats. *Proc. Am. Assoc. Cancer Res.* **21**, 72 (1980).
170. H. P. Witschi, P. J. Hakkinen, and J. P. Kehrer, Modification of lung tumor development in A/J mice. *Toxicology* **21**, 37–45 (1981).
171. Integrated Risk Information System (IRIS) [database], Printout of carcinogenicity assessment for lifetime exposure to manganese as verified 5/25/88, U.S. Environmental Protection Agency, Office of Health and Environmental Assessment, Environmental Criteria and Assessment Office, Cincinnati, OH, 1988. Available online from: TOXNET, National Library of Medicine, Bethesda, MD.
172. L. Davidsson et al., Manganese retention in man: A method for estimating manganese absorption in man. *Am. J. Clin. Nutr.* **49**, 170–179 (1989).
173. M. Aschner, K. E. Vrana, and W. Zheng, Manganese uptake and distribution in the central nervous system (CNS). *Neurotoxicology* **20**, 173–180 (1999).
174. D. B. Drown, S. G. Oberg, and R. P. Sharma, Pulmonary clearance of soluble and insoluble forms of manganese. *J. Toxicol. Environ. Health* **17**, 201–212 (1986).
175. D. Mergler and M. Baldwin, Early manifestations of manganese neurotoxicity in humans: An update. *Environ. Res.* **73**, 92–100 (1997).

176. A. Smargiassi, and A. Mutti, Peripheral biomarkers and exposure to manganese. *NeuroToxicology* **20**, 401–406 (1999).
177. M. E. Anderson, J. M. Gearheart, and H. J. Clewell, III, Pharmacokinetic data needs to support risk assessments for inhaled and ingested manganese. *NeuroToxicology* **20**, 161–172 (1999).
178. G. Gianutsos et al., Brain manganese accumulation following systemic administration of different forms. *Arch. Toxicol.* **57**, 272–275 (1985).
179. J. Komura and M. Sakamoto, Subcellular and gel chromatographic distribution of manganese in the mouse brain: Relation to the chemical form of chronically-ingested manganese. *Toxicol. Lett.* **66**, 287–294 (1993).
180. G. Oberdoerster and G. Cherian, Manganese. In T. W. Clarkson et al., eds., *Biological Monitoring of Toxic Metals*, Plenum, New York, 1998, pp. 283–301.
181. H. Tjülve and J. Henriksson, Uptake of metals in the brain via olfactory pathways. *NeuroToxicology* **20**, 181–196 (1999).
182. H. Tjülve et al., Uptake of manganese and cadmium from the nasal mucosa into the central nervous system via olfactory pathways in rats. *Pharmacol. Toxicol.* **79**, 347–356 (1996).
183. G. Gianutsos, G. R. Morrow, and J. B. Morris, Accumulation of manganese in rat brain following intranasal administration. *Fundam. Appl. Toxicol.* **37**, 102–105 (1997).
184. Mineral Gallery, *The Mineral Rheniite*, Amethyst Galleries, Inc., 1998. Available: http://mineral.galleries.com/
185. R. Colton, *The Chemistry of Rhenium and Technetium*, Interscience, New York, 1965. (Interscience monographs on chemistry, inorganic chemistry section).
186. Rembar Company, Inc., *Rhenium*, Rembar Company, Inc., Dobbs Ferry, NY, 1998. Available: www.rembar.com/rhen.htm
187. J. A. Biaglow, *Rhenium Material Properties*, NASA Tech. Memor. 107043, National Aeronautics and Space Administration, Washington, DC. Available from: NTIS, Springfield, VA(N96-13042/2).
188. R. D. Peacock, *The Chemistry of Technetium and Rhenium*, Top. Inorg. Gen. Chem., Monogr. No. 6, Elsevier, New York, 1966.
189. B. W. Wessels and R. D. Rogus, Radionuclide selection and model absorbed dose calculations for radiolabeled tumor associated antibodies. *Med. Phys.* **11**, 638–645 (1984).
190. K. Hashimoto and K. Yoshihara, Rhenium complexes labeled with [186, 188]Re for nuclear medicine. *Topi. Curr. Chemi.* **176**, 275–292 (1996).
191. R. A. Holmes, Radiopharmaceuticals in clinical trials. *Semin. Oncol.* **20** (Suppl. 2), 22–26 (1993).
192. M. H. Goldrosen et al., Biodistribution, pharmacokinetic, and imaging studies with [186]Re-labeled NR-LU-10 whole antibody in LS174T colonic tumor-bearing mice. *Cancer Res.* **50**, 7973–7978 (1990).
193. H. B. Breitz, D. R. Fisher, and B. W. Wessels, Marrow toxicity and radiation absorbed dose estimates from rhenium-186-labeled monoclonal antibody. *J. Nucl. Med.* **39**, 1746–1751 (1998).
194. M. Juweid et al., Pharmacokinetics, dosimetry and toxicity of rhenium-188-labeled anti-carcinoembryonic antigen monoclonal antibody, MN-14, in gastrointestinal cancer. *J. Nucl. Med.* **39**, 34–42 (1998).
195. D. Gbel et al., Chronische Polyarthritis und Radiosynoviorthese: Eine prospektive, kontrollierte Studie der Injektionstherapie mit Erbium-169 und Rhenium-186 [Radio-

synoviorthesis in rheumatoid arthritis: A prospective study of injectiontherapy with rhenium-186 and erbium-69]. *Z. Rheumatol.* **56**, 207–213 (1997).
196. J. R. Dilworth and S. J. Parrott, The biomedical chemistry of technetium and rhenium. *Chem. Soc. Rev.* **27**, 43–55 (1998).
197. S. V. Suvorov, Toxic effect of rhenium and its compounds in rats. *Chem. Abstr.* **71**, 53247h (1969).
198. J. M. Haguenoer and D. Furon, in J. M. Haguenoer, and D. Furon, eds. *Toxicologie et hygiene industrielles*. Tome 1, Part 1, Vol. 1, Paris, 1981, pp. 475–477.
199. P. Jacquet, Rhenium. In H. G. Seiler, and P. Sigel, eds., *Handbook on Toxicity of Inorganic Compounds*, Dekker, New York, 1988 pp. 557–560.
200. P. W. Durbin, K. G. Scott, and J. G. Hamilton, The distribution of radioisotopes of some heavy metals in the rat. University of California Press, Los Angeles. [*Univ. Calif., Berkeley, Publ. Pharmacol.* **3**, 1–34 (1957)].
201. J. F. Kopp and D. McKee, Methods for Chemical Analysis of Water and Wastes, Report No. EPA-600/4-79-020, U.S. Environmental Protection Agency, Environmental Monitoring and Support Laboratory, Cincinnati, OH, 1983, Available from: NTIS, Springfield, VA (PB84-128677).
202. L. C. Hurd, J. K. Colehour, and P. P. Cohen, Toxicity study of potassium perrhenate. *Proc. Soc. Exp. Biol. Med.* **30**, 926–928 (1933).
203. F. Maresh, M. J. Lustok, and P. P. Cohen, Physiological studies of rhenium compounds. *Proc. Soc. Exp. Biol. Med.* **45**, 576–579 (1940).
204. T. J. Haley and F. D. Cartwright, Pharmacology and toxicology of potassium perrhenate and rhenium trichloride. *J. Pharm. Sci.* **57**, 321–323 (1968).
205. S. V. Suvorov, Renii i ego soedineniya. In *Novye dannye po toksikologii redkikh metallov i ikh soedinenii [New Data on the Toxicology of Rare Metals and their compounds]*, Izdatel'stvo Meditsina, Moscow, 1967, pp. 45–50.
206. E. J. Baumann et al., Behavior of thyroid toward elements of the seventh peroidic group. II. Rhenium. *Proc. Soc. Exp. Biol. Med.* **72**, 502–506 (1949).
207. C. J. Shellabarger, Studies on the thyroidal accumulation of rhenium in the rat. *Endocrinology (Baltimore)* **58**, 13–58 (1955).
208. P. A. Bardfeld and K. Shulman, Transport and distribution of Rhenium-188 in the central nervous system. *Exp. Neurol.* **50**, 1–13 (1976).
209. H. R. Maxon et al., Re-186(Sn) HEDP for treatment of multiple metastatic foci in bone: Human biodistribution and dosimetric studies. *Radiology* **166**, 501–507 (1988), published erratum: Ibid, **167**(2), 582 (1988).
210. J. M. H. de Klerk et al., Pharmacokinetics of rhenium-186 after administration of rhenium-186-HEDP to patients with bone metastases. *J. Nucl. Med.* **33**, 646–651 (1992).
211. H. G. Nadler, Rhenium and rhenium compounds. In W. Gerhartz, and Y. S. Yamamoto, eds., *Ullmann's Encyclopedia of Industrial Chemistry*, 6th ed., Wiley-VCH Publishers, Weinheim, 1998, electronic release.
212. Rembar Company, Inc., (1997) *Material Safety Data Sheet [Rhenium]*, Rembar Company, Inc., Dobbs Ferry, NY, 1997. Available: www.rembar.com/safe-re.htm

CHAPTER FORTY

Iron and Cobalt

L. Faye Grimsley, MSPH, CIH

The following chapter discusses iron and cobalt and other selected compounds that exist with these specific elements. Elemental iron has been known since prehistoric times. Around 1200 B.C., iron was obtained from its ores; this achievement marks the beginning of the Iron Age. Even with the development of other materials, iron and its alloys remain crucial in the economies of modern countries. Iron is also critical to life. It is an essential element and a component of hemoglobin. Cobalt was known to be used by early civilizations. Minerals containing cobalt were of value to early Egyptians and Mesoptamia for coloring glass deep blue (1).

1.0 Iron

1.0.1 CAS Number: [7439-89-6]

1.0.2 Synonyms: Steel; Fe^{2+} ion; Fe^{3+} ion; $iron^{3+}$; stainless steel; iron powder

1.0.3 Trade Names: NA

1.0.4 Molecular Weight: 55.85

1.0.5 Molecular Formula: Fe

1.1 Chemical and Physical Properties

1.1.1 General

Iron is a silver-white solid metal of Group VIII, the transition elements of the periodic table. The chemical symbol, Fe, is from *ferrum*, the Latin word for iron.

Patty's Toxicology, Fifth Edition, Volume 3, Edited by Eula Bingham, Barbara Cohrssen, and Charles H. Powell. ISBN 0-471-31934-1 © 2001 John Wiley & Sons, Inc.

Table 40.1. Chemical and Physical Characteristics of Iron and Some of Its Salts

Form	At. or Mol. Wt.	Sp. Gr.	M.P. (°C)	B.P. (°C)	Solubility
Iron, Fe	55.85	7.86	1535	2750	Insol. water; sol. acids
Ferrous oxide, black, FeO	71.85	5.7	1420	—	Insol. water; sol. acid; insol. alcohol, alkalis
Iron oxide, magnetite, red, F_3O_4	231.54	5.19	Dec. 1538	—	Insol. water; sol. conc. acid; insol. alcohol, ether
Ferric chloride, $FeCl_3$	162.21	2.898 (25°C)	306	Dec. 315	5.35 kg/L (100°C); v. sol. EtOH, MeOH, ether
Ferric sulfate, $Fe_2(SO_4)_3$	399.87	3.097 (18°C)	—	—	Sl. sol. cold water; dec. hot water; insol. H_2SO_4
Ferrous sulfate, $FeSO_4 \cdot H_2O$	169.96	2.97 (25°C)	—	—	Sl. sol. cold water
Ferrocene, $C_5H_5FeC_5H_6$	186.04	—	172.5	Subl.	Insol. water; sol. EtOH, ether, MeOH
Iron carbonyl, $Fe(CO_5)$	195.9	1.46	−20	103	Insol. water, dilute acids; sol. most organic acids

Iron is the fourth most abundant element (5.1%) in the earth's crust. The molten core of the earth is primarily elemental iron. Iron occasionally occurs in its pure form; however, it is abundant in combination with other elements as oxides, sulfides, carbonates, and silicates. Other iron compounds discussed in this chapter include iron oxide CAS# [1309-37-1], ferrocene CAS# [102-54-5], iron pentacarbonyl CAS# [13463-40-6], and iron-dextran CAS# [9004-66-4]. Chemical and physical characteristics of iron and some of its compounds are listed in Table 40.1.

The physical properties of iron, the metal, are profoundly affected by impurities and by changes in temperature and treatment. Iron is superior to all other elements in magnetic properties. Iron, in an almost pure state, loses its magnetism when removed from an electric field; when iron contains small amounts of carbon, cobalt, or nickel, the retention of magnetism is increased. When heated to 770°C, iron loses its magnetism; on cooling, it retains this property. Iron undergoes a variety of structural changes (transformations) on heating that form the basis of the heat treatment of ferrous metals.

The principal compounds of iron are ferrous (Fe^{2+}) and ferric (Fe^{3+}). In general, ferrous and ferric forms are mutually interconvertible. The oxidation potential against the normal hydrogen electrode for the ferrous form is −0.43 V, and for the ferric form, +0.77 V. Ferrous compounds are more stable than ferric when ionized, less stable when covalent.

A large proportion of iron salts are water soluble; exceptions are carbonates, oxides, hydroxides, phosphates, sulfides, and ferrous fluoride. Iron of both valences tends to form complexes in which the most common coordination number is 6. Iron has a strong tendency to combine with oxygen, as in the form of hydroxyl groups, with resultant stable compounds, especially as chelates. Iron compounds exhibit marked catalytic activity in the promotion of oxidations, which are of both chemical and biological importance. Iron forms several carbonyls; their properties and uses are discussed.

An interesting aspect of iron chemistry is the array of compounds that bond to carbon. Cementite, Fe_3C, is a component of steel. The cyanide complexes of both ferrous and ferric iron are very stable and are not strongly magnetic, in contrast to most iron coordination complexes. The cyanide complexes form colored salts, including Prussian blue, $KFe_2(CN)_6$, made from ferric iron and potassium ferrocyanide. The compound Turnbull's blue, made from ferrous iron and potassium ferricyanide, is considered identical to Prussian blue.

Iron forms a large group of materials known as ferroalloys that are important as addition agents in steelmaking. Iron is also a major constituent of many special-purpose alloys developed for characteristics related to magnetic properties, electrical resistance, heat resistance, corrosion resistance, and thermal expansion.

Among the better-known types of Fe alloys are those with carbon, of which the principal ones are wrought iron, cast iron, and steel. Good wrought iron contains no more than 0.035% C, but also contains 0.075–0.15% Si, 0.1–0.25% P, less than 0.02% S, and 0.06–0.1% Mn, not all of which are alloyed with the iron.

Cast iron contains 2–4% C and varying amounts of silicon, phosphorus, sulfur, and manganese, to obtain a wide range of physical and chemical properties. Alloying elements such as silicon, nickel, chromium, molybdenum, copper, and titanium may be added in amounts varying from a few tenths to 30% or more.

Steel is a generic name for a large group of Fe–C alloys in which the carbon content is about 2%. To this basic steel, other alloying elements may be added, the more common types of which are aluminum, chromium, cobalt, Cr–Ni, Cr–Al, manganese, nickel, silicon, and tungsten, each of which has particular uses arising from its special properties.

Several iron oxide forms are used as paint pigments, polishing compounds, magnetic inks, and coatings for magnetic tapes. The soluble salts are variously used as dyeing mordants, catalysts, pigments, fertilizer, feeds, and disinfectants, and in tanning, soil conditioning, and treatment of sewage and industrial wastes.

The minimum ignition temperatures for iron dust clouds range from 470 to 780°C; for layered dust, the range is 220 to 520°C (1).

Iron pentacarbonyl, $Fe(CO)_5$, like nickel carbonyl, is insoluble in water and unreactive in dilute acids. It may ignite spontaneously in air. Concentrated reducing acids yield ferrous salts, as do gaseous halogens. Iron pentacarbonyl is a strong reducing agent, changing ketones to alcohols, benzil to benzoin, and nitrobenzene to aniline.

Iron pentacarbonyl has an ignition temperature of 320°C; the minimal explosive concentration is 105 oz/ft^3; 10% oxygen is the limiting concentration to prevent ignition (2).

Although information on storage and handling has been given specifically for $Fe(CO)_5$, it can be assumed that the information applies in like manner to all industrial metal carbonyls. Because the vapors of $Fe(CO)_5$ form explosive mixtures with air, this chemical should be stored under CO, CO_2, or N_2; and because of its high toxicity, handling of this substance should be done in well-ventilated hoods. The danger of spontaneous ignition can be reduced by the addition of hydrocarbons, their halogen derivatives, or alcohol. Workrooms should be provided with good general ventilation, and only persons trained in handling extra hazardous materials should be employed for this work.

1.2 Production and Use (1, 3)

Iron ore reserves are found worldwide. Areas with more than 1 billion metric tons of reserves include Australia, Brazil, Canada, the United States, Venezuela, South Africa, India, the former Soviet Union, Gabon, France, Spain, Sweden, and Algeria. The ore exists in varying grades, ranging from 20 to 70% iron content. North America has been fortunate in its ore deposits. There are commercially usable quantities in 22 U.S. states and in six Canadian provinces. In the United States the most abundant supplies, discovered in the early 1890s, are located in the Lake Superior region around the Mesabi Range. Other large deposits are found in Alabama, Utah, Texas, California, Pennsylvania, and New York. These deposits, particularly the Mesabi Range reserves, seemed inexhaustible in the 1930s when an average of 30 million tons of ore was produced annually from that one range. The tremendous demand for iron ore during World War II virtually tripled the output of the Mesabi Range and severely depleted its deposits of high-grade ore. The major domestic (U.S.) production is now from crude iron ore, mainly taconite, a low-grade ore composed chiefly of hematite [$FeO(OH) \cdot H_2O$] and silica found in the Great Lakes region.

After the war an intensive search revealed large quantities of rich ore, acceptable for blast-furnace use, in newly discovered deposits. Most of these discoveries involved reserves located close to the surface, allowing the use of open-pit mining rather than the more costly underground mining that had been necessary to reach many of the older reserves. In addition, new ore upgrading techniques were developed to exploit the large reserves of low-grade ores such as taconites and jaspers.

These techniques include sintering and pelletizing. Sintering is used when ore and other iron-bearing materials are too fine to be charged directly into the furnace. These materials are agglomerated with a mixture of coal and coke fines, or powders, which, when ignited, provide the heat for the sinter process. The result is a porous, clinker-like mass that enhances the upward flow of hot gases through the blast furnace burden.

Pelletizing is used to increase the iron content of low-grade (20–30% iron) ores. After being crushed, screened, and concentrated, the ore fines are formed into small balls or pellets with an iron content of 60% or more. The pellets are then hardened by heating to increase their strength and durability for subsequent processing. Thus ores that were once considered unsuitable now supply a substantial portion of the industry's requirements.

Perhaps the most important alloy of iron is steel, which contains up to approximately 2% carbon. Steels that contain about 0.25% carbon are called mild steels; those with about 0.45% carbon are medium steels; and those with 0.60–2% carbon are high-carbon steels. Within this range, the greater the carbon content, the greater the tensile strength of the steel. The hardness of steel may be substantially increased by heating the metal until it is red hot and then quickly cooling it, a process known as quench hardening. An important component of many steels is cementite, a carbon–iron compound. Mild steels are ductile and are fabricated into sheets, wire, or pipe. The harder medium steels are used to make structural steel. High-carbon steels, which are extremely hard and brittle, are used in tools and cutting instruments.

Wrought iron, which is nearly pure iron, has a lower carbon content than steel. Because of its low carbon content (usually below 0.035%), it is forgeable and nonbrittle. Iron of high carbon content (3–4%), obtained when pig iron is remelted and cooled, is called cast iron. If cast iron is cooled quickly, hard but brittle white cast iron is formed; if it is cooled slowly, soft but tough gray cast iron is formed. Because it expands while cooling, cast iron is used in molds.

The addition of other materials in alloys — for example, manganese or silicon — also increases the hardness of steel. The inclusion of tungsten permits high-speed drills and cutting tools to remain hard even when used at high temperatures. The inclusion of chromium and nickel improves the corrosion resistance of the steel and, within certain limits of composition, is called stainless steel. A common stainless steel contains 0.15% C, 18% Cr, and 8% Ni. It is used in cooking utensils and food-processing equipment. The inclusion of silicon, ranging from 1 to 5%, results in an alloy that is hard and highly magnetic. An alloy with cobalt is used for permanent magnets.

In the United States, steel ranks among the 10 largest industries. Steel producers fall into two major categories. Integrated steel makers convert iron ore into steel through a lengthy process that employs a blast furnace to produce iron from iron ore, and a basic oxygen or open hearth furnace to transform the iron into steel. Nonintegrated steelmakers melt steel scrap in electric arc furnaces to produce liquid steel in facilities that are sometimes referred to as minimills. Given the very large size of many nonintegrated steel facilities, however, the term "scrap-based mill" is also used to describe a steel plant that does not convert iron ore to iron; and "ore-based mill" has become another term to describe an integrated steelmaker.

The rapid expansion of foreign steel industries created unprecedented competition for the U.S. industry, which must increase its investment in new technologies to reduce costs, improve steel quality, and meet more demanding performance specifications. However, foreign steel, much cheaper than domestic steel, resulted in many older mills closing. The reduction of demand for domestic steel and the reduction of man-hours required to produce steel in modernized plants have reduced the number of workers exposed in this industry.

1.3 Exposure Assessment

Mining and handling of iron ores provide exposure to dusts of SiO_2 and iron oxides. Carbon monoxide is a hazard in the operation of blast furnaces for the production of pig iron. The use of fluorspar (CaF_2) in steelmaking gives rise to gases containing SiF_4 and other fluorine-containing substances. The manufacture of alloy steels introduces hazards attendant on the use of metals such as chromium, manganese, nickel, vanadium, tungsten, molybdenum, and copper. "Pickling" of iron containing arsenic and phosphorus liberates arsine and phosphine. Certain grades of ferrosilicon used in steelmaking decompose with explosive violence on contact with moist air, evolving various toxic gases such as acetylene, H_2S, SiH_4, AsH_3, and PH_3. Fatal intoxications have occurred from such accidents during transportation, particularly at sea (4).

Because iron is essential to health, iron supplements are frequently used in the treatment of iron deficiency or iron malabsorption syndromes. Iron dextran is a complex of ferric hydroxide with dextran. It is injected to treat iron-deficiency anemia in humans and in baby pigs. Exposure occurs in manufacturing and repacking, and use is limited. Slightly more than 1000 workers may be also exposed; about half are women (5). A great many more workers are exposed in the manufacture of oral iron preparations.

Iron in its various oxidation states readily combines with many carbon compounds to form organometallic compounds. Finely divided iron reacts with carbon monoxide under pressure to form the yellow liquid iron pentacarbonyl, $Fe(CO)_5$. This transition-metal carbonyl, like many others, contains the metal in a zero oxidation state. The compound is the starting material for iron compounds in unusually low oxidation states. On decomposition, iron pentacarbonyl yields pure iron. Iron pentacarbonyl is used as a gasoline additive (0.2%) in Europe, similar to the use of tetraalkyl lead in the United States.

A new type of organometallic compound was discovered in 1951. If ferrous chloride is reacted with cyclopentadiene in the presence of a strong organic base, the orange crystalline compound ferrocene is the product. This compound, which has a highly stable structure, is called a "sandwich" compound because the iron atom is strongly held between the two flat C_5H_5 rings. In this case, it is not useful to attempt to assign an oxidation state to iron. The characterization of this compound has led to extensive transition metal organometallic chemistry. Ferrocene (dicyclopentadienyl iron) is a relatively volatile, organometallic compound used as a chemical intermediate, a catalyst, and as an antiknock additive in gasoline.

1.3.1 Air

Collection on a mixed cellulose ester filter and analysis by inductively coupled plasma (ICP) is the National Institute of Occupational Health and Safety (NIOSH) method 7300 for iron oxide fume (6).

1.4 Toxic Effects

1.4.1 Experimental Studies

1.4.1.1 Acute and Chronic Toxicity. Ferrocene has been suggested as a therapeutic agent for anemia related to malabsorption of iron, as well as a gasoline additive. There are no published data with regard to adverse effects resulting from occupational exposure. However, F344/N rats and B6C3F1 mice were exposed to 0, 2.5, 5.0, 10, 20, and 40 mg ferrocene vapor/m^3, 6 h/day for 2 wk. During these exposures, there were no mortalities and no observable clinical signs of ferrocene-related toxicity in any of the animals. At the end of the exposures, male rats exposed to the highest level of ferrocene had decreased body-weight gains relative to the weight gained by control rats. The body-weight gains for all groups of both ferrocene and control female rats were similar. Male mice exposed to the highest level of ferrocene also had decreased body-weight gains, relative to controls. The female mice had relative decreases in body-weight gains at the three highest exposure levels. Male rats had a slight decrease in relative liver weights at the highest level of exposure, whereas no relative differences in organ weights were seen in female rats. Male mice had exposure-related decreases in liver and spleen weights, and an increase in thymus weights, relative to controls. For female mice, decreases in organ weights occurred in the brain, liver, and spleen. No exposure-related gross lesions were seen in any of the rats or mice at necropsy (7).

1.4.1.4 Reproductive and Developmental. Studies with injectable iron compounds have indicated that high doses given intravenously to pregnant rats may result in teratogenic changes (hydrocephalus, anophthalmia). These teratogenic effects can be reduced by deferoxamine (8).

1.4.2 Human Experience

1.4.2.1 General Information

1.4.2.2 Clinical Cases

1.4.2.2.1 Acute Toxicity. Ingestion of iron-containing tablets by children is a frequent occurrence. The estimated toxic dose for a 10-kg child is 20 mg Fe/kg. According to Ellenhorn and Barceloux (9), 5000 cases of iron poisoning occur in the United States each year. One case of acute industrial iron poisoning has been reported. In this case a worker fell into a vat of $FeCl_3$ (10).

The first phase of acute oral iron intoxication is gastrointestinal irritation and damage. Vomiting may occur at this phase. Central nervous system depression, as well as cardiovascular symptoms, such as pallor, tachycardia, and hypotension, may occur. Following the initial phase, the patients may appear to recover. However, in 12–48 h after the ingestion, life-threatening symptoms can appear. These include gastrointestinal perforation, coma, convulsions, vasomotor collapse, cyanosis, and pulmonary edema. Hepatorenal failure may develop. Most deaths occur during this phase. In the prolonged recovery, pyloric constriction and gastric fibrosis may occur (8).

Signs and symptoms of overexposure to $Fe(CO)_5$ resemble those of $Ni(CO)_4$ immediately upon exposure, giddiness and headache, occasionally accompanied by dyspnea and vomiting. Removal from exposure reverses the symptoms, but dyspnea returns in 12 to 36 h, accompanied by fever, cyanosis, and cough. Death usually occurs in 4–11 d from exposure to lethal concentrations. Pathological changes consist of pulmonary hepatization, vascular injury, and degeneration of the central nervous system (1).

1.4.2.2.2 Chronic and Subchronic Toxicity. Chronic oral iron intoxication is relatively rare, but can lead to hemosiderosis or hemochromatosis. Hemosiderosis is a condition in which there is a generalized increase in the iron content in the body tissues, particularly the liver and spleen. Hemochromatosis is marked by the accumulation of iron, as in the Kupffer cells of the liver and in the reticuloendothelial cells of the spleen and bone marrow. This is accompanied by fibrotic changes in the affected organ, most often the liver.

Hemosiderosis has been reported in the Bantu of Africa. This may be due to the use of iron pots for cooking, the nature of the diet, and the use of beer brewed in ironware. "Bantu siderosis" occurs more frequently in men than in women and may be a geographic cluster of primary hemochromatosis (8).

Primary hemochromatosis is a genetically determined autosomal recessive disorder occurring most often in men, characterized by the excessive accumulation of body iron (11). The disorder is determined by a locus closely linked to the HLA loci on the short arm of chromosome 6. There is a recessive mode of transmission. The gene frequency may be as high as 0.05 in some parts of the world. HLA typing makes it possible to identify family members who are homozygous for idiopathic hemochromatosis, and measurement of transferrin saturation and serum ferritin concentration will identify those with iron overload (12). Hypogonadism of either testicular or central origin is a frequent complication (13).

Pulmonary siderosis results from inhalation of iron dust or fumes. It falls into the group of pneumoconioses in which the pulmonary reaction is minimal, despite a heavy dust load. Because fibrosis is not caused by inhalation of iron dust, the clinical course is benign; and pulmonary function tests and blood gases are within normal limits (14).

Marazzini et al. (15) showed an increase of bronchial obstruction due to exposure in an iron foundry. In a 100-subject sample, all working in the iron foundry were affected only by small airway obstruction. Thirty months later, 99 of these subjects were reexamined and the present airway condition determined. In 43 subjects there were abnormal results of the tests, indicating total airway obstruction after 30 months. Even in the subsample of nonsmokers, a deterioration had occurred.

In 1967, 240 workers in the Kiruna, Sweden, iron mine were examined with regard to lung function and respiratory symptoms. Seventeen years later, 167 of these workers were reexamined using a structured interview that covered respiratory symptoms, smoking habits, and workplace conditions; lung function tests, including dynamic spirometry and closing volume, were also analyzed. The prevalence of chronic bronchitis in the latter study was 9.6%. There was a strong relationship between chronic bronchitis and smoking, but no relationship between chronic bronchitis and working underground in the mine. Only three persons had chronic obstructive lung disease. In the active mine workers, dynamic spirometry results showed no difference between smokers and nonsmokers or between underground and surface workers. Thus the authors reported no excess of chronic obstructive lung disease or lung function disturbances in the mine workers studied. This may reflect a self-selection process whereby the workers with airway obstruction due to smoking or underground exposure have left underground work and, also, the company. Underground workers with chronic mucous hypersecretion, on the other hand, have not felt motivated to leave underground work because of this. Some, however, may have stopped smoking, but not necessarily because of the hypersecretion (16).

1.4.2.2.3 Pharmacokinetics, Metabolism, and Mechanisms. The oral absorption of iron is largely limited by physiological homeostatic mechanisms that regulate the intake based on need. The intestinal mucosa is the major site at which the absorption is limited, but hepatic and pancreatic secretions may influence the absorption. However, in cases of acute iron poisoning the gastric mucosa is often disrupted. The iron transport system is overloaded, and this results in circulating free iron. In the normal homeostatic mode the divalent iron is absorbed into the gastric mucosa, where it is converted to the trivalent form.

The toxicokinetics of injectable iron and organo-iron compounds, like ferrocene, are not affected by the homeostatic gastrointestinal control of iron absorption. The trivalent iron attached to ferritin passes into the bloodstream and is converted into transferrin. Transferrin is transported to the spleen or liver, where it is stored as ferritin or hemosiderin. Under normal conditions the body burden of iron is about 4 g. Hemoglobin contains the greatest amounts of body iron (67%), and this largely in the red blood cells. Twenty-seven percent of the total body iron is in the liver as ferritin or in pathological conditions as hemosiderin. Because iron is so important in physiological function, the body tends to conserve iron.

The major mechanisms for the excretion of iron are desquamation of the gastrointestinal tract and blood loss. However, the iron–deferoxamine formed as the result of administering the specific iron chelator, deferoxamine, is excreted in the urine (8).

1.4.2.2.5 Carcinogenesis. Both NTP and IARC have determined that iron dextran may reasonably be anticipated to cause cancer in humans. This determination is based on the finding of injection-site tumors, particularly in rats after subcutaneous injections of iron dextran. Additionally, a few human cases of injection-site tumors arising after treatment with iron dextran have been reported (17). The nature of these reported tumors suggests that they may not have been due to iron dextran. However, the finding of injection-site tumors in experimental animals alone cannot be considered indicative of an occupational cancer hazard; there is virtually no information to suggest that exposure to iron or iron compounds by any route except intramuscular or subcutaneous injection poses a cancer hazard (8).

1.4.2.3 Epidemiology Studies

1.4.2.3.2 Chronic and Subchronic Toxicity. A retrospective cohort mortality study was conducted by Andelkovich et al. (18) among 8147 men and 627 women employed in a gray iron foundry for at least 6 months between 1950 and 1979. More than 1700 deaths occurred during a 35-year period of observation. Standardized mortality ratios (SMRs) for all causes were close to expected values, based on the U.S. general population as the standard. The mortality of nonwhite men was significantly increased for lung cancer (SMR = 132) and ischemic heart disease (SMR = 126). Other moderate, but nonsignificant, excesses were noted among nonwhite men for cancers of the stomach, pancreas, and prostate, for diabetes mellitus, and for pulmonary emphysema, and among white men for cancers of the lung and stomach, gastric and duodenal ulcers, pulmonary emphysema, and suicide. Small mortality increases were observed in both racial groups for cerebrovascular disease. The lack of a trend with time since hire and duration of foundry employment suggests that lung cancer mortality may not be associated with exposure to the foundry environment. Utilizing indirect measures of smoking, it appears that virtually all excess lung cancer deaths among whites, and at least some of the excess among nonwhites, could be explained by smoking habits. Similarly, smoking may have been responsible for the mortality excesses from emphysema, cerebrovascular diseases, and ischemic heart disease.

Underground hematite mining has been associated by IARC (17) with cancer among workers. It has been suggested that this may be due to excessive exposure to radon. In a retrospective cohort mortality study of 10,403 Minnesota iron ore (hematite) miners no excesses of lung cancer mortality were present among either underground (SMR = 100) or aboveground (SMR = 88) miners. Yugoslav-born miners incurred a twofold significant excess mortality for lung cancer that did not appear to be associated with their mining exposures. Significant excesses in mortality due to stomach cancer were found for both underground (SMR = 167) and aboveground (SMR = 181) miners as compared with U.S. white males. However, except among Finnish-born miners, these excesses disappeared when comparisons were made with the appropriate country rate. The authors (19) concluded that the apparent absence of significant radon exposure, a strict smoking prohibition underground, an aggressive silicosis control program, and the absence of underground diesel fuel use may explain why these underground miners did not appear to incur the lung cancer risk reported in other studies.

In contrast, a cohort mortality study was conducted with regard to a pyrite mine located in central Italy, where there was exposure to radon. The concentration of free silica in the dust was less than 2%. The cohort was determined from company files and included 1899 subjects. Mortality was studied for the years 1965–1983. The loss to followup was less than 2%. The SMR for all causes and all neoplasms was 97 and 107, respectively. That for lung cancer and for nonmalignant respiratory diseases was 131 and 173, respectively. The investigators (20) estimated that the extra cases of lung cancer attributable to radon daughters numbered 13 per 106 person-years and working level month in the whole cohort. The extra cases of lung cancer were 21.3 per 106 person-years in the subcohort with 10–25 years of exposure.

Mortality during the years 1947–1983 was studied by Cooper et al. (21) in 3444 men employed during the years 1947–1958 for at least 3 months in Minnesota taconite mining operations. Taconite is a low-grade iron ore consisting of iron, quartz, and numerous silicates. Taconite from the eastern part of the Mesabi Iron Range contains the amphibole silicate cummingtonite–grunerite, which is a mineral relative of amosite asbestos. During 86,307 person-years of observation, there were 801 deaths for a standardized mortality ratio of 88 (U.S. white male rates) or 98 (Minnesota rates). The 41 deaths from respiratory cancer were fewer than expected, the SMR being 61 (U.S. rates) and 85 (Minnesota rates). There were 25 respiratory cancers 20 or more years after first taconite employment, for an SMR of 57 (U.S. rates). SMRs for colon cancer, kidney cancer, and lymphopoietic cancer were elevated, but below the level of statistical significance. There was one death from pleural mesothelioma 11 years after first taconite employment in a man with a long prior employment as a locomotive operator. The pattern of deaths did not suggest asbestos-related disease in taconite miners and millers.

1.5 Standards, Regulations, or Guidelines of Exposure

Although iron dextran is classified as "reasonably expected to be carcinogenic" by NTP and a B2 carcinogen by IARC, its harmful exposure in workers is limited. The TLV for iron oxide fume (Fe_2O_3) is 5 mg/m^3, and the TLV committee also classifies it as a B2 carcinogen. The OSHA PEL for iron oxide is 10 mg/m^3 as total particulate. The TLV for iron salts is 1 mg Fe/m^3; the Federal OSHA PEL was vacated in 1989 (22). The TLV was selected to prevent the development of X-ray changes following long-term exposure to iron oxide dust and fume, whereas the TLV for the iron salts was recommended to reduce the likelihood of respiratory irritation and skin irritation (23).

The TLV for iron Fe(CO)$_5$, as Fe, is 0.23 mg/m^3 as an 8-h TWA with a short-term exposure limit (STEL) of 0.45 mg Fe/m^3, whereas the Federal OSHA PEL had previously been set at 0.1 mg/m^3 as Fe with a STEL of 0.2 mg Fe/m^3; currently there is not a Federal OSHA PEL because the 1989 PELs were vacated (22). The TLV is believed to be "more than adequate to protect against acute and chronic systemic effects." A previous TLV of 0.01 mg/m^3 was recommended because of acute toxicity and suspected carcinogenic potential (23).

2.0 Cobalt

2.0.1 CAS Number: [7440-48-4]

2.0.2 Synonyms: C.I. 77320, cobalt-59, aquacat, and super cobalt

2.0.3 Trade Names: NA

2.0.4 Molecular Weight: 58.93

2.0.5 Molecular Formula: Co

2.1 Chemical and Physical Properties

2.1.1 General

Cobalt is a hard, silver metal with a blue sheen. Cobalt is a transition element in Group VII of the periodic table. The name cobalt is derived from the German *kobald* (a malicious underground goblin or demon). Other cobalt compounds discussed in this chapter include cobaltous oxide [CAS# *1307-96-6*], cobalt chloride [CAS# *23670-59-9*], cobalt sulfate [CAS# *10124-43-3*], cobaltous nitrate [CAS# *10141-05-6*], cobalt carbonyl [CAS# *10210-68-1*], and cobalt hydrocarbonyl [CAS# *16842-03-8*]. Physical and chemical properties of cobalt and some of its compounds are listed in Table 40.2.

Cobalt is a hard magnetic metal, resembling nickel in appearance, but with a pinkish tinge. The metal crystallizes in two allotropic forms: alpha, a close-packed hexagonal, and beta, a face-centered cubic. The magnetic permeability of cobalt averages less than two-thirds that of iron; but when it is alloyed with iron and nickel, exceptional magnetic

Table 40.2. Physical and Chemical Properties of Selected Cobalt Compounds

Form	At. or Mol. Wt.	Sp. Gr.	M.P. (°C)	B.P. (°C)	Solubility
Cobalt, Co	58.93	8.9	1495	2870	Insol. water; sol. acid
Cobalt carbonyl, Co$_2$(CO)$_4$	229.90	—	50 (dec.)	—	—
Cobalt hydrocarbonyl, (HCo(CO)$_4$	171.98	—	−26.2	20 (dec)	—
Cobaltous oxide, CoO	74.93	6.45	1835	—	Insol. water; sol. acid
Cobaltic oxide, Co$_2$O$_3$	165.86	5.18	895	—	—
Cobaltic–cobaltous oxide, Co$_3$O$_4$	240.82	6.07	—	—	Insol. water; sol. acids, alkalies

properties are developed. Cobalt is a relatively unreactive metal; it does not oxidize in dry or moist air at ordinary temperatures, and at red heat oxidation is superficial. Cobalt reacts with most acids, but becomes passive in concentrated nitric acid. Cobalt is not attacked by alkalis, either in solution or when fused, but it combines with halogens when heated. When reduced from the oxide in a fine powder form, cobalt is pyrophoric. In NH_3, a nitride forms, which decomposes at higher temperatures. Under pressure at 150°C, cobalt forms the characteristic orange crystals of tetracarbonyl $[Co(CO)_4]_2$.

Cobalt carbonyl, a solid at room temperature, has a vapor pressure lower than that of iron or nickel carbonyl. The vapor pressure of cobalt hydrocarbonyl is very high.

Cobaltous oxide, CoO, varies in color from olive green to red, depending on particle size; usually, however, it is dark gray. It is the principal constituent of the "gray" cobalt oxide used in commerce; the other chief compound is Co_3O_4.

Cobaltic oxide, Co_2O_3, forms when cobalt compounds are heated at low temperatures in excess air; higher temperatures convert Co_2O_3 to Co_3O_4. Oxidation of cobaltous (Co^{2+}) salts in acid or alkaline solutions gives hydrated cobaltic oxide. This oxide is amphoteric and forms complexes with metal oxides, for example, $ZnO \cdot Co_2O_3$. The "black" cobaltic oxide of commerce consists chiefly of this oxide, together with a small amount of Co_3O_4.

Cobaltic–cobaltous oxide, Co_3O_4, is the stable oxide of cobalt. It is reducible, however, to cobalt by carbon, carbon monoxide, and hydrogen. The chlorides of cobalt and their hydrates compose a large group of cobalt compounds that are able to complex with NH_3 to form complex amines; $CoCl_2$ hydrolyzes in aqueous solution to the extent of 0.11% at 0.062 M and 0.17% at 0.031 M. Cobaltous salts are pink or red; the corresponding Co^{3+} salts are commonly so unstable they cannot exist under normal conditions. However, CoF_3 is used as a catalyst in the production of fluorocarbons and in the cracking of gasoline.

Cobalt forms three types of complexes: amines, $[Co(NH_3)_6]Cl_3$ and $[Co(NH_3)_6Cl]Cl_2$; complex nitrites, $K_3[Co(NO_2)_6]$; and complex cyanides, $K_4[Co(CN)_6]$ and $K_3[Co(CN)_6]$. Solutions of the cobalt amines show none of the reactions of cobalt. Cobalt also chelates with certain organic molecules that possess oxygen-carrying properties, such as ethylenediamine.

Only certain forms of cobalt metal are pyrophoric. The form prepared by reducing the oxides in H_2 is pyrophoric; when cobalt oxide is reduced with NH_3 so that it contains 14–16% O_2, it glows when exposed to air. Pyrophoric cobalt is a black powder that burns brilliantly when it is in contact with O_2 or air. Cobalt hydrocarbonyl is a flammable and toxic gas.

2.2 Production and Use (1, 3)

Cobalt is a relatively rare element, composing only 0.001% of the earth's crust, as compared with 0.02% Ni. Important cobalt-containing minerals are the arsenides, sulfides, and oxides. The principal arsenides of cobalt are smaltite ($CoAs_2$), skuterudite ($CoAs_3$), and cobaltite (CoAsS). The principal sulfide minerals are carrolite ($CuCo_2S_4$) and linnaeite (Co_3S_4). The principal oxide minerals are asbolite (an impure mixture of manganese and other oxides), neterogenite (a hydrated oxide usually containing copper and, occasionally, nickel and iron), sphaerocobaltite ($CoCO_3$), and erythrite ($3CoO \cdot As_2O_5 \cdot 8H_2O$). A number of other less known materials of cobalt exist, but in insufficient quantity to be mined.

World sources of the metal and the oxide are chiefly from Zaire, Belgium–Luxembourg, Norway, and Finland, in that order, with Zaire furnishing 58% of the world's supply.

Practically all cobalt produced is a by- or co-product of other metals, chiefly copper; accordingly, a description of the mining process is omitted. The processes used in extracting cobalt from its ores vary according to the type of ore and locations of the ore deposit.

Arsenical ores are concentrated by hand sorting, gravity separation, or froth flotation, and are smelted in a blast furnace with coke and limestone to a speiss (an impure mixture of iron, cobalt, and nickel arsenides). The speiss is ground, roasted with salt, and leached with water. Insoluble chlorides remaining after the leaching process are ground with sulfuric acid, washed, and filtered, and the washings are added to the liquid from the leaching step. The combined solution is oxidized and then neutralized with lime.

Basic ferric arsenate precipitates and is removed, leaving a solution containing cobalt and nickel. The addition of successive portions of sodium hydroxide and sodium hypochlorite precipitates cobalt as the hydroxide, which is initially pure but finally admixes with nickel hydroxide. The cobalt precipitate is dried, ground, and formed into pellets, which are reduced by heating with charcoal to cobalt metal.

About 80% of the world's cobalt output is used in the metallic state. Alloys that retain their strength and other desirable properties at high temperatures are widely used in jet aircraft, gas turbines, and other equipment that operates at high temperatures. Most of these alloys contain 20–65% Co, together with nickel, chromium, molybdenum, tungsten, and other elements. Large quantities of cobalt are used in the production of magnets. The best commercial magnet steel contains 35% Co, together with some tungsten and chromium. The Alnico magnet alloys usually contain 6–12% Al, 14–30% Ni, 5–35% Co, and the balance as iron.

Cobalt is also an important constituent of other permanent magnet alloys, such as Vicalloys, Cunicos, and Remalloy or Comol, and of soft magnet alloys like the Perminvar and Permindur types. It is found in magnets made from Fe–Co powder and from cobalt ferrites.

Iron-base and cobalt-base alloys containing about 6–65% Co, together with chromium, tungsten, and other alloying elements, are very hard and resistant to abrasion and corrosion. They are used extensively for cutting tools and hard facing.

Cobalt is employed as the binder for tungsten carbide (WC). Amounts of 3–25% provide the toughness and shock resistance required to make the hard carbide of practical value in drill bits and machine tools.

A dental and surgical alloy, Vitallium, containing essentially 65% Co, 30% Cr, and 5% Mo or W, is not attacked by body liquids and does not irritate tissues.

The artificially produced radioisotope ^{60}Co may be used in place of X-rays or radium for the inspection of materials to reveal internal structure, flaws, or foreign objects.

Cobaltous chloride, nitrate, and sulfate are all formed by the interaction of the metal, oxide, hydroxide, or carbonate with the corresponding acid. There are three main oxides of cobalt: gray cobaltous oxide, CoO; black cobaltic oxide, Co_2O_3, formed by heating compounds at a low temperature in an excess of air; and cobaltic–cobaltous oxide, Co_3O_4, the stable oxide, formed when salts are heated in air at temperatures that do not exceed 850°C.

In the glass and ceramic industries small quantities of cobalt oxide are used to neutralize the yellow tint resulting from the presence of iron in glass, pottery, and enamels. Larger quantities are used to impart a blue color to these products. Cobalt oxide is used in enamel coatings on steel to improve the adherence of the enamel to the metal.

Carbonyls are prepared by direct combination of metal, generally in finely divided form, with carbon monoxide. This is the basis of the Mond process used since 1890 in industry. Metal hydrocarbonyls may be prepared by acidification of a suitable organic base salt of the metal carbonyl. Thus cobalt hydrocarbonyl forms by adding sulfuric acid to pyridinium cobalt carbonyl. Cobalt hydrocarbonyl may be used as a catalyst in organic reactions. There is no mention of cobalt carbonyl use for any industrial purpose.

Cobalt linoleates, naphthenates, resinates, and ethylhexoates are excellent driers for paints, varnishes, and inks. Cobalt catalysts are used for many industrial reactions.

Cobaltous sulfate is sometimes added to nickel plating baths to improve smoothness, brightness, hardness, and ductility of deposits.

Cobalt compounds, such as the chloride, are added in very small amounts to livestock feeds, salt licks, and fertilizers in many parts of the world where a cobalt deficiency exists in the soil and natural vegetation.

There are many coordination compounds of Co^{3+}, such as the cobalt amines, $[Co(NH_3)_6]X_3$. An important, naturally occurring cobalt coordination compound is vitamin B_{12}, the anti-pernicious anemia factor.

Hazardous exposures to cobalt fume and dust from powder falls in the electric furnace and fume from melting and pouring of cobalt metal prior to pelleting may be sustained in the milling of cobalt.

In the production of cemented tungsten carbides (carballoy), exposures are to dust and fume of cobalt, in combination with dusts of WC, TiC, and TaC. Weighed charges of cobalt metal powder, tungsten metal powder, and lampblack, together with small additions of tantalum and titanium, are ground in ball mills. The charging and emptying of the containers cause dust exposures. After pressing, the material is put through a presintering process, following which it is cut and ground. This also presents a dust exposure. The material is given a final sintering, and the tips are brazed into holders (e.g., drills, lathe tools, sawteeth); some fume may be produced in these operations. The tools are then given a final (wet) grinding.

Cobalt is a common trace element in food. It is a component of vitamin B_{12} and is, therefore, an essential element.

2.3 Exposure Assessment

In 1971 McDermott (24) reported measurements of cobalt air levels in press, machine tool, and various operations associated with WC manufacture in seven plants in Michigan. Cobalt dust levels (particle size < 2 μm) in 121 of 173 samples were below the acceptable air limit of 0.1 mg/m^3. Levels exceeding the limit were found in the machine tool operations, grinding equipment cleaning, and screening. One worker, a grinder, developed symptoms of the disease in 1 month, but the others had exposures from 4 to 28 years (average 12.6 years) before developing symptoms.

Determination of a worker's exposure to airborne cobalt metal, dust, or fume (as Co) is made using a mixed cellulose ester filter (MCEF), 0.8 μm. NIOSH method 7027 recommends an atomic absorption spectrometry procedure for analysis of a sample collected on a particulate filter and digested with acid (25). For cobalt, when a flame of oxidizing air–acetylene is used, the method has a sensitivity of 0.15 μg/mL, with a range of 0.15–8 μg/mL.

Atomic absorption spectroscopy had previously been introduced by Slavin et al. (26) for the determination of cobalt in blood and urine. They claimed results as good as those by chemical methods, but with less required time and a smaller sample (ca. 2–4 mL for blood and urine).

Hubbard et al. (27) described a chemical method for the determination of cobalt in biological material. Small amounts (often less than 1 μg) are extracted with sodium diethyldithiocarbamate. The method has a sensitivity of 0.1 μg and a range of 0–5 μg/10 mL solution.

With its advantages of speed, sensitivity, and assurance of the identity of the element being determined, a spectrochemical method for ascertaining nanogram amounts in biological fluids has been reported by Kaibel et al. (28). The analysis is performed using an ion-exchange concentration method to separate nanogram amounts of cobalt ions from extraneous elements, followed by a copper spark procedure, which improves the limit of detection.

2.4 Toxic Effects

2.4.1 Experimental Studies

2.4.1.1 Acute Toxicity

2.4.1.2 Chronic and Subchronic Toxicity.
After 3 years, chronic inhalation by animals of a Co-metal blend used in industry (containing 46% WC, 28% TiC, 8% TaC, 6% Co, and 2.5% SiO_2 at a level of 20 mg Co/m^3) resulted in focal fibrotic lesions, hyperplasia of the bronchial epithelium, and developing granulomas in areas of dust deposition. These symptoms appeared to simulate those reported in industrial workers (1; H. E. Stokinger et al., unpublished results). Daily inhalation of cobalt metal fume, which was approximately equal parts of Co, CoO, and Co_3O_4 at 1 mg of Co/m^3 for 2 years, failed to elicit these pulmonary reactions. Delahant (29) also failed to find Co_3O_4 toxic by intratracheal injection, even at overwhelming doses of 150 mg/600 g guinea pig given in three equal doses. Cobalt metal, however, produced pneumonitis, pleural effusion, and pericarditis at these dosages.

2.4.1.3 Pharmacokinetics, Metabolism, and Mechanisms.
The degree of gastrointestinal absorption of cobalt and its salts depends on the dose. Very small doses on the order of a few micrograms per kilogram are absorbed almost completely; larger doses are less well absorbed. Copp and Greenberg (30), administering 10 μg of ^{60}Co to rats, found that more than 30% was excreted in the urine when given orally, and more than 90% when injected. Although the urinary values obtained by the authors appear somewhat high, their tissue distribution data are in conformity with others; the glandular organs, particularly the pancreas, accumulated the largest amounts. Of the injected cobalt, approximately equal

quantities appeared in the bile and feces. Sheline and Chaikoff (31), using intravenous radiocobalt in dogs, found 0.1–0.3% in the pancreatic juice in 2 or 3 days, whereas they found 5% in the bile at the same time. Their results and the later work of Goldner et al. (32) indicate that, possibly, cobalt in the pancreas is bound to cellular components, with lesser amounts in the pancreatic juice.

The 24-h deposition and clearance in hamsters exposed by inhalation and gavage to CoO showed that 87% was distributed throughout the body from an inhaled dose of 784 µg and 11.3%, from a 5-mg dose by gavage. The greatest amounts from both routes (60 and 11%, respectively) remained in the gastrointestinal tract. The carcass retained the next largest amounts, 23% from inhalation and 0.34% by gavage; the lung, 3.3% by inhalation and less than 0.06% by gavage. The liver and kidneys had small fractional percentages of the administered dose (33).

Cobalt-58 retention studies conducted during a 20-d period in young and old rats of both sexes by Strain et al. (34) showed that the uptake of the radioisotope was relatively high in the aorta, kidney, and liver, and relatively low in blood, femur, and hair (less than 0.20% of dose). The uptake was age-dependent only in the aorta. There were no biologically significant differences between ^{58}Co retentions in the two sexes. Elastin retained more ^{58}Co than either collagen or keratin; and aortic tissue uptake increased 2.5–3 times in old rats, a finding that correlates well with age changes in amino acid composition of aortic tissue.

Because the erythropoietic action of cobalt requires the interaction with copper and iron, the retention by rats of low (0.003 and 0.08 ppm) dietary levels of cobalt with and without iron and copper was studied by Houk et al. (35). Cobalt retention on the low cobalt diet was 30–42%, whereas retention on the higher dietary level was only 3.3–4.9%. Iron and copper did not affect cobalt retention, nor did cobalt markedly affect iron and copper retention.

Cobalt injected in rabbits as the citrate is more rapidly eliminated; 50 and 65% urinary elimination of the total doses of 10 and 13 mg, respectively, is excreted in 24 h (36).

Evidence indicates that (1) cobalt can exert a variety of physiological activities, and (2) these activities are manifested at various levels of tissue cobalt concentration. The cobalt content of fresh tissues of normal, unexposed dogs, rabbits, and rats ranged from a few milligrams (bone) to a few tenths of a milligram (thyroid and adrenal glands) when determined by a highly sensitive spectrochemical procedure developed by Keenan and Kopp (37). The pancreas, kidney, and lung showed intermediate amounts.

Aside from the indirect action of cobalt as part of vitamin B_{12}, one of whose actions is the regulation of sulfhydryl concentration (38) in tissue, the intimate mechanism of many activities of cobalt is as yet undisclosed. However, the overall effect of many of these activities of cobalt is to stimulate rather than to depress physiological action. The earliest sign of increased intake of cobalt in animals is the production of increased amounts of serum alpha globulins (1; H. E. Stokinger et al., unpublished results). Because neuraminic acid is principally associated with the α-globulins of the serum, their increased production is a direct result of cobalt action. This action of cobalt is not uniform in all animals, indicating that a sensitization mechanism is involved; and the action tends to regress to normal, despite continued cobalt intake. At slightly higher levels of cobalt intake (1–5 mg Co/kg as soluble salt by mouth), polycythemia develops (39) by a mechanism that directs a stimulating action on red bone marrow and, possibly, on extramedullary hematopoietic tissue in other organs. The polycythemia is regularly accompanied by hemoglobinogenesis (40), provided iron and copper are present in adequate amounts. It now appears probable that the cobalt-stimulated erythropoietic factor (41, 42) is part of the mechanism by which increased red blood cells are formed following the administration of cobalt. At higher doses cobalt specifically destroys the alpha cells in the pancreatic islets of the rabbit (32) and is accompanied by temporary elevation of blood sugar levels.

2.4.1.7 Other: Neurological, Pulmonary, Skin Sensitization. On the basis that cobalt toxicity is a sensitivity reaction, Kerfoot et al. (43) found an early appearance (3 mo) of pulmonary disease in miniswine exposed daily to cobalt metal dust at 0.1 mg/m^3, after having been given a sensitizing dose. Pulmonary disease was manifested by a marked decrease in pulmonary compliance and an increase in amounts of collagen in the central areas of the pulmonary alveolar septa. The cobalt dust used in the experiment was an equal mixture of alpha and beta variety, with a size range from 0.4 to 3.6 μm in diameter. Following a 1-wk sensitizing dose, wheezing by the animals, which occurred during the fourth week of exposure, was taken as evidence of sensitivity.

Similarly, an early developing pneumoconiosis occurred in hamsters exposed daily to CoO; the lesions were characterized by interstitial pneumonitis and diffuse granulomatous pneumonia, fibrosis of the alveolar septa, bronchial and bronchiolar epithelial hyperplasia, squamous metaplasia, emphysema, and/or atelectasis of variable degree (44).

2.4.2 Human Experience

2.4.2.1 General Information. Because of the instability of Co^{3+} salts in aqueous media, physiological response data are available mainly on Co^{2+} forms. Cobalt is an essential trace element for animals and humans, and as such, the body has built in procedures for metabolizing moderate amounts of cobalt substances. It is an important constituent of vitamin B$_{12}$ and certain enzymes, and is associated with the production of erythropoietin, the red cell stimulating factor. Common plants such as lettuce, beets, cabbage, spinach, and sweet potatoes are sources of dietary cobalt, containing from a few hundredths parts per million (sweet potatoes) to 0.7 ppm (spinach) on a moisture-free basis. The cobalt content of plants varies somewhat with the region in which they are grown (1).

2.4.2.2 Clinical Cases

2.4.2.2.1 Acute Toxicity. Overexposure to cobalt metal fume and dust has been reported to cause irritation of the upper respiratory tract. (1)

2.4.2.2.2 Chronic and Subchronic Toxicity. It has previously been shown that long-term oral exposure to cobalt can cause goiter and myxedema. The effect of industrial cobalt exposure on thyroid volume determined by ultrasonography and function was assessed for 61 female plate painters exposed to cobalt blue dyes in two Danish porcelain factories and 48 unexposed workers. The cobalt blue dyes were used in one of two forms, cobalt aluminate (insoluble) and cobalt–zinc silicate (semisoluble). Only the workers exposed to semisoluble cobalt had a significantly increased urinary cobalt content (1.17 versus 0.13 µg/mmol). These subjects also had increased levels of serum thyroxine (T4) and free thyroxine (FT4I). Unaltered serum thyroid stimulating hormone and 3,5,3′-triiodothyronine (T3) were marginally reduced. The thyroid volume tended to be lower ($p = .14$). The group exposed to insoluble cobalt did not differ significantly in any thyroid-related parameters. No correlation between urinary cobalt and FT4I or thyroid volume was found (45).

Cobalt unexpectedly caused severe lesions in cardiac muscle, hypothyroidism, and thyroid hyperplasia in excessive drinkers of beer that had $CoSO_4$ added as a foam stabilizer (46, 47). The unusual type of myocardiopathy recognized in 1965 and 1966 in Quebec (Canada), Minneapolis (Minnesota), Leuven (Belgium), and Omaha (Nebraska) was associated with episodes of acute heart failure (e.g., 50 deaths among 112 beer drinkers). The clinical course of a typical patient was dyspnea with abdominal pain and edema, lasting 1–2 wk. Extreme cardiomegaly was associated with low blood pressure and pulse, with peripheral cyanosis common. Early deaths occurred within 72 h of hospital admission, which in the Omaha group amounted to 11 of 28 (46). The subsequent clinical course of the disease, however, showed distinct differences in the various locations. Among the Canadian group, almost without exception, patients who recovered initially were free of chronic sequelae, quite different from the 33% chronically ill in the Omaha study (48). This presumably was due to differing beer intakes. In the Omaha study 20 of 34 hospitalized patients with myocardiopathy, survivors of a group of 60, regained normal cardiac status and had good exercise tolerance, normal heart size, and minimal electrocardiogram changes. Six of the 34 had recurrent or chronic cardiac failure. Four patients had neurological and mental deterioration, and 2 of the 34 died suddenly after leaving the hospital.

A cobalt-induced lipoic acid deficiency may have been the cause of the myocardiac lesions. In another study cobalt-damaged rat heart mitochondria showed strongly depressed oxygen uptake with 2-oxoglutarate as substrate (49). The patients' recovery after cobalt had been removed from the beer and the demonstration of similar myocardiopathies developing in rabbits given $CoCl_2$ (50) also offer convincing evidence that cobalt was the etiologic agent.

A detailed study of the thyroid pathology in 14 heavy beer drinkers who died from severe myocardiopathy in Quebec found 11 thyroids with follicular distortion, colloid depletion, and numerous cellular changes (47). Hypothyroidism and thyroid hyperplasia had previously been reported in patients treated with $CoCl_2$ for anemia (51). The mechanism appears to be that soluble Co^{2+} markedly interferes with the uptake of iodine by the adult thyroid (52).

2.4.2.2.3 Pharmacokinetics, Metabolism, and Mechanisms. Average values (to be considered with the wide variations in individual intake) for normal cobalt balance in humans as reported by Schroeder (53) are as follows: daily food intake, 300 (140–580 µg/d); water, 6 (0–10); and air, 0.1. For output, the values are for urine, 260 (120–330 µg/d); for feces, 40 (23–60); and for sweat and hair, 6. Forbes et al. (54) reported that the cobalt values for 10 tissues of one human cadaver ranged from 0.01 ppm Co for fat, nerve, muscle, and gastrointestinal tract to 0.06 ppm for liver, skin, skeleton, and heart were next highest. Endocrine organs were not analyzed.

Lung, liver, and kidney tissue concentrations of chromium, cobalt, and lanthanum by neutron activation analysis from 66 autopsied copper smelter workers were compared with 14 controls. The mean exposure time for the smelter workers was 30 years; the mean time to date of death after exposure stopped was 7.4 years. A fourfold increase of chromium ($p = .001$) and a twofold increase of cobalt ($p \leq .001$) and lanthanum ($p = .013$) in lung tissue was found for smelter workers, compared to controls. Of the smelters, nearly one-third died from malignancies (approximately 10% from respiratory cancer and approximately 45% from cardiovascular disease). In the control group nearly 80% died from cardiovascular diseases, but none from cancer. In lung tissues the concentration of chromium, cobalt, and lanthanum did not decrease markedly after exposure had ended, indicating a long biologic half-time (55).

Lasfargus et al. (56) compared cobalt, WC (tungsten–carbide), and a mixture of WC and cobalt. They found that the amount of cobalt excreted in the urine was higher after the intratracheal instillation of the mixture of WC and cobalt than that following either of the others. These investigators advised that the results signified a greater availability of cobalt when combined with WC.

The metabolic fate of 13 mg Co, as $CoCl_2$ intravenously injected in humans, when determined spectrographically, resulted in a 10-fold increase in urinary output, but only a 17-fold increase in fecal excretion during the first week after injection (36). Slightly less than 3 mg of Co injected was recovered in the excreta during 1 wk, indicating a rather slow elimination of injected cobalt. The normal weekly urinary output amounted to 0.21 mg Co (1.6%), and the fecal output, 1.04 mg (8%). Normal urinary excretion of cobalt represented 20% of the total output.

At the enzyme level, Co^{2+} acts as a relatively specific cofactor for a number of body enzymes. Divalent cobalt activates arginase (57), which subsequently liberates ammonia to control the acid–base balance of the body. Cobalt (2+) causes (58) an increase in phosphate turnover in both RNA and DNA and accelerates the hydrolytic rate of certain enzymes for peptide derivatives (59). Certain cobalt amine complexes (60) activate the serum inhibitor of hyaluronidase. Recently, Wiberg (49) stated that a lipoic acid deficiency caused by an excessive intake of $CoSO_4$ may be the immediate cause of the myocardial lesions seen in cobalt beer drinkers.

Cobalt acts synergistically with antibiotics (61), both *in vitro* and *in vivo* (mice). If antibiotics also synergize the action of cobalt, this could possibly explain the enhanced sensitivity to cobalt seen in some exposed individuals.

2.4.2.2.7 Other: Neurological, Pulmonary, Skin Sensitization, etc. A dermatitis of the allergic sensitivity type has been described by Schwartz et al. (62). Distribution of the eruption was most marked at points of friction and seemed to be related to the abrasive nature of the dust (cobalt-cemented tungsten carbide, WC). The same type of allergic dermatitis was described by Schwartz as occurring among cobalt-alloy workers and also among Finnish pottery workers handling cobalt clay (63).

An appreciable incidence (16.6%) of sensitivity to cobalt in cement worker's eczema was found in patch testing 246 workers in Finland (64). Cobalt content was 2 ppm in two factories that supply 80% of Finnish cement, 30 ppm in a third plant. The results suggested a positive correlation between the incidence of eczema and cobalt content. A review of 14 other reports on European cement workers showed an average incidence of about 25% among those patch tested between the years 1952 and 1963. Cobalt content ranged between 4.4 and 100 ppm.

Alveolitis progressing to lung fibrosis has been reported among workers exposed to a mixture of cobalt and WC in the hard metal industries, but it rarely occurs among workers exposed only to cobalt dust, as in cobalt production (56).

2.4.2.3 Epidemiology Studies

2.4.2.3.2 Chronic and Subchronic Toxicity. Associated with the manufacturing and grinding of WC in the cobalt-cemented WC industry both in the United States and in Europe is a pneumoconiosis. Miller et al. (63) reported three cases of "peculiar pulmonary reactions with hyperglobulinemia" among workers in the U.S. WC tool industry, who became asymptomatic upon removal from the dust exposure. Lundgren and co-workers (65, 66) reported a fatal case, which they describe on the basis of X-ray and pathological evidence as a chronic interstitial pneumonitis with pulmonary insufficiency, in a worker exposed to cobalt-cemented WC and TiC. Bech et al. (67) have summarized reports of 150 hard-metal disease cases up to 1962, mostly from western Europe; they also report seven additional cases. In 1968 and 1970 incidents were reported from Australia and Switzerland (68, 69). In those few reports in which cobalt dust concentrations were measured, exposure levels exceeded 0.1 mg/m^3 by at least 10-fold.

Coates and Watson (70) reported in considerable detail 12 additional cases from Michigan (U.S.) in workers manufacturing or grinding cobalt-cemented WC. They characterized the disease as a progressive, diffuse, interstitial pneumonia with a nonproductive cough and dyspnea on exertion, and pathologically, with various amounts of interstitial fibrosis, infiltration of mononuclear and mast cells, and desquamated histiocytes. Eight of the patients who had been followed since 1954 died. Coates and Watson also called attention to another type of respiratory disease affecting these workers, a form of "sensitization" characterized by cough, wheezing, and shortness of breath that developed at work, but was relieved by removal from it. Preexisting allergic background or existing lung disease had been considered a contributor, but this condition does not progress to the interstitial form.

For 40 years cases of interstitial pneumonia and bronchial asthma have been described in hard metal workers (i.e., alloys of WC and cobalt). Van den Eeckhout et al. (71) reported comparable pulmonary lesions in diamond industry workers who were exposed to cobalt not associated with WC. The exposure was caused by the diamond cobalt disks used for polishing diamonds. The disks had as the hard element microdiamonds cemented in an alloy of pure cobalt. The hard metals, on the other hand, consisted of cobalt and tungsten carbide. Forty-seven diamond cutters (i.e., nearly 1% of those exposed) presented with bronchopulmonary pathology due to cobalt. Nineteen had a fibrosing alveolitis sometimes documented by a pulmonary biopsy and, more often, by a broncho-alveolar lavage that revealed characteristic multinucleated giant cells. Thirteen had occupational asthma, often established by specific inhalation provocation tests to cobalt or by lung function measurements at the place of work. Two patients had mixed forms, and in 13 a probable diagnosis was suggested. The pathogenesis of cobalt might be explained by cytotoxic action such as that demonstrated in animal experiments. Results suggest either a sensitizing or an allergic action. Tungsten carbide does not produce pulmonary lesions, but its association with cobalt intensifies the effects of the latter.

Eight asthmatic patients who had no history of asthma before starting work in a hard metal plant and eight control subjects (three atopic, three nonatopic asthmatic, and two normal volunteers) without a history of exposure to hard metal dust were subjected to provocation tests, skin tests, radioallergosorbent tests (RAST), and Farr tests with cobalt. Four of the eight patients were atopic, and seven showed marked bronchial responsiveness to methacholine. Patch and intradermal skin tests with cobalt chloride ($CoCl_2$) could not distinguish the patients from the control subjects. All patients had positive reactions to $CoCl_2$ in the provocation tests; two developed immediate asthmatic reaction; four, late asthmatic reaction; and two, dual asthmatic reaction, whereas the control subjects showed no reaction. Evidence of specific IgE antibodies to cobalt-conjugated human serum albumin (Co-HSA) was displayed by four patients (RAST score greater than 2), based on comparison of serum samples from 60 asthmatic patients and 25 asymptomatic workers in the same plant (72). These findings in workers strongly imply that cobalt hypersensitivity plays a major role in the development of hard-metal-induced asthma.

A cross-sectional study of 1039 WC production workers was carried out by Sprince et al. (73). The purposes were (1) to evaluate the prevalence of interstitial lung disease and work-related wheezing, (2) to assess correlations between cobalt exposure and pulmonary disease, (3) to compare lung disease in grinders of hard carbide versus nongrinders, and (4) to evaluate the effects of new and previous threshold limit values for cobalt of 50 and 100 $\mu g/m^3$. Time-weighted-average cobalt concentrations were determined throughout the production process. Work-related wheeze occurred in 113 participants (10.9%).

Interstitial lung disease occurred in seven workers (0.7%). The relative odds of work-related wheeze were 2.1 times for present cobalt exposures exceeding 50 $\mu g/m^3$, compared with exposures equal to or less than 50 $\mu g/m^3$. Interstitial lung disease was found in three workers with very low average lifetime exposures (< 8 $\mu g/m^3$) and shorter latencies. Grinders of hard carbide had lower mean single-breath cobalt-diffusing capacity than nongrinders, even though their cobalt exposures were lower.

No mode of action has been proposed for the production of bronchial adenomatosis from cobalt metal dust, other than its irritative nature (74), or for an adenocarcinoma in the lung and a spindle-cell sarcoma in the rat (75). Later attempts through 1973 using $CoCl_2$ alone failed to elicit tumors of any type in laboratory animals (76), although an occasional tumor resulted when adjuvants were added (77). In any case, these weakly suggestive findings of animal tumorigenesis seem to find no parallel as yet in humans. High neoplasm rates have not been reported from cobalt mining areas: Canada (Cobalt City), Belgian Congo (Katanga), Norway (Skuterud), France (Allemont), or Czechoslovakia (Dobschina) (74). However, cobalt has recently been classified as possibly carcinogenic to humans by the IARC (78), based on the experimental animal work of Deutsche Forschungsgemeinschaft (79). Beyersmann and Hartwig (80) reviewed the genotoxicity of cobalt. They concluded that cobalt may interfere with the repair process for DNA and elicit possible carcinogenic effects through this mechanism.

2.4.2.3.5 Carcinogenesis. There is little epidemiologic support for the classification of cobalt as a possible human carcinogen. However, Mur et al. (81) studied the mortality between 1950 and 1980 of a cohort of 1143 workers in an electrochemical plant producing cobalt and sodium. Although mortality of the whole cohort was significantly lower than in the French population for all causes of death (SMR = 0.77), and especially for deaths from circulatory system diseases (SMR = 0.59), among cobalt production workers there was an increase of death from lung cancers (SMR = 4.66, 4 cases). The number of cases was small, and smoking was a possible confounding factor.

2.4.2.3.7 Other: Neurological, Pulmonary, Skin Sensization, etc. Jordan et al. (82) examined memory functioning on the Wechsler Memory Scale—Revised in a group of adult WC workers with hard metal disease and a group of matched controls. The hard-metal-exposed group of workers showed memory deficits related to difficulties in attention and verbal memory, with an apparent sparing of visual–spatial memory.

There is no therapy for cobalt poisoning at the present time. The depressing effect on blood pressure, observed when $Co(NO_3)_2$ is given intravenously to mice, can be diminished to a slight extent by BAL, according to Dalhamm (83). CaEDTA is more promising, for it is a reliable antidote against a lethal 20 mg/kg dose in the rabbit, before, during, and up to 30 min after administration of cobalt (84). Also, CaEDTA can rapidly remove Co from the body and prevents the polycythemic effect of cobalt, at least in rats (85). Dicobalt edeate (Co_2 EDTA) is used in Britain and France to treat cyanide poisoning. It acts more rapidly than nitrites, but concern about cobalt toxicity has led to its use as a second-line antidote (9).

2.5 Standards, Regulations, and Guidelines of Exposure

The OSHA and ACGIH air standards for cobalt (metal dust and fumes) are 0.2 and 0.02 mg Co/m^3, respectively, whereas the NIOSH REL is 0.05 mg Co/m^3 (22). A previous TLV for cobalt metal, dust, and fume of 0.05 was recommended by the TLV Committee of the ACGIH in 1976. They revised the TLV downward from its previous level of 0.1 mg/m^3 on the basis of the experimental results of Kerfoot et al. (43), which showed a marked decrease in pulmonary compliance in miniswine at the former level, and to provide a greater margin of protection against sensitization.

Germany allows 0.5 mg Co/m^3 during the handling of cobalt powder, but only 0.1 mg Co/m^3 for other uses of cobalt. These airborne concentrations result in whole blood concentrations of 25 or 5 µg Co/L and urinary concentrations of 300 or 60 µg Co/L. The normal blood concentration of cobalt is about 0.1 µg/L (80).

For cobalt carbonyl [$Co_2(CO)_4$] and cobalt hydrocarbonyl [$HCo(CO)_4$], NIOSH and ACGIH recommend on REL of 0.1 mg Co/m^3. The TLV is based on analogy to nickel carbonyl. The few studies referred to in the documentation (23) are only acute studies and are not suggestive of the marked toxicity implied by this low TLV.

BIBLIOGRAPHY

1. H. E. Stokinger, in G. Clayton and F. Clayton, eds., *Patty's Industrial Hygiene and Toxicology*, 3rd ed., Vol. 2A, Wiley, New York, 1981, pp. 1493–2060.
2. S. Martis, *Mechanical Engineering Handbook,* 6th ed., 1958, p. 41.
3. *Kirk-Othmer's Concise Encyclopedia of Chemical Technology*, 3rd ed., Wiley-Interscience, New York, 1985.
4. H. Hognested, *Med. Rev.* **48**, 409 (1931).
5. National Institute for Occupational Safety and Health (NIOSH), *National Occupational Exposure Survey*, NIOSH, Cincinnati, OH, 1990.
6. National Institute for Occupational Safety and Health (NIOSH), *NIOSH Manual of Analytical Methods*, 4th ed., NIOSH, Cincinnati, OH, 1994.
7. J. D. Sun et al., *Fundam. Appl. Toxicol.* **17**, 150 (1991).
8. P. B. Hammond and R. P. Beliles, in L. Casarett and J. Doull., eds., *Toxicology: The Basic Science of Poisons*, 2nd ed., Macmillan, New York, 1980, p. 409.
9. M. J. Ellenhorn and D. G. Barceloux, *Medical Toxicology*, Elsevier, New York, 1989.
10. E. J. Doolin, *J. Trauma* **20**, 518 (1980).
11. M. Rabinovitz et al., *Hepatology* **16**, 145 (1992).
12. M. Worwood, *J. Inherited. Metab. Dis.* **6**(Suppl. 1), 63 (1983).
13. T. M. Kelly et al., *Ann. Intern. Med.* **101**, 629 (1984).
14. A. H. Rubin and I. Bruderman, *Harefuah* **122**, 428 (1992).
15. L. Marazzini et al., *Bull. Eur. Physiopathol. Respir.* **13**, 219 (1992).
16. H. S. Jorgensen et al., *J. Occup. Med.* **30**, 953 (1988).
17. National Toxicology Program (NTP), *Sixth Annual Report on Carcinogens*, NTP, Research Triangle Park, NC, 1994.
18. D. A. Andelkovich et al., *J. Occup. Med.* **32**, 529 (1990).
19. A. B. Lawler et al., *J. Occup. Med.* **30**, 507 (1985).
20. G. Battista et al., *Scand. J. Work Environ. Health* **14**, 280 (1988).
21. W. C. Cooper et al., *J. Occup. Med.* **30**, 506 (1988).
22. American Conference of Governmental and Industrial Hygienists (ACGIH), *Guide to Occupational Exposure Values*, ACGIH, Cincinnati, OH, 1998.
23. American Conference of Governmental and Industrial Hygienists (ACGIH), *Documentation of Threshold Limit Values and Biological Exposure Indices*, 5th ed., ACGIH, Cincinnati, OH, 1986.
24. F. T. McDermott, *Am. Ind. Hyg. Assoc. J.* **32**, 188 (1971).
25. National Institute for Occupational Safety and Health (NIOSH), *Manual of Analytical Methods*, 4th ed., NIOSH, Cincinnati, OH, 1994.
26. W. Slavin et al., *At. Absorp. Newsl.* **17**, 7 (1978).
27. D. M. Hubbard et al., *Arch Environ. Health* **13**, 190 (1966).
28. A. M. Kaibel et al., *Appl. Spectrosc.* **22**, 183 (1968).
29. A. B. Delahant, *Arch. Ind. Health* **12**, 116 (1955).
30. D. H. Copp and D. M. Greenberg, *Proc. Natl. Acad. Sci. U.S.A.* **153** (1941).
31. G. E. Sheline and I. L. Chaikoff, *Am. J. Physiol.* **145**, 285 (1946).

32. M. G. Goldner, B. W. Volk, and S. Lazarus, *Metab. Clin. Exp.* **1**, 544 (1952).
33. A. P. Wehner and D. K. Craig, *Am. Ind. Hyg. Assoc. J.* **33**, 146 (1972).
34. W. H. Strain et al., *J. Nucl. Med.* **6**, 831 (1965).
35. A. E. Houk, A. W. Thomas, and H. C. Sherman, *J. Nutr.* **31**, 609 (1946).
36. N. L. Kent and R. L. McCance, *Biochem. J.* **35**, 877 (1941).
37. R. G. Keenan and J. F. Kopp, *Anal. Chem.* **28**, 185 (1956).
38. D. K. Kasbekar et al., *Biochem. J.* **72**, 374 (1959).
39. G. Brewer, *Am. J. Physiol.* **128**, 345 (1940).
40. F. A. Underhill et al., *J. Biol. Chem.* **91**, 13 (1931).
41. E. Goldwasser et al., *Science* **125**, 1085 (1958).
42. T. E. Brown and H. A. Meineke, *Proc. Soc. Exp. Biol. Med.* **99**, 435 (1958).
43. E. J. Kerfoot et al., *Am. Ind. Hyg. Assoc. J.* **36**, 17 (1975).
44. A. P. Wehner et al., *Am. Ind. Hyg. Assoc. J.* **38**, 338 (1977).
45. E. Prescott et al., *Scand. J. Work Environ. Health* **18**, 101 (1992).
46. J. F. Sullivan et al., *Ann. Intern. Med.* **70**, 277 (1969).
47. P. E. Roy et al., *Am. J. Clin. Pathol.* **50**, 234 (1968).
48. P. H. McDermott et al., *J. Am. Med. Assoc.* **198**, 253 (1966).
49. G. S. Wiberg, *7th Annu. Meet. Soc. Toxicol.*, Washington, DC, 1968, Abstr. 60.
50. J. L. Hall and E. B. Smith, *Arch. Pathol.* **86**, 403 (1968).
51. J. P. Kriss et al., *J. Am. Med. Assoc.* **157**, 117 (1955).
52. M. Roche and M. Layrisse, *J. Clin. Endocrinol.* **16**, 831 (1956).
53. H. A. Schoeder et al., *J. Chronic Dis.* **20**, 869 (1967).
54. R. M. Forbes et al., *J. Biol. Chem.* **209**, 857 (1954).
55. L. Gerhardsson et al., *Sci. Total Environ.* **37**, 233 (1984).
56. G. Lasfargus et al., *Toxicol. Appl. Pharmacol.* **112**, 41 (1992).
57. L. Hellerman and M. E. Perkins, *J. Biol. Chem.* **112**, 175 (1935).
58. H. B. Levy et al., *Arch. Biochem.* **24**, 199 (1949).
59. K. R. Rao et al., *J. Biol. Chem.* **198**, 507 (1952).
60. M. B. Matthews et al., *Arch. Biochem. Biophys.* **35**, 93 (1952).
61. R. Pratt, J. Dufrenoy, and L. A. Strait, *J. Bacteriol.* **55**, 75 (1948).
62. L. Schwartz et al., *Occupational Diseases of the Skin,* 2nd ed., Lea & Febiger, Philadelphia, PA, 1947, p. 176.
63. C. W. Miller et al., *Arch. Ind. Hyg. Occup. Med.* **8**, 453 (1953).
64. V. Pirila and H. Kajanne, *Acta Derm. -Venereol.* **45**, 9 (1965).
65. K. D. Lundgren and H. Ohman, *Virchows Arch. A: Pathol. Anat. Physiol.* **325**, 259 (1954).
66. K. D. Lundgren and A. Swensson, *Acta Med. Scand.* **115**, 20 (1953).
67. A. D. Bech et al., *Br. J. Ind. Med.* **19**, 239 (1962).
68. M. Joseph, *Australas. Radiol.* **12**, 92 (1968).
69. E. Reber and P. Burckhardt, *Respiration* **27**, 120 (1970).
70. E. O. Coates, Jr. and J. H. L. Watson, *Ann. Intern. Med.* **75**, 709 (1971).
71. A. V. van den Eeckhout et al., *Rev. Mal. Respir.* **6**, 201 (1989).

72. T. Shirakawa et al., *Chest* **95**, 29 (1989).
73. N. L. Sprince et al., *Am. Rev. Respir. Dis.* **138**, 1220 (1988).
74. G. W. H. Schepers, *Arch. Ind. Health* **12**, 127 (1955).
75. H. R. Schinz, *Schweiz. Med. Wochenschr.* **39**, 1070 (1942).
76. U.S. Department of Health, Education and Welfare, Public Health Services, National Institutes of Health, *Survey of Compounds Which Have Been Tested for Carcinogenic Activity*, U.S. Government Printing Office, Washington, DC, 1972–1973.
77. J. C. Heath, *Br. J. Cancer* **10**, 668 (1956).
78. International Agency for Cancer (IARC), *Monographs on the Evaluation of Carcinogenic Risks to Humans*, vol. 52, IARC, Lyon, France, 1991, p. 363.
79. Deutsche Forgschungsgemeinschaft, in S. Henschler and G. Lehnert, eds., *MAKWerte, Toxicologisch-arbeitmediziische Bergrundungen*, VCH-Verlagsgese., Weinheim, 1989.
80. D. Beyersmann and A. Hartwig, *Toxicol. Appl. Pharmacol.* **115**, 137 (1992).
81. J. M. Mur et al., *Am. J. Ind. Med.* **11**, 75 (1987).
82. C. Jordan et al., *Toxicol. Lett.* **54**, 241 (1990).
83. T. Dalhamm, *Acta Pharmacol. Toxicol.* **55**, 75 (1953).
84. M. Munari and V. Tinazzi, *Folia Med. (Naples)* **39**, 260 (1956).
85. J. T. Post, *Proc. Soc. Exp. Biol. Med.* **90**, 245 (1955).

CHAPTER FORTY-ONE

Nickel, Ruthenium, Rhodium, Palladium, Osmium, and Platinum

Slawomir Czerczak, Ph.D. and Jan P. Gromiec, Ph.D.

A NICKEL AND ITS COMPOUNDS

1.0 Nickel

1.0.1 CAS Number: [7440-02-0]

1.0.2 Synonyms: CI 77775; Nickel 200; Nickel 201; Nickel 205; Nickel 279; Alnico; NP 2

1.0.3 Trade Names: Monel; Iconel; Icoloy; Nimonic; Hastelloy; Udimet; Mar M; René 41; Waspaloy; Raney nickel

1.0.4 Molecular Weight: 58.69

1.0.5 Molecular Formula: Ni

1.1 Chemical and Physical Properties

Nickel (1–3) is a transition element in group VIII of the periodic system belonging with palladium and platinum to the 10 (nickel) triad. It is a silver-white metal with characteristic gloss and is ductile and malleable. It occurs in two allotropic forms. The specific density of nickel is 8.90 g/cm^3, melting point 1455°C, and boiling point 2730°C (Table 41.1). Nickel is not soluble in water, but it does dissolve in dilute oxidizing acids. It is resistant to lyes.

Nickel is passivated by treatment with concentrated nitric acid and aqua regia. Chemically, nickel is similar to iron and cobalt. Its valence is variable, and it may occur at various oxidation states, from $-$I to +IV, but only compounds at oxidation state +II are important. Ni^{2+} ions are stable in solution; there are numerous, both simple and complex, Ni(II) compounds. In the majority of its complex compounds, the coordination number of nickel is six, and Ni forms octahedral configuration of green color. In aqueous solutions, nickel occurs as a hexahydrate ion [Ni(H$_2$O)$_6$]$^{2+}$.

Natural nickel is a mixture of its five stable isotopes; besides, there are also seven unstable isotopes (Table 41.2) (4).

1.2 Production and Use (5)

Nickel is obtained by processing sulfide and laterite ore concentrates using pyrometallurgic and hydrometallurgic processes. The resultant nickel matte obtained by roasting and smelting is subjected to further cleaning by electro-, vapo-, and hydrometallurgic refining methods. Some portion of the matte is roasted to obtain commercial nickel oxide agglomerate. Pure, 99.9% nickel can be obtained by electrolytic refining process.

The most pure, 99.97%, nickel is obtained by vapometallurgy. In this process, known also as the *Mond method*, nickel and copper sulfide blend is converted to oxides, and then reduced by heating with water gas at 350–400°C. The resultant active form of nickel is treated with carbon monoxide to give volatile nickel carbonyl [Ni(CO)$_4$]. The latter reaction is reversible; heating results in pure nickel and carbon monoxide.

Nickel has been used predominantly as a component of alloys. The advantages of nickel-containing alloys include resistance to corrosion and high temperatures, high hardness, and strength. Copper/nickel alloy containing 63–77% nickel (Monel alloy) is used in industrial water supply/effluent systems, shipbuilding, petrochemical industry and for the manufacture of heat exchangers, pumps and welding electrodes. The coins, known in the United States as "nickels," are made of the alloy containing 75% copper and 25% nickel. Other commercial nickel alloys, such as nickel/chromium (Nichrom), nickel/iron/chromium (Iconel), or nickel/chromium/iron/molybdenum (Hastelloy), are resistant to elevated temperatures, oxidation and corrosion when exposed to acids and salts. Nickel/silver and nickel/zinc/copper alloys, due to their white color, are used for anticorrosive plating of cutlery and tools; the alloys are also used to manufacture electric contacts. Nickel/aluminium 50/50% (Raney) alloy is employed in the process of catalytic hydrogenation. Stainless steel usually contains 8–10% nickel, but sometimes nickel content may be as high as 25–30%. Typically, steel alloys contain approximately 0.3 to 5% nickel. Because of its magnetic properties, nickel, in the form of alloys with aluminium and iron, is used to manufacture permanent magnets. Nickel is also used in alkaline (cadmium/nickel) electric cells. Almost as much as 40% nickel production is used to manufacture stainless and heat-resistant steel grades, 21% for nonferrous alloys, 17% for electroplating, 12% for special-purpose alloys. The remaining 10% of nickel production is used in other applications (in the smelting, chemical, electric cell and ceramics industries).

Table 41.1. Physical Properties of Nickel and Nickel Compounds (1)

Compound	Molecular Formula	Molecular Weight	Boiling Point (°C)	Melting Point (°C)	Specific Gravity	Solubility in Water (g/100) mL	Refractive Index (20°C)	Vapor Pressure (mm Hg)
Nickel	Ni	58.69	2730	1455	8.90	Insoluble	ND[a]	—
Nickel oxide	NiO	74.69	—	1960	6.67	Insoluble	2.1818 (red)	—
Nickel carbonate	$NiCO_3$	118.70	—	Decomposes	—	0.0093 (25°C)	ND	—
Nickel subsulfide	Ni_3S_2	240.19	—	790	5.82	Insoluble	ND	—
Nickelocene	$Ni(C_5H_5)_2$	188.90	—	171–173	n.d.	Insoluble	—	—
Nickel carbonyl	NiC_4O_4	170.74	43	−25	1.32^{17}	0.18 (9.8°C)	ND	400 at 25.8°C
Nickel chloride	$NiCl_2$	129.60	Sublimes at 973	1001	3.55	64.2 (20°C)	ND	—
Nickel hydroxide	NiH_2O_2	92.70	—	Decomposes at 230	4.15	0.013	—	—
Nickel sulfate	$NiSO_4$	154.75	—	Decomposes at 848	3.68	29.3 (0°C)	ND	—

[a]No data.

Table 41.2. Main Isotopes of Nickel (4)

Nuclide	Molecular Weight	Natural Share (%)	Half-life $T_{1/2}$	Decomposition Energy (MeV)	Nuclear Spin
^{58}Ni	57.935	68.27	Stable	—	0+
^{59}Ni	58.934	0	7.6×10^4 L	EC(1.072); no gamma	3/2 −
^{60}Ni	59.931	26.10	Stable	—	0+
^{61}Ni	60.931	1.13	Stable	—	3/2 −
^{62}Ni	61.928	3.59	Stable	—	0+
^{63}Ni	62.930	0	100 L	Beta (0.065); no gamma	1/2 −
^{64}Ni	63.928	0.91	Stable	—	0+

1.3 Exposure Assessment

The techniques used most frequently for nickel quantitative analysis include atomic absorption spectroscopy (AAS), inductively coupled plasma–atomic emission spectroscopy (ICP-AES), and voltammetry.

Use of suitable procedures intended to prevent sample contamination during sampling, sample storage, preparation, and analyzing constitutes an important stage of the analysis, particularly when trace (mg/kg) or submicrometer (μg/kg) quantities are to be determined (6).

1.3.1 Air

Inductively coupled plasma–atomic emission spectroscopy (ICP-AES) is the analytical method recommended by NIOSH to determine airborne nickel (7).

Airborne nickel particles are collected on a cellulose membrane filter, mineralized with concentrated nitric and perchloric acid, and analyzed by ICP-AES. The working concentration range is 0.005–2.0 mg/m^3, and the estimated lod is 1 μg/sample. The procedure, which is also suitable for the determination of other elements suspended in the air, cannot be used to identify individual nickel compounds.

The methods enabling determination of different nickel compounds take advantage of their different solubility. A method of sequential selective leaching has been developed to determine the amount of nickel in four phase categories of a dust sample: soluble nickel, sulfidic nickel, metallic nickel, and refractory nickel oxides (8). Wong and Wu (9) used the adsorptive stripping voltammetry method to determine different forms of nickel in the air at a nickel manufacturing facility. Although it is important to characterize the individual nickel compounds, especially as components of complex mixtures, methods that determine nickel speciation are difficult and not in widespread use.

1.3.2 Background Levels

1.3.2.1 Determination of Nickel in the Biological Material. In the samples of biological material, such as tissues and body fluids, nickel concentrations are routinely determined by AAS and ICP-AES techniques. Before analyzing, the sample must be acid-mineralized and concentrated.

In the NIOSH recommended method (10) for the determination of nickel in blood or tissues, the sample, after being mineralized in a 3 : 1 : 1 mixture of the nitric, perchloric, and sulfuric acids, is analyzed by ICP-AES. The working concentration range is from 0.01–10 mg/100 g blood or 0.002–2 mg/g tissue, and the determination limit is 1 μg/g tissue.

The same ICP-AES technique with pretreatment of the sample with polydithiocarbamate resin is recommended by NIOSH for the determination of nickel in urine (11). The working concentration range is 0.25–200 μg/sample, and the estimated load limit is 0.1 μg/sample. The method is useful in particular for the determination of nickel in the urine of workers exposed to several metals simultaneously but is not compound-specific.

Sunderman (12) has developed a method of nickel determination by electrothermal atomic absorption spectroscopy (EAAS), which is suitable for the determination of nickel both in body fluids (serum, plasma, blood, urine, saliva), solid excreta (feces, hair), and tissues (e.g. lung, liver, kidney, heart); the determination limit of the method is 0.23 μg Ni/L body fluid and 0.4 μg/kg tissue.

A simplified method of nickel determination in human blood serum by the EAAS technique with Zeeman correction has been reported by Andersen et al. (13) (the estimated load is ca. 0.09 μg/L).

The recent advances in voltammetry has made it one of the most sensitive methods of nickel determination in the biological material. The limit estimated load below 0.002 μg/kg has been achieved by using differential pulse anodic stripping voltammetry (DPASV) after absorption of nickel–dimethylglyoxyme chelate (14).

Inductively coupled plasma mass spectrometry (ICP-MS) and ion-exchange chromatography are also used to determine nickel in the biological material (15–17).

1.3.3 Workplace Methods

NIOSH Analytical method 7300 is recommended for determining workplace exposures (17a).

1.4 Toxic Effects

1.4.1 Experimental Study

1.4.1.1 Acute Toxicity. Acute toxicity data of nickel metal are presented in Table 41.3 (18–20).

As seen in Table 41.3, the acute toxicity of metallic nickel given intravenously (IV), or intratracheally is considerably higher than in oral administration. Dogs tolerate nickel administered per os (orally) even at 3 g/kg body weight.

Exposure to metallic nickel inhibits the function of the immune system. A single intramuscular (IM) injection of 20 mg nickel powder resulted in prolonged reduction of natural-killer cell activity in the peripheral blood of rats. Within 8–18 weeks since the injection, the reduction of the activity was as high as 50–60% compared to control (21).

1.4.1.2 Chronic and Subchronic Toxicity. Hueper (22) exposed animals to metallic nickel dust at 15 mg/m^3. Nasal sinus inflammation and ulceration was observed in rats, exposed by inhalation for longer than 1.5 years; the guinea pigs and rats developed symptoms of lung irritation. A frequent symptom of the exposure was accumulation

Table 41.3. Summary of Toxic Effects of Nickel Metal

Species	Exposure route	Toxic effect	Reference
Rat	Oral	LD_{50} 9 g/kg	18
Rat	Intratracheal	LDL_0 12 mg/kg	19
Rat	Intramuscular	LDL_0 250 mg/kg LDL_0	19
Dog	Intravenous	LDL_0 10 mg/kg	19
Guinea pig	Oral	LDL_0 5 mg/kg	19
Dog	Oral	Tolerated 1–3 g/kg	20

of mesothelioma cells being formed. There was an increase in the phagocytic activity of the alveolar macrophages of rabbits exposed for 4 weeks to metallic nickel dust at 0.5–2.0 mg/m^3 (23).

Lundborg and Camner (24) exposed rabbits to nickel dust (0.1 mg/m^3 for 4–6 months) and found a considerable reduction in lysosomal enzyme activity versus controls. Inhalation exposure of male rabbits to 1 mg/m^3 nickel dust (6 h/day, 5 days/week) for 3 and 6 months resulted in two- or threefold increase in the density of type II alveoli in the lungs. The 6-month exposure resulted in focal pneumonia (25, 26).

Similar lesions, resembling hypoproteinosis, were observed in rabbits following 4-month exposure to metallic nickel dust (23).

1.4.1.3 Pharmacokinetics, Metabolism, and Mechanisms

1.4.1.3.1 Adsorption. Nickel and its inorganic compounds can be absorbed in the gastrointestinal (GI) tract and in the respiratory tract. The quantity of nickel absorbed is dependent not only on the quantity inhaled, ingested, or otherwise taken in but also on the physical and chemical properties of the relevant nickel compounds. Solubility is an important factor, regardless of the route of administration. Soluble nickel salts easily dissociate in the aqueous medium; resulting metal ions can penetrate through cellular membranes. Adsorption of nonsoluble nickel compounds is relatively small.

1.4.1.3.1.1 INHALATION EXPOSURE. The inhaled nickel particles, depending on their sizes, are deposited in the upper and lower respiratory tract; the smaller particles penetrate deeper into the respiratory system and therefore their relative adsorption is higher. The half-lives of nickelous oxide elimination from rat lungs ranged from 7.7 and 12.0 to 21 months, depending on mean aerodynamic diameter of the particles, which was 0.6; 1.2 and 4.0 µm, respectively (27, 28).

The rate of lung deposit elimination depends on the solubility of the deposits; it is low for the metallic nickel and nickel oxide dusts, and higher for the soluble nickel salts. The solubility of nickel compounds seems to be the most reliable index for the assessment of the rate at which the dust retained in the alveoli is absorbed to the blood. The half-life of nickel in the lungs of rats exposed by inhalation was approximately 30 h for nickel sulfate (29), 4 and 6 days for nickel subsulfide, and 120 days for nickelous oxide (30).

In rats given intratracheally a dose of 1 mg nickel in the form of soluble nickel chloride ($^{63}NiCl_2$), the major part of the dose was found in the kidneys (53%) and lungs (30%); the remaining portion of the dose was distributed between the adrenals, liver, spleen, heart, and testis (31). As renal clearance during 3 days was quick, the lungs were the organ in which ^{63}Ni content was the highest. Lung clearance during 6 h was 27%, which means that 70% of the retained material were absorbed. Nickel absorption following intratracheal administration of soluble nickel chloride was confirmed by Carvahlo and Ziemer (32).

1.4.1.3.1.2 ORAL EXPOSURE. Nickel absorption from the normal diet is poor. Tests on rats and dogs show that from 1–10% nickel given with food or intragastrically is absorbed from the GI tract (33).

Nickel absorption was higher for more readily soluble nickel salts: 0.01% of the administered dose for nickel administered in the form of nickel oxide, 0.09% for metallic nickel, 0.47% for nickel subsulfide, 11.12% for nickel sulfate, and 9.8% for nickel chloride (34).

1.4.1.3.1.3 DERMAL EXPOSURE. Animal studies have revealed that nickel can be absorbed through the skin (35, 36). Radioactive nickel sulfate, applied to shaved skin of rabbit or guinea pig was absorbed after 24 h and could be found in urine (35). Only a small percentage of radioactive nickel chloride was absorbed after 4–24 h since the application when it had been applied to the skin of guinea pigs. Levels as low as 0.005 and 0.51 of the dose were found in the blood and urine, respectively (36). The major part of nickel remained in the skin, mainly in its keratinized portions.

Elevated nickel levels observed in the liver and kidneys of the guinea pigs dermally exposed to nickel sulfate for 15 or 30 days also confirm possibility of dermal nickel absorption (37).

1.4.1.3.2 Distribution

1.4.1.3.2.1 INHALATION EXPOSURE. Results of studies on rats and mice show that less readily soluble nickel compounds are retained in the lungs for longer periods than the more easily soluble ones (28, 38, 39). Nickel retention varied from 6-fold (mice) to 10-fold (rats) higher in the animals exposed to slightly soluble nickel subsulfide than in the animals exposed to the more readily soluble nickel sulfate (38,39). The load on the lungs increased with duration and intensity of the exposure (40, 41).

After 3–4 days since the termination of the inhalation exposure of hamsters to high concentrations of nickel oxide (10–190 mg/m^3), as much as 20% of the inhaled oxide remained in the lungs. After 10 days following the termination of the exposure, the lungs retained 75%, and after 100 days 40% of the initial load (42).

Compared with the long retention of nickel oxide in the lungs, the more readily soluble nickel chloride is quickly removed from the organism of rats exposed intratracheally to a single dose at 1 mg/kg body weight (31). Six hours after the exposure, the highest nickel concentration was found in the kidneys; lower quantities were detected in the lungs, adrenals, pancreas, spleen, heart, and testis.

1.4.1.3.2.2 ORAL EXPOSURE. Owing to short-term and prolonged oral exposure of animals to different soluble nickel salts, nickel is accumulated primarily in the kidneys (34, 43, 44).

Considerable nickel levels were also detected in the liver, heart, and lungs (33, 43) and in the peripheral nerves and brain (44).

Nickel administered intragastrically to pregnant female rats can penetrate the placenta to the fetal blood and amniotic fluid (45).

1.4.1.3.2.3 DERMAL EXPOSURE. Nickel is absorbed through the skin of rabbits and guinea pigs (35, 36). After nickel chloride (^{63}NiCl$_2$) had been applied to the shaved skin of guinea pigs, radioactive nickel was detected in blood serum and urine (36).

After 24 h following its application to the depilated skin of rabbit and guinea pig, radioactive nickel ^{57}Ni (in the form of hexahydrate sulfate) was detected mainly in blood and kidneys; smaller quantities were also found in the liver (35).

1.4.1.3.3 Excretion

1.4.1.3.3.1 INHALATION EXPOSURE. Animal studies have revealed that the route of elimination after intratracheal administration of various nickel compounds depends on their solubility in water. When nickel chloride or nickel sulfite is administered to rats, approximately 70% of the dose is excreted with urine during 3 days (32, 46, 47). After 21 days, 96.5% of the dose is eliminated (32). Intratracheal administration of less readily soluble nickelous oxide or nickel subsulfide leads to nickel elimination with feces; this is associated with the mucociliary cleaning processes and swallowing.

An approximately equal proportion (90%) of the original dose of nickel oxide given to rats and of nickel subsulfide given to mice was eliminated during 35 days, whereas only 60% of the less soluble and more difficult-to-absorb nickel subsulfide was eliminated after 90 days (46, 48).

1.4.1.3.3.2 ORAL EXPOSURE. The major portion of the nickel dose given to animals with food is excreted with the feces (49). After one day since administration of nickel chloride to rats, 94 97% of the dose was removed with the feces, and 3–6% with urine (50). Only 1–3% of the dose of nickel sulfate given to dogs for 2 years in the diet was eliminated with urine (33).

1.4.1.3.3.3 DERMAL EXPOSURE. No data are available from studies on the elimination of nickel dermally administered to the organism.

1.4.1.3.3.4 KINETIC MODEL OF NICKEL METABOLISM. The kinetics of nickel metabolism in the organism was studied after injection of ^{63}Ni isotope in the form of nickel chloride and other soluble salts. Nickel concentration was determined in blood plasma relative to time since the injection, and then concentrations of nickel eliminated with urine and feces were assessed. Nickel is metabolized according to a typical two-compartment model (51). As a result of a study on female and male Wistar rats and rabbits exposed IV to ^{63}NiCl$_2$ at 17 µg/rat and 816 µg/rabbit, it has been found that during the first day following the injection, 68 and 78% of the dose, respectively, were eliminated with urine (52). Sunderman et al. (53) exposed rats to nickel isotope by IP injection of 2173 µg ^{63}Ni chloride. They reported a phase of quick clearance from blood plasma and serum during the first 2 days, followed by a phase of much slower clearing during the next 3–7 days.

The two-compartment model of nickel elimination was confirmed by the results of studies on volunteers, who were given nickelous sulfate at 12, 18, and 50 µg nickel/kg in drinking water or food; as much as 76% of the dose given in water and 102% of the dose given with food was eliminated with feces during 4 days. Half-life for the elimination of the absorbed nickel dose was 28 ± 9 h (17–48-h range) (54).

1.4.1.3.3.5 SUSPECTED MECHANISM OF CARCINOGENESIS. The mechanism of the carcinogenic activity of nickel remains unexplained. The ability of various nickel compounds to produce cancer is supposed to be associated with the process of their phagocytosis (55, 56) and with ability to induce erythrocytosis (57). Results of *in vitro* and *in vivo* tests revealed that nickel affected DNA and proteins and could produce such effects as DNA–protein crossbinding (probably heterochromatins), DNA thread fragmentation resulting in DNA damage, and other cytotoxic effects leading to changed gene expression of the surviving cells (58–63). Nickel influences DNA synthesis, inhibits DNA synthesis (which has been proved in tests on many organisms), and inhibits DNA repair processes (64, 65). Impaired repair mechanism may be an essential synergistic factor in the carcinogenic activity of nickel and other compounds (56, 66).

It has been shown that the Ni^{3+}/Ni^{2+} redoxy potential may, in the presence of certain peptides *in vitro*, enhance formation of free oxygen radicals (67, 68). Nickel, as metal ion, catalyzes the process of molecular oxygen transformations. The case is similar with other chemical compounds stimulating the phagocyting cells to produce oxygen radicals that damage DNA, proteins and lipids. Those phenomena may result in carcinogenic lesions (69).

According to Sunderman and Barber (70), the model of interaction between Ni^{2+} and DNA involves substitution of Zn^{2+} at the locus (identified in other protooncogenes) of binding that ion in the DNA. As the radii of Ni^2 and zinc ion are similar, the replacement of those metals may affect protein configuration and stability of the structures associated with DNA, which act on expression and specific free-radical reactions, causing DNA splitting formation of DNA-protein crossbinding and disturbed mitosis.

1.4.1.4 Reproductive and Developmental. Few data are available on the action of metal nickel on the reproductive system and fetal development. Chang et al. (71) observed that, after inserting a nickel wire into one corner of the uterus of female rats at day 3 of gestation, the number of implantations was reduced and the number of resorptions decreased compared to the opposite corner of the uterus, where no wire was present.

Addition of metallic nickel powder to the medium used to culture hen embryo myeloblasts (20 to 40 μg Ni/L medium) inhibited cell differentiation at only several mitoses visible after 5-day incubation. Cytodegradation-related reduction of mitotic cycles lead to a decrease in the size of the colony. At 80 μg/L there was an extensive degeneration of the culture and complete inhibition of the mitoses for 5 days (72).

1.4.1.5 Carcinogenesis. The carcinogenic activity of nickel and its alloys was studied on mice, rats, and guinea pigs exposed by inhalation; on rats exposed by intratracheal administration; on rats and hamsters receiving intramuscular injections; on rats, mice, and rabbits receiving IV injections; and also on rabbits and rats receiving intrathigh implants.

Hueper (22) and Hueper and Payne (73) studied the effects of exposure to metallic nickel suspended in air. Hueper (22) exposed inhalationally female C57Bl mice, Wistar rats (both sexes), female black Bethesda rats, and guinea pigs (both sexes) to metallic nickel dust containing 99% pure nickel (particle size below 4 μm at 5 mg/m^3 (6 h/day, 4–5 days/week for 21 months). Lung cancer was not observed in the exposed mice. Fifteen of the fifty rats (both strains) subjected to histological examinations had benign adenomas. Lung mesotheliomas were detected during zootomy in the guinea pigs. An anaplastic intra-alveolar carcinoma was found in one guinea pig.

In a study by Hueper and Payne (73) during which rats were exposed by inhalation to nickel metal powder blended with sulfur dioxide and chalk powder, lung cancer was not detected in any of the animals, but squamous metaplasia and peribronchial adenomatoses were observed. Kim et al. (74) exposed Wistar rats by inhalation to nickel metal dust at 31 mg/m^3 (6 h/day, 5 days/week) for 21 months. Two rats of the exposed group developed lung cancer. One rat of the control group also had lung cancer.

Results of studies in which experimental animals were exposed intratracheally to nickel confirm its carcinogenic potential. Lung cancers (adenocarcinomas and squamous carcinomas) were observed in rats (75) and hamsters (76) intratracheally exposed to nickel powder. It has been proved that IM administration of nickel powder causes development of sarcomas at the place of administration (77–79). Hamsters also develop sarcomas at the place of IM nickel powder administration, but the frequency was low (2 of 50 animals developed sarcomas) (80). Nickel powder given intrapleurally to the Fisher 344 rats produced mesotheliomas (81).

Nickel metal powder produced sarcomas in rats, but not in mice (82). Neoplastic tumors (sarcomas, mesotheliomas, and other cancer types) were observed after intraperitoneal administration of metallic nickel to rats (83, 84).

1.4.1.6 Genetic and Related Cellular Effects Studies. Nickel metal did not induce chromosomal aberrations in cultured human lymphocytes (85); however, dose-dependent increase of transformations in Syrian hamster embryo cells was observed (86).

Nickel metal inhibited CHO cell S mitosis phase in Chinese hamster (87). Hansen and Stern (88) found that nickel metal caused transformation of BHK21 cells. Cell proliferation in liquid agar was assessed. At the toxic doses, nickel metal and crystalline nickel subsulfide showed similar activity for cell transformation, and the 200 μg/mL nickel metal dose was equivalent to 10 μg/mL nickel subsulfide.

1.4.2 Human Experience

1.4.2.1 General Information. Information on the acute and chronic poisonings by nickel metal in people is limited and, in the majority of cases, refers to effects of the combined exposure to dusts or fumes comprising mixtures of metallic nickel, and its oxides and salts. Contact hypersensitivity to nickel and its salts, however, is quite well documented.

1.4.2.2 Clinical Cases

1.4.2.2.1 Acute Toxicity. Death due to adult respiratory distress syndrome 13 days after inhalation exposure (90 min) to high concentrations of respirable metallic nickel dust was reported by Rendall et al. (89). Simulated exposure conditions showed that air nickel concentration could be as high as 382 mg/m^3. Several days after the exposure, the patient's urine nickel concentration was 700 µg/L [vs. 0.1–13.3 µg/L in the urine of nonexposed people (90)]. Histological examination revealed damaged alveolar walls, lung swelling, and necrosis of the renal tubules.

Sandström et al. (91) reported a case where 13 people were poisoned (including one lethal outcome) as a result of inhalation exposure to respirable nickel dust in 1943.

Symptoms of acute poisoning, such as nausea, vomiting, weakness, headache, and palpitations, were reported in 23 dialyzed patients as a result of contamination of the dialysate by nickel from a nickel-plated container. After the dialysis had been terminated, the symptoms quickly disappeared (92).

1.4.2.2.2 Chronic and Subchronic Toxicity. Prolonged inhalation exposure to nickel dusts and fumes produces pulmonary lesions, including fibrosis. Zislin et al. (93) studied the respiratory function in 13 workers exposed to nickel dust for periods ranging from 12.9 to 21.7 years. The patients had lower pulmonary residual capacity and higher breathing rate; radiography revealed diffuse fibrosis diagnosed as nickel-induced pneumoconiosis.

Effects of nickel dust on the respiratory system were also studied in welders. In 1955–1979 Zober (94, 95) found 47 histopathologically confirmed cases of pulmonary fibrosis in electric arc welders. Although the pathology pointed to a number of causes, welding fumes were assumed to be the cause of the fibrosis in 20 cases. Zober (96) examined 40 welders (of whom 22 were nonsmokers and 18 were tobacco smokers), who welded chromium- and nickel-containing elements using inert-gas arc. Nickel concentration in the ambient air was not higher than $0.5 \, \text{mg/m}^3$, with the exception of one process, where nickel concentrations was as high as $1.2 \, \text{mg/m}^3$. The reaction of the respiratory system, including bronchitis and rales in the lungs, was stronger in the tobacco-smoking welders.

Metallic nickel and its alloys may cause contact dermatitis. Dermatoses and other dermal lesions due to nickel were observed in both occupationally exposed (97–99) and in the general (100–104) populations. Allergy to nickel is the most frequent cause of contact dermatoses in women and one of most frequent in men; approximately 10–15% of the female, and 1–2% of the male populations are hypersensitive to nickel (100, 101). The high incidence of allergy to nickel in women may be attributed to the fact that many women wear metal jewelry (earrings in particular); the allergy may be also due to wearing metal buttons, watches, or zippers (105, 106).

Allergic reactions or urticarious and eczematous dermatitis may occur in nickel-sensitive people following application of artificial prosthesis made of nickel-containing alloy (107).

1.4.2.2.3 Pharmacokinetics, Metabolism, and Mechanisms. Nickel exposure–related occupational health risks result primarily from the inhalation of nickel or its salts. Therefore, deposition, retention, and removal processes occurring in the lungs are essential. Other types of human nickel exposure include oral and dermal contact routes.

1.4.2.2.3.1 Absorption

Inhalation Exposure. The inhaled particles both of metallic nickel and its salts settle in the upper and lower portion of the respiratory tract. The deposition of the particles in the respiratory tract is dependent on particle size; particles $> 2\,\mu m$ settle in the upper respiratory tract, and smaller particles and aerosols travel to the lower portion of the respiratory tract, where less intense airflow and system geometry enhance sedimentation, diffusion, and electrostatic precipitation of the particles.

Approximately 35% of inhaled nickel is absorbed in humans from the respiratory system to blood (108, 109). The other portion of the inhaled nickel is removed by the mucociliary mechanism and expectorated or swallowed. Some small portion of the inhaled nickel is deposited in the lung and lymphatic tissues (90). Nickel was detected in urine of the exposed workers (110, 111). Nickel levels detected in the urine of workers exposed to soluble nickel compounds (nickel chloride, nickelous sulfate) were higher than those workers of exposed to the nonsoluble nickel chemicals (nickelous oxide, nickel subsulfide), indicating that soluble nickel compounds were more quickly absorbed from the respiratory system.

Oral Exposure. Following intake of nickelous sulfate with drinking water, the portion of nickel absorbed from the GI tract is 40 times higher (27%) than when the nickel is digested with food (0.7%) (54). The ability of the organism to absorb nickel fed to empty stomach, in terms of blood serum nickel level, is higher after intake of nickelous sulfate in drinking water (peak concentration $80\,\mu g/L$ after 3 h). The biological availability of nickel varies with the particular type of the drink; it was highest for the refreshing drinks, and lowest for milk, coffee, tea, and orange juice. Ethylenediaminetetraacetic acid added to the diet reduced nickel biological availability (112). Nickel intake index was determined from the results of examination of eight people on normal diets. Its value is 4.3% of the ingested nickel dose (113). Abundant gastric content remarkably reduces nickel absorption.

Dermal Exposure. Nickel salts are able to penetrate human skin. As much as 55–77% of the radioactive nickelous sulfate applied to human skin is absorbed during 24 h; the majority is absorbed within several hours since the application (114). Examinations of human skin specimen revealed that only 0.23% of the applied dose penetrated the skin after 144 h when nickel chloride was applied to the skin without any dressing; the corresponding value for nickel chloride applied under dressing was 3.5%. The nickelous ion of nickelous chloride solution penetrated the skin about 50 times as quickly as the nickelous ion of sulfate solution (115).

1.4.2.2.3.2 Distribution and Accumulation in the System

Inhalation Exposure. Nickel concentrations were determined in tissue and organ samples collected during autopsy from the general population and from the occupationally and environmentally exposed populations. Nickel concentration was higher in the exposed people (116–118). Lungs were the main place of its deposition in both nonexposed and exposed people. Nickel levels in liver and kidneys were considerably lower than those in the lungs (119). Nickel was also detected in the nasal mucosa of workers occupationally exposed to insoluble nickel forms (120).

Oral Exposure. Data on nickel deposition in the tissues of people who ingested nickel compounds are not available. Peak blood serum nickel concentration was found after 2.5–3 h following ingestion of nickelous sulfate (54, 113). Nickel appeared in blood serum following accidental intake of nickelous sulfate-contaminated water. The half-time of its elimination was 60 h (121).

Dermal Exposure. Data on the distribution of nickel applied to human skin are not available.

1.4.2.2.3.3 ELIMINATION

Inhalation Exposure. Independent of the route of administration, the absorbed nickel is excreted with urine (110, 120, 122). In the urine of exposed workers, nickel is found already at the beginning of the work shift which indicates its fast elimination from the system. Increased nickel excretion with urine was also observed at the end of the work week (123). Nickel is also eliminated with feces. This is probably associated with swallowing of the nickel removed by mucociliary clearance mechanism from the respiratory system (122). Higher nickel concentrations were found in the urine of the workers exposed to the soluble, more readily absorbable nickel compounds than of those exposed to the insoluble compounds (120).

Oral Exposure. The major portion of the ingested nickel in humans is removed with urine. Nickel absorption in the GI tract is lower when nickel compounds are given with food. As much as 27% nickel is absorbed from human GI tract when nickelous sulfate is given with drinking water, whereas only 0.7% is absorbed when the compound is given with food (54). Following 4-day medical treatment, 26% of the dose received with drinking water was removed with urine, and 76% with feces; when received with food, 2% of the dose was removed with urine and 102% with feces (54).

Dermal Exposure. Reports from studies on the elimination of nickel absorbed transcutaneously are not available.

1.4.2.2.4 Reproductive and Developmental. Data on the effects of nickel metal on human reproduction and offspring development are not available.

1.4.2.2.5 Carcinogenesis. An International Agency for the Research on Cancer (IARC) expert team has prepared a review of the epidemiological studies on nickel carcinogenicity (124). The reviewed studies referred to workers of nickel refining plants, nickel mines, and nickel processing plants in various countries. The employees of nickel processing plants are exposed to dusts of metallic nickel, nickel oxides, and nickel sulfites, and to soluble nickel salts occurring at various proportions and having different grain sizes.

The results of many studies are influenced by interfering factors, such as simultaneous exposure to other carcinogens (arsenic, cadmium) or irritating gases (hydrogen sulfide, ammonia, chlorine, sulfur dioxide) (124, 125).

1.4.2.2.5.1 NICKEL REFINING PLANTS AND MINES. First epidemiological studies performed in a nickel refinery in 1939 by Hill and published by Morgan (127) demonstrated that the O/E morbidity for lung cancer was 16/1, while the corresponding value for nose cancer was 22/1. Morgan (127) and Doll (128) found 161 deaths from lung cancer and 61 deaths from nose cancer in a group of 9340 examined workers (the studies referred to the period

1902–1957). Increased mortality was observed in the workers employed before 1924. The decreased death rate was said to be associated with improved working conditions such as use of protective masks and improved ventilation, and also with the use of pure sulfuric acid (not contaminated by arsenic). As much as 39 cases of nose cancer and 113 cases of lung cancer were detected in a cohort of 819 workers employed in the Clydach refinery for at least 5 years between 1920 and 1944, and those employed in 1934 (129). The O/E ratio for lung cancer was 10.5 in the workers employed in 1915–1919, 1.8 in those employed in 1925–1929, and 1.1 in the workers employed during 1930–1944.

In an analysis of death causes in a group of 297 workers exposed during 1937–1960 primarily to soluble nickel salts, nasal sinus cancer cases were not detected. However, the number of lung cancer cases was 13, rather than the 7.54 expected. The differences between these values and the data for England and Wales were not statistically significant (130).

In a cohort of 968 workers employed until 1981, when 159 deaths from lung cancer and 56 from nose cancer were recorded. A relationship was found to occur between the cancer incidence and the exposures at workplaces in the following refining plant departments: the calcining furnace area, the calcining/crushing area, the copper sulfate area, and the furnace area. The risk of development of the two cancer types increased with the duration of employment (131).

A cohort of 1916 workers employed in the Falconbridge refinery at Kristiansand (Norway) for minimum 3 years before 1961 was analyzed by Pedersen et al. (132). The risk of cancer development was assessed against worker job. The risk of lung cancer was the highest for the workers employed at the electrolysis department (O/E $26/3.6 = 7.2$); the risk of nasal cancer was also very high (O/E $6/0.2 = 30$). Nasal cancer risk was the highest at the roasting and smelting department (O/E $5/0.1 = 50$). The risk of lung cancer for the employees of that department was O/E $12/2.5 = 4.8$. Nickel concentration in the lungs of 15 workers employed at the department was 330 ± 380 μg/g dry mass [arithmetic mean ± standard deviation (SD)]. During autopsy, nickel concentration of 150 ± 280 μg/g was detected in the lungs of 24 electrolysis department workers; nickel concentration in the lungs of people not employed in the refinery was 0.76 ± 0.39 μg/g (117).

Andersen (133) built up a cohort of 1237 workers employed for at least one year who started working in 1968 or later. Seven cancer cases were observed in the workers starting their work in 1968–1972 against one expected. In 1979, nickel concentration in the electrolysis department was $0.1-0.5$ mg/m^3, and in the annealing department it ranged from 0.1 to 1.0 mg/m^3. The concentrations calculated for the period 1968–1977 were $0.5-2.0$ mg/m^3 and $2.0-5.0$ mg/m^3, respectively (120). Dysplasia of the epidermis was recorded from the results of autopsy in 318 active and 15 retired workers. As much as 14 cases (14.4%) of epithelial dysplasia were found in 97 workers of the roasting/smelting department, 16 cases (11%) in the workers of the electrolysis department, 98 cases (10%) in other workers, and 7 cases (47%) in the group of the retired workers. Nasal cancer was found in two co-workers from the roasting/smelting department. Despite changed exposure intensity due to the modernization of the industry, the risk of lung cancer among workers employed in 1930–1960 was not changed (134).

Nickel sulfide ores have been mined and processed by the International Nickel Company (INCO), Ontario. The analysis of a cohort of 2355 refinery workers (years

1930–1960) revealed 7 cases (at 0.19 expected) of death from sinus cancer [standardized mortality rate (SMR) = 3684] and 19, at 8.45 expected, from lung cancer (SMR = 224). Extending the time of employment (years 1930–1975) did not affect the risk of development of nasal cancer — 16 observed deaths at 0.166 expected (SMR = 9638) and lung cancer — 37 observed at 12.71 expected (SMR = 291). A cohort of 54,724 INCO employees was examined for the causes of mortality during the period 1950–1976. Among the workers employed at the ore sintering department, SMR for nose cancer was 2174 (O/E = 2/0.09), for lung cancer the corresponding value was 463 (O/E = 42/9.08). In other departments, SMR for nose cancer was 144 (O/E = 3/2.08), and for lung cancer SMR was 108 (O/E = 222/204.98) (135).

In a cohort of 11,494 workers of Falconbrige (Sudbury, Ontario) nickel mine using other sintering method, cases of nasal cancer were not observed. There was an insignificant increase in the incidence of lung cancer (SMR = 123, O/E = 46/37.5). Increased deaths from larynx tumors were recorded in the miners (SMR = 261, O/E = 5/1.92) (136).

In the Sherritt Gordon and Fort Saskatchewan (Alberta) mines, the employees were exposed during hydrorafination to nickel oxide ore concentrate, metallic nickel, and its soluble compounds. In a cohort of 720 workers, respiratory cancers were not recorded. Two cases of renal cancer were observed in the workers employed for 11 and 16 years, respectively (137).

In the Huntington (West Virginia) refinery operated from 1922 to 1947, mortality was analyzed in 1988 for workers employed for periods not shorter than one year (138). Air nickel concentration ranged from 20 to 350 mg/m^3 near the ore crushing area and from 5 to 15 mg/m^3 in the sintering department, SMR was 2443.5 (O/E = 2/0.08).

As much as 97 cases of lung cancer were found in workers employed in a New Caledonia refinery processing nickel oxide ore. The incidence of tumors was 3–7 times higher than in other Pacific regions and 3 times higher in the employees of the refinery than in people not employed in the refinery, independent of their age and tobacco smoking habit (139). Those data were questioned because of problems with selecting proper controls (140).

Increased incidence of lung cancer (180% higher than in the population of the town) and gastric carcinoma was recorded in nickel refineries in the region of the Ural Mountains in the former USSR (141, 142).

Andersen et al. (143) investigated the relation between occupational hazards among nickel refinery workers and their exposure to different forms of nickel over time and the interaction between smoking and total exposure to nickel.

The cohort consisted of 379 workers with firs employment 1916–1940 and at least three years of employment and 4385 workers with at least one year of employment 1946–1983. Data on smoking (ever or never) where available for almost 95% of the cohort. Two analyses were used: indirect standardization from observed and expected numbers and Poisson regression.

During the follow up 1953–1993, 203 new cases of lung cancer were observed against 68 expected [standardized incidence ratio (SIR) 3 × 0.95% confidence interval (95% CI) 2 × 6–3 × 4] and 32 cases of nasal cancer were observed against 1 × 8 expected (SIR 18 × 0, 95% CI 12–25). The Poisson regression analysis showed an excess risk of lung cancer in association with exposure to soluble forms of nickel, with a threefold increase in relative risk (RR) ($P < 0 \times 0.001$) and a multiplicative effect of smoking and exposure to nickel. The RRs were 1×1 (95% CI $0 \times 2-5 \times 1$) for exposed workers who had never smoked and 5×1 (95% CI $1 \times 3-20 \times 5$) for exposed workers who smoked. It is not possible to state with certainty which specific nickel compounds are carcinogenic, but a significant excess risk was found for workers exposed to soluble nickel alone or in combination with other forms of nickel. The study suggests a multiplicative effect of smoking and nickel exposure.

1.4.2.2.5.2 NICKEL PROCESSING PLANTS. Workers of the Herebord (England) nickel processing plants employed at the production of nickel alloys were exposed to metallic nickel and nickel oxides. Air nickel concentrations ranged from 0.04 to 0.84 mg/m^3, depending on the processing method. In the studied cohort of 1925 workers employed for at least 5 years from 1953 to 1978, nose cancer cases were not recorded, and SMR value for lung cancer was not increased (144). Other studies involved large cohorts. These consisted of nickel alloy smelting plant workers and included 28,261 people (145). Proportional mortality analyses were also performed, based on 3323 (146) and 851 (147) death cases in those groups. The results also failed to confirm statistically significant increase in the risk of developing nose, lung, sinus, throat, or kidney cancer.

In France, a study included a cohort of 269 workers employed during 1952–1982 at the production of iron/chromium alloys and stainless steel. The results confirmed statistically significant increase in the incidence of lung cancer (SMR = 2.04 vs. SMR = 0.32 for the nonexposed group) (148).

A cohort of 508 workers of an electroplating plant was classified into groups according to the duration of the employment. A statistically significant increase of the number of deaths from stomach cancer (SMR = 623, O/E 4/060) and from non-cancer respiratory diseases (SMR = 286, O/E = 8/2.8) was reported (149).

Nickel present in concentrations 0.1 to 1.0 mg/m^3 in the atmosphere of a uranium enrichment plant operated 1948–1969 did not cause any increase in the risk of respiratory system cancers. The study was performed in a cohort of 814 people (150).

Studies on a cohort of 3025 employees of a cadmium/nickel rechargeable cell plant revealed SMR increase: for the workers employed for the first time in 1923–1946 SMR = 123 (O/E = 52/42.4), and for the workers employed in 1947/75 SMR = 137 (O/E = 35/25.6). This could not be attributed solely to nickel since exposure to both nickel and cadmium was considerable (151).

1.4.2.2.6 Genetic and Related Cellular Effects Studies. Few data are available on the genotoxic activity of nickel and its compounds in the exposed humans.

Waksvik and Boysen (152) described 2 groups of workers employed at nickel refining, in whom an increased level of chromosome aberrations, was observed. Increased incidence of sister chromatid exchange was not observed. One group of the workers was employed at nickel ore crushing, sintering, and smelting department, where they were exposed mainly to nickel oxide and sulfide; the other group was employed at the electrolysis department, where exposures to nickel chloride and sulfate prevailed.

Waksvik et al. (153) studied also nine former workers, retired for 8 years, and who had been exposed to nickel for ~ 25 years, and whose blood plasma nickel level continued to be as high as 2 μg/L; the number of observed chromatid breaks was also small. Dang et al. (154, 155) found increased number of sister chromatid exchanges and chromosome aberrations (gaps, breaks, fragmentations) in seven workers employed at electroplating.

These data, although limited, suggest that inhalation exposure to nickel produces genotoxic effects in the exposed people.

1.5 Standards, Regulations, or Guidelines of Exposure

Recommended exposure limits are listed in Table 41.4 (156–159).

1.6 Environmental Impact

Nickel is emitted to the environment both from the natural (oceans, soil, volcanic dusts, ashes from forest fires) and anthropogenic (nickel mining and refining, fuel oil burning, steel an nickel-alloymaking) sources. The human-related sources supply 5 times as much nickel to the atmosphere as the natural ones (108). The environmental nickel circulates in the atmosphere, hydrosphere, and biosphere a result of such processes as settling and sedimentation and, to a much lesser extent, biotransportation by living organisms.

Data on the chemical form of the atmospheric nickel coming from the natural emission sources are not available. Nickel oxides and sulfates dominate in the fly ashes emitted from the anthropogenic sources (5).

Nickel enters the hydrosphere as a result of airborne dust fallout, from communal and industrial waste dumped into the rivers, by washing of ground surface, and as a result of the natural erosion of soil and rocks. Nickel ore refining plants and nickel-polluted sewage sludge dumping sites are the main sources of nickel emission. Nickel contained in the soil may, depending on its structure and chemical properties, show a considerable mobility throughout the layer (160).

Nickel bioavailability depends on its solubility. Most of nickel compounds are water-soluble at pH < 6.5. Thus, acid rains enhance nickel mobility in the soil and increase nickel concentration in the underground water, thus enhancing nickel intake by the living organisms and increasing its toxicity to microorganisms, plants, and animals (161).

Terrestrial and water plants absorb and accumulate nickel. Nickel content of the terrestrial plants ranges from 0.05 to 5 mg/kg dry mass; concentrations higher than ~50 mg/kg dry mass is usually toxic to them (162). Because of their essential role in the food chain of many animals, algae deserve special attention among the water plants. Nickel level in the algae collected from the contaminated areas was as high as 150.9 mg/kg dry mass, which was 10 times higher than the normal level (163).

Systemic sensitivity to nickel varies considerably. Generally speaking, microorganisms are less susceptible to its toxic activity than the higher ones. Nickel at 5–30 mg/L inhibited growth of many bacteria; yeast (*Aspergillus niger*) growth was inhibited at 1–40 mg/L, but some fungi reacted to nickel only at concentrations as high as 1000 mg/L (164, 165).

Water algae growth inhibition was observed at 0.05–5 mg/L (166). Nickel toxicity in the water invertebrates varies according to species abiotic factors, such as pH, hardness, temperature, and salinity of the water (167, 168). The LC_{50} value after 64 h determined for

Table 41.4. Exposure Limits for Nickel Metal (All Limits in mg/m^3)[a]

Exposure Limits	OSHA PEL (156)	NIOSH Exposure Limit (156)	ACGIH TLV (156)	AIHA WEEL	ANSI Standard	German MAK (157)	Swedish Standard (158)	British Standard (158)	Polish Standard (159)
Time-weighted average	1	0.015	1.5	—	—	—	0.5	0.5	0.24[b]
Short-term exposure limit	—	Ca	Ca A5 (i)	—	—	A$_1$(as inhalable dust)	S	—	—
	—	—	—	—	—	—	—	—	—
Ceiling limit	—	—	—	—	—	—	—	—	—
Biological limits (if available)	—	—	—	—	—	—	—	—	—

[a]*Carcinogen designations*: NIOSH—carcinogen defined with no further categorization; ACGIH—TLV A5, note suspected to be human carcinogen; German—MAK A1, capable of inducing malignant tumors as shown by experience with humans; (i)—Inhalable fraction; S—The substance is sensitizing.
[b]Standard for nickel and its compounds, excluding nickel carbonyl.

Daphnia magna was 0.32 mg nickel per liter at 25°C (169). Exposure of *D. magna* to nickel sulfate at concentrations ranging from 5 to 10 μg nickel/L for the life of three generations resulted in its extinction (170). Nickel threshold concentration for the inhibition of *Chironomus riparis* larva growth is 1.1 mg/L (171).

Just as in the invertebrates, the susceptibility of fish to nickel varies from species to species and is dependent largely on water hardness. Young individuals are poisoned much quicker than adults (172). The LC_{50} value after 48 h exposure of rainbow trout (*Salmo gairdneri*) was 80 mg/L for hard water and 20 mg/L in soft water (173). Studies by Calamari et al. (174) suggest that nickel ability to accumulate in the tissues of fish is small.

Few data are available regarding nickel effects on the terrestrial animals. Experiments on earthworms showed that they were relatively insensitive to nickel if the environment was rich in microorganisms and organic matter (14-day LC_{50} for *Eisenia foetida* was determined to be as high as 757 mg/kg soil (175).

Nickel is not considered to be a global environmental pollutant; nevertheless, gradual environmental changes, including reduced diversity of species, have been observed, to occur in the areas around the sources of nickel and other trace-element pollutants (176–178).

2.0 Nickel Oxide

2.0.1 CAS Number: *[1313-99-1]*

2.0.2 Synonyms: Bunsenite; C.I. 77777; green nickel oxide; mononickel oxide; nickel(II) oxide; nickelous oxide; nickel monoxide; nickel oxide sinter 75; nickel protoxide; nickel(II) oxide, green; nickel protoxide; nickel(II) oxide, black; nickel(II) oxide (1 : 1); mononickel oxide; nickel oxide–tungsten oxide.

2.0.3 Trade Names: NA

2.0.4 Molecular Weight: 74.69

2.0.5 Chemical Formula: NiO

2.0.6 Molecular Structure: Ni=O

2.1 Chemical and Physical Properties

Table 41.1 shows the physical properties of nickel oxide. Nickel oxide (1, 2) is insoluble in water; its solubility in acids and other properties depend on of its preparation process. The crystalline form of nickel oxide depends on its calcination temperature.

This compound occurs as either black or green solid. Black nickel oxide is chemically active and forms simple salts in the presence of acids, while green nickel oxide is passive and heat-resistant.

2.2 Production and Use

Green nickel oxide is prepared by roasting powdered nickel and water at a temperature of 1000°C (180). Black nickel oxide is made either of nickel hydrogen carbonate or nickel nitrate calcinated at a temperature of 600°C.

Nickel oxide is an important raw material in the processes of smelting and alloymaking. It can be also used as a catalyst or a stain to color glass and china (porcelain) ware (180).

2.3 Exposure Assessment

Nickel oxide content in air and biological materials is determined by the ICP-AES and AAS methods (See Sections 1.3.1 and 1.3.2).

2.4 Toxic Effects

2.4.1 Experimental Studies

2.4.1.1 Acute Toxicity. Table 41.5 gives lethal doses of nickel oxide.

Nickel oxide is characterized by rather low acute toxicity. A single administration of nickel oxide to rat trachea (3, 30, 300 µg/kg body weight (bw)) caused no effects within 7 days following the administration (181). Similarly negative results were obtained by Toya et al. (182) after a single injection of 13 mg NiO/kg body weight into rat trachea. All exposed animals survived the experiment, and neither macroscopic nor microscopic changes were observed in their lungs. The value of LD_{50} determined by these authors for nickel fumes (1.2–3.2% Ni_2O_3 and 96.8–98.8% NiO) administered intratracheally was 38.2 mg/kg body weight (182).

2.4.1.2 Chronic and Subchronic Toxicity. Prolonged inhalation exposure of hamsters, mice, and rats to nickel oxide caused this compound to accumulate in their lungs (41, 183).

Wehner et al. (183) exposed hamsters to nickel oxide at 53 mg/m^3 for their whole lifespan and found that nickel oxide molecules were accumulated in alveoli. During the initial exposure stage, emphysema was also observed and later on pneumonia was gradually developed. On the other hand, the animals' lifespan was not shortened.

Repeated inhalation exposure to nickel oxide caused chronic inflammatory conditions in rats' pulmonary alveoli and chronic interstitial pneumonia in mice. The growth in alveolar macrophages with NiO molecule accumulation was observed in both species (41).

Pneumonia and increased alveolar macrophage counts were observed in rats and mice exposed to nickel oxide with a concentration of 24 mg/m^3, 6 h/day, 5 day/week, for 12 days. Mouse lungs were less injured. Some rats showed also an olfactory epithelium atrophy, whereas athymia and lymphatic gland atrophy were observed in both species (184). A sixfold pulmonary weight increase and alveolar proteinosis as well as shorter life were observed in rats that had died during their lifelong exposure to 60 or 200 µg/m^3 nickel

Table 41.5. Acute Toxicity of Nickel Oxide

Species	Administration	Lethal dose	Ref.
Rat	Oral	LDL_0 5 g/kg	18
Rat	Intramuscular	LDL_0 180 mg/kg	19
Dog	Intravenous	LDL_0 7 mg/kg	19
Cat	Intravenous	LDL_0 10 mg/kg	19

oxide aerosol (185). Wistar rats exposed to green nickel oxide at 9.3 or 1.2 mg/m^3 showed no histopathological changes after 12 months of exposure (27).

Prolonged inhalation exposures of Wistar female rats for 120 days (2880 h) to nickel oxide aerosol (generated from nickel acetate at 550°C) at concentrations of 200, 400, and 800 μg/m^3 resulted in characteristic and dose-related decrease in renal and hepatic weights, a decrease in blood cell count, a drop in alkaline phosphatase activity in serum, an increase in lung weight and the number of leucocytes, and an increase in the average volume of blood cells (186).

2.4.1.3 Pharmacokinetics, Metabolism, and Mechanisms. See Section 1.4.1.3.

2.4.1.4 Reproductive and Developmental. Data on the effect of nickel oxide on reproduction are not available. No embryotoxic effects were observed in Wistar rats exposed to 1.6 mg/m^3 nickel oxide aerosol on days 1–20 of pregnancy (186).

2.4.1.5 Carcinogenesis. The carcinogenic effects of nickel oxide was examined in rats (187–189), hamsters (183) and mice (189) exposed by inhalation; in rats and mice exposed by intramuscular injection (190, 191); and in rats exposed to the chemical by intratracheal (192), intrapleural (193), and intraperitoneal (84) injections. The carcinogenic effects of nickel oxide on the experimental animals are listed in Table 41.6.

The results of the preceding studies on experimental animals (rats and mice) indicate that nickel oxide is carcinogenic at the locus of administration after intramuscular, intratracheal, and intrapulmonary injections. Rats' responses to inhalation exposures were not unambiguous, as the results of two experiments were negative (187, 188), whereas the results of experiments performed by NTP (189) indicate that nickel oxide increases the incidences of either benign or malignant pheochromocytoma of the adrenal medulla in both female and male rats. Mice proved to be less susceptible to the carcinogenic effects of nickel oxide (189).

The carcinogenic effects of nickel oxide were also observed after intratracheal administration (192). The inhalation exposure of Syrian hamsters did not produce carcinogenic effects (183).

2.4.1.6 Genetic and Related Cellular Effects Studies. Nickel oxide was not mutagenic in the *Bacillus subtilis* test (194), but it caused transformation of Syrian hamster embryo cells (55, 86) and chromosome aberrations in the culture of human lymphocytes (152).

Nickel oxide tended to prevent Chinese hamster CHO cells from undergoing the S phase of mitotic division (87).

2.4.2 Human Experience

2.4.2.1 General Information. Information on the toxic effects of nickel oxide on humans has been obtained from studies on workers engaged in, among other things, nickel refining processes and manufacture and welding stainless-steel and nickel-containing alloys. Nickel oxide along with nickel sulfide, sulfate, and carbonate as well as nickel metal occur in fumes from the refining processes (5). It is estimated that 50–95% nickel present in dusts and fumes emitted during alloy smelting (Inconel Alloy 600, Incoloy Alloy 800)

Table 41.6. Carcinogenic Effects of Nickel Oxide

Species	Administration	Dose	Experimental Conditions	Effects	Ref.
Male Wistar rats	Inhalation	0.6 or 0.8 mg/m^3	6 h/day, 5 days/week for 1 month followed by observation for 20 months	1 adenocarcinoma in the group exposed to lower concentrations	187
Male Wistar rats	Inhalation	0.06 mg/m, 0.2 mg/m^3	Continuous exposure for 18 months followed by observation for 1 year	No carcinogenic effects	188
F344/N rats	Inhalation	0.062, 1.25, or 2.5 mg/m^3	6 h/day, 5 days/week for 104 weeks	Alveolar/bronchiolar adenomas or carcinomas in the group of males: 1/54 (2%), 1/53 (2%), 6/53 (11%), 4/52 (2%), respectively, to the dose in the female group: 1/53 (2%), 0/53 (11%), 5/54 (9%), respectively, mild or malignant chromaffin tumors of adrenal glands in the male group: 27/54 (50%), 24/52 (46%), 27/53 (51%), 35/52 (67%), respectively in the female group: 4/51 (8%), 7/52 (13%), 6/53 (11%), 18/53 (34%), respectively	189
B6C3F$_1$ mice	Inhalation	0, 1.25, 2.5, or 5 mg/m^3	6 h/day, 5 days/week for 104 weeks	Alveolar/bronchiolar adenomas or carcinomas in the male group: 9/57 (6%), 14/67 (21%), 15/66 (23%), 14/69 (20%), respectively in the female group: 6/64 (9%), 15/66 (23%), 12/63 (19%), 8/64 (12%), respectively	189
Male Wistar rats	Intratracheal	5 or 15 mg	10 installations in weekly intervals, observation for 124 weeks	Pulmonary cancer in both exposed groups: 10/37 (27%) and 12/38 (31.6%), respectively, including 4 adenocarcinomas and 16 squamous carcinomas	192

Male Syrian golden hamsters	Inhalation	53.2 mg/m^3	7 h/day, 5 days/week whole life span	1/51 osteosarcoma, no pulmonary cancer was observed	183
Mice, 50 Swiss 52 C3G	intramuscular	5 mg	Single administration, observation up to 476 days	Local sarcomas (mainly fibrosarcomas) were observed among 33 Swiss mice and 23 C3H mice	191
Fisher 344 rats	Intramuscular	14 mg	Single administration, observation for 104 weeks	14/15 local sarcomas (mainly striated myosarcomas)	190
Female Wistar rats	Intraperitoneal	500 mg	Twice	46/47 new growths in abdominal cavity (sarcomas, mesotheliomas, or carcinomas)	192
Female Wistar rats	Intraperitoneal	25 or 100 mg	Single administration, then observation for 30 months	Sarcomas and mesotheliomas in abdominal cavity, 12/34 and 15/36, respectively, to the dose	84
Male Wistar rats	Intrapleural	10 mg	Single administration, then observation for 30 months	31/32 tumors at the administration locus (mainly striated myosarcomas)	193

occur in the form of oxides. Nickel oxide is the main component of fumes emitted during stainless-steel welding (5). Since the exposure to nickel compounds is often accompanied by an exposure to other metal compounds (chromium, iron), it is very difficult to establish the causal relationship between the exposure to an individual nickel compound and its toxic effects.

2.4.2.2 Clinical Cases

2.4.2.2.1 Acute Toxicity. The available literature contains no data concerning acute poisoning with nickel oxide among humans.

2.4.2.2.2 Chronic and Subchronic Toxicity. Toxic effects of chronic inhalation exposure to nickel oxide relate to the respiratory system.

Symptoms of pneumoconiosis determined by radiography were found among several of the 212 examined steelwork employees who had been engaged as pouring gate cleaners for 16 years. The total concentration of dusts at the workplaces ranged from 1.3 to 294.1 mg/m^3, and the dusts comprised iron, chromium, and nickel oxides at 6:1:1 ratio. The calculated nickel oxide concentration ranged from 0.15 to 34 mg/m^3. Two workers showed a distinct impairment of the respiratory system function, while symptoms of pneumoconiosis were observed in five workers (195). Fibrogenic changes in the lungs of electric-arc welders have been reported by Zober (94, 95) (see also Section 1.4.2.2.2.). These observations have not been confirmed by studies.

Muir et al. (196) reviewed X rays of 745 former sinter workers exposed to high concentrations of airborne dusts containing concentrations of nickel as high as 100 mg/m^3 (as nickel subsulfide and nickel oxide) and found no evidence of significant inflammatory or fibrogenic responses in the lung of the exposed workers. Symptoms of asthma that developed during a provocative test using dusts emitted during steel welding in stainless-steel welders who had reported respiratory system complications were described by Keskinen et al. (197).

2.4.2.2.3 Pharmacokinetics, Metabolism, and Mechanisms. See Section 1.4.2.2.3.

2.4.2.2.4 Reproductive and Developmental. The available literature contains no data concerning the effects of nickel oxide on human reproduction and fetal development.

2.4.2.2.5 Carcinogenesis. An epidemiological review by Doll et al. (198) suggests exposure to nickel oxides may result in an increased risk of lung and nose cancer. An increase in the lung cancer risk was observed among workers employed at the Falconbridge (Norway) roasting and smelting divisions where nickel oxide was considered the main hazard factor, although both the increased risk and the relation between the exposure duration and the risk were not clear. There is also evidence that lung cancer risk among workers of the roasting and smelting divisions decreased when exposure to nickel oxides was reduced by modifying the production process. The relationship between the lung and nasal cancer risks and exposure to nickel oxide has been confirmed by Andersen et al. (143).

A significantly increased risk of lung and nose cancer was observed among workers employed at the Clydach copper department, who were exposed to high concentrations of nickel oxide (> 10 mg/m^3). Unfortunately, it was not possible to definitely assess the increase of this risk because other soluble nickel salts were also present in the air (198).

According to the experts of the International Agency for Research on Cancer (IARC), there is a sufficient evidence that a combined exposure to nickel oxides and sulfides emitted during nickel refining is carcinogenic to humans (124). See also Section 1.4.2.2.5.

2.4.2.2.6 Genetic and Related Cellular Effects Studies. Waksvik and Boysen (152) have described two groups of workers exposed to nickel compounds who showed an increased level of chromosome aberrations, mainly gaps. Sister chromatid exchange (SCE) was not observed. The workers of one group were employed at a nickel ore crushing, roasting, and smelting division, where they were exposed mainly to nickel oxide and sulfide, while the other group operated electrolysis processes and was exposed to nickel chloride and sulfate. Waksvik et al. (153) examined nine retired workers (retired for 4–5 years) of a nickel refining plant who had been exposed to nickel oxide and sulfides or nickel chloride and sulfate aerosols, during employment and found that the nickel concentration in the workers' plasma was still 2 µg/L. An increased number of chromosome breaks and gaps was observed, but no differences were found in the incidence of SCE as compared with the reference group.

2.5 Standards, Regulations, or Guidelines of Exposure

Recommended exposure limits are listed in Table 41.7.

3.0 Nickel Carbonate

3.0.1 CAS Number: [3333-67-3]

3.0.2 Synonyms: Basic nickel carbonate; carbonic acid, nickel salt; nickelous carbonate; nickel monocarbonate; nickel(II) carbonate; nickel(II)carbonate basic; carbonic acid, nickel(2+) salt (1 : 1); nickel carbonate, 99.998%

3.0.3 Trade Names: NA

3.0.4 Molecular Weight: 118.70

3.0.5 Molecular Formula: NiCO$_3$

3.0.6 Molecular Structure:

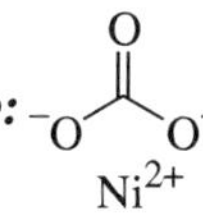

3.1 Chemical and Physical Properties

Table 41.1 shows the physical properties of nickel carbonate (1, 2). Nickel carbonate is a light green crystalline substance, which is almost insoluble (0.093 g/L) in water (25°C), nonsoluble in hot water, and soluble in acids. Nickel carbonate is available primarily as basic nickel carbonate (NiCO$_3$ · 2Ni(OH)$_2$ · 4H$_2$O), which is not soluble in water and

Table 41.7. Exposure Limits for Nickel Oxide as Ni (All Limits in mg/m^3)^a

Exposure Limits	OSHA PEL (156)	NIOSH Exposure Limit (156)	ACGIH TLV (156)	AIHA WEEL	ANSI Standard	German MAK (157)	Sweden Standard (158)	British Standard (158)	Polish Standard (159)
TWA	1^b	0.015^b Ca	0.2 A1 (i)	—	—	—	0.1	1^b	0.25^c
STEL	—	—	—	—	—	—	—	3^b	—
Ceiling limit	—	—	—	—	—	—	—	—	—
Biological limits (if available)	—	—	—	—	—	—	—	—	—

^{a}*Carcinogen designations*: NIOSH — Ca, carcinogen defined with no further categorization; ACGIH — TLV A1, not suspected as a human carcinogen; German — MAK A1, capable of inducing malignant tumors as shown by experience with humans; (i) — inhalable fraction.
^{b}Standard for insoluble nickel compounds.
^{c}Standard for nickel and its compounds excluding nickel carbonyl.

soluble in ammonia and dilute acids. In the natural environment, nickel carbonate tetrahydrate can be found as zaratite.

3.2 Production and Use

Pure nickel carbonate is produced by oxidizing nickel powder in ammonia and carbon dioxide. Nickel carbonate is used to manufacture nickel catalysts, colored glass, nickel pigments, nickel oxide, and nickel powder, and also as a neutralizing agent in nickel electroplating (180).

3.3 Exposure Assessment

See Section 1.3.

3.4 Toxic Effects

3.4.1 Experimental Studies

3.4.1.1 Acute Toxicity. Oral LD_{50} for the rat is 1044 mg/kg (18).

3.4.2.1 Chronic and Subchronic Toxicity. No data have been found in the available literature concerning chronic or subchronic exposure to nickel carbonate in the experimental animals.

3.4.1.3 Pharmacokinetics, Metabolism, and Mechanisms. See Section 1.4.1.3.

3.4.1.4 Reproductive and Developmental. No data have been found in the available literature concerning the effect of nickel carbonate on the reproduction and fetal development in the experimental animals.

3.4.1.5 Carcinogenesis. Intramuscular implantation of 7 mg nickel carbonate resulted in the development of sarcoma in 4 of 35 exposed rats (199), whereas IP injections did not result in any significant increase in the number of tumors in the exposed animals (84).

Although inhalation is the route of concern for human exposure, there has been no inhalation study in animals using nickel carbonate.

3.4.1.6 Genetic and Related Cellular Effects Studies. DNA lesions in nuclei isolated from rat tissues following IP injection of nickel carbonate were observed (59, 200, 201). Kidney, liver, lung, and thymus gland nuclei were examined at 3 and 20 h after treatment for the presence of DNA single-strand breaks and crosslinks. Single-strand breaks were detectable in lung and kidney nuclei, and both DNA–protein and DNA interstrand crosslinks were detectable in kidney nuclei. No DNA damage was observed in liver or thymus gland nuclei. A dose response to both single-strand breaks and crosslinks was observed in kidney nuclei. Time course studies revealed that maximum DNA damage in kidney nuclei occurred at 2–4 h following injection and also revealed the presence of an active repair process in these nuclei. Repair-resistant DNA–protein crosslinks were observed to persist through 48 h. Tissue and intracellular nickel concentrations as measured by electrothermal atomic absorption spectroscopy were observed to correlate with the levels of DNA damage and repair. A dose–response relation to the concentration of nickel in tissues and nuclei was observed (201).

3.4.2 Human Experience

No data have been found in the available literature concerning acute or chronic exposure to nickel carbonate in humans and its effects on reproduction and fetal development. Epidemiology studies intended to explain the carcinogenic effects of nickel carbonate alone in humans are also missing.

Considering that nickel carbonate is a component of nickel refining fumes, the effects of exposure occurring during the refining processes may be attributable also to nickel carbonate (see Section 1.4.2.2.5).

Data on the genotoxic effects of nickel carbonate in humans are not available in the literature, either.

3.5 Standards, Regulations, or Guidelines of Exposure

Recommended exposure limits are listed in Table 41.8.

4.0 Nickel Subsulfide

4.0.1 CAS Number: [12035-72-2]

4.0.2 Synonyms: Nickel sesquisulfide; nickel sulfide (3:2), trinickel disulfide; Heazlewoodite; khislevudite; nickel sulfide; α-nickel sulfide (3–2); nickel tritadisulfide

4.0.3 Trade Names: NA

4.0.4 Molecular Weight: 240.25

4.0.5 Molecular Formula: Ni_3S_2

4.1 Chemical and Physical Properties

Table 41.1 shows the physical properties of nickel subsulfide (1, 2). Nickel subsulfide occurs in two crystalline forms that are stable within different temperature ranges. Its low temperature form α-Ni_3S_2 is subject to phase transition at a temperature of 556°C into a high temperature form β-Ni_3S_2 (202). Heazlewoodite, a mineral occurring in the nature, is a variety of α-Ni_3S_2. Nickel subsulfide is insoluble in water and very resistant to the action of acids and alkalis.

4.2 Production and Use

Nickel subsulfide can be prepared by a direct reaction between nickel and sulfur at a temperature over 300°C. This process makes it possible to prepare the purest preparation in the form of a brittle melt or monocrystals (202).

Table 41.8. Exposure Limits for Nickel Carbonate (all limits in mg/m^3)[a]

Exposure Limits	OSHA PEL (156)	NIOSH Exposure Limit (156)	ACGIH TLV (156)	AIHA WEEL	ANSI Standard	German MAK (157)	Sweden Standard (158)	British Standard (158)	Polish Standard (159)
TWA	1[b]	0.015[b]	0.2	—	—	—	0.1	1[b]	0.25[c]
	—	Ca	A1	—	—	A1	—	—	—
	—	—	(i)	—	—	—	—	—	—
STEL	—	—	—	—	—	—	—	3[b]	—
Ceiling limit	—	—	—	—	—	—	—	—	—
Biological limits (if available)	—	—	—	—	—	—	—	—	—

[a]*Carcinogen designations*: NIOSH — Ca, carcinogen defined with no further categorization; ACGIH — TLV A1, not suspected as a human carcinogen; German — MAK A1, capable of inducing malignant tumors as shown by experience with humans; (i) — inhalable fraction.
[b]Standard for insoluble nickel compounds.
[c]Standard for nickel and its compounds excluding nickel carbonyl.

The largest industrial source of nickel subsulfide is the metallurgic process of nickel matte smelting. Nickel matte consists of nickel subsulfide and some quantities of iron and copper sulfides. It is an intermediate product in nickel metallurgy regardless of the type of the processed ores.

Nickel subsulfide is used as a catalyst in refining crude oil containing a considerable amount of sulfur (202).

4.3 Exposure Assessment

The content of nickel subsulfide in air and biological materials is determined by the ICP-AES and AAE methods (see Sections 1.3.1 and 1.3.2).

4.4 Toxic Effects

4.4.1 Experimental Studies

4.4.1.1 Acute Toxicity. The values of lethal doses LD_{50} for nickel subsulfide in oral administration to rats are 5.0 g/kg (18). The intratracheal administration of nickel subsulfide dust (1.8-µm-diameter particles) to BALB C mice at 12 µg/mouse causes acute changes in the form of pulmonary hemorrhage, which were particularly evident 3 months after administration. After a lapse of 20 h to 7 days since the moment of receiving nickel subsulfide, the rinsings from lungs showed an increased number of cells with nuclei of irregular shapes (203). After a single administration of subsulfide to the rat's trachea (3.2, 32, or 320 µg/kg bw), multifocal pulmonary alveolitis with type II alveolar cell hypertrophy was observed on day 7 (181).

4.4.1.2 Chronic and Subchronic Toxicity. Inhalation exposures of rats and mice to nickel subsulfide for 12 days (0.6–10 mg/m^3) brought about pathologic changes in their respiratory system with grave pulmonary lesions, including necrotic pneumonia (38, 39).

Pulmonary emphysema was developed in the rats exposed to 5 or 10 mg/m^3 nickel subsulfide; fibrosis was observed among mice at a nickel sulfide concentration of 5 mg/m^3. The exposed rats and mice showed pathological changes such as breathing difficulties, emaciation, dehydration, and reduced body weight gain (184).

Ottolenghi et al. (204) found abscesses and metaplasia in the lungs of rats after inhalation exposure to the subsulfide for 78 weeks.

Inhalation by rats and mice of 0.11–1.8 mg/m^3 nickel sulfide dust for 13 weeks caused lesions in their lungs: chronic inflammatory state, fibrosis, and proliferation of alveolar macrophages. Rats were more sensitive to that compound than mice (40).

4.4.1.3 Pharmacokinetics. See Section 1.4.1.3.

4.4.1.4 Reproductive and Developmental. Few data only are available on embryotoxic and teratogenic effects of nickel subsulfide, including its effect on reproduction.

Nickel subsulfide administered intramuscularly to pregnant rats (80 mg/kg bw) caused a decrease in the number of live fetuses in the litter. No congenital defects were observed (205). Sunderman et al. (206) examined the effect of nickel sulfide injected to female rat kidneys on the offspring. A dose of 30 mg/kg given 1 week before delivery caused an intensive polycythemia in the pregnant females. Such changes were not found in the offspring, but a considerable decrease in the average body weight was observed among the newborns after 2 and 4 weeks since their birth.

4.4.1.5 Carcinogenesis. Nickel subsulfide is a nickel compound that has been most extensively examined for its possible carcinogenic effects. The carcinogenic properties of nickel subsulfide were examined under conditions of inhalation and intratracheal exposures as well as after injections into various tissues, mainly in rats, mice, and hamsters. The results of the examinations listed in Table 41.9 (207–214) shows that all routes of nickel subsulfide administration resulted in tumor development. Ni_3S_2 administered IM and into the eye caused the highest neoplasm morbidity. After IM and inhalation administration, a dose–effect relationship was observed in the rats (207, 209). The tumors formed as a result of IM administration included mainly striated myosarcomas, fibrosarcomas, and undifferentiated sarcomas. Rats were more susceptible to the induction of sarcomas and pulmonary tumors by nickel subsulfide than mice or hamsters.

4.4.1.6 Genetic and Related Cellular Effects Studies. Swierenga and McLean (215) examined the genotoxic effects of nickel subsulfide in the T51B rat liver epithelial cell culture and found that this compound caused an increase in the number of mutations at the hypoxanthine–guanine phosphoriboxyl transferase locus.

Nickel subsulfide caused DNA strand breaks in the culture of human embryo lung fibroblast cell line (MRC-5 cells) (216). Nickel subsulfide induced an increase in the number of sister chromatid exchanges (SCEs) in human lymphocytes (217, 218) and morphological transformations in the Syrian hamster embryo SHE cells *in vitro* (86,219) and in mouse C3H10T cells (218).

4.4.2 Human Experience

4.4.2.1 General Information. The data on the toxic effects of nickel sulfide among humans have been obtained from the examinations carried out among workers employed in nickel refining plants. In addition to nickel oxide, sulfate, carbonate and nickel metal, nickel sulfide is present in the fumes emitted during nickel refining (5).

4.4.2.2 Clinical Cases

4.4.2.2.1 Acute Toxicity. No data on acute poisoning with nickel subsulfide among humans have been found in the available literature.

4.4.2.2.2 Chronic and Subchronic Toxicity consequences of chronic exposure to nickel subsulfide. The carcinogenic effects are discussed in Section 4.4.2.2.5.

4.4.2.2.3 Pharmacokinetics, Metabolism, and Mechanisms. See Section 1.4.2.2.3.

4.4.2.2.4 Reproductive and Developmental. No data on the effects of nickel subsulfide on the human reproduction and development have been found in the available literature.

Table 41.9. Carcinogenic Effects of Nickel Subsulfide

Species	Administration	Dose	Experimental conditions	Effects	Ref.
Fisher 344 rat males (122) females (109)	Inhalation	0.97 mg/m^3	6 h/day, 5 days/week for 78 weeks; observation for 30 weeks	14/228 malignant pulmonary tumors (10 adenocarcinomas, 3 squamous cell carcinomas, 1 fibrosarcoma); 15/228 mild pulmonary tumors	204
Fisher 344 rat 106 animals per group	Inhalation	0; 0.1 or 0.7 mg Ni/m^3 (as Ni$_3$S$_2$)	6 h/day, 5 days/week for 2 years	Increase in pulmonary adenomas and carcinomas 2/106, 12/106, and 20/106, respectively, by dose	207
Wistar female rat (47, 45, and 40)	Intratracheal	0.063, 0.125, or 0.25 mg	Once a week for 15 weeks	Malignant pulmonary tumors: 7/47, 13/45, and 12/40 respectively, by dose, including 12 adenocarcinomas, 15 squamous cell carcinomas, and 5 mixed neoplasms; no pulmonary tumors were observed in the control group	192
Mouse Swiss (45) C3H (18)	Intramuscular injections	5 mg	Single administration	Local sarcomas 27/45 and 9/18	191
B6C3F$_1$ mouse	Inhalation	0.04 or 0.8 mg Ni/m^3 (as Ni$_3$S$_2$)	6 h/day, 5 days/week for 2 years	Observed tumors: 22/109 (control), 7/118 and 8/118, respectively, to dose, which indicates no carcinogenic activity of nickel subsulfide in mice	207
Wistar rat (32)	IM injections	20 mg	Single administration	Local sarcomas were observed in 25 rats after 21 weeks	191
Fisher 344 rat	Subcutaneous or IM injections	3.3 mg or 10 mg	Single administration	Local sarcomas: after subcutaneous injection 37/39 and 37/40, respectively, to the dose; after IM injection 38/39 and 39/40, respectively, to the dose, mainly striated myosarcomas; no tumors were found in the reference group	208

Fisher 344 rat male (30)	IM injections	0.6, 1.2, 2.5, or 5 mg	Single administration	Local sarcomas 7/30, 23/30, 28/30, 29/30, respectively, to the dose, showing dose-dependent increase in the cases; no local sarcomas were observed in the reference group	209
Syrian hamster	IM injections	5 or 10 mg	Single administration	Local sarcomas in the animals exposed to the lower dose 4/15; to the higher dose 12/17; latency period 10 and 11 months, respectively; no local tumors were observed in the reference group	210
Wistar rat females (50)	IP injections	25 mg	Single administration	Tumors in the abdominal cavity (sarcomas, mesotheliomas, and carcinomas) in 27/42 animals; Average latency period: 8 months	192
Wistar rat males	Injection to kidneys	0.6, 1.2, 2.5, 5, or 10 mg	Single administration	Kidney cancers in 18/24 rats at 10 mg and 5/18 rats at 5 mg. No tumors at 0.6, 1.2 or 2.5 mg, or in the controls	211
Fisher 344 rat males (19)	To testicle	10 mg	Single administration to right testicle; observation for 20 months	Local sarcomas among 16/19 rats (10 fibrosarcomas, 3 malignant histocytomas, and 3 striated myosarcomas); remote metastases were observed in four rats; no tumors in the controls	212
Fisher 344 rat males groups of 20 individuals	IM, SC, intraarticular, or retroperitoneal fat injections	5 mg	Single administration	Malignant tumors in soft tissues after IM (19/20), SC (18/19), intraarticular (16/19), and retroperitoneal fat (9/20) injections; no tumors in the controls	213
Japanese newt (*Cynops pyrrhogaster*)	To eye	400–1100 µg per newt	Single administration	Malignant tumors of the melanoma type among 7/8 newts (striated myosarcomas, histiocytosis, and fibrosarcomas)	214

4.4.2.2.5 Carcinogenesis. The assessment of the relationship between the exposure to nickel subsulfide and increased risk of cancer in humans is difficult because Ni_3S_2 exposure is usually combined with exposures to some other nickel compounds.

Undoubtedly high pulmonary and nasal cancer risks occurred among workers of refining plants that processed nickel subsulfide ores at INCO, Ontario and in Clydach, Wales, where the exposure to nickel sulfides was very high, but not accompanied by high concentrations of nickel oxides and soluble compounds (131, 135, 198).

A review of the epidemiological studies by Doll et al. (198) shows that the role played by the exposure to nickel sulfide in inducing pulmonary and nasal cancers is not clear, but there are data suggesting that this compound increases the risk of these types of tumors.

According to the IARC, there is a sufficient evidence that the combined exposure to nickel sulfides and oxides emitted during nickel refining processes is carcinogenic to humans (124). See also Section 1.4.2.2.5.

4.4.2.2.6 Genetic and Related Cellular Effects Studies. Waksvik and Boysen (152) described two groups of workers exposed to nickel compounds who showed an increased level of chromosome aberrations, mainly breaks. SCE was not observed. Workers of one group were employed at the nickel ore crushing, roasting, and smelting division, where they were exposed mainly to nickel oxide and subsulfide, while the other group was exposed to nickel chloride and sulfate used during the electrolytic processes. Waksvik et al. (153) included also in their examination nine retired (for 4–15 years) workers of a nickel refining plant who were exposed to nickel oxides and sulfides or aerosols of nickel chloride and sulfate during employment; the authors found that nickel concentration in these workers' plasma was still 2 µg/L. Increased number of chromosome breaks and gaps, but no differences in SCE were found, compared with the control.

4.5 Standards, Regulations, and Guidelines of Exposure

Recommended exposure limits are listed in Table 41.10.

5.0 Nickelocene

5.0.1 CAS Number: [1271-28-9]

5.0.2 Synonyms: Dicyclopentadienylnickel, di-*p*-cyclopentadienylnickel; nickel, compound with *p*-cyclopentadienyl (1–2); bis($N^5$2,4-cyclopentadien-1-yl)nickel

5.0.3 Trade Names: NA

5.0.4 Molecular Weight: 188.9

5.0.5 Molecular Formula: $Ni(C_5H_5)_2$

5.1 Chemical and Physical Properties

Table 41.1 gives the physical properties of nickelocene. Nickelocene is a solid crystalline substance of dark green color, insoluble in water or carbon tetrachloride, slightly soluble in alcohol and liquid ammonia, and readily soluble in nonpolar organic solvents (180).

Table 41.10. Exposure Limits for Nickel Subsulfide as Ni (All Limits in mg/m^3)

Exposure Limits	OSHA PEL (156)	NIOSH Exposure Limit (156)	ACGIH TLV (156)	AIHA WEEL	ANSI Standard	German MAK (157)	Sweden Standard (158)	British Standard (158)	Polish Standard (159)
TWA	1[b]	0.015[b] Ca	0.1 A1	—	—	—	0.01	1[b]	0.25[c]
STEL	—	—	—	—	—	—	—	3[b]	—
Ceiling limit	—	—	—	—	—	—	—	—	—
Biological limits (if available)	—	—	—	—	—	—	—	—	—

[a] *Carcinogen designations*: NIOSH — Ca, carcinogen defined with no further categorization; ACGIH — TLV A1, confirmed human carcinogen.
[b] Standard for insoluble nickel compounds.
[c] Standard for nickel and its compounds excluding nickel carbonyl.

5.2 Production and Use

Nickelocene is obtained through reacting nickel halides with sodium cyclopentadienide. It is used as a catalyst or as a complexing agent (180).

5.3 Exposure Assessment

See Section 1.3.

5.4 Toxic Effects

5.4.1 Experimental Study

5.4.1.1 Acute Toxicity. Median lethal dose (LD_{50}) values are 490 mg/kg oral and 50 mg/kg IP for the rat; 600 mg/kg oral and 86 mg/kg IP for the mouse (220).

Nickelocene was given IP in single doses of 12.5–150 mg/kg and by gavage in single doses of 200–800 mg/kg to rats and mice. Initial signs of toxicity occurred 2–3 h after injection in both species and consisted of diarrhea, respiratory difficulty, and lethargy. Initial fatalities occurred 6–10 h after treatment. The animals surviving more than 48 h were still alive on day 10. In the orally treated group, mortalities occurred 24–72 h after treatment. Animals that survived for 10 days were also alive on day 35, when all animals were sacrificed and necropsied. No outstanding pathological changes were found. Mice tolerated slightly higher amounts on nickelocene than did rats. In mice, nickelocene was 7 times more toxic when given IP than orally, and, in rats, the compound was 10 times more toxic intraperitoneally than orally (220).

5.4.1.2 Chronic and Subchronic Toxicity. No data have been found in the available literature concerning chronic or subchronic exposure to nickelocene in the experimental animals.

5.4.1.3 Pharmacokinetics, Metabolism, and Mechanisms. see Section, 1.4.1.3.

5.4.1.4 Reproductive and Developmental. No data have been found in the available literature concerning the effects of chronic or subchronic exposure to nickelocene on reproduction or fetal development in experimental animals.

5.4.1.5 Carcinogenesis. Intramuscularly injected nickelocene induced tumors in rats and hamsters. Fisher 344 rats were exposed to total doses of 144 or 300 mg nickelocene by repeated IM injections. Local sarcomas were observed in 18 of 50 animals exposed to the lower dose, and in 21 of 50 rats receiving the higher dose. In the controls, local tumors were not detected. In hamsters exposed IM to a total dose of 40 mg nickelocene, local tumors were not observed, but the animals did develop tumors following a single injection of 25 mg nickelocene; fibrosarcoma was observed in 1 of 13 exposed female and in 3 of 16 exposed male hamsters (80).

In the opinion of IARC experts, the proof for the carcinogenic activity of nickelocene in animals is limited (124).

5.4.1.6 Genetic and Related Cellular Effects Studies.
Nickelocene did not cause mutations in tests on *Salmonella typhimurium* TA100, TA1535, TA1537, or TA98 strains (221).

No data have been found on the genotoxic activity of nickelocene in the eukaryotic organisms.

5.4.2 Human Experience

No data have been found in the available literature concerning the effects of chronic or subchronic nickelocene poisonings on reproduction or fetal development in humans. Data on the genotoxic activity of nickelocene in humans are not available, either.

Epidemiological studies mentioning exposures to nickelocene are not specific to this compound. No conclusion is possible about the carcinogenic potential of this material to humans.

5.5 Standards, Regulations, or Guidelines of Exposure

Recommended exposure limits are listed in Table 41.11.

6.0 Nickel Carbonyl

6.0.1 CAS Number: [13463-39-3]

6.0.2 Synonyms: Nickel tetracarbonyl; tetracarbonylnickel; (T-4)-nickel tetracarbonyl; nickel carbonyl, (T-4)-

6.0.3 Trade names: NA

6.0.4 Molecular Weight: 170.74

6.0.5 Molecular Formula: $Ni(CO)_4$

6.0.6 Molecular Structure:
$$O\equiv C - \underset{\underset{\underset{O}{\overset{\|}{C}}}{\overset{\overset{O}{\overset{\|}{C}}}{|}}}{Ni} - C\equiv O$$

6.1 Chemical and Physical Properties

Table 41.1 gives the physical properties of nickel carbonyl.

6.1.1 General

Nickel carbonyl is a colorless liquid insoluble in water, unreactive with aqueous acids and alkalis, but soluble in organic liquids. Air mixtures may explode at 20°C and at partial pressure of 15 mm Hg; O_2 rapidly decomposes the vapor to an amorphous hydrous nickel carbonate oxide. H_2SO_4 liberates CO and nickelous salts; CS_2 yields NiS and carbon. Thermodynamic data reviewed by Spice et al. (222) show that unless the pressure of CO approximates 1 atm, $Ni(CO)_4$ is almost completely dissociated; at a partial pressure of 1000 ppm CO at 25°C, the equilibrium concentration of $Ni(CO)_4$ is 0.02 ppm.

Table 41.11. Exposure Limits for Nickelocene (in mg/m^3)

Exposure Limits	OSHA PEL	NIOSH Exposure Limit	ACGIH TLV	AIHA WEEL	ANSI Standard	German MAK	Sweden Standard	British Standard	Polish Standard (159)
TWA	—	—	—	—	—	—	—	—	0.25[a]
STEL	—	—	—	—	—	—	—	—	—
Ceiling limit	—	—	—	—	—	—	—	—	—
Biological limits (if available)	—	—	—	—	—	—	—	—	—

[a]Standard for nickel and its compounds, excluding nickel carbonyl.

Decomposition is not instantaneous, however, for Kincaid et al. (223), working with air mixtures of Ni(CO)$_4$ between 2 and 350 ppm, reported about 5% dissociated in 50 and 30% dissociated in a chamber with an air change every 4 min. Ni(CO)$_4$ is flammable and burns with a yellow flame. It may decompose violently when heated at 60°C in the presence of air or oxygen. A concentration of 10 ppm in the atmosphere is sufficient to impart luminosity to alcohol or CO flames; this may be used as a semiquantitative test (224).

6.2.2 Odor and Warning Properties

Nickel carbonyl has a mild, nonpenetrating odor, often described as "sooty" or "musty," detectable at 500–3000 ppm (224).

6.2 Production and Use

Nickel carbonyl is produced in a reaction of carbon monoxide and nickel metal. It may also be formed as a by-product in the industrial processes using nickel catalysts, such as coal gasification, crude-oil refining, and hydrogenation reactions (225). Conditions for its formation occur in those processes where carbon monoxide is in contact with an active form of nickel under conditions of elevated pressure at 50–150°C (224).

Nickel carbonyl is used in nickel vapoplating processes in the metallurgical and electronics industry, and in the catalytic methyl- and ethylacrylate monomer synthesis. For many years it had been used to produce pure nickel by the Mond process, which has been considered to be outdated since around 1970 (3).

6.3 Exposure Assessment

6.3.1 Air

Volatile nickel carbonyl present in the ambient air can be determined by colorimetry. Nickel carbonyl is absorbed as Ni(II) in an alcoholic iodine solution, transferred to chloroform, and determined at 425 nm as a colored complex with α furildioxine. The limit of quantitation of the method is 1 µg/m^3 (226). The more sensitive chemiluminescence method enables nickel determination already at 0.2 µg/m^3 (227).

6.3.2 Workplace Methods

In the NIOSH-recommended method of nickel carbonyl determination, the air is passed through low Ni charcoal sorbent tube followed by a mixed cellulose ester membrane filter. After desorption with nitric acid, nickel is determined in the samples by atomic absorption spectrometry (AAS). The limit of quantitation of the method is 0.01 µg Ni per sample (228).

6.3.3 Background Levels

Sunderman et al. (229) have developed a very sensitive and rapid method for determining Ni(CO)$_4$ in air that uses a gas chromatographic technique, and have used it for the determination in blood and breath. The sampling procedure consists of trapping Ni(CO)$_4$-containing air in absolute ethyl alcohol at −78°C. For blood specimens, Ni(CO)$_4$ is extracted by vacuum, trapped in cold ethyl alcohol, and maintained at −78°C until

Table 41.12. Acute Toxicity of Nickel Carbonyl in Experimental Animals (Criteria)

Species	Administration	Lethal Dose (Duration of Exposure)	Ref.
Rabbit	Inhalation	$LC_{80} = 1.4$ mg/L (50 min)	231
Dog	Inhalation	$LC_{80} = 2.7$ mg/L (75 min)	231
Mouse	Inhalation	$LC_{50} = 0.067$ mg/L (30 min)	223
Rat	Inhalation	$LC_{50} = 0.24$ mg/L (30 min)	223
Cat	Inhalation	$LC_{50} = 0.19$ mg/L (30 min)	223
Rat	Inhalation	$LC_{50} = 0.1$ mg/L (20 min)	232
Rat	IV, SC, IP	$LD_{50} = 22$ mg/kg $LD_{50} = 21$ mg/kg $LD_{50} = 13$ mg/kg	233
Rat	Intravenous	$LD_{50} = 65$ mg/kg	234

injected into the chromatographic column. The most distinct and symmetrical peaks of $Ni(CO)_4$ are obtained by using Carbowax 20M. Peak heights to concentration were linear over a range of 0.0125–0.1 g/L (230).

6.4 Toxic Effects

6.4.1 Experimental Studies

6.4.1.1 Acute Toxicity. Table 41.12 gives lethal concentrations of nickel carbonyl for several experimental animal species (231–234).

Lethal doses for the examined animals ranged from LC_{50} 0.067 mg/L for mouse exposed during 30 min to LC_{80} 2.7 mg/L air for dog exposed by inhalation during 75 min. LD_{50} levels for rats exposed by other routes were 13–65 mg/kg.

The symptoms of acute inhalation exposure included dyspnea, cyanosis, fever, apathy, anorexia, vomiting, diarrhea, and, in some animals, hind-paw palsy.

Lungs are the target organs for nickel carbonyl in the exposed animals, and the harmful effects of $Ni(CO)_4$ at high concentrations are manifested quickly. Pneumonia and pneumoedema were observed one hour after 30-min inhalation exposure to 240 mg/m^3 $Ni(CO)_4$. Several days after the exposure, there was a heavy edema inside the alveoli with local bleeding and cell degeneration (235).

Damage in other organs following the acute exposure is milder than in the lungs. Local bleeding, hyperemia, and moderate inflammatory conditions were observed in the brain, liver, kidneys, spleen, and pancreas (233, 235). A 15-min inhalation exposure to 0.2–1.1 mg/L nickel carbonyl caused acute hyperglycemia (236). Proteins and aminoacids found in urine of the animals exposed to $Ni(CO)_4$ points to the nephrotoxic activity of this compound (237).

According to Oskarsson and Tjälve (238), who studied mice exposed to labeled nickel carbonyl, the target organs in the acute poisonings were lungs, central nervous system (CNS), adrenals, and heart.

6.4.1.2 Chronic and Subchronic Toxicity. Extensive inflammatory lesions of the lungs, contiguous pericarditis, and suppurative lesions of the thoracic walls were observed in the rats exposed by inhalation to nickel carbonyl at 0.03–0.06 mg/L (90 min 3 times a week for 52 weeks). Squamous-cell metaplasia was present in bronchiectatic walls of several rats (239).

6.4.1.3 Pharmacokinetics, Metabolism and Mechanisms. Absorption, distribution, and excretion of nickel carbonyl was studied in a number of experiments on dogs, cats, rats, mice, and rabbits (230–232, 240–242) with special stress on inhalation exposure, which is the most frequent route of human occupational exposure; however, IV and IP exposures were also considered.

In the inhalation experiments, animals received a single dose of 200–3050 mg Ni/m^3 air during 2–15 min.

After 10-min inhalation exposure of mice to nickel carbonyl at 3050 mg/m^3, high nickel concentrations were detected in the lungs, brain, spinal cord, heart, diaphragm, adrenals, kidneys and bladder from 5 min to 24 h following the exposure (241).

Sunderman and Selin (230) report that 24 h after ^{63}Ni(CO$_4$) inhalation, ^{63}Ni distribution in rat body was as follows: intestine and blood, 50%; muscles and adipose tissue 30%; bones and connective tissue, 16%; brain and spinal cord, 4%. Following IV administration, ^{63}Ni distribution differed considerably, and was as follows: muscles and adipose tissue, 41%, bones and connective tissue, 31%, intestine and blood, 27%, brain and spinal cord, 1%.

Different nickel distribution in the organisms of the exposed animals depending on administration route were observed also by Oskarsson and Tjälve (241), who found that nickel concentrations in the brain, spinal cord, heart, and diaphragm were considerably lower following IV compared with inhalation exposure. After 24 h following IV administration of labeled nickel carbonyl, radioactive nickel was found to be present in the lungs, liver, kidneys, and blood serum of the exposed mice.

According to Sunderman and Selin (230), lungs were the main organ eliminating the absorbed nickel carbonyl; 38% of the received ^{63}Ni(CO)$_4$ dose was found in the air exhaled by rats during 6 h following the IV exposure, whereas 31% of the absorbed nickel carbonyl was excreted with the urine during 4 days. Results of a number of studies on nickel carbonyl metabolism (230, 240) indicated that non-modified nickel carbonyl was present in the blood several hours after the exposure, and that could permeate alveoli in both directions without becoming decomposed.

Kasprzak and Sunderman (240) suggested that nickel carbonyl that has not been removed from the lungs by exhalation, is subject to slow intracellular decomposition [Ni(CO)$_4$ − > Ni^0 + 4 CO]. The released Ni° is oxidized to Ni^{2+}, which may be bound by nucleic acids and proteins, or transported to blood plasma where it is bound to albumins. Finally, Ni^{2+} is excreted with the urine and feces. Carbon monoxide released from nickel carbonyl is bound to hemoglobin to be exhaled from the lungs. Those suggestions have been confirmed by Oskarsson and Tjälve (241), who studied the distribution of nickel carbonyl (containing labeled nickel and carbon) [^{63}Ni(CO)$_4$ and Ni(^{14}CO)$_4$] given IV to mice by whole-body autoradiography using liquid scintillation counter. The distribution of radioactivity in the animals that had received nickel carbonyl containing ^{14}C was limited to blood, suggesting formation of the ^{14}CO–hemoglobin bond. When the animals received nickel carbonyl containing ^{63}Ni, the highest ^{63}Ni concentrations were found in the lungs, followed by the brain, spinal cord, heart, diaphragm, brown fat, adrenals, and yellow body (242).

6.4.1.4 Reproductive and Developmental. Sunderman et al. conducted a series of experiments indicating that nickel carbonyl is teratogenic and embryotoxic to hamsters and rats (206, 243–245). Intravenous injection of nickel carbonyl at 11 mg/kg b.w. to pregnant female rats on gestation day 7 caused fetal deaths, reduced body weight in the surviving pups, and congenital defects, such as anophthalmia, microphthalmia, cystic disease of the lungs and hydronephrosis in 16% fetuses (206). No toxic effects were observed in the dams. Similar effects were observed in the rats following inhalation exposure of the pregnant dams to nickel carbonyl at 160 mg/m^3 on days 7 or 8 of gestation, or at 300 mg/m^3 on gestation day 7 (244). Inhalation exposure of hamsters to nickel carbonyl at 60 mg/m^3 for 15 min on days 4 or 5 of gestation resulted in impaired fetal viability and increased number of fetuses with congenital malformations (245).

Inhalation exposure of male rats to nickel carbonyl (50 mg/m^3 for 15 min) for 2–6 weeks before mating had no adverse effect either on their ability for fertilization or reproduction, whereas injection of 22 mg/kg nickel carbonyl for a similar period and fertilization of the females during week 5 after the exposure resulted in reduced number of neonates in the litter (206).

6.4.1.5 Carcinogenesis. The carcinogenic activity of nickel carbonyl has been tested on rats exposed by inhalation and IV injection (246–248). In an inhalation experiment, groups of 64 and 32 Wistar rats were exposed to nickel carbonyl at 30 and 60 mg/m^3, respectively, for 3 min/day, 3 days/week during one year. Four of the nine rats surviving 2 years of the experiment developed lung cancer. No lung tumors were observed in the control group of 41 rats (246). Sunderman and Donnelley (247) experimenting on 285 Wistar rats observed pulmonary adenocarcinoma with metastasis in one of the 35 rats that had survived 2 years after single 30-min inhalation exposure to nickel carbonyl at 600 mg/m^3. In a group of 64 rats exposed to nickel carbonyl (30 mg/m^3 for 30 min, 3 days/week for their whole lifespan), one case of adenocarcinoma with metaplasia was detected among the eight rats that had survived more than 2 years. No lung tumors were observed in the control group.

Increased incidence of cancer was observed also in the rats exposed to nickel carbonyl by repeated IV injections (248). The following malignant tumor types were observed: pulmonary lymphoma; nondifferentiated pulmonary, pleural, hepatic, pancreatic, uterine, and extraabdominal sarcoma; neck, ear concha, orbit fibrosarcoma; liver, breast, renal cancer; and a case of endothelial cell hemangioma. In the opinion of the IARC Working Group there is a limited evidence for the carcinogenicity of nickel carbonyl in the experimental animals (124).

6.4.1.6 Genetic and Cellular Effects Studies. Nickel carbonyl injected IV to rats at 20 mg/kg bw inhibited DNA synthesis in the liver and kidneys (249).

6.4.2 Human Experience

6.4.2.1 General Information. Nickel carbonyl is the most toxic nickel compound. It is considered to be one of most dangerous chemicals used in the industry, and is responsible for a morbidity and mortality comparable with that caused by hydrogen cyanide (250). As nickel carbonyl is very volatile, it is absorbed primarily by inhalation. It is an irritant to the respiratory tract, and causes damage to the lungs and brain.

6.4.2.2 Clinical Cases and Epidemiology Studies

6.4.2.2.1 Acute Toxicity. The clinical symptoms of nickel carbonyl poisoning have been described in detail by Sunderman (251), Vuopala et al. (252), Shi (253, 254), and Kurta et al. (250).

The effects of inhalation exposure to high concentrations of nickel carbonyl are manifested in two stages: direct and delayed. The direct effects are manifested mainly by neurological symptoms and upper respiratory irritation; the symptoms include headache and vertigo, nausea, vomiting, sleeplessness, and irritation persisting for 4–5 h. Following the 12-h to 5-day asymptomatic period, the delayed symptoms occur: squeezing thoracalgia, dry cough, dyspnea, cyanosis, tachycardia, GI sensations, sweating, and weakness (254).

The symptoms of nickel carbonyl poisoning are similar to those in viral pneumonia. Lung hemorrhage, edema, and pneumonia, accompanied by alveolar cell disturbances, bronchial epithelium cell degeneration, and fibers present in the endovesical exudate, were observed in the people who had died as a result of nickel carbonyl poisoning. The pathology of the pulmonary tissue lesions was similar to that observed in the experimental animals. Other damaged organs included liver, kidneys, adrenals, and spleen, where parenchymatous degeneration was observed. Brain swelling with point hemorrhaging was observed in people who had died as the result of nickel carbonyl inhalation (251, 252). The recovery is slow; usually the patient is able to resume some easy work after 2–4 months (254).

Results of pulmonary function tests revealed disturbed FVC and FEV values in the patients after 3–5 years since the poisoning (254). Determination of urine nickel level makes it possible to assess the magnitude of the exposure to nickel carbonyl. According to Sunderman and Sunderman (255), when urine nickel content is $< 10\,\mu g/100\,mL$, the exposure is said to be slight. A concentration above $50\,\mu g/100\,mL$ is indicative of a high exposure. Urine nickel levels in the general population not exposed to nickel range between 0.1 and $13.3\,\mu g/100\,mL$ (90).

Agitation, shortness of breath, thoracalgia, and paresthesia were recorded in a 46-year-old man dermally and inhalationally exposed to nickel carbonyl as a result of an industrial emergency situation. Urine nickel content determined 24 h after the accident was $172\,\mu g/100\,mL$ (250).

6.4.2.2.2 Chronic and Subchronic Toxicity. Symptoms of asthma and the Löffler syndrome were found in a chemist exposed for a long time to low concentrations of nickel carbonyl. In addition to pulmonary infiltration and eosinophilia, the markers of the Löffler syndrome, the patient developed eczematous dermatitis of the hands (256).

In a study by Shi et al. (257), a group of workers employed at the production of nickel carbonyl was compared with 40 controls. The exposures ranged from 0.007 to 0.52 mg $Ni(CO)_4/m^3$ for 2–20 years. Reduced monoamine oxidase (MAO) activity in the serum and EEG irregularities were observed in the people exposed to the highest concentrations. Excitation, sleeplessness, headache, vertigo, weakness, tightness in the chest, weakness, excessive idrosis, hair loss, hypomnesia, and sexual frigidity were also observed in the exposed people. The incidence and intensity of those symptoms increased with the duration of the exposure.

6.4.2.2.3 Pharmacokinetics, Metabolism, and Mechanisms. The respiratory system is the main route of nickel carbonyl intake in humans. Oral or dermal absorption is practically insignificant and has not been reported in the available literature. Nickel carbonyl absorption was studied in people deceased as the result of poisoning by determining nickel concentration in the individual organs (258–260), and also by testing blood, plasma, and urine nickel concentrations in workers employed at nickel refining plants (261–263).

As a result of nickel carbonyl inhalation, the highest nickel concentrations were detected in the kidneys, liver, and brain. Nickel carbonyl absorbed in the lungs is quickly removed therefrom, which means that the absorption and elimination is quick. The solubility of nickel carbonyl vapors in fats indicates that they are able to penetrate the phospholipid-coated alveoli to produce symptoms of acute poisoning.

6.4.2.2.4 Reproductive and Developmental. No data have been found in the available literature concerning the effects of nickel carbonyl on reproduction or fetal development in humans.

6.4.2.2.5 Carcinogenesis. No data have been found in the available literature concerning the carcinogenic activity of nickel carbonyl in humans.

6.4.2.2.6 Genetic and Related Cellular Effects Studies. Decheng et al. (264) studied the frequency of sister chromatid exchanges and chromosome aberrations in workers employed in the production of nickel carbonyl. No decrease in the number of SCEs or chromosome aberrations (gaps, breaks, fragmentation) was observed in workers exposed to nickel carbonyl in relation to the corresponding number noted in nonexposed workers.

6.5 Standards, Regulations, or Guidelines of Exposure

Recommended exposure limits are listed in Table 41.13.

7.0 Nickel Chloride

7.0.1 CAS Number: [7718-54-9]

7.0.2 Synonyms: Nickel dichloride; nickelous chloride; nickel(II) chloride; nickel(II) chloride, ultradry, anhydrous, 99.9% (metals basis)

7.0.3 Trade Names: NA

7.0.4 Molecular Weight: 129.60

7.0.5 Molecular Formula: NiCl$_2$

7.0.6 Molecular Structure: Ni^{2+} Cl^- Cl^-

Table 41.13. Exposure Limits for Nickel Carbonyl as Ni (All Limits in mg/m^3)[a]

Exposure Limits	OSHA PEL (156)	NIOSH Exposure Limit (156)	ACGIH TLV (156)	AIHA WEEL	ANSI Standard	German MAK (157)	Sweden Standard (158)	British Standard (158)	Polish Standard (159)
TWA	0.007	0.007	0.05 PPM	—	—	—	0.007	—	—
STEL	—	Ca	—	—	—	A2, H	—	0.24	—
Ceiling limit	—	—	—	—	—	—	—	—	—
Biological limits (if available)	—	—	—	—	—	—	—	—	—

[a] *Carcinogen designations*: NIOSH—Ca, carcinogen defined with no further categorization; German—MAK A2: Unmistakably carcinogenic in animal experimentation only; H—danger of cutaneous absorption.

7.1 Chemical and Physical Properties

Table 41.1 gives the physical properties of nickel chloride. Nickel chloride occurs in the form of yellow deliquescent flakes, which are soluble in water, alcohol, and ammonia (1). The hexahydrate form of that compound ($NiCl_2 \cdot 6H_2O$) occurs as green monoclinic crystals readily soluble in water and alcohol (1).

7.2 Production and Use

Nickel chloride (hexahydrate) is obtained by reacting metal nickel powder or nickel oxide with hot, dilute hydrochloric acid (180).

The anhydrous salt is used as the ammonia absorbent in gasmasks and as a semiproduct in the nickel catalyst manufacturing processes. The hexahydrate salt is used in electroplating (2, 180).

7.3 Exposure Assessment

See Section 1.3.

7.4 Toxic Effects

7.4.1 Experimental Study

7.4.1.1 Acute Toxicity. Lethal dose values for nickel chloride given IP are 6–9.3 mg/kg for female Wistar rats (265), 11 mg/kg for rats, and 48 mg/kg for mice (266), whereas with the IM injection, the lethal dose value for female pregnant rats is 71 mg/kg (205). Ulcerative gastritis and enteritis were observed in rats that died after they had received a high intragastrical dose of nickel chloride (267). Nickel chloride injected IP produces rapid body temperature drop in rats and mice. At 250 µmol/kg bw the temperature drops by $3.0 \pm 0.5°C$, with the minimum after 1.5 h, and returns to the values observed in the control animals after 4 weeks. Nickel is suspected of directly affecting the autonomous thermoregulatory mechanism (268).

Atrophy of thymus was observed in rats following a single subcutaneous injection of nickel chloride (269); however, after the IP injection the animals developed aminoaciduria and proteinuria associated with the morphological lesions in renal glomeruli (270). The acute nephrotoxic activity resulting from a single IP injection of nickel chloride to rats was also reported by Sunderman et al. (271).

7.4.1.2 Chronic and Subchronic Toxicity. Nickel chloride given to rats orally caused ataxia, irregular breath, hypothermia, sialosis, diarrhea, and lethargy (267).

Exposure of rats to aerosol of soluble nickel chloride at 109 µg/m^3 caused bronchial and bronchiolar epithelium growth accompanied by peribronchial leucocyte infiltration (272). Tuberous macrophage agglomerations and increased numbers of type II alveolar cells were observed in the pulmonary parenchyma of rabbits exposed by inhalation to nickel chloride at 0.2–0.3 mg/m^3 for 1–8 months. After 1 month of exposure, the macrophage phagocyting activity was normal, but it was reduced after 3 months (26, 273).

The toxic effect in the alveolar macrophages was also detected following a single subcutaneous injection of nickel chloride to rats. The alveolar macrophages displayed morphological and biochemical symptoms of activation, functional defects, and lipoperoxidation (274).

7.4.1.3 Pharmacokinetics, Metabolism, and Mechanisms. See Section 1.4.1.3.

7.4.1.4 Reproductive and Developmental.
Effects of nickel chloride on reproduction and fetal development have been widely described. Increased numbers of deformed sperm cells were observed in male mice that had received a single dose of nickel chloride equivalent to 43 mg Ni/kg (275) with food; in female mice receiving nickel chloride with drinking water (160 mg Ni/kg per day on days 2–17 of gestation) increased numbers of spontaneous abortions, reduced body weight of the surviving pups, and reduced body weight of the dams were noted (276).

Sunderman et al. (205) studied nickel chloride embryo- and fetotoxicity on Fischer 344 rats. A significant drop in the mean number of live fetuses in the litter, on gestation day 8, reduced fetal body weight on gestation day 20, and a reduction in body weight of 4–8-week pups were observed following a single IM injection of nickel chloride at 12 or 16 mg Ni/kg. Nickel chloride was injected IM (repeated injections) at 1.5 or 2.0 mg/kg. With the higher dose, a significant increase of intrauterine mortality was observed on days 6–10 of gestation. The exposure did not result in a decrease of mean body weight of the pups, nor did it produce skeletal or intestinal defects.

Nickel chloride given to rats with food or drinking water increased the incidence of spontaneous abortions and also increased the frequency of prenatal and early postnatal mortality (277).

Pregnant CD-1 mice received single IP injections of nickel chloride at 1.2, 2.3, 3.5, 4.6, 5.7, and 6.9 mg Ni/kg on days 7–11 of gestation. A dose-related increase in fetal mortality and increased incidence of developmental malformations were observed (278).

Results of *in vitro* tests were used to assess the effect of nickel chloride on early embryogenesis (279). The two-, four-, and eight-cell mouse embryos were placed in the medium containing nickel chloride hexahydrate at concentrations from 10 to 1000 µmol/L. At 10 µmol/L, the compound adversely affected the development of the two-cell embryo, whereas eight-cell embryo development was affected by the 300-µmol/L solution. Storeng and Jonsen (280) studied the effects of single IP injection (20 mg/kg) of nickel chloride hexahydrate to mice on days 1–6 of gestation. Significantly reduced ovarian implantation frequency was noted on gestation day 19 in pregnant females that had received the nickel salt on gestation day 1. The number of pups in the litter was significantly smaller in the females exposed on days, 1, 3 and 5 of pregnancy. Increased numbers of fetal abnormalities (hematomas, cranial anostosis) were also observed in the exposed females (compared with the controls).

Nickel chloride injected at 0.02 and 0.7 mg into the hen eggs on days 1, 2, and 3 of incubation produced numerous developmental effects observable on the day 8 of incubation (281). The defects were cranial anostosis; abnormal intestine position; short twisted neck; microphthalmia; hemorrhage; and reduced body size. The developmental defects appeared usually on day 2 of incubation.

In the FETAX (frog embryo teratogenesis assay: *Xenopus*), nickel chloride was potent teratogen in *Xenopus*, causing concentration-related increase in the incidence of ophthalmic, skeletal, gut, craniofacial, and cardiac anomalies (282). The studies by Hauptman et al. (283) confirmed the teratogenic activity of nickel chloride in the frogs.

7.4.1.5 Carcinogenesis. The few data available on nickel chloride carcinogenicity indicate that its carcinogenic potential is small. None of the 35 rats exposed to nickel chloride by IM implantation did develop cancer (199). In the rats receiving the compound IP, four animals (of the 32 exposed) developed mesotheliomas and sarcomas (84). Reports on the carcinogenic effects associated with the inhalation or oral exposures are not available.

7.4.1.6 Genetic and Related Cellular Effects Studies. Nickel chloride does not show mutagenic properties in bacterial tests (284, 285); however, positive results were obtained in the eukaryotic cell tests (Chinese hamster ovary cells, mouse lymphoma cells) (286, 287).

A dose-dependent (up to four-fold vs. control) increase in the number of point mutations was observed in a study on the effect of nickel chloride on L5178Y mouse lymphoma cells (288). Association of that effect with chromosome damage observed during the same study could not be excluded.

Swierenga and McLean (215) assessed the genotoxic activity of nickel chloride (in the form of aqueous solution and aqueous suspension) in rat liver cell (T51B) culture. In all cases, the authors found increased number of mutations at the hypoxantine–guanine phosphoriboxyl transferase (HGPRT) locus.

In a study on chromosome aberrations in mouse breast cancer FM3A cell culture, the authors compared the effects of nickel chloride, nickel acetate, and nickel sulfide (289). They demonstrated that the inhibiting potential of those compounds on protein, RNA, and DNA synthesis was similar. Chromosome aberrations were evidenced by gaps and fragmentations. Crystalline nickel subsulfide and nickel chloride produced similar changes in the chromosomes of the cells of Chinese hamster ovary (CHO): breaks, gaps, and exchanges. In both cases, the changes in the chromosomes were associated with the heterochromatic centromere region. Nickel chloride caused a significant increase in the incidence of sister chromatid exchange (SCE) in human lymphocytes (218).

7.4.2 Human Experience

7.4.2.1 General Information. No reports could be found in the available literature on the effects of acute or chronic exposure to nickel chloride alone. Humans may be exposed to nickel chloride during nickel refining or electroplating; usually it is a combined exposure, where nickel chloride is accompanied by other nickel compounds, for example, nickel sulfate, carbonate, oxide or subsulfide (5). A considerable number of reports are available on the effects of soluble nickel salts used in the electroplating shops. The electroplating baths usually include nickel sulfate, nickel chloride and boric acid; the most abundant constituent was nickel sulfate (290).

7.4.2.2 Clinical Cases

7.4.2.2.1 Acute Toxicity. In an electroplating shop 32 workers drank, by accident, water containing nickel sulfate and chloride (1.63 g Ni/L). As much as 20 of them quickly developed symptoms of acute poisoning (nausea, vomiting, abdominal ailments, diarrhea, vertigo, fatigue, coughing, shortness of breath), which typically persisted for several hours, and only in seven cases continued for 1 or 2 days. Nickel dose absorbed by the patients was assessed to be 0.5–2.5 g. In 15 exposed patients examined on the next day, blood serum nickel level was 12.8–1340 µg/L (286 µg/L mean); the control levels in the other workers of the shop were 2.0–6.5 µg/L (4.0 µg/L mean) for blood serum nickel and 22–70 µg/L (50 µg/L mean) for the urine nickel. Laboratory examinations revealed a temporary increase in blood reticulocyte level (in 7 patients), urine albumins (3 patients) and blood serum bilirubin (2 patients) (121).

7.4.2.2.2 Chronic and Subchronic Toxicity. The effects of chronic exposure to soluble nickel salts (sulfate and chloride) are discussed in Section 9.4.2.2.2.

7.4.2.2.3 Pharmacokinetics, Metabolism, and Mechanisms. See Section 1.4.2.2.3.

7.4.2.2.4 Reproductive and Developmental: NA

7.4.2.2.5 Carcinogenesis. Results of cohort studies performed at the Electrolysis Department of the Falconbridge (Norway) nickel refining plant suggest a relationship between exposure to soluble nickel salts and lung cancer (133, 134). Increased risk of lung cancer in workers exposed to soluble nickel compounds was suggested earlier by Doll (128). No increase in the risk of pulmonary cancer was detected among the electrolysis workers of Port Colborne (135) and of a Finnish nickel refining plant (291), probably because concentrations of the soluble nickel compounds were considerably lower. The results of the Falconbridge study provided also a proof that combined exposure to high concentrations of nickel salts and nickel oxide is associated with higher risk of nose cancer (133). See also Section 1.4.2.2.5.

7.4.2.2.6 Genetic and Related Cellular Effects Studies. Deng et al. (154, 155) observed an increased number of SCEs and chromosome aberrations (gaps, breaks, fragmentations) in electrolysis workers. Waksvik and Boysen (152) described two groups of workers exposed to nickel compounds in whom increased level of chromosome aberrations (mainly gaps) was observed. SCE were not detected. One group was employed at ore crushing, roasting, and smelting and was exposed primarily to nickel oxide and sulfide; the other worked at the electrolytic process and was exposed to nickel chloride and sulfate. Waksvik et al. (153) included in their study a group of nine former nickel refining plant workers retired for 4–15 years, who were exposed on the job to nickel oxide or sulfide, or to nickel chloride or sulfate aerosols, and whose blood plasma nickel level continued to be as high as 2 µg/L. An increased number of chromosome gaps and breaks was observed compared to the control, but the incidence of SCE did not differ from that observed in the controls.

7.5 Standards, Regulations, or Guidelines of Exposure

Recommended exposure limits are listed in Table 41.14.

8.0 Nickel Hydroxide

8.0.1 CAS Number: [12054-48-7]

8.0.2 Synonyms: Nickel dihydroxide; nickelous hydroxide; nickel(II)hydroxide; nickel hydroxide, 99.999%

8.0.3 Trade Names: NA

8.0.4 Molecular Weight: 92.7

8.0.5 Molecular Formula: Ni(OH)$_2$

8.0.6 Molecular Structure: $\begin{matrix} & Ni^{2+} & \\ OH^- & & OH^- \end{matrix}$

8.1 Chemical and Physical Properties

Table 41.1 gives the physical properties of nickel hydroxide. Nickel hydroxide occurs in either crystalline or amorphous form. It is not water-soluble, but it is soluble in acids. Nickel hydroxide forms complex compounds with ammonia (1, 2).

8.2 Production and Use

Nickel hydroxide is obtained by treating nickel sulfate solution with sodium hydroxide, or by hot alcohol extraction of the precipitate formed as a result of the reaction of nickel nitrate with potassium hydroxide (180). Nickel hydroxide is used for the manufacture of nickel–cadmium electric cells (292), and as an intermediate product during the manufacture of nickel catalysts (180).

8.3 Exposure Assessment

See Section 1.3.

8.4 Toxic Effects

8.4.1 Experimental Study

8.4.1.1 Acute Toxicity. The oral LD$_{50}$ for the rat is 1600 mg/kg (18).

8.4.1.2 Chronic and Subchronic Toxicity. No data have been found in the available literature concerning chronic or subchronic exposure to nickel hydroxide in the experimental animals.

8.4.1.3 Pharmacokinetics, Metabolism, and Mechanisms. See Section 1.4.1.3.

Table 41.14. Exposure Limits for Nickel Chloride as Ni (All Limits in mg/m^3)[a]

Exposure Limits	OSHA PEL (156)	NIOSH Exposure Limit (156)	ACGIH TLV (156)	AIHA WEEL	ANSI Standard	German MAK (157)	Sweden Standard (158)	British Standard (158)	Polish Standard (159)
TWA	1[b]	0.015[b]	0.1[b]	—	—	—	0.1[b]	0.1[b]	0.25[c]
	—	Ca	A4	—	—	S	—	—	—
	—	—	(i)	—	—	—	—	—	—
STEL	—	—	—	—	—	—	—	0.3[b]	—
Ceiling limit	—	—	—	—	—	—	—	—	—
Biological limits (if available)	—	—	—	—	—	—	—	—	—

[a]*Carcinogen designations*: NIOSH — Ca, Carcinogen defined with no further categorization; ACGIH — TLV A4, not classifiable as a human carcinogen; S — respiratory allergen; (i) — inhalable fraction.
[b]Standard for soluble nickel compounds.
[c]Standard for nickel and its compounds, excluding nickel carbonyl.

8.1.1.4 Reproductive and Developmental. No data have been found concerning the effects of nickel hydroxide on reproduction or fetal development in experimental animals.

8.4.1.5 Carcinogenesis. Only limited data are available on the carcinogenic activity of nickel hydroxide. Gilman detected local sarcomas in rats exposed by IM injections of nickel hydroxide (293).

During a 2-year experiment, local sarcomas developed in 5 of 20 rats injected IM with nickel hydroxide air-dried gel and in 3 of 20 rats similarly injected with crystalline industrial nickel hydroxide. IM injections of freshly precipitated colloidal nickel hydroxide did not produce tumors (294).

No study in which nickel hydroxide was administered by inhalation has been reported. In the opinion of IARC experts, nickel hydroxide is carcinogenic to animals (124).

8.4.1.6 Genetic and related Cellular Effects Study. Data on the genotoxic effects of nickel hydroxide in humans could not be found in the available literature.

8.4.2 Human Experience

No data have been found in the available literature concerning acute or chronic nickel hydroxide poisoning in humans or its effects on reproduction and fetal development. Data on the genotoxic activity of nickel hydroxide in humans are not available, either.

Epidemiology studies mentioning exposures to nickel hydroxide are not specific to this compound. No conclusion is possible about the carcinogenic potential of this material to humans.

8.5 Standards, Regulations, or Guidelines of Exposure

Recommended exposure limits are listed in Table 41.15.

9.0 Nickel Sulfate

9.0.1 CAS Number: *[7786-81-4]*

9.0.2 Synonyms: Nickel monosulfate, nickelous sulfate; sulfuric acid nickel salt; sulfuric acid, nickel (2+) salt; nickel(II) sulfate; sulfuric acid, nickel salt; sulfuric acid, nickel(2+) salt (1 : 1); nickelsulfate (1 : 1); nickel sulfate, 97%- Carc.

9.0.3 Trade Names: NA

9.0.4 Molecular Weight: 154.75

9.0.5 Molecular Formula: $NiSO_4$

9.0.6 Molecular Structure:
$$O=\overset{\overset{O}{\|}}{\underset{\underset{O^-}{|}}{S}}-O^- \quad Ni^{2+}$$

Table 41.15. Exposure Limits for Nickel hydroxide as Ni (All Limits in mg/m^3)[a]

Exposure Limits	OSHA PEL (156)	NIOSH Exposure Limit (156)	ACGIH TLV (156)	AIHA WEEL	ANSI Standard	German MAK	Sweden Standard	British Standard (158)	Polish Standard (159)
TWA	1[b]	0.015[b]	0.2[b]	—	—	—	—	0.1[b]	0.25[c]
	—	Ca	A1	—	—	—	—	—	—
	—	—	(i)	—	—	—	—	—	—
STEL	—	—	—	—	—	—	—	0.3[b]	—
Ceiling limit	—	—	—	—	—	—	—	—	—
Biological limits (if available)	—	—	—	—	—	—	—	—	—

[a] *Carcinogen designations*: NIOSH—Ca, carcinogen defined with no further categorization; ACGIH—TLV A4, not classifiable as a human carcinogen; (i)—inhalable fraction.
[b] Standard for insoluble nickel compounds.
[c] Standard for nickel and its compounds, excluding nickel carbonyl.

247

9.1 Chemical and Physical Properties

Table 41.1 shows the physical properties of nickel sulfate. Anhydrous nickel sulfate occurs in the form of yellow crystals, soluble in water and insoluble in ethanol, ether, and acetone. Nickel sulfate hexahydrate occurs in two crystalline forms: blue and bluish-green tetragonal, which at 53.3°C is transformed into the green transparent form β (1–3).

9.2 Production and Use

Nickel sulfate can be prepared by dissolving nickel oxide or hydroxide in sulfuric acid. It is used as a main component of the baths in electroplating processes and as a raw material to make catalysts (177).

9.3 Exposure Assessment

See Sections 1.3.1 and 1.3.2.

9.4 Toxic Effects

9.4.1 Experimental Studies

9.4.1.1 Acute Toxicity. The lethal doses of nickel sulfate ($NiSO_4 \cdot 6H_2O$) are given in Table 41.16 (295).

High doses of nickel sulfate administered orally caused about acute irritations of the GI tract (18). Lethal dose administered intragastrically to rats caused a decrease in the rectal body temperature, excessive salivation, and convulsions. Histopathological examinations showed degradation of the liver and kidneys (295). A single instillation of nickel sulfate to rat (Fisher 344) trachea caused inflammation of pulmonary alveoli and changes in enzymatic activities measured in the bronchoalveolar lavage (BAL) fluid (181).

9.4.1.2 Chronic and Subchronic Toxicity. The inhalation exposure of Fisher 344/N rats and B6C3F$_1$ mice to hydrated nickel sulfate at concentrations from 0.8 to 13 mg/m^3 (6 h/day for

Table 41.16. Acute Toxicity of Nickel Sulfate

Species	Administration	Dose (mg/kg bw)		Ref.
Mouse	Intraperitoneal	LD$_{50}$	34	19
Guinea pig	Subcutaneous	LD$_{100}$	62	19
Rabbit	Subcutaneous	LD$_{100}$	500	19
Rabbit	Intravenous	MLD	35.8	19
Cat	Intravenous	MLD	71.5	19
Dog, rat	Subcutaneous	LD$_{100}$	500	19
Dog, rat	Intravenous	MLD	89.5	19
Rat	Oral	LD$_{50}$	300	18
Rat	Intragastrical	LD$_{50}$	500	295

12 days, 5 days per week, particle diameter 2 μm) resulted in death of all mice at concentrations of ≥ 1.6 mg/m³ or higher and of some rats at 13 mg/m³. Changes in lungs and nasal cavity were observed in both mice and rats when nickel sulfate concentration was 0.8 mg/m³. The changes included pneumonia with necrosis, chronic pneumonia, and degenerative lesions in bronchiolar epithelium and oflactory epithelium atrophy as well as the growth of bronchial and mediastinal glands (38, 184).

Histological changes in the rats' oflactory epithelium resulting from the inhalation exposure to nickel sulfate (0.635 mg Ni/m³, 6 h/day for 16 days) are described by Evans et al. (296). These changes disappeared within 22 days since the moment when the exposure ceased, indicating that the effects of short-lasting exposure to nickel sulfate are reversible. The injections of nickel sulfate to the male rat's peritoneal cavity (90 days, 3 mg/kg bw) caused proximal tubule necrosis and cell infiltrates around the portal vein as well as necrosis foci in the liver. The proliferation of biliary canaliculi, Kupffer's cell growth, and degenerative changes in some seminal canaliculi and internal wall of the heart were also observed (297).

Studies on long-term effects of nickel sulfate on rat's skin (doses 40, 60, and 100 mg/kg bw in a volume of 0.25 mL) applied once a day for 30 days have revealed epidermis atrophy and cornification zones, misarranged epidermal cells, and excessive keratinization. The results of histopathological examinations have also confirmed liver injury with necrotic foci (298). All mice receiving nickel sulfate in drinking water at concentrations 1, 5, or 10 g/L for 180 days survived (43).

No increased mortality was observed among rats and dogs fed for 2 years with fodder containing nickel sulfate up to 188 mg/kg per day (rats) and 62.5 mg/kg per day (dogs) (33). As dogs vomited for the first 3 days of the experiment, the daily nickel sulfate dose was reduced to 37.5 mg/kg and then, after 2 weeks, it was increased up to 62.5, after which GI disorders were not observed. This experiment indicates that high doses of nickel sulfate may irritate the GI tract and that adaptation to high levels of dietary nickel is possible.

9.4.1.3 Pharmacokinetics, Metabolism, and Mechanisms. See Section 1.4.1.3.

9.4.1.4 Reproductive and Developmental. An increased frequency of deformed sperm cells was observed among mice after a single oral administration of nickel sulfate at 28 mg Ni/kg (275).

During the studies on the nickel sulfate effects on male gonads and epididymis performed with Fisher rats, histopathological changes in gonads and appendages were observed. Contraction of the canaliculi and total degeneration of sperm cells was observed 18 h after intradermal exposure to a single dose of 0.04 μmol NiSO$_4$/kg bw (299). Oral administration of nickel sulfate to rats in a dose of 25 mg/kg for 120 days resulted in their sterility (300).

Nickel sulfate was applied onto male rats' skin at daily doses of 40, 60, and 100 mg Ni/kg bw for 15 and 30 days. The application of 60 mg Ni/kg for 30 days damaged the canaliculi and caused degeneration of sperm cells. With the 100-mg/kg dose, the reaction was stronger. No changes were observed in testicles after a dose of 40 mg Ni/kg had been administered for 30 days, and also when the exposure had lasted for 15 days regardless of the applied dose (298).

The administration of nickel sulfate to rats in their fodder or drinking water brought about an increased frequency of spontaneous and increased prenatal and early postnatal mortality (277).

No reports on teratogenic effects of nickel sulfate were found.

9.4.1.5 Carcinogenesis. Carcinogenic effects of nickel sulfate on mice and rats were examined using various routes of exposure. Nickel sulfate did not cause any cancerogenic changes in rats after IM administration (191,294) or in rats and mice exposed by inhalation (301). After repeated IP administration, tumors such as mesotheliomas and sarcomas were observed in 6 of 30 exposed rats (84). No reports are available on carcinogenic effects resulting from per os administration.

9.4.1.6 Genetic and Related Cellular Effects Studies. Nickel sulfate failed to induce point mutations in *Escherichia coli* or *Salmonella typhimurium* bacteria (302) or in Chinese hamster V79 cell cultures (303,304), but it did induce point mutations in mouse L5178Y lymphoma cell cultures (305) and in CHO AS52 cell culture (306).

Nickel sulfate induced an increase in the number of SCEs and chromosome aberrations in hamster cell cultures (307,308) as well as in the culture of human peripheral blood lymphocytes (307). Nickel sulfate showed a dose-dependent capability to induce morphological transformations in Syrian hamster embryo cells (SHE) *in vitro* (219). During *in vivo* examinations, nickel sulfate showed positive results in the *Drosophila* sex-linked recessive lethal (SLRL) assay. $NiSO_4$ induced SLRL to the extent dependent on the dose, but significant sex chromosome losses were detected only among individuals exposed to the highest concentrations (309).

9.4.2 Human Experience

9.4.2.1 General Information. Effects of an acute exposure to nickel sulfate were observed in incidental cases of nickel sulfate intake. The poisoning resulted first in disturbances of the alimentary and nervous systems (121).

Effects of prolonged exposure to nickel sulfate were observed among workers employed in the commercial processes of electroplating and nickel electrorefining. Operation of those processes involved a combined inhalation of nickel sulfate, which was the main component of the bath, and nickel chloride.

At the end of the nineteenth century, nickel sulfate was used as an oral medicine in the treatment of rheumatism and epilepsy. Daily doses of nickel sulfate of ≤ 500 mg were well tolerated by patients (310).

9.4.2.2 Clinical Cases

9.4.2.2.1 Acute Toxicity. In the literature on acute toxicity, several cases of poisoning with nickel sulfate can be found. Daldrup et al. (311) described a case of lethal poisoning of a 2.5-year-old girl who swallowed 10 or 15 g crystalline nickel sulfate ($\sim$2.2–3.3 g nickel). The symptoms included loss of consciousness, erythema, wide-open pupils not responding to light, tachycardia, and pulmonary rales. The child died from cardiac arrest

8 h after the poisoning. The autopsy showed acute, hemorrhagic gastritis and an elevated level of nickel in blood, urine, and liver.

Sunderman et al. (121) described the effects of incidental intake of water polluted with nickel sulfate and chloride. The poisoned people showed symptoms of alimentary system disorders (nausea, abdominal pain, diarrhea, and vomiting) and nervous system disturbances (headache, vertigo, and fatigue).

Nervous system disturbances manifested by temporary homonymous hemianopsia appeared in one of the volunteers participating in an experiment on nickel absorption and excretion in humans. Sight disturbances appeared 7 h after drinking water containing a single dose of nickel sulfate (0.05 mg Ni/kg) and persisted for 2 h. No side effects were observed among the remaining volunteers who received lower doses (54).

9.4.2.2.2 Chronic and Subchronic Toxicity. The consequences of chronic exposure among workers of nickel refining and electroplating plants included rhinitis, perforation of nasal septum, asthma, and contact dermatitis.

Tatarskaya (312) examined 486 workers of a nickel refining plant, exposed mainly to nickel sulfate during the electrorefining processes. Nasal mucositis was observed among 10–16% of the examined workers; 5.3% of them showed chronic inflammation. Erosions of nasal septum were found in 13%, perforations in 6.1%, and ulceration in 1.4% of the examined workers. Hyposmia and anosmia were found among 30.6 and 32.9% of the examined workers, respectively. Kucharin (313) examined 302 workers exposed for at least 10 years to nickel sulfate and sulfuric acid vapors at 0.02–4.5 mg/m^3. Clinical and X-ray examinations of the workers showed sinusitis among 83% of them. Severe injuries, such as nasal septum erosions (41.4%) and perforations (5.6%), were also found. Anosmia was found among 46% workers who developed sinusitis. In a nickel refining plant where hydrometallurgic processes were used, pathological changes within the nasopharynx were found among 37 of 151 examined workers (24%). Nasal septum erosions were observed among 14 workers (9.3%). The value of exposure to soluble nickel salts in terms of their concentrations ranged from 0.035 to 1.65 mg/m^3 during 1966–1970 (314).

The results of examinations of workers exposed to soluble nickel compounds (mainly nickel sulfate and chloride) for a long time (15–25 years) indicated moderate nephrotoxic activity of nickel sulfate (315).

Asthmatic lung diseases among workers of a nickel electroplating plant who were exposed to soluble nickel compounds has been reported (316–320). Cirla et al. (317) examined 12 workers of a nickel electroplating shop who reported respiratory system symptoms. Allergic asthma could be provoked among 6 workers by nickel sulfate aerosol inhalation. Novey et al. (318) described workers of an electroplating plant (chromium and nickel) who developed acute asthma in response to a provocative inhalation of nickel and chromium sulfates. Radioimmunological testing with a built-in provocative agent resulted in specific antiagent IgE antibodies forming.

Occupational asthma among seven workers of an electroplating plant exposed to soluble chromium and nickel salts has been also reported by Bright et al. (319).

Nickel sulfate shows an irritating effect on human skin: 5% aqueous solution of nickel sulfate applied onto the back (occlusion) caused skin irritation in some people (321). Unoccluded patch with nickel sulfate solution was applied onto a forearm once a day for 3 days, and it was found that the threshold concentration of the aqueous nickel sulfate solution causing skin irritation was 20% on uninjured skin and 0.13% on scarified skin (322).

Soluble salts of nickel are potential allergenic agents for the skin. Contact dermatitis caused by nickel sulfate and chloride is quite frequent among workers of electroplating plants (18, 97, 98). The resultant symptoms include itching eczema, known as "nickel itch," mainly on the hands and forearms. The nickel itch is stronger in summer when higher temperature and humidity in production halls enhance perspiration. If the exposure continues, eczema turns into a chronic phase during which hyperkeratosis and callous eczema are observed. Eczema often appears in the form of dermal eruptions over adjacent skin surfaces and, in the most extreme cases, it may spread over the whole surface of the body (97).

9.4.2.2.3 Pharmacokinetics, Metabolism and Mechanisms. See Section 1.4.2.2.3.

9.4.2.2.4 Reproductive and Developmental. Data on the effects of nickel sulfate on human reproduction and fetal development are very scanty and confined to a single study among women employed in nickel refining works in the Arctic region of Russia (323). Spontaneous and threatening abortions were reported in 16% and 17% of all pregnancies in nickel-exposed women, compared with 9 and 8%, respectively, in the controls. Structural malformations were found in about 17% of live-born infants with nickel-exposed mothers compared with about 6% in the reference group. The contribution of heavy lifting and possible heat stress to these effects is not known.

9.4.2.2.5 Carcinogenesis. According to the IARC assessment, there is sufficient evidence that the exposure to nickel sulfate is carcinogenic to humans (124). See also Sections 1.4.2.2.5 and 7.4.2.2.5.

9.4.2.2.6 Genetic and Related Cellular Effects Studies. Deng et al. (154, 155) found an increased number of SCE and chromosome aberrations (breaks, gaps, fragmentation) among workers employed at electroplating. Waksvik and Boysen (152) described two groups of workers exposed to nickel compounds who showed an increased level of chromosome aberrations, mainly gaps. SCE was not observed. The workers of the first group were employed at the nickel ore crushing, roasting and smelting division, where they were exposed mainly to nickel oxide and sulfide; the other group attended the electrolytic processes and was exposed to nickel chloride and sulfate. Waksvik et al. (153) examined also nine retired (for 4–15 years) workers of a nickel refining plant who had been occupationally exposed to nickel oxides and sulfides or nickel chloride and sulfate aerosols, and found that nickel concentration in their plasma was still 2 µg/L. The number of chromosome breaks and gaps was higher, but no differences in the incidence of SCE were found compared with the reference group.

9.5 Standards, Regulations, or Guidelines of Exposure

Recommended exposure limits are listed in Table 41.17.

Table 41.17. Exposure Limits for Nickel Sulfate as Ni (All Limits in mg/m^3)[a]

Exposure Limits	OSHA PEL (156)	NIOSH Exposure Limit (156)	ACGIH TLV (156)	AIHA WEEL	ANSI Standard	German MAK (157)	Sweden Standard (158)	British Standard (158)	Polish Standard (159)
TWA	1[b]	0.015[b]	0.1[b]	—	—	—	0.1[b]	0.1[b]	0.25[c]
		Ca	A4			S			
			(i)						
STEL	—	—	—	—	—	—	—	—	—
Ceiling limit	—	—	—	—	—	—	—	0.3[b]	—
Biological limits (if available)	—	—	—	—	—	—	—	—	—

[a]*Carcinogen designations*: NIOSH — Ca, carcinogen defined with no further categorization; ACGIH — TLV A4, not classifiable as a human carcinogen; S — respiratory allergen; (i) — inhalable fraction.
[b]Standard for soluble nickel compounds.
[c]Standard for nickel and its compounds, excluding nickel carbonyl.

B RUTHENIUM AND ITS COMPOUNDS

10.0 Ruthenium

10.0.1 CAS Number: [7440-18-8]

10.0.2 Synonyms: Royer@R Ruthenium Catalyst Beads

10.0.3 Trade Names: NA

10.0.4 Molecular Weight: 101.07

10.0.5 Molecular Formula: Ru

10.1 Chemical and Physical Properties

Ruthenium, a transition element, belongs to group VIII (iron) of the periodic classification and to the light platinum metals triad. It is a hard and brittle metal that resembles platinum. It crystallizes in hexagonal form and occurs in the form of seven stable isotopes: 96 (5.46%), 98 (1.87%), 99 (12.63%), 100 (12.53%); 101 (17.02%), 102 (31.6%), and 104 (18.87%). There are also several radioactive isotopes — 93, 94, 95, 97, 103, 105, 106, 107, and 108 — of which the 106 isotope characterized by strong β radiation and has a half-life of 368 days; since it is produced in large quantities in the nuclear reactors, it deserves special attention. Ruthenium, is the rarest of the platinum group elements (abundance in the earth's crust ~0.0004 ppm). In chemical compounds, it occurs at oxidation states from +2 to +8; the most frequent is +3 in ruthenium compounds. Ruthenium is resistant to acids and aqua regia, it is not oxidized in the air at room temperature, and in the form of powder it reacts with oxygen at elevated temperatures. It is dissolved in molten strong alkalis and reacts with alkaline metal peroxides and perchlorides. Ruthenium powder reacts with chlorine above 200°C and with bromine at 300–700°C (324, 325).

Ruthenium compounds are usually dark brown (ranging from yellow to black). Ruthenium forms alloys with platinum, palladium, cobalt, nickel, and tungsten.

Table 41.18 gives information on the molecular formula and weight, physical and chemical properties, and solubility of ruthenium and some of its compounds.

10.2 Production and Use

Elemental ruthenium occurs in native alloys of iridium and osmium (irridosmine, siskerite) and in sulfide and other ores (pentlandite, laurite, etc.) in very small quantities that are commercially recovered.

The element is separated from the other platinum metals by a sequence involving treatment with aqua regia (separation of insoluble osmium, rhodium, ruthenium, and iridium), fusion with sodium bisulfate (with which rhodium reacts) and fusion with sodium peroxide (dissolution of osmium and ruthenium). The resulting solution of ruthenate and osmate is treated with ethanol to precipitate ruthenium dioxide. The ruthenium dioxide is purified by treatment with hydrochloric acid and chlorine and reduced with hydrogen gas to pure metal.

Table 41.18. Physical and Chemical Properties of Ruthenium and Its Compounds

Compound	Molecular Formula	Molecular or Atomic, Weight	Boiling Point (°C)	Melting Point (°C)	Specific Gravity	Solubility in Water
Ruthenium	Ru	101.07	3900	2310	12.45	Insoluble
Ruthenium(IV) oxide	RuO$_2$	133.07	—	—	6.97	—
Ruthenium(VIII) oxide	RuO$_4$	165.07	40	25.4	—	2%
Ruthenium(III) chloride	RuCl$_3$	207.43	—	—	3.11	Insoluble
Ruthenium chloride hydroxide	RuCl$_3$OH	224.43	—	—	—	Insoluble
Ruthenium oxychloride ammoniated (Ruthenium Red)	[(NH$_3$)$_5$RuORu(NH$_3$)$_4$ORu(NH$_3$)$_5$]Cl$_6$	786.35	—	—	—	Soluble

Ruthenium is recovered from exhausted catalytic converters or, in a similar manner, from the waste produced during platinum and nickel ore processing.

Ruthenium is used in electronics and electrical engineering, and also in the chemical industry. Ruthenium metal is used as a catalyst in the oxidizing reactions and in the synthesis of long-chain hydrocarbons. Because of its catalytic activity, it is also used in the catalytic converters for motor car engines. Ruthenium is used to increase the hardness of platinum alloys designed to make electric contacts, to make resistance wires, circuit breakers, and other components. It is also employed as a substitute for platinum in jewelry and to make the tips of fountain pen nibs.

The commercial use of ruthenium salts is insignificant; saturated solutions of platinum and ruthenium salts are used to plate the surfaces of titanium electrodes employed in the production of chlorine and chlorates (326).

Ruthenium tetrachloride is used in chemical reactions as a strong oxidant.

Certain derived ruthenium(III) complexes are used in cancer therapy to prevent metaplasia (327, 328) or to inhibit tumor cell growth (329). Ruthenium 106 is also used for that purpose (330, 331). Ruthenium(III) complexes may be also applied to treat diseases resulting from exposure to nitric oxide (332). Ammoniated ruthenium oxychloride (Ruthenium Red) has been used as staining agent in microscopy.

10.3 Exposure Assessment

10.3.1 Air

Neither NIOSH nor OSHA-recommended methods for the determination of ruthenium or its compounds in the workplace air are available.

Air–acetylene flame atomic absorption spectrometry (FAAS) can be used to determine ruthenium and its compounds in the air. Cellulose ester membrane filters may be used for air sample collection. Samples are mineralized in concentrated acids and dissolved in diluted hydrochloric acid, and the absorbance of the resulting solution is measured at 349.9 nm wavelength. The standard solution is prepared from ruthenium(III) chloride trihydrate (333). Inductively coupled–plasma atomic emission spectrometry (ICP-AES) and neutron activation analysis (NAA) can be also used to determine ruthenium compounds (324).

10.3.2 Background Levels: NA

10.3.3 Workplace Methods: NA

10.3.4 Community Methods: NA

10.3.5 Biomonitoring/Biomarkers

10.3.5.1 Blood. Neutron activation analysis (NAA) may be used to determine ruthenium in dry bone, dry muscle, and hair (334, 335).

10.3.5.2 Urine. Radiometric methods are used to determine the ruthenium 106 radioactive isotope in urine. After urine sample has been mineralized by nitric acid, ruthenium is oxidized with potassium permanganate to ruthenium tetroxide. The resultant RuO_4 is distilled off, absorbed in hydrochloric acid, and precipitated by sodium carbonate, and the activity of the sample is determined by radiometry. The sensitivity of the method is 1.5 pCi ^{106}Ru/L urea (336).

In the nondistillation method ruthenium, after oxidation to tetroxide with potassium periodate, is absorbed in polyethylene powder and heated for 2.5 h in an oil bath. After the polyethylene powder containing the ruthenium tetroxide has been filtered off and washed with water, it is transferred to an aluminum planchet and slowly heated until the powder is melted. The activity of the sample is determined by radiometry. The detection limit of the method is 0.59 eq/L (337).

10.4 Toxic Effects

10.4.1 Experimental Studies

10.4.1.1 Acute Toxicity. No data are available in the literature on the acute toxicity of ruthenium metal.

The toxicity of ruthenium compounds increases with their solubility in water. Hypotaxia, clonic convulsions, forced respiration, disturbed breathing rhythm, and violent spasms were observed in the experimental animals at high doses of ruthenium chloride hydroxide and ruthenium dioxide. The animals died within several days (338, 339).

Organic ruthenium compounds exhibit significant toxicity resulting in rapid death of the test animals. Compounds containing bisbipyridyl groups produced a mean time to death of 10 min, in contrast to 4–7 days for the other compounds. The difference between toxic action of organic and inorganic ruthenium compounds is similar to that between alkyl lead compounds and its inorganic salts (340).

The values of LD_{50} for different ruthenium compounds are shown in Table 41.19 (341).

The value of 9.5 mg/m^3 has been assumed as to be threshold of the single-dose acute toxicity for ruthenium chloride hydroxide (338); the assumed threshold values of acute toxicity for ruthenium dioxide were 34.0 mg/m^3 for inhalation and 1000 mg/kg for intragastric exposure, respectively (339).

10.4.1.2 Chronic and Subchronic Toxicity. Data on the chronic and subchronic toxicity of ruthenium compounds are very scanty.

Chronic inhalation exposure of mice and guinea pigs to ruthenium dioxide at 21.4 mg/m^3 resulted in reduced motor activity, increased rectal temperature, and higher oxygen demand and carbon acid excretion, observed in the animals during the third month. After 4 months of exposure, there was an increase in the peripheral blood erythrocyte count, accompanied by altered hepatic function [manifested by lower total blood serum protein content, increased alkaline phosphatase activity, and higher blood urea nitrogen level (BUN)]. Changes in the upper respiratory tract and lungs (profound atrophy of bronchial mucosa, atrophy of the cilia in the ciliated epithelium, lung diffuse interstitional sclerosis, and granular dystrophy in the liver and kidneys) were detected in the sacrificed animals. No changes were noted in the animals exposed to 2.12 mg/m^3 ruthenium dioxide (339).

Table 41.19. LD$_{50}$ Values for Different Ruthenium Compounds

Chemical Name	CAS Number	Species	Exposure Route	LD$_{50}$ (mg/kg)	Ref.
Ruthenium chloride	[10049-08-8]	Rat	IP	360	341
			IP	108	340
Ruthenium chloride hydroxide	[16845-29-7]	Rat	Orl	1250	338
		Mouse	Oral	462.5	
			IP	225	
Ruthenium dioxide	[12036-10-1]	Rat	Oral	4580	339
		Mouse	Oral	5570	
			IP	3050	
Pentaminenitrosylruthenium(II) chloride	[15611-80-0]	Mouse	IP	8.9	340
Chloronitrobis(2,2-dipyridyl)ruthenium(II)	Unknown	Mouse	IP	55	340
Dichlorobis(2,2-dipyridyl)ruthenium(II)	[71230-28-9]	Mouse	IP	63	340
Potassium pentachlornitrosyl ruthenate(II)	[14854-54-7]	Mouse	IP	127	340

The different ruthenium complexes used in cancer therapy were administered to rats at 10% of the LD$_{50}$ twice a week for up to 5 weeks. No renal functional changes (excretion of water, protein, *p*-aminohippurate, and osmolytes) were observed (342).

10.4.1.3 Pharmacokinetics, Metabolism, and Mechanisms

10.4.1.3.1 Adsorption. Ruthenium and its compounds can be absorbed into the system in the respiratory tract, through skin, or per os; inhalation is the most important route in the occupational exposure. Absorption rate depends mainly on the physical and chemical characteristics of the compound and on the route of administration. The pharmacokinetics and metabolism have been determined only from single exposures to ruthenium radioisotopes in animals. No attempts have been made to demonstrate the applicability of the results to long-term exposures.

Whole-body retention of ^{106}Ru in the form of ^{106}RuCl$_3$ was measured in mice, rats, monkeys, and dogs by Furchner et al. (343). The animals received single oral, or single IV or IP injections of the compound. Retention from oral administration was considerably smaller than that from IP injection. The highest concentrations were found in the kidney, whereas bone maintained about 6% of the body burden over a period of 8 months.

Ruthenium-106 as a nitrosyl-ruthenium was absorbed completely from the GI tract of the rat in 2 h. About 20% of the dose was absorbed into the first section of the small-intestine wall after 1 h; by 6 h most of the ^{106}Ru was in the cecum or in the feces, and by 24 h the gut was clear. In fasted rats, absorption of ^{106}Ru was considerably increased in the first two sections of the small intestinal wall (344).

Following the inhalation exposure of rats to ^{106}Ru in the form of ruthenium tetroxide, 54% of activity was found in the upper respiratory tract, 45% amounted to external contamination, and 0.1% was in the pulmonary area (345). Similar retention pattern was observed for ruthenium tetroxide exposed beagle dogs. Initial deposition was primarily in the nasopharyngeal and tracheobronchial regions. Effective whole-body retention of ^{106}Ru followed a three-component exponential model with 99% of the initial body burden rapidly cleared with an effective half-life of 1.2 days; 0.7% with a half-life of 14 days and 0.3% with a half-life of 170 days (346).

In human exposure incidents to volatile ^{103}Ru compounds, probably ^{103}RuO$_4$ average effective retention half-life was 26.6 days (347) or 11.3 days (348).

Following a 6-h application of ^{106}RuCl$_3$ to the skin of the rats, the biological half-life was 12 h for the first 35% of the dose and 6 days for remaining 65%. The percentage of ^{106}Ru deposited in the internal organs was 1.37 (349).

10.4.1.3.2 Distribution. Distribution of ruthenium in the organism depends mostly on the route of exposure. Following inhalation, a majority of the dose is retained in the upper respiratory tract and trachea. The presence of ^{106}Ru in the GI tract at 2 h and 5 days after exposure was due mainly to clearance of ruthenium deposited primarily in the upper respiratory tract. An average of 4.9% of ^{106}Ru was found in the lungs after 2 h. Relative tissue ^{106}Ru concentrations 224 days after exposure were, in descending order: nasopharynx and tracheobronchial region, thoracic lymph nodes, lungs, kidneys, liver, GI tract and pelt, soft tissues, skeleton, and blood (346).

Eight months after oral administration of ruthenium chloride, the highest concentrations were found in the kidney, and bone maintained about 6% of the body burden (343).

10.4.1.3.2 Excretion. Excretion of ruthenium depends on the route of exposure; orally administered ruthenium is excreted primarily in feces and those injected IV or IP in urine. During the first week after the exposure the urinary-to-fecal excretion ratios (U/F) of ^{106}Ru were 2.34 (monkeys) or 7.52 (dogs) for IV; 1.65 (mice) or 2.46 (rats) for IP; and 0.034 (mice), 0.009 (rats), or 0.017 (monkeys) for oral administration (343, 345).

10.4.1.4 Reproductive and Developmental. Inhalation exposure of the experimental animals to ruthenium dioxide at 2.12 mg/m^3 did not produce any discernible effects in the reproductive or chromosomal systems (339).

10.4.1.5 Carcinogenesis: NA

10.4.1.6 Genetic and Related Cellular Effects Studies. A ruthenium complex caused mutations in *Salmonella typhimurium* strains TA100 and TA98 by frameshift and base-pair substitution. At a concentration of 400 μmol ruthenium continued to produce increased numbers of revertants. In the Comptest, ruthenium was found to be less toxic than platinum to *Bacillus-subtilis* (less ability to induce the so-called SOS system) (350).

10.4.1.7 Other: Neurological, Pulmonary, Skin Sensitization

10.4.1.7.1 Sensitization. Intravenous injection of ruthenium hydrochloride to rabbits was found to induce sensitization on day 5 after administration. The sensitization was confirmed by the basophil degranulation test, neutrophil damage index, neutrophil alteration, leucocyte agglomeration, drop skin, and skin fenestra tests. Ruthenium hydrochloride sensitizing properties were less pronounced than those of platinum, palladium, and rhodium compounds (351).

Ruthenium hydroxychloride applied as a 15% water solution to the skin exhibited weak sensitizing properties (338).

10.4.1.7.2 Irritation. A preparation containing 50% ruthenium dioxide applied to guinea pig skin did not cause irritation. Ruthenium dioxide instilled to the conjunctival sac of the rabbit produced blepharospasm. Iridosis, hyperemia of the mucosa, and other symptoms of irritation disappeared within 24 h (339).

Introduction of powdered ruthenium hydroxychloride in a quantity of 50 mg to rabbit conjunctival sac resulted in chronic blepharospasm. Putrescent conjunctivitis accompanied by profuse secretions developed within 24 h; after several days, all tissues of the eye were affected (338).

Becuase of its strong oxidizing activity, ruthenium tetroxide is irritant to the upper respiratory mucosa.

10.4.2 Human Experience

10.4.2.1 General Information. Exposure to ruthenium and its compounds is associated mainly with the working environment and may occur during ruthenium ore mining, refining, and processing. During mining, ruthenium occurs mainly as metal or in the form of insoluble compounds. During refining, the workers may be exposed to the soluble ruthenium compounds (chlororuthenium acid salts). In the situations mentioned, ruthenium occurs in the working environment usually in combination with other platinum group metals.

When used as a catalyst, ruthenium is employed in the form of chloride or hydroxychloride.

No data are available in the literature on health effects of ruthenium metal due to occupational exposure; the data on exposure to ruthenium compounds are very scanty.

10.4.2.3 Epidemilogy Studies

10.4.2.3.1 Acute Toxicity: NA

10.4.2.3.2 Chronic and Subchronic Toxicity. Among 16 women employed in surface coating of titanium anodes with ruthenium and platinum salts for 2–10 months, nasal ulceration was found in eight cases, including one case of perforation of the septum. The women were complaining of coughing and throat irritation. No platinum or ruthenium concentrations in workplace air were reported. The ulceration was attributed to ruthenium since no such effects were found in persons exposed to platinum salts only (326).

Detailed medical examinations of 17 people exposed to ruthenium compounds during the manufacture and use of ruthenium compounds (for 1–5 years) did not show any functional changes of the CNS or blood cell count irregularities. Elevated arterial blood pressure was found in 6 people. Allergic eczema was detected in 9 people. Workplace air concentration of ruthenium hydroxychloride ranged from 0.02–5.5 mg/m^3, while for ruthenium dioxide the corresponding values were 0.02–13 mg/m^3. The concentration of hydrochloride in the workroom air was also high (352).

10.4.2.3.3 Pharmacokinetics, Metabolism, and Mechanisms. Little quantitative information is available on biological behavior and pharmacokinetics of ruthenium compounds in humans. In general, simple salts and oxides of ruthenium are poorly absorbed after oral administration, whereas more compounds are absorbed to a much greater extent. Inhaled vapor and aerosol are deposited mainly in the upper respiratory tract.

When shellfish contaminated with radioruthenium compounds were fed to human volunteers, fecal excretion accounted for 95% of the dose in 2 days; about 1% was eliminated in the prolonged phase with a half-life of 30 days (353).

In accidental human exposures to volatile ruthenium compounds (103 or 106 isotopes), the effective retention half-life ranged from 11.3 to 26.6 days for the first phase of elimination (347, 348).

10.5 Standards, Regulations, or Guidelines of Exposure

No occupational exposure limits for ruthenium and its compounds have been recommended by ACGIH or adopted by OSHA.

The maximum allowable concentration (MAC) for ruthenium dioxide in workplace air, 1 mg/m^3, has been established in the former Soviet Union in 1979.

For ruthenium hydroxychloride, an occupational exposure limit of 0.1 mg/m^3 was proposed in the former Soviet Union (338) but it has not been formally established.

11.0 Radioactive Ruthenium

11.0.1 CAS Numbers: *[13967-48-1]* (Ruthenium-106); *[13968-53-1]* (Ruthenium-103)

11.0.2 Synonyms: Ruthenium-106; ruthenium-103; ^{106}Ru; ^{103}Ru; radioruthenium

11.0.3 Trade Name: NA

11.0.4 Molecular Weight: Atomic Weights 106 (^{106}Ru); 103 (^{103}Ru); for molecular weights of radioactive ruthenium compounds, see ruthenium entry in Table 41.18

11.1 Chemical and Physical Properties

Ruthenium-106 emits β emissions (energy 3.5 MeV, 78%) and γ emissions (energy 0.62 MeV 22%); physical half-life 368 days with rhodium-106 as the daughter product. Other radioactive isotopes are of less practical importance because of relatively short half-lives: 39 days and 4.5 h for ruthenium-103 and ruthenium-105, respectively. Radioactive ruthenium may be present in various chemical forms. For other chemical and physical properties, see Section 10.1.

11.2 Production and Use

Radioactive ruthenium is produced in large quantities in irradiated nuclear reactor fuel, which is one of the most abundant nucleotide products of atomic fission. At elevated temperatures, such as those that might occur in nuclear fuel reprocessing, radioactive ruthenium would be emitted mostly in the form of volatile oxides — ruthenium tetroxide or ruthenium dioxide.

Ruthenium-106 has been used in plaque radiotherapy of uveal melanoma and for the management of uveal metastasis (330, 331).

11.3 Exposure Assessment

Radiometric methods are used to measure radioactive ruthenium in tissues, blood, urine, and environmental media.

11.4 Toxic Effects

11.4.1 Experimental Studies

11.4.1.1 Acute Toxicity. Acute toxicity of radioactive ruthenium is due to β-radiation emission rather than to the metal itself.

Ruthenium-106/rhodium-106, with average energy 1.4 MeV when administered in insoluble form to rats (gavaged) and dogs (with food), caused severe damage to the GI tract, resulting in death. The calculated LD_{50} values were 1.5, 18, and 9 mCi/kg in suckling, weanling, and adult rats, respectively, and 3.5 mCi/kg in dogs, which corresponded to an absorbed dose of ~3300 rad. The greatest damage was found in the ileum of suckling rats and large intestine of adult rats (354).

11.4.1.2 Chronic and Subchronic Toxicity. In several studies of chronic and subchronic exposure of experimental animals to ^{106}Ru compounds by inhalation or oral administration, no effects other than tissue distribution of radioactivity were reported (343–346, 355).

11.4.1.3 Pharmacokinetics, Metabolism, and Mechanisms. See: Section 1.4.1.3.

11.4.1.4 Reproductive and Developmental. Development defects have been observed in experimental animals exposed to ionizing radiation, but no information related directly to radioactive ruthenium exposure was found.

11.4.1.5 Carcinogenesis. Ionizing radiation is potentially carcinogenic.

Squamous-cell carcinomas of bronchial origin have been produced in lungs of rats by implanting intrabronchial plates containing 5 mCi of ^{106}Ru. The rats were sacrificed serially at intervals of 122–356 days; the median time to tumor was 330 days (356). Tumor incidence was strongly dose related (356, 357).

11.4.1.6 Genetic and Related Cellular Effects Studies. Exposure of cells to ionizing radiation can cause gene mutations to occur in excess of the spontaneous mutation rate, but no information related directly to radioactive ruthenium exposure was found.

11.4.2 Human Experience

11.4.2.1 General Information. Human exposure to radioactive ruthenium may be limited to accidental situations in nuclear power plants or ruthenium plaque radiotherapy for melanoma treatment only. In both cases inhalation is the most important exposure route and radioruthenium would be released in the form of volatile vapors (mainly ruthenium tetroxide and dioxide) or aerosols. Gastrointestinal or skin absorption of radioactive ruthenium constitutes a minor hazard to humans and is less likely to appear.

No adverse health effects were reported in humans who experienced inhalation exposure incidents.

11.4.2.2 Clinical Cases

11.4.2.2.1 Acute Toxicity. In the reported cases of accidental inhalation of ^{103}RuO$_4$ (348) "an extremely volatile ^{103}Ru compound" (probably ^{103}RuO$_4$) (24) or volatile ^{106}Ru compound (probably ^{106}RuO$_4$) (358), ruthenium was deposited predominantly in the upper respiratory tract; the rate of ruthenium elimination was determined primarily by the rate of removal from the nasopharynx and oral cavities, but no health effects were described.

11.4.2.2.2 Chronic and Subchronic Toxicity. No health effects were observed when shellfish contaminated with radioruthenium compounds were fed to human volunteers (353).

11.4.2.2.3 Pharmacokinetics, Metabolism, and Mechanisms. For adsorption, distribution, and excretion of radioactive ruthenium compounds, see Section 10.4.1.3.

The mechanism of toxic action of radioactive ruthenium is that of β-radiation action. The main physical effect of exposure to ionizing radiation is an energy absorption by the object undergoing irradiation. If the object is a living organism, this may produce effects capable of significantly affecting its functions. The type of these effects depend on the total energy received by the object, namely on the absorbed dose. Relatively small doses may produce somatic effects only, such as carcinogenesis or gene mutations. At high doses above 1 Gy, effects may be deterministic; the number of damaged cells is so high that the functions of the organism are totally disturbed resulting in death.

11.4.2.2.4 Reproductive and Developmental. No information related to radioactive ruthenium exposure was found.

11.4.2.2.5 Carcinogenesis. Ionizing radiation has the potential for being carcinogenic, but no information on carcinogenic effects of human exposure to radioactive ruthenium was found.

11.4.2.2.6 Genetic and Related Cellular Effect Studies. Ionizing radiation can cause gene mutations and cellular damage.

Ruthenium-106 plaque radiotherapy has been used for the management of uveal metastases as more effective than external beam radiotherapy. Apart from a high degree of tumor control, the most frequent radiation-related complications were radiation cataract, vitreous hemorrhage, neovascular glaucoma, radiation retinopathy, and radiation papillomathy (331, 359).

Table 41.20. Guidelines for Exposure to Ionizing Radiation (360, 361)

Type of Exposure	Guideline
Effective dose	
In any single year	50 mSv (millisievert)[a]
Averaged over 5 years	20 per year
Annual equivalent dose to:	
Lens of the eye	150 mSv
Skin	500 mSv
Hands and feet	500 mSv
Embryofetus exposures once the pregnancy is known	
Monthly equivalent dose[b]	0.5 mSv
Dose to the surface of women's abdomen (lower trunk)	2 mSv for the remainder of the pregnancy
Intake of radionuclide	1/20 of annual limit on intake (ALI)
Radon daughters	4 working-level months (WLM)

[a] 10 msV = rem (quantity of ionizing radiation whose biological effect equals that produced by one X-ray roentgen).
[b] Sum of internal and external exposure but excluding doses from natural sources as recommended by NCRP

11.5 Standards, Regulations, or Guidelines of Exposure

The ACGIH Physical Agents Committee has adopted the occupational exposure guidance of the International Commission on Radiological Protection (ICRP) for ionizing radiation. The guiding principle of radiation is to avoid all unnecessary exposures. All exposures must be kept as low as reasonably achievable (ALARA), taking into account economic and social factors. The radiation doses that should not be exceeded are given in Table 41.20 (360, 361).

According to ICPR guidelines (361), the estimated average doses to the relevant critical groups of members of the public that are attributable to practices shall not exceed the following limits:

- An effective dose of 1 mSv in a year
- In special circumstances, an effective dose of ≤5 mSv in a single year, provided the average dose over 5 consecutive years does not exceed 1 mSv per year
- An equivalent dose to the lens of the eye of 15 mSv in a year
- An equivalent dose to the skin of 50 mSv in a year

The quoted guidelines apply to ionizing radiation from all possible sources and cover occupational and public exposure to radioactive exposure, respectively.

11.6 Studies on Environmental Impact

Ground contamination with ^{103}Ru and ^{106}Ru in the northern hemisphere of the earth due to radionuclides resulting from nuclear tests in the atmosphere in 1945–1980 was 30.1 and 24.2 kBq/m^2, respectively (362).

In Warsaw (Poland) 2 days after the Chernobyl nuclear power plant accident in April 1986, air concentration of ^{103}Ru at the ground surface was 10.2 Bq/m^3 and that of ^{106}Ru was 2.0 Bq/m^3. The concentrations dropped to 0.1 and 0.028 Bq/m^3, respectively, on day 5 after the accident (363).

C RHODIUM AND ITS COMPOUNDS

12.0a Rhodium

12.0.1a CAS Number: [7440-16-6]

12.0.2a Synonyms: Rhodium metal; elemental rhodium

12.0.3a Trade names: Rayer rhodium catalyst heads, rhodium on alumina

12.0.4a Molecular Weight: 102.9055

12.0.5a Molecular Formula: Rh

12.0b Rhodium Chloride

12.0.1b CAS Number: [10049-07-7]

12.0.2b Synonyms: Rhodium trichloride; *rhodium chloride* (III); rhodium chloride (RhCl$_3$)

12.0.3b Trade Names: NA

12.0.4b Molecular Weight: 209.26

12.0.5b Molecular Formula: Cl$_3$Rh

12.0.6b Molecular Structure:

$$\begin{array}{c} Cl^- \\ Rh^{3+} \\ Cl^- \quad Cl^- \end{array}$$

12.1 Physical and Chemical Properties

Table 41.21 gives data on the physical and chemical properties of rhodium and some of its compounds.

Table 41.21. Physical and Chemical Properties of Rhodium and Rhodium Chloride

Chemical Name	Chemical Formula	Molecular Weight	Melting Point (°C)	Boiling Point (°C)	Density (g/m^3)	Solubility
Rhodium	Rh	102.91	1966	3727	12.42	Insoluble
Rhodium chloride	RhCl$_3$	209.26	—	—	—	—

Rhodium is a transition element belonging to the cobalt group and to the light platinum triad at the same time. There is only one stable isotope: ^{103}Rh. Rhodium, in the elemental state, is a quite soft, forgeable, silver-white metal. It occurs in nature extremely rarely (abundance: 1×10^{-70} % by wt) in the form of alloys with other platinum metals (*e.g.*, in crude platinum) or accompanies gold. Because it is a very precious and expensive metal, rhodium is resistant to the action of cold chlorine and fluorine and insoluble in acids and aqua regia. In compounds, it assumes an oxidation state of +3 (RhO_3; $RhCl$; $Rh(SO_4)_3 \cdot 12H_2O$). In air, it occurs in the form of gray fume or dust (364).

12.2 Production and Use

Pure rhodium is prepared by the reduction of its ammonium salt (dichloropentaaminorhodium) (365).

Rhodium is used for the manufacture of thermocouples (in the form of platinum–rhodium alloy: 10% Rh and 90% Pt), laboratory vessels (crucibles), catalysts (as an additive to Pt and Pd), spinnerets for synthetic and glass fibers, surgical tools (Ph, Pt, and Ir alloys), and electroplating (364). Besides, rhodium is used in jewelry (365), $RhCl_3$ is capable of controlling some viruses (366). Anticarcinogenic activity of some rhodium compounds has also been confirmed (367, 368).

12.3 Exposure Assessment

12.3.1 Air

12.3.2 Background Levels: NA

12.3.3 Workplace Methods: NA

12.3.4 Community Methods: NA

12.3.5 Biomonitoring/Biomarkers

No toxic rhodium and rhodium compounds levels have been determined either for blood or urine. As the exposure of animals to rhodium results in respiratory function disorders, it seems useful to monitor the pulmonary function in the case of rhodium poisoning. Considering that central nervous system disorders have been observed among animals exposed to rhodium, it seems advisable to monitor this system in the case of rhodium poisoning in humans (366).

12.4 Toxic Effects

12.4.1 Experimental Studies

12.4.1.1 Acute Toxicity. Both fumes and dusts of rhodium metal are considered nontoxic. Data on the toxicity of soluble rhodium salts are very scarce. Intravenous injections of $RhCl_3$ to rabbits and rats caused coma and respiratory disorders. A slight reduction in animal body weight was observed. Histopathological examinations showed no changes in their internal organs, and their death, in the author's opinion, was due to the toxic effects in the central nervous system (369). (See also Tables 41.22 and 41.23).

Table 41.22. Rat Mortality Caused by IV Injection of RhCl$_3$ (369)

Dose (mg/kg)	Survivor/Dead Ratio	LD$_{50}$
125	10/10	—
150	9/10	—
175	7/10	—
200	4/10	198 mg/kg (184–212)
225	3/10	—
250	1/10	—
275	0/10	—

Table 41.23. Rabbit Mortality Caused by IV Injection of RhCl$_3$ (369)

Dose (mg/kg)	Survivor/Dead Ratio	LD$_{50}$
46.4	4/4	—
100.0	4/4	—
215.0	2/4	215 mg/kg (138–336)
464.0	0/4	—

The animals died within 12 h after RhCl$_3$ injection. Mice exposed to RhCl$_3$ developed respiratory disorders (370).

Table 41.24 gives the values of LD$_{50}$ for some rhodium compounds (371).

12.4.1.2 Chronic and Subchronic Toxicity: NA

12.4.1.3 Pharmacokinetics, Metabolism, and Mechanisms

12.4.1.3.1 Adsorption. Rhodium metal is absorbed by the organism only to an insignificant extent because of its low solubility. Organic rhodium salts are absorbed from the GI tract. However, under conditions of occupational exposure, rhodium and its compounds may also be absorbed by inhalation (366).

Table 41.24. LD$_{50}$ Values of Some Rhodium Compounds

Compound	Species	Administration	LD$_{50}$ (mg/kg)	Ref.
Rhodium trichloride	Rat	IV	198	369
	Rabbit	IV	215	369
Rhodium(I) carbonyl acetylacetate	Rat	Oral	50–200	371
	Mouse	IP	18	371

12.4.1.3.2 Distribution. After administration of Rh to mice in drinking water for their whole lifespan, considerable quantities of the element were found in their kidneys, heart, and spleen (372). Studies on dogs have revealed that 87–85% RhO_2 isotope is absorbed by inhalation remain in their lungs and the remaining quantity is deposited in the lymph nodes (366).

12.4.1.3.3 Excretion. Rhodium is excreted from the organism mainly in urine in two stages. The initial stage is fast; about 45% of the absorbed dose is excreted during the first day, and is followed by further gradual, slow excretion (373).

12.4.1.3.4 Mechanisms. It seems that rhodium, like platinum, reacts with DNA and RNA (366).

12.4.1.4 Reproductive and Developmental: NA

12.4.1.5 Carcinogenesis.
Schroeder and Nason (372) administered 5 ppm $RhCl_3$ in drinking water to mice for their lifespan. Tumors were found both in the control and exposed groups. As much as 28.8% malignant tumors were found in the exposed group; the corresponding figure for the control group was 13.8%. The authors concluded that the examined rhodium doses showed a slight carcinogenic activity.

12.4.1.6 Genetic and Related Cellular Effects Studies: NA

12.4.1.7 Other: Neurological, Pulmonary, Skin Sensitization

12.4.1.7.1 Irritating Effect. Soluble rhodium salts cause eye irritation. A 0.1 M solution of $RhCl_3$ instilled for 10 min to a rabbit's eye with previously removed corneal epithelium brought about opacification of the cornea, which continued for 2–3 weeks. The cornea assumed an orange color, which changed to yellow after 8 weeks (370, 374). Rhodium(I) carbonyl acetylacetonate instilled at the dose of 0.1 g to the rabbit's eye caused conjuctivitis and a positive reaction of both cornea and iris (365).

12.4.1.7.2 Allergic Effects. Magnussen and Kligmar (15) (375) report that rhodium(I) acetylacetonate caused allergic reaction in guinea pigs. During the same experiments, negative results were obtained for $[Rh(NH_3)Cl_5]Cl_2$ and Na_3RhCl_6.

12.4.2 Human Experience

Literature data on acute or chronic toxic effects of rhodium in humans are not available. The only data refer to allergic responses. Patch tests have confirmed the sensitizing activity of hexachlororhodiate in a 47-year-old woman employed at a goldsmith's workshop (376). Similar examinations showed allergenic activity of rhodium sulfate (377).

12.5 Standards, Regulations, or Guidelines of Exposure

Recommended exposure limits are listed in Tables 41.25, 41.26 and 41.27.

Table 41.25. Exposure Limits for Rhodium Metal, as Rh

Exposure Limits	CAS Number	OSHA PEL (mg/m^3)	NIOSH Exposure Limit (mg/m^3)	ACGIH TLV (mg/m^3)	AIHA WEEL	ANSI Standard	German MAK	Sweden Standard	British Standard	Polish Standard
TWA	[7440-16-6]	0.1	0.1	1 A4[a]	—	—	—	—	—	—
STEL	—	—	—	—	—	—	—	—	—	—
Ceiling limit	—	—	—	—	—	—	—	—	—	—
Biological limits (if available)	—	—	—	—	—	—	—	—	—	—

[a] A4—not classifiable as a human carcinogen.

Table 41.26. Exposure Limits for Rhodium Insoluble compounds, as Rh

Exposure Limits	CAS Number	OSHA PEL (mg/m^3)	NIOSH Exposure Limit (mg/m^3)	ACGIH TLV (mg/m^3)	AIHA WEEL	ANSI Standard	German MAK	Sweden Standard	British Standard	Polish Standard
TWA	—	0.1	0.1	1 A4[a]	—	—	—	—	—	—
STEL	—	—	—	—	—	—	—	—	—	—
Ceiling limit	—	—	—	—	—	—	—	—	—	—
Biological limits (if available)	—	—	—	—	—	—	—	—	—	—

[a] A4—not classifiable as a human carcinogen.

Table 41.27. Exposure Limits for Rhodium, Soluble Compounds, as Rh

Exposure Limits	CAS Number	OSHA PEL (mg/m^3)	NIOSH Exposure Limit (mg/m^3)	ACGIH TLV (mg/m^3)	AIHA WEEL	ANSI Standard	German MAK	Sweden Standard	British Standard	Polish Standard
TWA	—	0.001	0.001	0.01 A4[a]	—	—	—	—	—	—
STEL	—	—	—	—	—	—	—	—	—	—
Ceiling limit	—	—	—	—	—	—	—	—	—	—
Biological limits (if available)	—	—	—	—	—	—	—	—	—	—

[a] A4—not classifiable as a human carcinogen.

D PALLADIUM AND ITS COMPOUNDS

13.0a Palladium

13.0.1a CAS Number: [7440-05-3]

13.0.2a Synonyms: Palladium black; palladium element; Palladex 600; Pd; palladium on barium carbonate catalyst; palladium on barium sulfate catalyst

13.0.3 Trade Names: NA

13.0.4a Molecular Weight: 106.42

13.0.5a Molecular Formula: Pd

13.0b Palladium(II) Chloride

13.0.1b CAS Number: [7647-10-1]

13.0.2b Synonyms: Palladium chloride; palladous chloride; palladium dichloride; dichloropalladium; paladium(ous) chloride

13.0.3b Trade Names: NA

13.0.4b Molecular Weight: 177.33

13.0.5b Molecular Formula: $PdCl_2$

13.0.6b Molecular Structure: Pd^{2+}
Cl^- Cl^-

13.0c Potassium Tetrachloropalladate(II)

13.0.1c CAS Number: [10025-98-6]

13.0.2c Synonyms: Potassium chloropalladate; dipotassium tetrachloropalladate; potassium chloropalladite; potassium palladium chloride; potassium palladous chloride; potassium tetrachloropalladate

13.0.3c Trade Names: NA

13.0.4c Molecular Weight: 326.4

13.0.5c Molecular Formula: K_2PdCl_4

13.0.6c Molecular Structure: $K^+\ Cl^-\ \begin{smallmatrix}Cl\\ \diagdown\\ Pd\\ \diagdown\\ Cl\end{smallmatrix}\ K^+\ Cl^-$

13.0d Ammonium Tetrachloropalladate(II)

13.0.1d CAS Number: [13820-40-1]

13.0.2d Synonyms: NA

13.0.3d Trade Names: NA

13.0.4d Molecular Weight: 284.31

Table 41.28. Physical and Chemical Properties of Palladium and Some of Its Compounds

Chemical name	Chemical Formula	Molecular Weight	Boiling Point (°C)	Melting Point (°C)	Density (g/cm^3)	Solubility in Water
Palladium	Pd	106.42	3167	1555	12.02	Insoluble
Palladium (II) chloride	PdCl$_2$	177.3	—	501(dec)	4.0	Soluble
Potassium tetrachloride palladate (II)	K$_2$PdCl$_4$	326.4	—	105	—	—
Ammonium tetrachloride palladate (II)	(NH$_4$)$_2$PdCl$_4$	284.31	—	—	—	—

13.0.5d Molecular Formula: (NH$_4$)$_2$PdCl$_4$

13.0.6d Molecular Structure: $\text{NH}_4^+ \text{Cl}^- \quad \overset{\text{Cl}}{\underset{\text{Cl}}{\text{Pd}}} \quad \text{NH}_4^+ \text{Cl}^-$

13.1 Physical and Chemical Properties

Table 41.28 gives the data on chemical identification, physical and chemical properties, and solubility of palladium and some of its compounds.

Palladium, a transition element belonging to Group III in the periodic table (nickel group) and light platinum metals, is a medium-hard, moderately forgeable, ductile silver-white metal. In its compounds, palladium usually assumes oxidation state +2 and +4, forming bivalent and tetravalent salts.

Palladium occurs in the form of six isotopes: ^{102}Pd (1.0%), ^{104}Pd (11%), ^{105}Pd (22.2%), ^{106}Pd (27.3%), ^{108}Pd (26.7%), and ^{110}Pd (11.8%). Insignificant quantities of palladium can be found in the lithosphere in the form of native palladium and allopalladium as well as PdPt and PdAu alloys and as a contaminant of silver and nickel ores. Red heating of palladium results in a volatile palladium oxide, PdO$_2$. At elevated temperatures palladium can react with fluorine or chlorine to form palladium dihalides. It also reacts with sulfuric and nitric acids and slightly with hydrochloric acid. When palladium is heated in the presence of sulfur or phosphorus, palladium sulfides or phosphides are formed, respectively.

A characteristic feature of palladium is its high hydrogen absorption, which allows for use its in the form of palladium sponge or palladium black as a catalyst in reduction processes. Contrary to other platinum metals, palladium is considerably less resistant to chemicals. At elevated temperatures palladium reacts with oxygen, fluorine, chlorine, sulfur, and selenium. Palladium dust may constitute fire and explosion hazards (364, 378, 379). Palladium compounds show different water solubility. Palladium is soluble only in nitric and sulfuric acids as well as in aqua regia; its compounds such as PdCl$_2$, K$_2$(PdCl$_4$), and (NH$_3$)$_2$PdCl$_2$ are water-soluble.

13.2 Production and Use

Pure palladium is prepared by reducing its salts with hydrogen (364).

Palladium is not used extensively. In the form of alloys with gold, silver, and copper, palladium is used in dentistry, in jewelry as a "white gold," in making electrical contacts, and in the production of resistance wires with a very low temperature resistance coefficient. Palladium alloys are used to make mechanical parts of watches and clocks and to make mirrors for astronomic instruments. Palladium alloys are also used in the manufacture of electrical transmitters and switches in telecommunication. The colloidal form of palladium was used in the treatment of tuberculosis, urinary diathesis, and obesity.

Palladium and platinum have been used since 1974 as catalyst to oxidize carbon monoxide to carbon dioxide in automobile engines and in hydrocarbons conversion. This use of palladium results in palladium being emitted as solid particles with automotive exhaust gases (380).

13.3 Exposure Assessment

13.3.1 Air: NA

13.3.2 Background Levels: NA

13.3.3 Workplace Methods: NA

13.3.4 Community Methods: NA

13.3.5 Biomonitoring/Biomarkers

The concentration of palladium in biological samples such as tissues and systemic fluids is determined by absorption atomic spectroscopy (AAS) (381, 382). The limit of quantitation is 0.01 µg Pd in 5 g blood and 0.003 µg/g Pd in 50 g urine.

13.4 Toxic Effects

13.4.1 Experimental Studies

Palladium metal is practically nontoxic. The acute effects of palladium compounds depend on the type, dose and administration of the compound. In general, the effects are stronger after IV or IP administration than orally. Water-soluble palladium compounds, namely, those soluble in systemic fluids, show stronger toxic activity than do the insoluble ones (383).

In acute poisoning, the toxic activity of water-soluble palladium compounds was dependent on the administration route.

After IV injections, the experimental animals died within 5–7 min (384). Before the animals died, clonic and tonic convulsions were observed. A 7% weight reduction and an 80% reduction in fodder consumption were found in animals after the IP administration of $PdCl_2$. In addition, a decrease in water consumption accompanied by decreased urine excretion and albuminuria were found. The intraperitoneal administration of $PdCl_2$ to animals resulted in their death within 24 h and the histopathological examinations confirmed the peritonitis with numerous intestine adhesions and "chemical burn"–type changes (384).

In animals that died from intragastric administration of a $PdCl_2$ dose > 129 g/kg, necrotic changes in the mucous membrane of the GI tract, hepatocyte dystrophy, and renal

Table 41.29. LD$_{50}$ Values of Some Palladium Compounds

Compound	Species	Administration	LD$_{50}$	Ref.
PdCl$_2$	Rat	IV	3 mg/kg	384
	Rat	IV	5 mg/kg	
	Rabbit	IV	5 mg/kg	
	Rat	IV	6 mg/kg	
	Rat	IP	70 mg/kg	
	Rat	Per os	200 mg/kg	
	Rat	Per os	2704 mg/kg	383
PdCl$_2$H$_2$O	Rat	IP	0.4–0.6 mmol/kg	385
	Rat	Per os	2.7 mmol/kg	386
PdSO$_4$	Rat	IP	0.6 mmol/kg	385
	Rat	Per os	7 mmol/kg	385
PdO	Rat	Per os	40 mmol/kg	
Na$_2$PdCl$_4$3H$_2$O	Mouse	IP	122 mg/kg	387
K$_2$PdCl$_4$	Mouse	IV	6.4 mg/kg	388
	Mouse	IP	153 mg/kg	

tubule damage were found (383). Table 41.29 gives LD$_{50}$ values for some palladium compounds (383–388).

Table 41.30 gives the effects of acute toxicity of selected palladium compounds.

13.4.1.2 Chronic and Subchronic Toxicity. The main toxic effects observed in experimental animals after prolonged exposure to palladium compounds included decreased body weight, increased weight of internal organs such as spleen and heart, and decreased weight of liver and kidneys. Panova and Veselov (389) have examined the chronic toxicity of PdCl$_2$(NH$_3$)$_2$. This slightly soluble palladium salt is an intermediate product in the industrial process of palladium preparation (Table 41.31) (390–392). In the inhalation experiments, the exposure of animals to the dust of this salt at 18.35 and 5.4 mg/m^3 for 5 months resulted in liver and kidney function disorders. The extent or intensity of the changes and time of their appearance depended on the concentration. After the exposure to PdCl$_2$(NH$_3$)$_2$ at 5.4 mg/m^3, the toxic effect appeared during the final days of the experiment and disappeared within 30 days after the experiment was discontinued. The concentration 18.35 mg/m^3 caused permanent changes in the organisms of the exposed animals.

13.4.1.3 Pharmacokinetics, Metabolism, and Mechanisms

13.4.1.3.1 Adsorption. Palladium and its compounds may be absorbed by inhalation or per os. The absorption rate depends mainly on the way of administration and physical and chemical properties of the compound.

The absorption of palladium from the GI tract is insignificant. A single intragastrical dose of ^{103}Pd isotope given to male Charles River CD-1 rats was rapidly excreted from

Table 41.30. Acute Effects of Some Palladium Compounds

Compound	Species	Administration Dose (Concentration)	Effects Observed	Ref.
PdCl$_2$(II)	Rat, rabbit	IV	Death within 5–10 min; observation of survivors for 40 days: decreased water consumption and urine excretion; clonic and tonic convulsions	384
PdCl$_2$(II)	Rat, rabbit	IP	A 7% reduction in body weight; a 25% decreases in water consumption and urine excretion; albuminuria; peritonitis and "chemical burn" of internal organs were observed in animals that died within 24 h	384
PdCl$_2$ (II)	Rabbit	IV 0.5, 0.7, and 1.7 mg/kg	Weakness, apathy, decreased fodder, and water consumption; doses of 1.7 mg/kg were lethal to all animals	386
PdCl$_2$ (II)	Rat	SC 4–24 mg/kg	No toxic effects	386
PdSO$_4$, Pd(NO$_3$)$_2$, PdCl$_2$, (NH$_4$)PdCl$_4$, K$_2$PdCl$_4$	Rat	IV to 2 mg/kg for 40 s	Heart dysrhythmia with predominant additional systoles and fibrillation	388
PdCl$_2$(NH$_3$)$_2$	Rat	Inhalation 10.2 — 687.8 mg/m^3	Zero mortality; toxic effects — disturbed carbohydrate metabolism, proteometabolism, and lipometabolism evidenced by increased total protein, glucose, and cholesterol content and decreased blood serum urea level; acute toxicity threshold ~82.2 mg/m^3	383

their organism with feces. The total retention was lower than 0.5%. After 24 h, insignificant quantities of PdCl$_2$ were found only in liver and kidneys; after 104 days no palladium chloride was found in any analyzed tissue or organ (384).

A considerably higher absorption and retention were observed after IV administration of PdCl$_2$. The highest concentration of ^{103}PdCl$_2$ isotope was found in kidneys, spleen, liver, adrenals, lungs, and bones (384).

The administration route was a significant factor affecting the retention of palladium and its compounds in animals' organs. In an experiment during which rats received a single dose of PdCl$_2$ by various routes, the retention of ^{103}Pd isotope was lowest when administered orally where as the highest retention values were observed after intravenous administration. The routes of administration are arranged according to their corresponding retention values in the following descending order:

Intravenous > intratracheal > inhalation > per os

Table 41.31. Subchronic and Chronic Effects of Palladium and Some of Its Compounds

Compound	Species	Dose or concentration	Administration	Effects observed	Ref.
Pd (colloidal solution)	Rabbit	5 mg 1% solution/day (0.05 mg Pd/day)	Subcutaneous for 2 months	No toxic effects	390
Pd metal	Rat	50 mg/kg	Intragastrcal for 6 months	Lower body weight gain, shorter prothrombin time, decreased serum urea, and β-lipoprotein content; lower urine density and urine chloride content	383
Pd metal	Rat	5 ppm/day	In drinking water for whole lifespan	Increased number of malignant pulmonary tumors classified as lymphomas	391
$PdCl_2$	Rat	Saturated aqueous solution	In drinking water for 1 week	Lower body weight gain, increased spleen, heart, and testicle weight, lower liver and kidney weight	385
$PdCl_2$	Rat	3166 mg/kg	In food for 4 weeks	Body weight gain reduced by 25% compared with the control group	385
K_2PdCl_4	Rat	470 ppm/day	In drinking water for 23 days	Body weight gain reduced by 14.7% and water intake reduced by 32.8%; no pathological changes were observed in internal organs	384
K_2PdCl_4	Rat	235 ppm/day	In drinking water for 23 days	Lower body weight gain, reduced water intake; no pathological changes in internal organs	384
K_2PdCl_4	Rat	194 ppm/day	In drinking water for 33 days	No toxic effects	384

PdCl$_2$(NH$_3$)$_2$	Rat	18.35 mg/m^3 5 h/day for 5 months	Inhalation for 5 months	Atonia, reduced kinetic activity, reduced body weight gain, increased total serum protein content; after 6 days, reduced blood cholinesterase activity throughout the experiment; reduced liver weight, increased kidney and heart weight; reduced urine excretion, increased blood urea content and reduced urine urea content	389
Palladium hydrochloride	Rabbit	5.4 mg/kg per day, 2% aqueous solution	Application to shaved skin for 56 days	Dermatitis after 7–8 days, which disappeared on day 35 of the experiment; lower motor activity and 18% reduction in body weight gain were observed after 10 days; mucous and purulent secretion from the nose	392

In male Charles River CD-1 rats receiving 25 μg PdCl$_2$ in 0.2 mL saline, the retention in the animals' organs after 3 days was 0.4% of the initial charge and the absorption was less than 0.5% of the initial dose (384).

13.4.1.3.2 Distribution. After a single exposure of Charles River CD-1 rats to PdCl$_2$ orally, ^{103}Pd isotope after 24 h was confirmed to be present in the liver and kidneys only, where as Pd concentration was considerably higher in kidneys than in liver. After IV administration, ^{103}Pd isotope was found in all tissues examined. The highest concentrations were in kidneys, spleen, liver, adrenals, lungs, and bones (384). After a lapse of 104 days, in the group of rats receiving a single intragastrical dose of PdCl$_2$, no isotope was found in any of the organs examined. In the animals receiving PdCl$_2$ IV, the highest concentrations of ^{103}Pd were found in spleen, kidneys, liver, lungs, and bones. After intratracheal administration, the highest concentrations of ^{103}Pd were observed in lungs, followed by kidneys, spleen, bones, and liver (384).

The presence of ^{103}Pd isotope was also examined in the internal organs of female rats receiving PdCl$_2$ IV on 16th day of gestation. The concentration of ^{103}Pd was assessed in various organs of the dams and in fetuses 24 h after the administration (Table 41.32).

The low concentration of ^{103}Pd isotope in fetal bodies indicates that Pd does not easily penetrate the placenta (384).

13.4.1.3.3 Excretion. Palladium is eliminated from organisms of the experimental animals mainly via feces and urine. After a single acute exposure, most of the absorbed dose is excreted within 48 h to 5 days, although in the organs where Pd is accumulated, it may remain for 30 days after termination of the exposure (383).

Excretion rates of palladium and its salts depend on the route of administration and the physical and chemical properties of the compounds (384). Intragastrically received palladium is excreted from the organism with feces, and only trace quantities are excreted with urine. When administered intravenously, almost equal amounts of palladium are excreted with feces and urine, whereas in the final stage palladium is excreted mainly with urine.

13.4.1.3.4 Mechanisms. Studies *in vitro* and *in vivo* have shown that palladium and its compounds inhibit the activity of many enzymes.

Table 41.32. Concentrations of ^{103}Pd in Organs of Pregnant Female Rats and Fetal Bodies (384)

Tissue	Average Value (g tissue)
Kidneys	588.479
Liver	319.153
Ovary	29.625
Lungs	29.21
Bones	18.351
Blood	3.654
Placenta	58.321
Fetal liver	1.429
Fetus	757

The effect of PdCl$_2$ on the activity of CPK–MM creatine kinase of rabbit's muscles and human serum was examined (393) and a dose-dependent inhibition of the enzymatic activity, accompanied by a considerable increase in the electrophoretic enzyme mobility toward anode, were noted. Analysis of the Pd–CPK–MM bond has shown that the cation forms an extremely stable bond with the enzyme sulfhydryl groups, resulting in defective energy metabolism in the organism.

Palladium inhibits considerably the activity of the following enzymes: prolyl hydroxylase, creatine kinase, aldolase, succinic dehydrogenase, carbonic anhydrase, and alkaline phosphatase (394). According to the authors, these data indicate that Pd (+2) may disturb the energy metabolism in the organism, the acid–base equilibrium, electrolytic equilibrium, and the metabolism in the osseous tissue. These authors also believe that exposure to Pd(+2) during the developmental period may disturb fetal growth and development.

PdCl$_2$ given to rats in drinking water for 8 days decreased the activity of aniline hydroxylase and aminopyrine demethylase (395).

Palladium nitrate administered IP to rats at 56 or 113 µM/kg per day for 2 days caused a reduction in the content of aminopyrine demethylase and cytochrome P450 in isolated hepatic microsomes (395).

In the initial period of the chronic exposure to Pd(NO$_3$)$_2$, enzymatic activity is inhibited, but after some time this activity returns to its initial level. It is believed that, during chronic exposure, a protein structure is formed which combines and inactivates Pd (395).

In vivo palladium inhibits the addition of thymidine to a DNA molecule in spleen, liver, and testicles and, to a lesser extent, in kidneys (396).

13.4.1.4 Reproductive and Developmental. No findings on the effects of palladium and its salts on the processes of reproduction and progeny development are available in the literature.

13.4.1.5 Carcinogenesis. Examinations of rats and mice exposed to palladium at 5 ppm/day in drinking water for the whole lifespan have shown an increases in the number of pulmonary lymphomas (29.9%) compared with the control group (16.3%) (391).

13.4.1.6 Genetic and Related Cellular Effects Studies. Studies *in vitro* have not shown any mutagenic effects of palladium (397) or its salt: K$_2$(PdCl$_4$) (398). No information is available on *in vivo* studies.

13.4.1.7 Other: Neurological, Pulmonary, Skin Sensitization

13.4.1.7.1 Irritating Effects. Examinations of local irritating effects on rabbits and rats have shown that palladium salts, such as PdCl$_2$(NH$_3$)$_2$ and palladium hydrochloride, show irritating effects on the mucous membranes of the upper respiratory tract and eyes. High concentrations in air result in keratitis and conjunctivitis, whereas low concentrations cause conjunctivitis (383). Roshchin et al. (383) have determined threshold concentrations of PdCl$_2$(NH$_3$)$_2$ and its irritating effects on respiratory tracts and eyes, which are 65 and 50 mg/m^3, respectively.

Table 41.33. Irritating and Cytotoxic Effects of Palladium and Its Salts on Skin

Compound	Undamaged skin	Chafed skin	Recommendations
$Pd(NH_3)_2Cl_2$	No irritation	No cytotoxic effect	No risk for human skin
PdO	No irritation	No cytotoxic effect	No risk for human skin
K_2PdCl_4	No irritation	Slightly cytotoxic	No risk for undamaged human skin, caution recommended in the case of damaged skin.
$PdCl_2$	No irritation	Slightly cytotoxic	No risk for undamaged human skin, caution recommended in the case of damaged skin.
$K_2(PdCl_6)$	No irritation	cytotoxic	No risk for undamaged human skin, avoid contact with damaged skin.
$(C_3H_5PdCl)_2$	Strongly irritating	Strongly cytotoxic	Avoid contact with skin
$(NH_4)_2PdCl_2$	Strongly irritating	Strongly cytotoxic	Avoid contact with skin
$(NH_4)_2PdCl_6$	Strongly irritating	Strongly cytotoxic	Avoid contact with skin

The irritating effect of palladium compound dusts depends on their solubility in water and, consequently, in body fluids. Dusts of slightly soluble compounds remain for a long time on the surface of mucous membranes of eye, resulting in their considerable irritation (383).

Examination of the irritating and cytotoxic effects of palladium and its selected compounds on rabbit's skin made it possible to prepare hygienic recommendations for people occupationally exposed to that metal or its compounds (380) (Table 41.33).

13.4.1.7.2 Allergic Effects. Results of studies have shown that palladium compounds may cause bronchial spasm, anaphylactic shock, and increase in histamine concentration in serum in experimental animals. Some palladium salts in doses of 0.313–12.5 µM/kg were injected in cats IV, which resulted in anaphylactic effects with bronchial spasm, a drop in blood pressure, decreased hematocrit, and increased concentration of histamine in serum. The threshold doses for these effects range from 0.313 to 3.13 µM/kg, depending on the compound (399).

Tests performed with guinea pigs have shown that haptene alone (K_2PdCl_4) as well as haptene with a carrier (K_2PdCl_4 + egg albumin) exhibit allergenic properties. In the provocative tests, haptene plus carrier caused anaphylactic shock, but haptene alone caused only increased serum histamine concentrations (399).

13.4.2 Human Experience

There are no data concerning poisoning due to occupational exposures to palladium or its compounds. Reports on toxic effects, including mainly allergenic or irritating effects, refer to the therapeutic use of palladium, particularly in dentistry.

A patient with a metal dental bridge complained of itching and pain in the oral cavity. She suffered also from recurrent blisters, ulceration, and infection of the oral mucous membrane. Skin tests carried out by Koch (400) on this patient showed a strong and persistent allergic response to PdCl$_2$ (1% pet), a positive allergic reaction to (NH$_4$)$_2$PdCl$_2$ (0.25% pet) and palladium metal. The histopathologic examination of food from the test spot showed that (NH$_4$)$_2$PdCl$_2$ caused hemorrhagic eczematous changes, while PdCl$_2$ produced contact eczema and lichenoid changes.

Eczematous symptoms were observed among 8% patients subjected to skin patch tests for PdCl$_2$ (401). Positive results of skin tests for PdCl$_2$ were obtained among school youths and adults receiving dental treatment using palladium alloys (402). Among 417 examined girls, 44 (11%) showed allergic response to PdCl$_2$, whereas only 4 (1%) boys (of 283 examined) showed a positive response. In the adult group, 7% showed a positive allergic response to PdCl$_2$.

13.5 Hygienic Standards, Regulations, or Guidelines of Exposure

No hygienic standards have been established as yet for palladium and its salts.

E OSMIUM AND ITS COMPOUNDS

14.0 Osmium

14.0.1 CAS Number: [7440-04-2]

14.0.2 Synonyms: Metallic Osmium

14.0.3 Trade Names: NA

14.0.4 Molecular Weight: 190.2

14.0.5 Molecular Formula: Os

14.1 Chemical and Physical Properties

14.1.1 General

Osmium, a transition element belongs to the odd series 8$_1$ (iron) family, and at the same time to the heavy platinum metals. It has seven stable isotopes: ^{184}Os (0.02%), ^{186}Os (1.6%), ^{187}Os (1.6%), ^{188}Os (13.3%), ^{198}Os (16.1%), ^{190}Os (26.4%), and ^{192}Os (41.0%). It is found in water in the form of minerals: iridosmium (syerskite) and osmiridium (newianskite). In chemical compounds, it occurs at oxidation states +2, +3, +4, +6, and +8. Osmium easily forms alloys with other platinum group metals and with iron, cobalt, and nickel (403–406).

Osmium is a very hard and brittle gray-blue metal. It forms hexagonal crystals.

>   Melting point: ∼2700°C.
>   Boiling point: ∼5500°C.
>   Density at 20°C: 22.61 g/cm^3.
>   Solubility: Insoluble in water, slightly soluble in fuming nitric acid, and aqua regia, insoluble in ammonia.

Reactivity: Osmium is a chemically passive metal; finely divided, it is slowly oxidized in the air at room temperature to tetroxide; above 100°C it is attacked by fluorine to form fluorides: OsF_4, OsF_6; it reacts with chlorine at high temperatures to form chlorides; when osmium is heated, osmium sulfide (OsS_2) is easily formed; the majority of osmium compounds have a brown or brownish color.

14.1.2 Odor and Warning Properties

Powdered osmium has an acrid and pungent odor because at room temperature it is oxidized to osmium tetroxide (see discussion of osmium tetroxide in Section 15.1.2).

14.2 Production and Use

Osmium is obtained in the reaction of osmium tetroxide (OsO_4) reduction with carbon at red heat temperature (404). Osmium is also obtained from osmiridium. After separation from other metals with aqua regia, osmiridium is distilled in chlorine stream with formation of osmium tetrachloride ($OsCl_4$), from which the metal is recovered by reduction (404).

Osmium is used in the chemical industry as a catalyst in ammonia synthesis and in the hydrogenation of organic compounds (403). In the alloy with iridium, it is used to manufacture the tips of golden nibs for the fountain pens, compass needles, and engraving needles for use in jewelry. It is also used to make the bearings of small, high precision devices, for example, watches (404).

14.3 Exposure Assessment

14.3.1 Air

A NIOSH method does not exist for specifically measuring osmium in air. However, NIOSH method 7300 for measuring platinum should be applicable for osmium. A method has been published for measuring platinum in the air involving inductively coupled plasma—atomic emission spectrometry (ICP-AES) (7).

A similar method (ICP-AES) is recommended for analysis of osmium in biological materials (NIOSH (methods 8005 and 8310) (10, 11). Matusiewicz and Barnes (407) used ICP-AES with electrothermal vaporization for measuring osmium in serum and urine (range of quantitation from 1.0 to 19 ng/mL).

14.4 Toxic Effects

14.4.1 Experimental Studies

No data have been found in the relevant literature concerning the toxic effects of osmium metal in experimental animals. However, it oxidizes at relatively low temperatures to the volatile osmium tetroxide, which shows strong irritating activity to eyes, respiratory tract and skin (see osmium tetroxide discussion in Section 15.4.1).

14.4.1.3 Pharmacokinetics, Metabolism, and Mechanisms. Little is known about the absorption, excretion, and retention of osmium, but it has been suggested that, like many other trace elements, it is an essential nutrient in animal metabolism possibly in connection with the oxidation–reduction processes of the organism. Results of the studies are, however, equivocal (408, 409).

14.4.1.4 Reproductive and Developmental. Data have not been found.

14.4.1.5 Carcinogenesis. Data have not been found.

14.4.1.6 Genetic and Related Cellular Effects Studies. Data have not been found (see also osmium tetroxide discussion in, Section (15.4.1.6).

14.4.2 Human Experience

14.4.2.1 General Information. Osmium exposure in the occupational environment occurs during the roasting, smelting, and refining processes. Osmium metal itself and its natural and synthetic alloys are probably not harmful to the human organism. Already at relatively low temperatures, however, both osmium and its alloys are oxidized to volatile osmium tetroxide, which is toxic and strongly irritating (410, 411) (see also osmium tetroxide discussion in, Section 15.4.2).

14.4.2.3 Epidemiology Studies. No data have been found in the available literature that would point to harmful effects of osmium metal in humans.

14.5 Standards, Regulations, or Guidelines of Exposure

See osmium tetroxide discussion in, Section 15.5.

14.6 Studies on Environmental Impact

No analytical information on osmium contamination of the biosphere even in the vicinity of likely synthetic human made sources, or on the magnitude of synthetic contamination and its influence on the biosphere, is available (411).

The fate of osmium in wastewater is not known. The majority of osmium disposed of by pouring down the drain will be removed in the sludge. Soluble forms remaining in the water could be converted to osmium tetroxide during the chlorination operation and volatilized into the air. However, the majority of the soluble osmium should pass through the chlorination process unaffected. The small percentage present as osmium tetroxide will react rapidly and completely with residual organic matter in the receiving stream and be converted to osmium metal or to osmium dioxide and settle out in the sediment of the water course. The extremely low levels of osmium expected to be found in wastewater streams should pose no hazards to humans or the human environment (411).

The small amount of osmium metal disposed of as solid waste is resistant to chemical attack and, because of its wide dispersal into the environment, should pose no health hazard. The only situation that could create a hazard consists of incineration of wastes containing osmium metal. Under incineration conditions, the osmium will be converted to volatile osmium tetroxide. However, the quantities that would normally be found in solid waste to be incinerated are probably too small to pose a hazard (411).

15.0 Osmium Tetroxide

15.0.1 CAS Number: [20816-12-0]

15.0.2 Synonyms: Osmic acid; osmium oxide; osmium tetroxide; perosmic acid anhydride; perosmic oxide; perosmic acid; osmic acid anhydride; osmium (IV) oxide; osmium oxide, (T-4)-; osmium oxide OsO_4 (T-4); Osmium (VIII) tetroxide; osmium tetraoxide; osmium(VIII)oxide; Osmium Tetroxide, Solution

15.0.3 Trade Names: NA

15.0.4 Molecular Weight: 254.23

15.0.5 Molecular Formula: OsO_4

15.0.6 Molecular Structure:
$$O=\overset{\overset{O}{\|}}{\underset{\underset{O}{\|}}{Os}}=O$$

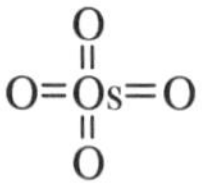

15.1 Chemical and Physical Properties

15.1.1 General

Osmium tetroxide is inflammable colorless to light-yellow crystalline solid body or crystalline mass of acrid, irritating odor and burning taste.

> Melting point: 40.6°C
> Boiling point: 130°C
> Density at 20°C: 4.91
> Vapor pressure at 27°C: 11 mm Hg; easily evaporates from aqueous solutions
> Solubility: Water solubility: 7.24 g/100 g; soluble in ethanol, benzene, ethyl ether

Osmium tetroxide is a strong oxidant. Numerous organic substances reduce it to black osmium dioxide (OsO_2) or to osmium metal (403–406).

15.1.2 Odor and Warning Properties

Osmium tetroxide has an acrid, irritating odor similar to that of bromine or chlorine. Odor threshold is 0.0019 ppm (412).

15.2 Production and Use

Osmium tetroxide is obtained by heating, at 300–400°C, finely divided osmium metal in the stream of air or oxygen (403). Commercially, it is received during osmium smelting and platinum annealing (404). Osmium tetroxide may be also produced by oxidizing osmium with aqua regia or nitric acid (404). It is often formed already at room temperature from osmium metal powder.

Osmium tetroxide is used in histopathological laboratories to stain the adipose tissue and as a stabilizing agent in scanning electron microscopy (403, 411, 412). In the chemical industry it is used as a catalyst in the organic synthesis, particularly as the oxidizing agent in olefin-to-glycol conversion (403, 412). In the past, osmium tetroxide in the form of aqueous solution was used in forensic medicine to examine fingerprints (413, 414). Osmium tetroxide is also used in medicine to treat rheumatoid arthritis (415–417).

15.3 Exposure Assessment

15.3.1 Air

See discussion of osmium in Section 14.3.

A specific method for the determination of osmium tetroxide in workplace air for the purpose of occupational environment monitoring has been developed in Poland. The method involves passing a known volume of air through a membrane filter impregnated with sodium carbonate in order to deposit osmium tetroxide aerosol thereon, and then washing it out with sulfuric acid and extracting with chloroform. A blue-violet complex is formed in the reaction of osmium tetroxide with diphenylcarbazide, which is then determined by spectrophotometry.

The limit of quantitation of the method (at 5000 L air sample and 25 L/min airflow) is 0.0006 mg/m^3 (418).

15.4 Toxic Effects

15.4.1 Experimental Studies

15.4.1.1 Acute Toxicity. Osmium tetroxide is very irritating to eyes, skin, and the respiratory tract. It may severely damage eyes, lungs, and kidneys in experimental animals (411). The IP osmium tetroxide LD$_{50}$ was 14.1 mg/kg for the rat, and 13.5 mg/kg for the mouse (20). Acute inhalation LCL$_{50}$ for rats and mice following 4-h exposure was 400 mg/m^3 (40 ppm) (320). With per os exposure, the LD$_{50}$ for mice was 162 mg/kg (419).

A 30-min exposure of rabbits to osmic acid vapors in airtight glass box in which ampoules were placed containing 250, 500, and 1000 mg osmic acid produced acute irritation of the mucous membranes and semicoma. The animals exposed to 1000 mg survived, on the average, for 30 h; for those exposed to 250 mg, the mean survival period was 4 days. The animals died of bronchopneumonia. Necropsy showed dark red consolidation with scattered purple areas in the lungs. The bronchi were filled with pus. The kidneys showed cloudy swelling and granulation of the tubular epithelium (420).

In another study, degenerative changes and hyperemia, mainly of the lungs, but also of liver, spleen, kidneys, and adrenals were detected in rabbits exposed for 24–48 h to vapors evolved from 125 mg OsO$_4$ in aqueous solution (421).

Osmium tetroxide given intragastrically to dogs and rabbits caused vomiting in the dogs, and considerable weakness in the rabbits, to which another injection 4 days later was lethal. Tissues contacting the osmium tetroxide were blackened (the usual effect when osmium tetroxide is reduced by organic matter (422).

The IV injections produced in the experimental animals extensive exudate in the lungs, palsy, and death (422). Subcutaneous and IM injections of 1% osmic acid solution did not produce any serious effects in the animals, but small amounts injected to the lungs resulted in instant death (423).

Intense osteocartilaginous lesion occurred in rabbits after injecting 0.1–0.5 mL 1% osmic acid to the knee joint, and the effect grew stronger when the dose and animal age increased. The lesion was less intense in rabbits with carrageenin-induced arthritis than in healthy animals (424).

Injection of 1 mL 1% osmium tetroxide solution to the knee joint of an adult rabbit caused widespread chondrocyte necrosis within 1 week and disintegration of cartilage surface within 3–7 months (425).

One or two droplets of 1% solution instilled to rabbit conjunctival sac caused serious damage of the eye. At 24 h after exposure there was eyelid swelling, all portions of conjunctiva turned brown, and after 48 h the rabbits developed persistent corneal opacity and superficial vascularization (420).

15.4.1.2 Chronic and Subchronic Toxicity. Toxic effects of osmium tetroxide on bone marrow of guinea pigs exposed to vapors from ampoules containing 50 mg osmium tetroxide 8 h/day for approximately 60 days were examined. The guinea pigs suffered from a chronic anemia with an initial hyperactivity of the bone marrow (426).

Marked sclerosis and collapse of lung tissue, bronchial compression, congestion and degeneration of the liver, sclerosis of the spleen, and fatty degeneration of the kidneys and adrenals were observed in rabbits chronically exposed 8 h/day to vapors of the aqueous solution containing 50 mg OsO_4 and killed after 45–60 days (421).

15.4.1.3 Pharmacokinetics, Metabolism, and Mechanisms. Osmium tetroxide is absorbed mainly by inhalation. Lungs and airways retain the major part of the inhaled OsO_4 vapors. Serious symptoms of irritation and dark color of pulmonary mucosa may indicate that, on contact with the tissues in the organic medium, OsO_4 is reduced to osmium metal (427).

Neither GSH nor ALA-dehydratase inhibition has been observed in guinea pigs in acute and chronic poisoning; this phenomenon is no doubt explained by the fact that OsO_4 causes serious damage to the mucous membranes of the airways where it is fixed without penetrating the alveolar barrier and consequently without entering the blood circulation (428). An *in vitro* inhibition of GSH and ALA-dehydratase of the red blood cells by scalar dilutions of OsO_4 has been observed.

No information is available that would suggest osmium accumulation in animal tissues or its dermal absorption.

Osmium tetroxide is a direct irritant to eyes, skin, and airways. It may damage kidneys, but the mechanism of renal damage is not known (417).

15.4.1.4 Reproductive and Developmental. A single intratesticular injection OsO_4 0.08 mmol/kg in rats caused of degeneration of the seminiferous epithelium and the interstitium (429). A single subcutaneous (SC) injection had no effect on the weight of the testis in rats but daily administration OsO$_4$ reduced the weight in mice. No necrotic changes were seen in the testis (429).

A single SC injection was without any effect on the residual spermatozoa in the ductus deferents of rats. Similar examination of spermatozoa was not done in mice after daily SC administration of the OsO$_4$ (429).

15.4.1.5 Carcinogenesis. No data could be found in the available literature on carcinogenic activity of osmium tetroxide.

15.4.1.6 Genetic and Related Cellular Effects Studies. Osmium tetroxide shows a mutagenic activity. Positive results have been obtained in the recassay system with *Bacillus subtilis* (194, 430). OsO$_4$ induced DNA lesions and DNA-repair replication in mammalian cells (431) and yielded positive results in DNA-repair systems in *Escherichia coli* (432, 433).

15.4.2 Human Experience

15.4.2.1 General Information. Osmium tetroxide vapors are toxic and highly irritant even at low concentrations. OsO$_4$ inhalation by humans results in severe irritation of the eyes, mucous membranes of the nose, throat, and bronchia; headache; and cough (427). Strong eye irritation is the first main symptom of OsO$_4$ exposure. Abundant lacrimation, conjunctivitis, a gritty feeling in the eyes, and the appearance of a halo around lights are frequently reported (420, 427). Prolonged exposure may damage eyelids, and cornea, and may result in disturbed vision (420, 427).

On direct contact with skin, OsO$_4$ causes serious contact dermatitis and blisters; sometimes the skin may assume a green or black hue (411, 413, 414, 434, 435).

The acute inhalation exposure may cause bronchial pneumonia, lung swelling, and degenerative lesions in the kidneys (427). No chronic or cumulative effects were noted.

15.4.2.2 Clinical Cases

15.4.2.2.1 Acute Toxicity. A metallic taste in the mouth was reported after a 10-min exposure to OsO$_4$ vapor by a scientist who decided to expose himself to that chemical. After 30 min he experienced a burning sensation in the eyes, lacrimation, reddening, and labored respiration (420).

Patients who received intraarticular injection of osmic acid also complained of strong metallic aftertaste (436).

McLaughlin et al. (427) referred to a human fatality resulting from inhalation of osmium tetroxide in 1874. The concentration of the vapor was not reported. The osmium tetroxide vapor caused capillary bronchitis, which brought about the death of the worker. The autopsy revealed frank pulmonary edema and other findings similar to those in animals exposed to the vapor of osmium tetroxide (427).

Dreisbach estimated the fatal dose of orally ingested osmic acid as 1 g, with an acid-like corrosive effect ranked 4 on a scale of 1–4. The rank of 4 denotes complete destruction of skin or mucous membrane (434).

Osmiridium refining plant workers (seven cases) exposed sometimes for several years to osmium tetroxide at 133–640 mg/m^3 in the fine spray produced in dissolving osmiridium ores in aqua regia, complained of very severe acute eye irritation manifested by pain, profuse lacrimation, and a gritty feeling in the eyes. The workers saw large halos around lights and were unable to read or see motion pictures during the height of the symptoms, which generally subsided within 24 h. Some of the workers developed conjunctivitis, cough, and headache. No chronic and cumulative effects were noted (427).

The symptoms of eye irritation were also observed in histologists who used 1–2% osmic acid to stain fat. The eye irritation was often accompanied by frontal or orbital headache (427). Severe irritation of nose, throat, and bronchi persisted for 12 h following single inhalation exposure of volunteers who inhaled vapors from a container filled with osmic acid (427).

Viotti et al. (437) were presented two cases of professional osmium tetroxide poisoning in medical staff. The presence in a case modifications of respiratory function tests, indicating a bronchitis with tendency to chronicity, was referred.

Osmium tetroxide is used in the treatment of rheumatoid arthritis. In 101 patients treated for knee joint hydrops with intraarticular osmic acid injection, increased incidence of proteinuria, and hematuria were observed during the first day following the injection. Temporary glycosuria was found in three of the patients. Permanent renal damage, however, was not detected (438).

In nonallergic patients, 1% solution of the osmic acid used for chemical synovectomy of arthritic joints was without harmful systemic effects (439). Nevertheless, research revealed that osteoarthrosis may develop as a side effect of osmic acid used to treat rheumatoidal arthritis (440).

Several cases of dermatitis, eczema, or urticaria have been traced back to local contact with osmic acid solutions during osmium extraction from osmiridium (427) or to using OsO$_4$ solution to examine fingerprints (414). Dermatitis was also observed during injection of the osmic acid into the knee joint (441, 442).

An acute immune response occurred in a patient with rheumatoidal arthritis following the second osmic acid injection, which was made after 4 years after the first one. After acute phase of the response, renal and hepatic lesions were detected within one week since the injection (443).

15.4.2.2.2 Chronic and Subchronic Toxicity. No chronic or cumulative effects were noted.

15.4.2.2.3 Pharmacokinetics, Metabolism, and Mechanisms. No data were found in the available literature concerning absorption, distribution, and elimination of osmium tetroxide in people occupationally exposed to OsO$_4$.

Examinations of the patients treated with 1% osmic acid solution used for chemical synovectomy of arthritic joints revealed that most of the injected osmium was excreted in urine with none accumulating in the contralateral knee, the regional lymph nodes, the liver, or the heart (439).

15.4.2.2.4 Reproductive and Developmental. No data were found.

15.4.2.2.5 Carcinogenesis. At the time of this review no data were available to assess the carcinogenic potential of this agent.

15.4.2.2.6 Genetic and Related Cellular Effects Studies. No data were found.

15.4.2.3 Epidemiology Studies. No data were found.

15.5 Standards, Regulations, or Guidelines of Exposure

ACGIH, considering very strong irritating activity in humans and no other harmful effects after 6-h exposure to osmium tetroxide at 0.001 mg/m^3, as well as taking into account the reported vision disturbances associated with exposures to 0.1–0.6 mg/m^3, has recommended a threshold limit value time-weighted average (TLV TWA) of 0.0002 ppm OsO$_4$ (equivalent to 0.0016 mg Os/m^3) and a TLV STEL (short-term exposure limit) of 0.0006 ppm OsO$_4$ (equivalent to 0.0047 mg Os/m^3) (412, 444).

Table 41.34 shows the values of occupational exposure limits for OsO$_4$ in some other countries.

F PLATINUM AND ITS COMPOUNDS

16.0 Platinum

16.0.1 CAS Number: [7440-06-4]

1.0.2. Synonyms: Platin; platinum black; platinum sponge; liquid bright platinum

16.0.3 Trade Names: Royer® platinum catalyst beads

16.0.4 Molecular Weight: 195.08

16.0.5 Molecular Formula: Pt

16.1 Physical and Chemical Properties

1.1 General

Platinum, an intermediate element belonging to group VIII (nickel, palladium, platinum) of the periodic table and at the same time to the heavy platinum group, is a relatively soft, very malleable, ductile, silver-white metal of very high melting point and high density. It occurs mainly in the form of stable isotopes: ^{190}Pt (0.01%), ^{192}Pt (0.08%), ^{194}Pt (32.9%), ^{195}Pt (33.8%), ^{196}Pt (25.2%), and ^{198}Pt (7.2%). It is rare in the earth crust (abundance 2×10^{-6}% by weight). Nevertheless, platinum is the most abundant element of the heavy platinum group. Platinum is usually at the 0, +2, or +4 oxidation state. The highest oxidation state for platinum is +6. Platinum is characterised by a high degree of nobility; it is not affected by atmospheric conditions (even at elevated temperatures) or acids (except aqua regia). When platinum is treated with aqua regia, chloroplatinic acid (H$_2$PtCl$_6$, dark red crystalline solid body) is formed. Platinum can react with molten hyperoxides, with

Table 41.34. Exposure Limits for Osmium Tetroxide (All Limits in mg/m^3)

Exposure limits	CAS Number	OSHA PEL (444)	NIOSH Exposure Limited (444)	ACGIH TLV (444)	AIHA WEEL	ANSI Standard	German MAK (157)	Sweden Standard (158)	British Standard (158)	Polish Standard (159)
TWA	[20816-12-0]	0.002	0.002	0.0016	—	—	0.0021, I^a	—	0.002	—
STEL	—	0.006	0.006	0.0047	—	—	—	—	—	—
Ceiling limit	—	—	—	—	—	—	—	—	—	—
Biological limits (if available)	—	—	—	—	—	—	—	—	—	—

^{a}Peak limitation category: I (local irritants).

hydroxides of the lithium group metals, and, at elevated temperatures, with fluorine. Metallic platinum extensively absorbs hydrogen, oxygen, and other gases, displaying very high catalytic activity (particularly in the form of sponge and platinum black); it is plastic at normal and high temperatures, and can be rolled and forged. In some applications it must be hardened by adding other platinum group metals and copper. Platinum can be bound to a number of ligands (ions or neutral molecules) that may have more than one binding site to form neutral or ionic salts or complexes.

Platinum compounds differ in their water solubility. Metallic platinum and platinum oxides are not water-soluble, unlike complex salts, such as potassium and sodium hexachloroplatinate(IV), which are soluble in water. Tetrachloroplatinates (II) are easier to dissolve than the respective hexachloroplatinates(IV) (2, 3, 404, 406, 445).

Table 41.35 gives information on the chemical identification, physical and chemical properties, and solubility of platinum and its salts.

16.2 Production and Use

Platinum is obtained mainly from copper and nickel ores, and platinum alloys and by recovery from the catalyst and other waste. The main stages of platinum production include extraction of the precious-metal concentrate from the ore followed by separation through a complex refining process, during which the concentrate is dissolved in aqua regia, and the platinum is precipitated in the form of ammonium(IV) hexachloroplatinate. The precipitate is then calcinated at 600–700°C to give platinum sponge, which is then hardened by melting at high temperatures, such as in the electric arc. The resultant gray platinum sponge contains 99.95–99.9% pure metal. Another method of platinum production involves its reduction to metal from the aqueous platinum salts by zinc, magnesium, iron, or aluminum. Similar procedures are used to recover platinum from the catalytic converters and other waste. The hexachloroplatinic(IV) acid produced by treating platinum with aqua regia is an important chemical platinum compound used to obtain various platinum salts (2, 404).

Platinum has been widely used in various industries, such as chemical, ceramic, electronic, automotive, petroleum. It is also used in medicine, dental surgery, and for jewelry manufacture (2, 404, 445, 446). Pure platinum and its alloys are used to produce special-purpose chemical apparatus, laboratory equipment (crucibles, evaporating dishes, platinum wire nets, electrodes), spinning dies for spinning chemical and glass fibers, and electric contacts. Platinum/iridium alloys are used to make length and weight standards. The industrial application of platinum is associated mainly with its catalytic activity. Platinum catalysts were used in crude oil reforming to increase the octane number of the liquid fuels. Before it had been used for the catalytic converters in automobile engines, platinum was used predominantly as a catalyst in curing of plastics, dehydrogenation, isomerization, and oxidation in the production of sulfuric, nitric, and organic acids and amines (446). In automobile engines, platinum catalyses the oxidation of carbon monoxide to carbon dioxide and the transformation of nitric oxide into nitrogen and water (446).

Platinum is used to make surgical instruments and implants.

The industrial application of platinum is not limited to its pure metal or alloy forms; it is used also in the form of chemical compounds to electroplate metal surfaces (airplane parts,

Table 41.35. Physical and Chemical Properties of Platinum and Platinum Compounds

Chemical Name	Molecular Mass	Melting Point (°C)	Boiling Point (°C)	Density (g/m^3 20°C)	Solubility in Water
Platinum Pt	195.08	1773.5	3827 (±100)	21.45	Insoluble
Platinum(II) chloride (PtCl$_2$)	265.00	581 (decomposes)	—	6.00	Insoluble
Platinum (IV) chloride (PtCl$_4$)	336.89	370 (decomposes)	—	4.30	Readily soluble
Platinum(II) oxide, Platinum monoxide (PtO)	211.08	550 (decomposes)	—	14.10	Insoluble
Platinum (IV) oxide, Platinum dioxide (PtO$_2$)	227.08	450	—	11.80	Insoluble
Platinum (IV) sulfate tetrahydrate Pt(SO$_4$)$_2$·4H$_2$O	459.29	—	—	—	Soluble
Potassium hexachloroplatinate(IV) (K$_2$PtCl$_6$)	485.99	250 (decomposes)	—	3.50	Slightly soluble
Potassium tetrachloroplatinate(II) (K$_2$PtCl$_4$)	415.09	—	—	3.5	Soluble

electrodes, turbine vanes, wires) or in jewelry. Besides, platinum salts are used for the production of catalysts [*e.g.*, potassium hexachloroplatinate, $K_2(PtCl_6)$] (2, 404, 445, 446).

Some complex platinum compounds, such as cisplatinum or carboplatinum, are used as drugs in the treatment of cancer (mainly cancer of ovary, testis, lungs, bladder, and Hodgkin's disease) (447–449) (see discussions of *cis*-diaminodichloroplatinum(II) in Section 2.2 and *cis*-diamine(1,1-cyclobutanedicarboxylato)platinum(II) in Section 21.2).

16.3 Exposure Assessment

16.3.1 Workplace Methods

The NIOSH-recommended method for the determination of total platinum in air samples is induction coupled plasma–atomic emission spectroscopy (ICP-AES). In this method (NIOSH 7300), airborne platinum particles are collected on a cellulose membrane filter, which is then mineralized by concentrated nitric and perchloric acids, and the resultant samples are analyzed by the ICP-AES technique. The determinable concentration range of the method is $0.005-2.0$ mg/m^3 (in a 500-L air sample). The estimated limit is 1 µg per sample (7).

Flame atomic absorption spectrometry (FAAS) with air–acetylene flame can be used to determine workplace air platinum content during the monitoring of the sanitary conditions. The principle of the method is as follows. A known volume of air is passed through a membrane filter; the filter is mineralized and the solution of the mineralizate in dilute hydrochloric acid is prepared. Platinum in the solution is determined using FAAS by measuring the absorbance at 265.9 µm wavelength. The estimated limit of the method is 0.25 mg/m^3 (450).

NIOSH method S191, enables determination of soluble platinum compounds and metallic platinum together with nonsoluble platinum compounds. The aerosol fraction is collected on a mixed cellulose filter and then wet-ashed by nitric acid. Soluble platinum salts are collected in the solution of nitric and perchloric acid, and insoluble platinum salts and metallic platinum are collected in the solution of the nitric and hydrochloric acid. The resultant solutions are then analyzed for their platinum content by graphite flask atomic absorption spectrometry (GFAAS). The method was checked with potassium hexachloroplatinate(IV) at $0.00079-0.0031$ mg/m^3 concentration range. The estimated limit of the method is 0.00014 mg/m^3 using the 720-L sample (451).

Air platinum can also be determined by induction coupled plasma–mass spectroscopy (ICP-MS), estimated limit 0.01 µg/m^3 for 15 min sample (452).

16.3.2 Biomonitoring/Biomarkers

In biological material samples, such as tissues and systemic fluids, platinum concentration is routinely determined by AAS and ICP-AES. Before proceeding with the analysis, the samples must be mineralized in acids. In the method recommended by NIOSH (8005) for the determination of platinum in blood and tissues, the sample, after being mineralized by an nitric/perchloric acid mixture, is dissolved in 10% sulfuric acid and analyzed by IPC-AES. The estimated limit of the method is $0.01-10$ mg/100 g blood or $0.02-2$ µg/g tissue (10).

The same ICP-AES technique is recommended by NIOSH method 8310 to determine urine platinum concentrations. Polydithiocarbamate resin is used as the extraction medium, and the nitric and perchloric acids are used for the mineralisation of the sample. The determinable concentration range is 0.25–200 µg/sample. The estimated limit of the method is 0.1 µg/sample (11).

Some other analytical methods are also used to determine platinum levels in the biological material, such as adsorptive voltammetry (AV) (453–456), derivative pulse polarography (DPP) (457), radiochemical neutron activation analysis (RNAA) (458), (IPC-MS) (459), and instrumental neutron activation analysis (NAA) (460).

16.4 Toxic Effects

16.4.1 Experimental Studies

16.4.1.1 Acute Toxicity. Metallic platinum in the form of very fine (1–5-µm-diameter) powder given intragastrically to rats produced weak necrotic lesions in the epithelium of the GI tract, hepatocyte dystrophy, and swelling in the epithelium of the renal tubules (461). The acute toxic effect of platinum depends on type of compound, dose, and administration route. Generally, the effect is stronger after IV and IP administration than after oral exposure. The water-soluble platinum compounds are more toxic than the nonsoluble ones (462). The toxicity of platinum compounds received per os is, in descending order:

$$PtCl_4 > Pt(SO_4)_2 \cdot 4H_2O > PtCl_2 > PtO_2 \quad \text{(see Table 41.36)}$$

In acutely-poisoned experimental animals, the platinum compounds produced vomiting, diarrhea, clonic convulsions, impaired respiration, and cyanosis (463).

Symptoms of nephrotoxicity were found in F344 rats given hexachloroplatinic acid IP at 40–50 mg/kg body weight. The rats died from renal insufficiency, hypocalcemia, hypokalemia, and necrotic lesions of renal tubules and cortex. Lesions were also found in thymus (464).

Table 41.36. LD_{50} Values for Some Platinum Salts Given to Rats

Compound	CAS Number	Exposure	LD_{50} (mg/kg bw)	Ref.
Platinum(II) chloride	[10025-65-7]	Per os	>2000	462, 467
Platinum(II) chloride	[10025-65-7]	Per os	3423	461
Platinum(II) chloride	[10025-65-7]	IP	670	462, 467
Platinum(IV) chloride	[13454-96-1]	Per os	240	462, 467
Platinum(IV) chloride	[13454-96-1]	Per os	276	461
Platinum(IV) chloride	[13454-96-1]	IP	38	462, 467
Platinum(IV) chloride	[13454-96-1]	IV	26.2	384
Platinum(IV) chloride	[13454-96-1]	IV	41.4	384
Platinum(IV) oxide	[1314-15-4]	Per os	>8000	462, 467
Platinum sulfate(IV) · 4H$_2$O	[69102-79-0]	Per os	1010	462, 467
Platinum sulfate(IV) · 4H$_2$O	[69102-79-0]	IP	310	462, 467
Potassium tetrachloroplatinate (II)	[10025-99-7]	Per os	50–200	446

Acute toxicity symptoms in the form of violent asthmatic bronchospasm resulting in death within 3 min were observed in guinea pigs given a single dose of 20 mg/kg bw sodium hexachloroplatinate IV Table 41.36 (465, 466). Specifies LD$_{50}$ values for rats (467).

16.1.2 Chronic and Subchronic Toxicity. The main effects observed in rats after subchronic oral exposure to platinum compounds (given with food or drinking water for about 4 weeks) were reduced body weight gain, disturbed renal function, and altered blood cell count (see Table 41.37) (385, 395, 468–470).

16.4.1.3 Pharmacokinetics, Metabolism, Mechanisms. Platinum and its compounds are absorbed by the organism by inhalation or per os; the main route in the occupational environment is inhalation.

The rate of absorption of the platinum compounds depends on their physical and chemical characteristics and on the absorption route (384, 470, 471).

Following single inhalational 48-min exposure of male Charles River CD-1 rats to various platinum compounds; including metallic platinum, platinum(IV) oxide, platinum(IV) chloride, and platinum(IV) sulfate at 5–8 mg/m^3, the bulk of the ^{191}Pt isotope was quickly removed from the organism, and the rate of ^{191}Pt absorption from the lungs of the rats depended on the water-solubility of the compounds. The water-soluble platinum(IV) sulfate was removed from the lungs quicker than metallic platinum or platinum(IV) oxide (471). Table 41.38 gives the percentages of the initial lung platinum content after inhalational exposure to the platinum compounds specified above.

The route of administration was also an important factor in determination of platinum retention by the organs of experimental animals. In an experiment during which Charles River CD-1 rats were exposed by various routes to a single dose of platinum(IV) chloride, ^{191}Pt retention decreased in the following order:

$$\text{Intravenous} > \text{intratracheal} > \text{inhalation} > \text{per os administration}$$

A comparison of the absorption after the inhalation and per os administration revealed that the absorption is more intense in the airways (470).

In male Charles River CD-1 rats exposed by inhalation to platinum(IV) chloride at 5 mg/kg for 48 min (1.0 µm particle diameter), systemic retention after 24 h was about 41% of the original load, whereas after 10 days the organism retained only 1% of the active component (471). After single intragastrical administration of platinum(IV) chloride, only 1% of the original dose were absorbed for 3 days (384). Transcutaneous absorption of platinum(IV) sulfate was not observed. Platinum was not detected in urine, blood serum, or spleen of guinea pigs repeatedly exposed to 0.19 or 0.25 g platinum(IV) sulfate by applying it to the skin of the animals (472).

After 48-min inhalation exposure of Charles River CD-1 rats to metallic (^{191}Pt) platinum at 7–8 mg/m^3, the isotope was found to be present in all tissues. During the first day following the exposure, 93.5% of the total radioactivity were detected in the lungs, and 3.9% in the trachea. Analysis of the radioactivity in other tissues revealed higher ^{191}Pt levels in the kidneys and bones, suggesting an accumulation in those organs (Table 41.39).

Table 41.37. Subchronic of Toxic Effects of Some Salts Given to Rats

Compound	CAS Number	Approximate dose	Exposure method	Observed effects	Ref.
Pt (metal powder, 0.5–150 μm particle size)	[7440-06-4]	≤50 mg/kg	With food for 4–12 weeks	No toxic effects, no disturbed food intake, no altered blood cell count	468
Platinum(II) chloride	[10025-65-7]	≤50 mg/kg (total dose 21 mg Pt/rat)	With food for 4 weeks	No toxic effects, unchanged body weight gain	469
Platinum(IV) chloride	[13454-96-1]	≤50 mg/kg (total dose 21 mg Pt/rat)	With food for 4 weeks	Lower erythrocyte count, lower haematocrit value, higher urine creatinine levels	469
Platinum(IV) chloride	[13454-96-1]	550 mg/L (total dose ≈ 250 mg Pt/rat)	With drinking water for 29 days	Body weight gain reduced by 20%, renal weight gain lower by 6–10%	385
Platinum(IV) oxide	[1314-15-4]	6800 mg/kg (total dose 4900 mg Pt/rat)	With food for 4 weeks	No toxic effects, unchanged body weight gain	395
Platinum(IV) sulfate · 4H$_2$O	[69102-79-0]	750 mg/L (total dose ≈ 60 mg Pt/rat)	With drinking water for 8 days	Reduced body weight gain, lower aniline hydroxylase activity, no changes in the organs	395
Potassium tetrachloroplatinate(II)	[10025-99-7]	235 or 470 mg/L	With drinking water for 23 days	Reduced body weight gain, lower water intake	470

Table 41.38. Percentage of the Initial Platinum Content in the Lungs after Inhalational Exposure to Metallic Platinum, Platinum(IV) Sulfate and Platinum(IV) Oxide (471)

Days	%Initial Pt Content		
	Platinum	Platinum(IV) oxide	Platinum(IV) sulfate
1	63.0	57.2	73.7
2	49.5	60.9	43.4
4	41.3	49.0	20.4
8	42.9	28.6	—
16	28.0	17.9	4.4

Table 41.39. Distribution of Radioactive ^{191}Pt in Rat Organs after Inhalation Exposure to Metallic Platinum (48 min, 7–8 mg/m^3) (471)

	Pulses (g wet mass)			
Days after Exposure:	1	2	4	8
Trachea	1,909	2,510	738	343
Lungs	45,462	28,784	28,280	23,543
Kidneys	750	1,002	906	823
Liver	52	46	37	17
Bones	281	258	231	156
Blood	61	43	30	12
Spleen	39	73	23	5
Heart	37	58	23	5
Muscles	22	10	28	0
Brain	5	3	1	0

After IV administration of ^{191}Pt-labeled platinum(IV) chloride to male rats, the isotope was found present in all tested tissues: liver, spleen, adrenals, and kidneys. Low level radioactivity was detected in the adipose tissue. High isotope levels were detected in kidneys (6.7% dose/g in the first day, and 1.2% dose/g on day 14. Those results point to an accumulation in that organ. The lowest isotope concentrations were found in the brain, suggesting a limited Pt transport through the blood–brain barrier (30,35). In rats given a single oral PtCl$_4$(IV) dose, the highest concentration of the ^{191}Pt isotope was found in kidneys and liver, but the levels were not elevated in other organs (470).

Platinum uptake by fetuses was only slight when pregnant rat females received ^{191}Pt-labeled platinum(IV) chloride IV on day 18 of gestation. After 24 h following exposure, only 0.01% of the isotope dose/g was found in the fetuses. The quantity of the isotope in the placenta was relatively high, 0.92% of the dose/g, but it was lower than that observed in the liver and kidneys of the mother rats (1.44 and 4.22% dose/g, respectively) (470).

Platinum-191 isotope was also found in the uterus and placenta of female rats receiving platinum(IV) chloride with food at various concentrations (to 100 mg/kg diet) during the 4 weeks preceding the pregnancy and up to day 20 of gestation. The highest concentration (~80–90%) of platinum was found in the amniotic fluid. Besides, ^{191}Pt was found to be present in the milk of the rat mothers maintained during lactation on a diet containing platinum(IV) chloride. Platinum levels in the bodies of the offspring were highest at the end of the lactation period (473).

Platinum was also detected in erythrocytes of female Spraque–Dawley rats receiving platinum(IV) chloride intragastrically; platinum concentration was higher in the blood than in the plasma (474).

Among the rats given intragastrically platinum(IV) sulfate for 4 weeks, Pt content in the organs of the rats receiving the highest (716 mg Pt/rat) dose was as follows: kidneys, 43.4 mg Pt/kg wet weight; liver, 3.5 mg Pt/kg; spleen, 3.2 mg Pt/kg; blood, 1.6 mg Pt/kg; testis, 1.1 mg Pt/kg; brain, 0.33 mg/kg. Those data show that highest Pt concentration was found in the kidneys, and the lowest in the brain (475).

After intragastric administration of platinum(IV) sulfate to Swiss mice at 144-213 mg/kg bw, platinum levels in the blood were several times higher than those in the brain. Systemic clearance was slower than in the rats. Besides, the authors noted stronger effect of the higher doses on the absorption (476).

Platinum is removed from the organism mainly with feces and urine. Lung clearance is slower than removal from the gastrointestinal tract.

After 48-min inhalational exposure of male Charles River CD-1 rats to platinum(IV) chloride at 5 mg/m^3, platinum(IV) sulfate (5–7 mg/m^3), platinum(IV) oxide (7–8 mg/m^3), and metallic platinum (7–8 mg/m^3), the bulk of the radioactive ^{191}Pt isotope was removed during the first day with the feces. Some slight radioactivity was found in urine (470). The values of ^{191}Pt retention in the organism of the rats in terms of the percentage of the original content 24 h after exposure to the above mentioned compounds were 41, 33, 31, and 20% respectively. After 10 days, the organism retained about 7–8% of the activity with the exception of platinum(IV) chloride, for which the corresponding value was 1%. Pt elimination from the lungs comprised two phases. The first (quick) phase took ca. 24 h, and was followed by slow phase with elimination half-period of ca. 8 days (470).

Removal of platinum (IV) sulfate from the lungs was quicker than elimination of the remaining test compounds, probably because it was more easily water-soluble (470).

After intravenous administration of 191 Pt-labelled platinum (IV) to male Charles River CD-1 rats, the absorbed platinum was removed with urine and feces, the major part being eliminated with urine. Total retention in animal body was 65% after 3 days, and ca. 14% after 28 days (470).

1.4.1.3.4 Mechanisms of Action. In vitro studies on the mechanism of action of platinum compounds show that, generally speaking, they become attached e.g. to albumin or transferrin and, probably, bonding power of the platinum/ligand bond, reactivity of the resultant complex of platinum with proteins or other carriers and the stability of the combination are essential for the sensitizing potential of platinum complexes. Inert or strong-bond ligand compounds are immunologically inactive (477). It has been also demonstrated that platinum complexes are attached to protein nitrogen or sulfur (478). Platinum attached to aminoacids may inhibit the activity of enzymes, e.g. of maleate dehydrogenase (479).

Platinum compounds may also be attached to DNA. This refers in particular to compounds used as antitumour drugs, such as cisplatinum and carboplatin (480–485) (see: cis-Diamminedichloroplatinum II, Section 6.4. and cis-Diammine(1,1-Cyclobutanedicarboxylato)platinum (II) Section 5.4.).

Experiments on animals have revealed that platinum compounds may produce asthmatic symptoms, such as bronchospasms, anaphylactic shock, and increased blood plasma histamine levels (monkey, dog, guinea pig, rat) as early as on the first contact due to the irritating or pharmacologic action (465, 486).

When guinea pigs received intravenously sodium (IV) hexachloroplatinate at 1 to 2 mg/kg, they developed a strong bronchospasm comparable to that resulting from the exposure to histamine dihydrochloride at 5 µg/kg. After several exposures, the symptoms disappeared. The animals pretreated with antihistamine drugs did not develop bronchospasm (465). The bronchospasmogenic activity of hexachloroplatinate has been confirmed by Biagini et al. (486), who assessed the pulmonary function of monkeys' lungs following provocation by aerosols containing growing levels of metacholine and, several weeks later, of sodium hexachloroplatinate. The authors noted respiratory system hyperreactivity with symptoms of expiratory resistance and smaller forced expiratory volume. In the opinion of the authors, the observed bronchospasm was due either to the direct action on the smooth muscles or to release of mediators causing bronchospasms without stimulating the irritation receptors.

Although the results of numerous studies have confirmed that platinum compounds induce immunological reactions of the respiratory system and skin, the exact mechanism of the allergy has not been yet completely explained.

16.4.1.4 Reproductive and Developmental Effects. The number of reports from experiments on the effect of metallic platinum and its compounds on reproduction or its embryotoxic and teratogenic properties is limited.

Metallic platinum (Pt°) in the form of wire or leaf is thought to be biologically inert, and its detrimental effect on the implantation is probably due to the presence of a foreign body in the uterus (71).

A 83% reduction in the number of implantation sites was found to occur in female rats after platinum wire had been placed in the uterus 3 days after the fertilization (71). A similar, 37% reduction in the number of implanted ovuli was found in female rabbits after platinum leaf had been placed in the uterus (487).

When metallic platinum Pt^+ was given with food to female Spraque–Dawley rats during 4 weeks and up to day 20 of gestation at 0.1, 0.5, 1.0, 50, and 100 mg Pt/kg food, neither mean (wet) weight of the fetuses nor the mean number of normal resorbed fetuses was dependent on mothers' platinum intake (488).

16.4.1.5 Carcinogenesis. No reports are available in the literature from the experimental tests of the carcinogenic activity of platinum and its compounds, except for cisplatinum and some other compounds of similar structure. Sufficient proof of the carcinogenic activity in animals exists for cisplatinum [see *cis*-diamminedichloroplatinum(II) discussion in (489–491), Section 21.4.1.5.].

16.4.1.6 Genetic and Related Cellular Effects Studies. No data have been found in the literature concerning the genotoxic activity of platinum metal.

The bulk of the data on the genotoxicity of platinum salts has been obtained from *in vitro* studies and, generally speaking, data from *in vivo* studies are not available. Platinum salts, such as platinum(IV) chloride or platinum(IV) sulfate showed mutagenic activity in several tests with bacteria, yeast, Chinese hamster CHO-S, CHO-AUXB1, or V9 cells, and the L5178Y cells of mouse lymphoma (194, 492–497).

No mutagenic activity of potassium(II) tetrachloroplatinate was observed in the *Drosophila melangoaster* recessive lethal mutations test (498) or in the test of micronucleated erythrocyte induction in mouse bone marrow cells (499).

16.4.1.7 Other: Neurological, Pulmonary, Skin Sensitization

16.4.1.7.1 Irritating Activity. No reports are available on the irritating activity of metallic platinum on skin or eyes. Studies of the local irritating activity in white rabbits in which aqueous solutions of platinum compounds were applied to nondamaged and chafed skin and to eyes revealed varying irritating action of the individual compounds to rabbit skin and eyes (500, 501). The reactions were assessed after 24, 48, and 72 h. Platinum(IV) chloride showed irritating activity on the skin (380). Potassium tetrachloroplatinate(II) showed irritating action to eyes (501).

16.4.1.7.2 Sensitizing Activity. No reports are available showing that platinum metal or dust displayed allergic activity.

Studies on possible sensitizing activity of platinum compounds, mainly hexachloroplatinates, to the respiratory tract and skin were performed on various experimental animal species exposed by various routes (465, 466, 472). The results indicate that platinum compounds may cause bronchospasm, anaphylactic shock, and increased histamine levels in blood plasma already on first contact during irritation test or pharmacological treatment.

Itching of mouth and feet, lower temperature of the extremities, and increased blood serum histamine concentrations were observed in rats receiving intravenously 40 mg/kg sodium(IV) hexachloroplatinate (466).

Very high increase of blood histamine concentration, from 20 to 1000 µg/kg during 2–5 min, was detected in a dog that died following IV injection of 30 mg/kg sodium(IV) hexachloroplatinate. Increased histamine concentration was not observed in another dog that received 10 mg/kg of the compound (466).

Acute asthmatic seizure occurred in guinea pigs exposed by inhalation to sodium(IV) hexachloroplatinate aerosol or after a single 10–20 mg/kg IV dose of that compound. Bronchospasm occurred after a single 1–2 mg/kg IV dose of sodium(IV) hexachloroplatinate, but the symptoms were not observed after subsequent doses (465, 466).

Hyperreactivity of the respiratory system with symptoms of the pulmonary flow resistance $(R_I)(R_L)$ and reduced forced expiratory volume (FEV$_{0.5}$/FVC) was found in

Cynomolgus monkeys exposed to aerosol containing sodium(IV) hexachloroplatinate at concentrations of ≤ 62.5 mg/mL 2 weeks after since termination of repeated inhalation exposure to that compound at 216 µg/m^3 (4 h/day, 2 days/week for 12 weeks) (486).

The bronchospasmic activity of the hexachloroplatinate was confirmed by the same authors in another study (486). *Cynomolgus* monkeys subjected to provocation with aerosol containing growing concentrations of metacholine (0.5, 1.05, and 6.25 mg/mL) several weeks afterward received sodium(IV) hexachloroplatinate (0.5, 2.5, 25, and 50 mg/mL). The results revealed concentration-related increase in pulmonary flow resistance (R_L), lower dynamic lung compliance (C_L dyn), and reduced maximum forced expiratory volume (MEFV) (502).

Allergic reaction was not observed when platinum(IV) sulfate was repeatedly injected (IV or SC) in the form of 0.05–0.35 mg/mL solutions to albino rabbits, guinea pigs, and white mice, or when platinum(IV) sulfate-containing paste (0.1–0.25 g) was applied to the skin (472).

16.4.1.7.3 Immunotoxicity. Metallic platinum seems to be nonallergic. On the other hand, halogenated platinum salts, such as potassium tetrachloroplatinate(II) and potassium hexachloroplatinate(IV), are very strong sensitizers.

The immunogenic activity of halogenated platinum compounds was assessed by various tests, such as PCA (passive cutaneous anaphylaxis test), RAST (radioallergosorbent test), PLN (popliteal lymphatic node test), LLNA (local lymphatic node assay), and GMT (guinea pig maximization test) on various animal species (rats, mice, guinea pigs) using various routes of administration (503–507).

Induction of IgE antibodies detected by the PCA and RAST tests occurred after exposure to complex platinum salts (hexachloroplatinates) in hen egg albumin (together with adjuvant), but not in the animals receiving free compounds, even when administered intratracheally (503).

Studies on the differences in the intensity of response to hexachloroplatinates between various varieties of mice using the PLN test showed no reaction in the mice without lymphocytes T. A dose-dependent activation of the node (increased cell mass and count) was observed in C57/BL/6 mice. The highest intensity of the symptoms was observed at 2–4 mg/kg bw (504, 505).

16.4.2 Human Experience

16.4.2.1 General Information. Exposure to platinum compounds refers primarily to the occupational environment and may occur during platinum ore mining, refining, and processing, and also during production and application of drugs [see *cis*-diaminedichloroplatinum(II), discussion in Section 21.4.2.1 and *cis*-diammine(1,1-cyclobutanedicarboxylato)platinum(II), discussion in Section 21.4.2.1]. During the mining, platinum is present mainly in the metallic form or in the form of compounds very difficult to dissolve in water. During the refining, the workers are exposed mainly to sodium, potassium, or ammonium salts of hexachloroplatinic(IV) and tetrachloroplatinic(II) acids. Generally speaking, metallic platinum is nontoxic, but allergic contact dermatitis may occur in oversensitive people. Skin lesions between fingers were observed in people wearing platinum rings (508).

Platinum compounds particularly responsible for the symptoms of hypersensitivity in humans include hexachloroplatinic(IV) acid and its chlorinated complex salts (465, 509–512).

Occupational exposure to platinum compounds is a recognized cause of allergic reactions of the skin and respiratory system. Hypersensitivity symptoms include urticaria, eczema (usually at the places exposed to platinum compounds) and lacrimation, eye burning, rhinitis, coughing, tight thorax, wheezing, and shallow breath. The cutaneous and respiratory symptoms quoted above were formerly referred to as "platinosis" (513).

About 60% people employed in the industry under conditions of exposure to platinum compounds may show symptoms of bronchial asthma. The latency period from the first contact with platinum compounds to first symptoms of hypersensitivity is from several weeks to several years. When the exposure is continued, the symptoms become more intense, and the hypersensitive people always react to platinum compounds in the air. After removal from the workplaces involving contact with platinum compound, the symptoms usually disappear, but cases have been reported where patients suffered asthmatic seizures at night as long as several weeks after cessation of the exposure (465, 477, 502, 513–519). Nonspecific symptoms of respiratory hypersensitivity have been reported (520). Besides, tobacco smokers were demonstrated to be more prone to develop the allergic symptoms. Tobacco smoking increases the likelihood of developing an allergy to platinum salts (516, 519–521). The general opinion is that tobacco smoking results in increased epithelial permeability, and it seems reasonable to suppose that the combined exposure to the irritating agents (chlorine, ammonia, ozone) may intensify the effects of the exposure to platinum salts (516).

It is difficult to determine the dose–effect or dose–response relationship because of the very wide range of the concentrations of platinum compounds in the ambient air of the workplaces and changes in workers' employment during the period of their occupational activity. Nevertheless, several authors have proved that the risk of developing allergy to platinum salts depended on exposure intensity and was higher in groups of people exposed to higher concentrations (502, 516, 517, 520, 522, 523).

No reports on other harmful health effects of occupational exposure to metallic platinum and its insoluble compounds are available in the literature.

16.4.2.2 Clinical Cases

16.4.2.2.1 Acute Toxicity. There are no data on the acute toxicity of metallic platinum. A few reports are available on the acute poisonings by platinum compounds in humans.

Nausea, vomiting, diarrhea, and leg cramping occurred within 2 h in a man who, in a suicidal attempt, ingested a 10-mL solution containing 600 mg potassium(II) tetrachloride. Medical examination revealed acute renal damage with hypouresia, fever, enteritis, metabolic acidosis, slight leukocytosis, and eosinophilia. The initial concentration of platinum in blood serum was 254 µg/dL, and in urine, 4200 µg/L. All symptoms of the poisoning disappeared within 6 days (524).

16.4.2.2.2 Chronic and Subchronic Toxicity. Information on the actual and possible effects of occupational exposure to platinum compounds includes data mainly on

- Hypersensitivity-related chronic irritation of the upper airways
- Bronchial asthma
- Allergic contact dermatitis

In a pilot study involving 21 male workers exposed to metallic platinum dust during its recovery from automotive catalytic converters for periods ranging from 6 weeks to 12 months, no symptoms of allergy to platinum were detected by relevant tests. The platinum levels in workplace air were 1.7–6.0 µg/m^3 (525).

Respiratory and dermal allergic reactions occurred in a chemist who had been exposed for several years to hexachloroplatinate in testing the purity of platinum and other precious metals. The specific hypersensitivity was detected by intradermal application of increasing doses of the allergen. The symptoms observed in the chemist suggested the serum-sickness reaction (509).

Complaints of disturbed respiratory function in people exposed to platinum compounds have been reported for many years.

Examinations of workers employed in 40 photographic laboratories and exposed to complex platinum salts revealed irritation of nose and throat accompanied by violent lacrimation and cough in nine employees. Bronchial irritation accompanied by difficult breathing occurred in the case of exposure to potassium(II) chloroplatinate–treated paper; in addition, skin lesions were observed (526).

Allergic rhinitis and allergic bronchitis were noted among the employees of British petroleum refineries. The symptoms were very severe in the workers employed at crushing platinum salts, where particle concentration ranged from 0.9 to 1700 µg/m^3. Dermal lesions (urticaria, or slight erythematous dermatitis in the hands, arms, and neck) were found in 13 employees, while manifestations of bronchial asthma (dripping from the nose, tight thorax, wheezing and shallow breath, coughing, and cyanosis) were detected in 52 of 91 workers. The respiratory symptoms were particularly pronounced in people exposed to platinum salt dusts and aerosols of the aqueous solutions of those salts. The latent period from the first contact to the first symptoms ranged from several months to 6 years (514).

Symptoms of "platinosis" were detected in 21 workers employed in a U.S. platinum refining plant. As much as 60% of the workers reported burning sensation in the eyes, tight larynx and thorax, dry cough, skin itching, and dermatosis. In 8 of 19 people (40%) skin tests were positive (513).

Similar allergic symptoms were found in employees of a French refinery. As much as 35 of 51 refinery workers reported respiratory and/or dermal symptoms, primarily at the beginning of the shift or at night (465).

16.4.2.2.3 Pharmacokinetics, Metabolism, and Mechanisms

16.4.2.2.3.1 ABSORPTION, DISTRIBUTION, AND EXCRETION. No detailed data can be found in the available literature on platinum pharmacokinetics in people occupationally exposed to metallic platinum or its salts. The results of platinum content determinations in samples collected for autopsy from people of unknown exposure to metals were questionable, sometimes due to doubtful reliability of the employed analytical procedures (527). Studies by Vaughan and Florence (528) revealed that, in humans, the major part of platinum salts ingested with food was absorbed in the alimentary tract, to be later excreted with urine (~42% of the dietary platinum) (528).

Studies by Schierl et al. (456) on urine platinum concentrations and the kinetics of platinum elimination with urine in 34 workers of a refinery revealed that, after the workshift, urine platinum levels were as high as 6270 ng/g creatinine, which was about 1000 times higher than the mean level in the nonexposed people. The elimination reached its maximum after nearly 10 h following the inhalation exposure to platinum-containing dust. The elimination referred to the first half-life was ~50 h, but it was proved that the built-in portion of platinum was stored for longer periods. The quantity of eliminated platinum correlated with the exposure levels monitored by the individual dosimetry (456).

16.4.2.2.3.2 MECHANISM OF ACTION. The effects of platinum salts on the skin and respiratory system in humans are thought to result from their allergic activity, although the exact mechanism of the activity is not clear. The symptoms occur in some people only after some period of exposure. Both atopic and nonatopic people may become sensitized. The allergic effects, such as bronchial asthma, may be explained by the general theory of the immunological mechanisms. The reaction mechanism seems to be an IgE-mediated response, as shown by dermal prick and *in vitro* tests (histamine release from leucocytes) and *in vivo* assays, such as the PCA test (466, 477, 502, 509, 510, 529, 530).

Platinum salts of relatively low molecular mass are thought to form, as heptenes, a complete antigen with serum protein. Complex platinum salts show activities of full allergens; this is particularly true for hexachloroplatinic acid and its ammonium salt (480, 531, 532).

Immunological reactions to platinum salts were detected many times on application of skin-prick tests (highly specific biological monitors of hypersensitivity) generally using sodium, ammonium, and potassium(IV) hexachloroplatinates as the test substances. In some instances, pulmonary hyperreactivity, diagnosed after provocation test or determined as work-related symptoms, preceded the positive results of skin-prick tests. This may suggest that the pulmonary hyperreactivity is due to pharmacological action rather than to allergy (477, 502, 517, 522, 533).

Heat-stable IgE antibodies were detected also by the PCA test on monkey skin (502).

Platinum salt-specific IgE antibodies were detected by the RAST tests in blood sera of workers exposed to platinum salts. Nonspecific intensification of immunological IgE response was proposed as a possible mechanism of the sensitization because very high total IgE concentrations were detected in the sera of the workers exposed to platinum salts, and most of the atopic individuals had been eliminated during the preliminary tests (502, 522, 523, 534, 535).

16.4.2.2.4 Reproductive and Developmental. No data were found in the available literature on the teratogenic activity of platinum and its salts on human reproduction.

Only weak inhibition of human sperm viability was observed during *in vitro* studies when the sperm was incubated with a strip of metallic platinum for 2–5 h (536). Slight spermicidal effects were observed when human spermatozoa were incubated *in vitro* with metallic platinum for 3 h (537).

16.4.2.2.5 Carcinogenesis. No reports are available on the carcinogenic activity of platinum in people occupationally exposed to metallic platinum or its salts.

16.4.2.2.6 Genetic and Related Cellular Effects Studies. There are no data on the mutagenic or genotoxic activity of metallic platinum or platinum salts in humans. Induction of micronucleated erythrocytes, increased chromosome aberrations, and sister chromatid exchange (SCE) were recorded following exposure to cisplatinum studies [see discussion of *cis*-diamminedichloroplatinum(II) in Section 21.4.2.2.6).

16.4.2.3 Epidemiology Studies. Cross-sectional surveys (502, 516, 517, 523, 535), cohort studies (530, 534, 538), historical prospective cohort studies (521, 539), and a prospective cohort study (519) relate mainly to the hazards that occur in platinum refining and processing plants and the resultant bronchial asthma, dermal hypersensitivity, and nonspecific respiratory diseases.

Prospective studies conducted until 1980 of a cohort of 91 workers employed in a U.K. platinum refining plant who started working there during 1973/74 revealed respiratory symptoms in 49, and positive skin-prick test for platinum compounds in 22. Positive results of the skin tests were more frequent in tobacco smokers. The risk of developing respiratory symptoms in the smokers was twice that in the nonsmokers (521).

Effects of tobacco smoking and exposure intensity on the development of allergy to platinum compounds were studied in the workers of a Republic of South Africa platinum refining plant during 1986/87. The tests were performed on fresh-employed workers without atopic symptoms. After 24 months, 32 people (41%) were examined for hypersensitivity to platinum salts. The positive results of skin-prick tests were obtained in 22 (28%) of the cases, and in 10 (13%) the results were negative in spite of evident symptoms of hypersensitivity. The risk of developing hypersensitivity was eightfold higher for the smokers and sixfold higher for high exposures (519).

Dermatitis and bronchial asthma, the symptoms associated with exposure to chloroplatinates, were detected in the workers of Japanese workers employed in the production of platinum-coated oxygen detectors. The ambient-air platinum concentrations were 0.14–1.83 µg/L. Bronchial asthma was present in 2 of 16 workers who were prick-tested with a 1% solution of chloroplatinate and after an environmental provocation test. The main symptoms detected during the medical examinations were contact dermatitis in 11 (78.6%), throat irritation in 6 (42.9%), and rhinostegnosis in 2 (14.3%) of the people examined and also frequent sneezing, coughing, and expectoration. Although platinum concentration in the ambient air was 2 µg/m^3, the workers were directly exposed to dry ammonium(IV) chloroplatinate powder at relatively high concentrations. The results of the pulmonary function and blood count tests were normal (540).

Results of a 1981 study involving 107 of 123 occupationally active and 29 former workers (removed from their workposts because of the suspected allergy to platinum compounds) employed in a U.S. plant where platinum was retrieved from the metal waste and automotive catalytic converters revealed that the incidence of the allergic symptoms was associated with the level of the exposure, and the increase of the risk of the positive reaction to the skin-prick tests was assessed by the authors to be 13% when the concentration of the platinum salts in the ambient air was increased by 1 µg/m^3. Positive results of the prick test were found in 67% workers exposed to complex platinum salts at 27.1 μg/m^3, and in 14% of those exposed at 10.7 μg/m^3. The reaction to skin prick test was significantly correlated with the rhinitis, symptoms of asthma, dermatitis, positive reaction to the cold-air test, after tobacco smoking and sensitization to allergens found in the ambient air had been taken into account. At the same time, higher blood plasma IgE levels were noted. Hypersensitivity to platinum salts was not related to the tendency for atopy. It was very interesting to note a high prevalence of the symptoms of hypersensitivity (e.g., positive cold-air test, FEV$_1$/FVC < 70%) among the former workers no longer exposed to platinum salts, who showed persistent positive skin prick-test as long as 5 years since they had been removed from the exposure (516, 517).

Persistent positive reaction to the prick tests and nonspecific hyperreactivity of the respiratory system were found in a study on 24 people (15 actual and former smokers) in Germany. In the majority of the people who developed the symptoms of asthma caused by exposure to platinum salts, the nonspecific hypersensitivity of the respiratory system persisted for a long time (1–27 months) after removal from the exposure (520).

Results of other studies performed since the late 1980s, when the concentration levels of platinum salts were reduced, indicate that the allergy may still constitute a serious occupational problem. In a study conducted in Germany, 65 workers of a platinum processing department were tested for frequency of allergic respiratory diseases. Mean duration of the occupational exposure was 8–9 years (complete range 1–40 years). Rhinitis, conjunctivitis, coughing, expectoration, and occupational exposure–related dyspnea were found in 23% of the workers. The symptoms were more frequent in the people exposed to higher concentrations (522).

16.5 Standards, Regulations, or Guidelines of Exposure

The ACGIH-recommended TLVTWA value for metallic platinum is 1.0 mg/m^3, because it has been proven that exposure to the metal dust did not give rise to allergic diseases (444). Occupational exposure to soluble platinum salts, on the other hand, may result in progressive allergic reactions responsible for the incidence of asthmatic symptoms and dermal allergies. The limited data from the determination of the air concentrations point to the necessity of keeping soluble platinum salt concentrations in the ambient air of the workplaces at very low levels in order to protect the workers from the risk of respiratory tract irritation, asthma, and allergic dermatitis. The TLV TWA value recommended by ACGIH for platinum salts is, in terms of metallic platinum, 0.002 mg/m^3; keeping the concentration within this limit is expected to minimise the risk of asthma or other allergic reactions. The rationale for the standards set out by ACGIH was based on data from works published until 1975 and was not supplemented with new data when the standards were reviewed in 1992 (541).

The IPCS documents (446) suggest that short-lasting exposures to high platinum salt concentrations may constitute a serious problem due to allergies. As unequivocal proof for the relationship between the intensity of the exposure and the development of allergies is not available, it does not seem reasonable to reduce the TLV TWA value. Nevertheless, the IPCS authors suggest to change the classification of the generally recommended value from $0.002\,mg/m^3$ 8-h time-weighted average (TLV TWA) to the ceiling concentration (TLV-C) (446).

Such an approach has been adopted in Germany, where the soluble platinum salts have been classified into the group of substances for which it is not possible at present to establish MAK values. However, it is recommended not to exceed the maximum value of $0.002\,mg/m^3$ (157).

Nevertheless, results of recent studies show that the $0.002\,mg/m^3$ value adopted in hygienic standards of many countries fails to offer sufficient protection against allergies (523, 530).

The analytical difficulties indicate that it is not feasible to use biological monitoring as a routine method for assessing exposure to platinum and its soluble compounds.

Tables 41.40 and 41.41 specifies current values of the hygienic standards in various countries.

16.6 Studies on Environmental Impact

Data on environmental platinum concentrations are limited, because the concentrations are so low that it is difficult to determine them using currently available analytical methods. Nevertheless, the widespread use in the automobile engines of the catalytic converters containing platinum, which is exhausted with the combustion gases to the atmosphere, constitutes an environmental hazard. Therefore, the knowledge of platinum concentrations in the air, particularly at the places with heavy road traffic, is necessary.

Platinum concentrations in the ambient air estimated from the emission data with the aid of dispersion models are probably 10,000 times lower than the occupational exposure limit of $1\,mg/m^3$ for the metallic platinum (total dust). Platinum concentrations in the air samples collected along heavy-traffic roads in the United States before the catalytic converters had been introduced were below the detection limit of $0.05\,pg/m^3$ (542). Platinum content determined in parking-lot dust in Sweden increased from 2 ng/g in 1984 to 10 ng/g in 1991. The increase was associated with increased number of the cars provided with the catalytic converters containing platinum (543).

Platinum concentrations in the soil and dust in the areas of heavy and low traffic in the United Kingdom ranged from < 0.30 to 40.1 ng/g. The highest concentrations were found in the areas of heavy traffic, whereas platinum levels of the samples collected from the low traffic roads and streets were represented by the lowest values of the specified range (544).

Determinations of platinum levels in the air over heavy-traffic streets of two large Polish cities (Katowice and Częstochowa) performed by heated graphite electrode atomic absorption spectrometry (GFAAS) gave a mean value of $6.69\,pg/m^3$ (range 2.35–$13.56\,pg/m^3$) (545).

Although definite data on the environmental concentrations are not available, it is quite likely that platinum and its compounds are present in the environment at very low concentrations that do not pose any risk to the organisms found in the natural environment.

Nevertheless, it has been demonstrated (primarily in laboratory tests) that fish and the aquatic and terrestrial plants may be subject to detrimental effects of platinum compounds when exposed to their relatively high concentrations.

Table 41.40. Exposure Limits for Metallic Platinum (All Limits in mg/m^3)

Exposure Limits	CAS Number	OSHA PEL (444)	NIOSH Exposure Limits (444)	ACGIH TLV (444)	AIHA WEEL	ANSI Standard	German MAK[a]	Sweden Standard	British Standard (158)	Polish Standard
TWA	[7440-06-4]	—	1.0	1.0	—	—	—	—	5.0	—
STEL		—	—	—	—	—	—	—	—	—
Ceiling limit		—	—	—	—	—	—	—	—	—
Biological limits (if available)		—	—	—	—	—	—	—	—	—

[a]No substances for MAK can be established at present.

Table 41.41. Exposure Limits for Soluble Salts of Platinum as Pt (All Limits in mg/m^3)

Exposure Limits	OSHA PEL (444)	NIOSH Exposure Limits (444)	ACGIH TLV (444)	AIHA WEEL	ANSI Standard	German MAK[a] (157)	Sweden Standard	British Standard (158)	Polish Standard
TWA	0.002	0.002	0.002	—	—	0.002[b] (S)[c]	—	0.002	—
STEL	—	—	—	—	—	—	—	—	—
Ceiling limit	—	—	—	—	—	—	—	—	—
Biological limits (if available)	—	—	—	—	—	—	—	—	—

[a]No substances for MAK can be established at present (157).
[b]Peak Concentration of 0.002 mg/m^3 should not exceeded (157).
[c]Danger of sensitization, also respiratory allergen (157).

Hexachloroplatinic(IV) acid at 250, 500, and 750 µg/L under laboratory conditions inhibited growth of algae (*Euglena gracilis*) (546).

The LC$_{50}$ value for *Daphnia magna* algae after a 3-week exposure to hexachloroplatinic(IV) acid was 520 µg Pt/l (547).

Studies on the effect of platinum on terrestrial plants were performed using the soluble chlorides. Tetrachloroplatinic(II) acid at 0.5 mg Pt/l in terms of pure platinum stimulated the growth of the *Setaria verticillata* grass. At higher concentrations, phytotoxic effects were observed, such as root dwarfishness and leaf chlorosis (548).

LC$_{50}$ values for the coho salmon (*Oncorhynchus kisutch*) exposed to tetrachloroplatinic(II) acid for 24, 48, and 96 h were 15.5, 5.2, and 2.5 mg Pt/l, respectively. The swimming activity and operculum movements were disturbed after exposure to 0.3 mg/L. However, the concentrations of 0.03 and 0.1 mg/L did not produce any harmful effects (549).

Platinum present in dust, mainly in the vicinity of roads, may pollute water, deposits, and soil. At present, however, there is no proof that environmental platinum produces allergic reactions in humans. Allergy to platinum is limited to a small number of its compounds containing reactive ligand systems, while the platinum is thought to be released from the catalytic converters in the metallic form or in the form of oxides, characterised by very weak sensitizing potential (544).

17.0 Platinum(IV) Chloride

17.0.1 CAS Number: [13454-96-1]

17.0.2 Synonyms: Platinum tetrachloride; tetrachloroplatinum; platinum chloride, (SP-4-1)-

17.0.3 Trade Names: NA

17.0.4 Molecular Weight: 336.89

17.0.5 Molecular Formula: PtCl$_4$(IV)

17.0.6 Molecular Structure: Cl$^-$ Pt^{2+} Cl$^-$
Cl^-
Cl^-

17.1 Chemical and Physical Properties

PtCl$_4$ occurs in the form of brown-red crystals (2).
See platinum entry in Table 41.35.

17.2 Production and Use

PtCl$_4$ is formed directly from the elements or by heating hexachloroplatinic(IV) acid (3).

17.3 Exposure Assessment

See discussion of Platinum in Section 16.3.

17.4 Toxic Effects

17.4.1 Experimental

17.4.1.1 Acute Toxicity. The values of LD_{50} for the rat after oral, IP, and IV administration are given in Table 41.36 (see platinum discussion in Section 16.4.1.1). The values in Table 41.36 have been obtained from documents presented in an IPCS work (446). No other data on acute $PtCl_4$ toxicity in the animals are available.

17.4.1.2 Chronic and Subchronic Toxicity. Some results of the studies on the subchronic toxicity of platinum(IV) chloride to rats are presented in Table 41.37 (see platinum discussion in Section 16.4.1.2.).

The results of Table 41.37 show that dietary platinum(IV) chloride given to male Spraque–Dawley rats in increasing doses of ≤ 50 mg/kg diet (total dose 21 mg Pt/rat) caused dose-related reduction of erythrocyte count and hamatocrit values (by ∼13% at the highest doses) and a significant increase of blood plasma creatinine level (469).

No changes were noted in the blood cell count values (hemoglobin content, hematocrit value, erythrocyte count, and volume) in Spraque–Dawley female rats receiving dietary platinum(IV) at 0.1, 0.5, 1.0, 50, and 100 mg/kg for 4 weeks before pregnancy and until day 12 of the pregnancy (474).

17.4.1.4 Reproductive and Developmental. Dietary platinum(II) chloride and platinum(IV) chloride given to feeding mother Spraque–Dawley rats at ≤100 mg/ Pt/kg did not affect either the weight of the offspring or blood cell counts of the mothers and offspring (473).

Spermatogenesis stoppage at the level of primary spermatocyte or spermatogonium was observed in male Swiss mice given platinum(IV) chloride SC for 30 days at a total dose of 27 mg/kg bw (429).

A reduction of testicle mass was observed in rats receiving a single intratesticular dose (27 mg/kg bw) of platinum(IV) chloride. Two days after exposure to $PtCl_4$, total testicular necrosis and spermatogonium destruction were observed (429).

Platinum(IV) chloride given intragastrically with food to rats for 4 weeks preceding gestation and until day 20 of gestation at 100 mg Pt/kg food did not affect the weight of the offspring or the number of fetuses. Increased numbers of congenital defects in the offspring were not observed, either (473).

17.4.1.5 Carcinogenesis: NA

17.4.1.6 Genetic and Related Cellular Effects Studies. Platinum(IV) chloride shows mutagenic activity, which has been demonstrated in both *in vitro* and *in vivo* tests.

Platinum(IV) chloride induced mutations in TA98 *Salmonella typhimurium* strains without the exogenous metabolizing system and showed strong genotoxic activity for *Bacillus subtilis* in the rec test (194).

At 0.3 mmol/L it caused inhibition of DNA, RNA, and ribosome synthesis in the cells of F51 *Saccharomyces* yeast (493).

A dose-dependent increase in the frequency of 8-azoguanine-resistant mutations was detected in CHO-S cells after 20-h exposure to platinum(IV) chloride and platinum(IV) sulfate (496).

Exposure to platinum(IV) chloride at 15 μmol/L caused a sevenfold increase of gene mutation frequency in Chinese hamster V9 cell locus HGPRT compared to controls (492).

Platinum(IV) chloride also caused increased frequency of mutations in L5178Y cell locus TK of mouse lymphoma and significantly augmented Syrian hamster SHE cell transformation (550)

In the tests using CHO-AUXB1 cells, platinum(IV) chloride induced dose-dependent increase in the incidence of spontaneous revertants (495, 497).

Platinum(IV) chloride caused a significantly elevated genotoxicity in the cytokinesis-block micronucleus test (MNT) with human lymphocytes and in bacterial SOS chromotest (551).

In the *in vivo* tests on *Drosophila melanogaster*, a significant increase in sex-related recessive lethal mutations was detected after exposure to 1.5×10^{-3} M platinum(IV) chloride solution (552).

17.4.1.7 Other: Neurological, Pulmonary, Skin Sensitization

17.4.1.7.1 Irritating Activity. Studies of the local irritant activity in white rabbits in which aqueous solutions of $PtCl_4$ were applied dermally to revealed irritating action to both chafed and nondamaged skin. The results were rated 2 on a 1–4 scale; there was reddening and swelling of the skin (380).

17.4.2 Human Experience

Data not available. (See discussion of platinum in Section 16.4.2.)

17.4.2.1 General Information: NA

17.4.2.2 Clinical Cases: NA

17.4.2.3 Epidemiology Studies. Data not available. (See discussion of platinum in Section 16.4.2.3.)

17.5 Standards, Regulations, and Guidelines of Exposure

Table 41.41 specifies the ACGIH, OSHA, and NIOSH hygienic standards for soluble platinum salts, and those currently in force in some other countries (see Section 16.5).

17.6 Studies on Environmental Impact

(See Section 16.6.)

18.0 Platinum(IV) Oxide

18.0.1 CAS Number: [1314-15-4]

18.0.2 Synonyms: Platinum oxide, platinic oxide, platinum(IV)oxide-hydrate, Adams catalyst

18.0.3 Trade Names: NA

18.0.4 Molecular Weight: 227.08

18.0.5 Molecular Formula: PtO$_2$

18.0.6 Molecular Structure:
$$\underset{O}{\overset{O}{\underset{\|}{\overset{\|}{Pt}}}}$$

18.1 Chemical and Physical Properties

PtO$_2$ occurs as black powder in the hydrated and anhydrous form (2, 3, 406). (See platinum entry in Table 41.35.)

18.2 Production and Use

PtO$_2$ is obtained by reduction of chloroplatinic acid with formaldehyde or by fusing chloroplatinic acid with sodium nitrate (2). It is used as a catalyst, and also to reduce the double and triple bonds of the carbonyl, nitrous and nitrile groups (2).

18.3 Exposure Assessment

See Section 16.3.

18.4 Toxic Effects

18.4.1 Experiments

18.4.1.1 Acute Toxicity. LD$_{50}$ value for the rat after intragastrical exposure is > 8000 mg/kg. No other data are available on acute toxicity (462). (See also platinum entry in Table 41.36.)

18.4.1.2 Chronic and Subchronic Toxicity. Tests on Spraque–Dawley rats maintained for 4 weeks on a diet containing, in terms of platinum, 6800 mg/kg (total dose 4000 mg Pt/rat) did not show abnormalities in body weight gain or toxic effects in the test animals (395). (See also platinum entry in Table 41.37.)

18.4.1.3 Pharmacokinetics, Metabolism, and Mechanisms. No data are available in the literature on the intake, distribution, and elimination of platinum(IV) oxide in animals.
   For mechanism of action, see Section 16.4.1.3.

18.4.2 Human Experience

Data not available. (See also Section 16.4.2.)

18.4.2.1 General Information: NA

18.4.2.2 Clinical Cases: NA

18.4.2.3 Epidemiology Studies. Data not available. (See Section 16.4.2.3.)

18.5 Standards, Regulations, or Guidelines of Exposure

Data not available.

18.6 Studies of Environmental Impact

(See Section 16.6.)

19.0 Platinum(IV) Sulfate Tetrahydrate

19.0.1 CAS Number: [69102-79-0]

19.0.2 Synonyms: Platinum sulfate tetrahydrate

19.0.3 Trade Names: NA

19.0.4 Molecular Weight: 459.29

19.0.5 Molecular Formula: $Pt(SO_4)_2 \cdot 4H_2O$

19.1 Chemical and Physical Properties

This compound occurs in the form of yellow flakes (406). See Platinum entry in Table 41.35.)

19.2 Production and Use

Data not available.

19.3 Exposure Assessment

(See Section 16.3.)

19.4 Toxic Effects

19.4.1 Experiments

19.4.1.1 Acute Toxicity. The values of LD_{50} for the rat after oral and IP administration are given in Table 41.36, (see section 16.4).

19.4.1.2 Chronic and Subchronic Toxicity. $Pt(SO_4)_2 \cdot 4H_2O$ given to Sprague–Dawley rats per os in drinking water at ~750 mg/L (1.63 mM/L; 60 mg/rat) resulted in a significantly lower body weight gain during the first week of exposure, but did not produce significant alterations in the mass of the individual organs. Lower aniline hydroxylase activity was noted in the isolated liver microsomes (385, 395). (See also platinum entry in Table 41.37.)

19.4.1.3 Pharmacokinetics, Metabolism, and Mechanisms. See Section 16.4.1.3.

19.4.1.4 Reproductive and Developmental. Effects of platinum(IV) sulfate on the offspring development was studied in Swiss IRC mice following a single 200 mg/kg intragastric dose on day 7 or 12 of gestation and on day 2 after birth. After birth, the offspring was assigned in a crosswise manner to the exposed and nonexposed females. Rate of growth and gross activity of the neonates were assessed. On day 60–65 postpartum open-field behavior (ambulation and rearing), rotarod performance, and passive avoidance learning were investigated in the adult offspring. Reduced body weight of the offspring was the main effect of the exposure of the mothers to platinum(IV) sulfate on day 12 of gestation. The effect persisted until day 45 after birth. The route by which platinum(IV) sulfate was administered to the assigned mothers also significantly affected offspring mass on day 45 after birth. Irrespective of the exposure during the fetal period, the weight of the offspring fed by mothers exposed to platinum sulfate during pregnancy was smaller than that of the offspring fed by the control mothers. The offspring fed by mothers exposed to platinum(IV) sulphate 2 days after birth showed smaller activity than that fed by the control mothers (553).

19.4.1.5 Carcinogenesis. Data not available.

19.4.1.6. Genetic and Related Cellular Effects Studies. $Pt(SO_4)_2$ induced dose-dependent (0–150 µmol/L) increase in spontaneous revertant frequency in CHO-AUXB1 cells constituting the FPGS (folylpolyglutamate synthetase) gene locus after 20–22 h of exposure (497). At 550 µgM, it produced mutations in the cultured CHO-S cell line after exposure continued for 4 months (554).

19.4.1.7 Other: Neurological, Pulmonary, Skin Sensitization

19.4.1.7.1 Sensitizing Activity. Allergic reaction was not observed when platinum(IV) sulfate was repeatedly injected (IV or SC) in the form of 0.05–0.35 mg/mL solutions to albino rabbits, guinea pigs, and white mice, or when platinum(IV) sulfate was applied to the skin (472).

19.4.2 Human Experience

Data not available. (See also Section 16.4.2.)

19.4.2.1 General Information: NA

19.4.2.2 Clinical Cases: NA

19.4.2.3 Epidemiology Studies. Data not available. (See also Section 16.4.2.3.).

19.5 Standards, Regulations, or Guidelines of Exposure

Table 41.41 specifies the ACGIH, OSHA, and NIOSH hygienic standards for soluble platinum salts, and those currently in force in some other countries (see: Section 16.5).

19.6 Studies on Environmental Impact

(See Section 16.6.)

20.0 cis-Diammine(1,1-Cyclobutanedicarboxylato)Platinum(II)

20.0.1 CAS Number: [41575-94-4]

20.0.2 Synonyms: Carboplatin; 1,1-cyclobutanedicarboxylatodiammine platinum(II); *cis*-(1,1-cyclobutanedicarboxylato)diammineplatinum(II); diamine(1,1-cyclobutanedicarboxylato)platinum(II); (SP-4-2)-Diammine[1,1-cyclobutanedicarboxylato(2-)-*O · O*]platinum; CBDCA; JM-8; NSC-241240; platinum, diammine(1,1-cyclobutanedicarboxylato(2-)-*O,O'*)-,(SP-4-2); Paraplatin

20.0.3 Trade Names: NA

20.0.4 Molecular Weight: 371.25

20.0.5 Molecular Formula: $C_6H_{12}N_2O_4Pt$

20.0.6 Molecular Structure:

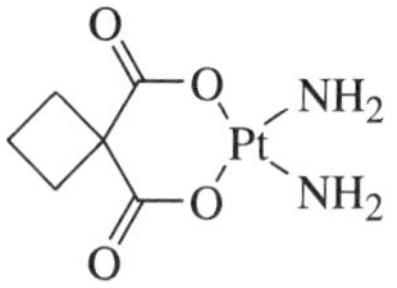

20.1 Chemical and Physical Properties

20.1.1 General

White crystals, soluble in water. Carboplatin has high stability in infusion fluids in the absence of chloride. Incompatible with aluminium. Other data have not been found (2, 447, 448).

20.2 Production and Use

Data on carboplatin production have not been found.

Carboplatin is used in chemotherapy to treat cancer, and more particularly to treat cancer of ovary, embryonal carcinoma of the testis, microcellular carcinoma of the lung, neuroblastoma, and squamous-cell carcinomas of the head and neck (447–449).

20.3 Exposure Assessment

See *cis*-diaminodichloroplatinum(II) discussion in Section 21.3.

The following methods are used to analyze carboplatin as platinum(II) in biological material: high pressure liquid chromatorgraphy (HPLC) (555), HPLC with differentaial pulse polarographic detection (556), and atomic absorption spectrometry (AAS) (557).

20.4 Toxic Effects

20.4.1 Experimental Studies

20.4.1.1 Acute Toxicity. LD$_{50}$ values after IP and IV administration of carboplatin in male (C57BL/RijXCBA/Rij) F$_1$ hybrid mice were 150.0 and 140 mg/kg bw, respectively. The LD$_{50}$ value after IV administration of carboplatin in Spraque–Dawley and WAg/Rij rats of both sexes was 85 mg/kg bw (558).

20.4.1.2 Chronic and Subchronic Toxicity. Carboplatin administered 4 times within 2 weeks at a maximum dose of 152 mg/kg bw in rats caused nephrotoxic lesions revealed by histopathological tests; however, functional tests did not reveal increased serum BUN or creatinine levels (558).

Blood serum urea nitrogen and creatinine concentrations remained within normal values in three dogs treated with carboplatin. Carboplatin did not cause any vomiting when administered at 12 mg/kg bw (288 mg/m^2) in three IV bolus injections at intervals of 3 weeks. Serum concentrations of electrolyte, liver function enzyme values remained within normal limits. Carboplatin caused a slightly reactive bone marrow, but the number and size of megakaryocytes were normal (559).

Dose-limiting neutropenia and thrombocytopenia were significant in cats given carboplatin at 200 or 250 mg/m^2. Weight loss, changes in appetite, and evidence of respiratory difficulty, as well as vomiting, diarrhea, or lethargy were not observed at any time during the 28-day period. At the highest dosage (250 mg/m^2), the neutrophil nadir (560 ± 303 neutrophils/μL) was observed on day 17 and the platelet count nadir (96.500 ± 11.815 platelets/μL) was observed on day 14 after carboplatin administration. Carboplatin appears to be safe when given IV in a single bolus at a dosage of 200 mg/m^2 to clinically normal cats (560).

Fourteen dogs with histologically confirmed transitional-cell carcinoma of the urinary bladder were treated with 300 mg/m^2 carboplatin every 3 weeks. The dogs were monitored for hematological toxicity with a complete blood counts (CBC) and platelet count performed immediately before and 10–14 days after carboplatin treatment. Toxicity included thrombocytopenia with or without neutropenia in seven dogs and gastrointestinal toxicity in six dogs (561).

A number of studies on the side effects of carboplatin in experimental animals have revealed that carboplatin shows a moderate nephrotoxic (562–564) and neurotoxic (565, 566) and ototoxic (567–571) activity, although its myelotoxic activity is strong (572, 573). Usually, the thrombocytopenia is more intense than the leutopenia (560, 561) [see discussion of *cis*-diamine(1,1-cyclobutanedicarboxylato)platinum(II) in Section 20.4.1.7].

20.4.1.3 Pharmacokinetics, Metabolism, and Mechanisms. Plasma concentrations of total platinum after carboplatin injection decayed biphasically with a rapid initial phase and prolonged second phase in rats that were injected IV with carboplatin 100 mg/kg and were killed at various times within 7 days after the injection. The initial and second half-life for carboplatin was 39 min and 9.2 h, respectively. Platinum in the whole plasma was detectable for up to 24 h after carboplatin injection. Approximately 90% of total platinum was ultrafilterable for the first 30 min following carboplatin injection. The proportion of ultrafilterable platinum in total platinum after carboplatin injection decreased slowly and accounted for 53% at 8 h. Ultrafilterable platinum was detectable for 8 h after carboplatin injection. Platinum in the kidney was detectable 5 days after carboplatin injection, and platinum levels in the cortex were almost the same as in the medulla (574).

Half-lives for the protein binding of carboplatin in plasma from dogs 39 h (575) and mice 45 h (576) were comparable with the half-life in human plasma.

The major antineoplastic mechanism of action for carboplatin is the production of crosslinks within and between DNA strands. As with cisplatin, carboplatin mechanism of action is thought to be related to platinum-DNA adduct formation. Because of the lesser reactivity of carboplatin, a larger dose is required to obtain binding to DNA equal to that of cisplatin. The difunctional DNA adducts of both compounds are similar (*i.e.* after reacting with $2Cl^-$ groups or the whole dicarboxylate group of carboplatin) (481, 484).

20.4.1.4 Reproductive and Developmental. In a series of reproductive and developmental studies in male and female rats, animals were given carboplatin up to 4 mg/kg per day (577–579). This treatment produced suppression of body weight in the adults and other general signs of toxicity but did not appear to impair fertility. Fetal mortality was increased, and there were decreases in intrauterine growth and skeletal ossification, consistent with general toxicity, but no increase in birth defects. Early behavior and reproductive capability of the F_1 progeny (which had been exposed in utero) were also unaffected. In a subsequent study of the same group (580) the dose was extended upward to 6 mg/kg per day with findings of an increase in congenital anomalies, including gastroschisis, ventriculomegaly, and skeletal anomalies. Exposure during day 6 of pregnancy appeared necessary for the induction of malformations.

The carboplatin was toxic to rat Leydig cells in culture; however, carboplatin appears to be less toxic than cisplatin (580). The drug-induced suppression of testosterone production by these cultured cells could be reversed by the addition of human chorionic gonadotropin to the cultures (581). The testicular toxicity of carboplatin has also been studied in mouse and rat experiments (582,583). On the basis of data collected from *in vivo* studies, one group of investigators concluded that carboplatin impairs spermatogenesis by damaging spermatogonia and Sertoli cells rather than selectively impairing Leydig cell function (583).

Kai et al. (577) gave carboplatin IV to rats on days 7–17 of gestation. At the highest dose (4 mg/kg bw,) maternal toxicity and lowered fetal weight occurred. Postnatal studies of behavior indicated no adverse affects.

20.4.1.5 Carcinogenesis. Data regarding the carcinogenicity of carboplatin are unavailable.

20.4.1.6 Genetic and Related Cellular Effects Studies. A limited amount of mutagenicity data suggest that carboplatin has mutagenic consequences similar to those of cisplatin (584). Carboplatin caused a significantly elevated genotoxicity in bacterial SOS chromatest with *Escherichia coli* (551, 585).

In Chinese hamster ovary (CHO) cells carboplatin showed a significant increase in chromosome aberrations at the doses assayed. The results suggest that carboplatin is a DNA-damaging drug with similar behavior as an alkylating agent (586).

The kinetics of bifunctional cisplatin and carboplatin adducts formation were studied with DNA *in vitro* and in cultured Chinese hamster ovary (CHO) cells. The data indicate that after 12 h postincubation, when all bifunctional adducts are formed, the total amount of the various bifunctional adducts after cisplatin treatment (37.5 ± 4.5 fmol/μg DNA) was in the same range as that after carboplatin (32.8 ± 6.3 fmol/μg DNA) (587).

The formation and persistence of platinum-DNA adducts were studied with immuno(cyto)chemical methods in male and female Sprague–Dawley rats treated with a single IP dose (80 mg/kg) of carboplatin. The levels of the various DNA adducts were measured in liver, kidney spleen, testis, and combined ovary/uterus in samples collected at 8 and 48 h after carboplatin administration. At both 8 and 48 h the highest platination levels were observed in the kidney. At 8 h after administration of carboplatin, the relative occurrence of the bifunctional adducts Pt-GG (34%), Pt-AG (27%), and G-Pt-G (32%) was similar in all tissues (588).

Carboplatin treatment (75–300 mg/kg bw) resulted in a dose-dependent increase in the frequency of micronuclei in Erlich ascites (EAT) tumor cells in mice. Carboplatin treatment also decreased the mitotic index. The maximal effect of the drug was obtained in all the treatments after 48 h (589).

The dose–response correlation and time course of micronucleated polychromatic erythrocytes (mPCE) in cat peripheral blood induced by various doses (150–250 mg/m^2) of carboplatin *in vivo* were investigated by Hahn et al. (560). The data indicate that carboplatin produced a significant ($p > .05$) dose-dependent increase in number of mPCE over baseline values: however, the times following carboplatin administered when mPCE were first observed differed significantly ($p > .05$) between the three carboplatin dose group (150 mg/m^2, 200 or 250 mg/m^2). The peak number of mPCE occurred on days 7, 14, and 17.5 following administration of carboplatin dose of 150, 200, and 250 mg/m^2, respectively.

Carboplatin induced chromosomal damage in two types of nonneuronal cells: fibroblasts and Schwann cells migrating from rat fetal dorsal root ganglia (DRG) in explant cultures. Dose–response curves for micronuclei in fibroblasts revealed normal distribution with the maximum at 100, 25, and 5 ηmol/L after 24, 48, and 72 h of carboplatin treatment, respectively. The maximum number of micronuclei in Schwann cells was obtained at 25 and 12.5 μmol/L at the same exposure time. Micronucleation of fibroblasts represented 293, 382, and 376% of control values and in Schwann cells 366, 819, and 1667% respectively (590).

Carboplatin induced formation of micronuclei (MN) in SCC VII murine cells. MN frequency was almost dose-dependent at lower concentrations, but at the highest concentration the micronucleus frequency was rather lower (591).

20.4.1.7 Other: Neurological, Pulmonary, Skin Sensitization

20.4.1.7.1 Nephrotoxicity. Male Wistar rats were administered single IV doses of 65 mg/kg carboplatin and observed for 4 days. Blood urea nitrogen (BUN), creatinine, glucose, and fractional electrolytes excretions were not significantly altered in carboplatin-treated animals. Carboplatin increased GGT (-glutamyltranspeptidase) excretion (approximately twofold). No functional changes as a consequence of direct nephrotoxicity were seen following carboplatin treatment. No renal lesions were detected by light or electron microscopy in the carboplatin-treated rats (563).

Intravenous carboplatin at 20 mg/kg in rats showed no significant change in effective renal plasma flow (ERPF), GFR, serum creatinine, or BUN, although the ERPF and GFR were slightly decreased 3 and 5 days posttreatment. No patological changes were observed (592).

The differential toxicity of cisplatin and carboplatin investigated in renal cortical slices over a 24 h period indicate that platinum compounds negatively affect cell function and viability. The concentration of slice-associated platinum following treatment with platinum-containing compounds increased with time and concentration. Inhibition of protein synthesis and loss of intracellular potassium correlated with increased total cellular platinum (564).

20.4.1.7.2 Neurotoxicity. Two different schedules of carboplatin administration (10 mg/kg and 15 mg/kg IP twice a week, 9 times) were evaluated in Wistar rats. Neurotoxicity was assessed for behavioral (tail-flick test), neurophysiological (nerve conduction velocity in the tail nerve), morphological, morphometrical, and analytical effects. Carboplatin administration induces dose-dependent peripheral neurotoxicity. Pain perception and nerve conduction velocity in the tail were significantly impaired, particularly after the high dose treatment. The dorsal root ganglia sensory neurons and, to a lesser extent, satellite cells showed the same changes as those induced by cisplatin, mainly affecting the nucleus and nucleolus of ganglionic sensory neurons. Moreover, significant amounts of platinum were detected in the dorsal root ganglia and kidney after carboplatin treatment. Carboplatin was neurotoxic, and the type of pathological changes it induces are so closely similar to those caused by cisplatin that neurotoxicity is probably induced by the same mechanism (566).

20.4.1.7.3 Ototoxicity. Studies in many animal species have indicated that carboplatin ototoxicity is expressed as damage to cochlear outer hair cells. The lesion is similar to, although less severe than, that resulting from cisplatin intoxication (593, 594).

Mount et al. (567) have demonstrated a species-specific ototoxicity of carboplatin in the chinchilla cochlea. The investigations of the vestibule by light and scanning electronmicroscopy indicate that sensory cell cilia became exfoliated or deformed in the crista, utricle, and in one instance in the saccule. In general, the pattern of damage is similar to that caused by other known ototoxic agents, including cisplatin.

Carboplatin induced ototoxicity in a guinea pig model, as determined by electrophysiological measurements and analysis of inner-ear outer hair-cell numbers. Delayed administration of sodium thiosulfate may provide a mechanism to reduce cochlear toxicity caused by carboplatin (568).

Carboplatin can damage the hair cells in the vestibular system; however, little is known about the time course of its vestibulotoxic effects. Ding et al. (571) examined the acute vestibulotoxic effects of carboplatin (50 mg/kg) in the chinchilla. The duration of the nystagmus response evoked by cold caloric stimulation was significantly reduced 6 hr following carboplatin treatment and showed a maximum, permanent reduction of approximately 50% by 24 h after injection. By 3 days after injection many type-I hair cells were filled with large vacuoles which often caused severe distortion of the plasma membrane. The results indicate that the vestibulotoxic effects of carboplatin occur quite rapidly and cause significant disruption of the mitochondria in hair cells and their afferent terminals (571).

20.4.1.7.4 Hematologic Effects. In rats a maximal tolerated dose of carboplatin (60 mg/kg i.v.) caused severe anemia, leucopenia and thrombocytopenia. These indices of hematological toxicity were also observed with a maximal tolerated dose of cisplatin (6.5 mg/kg i.v.), but reductions in blood cell counts were less than those observed with carboplatin, since red cell transfusion afforded protection to rats receiving a lethal dose of this compound (80 mg/kg i.v.). Anemia did not appear to be due to an increase in the susceptibility of carboplatin-exposed red cells to lysis, as concluded from results of osmotic fragility tests (573).

20.4.2 Human Experience

20.4.2.1 General Information. Carboplatin is used for chemotherapy and it may show some toxic side effects at the doses used in human treatment. Carboplatin is relatively well tolerated by the patients: vomiting or other gastric disturbances are less frequent than with cisplatin (595, 596).

Myelotoxicity constitutes the main side effect of carboplatin-based drugs. Myelosupression caused by carboplatin is dose-related and reversible and is usually characterized by thrombocytopenia, leukopenia and/or neutropenia. Platelets are usually more affected than leukocytes or erythrocytes (595–597). The risk of severe myelosuppression is increased in patients who have previously received cisplatin and/or radiation therapy and other cytostatics (598–600).

Compared to cisplatin, carboplatin shows a weaker nephrotoxic activity (601–603), which is more pronounced in patients with coexistent renal insufficiency (604). Decreased creatinine clearance (605) and glomerular filtration rate (GFR) (603) occurred in patients given carboplatin. It has been suggested that renal toxicity may be more likely at cumulative carboplatin doses of about 800 mg/m^2 or more (606).

Acute interstitial nephritis and renal failure were reported in two patients receiving intraperitoneal carboplatin as chemotherapy for advanced ovarian carcinoma (607).

The neurotoxic activity of carboplatin is weaker than that of cisplatin. Peripheral neuropathy occurs only in 6% patients receiving carboplatin (608). The incidence of neuropathy increases in patients previously treated with cisplatin. Mild paresthesia seems to be the most common manifestation of carboplatin neurotoxicity (595, 609, 610).

Clinical ototoxicity occurred in only 5 patients (19%) of 27 treated with 300 to 400 mg/m^2 total doses of carboplatin. Hearing loss tended to be cumulative with increasing dose and occurred most frequently in the higher frequencies (8000 Hz) (611).

Severe headaches, with concomitant bilateral vision loss, were associated with high-dose (3000 mg/m^2) carboplatin administration (595).

Fatal hepatotoxicity with transient elevation of liver enzymes has been reported (597) Alopecia, skin rash a flulike syndromic and local effects at the injection site have also been reported (608, 612).

Facial edema and painful swelling of hands and feet occured in patients with germ cell tumours which were treated with carboplatin (1500–2000 mg/m^2). No ulcerations or bullae formation were seen and changes resolved spontaneously (612).

Critical blindness developed in patients with impaired renal function receiving high-dose carboplatin (613).

Hypersensitivity to carboplatin has been reported in 2% of patients; reactions are similar to those seen with cisplatin and include rash, urticaria, erythema, pruritus, bronchospasm and hypotension (614, 615).

20.4.2.2.3 Pharmacokinetics, Metabolism and Mechanisms. Carboplatin is less reactive than cisplatin, and the drug is not bound to plasma protein to a significant extent. Protein binding is limited (616, 617). Carboplatin is usually administered intravenously and its pharmacokinetics are linear up to a dose of 2400 mg/m^2 (616). Following intravenous injection carboplatin exhibits a biphasic elimination and is excreted primarily in the urine. Approximately 77% of the administered carboplatin dose as measured by total platinum is excreted in the urine over the first 24 hours. Platinum slowly becomes protein bound and is subsequently excreted with a half-life of 5.4 days. Intraperitoneal administration has been used in cases of residual ovarian cancer. As a result of its higher hydrophilicity and higher molecular weight, carboplatin is cleared more slowly from the peritoneal cavity than cisplatin (6 and 15 mL/min, respectively) (616).

The major antineoplastic mechanism of action for carboplatin is the production of crosslinks within and between strands of DNA. *In vitro* and *in vivo* studies indicate that the differences in cytotoxicity between cisplatin and carboplatin may be related to the kinetics of their interaction with DNA (481) [see also *cis*-Diammine(1,1-cyclobutanedicarboxylato)platinum(II) discussion in Section 20.4.1.3.).

20.4.2.2.4 Reproductive and Developmental. Henderson et al. (618) report the antepartum use of cisplatin, followed by carboplatin, for an ovarian serous cystadenocarcinoma. During this treatment, serial sonographic assessments of fetal morphometric parameters and biophysical profiles with fetal heart rate monitoring were performed to document fetal well being. Examinations revealed that fetal exposure to cisplatin between 20 and 30 weeks of gestation and to carboplatin during gestation weeks 31–36 had no adverse effects on fetal development (618).

The result of studies by Horwich et al. (619) have revealed that carboplatin has activity equivalent to cisplatin in germ-cell tumors of the testis and is less toxic (618).

20.4.2.2.5 Carcinogenesis. Data regarding the carcinogenicity of carboplatin are unavailable.

20.4.2.2.6 Genetic and Related Cellular Effects Studies. Carboplatin caused a significantly elevated genotoxicity in the MNT cytokinesis–block micronucleus test.

Carboplatin caused a significant increase in the number of micronuclei at minimum doses of 0.5 μM (551).

In human lymphocytes carboplatin produced a sevenfold increase in the frequency of sister chromatid exchange (SCE) *in vitro* and a threefold increase in the number of cells with structural abnormalities compared with a control. Likewise, at this highest dose a significant increase was induced in the value of micronuclei (MN), and an important delay in the lymphocyte cycle progression was observed (586) [see also discussion of *cis*-diaminodichloroplatinum(II) in Section 21.4.2.2.6.].

20.4.2.3 Epidemiology Studies. Epidemiologic data on the harmful effects of carboplatin alone in the occupationally exposed people are not available [see also *cis*-diaminodichloroplatinum(II) discussion in, Section 21.4.2.3]

20.5 Standards, Regulations, or Guidelines of Exposure

Table 41.41 specifies the ACGIH, OSHA, and NIOSH hygienic standards for soluble platinum salts and the current corresponding values adopted in selected countries (see Section 16.5).

20.6 Studies on Environmental Impact

Data have not been found. [See also *cis*-diaminodichloroplatinum(II) discussion in Section 21.4.2.3].

21.0 *cis*-Diamminedichloroplatinum(II)

21.0.1 CAS Number: [15663-27-1]

21.0.2 Synonyms: Cisplatin; CDDP; cisplatinum; *cis*-platinum; DDP; *cis*-DDP; NSC 119875; Peyrone's salt, Peyrone's chloride; platinum diamminochloride cisplatyl; NSC-119875; platinol; (SP-4-2)-diamminodichloroplatinum; *cis*-platinous diamine dichloroplatin; CACP; CPDD; *cis*-platinous diamine dichloride; dCDP; *cis*-Pt(II); *cis*-diamminedichloroplatinum; DDPt; platiblastin; *cis*-dichlorodiamineplatinum(II); *cis*-diamminodichloroplatinum(II); *cis*-platinum(II) diamine dichloride; cisplatyl; CPDC; *cis*-ddp; neoplatin; platinex; PT-01; diamineedichloroplatinum; *cis*-dichlorodiaminoplatinum(II); *cis*-dichlorodiamineplatinum; *cis*-platinous diaminodichloride; 2'-deoxycytidine diphosphate; *cis*-diammine dichloroplatinum(II); *cis*-dichlorodiammine platinum(II); cisplatin [*cis*-diamminedichloroplatinum(II)]

21.0.3 Trade Names: NA

21.0.4 Molecular Weight: 300.5

21.0.5 Molecular Formula: $Cl_2H_6N_2Pt$; also $[NH_3]_2PtCl_2$

21.0.6 Molecular Structure:
$$\text{Cl}-\underset{\underset{\text{Cl}}{|}}{\overset{\overset{\text{NH}_2}{|}}{\text{Pt}}}-\text{NH}_2$$

21.1 Chemical and Physical Properties

21.1.1 General

Cisplatin [*cis*-diamminodichloroplatinum(II)] occurs in the form of yellow powder or orange-yellow crystals. It slowly changes to the trans form in aqueous solutions. When heated, it decomposes and releases toxic fumes of chlorine and nitrogen oxides. It quickly decomposes in the presence of bisulfite or metabisulfite. Cisplatin is commercially available as freeze-dried powder and in the form of aqueous solutions for injection.

Melting point: 270°C (decomposes)
Density: 3.738 g/cm^3
Solubility: slightly soluble in water (0.253 g/100 g at 25°C); insoluble in the majority of common solvents with the exception of *N,N*-dimethyl formamide and DMSO (2, 447, 448).

21.2 Production and Use

Cisplatin is obtained by the method described by Kauffman and Cowan (620), in which potassium(II) tetrachloroplatinate is treated with buffered aqueous ammonia solution. Pure cisplatin is obtained by recrystallization from dilute hydrochloric acid (489).

Cisplatin is a cytostatic agent and it is used to treat various cancer types, including cancer of ovary, testis, lung, head, neck, bladder, neuroblastoma, and nephroblastoma, and Hodgkin's disease and non-Hodgkin lymphoma (447–449).

21.3 Exposure Assessment

21.3.1 Air

The NIOSH-recommended methods for total air platinum determination are methods 7300 and S191 (see also, Section 16.3.1).

In samples of biological material, such as tissues and systemic fluids, cisplatin is determined as total concentration of platinum(II) by atomic absorption spectrometry (AAS) and induction plasma coupled–atomic emission spectrometry (IPC-AES). NIOSH recommends method 8005 (see Section 16.3.1) for the determination of total platinum in blood and tissues.

In addition the following methods are used to analyze cisplatin as platinum(II) in the biological material: high-pressure liquid chromatography (HPLC) (621); HPLC with a glassy carbon based wall-jet amperometric detection (622); HPLC with AAS detection (623), differential-pulse polarography (457), and differential-pulse voltametry (DPV) (624).

21.4 Toxic Effects

21.4.1 Experimental Studies

21.4.1.1 Acute Toxicity. Table 41.42 gives LD$_{50}$ values for cisplatin in experimental animals. (558, 625–629). A single IV injection of 2.5 mg/kg bw or 5 consecutive daily

Table 41.42. Acute Toxicity of Cisplatin

Species	Exposure Route	LD$_{50}$ (mg/kg bw)	Ref.
BALB/C female mice	IP	13	625
C57 BL/RijXCBA/Rij/F$_1$ hybrid mice	IP	14.2	558
C57 BL/RijXCBA/Rij/F$_1$ hybrid mice	IV	13.2	558
Swiss male mice	IV	12.32	626
Swiss female mice	IV	13.36	626
Spraque–Dawley rats	IP	12	627
F344 rats	IP	7.7	628
Spraque–Dawley rats of both sexes	IV	5	558
Guinea pigs	IP	9.7	629

injections 0.75 mg/kg bw each was the minimal lethal dose for dogs. Heavy morbidity or death within 5–17 days were observed in the exposed dogs; the poisoned animals had anorexia, vomiting, abdominal tenderness, diarrhea, dehydration, and body weight loss. The toxic symptoms included hemorrhagic enteritis, serious lesion of bone marrow and lymphatic tissue, necrosis of renal tubules, and pancreatitis (626).

Five daily IV injections of 2.5 mg/kg bw constituted the minimum lethal dose for the monkeys. Symptoms of poisoning in the monkeys were similar. Pathologic examinations revealed nephritis, enteritis, bone marrow hypocellularity, lymphatic tissue hypocellularity, myocarditis, and spermatogenic cell degeneration. Grave renal damage — manifested, among other things, by azotemia, hypochloremia and proteinuria — was the most serious symptom of cisplatin poisoning. The monkeys recovered within 55–124 days (626).

A single IP LD$_{50}$ 7.7-mg/kg bw dose (range 0.5–12 mg/kg bw) of cisplatin given to male F344 rats in physiological saline solution caused the animals to die within 2–7 days. Nephrotoxic activity was detected, manifested by increased blood urea nitrogen (BUN) and creatinine levels on days 4–5 and acute degenerative and necrotic lesions in the renal tubules. Diarrhea appeared on day 3 after poisoning. In addition, small-intestine inflammation, cecitis, colitis, and dehydration of bone marrow and thymus cells were detected (628).

Leukopenia with reduced neutrophil, lymphocyte circulating platelet counts, intestinal epithelium lesions, and renal tubule epithelium sloughing were observed in Spraque–Dawley rats that received a single IV dose LD$_{50}$ 12.0 mg/kg bw of cisplatin. The symptoms were strongest on days 2–4 after the injection (627).

An evident hearing loss was observed in guinea pigs exposed to a single IP dose of cisplatin, 9.7 mg/kg (629).

21.4.1.2 Chronic and Subchronic Toxicity. A study of the relationship between cisplatin doses and the time when organ lesions became evident was conducted on 18 guinea pigs injected IP with cisplatin at 2 mg/kg per day, 5 days/week until they died. Disturbed iron metabolism and hearing organ damage were observed in the animals receiving cisplatin at 10.2 and 10.8 mg/animal, respectively; salivary gland lesions were observed for the 15-mg/animal dose; renal lesions were observed at 19.5 mg/animal, whereas cisplatin at

19.7–33.6 mg/animal produced bone marrow damage. Minimal hepatic dysfunction was observed at 21.3 mg/animal. The results of the study show that the onset of the organ lesions in guinea pigs is dose-related and can be arranged in the following sequence: iron metabolism, hearing organ, salivary glands, kidneys, bone marrow, and liver (630).

A number of studies revealed that cisplatin in the chronic exposure causes serious toxic effects in the kidneys, nervous system, GI tract, and bone marrow of mice, rats, guinea pigs, dogs, and monkeys (see *cis-diaminedichloroplatinum (II)* discussion in Section 21.4.1.7).

21.4.1.3 Pharmacokinetics, Metabolism, and Mechanisms. Animal studies have revealed that cisplatin distribution in the individual organs and its serum levels and excretion are similar in different experimental animal species (mice, rats, dogs). Intravenously injected cisplatin was quickly distributed almost in all organs; highest levels were detected in the kidneys, liver, ovary, uterus, skin, and bone (Litterst, 1976). After 24 h, tissue/plasma drug ratios are greater than 1 in the other tissues; these persisted for at least a week in dogs and other species. Up to 4 weeks after a single dose, platinum was still detectable in kidneys, liver, skin and lungs (631).

The main route of cisplatin elimination from the system is excretion with urine through kidneys. About 50% of the dose was removed within 4 h, and 76% of the dose was removed within 48 h. As much as 90% of plasma platinum was bound to serum proteins within 2 h following IV injection of cisplatin (632).

Cisplatin appears to enter cells by diffusion. The chloride atoms may be displaced directly by reaction with nucleophils such as thiols; replacement of chloride by water yields a positively charged molecule and is probably responsible for formation of the activated species of the drug, which then reacts with nucleic acids and proteins. Cisplatin can react with DNA, forming both intrastrand and interstrand crosslinks (482, 483, 485). The N_7 of guanine is very reactive, and platinum crosslinks between adjacent guanine molecules on the same DNA strand: guanine–adenosine crosslinks are also readily formed. The formation of interstrand crosslinks is a slower process and occurs to a lesser extent. DNA adducts formed by cisplatin inhibit DNA replication and transcription and lead to breaks and miscoding. The covalent binding of protein to DNA has also been demonstrated. The cisplatin–DNA interaction is responsible for cisplatin anticarcinogenic activity. Selective killing of tumor cells is probably due to attack on the guanine- and cytosine-rich regions of DNA, producing damage that is repairable by normal cells (482, 483, 485).

21.4.1.4 Reproductive and Developmental. Cisplatin displays strong embryolethal and embryotoxic activity in experimental animals. Cisplatin induced atrophy of testis and prostate in dogs and monkeys (626), and in mice it caused spermatocyte and spermatoid necrosis (633).

At high doses (8 mg/kg) it had a damaging effect on the Sertoli cells in Spraque–Dawley rat testis (634).

Cisplatin shows strong embryotoxic activity in both rats and in mice when administered at doses normally used to treat human cancer. Teratogenic effects were observed at lower doses, but they were not statistically significant (635–641).

Cisplatin administered IP at 13 mg/kg bw to Swiss Webster mice at day 8 of gestation resulted in death of all fetuses of 10 dams, while the 8-mg/kg bw dose was lethal to 98% fetuses of 13 dams. After the dose of 3 mg/kg bw, 31% fetuses from 12 dams died. Examination of the surviving fetuses revealed growth retardation and developmental anomalies (635).

High embryotoxicity was noted in rats and mice receiving cisplatin at doses lower than those used in treatment of adult people. Embryal LD_{50} for Wistar rats was 2.88, 1.28, and 1.0 mg/kg bw on days 6, 8, and 11, respectively, following IP administration of cisplatin at 0.3; 1.0, 2.5, and 3.0 mg/kg bw on gestation days 6, 8, and 11. None of the doses administered on gestation day 14 produced a significant increase in the embryotoxicity. In the same experiment, embryal LD_{50} for Swiss Webster mice was 5.24 mg/kg. The mice received single IP injections of cisplatin at 0.3, 3.0, 6.0, 8.0, and 13.0 mg/kg on day 8 of gestation. An increase in the incidence of growth retardation or gross malformation was not discernible in the surviving fetuses with the number of dams used in this study. This embryolethality is gestation-stage-specific, and the highest mortality corresponds to the period of rapid DNA replication in early organogenesis (636).

A dose-dependent reduction in the number of fetuses in the litter, reduced fetal body weight, and marked retardation of skeletal ossification processes were found in mice receiving a single injection of cisplatin at 2.5, 5, 10, and 20 mg/kg bw on gestation days 8, 10, 12, 14, and 16. The most evident effects were noted when cisplatin was given during organogenesis, that is, on gestation days 8, 10, and 12, which points to a serious embryolethal and embryotoxic activity of cisplatin (638).

Studies on the effect of exposure of CD rat females to cisplatin given IP at 3 and 6 mg/kg bw during the preimplantation period (gestation day 3) on blastopathies and postimplantation embryotoxicity revealed reduced number of blastocyte cells after the 6-mg/kg bw dose, while increased numbers of micronuclei and micronucleated blastocytes were observed at both doses. The assessment of the postimplatation embryotoxicity was performed in the females that were sacrificed on gestation day 21 (640).

Teratogenic activity of cisplatin was tested on rats receiving cisplatin intravenously (0.375–6 mg/kg) each day during gestation days 5–16. The rats were sacrificed on gestation day 21. The results revealed high embryolethality among the fetuses of the dams that received cisplatin during the period from gestation days 6–9. Extreme malformations were observed in the fetuses of the dams exposed on days 5 and 8 of gestation. Malformed tails and limbs were found in the groups exposed on days 5 and 6. Cisplatin displayed teratogenic and embryolethal activity when administered prior to limbs and tail organogenesis (639).

21.4.1.5 Carcinogenesis. Although cisplatin shows anticarcinogenic activity, several tests on mice and rats demonstrated that cisplatin given IP several times to A/Jax mice increased the incidence of pulmonary adenoma in the exposed animals (up to 100% cases) (642).

Cisplatin administered IP caused a significant increase in the incidence of skin papillomas in female CD-1 mice that were additionally exposed to croton oil applied locally to the skin. The number of epidermoid carcinomas and both malignant and benign tumors in the internal organs was higher, but the increase was not statistically significant compared to controls (642).

Intraperitoneal administration of cisplatin to BD IX rats 3 times a week for 3 weeks at a dose of 1 mg/kg bw induced leukemia (643, 644).

As cisplatin is used to treat pregnant women for malignant ovarian and uterine tumors, studies on transplacental carcinogenic activity of cisplatin were performed in F344 rats and SENCAR mice (645, 646). The results revealed that a single IP injection of cisplatin to pregnant rats F344 on gestation day 18 produced cancer of kidney, liver, and nervous system in the offspring. The number of hepatocellular adenomas was significantly higher in the offspring transplacentally exposed to cisplatin (646).

Intraperitoneal administration to pregnant SENCAR mice cisplatin on gestation day 17, followed by local application of cisplatin to the offspring after week 4 of their life, caused development of skin papillomas. Although cisplatin alone (without TPA promotion) did not cause skin cancer, transplacental cisplatin administration induced the development of thymus lymphoma, lung cancer, and proliferative renal lesions in the offspring. The results show that cisplatin can initiate and/or induce preneoplastic and neoplastic lesions in the tissues of transplacentally exposed offspring (645). Table 41.43 presents the conditions and results of the experiment.

21.4.1.6 Genetic and Related Cellular Effects Studies.
Cisplatin shows a strong mutagenic and genotoxic activity (490).

Cisplatin induced reverse mutations in *Salmonella typhimurium* strains TA100 and TA98, and in *Escherichia coli* (647–650). It also induced mutations and gene conversion in *Saccharomyces cerevisiae* yeast cells (651) and mutations in *Neurospora* (652).

In *Drosophila melanogaster* it induced aneuploids, dominant lethal mutations, and sex-related recessive lethal mutations (552, 653).

In CHO or V79 cells of Chinese hamster, cisplatin induced a dose-related increase in 8-azaguanine-resistant and thioguanine-resistant mutations (495).

Increased frequency of sister chromatid exchanges and chromosome aberrations was observed *in vitro* in Chinese hamster CHO cells (654, 655).

During *in vivo* tests, cisplatin induced chromosome aberrations in bone marrow and spermatocytes of mice (633, 656).

Increased numbers of chromosome aberrations were observed in bone marrow cells of rats receiving cisplatin IP at 6.0 mg/kg, which were killed 6, 12, 18, 24, and 48 h after the injection. A significant increase was detected after 6–24 h; the majority were of the break and gap type (657).

A significant genotoxic activity was demonstrated in *E. coli* bacterial cells in the SOS chromotest (551,585,658).

Cisplatin induces primarily interstrand DNA–DNA crosslinks (485, 659).

Cisplatin induces also gene conversion (in meiotic stage cells) during the meiosis in spermatids of the transgenic mice (660).

In studies on the transplacental mutagenicity in SENCAR mice it has been demonstrated that cisplatin may cause mutations in the skin of the fetuses (661).

The DNA adducts were found to be present in kidneys, liver, lungs, and brain of the dams and of the fetuses of the rats exposed transplacentally on gestation day 18 to cisplatin at 5, 10, and 15 mg/kg bw, which caused cancer in the fetuses. These results confirm the

Table 41.43. Carcinogenic Effects of Cisplatin

Species, Sex, Number (N), Age	Administration	Dose	Conditions of experiment	Effects	Ref.
A/Jax female mice, N 10 or 20, 8 weeks	IP (in 0.85% NaCl)	3.25 mg/kg bw (total dose 108 μmol/kg bw)	10 injections 1 × week; observation 8 months	100% pulmonary adenomas; 14.2% adenomas/mouse	642
	IP (in 0.85% NaCl)	1.62 mg/kg bw (total dose 103 μmol/kg bw)	19 injections 1 × week; observation 8 months	100% pulmonary adenomas; 15.8%adenomas/mouse	
	IP (in trioctanoin)	3.25 mg/kg bw (total dose 108 μmol/kg bw)	10 injections 1 × week; observation 8 months	100% pulmonary adenomas; 10.4% adenomas/mouse	
	IP (in trioctanium)	1.62 mg/kg bw (total dose 54 μmol/kg bw)	10 injections 1 × week; observation 8 months	94% pulmonary adenomas; 5.4% adenomas/mouse	
	IP (in trioctanoin)	3.25 mg/kg bw (total dose 54 μmol/kg bw)	5 injections 1 × week; observation 8 months	100% pulmonary adenomas; 7.2% adenomas/mouse	
	IP (0.85% NaCl alone)	6.5 ml/kg bw	19 injections; observation 8 months	67%pulmonary adenomas; 15.8% adenomas/mouse	
	IP (trioctanoin alone)	5 ml/kg bw	10 injections; observation 8 months	26%pulmonary adenomas; 0.5% adenomas/mouse	
CD-1 female mice, N 40, 8 weeks	IP (in 5 mL 0.85% NaCl)	1.62 mg/kg bw	1 × week for 16 weeks; observation 52 weeks	After 41 weeks; no skin tumors; After 52 weeks, 1 epidermoid carcinoma in the external ear 2 thymic lymphomas 1 pulmonary adenoma 3 mammary adenocarcinomas 1 subcutaneous fibroliposarcoma	642

CD-1 female mice, N 40	IP (in 5 mL 0.85% NaCl and croton oil to the skin)	1.62 mg/kg bw	1 × week for 16 weeks and application 0.15 mL 0.6% croton oil to shaved portion of skin 2 ×/week (1 and 4 days after injections of cisplatin); observation 52 weeks	After 41 weeks; 15/30 skin papillomas; 3.2 papillomas/mouse; after 52 weeks: 3 epidermoid carcinomas 1 thymic lymphomas 1 pulmonary adenoma
	IP 0.85% NaCl alone and application of croton oil on the skin	—	Injections of NaCl and application of croton oil on the skin	1 pulmonary carcinoma 1 reticulum cell sarcoma of the spleen
BDIX rats, N 50	IP (in 0.9% NaCl)	1 mg/kg bw	3 × week for 3 weeks; observation 455 days	455 days after the first injection, 33 animals died; 13 of them developed malignancies, including 12 cases of leukemia and 1 renal fibrosarcoma	643
BDIX rats, N 25	IP (0.9% NaCl alone)	1 mg/kg bw	3 × week for 3 weeks; observation 455 days	No malignancies	
F344/NCr female rats	IP (in 2.5% NaCl)	5 mg/kg bw	Single dose to pregnant females on gestation day 18; observation until 79 weeks of pup life	Males: 2/19 (10.5%) renal cell adenomas Males and females: 9/40 (22.5%) hepatocellular adenomas 3/40 (7.5%) pulmonary tumors Nervous system tumors; 2/40 brain; (gliomas 5%) 1/40 peripheral nervous system (Schwannomas 2.5%)	646
	IP (in 2.5% NaCl alone)	2 mL/kg bw	Single dose to pregnant females on gestation day 18	Males: 1/20 (5%) pelvic transitional cell papillomas 1/36 (2.8%) hepatocellular adenoma 1/36 (2.8%) pulmonary tumor	

Table 41.43. (*Continued*)

Species, Sex, Number (N), Age	Administration	Dose	Conditions of experiment	Effects	Ref.
SENCAR female mice	IP (in 2.5% NaCl)	7.5 mg/kg bw	Single dose to pregnant females on gestation day 17; since 4 week of life, the pups received local dose of 2 mg TPA[a] 2 × week for 20 weeks; observation until 25 week of pup life	18/37 (48.7%) papillomas of skin	645
	IP (in 2.5% NaCl)	7.5 mg/kg bw	Single dose of cisplatin alone on gestation day 17; observation until 25 week of pup life	No skin tumors; the following were detected in the offspring: thymic lymphomas, lung tumors, proliferative kidney lesions	
	Local TPA application in acetone	—	Observation until 25 week of pup life	4/40 (10%) papillomas of skin	

[a]12-O-Tetradecanoylphorbol-13-acetate.

hypothesis that genotoxic mechanism play an important role in the induction by cisplatin of tumors in the offspring (662).

Intense formation of cisplatin–DNA adducts in the brain and liver mitochondria of the fetuses was found to occur in the offspring of rats transplacentally exposed to cisplatin. The results suggest that the mitochondrial DNA in some organs may be a particular target place for the genotoxic activity of cisplatin (663).

21.4.1.7 Other: Neurological, Pulmonary, Skin Sensitization

21.4.1.7.1 Nephrotoxicity. The nephrotoxic activity detected as a result of numerous biochemical investigations, functional tests, and morphological examinations is the main toxic effect of cisplatin (664–670).

Increased BUN level, reduced insulin concentration, increased K^+ elimination, and glycosuria were found in male Fischer 344 rats following single intravenous injection of cisplatin at 5, 10, or 15 mg/kg bw (664).

A fourfold increase in N-acetyl-β-D-glucosaminidase (NAG) elimination with urine occurred following single IP administration of cisplatin at 1.5 or 5 mg/kg bw to Wistar rats. The S_3 segments of the proximal tubule were the main location of the nephrotoxic effects (665).

Cisplatin produced significant chronic effects in the structure and function of kidneys following single IP administration of cisplatin at 6 mg/kg bw to Fischer 344 rats. Examinations performed 15 months after the exposure revealed increased platinum concentration in the renal cortex and increased number of atrophic or hyperplastic proximal renal tubules as compared to the control rats. In addition, papillary hyperplasia was visible after 15 months, and involved the epithelium lining the renal papilla (666, 667).

Changes in the activity of glutathione S-transferase (GST) isoenzymes were observed in rat kidneys following cisplatin exposure. The activity of the renal GST-α dropped to 33.4% compared with the control levels, and the GST-μ activity increased 1.9-fold after cisplatin administration. The levels of activity of total GST and GST-μ in urine correlated with BUN levels, which closely parallelled the course of nephrotoxicity after cisplatin administration (668).

In male (Swiss OF11) mice receiving subcutaneously cisplatin at 20 mg/kg bw, histopathological examinations of the kidneys revealed 10, 20, 40, and 50% proximal tubule damage after 7, 24, 48, and 72 h, respectively (670).

Fourfold BUN increase, threefold increase in glucose and fractional electrolyte excretions, and reduced creatinine clearance on day 4 after the poisoning were observed in male Wistar rats given a single IV injection of cisplatin at 6.5 mg/kg bw Besides, cisplatin produced the following increases of elimination with urine: sixfold for LDH (lactate dehydrogenase); twofold for NAG (N-acetyl-β-glucosaminidase); twofold for GGT (glutamyltranspeptidase). Cisplatin induced also changes resulting from direct nephrotoxicity and lowered: GFR (glomerular filtration rate) (84%), ERPF (estimated renal plasma flow) (97%), ERBF (estimated renal blood flow) (96%), and ERTS (estimated renal tubular secretion) (95%), whereas the FF (filtration fraction) was increased fivefold. All rats receiving cisplatin developed proximal tubular necrosis in the outer stripe of the outer medulla, extending multifocally into inner cortical medullary rays (563).

After *in situ* perfusion, cisplatin lowered (45%) Na$^+$ retention and increased (117%) Ca$^+$ retention in the kidneys of Spraque–Dawley rats, which points to a nephrotoxic activity of this drug (669).

Protein synthesis inhibition and intracellular potassium loss correlated with the increase in total cellular platinum content, indicating that platinum compounds adversely affect the function and viability of renal cortex cells (564).

21.4.1.7.2 Neurotoxicity. Neurotoxic effects of cisplatin in the experimental animals include peripheral neuropathy, retrobulbar neuritis, and neurosensory hearing loss. Cisplatin-induced neuropathy is mainly of the sensory type with early deterioration of the vibratory sensibility in toes, diminished propriocepsis, and loss of ankle jerks. Neurophysiological examinations revealed lower sensory but normal motor-nerve conduction velocities with occasional signs of denervation (671, 672).

Following IP administration of cisplatin to Wistar rats 2 times a week at 1 mg/kg bw up to a cumulative dose of 19.0 mg/kg bw, a reduction of H-reflex-related sensory-nerve conduction velocity was recorded. After the termination of the exposure, the H-reflex-related sensory-nerve conduction velocity slowly improved (671).

Similar effects, including slowing of the H-related sensory-nerve conduction velocity, were obtained when Wistar rats were exposed to cisplatin at 2 mg/kg bw 2 times a week. Administration of the neurotropic peptide ORG 27b6 canceled the neurotoxic effects of cisplatin (672, 673).

Administration of cisplatin alone to Wistar rats at 2 mg/kg per/week in nine cycles resulted in morphological changes, increased platinum concentration in the dorsal root ganglia (DRG), and reduced sensory-nerve conduction velocity in the tail nerve. Glutathione (GSH) given to the rat effectively reduced the neurotoxic effects produced by cisplatin (674).

Dorsal root ganglia neuropathy was found in Wistar rats as a result of chronic exposure (2 mg/kg twice per/week in nine injections) to cisplatin. The examinations confirmed that dorsal root ganglia cells were the primary target of cisplatin-induced neurotoxicity. Milder alterations could be detected in peripheral nerves (675).

21.4.1.7.3. Ototoxicity. Cisplatin induces sensorineural hearing loss with deafness for high frequency tones preceding that for low frequencies. Disappearance of the ear-twitch response correlated with histopathological lesions in the organ of Corti (629, 676–680).

Permanent deafness and histopathological lesions with pronounced loss of outer hair cells in the lower turns of the organ of Corti were detected in guinea pigs that had received 8–40 IP cisplatin injections at 1 mg/kg (5 doses/week) or 1.5 mg/kg (10–15 doses). Single doses of 6, 9, 12, or 18 mg/kg bw cisplatin produced permanent hearing loss as early as day 3 and a scattered pattern of outer hair-cell loss on day 4 with cytological changes similar to but severe than those observed for multiple doses (629).

Functional and morphological cochlear lesions were detected in guinea pigs that had received cisplatin twice at 7.5 mg/kg or 10 times at 1.5 mg/kg. Drastic and quick hearing loss occurred after second injection in the acute experiment, while in the chronic experiment the threshold shifts were observed 8 and 9 days after the poisoning. The morphological changes in the outer hair cells were more pronounced in the acute than in the chronic poisoning. Partial damage in the stria vascularis was observed only after acute treatment (677).

Investigations of the differences in the ototoxic activity of cisplatin and transplatin in guinea pigs exposed twice at 5-day intervals to 7.5 and 30 mg/kg, respectively, revealed severe losses of outer hair cells essentially at the basal and second turns after administration of cisplatin. Transplatin did not induce detectable functional or morphological changes (680).

Cisplatin given in a series of five injections induced widespread loss of outer hair cells along much of the cochlea in the hamster, especially in the vasal and middle turns, with an average survival of only 56% of the outer hair-cell population. In contrast, inner hair cells resisted cisplatin ototoxicity in the hamster (678).

21.4.2 Human Experience

21.4.2.1 General Information.
Cisplatin applied to humans at the therapeutic doses in chemotherapy of some tumors may produce certain chronic toxic effects primarily in kidneys, nervous system, GI tract, and bone marrow. Cisplatin is usually administered intravenously at 3- or 4-week intervals. It is usually injected at the single daily dose of 50–120 mg/m^2 for 5 consecutive days at 3-week intervals (447, 448).

Dysopsia, hypertension, convulsions, loss of consciousness, and hypomagnesemia were diagnosed in a 70-year-old patient treated with cisplatin for upper-limb osteosarcoma (681). Strong nausea, vomiting, alopecia, and GI disorders are observed in the majority of cisplatin-treated patients (609, 682, 683). Nausea and vomiting were noted in 72% of a group of 49 patients treated for endometrial carcinoma with cisplatin injected IV at 50 mg/m^2 every 3 weeks (682).

Cisplatin is strongly nephrotoxic, and the symptoms of nephrotoxicity are evident in a considerable majority of cisplatin-treated patients. The manifestations of nephrotoxicity include damaged nephrons (of distal tubules in particular), reduced renal flow and filtration, increased concentrations of urea and creatinine, reduced creatinine clearance, azotemia, hyperuricemia, and hypomagnesemia. Pathological lesions occur in segment S_3 of the proximal tubule and situated in the outer stripe of outer medulla (684–688). A significant increase in serum creatinine level, 78 ± 21 mmol/L and 88 mmol/L ± 3 before and after chemotherapy, respectively, (P < .005) and significantly lower glomerular filtration rate (GFR) (92 ± 5 mL/min) estimated renal plasma flow (ERPF) (362 ± 21 mL/min) compared with the control (110 ± 3 and 436 ± 24 mL/min), respectively, were detected in 35 patients treated with cisplatin at the cumulative dose of 603 ± 37 mg/m^2. The mean enzymuria and the renal size remained within the normal range (684).

The neurotoxic activity of cisplatin constitutes a serious side effect associated with cisplatin treatment. The neurotoxic effects include peripheral neuropathies, the risk of which is increased after a cumulative dose of 300 mg/m^2 is exceeded. Optical neuritis, papilloedema, cerebral blindness, seizures, and Raynod-like phenomena were observed in people treated with cisplatin for various tumors. The symptoms of the neuropathy are reversible, but they may persist for a year or longer. Neurologic disturbances refer primarily to sensory fibers in combination with vibratory sensation (683, 689–693). The symptoms of neuropathy occurred in 12 of 19 patients who received cisplatin at a cumulative dose of 360 mg/m². Decreased vibration sensation at the ankles and depressed ankle reflexes were the early manifestations (694). Symptoms of epileptic seizures occurred within 3 months in 8 (9.8%) patients treated for 4 years with cisplatin for germ cells tumors or ovarian carcinoma (689).

Ototoxic side effects of cisplatin found in patients treated with cisplatin are manifested by tinnitus, hearing loss in the high and speech frequency ranges and partial deafness or vestibular toxicity (695–698). Serious ototoxic disturbances occur in children treated with cisplatin for tumors. Serious ototoxic symptoms appeared in 46% patients (37 people were examined) who received cisplatin IV at 50 mg/m² every 3 or 4 weeks. Hearing loss in the speech frequency range was found in 14% patients (695). Examinations of temporal bones in cisplatin-treated patients revealed degenerative lesions in the outer hair cells in the lower turns of the cochlea and in the spiral ganglion and cochlear nerve (696).

Very heavy depression of bone marrow may appear following cisplatin administration at higher doses. Mild-to-moderate myleosuppression may occur with transient leukopenia, thrombocytopenia, and anemia. Nadirs in platelet and leucocyte counts occur between days 18 and 23, and most patients recover by day 39 (447, 682, 683). Myleosuppressia occurred in 9 of 20 examined patients receiving cisplatin at high doses of 200 mg/m² (mean treatment time was 44 weeks). Anemia was observed in 5 patients, and 6 developed thrombocytopenia (683). Mild leukopenia occurred in 31% patients as a side effect of cisplatin administered at 50 mg/m² to treat endometrial carcinoma (682).

Circulatory disorders and anaphylactic reactions were observed during cisplatin treatment. Allergic reactions were manifested by face swelling, dyspnea, tachycardia, lower arterial blood pressure, bronchospasm, hyperidrosis, pruritus, and skin erythema. The anaphylactic response occurred in a man treated IV with cisplatin for transitional-cell carcinoma of bladder (699). Two cisplatin-treated patients developed serious anaphylactic symptoms. The latter were preceded by mild allergic reactions that had been overlooked in previous treatment courses (700).

Heavy allergic exfoliative dermatitis occurred in patients treated with several doses of cisplatin for hand ischemia and necrosis (701).

A number of surveys and original works are also available in which the authors report the results of studies on the toxic effects of anticarcinogenic drugs in medical workers (including nurses, physicians, pharmacists, auxiliary personnel, chemists) occupationally exposed to the cytostatic drugs (702–706). Many of the reports point to the possible respiratory and dermal adsorption of the cytostatic drugs during preparation of the injectied solutions. It is a common practice to inject drug mixtures and, as it is not possible to distinguish between the effects of the individual drugs, only combined effects can be assessed.

A study by Sotaniemi et al. (707) revealed that the exposure associated with the manipulation of cytostatic drugs containing, among other things, cisplatin may cause hepatic damage and, after some time, lead to irreversible fibrosis, whereas the intensity and duration of the exposure may affect the extent of the hepatic lesions. Three cases of hepatic lesion were detected in nurses exposed to cytostatic drugs for 6, 8, and 16 years. The nurses had elevated levels of serum alanine transferase (ALAT) and alkaline phosphatase, and hepatitis with partial necrosis. The hepatic damage was found in those nurses who handled the cytostatic drugs twice as often as the other nurses (707).

Complaints about pruritus, nasal mucosa ulcer, and nausea were noted among hospital personnel exposed to the cytostatic drugs (708).

21.4.2.2.1 Acute Toxicity: NA

21.4.2.2.2 Chronic and Subchronic Toxicity: NA

21.4.2.2.3. Pharmacokinetics, Metabolism, and Mechanisms. Cisplatin injected IV to people has a two-phase half-life; the initial one is 25–49 min, and the final one is 58–74 h. High cisplatin concentrations were found in the kidneys, liver, intestines, and testis. Cisplatin permeates with difficulty to the cerebrospinal fluid. Over 90% of blood platinum is covalently associated with plasma protein. Plasma levels of sequential courses of cisplatin increased, indicating accumulation. Cisplatin is eliminated mainly with urine. Only a small portion of cisplatin is eliminated by kidneys during the first 6 h. As much as 25% is eliminated during 24 h, and up to 43% of the received dose is recovered in urine within 5 days. In the tissues, platinum may be detected several months after the administration. When given by infusion (instead of rapid injection), the plasma half-life is shorter and the quantity of removed cisplatin is greater. The excretion with feces is supposed to be insignificant, although cisplatin was detected in the gallbladder. Platinum was also found to be present at 0.9 mg/mL in the milk of women subjected to chemotherapy (447, 709–712).

[For mechanisms, see discussion of *cis*-diaminedichloroplatinum(II) in Section 21.4.1.3.]

21.4.2.2.4 Reproductive and Developmental. The majority of data on the incidence of reproductive disorders, embryotoxic and teratogenic effects associated with exposure to cytostatic drugs, have been obtained from experiments on animals [see *cis*-diaminedichloroplatinum(II) discussion in Section 21.4.1.4] and from studies on the side effects of those drugs in patients subjected to chemotherapy (702, 713–715) or from examinations of medical personnel who might be occupationally exposed to the cytostatic drugs (716–718).

An increase in the incidence of spontaneous abortions and congenital defects was noted in women who were treated with cytostatic drugs for cancer during embryogenesis; there was no such risk when the chemotherapy was applied after the third trimester of pregnancy. Among 13 women treated with cytostatic drugs containing cisplatin during the first trimester of pregnancy, lower birth weight compared to a selected control was noted in two of five mothers who continued the pregnancy until child birth (2227 ± 558 g vs. 3519 ± 272 g) ($p < .001$). Spontaneous abortions occurred in 4 women. Women subjected to chemotherapy at the third trimester gave birth to healthy infants (713).

The assessment of the interfertility indices in patients with germ-cell tumors who had been treated with cisplatin, etoposide, VP16, and bleomycin revealed persistent sperm anomalies (oligospermia) (715).

Studies on the effects of occupational exposure of medical personnel to cytostatic drugs on reproductive function abnormalities revealed that the exposure may result in the increased incidence of spontaneous abortions, stillbirths, lower birth weight, and malformations (716–719).

The risk of spontaneous abortion among French nurses routinely preparing cytostatic drugs in chemotherapy departments was 1.7, which was statistically significant. In the group of the exposed women, the frequency of spontaneous abortions was 25.9% and the corresponding value for the controls was 15.1%. The occupational exposure was determined in relation to the cytostatic drugs in general as it was not possible to determine the effects of the individual agents or time of the exposed employment. Mean weekly number of the infusions prepared by the nurses was 18.6 (717).

The birth weight of infants of the mothers exposed to cytostatic drugs was by 85 g lower than the weight of infants of nonexposed mothers. The difference was not statistically significant (718).

21.4.2.2.5 Carcinogenesis. The International Agency for the Research on Cancer (IARC) has classified cisplatin as probably carcinogenic to humans (group 2A) considering the adequate proof of its carcinogenic activity in experimental animal [see *cis*-diaminedichloroplatinum(II) discussion in Section 21.4.2.1.5) and the inadequate proof of carcinogenic activity in humans (489–491).

No epidemiological data are available in the literature on the carcinogenic activity of cisplatin in people treated with cisplatin alone or occupationally exposed to cisplatin. Several cases of secondary leukemia have been reported in the literature in patients treated for long-timer cisplatin exposure, often in combination with alternative treatment using ionizing radiation or other carcinogens, such as alkylating agents or other cytostatic drugs (720–725).

In the opinion of IARC, however, those few reported cases of the secondary leukemia do not constitute an adequate proof of the carcinogenic activity of cisplatin in humans (491).

21.4.2.2.6 Genetic and Related Cellular Effects Studies. The ability of the cytostatic drugs (including cisplatin) to damage human chromosomes has been demonstrated as a result of *in vitro* tests (726, 727) and *in vivo* studies on patients treated for cancer (728, 729) or on people occupationally exposed to cytostatic drugs, usually medical personnel (730–735).

Dose- and exposure-time-related induction of micronuclei in two cellular lines of human skin fibroblasts was observed during *in vitro* tests, in which the fibroblasts were treated with cisplatin at concentrations of 2–80 mmol/L for 2, 24, and 48 h (726).

In the human lymphocytes *in vitro* cisplatin induced generation of micronuclei, as demonstrated in the cytokinesis–block micronuleus test (551).

Experiments on peripheral blood lymphocytes of healthy people, which were *in vitro* exposed to cisplatin revealed a significant dose-related increase in sister chromatid exchanges (SCE) (727).

A significant increase in the frequency of micronuclei in peripheral blood binucleated lymphocytes was detected in patients treated with cisplatin in combination with other cytostatic drugs. Lymphocytes containing chemically induced chromosome lesions were noted after 9.3 years following termination of chemotherapy. The authors conclude that the implications of those results should be taken into account in view of the increased risk of developing secondary tumors (728).

An analysis of blood samples collected from patients with reproductive cell cancer who had been treated with cisplatin or carboplatin in combination with other cytostatic drugs and tested for the presence of seven different biological markers [platinum–protein adducts, platinum–DNA adducts, sister chromatid exchange (SCE), micronuclei (MN), and somatic gene mutation at the hypoxanthine phosphoribosyl transferase (HPRT) locus and the glycophorin A(GPA) loci (NO and NN)] revealed that Pt-protein adducts, Pt-DNA adducts and SCE were remarkably higher compared to baseline samples and very highly correlated with effect of treatment and remained elevated 3–6 months after termination of the treatment (729).

Statistically significant increase of SCE frequency in peripheral blood lymphocytes was noted in the nurses occupationally exposed to cytostatic drugs compared to the administration workers (730) or to nurses employed in other hospital wards (731).

A statistically significant increase in SCE compared to the control was observed in peripheral blood lymphocytes of 17 nurses exposed to cytostatic drugs (including cisplatin) who worked in an oncology ward without using suitable personal protective equipment ($p < .001$). The increase was also statistically significant when compared with a group of nurses provided with suitable protective means during performing their daily duties who also experienced SCE, which was, however, statistically insignificant (734).

Studies of the genetic effects of the occupational exposure to cytostatic drugs, including cisplatin, conducted among 20 nurses and blue-collar workers employed in various hospital wards and the controls (11 healthy people), showed statistically significant differences in the number of translocations (FG/100 = 2.25 ± 1.50 vs. 0.66 ± 0.21, $p < .001$) and unstable chromosome aberrations between the subgroups of the nurses and the blue-collar workers, whereas a high statistically significant difference was found to occur between all medical personnel and the control. Significant differences were not detected in the number of stable and unstable aberrations between the exposed nurse group and the blue collar group; high statistically significant differences ($p < .01$) were demonstrated to occur between all medical personnel and the controls (735).

Increased frequency of chromosome aberrations ($3.3 \pm 0.1/100$ cells) in peripheral blood lymphocytes was found in a group of 106 people exposed to cytostatic drugs, who worked without personal protective equipment as compared to the group of nonexposed controls, whose aberration frequency was $0.6 \pm 0.1/100$ cells. The frequency of the chromosome aberrations did not correlate with the age of the examined people, duration of exposure, or tobacco smoking (732).

Studies on the mutagenic activity of urine collected from people occupationally exposed to cytostatic drugs did not show any significant correlation between the exposure and the mutagenic activity of the urine assessed by the Ames test in *Salmonella typhimurium* TA98/S9 mix. Only for three people of the exposed group the urine samples collected at the end of the working shift displayed a mutagenic activity in TA102/S9 mix (736).

21.4.2.3 Epidemiology Studies. No epidemiological data have been located in the literature on the toxic activity of cisplatin alone in occupationally exposed people. The case-control data reported in the literature relate primarily to the side effects of prolonged treatment of cancerous patients with cytostatic drugs, which is in most cases accompanied by simultaneous application of ionizing radiation, alkylating agents, or other cytostatic drugs (682,693). The results of those studies indicate that the strong nephrotoxic activity, strong neurotoxic activity manifested mainly by peripheral neuropathy and hearing loss [see *cis*-diamminedichloroplatinum(II) discussion in Section 21.4.2.1.], and harmful effect on the infants of women undergoing chemotherapy during the first trimester of the pregnancy (713, 714) constitute the main side effects of cisplatin.

Epidemiological case-control studies on the effects of embryotoxic teratogenic activity of cytostatic drugs in medical personnel (nurses in particular) are made difficult because the number of cases is small, no data are available for comparison, the exposed people suffer from other diseases, and, finally, because of the combined exposure. Although the results of those studies are equivocal, cautiousness suggests that the possibility of such effects occurring in pregnant women is taken into account (716–718, 723) [see *cis*-diamminedichloroplatinum(II) discussion in Section 21.4.2.2.4].

21.5 Standards, Regulations, or Guidelines of Exposure

The values of hygienic standards by ACGIH, OSHA, and NIOSH and those issued in the individual countries for the soluble platinum salts are given in Table 41.41 (see Section 16.5).

21.6 Studies on Environmental Impact

Considering the high melting point and applications, it is not reasonable to suppose that cisplatin is present in the atmospheric air. Nevertheless, it is produced and widely used as an anticarcinogenic drug and therefore may be present in the environment during its manufacture and use. Cisplatin has an aqueous solubility of 2.530 mg/L and the compound is stable toward ionization. Therefore, the complex should leach through soil unless it is converted into some insoluble simple salt such as platinum hydroxide or platinum monosulfide or converted into ionic species by biotic and abiotic reaction. No data regarding biotic or abiotic speciation of cisplatin in soil are available (736).

The aqueous solubility of cisplatin suggests that entering water it will be present in the aqueous solution. Cisplatin is known to slowly transform to *trans*-platin in water (489). Some of the *trans*-platin should be adsorbed to suspended solids and sediments in water, but the majority of the compound should stay in the aquatic phase unless biotic or abiotic processes convert it into an ionic salt or insoluble precipitate—SM insoluble precipitate. The ionic salt could be adsorbed by clay materials by ion-exchange process and precipitate out in the sediment (489, 736).

22.0. Potassium Hexachloroplatinate(IV)

22.0.1 CAS Number: *[16921-30-5]*

22.0.2 Synonyms: Platinic potassium chloride; potassium platinic chloride; potassium chloroplatinate; platinum potassium chloride; Platinate(2-), hexachloro-, dipotassium, (OC-6-11)-; Platinium (IV) potassium chloride; Potassium hexachloroplatinate(IV); potassium platinum(IV) hexachloride; platinum (IV) Potassium chloride

22.0.3 Trade Names: NA

22.0.4 Molecular Weight: 485.18

22.0.5 Molecular Formula: K_2PtCl_6

22.0.6 Molecular Structure

>Melting point (°C): 250
>Specific gravity: 3.499

22.1 Chemical and Physical Properties

Occurs in the form of orange-yellow or yellow crystals (3). (See also platinum entry in Table 41.35)

22.2 Production and Use

K_2PtCl_6 is used in photography (2). (See also Section 16.2.)

22.3. Exposure Assessment

(See Section 16.3.)

22.4 Toxic Effects

22.4.1 Experimental Studies

22.4.1.1 Acute Toxicity. Data not available. (See also Section 16.4.)

22.4.1.2 Chronic and Subchronic Toxicity. Data not available.

22.4.1.3 Pharmacokinetics, Metabolism, and Mechanisms. Data on absorption, distribution, and elimination of K_2PtCl_6 experimental animals are not available. *In vivo* tests revealed that a single subcutaneous 60-mg/kg bw dose of $K_2[PtCl_6]$ given to Sprague–Dawley rats affected with the heme biosynthesis and reduced the activity of the enzymes controlling heme biosynthesis in kidneys (δ-aminolevulinic acid dehydratase, uroporphyrinogen I synthetase, ferrochelatase), and lowered total porphyrin content in the kidneys (during 24 h) (737) (For mechanisms of action, see Section 16.4.1.3.4.)

22.4.1.4 Reproductive and Developmental Effects: NA

22.4.1.5 Carcinogenesis: NA

22.4.1.6 Genetic and Related Cellular Effects Studies. K_2PtCl_6 at 10 μmol/L showed weak mutagenic activity in CHO-S cells (495). Other authors have demonstrated that K_2PtCl_6 at concentrations of ≤ 220 μmol/L induced mutations in CHO-S cells following repeated 5-month exposure (554).

A dose-dependent increase in the frequency of spontaneous revertants in CHO-AUXB1 cells (at a concentration where the cell survival remains high) was induced after 20–22 h with K_2PtCl_6 (497).

22.4.2 Human Experience

22.4.2.1 General Information. Many authors have demonstrated that exposure to soluble complex platinum salts, such as tetra- and hexachloroplatinates, which are present in the occupational environment during platinum refining and processing, may be detrimental to the health of the workers, who may develop symptoms of respiratory tract and dermal allergy.

Exposure to complex platinum salts may produce allergic reactions, such as conjunctivitis, rhinorrhea, sneezing, coughing, asthma, urticaria, and eczema (477, 509, 512–515). About 60% people employed in the industry under conditions of exposure to platinum compounds may show the symptoms of bronchial asthma (512–515).

(See also Section 16.4.1.2.)

22.4.2.2 Clinical Cases

22.4.2.2.1 *Acute Toxicity.* Data not available.

22.4.2.2.2 *Chronic and Subchronic Toxicity.* Data not available. (See Section 16.4.2.2.2.)

22.4.2.2.3 *Pharmacokinetics, Metabolism, and Mechanisms.* Data not available. (See Section 16.4.2.3.)

22.4.2.3 Epidemiology Studies. Epidemiology studies are concerned primarily with hazards occurring in platinum refining and processing plants, including exposure to complex platinum salts—the primary source of dermal and respiratory hypersensitivity, asthma, and nonspecific diseases of the respiratory system (502, 515, 517, 519, 521, 522, 530, 534, 539).

In a cross-sectional study, 65 workers in the chemical industry with exposure to platinum salts were investigated with regard to the prevalence to allergic respiratory tract diseases. A respiratory questionnaire, a skin-prick test with Cl_6K_2Pt and environmental allergens, determination of total IgE, platinum-specific IgE and histamine release in basophilic granulocytes, and lung function tests were applied before and after a Monday shift and after a Friday shift. Work-related symptoms of respiratory allergy were present in 23% of all workers, but were significantly more frequent in the most exposed group in the platinum refinery (52.4%). Of all workers, 18.7% had a positive skin-prick test with platinum salt. As compared to the other workers, the workers with work-related symptoms of respiratory allergy had significantly more positive skin-prick tests (64.3%) and a higher total IgE and platinum-specific IgE; they did not, however, show higher histamine release. In the course of the week, a significantly fall in lung function, namely in FEV_1 and FEF_{25}, was recorded in the group of workers with work-related symptoms (522).

(See also Section 16.4.2.3.)

NICKEL, RUTHENIUM, RHODIUM, PALLADIUM, OSMIUM, AND PLATINUM

22.5 Hygienic Standards, Regulations, or Guidelines of Exposure

Table 41.41 specifies the ACGIH, OSHA, and NIOSH hygienic standards for soluble platinum salts, and those currently in force in some other countries.

(See also Section 16.5.)

22.6 Studies on Environmental Impact

(See Section 16.6.)

23.0 Potassium Tetrachloroplatinate(II)

23.0.1 CAS Number: [10025-99-7]

23.0.2 Synonyms: Platinous potassium chloride; potassium platinochloride; potassium chlorplatinate; platinate(2-), tetrachloro-, dipotassium, (SP-4-1)-; potassium tetrachloroplatinate(II); potassium tetrachloroplatinate; bipotassium tetrachloroplatinate; chloroplatinic acid, dipotassium salt; potassium chloroplatinite; potassium platinous chloride; potassium tetrachloroplatinite

23.0.3 Trade Names: NA

23.0.4 Molecular Weight: 415.09

23.0.5 Molecular Formula: K_2PtCl_4

23.0.6 Molecular Structure:
$$\begin{array}{c} Cl \\ | \\ Cl-Pt-Cl \\ | \\ Cl \end{array} \quad K^+ \quad K^+$$

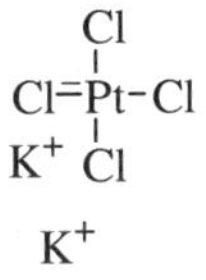

23.1 Chemical and Physical Properties

Cl_4K_2Pt occurs in the form of ruby-red crystals (3). (See platinum entry in Table 41.35.)

23.2 Production and Use

Cl_4K_2Pt is used in photography (2). (See also Section 16.2.)

23.3 Exposure Assessment

See: Platinum — Section 1.3.

23.4 Toxic Effects

23.4.1 Experimental Studies

23.4.1.1 Acute Toxicity. The LD_{50} for the rat after per os administration of potassium tetrachloroplatinate(II) is 50–200 mg/kg bw (446).

No other data on potassium tetrachloroplatinate(II) acute toxicity are available. See also platinum entry in Table 41.36.)

23.4.1.2 Chronic and Subchronic Toxicity. Data not available. (See also Section 16.4.1.2.)

23.4.1.3 Pharmacokinetics, Metabolism, and Mechanisms. Data on $K_2[PtCl_4]$ absorption, distribution, and elimination are not available. *In vivo* tests revealed that a single subcutaneous 52 mg/kg bw dose of K_2PtCl_4 affected the enzymes controlling the heme pathway in liver and kidneys of Sprague–Dawley rats (737). See also Section 16.4.1.3.)

23.4.1.4 Reproductive and Developmental Effects. Data not available.

23.4.1.5 Carcinogenesis. Data not available.

23.4.1.6 Genetic and Related Cellular Effects Studies. K_2PtCl_4 at 0.8–100 µmol/plate showed weak mutagenic activity in the Ames test with TA98 and TA100 *Salmonella typhimurium* after metabolic S9 mix activation (738). It weakly induced dipolidal spores in *Saccharomyces cerevisiae* at the most efficient dose 42 µg/ml (739). It produced dose-dependent increase in the frequency of spontaneous revertants in CHO-AUXB1 cells at 0–103 µmol/L (497) but did not induce increase in the frequency of sex-related recessive mutations in *Drosophila melanogaster* (498). It did not produce mutagenic effects in bone marrow cells of mice and rats (499). It caused a significantly elevated genotoxicity in cytokinesis–block micronucleus tests (MNT) with human lymphocytes and the bacterial SOS chromotest (551).

23.4.1.7 Other: Neurological, Palmonary, Skin Sensitization

23.4.1.7.1 Irritating Activity. K_2PtCl_4 did not show local irritating activity after 4-h contact with rabbit skin (500); however, it did show irritating action on the eyes (501).

23.4.1.7.2 Sensitizing Activity. Results of studies on potential sensitizing activity of complex platinum salts show that these compounds may produce anaphylactic shock, and increased blood plasma histamine levels in experimental animals (465, 466, 472).

Reports on the potential sensitizing or immunotoxic activity of K_2PtCl_4 are not available in the literature.

(See also Section 16.4.1.7.2.)

23.4.2 Human Experience

23.4.2.1 General Information. Many authors have demonstrated that exposure to soluble complex platinum salts, such as tetra- and hexachloroplatinates, which are present in the occupational environment during platinum refining and processing, may be detrimental to the health of the workers who may develop symptoms of respiratory tract and dermal allergy. Exposure to complex platinum salts may produce allergic reactions, such as conjunctivitis, rhinorrhea, sneezing, coughing, asthma, urticaria, and eczema (465, 477,

509, 512–515). About 60% of people employed in the industry under conditions of exposure to platinum compounds may show the symptoms of bronchial asthma (465, 512–515). (See also Section 16.4.2.)

23.4.2.2 Clinical Cases

23.4.2.2.1 Acute Toxicity. A case of a young photographer has been reported who, in a suicide attempt, drank a 10-mL solution containing 600 mg K_2PtCl_4. Within 2 h the patient developed nausea, vomiting, diarrhea, leg cramps, symptoms of acute renal damage, hepatitis, enteritis, leucocytosis, eosinophilia, and mild metabolic acidosis. All symptoms of the poisoning disappeared within 6 days (524).

23.4.2.2.2 Chronic and Subchronic Toxicity. Data not available. (See also Section 16.4.2.2.2.)

23.4.2.2.3 Pharmacokinetics, Metabolism, and Mechanisms. Data not available. (See also Section 16.4.2.2.3.)

23.4.2.3 Epidemiology Studies. Epidemiology studies are concerned primarily with hazards occurring in platinum refining and processing plants, including exposure to complex platinum salts — the primary source of dermal and respiratory hypersensitivity, asthma, and nonspecific diseases of the respiratory system (502, 515, 517, 519, 521, 522, 530, 534, 539).
(See also Section 16.4.2.3.)

23.5 Standards, Regulations, or Guidelines of Exposure

Table 41.41 specifies the ACGIH, OSHA, and NIOSH hygienic standards for soluble platinum salts, and those currently in force in some other countries. (See also Section 16.5.)

23.6 Studies on Environmental Impact

(See Section 16.6.)

BIBLIOGRAPHY

1. R. C. West, ed., *CRC Handbook of Chemistry and Physics*, 66th ed., CRC Press, Boca Raton, FL, 1985, pp. B-118, B-119, D-194.
2. S. Budavari, ed., *The Merck Index: An Encyclopedia of Chemicals, Drugs, and Biologicals*, 12th ed., Merck & Co, Whitehouse Station, NY, 1996, pp. 1116–1118.
3. J. D. Lee, *A New Concise Inorganic Chemistry*, 3rd ed. Van Nostrand-Reinhold, UK, 1977. Polish edition by Wydawnictwo Naukowe PWN Sp. z o. o., Warszawa, 1994, pp. 336–374 (in Polish).
4. J. Emsley, *Oxford Chemistry Guides: The Elements*, 2nd ed. Polish edition by Wydawnictwo Naukowe PWN, Sp. z o. o., Warszawa, 1997, pp. 138–139 (in Polish).

5. J. S. Warner, Occupational exposure to airborne nickel in producing and using primary nickel products. In F. W. Sunderman, Jr., ed., *Nickel in the Human Environment*, IARC Sci. Publ. No. 53, IARC, Lyon, France, 1984, pp. 419–437.
6. K. W. Boyer and W. Horwitz, Special considerations in trace element analysis of foods and biological materials. In I. K. O'Neill, P. Schuller, and L. Fishbein, eds. *Environmental Carcinogens: Selected Methods of Analysis*, IARC Sci. Publ. No. 71, IARC, Lyon, France, 1986, pp. 191–220.
7. National Institute for Occupational Safety and Health (NIOSH), *Manual of Analytical Methods*, 4th ed., Method 7300, U. S. Department of Health and Human Services, Washington, DC, 1994.
8. V. J. Zatka, *Chemical Speciation of Nickel Phases in Industrial Dust*, Method 90-05-03, Inco, Ltd. J. Roy Gordon Laboratory, Sheridan Park, Mississanga, Ontario, Canada, 1990.
9. J. L. Wong and T. G. Wu, Speciation of airborne nickel in occupational exposure. *Environ. Sci. Technol.* **25**, 306–309 (1991).
10. National Institute for Occupational Safety and Health (NIOSH), *Manual of Analytical Methods*, 4th ed., Method 8005, U. S. Department of Health and Human Services, Washington, DC, 1994.
11. National Institute for Occupational Safety and Health (NIOSH), *Manual of Analytical Methods*, 4th ed., Method 8310, U.S. Department of Health and Human Services, Washington, DC, 1994.
12. F. Sunderman, Jr., Determination of nickel in water body fluids, tissues and excreta. In J. K. O'Neill, P. Schuller, and L. Fishbein, eds., *Environmental Carcinogens: Selected Methods of Analysis*, IARC Sci. Publ. No. 71, IARC, Lyon, France, 1986, pp. 319–334.
13. J. R. Andersen, B. Gammelgaard, and S. Reimert, Direct determination of nickel in human plasma by Zeeman-corrected atomic absorption spectrometry. *Analyst (London)* **111**, 721–722 (1986).
14. M. Stoeppler, Analytical chemistry of nickel. In F. W. Sunderman, Jr., ed., *Nickel in the Human Environment*, IARC Sci. Publ. No. 53, IARC, Lyon, France, 1989, pp. 459–468.
15. A. Alimonti et al., Determination of chromium and nickel in human blood by means of inductively coupled plasma mass spectrometry. *Anal. Chim. Acta* **306**, 35–41 (1995).
16. M. Patriarca et al., Determination of selected nickel isotopes in biological samples by inductively coupled plasma mass spectrometry with isotope dilution. *J. Anal. At. Spectrom.*, **11**, 297–302 (1996).
17. A. Sturaro et al., Simultaneous determination of trace metals in human hair by dynamic ion-exchange chromatography. *Anal. Chim. Acta* **274**, 163–170 (1993).
17a. NIOSH, *Manual of Analytical Method*, 4th ed., 1994.
18. E. Mastromatteo, Nickel. *Am. Ind. Hyg. Assoc. J.* **47**, 589–601 (1986).
19. B. Venugopal and T. D. Luckey, *Metal Toxicity in Mammals. 2. Chemical Toxicity of Metals and Metalloids*, Plenum, New York, 1978, pp. 289–297.
20. F. W. Sunderman, Jr., Hazards from exposure to nickel: A historical account. In E. Nieboer and J. O. Nriagu, eds., *Nickel and Human Health*, Wiley, New York, 1992, pp. 1–20.
21. J. G. Judde et al., Inhibition of rat natural killer cell function by carcinogenic nickel compounds: Preventive action of manganese. *J. Natl. Cancer. Inst.*, **78**, 1185–1190 (1987).
22. W. C. Hueper, Experimental studies in metal carcinogenesis. IX. Pulmonary lesions in guinea-pigs and rats exposed to prolonged inhalation of powdered metallic nickel. *Arch. Pathol.*, **65**, 600–607 (1958).

23. P. Camner, A. Johansson, and M. Lundberg, Alveolar macrophages in rabbits exposed to nickel dust. *Environ. Res.* **16**, 226–235 (1978).
24. M. Lundborg and P. Camner, Decreased level of lysozyme in rabbit lung lavage fluid after inhalation of low nickel concentrations. *Toxicology* **22**, 353–358 (1982).
25. A. Johansson, P. Camner, and B. Robertson, Effects of long-term nickel dust exposure on rabbit alveolar epithelium. *Environ. Res.* **25**, 391–403 (1981).
26. P. Camner et al., Toxicology of nickel. In F. W. Sunderman, Jr., ed., *Nickel in the Human Environment*, IARC Sci. Publ. No. 53, IARC, Lyon, France, 1984, pp. 267–276.
27. I. Tanaka et al., Biological half time of deposited nickel oxide aerosol in rat lung by inhalation. *Biol. Trace Elem. Res.* **8**, 203–210 (1985).
28. I. Tanaka et al., Biological half-time in rats exposed to nickel monosulfide (amorphous) aerosol by inhalation. *Biol. Trace Elem. Res.* **17**, 237–246 (1988).
29. S. Hirano et al., Pulmonary clearance and inflammatory potency of intratracheally instilled or acutely inhaled nickel sulphate in rats. *Arch. Toxicol.* **68**, 548–554 (1994).
30. J. M. Benson et al., Fate of inhaled nickel oxide and nickel subsulfide in F344/N rats. *Inhalation Toxicol.* **6**, 167–183 (1994).
31. J. J. Clary, Nickel chloride-induced metabolic changes in the rat and guinea-pig. *Toxicol. Appl. Pharmacol.*, **31**, 55–65 (1975).
32. S. M. Carvalho and P. L. Ziemer, Distribution and clearance of ^{63}Ni administered as ^{63}NiCl$_2$ in the rat: Intratracheal study. *Arch. Environ. Contam. Toxicol.* **11**, 245–248 (1982).
33. A. M. Ambrose et al., Long-term toxicologic assessment of nickel in rats and dogs. *J. Food Sci. Technol.*, **13**, 181–187 (1976).
34. S. Ishimatsu et al., Distribution of various nickel compounds in rat organs after oral administration. *Biol. Trace Elem. Res.* **49**(1), 43–52 (1995).
35. O. Norgaard, Investigations with radioactive nickel, cobalt and sodium on the resorption through the skin in rabbits, guinea-pigs and man. *Acta Derm.-Venereol.* **37**, 440–445 (1957).
36. G. K. Lloyd, Dermal absorption and conjugation of nickel in relation to the induction of allergic contact dermatitis: Preliminary results. In S. S. Brown and F. W. Sunderman, Jr., eds, *Nickel Toxicology*, Academic Press, London, 1980, pp. 145–148.
37. A. K. Mathur and B. N. Gupta, Dermal toxicity of nickel and chromium in guinea pigs. *Vet. Hum. Toxicol*, **36**(2), 131–132 (1994).
38. J. M. Benson et al., Comparative inhalation toxicity of nickel subsulfide to F344/N rats and B6C3F$_1$ mice exposed for twelve days. *Fundam. Appl. Toxicol*, **9**, 251–265 (1987).
39. J. M. Benson et al., Comparative inhalation toxicity of nickel subsulfide to F344/ rats and B6C3F$_1$ mice exposed for twelve days. *Fundam. Appl. Toxicol.* **10**, 164–178 (1988).
40. J. K. Dunnick et al., Lung toxicity after 13-week inhalation exposure to nickel oxide, nickel subsulfide, or nickel sulfate hexahydrate in F344/N rats and B6C3F$_1$ mice. *Fundam. Appl. Toxicol.* **12**, 584–595 (1989).
41. J. M. Benson, I. Y. Chang, and Y. S. Cheng, Particle clearance and histopathology in lungs of F344/N rats and B6C3F$_1$ mice inhaling nickel oxide or nickel sulfate. *Fundam. Appl. Toxicol.* **38**, 232–244 (1995).
42. A. P. Wehner and D. K. Craig, Toxicology of inhaled NiO and CoO in Syrian golden hamsters. *Am. Ind. Hyg. Assoc. J.* **33**, 146–155 (1972).

43. M. P. Dieter et al., Evaluation of tissue disposition, myelopoietic, and immunologic responses in mice after long-term exposure to nickel sulfate in the drinking water. *J. Toxicol. Environ. Health* **24**, 356–372 (1988).
44. K. Borg and H. Tjalve, Uptake of $^{63}Ni_2^+$ in the central and peripheral nervous system of mice after oral administration: Effects of treatment with halogenated 8-hydroxyquinolines. *Toxicology* **54**, 59–68 (1989).
45. E. Szakmáry et al., Offspring damaging effect of nickel in rat, mouse and rabbit. *Cent. Eur. J. Occup. Environ. Medi.*, **2**(3), 277–287 (1996).
46. J. C. English et al., Toxicokinetics of nickel in rats after intratracheal administration of a soluble and insoluble form. *Am. Ind. Hyg. Assoc. J.* **42**(7), 486–492 (1981).
47. M. A. Medinsky, J. M. Benson, and C. H. Hobbs, Lung clearance and disposition of the ^{63}Ni in F344/N rats after intratracheal instillation of nickel sulfate solutions. *Environ. Res.* **43**, 168–178 (1987).
48. R. Valentine and G. L. Fisher, Pulmonary clearance of intratracheally administered $^{63}Ni_3S_2$ in strain A/J mice. *Environ. Res.* **34**, 328–334 (1984).
49. G. D. O'Dell et al., Effect of dietary nickel level on excretion and nickel content of tissues in male calves. *J. Anim. Sci.* **32**(4), 769–733 (1971).
50. W. Ho and A. Furst, Nickel excretion by rats following a single treatment. *Proc. West. Pharmacol. Soc.* **16**, 245–248 (1973).
51. C. Onkelinx and F. W. Sunderman, Modeling of nickel metabolism. In J. O. Nriagu, ed., *Nickel in the Environment*, Wiley, New York, 1980, pp. 525–545.
52. C. Onkelinx, J. Becker, and F. W. Sunderman, Compartmental analysis of the metabolism of $^{63}Ni(II)$ in rats and rabbits. *Res. Commun. Chem. Pathol. Pharmacol.* **6**(2), 663–676 (1973).
53. F. W. Sunderman, Jr. et al., Effects of triethylenetetramin upon the metabolism and toxicity of $^{63}NiCl_2$ in rats, *Toxicol. Appl. Pharmacol.* **38**, 177–188 (1976).
54. F. W. Sunderman, Jr. et al., Nickel absorption and kinetics in human volunteers. *Proc. Soc. Exp. Biol. Med.* **191**(5), 5–11 (1989).
55. M. Costa et al., Phagocytosis, cellular distribution, and carcinogenic activity of particulate nickel compounds in tissue culture. *Cancer Res.* **41**(7), 2868–2876 (1981).
56. M. Costa, Perspectives on the mechanism of nickel carcinogenesis gained from models of in vitro carcinogenesis. *Environ. Health Perspect* **81**, 73–76 (1989).
57. F. W. Sunderman, Jr. and S. M. Hopfer, Correlation between the carcinogenic activities of nickel compounds and their potencies for stimulating erythropoiesis in rats. In B. Saker, ed., *Biological Aspects of Metals and Metal-related Diseases*, Raven Press, New York, 1983, pp. 171–181.
58. R. B. Ciccarelli and K. E. Wetterhahn, In vitro interaction of $^{63}nickel(II)$ with chromatin and DNA from rat kidney and liver nuclei. *Chem.-Biol. Interact.* **52**(3), 347–360 (1985).
59. R. B. Ciccarelli, T. H. Hampton, and K. W. Jennette. Nickel carbonate induces DNA-protein crosslinks and DNA strand breaks in rat kidney. *Cancer Lett.* **12**(4), 349–354 (1981).
60. S. H. Robison and M. Costa, The induction of DNA strand breakage of nickel compounds in cultured Chinese hamster ovary cells. *Cancer Lett.* **15**, 35–40 (1982).
61. S. R. Patierno and M. Costa, DNA-protein cross-links induced by nickel compounds in intact cultured mammalian cells. *Chem.-Biol. Interact.* **55**, 75–91 (1985).
62. M. Costa et al., Molecular mechanisms of nickel carcinogenesis. *Sci. Total Environ.* **148**, 191–199 (1994).

63. M. Costa, Model for the epigenetic mechanism of action of nongenotoxic carcinogens. *Am. J. Clin. Nutr.* **61** (Suppl.), 666S–669S (1995).

64. K. S. Kasprzak and L. Hernandez, Enhancement of hydroxylation and deglycosylation of 2'-deoxyguanosine by carcinogenic nickel compounds. *Cancer Res.* **49**, 5964–5968 (1989).

65. A. Hartwig, I. Kruger, and D. Beyersmann, Mechanisms in nickel genotoxicity: The significance of interactions with DNA repair. *Toxicol. Lett.* **72**, 353–358 (1994).

66. Y. Kurokawa et al., Promoting effect of metal compounds on rat renal tumorigenesis. *J. Am. Coll. Toxicol.* **4**, 321–330 (1985).

67. M. Misra, R. E. Rodriguez, and K. S. Kasprzak, Nickel induced lipid peroxidation in the rat: Correlation with nickel effect on antioxidant defense systems. *Toxicology* **64**, 1–17 (1990).

68. J. Torreilles and M. C. Guerin, Nickel(II) as a temporary catalyst for hydroxyl radical generation. *FEBS Lett.* **272**, 58–60 (1990).

69. F. W. Sunderman, Jr., Recent advances in metal carcinogenesis. *Ann. Clin. Lab. Sci.* **14**, 2 (1984).

70. F. W. Sunderman and A. M. Barber, Finger loops, oncogens, and metals. *Ann. Clin. Lab. Sci.* **18**, 267–288 (1988).

71. C. C. Chang, H. J. Tatum, and F. A. Kincl, The effect of intrauterine copper and other metals on implantation in rats and hamsters. *Fertil. Steril.* **21**, 274–278 (1970).

72. M. Daniel, M. Edwards, and M. Webb, The effect of metal-serum complexes on differentiating muscle in vitro. *Br. J. Exp. Pathol.* **55**, 237–244 (1974).

73. W. C. Hueper and W. W. Payne, Experimental studies in metal carcinogenesis. *Arch. Environ. Health* **5**, 445–462 (1962).

74. M. K. Kim, A. M. Fischer and R. J. Mackay, *Pulmonary Effects of Metallic Dusts—Nickel and Iron*, University of Toronto, School of Hygiene, Department of Physiological Hygiene, Toronto, 1969.

75. F. Pott et al., Carcinogenicity studies on nickel compounds and nickel alloys after intraperitoneal injection in rats. *4th Int. Conf. Nickel Metab. Toxicol.*, Espoo, Finland, 1988, Helsinki, Institute of Ocupational Health, Abstr. p. 42.

76. H. Muhle, B. Bellmann, and S. Takenaka, Chronic effects of intratracheally instilled nickel containing particles in hamsters. *4th Int. Conf. Nickel Metab. Toxicol.* Espoo, Finland, 1988, Helsinki, Institute of Occupational Health, Abstr. p. 41.

77. J. C. Heath and M. R. Daniel, The production of malignant tumors by nickel in the rat. *Br. J. Cancer* **18**, 261–264 (1964).

78. F. W. Sunderman, Jr. and R. M. Maenza, Comparisons of carcinogenicities of nickel compounds in rats. *Res. Commun. Chem. Pathol. Pharmacol.* **14**, 319–330 (1976).

79. F. W. Sunderman, Jr., Carcinogenicity of nickel compounds in animals. In F. W. Sunderman, Jr. ed., *Nickel in the Human Environment*, IARC Sci. Publ. No. 53, IARC, Lyon, France, 1984, pp. 127–142.

80. A. Furst and M. C. Schlauder, The hamster as a model for metal carcinogenesis. *Proc. West. Pharmacol. Soc.* **14**, 68–71 (1971).

81. A. Furst, D. M. Cassetta, and D. P. Sasmore, Rapid induction of pleural mesotheliomas in the rat. *Proc. West. Pharmacol. Soc.* **16**, 150–153 (1973).

82. W. C. Hueper, Experimental studies in metal cancerogenesis. IV. Cancer produced by parenterally introduced metallic nickel. *J. Natl. Cancer Inst. (U. S.)* **16**, 55–67 (1955).

83. A. Furst and A. D. Cassetta, Carcinogenicity of nickel by different routes. *Proc. Am. Assoc. Cancer Res.* **14**, 31, Abstr. 121 (1973).

84. F. Pott, et al., Carcinogenicity of nickel compounds and nickel alloys in rats by intraperitoneal injection. In E. Nieboer and J. O. Nriagu, eds., *Nickel and Human Health*, Wiley, New York, 1992, pp. 491–502.

85. G. R. Paton and A. C. Allison, Chromosome damage in human cell cultures induced by metal salts. *Mutat. Res.* **16**, 332–336 (1972).

86. M. Costa, M. P. Abbracchio, and J. Simmons-Hansen, Factors influencing the phagocytosis, neoplastic transformation and cytotoxicity of particulate nickel compounds in tissue culture systems. *Toxicol. Appl. Pharmacol.* **60**, 313–323 (1981).

87. M. Costa, J. D. Heck, and S. H. Robison, Selective phagocytosis of crystalline metal sulfide particles and DNA strand breaks as a mechanism for the induction of cellular transformation. *Cancer Res.* **42**, 2757–2763 (1982).

88. K. Hansen and R. M. Stern, Toxicity and transformation potency of nickel compounds in BHK cells in vitro. In F. W. Sunderman, Jr., ed., *Nickel in the Human Environment*, IARC Sci. Publ. N. 53, IARC, Lyon, France, 1984, pp. 193–200.

89. R. E. G. Rendall, J. I. Phillips, and K. A. Renton, Death following exposure to fine particulate nickel from a metal arc process. *Ann. Occup. Hyg.* **38**(6), 921–930 (1994).

90. F. W. Sunderman, Biological monitoring of nickel in humans. *Scand. J. Work. Environ. Health* **19** (Suppl. 1), 34–38 (1993).

91. A. I. M. Sandström, S. G. I. Wall, and A. Taube, Cancer incidence and mortality among Swedish smelter workers. *Br. J. Ind. Med.* **46**, 82–89 (1989).

92. J. D. Webster et al., Acute nickel intoxication by dialysis. *Ann. Intern. Med.* **92**, 631–633 (1980).

93. D. M. Zislin et al., Residual air in a complex evaluation of the respiratory system function in initial and suspected pneumonconiosis. *Gig. Tr. Prof. Zabol.* **13**, 26–29 (1969).

94. A. Zober, Symptoms and findings at the bronchopulmonary system of electric arc welders. II. communication: Pulmonary fibrosis. *Zentralbl. Bakteriol., Parasitenkol., Infektionskr. Hyg., Abt. I: Orig., Reihe B* **173**, 120–148 (1981).

95. A. Zober, Symptoms and findings at the bronchopulmonary system of electric arc welders. I. communication: epidemiology. *Zentralbl. Bakteriol., Parasitenkol., Infektionskr. Hyg., Abt. I: Orig., Reihe B* **173**, 92–119 (1981).

96. A. Zober, Possible dangers to the respiratory tract from welding fumes. *Schweissen Schneiden*, **34**, 77–81 (1982).

97. H. J. Raithel, K. H. Schaller, and H. Valentin, Medical and toxicological aspects of occupational nickel exposure in the Federal Republic of Germany—Clinical results (carcinogenicity, sensitization) and preventive measures (biological monitoring). In F. W. Sunderman, Jr. ed., *Nickel in the Human Environment*, IARC Sci. Publ. No. 53, IARC, Lyon, France, 1984, pp. 403–514.

98. L. Kanerva et al., Hand dermatitis and allergic patch test reactions caused by nickel in electroplaters. *Contact Dermatitis* **36**(3), 137–140 (1997).

99. M. Shah, F. M. Lewis, and D. J. Gawkrodger, Nickel as an occupational allergen. A survey of 368 nickel-sensitive subjects. *Arch. Dermatol.* **134**(10), 1231–1236 (1998).

100. L. Kanerva, T. Estlander, and R. Jolanki, Bank clerk's occupational allergic nickel and cobalt contact dermatitis from coins. *Contact Dermatitis* **38**(4), 217–218 (1998).

101. L. Peltonen, Nickel sensitivity in the general populations. *Contact Dermatitis* **5**, 27–32 (1979).
102. T. Menné, O. Borgan, and A. Green, Nickel allergy and hand dermatitis in a stratified sample of the Danish female population: An epidemiologic study including a statistic appendix. *Acta Derm.-Venereol.* **62**, 32–41 (1982).
103. L. K. Dotterud and E. S. Falk, Metal allergy in north Norwegian schoolchildren and its relationship with ear piercing and atopy. *Contact Dermatitis* **31**, 308–313 (1994).
104. J. Brasch and J. Geier, Patch test results in schoolchildren. Results from the Information Network of Departments of Dermatology (JVDK) and the German Contact Dermatitis Research Group (DKG) *Contact Dermatitis* **37**(6), 286–293 (1997).
105. C. Linden et al., Nickel release from tools on the Swedish market. *Contact Dermatitis* **39**(3), 127–131 (1998).
106. F. S. Larsen and F. Brandrup, Nickel dermatitis provoked by buttons in blue jeans. *Contact Dermatitis* **6**, 298 (1980).
107. R. Deutman et al., Metal sensitivity before and after total hip arthroplasty. *J. Bone Jt. Surg.* **59**(7), 862–965 (1997).
108. B. G. Bennett, Environmental nickel pathways to man. In F. W. Sunderman, Jr., ed., *Nickel in the Human Environment*, IARC Sci. Publ. No. 53, IARC, Lyon, France, 1984, pp. 487–495.
109. P. Grandjean, Human exposure to nickel. In F. W. Sunderman, Jr., ed., *Nickel in the Human Environment*, IARC Sci. Publ. No. 53, IARC Lyon, France, 1984, pp. 469–485.
110. J. Angerer and G. Lehnert, Occupational chronic exposure to metals. II: Nickel exposure of stainless steel welders—biological monitoring. *Int. Arch. Occup. Environ. Health* **62**, 7–10 (1990).
111. B. Baranowska-Dutkiewicz, R. Rózañska, and T. Dutkiewicz, Occupational and environmental exposure to nickel in Poland. *Pol. J. Occup. Med. Environ. Health* **5**(4), 335–343 (1992).
112. N. W. Solomons et al., Bioavailability of nickel in man: Effects of foods and chemically-defined dietary constituents on the absorption of inorganic nickel. *J. Nutr.* **112**(1), 39–50 (1982).
113. O. B. Christensen and V. Lagesson, Nickel concentration of blood and urine after oral administration. *Ann. Clin. Lab. Sci.* **11**, 119–125 (1981).
114. O. Norgaard, Investigation with radioactive Ni-57 into the resorption of nickel through the skin in normal and in nickel-hypersensitive persons. *Acta Derm.-Venereol.* **35**, 111–117 (1995).
115. A. Fullerton, et al., Permeation of nickel salts through human skin in vitro. *Contact Dermatitis* **15**, 183–177 (1986).
116. H. J. Raithel et al., Analyses of chromium and nickel in human pulmonary tissue. *Int. Arch. Occup. Environ. Health* **61**, 507–512 (1989).
117. I. Andersen and K. B. Svenes, Determination of nickel in lung specimens of thirthy-nine autopsied nickel workers. *Int. Arch. Occup. Environ. Health* **61**, 289–295 (1989).
118. H. Kollmeier et al., Age, sex, and region adjusted concentrations of chromium and nickel in lung tissue. *Br. J. Ind. Med.* **47**, 682–687 (1990).
119. W. N. Rezuke, J. A. Knight, and F. W. Sunderman, Jr., Reference values for nickel concentrations in human tissues and bile. *Am. J. Ind. Med.* **11**, 419–426 (1987).
120. W. Torjussen and J. Andersen, Nickel concentrations in nasal mucosa, plasma and urine in active and retired nickel workers. *Ann. Clin. Lab. Sci.* **9**, 289–298 (1979).
121. F. W. Sunderman, Jr. et al., Acute nickel toxicity in electroplating workers who accidently ingested a solution of nickel sulfate and nickel chloride. *Am. J. Ind. Med.* **14**, 257–266 (1988).

122. E. Hassler et al., Urinary and fecal elimination of nickel in relation to air-borne nickel in battery factory. *Ann. Clin. Lab. Sci.* **13**, 217–224 (1983).

123. I. Ghezzi et al., Behavior of urinary nickel in low-level occupational exposure. *Med. Lav.* **80**, 244–250 (1989).

124. International Agency for Research on Cancer, (IARC), *Monographs on the Evaluation of Carcinogenic Risks to Humans. Chromium Nickel and Welding.* Vol. 49. IARC, Lyon, France, 1990, pp. 257–445.

125. A. Aitio, *The Nordic Expert Group for Criteria Documentation of Health Risks from Chemicals. 119. Nickel and Nickel Compounds* Arbete och Halsa, Arbetslivsinstitutet, Solna, Sverige, 1996.

126. J. C. Bridge, *Annual Report of the Chief Inspector of Factories and Workshops for the Year 1932*, HM Stationery Office, London, 1933, pp. 103–109.

127. J. G. A. Morgan, Some observations on the incidence of respiratory cancer in nickel workers. *Br. J. Ind. Med.* **15**, 224–234 (1958).

128. R. Doll, Cancer of the lung and nose in nickel workers. *Br. J. Ind. Med.* **15**, 217–223 (1958).

129. R. Doll, L. G. Morgan, and F. E. Speizer, Cancer of the lung and nasal sinuses in nickel workers. *Br. J. Cancer.* **24**(4), 623–632 (1970).

130. H. Cuckle, R. Doll, and L. G. Morgan, Mortality study of men working with soluble nickel compounds. In S. S. Brown and F. W. Sunderman, Jr., eds., *Nickel Toxicology*, Proc. 2nd Int. Conf. Nickel Toxicol., Swansea, London, Academic Press, New York, 1980, pp. 11–14.

131. J. Peto et al., Respiratory cancer mortality of Welsh nickel refinery workers. In F. W. Sunderman, Jr., ed., *Nickel in the Human Environment*, IARC Sci. Publ. No. 53, IARC, Lyon, France, 1984, pp. 37–46.

132. E. Pedersen, A. C. Hogetveit, and A. Andersen, Cancer of respiratory organs among workers at a nickel refinery in Norway. *Int. J. Cancer* **12**(1), 32–41 (1973).

133. A. Andersen, Recent follow-up of respiratory cancer in a Norwegian nickel refinery. *4th Int. Conf. Nickel Metab. Toxicol.*, Espoo, Finland, 1988, Institute of Occupational Health, Helsinki, Abstr. p. 49.

134. K. Magnus, A. Andersen and A. C. Hogetveit, Cancer of respiratory organs among workers at a nickel refinery in Norway. *Int. J. Cancer* **30**(6), 681–685 (1982).

135. R. S. Roberts et al., Cancer mortality associated with the high-temperature oxidation of nickel subsulfide. In F. W. Sunderman, Jr., ed., *Nickel in the Human Environment*, IARC Sci. Publ. No. 53, IARC Lyon, France 1984, pp. 23–35.

136. H. S. Shannon et al., A mortality study of Falconbridge workers. In F. W. Sunderman, Jr., ed., *Nickel in the Human Environment*, IARC Sci. Publ. No. 53, IARC, Lyon, France, 1984, pp. 117–124.

137. R. Egedahl and E. Rice, Cancer incidence at a hydrometallurgical nickel refinery. In F. W. Sunderman, Jr., ed., *Nickel in the Human Environment*, IARC Sci. Publ. No. 53, IARC, Lyon, France, 1984, pp. 47–55.

138. P. E. Enterline and G. M. Marsh, Mortality among workers in a nickel refinery and alloy manufacturing plant in West Virginia. *J. Natl. Cancer Inst.* **68**, 925–933 (1982).

139. R. Lessard et al., Lung cancer in New Caledonia, a nickel smelting island. *J. Occup. Med.* **20**(12), 815–817 (1978).

140. J. Meininger, P. Raffinot, and G. Troly, Cancer in nickel processing workers in New Caledonia. *Science* **215**, 424–425 (1982).
141. A. V. Saknyn and N. K. Shabynina, Some statistical data on the carcinogenous hazards for workers engaged in the production of nickel from oxidized ores. *Gig. Tr. Prof. Zabol.* **14**(11), 10–13 (1970).
142. A. V. Saknyn and N. K. Shabynina, Epidemiology of malignant neoplasms in nickel plants. *Gig. Tr. Prof. Zabol.* **9**, 25–28 (1973).
143. A. Andersen et al., Exposure to nickel compounds and smoking in relation to incidence of lung and nasal cancer among nickel refinery workers. *Occup. Environ. Med.* **53**, 708–713 (1996).
144. J. E. Cox et al., Mortality of nickel workers: Experience of men working with metallic nickel. *Br. J. Ind. Med.* **38**(3), 235–239 (1981).
145. C. K. Redmond, Site-specific cancer mortality among workers involved in the production of high nickel alloys. In F. W. Sunderman, Jr., ed., *Nickel in the Human Environment*, IARC Sci. Publ. No. 53, IARC, Lyon, France, 1984, pp. 73–86.
146. R. G. Cornell, Mortality patterns among stainless-steel workers. In F. W. Sunderman, Jr., ed., *Nickel in the Human Environment*, IARC Sci. Publ. No. 53, IARC, Lyon, France, 1984, pp. 65–71.
147. R. G. Cornell and K. R. Landis, Mortality patterns among nickel/chromium alloy foundry workers. In F. W. Sunderman, Jr., ed., *Nickel in the Human Environment*, IARC Sci. Publ. No. 53, IARC, Lyon, France, 1984, pp. 87–93.
148. J. J. Moulin et al., Mortality study among workers producing ferroalloys and stainless steel in France. *Br. J. Ind. Med.* **47**, 537–543 (1990).
149. D, Burges, Mortality study of nickel platers. In S. S. Brown and F. J. Sunderman, Jr., eds., *Nickel Toxicology*, Proc. 2nd Int. Conf. Nickel Toxicol., Swansea, London, Academic Press, New York, 1980, pp. 15–18.
150. J. H. Godbold, Jr. and E. A. Tompkins, A long-term mortality study of workers occupationally exposed to metallic nickel at the Oak Ridge Gaseous Diffusion Plant. *J. Occup. Med.* **21**(12), 799–806 (1979).
151. T. Sorahan, Mortality from lung cancer among a cohort of nickel cadmium battery workers: 1946-84. *Br. J. Ind. Med.* **44**, 803–809 (1987).
152. H. Waksvik and M. Boysen, Cytogenetic analyses of lymphocytes from workers in a nickel refinery. *Mutat. Res.* **103**, 185–190 (1982).
153. H. Waksvik, M. Boysen, and A. C. Hogetveit, Increased incidence of chromosomal aberrations in peripheral lymphocytes of retired nickel workers. *Carcinogenesis (London)* **5**(11), 1525–1527 (1984).
154. C. Deng et al., Cytogenetic effects of electroplating workers. *Acta Sci. Circumstantiae*, **3**, 167–171 (1983) (in Chinese with English abstract).
155. C. Deng et al., Chromosomal aberrations and sister chromatid exchanges of peripheral blood lymphohocytes in Chinese electroplating workers. *J. Trace Elem. Exp. Med.* **7**, 57–62 (1988).
156. American Conference of Governmental Industrial Hygienists (ACGIH), *Threshold Limit Values for Chemical Substances and Physical Agents and Biological Exposure Indices*, ACGIH, Cincinnati, OH, 1998.
157. Deutsche Forschungsgemeinschaft (DFG), *List of MAK and BAT Values 1996*, Rep. No 32. VCH Verlagsges., Weinheim, 1996 (in English).
158. International Programme on Chemical Safety (IPCS), Data-Base, (INCHEM) 1999.

159. Maximum Allowable Concentrations and Intensities of Harmful Agents in Work Environment. Ordinance of the Minister of Labour and Social Policy of the Republic of Poland of 17.06.1998. Dz. Ustaw No 179 p02.513, 1998 (in Polish).
160. H. Heinrichs and R. Mayer, Distribution and cycling of nickel in forest ecosystems. In J. O. Nriagu, ed., *Nickel in the Environment*, Wiley, New York, 1980, pp. 431–455.
161. F. W. Sunderman, Jr. and A. Oskarsson, Nickel. In Merian, ed., *Metals and Their Compounds in the Environment*, VCH Verlagsges., Weinheim, 1988, pp. 1–19.
162. National Academy of Sciences (NAS), *Nickel*, NAS, Washington, DC, 1975.
163. D. W. Jenkins, Nickel accumulation in aquatic biota. In J. O. Nriagu, ed., *Nickel in the Environment*, Wiley, New York, 1980, pp. 273–337.
164. H. Babich and G. Stotzky, Nickel toxicity to microbes: Effect of pH and implications for acid rain. *Environ. Res.* **29**, 335–350 (1982).
165. H. Babich and G. Stotzky, Nickel toxicity to fungi: Influence of some environmental factors. *Ecotoxicol. Environ. Saf.* **6**, 577–589 (1982).
166. D. F. Spencer, Nickel and aquatic algae. In J. O. Nriagu, ed., *Nickel in the Environment*, Wiley, New York, 1980, pp. 339–347.
167. L. Brkovic-Popovic and M. Popovic, Effects of heavy metals on survival and respiration rate of tubificid worms: Part 1. Effects on survival. *Environ. Pollut.* **13**, 65–72 (1977).
168. V. Bryant et al., Effect of temperature and salinity on the toxicity of nickel and zinc to two estuarine invertebrates (*Corophium volutator, Macoma balthica*). *Mar. Ecol., Prog. Ser.* **24**(1–2), 139–153 (1985).
169. B. G. Anderson, The apparent thresholds of toxicity to Daphnia magna for chlorides of various metals when added to Lake Erie water. *Sewage Works J.* **16**, 96–113 (1950).
170. L. P. Lazareva, Changes in biological characteristics of *Daphnia magna* from chronic action of copper and nickel at low concentrations. *Gidrobiol. Zh.* **21**, 59–62 (1985).
171. C. Powlesland and I. C. George, Acute and chronic toxicity of nickel to larvae of *Chronomus riparis* (Meigen). *Environ. Pollut.* **42**(1), 47–64 (1986).
172. W. J. Birge and J. A. Black, Aquatic toxicology of nickel. In J. O. Nriagu, ed., *Nickel in the Environment*, Wiley, New York, 1980, pp. 349–406.
173. V. M. Brown, The calculation of the acute toxicity of mixtures of poisons to rainbow trout. *Water Res.* **2**, 723–733 (1968).
174. D. Calamari, G. F. Gaggino, and G. Pacchetti, Toxokinetics of low levels of Cd, Cr, Ni and their mixture in long-term treatment of *Salmo gairdneri*, Rich. *Chemosphere* **11**(1), 59–70 (1982).
175. E. F. Neuhauser, R. C. Loehr, and M. R. Malecki, Contact and artificial soil tests using earthworms to evaluate the impact of wastes in soil. In J. K. Petros, W. J. Lacy, and R. A. Conway, eds., *Hazardous and Industrial Solid Waste Testing*, 4th Symp., (AST MSTP) 886. American Society for Testing and Materials, Philadephia, PA, 1985, pp. 192–203.
176. N. D. Yan and R. Strus, Crustacean zooplankton communities of acidic, metal-contaminated lakes near Sudbury, Ontario. *Can. J. Fish. Aquat. Sci.* **37**(12), 2282–2294 (1980).
177. N. D. Yan et al., Richness of aquatic macrophyte floras of soft water lakes of differing pH and trace metal content in Ontario, Canada. *Aquat. Bot.* **23**(1), 27–40 (1985).
178. L. D. Gignac and P. J. Beckett, The effect of smelting operations on peatlands near Sudbury, Ontario. Canada. *Can. J. Bot.* **64**(6), 1138–1147 (1986).
179. F. W. Sunderman, Jr. et al., Physicochemical characteristics and biological effects of nickel oxides. *Carcinogenesis (London)* **8**(2) 305–313 (1987).

180. D. H. Antonsen, Nickel compounds. In H. F. Mark, ed., *Kirk-Othmer Encyclopedia of Chemical Technology*, 3rd ed., Vol. 15, Wiley, New York, 1981, pp. 801–819.

181. J. M. Benson et al., Comparative acute toxicity of four nickel compounds to F344 rat lung. *Fundam. Appl. Toxicol.* **7**, 340–347 (1986).

182. T. Toya et al., Lung lesions induced by intratracheal instillation of nickel fumes and nickeloxide powder in rats. *Ind. Health* **35**, 69–77 (1977).

183. A. P. Wehner et al., Chronic inhalation of nickel oxide and cigarette smoke by hamsters. *Am. Ind. Hyg. Assoc. J.* **36**, 801–810 (1975).

184. J. K. Dunnick et al., Comparative toxicity of nickel oxide, nickel sulfate hexahydrate, and nickel subsulfide after 12 days of inhalation exposure to F344/N rats and B6C3F$_1$ mice. *Toxicology* **50**, 145–156 (1988).

185. S. Takenaka, D. Hochrainer, and H. Oldiges, Alveolar proteinosis induced in rats by long-term inhalation of nickel oxide. In S. S. Brown and F. W. Sunderman, Jr., eds., *Progress in Nickel Toxicology*, Proc. 3rd Int. Conf. Nickel Metab. Toxicol. Paris, 1984, Blackwell, Oxford, UK, 1985, pp. 89–92.

186. C. H. Weischer, W. Kördel, and D. Hochrainer, Effects of NiCl$_2$ and NiO in Wistar rats after oral uptake and inhalation exposure respectively. *Zentralbl. Bakteriol., Parasitenkol., Infektionsk. Hyg., Abt. I: Orig., Reihe B.* **171**, 336–351 (1980).

187. A. Horie et al., Electron microscopical findings, with special reference to cancer in rats caused by inhalation of nickel oxide. *Biol. Trace Elem. Res.* **7**, 223–239 (1985).

188. U. Glaser et al., Long-term inhalation studies with NiO and As$_2$O$_3$ aerosols in Wistar rats. *Int. Congr. Sci. — Excerpta Med.* **676**, 325–328 (1986).

189. National Toxicology Program (NTP), *Technical Report on the Toxicology and Carcinogenesis Studies of Nickel Oxide (CAS No 1313-99-1) in F344/N Rats and B6C3F$_1$ Mice* (Inhalation Studies), NTP TR 451, NIH Publ. No. 96-3367, U. S. Department of Health and Human Services, Public Health Service, National Institute of Health, Washington, DC, 1996.

190. F. W. Sunderman, Jr. and K. S. McCully, Carcinogenesis tests of nickel arsenides, nickel antimonide, and nickel telluride in rats. *Cancer Invest.* **1**, 469–474 (1983).

191. J. P. W. Gilman, Metal carcinogenesis. II. A Study of the carcinogenic activity of cobalt, copper, iron, and nickel compounds. *Cancer Res.* **22**, 158–162 (1962).

192. F. Pott et al., Carcinogenicity studies on fibres, metal compounds and some other dusts in rats. *Exp. Pathol.* **32**, 129–152 (1987).

193. V. Skaug et al., Tumor induction in rats after intrapleural injection of nickel subsulfide and nickel oxide. In S. S. Brown and F. W. Sunderman, Jr., eds., *Progress in Nickel Toxicology*, Blackwell, Oxford, UK, 1985, pp. 37–41.

194. N. Kanematsu, M. Hara, and T. Kada, Rec assay and mutagenicity studies on metal compounds. *Mutat. Res.* **77**, 109–116 (1980).

195. J. G. Jones and C. G. Warner, Chronic exposure to iron oxide, chromium oxide, and nickel oxide fumes in metal dressers in a steelworks. *Br. J. Ind. Med.* **29**, 168–177 (1972).

196. D. C. F. Muir et al., Prevalence of small opacities in chest radiographs of nickel sinter plant workers. *Br. J. Ind. Med.* **50**, 428–431 (1993).

197. H. Keskinen, P. L. Kalliomaki, and K. Alanko, Occupational asthma due to stainless steel welding fume. *Clin. Allergy* **10**, 151–159 (1980).

198. R. Doll, author-in-chief, Report of the International Committee on Nickel Carcinogenesis in Man. *Scand. J. Work Environ. Health* **16**(1) (Spec. Issue), 1–82 (1990).

199. W. W. Payne, Carcinogenicity of nickel compound in experimental animals. *Proc. Am. Assoc. Cancer Res.* **5**, 50 (abstr.) (1964).
200. R. B. Ciccarelli and K. E. Wetterhahn, Nickel distribution and DNA lesions induced in rat tissues by the carcinogen nickel carbonate. *Cancer Res.* **42**(9), 3544–3549 (1982).
201. R. B. Ciccarelli and R. B. Wetterhahn, Molecular basis for the activity of nickel. In F. W. Sunderman, Jr., ed., *Nickel in the Human Environment*, IARC Sci. Publ. No. 53, IARC, Lyon, France, 1984, pp. 201–213.
202. K. S. Kasprzak, *Nickel subsulfide—Ni_3S_2*: Chemistry, Applications, Carcinogenicity, Politechnika Poznańska, Rozprawy, No. 88, Wydawnictwo Politechniki Poznańskiej, Poznań, 1978 (in Polish).
203. G. L. Finch, G. L. Fisher, and T. L. Hayes, The pulmonary effects and clearance of intratracheally instilled Ni_3S_2 and TiO_2 in mice. *Environ. Res.* **42**, 83–93 (1987).
204. A. D. Ottolenghi et al., Inhalation studies of nickel sulfide in pulmonary carcinogenesis of rats. *J. Natl Cancer Inst. (U.S.)* **54**(5), 1165–1171 (1974).
205. F. W. Sunderman, Jr. et al., Embryotoxicity and fetal toxicity of nickel in rats. *Toxicol. Appl. Pharmacol.* **43**, 381–390 (1978).
206. F. W. Sunderman, Jr. et al., Embryotoxicity and teratogenicity of nickel compounds. In T. W. Clarkson, G. F. Nordberg, and P. R. Sager, eds., *Reproductive and Developmental Toxicity of Metals*, Plenum, New York, 1983, pp. 399–416.
207. National Toxicology Program (NTP), *Technical Report on the Toxicology and Carcinogenesis Studies of Nickel Subsulfide (CAS No. 12035-72-2) in F344/N Rats and B6C3F$_1$ Mice (Inhalation Studies)*. NTP-TRS No. 453, U. S. Department of Health and Human Services, Public Health Service, National Institute of Health, Washington, DC, 1996.
208. M. M. Mason, Nickel sulfide carcinogenesis. *Environ. Physiol. Biochem.* **2**, 137–141 (1972).
209. F. W. Sunderman, Jr. et al., Effects of manganese on carcinogenicity and metabolism of nickel subsulfide. *Cancer Res.* **36**, 1790–1800 (1976).
210. F. W. Sunderman, Jr., Organ and species specificity in nickel subsulfide carcinogenesis. In R. Langenbach, S. Nesnow, and J. M. Rice, eds., *Organ and Species Specifity in Chemical Carcinogenesis*, Plenum, New York, 1983, pp. 107–126.
211. F. W. Sunderman, Jr. et al., Induction of renal cancer in rats by intrarenal injection of nickel subsulfide. *J. Environ. Pathol. Toxicol.* **2**, 1511–1527 (1979).
212. I. Damjanov et al., Induction of testicular sarcomas in Fisher rats by intratesticular injection of nickel subsulfide. *Cancer Res.* **38**, 268–276 (1978).
213. M. Shibata et al., Induction of soft tissue tumors in F344 rats by subcutaneous, intramuscular, intraarticular, and retroperitoneal injection of nickel sulphide (Ni_3S_2). *J. Pathol.* **147**, 263–274 (1989).
214. O. Mitsumasa, Induction of ocular tumor by nickel subsulfide in the Japanese common newt, *Cynops pyrrhogaster*. *Cancer Res.* **47**, 5213–5217 (1987).
215. S. H. H. Swierenga and J. R. McLean, Further insights into mechanisms of nickel-induced DNA damage: Studies with cultured rat liver cells. In S. S. Brown and F. W. Sunderman, Jr., eds., *Progress in Nickel Toxicology*, Blackwell, Oxford, UK, 1985, pp. 101–104.
216. Z. X. Zhuang et al., DNA strand breaks and poly (ADP-ribose) polymerase activation induced by crystalline nickel subsulfide in MRC-5 lung fibroblast cells. *Hum. Exp. Toxicol.* **15**(11), 891–897 (1996).

217. H. C. Wulf, Sister chromatid exchanges in human lymphocytes exposed to nickel and lead. *Dan. Med. Bull.* **27**, 40–42 (1980).
218. J. J. K. Saxholm, A. Reith, and A. Brøgger, Oncogenic transformation and cell lysis in C3H/10T1/2 cells and increased sister chromatid exchange in human lymphocytes by nickel subsulfide. *Cancer Res.* **41**, 4136–4139 (1981).
219. J. A. Di Paolo and B. C. Casto, Quantitative studies of in vivo morphological transformation of Syrian hamster cells, by inorganic metal salts. *Cancer Res.* **39**, 1008–1013 (1979).
220. R. T. Haro and A. Furst, Studies on the acute toxicity of nickelocene. *Proc. West. Pharmacol. Soc.* **11**, 39–42 (1968).
221. S. Haworth et al., Salmonella mutagenicity, test results for 250 chemicals. *Environ. Mutagen.* **5** (Suppl. 1), 3–142 (1983).
222. J. E. Spice, L. A. K. Staveley, and G. A. Harrow, The heat capacity of nickel carbonyl and the thermodynamics of its formation from nickel and carbon monoxide. *J. Chem. Soc.* **55**, 100–104 (1995).
223. J. E. Kincaid, J. S. Strong, and F. W. Sunderman, Nickel poisoning. I. Experimental study of the effects of acute and subacute exposure to nickel carbonyl. *Arch. Ind. Hyg. Occup. Med.* **8**, 48–60 (1953).
224. National Institute for Occupational Safety and Health (NIOSH), *Special Occupational Hazard Review and Control Recommendations for Nickel Carbonyl*, Publ. No. 77-184, Department of Health, Education, and Welfare, Public Health Service, Center for Disease Control, Division of Criteria Documentation and Standards Development, Rockville, MD, 1977.
225. J. M. Stellman, ed.-in-chief, *Encyclopedia of Occupational Health and Safety*, 4th ed., International Labour Office, Geneva, 1998, pp. 63.32–63.34.
226. D. H. Stedman, Determination of nickel carbonyl in air by colorimetry. In J. K. O'Neill, P. Schuller, and L. Fishbein, eds., *Environmental Carcinogens: Selected Methods of Analysis*. IARC Sci. Publ. No. 71, IARC, Lyon, France, 1986, pp. 261–267.
227. D. H. Stedman, Determination of nickel carbonyl in air by chemiluminescence. In J. K. O'Neill, P. Schuller, and L. Fishbein, eds., *Environmental Carcinogens: Selected Methods of Analysis*. IARC Sci. Publ. No. 71, IARC, Lyon, France, 1986, pp. 269–273.
228. National Institute for Occupational Safety and Health (NIOSH), *Manual of Analytical Methods*, 4th ed., Method 6007, U. S. Department of Health and Human Services, Washington, DC, 1994.
229. F. W. Sunderman, Jr., N. O. Roszel, and R. J. Clark, Gas chromatography of nickel. *Arch. Environ. Health* **16**, 836–843 (1968).
230. F. W. Sunderman, Jr. and C. E. Selin, The metabolism of nickel-63 carbonyl. *Toxicol. Appl. Pharmacol.* **12**, 207–218 (1968).
231. H. W. Armit, The toxicology of nickel carbonyl. Part II. *J. Hygi.* **8**, 565–600 (1908).
232. J. Ghiringhelli and M. Agamennone, The metabolism of nickel in animals experimentally poisoned with nickel carbonyl. *Med. Lav.* **48**, 187–194 (1957) (in Italian).
233. R. L. Hackett and F. W. Sunderman, Jr., Acute pathological reactions to administration of nickel carbonyl. *Arch. Environ. Health* **14**(4), 604–613 (1967).
234. R. L. Hackett and F. W. Sunderman, Jr., Pulmonary alveolar reaction to nickel carbonyl: Ultrastructural and histochemical studies. *Arch. Environ. Health* **16**(3), 349–362 (1968).
235. F. W. Sunderman, A pilgrimage into the archives of nickel toxicology. *Ann. Clin. Lab. Sci.* **19**(1), 1–16 (1989).

236. E. Horak et al., Effect of nickel chloride and nickel carbonyl upon glucose metabolism in rats. *Ann. Clin. Lab. Sci.* **8**, 476–482 (1978).
237. E. Horak and F. W. Sunderman, Jr., Nephrotoxicity of nickel carbonyl in rats. *Ann. Clin. Lab. Sci.* **10**, 425–431 (1980).
238. A. Oskarsson and H. Tjälve, Binding of ^{63}Ni by cellular constituents in some tissues of mice after the administration of ^{63}NiCl$_2$ and ^{63}Ni(CO)$_4$. *Acta Pharmacol. Toxicol.* **45**, 306–314 (1979).
239. F. W. Sunderman et al., Nickel poisoning. IV. Chronic exposure of rats to nickel carbonyl; A report after one year of observation. *Arch. Ind. Health* **16**, 480–484 (1957).
240. K. S. Kasprzak and F. W. Sunderman, Jr., The metabolism of nickel carbonyl—^{14}C. *Toxicol. Appl. Pharmacol.* **15**(2), 295–303 (1969).
241. A. Oskarsson and H. Tjälve, The distribution and metabolism of nickel carbonyl in mice. *Br. J. Ind. Med.* **36**(4), 326–335 (1979).
242. H. Tjälve, S. Jasim, and A. Oskarsson, Nickel mobilization by sodium diethyldithiocarbamate in nickel-carbonyl-treated mice. In F. W. Sunderman, Jr., ed., *Nickel in the Human Environment*, IARC Sci. Publ. No. 53, IARC, Lyon, France, 1984, pp. 311–320.
243. F. W. Sunderman, Jr., P. R. Allpass, and J. M. Mitchell, Ophthalmic malformations in rats following prenatal exposure to inhalation of nickel carbonyl. *Ann. Clin. Lab. Sci.* **8**, 499–500 (1978).
244. F. W. Sunderman, Jr. et al., Eye malformations in rats: Induction by prenatal exposure to nickel carbonyl. *Science*, **203**, 550–553 (1979).
245. F. W. Sunderman, Jr. et al., Teratogenicity and embryotoxicity of nickel carbonyl in Syrian hamsters. *Teratog. Carcinog. Mutagen.* **1** (2), 223–233 (1980).
246. F. W. Sunderman, Jr. et al, Nickel poisoning. IX. Carcinogenesis in rats exposed to nickel carbonyl. *Arch. Environ. Health* **20**, 36–41 (1959).
247. F. W. Sunderman and A. J. Donnelly, Studies of nickel carcinogenesis: Metastasizing pulmonary tumors in rats induced by the inhalation of nickel carbonyl. *Am. J. Pathol.* **46**, 1027–1041 (1965).
248. T. J. Lau, R. L. Hackett, and F. W. Sunderman, Jr., The carcinogenicity of intravenous nickel cabonyl in rats. *Cancer Res.* **32**, 2253–2258 (1972).
249. G. Hui and F. W. Sunderman, Jr., Effects of nickel compounds on incorporation of [^{3}H]-thymidine into DNA in rat liver and kidney. *Carcinogenesis (New York)* **1**, 297–304 (1980).
250. D. L. Kurta, B. S. Dean, and E. P. Krenzelok, Acute nickel carbonyl poisoning. *Am. J. Emerg. Med.* **11**, 64–66 (1993).
251. F. W. Sunderman, Nickel poisoning. In F. W. Sunderman, Jr., ed., *Laboratory Diagnosis of Diseases Caused by Toxic Agents*, Warren H. Green, St. Louis, MO, 1970, pp. 387–396.
252. U. Vuopala et al., Nickel carbonyl poisoning. Report of 25 cases. *Ann. Clin. Res.* **2**, 214–222 (1970).
253. Z. Shi, Acute nickel carbonyl poisoning: A report of 179 cases. *Br. J. Ind. Med.* **43**, 422–424 (1986).
254. Z. Shi, Nickel carbonyl: Toxicity and human health. *Sci. Total Environ.* **148**, 293–298 (1994).
255. F. W. Sunderman and F. W. Sunderman, Jr., Nickel poisoning VIII. Dithiocarb: A new therapeutic agent for persons exposed to nickel carbonyl. *Am. J. Med., Sci.* **236**, 26–31 (1958).
256. F. W. Sunderman and F. W. Sunderman, Jr., Löfflers syndrome associated with nickel sensitivity. *Arch. Int. Med.* **107**, 405–408 (1961).

257. Z. Shi, A. Lata, and Han Yuhua, A study of serum monoamine oxidase (MAO) activity and the EEG in nickel carbonyl workers. *Br. J. Ind. Med.* **43**, 425–426 (1986).
258. F. W. Sunderman and J. F. Kincaid, Nickel poisoning II. Studies on patients suffering from acute exposure to vapors of nickel carbonyl. *J. Am. Med. Assoc.* **155**(10), 889–894 (1954).
259. H. J. Ludewigs and A. M. Thiess, Knowledge in occupational medicine of nickel carbonyl poisoning. *Zentralbl. Arbeitsmed. Arbeitsschutz* **20**, 329–339 (1970).
260. F. W. Sunderman, The treatment of acute nickel carbonyl poisoning with sodium diethyl dithiocarbamate. *Ann. Clin. Res.* **3**(3), 182–185 (1971).
261. J. F. Kincaid et al., Nickel poisoning. III. Procedures to detection, prevention and treatment of nickel carbonyl exposure including a method for the determination of nickel in biologic materials. *Am. J. Clin. Pathol.* **26**, 107–119 (1956).
262. S. N. Sorinson, A. P. Kornilova, and A. M. Artemeva, Concentrations of nickel in blood and urine of workers in the nickel carbonyl industry. *Gig. Sanit.* **23**(9), 69–72 (1958).
263. H. Hagedorn-Gotz, G. Kuppers, and M. Stoppler, On nickel contents in urine and hair in a case of exposure to nickel carbonyl. *Arch. Toxicol.* **38**(4), 275–285 (1977).
264. C. Decheng et al., Cytogenic analysis in workers occupationally exposed to nickel carbonyl. *Mutat. Res.* **188**, 149–152 (1987).
265. A. Mas, D. Holt, and M. C. Webb, The acute toxicology and teratogenicity of nickel in pregnant rats. *Toxicology*, **35**, 47–57 (1985).
266. National Research Council (NRC), *Medical and Biological Effects of Environmental Pollutants, Nickel*, Committee on Medical and Biological Effects of Environmental Pollutants, National Academy of Sciences, Washington, DC, 1975.
267. American Biogenics Corporation, *Ninety Day Gavage Study in Albino Rats using Nickel.*, Final report submitted to U. S. Environmental Protection Agency, Office of Solid Waste, Submitted by Research Triangle Institute and American Biogenics Corporation, 1988.
268. C. Watanabe et al., Modification by nickel of instrumental thermoregulatory behavior in rats. *Fundam. Appl. Toxicol.* **14**(3), 578–588 (1990).
269. Y. A. Knight et al., Acute thymic involution and increased lipoperoxides in thymus of nickel chloride-treated rats. *Res. Commun. Chem. Pathol. Pharmacol.* **55**, 291–302 (1987).
270. P. H. Gitlitz, F. W. Sunderman, and P. J. Goldblatt, Aminoaciduria and proteinuria in rats after single intraperitoneal injection of Ni(II). *Toxicol. Appl. Pharmacol.* **34**, 430–440 (1975).
271. F. W. Sunderman, Jr. and E. Horak, Biochemical indices of nephrotoxicity, exemplified by studies of nickel neophropathy, In S. S. Brown and D. S. Davies, eds., *Organ-Directed Toxicity: Chemical Indices and Mechanisms*, Pergamon, Oxford, 1982.
272. E. W. Bingham et al., Responses of alveolar macrophages to metals. I. Inhalation of lead and nickel. *Arch. Environ. Health* **25**, 406–414 (1972).
273. M. Lundborg and P. Camner, Lysozyme levels in rabbit lung after inhalation of nickel, cadmium, cobalt and copper chlorides. *Environ. Res.* **34**, 335–342 (1984).
274. F. W. Sunderman, Jr. et al., Toxicity of alveolar macrophages in rats following parenteral injection of nickel chloride. *Toxicol. Appl. Pharmacol.* **100**, 107–118 (1989).
275. R. C. Sobti and R. K. Gill, Incidence of micronuclei and abnormalities in the head of spermatozoa caused by the salts of a heavy metal nickel. *Cytologia.* **54**, 249–254 (1989).
276. E. Berman and B. Rehnberg, *Fetotoxic Effects of Nickel in Drinking Water in Mice*, EPA 600/1-83-007, NTIS PB83–225383, U. S. Environmental Protection Agency, Washington, DC, 1983.

277. V. G. Nadeenko et al., Embryotoxic effect of nickel entering the body via drinking water. *Gig. Sanit.* **6**, 86–88 (1979).
278. C. C. Lu, N. Matsumoto, and S. Iijima, Placental transfer and body distribution of nickel chloride in pregnant mice. *Toxicol. Appl. Pharmacol.* **53**(3), 409–413 (1981).
279. R. Storeng and J. Jonsen, Effects of nickel chloride and cadmium acetate on the development of preimplantation mouse embryos *in vitro*. *Toxicology* **17**, 183–187 (1980).
280. R. Storeng and J. Jonsen, Nickel toxicity in early embryogenesis in mice. *Toxicology*, **20**(1), 45–51 (1981).
281. S. H. Gilani and M. Marano, Congenital abnormalities in nickel poisoning in chick embryos. *Arch. Environ. Contam. Toxicol.* **9**(1), 17–22 (1980).
282. S. M. Hopfer et al., Teratogenicity of Ni^{2+} in *Xenopus laevis*, assayed by the FETAX procedure. *Biol. Trace Elem. Res.* **29**(3), 203–216 (1991).
283. O. Hauptman et al., Ocular malformations of *Xenoplus laevis* exposed to nickel during embryogenesis. *Ann. Clin. Lab. Sci.* **23**(6), 397–406 (1993).
284. M. Green and B. A. Bridges, Use of simplifield fluctation test to detect low levels of mutagens. *Mutat. Res.* **38**, 33–42 (1976).
285. N. W. Biggart and M. Costa, Assessment of the uptake and mutagenicity of nickel chloride in Salmonella tester strains. *Mutat. Res.* **175**, 209–215 (1986).
286. N. W. Biggart and E. Murphy, Jr., Analysis of metal-induced mutation altering the expression or structure of retroviral gene in a mammalian cell line. *Mutat. Res.* **198**, 115–130 (1988).
287. A. Hartwig and O. Beyersmann, Enhancement of UV mutagenesis and sister-chromatid exchanges by nickel ions in V79 cells: Evidence for inhibition of DNA repair. *Mutat. Res.* **217**, 65–73 (1989).
288. D. E. Amacher and S. C. Pailet, Induction of trifluorothymidine-resistant mutants by metal ions in L5178Y/TK +/− cells. *Mutat. Res.* **78**(3), 279–288 (1980).
289. M. Nishimura and M. Umeda, Induction of chromosomal aberrations in cultured mammalian cells by nickel compounds. *Mutat. Res.* **68**, 337–349 (1979).
290. P. J. Tsai et al., Worker exposure to nickel-containing aerosol in two electroplating shops: Comparison between inhalable and total aerosol. *Appl. Occup. Environ. Hyg.* **11**(5), 484–492 (1996).
291. S. Karjalainen, R. Kerttula, and E. Pukkala, Cancer risk among workers at a copper/nickel smelter and nickel refinery in Finland. *Int. Arch. Occup. Environ. Health*, **63**(8), 547–551 (1992).
292. R. G. Adams, Manufacturing process, resultant risk profiles and their control in the production of nickel-cadmium (alkaline) batteries. *Occup. Med.* **42**(2), 101–106 (1992).
293. J. P. W. Gilman, Muscle tumourigenesis. *Can. Cancer Conf.* **6**, 209–223 (1966).
294. K. S. Kasprzak, P. Gabryel, and K. Jarczewska, Carcinogenecity of nickel(II) hydroxides and nickel(II) sulfate in Wistar rats and its relation to the in vitro dissolution rates. *Carcinogenesis (London)* **4**, 275–279 (1983).
295. L. V. Kosova, Toxicity of nickel sulfate. *Gig. Tr. Prof. Zabol.* **6**, 48–49 (1979).
296. J. E. Evans et al., Behavioral, histological, and neurochemical effects of nickel(II) on the rat olfactory system. *Toxicol. Appl. Pharmacol.* **130**, 209–220 (1995).
297. A. K. Mathur et al., Biochemical and morphological changes in some organs of rats in nickel intoxication. *Arch. Toxicol.* **37**, 159–164 (1977).
298. A. K. Mathur et al., Effect of nickel sulphate on male rats. *Bull. Environ. Contam. Toxicol.* **17**, 241–247 (1977).

299. M. J. Hoey, The effects of metallic salts on the histology and functioning of the rat testis. *J. Reprod. Fertil.* **12**, 461–471 (1966).

300. W. Waltschewa, M. Slatewa, and I. Michailow, Testicular changes due to long-term administration of nickel sulphate in rats. *Exp. Pathol.* **6**(3), 116–120 (1972) (in German).

301. National Toxicology Program (NTP), *Technical Report on the Toxicology and Carcinogenesis Studies of Nickel Sulfate Hexahydrate (CAS No. 10101-97-0) in F344/N Rats and B6C3F1 Mice (Inhalation Study)*. NTP TR 454, NIH Publ. No. 96-3370, U. S. Department of Health and Human Services, Public Health Service, National Institute of Washington, DC, Health, 1996.

302. A. Arlauskas et al., Mutagenicity of metal ions in bacteria. *Environ. Res.* **36**(2), 379–388 (1985).

303. N. T. Christie, The synergistic interaction of nickel(II) with DNA damaging agents. *Toxicol. Environ. Chem.* **22**, 51–59 (1989).

304. Y. W. Lee et al., Mutagenicity of soluble and insoluble nickel compounds at the gpt locus in G12 Chinese hamster cells. *Environ. Mol. Mutagen.* **21**, 365–371 (1993).

305. D. B. McGregor et al., Responses of the L5178Y tk$^+$/tk$^-$ mouse lymphoma cell forward mutation assay: III. 72 coded chemicals. *Environ. Mol. Mutagen.* **12**, 85–154 (1988).

306. G. G. Fletcher et al., Toxicity, uptade, and mutagenicity of particulate and soluble nickel compounds. *Environ. Health Perspect.* **102**(Suppl. 3), 69–79 (1994).

307. M. L. Larramendy, N. C. Popescu, and J. A. DiPaolo, Induction by inorganic metal salts of sister chromatid exchanges and chromosome aberrations in human and Syrian hamster cell strands. *Environ. Mutagen.* **3**, 597–606 (1981).

308. H. Ohno, F. Hanaoka, and M. Yamada, Inducibility of sister chromatid exchanges by heavy metal ions. *Mutat. Res.* **104**, 141–145 (1982).

309. R. Rodriguez-Arnaiz and P. Ramos, Mutagenicity of nickel sulphate in Drosophila melanogaster. *Mutat. Res.* **170**, 115–117 (1986).

310. J. M. Da Costa, Observations on the salts of nickel, especially the bromide of nickel. *Med. News* **43**, 337–338 (1883).

311. T. Daldrup, K. Haarhoff, and S. C. Szathmary, Toedliche nickel — sulfate — intoxikation. *Ber. Gerichtl. Med.* **41**, 141–144 (1983).

312. A. A. Tatarskaya, Occupational disease of upper respiratory tract in persons employed in electrolytic nickel refinning departments. *Gig. Tr. Prof. Zabol.* **6**, 35–38 (1960).

313. G. M. Kucharin, Occupational disorders of the nose and nasal sinuses in workers in an electrolytic nickel refining plant. *Gig. Tr. Prof. Zabol.* **14**, 38–40, (1970).

314. O. V. Sushenko and K. E. Rafikova, Questions of work hygiene in hydrometallurgy of copper, nickel and cobalt in a sulfide ore. *Gig. Tr. Prof. Zabol.* **16**, 42–45 (1972).

315. A. Vyskocil et al., Biochemical renal changes in workers exposed to soluble nickel compounds. *Hum. Exp. Toxicol.* **13**(4), 257–261 (1994).

316. J. Dolovich, S. L. Evans, and E. Nieboer, Occupational asthma from nickel sensitivity: I. Human serum albumin in the antigenic determinant. *Br. J. Ind. Med.* **41**(1), 51–55 (1984).

317. A. M. Cirla et al., Nickel-induced occupational asthma -immunological and clinical aspects. In S. S. Brown and F. W. Sunderman, Jr., eds., *Progress in Nickel Toxicology*. Proc. 3rd Int. Cong. Nickel Metab. Toxicol, Paris, UK, 1984, Blackwell, Oxford, pp. 165–168.

318. H. S. Novey, M. Habib, and I. D. Wells, Asthma and IgE antibodies induced by chromium and nickel salts. *J. Allergy Clin. Immunol.* **72**, 407–412 (1983).

319. P. Bright et al., Occupational asthma due to chrome and nickel electroplating. *Thorax* **52**(1), 28–32 (1997).
320. G. T. Block and M. Yeung, Asthma induced by nickel. *J. Am. Med. Assoc.* **247**(11), 1600–1602 (1982).
321. K. Kalimo and K. Lammintausta, 24 and 48h allergen exposure in patch testing. *Contact Dermatitis* **10**, 25–29 (1984).
322. P. J. Frosch and A. M. Kligman, The chamber-scarification test for irritancy. *Contact Dermatitis* **2**, 314–324 (1976).
323. V. P. Chashschin, G. P. Artunina, and T. Norseth, Congenital defects, abortion and other health effects in nickel refinery workers. *Sci. Total Environ.* **148**, 287–291 (1994).
324. P. L. Goering, Platinum and related metals: Palladium iridium, osmium, rhodium, and ruthenium. In J. B. Sullivan, Jr. and G. R. Krieger, eds., *Hazardous Materials Toxicology: Clinical Principles of Environmental Health*, Williams & Wilkins, Baltimore, MD, 1992, pp. 874–881.
325. *Chemie*, vol. 2. F. A. Brockhaus Verlag, Leipzig, 1987, pp. 1001–1992.
326. S. Harris, Nasal ulceration in workers exposed to ruthenium and platinum salts. *J. Soc. Occup. Med.* **25**, 133–134 (1975).
327. G. Sava et al., Pharmacological control of lung metastases of solid tumours by a nover ruthenium complex. *Clin. Exp. Metastasis* **16**, 371–379 (1998).
328. G. Sava et al., Effects of ruthenium complexes on experimental tumors: Irrelevance of cytotoxicity for metastasis inhibition. *Chem.-Biol. Interact.* **95**(1–2), 109–126 (1995).
329. M. Carballo et al., A newly synthesized molecule derived from ruthenium cation, with antitumor activity, activates NADPH oxidase in human neutrophils. *Biochem. J.* **328**, 559–564 (1997).
330. S. Seregard et al., Results following episcleral ruthenium plaque radiotherapy for posterior uveal melanoma. The Swedish experience. *Acta Ophthalmol. Scand.* **75**(1), 11–16 (1997).
331. C. L. Shields et al., Plaque radiotherapy for the management of uveal metastasis. *Arch. Ophthalmol.* **115**(2), 203–209 (1997).
332. S. P. Fricker et al., Ruthenium complexes as nitric oxide scavengers: A potential therapeutic approach to nitric oxide-mediated diseases. *Br. J. Pharmacol.* **122**(7), 1441–1449 (1997).
333. Flame Atomic Absorption Spectrometry. Analytical methods, *Varian Publ.* **85-100009-00** p 53 (1989).
334. H. J. M. Bowen and P. A. Cawse, *The Determination of Inorganic Elements in Biological Tissue by Activation Analysis*, Rep. No AERE-R-4309, U. K. Ato. Energy Authority, Wantage, 1963.
335. M. S. Nagra et al., A study of trace elements in scalp hair and fingernails of industrial workers of Ontario. *J. Radioanal. Nucl. Chem.* **162**(2), 283–288 (1992).
336. Z. Holgye, Determination of radioactive ruthenium in urine. *Prac. Lek.* **26**(4), 127–129 (1974).
337. M. V. R. Prasad, D. S. Suryanarayana, and R. K. Jeevanram, A simple non-distillation method for the estimation of Ru-106 in urine. *Appl. Radiat. Isot.* **45**(1), 35–40 (1994).
338. T. A. Akinfieva, Action of ruthenium chloride hydroxide on organism. *Gig. Tr. Prof. Zabol.* **23**(6), 54–55 (1979).
339. T. A. Akinfieva, Establishing a TLV for ruthenium dioxide in workroom air. *Gig. Tr. Prof. Zabol.* **25**(1), 46–47 (1981).

340. H. Kruszyna et al., Ruthenium compounds: Vascular smooth muscle relaxation by nitrosyl derivatives of ruthenium and iridium. *J. Toxicol. Environ. Health* **6**, 757–773 (1980).

341. *Environmental Quality and Safety Supplement*, Academic Press. New York, 1975.

342. L. Kersten et al., Comparative nephrotoxicity of some antitumour-activate platinum and ruthenium complexes in rats. *J. Appl. Toxicol.* **18**(2), 93–101 (1998).

343. J. E. Furchner, C. R. Richmond, and G. A. Drake. Comparative metabolism of radionuclides in mammals. VII. Retention of ^{106}Ru in the mouse, rat, monkey and dog. *Health Phys.* **21**(9), 355–365 (1971).

344. R. S. Bruce, T. E. F. Carr, and M. E. Collins, Studies in the metabolism of carrier-free radioruthenium. III. The behaviour of nitosylruthenium in the gastrointestinal tract. *Health Phys.* **8**, 397–406 (1962).

345. G. E. Runkle et al., Metabolism and dosimetry of inhaled RuO_4 in Fischer-344 rats. *Health Phys.*. **39**, 543–553 (1980).

346. M. B. Snipes, Metabolism and dosimetry of ^{106}Ru inhaled as ^{106}RuO$_4$ by beagle dogs. *Health Phys.* **41**(8), 303–317 (1981).

347. W. M. Pusch, Determination of effective half-life of ^{103}Ru in man after inhalation. *Health Phys*, **15**, 515 (1968).

348. C. E. Weber and J. W. Harvey, Accidental human inhalation of ruthenium tetroxide. *Health Phys.* **30**, 352 (1976).

349. L. G. Barsegjan, O. V. Klykov, and D. P. Osanov, Penetration of ^{95}Zr, ^{95}Nb and ^{106}Ru into the body by skin contamination. *Gig. Sanit.* **46**(1), 32–34 (1981).

350. R. E. Yasbin, C. R. Matthews, and M. J. Clarke, Mutagenic and toxic effect of ruthenium. *Chem.-Biol. Interact.* **31**(3), 355–365 (1980).

351. A. F. Kolpakova and F. J. Kolpakov, Comparative study of the sensitizing action of platinum-group metals. *Gig. Tr. Prof. Zabol.* **27**(7), 22–24 (1983).

352. T. A. Akinfieva et al., Working conditions and health status of workers in contact with certain ruthenium compounds. *Gig. Tr. Prof. Zabol.* **25**(4), 44–45 (1981).

353. N. Yamagata et al., Update and retention experiments of radioruthenium in man. *Health Phys.* **16**, 159–166 (1969).

354. M. F. Sullivan et al., Acute toxicity of β-emitting radionuclides that may be released in a reactor accident and ingested. *Radiat. Res.* **73**(1), 21–36 (1978).

355. J. E. Ballou and R. C. Thompson, *Physiological Parameters for Assessing the Hazard of Exposure to Ruthenium Radioisotopes. II. Chronic Exposure Studies*, Rep. No. HW-46409, Hanford Atomic Products Operation, General Electric, Richland, WA. (1956).

356. S. Laskin et al., Tissue reactions and dose relationships in rats following intrapulmonary β-radiation. *Health Phys.* **10**(12), 1229–1223 (1964).

357. S. Laskin, M. Kuschner, and N. Nelson, Carcinoma of the lung in rats exposed to the β-radiation of intrabronchial ruthenium-106 pellets. I. Dose-response relationships. *J. Natl. Cancer Inst. (U.S.)* **31**(2), 219–226 (1963).

358. H. Howells et al., In vivo measurement and dosimetry of ruthenium-106 oxide in the lung. In *Handling of Radiation Accidents*, Proc. Symp. Handling Radiat. Accidents, IAEA, Vienna, 1977, p. 83.

359. P. Summanen et al., Radiation related complications after ruthenium plaque radiotherapy of uveal melanoma. *Br. J. Ophthalmol.* **80**(8), 732–739 (1996).

360. American Conference of Governmental and Industrial Hygienists (ACGIH), *Threshold Limit Values for Chemical Substances and Physical Agents*, ACGIH Worldwide, Cincinnati, OH, 1998, pp. 132–133.

361. International Atomic Energy Agency (IAEA), *International Basic Safety Standards for Protection against Ionizing Radiation and for the Safety of Radiation Sources*, IAEA, Vienna, 1996.

362. United Nations Scientific Committee on the Effects of Atomic Radiation, *Ionizing Radiation Sources and Biological Effects*, United Nations, New York, 1982.

363. J. Jagielak et al., *Changes of Radiologic Situation in Poland During 10 years Period Following the Chernobyl Accident*, Polish State Environmental Inspectorate, Warsaw, 1996 (in Polish).

364. *Technical Encyclopedia: Chemistry*, WNT 600–601. 1972.

365. E. Browning, *Toxicity of Industrial Materials* 2nd ed., Appleton Century-Crofts, New York, 1969, p. 278.

366. G. D. Clayton and F. E. Clayton, eds., *Patty's Industrial Hygiene and Toxicology*, 3rd ed., Vol. 2A, Wiley, New York, 1981.

367. S. Zyngier, Antitumor effects of rhodium (II) citrate in mice bearing Ehrlich tumors. *Braz. J. Med. Biol. Res.* **22**, 397 (1989).

368. D. G. Craciunescu, Pharmacological and toxicological studies on new Rh (I) organometallic complexes. *In Vivo* **5**, 329 (1991).

369. R. R. Landolt, H. W. Berk, and H. T. Russel, Studies on the toxicity of rhodium trichloride in rats and rabbits. *Toxicol. Appl. Pharmacol.* **21**, 589–590 (1972).

370. N. H. Proctor, J. P. Hughes, and M. L. Fishman, *Chemical Hazards of the Workplace*, 2nd ed., Lippincott, Philadelphia, PA, 1988.

371. B. J. Chase; *Material Safety Report*, N5 81–84, Johnson Matthey Research Center, Reading, UK, 1981.

372. H. A. Schroeder and M. Nason, Interactions of trace metals in mouse and rat tissues; Zinc, chromium, copper, and manganese with 13 other elements. *J. Nutr.* **160**(2), 198–203 (1976).

373. B. Venugopal and T. D. Luckey, *Metal Toxicity in Mammals* Vol. 2. Plenum, New York, 1978, p. 299.

374. W. M. Grant, *Toxicology of the Eye*, 3rd ed., Thomas, Springfield, IL, 1986.

375. B. Magnusson and A. M. Kligmann, Usefulness of guinea pig tests for detection of contact sensitizers. *Adv. Mod. Toxicol.* **4**, 551–560, (1977).

376. P. B. Bedello, M. Goitre, and G. Roncarozo, Contact dermatitis to rhodium. *Contact Dermatitis* **17**, 111–112 (1987).

377. De La Cuadra and N. Grau-Massaries, Occupational contact dermatitis from rhodium and cobalt. *Contact Dermatitis* **25**, 182–184 (1991).

378. *The Merck Index*: In *Encyclopedia of Chemicals, Drugs and Biologicals* Merck & Co., Rahway, NJ: 1989.

379. *MAK—Werten*. Toksikol.—arbeitsmed. Berundungen. Bacol III, Verlag Chemie, Weinheim, 1986.

380. K. J. Campbell et al., Dermal irritancy of metal compounds. *Arch. Environ. Health.* **30**, 168–170 (1975).

381. M. F. Pera and H. C. Harder, Analysis for platinum in biological material by flameless atomic absorption spectrometry. *Clin. Chem.* (Winston-Salem, N.C.) **23**, 1245 (1977).

382. A. H. Jones, Determination of platinum and palladium in blood and urine by flameless atomic absorption spectophotometry. *Anal. Chem.* **48**, 1472 (1976).
383. A. W. Roshchin, V. G. Veselov, and A. J. Panova, Toxicology of platinum and platinum-group metals. *Gig. Tr. prof. Zabol.* **9**(4–9) (1979).
384. W. Moore et al., Prelimiary studies on the toxicity and metabolism of palladium and platinum. *Environ. Health Perspect.* **10**, 63–71 (1975).
385. D. J. Holbrock et al., Studies on the evaluation of the toxicity of various salts of lead, manganese, platinum, and palladium. *Environ. Health Perspect.* **10**, 95–101 (1975).
386. S. F. Meec, G. C. Harrold, and C. P. McCord, The physiologic properties of palladium. *Ind. Med. Surge.* **12**, 447 (1943).
387. M. M. Jones, E. Schoenheit, and A. D. Wecever, Pretreatment and heavy metal LD_{50} values. *Toxicol. Appl. Pharmacol.* **49**(41), 41–44 (1979).
388. M. J. Wiester, Cardiovascular actions of palladium compounds in the unanesthetized rat. *Environ. Health Perspect.* **12**, 41 (1975).
389. A. J. Panova and V. G. Veselov, Toxicity of chlorpalladosamine in the chronic inhalatory exposure of experimental animals. *Gig. Tr. Prof. Zabol.* **11**(45) (1978).
390. M. Kaufmann, *Muench. Med. Wochenschr.* **60**, 525 (1913).
391. H. A. Schroeder and M. Mitchener, Scandium, chromium (IV), gallium, yttrium, rhodium, palladium, indium in mice: Effects on growth and life span. *J. Nutr.* **101**, 1431 (1971).
392. F. J. Kolpakov, A. F. Kolpakova, and V. J. Prochorenko, Toxic and sensitizing properites of palladium hydrochloride. *Gig. Tr. Prof. Zabol.* **4**, 52–54 (1980).
393. T. Z. Liu and K. Bashi, Inhibition of creatine kinase activity and alterations in electrophoretic mobility by palladium ions. *J. Environ. Pathol. Toxicol.* **2**(3), 907–916 (1979).
394. T. Z. Liu, S. D., Lee and R. S. Bhatangar, Inhibition of creatine kinase activity and alterations in electrophoretic mobility by palladium ions. *Toxicol. Lett.* **4**(6), 469–473 (1979a).
395. D. J. Holbrook et al., Effects of platinum and palladium salts on parameters of drug metabolism in rat liver. *J. Toxicol. Environ. Health* **1**(6), 1067–1079 (1976).
396. R. F. Fisher et al., Effect of platinum and palladium salts in thymidine incorporation into DNA of rat tissues. *Environ. Health Perspect.* **12**(57), 57–62 (1975).
397. J. Bunger, J. Storch, and K. Stadler, Cyto- and genotoxic effects of coordination complexes of platinum, palladium and rhodium *in vitro*. *Int. Arch. Occup. Environ. Health* **69**(1), 33–38 (1996).
398. U. Yuriko and M. Masatoshi, Mutagenic activity of some platinum and palladium complexes. *Mutat. Res.* **298**, 269–275 (1993).
399. V. A. Tomilets and J. A. Zakharova, Anaphylactic and anaphulactoid properties of complex palladium compounds. *Farmakol. Toksikol.* **42**(170), 170–173 (1979).
400. P. Koch and H. P. Baum, Contact stomatitis due to palladium and platinum in dental alloys. *Contact Dermatitis.* **34**, 253–257 (1996).
401. L. Kanerva et al., Allergic patch test reactions to palladium chloride in schoolchildren. *Contact Dermatitis.* **34**, 39–42 (1996).
402. B. Kränke and W. Aberer, Multiple sensitivities to metals. *Contact Dermatitis.* **34**, 225 (1996).
403. S. Budavari, ed., *The Merck Index: An Encyclopedia of Chemicals, Drugs, and Biologicals*, 12th ed., Merck & Co., Whitehouse Station, New York, 1996.
404. H. Chmielewski, ed., *Encyklopedia Techniki*, Chemia, 3rd ed., Wydawnictwo Naukowo-Techniczne, Warszawa, 1972, p. 595 (in Polish).

405. J. D. Lee, *A New Concise Inorganic Chemistry*, 3rd ed., Van Nostrand-Reinhold, UK; Polish edition by Wydawnictwo Naukowe PWN Warszawa, 1994, pp. 357–358 (in Polish).
406. R. C. West, ed., *CRC Handbook of Chemistry and Physics, (1985-1986)*, 66th ed., CRC Press, Boca Raton, FL, 1985, p. B-121.
407. H. Matusiewicz and R. M. Barnes, Determination of metal chemotherapeutic agents in human body fluids using inductively coupled plasma atomic-emission spectrometry with electrothermal vaporization. *Acta Chim. Hung.* **125**, 777–784 (1988).
408. J. Bunyan, F. E. Edwin, and J. Green, Protective effects of trace elements other than selenium against dietary necrotic liver degeneration. *Nature (London)* **181**, 1801 (1958).
409. K. Schwarz, E. E. Roginski, and C. M. Foltz, Ineffectiveness of molybdenum, osmium, and cobalt in dietary necrotic liver degeneration. *Nature (London)* **183**, 472 (1959).
410. Osmium and its compounds. *Am. Ind. Hyg. Assoc. J.* **29**(6), 621–623 (1968).
411. J. C. Smith, B. L. Carson, and T. L. Ferguson, Osmium: An appraisal of environmental exposure. *Environ. Health Perspect.* **8**, 201–213 (1974).
412. American Conference of Governmental and Industrial Hygienists (ACGIH), *Documentation of the Threshold Limit Values*, 6th ed., ACGIH Cincinnati, OH, 1996.
413. L. T. Fairhall, The toxicology of the newer metals. *Br. J. Ind. Med.* **3**(4), 207–212 (1946).
414. D. Hunter, Poisoning by the newer metals: Beryllium, cadmium, osmium and vanadium. *Arch. Hig. Rada*, **1**(2), 113–121 (1950).
415. M. Nissila, Osmic acid treatment for rheumatoid synovitis. *Ann. Clin. Res.* **7**(3), 202–204 (1975).
416. H. Sheppeard and D. J. Ward, Intra-articular osmic acid in rheumatoid arthritis: Five year's experience. *Rheumatol. Rehabil.* **19**(1), 25–29 (1980).
417. E. Honkanen et al., Membranous glomerulo-nephritis in rheumatoid arthritis not related to gold or D-penicilamine therapy: A report of four cases and review of their literature. *Clin. Nephrol.* **27**(2), 87–93 (1987).
418. W. Matczak, Osmium tetroxide. In *Podstawy i Metody Oceny Srodowiska Pracy*, Vol. **16**, Centralny Instytut Ochrony Pracy, Warszawa, 1997 (in Polish).
419. Shell Chemical Company, New York, 1961, p. 10.
420. F. R. Brunot, The toxicity of osmium tetroxide (osmic acid). *J. Ind. Hyg.* **15**(136) (1933).
421. A. Masturzo, Ricerche anatomo-patologiche nella intossicazione spermentale da osmio. *Folia Med. (Naples)* **33**, 546 (1950).
422. C. Gmelin, Experiments to determine the action of titanium and osmium on the human body. *Edinburgh Med. J.* **3**, 324 (1827), in Smith et al. (411)
423. J. B. Bardirux, Osmic acid from the therapeutic point of view. Thesis, Paris, 1898, in Smith et al. (411)
424. C. J. Menkes et al., Effects of articular injections of osmic acid in rabbits. Influence on bone growth. *Rev. Rhum. Mal. Osteo-Articulaires* **36**(8), 513–521 (1972).
425. N. Mitchell, C. Laurin, and N. Schepard, The effect of osmium tetroxide and nitrogen mustard on normal articular cartilage. *J. Bone J. Surg. Br.* Vol. **55**(4), 814–821 (1973).
426. A. Maturzo, Sangue periferico e mielogramma nella intossicazione sperimentale da osmiu. *Folia Med. (Naples)* **34**, 27 (1951).
427. A. J. G. McLaughlin, R. Milton, and K. M. A. Perry, Toxic manifestations of osmium tetroxide. *Br. J. Ind. Med.* **3**(3), 183–186 (1946).

428. V. Ardoino et al., Studi sull intossicazione sperimentale da tetrossido di osmio. *Lav. Med.* **23**(6), 247–252 (1969).
429. V. P. Kamboj and A. B. Kar, Antitesticular effect of metallic and rare earth salts. *J. Reprod. Fertil.* **7**, 21–28 (1964).
430. T. Kada, K. Hirano, and Y. Shirasu, Screening of environmental chemical mutagens by the recassay system with *Bacillus Subtilis.* **6**, 149–173 (1980).
431. S. H. Robinson, O. Cantoni, and M. Costa, Analysis of metal-induced DNA lesions and DNA-repair replication in mamman cells. *Mutat Res.* **131**, 173–181 (1984).
432. H. W. Thielmann and H. Gersbach, The nucleotide - permeable *Escherichia coli* cell, sensitive DNA repair indicator for carcinogens, mutagens and antitumor agents binding covalently to DNA. *Z. Krebsforsch. Klin. Onkol.* **192**, 177–214 (1978).
433. B. F. Demple, DNA repair systems in *Escherichia coli* that recognize distortion or specific lesions. *Diss. Abstr. Int. B.* **42**, 2654 (1982).
434. R. H. Dreisbach, *Handbook of Poisoning*, 4th ed., Lange Med. Publ., Los Altos, CA, 1963.
435. J. J. Hostynek et al., Metals and skin. *Crit. Rev. Toxicol.* **23**(2), 206 (1993).
436. A. F. Bakken and P. Blichfeldt, Post-Yerisnotic Reiter's disease in a physican treated with osmium tetroxide. *Scand. J. Rheumatol.* **5**(3), 174–176 (1976).
437. G. Viott, M. Valbonesi, and V. Ardoino, Two cases of professional poisoning with OsO_4. *Lav. Med.* **23**(2), 77–82 (1969).
438. M. Nissila, H. Isomaki, and S. Jalava, Reversible renal side effects of intra-articular osmic injection. *Scand. J. Rheumatol.* **7**(2), 79–80 (1978).
439. M. Oka, A. Rekonen, and A. Rutosi, The fate of intra-articulary injected osmium tetroxide. *Acta Rheumatol. Scand.* **15**, 35 (1969).
440. C. Cruz-Esteban and W. S. Wilke, Non-surgical synovectomy. *Baillieres Clin. Rheumatol.* **9**(4), 787–801 (1995).
441. A. Kajander and A. Ruotsi, The effect of intra-articular osmic acid or rheumatoid knee joint affections. *Ann. Med. Intern. Fenn.* **56**, 87 (1967).
442. P. Jean et al., Skin necrosis following chemical synovectomy with osmium tetroxide. *Therapi.* **41**(5), 357–359 (1986).
443. Y. Collan, C. Servo, and J. Winblad, An acute immune response to intra-articular injection of osmium tetroxide. *Acta Rheumatol. Scand.* **17**, 236–242 (1971).
444. American Conference of Governmental Industrial Hygienists (ACGIH) (1999) *Threshold Limit Values for Chemical Substances and Physical Agents and Biological Exposure Indices.* Cincinnati, OH, 1999.
445. R. P. Beliles, Platinum-group metals: Platinum, Pt; Palladium, Pd; Iridium, Ir; Osmium, Os; Rhodium, Rh; Ruthenium, Ru. In B. D. Clayton and F. E. Clayton, eds., *Patty's Industrial Hygiene and Toxicology*, 4th ed., Vol. 2C, Wiley, New York, 1994, pp. 2183–2201.
446. Internal Programme on Chemical Safety (IPCS), *Environmental Health Criteria 125: Platinum*, World Health Organization, Geneva, 1991.
447. J. E. F. Reynolds, ed., *Martindale: The Extra Pharmacopeia*, 31st ed., Royal Pharmaceutical Society, London, 1996.
448. J. R. Podlewski and A. Chwalibogowska-Podlewska, eds. *Leki Współczesnej Terapii*, 13th ed., Wydawnictwa Fundacji Büchnera, Warszawa, 1998 (in Polish).
449. J. G. Hardman, et al., eds. *Goodman and Gilman's Pharmacological Basis of Therapeutics*, 9th ed., International Edition, McGraw-Hill, Health Professions Division, New York, 1996.

450. E. Gawêda, Platyna. In: *Podstawy i Metody Oceny Srodowiska Pracy*, Vol. 17, Centralny Instytut Ochrony Pracy, Warszawa, 1997 (in Polish).

451. National Institute for Occupational Safety and Health (NIOSH), *Manual of Analytical Methods*, 2nd ed., Method S191, U. S. Department of Health and Human Services, Washington, DC, 1981.

452. A. D. Maynard et al., Measurement of short-term exposure to airbone soluble platinum in the platinum industry. *Ann. Occup. Hyg.* **41**(1), 77–94 (1997).

453. J. Messerschmidt, et al., Adsorptive valtammetric procedure for the determination of platinum baseline, levels in human body fluids. *Fresenius' J. Anal. Chem.* **343**, 391–394 (1992).

454. T. Minami, M. Ichii, and Y. Okazaki, Comparison of threee different methods for measurment of tissue platinum level. *Biol. Trace Elem. Res.* **48**, 37–44 (1995).

455. O. Nygren, et al., Determination of platinum in blood by adsorptive voltammetry. *Anal. Chem.* **62**, 1637–1640 (1990).

456. R. Schierl et al., Urinary excretion of platinum from platinum industry workers. *Occup. Environ. Med.* **55**(2), 138–140 (1998).

457. J. Gasparic et al., Determination of platinum in bioloigcal material by differential-pulse polarography following wet mineralization. *Chem. Listy.* **84**(10), 1098–1104 (1990).

458. L. Xilei, K. Heydorn, and B. Reitz, Limit of detection for the determination of platinum in biological material by RNAA using electrolytic separation of gold. *J. Radioanal. Nucl. Chem.* **160**(1), 85–99 (1992).

459. P. Schramel, I. Wendler, and J. Angerer, The determination of metals (antimony, bismuth, lead, cadmium, mercury, palladium, platinum, tellurium, thallium tin, tungsten in urine samples by inductively coupled plasma — mass spectrometry. *Int. Arch. Occup. Environ. Health* **69**(3), 219–223 (1997).

460. G. W. Haverland and L. I. Wiebe, Determination of platinum biological tissue by instrumental neutron — activation analysis. *Appl. Radiat. Isot.* **42**(8), 775–776 (1991).

461. A. V. Roshchin, V. G. Veselov, and A. I. Panova, Industrial toxicology of metals of the platinum group. *J. Hyg. Epidemiol. Microbiol. Immunol.* **28**, 17–24 (1984).

462. D. J. Holbrook, *Assessment of Toxicity of Automotive Metallic Emissions* Vol. 1, EPA/600/1-76/010a, University of North Carolina, Chapel Hill, 1976, (in Ref. 446).

463. A. G. Degussa, *Ammonium — tetrachloroplatinate (II) — Acute Toxicity after Single Oral Administration in Rats*, unpublished rep. No. 863910, Hanan Degussa, Germany.

464. J. M. Ward et al., Comparative nephrotoxicity of platinum chemotherapeutic agents. *Cancer Treat. Rep.* **69**, 1675–1678, (1976).

465. J. L. Parrot et al., Platinum and platinosis. *Arch. Environ. Health* **19**, 685–691 (1969).

466. A. Saindelle and A. Ruff, Histamine relase by sodium chloroplatinate. *Br. J. Pharmacol.* **35**, 313–321 (1969).

467. D. J. Holbrook, *Assessment of Toxicity of Automotive Metallic Emissions* Vol. 2, EPA/600/1-76/010b, University of North Carolina, Chapel Hill, 1976.

468. R. Bader, A. M. Reichlmayr-Lais, and M. Kirchgessner, Effecte von alimentärem metallischen Platin bei wachsenden Ratten in Abhängigkeit von der Applikationsdaure und der Partikelgrösse. *J. Anim. Physiol. Anim. Nutr.* **67**, 181–187 (1992) (in German).

469. A. M. Reichlmayr-Lais, M. Kirchgessner, and B. Bader, Dose-response relationships of alimentary $PtCl_2$ and $PtCl_4$ in growing rats. *J. Trace Elem. Electrolytes Health Dis.* **6**, 183–187 (1992).

470. W. Moore et al., Biological fate of a single administration of ^{191}Pt in rats following different routes of exposure. *Environ. Res.* **9**, 152–158 (1975).
471. W. Moore et al., Whole body retention in rats of different ^{191}Pt compounds following inhalation exposure. *Environ. Health Perspect.* **12**, 35–39 (1975).
472. J. Taubler, *Allergic Reponse to Platinum and Palladium Complexes. Determination of No-effect Level*, EPA-600/1/77-039, NTIS Access. No. PB 271 659, University of North Carolina, Chapel Hill, 1977.
473. M. Kirchgessner and A. M. Reichlmayr-Lais, Pt-Gehalte in Milch und Nachkommen von Ratten nach Applikation von Platin in Form von PtCl$_2$ und PtCl$_4$ während der Laktation. *J. Anim. Physiol. Anim. Nutr.* **68**, 151–155 (1992) (in German).
474. A. Bogenrieder, A. M. Reichlmayr-Lais, and M. Kirchgessner, Pt-Retention in maternalen Geweben nach unterschiedenie hoher PtCl$_4$- und Pt^0-Ingestion. *J. Anim. Physiol. Anim. Nutr.* **69**, 143–150 (1993) (in German).
475. D. J. Holbrook, Jr., *Content of Platinum and Palladium in Rat Tissue: Correlation of Tissue Concentration of Platinum and Palladium with Biochemical Effects*, EPA 600/1-77/051, University of North Carolina, Chapel Hill, 1977.
476. B. A. Lown et al., Tissue organ distribution and behavioral effects of platinum following acute and repeated exposure of the mouse to platinum sulfate. *Environ. Health Perspect.* **34**, 203–212 (1980).
477. M. J. Cleare et al., Immediate (type I) allergic responses to platinum compounds. *Clin. Allergy* **6**, 183–195 (1976).
478. National Academy of Sciences (NAS), *Platinum-Group Metals*, EPA-600/1/77-040. NTIS Pb 600/1-77-040, National Research Council, Washington, DC, 1977.
479. J. E. Teggins and M. E. Friedman, The inhibition of malate dehydrogenase by chlorammine - platinum complexes. *Biochim. Biophys. Acta* **350**, 272–276 (1974).
480. G. Kazanzis, Role of cobalt iron, lead, manganes, mercury, platinum, selenium and titanium in carcinogenesis. *Environ. Health Perspect.* **40**, 143–161 (1981).
481. R. J. Knox et al., Mechanism of cytotoxicity of anticancer platinum drugs: Evidence that *cis*-diamminedichloroplatinum (II) and *cis*-Diammine-(1,1-Cyclobutanedicarboxylato)platinum (II) differ only in the kinetics of their interaction with DNA. *Cancer Res.* **46**, 1972–1979 (1986).
482. R. Oliñski and T. H. Zastawny, Reactions with DNA of antitumour drug cis-Diamminedichloroplatinum (II). *Postepy Biochem.* **37**(1), 41–48 (1991).
483. T. H. Zastawny and R. Oliñski, Molecular and cellular basis of cis-DDP-DNA interaction. *Postepy Hig. Med. Dosw.* **47**(2), 103–123 (1993).
484. W. DeNeve et al., Discrepancy beetwen cytotoxicity and DNA interstrand crosslinking of carboplatin and cisplatin in vivo. *Invest. New Drugs.* **8**, 17–24 (1990).
485. C. Perez, M. Leng, and J. M. Malinge, Rearrangement of interstrand cross-links into interstrand cross-links in cis,-Diamminedichloroplatinum (II)-modified DNA. *Nucleic Acids Res.* **25**(4), 896–903 (1997).
486. R. E. Biagini et al., Pulmonary hyperactivity in cynomolgus monkeys (*Macaca fasicularis*) form nose-only inhalation exposure to disodium hexachloroplatinate, Na$_2$PtCl$_6$. *Toxicol. Appl. Pharmacol.* **69**, 377–384 (1983).
487. A. J. Tobert and D. R. Davies, Effect of copper and platinum intrauterine devices on endometrial morphology and implantation in the rabbit. *J. Reprod. Fertil.* **50**, 53–59 (1977).

488. A. Bogenrieder, A. M. Reichlmayr-Lais, and M. Kirchgessner, Einfluss von Alimentärem PtCl$_4$ und Pt^0 auf Wachstum, Hämatologische Parameter und auf Reproduktionsleistung. *J. Anim. Physiol. Anim. Nut.* **68**, 281–288 (1992) (in German).

489. International Agency for Research on Cancer (IARC), Cisplatin. In *IARC Monographs on the Evaluation of the Carcinogenic Risk of Chemicals to Humans*, Vol. 26, IARC, Lyon, France, 1981, pp. 151–164.

490. International Agency for Research on Cancer (IARC), Cisplatin. In *Genetic and Related Effects: An Updating of Selected IARC Monographs from Volumes 1 to 42*, IARC, Lyon, France, 1987, Suppl. 6, pp. 178–181.

491. International Agency for Research on Cancer (IARC), Cisplatin. In *IARC Monographs on the Evaluation of the Carcinogenic Risk of Chemicals to Humans. Overall Evaluations of Carcinogenicity: An Update of IARC Monographs*, Vols. 1 to 42. IARC, Lyon, France, 1987, Suppl. 7, pp. 170–171.

492. N. Kanematsu et al., Mutagenicity of cadmium, platinum and rhodium compounds in cultured mammalian cells. *J. Gifu Dent. Soc.* **17**, 575–581 (1990).

493. R. L. Hoffmann, The effect of cisplatin and platinum (IV) chloride on cell growth, RNA, protein, ribosome and DNA synthesis in yeast. *Toxicol. Environ. Chem.* **17**, 139–151 (1988).

494. R. T. Taylor, J. A. Happe, and R. Wu, Methycobalamin methylation of chloroplatinate: Bound chloride, valence state and relative mutagenicity. *J. Environ. Sci. Health*, Part A **A13**, 707–723 (1978).

495. R. T. Taylor et al., Platinum-induced mutations to 8-Azaguanine resistance in Chinese hamster ovary cells. *Mutat. Res.* **67**, 65–80 (1979).

496. R. T. Taylor, J. A. Happe, and R. Wu, Platinum tetrachloride: Mutagenicity and methylation with methycobalamin. *J. Environ. Sci. Health* Part A **A14**, 87–109 (1979).

497. R. T. Taylor, R. Wu, and M. L. Hanna, Induces reversion of a Chinese hamster ovary triple auxotroph. *Mutat. Res.* **151**, 193–308 (1985).

498. Life Science Research, *Potassium Chloroplatinate (II): Assessment of its Mutagenic Potential in Drosophila melanogaster, Using the Sex-linked Recessive Lethal Test, Eye*, unpublished Rep. No. 81/JOM012/474, Life Science Research, Suffolk, (1981) (in Ref. 446).

499. Life Science Research, *Potassium Chloroplatinate (II): Assessment of Clastogenic Action on Bone Marrow Erythrocytes in the Micronucleous Test, Eye*, unpublished Rep. No. 81/JOM013/241, Life Science Research, Suffolk, (1981).

500. Johnson Matthey, *OECD Skin Irritation Test: Determination of the Degree of Primary Cutaneous Irritation Caused by Potassium Tetrachloroplatinate (II)-MS 322 in the Rabbit*, Experiment No. 631/9111, (unpublished report), Johnson Matthey Research Centre, Reading UK, 1981.

501. Johnson Matthey, *OECD Eye Irritation Test: Determination of the Degree of Ocular Caused by Potassium Tetrachloroplatinate (II)-MS 322 in the Rabbit*, Experiment No. 278/8112, (unpublished report) Johnson Matthey Research Centre, Reading, UK, 1981 (in Ref. 446).

502. R. E. Biagini, Pulmonary responsiveness to methacholine and disodium hexachloroplatinate (Na$_2$PtCl$_6$) aerosols in cynomolgus monkeys (*Macaca fascicularis*). *Toxicol. Appl. Pharmacol.* **78**, 139–146 (1985).

503. R. D. Murdoch and J. Pepys, Immunological Responses to complex salts of platinum. *Clin. Exp. Immunol.* **57**, 107–114 (1984).

504. H. C. Schuppe et al., T-cell-dependent poplietal lymph node reactions to platinum compounds in mice. *Int. Arch. Allergy Immunol.* **97**, 308–314 (1992).

505. H. C. Schuppe et al., Specific immunity to platinum compounds in mice. *J. Invest. Dermatol.* **98**, 517 (1992).

506. H. C. Schuppe et al., Immunostimulatory effects of platinum compounds: Correlation between sensitizing properties in vivo and modulation of receptor-mediated endocytosis in vitro. *Int. Arch. Allergy Immunol.* **112**(2), 125–132 (1997).

507. D. A. Basketter and E. W. Scholes, Comparison of the local lymph node assay with the guinea-pig maximization test for the detection of a range of contact allergens. *Food Chem. Toxicol.* **30**, 65–69 (1992).

508. C. Sheard, Contact dermatitis from platinum and related metals. *AMA Arch. Dermatol.* **71**, 357–360 (1955).

509. G. M. Levene and C. D. Calnan, Platinum sensitivity: Treatment by specific hyposensitization. *Clin. Allergy* **1**, 75–82 (1971).

510. P. J. Linnett, Platinum salt sensitivity. *J. Mine Med. Off. Assoc. S. Afr.* **63**, 24–28 (1987).

511. P. Örback, Allergy to the complex salts of platinum. *Scand. J. Work Environ. Health* **8**, 141–145 (1982).

512. B. Liechti, Asthme au platine. *Arch. Mal. Prof. Med. Trav. Secur.* **46**, 541–542 (1985).

513. A. E. Roberts, Platinosis. *Arch. Ind. Hyg. Occup. Med.* **4**, 549–559 (1951).

514. D. Hunter, R. Milton, and K. M. A. Perry, Asthma caused by the complex salts of platinum. *Br. J. Ind. Med.* **2**, 92–98 (1945).

515. J. Pepys and B. J. Hutchcroft, Bronchial provocation test in etiologic diagnosis and analysis of asthma. *Am. Rev. Respir. Dis.* **112**, 829–859 (1975).

516. D. B. Baker et al., Cross-sectional study of platinum salts sensitization among precious metals refinery workers. *Am. J. Ind. Med.* **18**, 653–664 (1990).

517. M. S. Brooks et al., Cold air challenge and platinum skin reactivity in platinum refinery workers. *Chest* **97**, 1401–1407 (1990).

518. J. J. Hostynek et al., Metals and the skin. *Crit. Rev. Toxicol.* **23**, 171 (1993).

519. A. E. Calverley et al., Platinum salt sensitivity in refinery workers: Incidence and effects of smoking and exposure. *Occup. Environ. Med.* **52**, 661–666 (1995).

520. R. Merget et al., Nonspecific and specific bronchial responsiveness in occupational asthma caused by platinum salts after allergen avoidance. *Am. J. Respir. Crit. Care. Med.* **150**, 1146–1149 (1994).

521. K. M. Venables et al., Smoking and occupational allergy in workers in a platinum refinery. *Br. Med. J.* **299**, 939–942 (1989).

522. U. Bolm-Audorff et al., Prevalence of respiratory in a platinum refinery. *Int. Arch. Occup. Environ. Health* **64**, 257–260 (1992).

523. R. Merget et al., Asthma due to the complex salts of platinuma cross-sectional survey of workers in a platinum refinery. *Clin. Allergy* **18**, 569–580 (1988).

524. A. D. Woolf and T. H. Ebert, Toxicity after self-poisoning by ingestion of potassium chloroplatinite. *Clin. Toxicol.* **29**, 467–472 (1991).

525. A. Weber et al., Objektivierung und Quantifizierung, einer Beruflichen Platinbelastung beim Umgang mit Platinhaltigen Katalysatoren. *Verh. Dtsch. Ges. Arbeitsmed.* **31**, 611–614 (1991).

526. S. R. Karasek and M. Karasek, *The Use of Platinum Paper*. Report of (Illinois) Commission on Occupational Diseases to His Excellency Governor Charles S. Deneen, Warner Printing Company, Chicago, IL, 1911, (in Ref. 446).

527. F. V. P. Duffield et al., Determination of human body burden baseline data of platinum through autopsy tissue analysis. *Environ. Health Perspect.* **15**, 131–134 (1976).

528. G. T. Vaughan and T. M. Florence, Platinum in the human diet, blood, hair and excreta. *Sci. Total Environ.* **111**, 47–58 (1992).

529. R. D. Murdoch and J. Pepys, Enhancement of antibody production by mercury and platinum group metal halide salts. *Int. Arch. Allergy Appl. Immun.* **80**, 405–411 (1986).

530. R. Merget, et al., Quantitative skin prick and bronchial provocation test with platinum salt. *Br. J. Ind. Med.* **48**, 830–837 (1991).

531. W. Zachgo, R. Merget, and G. Werninghaus, Bestimmungsverfahren fur spezifisches Immunoglobulin E gegen wiedermolekulare Substanzen (Platinsalz). *Atemwegs- Lungenkr.* **11**, 267–268 (1985).

532. D. L. Gauggel, K. Sarlo, and T. N. Asquith, A proposed screen for evaluation low-molecular-weight chemicals as potential respiratory allergens. *J. Appl. Toxicol.* **13**, 307–313 (1993).

533. M. B. Dally et al., Hypersensitivity to platinum salts: A population study. *Am. Rev. Respir. Dis., Suppl.* **30**, 121 (1980).

534. O. Cromwell et al., Specific IgE antibodies to platinum salts in sensitized workers. *Clin. Allergy* **9**, 109–117 (1979).

535. R. D. Murdoch, J. Pepys, and E. G. Hughes, IgE antibody responses to platinum group metals: A large scale refinery survey. *Br. J. Ind. Med.* **43**, 37–43 (1986).

536. E. Kessern and F. Leon, Effect of different solid metals and metallic pairs on human sperm motility. *Int. J. Fertil.* **19**, 81–84 (1974).

537. M. K. Holland and J. G. White, Heavey metals and spermatozoa. 1. Inhibition of the motility and metabolism of spermatozoa by metals related to copper. *Fertil. Steril.* **34**, 483–489 (1980).

538. J. Pepys, C. A. C. Pickering, and E. G. Hughes, Asthma due to inhaled chemical agents-complex salts of platinum. *Clin. Allergy* **2**, 391–396 (1972).

539. E. G. Hughes, Medical surveillance of platinum refinery workers. *J. Soc. Occup. Med.* **30**, 27–30 (1980).

540. S. Shima et al., Bronchial asthma due to inhaled chloroplatinate. *Jpn. J. Ind. Health* **26**, 500–509 (1984).

541. American Conference of Governmental and Industrial Hygienists (ACGIH), *Documentation of the Threshold Limit Values*, 6th ed., ACGIH, Cincinnati, OH, 1996.

542. D. E. Johnson, J. B. Tillery and R. J. Prevost, Levels of platinum, palladium, and lead in populations of southern california. *Environ. Health Perspect.* **12**, 27–33 (1975).

543. C. Wei and G. M. Morrison, Platinum concentrations in dust of a car park. *Sci. Total Environ.* **146/147**, 169–174 (1993).

544. M. E. Farago et al., Platinum concentration in urban road dust and soil and in blood and urine in the United Kingdom. *Analyst (London)* **123**(3), 451–454 (1998).

545. A. Jendryczko et al., Stêzenie platyny w powietrzu arterii komunikacyjnych Katowic i Czêstochowy. *Bromatol. Chem. Toksykol.* **30**(3), 269–271 (1997).

546. G. D. Barnes and L. D. Talbert, The effect of platinum on population and absorbance of *Euglena gracilis*, Klebs utilizing a method with atomic absorption and coulter counter analysis. *J. Miss. Acad. Sci.* **29**, 143–150 (1984).

547. K. E. Biesinger and G. M. Christensen, Effects of various metals on survival, growth, reproduction, and metabolism of *Daphnia magna*. *J. Fish. Res. Board Can.* **29**, 1691–1700 (1972).

548. M. E. Farago and P. J. Persons, The effect of platinum, applied as potasssium tetrachloroplatinate, on *Setaria verticillata* (L) *P. Beav.* and its growth on flotation tailings. *Environ. Technol. Lett.* **7**, 147–154 (1986).

549. P. F. Ferreira and R. E. Wolke, Acute toxicity of platinum to coho salmon *(Oncorhynchus kisutch)*. *Mar. Pollut. Bull.* **10**, 79–83 (1979).

550. B. C. Casto, J. Meyers, and J. A. DiPaolo, Enhancement of viral transformation for evaluation of the carcinogenic or mutagenic potential of inorganic metal salts. *Cancer Res.* **39**, 193–198 (1979).

551. T. Gebel, Genotoxocity of platinum and palladium compounds in human and bacterial cells. *Mutat. Res.* **389**, 183–190 (1997).

552. R. C. Woodruff et al., The mutagenic effect of platinum compounds in *Drosophila melanogaster. Environ. Mutagen.* **2**, 133–138 (1980).

553. R. B. D'Agostino et al., Effects on the development of offspring of female mice exposed to platinum sulfate or sodium hexachloroplatinate during pregnancy or lactation. *J. Toxicol. Environ. Health* **13**, 879–891 (1984).

554. B. L. Smith, M. L. Hanna, and R. T. Taylor, Induced resistance to platinum in Chinese hamster ovary cells. *J. Environ. Sci. Health, Part A* **A19**, 267–298 (1984).

555. R. C. Gaver and G. Deeb, High-performance liquid-chromatographic procedures for the analysis to carboplatin in human plasma and urine. *Cancer Chemother. Pharmacol.* **16**(3), 201–206 (1986).

556. F. Elfernik, W. F. J. Van-Der-Vijgh, and H. M. Pinedo, On-line differential pulse polarographic detection of carboplatin in biological samples after chromatographic separation. *Anal. Chem.* **58**(11), 2293–2296 (1986).

557. L. J. C. Van-Warmerdam et al., Validated method for the determination of carboplatin in biological fluids by Zeeman atomic-absorption spectrometry. *Fresenius J. Anal. Chem.* **351**(8), 777–781 (1995).

558. P. Lelieveld et al., Preclinical studies on toxicity, antitumour activity and pharmacokinetics of cisplatin and three recently developed derivatives. *Eur. J. Cancer Clin. Oncol.* **20**(8), 1087–1104 (1984).

559. P. Lelieveld, W. J. Van-Der-Vijgh, and D. Van-Velzen, Preclinical toxicology of platinum analogues in dogs. *Eur. J. Cancer Clin. Oncol.* **23**(8), 1147–1154 (1987).

560. K. A. Hahn et al., Hematologic and systemic toxicoses associated with carboplatin administration in cats. *Am. J. Vet. Res.* **58**(6), 677–679 (1997).

561. R. Chun et al., Phase II. Clinical trial of carboplatin in canine transitional cell carcinoma of the urinary bladder. *J. Vet. Intern. Med.* **11**(5). 279–283 (1997).

562. S. K. Aggarwal and J. M. Fadool, Cisplatin and carboplatin induced changes in the neurohypophysis and parathyroid, and their role in nephrotoxicity. *Anticancer Drugs* **4**(2), 149–162 (1993).

563. G. H. Wolfgang et al., Comparative nephrotoxicity of a novel platinum compound, cisplatin and carboplatin in male Wistar rats. *Fundam. Appl. Toxicol.* **22**(1), 73–79 (1994).

564. M. E. Leibbrant and G. H. Wolfgang, Differential toxicity of cisplatin, carboplatin and CI-973 correlates with cellular platinum levels in rat renal cortical slices. *Toxicol. Appl. Pharmacol.* **132**(2), 245–252 (1995).

565. K. Jirsova and W. Mandys, Differences in the inhibition of neurotic outhrowth in organotypic cultures of rat foetal dorsal root ganglia treated with cisplatin and carboplatin: A comparative study. *Folia Histochem. Cytobiol.* **35**(4), 215–219 (1997).

566. G. Cavaletti et al., Carboplatin toxic effects in the peripheral nervous system of the rat. *Ann. Oncol.* **9**(4), 443–447 (1998).

567. R. J. Mount et al., Carboplatin ototoxicity in the chinchilla: Lesions of the vestibular sensory epithelium. *Acta Oto-Laryngol. Suppl.* **519**, 60–65 (1995).

568. E. A. Neuwelt et al., In vitro and animal studies of sodium thiosulfate as a potential chemoprotectant against carboplatin-induced ototoxicity. *Cancer Res.* **56**(4), 706–709 (1996).

569. M. Wake et al., Selective inner hair cell ototoxicity induced by carboplatin. *Laryngoscope* **104**(4), 488–493 (1994).

570. M. Wake et al., Recording from the inferior colliculus following cochlear inner hair cell damage. *Acta Oto-Laryngol.* **116**(5), 714–720 (1996).

571. D. Ding, J. Wang, and R. J. Salvi, Early damage in the chinchilla vestibular sensory epithelium from carboplatin. *Audiol. Neuropatol.* **2**(3), 155–167 (1997).

572. M. Treskes and W. J. Van-Der-Vijgh, WR2721 as a modular of cisplatin and carboplatin induced side effects in comparison with other chemoprotective agents: A molecular approch. *Cancer Chemother. Pharmacol.* **33**(2), 93–106 (1993).

573. Z. H. Siddik et al., The comparative pharmacokinetics of carboplatin and cisplatin in mice. *Cancer Chemother. Pharmacol.* **36**, 1925–1932 (1987).

574. T. Yasumasu et al., Comparative study of cisplatin and carboplatin on pharmacokinetics. *Pharmacol. Toxicol.* **70**(2), 143–147 (1992).

575. R. C. Gaver, A. M. George and G. Deeb, In vitro stability, plasma protein binding and blood cell paritioning of ^{14}C-carboplatin. *Cancer Chemother. Pharmacol.* **20**, 271–276 (1987).

576. Z. H. Siddik et al., Comparative distribution and excretion of carboplatin and cisplatin in mice. *Cancer Chemother. Pharmacol.* **21**, 19–24 (1988).

577. S. Kai et al., Reproduction studies of carboplatin. 1. Intravenous administration to rats prior to and in the early stages of pregnancy. *J. Toxicol. Sci.* **13**(Suppl. 2), 23–24 (1988).

578. S. Kai et al., Reproduction studies of carboplatin. 2. Intravenous administration to rats during the period of fetal organogenesis. *J. Toxicol. Sci.* **13**(Suppl. 2), 35–61 (1988).

579. S. Kai et al., Reproduction studies of carboplatin. 3. Intravenous administration to rats during the perinatal and lactation periods. *J. Toxicol. Sci.* **13**, 63–81 (1988).

580. S. Kai et al., Teratogenic effects of carboplatin an oncostatic drug, administered during early organogenetic period in rats. *J. Toxicol. Sci.* **14**, 115–130 (1989).

581. H. Azouri, J. M. Bidart, and C. Bohuon, *In vivo* toxicity of cisplatin of carboplatin on the leydig cell function effect of the human choriogonadotropin. *Biochem. Pharmacol.* **38**(4), 567–571 (1989).

582. P. Kopf-Maier, Effects of carboplatin on the testis, a histological study. *Cancer Chemother. Pharmacol.* **29**, 227–235 (1992).

583. H. Fuse et al., Effect of carboplatin on rat spermatogenesis. *Urol. Int.* **56**(4), 219–223 (1996).

584. B. J. Sanderson, L. R. Ferguson, and W. A. Denny, Mutagenic and carcinogenic properties of platinum-based anticancer drugs. *Mutat. Res.* **355**(1–2), 59–70 (1996).

585. T. L. Overbeck, J. M. Knight, and D. J. Beck, A comparison of the genotoxic effects of carboplatin and cisplatin in *Escherichia coli. Mutat. Res.* **362**(3), 249–259 (1996).

586. C. M. Gonzales, M. Mudry, and I. Larripa, Chromosome damage induced by carboplatin (CBDCA). *Toxicol. Lett.* **76**(2), 97–103 (1995).

587. A. M. Fichtinger-Schepman et al., Cisplatin and carboplatin—DNA adducts: Is PT-AG the cytotoxic lession *Carcinogenesis (London)* **16**(10), 2447–2453 (1995).

588. F. A. Blommaert et al., The formation and persistence of carboplatin—DNA adducts in rats. *Cancer Chemother. Pharmacol.* **38**(3), 273–280 (1996).

589. E. Quintana et al., Carboplatin treatment induces dose-dependent increases in the frequency of micronuclei in Ehrlich ascites tumor cells. *Mutat. Res.* **322**(1), 55–60 (1994).

590. K. Jirsova and V. Mandys, Carboplatin-induced micronuclei formation in non-neuronal cells of rat foetal dorsal root ganglia cultured in vitro and comparison with another anticancer drug-cisplatin. *Sb. Lek.* **97**(3), 331–342 (1996).

591. B. Jeremic, J. Sibamoto, and M. Abe, Significance of formation of micronuclei in SCC VII murine cells treated with various chemotherapeutic agents. *Srp. Arh. Celok. Lek.* **124**(7–8), 169–174 (1996).

592. A. Laznickova et al., Effect of oxoplatinum and CBDCA on renal functions in rats. *Neoplasma* **36**, 161–169 (1989).

593. S. Takeno et al., Cochlear function after selective inner hair cell degeneration induced by carboplatin. *Hear. Res.* **75**(1–2), 93–102 (1994).

594. S. Takeno et al., Induction of selective inner hair cell damage by carboplatin. *Scanning Microsc.* **8**(1), 97–106 (1994).

595. D. J. Stewart et al., Phase I study of intracarotid administration of carboplatin. *Neurosurgery* **30**(4), 512–516 (1992).

596. A. H. Calvert et al., Phase I studies with carboplatin at the Royal Marsden Hospital. *Cancer Treat. Rev.* **12**(Suppl. A), 51–57 (1985).

597. R. H. Hruban et al., Fatal trombocytopenia and liver failure associated with carboplatin therapy. *Cancer Invest.* **9**(3), 263–268 (1991).

598. J. E. Smith and B. D. Evans, Carboplatin (JM8) as a single agent and in combination in the treatment of small cell lung cancer. *Cancer Treat. Rev.* **12**(Suppl. A), 73–75 (1985).

599. R. F. Ozols et al., High-dose cisplatin and high dose carboplatin in refractory ovarian cancer. *Cancer Treat. Rev.* **12**(Suppl. A), 59–65 (1985).

600. J. Chauvergne et al., Carboplatin and etoposide combination for the treatment of recurrent epithelial ovarian cancer. *Bull. Cancer* **83**(4), 315–323 (1996).

601. D. T. Sleijfer et al., Acute and cumulative effects of carboplatin on renal function. *Br. J. Cancer* **60**(1), 116–120 (1989).

602. T. L. Cornelison and E. Reed, Nephrotoxicity and hydration management for cisplatin, carboplatin and ormaplatin. *Gynecol. Oncol.* **50**(2), 147–158 (1993).

603. P. O. Mulder et al., Renal dysfunction following high-dose carboplatin treatment. *J. Cancer Res. Clin Oncol.* **114**(2), 212–214 (1988).

604. M. J. Egorin et al., Pharmacokinetics and dosage reduction of cis-Diamine(1,1-cyclobutane-dicarboxylato)platinum in patients with impaired renal function. *Cancer Res.* **44**, 5432–5438 (1984).

605. E. Reed and J. Jacob, Carboplatin and renal dysfunction. *Ann. Intern. Med.* **110**(5), 409 (1989).

606. M. E. Gore, A. H. Calvert, and L. E. Smith, High dose carboplatin in the treatment of lung cancer and mesothelioma: A phase I dose escalation study. *Eur. J. Cancer Clin. Oncol.* **23**(9), 1391–1397 (1987).

607. B. R. McDonald et al., Acute renal failure associated with the use of intraperitoneal carboplatin: A report of two cases and review of the literature. *Am. J. Med.* **90**(3), 386–391 (1991).

608. R. Canetta, M. Rozencweig, and S. K. Carter, Carboplatin: The clinical spectrum to data. *Cancer Treat. Rev.* **12**(Suppl. A), 125–136 (1985).

609. M. J. McKeage, Comparative adverse effect profiles of platinum drugs. *Drug. Saf.* **13**(4), 228–244 (1995).

610. L. R. Kelland, New platinum antitumor complexes. *Crit. Rev. Oncol. Hematol.* **15**(3), 191–219 (1993).

611. J. C. Kennedy et al., Carboplatin is ototoxic. Cancer Chemother. *Pharmacol.* **26**(3), 232–234 (1990).

612. J. Beyer et al., Cutaneous toxicity of high-dose carboplatin, etoposide and fosfamide followed by autologous stem cell reinfusion. *Bone Marrow Transplant.* **10**(6), 491–494 (1992).

613. M. E. O'Brien et al., Blindness associated with high-dose carboplatin. *Lancet* **339**, 558 (1992).

614. K. S. Tonkin, P. Rubin, and L. Levin, Carboplatin hypersensitivity case reports and review of the literature. *Eur. J. Cancer* **29A**(9), 1356–1357 (1993).

615. H. H. Windom et al., Anaphylaxis to carboplatin a new platinum chemotherpaeutic agent. *J. Allergy Clin. Immunol.* **90**(4), 681–683 (1992).

616. W. J. Van-Der-Vijgh, Clinical pharmacokinetics of carboplatin. *Clin. Pharmacokinet.* **21**(4), 242–261 (1991).

617. H. Calvert, I. Judson, and W. J. Van-Der-Vijgh, Platinum complexes in cancer medicine: Pharmacokinetics and pharmacodynamics in relation to toxicity and therapeutic activity. *Cancer Surv.* **17**, 189–217 (1993).

618. C. E. Henderson et al., Platinum chemotherapy during pregnancy for serous cytadenocarcinoma of the ovary. *Gynecol. Oncol.* **49**(1), 92–94 (1993).

619. A. Horwich, M. Mason, and D. P. Dearnaley, Use of carboplatin in germ cell tumors of the testis. *Semin. Oncol.* **19**(1) (Suppl. 2), 72–77 (1992).

620. C. B. Kauffman and D. O. Cowan, Cis-and trans-Dichlorodiammineplatinum (II). *Inorg. Synth.* **7**, 239–245 (1963).

621. H. H. Farrish et al., Validation of a Liquid Chromaotgraphy post-columin derivatization assay for determination of cisplatin in plasma. *J. Pharm. Biomed. Anal.* **12**(2), 265–271 (1994).

622. K. Digua et al., Determination of cisplatin in human plasma by HPLC with a glassy carbon-based wall-jet amperometric detection. *J. Liq. Chromatogr.* **15**(18), 3295–3313 (1992).

623. M. Kinoshita et al., High-performance liquid chromatographic analysis of unchanged cis-Diamminedichloroplatinum (II) (cisplatin) in plasma and urine with post-column derivatization. *J. Chromatogr. Biomed. Appl.* **94**(2), 462–467 (1990).

624. A. M. O. Brett et al., Electrochemical determination of carboplatin in serum using a DNA-modified glassy-carbon electrode. *Electroanalysis* **8**(11), 992–995 (1996).

625. T. A. Connors et al., New platinum complexes with anti-tumour activity. *Chem.-Biol. Interact.* **5**, 415–424 (1972).

626. U. Schaeppi et al., cis,-Dichlorodiammineplatinum (II) (NSC-119-875) preclinical toxicologic evaluation of intravenous injection in dogs, monkeys and mice. *Toxicol. Appl. Pharmacol.* **25**, 230–241 (1973).

627. R. J. Kociba and S. D. Sleight, Acute toxicologic and pathologic effects of *cis*-Diamminedichloroplatinum (NSC-119875) in the male rat. *Cancer Chemother. Rep.* **55**(Part 1), 1–8 (1971).

628. J. M. Ward and K. A. Fallvie, The nephrotoxic effects of *cis*-Diamminechloroplatinum (II) (NSC-119875) in male F344 rats. *Toxicol. Appl. Pharmacol.* **38**, 535–547 (1976).

629. R. W. Fleischman et al., Ototoxicity of *cis*-Dichlorodiammineplatinum (II) in the guinea pig. *Toxicol. Appl. Pharmacol.* **33**, 320–332 (1975).

630. Y. Chiba and Y. Kano, Dose effects on vital parameters of guinea pig continuous administration of cisplatin. *Gan to Kagaku Ryoho* **21**(4), 525–530 (1994).

631. C. L. Litterst, A. F. Leroy, and A. M. Guarino, Disposition and distribution of platinum following parenteral administration of cis-Dichlorodiammineplatinum II to animals. *Cancer Treat. Rep.* **63**, 1485–1492 (1979).

632. C. L. Litterst, I. J. Torres, and A. M. Guarino, Plasma levels and organ distribution of platinum in the rat, dog and fish shark following single intravenous administration of cis-Dichlorodiamineplatinum II. *J. Clin. Hematol. Oncol.* **7**, 169–179 (1976).

633. M. L. Meistrich et al., Damaging effects of fourteen chemotherapeutic drugs on mouse testis cells. *Cancer Res.* **42**(1), 122–131 (1982).

634. A. Nambu and Y. Kumamoto, Studies of spermatogenic damages induced by anti-cancer agent and anti-androgenic agents in rats' testes. *Nippon Hinyokika Gakkai Zasshi* **86**(7), 1221–1230 (1995).

635. R. Lazar, P. C. Conran, and I. Damjanov, Embryotoxicity and teratogenicity of *cis*-Diamminedichloroplatinum. *Experientia*, **35**, 647–648 (1978).

636. K. A. Keller and S. K. Aggarwal, Embryotoxicity of cisplatin in rats and mice. *Toxicol. Appl. Pharmacol.* **69**(2), 245–256 (1983).

637. M. L. Bajt and S. K. Aggarwal, An analysis of factors responsible for resorption of embryos in cisplatin-treated rats. *Toxicol. Appl. Pharmacol.* **80**, 97–107 (1985).

638. P. Kopf-Maier, P. Erkenswick, and H. J. Merker, Lack of severe malformation versus occurance of marked embryotoxic effects after treatment of pregnant mice with cis-platinum. *Toxicology*, **34**(4), 321–331 (1985).

639. R. Muranaka et al., Teratogenic characteristic by single dosing of antineoplastic platinum complexes in rats. *Teratology* **44**(6), 7B–8B (1991).

640. E. Giavani et al., Induction of micronuclei and toxic effects in embryos of pregnant rats treated before implantation with anticancer drugs: Cyclophosphamide cis-platinum, adriamycin. *Teratog., Carcinog., Mutagen.* **10**(5), 417–426 (1990).

641. Z. Zemanova et al., Embryotoxicity and nephrotoxicity of cisplatin. *Reprod. Toxicol.* **6**(2), 190 (1992).

642. W. R. Leopold, F. C. Miller, and J. A. Miller, Carcinogenicity of antitumour *cis*-platinum (II) coordination complexes in the mouse and rat. *Cancer Res.* **39**, 913–918 (1979).

643. S. R. Kempf and S. Ivankovic, Carcinogenic effect of cisplatin [*cis*-Diaminedichloroplatinum (II), CDDP] in BD IX rats. *J. Cancer Res. Clin. Oncol.* **111**(2), 133–136 (1986).

644. S. R. Kempf and S. Ivankovic, Chemotherapy-induced malignancies in rats after treatment with cisplatin as single agents and in combination: Preliminary results. *Oncology*, **43**(3), 187–191 (1986).

645. B. A. Diwan et al., Transplacental carcinogenicity of cisplatin: Initiation of skin tumors and induction of other preneoplastic and neopalstic lesions in SENCAR mice. *Cancer Res.* **53**(17), 3874–3876 (1993).

646. B. A. Diwan et al., Transplantacental carcinogenesis by cisplatin in F344/NCr rats: Promotion of kidney tumors by postnatal administration of sodium barbital. *Toxicol. Appl. Pharmacol.* **132**(1), 115–121 (1995).

647. C. Monti-Bragadin, M. Tamaro, and E. Banfi, Mutagenic activity of platinum and ruthenium complexes. *Chem. Biol. Interact.* **11**, 469–472 (1975).

648. K. S. Andersen, Platinum (II) complexes generate frameshift mutations in test strains of *Salmonella thyphimurium*. *Mutat. Res.* **67**, 209–214 (1979).

649. D. Cunningham, J. T. Pembroke, and E. Stevens, Cis-platinum (II) diamminedichloride-induced mutagenesis in *E. coli* K12: Crowding depresion of mutagenesis. *Mutat. Res.* **84**, 273–282 (1981).

650. L. J. Bradley et al., Mutagenicity and genotoxicity of the major DNA adduct of the antitumor drug cis-Diamminedichloroplatinum (II). *Biochemistry* **32**(3), 982–988 (1993).

651. M. A. Hannan, S. G. Zimmer, and J. Hazle, Mechanisms of cisplatin (*cis*-Diamminedichloroplatinum II) — induced cytotoxicity and genotoxicity in yeast. *Mutat. Res.* **127**, 23–30 (1984).

652. S. Pope, J. M. Baker, and J. H. Parish, Assay of cytotoxicity and mutagenicity of alkylating agents by using neurospora spheroplasts. *Mutat. Res.* **125**, 43–53 (1984).

653. R. K. Brodberg, R. F. Lyman, and R. C. Woodruff, The induction of chromosome aberrations by cis-Platinum (II) diamminechloride in drosophila melanogaster. *Environ. Mol. Mutagen.* **5**, 285–297 (1983).

654. M. O. Bradley, I. C. Hsu, and C. C. Harris, Relationships between sister chromatid exchange and mutagenicity toxicity and DNA damage. *Nature (London)*, **282**, 318–320 (1979).

655. D. Turnbull et al., Cisplatinum(II)diammine dichloride cause mutation, transformation and sister-chromatid exchanges in cultured mammalian cells. *Mutat. Res.* **66**, 267–275 (1979).

656. P. Tandon and A. Sodhi, Cis-dichlorodiammine platinum (II) induced aberrations in mouse bone-marrow chromosomes. *Mutat. Res.* **156**, 187–193 (1985).

657. M. J. Edelweiss et al., Clastogenic effect of cisplatin on Wistar rat bone marrow cells. *Braz. J. Med. Biol. Res.* **28**, 679–683 (1995).

658. H. Lantzch and T. Gebel, Genotoxicity of selected metal compounds in the SOS chromotest. *Mutat. Res.* **389**, 191–197 (1997).

659. K. J. Yarema, S. J. Lippard, and J. M. Essigmann, Mutagenic and genotoxic effects of DNA adducts formed by anticancer drug cis-Diamminedichloroplatinum (II). *Nucleic Acids Res.* **23**(20), 4066–4072 (1995).

660. W. M. Hannemon, K. J. Schimenti, and J. C. Schimenti, Molecular analysis of gene conversion in spermatids from transgenic mice. *Gene* **200**(1–2), 185–192 (1997).

661. E. F. Munoz et al., Transplacental mutagenicity of cisplatin: H-ras codon and 13 mutations in skin tumors of SENCAR mice. *Carcinogenesis (London)*, **17**(12), 2741–2745 (1996).

662. A. J. Giurgiovich et al., Cisplatin - DNA adduct formation in maternal and fetal rat tissues after transplacental cisplatin exposure. *Carcinogenesis (London)*, **17**(8), 1665–1669 (1996).

663. A. J. Giurgiovich et al., Elevated mitochondrial cisplatin — DNA adducts levels in rat tissues after transplacental cisplatin exposure. *Carcinogenesis (London)*, **18**(1), 93–96 (1997).

664. R. S. Goldstein and G. H. Mayor, The nephrotoxicity of cisplatin. *Life Sci.* **32**, 685 (1983).

665. C. Cojocel et al., Renal protein degeneration: A biochemical target of specific nephrotoxicants. *Fundam. Appl. Toxicol.* **3**(4), 278–284 (1983).

666. R. E. Bulger and D. C. Dobyan, Proliferative lesions found in rat kidneys after a single dose of cisplatin JNCI, *J. Natl. Cancer Inst.* **73**(5), 1235–1242 (1984).

667. D. C. Dobyan, Long-term consequence of *cis*-platinum-induced renal injury: A structural functional study. *Anat. Rec.* **212**(3), 239–245 (1985).

668. Y. Sadzuka, Y. Shimizu, and Y. Takino, Role of glutathione S-transferase isoenzymes in cisplatin-induced nephrotoxicity in the rat. *Toxicol. Lett.* **70**(2), 211–222 (1994).

669. A. B. Bikkazi et al., Comparative nephrotoxic effects of *cis*-Platinum (II), *cis*-Palladium (II) and cis-Rhodium (III) metal coordination compounds in rat Kidneys. *Comp. Biochem. Physiol. Pharmacol. Toxicol. Endocrinol.* **111**(3), 423–427 (1995).

670. M. Ban, D. Hettich, and N. Huguet, Nephrotoxicity mechanism of *cis*-Platinum(II)diammine dichloride in mice. *Toxicol. Lett.* **71**(2), 161–168 (1994).

671. P. D. Koning et al., Evaluation of cis-Diamminedichloroplatinum (II) (cisplatin) neurotoxicity in rats. *Toxicol. Appl. Pharmacol.* **89**, 81–87 (1987).

672. F. P. Hamers et al., Putative neurotorphic factors in the protection of cisplatin-induced peripheral neuropathy in rats. *Toxicol. Appl. Pharmacol.* **111**(3), 514–522 (1991).

673. R. G. Van-Der-Hoop et al., Protection against cisplatin induced neurotoxicity by ORG 2766: Histological and electrophysiological evidence. *J. Neurol. Sci.* **126**(2), 109–115 (1994).

674. G. Cavaletti et al., Protective effects of glutathione on cisplatin neurotoxicity in rats. *Int. J. Radiat. Oncol. Biol. Phys.* **29**(4), 771–776 (1994).

675. I. Barajon et al., Neuropeptides and morphological changes in cisplatin-induced dorsal root ganglion neuropathy. *Exp. Neurol.* **138**(1), 93–104 (1996).

676. R. Ravi, S. M. Somani, and L. P. Rybak, Mechanism of cisplatin ototoxicity: Antitoxidant system. *Pharmacol. Toxicol.* **76**(6), 386–394 (1995).

677. T. Saito and J. M. Aran, Comparative ototoxicity of ciplatin during acute and chronic treatment. *J. Oto-Rhino-Laryngol. Relat. Spectrom.* **56**(3), 315–320 (1994).

678. J. A. Kaltenbach et al., Comparison of five agents in protecting the cochlea against the ototoxic effects of cisplatin in the hamster. *Otolaryngol. Head Neck Surg.* **117**(5), 493–500 (1997).

679. C. Stengs et al., Cisplatin-induced ototoxicity: Electophsiological evidence of spontaneous recovery in the albino guinea pig. *Hear. Res.* **111**(1–2), 103–113 (1997).

680. T. Saito et al., Similar pharmacokinetics and differential ototoxicity after administration with cisplatin and transplatin in guinea pigs. *Acta Otolaryngol.* **117**(1), 61–65 (1997).

681. Y. Ito et al., Cisplatin neurotoxicity presenting as reversible posterior leukoencephalopathy syndrome. *Am. J. Neuroradiol.* **9**(3), 415–417 (1998).

682. J. T. Thigpen et al., Phase II trial of cisplatin as first-line chemotherapy in patients with advanced or recurrent endometrial carcinoma: A gynecologic oncology group study. *Gynecol. Oncol.* **33**(1), 68–70 (1989).

683. S. Kehoe et al., Single agent high-dose cisplatin (200 mg/m^2) treatment in ovarian carcinoma. *Br. J. Cancer* **66**(4), 717–719 (1992).

684. G. Brillet et al., Long-term renal effect of cisplatin in man. *Am. J. Nephrol.* **14**(2), 81–84 (1994).

685. K. B. Meyer and N. E. Madias, Cisplatin nephrotoxicity. *Miner. Electrolyte Metab.* **20**(4), 201–213 (1994).

686. A. J. Verplanke et al., Comparison of renal function parameters in the assessment of *cis*-platin induced nephrotoxicity. *Nephron* **66**(3), 267–272 (1994).

687. D. Sheikh-Hamad, K. Timmins, and Z. Jalali, Cisplatin induced renal toxicity: Possible reversal by N-acetylcysteine treatment. *J. Am. Soc. Nephrol.* **8**(10), 1640–1644 (1997).

688. G.Ariceta et al., Acute and chronic effects of cisplatin therapy on renal magnesium homeostasis. *Med. Pediat. Oncol.* **28**(1), 35–40 (1997).

689. G. M. Mead et al., Epileptic seizures associated with cisplatin administration. *Cancer Treat. Rep.* **66**(9), 1719–1722 (1982).

690. M. Ashraf et al., Cis-platinum-induced hypomagnesemia and peripheral neuropathy. *Gynecol. Oncol.* **16**(3), 309–318 (1983).

691. R. W. Gregg et al., Cisplatin neurotoxicity: The relationship between dosage, time and platinum concentration in neurologic tissues and morphologic evidence of toxicity. *J. Clin. Oncol.* **10**(5), 795–803 (1992).

692. G. Cavaletti et al., Long-term peripheral neurotoxicity of cisplatin in patients with successfully treated epithelial ovarian cancer. *Anticancer Res.* **14**(3B), 1287–1292 (1994).

693. S. D. Fossa et al., Clinical and biochemical long-term toxicity after postoperative *cis*-platin-based chemotherapy in patients with low-stage testicular cancer. *Oncology* **52**(4), 300–305 (1995).

694. K. K. Fu, E. F. Kai, and C. K. Leung, Cisplatin neuropathy: A prospective clinical and electophysiological study in patients with ovarian carcinoma. *J. Clin. Pharmacol. Ther.* **20**(3), 167–172 (1995).

695. R. L. Brown et al., Audiometric monitoring of *cis*-platinum ototoxicity. *Gynecol. Oncol.* **16**(2), 254–262 (1983).

696. M. Strauss et al., *Cis*-platinum ototoxicity: Clinical experience and temporal bone histopathology. *Laryngoscope*, **93**(12), 1554–1559 (1983).

697. I. Ilveskoski et al., Ototoxicity in children with malignant brain tumors treated with the "8 in 1 chemiotherapy protocol." *Med. Pediatr. Oncol.* **27**(1), 26–31 (1996).

698. C. Bokemeyer et al., Analysis of risk factors for cisplatin-induced ototoxicity in patients with testicular cancer. *Br. J. Cancer.* **77**(8), 1355–1362 (1998).

699. R. J. Babaian, Toxicity associated with intravesical cisplatinum in a patient with superficial bladder cancer. *J. Urol.* **130**(5), 974 (1983).

700. A. Goldberg et al., Anaphylaxis to cisplatin: Diagnosis and value of pretreatment in prevention of recurrent allergic reactions. *Ann. Allergy* **73**(3), 271–272 (1994).

701. T. C. Lee, C. C. Hook, and H. J. Long, Severe exfoliative dermatitis associated with hand ischemia during cisplatin therapy. *Mayo Clin. Proc.* **69**(1), 80–82 (1994).

702. L. Fisshbein, Perspectives on occupational exposure to antineoplastic drugs. *Arch. Geschwulstforch.* **57**, 219–248 (1987).

703. P. Siedlecki, Toksycznosc Cytostatyków dla personelu. *Nowotwory*, **37**, 56–59 (1987).

704. M. Sorsa, K. Hemminki, and H. Vainio, Occupational exposure to anticancer drugs — potential and real hazards. *Mutat. Res.* **156**, 135–149 (1985).

705. P. J. M. Sessink et al., Occupational exposure to antineoplastic and parameters for renal dysfunction. *Int. Arch. Occup. Environ. Health* **69**(3), 215–218 (1997).

706. A. S. Ensslin et al., Biological monitoring of hospital pharmacy personnel occupationaly exposed to cytostatic drugs: Urinary excretion and cytogenetics studies. *Int. Arch. Environ. Health* **70**(3), 205–208 (1997).

707. E. A. Sontaniemi, Liver damage in nurses handling cytostatic agents. *Acta Med. Scand.* **214**, 181–189 (1983).

708. L. S. Forst, Antineoplastic drugs as an occupational hazards in hospitals. *Ann. Intern. Med.* **103**, 473–476 (1985).

709. R. C. DeConti et al., Clinical and pharmacological studies with *cis*-Diamminedichloroplatinum (II). *Cancer Res.* **33**, 1310–1315 (1973).

710. A. W. Prestayko et al., *Cis*-platin (*cis*-Diamminedichloroplatinum (II)). *Cancer Treat. Rev.* **6**, 17–39 (1979).

711. J. B. Vermorken, W. J. F. Van-Der-Vijgh, and H. M. Pinedo, Pharmacokinetic evidence for an enterohepatic circulation in a patient treated with cis,-Dichlorodiamineplatinum (II). *Res. Commun. Chem. Pathol. Pharmacol.* **28**, 319–328 (1980).

712. D. F. Bajorin et al., Pharmacokinetics of *cis*-Diamminedichloroplatinum (II) after administration in hypertonic saline. *Cancer Res.* **46**, 5969–5972 (1986).

713. D. Zemlickis et al., Fetal outcome after in utero exposure to cancer chemotherpay. *Arch. Intern. Med.* **152**(3), 573–576 (1992).

714. D. Zemlickis et al., Cisplatin protein binding in pregnancy and the neonatal period. *Med. Pediatr. Oncol.* **23**(6), 476–479 (1994).

715. W. T. Stephenson et al., Evaluation of reproductive capacity in germ cell tumor patients following treatment with cisplatin, etoposide and bleomycin. *J. Clin. Oncol.* **139**, 2278–2280, (1995).

716. K. Hemminki, P. Kyyrönen, and M. L. Lindbohm, Spontaneous abortions and malformations in the offspring of nurses exposed to anaesthetic ases cytostatic drugs and other potential hazards in hospitals, based on registered information of outcome. *J. Epidemiol. Commun. Health* **39**(2), 141–147 (1985).

717. I. Stücker et al., Risk of spontaneous abortion among nurses handling antineoplastic drugs. *Scand. J. Work Environ. Health* **16**(2), 102–107 (1990).

718. I. Stücker, L. Manderean, and D. Hemon, Relationship between birthweight and occupational exposure to cytostatic drugs during or before pregnancy. *Scand. J. Work, Environ. Health* **19**(3), 148–153 (1993).

719. S. G. Selevan et al., A study of occupational exposure to antineoplastic drugs and fetal loss in nurses. *N. Engl. J. Med.* **19**(313), 1173–1178 (1985).

720. J. Pedersen-Bjergaard et al., Acute nonlymphocytic leukemia following treatment of testicular cancer and gastric cancer with combination chemotherapy not including alkylating agents: Report of two cases. *Am. J. Hematol.* **18**(4), 425–429 (1985).

721. J. M. Kaldor et al., Leukemia following chemotherapy for ovarian cancer. *N. Engl. J. Med.* **322**(1), 1–6 (1990).

722. M. H. Green, Is cisplatin a human carcinogen *J. Natl. Cancer Inst.* **84**(5), 306–312 (1992).

723. T. Skov et al., Leukemia and reproductive outcome among nurses handling antineoplastic drugs. *Br. J. Ind. Med.* **49**(12), 855–861 (1992).

724. C. Bokemeyr and H. J. Schmoll, Treatment of testicular cancer and the development of secondary malignancies. *J. Clin. Oncol.* **13**(1), 283–292 (1995).

725. R. Minero et al., Acute lymphoblastic leukemia in a girl treated for osteosarcoma. *Pediatr. Hematol. Oncol.* **12**(2), 185–188 (1995).

726. K. Jirsova and V. Mandys, Induction of micronuclei and granular chromatin condensation in human skin fibroblasts influenced by cisplatin (*cis*-DDP) in vitro. *Mutat. Res.* **310**(1), 37–44 (1994).

727. A. Kojima, T. Shinkai, and N. Saijo, Cytogenic effects of CPT-11 and its active metabolite, SN-38 on human lymphocytes. *Jpn. J. Clin. Oncol.* **23**(2), 116–122 (1993).

728. S. Osanto et al., Increased frequency of chromosomal damage in peripheral blood lymphocytes up to nine years following curative chemotherapy of patients with testicular carcinoma. *Environ. Mol. Mutagen.* **17**(2), 71–78 (1991).

729. F. P. Perera et al., Multiple biological markers in germ cell tumor patients treated with platinum-based chemotherpay. *Cancer Res.* **52**(13), 3558–3565 (1992).
730. H. Norppa et al., Increased sister chromatid exchange frequencies in lymphocytes of nurses handling cytostatic drugs. *Scand. J. Work, Environ. Health* **6**, 299–231 (1980).
731. G. Thiringer et al., Comparison of methods for the biomonitoring of nurses handling antitumor drugs. *Scand. J. Work, Environ. Health* **17**, 133–138 (1991).
732. I. Grummt, H. J. Grummt, and G. Schott, Chromosomal aberrations in peripheral lymphocytes of nurses and physicians handling antineoplastic drugs. *Mutat. Res.* **302**(1), 19–24 (1993).
733. W. A. Anwar et al., Chromosomal aberrations on micronucleus frequency in nurses occupationaly exposed to cytotoxic drugs. *Mutagenesis* **9**(4), 315–317 (1994).
734. W. Brumen and D. Horvat, Work environment influence on cytostatics-induced genotoxicity in oncologic nurses. *Am. J. Ind. Med.* **30**(1), 67–71 (1996).
735. J. Rubes et al., Cytogenetic analysis of peripheral lymphocytes in medial personnel by means of FISH. *Mutat. Res.* **412**(3), 293–298 (1998).
736. M. P. DeMeo et al., Monitoring exposure of hospital personnel handling cytostatic drugs and contaminated materials. *Int. Arch. Occup. Environ. Health* **66**(6), 363–368 (1995).
737. A. Oskarsson and B. A. Folwer, Alterations in renal heme biosynthesis during metal nephrotoxicity. *Ann. N. Acad Sci.* **514**, 268–277 (1987).
738. Y. Uno and M. Morita, Mutagenic activity of some platinum and palladium complexes. *Mutat. Res.* **298**, 269–275 (1993).
739. S. Sora and G. E. Mag, Induction of meiotic chromosomal malsegragation in yeast. *Mutat. Res.* **201**, 375–384 (1988).

CHAPTER FORTY-TWO

Uranium and Thorium

Melissa A. McDiarmid, MD, MPH and Katherine S. Squibb, Ph.D.

1.0 Uranium

1.0.1 CAS Number: *[7440-61-1]*

Uranium is a heavy, radioactive metal, the 92nd element in the periodic table, and a member of the actinide series. Its name and chemical symbol U are derived from the planet Uranus, discovered (1781) a few years before the element. A compound of uranium (uranium oxide) was discovered in the uranium ore pitchblende by M. H. Klaproth in 1789. Klaproth believed that he had isolated the element, but this was not achieved until 1841 when a French chemist, E. M. Peligot, reduced uranium tetrachloride with potassium in a platinum crucible to obtain elemental uranium.

Uranium is not as rare as once believed. Widely distributed in the earth's crust, uranium occurs to the extent of about 0.0004%, making the metal more plentiful than mercury, antimony, or silver. Before World War II, uranium was of interest only to the chemists and physicists who studied the element as they would any other substance. With the advent of the nuclear age, uranium now occupies a key position in nuclear weapons and energy.

1.1 Chemical and Physical Properties

The physical and chemical properties of uranium and some of its compounds are listed in Table 42.1 (1,2; P. W. Durbin, unpublished data). Natural uranium is a mixture of three isotopes: ^{234}U (0.0057%), ^{235}U (0.7204%), and ^{238}U (99.2739%), giving an atomic weight of 238.03. Uranium-235 undergoes fission with slow neutrons to release large amounts of energy. Uranium-238 absorbs slow neutrons to form ^{239}U, which in turn decays to fissile ^{239}Pu by emitting two beta particles. Of the three natural isotopes, the most abundant is

---

Patty's Toxicology, Fifth Edition, Volume 3, Edited by Eula Bingham, Barbara Cohrssen, and Charles H. Powell.
ISBN 0-471-31934-1 © 2001 John Wiley & Sons, Inc.

Table 42.1. Physical and Chemical Properties of Uranium and Some of Its Important Compounds[a]

Compound	Formula	At. or mcl. wt.	Sp gr (°C)	Mp (°C)[b]	Bp (°C)	Solubility	Aqueous solubility[c]	Transport-ability[d]
Uranium	U, ^{238}U, ^{235}U, ^{234}U	238.03	19.05 (25)	1132	3818	Insol. H_2O; sol. acids; insol. alkalies, alcohol		
Uranium dioxide	UO_2	270.03	10.96	2500	—	Insol. H_2O; sol. HNO_3, conc. H_2SO_4	insol.	Slightly
Uranyl oxide	UO_3	286.03	7.29	Dec.	—	Insol. H_2O; sol. mineral acids	sl.sol.0.01 to 1 g/L	Moderately
Triuranium octoxide	U_3O_8	842.09	8.3	Dec. 1300 to UO_2	—	Insol. H_2O; sol. HNO_3, H_2SO_4	insol.	Slightly
Uranium tetrafluoride	UF_4	314.02	6.7	960	—	Insol. cold H_2O; sol. conc. acids, alkalies; insol. dil. acids, alkalies	v.sl.sol.-<0.01 g/L	Moderately
Uranium hexafluoride	UF_6	352.02	4.68 (20.7)	69.2 (2 atm)	56.2	Dec. cold H_2O, alcohol, ether, CS_2, CCl_4, $CHCl_3$		Highly
Uranyl chloride	UO_2Cl_2	340.93	—	578	Dec.	3.2 kg/L(13°C); sol. ether, alcohol, amyl alcohol	≥200 g/L sol.	
Uranyl nitrate	$UO_2(NO_3)_2 \cdot 6H_2O$	502.13	2.807 (13)	60.2 Dec. 100	118	8 g/L(14°C), sol. mineral acids, alkalies, oxalates	1–200 g/L	Highly
Uranyl fluoride (Uranium oxyfluoride)	UO_2F_2	308.03		Dec. 300	—	Soluble H_2O	V.sol. >200 g/L	
Uranyl acetate, dihydrate	$UO_2(CHCOO)_2 \cdot 2H_2O$	424.15		loses $2H_2O$ 110	Dec. 275			Moderately
Uranium tetrachloride	UCl_4	379.84		590	792			Highly
Uranium sulfates								Highly
Uranium carbonates								Moderately
Uranium nitrates								Slightly
Uranium oxides								Slightly
Uranium hydrides								Slightly
Uranium carbides								

[a] Ref. 1.
[b] Dec. = decomposed.
[c] P. W. Durbin, unpublished data.
[d] Ref. 2; sol. = soluble.

^{238}U (with a half-life of 4.51×10^9 years), which occurs with ^{235}U (with a half-life of 7.00×10^8 years), and ^{234}U (with a half-life of 2.46×10^5 years). Fourteen isotopes of uranium, that range in mass from ^{227}U through ^{240}U were prepared by radioactive processes.

Except for traces of neptunium and plutonium, uranium is the heaviest atom found in nature. Uranium metal is strongly electropositive, reactive, ductile, and malleable, but a poor conductor of electricity. It exists in three crystalline modifications. Alpha-uranium exists at 25 to 668°C and is orthorhombic. The beta phase exists at 668 to 774°C and has a complex tetragonal structure. Gamma uranium exists at 774 to 1132°C and is body-centered cubic.

The unique nature of the room-temperature, alpha structure curtails solid solution of uranium with many other metals. Extensive solid solution without compound formation has been found only with molybdenum, niobium, titanium, and zirconium, but aluminum, beryllium, bismuth, cobalt, copper, gallium, gold, iron, lead, manganese, mercury, nickel, and tin form one or more intermetallic compounds with uranium. Chromium, magnesium, silver, tantalum, thorium, tungsten, vanadium, calcium, sodium, and some of the lanthanons form neither compounds nor solid solutions. Uranium alloys are of great interest in nuclear technology because the pure metal is chemically active and anisotropic and has poor mechanical properties. Alloys can also be useful in diluting enriched uranium for reactors and in providing liquid fuels.

In the dry state, uranium forms compounds in which the valence is 3+, 4+, 5+, or 6+. In aqueous media U^{3+} and U^{5+} are unstable. U^{3+} readily oxidizes, and U^{5+} disproportionates to U^{4+} and U^{6+}. The latter is the most stable form and exists as the oxygen-containing cation UO_2^{2+} (uranyl) in acid solution and in the body. The uranyl ion has a green-yellow fluorescence. Although uranium forms a great variety of compounds in which uranium is either a cation or an anion, the most industrially important compounds are the dioxide (UO_2), the trioxide (UO_3), the octoxide (U_3O_8), the tetrafluoride (UF_4), the hexafluoride (UF_6), and uranyl nitrate hexahydrate ($UO_2(NO_3)_2 \cdot 6H_2O$). The uranyl ion forms soluble complexes with various inorganic and organic anions (e.g., uranyl carbonate and uranyl proteinate), but the exact composition of the complexes cannot be specified.

There are only a few organometallic complexes of uranium. A uranyl protoporphyrin complex has been reported (3) in which the uranyl (UO_2) ion is so tightly bound as not to be nephrotoxic.

Of the four stable oxides of uranium, U_3O_8 and UO_2 are the chief intermediates in the production of uranium metal from its ores. Of the uranium halides, the green UF_4 is the intermediate in the preparation of UF_6, the most volatile uranium compound used in the isotopic separation of ^{235}U and ^{238}U. Because uranyl acetate is decomposed by light, it has been tried in solar batteries and is used chiefly in analytic chemistry. Uranyl nitrate is a severe fire and explosion risk when shocked or heated, or when in contact with organic materials. It is a source of UO_2 and is used in photography and in uranium glazes. Uranyl perchlorate is also a strong oxidizing agent and is available commercially.

Uranium is pyrophoric and reacts with carbon dioxide. A uranium-metal dust cloud ignites at ordinary temperatures. The explosive concentration is 60 oz/1000 ft^3. Complete coverage of uranium metal scrap with oil is essential to prevent fires; supervised burning of all finely divided uranium metal before accumulation occurs is essential. Graphite chips should be used for fighting fires; speed is essential. Combustibles should not be stored near $UO_2(NO_3)_2 \cdot 6H_2O$ because of excess nitric acid in the product. Cylinders of solid UF_6 should be warmed with extreme caution, when preparing release of the gaseous contents, to prevent fracturing the container.

1.1.1 General

To enhance its use in reactors and nuclear weapons, uranium undergoes an industrial enrichment process that increases the ^{235}U content from 0.7% found naturally to a content between 2 and 90%. ^{235}U is the only natural uranium isotope that can sustain the nuclear chain reaction required for reactors and weapons processes. Both the ^{235}U and ^{234}U concentrations are increased in the enrichment process and produce a by-product, UF_6, that is "depleted" of ^{235}U and ^{234}U. Depleted UF_6 is processed to DUF_4, and, due to its density and metallurgical qualities, it is used to produce metal for missiles and armament (4). The specific activity of DU is roughly (0.4 µCi/g) 60% that of natural uranium (0.7 µCi/g) due to the depletion of ^{234}U and ^{235}U. [See Table 42.2 (4).]

1.2 Production and Use

Uranium never occurs in its elemental state but, rather, is always combined with other elements in about 150 known minerals. There are several important uranium ores. Carnotite ($K_2O \cdot 2U_2O_3 \cdot V_2O_5 \cdot 3H_2O$), pitchblende (a mineral complex of $UO_3 \cdot UO_2$, PbO, Th, Y, etc.), tobernite $Cu(UO_2)_2P_2O_8 \cdot 12H_2O$, autunite, and a few others are uranium-bearing minerals of commercial interest. The largest quantities of uranium ore are mined from the Blind River area of Canada. Other significant uranium deposits have been found in South Africa, Australia, France, and the United States (Colorado, Utah, New Mexico, and Wyoming).

No deposits of concentrated uranium ore have been discovered. As a result, uranium must be extracted from ores containing less than 0.1% U. Because it is necessary to use low-grade ores, substantial and complex processing of these ores is required to obtain pure uranium. Usually it is necessary to preconcentrate the ore by grinding and flotation or similar processes. The preconcentrated ore is then leached to dissolve the uranium compounds. The resulting solution is filtered, adsorbed onto an ion-exchange resin, eluted,

Table 42.2. Natural Uranium Compared with the Depleted Uranium Used by DoD[a]

Material	Component by weight %				Radioactivity[c] (µCi/g)
	^{234}U	^{235}U	^{236}U	^{238}U	
U[b] found in nature	0.01	0.72	0	99.28	0.7
DU used by DOD	0.00	0.20	0.00	99.80	0.4

[a]Ref. 4.
[b]The weight percentages quoted for natural uranium vary slightly from source to source.
[c]Reported values for the radioactivity (specific activity) of depleted uranium vary depending primarily on the weight percentages of ^{234}U and ^{235}U (10 *CFR* 20). Although the exact ratio varies, the radioactivity of depleted uranium is always less than that of natural uranium.

URANIUM AND THORIUM

precipitated, and purified to yield uranyl nitrate, which may be converted chemically to other uranium compounds or reduced to yield the pure metal.

Uranium is distributed more abundantly on the earth's surface than the combined total of antimony, bismuth, cadmium, gold, mercury, and silver and averages about 3 g/ton of rock where it occurs. Basic rocks (basalts) contain less than 1 ppm U, whereas acidic rocks (granites) may have 8 ppm or more. Estimates of U content for sedimentary rocks are 2 ppm and for ocean water, 0.001 ppm. The total uranium content of the earth's crust to a depth of 15 miles is calculated as 1014 tons; the oceans may contain 1010 tons of uranium.

The United States, Gabon, and Niger have high-grade sandstone-type deposits of uranium ore. The Lake Elliott area of Canada has Precambrian quartz pebble deposits. Gold–uranium deposits are found in South Africa. The Lake Athabasca region of Canada and the Alligator River region in Australia have high-grade deposits. The annual natural uranium world requirement is expected to increase from three- to fivefold to meet the needs of nuclear power.

New Mexico and Wyoming in 1975 were the leading producer states. Together they accounted for three-quarters of the total domestic production of 12,300 tons U_3O_8 content; Colorado, Texas, Utah, and Washington together accounted for 3000 tons. About 55% of the total production was from open-pit mines, 43% from underground mines, and 2% from leaching and recovery. Open-pit mines totaled 23, underground, 121, and miscellaneous sources, 25 sites. In the United States, most underground uranium mines were shut down by the late 1980s. Worldwide, uranium mining continues, and production is documented in Canada, South Africa, and other African countries, as well as in Australia. The radon in underground uranium mines remains a significant occupational hazard (5).

The National Institute for Occupational Safety and Health (NIOSH) estimates for the number of U-exposed workers are derived from the Standard Industrial Classifications (SIC) codes that cover potential uranium exposure. These include

SIC 1094 Uranium, radium, and vanadium mining
SIC 3483 Ammunition, excluding small arms
SIC 2819 Industrial Inorganic Chemicals NEC (includes U slug manufacture)

The County Business Patterns (Bureau of Census) data for mid-March 1995 yield the following employment data:

SIC 1094 1,001 employees
SIC 3483 12,588 employees
SIC 2819 60,176 employees

County business pattern data do not discriminate among workers in the SIC code in trades without related exposures, so for SIC codes 3483 and 2819, the employment figures are likely to be a large overestimate of workers exposed to U.

Thus, the National Occupational Exposure Survey (NOES) determined that the only SIC code observed for ^{238}U exposure was SIC 7391 "Research and Development

Laboratories. NOES visited seven facilities and estimated that 5328 were exposed, including 720 women, during 1981–1983 (6).

A number of changes have been introduced into the production of uranium. Formerly, processing was by acid or carbonate leaching to produce a concentrate, after which (*1*) uranium was recovered by classified procedures to produce the orange UO_3, (*2*) UO_3 was converted to UF_4, and (*3*) part of the UF_4 was converted to UF_6 for the ^{235}U gaseous diffusion plants, and the remainder to uranium metal for plutonium manufacture. Now *in situ* leaching or solution mining, aided by ferrous ion oxidation and uranium extraction with *Thiobacillus ferroxidans* (7), reduces investment and construction lead time to produce uranium from low-grade ores. Uranium is leached with acidic ferric sulfate from relatively coarse mill-grade ore. The bacteria *T. ferroxidans* is introduced to convert the leach reaction product ferrous sulfate back to ferric sulfate, which is then recycled. A Japanese process purifies uranium in conventional sulfuric acid leach solution by solvent extraction, followed by stripping with chloride solution. The uranium is converted to the uranous state by electrolytic reduction in a cation-exchange membrane cell, then precipitated with HF as insoluble UF_4 "green cake." A French process is similar, although in one approach, the uranium in the chloride stage can be chemically reduced with SO_2 in the presence of copper and HF to form "green cake." Both processes produce an intermediate suitable for direct fluorination to UF_6. An Australian process that involves acid leaching and other solvents for extration is displayed in Fig. 42.1 (8). The chemistry involved in the process is as follows. The ore is crushed and ground to liberate the mineral particles. It is then leached with sulfuric acid

$$UO_3 + 2H^+ \Rightarrow UO_2^{2+} + H_2O$$

$$UO_2^{2+} + 3SO_4^{2-} \Rightarrow UO_2(SO_4)_3^{4-}$$

With some ores, carbonate leaching is used to form a soluble uranyl tricarbonate ion: $UO_2(CO_3)_3^{4-}$. This can then be precipitated with an alkali, e.g., as sodium or magnesium diuranate. Alkaline leaching is not done in Australia at present. Two methods have been used to concentrate and purify uranium: ion exchange and solvent extraction. Early operations in Australia used ammonium-type resins in polystyrene beads for ion exchange, but solvent extraction is now in general use. In solvent extraction, tertiary amines are used in a kerosene diluent, and the phases move countercurrently

$$2R_3N + H_2SO_4 \Rightarrow (R_3NH)_2SO_4$$

$$2(R_3NH)_2SO_4 + UO_2(SO_4)_3^{4-} \Rightarrow (R_3NH)_4UO_2(SO_4)_3 + 2SO_4^{2-}$$

The loaded solvents may then be treated to remove impurities. First, cations are removed at pH 1.5 using sulfuric acid and then anions are dealt with using gaseous ammonia. The solvents are then stripped in a countercurrent process using ammonium sulfate solution.

$$(R_3NH)_4UO_2(SO_4)_3 + 2(NH_4)2SO_4 \Rightarrow 4R_3N + (NH_4)_4UO_2(SO_4)_3 + 2H_2SO_4$$

Ammonium diuranate is precipitated by adding gaseous ammonia to neutralize the

URANIUM AND THORIUM

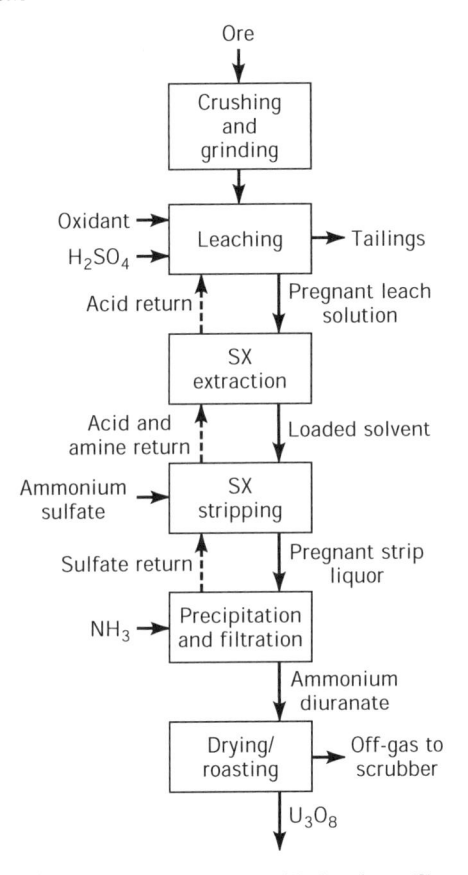

Figure 42.1. Uranium mill chemistry (8).

solution (though in earlier operations caustic soda and magnesia were used)

$$2\,NH_3 + 2\,UO_2(SO_4)_3^{4-} \Rightarrow (NH_4)2U_2O_7 + 4SO_4^{2-}$$

The diuranate is then dewatered and roasted to yield U_3O_8 product, which is the form in which uranium is marketed and exported.

Uranium's commercial importance stems from its use in nuclear weapons. It was first used during the World War II "Manhattan Project." In the subsequent Cold War years, the U.S. Atomic Energy Commission controlled the uranium market in the United States. Because of its density, uranium in its depleted form has also been used for military armament and weaponry. Small amounts of uranium are also used in chemicals, ceramics, glass, and photography.

1.3 Exposure Assessment

Hazardous exposures in the uranium industry begin in the mining process. Hazards are of two types, chemical and radiological; of the two, radiation is the more dangerous.

Effective ventilation control measures have reduced the radiation exposures in the larger mines, but far less satisfactory radiation-exposure conditions exist in small mines without the benefit of ventilation. In addition to the alpha-particle radiation hazard from uranium in the ore, the most hazardous elements are radon gas and its particulate daughters, RaA and RaC, all alpha emitters. Some mine waters are high in radon and thus are an additional exposure source and should not be used for wet drilling. In the mines some beta and gamma exposures from RaB, RaC, and Ra also occur but are of relatively minor importance. The chemical toxicity of uranium is similar to other heavy metals. Storage in the skeleton and excretion via the urine are accompanied by renal toxicity and are discussed further later.

Hazards in milling uranium to produce a concentrate were thought to be relatively minor because a wet process was used. However, some chronic health effects, including nonmalignant respiratory disease and renal tubular biochemical abnormalities, have been documented in these workers and are discussed subsequently. Somewhat greater exposures occur in the production of uranium metal from dusts of the intermediates, UO_3 and UF_4, and from the gaseous UF_6 (UO_2F_2 and HF) in accidents or leaks. Hazards in producing uranium metal briquettes or in hot-rolling uranium rods are relatively small. An evaluation of surface-contamination control for uranium-rolling operations has been made by both Blackwell (9) and Hyatt (10). Uranium metal is pyrophoric; chips from cleaning the briquette readily ignite. Radiation hazards from ^{235}U-enriched uranium are high and are recognized. An overall evaluation of various uranium fabrication procedures is given by Harris and Kingsley (11).

1.3.1 Air

A number of methods for determining uranium in air and biological samples, which were developed by the Manhattan Project at Rochester, New York (1) and quoted in the second edition of that work, are described here because these methods were used to determine the concentrations of U in animal inhalation exposure chambers and the body tissues and fluid levels in the exposed animals that are described in this text. However, there have also been significant advances in U determination techniques that are also described.

A simple, rapid colorimetric method that uses ferrocyanide to determine uranium in dust samples ranging upward from 80 mg is described by Cohenour and Davis in the monograph on uranium (1). The fluorophotometric method was commonly used in the past for small samples from air or samples of biological origin. It is suitable for determining uranium in the parts per billion range if preliminary protein isolation and electrolytic procedures are used (1). Polarographic determination of uranium is particularly valuable for analyzing trace amounts of the hexavalent form in the presence of tetravalent uranium because it requires no prior separation procedures. A spectrochemical method devised by Steadman for determining uranium in all types of samples is given in detail in the monograph on uranium (1).

Vernon et al. (12) recovered uranium quantitatively from seawater by spectrophotometric estimation with 8-hydroxyquinoline after passage through a chelating ion-exchange column, preceded by a trioctylamine extraction. This procedure allowed determination of 200 mg U in the presence of 10 mg each of 23 cations, including larger amounts of calcium, magnesium, ammonium, sodium, and potassium.

URANIUM AND THORIUM

According to Becker and LaFleur (13), neutron activation analysis is a good method for determining uranium in a wide variety of inorganic and biological samples, provided that a reactor is available. With a rapid radiochemical separation, the chemical yield for the complete analysis is 95 ± 3% for metals and 91 ± 4% for biological materials.

Hamilton in the United Kingdom (14) used measurements of the concentration of uranium in normal human tissues to estimate that the 70-kg, ICRP standard man contains a minimum of about 100 mg U and a maximum of about 125 mg. The concentration and tissue content of uranium calculated for the standard man, on which this estimate is partly based, is shown in Table 42.3 (15) and is determined by the delayed-type neutron activation analysis method for residents in the United Kingdom. The skeleton, comprising the analyzed skull, rib, femur, sternum, and vertebrae, has a concentration of 6.96 ng U/g. With the range of 3.22 to 12.10 ng U/g, skeletal concentration is from 14 to 20 times that of all other five analyzed tissues combined.

Much progress has been made recently in laboratory techniques to measure uranium in both environmental and biological samples. Kinetic phosphorescence analysis (KPA) is a computer- controlled, laser-induced, time-resolved luminescence technique with a built-in nitrogen ion laser that excites a stilbene-420 dye laser and stimulates uranium ions in reference and sample cells by a pulsed light source. Emission of photoluminescence during deexcitation of uranium ions back to ground state is captured and converted to an electronic signal by a photomultiplier tube connected to a dual scaler. Time gates of a specified millisecond duration are used after each laser pulse to measure photoluminescent intensity. For luminescent emissions, the logarithm of intensity is a linear function of time (16). A second technique is inductively coupled, argon-plasma, mass spectrometry (ICP-MS) that allows lower limits of detection and alleviates some laboratory preparation of specimens. It has also helped overcome interference and matrix problems (17).

1.3.2 Background Levels

The total body burden of 86.3 mg estimated from the six tissues listed in Table 42.3 (15), constitutes 82% of body mass. Assuming that the remaining body mass of 18% contains uranium of the same order of magnitude, Hamilton tentatively arrived at 100 to 125 mg U for total body burden (age not specified).

Table 42.3. Concentration of Uranium in Human Tissues[a]

Tissue	Average concentration (ng U/g wet wt.)	Range (ng U/g)	Total amount U in tissue standard man (μ)
Skeleton	6.96	3.22–12.1	69.6
Blood	0.84	0.12–1.41	6.0
Fat	0.60	0.52–0.68	5.7
Liver	0.25	0.21–0.28	4.5
Muscle	0.19	0.05–0.33	0.43
Heart	0.16	—[b]	0.048
Total			86.3

[a]Ref. 15.
[b]One sample.

The primary source of uranium was food of the order of 1 mg/day. This value from the United Kingdom agrees fairly well with 1.3, 1.4, and 1.8 mg U/day for New York, Chicago, and San Francisco, which Welford and Baird obtained (15). Table salt with its relatively high uranium content (40 ng/g) could add appreciable amounts depending on dietary habits, as could some water supplies (Long Beach, California, 4.8 to 8.6 mg/L). Far lower values of U (0.024 to 0.041 mg/L) were found in New York City tap water in 1967 (15).

Scattered values of normal uranium content in body tissues and fluids have been appearing since 1960. Welford et al. (18) reported that the range of urinary uranium levels in 26 nonexposed laboratory workers in the New York area was from 0.03 to 0.300 mg/L, and the average varied only 1.5- to 3-fold; that of bone was 0.02 mg U/g ash. The average concentration of uranium in adult lung tissue from the same area was 1 ng/g wet tissue, and that for whole blood, 57 ng/100 g (19), a value approximately that found in the United Kingdom of 84 ng/100 g, which had a 12-fold spread in values (Table 42.3). In the same year, 1970, samples of human bone, rib, and sternum from 47 residents of Kyoto, Osaka, Sapporo, and Tokyo showed an average concentration of 2 ng U/g wet bone (20), a value 3.5 times lower than that found in the United Kingdom.

Assuming no serious analytic errors, it is logical to attribute the differences to variations in locale. No differences, however, were found in geographic district, subject, age, or sex in the four areas of Japan that were sampled.

As part of The Third National Health and Nutrition Examination Survey (NHANESIII) conducted by The National Center for Health Statistics, Centers for Disease Control and Prevention (CDC), urinary uranium was measured in 500 U.S. residents to determine reference range concentrations using a magnetic-sector, inductively coupled, argon-plasma mass spectrometer (ICP-MS) (17). Mean urinary uranium was 11 ng/L or 10 ng/g creatinine. Means for the 95th percentile of the subjects were 34.5 ng/L or 34.9 ng/g creatinine. No significant differences in urinary uranium concentrations were detected among racial groups, between children and adults, or between males and females. Data for geographic differences were not reported.

1.3.3 Workplace Methods

The correlation between environmental air sampling for U and individual urinary bioassay results for exposed workers has not been reliably strong, although both have been used to estimate lung exposure doses. In a recent paper, historical air monitoring and urinary bioassay data were reanalyzed to examine correlation of lung dose estimates made from the two different data sources, and only minimal correlation was demonstrated. An additional finding of the study showed that area air sampling results varied significantly among and within areas at different sampling locations and over time due to variability of operations, controls, and operating techniques. The authors concluded that urinalysis is a better measure of individual dose (21).

1.3.5 Biomonitoring/Biomarkers

Because urinary uranium determinations are the primary method of biological monitoring of environmental and occupationally exposed populations, some attention to more recent background concentrations reported and methods available is in order.

URANIUM AND THORIUM

Using a neutron activation technique, Dang et al. reported average and geometric mean concentrations of urinary uranium at 12.8 and 9.4 ng/L, (range = 2.9–40 ng/L) respectively, in 27 healthy subjects, never occupationally exposed, who lived and worked in Bombay, India (22).

This agrees generally with the findings of Medley et al. (16) who determined that the urinary uranium excretion of six nonoccupationally exposed men from the Tri-Cities, Washington, area ranged from 7.0–72.9 ng in 24 hours and 4–58 ng/L using kinetic phosphorescence analysis (KPA).

Medley and colleagues also addressed the health physics practice of performing urinary bioassays on exposed workers using "simulated" or incremental 24-hour samples. These samples are defined by the National Council on Radiation Protection and Measurements (NCRP) as the last voiding before bedtime and "all samples until and including the first voiding upon rising the next morning" (23). This was traditionally assumed to constitute one-half of the daily urinary volume and thus one-half of the daily urinary excretion. Because of the variability of fluid intake, metabolism, and hydration, it is likely that the assumptions regarding representativeness of the simulated sample are enormous, and supporting evidence is not available. Medley and colleagues demonstrated that the simulated sample underestimated the total daily urinary uranium excretion in exposed workers by a factor of 2. Recently a group of investigators has recommended normalizing urinary uranium concentrations to urinary creatinine concentration to address subject hydration status and diurnal variability of uranium excretion (24). Although not classically observed in the health physics community, this practice of normalization to creatinine aligns the bioassay procedures of health physics with the wider toxicology community and may narrow the considerable variability often observed in uranium bioassays, even in nonexposed populations.

1.3.5.3 Other. The radiochemical neutron activation analytic technique was used in a 1991 report to explore uranium concentrations in hair, urine, and blood of nonexposed and occupationally exposed persons in Yugoslavia. For controls, typical levels found in hair, urine, and blood were 10 ng/g, 10 and 5 ng/L, respectively. In occupationally exposed persons, hair concentration shows some promise as an indicator of exposure (25). These concentrations seem markedly lower than those reported earlier and from other geographic areas. The investigators imply that the analytic technique previously used may be at fault.

Fingernail cystine may be a biological marker of effect. Mountain and Stokinger (26) found a good correlation on two occasions between the degree of uranium exposure and lowered fingernail cystine content among workers in a uranium production plant. The decreasing fingernail cystine values as uranium exposure increased presumably indicate response from an alpha emitter, ^{238}U. No subsequent mention of this end point is found in the literature.

Whole Body Radiation Counting. Although urinary uranium determination is the method of choice for evaluating systemic exposure to soluble U, lung measurements of radiation (gamma counts) is the best method for estimating exposures from inhaled insoluble U constituents. Several different detector types available range from the most sensitive, high-purity germanium detectors whose sensitivity for a 1-hour count is about

1.6 mg of natural U at a 95% confidence level, to phoswich detectors at about 4 mg and about 8 mg for the least sensitive NaI detectors. The sensitivity depends on the isotopic mixture of the U counted. Enriched material is more easily detected, and depleted uranium requires a high concentration for detection (27).

1.4 Toxic Effects

1.4.1 Experimental Studies

Based on animal studies, the toxicity of uranium is quite variable and depends on dose, exposure route, and specific uranium compound. Generally, for inhalation exposures, the more soluble compounds are less toxic to the lungs (partly because of decreased retention time) but more toxic systemically from absorption by the lungs and transport to distal organs. The more insoluble compounds tend to pose a greater hazard to the lungs. The pharmacology and toxicology of uranium compounds of industrial (atomic energy) interest are presented in a monograph of four volumes entitled *Pharmacology and Toxicology of Uranium Compounds*. Comprising about 2300 pages, for its time, this represented the most thorough and extensive study ever given a hazardous substance (1).

The maximal dosage that just failed to be lethal for rats in 30-day feeding tests was about 0.5% U compound in the diet for six soluble compounds tested and 20% U compound for three insoluble uranium compounds (UO_2, U_3O_8, and UF_4). No amount of insoluble uranium compounds acceptable to the rat was lethal, but levels of 1 to 4% soluble uranium compound produced 50% mortality in 30 days. Compared with toxicity by inhalation, toxicity by ingestion was far less and ranged from 30-fold less for UO_4 to 3300-fold less for UF_4 and U_3O_8 for the rat; for the dog, differences were even greater (1).

Acute intravenous toxicity of soluble uranium compounds (uranyl nitrate) is extremely high (1). The approximate LD_{50} for rabbits is 0.1 mg U/kg; for guinea pigs, 0.3 mg U/kg; for rats, 1 mg U/kg; and for mice, 10 to 20 mg U/kg.

The acute intraperitoneal toxicity of soluble uranium compounds for the rat is considerably less than the intravenous toxicity; approximate LD values ranged from 40 mg UO_2F_2/kg to 400 mg UCl_4/kg. About a twofold increase in toxicity with increasing age occurred (1).

All soluble uranium compounds are lethal when applied in a single dose to the skin of rabbits, either in various vehicles or, in some cases (UCl_4, UCl_5), without vehicle; the insoluble salts UO_4, UF_4, UO_2, and U_3O_8 were not lethal by the same route and caused no signs of poisoning (1).

Soluble uranium compounds may be lethal when placed in the conjunctival sac of the rabbit eye, either in a vehicle or without vehicle; insoluble salts resulted in no mortality. Uranyl oxide was remarkable in that it was uniformly fatal but failed to produce any local signs of poisoning (1)!

Dusts and mists of respirable particle size of UF_6, UO_2F_2, and UCl_4, as well as $UO_2(NO_3)_2 \cdot 6H_2O$, were generally fatal to most laboratory species exposed daily for 1 month at 20 mg of compound/m^3; 2.5 mg/m^3 was fatal to some species; 0.2 mg/m^3 was fatal to an occasional animal; and 0.05 mg/m^3 resulted in no histological damage to any species. At 20 mg/m^3, UF_4, UO_2, and "high-grade" ore were occasionally fatal to some species; 2.5 mg/m^3 was essentially nonfatal and caused mild or no renal damage. On a relative toxicity scale, UF$_6$ was the most toxic uranium compound, followed by UO$_2$F$_2$, UCl$_4$, and UO$_2$(NO$_3$)2·6H$_2$O. UO$_2$ and U$_3$O$_8$ were the least toxic (1).

Carnotite ore dust (20% U, 5% V) at a daily average concentration of 84 mg/m^3 and particle diameter (MMD) of 1.39 μm was lethal to rabbits (92% mortality), but resulted in only 10% mortality of rats in a 32-day intermittent daily exposure. Other criteria of injury were consistent with these findings. The degree and nature of the injury resembled those of UO$_3$ dust (28).

1.4.1.2 Chronic and Subchronic Toxicity. The kidney was classically and is still considered the critical organ for uranium toxicity. Tests of renal injury in the older literature include changes in urinary catalase, phosphatase, and protein that exhibit a sharp peak 2 to 5 days after acute injury. Catalase normally not present in urine appears in response to as little as 0.02 mg uranyl acetate/kg in the rabbit. Tests of altered functional capacity of the kidney used in studies in the early 1950s when chronic U toxicity was first being characterized included phenol red removal from blood and the urinary amino acid nitrogen/creatinine ratio, and various clearance tests showed maximal change 3 to 6 days after exposure to uranium. These functions return to normal despite continued exposure, provided that exposure is not so severe as to result in fatality. Only chloride clearance and amino acid/creatinine ratio reportedly remained elevated for long periods after severe exposure, indicating prolonged interference with renal tubular function (1).

There is a fairly extensive though aging literature on the chronic toxicity of U derived from the military and nuclear energy interest in it after the Second World War. The soluble uranium compounds UO$_2$F$_2$, UCl$_4$, and UO$_2$(NO$_3$)$_2$·6H$_2$O were tolerated by dogs for 1 year when incorporated in the diet at a level of 0.2 mg/(kg)(day); 10 g/(kg)(day) of UO$_2$ was tolerated. UO$_2$(NO$_3$)$_2$·6H$_2$O caused an adverse effect on growth at the end of 1 year at a level of 0.2 g/(kg)(day). Borderline effects resulted from UF$_4$ at 5 g/(kg)(day). Most of the dogs showed little change in nonprotein or urea nitrogen, although some showed abnormal urinary protein and sugar values. Renal cortical changes were histologically observable in the dogs that showed urinary changes. No uranium compound produced significant alteration in the cellular blood picture. In male rats 0.05% UO$_2$(NO$_3$)$_2$·6H$_2$O in the diet did not alter weight gain over a 2-year period; 0.1% caused just detectable growth depression for male rats and 0.5% for female rats. Two percent did not lessen life span. A tolerance to acute UO$_2$(NO$_3$)$_2$·6H$_2$O poisoning was said to be demonstrated in proportion to the chronic dosage fed (1). "Tolerance" is discussed subsequently in this chapter.

Soluble uranium compounds clearly differ from insoluble compounds in toxicity in 1-year dog inhalation studies: 0.2 mg of U/m^3 as UO$_2$(NO$_3$)$_2$·6H$_2$O, UF$_6$ or UCl$_4$ was toxic, whereas 10 mg U/m^3 as UO$_2$ was a tolerated concentration. The dog was the most sensitive species, and 1 mg U/m^3 was without detectable effect in any animal by any test employed (1).

Foulkes and associates (29–31) studied how soluble hexavalent uranium acts on the renal nephron by analyzing *in vitro* transport kinetics from U-poisoned rabbits. They found that, although inhibition of proximal tubular function could readily be demonstrated, the major site of inhibition was at the first step of the overall secretory process, the transfer of *p*-aminohypurate across the peritubular cell membrane. No damage was recognized on the luminal side of the cells, as detected by tests of physiological function (29). Damage to the nephron clearly occurs beyond the proximal convoluted tubule, previously believed to be the major site of attack of the uranyl ion on the rabbit nephron. This was shown by altered functions of the more distal parts of the nephron: (*1*), depression of free-water clearance during water diuresis, (*2*), decreased water conservation during hydropenia, and (*3*), dissipation of the normally high tissue sodium concentration in the renal medulla. These altered physiological functions point to inhibition of sodium resorption from the ascending limb of Henle's loop, all showing that damage to the nephron occurs beyond the proximal tubule (30). No gross changes in glomerular permeability could be demonstrated in uranium-poisoned rabbits injected with 0.2 mg U/kg as uranyl nitrate; renal clearance of high molecular weight dextrans relative to glomerular filtration rate remained unchanged (31).

Rats were injected intraperitoneally with uranyl fluoride (UO_2F_2) (cumulative dose: 0.66 or 1.32 mg U/kg body wt). Renal injury was characterized histologically by cellular and tubular necrosis of the pars recta of the proximal tubule (S2 and S3). There was less marked cellular injury apparent in the thick ascending limb of the loop of Henle and the collecting tubule when renal uranium concentrations were between 0.7 and 1.4 µg U/g wet kidney. The damage was most severe when the renal uranium burden was between 3.4 and 5.6 µg U/g. Injury repair was rapid, and completely reversed in 35 days. Abnormalities in renal function associated with injury included impaired tubular reabsorption, proteinuria, and enzymuria, which appeared temporally related to development of renal injury. The reversible renal injury occurs in the rat at concentrations of uranium in the kidney below the present Nuclear Regulatory Commission standard of 3 ug U/g kidney for renal injury in humans (32), suggesting the need to reassess this threshold.

To focus attention on other than the renal effects of uranium and on the possible radiological hazards from prolonged accumulation and storage of uranium in the respiratory tract, a 5-year inhalation toxicity study of natural UO_2 was done on monkeys, dogs, and rats (33). The concentration of UO_2, particle size about 1 µm MMD, averaged 5 mg U/m^3. Exposures were 6 h/day, 5 days/week, for periods up to 5 years. The two major sites of uranium accumulation, the lungs and tracheobronchial lymph nodes (TLN), accounted for more than 90% of the uranium found in the body without evidence of toxicity, as measured by body weights or mortality, blood nonprotein nitrogen levels, or the hematologic picture.

A rapid buildup of uranium occurred in the lungs and the TLN of the dog and monkey during the first year of exposure. After 1 year, the lungs contained approximately 2000 µg U/g in the dog and 3600 µg U/g in the monkey, near-maximal values. Unlike the lung, the uranium content of the TLN for both species continued to rise and reached maximal values of 50,000 to 70,000 µg U/g after 4 years of exposure. In kidney, femur, spleen, and liver, uranium concentrations were comparatively low; after 5 years of exposure, monkey spleen showed the highest concentration (350 µg U/g), whereas dog spleen showed the lowest (0.9 µg U/g).

Alpha-radiation dosages, calculated from the organ burdens of uranium, indicate that dose rates to the lungs and the lymph nodes of each species surpassed 0.03 rad/week during the first few months of exposure. At 5 years, dose rates to dog lung and TLN and to monkey lung and TLN were 1.8, 55, 3.3, and 64 rads/week, respectively. The integrated alpha radiation dose to dog and monkey lung after 5 years of exposure was estimated at 500 and 900 rads, respectively. At this time TLN values for both species were of the order of 10,000 rads.

Kidney injury did not occur at the exposure level of 5 mg U/m^3. Fibrotic changes that suggested radiation injury, however, were seen occasionally in the TLN of dogs and monkeys and in monkey lungs after exposure periods longer than 3 years in tissues whose estimated alpha doses were higher than 500 rads for the lung and 7000 rads for the TLN.

The lung and lymph node data obtained in this study show that the animal body can accumulate sufficient uranium from prolonged exposures to insoluble uranium dust at 5 mg/m^3 to create potential radiological hazards. The lung and TLN values were high enough, in fact, to anticipate radiation hazards in these tissues from exposures at or lower than the occupational TLV (200 μg U/m^3) recommended by the ACGIH or the MPC alpha (6×10^{-11} mCi/cm^3 = 180 mg U/m^3) suggested by the ICRP.

The deposit of uranium compounds in calcifying zones has been demonstrated in bone. The alveolar bone volume, total bone formation areas, and the volume density of bone in the alveolar apical third were significantly lower in rats given a single intraperitoneal uranyl nitrate dose of 2 mg/kg than in the controls (34).

1.4.1.3 Pharmacokinetics, Metabolism, and Mechanisms. Approximately 20% of U^{6+} in the bloodstream is deposited immediately in the kidney, followed by a 60% mobilization of the dose to the urine in 24 h. About 10 to 30% of the dose is deposited immediately in the bone, which is followed by a slow mobilization. No significant excretion of intravenous U^{6+} occurs via the gastrointestinal tract. Considerably less U^{6+} is deposited in the liver or in other soft tissues.

Animal studies have indicated (28) that uranium in the tetravalent state (U^{4+}) cannot enter the bloodstream, except by direct injection. Uranium in this form is first oxidized to the hexavalent state (U^{6+}), after which its absorption and fate are indistinguishable from that of U^{6+} from other sources.

The distribution and excretion of absorbed U^{6+} and U^{4+} are compared in Table 42.4 (1). Comparison of the kinetics of U^{6+} with that of U^{4+} following a single exposure shows that the more insoluble U^{4+} is less readily absorbed (appreciable fecal excretion versus essentially none by U^{6+}) and less readily excreted (10% in 24 h versus 50% for U^{6+}).

The relevance of these data for industrial exposures to soluble uranium is that, despite its solubility, more than 1.5 years are required to rid the body of accidental high doses, at which time the bone and kidney still show measurable uranium retention. It should be noted that variations in acid–base balance have a profound effect upon the amount of uranium excreted. Increased alkalinity increases urinary excretion, and acidity decreases it.

The manner in which uranium compounds are absorbed cannot be predicted from their water solubility, as Morrow (35) demonstrated with water-insoluble UO$_3$. Although UO$_3$ is soluble only to the extent of 1 mg% in water, its solubility in blood plasma is 40-fold greater because of its reaction with bicarbonate and the production of a soluble complex. An inhalation study by Morrow et al. of UO$_3$ enriched with ^{235}U in dogs, contrary to the

Table 42.4. Comparison of the Distribution and Excretion of Hexavalent and Tetravalent Uranium From Animals[a]

Tissue	Hexavalent U	Tetravalent U
Kidney	~20% of dose deposited immediately, followed by mobilization to urine	~10% of dose deposited immediately, followed by mobilization, possibly to urine
Liver	No significant changes	As much as 50% deposited, followed by mobilization, possibly to feces
Bone	10–30% deposited immediately, followed by slow mobilization	10–20% deposited, immediately followed by slow mobilization
Urine	Up to 50% excreted in 24 h, followed by continued excretion of small amounts	Up to 10% excreted in 24 h, followed by continued excretion of small amounts
Feces	No significant excretion at any time	Significant excretion on 2nd, 3rd, 4th days; continued excretion of small amounts

[a]Ref. 1.

long retention times in the lungs for insoluble U compounds, revealed a 4.7-day biological half-life for UO_3, indicating rapid removal of a soluble uranium compound. Moreover, systemic absorption amounted to more than 20% of the exposure burden, and urinary excretion accounted for more than 20% of the body burden.

The soluble uranyl ion UO_2^{2+} is the most commonly encountered species found in body fluids (36). It may be complexed with transferrin in plasma which is not filterable at the glomerulus. About 60% of the uranyl ion may complex with anionic constituents, primarily carbonate species that have sufficiently low molecular weight to be filterable at the glomerulus (36, 37). As the UO_2–bicarbonate complex in plasma dissociates, the weak U-transferrin complex dissociates to allow a continuous equilibrium between the U-transferrin and the low molecular weight uranium complexes which facilitates relatively rapid elimination of U through the glomerulus or to soft tissue and bone binding (36, 38). Estimates for humans are that 80% or more of injected uranyl nitrate may be filtered in the first 24 hours. The data are based on several studies of injected U in humans (39, 40).

The filtrate that passes through the glomerulus enters the proximal tubule that is lined with epithelium that has a brush border made up of microvilli that resorb components of the filtrate. The pH drop along the course of the tubule facilitates dissociation of the U-bicarbonate complex and in turn enhances the uranyl ion's ability to interact with other complexing species or to interact with (or be resorbed across) the brush border membrane (36, 41) most prominently in the distal third of the proximal tubule. Complexed uranium that is not absorbed may be eliminated to the bladder and thus is rapidly eliminated after injection (39, 40, 42). As with other weak acids (cations), the pH in the lumen and the presence of bicarbonate determine the rate of U elimination in urine.

Specific uranyl binding sites on brush border membranes are not known. However, the importance of phosphate and phospholipid binding sites for U has been repeatedly cited (43–45).

It was traditionally held that U does not enter cells to achieve its effects (46), and there are a number of studies documenting poor cell membrane penetration of U (47–49). However, in the last twenty years, a number of studies have documented intracellular accumulation of U in various cell types and animal species (37, 50, 51), in various microorganisms (52, 53), and in human cell lines (54).

U may be absorbed across the brush border membrane of renal proximal tubular cells through an energy-dependent process similar to that which occurs for cadmium and possibly mercury (55). However, a likely and more important mechanism is endocytosis and uptake of a U-protein complex attached to the cell membrane, as reported for gentamycin (56). Subsequently, a lysosome attaches to and releases its enzymatic contents into the endocytotic vesicle. These hydrolytic enzymes breakdown the vesicular contents allowing diffusion into cytosol and back into the tubular lumen by some form of exocytosis. Lysosomes of the renal tubule concentrate a number of heavy metals such as mercury, chromium, and copper, as well as U (55, 57, 58).

The dynamic process of U deposition on and removal from proximal tubular cell membrane results in changing fluctuations in the fraction of the U burden in the kidney after exposure (59, 60). In beagle dogs, 80% of absorbed U is excreted in the urine in the first 24 hours. Similar values were observed for rats and in human studies (40, 59).

A two-compartment model is postulated for U. The short term compartment is as described before and has a $T_{1/2}$ of 6 d and a much smaller, longer term compartment has a $T_{1/2}$ of 1500 d (61). The authors caution that the dose of U, as well as the turnover of U in bone, also influences U entry into kidney (62).

The mechanism of renal insult has evolved over the years and several key observations have clarified the current view. Unlike some other heavy metals, U does not bind sulfhydral groups (41, 63–65) although there may be some indirect effect on them.

Phospholipid constituents of cell membranes that generally regulate membrane properties and cell-membrane interactions and membrane-bound enzymes such as Na-K ATPase (55) may be a potential site of U action. The effects on lysosomes, mitochondria, and calcium metabolism and the insult to the microvilli of brush border membranes may also be potential sites of toxic action resulting in reduced Na and consequently reduced glucose and amino acid and protein resorption, even before extensive cell damage occurs. Many of the observations of animals and physiological models were based on usually acute and relatively high-dose U exposure. Sites of toxic action in the kidney may be altered by a chronic lower dose exposure. These considerations and the chemical toxicity of U in the kidney are discussed in detail in the excellent review by Legget (62).

The guidance level of 3 µg U/gram kidney traditionally applied as a safe limit that, if observed, would avoid kidney toxicity, is currently undergoing scrutiny. It was fairly recently upheld (66); however, others have voiced concern. Morrow and colleagues (67) showed that the renal injury threshold in rats exposed to UO_2F_2 is at or less than 1 µg/g wet kidney tissue. Leggett states that "the level is based on tests of chemical toxicity that appear to have been less sensitive, and on definitions of chemical toxicity that were often less stringent, than those applied by more recent investigators. Moreover, in the human studies considered in developing this guideline, it appears that the subjects may have experienced concentrations on the order of 3 µg U/gram only for very brief periods. In view of current information, it may be prudent to lower this long standing guidance, roughly an order of magnitude until more is known about the subtle physiological effects of small quantities of U in the kidney (62). Morrow (60) has also voiced similar concerns.

"Tolerance," initially described as decreased mortality from multiple nephrotoxic exposures (68), may really be a more complex response pattern of the kidney to multiple exposures (60). This view has been shaped by a number of observations. Proximal tubular cells that regenerate after U insult possess flattened brush borders without microvilli. Mitochondrial number is also reduced. There are also functional differences in "tolerant" animals, including high urine volumes (69) and decreased glomerular filtration rates (67). Therefore, these regenerated cells are not normal in either structure or full function.

A "reversibility" of nephrotoxicity was defined historically as a return to normal of some renal functional parameters, which also should be viewed with caution as "overly simplistic" due to the complexities of kidney response (60).

1.4.1.4 Reproductive and Developmental. The effect of uranium on reproductive performance has not been extensively investigated. Daily oral doses of 0, 0.05, 0.5, 5, and 50 mg/kg of uranium as uranyl acetate dihydrate were given to Swiss mice from day 13 of pregnancy until weaning of the litters at 21 days of age. There were significant decreases in the mean litter size on postnatal day 21 and in the viability and lactation indexes that were observed at the 50 mg/kg dose (70). Male Swiss mice were treated with uranyl acetate dihydrate at doses of 0, 10, 20, 40, and 80 mg/(kg)(day) given in drinking water for 64 days. They were mated with untreated females for 4 days. There was a significant decrease in the pregnancy rate at 10, 20, 40, or 80 mg/(kg)(day). The decreased rates of pregnancy did not follow a clear dose–response relationship. Body weights of the males were significantly depressed only in the 80 mg/(kg)(day) group. Testicular function/spermatogenesis was not affected by uranium at any dose. Interstitial alterations and vacuolization of Leydig cells were seen at 80 mg/kg/day (71).

1.4.1.5 Carcinogenesis. The older toxicology literature reported that sarcomas resulted in rats injected with metallic uranium in the femoral marrow and in the chest wall (72). It is unknown whether the sarcomas were due to metallocarcinogenic or radiocarcinogenic action.

Very recently, to characterize internalized DU's carcinogenic potential, Miller and colleagues exposed immortalized human osteoblastic cells (HOS) in cell culture to DU-uranyl chloride 24 hours at varying doses beginning at 10 µM. At this lowest dose and at all higher doses, HOS cells were transformed into a tumorigenic phenotype. These transformed cells were characterized by anchorage-independent growth, tumor formation in nude mice, expression of high levels of the K-ras oncogene, reduced production of the retinoblastoma (Rb) tumor-suppresser protein and elevated levels of sister chromatid exchanges. The authors compared DU's transforming ability with other known heavy metal carcinogens, nickel sulfate, and lead acetate, and found that DU-UO_2^{2+} has slightly greater transforming ability *in vitro* and induction of tumors in animals than nickel and lead (73).

The authors also addressed the radiological versus the chemical toxicity of the induced effect by modeling the number of all nuclei potentially hit by a DU-alpha particle and found that it is small (0.0014%). Therefore, they attribute the transforming ability to DU's heavy metal chemical toxicity rather than its radiological activity.

1.4.1.6 Genetic and Related Cellular Effects Studies. Uranyl nitrate (UO_2^{2+}) is genotoxic and cytotoxic in Chinese hamster ovary cells in culture at concentrations ranging from 0.01 to 0.3 mM (74). The toxic end points observed included decreased viability and depressed cell-cycle kinetics that increased micronuclei, sister chromatid exchanges and chromosomal aberrations. The authors suggest that these data provide a possible mechanism for the teratogenic effects of uranium in fetal mice, reported by Domingo and colleagues, that produce cleft palate and bipartite sternebrae (75).

1.4.2 Human Experience

The toxicity of U^{6+} in humans following intravenous administration was determined in five terminal patients who had brain tumors and were given uranyl nitrate (39, 49). The three patients who received the largest doses (0.17 and 0.23 mg/kg of U) showed an elevation of urinary catalase and albumin. Other evidences of renal toxicity were the appearance of casts in the urine and elevated nonprotein nitrogen. Evidence suggested the loss of renal capacity for sodium and chloride reabsorption and potassium secretion. Electrocardiograms, liver function studies, hematologic tests, and glucose tolerance tests did not reveal other evidences of toxicity. The autopsies of four of the patients did not reveal any unusual findings. The authors concluded that humans correspond most closely to the laboratory white rat in our tolerance to soluble uranium, a conclusion arrived at in a University of Rochester Medical School (39) study on terminal patients administered soluble uranium, whose tolerance was 10 times that of the rabbit, a species far more susceptible to uranium than the rat.

Except for accidental acute exposures, no evidence of either chemical or radiative chronic toxicity was obtained from any worker for any uranium compound during the many years of the atomic energy program. The biochemical measures available and routinely used in surveillance of workers during those years (the 1940s and forward), however, temper these negative observations.

The two accidental exposures involved the same compounds, gaseous UF_6 and the hydrolysis products UO_2F_2 and HF. In one accident the sudden rupture of a tank of UF_6 and steam lines resulted in the deaths of two persons in the path of the UF_6 cloud. One who was exposed for 5 min died 10 min later; the other, after a rapid escape, died 70 min later. The acute onset of death implicated the pulmonary and cardiac toxicity of HF. Live steam also complicated the exposure; some persons had third-degree burns. Of three other persons seriously injured, two were in the vicinity of the accident, and the third was outside nearby. Ten to fourteen days of hospitalization were required for recovery. The 13 other exposed individuals required only dispensary care. Detailed case findings are given in a monograph (Vol. 2 of Ref. 1).

In the second case of accidental exposure to UF_6, important findings were related to the eyes (chemical conjunctivitis and corneal necrosis), the respiratory tract (increased density of bronchovascular markings and hilar shadows), and the urinary tract (increased amount of urinary solids). In 5 days with treatment, the corneal epithelium had almost completely regenerated, and at time of discharge, visual acuity was normal. In 10 days the chest was clear; in 13 days the hemorrhages of the larynx had disappeared. The urinary signs cleared with improvement of the patient. Some mental derangement, restlessness, and nervous tension accompanied the early responses, but these disappeared in a week (see Vol. 2 of Ref. 1).

Two individuals were exposed to massive quantities of airborne uranium hexafluoride (UF$_6$) and its hydrolytic products following a World War II equipment rupture. In the light of current knowledge (1985), the excretion pattern for uranium exhibited by these patients is peculiar. Pulmonary edema occurred, but examination of these individuals 38 years later revealed no physical changes related to their uranium exposure, and no deposition of uranium could be detected (76).

1.4.2.2.2 Chronic and Subchronic Toxicity. Uranium's nephrotoxicity is the principal manifestation of chronic exposure in humans. Its action targets the proximal tubule primarily, as do other heavy metals, but may also involve glomerular and distal tubular insult (31). Compared to other heavy metals, such as cadmium, lead, and mercury, however, uranium's nephrotoxic effect is less intense (77). Although the nephrotoxicity of uranium is attributed primarily to its metallotoxicity (chemical toxicity), some authors have discussed that renal damage from exposure to high-LET alpha-emitting heavy metals, like uranium, is derived from combined chemical and radiotoxic effect (78).

A number of epidemiological studies of uranium workers have failed to find excess mortality from renal disease (79–81). Measures of renal function in uranium mill workers chronically exposed to insoluble uranium dioxide reported a dose-related perturbation in beta-2 microglobulin clearance and mild proteinuria and aminoaciduria, indicative of proximal tubular effects (82). Another study compared 39 uranium mill workers exposed to "yellow cake" (26–86% ammonium diuranate) and 36 cement plant workers similar to the uranium-exposed in age, sex, and race. The U-exposed groups' results revealed a significantly higher excretion of beta-2 microglobulin and five amino acids than the reference group (83). The length of time workers were in the yellow cake area of the plant was used as an exposure surrogate. Unfortunately, urinary uranium measures were not obtained, though historical urinary bioassay results from the early 1980s revealed relatively high measurements in the year before the study, a mean of 7.2 µg/L, and a median of 6 µg/L, although the U.S. Nuclear Regulatory Commissions (NRC's) action level at that time was 30 µg/L.

Of interest, in a recent study, kidney tissue sections obtained at autopsy of seven occupationally exposed uranium workers were compared to six unexposed controls in a blind fashion. Pathologists were unable to distinguish the uranium workers or to identify uranium-induced renal damage (84). Workers were exposed to low-level uranium concentrations, about one-tenth of the 3-µg/gm kidney tissue threshold traditionally held as a permissible level for kidney exposure. The authors suggest that uranium exposure at these concentrations is not associated with nephrotoxic effects.

A Canadian study examined uranium kidney effects in persons environmentally exposed through drinking water. Thirty subjects whose uranium in drinking water was ≥ 1 µg/L and half of the subjects whose water exceeded the Canadian Guideline of Health Canada of 100 µg/L were compared with 20 subjects whose drinking water uranium concentration was ≤ 1 µg L. All subjects' exposure was at least one year and the majority were exposed for many years. Urinary glucose was significantly different and positively correlated with uranium intake for males, females, and pooled data (85).

Increases in alkaline phosphatase and beta-2 microglublin, both proximal tubular markers, were correlated with uranium intake. Unfortunately, urinary uranium values were not reported.

A small cohort of Gulf War veterans involved in friendly fire incidents where DU shells (penetrators) were used is being followed prospectively to assess the health effects from inhalation, wound contamination, and systemic absorption of retained DU metal fragments. A group of 33 soldiers was first evaluated in 1993/1994. At that time, they had elevated concentrations of urinary uranium, and mean urine uranium excretion was significantly higher in soldiers with retained metal fragments compared to those without fragments (4.47 vs. 0.03 µg/g creatinine) (86). No evidence of a relationship between urine uranium and abnormal renal function could be demonstrated.

In a subsequent follow-up of the same cohort, 29 of the original 33 were examined in 1997 and their results compared to 38 non-DU exposed, but Gulf War deployed soldiers. The correlation between 1994 and 1997 24-h urinary uranium determinations was highly significant (Rsq = 0.8623) and urine uranium was again correlated with the presence of retained DU fragments. Exposed soldiers (with and without fragments) had 24-h urinary uranium results ranging from 0.01 to 30.74 µg/g creatinine, whereas the nonexposed group's results ranged from 0.01 to 0.047 µg/g creatinine (87). The persistence of elevated uranium excretion suggests ongoing mobilization from a storage depot and results in chronic systemic exposure. Again, no renal abnormalities were found but neurocognitive examinations demonstrated a statistical relationship between urine uranium levels and lowered performance on computerized tests assessing performance efficiency. Elevated urinary uranium was also statistically related to a high prolactin level (> 1.6 ng/mL; $p = .04$). Uranium was also detected in the semen of five of seventeen exposed veterans, but in none of five nonexposed veterans. The authors state that these findings are subtle and of uncertain clinical significance but nonetheless document elevated urinary uranium excretion and small, but measurable, biochemical effects on the neuroendocrine and central nervous systems seven years after first exposure.

1.4.2.3 Epidemiology Studies. The National Institute for Occupational Safety and Health (NIOSH) has engaged in a cooperative agreement with the Epidemiology and Cancer Control Program of the University of New Mexico Health Sciences Center to study the health of men and women who worked in uranium mills. The goal of the study is to describe changes in pulmonary and renal function as measured by spirometry and proximal tubular markers in former uranium mill workers.

1.4.2.3.5 Carcinogenesis. Evaluation of the radiation health hazards in all phases of uranium fabrication, as well as practical methods of control, is given by Harris and Kingsley (11). Hazards associated with rolling normal and enriched uranium are similarly treated by Hyatt (10) and by Blackwell (9). Data on uranium concentrations in air and in urine and medical findings are presented by Lippman (88) for two uranium refineries during a 2-year period. External radiation hazards, which are chiefly beta emissions, can be controlled if attention is paid to prescribed procedures. Control of exposures to uranium is gradually being tightened. No dermatitis problems relating to uranium itself have been found in a 2 1/2-year study of a plant that processes uranium ore to form UF_4 (89). EDTA was employed to aid removal of uranium (and thorium) in four accidentally exposed cases and had some value (90).

In uranium mining, compared with uranium fabrication, health hazards have resulted from largely uncontrolled exposure to radon daughters. The incidence of respiratory disease mortality, including that of cancer and pulmonary insufficiency in smoking and nonsmoking uranium miners in the United States, was first determined by Archer et al. in a 1973 publication (91) on 3366 whites and 780 nonwhites. As expected, in the early days of mining (1948 to 1967) when exposures to radon daughters were 50 to 80 times the permissible limit of 80 mCi/L, a significant excess of respiratory cancer deaths among white and Indian miners was found, 4.8 and 4.2, respectively. At the study's conclusion in January 1974, nonmalignant respiratory disease among the whites was approaching cancer in importance as a cause of death. In the group that had 5 to 9.9 years experience in underground mining, the ratio of observed to expected deaths among whites was in excess of twofold; for those who had 10 to 24.9 years experience, in excess of 3.6-fold; and for 25 years and more, 3.75-fold. Not stressed in the reports was a reduced incidence of vascular disease of the CNS (19 observed versus 32 expected) and heart disease (166 observed versus 218 expected), both significant at the 5% level.

Horacek et al. (92) indicated that the frequency of the two major (small-cell undifferentiated and epidermoid) histological types of cancer may be influenced by the level of radiation exposure and by the time course of accumulation of exposure in a different way. In 299 uranium miners (1976 to 1980) in Czechoslovakia, there were 52.8% epidermoid carcinomas, 32.8% small cell carcinomas, 5% adenocarcinomas, and 9.4% of other, mixed, undifferentiated carcinomas (93). These findings, in agreement with the data of Archer et al., eliminate the former assumption that radiation can induce an elevated frequency of only one histological type of lung cancer.

Lung cancer has been a rare disease among the Indians of the southwestern United States. In Navajo men with lung cancer who were admitted to the hospital from February 1965 to May 1979, 16 among a total of 17 patients with lung cancer were uranium miners, and one was not. The mean value of cumulative radon exposure for this group was 1139.5 working level months (WLMs). The predominant cancer type was the small-cell undifferentiated category (62.5%). The low frequency of cigarette smoking in this group supports the view that radiation is the primary cause of lung cancer among uranium miners and that cigarette smoking is a promoting agent (94).

Samet studied a cohort of underground uranium miners in New Mexico and found that exposures to radon progeny declined dramatically after 1950 when the annual exposure for the 50th percentile of workers was > 0.5 Jhm^{-3}; by the 1970s exposure had declined to less than the annual exposure maximum mandated by a Federal Standard in 1971 of 0.0145 Jhm^{-3} or four working level months (WLM) (95). He notes that the decline continued and was less than 0.001 Jhm^{-3} in the mid-1980s.

A total of 2574 workers from the Radium Hill uranium mine, which operated in eastern South Australia from 1952 to 1961, were identified from mine records. Exposures to radon daughters were estimated from historical records of radon gas concentrations in the mine and from individual job histories. Exposures of underground workers were low in comparison with other mines of that period (mean 7.0 WLM, median 3.0 WLM). Thirty-six percent of the cohort could not be traced beyond the end of employment at Radium Hill. Among those traced to the end of 1987, lung cancer mortality was increased relative to the Australian national population of the period (SMR = 194, 95%, CI 142 to 245). Compared with surface workers, lung cancer mortality markedly increased in the underground workers who had radon daughter exposures greater than 40 WLM (relative risk = 5.2, CI 1.8 to 15.1). From the available information, the authors (96) concluded, in contrast to other investigators, that this increase is unlikely to be caused by differences in smoking habits or other confounders.

A total of 65 lung cancer deaths was observed (34.24 expected) among a cohort of 8487 workers employed between 1948 and 1980 at a uranium mine in Saskatchewan, Canada. There was a highly significant linear relationship between dose and increased risk of lung cancer (97).

Among nonmining workers in the uranium processing industry, the rate of respiratory cancer is not as elevated as among miners, presumably because of decreased exposure to radon. In a retrospective cohort mortality study of 995 white males employed more than 30 days at a uranium processing facility in upstate New York between 1943 and 1949, increased standardized mortality ratios were observed for all causes (SMR = 118), laryngeal cancer (SMR = 447), all circulatory diseases (SMR = 118), arteriosclerotic heart disease (SMR = 119), all respiratory diseases (SMR = 152), and pneumonia (SMR = 217). No association was found with length of employment or work in the most hazardous areas of the plant (98).

In a cohort of 18,869 white males who were employed between 1943 and 1947 at a uranium conversion and enrichment plant in Oak Ridge, Tennessee, the SMR for lung cancer was 1.22. Workers in certain departments (especially chemical workers) were exposed to high average air levels of uranium dust. SMRs for various causes, including lung cancer, were not higher in 8345 workers employed in areas where uranium dust was present or in 4008 of these 8345 workers employed for one year or longer at the plant. Other causes of particular interest (i.e., bone cancer, leukemia, diseases of respiratory and genitourinary systems) did not exhibit high SMRs. The suggestive finding of the authors was an increased number of lung cancer deaths in a group of chemical workers hired at an age of 45 years or older (99).

On the other hand, there is a report from Russia that the mortality rates from acute leukemia among the male staff of industrial atomic reactors and the radiochemical industry for irradiated uranium increased in the first 5 years of exposure. However, in the following years, 1953 to 1987, the risk of death from this cause declined (100).

Generally, exposure response curves for nonsmokers were linear for both respiratory cancer and "other respiratory disease;" cigarette smoking by both whites and nonwhites elevates and distorts the linearity and raises respiratory cancer/1,000 person-years from 1.5 for nonsmokers at WLM of 2100 to 8.2 for those who smoked 1 to 19 cigarettes/day and to 13 for those who smoked more than 20 a day for the same WLM of 2100.

The linear curves for nonsmokers suggest a direct dose response for both respiratory cancer and "other respiratory disease" deaths. Plots of the response of cigarette smokers are markedly curvilinear for respiratory cancer, whereas those for respiratory disease mortality are essentially linear. Oddly, with respect to deaths from respiratory disease, light smokers fared worst. At 2100 WLM, light smokers had a respiratory disease/1,000 person-years rate of almost 7, compared with that of heavy smokers at half that value. The authors attributed the difference to the shielding of cells from low-penetrating alpha particles of radon daughters by production of extra mucus in the heavy smokers, which does not occur among light smokers.

When the overall summary of cancer induction periods was made, the mean induction–latent period for eliciting authenticated respiratory cancer for 15 white uranium miners who smoked 20 or more cigarettes/day was 13.7 years, contrasted with a mean of 20.5 years for four nonsmokers. An intermediate latent period of 17.1 years for light or nonsmokers shows a direct dose–time response for respiratory cancer induction attributable to smoking (91).

A cohort of 3469 males with at least 1 year of underground uranium mining experience in New Mexico was assembled, and mortality was followed through December 31, 1985 by Samet et al. (95). Significant increases were observed for lung cancer (SMR = 4.0, 95%, CI 3.1 to 5.1) and for external causes of death (SMR = 1.5, 95%, CI 1.3 to 1.7). The risk of lung cancer increased for exposure categories above 100 WLM. Data were consistent with a multiplicative interaction between smoking and exposure to radon progeny in an exponential relative risk model. The risk of lung cancer varied with age at observation. The odds ratios rose more steeply with exposure to radon progeny for those less than age 55 at observation.

Samet reports that New Mexico miners were employed only since the 1960s, when there was some governmental regulation of mining exposures and thus, the New Mexico miners have generally lower exposures than the miners of the Colorado Plateau.

A nested case-control study of 65 cases and 230 age-matched control New Mexican underground miners who had at least one year of experience was undertaken. The risk for lung cancer increased for cumulative exposures to Rn progeny of 0.35 Jhm^{-3} (100 WLM) or greater. The risk was greater for younger subjects, and results were consistent with a multiplicative interaction between exposure to Rn progeny and cigarette smoking (101).

A comparison of a Czechoslovak study of lung cancer frequency among uranium miners to 1973 (102), 70% of whom smoked cigarettes, with the 1973 American study (103) shows that there was a significant excess of respiratory cancer at a similar level of exposure in the lowest exposure categories, 100 to 149 WLM in the Czech study and 120 to 359 WLM in the American, for an approximately equal period of observation. In these groupings, the observed/expected ratio (O/E) for Czech miners was 2.5; for the American O/E for all groups of smokers, the range was 1.6 to 3.0. At the Czech category of 600 and over WLM, an O/E ratio of 8.1 was found, compared with an O/E of 8.2 for all groups of smoking American miners at 1800 WLM.

In 1976 the pathogenesis of respiratory cancers in the population of 2500 to 3500 American uranium miners was reported (91). Saccomanno et al. (104) reported on studies beginning in 1957. From repeated sputum samples and from biopsies, surgery, and autopsies of miner patients, the course of the development of bronchogenic epidermoid carcinoma from squamous cell metaplasia of the bronchi arises as small patches from normal, tall, columnar epithelium. Many years may elapse before these patches develop into carcinoma, for example, about 15 years for epidermoid carcinoma; but the average time is about 4 years from the marked atypia stage to an invasive tumor. On occasion, concomitant tumors, such as small-cell carcinoma, may develop very rapidly and terminate in death. On the other hand, an *in situ* lesion may remain in its early developmental state, identified only at autopsy.

A cohort mortality study of French uranium miners who experienced more than 2 years of underground mining with first exposure between 1946 and 1972 reported a statistically significant excess for lung and laryngeal cancer deaths. This cohort experienced relatively lower radon exposures. An SMR for lung cancer of 0.06% per WLM was observed (standard error: 0.4%) and an estimated intercept at O WLM of 1.68 (standard error: 0.4) with a mean cumulative exposure to radon of 70 WLM over a mean duration of 14.5 years (105).

Other considerations for radiologically mediated health effects in humans in exposures to natural or depleted U are thought to be unlikely due to their relatively low specific activity. These postulations have been supported by available animal and human data to date (106). Enriched U, however, may pose true carcinogenic and other noncarcinogenic hazards, although even in this case, no human cancers have been documented from enriched U exposure. The BEIR IV report surmises that if uranium were a human carcinogen, bone sarcomas would be the most likely cancer type seen (107). The report stated that "exposure to natural uranium is unlikely to be a significant health risk in the population and may well have no measurable effect." Therefore, the risk from depleted U would be even less.

1.4.2.2.6 Genetic and Related Cellular Effects Studies. Radon, a product of the U decay series, believed to be the principal constituent responsible for the carcinogenic effects seen in animals and humans exposed to U, causes lung cancer in animals and humans.

A number of cytogenetic studies of radon's effects have been examined in a variety of cell types and employing various end points. Cytogenetic damage of radon to cell nuclei has been demonstrated in several animal cell lines using micronuclei as the indicator (108–110).

Many studies have documented that alpha irradiation *in vitro* produces large or complete gene deletions as the predominant mutational event and loss of heterozygosity in neighboring chromosomal regions up to 34 Mb distance from the target gene (111). However, mutations in cancer-relevant genes found in lung tumor tissue from uranium miners are base-pair changes or frameshift mutations (112, 113).

The mutational spectrum from exposure to high levels of radon among uranium miners was investigated by Vahakangas et al. (112). They sequenced exons 5 to 9 of the p53 tumor suppresser gene and codons 12 to 13 of the K-ras protooncogene in 19 lung cancers from uranium miners exposed to radon and tobacco smoke. Mutations were not found in K-ras, but nine p53 mutations, including two deletions, were found in seven patients by direct DNA sequencing after polymerase chain reaction amplification of DNA from formalin-fixed, paraffin-embedded tissue. In tumors from five patients, the mutation produced an amino acid change and an increased nuclear content of p53 protein. The tumors, that had either a stop codon or frameshift deletion in the p53 gene, were negative by immunohistochemistry. None of the mutations was G:C to T:A transversion in the coding strand of the p53 gene, which is the most frequent base substitution associated with tobacco smoking; and none was found at the hot-spot codons described in lung cancer. The observed differences from the usual lung cancer mutational spectrum may reflect the genotoxic effects of radon.

In a subsequent publication, Taylor (113) and colleagues reported on a specific p53 mutational hot spot in radon-associated lung cancer wherein 16 (31%) of 52 large cell and squamous-cell cancers from U miners contained the same AGG to ATG transversion at codon 249, including cancers from 3–5 miners who were never smokers. The authors stated that this particular mutation had been previously reported in only one of 241 other p53 mutations from lung cancer and postulate that codon 249 is a marker for radon-associated lung cancer (113).

Immortalized human cells in culture have been malignantly transformed by a single 0.3-Gy dose of alpha particles (114). The inverse dose-rate effect described in rats and first observed with neutron-induced transformation (115) was also observed with alpha-induced transformation (116). The authors attribute their observation of increased transformation efficiency seen only at LETs between 50 and 120 keV/µm and only in cycling cells in log phase to a period of high sensitivity to cell transformation, and the effect is not seen above 120 keV/µm. The authors postulate that this is due to a reduction of the number of cells being hit, as previously posited by Brenner (117) and Endlich (118).

This inverse dose-effect rate has also been described in studies of lung cancer in miners where an increase in exposure rate resulted in a decrease in cancer development (119). This phenomenon has also been reported in studies of lung cancer occurrence in (Ref. 120) when similar radon doses are protracted over time (121).

The chemically mediated genotoxicity of U was examined in 20 occupationally exposed fuel production plant workers who had a wide range of exposures to soluble and insoluble U (group A) and in 73 fuel enrichment plant workers (group B) whose exposure was to soluble UF alone (122). Significant increases in both sister chromatid exchanges (SCE), a measure of point mutation, and chromosomal aberrations were observed in workers' peripheral blood lymphocytes compared to nonexposed controls. The radiation exposure in the work settings, "well within the ICRP protection limits," according to the authors, was not thought sufficient to explain the aberrations. Because SCE were elevated (not usually seen from a radiation exposure), a chemical rather than a radiological mechanism of insult was postulated.

1.5 Standards, Regulations, or Guidelines of Exposure

A variety of both mandatory and voluntary health-based exposure limits for uranium are derived from both its chemical and radiological toxicity [See Table 42.5 (106)]. Regulating bodies include international, national, and state organizations. Some of the pertinent regulations and guidelines on exposure limits are summarized here, but the reader is cautioned to consult other sources to ensure health protection and regulatory compliance.

Uranium is unusual among the elements because it presents both chemical and radiological hazards. Current federal and state regulations limit radiation workers' doses to a total effective dose equivalent (TEDE) of 5 rem/year and a committed dose equivalent to

Table 42.5 Uranium Exposure Limits[a,b,c]

Chemical Based Limits

	OSHA PEL		ACGIH'94		NIOSH'92 REL-TWA	DHHS IDLH	NRC
	8 h TWA	STEL	8 h-TWA	STEL			
			Soluble				
Inhalation	0.05 mg U/m^3		0.05 mg U/m^3	0.025 mg U/m^3	0.05 mg U/m^3	10 mg U/m^3	0.2 mg/m^3 per 40 h workweek
			Insoluble				
	0.2 mg U/m^3	0.6 mg U/m^3	0.05 mg U/m^3	0.025 mg U/m^3	0.2 mg U/m^3	10 mg U/m^3	

Radiation Based Limits

	ICRP			NRC			
	Individual short term	Individual chronic	Occupational ALI for inhalation (1993)		DAC (1993)		Ingestion Occupational ALI for oral ingestion (1991)
Occupational-whole body exposure							
5 rem/yr (50 mSv)	0.5 rem/yr (5mSv)	0.1 rem/yr (1 mSv)	U-230	4.0 × 10^{-1} μCi	U-230	2.0 × 10^{-10} μCi	U-230 4.0 μCi
			U-231	8 × 10^3 μCi	U-231	3.0 × 10^{-6} μCi	U-231 5.0 × 10^3 μCi
			U-232	2 × 10^{-1} μCi	U-232	9.0 × 10^{-11} μCi	U-232 2.0 μCi
			U-233	1.0 Ci	U-233	5.0 × 10^{-10} μCi	U-233 1.05 μCi
			U-234	1.0 Ci	U-234	5.0 × 10^{-10} μCi	U-234 1.0 × 10^1 μCi
			U-235	1.0 Ci	U-235	6.0 × 10^{-10} μCi	U-235 1.0 × 10^1 μCi
			U-236	1.0 Ci	U-236	5.0 × 10^{-10} μCi	U-236 1.0 × 10^1 μCi
			U-237	3 × 10^3 μCi	U-237	1.0 × 10^{-6} μCi	U-237 2.0 × 10^3 μCi
			U-238	1.0 μCi	U-238	6.0 × 10^{-10} μCi	U-238 1.0 × 10^1 μCi
			U-239	10^5 μCi	U-239	8.0 × 10^{-5} μCi	U-239 7.054 μCi
			U-240	4 × 10^3 μCi	U-240	2.0 × 10^{-6} μCi	U-240 1.0 × 10^3 μCi
			U natural	5.0 × 10^{-10} μCi	U natural	5.0 × 10^{-10} μCi	U natural 1.0 × 10^1 μCi

[a]Ref. 106.
[b]Ref. 124.
[c]PEL-permissible exposure limits, STEL-short term exposure limits, TWA-time-weighted average, REL-recommended exposure limit, IDLH-immediately dangerous to life and health, ICRP-International Commission on Radiological Protection, ALI = Annual limits on intake; DAC = derived air concentrations; Rem-unit of dose equivalent (1 Rem = 0.01 Sievert), Sievert(Sv)-unit of dose equivalent(absorbed dose, in gray, multiplied by a quality factor), Absorbed dose-one rad = 100 ergs per gram. In SI units–absorbed dose unit is the gray J/kg, Curie(Ci)- unit of radioactivity. One curie =3.7 × 10^{10} nuclear transformations per second, Microcurie(μCi) = one millionth of a Curie (3.7 × 10^4 disintegrations per second).

any organ of 50 rem/year (123–125). These limits apply to the sum of external and internal doses. The limits are upper limits, and the public health radiation protection proviso of limiting radiation exposures to as little as reasonably achievable (ALARA) is recommended.

The annual limits of intake (ALI) is the activity of a radionuclide that can be taken into the body by inhalation or ingestion without exceeding a committed effective dose equivalent (CEDE) of 5 rem/year or a committed dose equivalent to organ of 50 rem/year, whichever is more limiting. The TEDE is the sum of the CEDE and any penetrating external dose (10 *CFR* 20). If any external dose is present, the ALI must be reduced by a proportional amount to ensure that the dose limits are not exceeded. For example, if a worker received an external dose of 1 rem/year, the ALI would have to be reduced by 20% to ensure that the TEDE did not exceed 5 rem/year.

The derived air concentration (DAC), another limit used in radiation protection practice, is defined as the inhalation ALI divided by the volume of air that a worker is assumed to breathe in a year (2,400 m^3). Thus, if the air concentration is controlled so as not to exceed the DAC, a worker dose not take in more than an ALI, and the worker's dose will not exceed 5 rem CEDE or 50 rem committed dose equivalent to any organ (126).

For soluble uranium, whose ^{235}U enrichment is no greater than 5%, limits on intakes and air concentrations for radiation workers are based on the chemical toxicity of uranium because it is more potentially hazardous to health than the radiological hazard. For this case, the NRC's limit for a 40-hour workweek is 0.2 mg uranium per cubic meter or air average (10 CFR 20). OSHA regulations specify that an employer must use engineering and work practice controls, if feasible, to reduce exposures to or below an 8-h TWA of 0.05 mg per cubic meter of soluble uranium (127). For insoluble uranium, values of the ALI and DAC are based on the radiation dose limits discussed before.

Two radiation protection professional bodies of the ICRP (128) and the NCRP (129) have recommended lower worker dose limits. The ICRP recommends a limit on the total effective dose of 2 rem/year averaged over 5 years, with the additional provision that the dose not exceed 5 rem in any single year. The NCRP's recommendations are that workers' total accumulated dose should not exceed their age in years times 1 rem, and that the dose should not exceed 5 rem in any single year. These recommendations have not yet been incorporated into regulations.

Regulations for the general public are based on an annual TEDE of 0.1 rem/year, with provisions for a limit of 0.5 rem/year under special circumstances. Considering the lower limit for members of the public and their continuous exposure, the limits on air concentrations of radionuclides for the public are two orders of magnitude lower than the DACs for radiation workers. Regulations for specific applications limit the dose to the public to values < 0.1 rem/year. Under its responsibility for generally applicable environmental radiation standards, the EPA has issued regulations for the nuclear fuel cycle that limit the total-body dose of members of the public to 0.025 rem/year and a single organ (except the thyroid) dose to 0.050 rem/year (40 *CFR* 190). Also, based on the Clean Water Act, EPA has proposed a drinking water standard for naturally occurring uranium of 20 µg/L (130). These recommendations and guidelines are summarized from ATSDR (106).

2.0 Thorium and Thorium Compounds

Thorium, the second element in the actinide series, exists in the earth's crust as an unstable, radioactive element that undergoes decay by alpha emission and gives rise to a series of short-lived daughter products that ends in a stable isotope of lead. Thorium is used as a source of atomic fuel, in the production of incandescent mantles, as an alloying element with magnesium, tungsten and nickel, and in the past was used as a diagnostic agent for systemic radiological studies. Thorium is primarily a radioactive hazard in humans; however, its chemical toxicity must also be considered.

2.1 Chemical and Physical Properties

Thorium [CAS Number: *7440-29-1*], a naturally radioactive element in the actinide series, has an atomic number of 90 and an atomic mass of 232.04. This element, named for the Scandinavian god, thor, was discovered by Berzelius in 1828.

^{232}Th decays by alpha emission and has a half-life of 1.41×10^{10} years. Further decay by alpha, beta, and gamma emissions gives rise to a series of isotopes which includes ^{228}Th, ^{224}Ra, ^{212}Bi, ^{216}Po, ^{212}Po, and, finally, stable ^{208}Pb. The energies for the alpha emissions in this series range from 4.01 keV for ^{228}Th up to 8.79 keV for ^{212}Po (131).

The physical and chemical properties of thorium and its industrially important compounds are listed in Table 42.6 (1). Thorium is a white, relatively soft, ductile metal. The powdered metal burns readily in air and ignites at 270°C; ThH$_4$ ignites at 260°C. There is no limiting O$_2$ pressure that prevents thorium from igniting, for combustion will occur in a pure carbon dioxide atmosphere. A limiting concentration of 6% O$_2$ will prevent ThH$_4$ from ignition. The minimal explosive concentrations for thorium metal powder and ThH$_4$ are 75 and 80 oz/1000 ft^3, respectively.

Thorium resembles the rare earth metals in its analytic reactions and can be separated from the common metals as an oxalate or fluoride. Thorium forms a series of tetravalent salts, many of which exist as various crystalline hydrates. Water and aqueous alkalies have little effect on thorium metal, but it slowly reacts with acids to form soluble salts. The oxides of thorium are insoluble in water and alkalies, but dissolve in acids.

2.1.2 Odor and Warning Properties

No characteristics of thorium warn of its presence without specialized detection techniques. The thorium ion is colorless. Its presence can be measured by radiation detection devices.

2.2 Production and Use

Thorium is a relatively abundant constituent of the earth's surface and occurrs in more than 100 minerals. Usually found with uranium and certain rare earth elements, it is three times as abundant as uranium, about as abundant as beryllium and cobalt, and one-half as common as lead. The thorium minerals, thorianite [(Th,U)O$_2$] and thorite (ThSiO$_4$), do not

Table 42.6. Physical and Chemical Properties of Thorium and Its Industrially Important Compounds[a]

Form	At. or mol wt.	Sp gr.	Mp (°C)	Bp (°C)	Solubility
Thorium, Th	232.04	11.7	1845	4500	Insol. water; sol. HCl, H_2SO_4, aqua regia; sl. sol. HNO_3
Thorium dioxide, ThO_2	264.04	9.86	3050	4400	Insol. water, alkalies; dil. acids; sol. hot H_2SO_4
Thorium hydroxide, $Th(OH)_4$	300.02		Dec.[b]		Insol. water; sol. acid; insol alkalies, HF
Thorium fluoride, ThF_4	308.03	6.32 (24°C)	> 900		Sl. dec. dil. H_2SO_4, HCl; insol. conc. H_2SO_4
Thorium oxalate, $Th(C_2O_4)_2$	408.08	4.637 (16°C)	Dec.[b]		0.017 g/L (16°C); sl. sol. acid
Thorium nitrate tetrahydrate, $Th(NO_3)_4 \cdot 4H_2O$			Swells		V. sol. water, alcohol, acids
Thorium sulfate, $Th(SOC_4)_2$	426.16	4.225 (16°C)			Sol. water; insol. acid; v. sol. $NH_4C_2H_3O_2$

[a]Ref. 1.
[b]Dec. = Decompose.

occur in commercially significant quantities. Monazite is the most common and commercially important thorium-bearing mineral. Important deposits are in India, Brazil, Sri Lanka, South Africa, the former USSR, Scandinavia, the United States, and Australia. Thorium is obtained as a by-product of beach-sand mining for rare earth elements. The monazite sand, which is separated physically or mechanically from other sands, is essentially an orthophosphate of rare earth elements (Ce, La, Y, Th)PO_4, that contains from 3 to 10% ThO_2 or more.

During 1974, the United States exported thorium-bearing ore containing approximately 156,430 lb of ThO_2, the first significant export of thorium. Production was from a mine in Green Cove Springs, Florida, where monazite sand was recovered by dredging. This mine ceased production in 1995, however, due to decreased demand for thorium. In 1995, approximately 10 metric tons of thorium ore and concentrates from monazite were imported from Australia. In 1995 U.S. reserves of thorium oxide equaled 158,000 metric tons. Other countries that have significant stores of thorium include Brazil, Canada, India, Malaysia, Norway, and South Africa.

Thorium is extracted from monazite sand concentrates for metallurgical and other purposes by digestion with either hot, fuming sulfuric acid or caustic soda. The resultant mass is diluted with water which dissolves thorium, uranium, and rare earth metals, leaving unreacted monazite, silica, rutile (TiO_2) and zircon ($ZrSiO_4$). Neutralization of the liquor precipitates thorium phosphate, leaving behind uranium and most of the rare earth metals.

In 1974, U.S. domestic use of thorium was about 80 tons, about one-half of which was employed to produce nuclear fuels and for nuclear research. Principal nonenergy applications were in the production of Welsbach incandescent gaslight mantles, as a hardener in Th–Mg alloys, in thoriated tungsten electrodes, and for chemical catalytic uses. The use of thorium has declined since the 1970s, however, due to the development of nonradioactive substitutes for many applications. Consumption of thorium dropped in 1991 from 54 metric tons of thorium oxide to 17 metric tons in 1994.

Nuclear Fuel: Because thorium can be converted via neutron capture followed by beta-decay into fissionable uranium-233 in a breeder reactor, it can be used as a source of atomic fuel. Because partial substitution of thorium for uranium in nuclear fuel decreases the amount of bomb-grade plutonium produced by reactors, this use of thorium may increase in the future.

Reactor-grade thorium can be obtained from the impure concentrate extracted from monazite sand by a liquid–liquid extraction process using a mixture of water, tributyl phosphate, and nitric acid. Thorium is extracted into the water-immiscible tributyl phosphate phase, and impurities remain in the aqueous phase. Generally, the purified thorium is back-extracted into an aqueous solution and either crystallized from solution as the nitrate or precipitated as the oxalate. The oxide or other compounds of thorium can be prepared from these pure salts.

Metallurgy: Thorium metal is produced by reducing thorium fluoride with calcium and a zinc chloride booster. At sufficiently high temperatures, Zn–Th melts and settles to the bottom of the furnace pot, and the slag, composed of CaF_2, $CaCl_2$, and ZnF_2, is displaced to the top. The exclusion of air at this phase is necessary because thorium is highly reactive at elevated temperatures (660°C). The Zn–Th metal is cleaned by chipping and brushing, heated in a steel retort to 1150°C under vacuum, and the zinc is distilled off, leaving a thorium sponge. The sponge or powder is melted under vacuum or in an inert atmosphere as a consumable electrode made of compressed thorium metal. The metal consumed by the arc forms an ingot by dropping into a water-cooled copper crucible, ready for conventional methods of extrusion, rolling, forging, swaging, wiredrawing, and machining. Powder metallurgy techniques have been developed extensively for thorium.

Alloying with thorium imparts desirable heat-resistant properties to magnesium. Thus, Mg–Th alloys are used in constructing airplane engines, frames, and missiles. Manufacture occurs in two steps: a master alloy that contains roughly equal parts of thorium and magnesium is remelted; further magnesium is added to a maximal concentration of 4% thorium. The final product is handled by standard foundry practices for melting, casting, grinding, and welding.

Thoriated tungsten alloy containing 1 to 2% thoria (ThO_2) is produced by "doping" tungsten powder with $Th(NO_3)$ solution fired at 900°C, converting the nitrate to the oxide. Conversion to the metal oxide is done in a hydrogen reduction furnace, and the resulting thoriated tungsten oxide powder is pressed into bars and sintered at about 2500°C under a hydrogen atmosphere. During 1987 approximately 1350–1500 kg of thorium dioxide were used in the United States to manufacture thoriated tungsten electrodes which are used in very high voltage filaments (132).

Ceramics/Glass: Thorium oxide ceramics are fabricated by many of the common operations of ceramic manufacture (cold pressing, hot pressing, slip casting, extrusion, isostatic pressing, tamping, and injection molding) using the same types of binders, lubricating agents, and plasticizers necessary in fabricating alumina, beryllia, and similar materials. Glasses that contain thorium oxide are used as lenses for cameras and scientific instruments because of their high refractive index and low dispersion.

Gaslight Mantles: The manufacture of thorium lighting mantles is a small scale, hand operation that involves dipping a highly absorbent rayon webbing in a concentrated $Th(NO_3)_4$ solution containing about 1% Ce nitrate for luminosity and some aluminum and beryllium nitrates for added strength. The impregnated web is exposed to NH_3 fumes that convert the metal nitrates to insoluble hydroxides, rinsed to remove NH_4NO_3, and dried. For "hard" mantles, after fastening the rayon to a chrome mounting, it is ignited, leaving a skeleton of oxides. For "soft" mantles, the webbing is merely cut off and sewn at one end.

Catalyst: Thorium is used by chemical industries as a catalyst in converting ammonia to nitric acid, in the production of sulfuric acid, and in petroleum cracking.

Medical: Thorotrast (colloidal thorium dioxide) was introduced in the late 1920s as a diagnostic roentgenographic agent and its use was widespread until the 1950s when evidence of its carcinogenic and fibrogenic properties led to a ban. It was given intravenously to outline the vascular system, especially in the liver and spleen, where it was picked up by the reticuloendothelial cells of these organs. Thorotrast was also employed for ventriculograms of the brain and for mammograms of the lacteal duct system.

2.3 Exposure Assessment

Industrial exposures to thorium occur during the refining of monazite sand, the handling of various thorium salts, the fabrication of thorium ingots, the handling thorium salts in various industrial uses, the casting and machining of thorium alloy parts, and the production of glazed tiles. In addition, there is potential for exposure through fumes from welding with thoriated-tungsten electrodes and from fires and explosions caused by thorium metal powder. In 1990, refractory ceramics accounted for approximately 70% of domestic consumption, incandescent lamp mantles accounted for 26%, and ceramics and welding accounted for the additional 3% (133). In 1998, the National Occupational Exposure Survey (NOES) (7) estimated that 5,318 workers in the United States were potentially exposed to thorium (SIC code 7391).

Thorium can be measured by a variety of analytical techniques, including alpha spectrometry, neutron activation analysis, spectrochemical analysis, and inductively coupled, argon-plasma, mass spectrometry (ICP-MS) (134, 135). Each has its advantages and disadvantages. Alpha spectrometry requires large sample sizes and long counting times, whereas neutron activation requires relatively rare and expensive equipment. Successful application of many of these methods for low concentrations of thorium is difficult and requires preliminary concentration procedures involving ion exchange or coprecipitating agents. Interference by commonly encountered ions can be a serious problem, and recoveries are often of the order of only 80%. The most sensitive reagent for colorimetric determination of thorium in bone ash is the dye Arseno III, which can measure as little as 1 µg under suitable conditions (136). Interfering ions such as calcium and phosphate, however, require preliminary separation of the thorium. The development of newer methods such as the quadrupole and magnetic-sector ICP-MS, which provide higher sensitivity and lower detection limits, has helped to overcome these problems.

2.3.2 Background Levels

Thorium is a relatively abundant element, but natural background concentrations vary significantly in different geological strata. A study in Pakistan found that average thorium concentrations were highest in alluvium (shales and clays), granitic and metamorphic rocks, and intermediate in acidic rocks, metamorphic schist, and gneiss, ranging from 11 to 18 ppm thorium. Much lower thorium concentrations were found in ultrabasic rocks (0.004 ppm Th), and sandstones and basic rocks contain 5.5 and 2.2 ppm Th, respectively (137).

In a study of 500 United States residents, 39.6% of the individuals had detectable concentrations (> 0.85 ng/L) of thorium in their urine. Concentrations ranged up to 7.7 ng/L, with a 95th percentile concentration of 3.09 ng/L (17).

2.3.3 Workplace Methods

Workplace exposures to thorium ionizing radiation can be monitored by measuring γ-rays using either NaI scintillation detectors or cylinder ionizing chambers to determine radiation dose rates. Monitoring room air concentrations of Th involves collecting air particles by using a filtering device, followed by analysis of the filter for Th either by neutron activation or direct alpha spectrometry (131). Air concentrations of Th generated by grinding and welding thoriated-tungsten electrodes were measured in the personal breathing zones and area air samples of experienced welders (138). Concentrations detected during grinding operations ranged from 0.001 to 0.3 Bq m^{-3}, depending upon the grinding method used. During welding operations, air concentrations of Th were 0.000004 to 0.0003 Bq m^{-3}; these concentrations are well below the derived air concentration (DAC) of 0.04 Bq m^{-3} for solubility class Y.

2.3.4 Community Exposures

Glazed tiles made in China from zircon sand can be a significant source of ionzing radiation to the general public and workers in the tile industry because of high concentrations of radionuclides, including ^{232}Th, in zircon sand. Zircon sand from China contains 8,200 ± 900 Bq/kg of thorium, which is approximately 10-fold higher than zircon from other areas of the world and 100 times higher than normal building materials (139).

2.3.5 Biomonitoring/Biomarkers

The best methods available for quantitating past exposure to thorium depend on measurements of the daughter products of ^{232}Th. Thorium body burdens have been determined *in vivo* by measuring the external gamma rays emitted by daughter products of thorium stored in a subject's body. These measurements are conducted in a shielded room with low background radiation levels (140). Measurement of the radioactive gas, thoron (^{220}Rn), a daughter product of Th, in exhaled air provides another measure of past thorium exposure (141, 142). Tissue concentrations in biopsy or postmortem samples from Thorotrast patients have been determined from measurements of ^{228}Ac γ-rays using a NaI(Tl) well-type scintillation spectrometer (143).

2.4 Toxic Effects

2.4.1 Experimental Studies

2.4.1.1 Pharmacokinetics, Metabolism, and Mechanisms. The metabolism of Th involves consideration of this species as a chemical entity, as well as the metabolism of some of its principal decay products, ^{224}Ra, ^{228}Ra, ^{220}Rn, and ^{212}Pb as radiological entities. The general kinetic behavior of thorium compounds is governed by their ready hydrolysis to colloidal polymers. Only at extremely low concentrations of the order of radio tracer amounts does thorium exist in monomeric, ionic form. Accordingly, organ distribution and excretion are governed by the dose and the physical state when introduced into the body. Thorium may enter the body as an inorganic, unchelated salt, as a complex ion chelate, or as a colloidal suspension, as in Thorotrast.

2.4.1.3.1 Absorption. Thorium inorganic salts are characterized by very low absorption from the gastrointestinal tract; less than 0.001% is absorbed at doses of 500 to 800 mg/kg Th(NO$_3$)$_4$, and 0.05% at doses of 5 mg/kg (144). ICRP (1979) assumes a gastrointestinal transfer factor of 0.02% (61).

2.4.1.3.2 Distribution
Thorium salts: Tissue distribution and retention of thorium is highly dependent upon dose and route of exposure. A tracer dose of carrier-free ^{227}Th citrate is absorbed within 1 day after intramuscular injection, and two-thirds of the dose is found in the skeleton (145). On the other hand, intravenous administration of large amounts (9 to 61 mg/kg) of ^{230}Th citrate has a deposition pattern predominantly in the reticuloendothelial system, in keeping with colloidal formation of large doses of thorium salts (146). Following intravenous administration, a soluble thorium salt is slowly removed from the blood, and 30% of the dose remains after 6 h. Most of the thorium goes to the reticuloendothelial system, the liver, spleen, and bone marrow (147). When thorium salts are intratracheally injected, they remain in the lung; when injected in the muscle, they remain at the injection site (144). When introduced into the body as an inorganic salt, most of the thorium is hydroxylated and polymerized, and the resultant colloidal particles are engulfed by the phagocytic cells of the reticuloendothelial system, liver, spleen, and bone marrow. Very small polymeric particles may also attach to cell surfaces or soft tissues. Only about 0.5% of the body burden of ^{232}Th is translocated from the reticuloendothelial system to the skeleton.

Thorium dust: Retention of thorium from inhalation of ThO$_2$ dust for 1 year at approximately 50 times the maximum permissible concentration (MPC) at a particle diameter near 1 mm, left 98% of the thorium body burden in the lung and pulmonary lymph nodes, 2% in bone, and less than 0.1% in the rest of the body (148, 149). The relative immobility of tissue-deposited thorium is shown by finding 97.7% in the pulmonary lymph nodes, 2.1% in the lung, 0.003% in the femur, and fractions of this latter amount in the kidney, spleen, and liver 6 to 7 years after inhalation exposure had been terminated (148, 149).

Thorotrast: Thorotrast is a colloidal suspension of thorium stabilized in a dextran solution. After intravascular administration, thorium is accumlated by the liver, spleen, lymph nodes, bone, and other tissues of the reticuloendothelial system by macrophage phagocytosis (150). Thorotrast also deposits and remains in areas around injection sites. Special interest in the effects of Thorotrast on bone marrow cells led to a recent study with monkeys, which found that by one week after intravenous or arterial injection of Thorotrast, thorium was evenly distributed throughout the cellular red bone marrow, and less than 1% of the skeletal burden was present in the bone matrix. By 3-4 years after injection, bone marrow thorium was present in conglomerates resident in macrophages. These findings are similar to those found in human tissue samples from Thorotrast patients in which similar levels of deposits were observed throughout the red bone marrow (151). Radiation dose estimates from studies of Thorotrast cancer patients reported mean doses to the liver of 22.2-34.7 rad/yr, to the spleen of 67.8-137.8 rad/yr, and to the bone marrow of 15.9-36.6 rad/yr (143).

2.4.1.3.3 Excretion. Excretion of polymeric thorium from the reticuloendothelial system is slow and, as expected, predominantly fecal by way of bile from the liver. Monomeric thorium from tracer doses of radioactive thorium isotopes is excreted mainly in the urine and occurs in the first few days after administration. Two-thirds of the total excretion of 35% of intravenously injected, essentially carrier-free, ^{228}Th citrate in the rat occurred in the first day, and practically no excretion occurred after 30 days (152). The fecal to urinary ratio was 1.6, compared with a ratio of 45 at higher dose levels of thorium.

2.4.1.5 Carcinogenesis. There are a limited number of studies of the carcinogenicity of ThO$_2$ in animals. These have been reviewed by Cohen and co-workers (153). The primary tumors observed in exposed animals were spindle cell carcinomas and osteosarcomas. In genotoxicity tests using the nonmammalian *B. subtilis* rec assay, thorium tetrachloride was not mutagenic.

2.4.1.6 Genetic and Related Cellular Effect Studies. There are few investigations into the mechanisms by which thorium interacts chemically with body constituents. One such investigation (154) showed that thorium (as the nitrate) binds strongly to the (bovine) cortical bone glucoprotein, sialoprotein, which has a high-density distribution of carboxylic acid groups along the protein chain; 42% of the amino acids in bone sialoprotein are either aspartic or glutamic acids and are mainly responsible for binding highly charged thorium ions. Similarly, the chondroitin sulfate protein of cortical bone can bind thorium (155). An interaction of thorium with zinc in the rat prostate has also been reported (156).

2.4.2 Human Experience

2.4.2.3 Epidemiology Studies. Humans have been exposed to thorium through occupational exposure and through the medicinal use of Thorotrast as a contrasting agent. These are dealt with separately as they represent two very different types of exposure, differing in both chemical/physical forms of thorium and rates and durations of exposure.

Thorotrast

With increasing frequency since 1955, reports in the medical literature have indicated severe radiation damage and cancers of the blood vessels, liver, kidney, reticuloendothelial system and other organs as a result of Thorotrast or thorium X administration as a radiopaque medium. Epidemiological studies have shown correlations between Thorotrast exposure and increased incidence of liver cancer, hepatoma, leukemia, and liver cirrhosis (157,158). In some cases the latency was only 11 years but generally was of the order of 25–35 years. Among 71 autopsies of Thorotrast patients, there were 45 cases of malignant hepatic tumor, 12 cases of liver cirrhosis, 7 cases of blood disease, and 7 cases of other diseases (143). During the period 1949 to 1974, 45 cases of tumors of the renal pelvis alone, mostly squamous cell or transitional cell types, were found in patients injected with Thorotrast for pyelography in Belgium, France, Germany, and the Netherlands (159). The latent period of the tumors varied from 16 to 37 years (in 1974) after Thorotrast pyelography, and the mean was 27 years. In patients who received Thorotrast during transcervical carotid angiography, granulomas resulting from a fibrotic reaction caused by the extravasation of the Thorotrast at the injection site are most commonly encountered, although squamous cell or mucoepidermoid carcinomas of the maxillary sinus were reported (150).

No correlation was found between the amount of Thorotrast, the latency, and the development of malignancy. The extremely variable composition of Thorotrast may, in part, be responsible for the lack of correlation between the amount injected, latency, and malignancy development. Depending on the number of times thorium was chemically processed to remove radium and other members of the decay chain, the amount of ^{228}Th relative to ^{232}Th changed. Moreover, the composition of the injected Thorotrast can be markedly affected by adhesion of radium to glass. Because it is thought that the tumors result from the alpha emission of ^{232}Th decay series, ^{228}Th and ^{224}Ra, from beta emission of ^{228}Ra, and from gamma emission from ^{224}Ra, it is evident that varying composition could result in varying effects. A recent study of four primary cancers found at autopsy in a patient who had received Thorotrast showed the presence of multiple mutations in the p53 gene of the cancerous tissues and in noncancerous small intestine and liver tissues, suggesting that cells with p53 abnormalities have a survival advantage that allows them to expand clonally (160).

In contrast to Thorotrast exposed patients, the work experience with thorium and its compounds has been exceptionally good. Although the workforce was small, neither workers in plants who made incandescent Welsbach mantles (Th 99%, Ce 1% from the nitrates) for more than 70 years nor workers who refined thoria from monazite experienced any health effects attributable to chemical toxicity or radiation injury. Moreover, exposures during the earlier years were largely unregulated and unquestionably exceeded the MPC for ^{232}Th (2×10^{-12} mCi/cm^3). An industrial hygiene and medical survey of a thorium refinery in 1955 that had operated for 30 years revealed exposures well in excess of current standards. However, "no evidence of overt industrial diseases" was discovered (161). The employee force at the time was 84, of whom 60 worked with thorium and thoron; one-half had spent 10 years or more in the thorium process. Of 693 former employees for whom records existed, there was no indication of a single disease related to their work. In a study of thorium refinery workers conducted by Argonne National Laboratory, there was no correlation between *in vivo* measurements of body burden and complete blood count parameters (162).

In a study of mortality carried out for 3796 workers (3119 men and 677 women) employed in a U.S. thorium-processing plant between 1915 and 1973, 2620 (2161 men and 459 women) were still alive. The standardized mortality ratio (SMR) for the male workers was significantly increased for all cancers (SMR 1.23) and for lung cancer (SMR 1.36). For the female workers, the SMR was 0.53 for all cancers (163).

Generally speaking, few liver tumors have been reported among workers compared to Thorotrast patients. The correlation of some hepatic function tests with the body burden of radioactivity among workers, however, suggests a radiation and/or chemical effect (140). Serum aspartate aminotransferase and alkaline phosphatase levels were associated with radioactivity body burden and were significantly higher than those of the general population.

A case-control study of 112 New Jersey households in the vicinity of a thorium waste disposal site found a higher prevalence of birth defects (relative risk 2.1) and liver diseases (relative risk 2.3) among the exposed population than among the unexposed group. The numbers were quite small and the confidence intervals wide, so the investigators (164) drew no definite conclusions.

2.5 Standards, Regulations, or Guidelines of Exposure

Because radiation effects are the principal hazards from exposure to thorium, permissible concentration limits have been set by the ICRP (1979 and 1991) for ^{232}Th and ^{228}Th on the basis of their radioactivity (61, 165). The annual limit of activity intake by inhalation is 40 Bq according to the guidelines of ICRP 30. According to German guidelines, the limit is 30 Bq (166). The derived air concentration (DAC) of ^{232}Th established by the NRC (1988) (123) in 10 CFR 20 is 0.04 Bq/m^3 (1 × 10^{-12} µCi/mL) for solubility class Y.

BIBLIOGRAPHY

1. C. Voegtlin and H. Hodge, Eds., *Pharmacology and Toxicology of Uranium Compounds*, McGraw-Hill, New York, 1953.
2. L. M. Scott, H. C. Hodge, and J. B. Hursh, eds., *Uranium Plutonium Transplutonic Elements*, Springer-Verlag, New York, 1973, pp. 271–306.
3. R. E. Bases, *Science* **126**, 164 (1957).
4. AEPI, *Health and Environmental Consequences of Depleted Uranium Use by the U.S. Army*, US Army Health Policy Institute, Tech. Rep. 1995.
5. J. M. Samet, *Occup. Med.* **6**, 629–639 (1991).
6. D. S. Sundin and T. M. Frazier, *Am. J. Public Health* **79** (Suppl.), 32–37 (1989).
7. R. Guay et al., *Biotechnol. Bioeng.* **19**, 727–740 (1977).
8. UIC Uranium Mill Chemistry, 1998. Available at: *www.uic.com.au.briefing*.

9. C. D. Blackwell, *Am. Ind. Hyg. Assoc.* **20**, 92 (1959).
10. A. C. Hyatt, *J. Am. Ind. Hyg. Assoc.* **20**, 82 (1959).
11. W. B. Harris and I. Kingsley, *Arch. Ind. Hyg.* **19**, 540 (1959).
12. F. Vernon et al., *Anal. Chim. Acta* **87**, 491–493 (1976).
13. D. A. Becker and P. D. LaFleur, *Anal. Chem.* **44**, 1508 (1972).
14. E. I. Hamilton, *Health Phys.* **22**, 149–153 (1972).
15. G. A. Welford and R. Baird, *Health Phys.* **13**, 1312–1324 (1967).
16. D. W. Medley et al., *Health Phys.* **67**, 122–130 (1994).
17. B. G. Ting, et al., *Environ. Res.* **76**, 53–59 (1998).
18. G. A. Welford, *Am. Ind. Hyg. Assoc. J.* **21**, 68–70 (1960).
19. G. A. Welford *76th Annu. Bioassay Anal. Chem.Conf.*, Bethesda, MD, Oct. 8–9, 1970.
20. T. Nozaki, *J. Radioanal. Chem.* **6**, 33 (1970).
21. C. M. West et al., *Health Phys.* **69**, 481–486 (1995).
22. H. S. Dang et al., *Health Phys.* **62**, 562–566 (1992).
23. NCRP, *NCRP Rep.* **87** (1987).
24. Z. Karpas et al., *Health Phys.* **73**, 86–100 (1998).
25. A. R. Byrne and L. Benedik, *Sci. Total. Environ.* **107**, 143–157 (1991).
26. S. Mountain and H. Stokinger, *Am. Ind. Hyg. Assoc. J.* **23**, 8 (1962).
27. H. E. Palmer et al., *Health Phys.* **47**, 632–634 (1984).
28. H. B. Wilson, *Arch. Ind. Hyg. Occup. Med.* **7**, 301 (1953).
29. K. Nomiyama and E. C. Foulkes, *Toxicol Appl. Pharmacol.* **13**, 89–98 (1968).
30. F. J. Bowman and E. C. Foulkes, *Toxicol Appl. Pharmacol.* **16**, 391–399 (1970).
31. E. C. Foulkes, *Toxicol Appl. Pharmacol.* **20**, 380–385 (1971).
32. G. L. Diamond et al., *Fundam. Appl. Toxicol.* **13**, 65–78 (1989).
33. L. J. Leach et al., *Health Phys.* **18**, 599–612 (1970).
34. M. B. Guglielmotti et al., *J. Oral Pathol.* **14**, 565–572 (1985).
35. P. E. Morrow et al., *Health Phys.* **23**, 273–280 (1972).
36. P. W. Durbin, in R. H. Moore, ed., *Biokinetics and Analysis of Uranium in Man*, Proc. Colloq. Richland, WA, USUR-05 HEHF-47, Hanford Environmental Health Foundation, Richland, WA, 1984.
37. W. Stevens et al., *Radiat. Res.* **83**, 109–126 (1980).
38. K. A. Stevenson and E. P. Hardy, *Health Phys.* **65**, 283–287 (1993).
39. S. R. Bernard et al., *Proc. 1st Annu. Meet., Health Phys. Soc., 1957*.
40. S. Bernard and E. C. Struxness, *A Study of the Distribution and Excretion of Uranium in Man*, Oak Ridge National Laboratory, Oak Ridge, TN 1957, pp. 1–57.
41. A. L. Dounce and T. H. Lan, in C. Voegtlin, ed., *Pharmocology and Toxicology of Uranium Compounds*, McGraw-Hill, New York, 1949, pp. 759–888.
42. J. B. Hursh and N. L. Spoor, *Uranium Plutonium, Transplutonic Elements*, Vol 36, In H. C. Hodge et al., Eds., *Handbook of Experimental Pharmacology*. Springer-Verlag, Berlin, 1973, pp. 197–239.
43. F. Bernheim, *Microbios* **4**, 87–92 (1971).
44. Q. Y. Hu and S. P. Zhu, *Radiat. Environ. Biophys.* **29**, 161–167 (1990).

45. N. Friis and P. Myers-Keith, *Biotechnol. Bioeng.* **28**, 21 (1986).
46. P. W. Durbin and M. E. Wrenn, *Conf. Occup. Health Experience with Uranium*, Arlington, VA, April, 1975. U.S. Energy Research and Development Administration, Washington, DC, ERDA-93, pp 67–129.
47. A. Rothstein et al., eds., *Effects of Metals on Cells, Subcellular Elements and Macromolecules*, Thomas,Springfield, SL, 1970, 365–385.
48. T. Tyrakowski, *Acta Med. Pol.* **20**, 297–306 (1979).
49. A. J. Luessenhop et al., *Am. J. Roentgenol.* **79**, 83–100 (1958).
50. P. Galle, *J. Microsc. (Paris)* **19**, 17 (1974).
51. P. Galle, in P. Galle, ed., *Toxique Nucleaires*, Masson, France, 1982, pp. 224–240.
52. T. Horikoshi et al., *Radioisotopes* **28**, 485 (1979).
53. G. W. Strandberg et al., *Appl. Environ. Microbiol.* **41**, 237 (1981).
54. F. N. Ghadially et al., *Br. J. Exp. Pathol.* **63**, 227–234 (1982).
55. J. M. Weinberg, R. W. Schrier, and C. W. Gottschalk, eds., *Diseases of the Kidney*, Little, Brown, Boston, MA, 1988, pp. 1137–1195.
56. F. J. Silverblatt and C. Kuehn, *Kidney Int.* **15**, 335–345 (1979).
57. J. P. Berry et al., *J. Histochem. Cytochem.* **26**, 651–657 (1978).
58. A. Roth, *Virchows Arch.* **405**, 131 (1984).
59. P. E. Morrow et al., *Health Phys.* **43**, 859–873 (1982).
60. P. E. Morrow, *Biokinetics and Toxicology of Uranium*, Proc. Colloq. Richland, WA., USUR-05, HEHF-47, Harford Environmental Health Foundation, Richland, WA, 1984.
61. International Commission Radiological Protection (ICRP), *Ann. ICRP* **2**, 30 (1979).
62. R. W. Leggett, *Health Phys.* **57**, 365–383 (1989).
63. T. P. Singer et al., in A. Tannenbaum, ed., *Toxicology of Uranium*, McGraw-Hill, NY, 1951, pp. 208–245.
64. J. H. Schwartz and W. Flamenbaum, *Kidney Int., Suppl.* **6**, S123–S127 (1976).
65. J. H. Schwartz and W. Flamenbaum, *Am. J. Physiol.* **230**, 1582–1589 (1976).
66. A. Brodsky et al., *Bioassay Programs for Uranium*, HPS N13.22–1995, 1995, pp. 1–77.
67. P. E. Morrow et al., *Metabolic Fate and Evaluation of Injury to Rats and Dogs Following Exposure to the Hydrolysis Products of Uranium Hexafluoride*, NTIS/UREG/CR-2268, National Technical Information Services, Springfield, VA, 1982.
68. F. L. Haven, *J. Biol. Chem.* **175**, 737 (1948).
69. C. L. Yuile eds., *Uranium Plutonium Transplutonic Elements*, Springer-Verlag, New York, 1973, pp. 165–196.
70. J. L. Domingo et al., *Arch. Environ. Health* **44**, 395–398 (1989).
71. J. M. Llobet et al., *Fundam. Appl. Toxicol.* **16**, 821–829 (1991).
72. W. C. Hueper, *J. Natl. Cancer Inst. (U.S.)* **13**, 291–305 (1952).
73. A. C. Miller et al., *Environ. Health Perspect.* **106**, 465–471 (1998).
74. R. H. Lin et al., *Mutat. Res.* **319**, 197–203 (1993).
75. J. L. Domingo et al., *Toxicology* **55**, 143–152 (1989).
76. R. H. Moore and R. L. Kathren, *J. Occup. Med.* **27**, 753–756 (1985).
77. D. R. Goodman, P. L. Williams, and J.L. Burson, eds., *Industrial Toxicology Safety and Health Applications in the Workplace*, Van Nostrand-Rheinhold, New York, 1985.

78. M. E. Wrenn et al., *Am. Water Works Assoc. J.* **79**, 177–184 (1987).
79. V. E. Archer et al., *J. Occup. Med.* **15**, 11–14 (1973).
80. D. P. Brown and T. Bloom, *Mortality among Uranium Enrichment Workers*, Report to NIOSH, NTIS PB87-188991, National Technical Information Service, Springfield, VA, 1987.
81. H. Checkoway and D. Crawford-Brown, *J. Chronic Dis.* **40** (Suppl. 2), 191S-200S (1987).
82. G. Saccomanno, *Recent Results Cancer Res.* **82**, 43–52 (1982).
83. M. J. Thun et al., *Scand. J. Work Environ. Health* **11**, 83–90 (1985).
84. J. J. Russell et al., *Health Phys.* **70**, 466–472 (1996).
85. M. L. Zamora et al., *Toxicol. Sci.* **43**, 68 (1998).
86. F. J. Hooper et al., *Health Physics* **77**, 512–519 (1999).
87. M. A. McDiarmid et al., *Env. Res.* **82**, 168–180 (2000).
88. M. Lippmann, *Arch. Ind. Health* **20**, 211 (1959).
89. R. L. Kile and A. J. Quigley, *Arch. Dermatol.* **27**, 220 (1973).
90. W. N. Young and H. A. Tebroack, *Ind. Med. Surg.* **27**, 229 (1958).
91. V. E. Archer et al., *Ann. N. Y. Acad. Sci.* **271**, 280–293 (1976).
92. J. Horacek et al., *Cancer* **40**, 832–835 (1977).
93. Z. Vich and V. Pacina, *Neoplasma* **34**, 211–215 (1987).
94. L. S. Gottlieb and L. A. Husen, *Chest* **81**, 449–452 (1982).
95. J. M. Samet et al., *Health Phys.* **61**, 745–752 (1991).
96. A. Woodward et al., *Cancer Causes Control* **2**, 213–220 (1991).
97. G. R. Howe et al., *J. Natl. Cancer Inst.* **77**, 357–362 (1986).
98. E. A. Dupree et al., *Scand. J. Work Environ. Health* **13**, 100–107 (1987).
99. A. P. Polednak and E. L. Frome, *J. Occup. Med.* **23**, 168–178 (1981).
100. G. D. Baisogolov et al., *Vopr. Onkol.* **37**, 553–559 (1991).
101. J. M. Samet et al., *Health Phys.* **56**, 415–421 (1989).
102. J. Sevc et al., *Health Phys.* **30**, 433–437 (1976).
103. V. E. Archer et al., *Health Phys.* **25**, 351–371 (1973).
104. G. Saccomanno et al., *Ann. N. Y. Acad. Sci.* **271**, 377–383 (1973).
105. M. Tirmarche et al., *Br. J. Cancer* **67**, 1090–1097 (1993).
106. ATSDR, *Toxicological Profile for Uranium*, ATSDR, Agency for Toxic Substances and Disease Regestry, Atlanta, GA, 1999.
107. BEIR IV (Committee on the Biological Effects of Ionizing Radiation), *Health Risks of Radon and Other Internally Deposited Alpha Emitters*, National Research Council (NRC), National Academy Press. Washington, DC, 1988.
108. A. L. Brooks et al., *Int. J. Radiat. Biol.* **66**, 801–808 (1994).
109. A. L. Brooks et al., *Radiat. Res.* **144**, 114–118 (1995).
110. M. A. Khan et al., *Mutat. Res.* **334**, 131–137 (1995).
111. C. Y. Bao et al., *Mutat. Res.* **326**, 1–15 (1995).
112. K. H. Vahakangas et al., *Lancet* **339**, 576–580 (1992).
113. J. A. Taylor et al., *Lancet* **343**, 86–87 (1994).
114. T. K. Hei et al., *Carcinogenesis* **15**, 431–437 (1994).
115. C. K. Hill et al., *Nature (London)* **298**, 67–69 (1982).

116. R. C. Miller et al., *Radiat. Res.* **133**, 360–364 (1993).
117. D. J. Brenner et al., *Radiat. Res.* **133**, 365–369 (1993).
118. B. Endlich et al., *Int. J. Radiat. Biol.* **64**, 715–726 (1993).
119. X. Z. Xuan et al., *Health Phys.* **64**, 120–131 (1993).
120. J. Chameaud et al., *Recent Results Cancer Res.* **82**(11–20), 11–20 (1982).
121. R. F. Jostes, *Mutat. Res.* **340**, 125–139 (1996).
122. F. Martin et al., *Br. J. Ind. Med.* **48**, 98–102 (1991).
123. National Research Council (NRC), *Code of Federal Regulations*, 10 CFR 20, U. S. Nuclear Regulatory Commission, Washington, DC, 1988.
124. U. S. Department of Energy (DOE), *Code of Federal Regulations*, 10 CFR 835, DOE, Washington, DC, 1993.
125. U.S. Environmental Protection Agency (USEPA), *Limiting Values of Radionuclide Intake and Air Concentration and Dose Conversion Factors for Inhalation, Submersion and Ingestion*, EPA-520/1-88-020, USEPA, Washington, DC, 1988.
126. International Commission on Radiological Protection (ICRP), *Recommendations of the International Commission on Radiological Protection*, Publ. 26, ICRP, Washington, DC, 1977.
127. OSHA Occupational Safety and Health Administration (OSHA), *Code of Federal Regulations*, 29 CFR 1910.1000, OSHA, Washington, DC, 1989.
128. International Commission on Radiological Protection (ICRP), *Recommendations of the International Commission on Radiological Protection* (Adopted 1978), Publ. 60, ICRP, Washington, DC, 1991.
129. NCRP, *NCRP, Rep* **116** (1993).
130. U.S. Environmental Protection Agency (USEPA), *Radionuclides in Drinking Water Fact Sheet. National Primary Drinking Water Regulations for Radionuclides*, (proposed rule, June,1991), EPA-570/9-91-700. USEPA, Washington, DC, 1991.
131. H. Hotzl et al., *Health Phys.* **70**, 651–655 (1996).
132. NCRP, *NCRP Rep.* **95** (1987).
133. Mineral Commodity Summaries, January, 1996 at http://minerals.er.usgs.gov/minerals/pubs/msc/1996/thorium.txt.
134. S. D. Pleskach, *Health Phys.* **48**, 303–307 (1985).
135. B. G. Ting et al., *J. Anal. At. Spectrom.* **11**, 339–342 (1996).
136. H. G. Petrow and C. D. Strehlow, *Anal. Chem.* **39**, 265–267 (1967).
137. K. A. Butt, et al., *Health Phys.* **75**, 63–66 (1998).
138. E. M. Crim and T. D. Bradley, *Health Phys.* **68**, 719–722 (1995).
139. W. Deng, et al., *Health Phys.* **73**, 369–372 (1997).
140. I. Farid and S. A. Conibear, *Health Phys.* **44** (Suppl. 1), 221–230 (1983).
141. J. Rundo et al., *Environ. Res.* **18**, 94–100 (1979).
142. A. F. Stehney et al., *Health Status and Body Radioactivity of Former Thorium Workers*, Argonne National Lab Report NUREG/CR-1420, ANL 80-37, 1980.
143. Y. Kato et al., *Health Phys.* **44** (Suppl. 1), 273–279 (1983).
144. P. R. Salerno and P. A. Mattis, *J. Pharmacol. Exp. Ther.* **101**, 31 (1951).
145. B. J. Stover and D. S. Buster, *The Dosimetry and Toxicity of Thorotrast*, IAEA-106, pp. 53–64.

146. I. U. Boone et al., *Am. Ind. Hyg. Assoc. J.* **19**, 285 (1958).
147. J. K. Scott, *J Pharmacol. Exp. Ther.* **106**, 286 (1951).
148. H. C. Hodge, Proceedings Rochester Conference on Thorium Toxicity, 1960.
149. R. E. Albert, *Thorium — Its Industrial Hygiene Aspects.* Academic Press, New York, 1966.
150. F. Ung et al., *Ann. Otol., Rhinol., Laryngol.* **107**, 708–712 (1998).
151. J. A. Humphreys et al., *Health Phys.* **74**, 442–447 (1998).
152. R. G. Thomas, *Health Phys.* **9**, 153 (1983).
153. M. D. Cohen et al., in L. W. Chang, ed., *Toxicology of Metals*, CRC Press, Boca Raton, FL, 1996 pp. 253–284.
154. A. R. Peacocke and P. A. Williams, *Nature (London)* **211**, 1140–1141 (1966).
155. G. M. Herring, *Calcifi. Tissues, Proc. 4th Eur. Symp., Int. Congre. Ser.* 120 (1966).
156. A. R. Chowdhury, *Indian J. Exp. Biol.* **12**, 563–564 (1974).
157. M. Andersson et al., *Radiat. Res.* **134**, 224–233 (1993).
158. M. Andersson et al., *Radiat. Res.* **142**, 305–319 (1995).
159. R. L. Verhaak et al., *Cancer* **34**, 2061–2068 (1974).
160. K. S. Isamoto et al., *Hum. Pathol.* **29**, 412–416 (1998).
161. R. Albert, *Arch. Ind. Health* **11**, 234–242 (1955).
162. S. A. Conibear, *Health Phys.* **44** (Suppl. 1), 231–237 (1983).
163. Z. Liu et al., *Scand. J. Work Environ. Health* **18**, 162–168 (1992).
164. G. R. Najem and L. K. Voyce, *Am. J. Public Health* **80**, 478–480 (1990).
165. International Commission on Radiological Protection (ICRP), *Annual Limits on Intake of Radionuclides by Workers Based on the 1990 Recommendations*, Rep. 61, ICRP, Washington, DC, 1991.
166. *Bundesminister fur Umvelt, Naturschutz and Reaktorsicherheit (BMU), Verordnung uber den Schuts vor Schaden durch Ionisierende Strahlen,* (Strahlenschutsverdnung-StrlSchV), Report 34, Bundesgesetzblatt Teil I, Bonn, 1989.

CHAPTER FORTY-THREE

The Lanthanides, Rare Earth Metals

W. H. (Bill) Wells Jr., Ph.D., CIH and Vickie L. Wells, MS, CIH

The lanthanides (or lanthanons) are a group of 15 elements of atomic numbers from 57 through 71 in which scandium (atomic number 21) and yttrium (atomic number 39) are sometimes included. The lanthanide series proper is that group of chemical elements that follow lanthanum in its group IIIB column position of the periodic table. Their distinguishing atomic feature is that they fill the $4f$ electronic subshell. Actually, only those elements with atomic numbers 58–71 are lanthanides. Most chemists also include lanthanum in the series because, although it does not fill the $4f$ subshell, its properties are very much like those of the lanthanides. The elements scandium and yttrium are also known as the "rare earths" because they were originally discovered together with the lanthanides in rare minerals and isolated as oxides, or "earths." In comparison with many other elements, however, the rare earths are not really rare, except for promethium, which has only radioactive isotopes. Yttrium, lanthanum, cerium, and neodymium are all more abundant than lead in the earth's crust. All except promethium, which probably does not occur in nature, are more abundant than cadmium. The relative abundance and atomic numbers are tabulated in Table 43.1. The more common lanthanide compounds are listed in Sections 1.0a–1.0o.

Scandium is a silvery white metallic chemical element, the first member of the first transition-metal series in the periodic table. The name is derived from Scandinavia, where the element was discovered in the minerals euxenite and gadolinite. In 1876, L. F. Nilson prepared about 2 g of high purity scandium oxide. It was subsequently established that scandium corresponds to the element "ekaboron," predicted by Mendeleyev on the basis

Patty's Toxicology, Fifth Edition, Volume 3, Edited by Eula Bingham, Barbara Cohrssen, and Charles H. Powell.
ISBN 0-471-31934-1 © 2001 John Wiley & Sons, Inc.

Table 43.1. Atomic Numbers, Weights, and Concentrations of the Lanthanides, Cadmium, and Lead Included for Reference

Element	Atomic No.	Atomic Weight	Earth's Crust (ppm)
Scandium (Sc)	21	44.9559	5–6
Yttrium (Y)	39	88.9059	28–70
Cadmium (Cd)	48	112.41	0.1–0.2
Lanthanum (La)	57	138.9055	5–18
Cerium (Ce)	58	140.12	46
Praseodymium (Pr)	59	140.9077	6
Neodymium (Nd)	60	144.24	24
Promethium (Pm)	61	145	4.5×10^{-20}
Samarium (Sm)	62	150.4	6
Europium (Eu)	63	151.96	1
Gadolinium (Gd)	64	157.3	6
Terbium (Tb)	65	158.9254	0.9
Dysprosium (Dy)	66	162.50	4
Holmium (Ho)	67	164.9304	1
Erbium (Er)	68	167.26	2
Thulium (Tm)	69	168.9342	0.2
Ytterbium (Yb)	70	173.04	3
Lutetium (Lu)	71	174.97	0.8
Lead (Pb)	82	207.2	1.6×10^{-3}

of a gap in the periodic table. Scandium occurs in small quantities in more than 800 minerals and causes the blue color of aquamarine beryl.

Yttrium is one of four chemical elements (the others are erbium, terbium, and ytterbium) named after Ytterby, a village in Sweden that is rich in unusual minerals and rare earths. Yttrium is a metal with a silvery luster and properties closely resembling those of rare earth metals. It is the first member of the second series of transition metals. Yttrium is found in several minerals and is produced primarily from the ore material xenotime.

Lanthanum is a white, malleable metal; it is the first member of the third series of transition metals, and the first of the rare earths. Lanthanum is found with other lanthanides in the ore minerals monazite, bastnaesite, and xenotime, and in other minerals. It was discovered in 1839 by the Swedish chemist Carl G. Mosander. Scientists have created many radioactive isotopes of lanthanum.

1.0a Scandium Chloride

1.0.1a CAS Number: *[10361-84-9]*

1.0.2a Synonyms: NA

1.0.3a Trade Names: NA

1.0.4a Molecular Weight: 151.31

THE LANTHANIDES, RARE EARTH METALS

1.0.5a **Molecular Formula:** $ScCl_3$

1.0b Lanthanum Oxide

1.0.1b **CAS Number:** *[1312-81-8]*

1.0.2b **Synonyms:** Dilanthanum oxide, dilanthanum trioxide, lanthana, lanthania (La_2O_3), lanthanum sesquioxide, lanthanum trioxide, lanthanum (+3) oxide, lanthanum(III) oxide

1.0.3b **Trade Names:** NA

1.0.4b **Molecular Weight:** 325.09

1.0.5b **Molecular Formula:** La_2O_3

1.0c Lanthanum Nitrate

1.0.1c **CAS Number:** *[10099-59-9]*

1.0.2c **Synonyms:** NA

1.0.3c **Trade Names:** NA

1.0.4c **Molecular Weight:** 324.94

1.0.5c **Molecular Formula:** $La(NO_3)_3$

1.0d Cerium

1.0.1d **CAS Number:** *[7440-45-1]*

1.0.2d **Synonyms:** NA

1.0.3d **Trade Names:** NA

1.0.4d **Molecular Weight:** 140.12

1.0.5d **Molecular Formula:** Ce

1.0e Neodymium

1.0.1e **CAS Number:** *[7440-00-8]*

1.0.2e **Synonyms:** NA

1.0.3e **Trade Names:** NA

1.0.4e **Molecular Weight:** 144.24

1.0.5e **Molecular Formula:** Nd

1.0f Neodymium Oxide

1.0.1f CAS Number: [1313-97-9]

1.0.2f Synonyms: Dineodymium trioxide, neodymia, neodymium sesquioxide, neodymium trioxide, neodymium (+3) oxide, neodymium(III) oxide

1.0.3f Trade Names: NA

1.0.4f Molecular Weight: 336.48

1.0.5f Molecular Formula: Nd_2O_3

1.0g Samarium Oxide

1.0.1g CAS Number: [12060-58-1]

1.0.2g Synonyms: Disamarium trioxide, samaria, samarium sesquioxide, samarium trioxide, samarium (+3) oxide, samarium(III) oxide

1.0.3g Trade Names: NA

1.0.4g Molecular Weight: 348.72

1.0.5g Molecular Formula: Sm_2O_3

1.0h Europium Oxide

1.0.1h CAS Number: [1308-96-9]

1.0.2h Synonyms: Dieuropium trioxide, europia, europium sesquioxide, europium trioxide, europium (+3) oxide, europium(III) oxide

1.0.3h Trade Names: NA

1.0.4h Molecular Weight: 351.92

1.0.5h Molecular Formula: Eu_2O_3

1.0i Gadolinium

1.0.1i CAS Number: [7440-54-2]

1.0.2i Synonyms: NA

1.0.3i Trade Names: NA

1.0.4i Molecular Weight: 157.25

1.0.5i Molecular Formula: Gd

THE LANTHANIDES, RARE EARTH METALS 427

1.0j Gadolinium Oxide

1.0.1j **CAS Number:** *[12064-62-9]*

1.0.2j **Synonyms:** Digadolinium trioxide, gadolinia, gadolinium sesquioxide, gadolinium trioxide, gadolinium (+3) oxide, gadolinium(III) oxide

1.0.3j **Trade Names:** NA

1.0.4j **Molecular Weight:** 362.50

1.0.5j **Molecular Formula:** Gd_2O_3

1.0k Terbium Trinitrate

1.0.1k **CAS Number:** *[10043-27-3]*

1.0.2k **Synonyms:** Terbium (+3) nitrate; nitric acid, terbium (+3) salt

1.0.3k **Trade Names:** NA

1.0.4k **Molecular Weight:** 344.95

1.0.5k **Molecular Formula:** $Tb(NO_3)_3$

1.0l Dysprosium Oxide

1.0.1l **CAS Number:** *[1308-87-8]*

1.0.2l **Synonyms:** Didysprosium trioxide, dysprosium sesquioxide, dysprosia, dysprosium (+3) oxide, dysprosium(III) oxide

1.0.3l **Trade Names:** NA

1.0.4l **Molecular Weight:** 373.00

1.0.5l **Molecular Formula:** Dy_2O_3

1.0m Erbium Trioxide

1.0.1m **CAS Number:** *[12061-16-4]*

1.0.2m **Synonyms:** Dierbium trioxide, erbia, erbium sesquioxide, erbium (+3) oxide, erbium(III) oxide

1.0.3m **Trade Names:** NA

1.0.4m **Molecular Weight:** 382.52

1.0.5m **Molecular Formula:** $Er_2(O_3)_3$

1.0n Ytterbium

1.0.1n CAS Number: [7440-64-4]

1.0.2n Synonyms: NA

1.0.3n Trade Names: NA

1.0.4n Molecular Weight: 173.04

1.0.5n Molecular Formula: Yb

1.0o Yttrium

1.0.1o CAS Number: [7440-65-5]

1.0.2o Synonym: Yttrium-89

1.0.3o Trade Names: NA

1.0.4o Molecular Weight: 88.91

1.0.5o Molecular Formula: Y

1.1 Chemical and Physical Properties

The chemical and physical properties of the lanthanides are given in Table 43.2. The unique characteristic of the chemistry of the lanthanides is their similarity. The defining $4f$ electrons are deep within the lanthanide atoms and are shielded from the environment around the atom by their outer $6s$ and $5d$ electrons. Thus the f electrons do not cause strong variations in chemical properties. This is in contrast to the wide variation in properties shown by the transition metals, whose defining outer d electrons project from the atoms and ions and interact with surrounding materials.

All the lanthanides form ions in the characteristic group IIIB oxidation state of $+3$. Some of the elements exhibit other oxidation states, but they are less stable. There is evidence that the elements with empty, half-filled, and filled $4f$ orbital shell ions have very stable configurations; thus La (empty), Gd (half-filled), and Lu (filled) form only $+3$ ions. The most stable $+2$ and $+4$ ions are Eu^{2+}, Yb^{2+}, Ce^{4+}, and Tb^{4+}. The ceric ion, Ce^{4+}, is a good oxidizer.

The elements occur together in nature in large part due to their chemical similarity. The exception is promethium, which is radioactive and probably occurs naturally only in trace amounts, if at all. The elements are extremely difficult to separate. Modern ion-exchange and repeated fractional crystallization techniques have been developed that result in the availability of pure (99.99%) materials.

All the lanthanides are silvery white, very reactive metals with high melting points. Their $+3$ ion reduction electrode potentials are very similar. The metals react slowly with cold water, but rapidly with acids to produce hydrogen, but not readily with concentrated H_2SO_4. The metals are active reducing agents, ignite in air, and are pyrophoric on filing (mixed metals). They react with halogens, carbon, silicon, nitrogen, and sulfur at elevated

THE LANTHANIDES, RARE EARTH METALS

Table 43.2. Physical and Chemical Characteristics of the Lanthanides

Element	Atomic No.	Electronic Configuration	Atomic Weight	MP (°C)	Density	Radius of RE^{3+}(Å)[a]
Scandium (Sc)	21	$3d^1 4s^2$	44.9559	1541	2.989	0.83
Yttrium (Y)	39	$4d^1 5s^2$	88.9059	1552	4.472	1.06
Lanthanum (La)	57	$5d^1 6s^2$	138.9055	921	6.146	1.22
Cerium (Ce)	58	$4f^2 6s^2$	140.12	798	6.162	1.18
Praseodymium (Pr)	59	$4f^3 6s^2$	140.9077	931	6.769	1.16
Neodymium (Nd)	60	$4f^4 6s^2$	144.24	1016	7.007	1.15
Promethium (Pm)	61	$4f^5 6s^2$	145	1168	7.264	
Samarium (Sm)	62	$4f^6 6s^2$	150.4	1072	7.54	1.13
Europium (Eu)	63	$4f^7 6s^2$	151.96	817	5.24	1.13
Gadolinium (Gd)	64	$4f^7 5d^1 6s^2$	157.3	1313	7.868	1.11
Terbium (Tb)	65	$4f^9 6s^2$	158.9254	1357	8.253	1.09
Dysprosium (Dy)	66	$4f^{10} 6s^2$	162.50	1410	8.556	1.07
Holmium (Ho)	67	$4f^{11} 6s^2$	164.9304	1740	8.799	1.05
Erbium (Er)	68	$4f^{12} 6s^2$	167.26	1522	9.006	1.04
Thulium (Tm)	69	$4f^{13} 6s^2$	168.9342	1525	9.318	1.04
Ytterbium (Yb)	70	$4f^{14} 6s^2$	173.04	824	6.959	1.00
Lutetium (Lu)	71	$4f^{14} 5d^1 6s^2$	174.97	1663	9.849	0.99

[a] Radius of rare earth (+3) in angstroms.

temperatures. Reaction with hydrogen produces salt-like hydrides. The metals also absorb hydrogen without reacting, forming interstitial hydrides of the approximate composition M(1):H(2.8). The metals react with oxygen to yield oxides of the general formula M_2O_3 (except cerium, which yields CeO_2). The M_2O_3 oxides react with water to form insoluble hydroxides, $M(OH)_3$, which are not amphoteric (*i.e.*, capable of reacting chemically as either an acid or base). Insoluble carbonates, $M_2(CO_3)_3$, may be produced by the reaction of carbon dioxide with the oxides or hydroxides. The sulfates, nitrates, and chlorates of the +3 ions are soluble in water; the phosphates, fluorides, and oxalates are insoluble. The precipitation of these elements as oxalates serves as the basis of an analytical determination. Generally, the compounds are paramagnetic and highly colored. Lanthanides alloy with almost all metals. It is convenient to group the rare earths into two broad classes based on their properties and occurrence as the "light" or cerium group, and the "heavy" or yttrium group. The cerium group includes La, Ce, Pr, Nd, Sm, Eu, and Gd. The yttrium group includes Y, Tb, Dy, Ho, Er, Tm, Yb, and Lu. The cerium-group metals are soft; however, hardness increases with atomic number. Oxide-free metals are malleable, but oxide inclusions reduce this property. The metals are good heat conductors and moderate electrical conductors.

The lanthanides exhibit a prominent characteristic termed "lanthanide contraction." From La to Lu, the radius of lanthanide ions (+3) decreases as the atomic number increases. This characteristic is due to the attraction of 4*f* orbital electrons by the increasing positive charge of the nucleus with increasing atomic number. A decrease in basicity and increase in solubility are also associated with increasing atomic number. These physiochemical qualities seem to correlate with distribution in tissues; the larger lighter and less soluble ions tend to deposit in the liver, and the smaller, heavier, more soluble ions tend to deposit in bone. The ionic size of the lanthanides is similar to Ca^{2+}, and as such lanthanides have been used as Ca^{2+} probes in physiological and biochemical studies. Lanthanides, especially La, have been shown to be calcium antagonists, as well as elements that produce tissue calcification (1, 2). The chlorides, nitrates, and sulfates of lanthanides are soluble whereas their carbonates, phosphates, and hydroxides are insoluble. These solubility differences among ionic forms of the lanthanides also seem to influence their metabolic fate in biological systems. In general, the toxicity of the lanthanides decreases as the atomic number increases, probably because of the greater solubility and ionic stability of the heavier lanthanide ions (2).

The lanthanide salts precipitate readily at physiological pH range because their isoelectric point is pH < 7. Lanthanum and other rare earths form insoluble complexes with nucleic acids (3, 4). The stability of lanthanide complexes with EDTA has been determined at 20°C and at an ionic strength of $m = 0.1$ (3, 4).

Relatively little is known about the chemistry of scandium, even though it is not particularly rare. Its chemistry resembles that of aluminum in many ways. The metal reacts vigorously with water to liberate hydrogen when its oxide coating is removed. Like Al_2O_3, the oxide Sc_2O_3 is insoluble in water, but because of the larger size of the Sc^{3+} ion, Sc_2O_3 is basic rather than amphoteric. Like aluminum, scandium forms stable compounds only in the +3 oxidation state. It has 11 known isotopes, only one of which occurs in nature.

Yttrium is very similar to scandium. It is also an active metal. The oxide of yttrium, Y_2O_3, is a white powder, insoluble in water but soluble in acids. In its compounds, yttrium displays only +3 oxidation state.

1.2 Production and Use

Although many minerals contain rare earths, only three contain sufficient percentages to be of widespread commercial value. (*1*) *monazite*, an orthophosphate primarily of the light, or cerium group; (*2*) *bastnaesite*, a fluorocarbonate of the cerium group; and (*3*) *xenotime*, an orthophosphate of the heavy, or yttrium, group. The average rare earth content in these minerals is shown in Table 43.3. Other minerals, including allanite, cerite, gadolinite, and euxenite, contain extractable quantities of rare earths but are not in widespread commercial use.

The most important sources of monazite are placer and marine beach deposits in the form of heavy mineral sands found in Australia, Brazil, India, South Africa, and the United States (Florida and the Carolinas). The most significant bastnaesite deposits are found in Palabora, South Africa; Paotou, Inner Mongolia in the People's Republic of China; and San Bernadino, California (USA). In California (USA), Molycorp, Inc. operates local deposits and has been the primary world source, capable of supplying the industry for many years. Huge ore reserves are reported from the People's Republic of China, where an existing mining–distribution infrastructure is in place from a producing hematite mine (5). Xenotime is also found in placer and beach deposits with commercial quantities extracted with placer tin deposits in Southeast Asia, and from the monazite deposits in the U.S. Carolinas and Florida. The production of these yttrium-group xenotime ores is considerably less than that of monazite or bastnaesite.

THE LANTHANIDES, RARE EARTH METALS

Table 43.3. Average Rare Earth Content of the Major Ore Minerals (in percent)

Element	Monazite	Bastnaesite	Xenotime
La	23 ⎫	32 ⎫	⎫
Ce	46 ⎬ 92	50 ⎬ 98.7	⎬ 10.6
Pr	5 ⎪	4 ⎪	⎪
Nd	18 ⎭	13 ⎭	⎭
Sm	2.3 ⎫	0.5 ⎫	1.2 ⎫
Eu	0.7 ⎬ 4.7	0.1 ⎬ 0.75	0.01 ⎬ 4.8
Gd	1.7 ⎭	0.15 ⎭	3.6 ⎭
Tb	0.16 ⎫		1.0 ⎫
Dy	0.5 ⎪		7.5 ⎪
Ho	0.09 ⎪		2.0 ⎪
Er	0.13 ⎬ 0.9		6.2 ⎬ 84.6
Tm	0.01 ⎪	Almost absent	1.27 ⎪
Yb	0.06 ⎪		6.0 ⎪
Lu	0.006 ⎭		0.63 ⎭
Y	2	2	60.0
Th	10	0.5	

Monazite is processed in one of two ways depending on the desired product: a sulfuric acid digestion or a sodium hydroxide digestion followed by treatment with hydrochloric acid. The process involves cracking the ore, removing the thorium, and separating the lanthanides. The latter process is more commonly used because it has the advantage of recovering the phosphate as the useful by-product, trisodium phosphate, and yielding rare earth chlorides, which are a useful intermediate in further processing (6).

The Molycorp method for bastnaesite processing is to calcine the ore after concentration by flotation. It is then attacked by hydrochloric acid to solubilize most of the trivalent rare earths as chlorides. The residue containing the tetravalent components is filtered for recovery and contains 65–80% cerium oxide. This fraction, after further calcination, can be used directly as a glass polishing compound. The rare earth chloride fraction is used for processing into specific applications. Xenotime, also an orthophosphate, is usually treated by the sulfuric acid method described for monazite to yield the rare earth yttrium group (6).

Separation of the individual lanthanides and yttrium is difficult because their identical external electron shell configuration gives them very similar chemical properties. The development of ion-exchange and liquid–liquid extraction techniques allows separation processes to take advantage of their substantial difference in ionic sizes that result from their lanthanide contraction property. Kaczmarek describes the overall separation strategy as a combination of the following procedures: (*1*) fractional crystallization of amorphous salts, (*2*) selective oxidation or reduction, and (*3*) basicity differences, including (*a*) ion exchange and (b) solvent extraction. All of these procedures, with the exception of oxidation or reduction, are fractional, and require repeated steps to obtain desired purities. Selective oxidation is used to obtain relatively pure quantities of the +2 valent metals europium and samarium, both of which have significant commercial applications (7).

Scandium has only a few commercial uses because it is difficult to process. It is found as a minor element in many minerals but there is only one, thortveitite, which contains significant concentrations. This mineral is a silicate that was found in granite pegmatite in southern Norway. A similar mineral, befanamite, has been found in Madagascar (8). Scandium is most widely used by the lighting industry in the manufacture of "metal halide" lamps. The use of scandium iodide in these lamps produces a nearly perfect spectral match to natural daylight (9).

Yttrium is used as an alloying agent in specialty stainless steels that require high resistance to corrosion at temperature extremes. These alloys have been used in the nuclear industry, as the base metal for catalytic converters for automobile exhaust systems, and as protective coating on turbine engine blades. As an alloying agent with titanium, it improves ductility and ease of fabrication. It improves the strength of magnesium castings when used in combination with zirconium. The conductivity of aluminum transmission lines is improved by as much as 50% with as little as 100 ppm of yttrium in the alloy. In the electronics industry, yttrium is used as the matrix of europium-activated red phosphors that give the red hue in color television tubes (9, 10).

At one time, the only commercial use of the rare earths was as misch metal, an alloy consisting principally of cerium (47%), neodymium (19%), and lanthanum (27%), plus other rare earths. It is pyrophoric (catching fire in air) when finely divided and is used to make cigarette-lighter flints. Commercial usage of the rare earths continues to expand as quantities of pure materials become more economically available and with technical innovations that take advantage of newly discovered properties and applications. Their use in industry falls into five general categories: (*1*) metallurgical, (*2*) lighting and chemical catalysis, (*3*) glass and ceramics, (*4*) phosphors and electronics, and (*5*) pharmaceuticals. An extensive reference regarding the technology and uses of lanthanides is found in the *Handbook on the Physics and Chemistry of Rare Earths*, which in 1998 included about 25 annual volumes (11).

The most common and known use of lanthanides is in the form of mixed lanthanide metal (misch metal) and cerium metal. These metals are used in lighter flints, magnesium alloys, and some of the ferrous alloys. Cerium in the pure state is not very pyrophoric, but when slightly oxidized or alloyed with iron, it becomes readily pyrophoric. Cerium flashes at 150–180°C. Misch metal, containing cerium, becomes pyrophoric on filing and for this reason is used for lighter flints (12). In metallurgy, the lanthanides Ce, La, Nd, and Pr have been added to iron to improve the strength and ductility of cast-iron products by the removal of oxygen and sulfur, and by controlling the formation of graphite crystals. These metals have been added to steel to improve forming characteristics for cold-stamping applications. In higher concentration alloys they are used to form specialty stainless steels. Samarium is alloyed with cobalt in a 1 : 5 ratio to make strong permanent magnets. Samarium and europium are used in nuclear reactor control rods. Lanthanum is alloyed with nickel in 1 : 4 and 1 : 5 ratios to form solid-state hydrogen storage containers that have significant weight and pressure advantages over gas or liquid hydrogen storage containers.

About one-quarter of the lanthanide chemicals produced are used in carbon-arc lighting applications. Lanthanide-cored carbons are indispensable to the motion picture industry, in both studio lighting and in theater projection. U.S. Army, Navy, and Coast Guard searchlights also use these lanthanide-cored carbons. Another lighting application is in their use in incandescent gas mantles. A small cotton sock which has been saturated with a salt solution of La_2O_3 and ZrO_2 is raised in temperature in the hottest part of a flame until the metals give off visible light radiation. The salt mixture and surface area of the mantle substrate are optimized to give off the most light in the visible range. Later versions of this mantle, which contain thorium oxide to improve spectral characteristics, are still in use today.

The lanthanides are used widely in the chemical industry as process materials and as catalysts. Lanthanum is largely used as a component of zeolite catalysts for cracking crude petroleum into gasoline and other light-petroleum fractions. This catalytic property also is used in some catalytic converters for automobile exhaust systems. Lanthanides are also used as a reagent and as a phosphor in fluorescent lamps. Gadolinium is used in X-ray film emulsions to intensify exposed images, which allows lower exposure doses. Neodymium is used as the dopant in YAG (yttrium aluminum garnet) lasers.

Cerium oxide and some forms of specially prepared lanthanide oxides are widely used in the polishing of spectacle and optical instrument lenses and in the surface preparation of mirror glass, gemstones, and other glass specialties. This polishing process is dependent on both the abrasive nature and high melting temperature of the oxide.

Lanthanum increases the refractive index of glass and is used in manufacturing high quality lenses and in the manufacture of fiber-optic glass. Cerium is added to glass to decolorize and to provide UV light absorption characteristics without adding color to the glass. Neodymium, praseodymium, and erbium oxides are used in art glass to produce specific colors. When neodymium is used, it provides a delicate pink tint with violet reflections. The color varies with glass thickness, concentration of neodymium, and source of illumination. The color shifts from light pink in thin sections to a blue-violet in thicker pieces in a process called *dichroism*, which is used in art glasses and for special filters. Higher concentrations of neodymium are used to produce welding glasses that protect against the yellow flame color. Praseodymium gives a green color to glass similar to chrome glasses. Erbium produces a pale pink that cannot be obtained by other means.

In the electronics industry europium finds its greatest use as the material that produces the vivid red in color television screens. This is due to its narrow emission spectrum, which peaks near 620 nm. Bubble domain memory materials for the computer industry are made on a gadolinium gallium garnet (GGG) substrate. PLZT electrooptical devices are made from a ceramic material composed of lead, lanthanum, zirconium, and titanium. An example application is an electric welding helmet. When an arc is struck, sensors detect the intense light and remove voltage from the PLZT shutter to instantly darken the viewing lens. When the arc is stopped, the shutter reactivates and eliminates the necessity of raising and lowering the helmet; thus the mask can be kept in the down position, eliminating the possibility of eye burns from adjacent welding operations emission (13). Oxygen sensors, first used in automobile exhaust systems to minimize hydrocarbon and carbon monoxide emissions, contain zirconium in a yttrium oxide stabilization medium.

Several lanthanides are components of crystals that have been reported to demonstrate superconductivity.

As pharmaceuticals, the lanthanides have found applications as therapeutic agents. Samarium-153 ethylenediaminetetramethylene phosphoric acid ([^{153}Sm]EDTMP) is injected as a radiopharmaceutical to combine with metastatic bone cancer lesions to reduce pain resulting from skeletal carcinoma (14). Gadolinium, as gadopentetate dimeglumine, is used as a magnetic resonance imaging (MRI) contrast enhancing agent, which represented the first approved application of this technique in the United States (15).

1.3 Exposure Assessment

Characteristically, the analytic methodology of the lanthanides has progressed through the quantitative estimation by the arc spectrographic method, using a diffraction grating of sufficient dispersion to separate the complex spectra of the lanthanides (16), followed by flame photometry (17), and finally atomic absorption spectrometry (AAS) using nitrous oxide–acetylene flame (18). Yttrium, by this method for the determination in air, has a range of 0.529–2.211 mg/m^3 with a precision of 0.054, and a sensitivity of 10 mg (at 2 mg/mL for 1% absorption) in 5 mL final volume. Interferences of aluminum, potassium, and H_3PO_4, which cause depression of absorbance, can be overcome by the addition of NaCl to samples and standards. Lamps are presently available (Perkin-Elmer) for all lanthanides except cerium and lutetium (12). Most lamps require the nitrous oxide–acetylene flame.

With regard to yttrium, the only one of the rare earths that is regulated, the preferred NIOSH method (7300) involves collection on a particulate filter and acid digestion. Analysis is carried out by inductively coupled plasma spectroscopy (19).

The three favored methods for determination of lanthanides from biological specimens are neutron activation analysis (NAA), high pressure liquid chromatography (HPLC), and inductively coupled plasma–mass spectrometry (ICP-MS). Chiba et al. evaluated the comparability of these three methods by injecting mice with the chloride compounds of Y, La, Ce, Nd, Sm, Eu, Gd, Tb, Dy, Ho, Er, or Yb at a dose of 25 mg/kg body weight and sacrificing the animals after 20 h. Element ratios between the methods were calculated to compare results. The ratios for HPLC/ICP-MS ranged from 0.91 to 1.17; those for the NAA/ICP-MS, 0.76 to 1.33. The better agreement between the former comparison was attributed to similarity of sample preparation, rather than any deficiency between the methods. The authors concluded that the results obtained by the three methods agreed well within acceptable limits (20).

Allain et al. (21) used ICP-MS to determine the background concentrations of rare earths in the plasma and urine from healthy male volunteers from France. The concentrations of La, Ce, Gd, Tb, and Yb (the ions selected for analysis) were all < 0.3 µg/L, except for one urine that contained cerium at 1.5 µg/L. The authors indicate that their data are in agreement with general data about rare trace elements in biological samples.

1.4 Toxic Effects

As discussed in detail in Section 1.4.1, rare earths have been shown to have both acute and chronic toxic effects in animals. Specifically, rare earths prevent blood coagulation and adversely impact reproduction and development. Further, they are toxic to the gastrointestinal (GI) tract, liver, kidney, spleen, and lung tissues. They are irritating to the eye and to abraded skin. They also act as calcium agonists and impact a variety of enzymatic functions. Rare earth chlorides and oxides have been proven to cause bronchitis, pneumonitis, and granulomatous lesions in animals. However, there is no evidence that they cause pneumoconiosis or chronic pulmonary reactions in animals or are carcinogenic.

As discussed in detail in Section 1.4.2, rare earths have been shown to act as an anticoagulants in humans. However, the side effects are too severe for these agents to be used clinically. More importantly, a number of case studies demonstrate that exposure to rare earths can cause pneumoconiosis in humans.

1.4.1 Experimental Studies

Early research investigations on the biological and medical aspects of the lanthanides are recorded through 1955 in the proceedings of a conference sponsored by the Medical Division of the Oak Ridge Institute of Nuclear Studies (22). This evidence, which was later confirmed through extensive investigations of Haley and associates (23–28) and Graca et al. (29), was that the toxicity of the lanthanides, including yttrium, is low orally and high parenterally and varies greatly according to the associated nonmetallic component.

Under the general title of pharmacology and toxicology of the lanthanides, Haley and his associates studied 11 of the lanthanides (Pr, Nd, Sm, Gd, Tb, Dy, Ho, Er, Tm, Yb, and Lu). These investigations included acute intraperitoneal and oral toxicity of the chlorides in mice; chronic (12-week) oral toxicity in rats; determinations of ocular and dermal irritation, and effects on isolated ileum in rabbits, and pharmacological effects consisting of electrocardiographic changes and changes in acetylcholine, epinephrine, and histamine and effects on vagal stimulation.

Because separation and purification procedures before the 1950s yielded impure lanthanides, with resulting questionable toxicity data, only data obtained on lanthanides since the institution of newer purification methods are included. Data supplied by Haley et al. were obtained on lanthanides of 98 or 99% minimal purity and those by Graca et al. (29) on lanthanides from Ames, Iowa Laboratories, where the newer purification procedures originated. Thus the rat intraperitoneal (IP) LD_{50} of 99% pure $LaCl_3$, determined by Kyker and Cress (30) to be 187 mg compound/kg, was about half that reported by Cochran et al. (31) some years previously (350 mg/kg). The corresponding LD_{50} value for YCl_3 was 100 mg/kg, compared with 450 mg/kg according to Cochran et al. As pointed out by Kyker and also by Graca (29), however, precise acute toxicity determinations of the lanthanides are difficult because of their protein-precipitating capacity and the unusually great influence of the nonmetallic components. Toxicity values could be greatly modified, therefore, according to concentration and rate of the injected dose; animal strain differences may be another factor. Hart et al. (32) reported that when lanthanum and yttrium were administered to humans and animals, the uptake of lanthanum and yttrium by the tissues varied according to the form administered, that is, whether ionized or as an EDTA complex.

1.4.1.1 Acute Toxicity. The acute toxic response has been determined in laboratory animals for the lanthanides by IP, IV, and oral routes as nitrates, chlorides, oxides, and chloride–citrate complexes. Bruce et al. (33), reporting on all the lanthanides except lanthanum and promethium, found that for mice the approximate IP LD$_{50}$ values of the nitrates ranged from 225 to 480 mg compound/kg bw; for rats, the values ranged from 210 to 335 mg/kg. Oddly, the most and least toxic lanthanides for the mouse, Tb and Er, did not correspond to the most and least toxic for the rat (Eu and Lu), although administered by the same route, indicating differing metabolism for these elements in the two species. There were other discrepancies (Gd, Dy, and Ho), but generally the degree of toxicities was parallel in the two species.

As expected, toxicity by the IV route was far greater, ranging from 49.6 to 77.2 mg/kg for the male rat for the five lanthanides tested (Ce, Pr, Nd, Sm, Er), with a surprising increase in toxicity for the female rat (from 4.3 to 35.8 mg/kg). Again, the relative toxic degree was not always parallel in the two sexes, although cerium proved the most toxic for both sexes. Toxicities for IV and IP routes showed no relationship among the lanthanides tested (33).

Typical of polyvalent metals with their protein-precipitating potential and consequent poor absorption from the GI tract, the oral toxicity of the nitrates was of a different order of magnitude, with LD$_{50}$ values ranging from 2750 to 4200 mg/kg, 10 or more times that of the IP route. With the exception of the elements Eu, Gd, and Tb, for which the oral LD$_{50}$ values were greater than 5000 mg/kg, the nitrates roughly exhibited an increase in toxicity with increasing atomic weight (33).

All the oxides of the lanthanides were tolerated by female rats at 1000 mg/kg (33). Haley et al. (23–28) tested the toxicity of the chlorides of the lanthanides, all except those of La, Ce, Pm, Eu, and Y, and found the IP LD$_{50}$ values to range from 315 to 600 mg compound/kg; lutetium was the most and praseodymium and neodymium the least toxic, showing that with acute exposures the chlorides are about half as toxic as the nitrates. As with the nitrates, the chlorides' oral LD$_{50}$ levels were of a different order, ranging from 7.5 to 22.5 times lower than those by the IP route. Although a rough parallelism existed between IP toxicity and increasing atomic weight, no corresponding parallelism was found for the oral route.

Ji and Cui (34) studied a mixture of rare earth metal nitrates Ce, La, Nd, Pr, and Sm. They reported that in mice, rats, and guinea pigs, the oral LD$_{50}$ ranged from 1397 to 1876 mg/kg. They concluded that absorption from the GI tract was low.

Hirano and Suzuki (2) reviewed the mortality data for ionic and chelated forms of rare earths administered orally, IP, and IV. They concluded that the median lethal dose (LD$_{50}$) indicated the rare earths were not highly toxic for oral or IP administration. However, they questioned the validity of establishing LD$_{50}$ values for intravenous administration. They reported that the percent mortality peaked at 20–40 mg Pr(NO$_3$)$_3$/kg bw following IV injection in both mice and rats of both sexes; however, the lethality then decreased as the dose increased. They postulated that the colloid formation of ionic rare earths in blood at higher doses of rare earth chlorides or nitrates might be responsible for the bell-shaped dose–response curve. More study is needed to test this hypothesis.

Uniform signs and symptoms of acute poisoning reported by Haley et al. were ataxia, writhing, slightly labored and depressed respiration, arched back, stretching of limbs on walking, and lacrimation, all starting within 24 h and, depending on the lanthanide, peaking between 48 and 72 h. Gadolinium and samarium differed in that peak mortality was delayed until the fourth or fifth day. Graca et al. (35), while observing the effect on heart rate, blood pressure, and respiration in 135 dogs, administered 100 mg compound/kg in 10 mg/kg doses at 10-min intervals and found that the only specific effects were increases in prothrombin and coagulation times. Cardiovascular collapse and respiratory paralysis were considered by both investigators to be the cause of death.

1.4.1.2 Chronic and Subchronic Toxicity. Studies on the toxicity and safety of a mixture of rare earth metal nitrates Ce, La, Nd, Pr, and Sm used in agricultural operations were reported (34). Subchronic and chronic toxicity studies were done at different doses in monkeys (100 mg/kg) and rats (200 and 1800 mg/kg); biochemical and histopathological examination of tissues showed no abnormal or specific pathological changes. In chronic feeding studies with rats, the incidence of benign and malignant tumors in the test groups was lower than that in the control. Rat fetuses did not show any teratogenicity when the dams were orally fed up to 330 mg/kg of this nitrate mixture. Ames mutagenicity tests were negative at 50 mg/kg. The investigators (34) concluded that an oral dose of 60 mg/kg should be considered as a no-effect level with an allowable dietary intake of 0.6 mg/kg.

1.4.1.3 Pharmacokinetics, Metabolism, and Mechanisms. The distribution of all 15 lanthanides has been determined (22, 36) from their radioisotopes in a dozen tissues and body sites of rats for periods of 1–256 days for the longer-lived isotopes, following oral or intramuscular (IM) injection as citrate complexes. In general, absorption from parenteral injection was relatively complete in 4 days ($<6.5\%$ unabsorbed), whereas absorption from the GI tract of the four lanthanides ^{144}Ce, ^{152,154}Eu, ^{160}Tb, and ^{170}Tm was less than 1% of the administered dose. The low parenteral absorption of ^{140}La (22.5% unabsorbed in 4 days) was attributed to its low solubility at body pH and, in general, to the amount of stable carrier administered (36).

The skeleton was the most important site of deposition for all lanthanides, particularly the heavy ones; deposition ranged from 62% for gadolinium to 90% of the administered dose for lutetium. The skeleton was, however, initially a secondary site of deposition for the light lanthanides, whose major deposition site was the liver, ranging from 67% for lanthanum to 45% for samarium of the administered dose. For the transition lanthanides (Eu and Gd), deposition was nearly equal in liver and skeleton (3).

The kidney, which initially contained a greater concentration than any other soft tissue for the light and heavy lanthanides (1.6 and 0.9%/g), decreased in all cases to very low levels at 8 months (3).

In the blood, lanthanide levels, even at 1 day, never exceeded 0.02%/mL, and steadily decreased thereafter to unmeasurable amounts. Initial levels in muscle were about the same as in blood, but those in the skin were 3–4 times greater. The GI tract showed concentrations about 10 times those of blood and muscle initially, but concentrations were still measurable 8 months later for the lighter lanthanides, for which the GI tract is the main excretory route. Concentrations in the spleen depend on the route of administration. By IM injection they are low ($<0.5\%$/g wet tissue); but by vein, they are high if administered as halides or nitrates, from which colloids form to be deposited in the reticuloendothelial system (3).

Hirano and Suzuki (2) summarized inhalation studies that showed that significant amounts of inhaled $CeCl_3$, $Ce(OH)_3$, and Y (chemical form unknown) were translocated to the skeleton or liver in rats or hamsters. They also reported a half-life of 140 days for inhaled $Ce(OH)_3$.

Berthezène et al. (15) evaluated the safety aspects and pharmacokinetics of aerosolized gadopentetate dimeglumine in Sprague–Dawley rats. They determined that pulmonary clearance of aerosolized gadopentetate dimeglumine had a half-life of 2.16 h. They also determined that material was excreted via the kidneys and renal elimination was complete in 30 h. No acute hemodynamic effect, histological change, or induction of edema was demonstrated. Thus they concluded that inhalation is well tolerated. In reviewing these data, Hirano and Suzuki (2) concluded that gadopentetate dimeglumine was stable in the alveolar space and was hardly taken up by macrophages because of limited release of ionic Gd from the complex.

Galle et al. (37) exposed male Wistar rats to submicrometer aerosols generated from a 1% solution of cerium chloride ($CeCl_3$), chromium chloride ($CrCl_3$), uranyl nitrate $[UO_2(NO_3)_2 \cdot 6H_2O]$, and aluminum chloride ($AlCl_3$). Rats were exposed to a concentration of $10–30\,\mu g/cm^3$ for 5 h/day for 5 days. The ultrastructure of the cells in lung sections was examined using a Philips EM 300 microscope. The intracellular localization of the elements was determined using an electron probe Camebax Camerca microanalyzer equipped with thallium acid phthalate (TAP), pentaerythritol (PET), and lithium fluoride (LiF) crystals. A specially adapted electron microscope allowed the analysis of intracellular organelles. The study showed that inhaled cerium chloride accumulated in alveolar and tissue macrophages and alveolar walls. They attributed the mechanism involved to the high acid phosphatase activity of lysosomes and considered it as an *in vivo* Gomori reaction. They postulated that in the lung macrophage this mechanism of intralysosomal concentration and precipitation may prevent the diffusion of the toxic elements through the alveolar membrane.

Hirano et al. (38) investigated the metabolic behavior and pulmonary toxicity of intratracheally administered yttrium chloride (YCl_3). Yttrium chloride solution was intratracheally instilled in male Wistar rats at doses of 10, 20, 50, 100, and 200 µg Y/rat. The time-course and dose-related changes in distribution of Y between lung tissue and bronchoalveolar lavage fluid (BALF) and pulmonary inflammatory responses were investigated. It was determined that the pulmonary clearance of Y was very slow, with an estimated half-life of 168 days. Yttrium content in the supernatant of BALF did not exceed 5 µg Y/lung, even at a dose of 200 µg Y/rat.

Hirano et al. concluded that the alveolar surface fluid could retain at most 5 µg Y. Yttrium content in the pellet of BALF changed with the number of macrophages retrieved in BALF. Transmission electron microscopy and X-ray microanalysis suggested that Y was localized in lysosomes of alveolar and interstitial macrophages and in basement membranes. No detectable amount of Y was accumulated in the liver. Renal Y content increased with time, but only 0.1% of the initial dose was accumulated in the kidney, even at 162 days.

Suzuki et al. (39) investigated the distribution and health effects of lanthanum chloride ($LaCl_3$) intratracheally instilled in rats. Lanthanum chloride solution was intratracheally instilled in male Wistar rats at doses of 0.5, 1, 10, 20, 50, 100, and 200 µg La/rat. The distribution of La among tissues revealed that the metal remains mostly in the lung with a biological half-life of 244 days. The subcellular localization by transmission electron microscopy with an X-ray microanalyzer indicated that La localizes in macrophages as high electron-dense granular inclusions in lysosomes and cytoplasm and on the cell surface and basement membranes of type I pneumocytes among lung cells. La was not detectable in any femur samples, and was detected marginally in liver and kidney samples. La was detected at high concentration in the pulmonary hilum lymph node samples, which the authors interpreted as suggesting that La is transferred from the lung through the lymph nodes.

Yoneda et al. (40) investigated the effects of gadolinium chloride on rat lung. They performed the study in the same manner as those Hirano et al. performed for yttrium chloride and Suzuki et al. performed for lanthanum chloride. $GdCl_3$ solution was intratracheally instilled at doses of 10, 20, 50, and 100 μg Gd/rat. The results were quite similar to those obtained with yttrium chloride and lanthanum chloride. Gd in the lung tissue decreased very slowly with a biological half-life of 136 days at a dose of 50 μg Gd/rat. Gd content in the supernatant of bronchoalveolar lavage fluid (BALF) did not exceed 5 μg Gd/BALF even at a dose of 100 μg Gd/rat. These results suggest that intratracheally instilled Gd can be retained in epithelial lining fluid only to a limited extent as soluble forms, and is deposited in the lung tissue probably in insoluble forms that are metabolized very slowly. They also found that Gd caused a rapid and selective infiltration of serum calcium before acute lung toxicity.

Hirano and Suzuki (2) postulated that the differences in the extrapulmonary translocation of rare earths between the intratracheal instillation and inhalation studies may be due to absorption of rare earths from the upper airways or GI tract after being transported through the esophagus.

Excretion of the lanthanides is both urinary and fecal, depending on their position in the series; light lanthanides are excreted in the feces following accumulation in the liver. The heavier lanthanides are excreted primarily in the urine, and those in the middle of the series by both routes, about equally (41).

Dean et al. (42) concluded that intravenously administered gadolinium chloride ($GdCl_3$) forms insoluble carbonate and phosphate precipitates in the blood, which are taken up by the reticuloendothelial system. Barnhart et al. (43) also found that intravenously injected $GdCl_3$ was found primarily in the liver and spleen. Further, they demonstrated that it accumulated rapidly, reaching 72% of the injected dose in 2 h. They also found that Gd-DPTA was quickly accumulated in the kidney and excreted in urine, with only 2% of the injected dose remaining after two h.

The metabolism and toxicity of IV injected yttrium chloride (YCl_3) in male Wistar rats was studied by Hirano et al. (44). They concluded that IV injected YCl_3 resulted in the dose-dependent formation of colloidal material in blood plasma. The collodial plasma Y was accumulated predominantly in phagocytic cells in the liver and spleen. The liver Y was cleared slowly with a half-life of 144 days.

Hirano and Suzuki (2) concluded that chelated rare earths are excreted mainly via the urine after a transient accumulation in the kidney and that their whole-body half-lives are several hours. Rare-earth chlorides are taken up by the liver and spleen, and those rare earths are not easily excreted.

A study of the subcellular distribution in rat liver after IV administration of representative lanthanides, Ce, Pm, Tb, Ho, Yb, and Y revealed that in females the largest amounts of ^{144}Ce and ^{169}Yb were in the microsomal fraction; in males, in the supernatant for ^{144}Ce and in the mitochondria and supernatant for ^{169}Yb. The nuclear fraction was lowest for both isotopes in male and female rats (12).

The complexing capacity of the lanthanides, particularly with proteins, is unquestionably the basis for their characteristic distribution and excretion pattern, varied toxicity by different routes according to compound type, and lack of mobility from an injection site. This property is, in turn, a consequence of their high ionic charge, tri- or tetravalence. Lanthanum, for example, forms insoluble complexes with nucleic acids in a physiological pH range (45–48). This has been useful in analyzing and purifying nucleic acids and for electron-stain chromatin fibrils. Similarly, all salts of lanthanum, praseodymium, neodymium, and samarium precipitate fibrinogen at concentrations of 0.01–1%. Higher salt concentrations partially resolve the precipitates, and the chlorides of praseodymium, neodymium, and samarium completely resolve them. Human serum proteins are incompletely precipitated at 0.25–0.7% lanthanide salt and dissolve at concentrations >1.5% (49). Bamann (50), however, believed that the well-studied anticoagulant action of the lanthanides on blood is mediated through their capacity to act as phosphate acceptors and indirectly reduce prothrombin content of blood. The reasons for these actions are open to other interpretations.

1.4.1.4 Reproductive and Developmental. A single IP injection of lanthanum chloride (44 mg La/kg) into pregnant mice reduced the number of successful pregnancies and the average litter size. The most susceptible periods of pregnancy were preimplantation (days 4 and 6) and near-term period (days 14 and 16). Injection of lanthanum during the preimplantation period resulted in a cessation of pregnancy in 24–43% of females, and injection during near-term period produced the cessation of pregnancy in 36–46% of the females. The average litter size after injection of lanthanum during preimplantation or near-term periods was reduced to about 75% of the average litter size in the control animals. No external malformations were observed among fetuses, but the exposure of single-cell stage embryos to lanthanum chloride *in vitro* resulted in an increase in the fraction of embryos developing into blastocysts (51).

Pregnant female mice received either a single subcutaneous dose of cerium citrate (80 mg Ce/kg) or an equivalent (in citrate) dose of sodium citrate on day 7 or 12 of gestation or on day 2 postpartum. To separate effects of prenatal and postnatal exposure, a cross-fostering design was employed. The weight and gross activity of the neonates were assessed on day 8 or 13 postpartum. Open-field behavioral parameters, accelerating rotarod performance, and passive avoidance learning were assessed on days 60–65 postpartum. Maternal offspring retrieval latency was measured on day 3 postpartum. Analyses revealed that neonatal weight was reduced both in offspring exposed to cerium *in utero* and in the offspring of mothers receiving cerium during lactation. Cerium also appeared to affect maternal–offspring interaction: pups exposed prenatally to cerium were retrieved in less time than control pups. Except for an increased frequency of rearings in the open field of adult offspring exposed to cerium *in utero*, cerium exposure had no apparent effect on behavioral parameters, either in neonatal or adult offspring (52).

Chronic (12-week) oral toxicity to male and female rats at dietary levels of 0.01, 0.1, and 1% of lanthanide chloride uniformly had no influence on growth or hematological variables. At autopsy, tissues of eight internal organs showed no histological changes at the 0.01 or 0.1% levels; but some, gadolinium, terbium, thulium, and ytterbium, showed nonspecific liver damage (perinuclear vacuolization and granular cytoplasm) at 1%. Hutcheson et al. (53) evaluated the possible toxic effects of scandium, lanthanum, samarium, europium, dysprosium, terbium, thulium, and ytterbium oxides on growth, general development, reproduction, and lactation. Mice were fed different doses of these compounds for three generations. The amount of elements fed were 0, 1, 10, 100, and 1000 times the use amount. The use amounts were (in ppm.): Sc, 0.12; La, 0.40; Sm, 0.80; Eu, 0.036; Tb, 1.20; Dy, 1.20; Tm, 0.08; and Tb, 0.12. The use amount was one-fifth of the concentration required for activation analysis. Mortality and morbidity were negligible. No consistent growth rate changes were observed. The number of mice born showed no significant differences among treatment groups. Survival, growth rate, hematology, morphological development, maturation, reproduction, and lactational performance were comparable in mice fed the different levels of metal oxides to those mice fed the basal diet.

Rat fetuses did not show any teratogenicity when the dams were orally fed up to 330 mg/kg of a mixture of rare earth nitrates (34).

1.4.1.5 Carcinogenesis. Scant investigative attention has been given to the neoplastic potential of the stable isotopes of the lanthanides. According to the National Cancer Institute survey of compounds for carcinogenic activity for 1970 to 1971, only gadolinium and ytterbium have been investigated, and only sporadic testing has been done on some of the radioisotopes of lanthanum and cerium or their admixtures with other lanthanide radioisotopes.

Subcutaneous implantation of 118 CFW male and female mice with 200 mg each of gadolinium and ytterbium metal pellets by Ball et al. (54) resulted in a variety of neoplastic growths, 12 carcinomas at implant sites with pulmonary metastases in four, and 24 additional neoplasm, 16 of which, in female implanted mice, were diagnosed as mammary carcinomas, 9/30 (30%) from gadolinium and 7/28 (25%) from ytterbium. By comparison, sham-implanted female CFW mice developed 44 and 40% mammary cancers from gadolinium and ytterbium, respectively. Similarly, whereas sham-implanted female mice developed 25% bronchial adenomas, only 7% developed the same tumor from gadolinium; 25% of ytterbium controls developed the tumors versus 14% ytterbium-implanted. Occasional tumors at other sites — hemangiomas of the liver, ovarian, salivary, and vulvar adenoma, and malignant lymphoma — were too few in number in either group to permit comparison. They were about equal in incidence in both treated and in controls. These data indicate that both gadolinium and ytterbium are antitumor agents of mild degree in a species (CFW mice) that has a high spontaneous incidence of mammary carcinomas and bronchial adenomas in females and hemangiomas and lymphomas in males.

In chronic feeding studies with rats, the incidence of tumors and malignant tumors in test groups was lower than that in the control groups (34). Using radioisotopic mixtures of ^{140}La and other lanthanides, investigators in the former Soviet Union found a 2% incidence of lymphosarcomas in rats at 500 rad to bone marrow and 350 rad to the GI tract in 320–640 days (55). Other Soviet investigators reported five bronchogenic or alveolar lung carcinomas in rabbits and rats intratracheally injected with 25 mCi colloidal ^{144}CeF$_3$ (56). Magnusson (57) found five squamous cell carcinomas with two metastases among 23 rats inhaling an average of 170 mCi/kg; tests with ^{144}Ce by other investigators were negative. Hirano and Suzuki (2) concluded that rare earths have not been found to be carcinogenic in animals.

From these reports, it can be concluded that at least certain of the stable lanthanide isotopes have, if anything, mild antitumor activity, and that certain radioisotopic forms are mildly carcinogenic.

1.4.1.6 Genetic and Related Cellular Effects Studies.

The La^{3+} ion, long known to antagonize calcium binding in heart muscle, inhibits the calcium influx requirement for the human neutrophil chemotaxis, thus inactivating the contractile chemotactic process (58). In a related manner, if sialic acid is removed from cultured heart cells, lanthanum, normally restricted to cell surfaces, can now enter the cell and displace more than 80% of the cellular calcium, thus demonstrating a specific role of cell surface constituents in the control of calcium exchangeability in the heart (59).

In vitro, rare earth chlorides (EuCl$_3$, DyCl$_3$, HoCl$_3$, and ErCl$_3$) dose-dependently decrease the tonus and contractility of rabbit ileum in response to acetylcholine or nicotine (28). Triggle and Triggle (60) demonstrated that lanthanide cations strongly inhibit the phasic and tonic component of the response to muscarine agonist *cis*-2-methyl-4-dimethylaminomethyl-3-dioxolane methiodide (CD) in guinea pig ileal longitudinal smooth muscle. They found that lanthanide cations also inhibited the responses evoked by high K$^+$, but here the effect was mainly on the phasic response, and the tonic component was merely delayed. Other members of the lanthanide series, with the exception of cerium, were more effective than lanthanum in their ability to inhibit the CD response. Thulium (Tm^{3+}) was the most effective. They concluded that lanthanide cations combine with membrane sites specifically involved in Ca^{2+} translocation during excitation–contraction coupling.

Swamy et al. (61) evaluated the effect of La^{3+} and Tm^{3+} on the mechanical responses of rat vas deferens and found that they inhibited the noradrenaline and K$^+$-induced responses; complete inhibition was obtained at approximately 10^{-3} M Ln^{3+}. La^{3+} and Tm^{3+} were equally effective in inhibiting noradrenaline and K$^+$ responses. The phasic and tonic components of the noradrenaline response were equally sensitive to lanthanide cations, but the phasic component of the K$^+$ response was more sensitive than the tonic component. They theorized that the action of Ln^{3+} in the rat vas deferens are mediated through some kind of membrane stabilization rather than via a specific Ca^{2+} binding site, as postulated for the Ln^{3+} action in guinea pig ileal longitudinal muscle.

Ghosh et al. (62) reported that administering a single IP dose of 250 mg/kg bw of lanthanum chloride and neodymium chloride to chicks reduced the activity of certain enzymes bound to the erythrocyte membranes. Specifically they reported that activities of acetylcholinesterase, NADH dehydrogenase, Mg^{2+}-ATPase, and *p*-nitrophenyl phosphatase were reduced. Membrane bound glycosidases were also reduced. They also found that cholesterol and phospholipid residues were reduced, but the ratio of cholesterol to phospholipid remained constant. In addition, they reported that membrane sulfhydral groups were also significantly inhibited.

In summarizing enzymatic and cellular effects, Hirano and Suzuki (2) reported that IP administration of LaCl$_3$ to chicks decreased lipid peroxides and the contents of sulfhydryl groups while increasing the activities of glutathion peroxidase, glutathione reductase, glutathione-S-transferase, and catalase in bone marrow. In addition, there was a marked depression in the activities of neural Ca^{2+}-ATPase, Mg^{2+}-ATPase, and cholinesterase. They concluded that the depression of these enzymes may be related to inhibitory effects of La^{3+} on binding of Ca^{2+} to brain synaptosomal membrane.

Yajima et al. (63) reported forming adenosine 3'5'-cyclic monophosphate (cAMP) from adenosine triphosphate at pH 8 and 50°C using lanthanide ions. Pr^{3+} and La^{3+} were reported as being the most active. The authors concluded that lanthanide ions can function as the catalytic center of artificial adenylate cyclase.

Hirano and Suzuki (2) summarizing the work of Komiyama et al., reported that rare earth ions hydrolyze RNA dinucleoside monophates and phosphatidylinositol *in vitro* under physiological conditions. They also reported that Ce^{4+} hydrolyzes cAMP under physiological conditions.

1.4.1.7 Other: Neurological, Pulmonary, Skin Sensitization

1.4.1.7.1.1 Eye. Only terbium produced detectable damage to the cornea and iris after instillation of 0.1 mL of a 1 : 1 solution of the chloride into the rabbit eye, showing a maximal irritation index immediately; healing required 18 days. Several of the lanthanides produced multiple 1-mm-diameter conjunctival ulcers, particularly lutetium, praseodymium, neodymium, thulium, ytterbium, and erbium (23–28).

In rabbits, a topical application of a suspension of 500 mg/mL produced mild irritation of the skin and eye mucosa (34).

1.4.1.7.1.2 Gastrointestinal Tract. Haley reported pharmacological effects uniformly had a depressing effect on blood flow and pressure, intestinal tonus, and contractility of rabbit ileum (23–28).

Stineman et al. (64) orally administered doses of 1000 and 1163 mg Ce/kg bw to mice. At both doses, gastritis and enteritis were evident in all animals. The stomach and the duodenum manifested focal hemorrhages and necrosis of the mucosa. Homogenous conglomerates of eosinophilic substances were found within the necrotic mucosa. In addition, conspicuous polymorphonuclear cell infiltration was observed in the gastric and duodenal lamina propria and submucosa, especially in those areas adjacent to the necrotic zones.

1.4.1.7.1.3 Kidney. Hirano and Suzuki (2) summarized the work of Endre et al. (65) and reported that perfusing rat kidney for 30 min with a solution containing 3–5.5 mM of chelated Dy reduced urinary concentrating ability and increased renal vascular resistance.

Salonpaa et al. (66) investigated the effects of liver-damaging doses of cerium chloride (2 mg/kg) on the expression of Cyp2a-4 and Cyp2a-5 in the liver and kidney, which differ in the regulation of the Cyp2a-4/5 gene complex. A dose of 2 mg/kg cerium was administered IV to adult male DBA/2 (D2) and C57BL/6N (B6) mice. In D2 mice they found that COH activity peaked 4.4-fold over the control value at 4 days after treatment, and then returned to the normal level. The activities of 7-ethoxyresorufin O-deethylase (EROD) and testosterone 15α-hydroxylase were unchanged during the first 2 days, and then showed a slight reduction, coinciding with the increase in COH activity at days 3 and 4 after treatment. A comparable change was not seen in B6 mice. Northern blot analysis showed 7 and 6-fold increases in the level of Cyp2a-4/5 mRNA in the kidneys of D2 mice at 6 h and 1 day after the cerium injection. No increases occurred in the mRNA levels in the kidneys of B6 mice.

Maulik et al. (67) found that administering a single acute dose of lanthanum chloride to chicks altered the level of enzymes of the antioxidant defense system of the renal cortex fractions. There was significant decrease in the activities of glucose-6-phosphate dehydrogenase, glutathione reductase, glutathione peroxidase, and catalase of kidney epithelial cells. Glutathione and total thiol contents were decreased while lipoperoxidative reactions in the kidney cortex was significantly enhanced. They concluded that amelioration of lanthanum toxicity by methionine supplementation may be due to the methionine serving as a precursors of glutathione.

1.4.1.7.1.4 LIVER. Intravenous injection of the "light" lanthanides (cerium) produced fatty infiltration in the livers of rats (68), characterized by an increase in neutral fat esters; total hepatic cholesterol and phospholipid levels were unchanged. Cerium was associated almost wholly with the acid-soluble fraction of the liver, and not with the lipid. The action of cerium was sex-based because it appeared most pronounced in females; testosterone-treated and ovariectomized females showed marked reduction in the response. Neither choline nor methionine exerted a protective effect against cerium. Injection of $CeCl_3$ caused fatty liver in female rats, but not in male rats (57); the reason for this limitation remains unknown (2).

Stineman et al. (64) found that subcutaneous administration of Ce at a dose of 173 mg Ce/kg bw caused focal necrosis in the liver of mice. Regenerative changes characterized by a moderate number of mitotic figures in the hepatocytes were also observed.

Electron microscopic examination of the liver by Magnusson (57) showed changes in both the mitochondria and endoplasmic reticulum (ER). The mitochondria were enlarged, and changes in the ground substance and cristae were seen chiefly in female rats treated with cerium. Changes in the endoplasmic reticulum were manifested as a dilatation of the cisternae and a loss of ribosome; similar changes occurred with all the lanthanides tested in female as well as male rats. Associated with these changes was a lowering of the blood glucose level during the first 3 or 4 days; the lowest levels occurred in female rats injected with cerium.

Hirano and Suzuki (2) reported that dilation, disorganization, and degranulation of rough endoplasmic reticulum and proliferation of smooth endoplasmic reticulum occurred following the IV injection of $CeCl_3$ in rats. They concluded that the liver is the primary target organ of IV injected $CeCl_3$.

Magnusson (57) reported that IV injection of rare earth chlorides (YCl_3, $TbCl_3$, $HoCl_3$ and $YbCl_3$), except $CeCl_3$, caused focal necrosis with calcium deposition in rats.

Hirano et al. (44) administered yttrium chloride solution to male Wistar rats at doses of 0.1, 0.2, 0.5, 1, and 2 mg Y/rat to evaluate its metabolism and toxicity. They showed that some plasma calcium was translocated to the colloidal material and plasma calcium concentration was increased transiently following injection of YCl_3. The authors theorized that the increase in calcium concentration was due to resorption of bone. At doses of 1 mg Y/rat, the calcium content of the liver increased over 10-fold. Gross observation of the tissues indicated there were white blotchy necroses centriportally in the liver. Examination by transmission electron microscopy (TEM) equipped with an X-ray microanalyzer (XMA) revealed that calcium was deposited focally in the necrotic sites of the liver. As well as calcium deposition, the study indicated that at a dose of 1 mg Y/rat, 75% of the dose was accumulated in the liver at 7 h postinjection. The authors concluded that the liver is one of the primary target organs of intravenously injected YCl_3.

At the enzyme level, Cochran et al. (31) found that lanthanum stimulated the activity of the succinic dehydrogenase system, as did aluminum, a trivalent element; but both lanthanum and yttrium inhibited adenosine triphosphatase activity at 10^{-3} M. Horecke et al. (69) had previously shown that, like polyvalent aluminum and chromium, certain lanthanides had a promoting effect on the succinic dehydrogenase-cytochrome oxidase system.

Salonpaa et al. (66) summarized the enzymatic activities in mouse liver and kidney as follows:

Coumarin 7-hydroxylase (COH) is an activity catalyzed by the Cyp2a-5 gene product (P450COH) in mice. Cyp2a-4 mediates the 15 α-hydroxylation of testosterone and some other steroids in the mouse liver and kidney. Cyp2a-5 is predominantly expressed in the liver, while Cyp2a-4 is the main component in the kidney. Mouse hepatic Cyp2a-5 expression is often increased in conditions in which other P450 isoforms are repressed.

This is helpful in understanding the impact of cerium on the liver, as discussed below.

When $CeCl_3$ was injected IV into two strains of mice at doses of 0.5, 1.0, and 2.0 mg/kg, coumarin 7-hydroxylase (COH) was increased in a dose-dependent manner in DBA/2 (D2) animals after 24 h, whereas no change was seen in C57BL/6N (B6) animals. A significant increase in all other enzymes studied (cytochrome P450, ethoxycoumarin O-deethylase, and ethoxyresorufin O-deethylase) was seen in D2 mice injected with the highest dose of cerium. In B6 mice at 72 h cerium increased COH activity, as well as other enzymes, in a dose-dependent manner. In D2 mice the increase was seen only after the two lower doses; the highest dose caused severe morphological changes in the liver structure and a clear decrease in COH and other activities. This demonstrated that an IV dose of 2 mg/kg caused severe disintegration of the liver tissue in D2 mice. The changes consist of diffuse panlobular necrosis, disintegration of nuclei, cytoplasmic accumulation of fat droplets, and sinusoid congestion. These changes do not occur in the B6 mice at the same dose level. The differences are not due to unequal tissue distribution of cerium in the two mouse strains (70).

Salonpaa et al. (66) investigated the effects of liver-damaging doses of cerium chloride (2 mg/kg) on the expression of Cyp2a-4 and Cyp2a-5 in the liver and kidney, which differ in the regulation of the Cyp2a-4/5 gene complex. A dose of 2 mg/kg cerium was administered IV to adult male DBA/2 (D2) and C57BL/6N (B6) mice. As described above, Arvela et al. (70) had demonstrated that this dose resulted in severe liver damage in D2 mice, but not in B6 mice. Salonpaa et al. found that in D2 mice COH activity in the liver was increased 3.2 fold 2 days after the cerium was administered, and 3 and 4 days after cerium treatment, there was a dramatic decrease, to about 10% of the control value, in COH activity. Testosterone 15α-hydroxylase activity was steadily decreased by cerium, and also the Cyp1a-1-mediated 7-ethoxyresorufin O-deethylase (EROD) activity was substantially decreased after a slight initial increase. The sharp decrease in enzymatic activity coincided with the occurrence of overt liver damage. In the B6 mice there was only a 1.7-fold increase in COH activity 2 days after treatment. Further, no sharp decrease in the enzyme activities comparable to those in the D2 mice took place in the B6 mice.

Northern Blot analysis showed a 21-fold increase in the hepatic Cyp2a-4/5 mRNA in the D2 mice at day 2, whereas no increase occurred in the B6 mice. A polymerase chain-reaction-mediated analysis method utilizing a unique *Pst*I restriction site in the Cyp2a-5 cDNA was used to differentiate between the highly homologous Cyp2a-4 and Cyp2a-5 mRNAs. Cerium was found to increase the amount of hepatic Cyp2a-4 and Cyp2a-5 mRNA only in the D2 mice. This study showed that administration of a liver-damaging dose of cerium causes an increase in the expression of Cyp2a-4/5 in the liver of D2 mice, but not B6 mice. Salonpaa et al. concluded that the Cyp2a-4/5 complex is regulated in a different way in D2 and B6 mice, and that some association exists between the development of liver damage and COH induction. In reviewing these data, Hirano and Suzuki (2) concluded that hepatic injury caused by IV injection of $CeCl_3$ seemed to reduce P450 content and P450-related enzyme activities in rats and mice.

Hirano and Suzuki (2) also reported that IV injection of $Pr(NO_3)_3$ in rats decreased serum very low density lipoprotein (VLDL) and high density lipoprotein (HDL). They theorized that the decrease could be due to decreased hepatic secretions of these lipoproteins. Hirano and Suzuki (2) noted that the effect of IV injected $Pr(NO_3)_3$ on serum glutamic–oxaloacetic and glutamic–pyruvic transaminase activities was dose-dependent. Enzymatic activity increased significantly at doses of ≤ 20 mg/kg bw, indicating hepatic injury, but decreased at doses above this level. Hirano et al. (44) had previously shown that increasing the dose of YCl_3 significantly increased the formation of collodial rare earths in blood. Thus, Hirano and Suzuki (2) postulated that the IV injected $Pr(NO_3)_3$ was taken up by Kupffer cells, rather than by hepatocytes, at doses higher than a maximum lethality. They further concluded that the uptake of collodial rare earths by the Kupffer cells may have reduced the uptake of the rare earths by the hepatocytes, resulting in reduced hepatic injury.

Hirano and Suzuki (2) summarized data reporting a wide variety of hepatic biochemical changes following IV injection of rare earth chlorides and nitrates. This information is summarized in Table 43.4 (44, 57, 66, 70–79). Hirano and Suzuki concluded that, with the exception of the effect on RNA polymerase activity, the rare earths consistently increased hepatic triglyceride levels and increased leakage of hepatic enzymes into the blood.

Basu et al. (79) administered lanthanum chloride IP to chicks. They found that lanthanum chloride markedly altered the antioxidant defense system of the liver. In treated animals they observed significant elevation of lipid peroxidation with concomitant depression in the activities of glutathione reductase, and glutathione peroxidase and the levels of reduced and oxidized glutathione. L-Cysteine hydrochloride supplementation in treated animals could partially prevent the alterations produced by toxic doses of lanthanum.

Table 43.4. Hepatic and Liver-Associated Biochemical Changes Following IV Injection of Rare Earth (2)[a]

Effect	Compound	Dose	Animal	Ref.
s-GOT, s-GPT↑	CeCl₃, Ce(NO₃)₃	2–10 Ck/kg	Rat	(44, 70, 71)
	La(NO₃)₃	3–10 mg La/kg		
	YCL₃	1 mg Y/rat		
s-GOT, s-GPT↑	Pr(NO₃)₃	3–40 mg Pr(NO₃)₃/kg	Rat, mouse	(71–73)
s-SDH↑	Ce(NO₃)₃	3–10 mg RE/kg	Rat	(71)
	Pr(NO₃)₃			
	La(NO₃)₃			
s-OCT↑	CeCl₃	1.5–3 mg Ce/kg	Rat	(57, 74)
	PrCl₃	3 mg Pr/kg		
	LaCl₃	0.75 mg La/kg		
	YCl₃	9 mg Y/kg		
	TbCl₃	35 mg Tb/kg		
	HoCl₃	40 mg Ho/kg		
	YbCl₃	60 mg Yb/kg		
s-FFA↑	Pr(NO₃)₃	10 mg Pr(NO₃)₃/kg	Rat	(75)
s-VLD, s-HDL↓	Pr(NO₃)₃	10 mg Pr(NO₃)₃/kg	Rat	(76)
s-Triglyceride↓				
Triglyceride (liver)↑	CeCl₃	10 mg CeCl₃/kg	Rat	(77)
Triglyceride (liver)↑	Pr(NO₃)₃	7 mg Pr/kg	Rat, mouse	(71, 73, 75)
		10–20 mg Pr(NO₃)₃/kg		
ATP↓	CeCl₃	10 mg CeCl₃/kg	Rat	(77)
P450, AH, AD↓	Pr(NO₃)₃	7 mg Pr/kg	Rat	(71)
COH, EROD, Cyp2a-4/5-mRNA↑↓	CeCl₃	2 mg CeCl₃/kg	Mouse	(66, 70)
		0.5–2 mg Ce/kg		
RNA polymerase I↓	Pr(NO₃)₃	35 µmol/kg	Rat	(78)
	Nd(NO₃)₃			
RNA polymerase II↑	Gd(NO₃)₃	35 µmol/kg	Rat	(78)
	Dy(NO₃)₃			
	Er(NO₃)₃			
RNA polymerase II↓	Pr(NO₃)₃	35 µmol/kg	Rat	(78)
	Nd(NO₃)₃			
	Sm(NO₃)₃			
LPO↑	LaCl₃	250 mg LaCl₃/kg	Chick (lip)	(79)
GR↓	LaCl₃	250 mg LaCl₃/kg	Chick (lip)	(79)

[a] These studies indicate that cerium, and possibly other rare earths, are hepatotoxic agents.

1.4.1.7.1.5 LUNG. Chronic inhalation of mixtures of lanthanides high in fluorides (65% fluorides, 10% oxides, 31% C), with a calculated particle size of 1–2 mm, by guinea pigs at massive concentrations of 200–300 million particles per cubic foot daily for 3 years showed histological changes consisting of focal hypertrophic emphysema, regional bronchiolar stricturing, and subacute chemical bronchitis. Although pigment was focally retained, no fibrosis or granulomatosis developed (80). By comparison, guinea pigs exposed to the high oxide mixture (26.4% oxide, 39.6% fluorides, 17% C) had a less severe reaction. The most prominent feature was a cellular eosinophilia, and although most of the dust was trapped within focal atelectatic areas, no substantial chronic cellular reaction or fibrosis occurred around these deposits (54).

Ball and Van Gelder (81) exposed mice to gadolinium oxide aerosol for 20–120 days to evaluate the chronic effects of the exposure. They found that there was a trend toward shorter lifespans in exposed mice. Further, they observed an accumulation of macrophages in the lung, with focal areas of interstitial thickening around macrophages, particularly for mice exposed for 120 days. Some macrophages containing gadolinium oxide were undergoing degeneration, evidenced by pyknosis and karyolysis. There was little tendency toward pulmonary fibrosis, but there was calcification in the region of the alveolar basement membrane and elastic laminae of small pulmonary vessels.

The rare earth metals cerium, lanthanum, and neodymium were each evaluated in an *in vitro* cytotoxicity assay system using adult male rat pulmonary alveolar macrophages (82). Both the soluble chlorides and the insoluble metal oxides were studied. For comparison purposes, the cytotoxicities of cadmium chloride and cadmium oxide were also quantified in this test system. In general, regardless of the cytotoxicity parameter measured, that is, cell viability, lysosomal enzyme leakage, or changes in cell surface morphology, cadmium was more toxic to these cells than were the rare earth metals. Of the rare earth metals studied, only lanthanum chloride ($LC_{50} = 52$ mM), cerium chloride ($LC_{50} = 29$ mM), and neodymium oxide ($LC_{50} = 101$ mM) displayed significant cytotoxicity in this test system. Cadmium chloride exhibited an LC_{50} value of 28 mM, whereas the LC_{50} value for cadmium oxide was 15 mM. These findings support the conclusions of Haley (1) that rare earth metal fumes should be considered as cytotoxic to lung tissue.

As discussed in detail in Section 1.4.1.3, Hirano et al. (38) investigated the metabolic behavior and pulmonary toxicity of intratracheally administered yttrium chloride in male Wistar rats. Examination of the lung tissue showed that rats developed granulomatous lesions. Yttrium was observed as black, granular inclusions in alveolar macrophages. Transmission electron microscopy and X-ray microanalysis suggested that Y was localized in lysosomes of alveolar and interstitial macrophages, and basement membranes. Comparative dose–effect profiles of lactate dehydrogenase activity in BALF supernatant revealed that 1 mol of YCl_3 is equivalent to about 0.33 mol of cadmium compounds and about 3 mol of zinc oxide in the potency for acute pulmonary toxicity. On the basis of this study, Suzuki et al. (39) concluded that Y showed acute toxicity comparable to copper when the metal was administered intratracheally.

Suzuki et al. (39) investigated the distribution and health effects of lanthanum chloride ($LaCl_3$) intratracheally instilled in male Wistar rats. This study is discussed in more detail in Section 1.4.2.2.3 and parallels the study described above. The authors used lactate dehydrogenase (LDH), β-glucuronidase activities and P content in the supernatant fraction of BALF as indices for cell lysis, secretion of lysosomal enzymes, and surfactant content, respectively. This represented acute inflammatory responses in the bronchoalveolar tissues. They found that LDH and β-glucuronidase activities increased significantly at doses higher than 1 µg/rat. P content increased significantly at doses of 0.5 µg/rat. Using LDH activity as an indicator of acute toxicity, the authors compared La and Y. At a dose of

10 µg metal/rat, both Y and La induced 1.8-fold increases in LDH activity. Thus, the authors concluded that La and Y showed comparable acute pulmonary toxicity when their chloride salts were intratracheally instilled in rats. Both compounds were shown to cause pneumonitis and acute inflammation in lung tissue.

In a study that paralleled those of Hirano et al. (38) and Suzuki et al. (39), Yoneda et al. (40) investigated the effects of gadonlinium chloride when intratracheally instilled in rat lung. As described more fully in Section 1.4.1.3., Yoneda et al. found that pulmonary clearance of Gd was slow, with an estimated biological half-life of 136 days. They found that calcium content in BALF increased more rapidly than did other toxicological indices such as LDH activity, protein content, and inflammatory cell counts. Gd also showed acute lung toxicity. Their findings were very similar to those for YCl_3 and $LaCl_3$. Thus it is clear that, in animals, inhalation or intratracheal exposure to rare earths causes acute pneumonitis and inflammation in lung tissue.

1.4.1.7.1.6 SKIN. Of the 11 lanthanides tested by Haley et al., only erbium produced irritation to the intact rabbit skin; however, all uniformly and severely irritated abraded skin, producing within a day erythema and edema (maximum score), with scars 25–35 mm in diameter and complete healing only after 45 days, according to the Draize procedure. Skin reactions to intradermal injection of seven lanthanides (Pr, Nd, Eu, Dy, Ho, Er, Lu) as the chlorides into guinea pigs resulted in the formation of nodules 2 mm in diameter, which contained crystalline deposits (83). These deposits, resulting from 0.05–5 mg lanthanide administration, contained calcium (but no iron) and were surrounded by histiocytes and foreign-body giant cells with fibroblasts and granulation tissue extending into the area. No decrease in nodule size or indication of any resorption of the crystalline deposit occurred within the 45-day observation period (23–28).

In rabbits, a topical application of a suspension of 500 mg/mL produced mild irritation of the skin and eye mucosa (34).

1.4.1.7.1.7 SPLEEN. Stineman et al. (64) observed hypertrophy, reticuloendothelial hyperplasia, and hyperactive lymphoid follicles in the spleens of mice who had been treated with cerium, either orally or subcutaneously. As summarized by Hirano and Suzuki (2), Marciniak and Baltrukiewicz (74) reported that in mice $LaCl_3$ and $CeCl_3$ increased vascular permeability of the spleen.

Yttrium chloride solution was administered IV to male Wistar rats at doses of 0.1, 0.2, 0.5, 1, and 2 mg Y/rat to evaluate its metabolism and toxicity. The study demonstrated that some plasma calcium was translocated to the collodial material and plasma calcium concentration was increased transiently following injection of YCl_3. The authors theorized that the increase in calcium concentration was due to resorption of bone. At doses of 1 mg Y/rat, the calcium content of the spleen increased over 100-fold. Gross observation of the tissues indicated the presence of white blotchy necroses in the spleen. Examination by transmission electron microscopy (TEM) equipped with an X-ray microanalyzer (XMA) revealed that calcium was deposited focally in the necrotic sites of the spleen. In addition to calcium deposition, the study indicated that at a dose of 1 mg Y/rat, 20% of the dose was accumulated in the spleen at 7 h postinjection. The authors concluded that the spleen is one of the primary target organs of IV injected YCl_3 (44).

1.4.2 Human Experience

1.4.2.1 General Information. The pulmonary toxicity of inhaled lanthanides has been the subject of debate. The relative contributions of radioactive versus stable elements in the development of lanthanide-associated progressive pulmonary interstitial fibrosis has been questioned. Haley (1) examined the epidemiological and experimental data and concluded that significant pathogenic potential of inhaled lanthanides exists and is related to the type and physicochemical form of the material inhaled and to the dose and duration of exposure. Although contamination of the dust of lanthanides with radioactive materials may accelerate and enhance the pathological response, depending on the form and dose of radioactivity encountered, there is little evidence to suggest that the level of radioactive contamination of occupationally encountered lanthanide dusts is sufficient to be included as a risk factor for pulmonary disease. Haley (1) concluded that the pulmonary syndrome induced by stable rare earths includes progressive pulmonary fibrosis and should not be referred to as "benign pneumoconiosis."

Hirano and Suzuki (2) concluded that chronic exposure to rare earth dusts probably causes pneumoconiosis in humans.

1.4.2.2 Clinical Cases. A case of rare earth pneumoconiosis was described by Sabbioni et al. (84). A man working in a lithographic laboratory as a photoengraver and exposed to the smoke of cored carbon-arc lamps over a period of 46 years developed an interstitial pneumoconiosis. Neutron activation analysis (NAA) of eight rare earths (La, Ce, Nd, Sm, Eu, Tb, Yb, and Lu) in lung and lymph node biopsies showed an abnormally high amount of these elements in comparison to the corresponding values from 11 autopsied unexposed subjects. The estimated radiological dose due to the inhalation of natural thorium, as calculated from NAA of thorium in the biopsies, tends to exclude the effect of ionizing radiation in the pathogenesis of lung fibrosis. The findings strongly suggest that a relationship exists between the pneumoconiosis diagnosed and the occupational exposure to rare earth dusts.

Waring and Watling (85) reported the case of a movie projectionist with approximately 25 years of occupational exposure to carbon-arc-lamp fumes. The carbon-arc deposits were visible in histological sections as small granules within macrophages of the tracheobronchial lymph nodes and hepatic Kupffer cells. Electron microprobe analysis by energy-dispersive analysis of X rays showed the granules were composed of the rare earth elements cerium, lanthanum, and neodymium, which are the major constituents of carbon-arc rods. Tissue concentrations, as determined by inductively coupled plasma spectroscopy, were approximately 250–2000 times those of unexposed controls, and there was evidence of their redistribution throughout the reticuloendothelial system. There were no respiratory symptoms or radiographic or histological pulmonary changes attributable to the progressive accumulation of the rare earth elements; as such, the patient cannot be considered to have suffered from pneumoconiosis. However, the authors reviewed 21 published cases of rare earth pneumoconiosis, mainly in photoengravers exposed to carbon-arc fumes, and concluded that rare earth oxides are not innocuous dusts.

McDonald et al. (86) described a case of rare earth pneumoconiosis in a 68-year-old male patient with a 35-year history of optical lens grinding, which is associated with exposure to cerium oxide. The patient had ceased working as a lens grinder 13 years prior to seeking medical treatment. The patient initially presented with progressive dyspnea. Chest X rays revealed an interstitial lung pattern that was consistent with fibrosis. Open-lung biopsy showed interstitial fibrosis histologically indistinguishable from usual interstitial pneumonitis. Numerous birefringent particulates were observed in the interstitium using polarized light microscopy. Scanning electron microscopy with energy-dispersive X-ray analysis demonstrated numerous particulate deposits in the lung, the majority of which contained cerium alone or in combination with other elements. This case supports Haley's conclusion (1) that stable rare earths cause pneumoconiosis.

1.4.2.2.1 Acute Toxicity. Considerable study has been conducted in the past on the anticoagulant action of the lanthanides on blood for the prevention of thrombosis. Neodymium salts injected IV in total dosages of 250–500 mg have been reported to prolong the coagulation time of human blood (83) for 6 h; no ill effects were noted, but oral or subcutaneous administration was without anticoagulant effect. Beaser et al. (87), after having confirmed in rabbits the innocuousness of small anticoagulant doses of neodymium (10 mg/kg), critically evaluated the anticoagulant effects of neodymium, lanthanum, and cerium in 18 patients with single and repeated daily doses of 3–12.5 mg/kg. All the lanthanides tested increased the clotting time of blood to the point of incoagulability, which persisted in diminishing degree for 8 h. The toxic reactions of chills, fever, headache, muscle pains, abdominal cramps, hemoglobinemia, and hemoglobinuria were such as to exclude the use of lanthanides as anticoagulant agents. Determination of blood levels of neodymium showed that it was still present in considerable amounts during decline of the anticoagulant effect. Neodymium was not found in the urine of the treated patients, despite repeated analysis; available evidence suggested removal by the reticuloendothelial system of the liver and spleen. Local thrombophlebitis was observed in humans, and also reported in dogs, after injection of the lanthanide chloride. Hemoglobinemia was unexpected; it had not been seen previously in animals and does not occur when the lanthanides are added to blood *in vitro*. These findings provide guideposts in the medical control of exposures to lanthanides. Blood serum hemoglobin values would be the most sensitive indicator; values of $\sim 9\%$ would indicate overexposure. Initial appearance of or repeated headache, malaise, chills and fever, and nausea would be indicative symptoms.

1.4.2.2.2 Chronic and Subchronic Toxicity. As discussed in Section 1.4.2.2.1, Haley (1, 23–28) and Hirano and Suzuki (2) concluded that chronic exposure to rare earth dusts probably causes pneumoconiosis in humans.

1.4.2.2.3 Pharmacokinetics, Metabolism, and Mechanics. Lung, liver, and kidney tissue concentrations of lanthanum from 66 deceased copper smelter workers have been compared with 14 controls. Samples were taken in connection with ordinary autopsies at the local hospital. Neutron activation analysis was used. There was a twofold increase of lanthanum in lung tissue among smelter workers as compared to controls. Of the smelters, nearly 33% died from malignancies (approximately 10% of these from respiratory cancer) and approximately 45% from cardiovascular disease. In the control group nearly 80% died from cardiovascular diseases, but no malignancies were present. In lung tissues the lanthanum did not decline with time after exposure had ended, indicating a long biologic half-life. The causes of death could not be related to any single factor (88).

Samarium-153 ethylenediaminetetramethylene phosphonic acid (EDTMP) was developed as a radiopharmaceutical to treat metastatic bone cancer pain. In this study 2 mCi of [^{153}Sm]EDTMP was injected IV into five male patients. The study showed that the chelate was cleared rapidly, with only 5.17% of the activity remaining at 2 h postinjection and 2.09% of the activity remaining at 4 h postinjection. The complex clears through the kidneys, and approximately 50% of the injected dose is excreted into the urine in 8 h (14).

1.4.2.3 Epidemiology Studies. Weifang et al. (89) compared the incidence of arteriosclerosis in individuals living in three different regions of China. The three regions were described as areas A, B, and C. In areas A and B, the soil contains 40–90% exchange-state rare earth elements (REE) and there were numerous places where rare earths were mined. In area A the concentration of REE in drinking water averaged 15.9 µg/L water. In area B the concentration of REE in drinking water averaged 47.0 µg/L water. In area C, the soil contained less than 25% REE and there were no mining locations. In area C the concentration of REE in drinking water averaged only 0.92 µg/L water. The authors calculated the daily REE intake for each of the areas. Villagers in area A were estimated to take in 16 times more REE than area C villagers. Area B villagers were estimated to take in 45 times more REE than area C villagers.

Villagers aged 25–45 years with no history of angina or hypertension were selected from each geographic area. They were examined using an ophthalmofunduscope, and the extent of arteriosclerosis was determined. The arteriosclerosis of the fundus aculi was graded according to Scheie's method. The authors reported a significantly higher number of arteriosclerotic villagers in areas A and B than in area C. As all the villagers were poor and did not have access to a high lipid diet, which is a known risk factor for arteriosclerosis, the authors postulated that intake of REE could be responsible for the increased incidence of arteriosclerosis.

1.4.2.3.1.1 YTTRIUM. Although yttrium (see Section 1.0o for CAS number, *etc.*) reacts chemically in a manner similar to La and is included with the lanthanides, early studies indicated that toxicologically it might differ sufficiently to single it out from the lanthanides. The yttrium citrate–chloride complex was reported by Graca et al. (29), for example, to be the most acutely toxic to both guinea pigs and mice by the intraperitoneal route of all the lanthanides (with LD$_{50}$ values of 42 and 78 mg/kg, respectively, compared with a close member in the series, cerium, with respective LD$_{50}$ values of 82 and 150 mg/kg). Further, in a comparative investigation of the oxides of yttrium, cerium, and neodymium in rats intratracheally administered at a dose of 50 mg (0.222, 0.152, and 0.148 mM, respectively), Y$_2$O$_3$ showed the most pronounced changes in the lung of the three oxides tested (90). At 8 months, in the lung tissue in which Y$_2$O$_3$ was present, characteristic granulomatous nodules developed, consisting of crystalline deposits of the oxide and cellular elements. Nodules in the peribronchial tissue compressed and deformed several bronchi; the surrounding lung areas were emphysematous, the interalveolar walls were thin and sclerotic, and the alveolar cavities dilated. In contrast, Ce$_2$O$_3$ did not elicit serious changes, and produced neither diffuse nor nodular fibrotic processes. Although Nd_2O_3 granulomas resembled those induced by Y_2O_3, their further development differed. Their characteristic was a very poor formation of connective-tissue fibers, which were extremely weak around the granulomas, and sclerosis of the interalveolar walls was only moderate. It was concluded that, regardless of the similarity of their chemical and physical properties, Y_2O_3 exhibited specific toxicologic actions that were the most severe of the three oxides studied.

A still further toxicological difference is the lesser degree of skeletal deposition. Yttrium chloride, YCl_3, administered to rats IP on alternate days, for as many as 83 injections totaling 936 mg, did not accumulate large amounts of yttrium in femoral bone (91). Yttrium never exceeded 330 ppm of bone ash, and after it had been deposited to the extent of 150–200 ppm (corresponding to about 50 mg injected YCl_3), further accumulation proceeded at a very slow rate. With age, the ratio of yttrium concentration in spongy to compact bone approached 1. By comparison, skeletal deposition amounted to 20% for lanthanum, ranging to 70% for lutetium (Lu).

Conversely, Hirano and Suzuki et al. (38–40) more recently studied the effects of yttrium chloride, lanthanum chloride, and gadolinium chloride when intratracheally instilled in rat lung. The toxicities and action mechanisms of all three compounds were similar, which does not support the previous conclusion that Y differs toxicologically from the other lanthanides.

1.5 Standards, Regulations, or Guidelines of Exposure

A TLV (threshold level value) of $5 mg/m^3$ was recommended by the ACGIH TLV Committee in 1960 for yttrium, the only correlative member of the lanthanides for which a limit has been set. This limit was revised downward to $1 mg/m^3$ in 1964 when the ACGIH TLV Committee learned from a Soviet Union publication (90) that Y_2O_3, on intratracheal administration, resulted in severe lung damage (12). The ACGIH (92) also cites a study by Tebrock and Machle (93) in which exposure to a mixture of yttrium, europium, and vanadate produced irritation to the eyes and respiratory tract as a result of vanadium content, but the concentration of yttrium was $1.4 mg/m^3$. The OSHA PEL is also $1 mg/m^3$ (94).

No standards have been recommended for any of the other lanthanides because either suitable data for setting a standard, such as inhalation studies, or studies on individual lanthanides are lacking (12). However, because of the accumulating evidence of induction of fibrosis with the lanthanides and their expanding use, the exposure should probably be limited to $1 mg/m^3$.

1.6 Studies on Environmental Impact

Volokh et al. (95) reported on a study of REO pollution from a phosphorus fertilizer production facility in Russia. They studied air-pollution sediment that fell on soil or snow, uptake into agricultural plants, and human uptake as indicated by hair samples from plant operators and local residents. Samples were analyzed using X-ray fluorescence and neutron activation analysis. Results are reported for Sc, Cr, As, Se, Br, Sb, La, Ce, Sm, and Eu. They concluded that these elements were emitted from the plant in the form of a phosphogypsum dust with a diameter of $\leq 8\,\mu m$. Abnormally high concentrations were measured in soil and snow within 1 km of the plant and extended up to a radius of 8 km that was influenced by prevailing wind direction. Apples, beets, potatoes, and carrots showed an increased storage of ytterbium and europium. Hair samples of plant employees demonstrated storage of Sm, La, and Ce. Resident children's hair stored La, Sm, and Ce, but in lesser amounts than did the hair of plant workers. No values from a representative control group were provided for comparison, nor were any expressions regarding toxicity or adverse effects presented.

BIBLIOGRAPHY

1. P. J. Haley, Pulmonary toxicity of stable and radioactive lanthanides. *Health Phys.* **61**(6), 809–820 (1991).
2. S. Hirano and K. T. Suzuki, Exposure, metabolism, and toxicity or rare earths and related compounds. *Environ. Health Perspect.* **104**(Suppl. 1), 85–95 (1996).
3. D. Laszlo et al., *J. Natl. Cancer Inst.* (U.S.). **13**, 559 (1952).
4. E. J. Wheelwright et al., *J. Am. Chem. Soc.* **75**, 4196 (1953).
5. E. Greinacher, History of rare earth applications, rare earth market today: Overview. In K. A. Gschneidner, Jr., ed., *Industrial Applications of Rare Earth Elements*, American Chemical Society, Washington, DC, 1981, pp. 3–17.
6. T. K. S. Murthy and C. K. Gupta, Rare earth resources, their extraction and application. In E. C. Subbarao and W. E. Wallace, eds., *Science and Technology of Rare Earth Materials*, Academic Press, New York, 1980, pp. 3–23.
7. J. Kaczmarek, Discovery and commercial separations. In K. A. Gschneidner, Jr., ed., *Industrial Applications of Rare Earth Elements*, American Chemical Society, Washington, DC, 1981, pp. 135–166.
8. E. V. Kleber and B. Love, *The Technology of Scandium, Yttrium and the Rare Earth Metals*, Macmillan, New York, 1963, pp. 2–48.
9. K. E. Davies, Industrial applications of pure rare earth metals and related alloys. In K. A. Gschneidner, Jr., ed., *Industrial Applications of Rare Earth Elements*, American Chemical Society, Washington, DC, 1981, pp. 167–175.
10. J. R. McColl and F. C. Palilla, Use of rare earths in television and cathode ray phosphors. In K. A. Gschneidner, Jr., ed., *Industrial Applications of Rare Earth Elements*, American Chemical Society, Washington, DC, 1981, pp. 177–193.
11. K. A. Gschneidner, Jr. and L. Eyring, eds., *Handbook on the Physics and Chemistry of Rare Earths. Vol.* 1–25, Elsevier, Amsterdam, 1998.
12. H. E. Stokinger, in , G. Clayton and F. Clayton, eds., *Patty's Industrial Hygiene and Toxicology*, 3rd ed., Vol. 2A, Wiley, New York, 1981, pp. 1493–2060.
13. G. H. Haertling, PLZT electrooptic ceramics and devices. In K. A. Gschneidner, Jr., ed., *Industrial Applications of Rare Earth Elements*, American Chemical Society, Washington, DC, 1981, pp. 265–283.
14. A. Singh et al., Human pharmacokinetics of Samarium-153 EDTMP in metastatic cancer. *J. Nucl. Med.* **30**(11), 1814–1818 (1989).

15. Y. Berthezène et al., Safety aspects and pharmacokinetics of inhaled aerosolized gadolimium. *J. Magn. Reson Imaging* **3**, 125–130 (1993).
16. V. A. Fassel, *J. Opt. Soc. Am.* **39**, 187 (1949).
17. *Chem. Eng. News*, p. 41 (1959).
18. D. G. Taylor, manual coordinator, *NIOSH Manual of Analytical Methods*, NIOSH Publ. No. 77-157-A, 2nd ed., Vol 2, U.S. Department of Health, Education, and Welfare, Cincinnati, OH, 1977.
19. National Institute for Occupational Safety and Health (NIOSH), *NIOSH Pocket Guide to Hazardous Chemicals*, NIOSH, Cincinnati, OH, 1990.
20. M. Chiba et al., Comparative study of methods for determining lanthanide elements in biological materials by using NAA, HPLC postcolumn reaction, and ICP-MS. *Biol. Trace Elem. Res.*, **43–45**, 561–569 (1994).
21. P. Allain et al., Concentrations of rare earth Elements in plasma and urine of healthy subjects determined by inductively coupled plasma mass spectrometry. *Clin. Chem. (Winston-Salem, N.C.)* **36**(11), 2011–2012 (1990).
22. *Rare Earths in Biochemical and Medical Research*, ORINS-12, Institute or Nuclear Studies, Tech. Inf. Serv. Ext., Oak Ridge, TN, 1955.
23. T. J. Haley et al., *Br. J. Pharmacol.* **17**, 526 (1961).
24. T. J. Haley and H.C. Upham, *Nature (London)* **200**, 271 (1963).
25. T. J. Haley et al., *Toxicol. Appl. Pharmacol.* **5**, 427 (1963).
26. T. J. Haley et al., *Toxicol. Appl. Pharmacol.* **6**, 614 (1964).
27. T. J. Haley et al., *J. Pharm. Sci.* **53**, 1186 (1964).
28. T. J. Haley et al., Pharmacology and toxicology of dysprosium, holmium, and erbium chlorides. *Toxicol. Appl. Pharmacol.* **8**(1), 37–43 (1966).
29. J. G. Graca et al., *Arch. Environ. Health* **5**, 437 (1962).
30. G. C. Kyker and E. A. Cress, *Arch. Ind. Health* **16**, 475 (1957).
31. K. W. Cochran et al., *Arch. Ind. Hyg. Occup. Med.* **1**, 637 (1950).
32. H. E. Hart et al., *J. Lab. Clin. Med.* **46**, 182 (1955).
33. D. W. Bruce et al., *Toxicol. Appl. Pharmacol.* **5**, 750 (1963).
34. Y. J. Ji and M. Z. Cui, Toxicological studies on safety of rare earths used in agriculture. *Biomed. Environ. Sci.* **1**(3), 270–276 (1988).
35. J. G. Graca et al., *Arch. Environ. Health* **8**, 555 (1964).
36. P. W. Durbin et al., *U.S. At. Energy Comm. Rep.* **UCRL-3066** (1955).
37. P. Galle, J. P. Berry; and C. Galle, Role of aveolar macrophages in precipitation of mineral elements inhaled as soluble aerosols. *Environ. Health Perspect.* **97**, 145–147 (1992).
38. S. Hirano et al., Distribution, localization, and pulmonary effects of yttrium chloride following intratracheal instillation into the rat. *Toxicol. Appl. Pharmacol.* **104**(2), 301–311 (1990).
39. K. T. Suzuki et al., Localization and health effects of lanthanum chloride instilled intratracheally into rat. *Toxicology* **76**(2), 141–152 (1992).
40. S. Yoneda et al., Effects of gadolinium chloride on the rat lung following intratracheal instillation. *Fundam. Appl. Toxicol.* **28**, 65–70 (1995).
41. P. W. Durbin et al., *Proc. Soc. Exp. Biol. Med.* **91**, 78 (1956).
42. P. B. Dean et al., Comparative pharmacokinetics of gadolinium DTPA and gadolinium chloride. *Invest. Radiol.* **23**(Suppl. 1), S258–S260 (1988).

43. J. L. Barnhart et al., Biodistribution of GdCl₃ and Gd-DTPA and their influence on proton magnetic relaxation in rat tissue. *Magn. Reson. Imaging* **5**, 221–231 (1987).
44. S. Hirano et al., Metabolism and toxicity of intravenously injected yttrium chloride in rats. *Toxicol. Appl. Pharmacol.* **121**, 224–232 (1993).
45. T. Cassperson et al., *Trans. Faraday Soc.* **31**, 367 (1935).
46. J. N. Davidson and C. Waymouth, *Biochem. J.* **38**, 39 (1944).
47. E. Chargaff et al., *J. Biol. Chem.* **177**, 405 (1949).
48. F. Calvert et al., *Nature (London)* **162**, 305 (1948).
49. E. Vincke et al., *Z. Gesamet. Exp. Med.* **113**, 536 (1944).
50. E. Bamann, *Klin. Wochenschr.* **32**, 588 (1954).
51. J. W. Abramczuk, The effects of lanthanum chloride on pregnancy in mice and on preimplantation mouse embryos in vitro. *Toxicology* **34**(4), 315–320 (1985).
52. R. B. D'Agostino et al., Effects of in utero or suckling exposure to cerium (citrate) on the postnatal development of the mouse. *J. Toxicol. Environ. Health* **10**(3), 449–458 (1982).
53. D. P. Hutcheson et al. Studies of nutritional safety of some heavy metals in mice. *J. of Nutr.* **105**(6), 670–675 (1975).
54. R. A. Ball et al., Neoplastic sequelae following subcutaneous implantation of mice with rare earth metals. *Proc. Soc. Exp. Biol. Med.* **135**(2), 426–430 (1970).
55. N. H. Zapolskaya et al., *Gig. Sanit.* **36**, 55 (1971).
56. N. N. Kushakova and A. E. Ivanov, *Zh. Eksp. Biol. Med.* **54**, 79 (1962).
57. G. Magnusson, The behavior of certain lanthanoids in rats. *Acta Pharmacol. Toxicol.* **20**(Suppl. 3), 1–95 (1963).
58. M. M. Boucek and R. Snyderman, Calcium influx requirement for human neutrophil chemotaxis: Inhibition by lanthanum chloride. *Science* **193**, 905–907 (1976).
59. G. A. Langar et al., Sialic acid: Effect of removal on calcium exchangeability of cultured heart cells. *Science* **193**, 1013–1015 (1976).
60. C. R. Triggle and D. J. Triggle, An analysis of the action of cations of the lanthanide series on the mechanical responses of guinea-pig ileal longitudinal muscle. *J. Physiol. (London)* **254**(1), 39–54 (1976).
61. V. C. Swamy, C. R. Triggle, and D. J. Triggle, The effects of lanthanum and thulium on the mechanical responses of rat vas deferens. *J. Physiol. (London)* **254**(1), 55–62 (1976).
62. N. Ghosh, D. Chattopadhyay, and G. C. Chatterjee, Chicken erythrocyte membrane: Lipid profile and enzymatic activity under lanthanum chloride and neodymium chloride administration. *Indian J. Exp. Biol.* **29**, 226–229 (1991).
63. H. Yajima et al., Lanthanide ions for the first non-enzymatic formation of adenosine 3′, 5′-cyclic monophosphate from adenosine triphosphate under physiological conditions. *J. Biochem. (Tokyo).* **115**, 1038–1039 (1994).
64. C. H. Stineman et al., Cerium tissue/organ distribution and altercations in open field and exploratory behavior following acute exposure of the mouse to cerium (citrate). *J. Exp. Pathol. Toxicol.* **2**, 553–570 (1978).
65. Z. H. Endre, J. L. Allis, and G. K. Radda, Toxicity of dysprosium shift reagents in the isolated perfused rat kidney. *Magn. Reson. Med*, **11**, 267–274 (1989).
66. P. Salonpaa et al., Cerium-induced strain-dependent increase in Cyp2a-4/5 expression in the liver and kidneys of inbred mice. *Biochem. Pharmacol.* **44**, 1269–1274 (1992).

67. G. Maulik et al., Curative effect of methionine on certain enzymes of chick kidney cortex under lanthanum toxicity situation. *Indian J. Exp. Biol.* **30**, 1166–1169 (1992).
68. F. Snyder et al., *J. Lipid Res.* **1**, 125 (1959).
69. L. Horecke et al., *J. Biol. Chem.* **128**, 251 (1939).
70. P. Arvela et al., The cerium-induced liver injury and oxidative drug metabolism in DBA/2 and C57BL/6 mice. *Toxicology* **69**(1), 1–9 (1991).
71. O. Strubelt, C. P. Siegers, and M. Younes, The influence of silybin on the hepatotoxic and hypoglycemic effects of praseodymium and other lanthanides. *Arzneim.-Forsch*, **30**(II), 1690–1694 (1980).
72. B. Tuchweber et al., Effect of praseodymium nitrate on hepatocytes and Kupffer cells in the rat. *Can. J. Physiol. Pharmacol.* **54**, 898–906 (1976).
73. M. Conti, S. Malandrino, and M. J. Magistretti, Protective activity of silipide on liver damage in rodents. *Jpn. J. Pharmacol.* **60**, 315–321 (1992).
74. M. Marciniak and Z. Baltrukiewicz, Serum ornithine carbamoyltransferase (OCT) in rats poisoned with lanthanum, cerium, and praseodymium. *Acta Physiol. Pol.* **28**, 589–594 (1977).
75. E. Oberdisse, P. Arvela, and U. Gross, Lanthanon-induced hepatotoxicity and its prevention by pretreatment with the same lanthanon. *Arch. Toxicol.* **43**, 105–114 (1979).
76. O. Grajewski et al., Alterations of rat serum lipoproteins and lecithine-cholesterol-acyltransferase activity in praseodymium-induced liver damage. *Naunyn-Schmiedeberg's Arch. Pharmacol.* **301**, 65–73 (1977).
77. M. Salas et al., Effect of cerium on the rat liver. An ultrastructural and biochemical study. *Beitr. Pathol.* **157**, 23–44 (1976).
78. H. I. Sarkander and W. P. Brade, On the mechanism of lanthanide-induced liver toxicity. *Arch. Toxicol.* **36**, 1–17 (1976).
79. A. Basu et al., Effects of cysteine supplementation on lanthanum chloride induced alteration in the antioxidant defense system of chick liver. *Indian J. Exp. Biol.* **22**, 432–434 (1984).
80. G. W. H. Schepers, *Arch. Ind. Health* **12**, 306 (1955).
81. R. A. Ball and G. Van Gelder, Chronic toxicity of gadolinium oxide for mice following exposure by inhalation. *Arch. Environ. Health* **13**(5), 601–608. (1966).
82. R. J. Palmer, J. L. Butenhoff, and J. B. Stevens, Cytotoxicity of the rare earth metals cerium, lanthanum, and neodymium in vitro: Comparisons with cadmium in a pulmonary macrophage primary culture system. *Environ. Res.* **43**(10), 142–156 (1987).
83. H. Dyckerhoff and N. Goosens, *Z. Gesamte Exp. Med.* **106**, 181 (1939).
84. E. Sabbioni et al., Long-term occupational risk of rare earth pneumoconiosis. A case report as investigated by neutron activation analysis. *Sci. Total Environ.* **26**(1), 19–32 (1982).
85. P. M. Waring and R. J. Watling, Rare earth deposits in a decreased movie projectionist. A new case of rare earth pneumoconiosis? *Med. J. Aust.* **153**(11–12), 726–730 (1990).
86. J. W. McDonal et al., Rare earth (cerium oxide) pneumoconiosis: Analytical scanning electron microscopy and literature review. *Mod. Pathol.*, **8**(8), 859–865 (1995).
87. S. B. Beaser et al., *J. Clin. Invest.* **21**, 447 (1942).
88. L. Gerhardsson et al., *Sci. Total Environ.* **37**, 233 (1984).
89. Z. Weifang et al., Investigation on arteriosclerosis among population in a rare earth area in South China. *Biol. Trace Elem. Res.* **59**, 93–98 (1997).

90. Z. I. Israelson, ed., *Toxicology of Rare Metals*, Moscow, 1963; translated by Mrs. Irene Campbell as cited in G. Clayton and F. Clayton, eds., *Patty's Industrial Hygiene and Toxicology*, 3rd ed., Wiley, New York, 1981, pp. 195–208.
91. N. S. MacDonald et al., *J. Biol. Chem.* **195**, 837 (1952).
92. American Conference of Governmental Industrial Hygienists (ACGIH), *Documentation of the Threshold Limit Values and Biological Exposure Indices*, 6th ed., ACGIH, Cincinnati, OH, 1991.
93. H. E. Tebrock and W. Machle, Exposure to europium-activated yttrium orthovanadate: A cathodluminescent phosphor. *J. Occup. Med.* **10**(12), 692–696 (1968).
94. Occupational Safety and Heath Administration (OSHA), *Permissible Exposure Limit*, 29 CFR 1910.1000, Table Z-1. OSHA, Washington, DC, 1996.
95. A. A. Volokh et al., Phosphorus fertilizer production as a source of rare earth elements pollution of the environment. *Sci. Total Environ.* **95**, 141–148 (1990).

CHAPTER FORTY-FOUR

Phosphorus, Selenium, Tellurium, and Sulfur

Eula Bingham, Ph.D.

Phosphorus and sulfur are elements 15 and 16 in the periodic chart and selenium and tellurium are in the same group as sulfur.

A PHOSPHORUS

1.0 Phosphorus

1.0.1 CAS Number: [7723-14-0]

1.0.2 Synonyms: Red Phosphorus; phosphorus, red, amorphous; phosphorus (yellow); yellow phosphorus; elemental white phosphorus; WP; phosphorus, amorphous, red; Red nip; Bonide Blue Death Rat Killer, White Phosphorus; phosphorus (white); phosphorus (yellow or white); phosphorus atom; phosphorous, yellow/white

1.0.3 Trade Names: NA

1.0.4 Molecular Weight: 30.97

1.1 Chemical and Physical Properties

The properties of phosphorus and some selected phosphorus compounds are presented in Table 44.1. Yellow or white phosphorus ignites spontaneously in air at 34°C. It should be

---

Patty's Toxicology, Fifth Edition, Volume 3, Edited by Eula Bingham, Barbara Cohrssen, and Charles H. Powell.
ISBN 0-471-31934-1   © 2001 John Wiley & Sons, Inc.

Table 44.1. Physical and Chemical Properties of Selected Phosphorus Compounds

Phosphorus Compound	CAS#	Physical State	Mol. Wt.	Specific Gravity	MP (°C)	BP (°C)	Solubility
Phosphorus, P	[7723-14-0]	Yellow (white) wax-like solid or red (reddish) brown powder	30.97	1.82 (20°C)	44	280	0.0003 g in 100 mL of water; very soluble in carbon disulfide, ether, chloroform, and benzene
Phosphorus pentoxide (phosphoric anhydride), P_2O_5	[1314-56-3]	White fluffy powder	283.9	2.39	347		Very soluble in water; reacts violently with evolution of heat to form phosphoric acid, H_3PO_4
Phosphoric acid (orthophosphoric acid), H_3PO_4	[7664-38-2]	Clear, thick liquid or deliquescent crystals	98	1.83	42.3	213	Soluble in water and ethanol
Tetraphosphorus trisulfide (phosphorus sesquisulfide), P_4S_3	[1314-85-8]	Yellow crystals	220.12	2.03	172.5	407.5	Insoluble in cold water water; decomposes in hot water
Phosphorus pentachloride PCl_5	[10026-13-8]	Pale yellow, fuming solid	208.21		148	160 sublimes at 100	Decomposes in water to form phosphorus oxychloride, phosphoric acid, and HCl
Phosphorus oxychloride, $POCl_3$	[10025-87-3]	Clear colorless fuming liquid	153.35	1.645	1.25	105.8	Decomposes in water to form phosphoric acid and hydrogen sulfide
Phosphorus pentasulfide, P_2S_5 (generally exists as P_4S_{10})	[1314-80-3]	Light yellow crystals	222.29 or 444.5	2.09	286	513	Decomposes in water to form phosphoric acid and hydrogen sulfide
Phosphorus trichloride, PCl_3	[7719-12-2]	Colorless liquid	137.39	1.574 (21°C)	−112	75.5	Reacts with water to produce HCl and H_3PO_4
Phosphine, PH_3	[7803-51-2]	Colorless gas	34.04	1.17 (air = 1)	−133.5	87.4	26 mL of gas dissolves in 100 mL of water at 17°C

stored under water. Under this condition, however, it may form phosphoric acid. Stainless steel containers should be used to hold the corrosive material. White phosphorus fires can be controlled by using water or sand or by excluding air.

1.1.1 General

Phosphorus (P) is not found naturally in the elemental state. The name is from the Greek for "light-bring." Phosphorus is found in all life forms as phosphates. Phosphates are essential for energy transfer reactions. Of major importance is adenosine triphosphate (ATP) that is involved in nearly all metabolic reactions. Phosphates are an important ingredient of bone. The human skeleton contains about 1.4 kg as calcium phosphate. Phosphates also form part of a number of coenzymes and nucleic acids. Most of the available forms are phosphates.

1.1.2 Odor and Warning Properties

Acrid fumes in air.

1.2 Production and Use

Elemental phosphorous is produced as a by-product or intermediate in the production of phosphate fertilizer. Environmental contamination with phosphorus results from its manufacture into phosphorus compounds and during the transport and use of these compounds. In the manufacturing process, phosphate rock containing the mineral apatite (tricalcium phosphate) is heated, and elementary phosphorus is liberated as a vapor. Phosphorus is used to manufacture explosives, incendiaries, smoke bombs, chemicals, rodenticides, phosphor bronze, and fertilizer. The use of phosphate fertilizers results in increased nutrients in fresh water and is a major environmental pollution problem.

Phosphorus exists in several allotropic forms: white (or yellow), red, and black (or violet). The last is of no industrial importance. Elemental yellow phosphorus extracted from bone was used to make "strike-anywhere" matches. In 1845, the occupational disease "phossy jaw," a jaw bone necrosis, was recognized in workers who manufactured such matches. A prohibitive tax imposed in 1912 on matches made from yellow phosphorus led to the use of less toxic materials, red phosphorus and phosphorus sesquisulfide. The United States appears to have lagged behind European countries in that signatories of the Berne Convention of 1906 agreed not to manufacture or import matches made with yellow phosphorus. Occasional injuries continued to result from using yellow phosphorus to manufacture fireworks until 1926, when an agreement was reached to discontinue using yellow phosphorus for this purpose.

The world production of elemental phosphorus exceeds 1,000,000 metric tons. It is manufactured either in electric or blast furnaces. Both depend on silica as a flux for the calcium present in the phosphate rock. Nearly all of the phosphorus produced is converted into phosphoric acid or other phosphorus compounds (1).

Red phosphorus does not ignite spontaneously but may be ignited by friction, static electricity, heating, or oxidizing agents. Handling it in an aqueous solution helps prevent fires.

1.3 Exposure Assessment

1.3.1 Air

The separation of free phosphorus from many phosphorus compounds in the air is a major problem. Air samples may be obtained using an impinger containing xylene, depending on the sampling rate. The exit side of the impinger should be equipped with a filter to catch fumes that may pass through the xylene (2). The collected samples can be analyzed by gas (flame photometric detector) chromatography (3). NIOSH Analytical Method 7905 is recommended for determining workplace exposures and uses the methodology described here.

Alternative methods of collection using filtration procedures have been suggested. Likewise, other methods of analysis are available, based mostly on the conversion of phosphorus to phosphate, which is then treated with ammonium molybdate to form a colored complex (4).

1.4 Toxic Effects

Phosphorus (white-yellow) can be absorbed through the skin, respiratory tract, and gastrointestinal (GI) tract. Experimental investigations in rats show the highest retention 5 days after oral administration in the liver, skeletal muscle, GI tract, blood, and kidney. Phosphorus is converted to phosphates in the body. Urinary excretion, the chief mode of elimination, is largely as organic and inorganic phosphates (5).

1.4.2.3 Experimental Studies

1.4.2.2.3 Pharmacokinetics, Metabolism, and Mechanisms. Experimental studies in rats demonstrate the highest retention in the liver, skeletal muscle, gastrointestinal tract, blood and kidney after oral administration. Phosphorus is converted to phosphates in the body (6). Phosphorus (yellow) is absorbed from the respiratory and gastrointestinal tracts (7).

1.4.2.3.1 Acute Toxicity. The human lethal oral dose of phosphorus (white) is about 1 mg/kg body weight, and as little as 0.2 mg/kg may produce adverse effects (8). Acute oral phosphorus intoxication generally has two stages. In the initial phase, GI effects predominate and may include nausea, vomiting, and belching. The onset may be within 30 min after ingestion. Death from cardiovascular collapse can occur in about 12 h. A period of regression and apparent recovery that lasts about 2 days may occur. The second stage is characterized by the return of the GI distress plus signs of hepatic, renal, and cardiovascular problems, for example, jaundice, pitting edema, oliguria, high pulse rate, and low blood pressure. In either phase, smoking, luminescence, and a garlic odor of the vomitus and feces are characteristic but not diagnostic. The most common pathological finding in deaths has been fatty degeneration of the liver and kidneys (9).

Acute yellow phosphorus poisoning from rat poison consists of garlic odor, mucosal burns, and phosphorescent vomitus or feces that occurs from a few minutes to 24 h after ingestion and vary in occurrence. These initial symptoms are related to GI or central nervous system (CNS) effects. Mortaliy rates of 23% were associated with the early symptoms of vomiting or abdominal pain, whereas the mortality rate was 73% for victims who had the initial manifestations related to CNS effects, consisting of restlessness, irritability, drowsiness, stupor, or coma. The mortality rate for patients who had both GI- and CNS-related symptoms was 47% (10). These findings differ somewhat from the classical three stages of phosphorus poisoning. McCarron ct al. (10) recommend gastric lavage with a 1:5000 solution of potassium permanganate. They did not recommend treating with copper sulfate, which reacts to form insoluble copper phosphide, because the dose of copper sulfate is near the lethal dose (15 g) for that compound.

1.4.2.3.2 Chronic and Subchronic Toxicity. Chronic toxicity after exposure to phosphorus vapors has occurred in factory workers. Generally, the onset of signs occurs only after many years of exposure. Chronic poisoning after ingestion is rare. Cases of occupational exposure in which the onset of signs has occurred after the personnel have ceased working in a phosphorus-contaminated area have been reported. Early symptoms of chronic intoxication include gastrointestinal distress and sometimes a phosphorus odor (garlic-like) of the breath. Slight jaundice is not uncommon. The classical effect of chronic phosphorus intoxication is on the bone. Most typically this involves the jaw and leads to the so-called "phossy-jaw" or necrosis of the jaw (11).

Phosphorus necrosis of the jaw has been reported among workers in the chemical and fireworks industries (11–13); radiological evidence of bone damage has been observed as one component of phosphorus poisoning (14). Dental and medical surveillance of employees for 40 years at a phosphorus-producing and fertilizer manufacturing plant did not reveal any deterioration of health; the absence of "phossy jaw" or Lucifer's jaw in these employees probably emphasized the importance of scrupulous dental hygiene (15). In chronic phosphorus intoxication, hypophosphatemia blood levels, hypocalcemia, or increased involvement were generally not characteristic (12). Others state that "phossy jaw" may be accompanied by signs of liver and kidney damage (7, 16).

In chronic phosphorus intoxication, lowered potassium blood levels or increased chloride concentrations along with leukopenia and anemia have been reported. Hepatic and renal involvement are generally not characteristic (12). Hematologic changes have been produced in experimental animals as have the classical effects on bone. In addition, hepatic degeneration leading to cirrhosis from chronic oral administration of phosphorus has been produced in experimental animals (17, 18).

In general, red phosphorus is considered much less toxic than white. Experimental animals subjected to red phosphorus have shown liver and kidney effects similar to those reported for white phosphorus (19). However, exposure data are nonexistent for humans. The cellular basis for the toxic effects of phosphorus is probably its reducing properties, which may disturb intracellular oxidative processes.

1.4.2.3.7 Other: Neurological, Pulmonary, Skin Sensitization, etc. Skin contact with white phosphorus results in severe and painful burns. Usually the affected area turns grayish white and infection ensues. Treatment should be prompt to minimize deeper penetration. Ocular damage may result from direct contact. Animal studies have shown that liver and kidney damage may result from dermal application (20). Sudden deaths after skin contact with yellow phosphorus have been reported (21).

Acute inhalation of white phosphorus vapors or smoke that usually contain phosphorus pentoxide at concentrations of about 0.035 mg/L, has resulted in respiratory tract irritation. In at least one case, hepatic involvement was reported (22). In humans, short-term exposure, 10 to 15 min, at concentrations above 400 mg/m^3 has resulted in signs of respiratory tract irritation (23).

1.5 Standards, Regulation, or Guidelines of Exposure

Occupational Safety and Health Administration (OSHA) time-weighted average (TWA) permissible exposure limit for phosphorus (yellow) is 0.1 mg/m^3 (24). The American Conference of Governmental Industrial Hygienists (ACGIH) adopted TLV is 0.02 ppm and no standards for red phosphorus have been established.

Suitable skin and eye protection should be required. Pre-employment medical examination should include X ray studies of the teeth and jaw, and good dentition should be required for placement. Routine dental examination should be monthly if exposure is high or prolonged. Dental work, fillings, and extractions should be followed by exclusion from exposure for several months, and any suspicion of jaw injury should result in permanent removal from phosphorus exposure.

B PHOSPHORUS COMPOUNDS

2.0 Phosphorus Pentoxide

2.0.1 CAS Number: [1314-56-3]

2.0.2 Synonyms: Phosphorous pentaoxide; phosphoric; phosphoric anhydride; phosphoric pentoxide; phosphorus (V) oxide; Phosphorus oxide; diphosphorus pentaoxide; diphosphorus pentaoxide, phosphorous pentoxide (phosphorous anhydride)

2.0.3 Trade Names: NA

2.0.4 Molecular Weight: 283.889

2.0.5 Molecular Formula: P$_2$O$_5$

2.1 Chemical and Physical Properties

Properties of phosphorus pentoxide are listed in Table 44.1.

2.2 Production and Use

Phosphorus pentoxide or phosphoric anhydride (P$_2$O$_5$) is formed by burning yellow phosphorus in dry air or oxygen. It has great affinity for water and is used as a desiccating agent.

2.4 Toxic Effects

Phosphorus pentoxide is corrosive and irritating to mucosal surfaces, eyes, and skin. The resulting phosphoric acid is less harmful than sulfuric acid. See section 3.0 on phosphoric acid.

2.4.2.3 Epidemiology Studies. Pulmonary function was studied in phosphorus refinery workers. The study group consisted of 131 workers employed for 3 to 46 years at a refinery where phosphorus rock was processed to elemental phosphorus. They were typically exposed to phosphoric acid, phosphorus oxides, fluorides, and coal-tar-pitch volatiles (CTPVs). The subjects were followed for up to 7 years during which they received annual evaluations of pulmonary function. Annual changes in pulmonary function were analyzed by logistic regression techniques to assess any effects of occupational exposure. The maximum concentrations of phosphorus pentoxide, fluorides, and CTPVs in the refinery air were 2.23, 4.21, and 1.04 mg/m^3, respectively. Fifty-five subjects were current smokers, 39 were former smokers, and 37 were nonsmokers. Regression analysis of individual mean values of forced vital capacity (FVC), 1-second forced expiratory volume (FEV$_1$), and mild expiratory flow (FEF25-75) for predicted values on years of exposure indicated no statistically significant decreases among nonsmokers and former smokers. Significant decreases in FVC, FEV$_1$, and FEF25-75 were found among the smoking workers. Longitudinal and cross-sectional regression of the FVC, FEV$_1$, and FEF25-75 data for the smokers found no consistent exposure-related effects after adjusting for age and smoking. Occupational exposures in the refinery do not contribute to annual changes in pulmonary function in the workers. The observed deficits in smokers reflect the effects of smoking (25).

2.5 Standards, Regulations, or Guidelines of Exposure

The hygienic standard recommended in the American Industrial Hygiene Association (AIHA) Hygienic Guide is 1 mg P$_2$O$_5$/m^3 of air (26).

3.0 Phosphoric Acid

3.0.1 CAS Number: [7664-38-2]

3.0.2 Synonyms: white phosphoric acid; orthophosphoric acid; Sonac

3.0.4 Molecular Weight: 97.9951

3.0.5 Molecular Formula: H$_3$PO$_4$

3.1 Chemical and Physical Properties

The properties of phosphoric acid are given in Table 44.1.

3.1.2 Odor and Warning Properties

Odorless liquid.

3.2 Production and Use

Phosphoric acid (orthophosphoric acid), (H$_3$PO$_4$), is a weak acid. It is an intermediate in the manufacture of superphosphate fertilizers. In addition, it may be used in cleaning metals. With regard to this use, impurities in the metals may lead to the formation of phosphine.

3.3 Exposure Assessment

NIOSH Analytical method recommended for determining workplace exposures is #7903 for inorganic acids.

3.4 Toxic Effects

Irritation of the eyes, respiratory tract, and mucous membranes may occur as with any weak acid. There is no evidence that phosphorus poisoning results from contact with phosphoric acid.

3.4.2.3 Epidemiology Studies. Ambulatory and preening activity of rock doves was significantly decreased during the first day of exposure to phosphoric acid. The 1.0 mg/L aerosol did not affect spontaneous activity. No treatment-related effects on activity were seen in either species 6 days after exposure to either aerosol concentration (27). The effects of phosphoric acid aerosols on body weight and on feed and water consumption were studied in black-tailed prairie dogs (*Cynomys ludovicianus*) and rock doves (*Columbia livia*). Prairie dogs and rock doves were exposed to 1.0 or 4.0 mg/L phosphoric acid aerosols in an exposure chamber for approximately 80 minutes daily for 4 and 2 days, respectively . The authors concluded that exposure to phosphoric acid aerosol induces transient decreases in ingestive behaviors accompanied by decreases in body weight gain. The authors suggest that phosphoric acid causes transient irritation and ulceration of mucosal esophageal tissue (28).

3.5 Standards, Regulations, of Guidelines of Exposure

The ACGIH/OSHA TWA is 1 mg H_3PO_4/m^3, and the ACGIH short-term exposure limit (STEL) is 3 mg/m^3 (24). Eye and skin protection should be provided for personnel likely to be involved with accidental spills (29).

4.0 Tetraphosphorus Trisulfide

4.0.1 CAS Number: [1314-85-8]

4.0.2 Synonyms: Phosphorus sesquisulfide; trisulfurated phosphorus

4.0.4 Molecular Weight: 220.1

4.1 Chemical and Physical Properties

Some of the properties of tetraphosphorus trisulfide are given in Table 44.1.

4.2 Production and Use

Tetraphosphorus trisulfide, or phosphorus sesquisulfide (P_4S_3) is used in making matches and friction strips for safety-match boxes.

4.4 Toxic Effects

Phosphorus sesquisulfide is not sufficiently volatile to present a vapor hazard at the temperatures of occupied spaces. The dust or fume is an irritant to the eyes, the respiratory tract, and the skin. Its toxicity, however, is minor compared with that of yellow phosphorus; and serious ill effects other than eczema have not been reported (30).

4.5 Standards, Regulations, or Guidelines of Exposure

No hygienic standard has been suggested.

5.0 Phosphorus Pentachloride

5.0.1 CAS Number: [10026-13-8]

5.0.2 Synonyms: Phosphorous (V) chloride; pentachlorophosphorane; phosphoric perchloride

5.0.3 Trade Names: NA

5.0.4 Molecular Weight: 208.24

5.0.5 Molecular Formula: PCl_5

5.0.6 Molecular Structure:

$$\text{Cl} - \underset{\underset{\text{Cl}}{|}}{\overset{\overset{\text{Cl}}{|}}{P}}\begin{matrix}\text{\tiny{,,,}}\text{Cl}\\ \text{Cl} \end{matrix}$$

5.1 Chemical and Physical Properties

Some of the properties of phosphorus pentachloride are given in Table 44.1.

5.2 Production and Use

Phosphorus pentachloride (PCl_5) is made by reacting yellow phosphorus with chlorine. It is used in chemical manufacturing, and it produces phosphorus trichloride and chlorine when heated, and phosphorus oxychloride, phosphoric acid, and hydrochloric acid when decomposed in water.

5.3 Exposure Assessment

For workplace exposures, NIOSH Analytical Method 5257 is recommended (30a).

5.4 Toxic Effect

Phosphorus pentachloride has a pungent, unpleasant odor, and its vapor or fume is very irritating to all mucous surfaces including the lungs.

5.5 Standards, Regulations, or guidelines of Exposure

The threshold limit value (TLV) as a TWA has been set by ACGIH for PCl_5 at 0.1 ppm. The OSHA PEL and NIOSH REL is 1 mg/m^3. It is prudent to control for the possible breakdown products: phosphorus trichloride, chlorine, phosphorus oxychloride, phosphoric acid, and hydrochloric acid.

6.0 Phosphorus Oxychloride

6.0.1 CAS Number: [10025-87-3]

6.0.2 Synonyms: Phosphoryl chloride; phosphoryl trichloride; phosphoric trichloride; Phosphoryl Oxychloride; Phosphorous Oxychloride (Phosphoryl Chloride)

6.0.4 Molecular Weight: 153.33

6.0.5 Molecular Formula: $POCl_3$

6.0.6 Molecular Structure:

$$Cl-\underset{Cl}{\overset{\overset{O}{\|}}{P}}-Cl$$

6.1 Chemical and Physical Properties

The reaction of phosphorus oxychloride with water and alcohol generates heat. Properties of phosphorus oxychloride are given in Table 44.1.

6.1.2 Odor and Warning Properties

Corrosive material; volatile, colorless to pale yellow; strongly fuming liquid.

6.2 Production and Use

Phosphorus oxychloride ($POCl_3$) is used in manufacturing plasticizers, hydraulic fluids, and gasoline additives (26).

6.4 Toxic Effects

The main effects of phosphorus oxychloride inhalation are on the mucous membranes, and effects as severe as emphysema have been reported. The ACGIH (26) reports that cases of both chronic and acute intoxication have been observed; some instances of nephritis have occurred.

6.5 Standards, Regulations, of Guidelines of Exposure

The OSHA and the ACGIH TWAs for phosphorus oxychloride (POCl$_3$) are 0.1 ppm. The NIOSH REL is also this limit. The ACGIH limit is based largely on analogy to phosphorus trichloride.

7.0 Phosphorus Pentasulfide

7.0.1 CAS Number: [1314-80-3]

7.0.2 Synonyms: Reactive phosphorus pentasulfide; undistilled pentasulfide; phosphoric sulfide; sulfur phosphide; di-phosphorus pentasulfide; phosphorus sulfide; phosphorus (V) sulfide

7.0.3 Trade Names: NA

7.0.4 Molecular Weight: 222.29

7.0.5 Molecular Formula: P$_2$S$_5$

7.0.6 Molecular Structure:

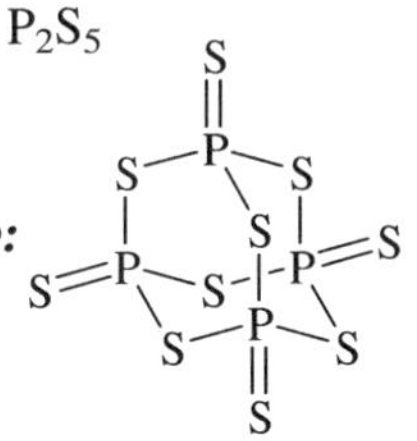

7.1 Chemical and Physical Properties

Some of the properties of phosphorus pentasulfide are included in Table 44.1.

7.2 Production and Use

This sulfide, (P$_2$S$_5$), generally exists as P$_4$S$_{10}$ and is prepared by fusing red phosphorus with sulfur. It is used in manufacturing safety matches and ignition chemicals and in chemical processes for introducing sulfur into organic compounds.

7.4 Toxic Effects

Irritation due to the decomposition products of phosphorus pentasulfide is the major toxicity.

7.5 Standards, Regulations, or Guidelines of Exposure

The ACGIH TWA-TLV and the NIOSH REL are 1 mg/m^3, and the STEL is 3 mg/m^3. OSHA has adopted the same standards (24).

8.0 Phosphorus Trichloride

8.0.1 CAS Number: [7719-12-2]

8.0.2 Synonyms: Phosphorus (III) chloride; chloride of phosphorus; PICI; phosphorus chloride

8.0.3 Trade Names: NA

8.0.4 Molecular Weight: 137.33

8.0.5 Molecular Formula: PCl$_3$

8.0.6 Molecular Structure:

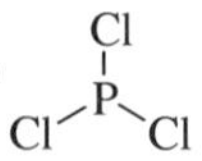

$$\underset{Cl\qquad Cl}{\underset{|}{Cl-P}}$$

8.1 Chemical and Physical Properties

Some of the properties of phosphorus trichloride are given in Table 44.1.

8.1.2 Odor and Warning Properties

Smells like hydrochloric acid, corrosive material, fuming liquid.

8.2 Production and Use

Phosphorus trichloride (PCl$_3$) is made by reacting yellow phosphorus with chlorine and is used in chemical manufacturing. It hydrolyzes to phosphoric acid and hydrochloric acid.

8.3 Exposure Assessment

For workplace exposures, NIOSH Analytical method 6402 is recommended (30b).

8.4 Toxic Effects

Phosphorus trichloride reacts with water exothermically and gives off hydrochloric acid and phosphoric acid gases. Phosphorus trichloride itself volatilizes at room temperature. As a vapor, it is irritating to the skin, eye, and mucous membranes. Severe acid burns can occur. The onset of acute pulmonary edema as a consequence of inhalation exposure may be delayed 2 to 6 h or 12 to 24 h in minor and moderate to severe exposures, respectively.

8.5 Standards, Regulations, or Guidelines of Exposure

The ACGIH and NIOSH recommends 0.2 ppm (1.1 mg of PCl$_3$/m^3) for the TWA and 0.5 ppm for the STEL. The OSHA TWA is 0.5 ppm (24).

9.0 Phosphine

9.0.1 CAS Number: [7803-51-2]

9.0.2 Synonyms: Hydrogen phosphide; phosphorus hydride; phosphorated hydrogen; phosphorus trihydride, Detia gas EX-B; Gas EX-B

9.0.4 Molecular Weight: 33.9975

9.0.5 Molecular Formula: H₃P

9.0.6 Molecular Structure:

$$\begin{array}{c} H \\ | \\ H-P-H \end{array}$$

9.1 Chemical and Physical Properties

Some of the properties of phosphine are included in Table 44.1.

9.1.2 Odor and Warning Properties

Colorless gas with a characteristic, fishy odor.

9.2 Production and Use

Phosphine, also known as phosphorated hydrogen or hydrogen phosphide, (PH₃), has no direct commercial use. However, it may be generated from aluminum or zinc phosphide and water for grain fumigation. It may be present in phosphorus as a polymer or generated at low rates under alkaline conditions and at a temperature of 85°C. The generation of acetylene from calcium carbide containing calcium phosphide as an impurity and metal processing procedures in which phosphides are formed are the most frequent sources of industrial hygiene problems with phosphine.

9.3 Exposure Assessment

9.3.1 Air

Airborne concentrations of phosphine can be collected in a fritted glass bubbler containing silver diethyldithiocarbamate. A complex is formed and analyzed spectrographically (31). Gas chromatography using a flame photometric (phosphorus mode) detector is also applicable (32).

OSHA Analytical Method #ID 180 is recommended to provide for workplace exposure assessments (31a).

9.4 Toxic Effects

9.4.1 Experimental Studies

Phosphine was studied in male and female F344 rats and B6C3F1 mice. In a 4-day pilot study, male rats and mice were exposed to 0, 1, 5, and 10 parts per million (ppm) of phosphine for 6 hours daily, five animals per group. Immediately after the last exposure, animals were sacrificed, and blood, lung, liver, heart, and kidneys were collected. In another 2-week study, male and female rats were exposed to 0, 1.25, 2.5, and 5 ppm of phosphine for 6 hours daily, 5 days a week. All mice exposed to 10 ppm phosphine were moribund after the last exposure. Statistically significant decreases in erythrocyte counts, hemoglobin, and hematocrits occurred in mice exposed to 1 and 10 ppm. Mice exposed to 10 ppm of phosphine showed significant increases in serum urea nitrogen alanine aminotransferase, and sorbitol dehydrogenase compared with controls. Moribund mice showed necrotic kidney lesions, hemorrhagic and necrotic liver lesions, and myocardial degeneration. No mortalities occurred in rats and mice exposed for 2 weeks. No microscopic evidence of treatment-related effects occurred in any tissue of rats and mice exposed to 5 ppm for 2 weeks. The authors concluded that phosphine inhalation lacks target organ specificity in rats and mice and lethality is the primary hazard of subchronic pH_3 exposure (33). The exposure of male F344/N rats and male B63F1-mice to multiple dosing with phosphine during an 11-day period to determine the *in vivo* cytogenetic effects was examined in this study. Exposures were to 1.25, 2.5, or 5 parts per million (ppm) of phosphine, 6 hours a day, on 9 days during an 11-day period. Phosphine inhalation caused no statistically significant increases in sister chromatid exchange (SCE) or chromosomal aberrations (CAs) in peripheral blood lymphocytes (PBLs) or micronuclei (MN) in peripheral blood PCEs or binucleated (BN) lymphocytes. All of the chromosomal aberrations observed were either simple chromatid or chromosomal deletions. Cytogenetic results were similar for the rats and mice. The only phosphine effect noted in the dominant lethal study was a slight decrease in the number of implants per female in mating interval six, which sampled sperm exposed primarily as spermatocytes. Based on the *in vivo* exposure of rats and mice, the authors concluded that PH_3 is at most a weak genotoxic agent and may sometimes cause slight increases in cytogenetic damage when exposures are near toxic levels (34).

9.4.2 Human Experience

Phosphine differs from arsine in that red blood cell hemolysis does not occur. When low concentrations are inhaled, headaches, dizziness, tremors, general fatigue, GI distress, and burning substernal pain may result. Toxic exposures to phosphine have been documented as a result of grain fumigation, attempted suicide, and ferrosilicon decomposition. A productive cough with fluorescent green sputum, acute dyspnea, and pulmonary edema may develop. Death may be preceded by tonic convulsions, which may ensue after apparent recovery. At higher concentrations, death may occur after 1/2 to 1 h of exposure at concentrations of 400 to 600 ppm. Serious effects may be produced by exposure to 5 to 10 ppm for several hours. Phosphine's characteristic decayed fish odor is barely detectable at concentrations of 1.5 to 3 ppm. The chronic effects produced by phosphine are essentially the same as those produced by phosphorus (35, 36).

Aluminum phosphide is used worldwide as a rodenticide to protect stored grain. It reacts with moisture to release the highly toxic gas phosphine. It normally dissipates into the air, leaving little residue. Thus the grain is fit for human consumption. Fifteen cases of aluminum phosphide oral intake poisoning have been reported (37). The clinical and pathological changes were in the GI tract and in the respiratory, cardiovascular, and central nervous systems. These changes have been reported after inhalation exposure to phosphine. Hypotension and hepatic toxicity were a usual feature of these poisonings. The time between ingestion and hospital admission averaged 5.3 h (0.5 to 16 h). The average time between ingestion and death in patients was 31 h (1 to 106 h). The average dose was 1.5 to 90 g (average 4.7 g). Assuming 65 kg of body weight, this amounts to 72 mg/kg and is greater than the 40.5 mg/kg LD_{50} for zinc phosphide in rats.

Of the 1,000,000 U.S. workers possibly exposed to phosphine, 10,000 of them may be exposed aboard ships as a result of generation of phosphine to fumigate grain. Unlike industrial workers, crew members and others living aboard ships may be exposed constantly and for long durations. Aboard-ship exposures may be intensified by the confined nature of the working conditions and life.

The wife of the captain and two children, plus 29 of 31 crew members on a grain carrier, became ill after inhaling phosphine generated from using aluminum phosphide. The first cases occurred 2 days after the fumigation started. The predominant symptoms were fatigue (86%), nausea (72%), headache (66%), vomiting, cough, and shortness of breath. Abnormal physical findings included paresthesia (59%), jaundice (52%), tremor (31%), ataxia, and diplopia. The younger of the two children, who were 2- and a 4-years old, died. Necropsy findings included focal myocardial infiltration, necrosis, pulmonary edema, and widespread small vessel injury. The older surviving daughter had ECG changes and echocardiographic changes indicating heart muscle damage accompanied by an increase of skeletal muscle serum creatinine phosphokinase. She recovered in 24 h. Except for nausea, vomiting, and paresthesia, values of parameters from physical, neurological, and clinical pathological evaluation of the mother were normal; she recovered 24 h after exposure ceased. Concentrations ranged from 30 ppm in the void space near the midship ventilator intake to 0.5 ppm in the living quarters (38). The severity of the effects in the younger child and recovery course of the surviving child compared with that of the mother suggest that children may be especially susceptible to the toxicity of phosphine at low levels. This may be related to the greater intake rate in the young because of a higher rate of metabolism.

9.5 Standards, Regulations, or Guidelines of Exposure

The OSHA TWA is 0.3 ppm (18, 24). The PEL of 0.3 ppm is below the odor threshold of 1.5 to 3 ppm. The ACGIH TLV and the NIOSH REL TWA-TLV (ACGIH) is 0.3 ppm (0.42 mg/m^3). The STEL (15 min) is 1 ppm (1.4 mg/m^3).

C SELENIUM

10.0 Selenium

10.0.1 CAS Number: [7782-49-2]

10.0.2 Synonyms: Selenium atom, selenium powder

10.0.3 Trade Names: NA

10.0.4 Molecular Weight: 78.96

10.0.5 Molecular Formula: Se

10.1 Chemical and Physical Properties

Properties of selenium and some of its compounds are listed in Table 44.2.

10.1.1 General

Selenium exists in three allotropic forms. When heated, the red amorphous powder cools to the vitreous form. A gray form of hexagonal structure is photoconducting and semiconducting. The third form is that of monoclinic red crystals.

10.1.2 Odor and Warning Properties

Appearance and odor vary depending upon the specific compound. Selenium burns in air with a bright blue flame and emits a horseradish-like odor.

10.2 Production and Use

Selenium (Se), a nonmetallic element of the sulfur group, is widely distributed in nature. It is obtained along with tellurium as a by-product of metal or refining, chiefly from copper. About sixteen tons are mined a year globally (39).

Because selenium is present in fossil fuels, up to 90% of the selenium content in ambient air is emitted during combustion of them. Air pollution concentrations averaged from 13 ng/mg^3 in urban areas to 0.38 in remote areas. The mass medium diameter was 0.92 mm. The worldwide emissions of 10,000 tons/year from natural sources exceeds the atmospheric emissions from anthropogenic sources (5100 tons). However, 41,000 tons is emitted into the aquatic ecosystems. The largest contributor is electric power generating plants that produce 18,000 tons; manufacturing processes account for 12,000 tons (39).

Selenium is an essential trace metal. Because of data suggesting that it may inhibit chemical carcinogenesis, it has been widely promoted as a dietary supplement (40). Selenium may replace sulfur and forms selenoproteins in plants and animal systems.

Most of the world's selenium today is provided by recovery from anode muds of electrolytic copper refineries. Selenium is recovered by roasting these muds with soda or sulfuric acid or by melting them with a soda and niter.

One of the important uses of selenium is in photoelectric cells. In addition, it is used in manufacturing rectifiers to convert alternating current to direct current. Its conductivity increases up to 1000 times on exposure to light. The greatest use is as a decolorizer for glass and ceramic. Cadmium selenide was the red pigment that was used for the ruby glass of automobile taillights. It is also used in various alloys and in rubber manufacture. Some compounds have been used as insecticides. Selenium dioxide is the most widely used selenium compound. It is produced by burning selenium in oxygen and is employed to produce of other selenium compounds (1). Approximately 1,000,000 lb. of selenium are used annually in the United States. Use in electronic and photocopier components accounts for 35%, glass manufacturing for 30%, and chemicals and pigments for 25% (41). Radionuclide imaging of the pancreas uses ^{75}Se-selenomethionine.

Toxic gases and vapors may be released in a fire involving selenium. Selenium can react violently with chromic oxide (CrO_3), lithium silicon (Li_6Si_2), nitric acid, nitrogen trichloride, oxygen, potassium bromate, silver bromate, and fluorine.

Table 44.2. Physical and Chemical Properties of Selenium and Selected Compounds

Form of Selenium	CAS Number	Physical State	Atomic or Mol. Wt.	Sp. Gr.	MP (°C)	Solubility
Selenium (Se)	[7782-49-2]	Red amorphous powder turning black on standing and vitreous on heating	78.96	4.3–4.8 (20°C)	217 (b.p. 688)	Soluble in H_2SO_4, ether, chloroform
Selenium dioxide (SeO_2)	[7446-08-4]	White crystalline powder	110.96	3.95	340 under pressure (subl. 315)	Readily sol. in hot or cold water to form selenious acid, H_2SO_3
Selenium trioxide (SeO_3)	[13768-86-0]	Yellowish white hygroscopic powder	126.96	3.6		Readily sol. in water to form selenic acid similar to H_2SO_4
Sodium selenite ($Na_2SO_3 \cdot 5H_2O$)		White powder	80.98		710	Freely sol. in water to form a slightly alkaline solution
Sodium selenate	[10102-18-8]	Colorless crystals, with or without 10 mol H_2O	188.95			Very sol. in water moisture sensitive
Hydrogen selenide (H_2Se)	[7783-07-5]	Colorless gas	80.98	—	−66	0.70 g/100 mL
Selenium hexafluoride (SeF_6)	[7783-79-1]	Gas	193.0	3.25 at −28°C	−39	Insoluble in water
Selenium oxychloride ($SeOCl_2$)	[7791-23-3]	Clear pale yellow liquid	165.87	2.42 at 22°C	176.4	Hydrolyzes in water to HCl and selenious acids

10.3 Exposure Assessment

10.3.1 Air

Particulates that contain selenium can be collected on filters treated with nitric acid to extract the selenium. Gases or vapors may be scrubbed through 40 to 48% hydrobromic acid containing 5 to 10% free bromine. Soda lime has been used to collect hydrogen selenide. Selenium dioxide can be collected in 10 mL of water with a midget impinger.

The use of an atomic absorption spectrophotometer with an argon hydrogen flame and an electrodeless discharge lamp for selenium to provide a characteristic selenium line at 196.0 nm is recommended (42). Gravimetric, volumetric, colorimetric, or spectrophotometric methods have been used for analysis.

To determine workplace exposures to selenium NIOSH Analytical Methods 7300 and S190 are recommended (30b, 42a).

10.4 Toxic Effects

Selenium is an essential element. It interacts with a wide variety of vitamins, xenobiotics, and sulfur-containing amino acids. Selenium reduces the toxicity of many metals such as mercury, cadmium, lead, silver, copper, and arsenic (29, 43).

Selenium and most of its compounds are readily absorbed by oral intake or by breathing. Dermal exposure generally does not result in elevated selenium blood concentration. After absorption, high concentrations are found in the liver and kidney. In humans, dimethylselenide is formed and may account for the garlic odor of the breath. Selenium crosses the placenta in rats, dogs, mice, and humans. Trimethylselenium is a predominant urinary metabolite, at least in rats. Urinary and fecal excretion of selenium accounts for about 50% of the total selenium output. At higher oral doses, excretion in the exhaled air becomes more important. In humans as in rats, elimination of selenium is triphasic. The half-life of the first phase in humans is about 1 day; the second phase has a $t_{1/2}$ of 8 to 9 days; and the third phase has a $t_{1/2}$ of 115 to 116 days. The first two elimination phases represent the fecal excretion of unabsorbed selenium and the urinary excretion of absorbed but unmetabolized material (43). In rats, selenium is highly concentrated in bile relative to plasma. The liver and kidney contain the highest concentrations of selenium. Selenium is removed predominantly by the urinary tract (44). In the liver, many selenium compounds are biotransformed to excretable metabolites. Identified metabolites are trimethylselenide in urine and dimethylselenide in breath (45).

In farm animals (cattle, sheep, hogs, and horses), toxicity from intake of feed containing excessive selenium has resulted in blind staggers and/or alkali disease, characterized by anorexia, emaciation, and collapse. At lower concentrations, loss of hair and hooves has occurred, and reproductive capacity was reduced. Pathology includes hepatic necrosis, nephritis, hyperemia, and ulceration in the upper GI tract. In experimental animals, liver injury, disturbance of the endocrine system, anemia, nephritis, myocarditis, and pancreatitis have been produced (46).

10.4.1 Experimental Studies

Organic selenoamino acids are among the most toxic forms of ingested selenium and are more toxic to primates than rodents. In a 30-day oral study, the maximum dose of

L-selenomethionine tolerated by female macaques (Macaca fascicularis) was 188 mg/(kg)(day). Signs of adverse effects included hypothermia, weight loss, dermatitis, and disturbance in menstruation. The addition of dietary supplements and fruits was necessary to prevent death at this dose (47).

10.4.1.2 Chronic and Subchronic Toxicity. Most concern has related to the possible effects of chronic oral intake and arises from results of long-term oral feeding studies in rodents. The U.S. governmental authorities have now accepted the fact that the use of selenical feed additives does not pose a carcinogenic hazard for humans. The industrial evidence is a little clearer with regard to this question. A series of oral rodent bioassays was performed on selenium compounds in the period 1943 to 1977. The results were mixed, some indicated carcinogenic potential (most frequently in the liver), and some showed a reduction of cancer.

10.4.1.4 Reproductive and Developmental. Both inorganic and organic selenium compounds cross the rodent, cat, sheep, and non human primate placenta (48, 49) Fetal retention of organic selenium is greater than that seen after equivalent exposure to inorganic selenium (50). Selenium accumulated in hamster embryos after doses that failed to elicit embryotoxicity (51). Although environmental levels of selenium have been associated with embryonic mortality and teratogenesis in birds, especially in the western United States, few investigators other than Robertson (52) have shown that it is a problem in humans or animals. Tarantal et al. (53) tested L-selenomethionine in cynomolgus monkeys. Embryonic toxicity occurred in three monkey fetuses at 300 mg/kg. All of the dams were compromised at this dose, and one of the three dams died.

10.4.1.5 Carcinogenesis. In 1975, the International Agency for Research on Cancer (IARC) considered data on sodium selenate, sodium selenite, and organic forms of selenium contained in plants and judged that the available data did not suggest that selenium is carcinogenic to humans (54). Since then, the National Toxicology Program (NTP) has shown that selenium sulfide, a material used in shampoo, but not absorbed through the skin, causes increased hepatocellular cancer in both rats and mice and alveolar/bronchial carcinomas and adenomas in mice when administered orally (55). Dermal administration to mice did not produce significantly more tumors in the treated than in the controls (56), although the incidence of lung tumors in the males was increased by a prescription dandruff shampoo containing 2.5% selenium sulfide (57). The latter could be the result of grooming by the test animals.

10.4.2 Human Experience

Inorganic (58, 59) or organic (60, 61) forms of selenium are rapidly absorbed from the human gut, and the bioavailability is 84 to 97%, respectively (59, 62). Following a single oral 200-μg dose of sodium selenite in six male and female volunteers, the terminal plasma elimination half-time was 200 to 285 hours and that for tissues was 115 to 285 days. After 12 days, 65% of the administered selenium remained in the body, and at 90 days, approximately 35% was still retained (62).

Selenium dioxide is the primary compound involved with most industrial exposure to the element because the oxide is formed when selenium is heated. The dioxide itself forms selenious acid with water or sweat, and the acid is an irritant. Selenium dioxide splashed in the eye or entering the eye because of airborne concentrations causes the development of a pink allergic-type reaction of the eyelids, which may become puffy. There is usually conjunctivitis of the palpebral conjunctiva (63). Two hours after a fatal suicidal ingestion of selenium dioxide, a 17-year-old male was admitted to a hospital with asystolia and apnea. Autopsy findings included congestion of the lungs and kidneys, diffuse swelling of the heart, and brain edema. The most striking finding was an orange-brown discoloration of the skin and all viscera, probably due to hemolysis. The selenium blood and tissues levels were 100 to 1000 times normal. The concentration in the pancreas was remarkably high. Tissue concentrations were predominantly elemental selenium or selenium disulfide, although the authors hypothesized that the discoloration was due to hemolysis caused by hydrogen selenide formed metabolizing of selenium (64).

A case of alopecia universalis due to occupational selenium intoxication was reported. The patient was a 31-year old male who had a 4-year history of working in a photocopy machine manufacturing facility. He worked in an area where a selenium alloy was used to coat photocopy drums. He reported noticing hair loss along with nail brittleness about 6 months after beginning work. He later lost all of the hair on his body, including eyebrows and lashes. Various treatments had been temporarily successful, but he lost all of his hair again after returning to work. His blood selenium level was 0.5 µg/mL, and his nail selenium level was 2.04 µg/g. Comparison of selenium values from this patient with those from other workers from the same factory, but not exposed to selenium, demonstrated increased levels in this patient, although they were below those reported for clinical selenosis. The authors concluded that the alopecia in this subject may have been caused by his exposure to selenium at work (65).

Acute selenium toxicity is primarily limited to inhalation of selenium dust or fumes, which are irritants. Oral doses of 1 to 2 mg can rapidly be fatal to animals. Two deaths caused by the ingestion of gun bluing solutions containing selenious acid have been reported. In one case the estimated dose was 10 to 20 mg/kg of selenious acid. Hypertension and garlicky breath (from dimethylselenide) were the major signs on hospital admission 1 h after ingestion. Autopsy findings on day 8 indicated death from respiratory problems, chiefly edema and hemorrhage. The patient's serum concentration and urinary excretion of selenium were 20 times normal. However, 24-h selenium levels of 100 mg have been associated with vomiting and other gastrointestinal problems. Other clinical findings of acute inorganic selenium toxicity included vomiting, diarrhea, labored breathing, weakness, unsteady gait, and coma. The histological evaluation may include pulmonary edema, hepatic necrosis, skeletal muscle degeneration, renal tubular hydropic degeneration, and swelling and disruption of myocardial mitochondria (39).

10.4.2.3 Epidemiology Studies. Symptoms and signs of occupational disease resulting from long-term industrial exposure to selenium were recognized early. These were bronchial irritation, gastrointestinal distress, nasopharyngeal irritation, and a persistent garlic odor on the breath (66). This odor, however, is not reliable and may be present when urinary excretion is as high as 0.1 mg/L. The normal value is about 0.05 mg/L, but can vary somewhat because of large differences in dietary content.

Human whole blood cultures were exposed to selenium dioxide and selenium in other valence states. The ability of the selenium compound to induce sister chromatid exchanges were, in decreasing order of their effectiveness, selenium, selenium dioxide, sodium selenide, sodium selenite, and sodium selenate (67).

Both experimental evidence and epidemiological studies support the contention that selenium increases dental caries. This aspect of selenium industrial toxicity has not been investigated.

10.4.2.3.2 Chronic and Subchronic Toxicity. A metallic taste has also been reported frequently in instances of chronic industrial exposure. Other signs include pallor, irritability, and excessive fatigue (63). There have been only occasional associations made to suggest that either industrial or environmental exposure of humans may result in the renal or hepatic lesions that have been produced experimentally by selenium (68, 69).

10.4.2.3.5 Carcinogenesis. Several studies have demonstrated an increased risk of malignant disease and mortality associated with reduced circulating selenium concentrations (76–81). In a 26-year period of selenium rectifier processing in Britain, the cancer deaths were comparable to the national average (63). In addition, epidemiological studies in geographic areas of low and high selenium (forage and blood concentrations) levels suggested an inverse relationship between selenium and cancer in humans (76, 82). NTP (83) assigned selenium sulfide to Group 2, substances reasonably anticipated to be a carcinogen.

10.4.2.3.6 Genetic and Related Cellular Effects Studies. An accidental exposure of female laboratory workers to selenite resulted in termination of several pregnancies by miscarriages and one bilateral clubfoot infant (52). Although results of investigations in the chick embryo and experience in farm animals support this possibility (70), experimental investigations in the hamster failed to show a teratogenic potential (71). Epidemiological investigations have actually suggested that neonatal deaths declined as the environmental test selenium concentration rose (72). Positive mutagenic effects in various experimental test systems are supported by reports of excess deaths before weaning, failure to breed, and an unusual sex ratio when selenium is incorporated into the drinking water of rats and mice for three generations, (70, 73–75).

10.4.2.3.7 Other: Neurological, Pulmonary, Skin Sensitization, etc. Dermal and oral exposure to some selenium compounds may present a risk of cancer. Medical examination should be made available annually to each employee who is exposed to selenium and its inorganic compounds at potentially hazardous levels. Persons with a history of asthma, allergies, or known sensitization to selenium, or with a history of other chronic respiratory disease, GI disturbances, disorders of liver or kidneys, or recurrent dermatitis would normally be at increased risk from exposure. Special consideration should be given to women of childbearing age, because the possibility that selenium may be teratogenic could place these women in a high-risk group. Because selenium causes liver damage and tumors in animals, a profile of liver function should be obtained. Whole blood selenium levels in individuals range from 1.33 to 7.5 mg/mL.

Normal levels in the United States are 0.1 mg/mL, but no biological exposure indexes are established.

The maximum contaminant level (MCL) of selenium in drinking water is 0.01 mg/L (40 CFR 141.11).

10.5 Standards, Regulations, or Guidelines of Exposure

Most of the toxicological information related to selenium itself is applicable to selenium salts. The analytical procedures are useful in determining the airborne concentration of most materials. The ACGIH TLV-TWA for selenium and its compounds is 0.2 mg Se/m^3. The current OSHA standard is 0.2 mg/m^3. IARC currently classifies selenium and its compounds as a Group 2B carcinogen (probably carcinogenic in humans, but having [usually] no human evidence), whereas they classify selenium sulfide as a Group 3 carcinogen (sufficient evidence in experimental animals). The NTP indicates that selenium sulfide is "reasonably anticipated to be a carcinogen with limited evidence in humans or sufficient evidence in animals" (24).

D SELENIUM COMPOUNDS

11.0 Hydrogen Selenide

11.0.1 CAS Number: [7783-07-5]

11.0.2 Synonyms: Hydroselenic acid; selenium hydride

11.0.4 Molecular Weight: 80.98

11.0.5 Molecular Formula: H$_2$Se

11.0.6 Molecular Structure: H–Se–H

11.1 Chemical and Physical Properties

See Table 11.2.

11.1.2 Odor and Warning Properties

Colorless gas with a very offensive odor.

11.2 Production and Use

Hydrogen selenide (H$_2$Se) may be formed by the reaction of acids or water with metal selenides or wherever nascent hydrogen is in contact with soluble selenide compounds. It has no commercial use.

11.4 Toxic Effects

The results of acute hydrogen selenide inhalation intoxication in humans include irritation of the respiratory tract mucous membranes, pulmonary edema, severe bronchitis, and bronchial pneumonia. Accidental inhalation of hydrogen selenide has produced upper respiratory tract irritation and wheezing, followed by progressive dyspnea and reduced expiratory flow rates in 18 h. The patients improved during 5 days, but some had pulmonary function changes that remained three years later (84). Similar signs have been observed in rats at concentrations of 1 to 4 µg/L (85). Five cases of subacute intoxication from less than 0.2 ppm of hydrogen selenide, probably generated from the use of selenious acid in an etching and printing operation, have been reported (86). Gastrointestinal distress, dizziness, increased fatigue, and a metallic taste in the mouth were reported. A single case has been reported in which exposure to a high concentration caused severe hyperglycemia, controllable by increasing doses of insulin (87).

11.5 Standards, Regulations or Guidelines of Exposure

The OSHA PEL TWA, NIOSH REL, and ACGIH (24) TLVs are 0.05 ppm (0.16 mg/m^3). Hydrogen selenide has an offensive odor somewhat resembling that of decayed horseradish. In the lower toxic range, this odor is not a dependable warning. Eye and nasal irritation are moderate.

12.0 Selenium Hexafluoride

12.0.1 CAS Number: *[7783-79-1]*

12.0.4 Molecular Weight: 192.96

12.0.5 Molecular Formula: SeF$_6$

12.1 Chemical and Physical Properties

See Table 44.2.

12.2 Production and Use

Selenium hexafluoride is used as a gaseous electric insulator (88). It may be prepared by passing gaseous fluorine over finely divided selenium in a copper vessel. Industrial selenium exposure is common for makers of arc light electrodes, electric rectifiers, and semiconductors (89).

12.4 Toxic Effects

Short-term exposures of experimental rodents to airborne concentrations of selenium hexafluoride produced signs of pulmonary edema. Exposure to 5 ppm for 4 h was the lowest effect level, and 1 ppm showed no effect. In a 5-day, 1-h exposure regimen, 5 and 1 ppm produced similar results (90).

12.4.1 Experimental Studies

12.4.1.1 Acute Toxicity. The acute toxicity of selenium hexafluoride was studied in a few animals (rabbit, guinea pig, rat, and mouse) at 100, 50, 25, 10, 5, and 1 ppm for 4 hours. Those exposures of ≥10 ppm were fatal. Inhalation of 5 ppm for 4 hours resulted in pulmonary edema, but 1 ppm was without observable effects. Repeated exposures of 1 hour each for 5 days at 5 ppm and 1 ppm resulted in signs of pulmonary injury (90).

12.4.2.3 Epidemiology Studies. No data on the toxicity of occupational or environmental exposure to selenium hexafluoride per se in humans were located in the open literature.

12.5 Standards, Regulations, of Guidelines of Exposure

The National Institute for Occupational Safety and Health (NIOSH) (29) indicated that 2 ppm is immediately dangerous to life or health (IDLH). The OSHA PEL TWA, the NIOSH REL, and ACGIH TWA-TLV are all 0.05 ppm measured as selenium (24).

13.0 Selenium Oxychloride

13.0.1 CAS Number: [7791-23-3]

13.0.2 Synonyms: Seleninyl chloride; selenium chloride oxide; selenium (IV) oxychloride; selenium dichloride oxide

13.0.4 Molecular Weight: 165.87

13.0.5 Molecular Formula: $SeOCl_2$

13.1 Chemical and Physical Properties

See Table 44.2.

13.2 Production and Use

Selenium oxychloride ($SeOCl_2$) is a powerful solvent, chlorinating agent, and resin plasticizer used in the chemical industry.

13.4 Toxic Effects

Selenium oxychloride is strongly vesicant and will rapidly destroy the skin upon contact unless immediately removed by washing. Less than 0.1 mL on the skin of rabbits has proved fatal within 24 h, and selenium was found in the blood and liver. The minimum lethal dose, when applied to the skin of a rabbit, is 7 mg/kg. This would be equivalent to approximately 0.2 mL applied to a man of average size. The application of less than 0.005 mL to the arm of a man caused a painful burn with swelling, and healing required a month. The vapors of selenium oxychloride are toxic, but their irritant and corrosive actions on the respiratory tract are not so great as might be supposed because the vapor readily decomposes in air. Its low vapor pressure also, limits the concentration possible in air (91).

13.5 Standards, Regulations, or Guidelines of Exposure

No health standard has been published.

14.0 Sodium Selenite

14.0.1 CAS Number: [10102-18-8]

14.0.2 Synonyms: Selenious acid; disodium salt; disodium selenite; disodium selenium trioxide; sodium selenite ($Na_2SeO_3 \cdot 5H_2O$)

14.0.4 Molecular Weight: 172.94

14.0.5 Molecular Formula: $Na_2SeO_3 \cdot 5H_2O$

14.0.6 Molecular Structure:

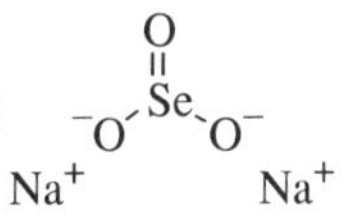

14.1 Chemical and Physical Properties

See Table 44.2.

14.4.1 Experimental Studies

14.4.1.1 Acute Toxicity. Oral LD_{50} values for sodium selenite of 4.8 to 7.0 mg selenium/kg body weight in rats have been reported (92).

14.4.1.2 Chronic and Subchronic Toxicity. Male Sprague–Dawley rats fed 2.5 to 10 ppm sodium selenite showed dose toxicity; all rats fed 10 ppm died within 29 days, however, rats given 2.5 ppm had normal growth. Rats fed 5 ppm exhibited reduced body weight gain, and dose-related selenium concentrations were found (93). Yorkshire–Hampshire pigs fed 0.1, 5, 10, 20, 45, or 100 ppm sodium selenite for up to 84 days had reduced body weight gain, emesis, anorexia, diarrhea, lethargy, dyspnea, and hypothermia (94).

14.4.1.4 Reproductive and Developmental. The teratogenicity of selenium was examined by using a postimplantation rat embryo culture. Wistar-rat embryos at day 9.5 of gestation were cultured for 48 hours in the presence of sodium selenite, sodium selenate, seleno-DL-methionine (SeMet) and seleno-DL-cystine (SeCys). These compounds were embryolethal at 20, 300, 1000, and 1000 (sic) micromolar (μM) selenium, respectively. Embryonic viability and growth were significantly reduced above the 20 μM level. Selenite significantly increased the incidence of abnormal embryos also, the abnormalities were mainly in the head region, including hypoplastic telencephalon, swollen rhombencephalon, deformed optic vesicle, and deformed foci vesicle. Selenate slightly reduced embryonic growth and significantly increased the incidence of abnormal embryos at 100 μM. Selenate was less embryotoxic than selenite. SeMet was far less embryotoxic than the inorganic selenium compounds. However, most of the abnormalities such as hypoplastic olfactory system, deformed optic vesicle, deformed foci vesicle, and fused bronchial arch were noted in the head region, as were the abnormalities observed on treatment with the inorganic selenium compounds. SeCys was the least embryotoxic among the compounds tested (95).

14.4.1.6 Genetic and Related Cellular Effects Studies. The clastogenic effects of sodium selenite (Na_2SeO_3) and sodium selenate ($NaSeO_4$) on bone marrow cells were examined. Male Swiss albino mice were gavaged with Na_2SeO_3 and Na_2SeO_4 doses of 7 to 28 mg/kg. Negative controls received distilled water, whereas positive controls received mitomycin-C. Twenty-four hours after exposure, the animals were sacrificed, and bone marrow cells were obtained and prepared on slides. The frequencies of total chromosomal aberrations, primarily chromatid gaps and chromatid breaks, increased with increasing concentrations of Na_2SeO_3 and $NaSeO_4$. The number of chromosomal breaks per cell and the percentage of damaged cells were significantly and dose-dependently increased in mice exposed to $NaSeO_3$ and $NaSeO_4$, compared to controls. Sodium selenite was slightly more toxic than sodium selenate. The authors concluded that further study is needed to elucidate the mechanisms of the clastogenic effects of selenium salts (96).

14.5 Standards, Regulations, or Guidelines of Exposure

No health standards have been published.

D TELLURIUM

15.0 Tellurium

15.0.1 CAS Number: [13494-80-9]

15.0.2 Synonyms: NA

15.0.3 Trade Names: NA

15.0.4 Molecular Weight: 127.60

15.0.5 Molecular Formula: Te

15.1 Chemical and Physical Properties

Elemental tellurium burns slowly in air; a finely divided suspension in air can be exploded. Reactions with zinc, chlorine, fluorine, and solid sodium are vigorous and can cause fires (97).

15.2 Production and Use

Elemental tellurium (Te) has some metallic properties, although it is classed as a nonmetal or metalloid. The name is derived from the Latin word for earth. Tellurium is occasionally found naturally, more often as telluride of gold, calaverite. The elemental form has a bright luster, is brittle, readily powders, and burns slowly in air. Tellurium exists in two allotropic forms, as a powder and in the hexagonal crystalline form (isomorphous) with gray selenium. The concentration in the earth's crust is about 0.002 ppm. It is recovered from anode muds during the refining of blister copper. It is also found in various sulfide ores along with selenium and is produced as a by-product of metal refineries. The United States, Canada, Peru, and Japan are the largest producers (1).

Tellurium's industrial applications include its use as a metallurgical additive to improve the characteristics of alloys of copper, steel, lead, and bronze. Increased ductility results from its use in steel and copper alloys. Addition of tellurium to cast iron is used for chill control, and it is a basic part of blasting caps (1). It is used in some chemical processes as a catalyst. Tellurium vapor is used in "daylight lamps." The use of tellurium, along with selenium, in semiconductors is expanding. Tellurium's use in pottery glazes is limited. Its major use is in vulcanizing of rubber.

15.3 Exposure Assessment

15.3.1 Air

Dusts and fumes can be collected on a cellulose membrane filter. The samples are then washed with perchloric and nitric acids to destroy other organic materials and the filter. A dilute solution of the tellurium in nitric acid is prepared and aspirated into the oxidizing air-acetylene flame of an atomic absorption spectrophotometer. This is good in the range of 0.0495 to 0.240 mg/m^3 (42).

Alternative methods of sampling for tellurium hydride or methyl telluride include absorption in concentrated hydrochloric acid containing 5 to 10% free bromine. The hydride can be collected in dilute sodium hydroxide. When it is collected in alkali (98) in the presence of air, polytellurides are formed. Alternative analytical methods include determinations as the iodotellurite complex or precipitation and determination as a hydrosol (99). NIOSH method 7300 (30b) is a recommended method for determining potential exposures to tellurium in the workplace.

15.4 Toxic Effects

Elemental tellurium is poorly absorbed, but its more soluble compounds may undergo some oral absorption. Soluble tellurium can be absorbed through the skin, although ingestion or inhalation of fumes presents the greatest industrial hazard. A metallic taste in the mouth may result from excessive absorption. The characteristic sign of tellurium absorption is the garlic-like odor attributed to dimethyltelluride in the breath and sweat. This may persist for many days after exposure. Urinary, fecal, and biliary excretion also occur. Urinary excretion is probably more important than respiratory excretion in eliminating absorbed tellurium (100, 101). The normal concentration of tellurium in the urine is 0.2 to 1.0 mg/mL. Tellurium is complexed to plasma proteins, and little is found in the red blood cells (102). In the nervous system, tellurium accumulates in the gray matter, not the white matter, when injected intracerebrally. The metal is found in phagocytic and ependymal cells and in lysosomes as fine needles (103). The whole-body retention model assumes a long half-life based on tellurium dioxide (104).

15.4.1 Experimental Studies

15.4.1.1 Acute Toxicity. The acute toxicity of tellurium has not been well explored, but it does not seem to be an industrial problem. In animals, repeated administration by the oral route has produced kidney and nerve damage in several species (105, 106). The appearance of endoneurial edema early in the evolution of tellurium neuropathy raises the possibility that a breakdown of the blood–nerve barrier (BNB) plays a role in the pathogenesis of tellurium-induced demyelination (107). Inclusion of 1.1% tellurium in the diet of developing rats causes a highly synchronous primary demyelination of peripheral nerves, which is followed closely by a period of rapid remyelination. The demyelination is related to the inhibition of squalene epoxidase activity, which results in a block in cholesterol synthesis and accumulation of squalene (108).

15.4.1.4 Reproductive and Developmental. Teratogenesis characterized by hydrocephalus has also been reported following the administration of tellurium to pregnant rats (109). Neonatal rats were exposed to tellurium, via the mother's milk, from the day of birth until sacrifice at 7, 14, 21, and 28 days of age. Light and electron microscopy revealed Schwann cell and myelin degeneration in the sciatic nerves at each age studied. These changes were similar to those described in weanling rats as a result of tellurium intoxication. In the CNS, hypomyelination of the optic nerves was demonstrated at 14, 21, and 28 days of age, accompanied by some evidence of myelin degeneration. These changes were also seen in the ventral columns of the cervical spinal cords, although less markedly, and were confirmed by quantitative methods (110).

15.4.2.3 Epidemiology Studies. Exposure of iron foundry workers to concentrations of 0.01 and 0.1 mg Te/m^3 for 22 months produced mild GI distress, the characteristic garlic odor, dryness of the mouth, metallic taste, and somnolence (100). A tellurium-containing catalyst from an industrial plant in the Netherlands caused a serious odor problem during biological treatment of wastewater from about 35 plants that produced organic and inorganic chemicals. Odor problems, not noticeable during the first years of wastewater plant operation, developed on the skin of the operators and at a distance of 0.5 to 1 km from the plant. It was determined that the odor-producing organic tellurium compound was produced under anoxic conditions in the denitrification step. Faint odor was detected after the adding of as little as 0.01 mg/L of tellurium tetrachloride. This odor was apparent under anoxic conditions earlier than under aerobic conditions. Strong odor was produced under anoxic conditions after 20 h at concentrations of 0.02, 0.05, and 0.10 mg/L. Under aerobic conditions, the higher concentrations produced a faint odor except for one trial. To avoid odor problems, a process was developed to reduce the tellurium content of the wastewater to 5 to 10% of the usual values before allowing it to leave the plant of origin (111).

15.5 Standards, Regulations, or Guidelines of Exposure

The OSHA/ACGIH (24) TWA and NIOSH REL limit is 0.1 mg/m^3. This is about 10 times greater than the concentration that will cause the garlic odor on the breath. Protective equipment (hands and respiratory) should be used in dust operations such as grinding. Local ventilation is recommended (112). Women of childbearing age should be informed of a possible adverse effect from tellurium on offspring when exposed in utero or by nursing.

E TELLURIUM COMPOUNDS

16.0 Tellurium Hexafluoride

16.0.1 CAS Number: [7783-80-4]

16.0.2 Synonyms: (OC-6-11) tellurium fluoride; tellurium (VI) fluoride

16.0.3 Trade Names: NA

16.0.4 Molecular Weight: 241.59

16.0.5 Molecular Formula: TeF$_6$

16.0.6 Molecular Structure:

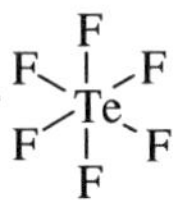

$$\begin{array}{c} F \\ F \diagdown | \diagup F \\ F \diagup \overset{|}{\underset{F}{Te}} \diagdown F \end{array}$$

16.1 Physical and Chemical Properties

16.1.1 General

Tellurium hexafluoride, TeF$_6$, is the only tellurium compound for which a separate permissible exposure level has been developed. Tellurium dioxide (TeO$_2$), potassium tellurite (K$_2$TeO$_3$), and hydrogen telluride (H$_2$Te) have significant potentials for industrial exposure. The oxide is found only at temperatures more than 450°C and is almost insoluble in water and body fluids, which limits the hazard from this compound. The only cases of poisoning by potassium tellurite have been associated with accidental poisoning.

16.1.2 Odor and Warning Properties

Repulsive odor.

16.2 Production and Use

Tellurium hexafluoride (TeF$_6$) is prepared by the direct fluorination of tellurium metal. It is a by-product of ore-refining (113).

16.3 Exposure Assessment

The recommended method for determining workplace exposures is NIOSH Analytical Method #S187 (112a).

16.4 Toxic Effects

Short-term exposure of experimental rodents to airborne concentrations of tellurium hexafluoride produced evidence of pulmonary edema. The animals exposed to 1 ppm for 4 h were adversely affected. Those exposed for 1 h to 1 ppm had an increased respiratory rate; however, repeated exposure to 1 ppm for 5 days produced no effect (114). Tellurium hexafluoride has caused headache, chest pain, and dyspnea (115, 116). It is considered toxic by inhalation and may produce pulmonary edema and death.

Two cases of excessive occupational exposure to tellurium hexafluoride were reported. Because both workers were also handling volatile liquid esters, some increased absorption and deposition of elemental tellurium in the skin may have occurred. The signs included garlic breath. An unusual feature was bluish black discoloration of the webs of the fingers and streaks on the face and neck. No permanent damage was noted (117).

16.5 Standards, Regulations, or Guidelines of Exposure

The OSHA ACGIH (24) TWA and exposure limit is 0.2 mg/m^3 (0.02 ppm). This limit is consistent with that for tellurium and its other compounds. The OSHA PEL, NIOSH REL and ACGIH TLV is 0.02 ppm (0.2 mg/m^3).

17.0 Hydrogen Telluride

17.0.1 CAS Number: [7783-09-7]

17.0.2 Synonyms: NA

17.0.3 Trade Names: NA

17.0.4 Molecular Weight: 129.62

17.0.5 Molecular Formula: H$_2$Te

17.1 Chemical and Physical Properties

See Table 44.3.

Table 44.3. Physical and Chemical Properties of Tellurium, and Selected Compounds

Tellurium Compounds	CAS Number	Physical State	Mol. Wt.	Sp. Gr.	MP (°C)	BP (°C)	Solubility
Tellurium (Te)	[13494-80-9]	Amorphous brownish black powder	127.61	6.24 at 20°C	450	1390	See text
Tellurium hexafluoride (TeF$_6$)	[7783-80-4]	Colorless gas	241.59		−37.6		Reacts slowly
Hydrogen Telluride (H$_2$Te)	[7783-09-7]	Colorless gas	129.63		−51	−4	See text

17.2 Production and Use

It is synthesized by the action of acid on aluminum telluride. It has no industrial uses.

17.1.2 Odor and Warning Properties

Offensive odor similar to hydrogen sulfide.

17.4 Toxic Effects

Hydrogen telluride is an irritant at relatively low concentrations. Headaches may follow an exposure to hydrogen telluride. Dry mouth and throat have also occurred after exposure to this compound. Hydrogen telluride causes pulmonary irritation and edema in animals. It produces symptoms similar to those of hydrogen selenide or arsine, including headaches, malaise, weakness, dizziness, and respiratory and cardiac symptoms. It is probably less toxic than arsine or hydrogen selenide, because it readily decomposes; however, exposure of guinea pigs to airborne concentrations caused pronounced hemolysis (116).

17.5 Regulations, Standards, or Guidelines of Exposure

No health standards have been developed.

F SULFUR

18.0 Sulfur

18.0.1 CAS Number: [63705-05-5]

18.0.2 Synonyms: NA

18.0.3 Trade Names: NA

18.0.4 Molecular Weight: 32.06

18.0.5 Molecular Formula: S

18.1 Chemical and Physical Properties

See Table 44.4.

18.2 Production and Use

Sulfur (S) occurs naturally as a yellow, water-insoluble solid. The name is from the Latin "sulphur." Early Greek physicians mention sulfur and the fumes from burning sulfur in religious ceremonies. Sulfur constitutes about 0.053% of the earth's crust and occurs in two allotropic crystalline forms, rhombic and monoclinic. Below 96°C, only the rhombic form is stable. Large sedimentary deposits of almost pure sulfur are mined in Texas and Louisiana. Sulfur can be extracted from crude oil in the refining process, as well as from stack gases resulting from coal combustion. Sulfur occurs in fossil fuels and in metal (Fe,

Table 44.4. Physical and Chemical Properties of Sulfur, Sulfuroxide, and Selected Compounds

Sulfur Compound	CAS Number	Physical state	Mol Wt.	Sp. Gr.	Density	MP (°C)	BP (°C)	Solubility
Sulfur (S)	[63705-05-5]	Yellow solid	32.06		1.96	119	444.6	Very soluble in carbondisulphide
Sulfur dioxide (SO$_2$)	[7446-09-5]	Colorless liquid or gas	64.06			−75.51	−10.06	10 g/100 ml at 20°C
Sulfur Trioxide (SO$_3$)	[7446-11-9]	Colorless liquid or crystals	80.07	1.92		16.8	44.8	Forms H$_2$SO$_4$ in water
Hydrogen sulfide (H$_2$S)	[7783-06-4]	Colorless gas	34.08			−85.4	−60.3	437 ml in 100mL of water
Carbon disulfide (CS$_2$)	[75-15-0]	Colorless to faintly yellow liquid	76.13	1.26		−110	46.2	Slightly soluble 0.1185 g/100 ml
Sulfur monochloride (S$_2$Cl$_2$)	[10025-67-9]	Red fuming oily liquid	135.0	1.7		−80	136	Soluble in carbondisulfide, benzene, ether, liquid decomposes
Thionyl chloride (SOCl$_2$)	[7719-09-7]	Pale yellow to red liquid	119.0	1.63		−104	76	Decomposes
Sulfuric acid (H$_2$SO$_4$)	[7664-93-9]	Colorless to dark brown oily dense liquid		1.84		−3	280	Soluble in water

Pb) ores. Exposure may occur in numerous operations related to the mining and recovery of sulfur. The recovery of sulfur as a by-product accounts for a larger portion of the world's production than the mined mineral. Sulfur is one of the most important raw materials, particularly in the fertilizer industry (1).

Organic sulfur compounds occur in garlic, mustard, onions, and cabbage and are responsible for the odor of skunks. Sulfur occurs in living tissue and is part of some amino acids. Unlike many other inorganic elements, sulfur itself is relatively nontoxic. Sulfur and some of its salts have been used medicinally. The consumption of sulfur is a measure of national industrial development and economic activity. Sulfur is most often used as a chemical reagent, rather than as part of a finished product (1).

18.3 Exposure Assessment

18.3.1 Air

Normally, airborne sulfur is collected and analyzed based on the individual sulfur compound.

18.4 Toxic Effects

Various sulfur preparations have been used therapeutically. For example, sodium thiosulfate has been used as a cathartic, in treating of parasitic diseases, and in treating some types of poisoning. Up to 12 g/day has been consumed without illness except for emesis or vomiting. Orally, colloid preparation of sulfur leads to urinary excretion of the sulfur load as sulfate in 24 h (118). Exposure to sulfur particulates produces tracheobronchitis, characterized by cough, sore throat, chest pain, and lightheadedness (119).

18.5 Standards, Regulations, or Guidelines

There are no airborne standards for elemental sulfur.

G SULFUR OXIDES

Sulfur oxides include major pollutants, and whereas other sulfur compounds are highly toxic, either type of compound can have a major impact on health and the quality of life.

19.0 Sulfur Dioxide

19.0.1 CAS Number: [7446-09-5]

19.0.2 Synonyms: Sulfurous oxide; sulfur oxide; sulfurous anhydride; sulfur dioxide; anhydride, 99.9%

19.0.3 Trade Names: NA

19.0.4 Molecular Weight: 64.06

19.0.5 Molecular Formula: SO_2

19.0.6 Molecular Structure: O=S=O

19.1 Chemical and Physical Properties

The physical and chemical properties of the oxides are listed in Table 44.4.

19.1.2 Odor and Warning Properties

Pungent odor detectable at 0.3 to 5 ppm, can be liquid at < 14°F.

19.2 Production and Use

Sulfur dioxide is formed whenever sulfur burns in air. It is the most widely encountered and best known irritant gas, because of its wide usage, and also because of its frequent occurrence as an undesired by-product in smelting sulfide ores, in paper manufacture, in the combustion of sulfur-bearing coals and petroleum fuels, and in the action of sulfuric acid on reducing agents. The predominant use of sulfur dioxide is in producing sulfuric acid. Sulfur dioxide is also used as a refrigerant (1).

Sulfur dioxide is one of the most prominent gases that contribute to atmospheric pollution in large cities and in areas surrounding smelters. Mobile sources (automobiles) were one of the main contributors, but catalytic converters have reduced emissions. The sulfur oxides, along with those of nitrogen, are the principal cause of acid rain, which adversely affects the ecology. The residence time for sulfur dioxide in ambient air is between 2.5 and 5 days (120).

Sweetening plants for petroleum products sometimes dispose of sulfide gases by burning them to sulfur dioxide and discharging the resulting gas from high stacks. The terrain, height of the stack, rate of gas discharge, and atmospheric conditions present variable factors that have made the success of the dilution method unpredictable and frequently disappointing. The gas rises vertically for some distance above the stack and then spreads laterally. The important factors in dispersing it are fog, wind direction and velocity, inversion, and turbulence. Sulfur dioxide in moist air or fog combines with the water to form sulfurous acid and is slowly oxidized to sulfuric acid. In the United States, coal-fired electric power plants generate about 70% of the sulfur dioxide. The Clean Air Act of 1990 called on utilities to reduce their annual emissions of SO_2 from about 17.9 million tons to 8.9 million tons by the year 2000.

Sulfur dioxide is an intermediate in the manufacture of sulfuric acid. It is also used to manufacture sodium sulfite and other chemicals. Large quantities are used in refrigeration, bleaching, fumigating, and preserving. It is used as an antioxidant in melting and pouring magnesium, where it is applied as the gas or is generated by adding powdered sulfur to the surface of the molten metal in the ladle and to the surface of the poured casting. Up to 0.5% sulfur dioxide is also used to prevent oxidation in controlled-atmosphere heat-treat ovens for magnesium (1).

Before or during inhalation, sulfur dioxide (SO_2) may react with water to form sulfurous acid (H_2SO_3) and be oxidized to form sulfur trioxide. The latter reacts rapidly to form sulfuric acid (H_2SO_4), which forms ammonium sulfate in the presence of ammonia. Sulfurous acid disassociates to sulfite and bisulfite ions.

19.3 Exposure Assessment

19.3.1 Air

Workplace sulfur dioxide may be collected on a charcoal tube. The maximum flow rate of 0.2 L/min is used until 24 liters are collected. Analysis is by ion chromatography (121). As an alternate, a maximum flow rate of 1.5 L/min is used and 400 liters are collected. The sulfur dioxide is desorbed using sodium bicarbonate solution and the analysis is again by ion chromatography (42). This is the method employed in NIOSH Analytical method #6004 (30a).

Automatic recorders for SO_2 are available and are widely used for area or atmospheric sampling. Passive dosimeters, detector tubes, and portable direct-reading meters, some pocket sized, can be used to augment approved workplace monitoring methods.

19.4 Toxic Effects

Only a small portion of inhaled sulfur dioxide penetrates the lower respiratory tract because sulfur dioxide is water soluble. However, lower respiratory tract penetration may occur during exercise. The major effects of sulfur dioxide are on the upper respiratory tract. It may cause edema of the lungs or glottis and can produce respiratory paralysis. Sulfur dioxide produces sulfurous acid on contact with the moist mucous membranes. This is a direct irritant and inhibits mucociliary transportation. Most of the inhaled sulfur dioxide is detoxified in the liver by a molybdenum-dependent sulfite oxidase pathway. Concentration rather than duration of the exposure is the more important determinant of histopathological damage (122).

The acute effects of sulfur dioxide and causal concentrations are listed here:

400 to 500 ppm	Immediate danger to life
100 ppm	NIOSH IDLH
	Maximum tolerated exposures for 30 to 60 min
20 ppm	Chronic respiratory symptoms
6 to 12 ppm	Nasal and throat irritation
3 ppm	Odor detected
0.3 to 1 ppm	Threshold for smell or taste

Individuals who have hyperactive airway disease, including asthma, may be particularly sensitive to the bronchospastic properties of sulfur dioxide. Mild asthmatics selected for methacholine sensitivity have significant bronchoconstriction as a group in response to short-term moderate exercise when breathing in 1.0 and 0.5 ppm sulfur dioxide (123). In addition, the bronchoconstriction induced decreases after short-term repeated exercise in the presence of elevated sulfur dioxide concentrations. Particulates and water droplets can take sulfur dioxide into the lower respiratory tract and thereby worsen cardiopulmonary diseases.

Exposures to sulfur dioxide (SO_2) have been associated with progressive, dose-dependent bronchoconstriction in sensitive individuals. The clinical significance of such changes remains poorly characterized. Witek et al. (124) studied subjective responses following exposure to low-level concentrations of SO_2 (less than 1 ppm) in a group of 10 healthy and 10 asthmatic subjects. The number and severity of complaints associated with SO_2 increased with concentrations in both healthy and asthmatic subjects. Asthmatics had progressive lower respiratory complaints such as wheezing, chest tightness, dyspnea, and cough, at increasing levels of SO_2, whereas healthy subjects complained more frequently of upper airway symptoms such as taste and odor at increasing levels of SO_2. Although exercise increased the frequency of lower airway symptoms in asthmatics, there was no increase in symptoms for healthy subjects.

Sulfur trioxide and sulfuric acid mists are strongly irritant, and inhaling concentrations of approximately 3 mg/m^3 causes a choking sensation. Persons accustomed to the exposure are unable to notice concentrations at this order of magnitude. Sulfur trioxide is irritating and corrosive to all mucous surfaces and causes inflammation of the upper respiratory tract and possible lung injury. Sulfuric acid also attacks tooth enamel.

19.4.2.2.1 Acute Toxicity. Acute exposure to sulfur dioxide has caused severe obstructive pulmonary disease that is unresponsive to bronchodilators and lasts for 3 months after exposure. Only on occasion have short-term acute exposures resulted in moderate to severe obstructive defects accompanied by persistent and productive cough (119). A chronic cough and sputum production are associated with exposure to sulfur dioxide in ambient concentrations in women and nonsmokers (125).

19.4.2.2.5 Carcinogenesis. Sulfur dioxide may act as a cancer promoter. The mortality of arsenic smelter workers was higher when they had also been exposed to sulfur dioxide (126). Additionally, rats exposed to 3.5 or 10 ppm of sulfur dioxide developed squamous cell carcinomas from inhalation of benzo[a]pyrene, but neither compound alone produced carcinomas under the conditions of this experiment (127).

19.4.2.3 Epidemiology Studies. Signs and symptoms of respiratory distress were reported among 100 refrigeration workers exposed to 20 to 30 ppm of sulfur dioxide. In this study of sulfur dioxide the exposure was almost purely sulfur dioxide (128). More recently, a high incidence of respiratory symptoms was reported in pulp mill workers exposed to 10 to 20 ppm of sulfur dioxide (129). However, a 10-year follow-up of workers exposed to concentrations ranging from 4 to 33 ppm did not reveal an increased prevalence of respiratory disease or deteriorating pulmonary function compared with the control group (130). Smelter workers exposed to sulfur dioxide at 2 ppm or lower developed pulmonary disease, and workers exposed to 1 ppm or more had accelerated loss of pulmonary function (131).

19.5 Standards, Regulations, or Guidelines of Exposure

The ACGIH TLV for sulfur dioxide is 2 ppm. A STEL of 5 ppm is recommended (13 mg/m^3). OSHA's PEL is 5 ppm (13 mg/m^3) NIOSH recommends a TWA of 2.0 ppm (5 mg/m^3). The NIOSH IDLH estimation is 100 ppm (29).

20.0 Hydrogen Sulfide

20.0.1 CAS Number: [7783-06-4]

20.0.2 Synonyms: Hepatic acid; Stink Damp; sulfuretted hydrogen; hydrosulfuric acid; sulfur hydride; sewer gas; sour gas

20.0.3 Trade Names: NA

20.0.4 Molecular Weight: 34.08

20.0.5 Molecular Formula: H_2S

20.0.6 Molecular Structure: $H{-}S{-}H$

20.1 Chemical and Physical Properties

See Table 44.4.

20.1.2 Odor and Warning Properties

Strong odor of rotten eggs detectable at 0.001 to 0.1 ppm. It is detectable by odor at about 1/400th of the lowest amount that can cause injurious effects.

20.2 Production and Use

It occurs as a by-product in industrial or natural processes wherever proteins decompose. It is encountered in mining, especially where sulfide ores are found; in excavating swampy or filled ground, and hence sometimes in wells, caissons, and tunnels; in natural gas; in the production and refining of petroleum; in the waters of certain natural springs; in volcanic gases; in the low-temperature carbonization of coal; in the manufacture of chemicals, dyes, and pigments; in the rayon industry; in the rubber industry; in tanneries; in the manufacture of glue; in the washings from sugar beets; and in sewer gases. Because hydrogen sulfide is soluble in water and oil, it may flow for a considerable distance from its place of origin to escape in unexpected areas.

Hydrogen sulfide is used in the manufacturing chemicals; in metallurgy; as an analytical reagent; as an agricultural disinfectant; as an intermediate for sulfuric acid, elemental sulfur, sodium sulfide, and other inorganic sulfides; as an additive in extreme-pressure lubricants and cutting oils; and as an intermediate for organic sulfur compounds (1). It is not registered as a pesticide in the United States (132).

A hygienic survey of hydrogen sulfide, methyl mercaptan and its derivatives, and sulfur dioxide in kraft mills and in sulfite mills revealed concentrations that varied from 0 to 20 ppm hydrogen sulfide, 0 to 15 ppm methyl mercaptan, and comparable amounts of dimethyl sulfide, and dimethyl disulfide up to 1.5 ppm. The greatest emissions were detected at chip chutes and evaporation vacuum pumps. Batch operations yielded clearly higher sulfur dioxide concentrations (up to 20 ppm) compared with a continuous ammonia-base digester. Furthermore, there was a strong correlation with season in the sulfite mills, higher concentrations were found in the winter when natural ventilation was poorer.

Hydrogen sulfide is a by-product of organic decomposition, the petroleum industry, tanning, rubber vulcanizing, and heavy water production. Deaths have been reported as a result of agitation of underground manure tanks, the addition of sulfuric acid to drains, HCl to wells, cleaning of propane tanks, and entry of both victims and rescuers into sewers and ship holds containing fish meal (119). The 1969 air pollution levels ranged from an average of 1 to 6 mg/m^3, and concentrations were as high as 300 mg/m^3 near industrial sources, largely from combustion of sulfur-containing materials. Nonindustrial sources of H_2S include volcanoes, bacteria, and plants (133).

20.3 Exposure Assessment

20.3.1 Air

Passive dosimeters, detector tubes, and portable direct-reading meters, some pocket size, can be used to augment approved workplace monitoring methods for hydrogen sulfide. An automatic detection and control system has been described.

The OSHA method (121) calls for collection on a silver nitrate-impregnated cellulose filter, 17 mm in diameter. The maximum flow rate is 0.2 L/min. The samples are analyzed by a polarographic analyzer system using differential pulse polarography. The limit of detection is 0.4 ppm (0.5 mg/m^3) for a 2-liter volume.

The NIOSH analytical method, #6013 uses charcoal tubes for collection with ammonia and hydrogen peroxide as reagents and analysis by ion chromatography.

20.4 Toxic Effects

As to health effects, exposed workers complained of headaches and a decrease in concentration capability more often than matched controls. The number of sick days was greater for the exposed workers than among the controls (134).

Many budding chemists have developed a casual disregard for the toxicity of hydrogen sulfide because of its general, and sometimes careless, use in the teaching of qualitative and quantitative analysis; and it is with great surprise that they later learn that the gas they used so consistently has a toxicity comparable to that of hydrogen cyanide.

Absorption of hydrogen sulfide is almost exclusively through the respiratory tract. Absorption through the skin has been demonstrated and discoloration of the skin reported. This is not a significant source of systemic poisoning, as indicated by the fact that the routine for gas mask approval testing by the U.S. Bureau of Mines has included wearing gas masks for 30-min periods in atmospheres containing 2% (20,000 ppm) hydrogen sulfide. During these tests, which include strenuous exercise, the subjects have noted slight skin irritation, but no systemic effects indicative of hydrogen sulfide absorption and no discoloration of the skin. When free sulfide exists in the circulating blood, a certain amount of hydrogen sulfide is excreted in the exhaled breath. This is sufficient to be detected by odor. Although the characteristic odor of gas is detectable in concentrations as low as 0.025 ppm, is distinct at 0.3 ppm, is offensive and moderately intense at 3 to 5 ppm, and is strong and marked, but not intolerable, at 20 to 30 ppm, the odor of higher concentrations does not become more intense; above about 200 ppm the disagreeable odor appears less intense. These perceptions are based upon initial inhalations, and the olfactory sense fatigues rapidly from continuous inhalation. The initial event in the absorption of H$_2$S is its disassociation into HS$^-$ anion (pKa = 7.08 at 18°C). The reverse reaction does not occur at normal blood pH or in histotoxic acidosis caused by H$_2$S. The greater portion of hydrogen sulfide, however, is excreted in the urine, chiefly as sulfate (particularly thiosulfate); but some is also excreted as sulfide. The blood sulfide has a short half-life of about 60 minutes. Thus this possible marker must be sampled shortly after exposures cease (135).

Hydrogen sulfide is also a potent eye and mucous membrane irritant, even at low concentrations (50 to 200 ppm). Pulmonary edema is often a clinical finding in persons who have been rendered unconscious by hydrogen sulfide exposure (136–139). In several of the reported fatalities, the individuals apparently died of acute respiratory distress syndrome due to pulmonary edema. Irritation of the eye results in initial lachrymation, loss of coronary reflex, and changes in visual acuity and perception (usually at concentrations in excess of 50 ppm), which may progress to inflammation and ulceration, and the possibility of permanent scarring of the cornea in severe cases. Inflammation of the cornea of the eye has been reported in workers exposed to as little as 10 ppm hydrogen sulfide for 6 to 7 h (140, 141). The blood sulfide concentrations, air concentrations, and effects are tabulated in Table 44.5.

Hydrogen sulfide is an irritant gas and exposure to concentrations between 70 and 700 ppm may irritate the mucous membranes of the eyes and of the respiratory tract. Pulmonary edema or bronchial pneumonia is likely to follow prolonged exposure to

Table 44.5. Exposure and Blood Sulfide Concentrations and Clinical Effects of Acute Hydrogen Sulfide Exposure (96, 98)

Concentration (mg/m$_3$)	Duration (min)	Blood Conc (mmol/l)[a]	Effect
7	480		TLV-TWA (notice of intended changes 1999)
16-32	360–420		Eye irritation
		1.3	Clear smell, worker immeadiately escaped, minor headache
38	15		OSHA ceiling
70	10		OSHA max peak above acceptable ceiling from 8 hr shift
150–300	2–30		Loss of smell
1350	<30		Marked systemic symptoms
		5.0	No smell, worker collapsed
		19	Worker found unconscious
1350	>60		Death
		53–117	Five fatal cases from oil refineries, tanneries, and sewage work
2250	15–30		

[a]In the nonfatal cases, the blood samples were taken within 30 min of the event. Maximum concentrations were possibly higher.

concentrations of the order of 250 to 600 ppm. These levels of exposure may cause such symptoms as headache, dizziness, excitement, nausea or GI disturbances, dryness and sensation of pain in the nose, throat, and chest, and coughing. Table 44.5 indicates responses to various concentrations of hydrogen sulfide in the atmosphere.

20.4.1 Experimental Studies

After inhalation of hydrogen sulfide for 1 min to 10 h at high concentrations, changes were noted in the brain, pancreas, liver, kidney, and small intestine of rats and guinea pigs. The formation of methemoglobinemia protects against H_2S toxicity experimentally, although oxygen therapy is recommended in humans (133). Detoxification of hydrogen sulfide occurs rapidly (85% of a lethal dose per hour in animal). The red blood cells and the liver mitochondria are the main sites. Hydrogen sulfide intracellular cytochrome oxidase by altering electron transport is the most likely mechanism of action (119).

20.4.1.2 Chronic and Subchronic Toxicity. Subchronic (90-day) vapor inhalation studies of hydrogen sulfide were conducted using Sprague–Dawley rats and Fischer-344 rats (142, 143). These animals were exposed to 0, 10.1, 30.5, and 80 ppm for 6 h/day, 5 days/week, for at least 90 days, simultaneously and in the same chambers as the mice in the principal study. Three groups of 15 males and 15 females of each strain of rats were used for each dose. In addition, a control group consisting of 15 of each sex was exposed only to clean air. When exposure terminated, the animals were examined as described for the critical study (144). A significant reduction in body weight gain was noted in all animals exposed to 80 ppm. In F-344 and male Sprague–Dawley rats, the effect on body weight was statistically significant at some times in all exposed groups, but mean body weights were never less than 93% of control. In female Sprague–Dawley rats, mean body weight in the 80 ppm group was < 90% of that of the control groups during most of the study. Brain weight was significantly reduced in the male Sprague–Dawley rats in the high-dose group and slightly, but not significantly, reduced in females (143), indicating a lowest observed adverse effect level (LOAEL) of 80 ppm. No clinical signs in rats were observed to be exposure-related. Examinations of neurological function yielded negative results. Blood volume, appearance, occult blood, specific gravity, protein, pH, ketone, and glucose values were all normal. Ophthalmoscopic examination, hematology, serum chemistry parameters, and urinalysis were also normal. Histopathological examination, which included four sections of the nasal turbinates, revealed no abnormalities in comparison with controls.

Male F-344 rats were exposed to 0, 10, 200, or 400 ppm nominal concentration (14, 279, and 557 mg/m^3) of hydrogen sulfide for 4 h (145), and samples were collected by broncho alveolar lavage and nasal lavage at 1, 20, and 44 h after exposure (four rats/exposure level). Increased number of cells in nasal lavage and increased protein and lactate dehydrogenase in both bronchoalveolar and nasal lavage fluid from rats exposed to 400 ppm were observed. Male F-344 rats (four rats/exposure level) were exposed to 0, 116, or 615 mg/m^3 for 4 h (146). Rats exposed to 615 mg/m^3 had marked perivascular and alveolar edema and bronchioles containing granular leukocytes, proteinaceous fluid, fibrin, and exfoliated cells. Necrosis of bronchiolar ciliated cells and hyperplasia of alveolar type II cells were observed in rats exposed to 615 mg/m^3. In rats exposed to 116 mg/m^3, only mild perivascular edema was observed. Nasal structures were not examined. Male F-344 rats were exposed to 0, 14, 280, or 560 mg/m^3 hydrogen sulfide for 4 h, and four levels of the nasal cavity were examined histologically at 1, 18, and 44 hours after exposure (147). Necrosis and exfoliation of respiratory and olfactory mucosal cells, but not squamous epithelial cells, were observed in rats exposed to 560 mg/m^3. No nasal lesions were seen in the controls or the two lower exposure levels.

Subchronic (90-day) inhalation studies were conducted using B6C3F1 mice (144). Three groups of 10 males and 12 female mice each were exposed to 0, 10.1, 30.5, and 80 ppm 6 h/day, 5 days/week for 90 days. A control group of mice was exposed only to clean air. Animals were examined using neurological function tests of posture, gait, facial muscle tone, and reflexes. Ophthalmological examination using a slit lamp was performed. The only exposure-related histopathological lesion was inflammation of the nasal mucosa in the anterior segments of the nose, which was observed in eight of nine male mice and in seven of nine female mice in the group exposed to 80 ppm. This lesion was also present in two high-dose mice that died during the course of the study. The lesion was generally minimal to mild and was located in the anterior portion of the nasal structures, primarily in the squamous portion of the nasal mucosa, but extended to areas covered by respiratory epithelium. This lesion was not observed in any animals in the other exposure groups. Thus for mice, 80 ppm is considered a LOAEL for nasal inflammation, and 30.5 ppm is a no observed adverse effect level (NOAEL).

A minimal submucosal lymphocytic cellular infiltrate was observed in the posterior section of the nasal structures. This lesion occurred with approximately equal frequency in control and exposed animals and was not considered exposure-related. Histopathological evaluation revealed no other abnormalities compared with controls. Significant reductions in body weight gain were noted in all exposure groups at various times during the study. Decreased weight gain in animals exposed to 80 ppm occurred consistently in both male (approximately 90% of control during the last 7 weeks of study) and female (< 90% of control during the last 3 weeks of study) mice, and this level is considered a LOAEL. Statistically significant changes in absolute kidney, liver, and spleen weight were also observed in the male rats exposed to 80 ppm, but no differences were apparent when organ weights were normalized to body weight. Neurological function examinations yielded negative results. Ophthalmoscopic examinations and clinical pathological values were also normal.

20.4.1.4 Reproductive and Developmental. No data on human developmental effects of inhaled hydrogen sulfide were found; but based on the limited information available in animals, hydrogen sulfide does not induce developmental effects. In a preliminary study, Saillenfait et al. (148) administered 0, 50, 100, or 150 ppm hydrogen sulfide 6 h/day to pregnant rats during gestational days 6 to 20. Maternal body weight gain was significantly reduced at 150 ppm, and fetal body weight was slightly (4 to 7%) reduced in all exposed groups. In dams exposed to 100 or 150 ppm, reduced absolute weight gain, increased implantations, and increased live fetuses were observed. In a follow-up experiment, 20 pregnant females were exposed to 100 ppm for 6 h/day on days 6 to 20 of gestation. Fetal weights, number of live and dead fetuses, number of implantation sites and resorptions, and external malformations were recorded. Viable fetuses were then prepared for soft tissue and skeletal examination. Neither maternal toxicity nor adverse effects on the developing embryo or fetus are observed. The preliminary study identifies a LOAEL of 50 ppm for maternal weight gain. Because of the larger number of animals in the main study, a NOAEL of 100 ppm for maternal effects and developmental effects was identified.

20.4.2 Human Experience

Hydrogen sulfide is an irritant gas, and exposure to concentrations between 70 and 700 ppm may irritate the mucous membranes of the eyes and the respiratory tract. Pulmonary edema or bronchial pneumonia is likely to follow prolonged exposure to concentrations of the order of 250 to 600 ppm. These levels of exposure may cause such symptoms as headache, dizziness, excitement, nausea or GI disturbances, dryness and sensation of pain in the nose, throat, and chest, and coughing. Table 44.5 lists responses to various concentrations of hydrogen sulfide in the atmosphere.

Among the subacute and chronic effects of exposure to hydrogen sulfide, eye irritation that results in conjunctivitis or "gas eyes" is the most common and, ranging from mild to severe with extent and intensity of exposure, may include itching and smarting, a feeling of sand in the eyes, marked inflammation and swelling, cloudy cornea, and destruction of the epithelial layer and scaling that results in blurring of vision. Exposure to light may increase the painful effect. Atmospheric concentrations above 50 ppm and up to 300 ppm are conducive to this condition.

20.4.2.2.1 Acute Toxicity. By far the greatest danger of inhaling of hydrogen sulfide is its acute effects. Whether the effects are acute or subacute and chronic depends on the concentration of the gas in the atmosphere. Death or permanent injury may also occur after very short exposure to small quantities of hydrogen sulfide (149). It acts directly upon the nervous system and results in paralyzing of respiratory centers. Contact with eyes causes painful conjunctivitis, sensitivity to light, tearing, and clouding of vision. Inhalation of low concentrations causes a runny nose with a sense of smell loss, labored breathing, and shortness of breath. Direct contact with skin causes pain and redness. Other symptoms of exposure include profuse salivation, nausea, vomiting, diarrhea, giddiness, headache, dizziness, confusion, rapid breathing, rapid heart beat, sweating, weakness, sudden collapse, unconsciousness, and death due to respiratory paralysis (150). Deaths occur rapidly on the site. Patients who have vital signs at hospital arrival usually survive, if severe hypoxic encephalopathy is absent (121). Histopathological evaluation most often reveals changes in the lung, brain, and heart. The thyroid and heart may also be target organs (133).

Concentrations of 700 ppm and above may result in acute poisoning, and although the gas is an irritant, the systemic effects from absorption of hydrogen sulfide into the bloodstream overshadow the irritant effects. These acute systemic effects result from the action of free hydrogen sulfide in the blood stream and occur whenever the gas is absorbed faster than it can be oxidized to pharmacologically inert compounds, such as thiosulfate or sulfate. Such oxidation occurs rapidly in humans or animals, and even following inhalation exposure to concentrations up to 700 ppm of hydrogen sulfide in the atmosphere, hydrogen sulfide does not appear in the exhaled breath. Relatively massive doses are required to overcome this protective activity of the body. Sodium sulfhydrate, (NaHS), solution injected intravenously into dogs rapidly disappears from the circulating blood when a rate equivalent to 0.1 to 0.2 mg of hydrogen sulfide per kilogram of body weight per minute is not exceeded. When the amount absorbed into the bloodstream exceeds that which is readily oxidized, systemic poisoning results. There is a general action on the nervous system, hyperpnea occurs shortly; and respiratory paralysis may follow immediately. This condition may be reached almost without warning because the originally detected odor of hydrogen sulfide may have disappeared as a result of olfactory fatigue. Unless the victim is removed to fresh air within a very few minutes and breathing is stimulated or induced by artificial respiration, death occurs. Unconsciousness and collapse occur within seconds in high concentrations; for that reason many persons have lost their lives attempting to save victims who collapsed from exposure. In such a case, holding the breath permits a brief stay in the atmosphere, whereas inhaling would cause almost immediate collapse.

Hydrogen sulfide is acutely toxic to humans as evidenced by the numerous reports of individuals fatally poisoned by accidental exposure (135, 139, 151–155). According to NIOSH, hydrogen sulfide is a leading cause of sudden death in the workplace (156). The odor threshold is reported as 25 ppb (0.035 mg/m^3). Levels in the 3 to 5 ppm range cause an offensive odor. At levels around 100 ppm, no odor is detectable, because olfactory sensation is lost, which results in loss of warning properties at lethal levels. In reports of acute poisoning, systemic intoxication can result from a single (one to two breaths) massive exposure to concentrations usually greater than 1000 ppm (153, 155). High levels of hydrogen sulfide inhaled act directly on the respiratory center, causing respiratory paralysis with consequent asphyxia and subsequent death (139, 151, 157). At levels between 500 and 1000 ppm, acute effects include symptoms of sudden fatigue, headache, dizziness, intense anxiety, loss of olfactory function, nausea, abrupt loss of consciousness, disturbances of the optic nerves, hypertension, insomnia, mental disturbances, pulmonary edema, coma, and convulsions and respiratory arrest, followed by cardiac failure and, often, death (136, 137, 140). Levels estimated at 250 ppm resulted in unconsciousness in three workers after several minutes of exposure (154). Cardiac effects in acute hydrogen sulfide intoxication have been reported in humans (138) and animals (158). If exposure is terminated promptly, recovery occurs quickly. However, neurological effects have reportedly persisted in survivors of high-level exposure (159).

Two case studies noted neuropsychological dysfunction characterized by cognitive impairment, deficits of verbal fluency, disorders of written language, and impairment of various memory, psychomotor, and perceptual abilities in individuals acutely exposed to hydrogen sulfide (160, 161). The persistent damage that has been observed after hydrogen sulfide exposure is not distinguishable from the effects of systemic anoxia or ischemia of the brain or heart, and no specific hydrogen sulfide chronic systemic toxicity has been defined (162).

20.5 Standards, Regulations, or Guidelines of Exposure

Based on epidemiological data (163) the ACGIH (26) has a notice of intended change of TLV-TWA of 5 ppm (7.0 mg/m^3) for hydrogen sulfide. However, citing evidence of eye injury, headaches, nausea, and insomnia after exposure to H$_2$S at low concentrations for several hours, NIOSH (156) adopted a ceiling occupational exposure limit of 10 ppm

(15 mg/m^3) with a 10-min maximum exposure to this concentration. The OSHA PEL is 20 ppm, a ceiling concentration TWA with a maximum peak above the acceptable ceiling concentrations for an 8 hr shift.

21.0 Carbon Disulfide

21.0.1 CAS Number: [75-15-0]

21.0.2 Synonyms: Carbon bisulfur; dithiocarbonic anhydride; alcohol of sulfur; carbon bisulfuret; carbon sulfide; weeviltox; sulfocarbonic anhyride

21.0.3 Trade Names: NA

21.0.4 Molecular weight: 76.13

21.0.5 Molecular Formula: CS_2

21.0.6 Molecular Structure: S=S

21.1 Physical and Chemical properties

See Table 44.4.

21.1.2 Odor and Warning Properties

Carbon disulfide has a foul, slightly ethereal (cabbage-like) odor that nevertheless does not offer adequate warning in the lower harmful concentrations.

21.2 Production and Use

Carbon disulfide (CS_2) or carbon bisulfide, is used to xanthate cellulose in preparing viscose; exposures exist in the xanthating process, and also during spinning and washing of the viscose. Carbon disulfide has been used in the rubber industry, as a solvent for sulfur or as a diluent for sulfur chloride in vulcanizing and as a solvent for rubber cement. It has been used as an insecticide and in the chemical industry as a solvent for phosphorus, fats, oils, resins, and waxes. It is used to manufacture optical glass. Carbon disulfide is also encountered in the destructive distillation of coal. Carbon disulfide is used to manufacture soil disinfectants and vacuum tubes and as a solvent for cleaning and extractions, especially in metal treatment and plating. It is a fumigant for grain, a corrosion inhibitor, and a polymerization inhibitor for vinyl chloride (1).

21.3 Exposure Assessment

21.3.1 Air

Carbon disulfide can be collected by using a sodium sulfide drying tube (270 mg) connected by a 20-mm section of tubing to the front section of a coconut shell charcoal tube (100/50 mg sections, 20/40 mesh. The maximum flow rate is 0.2 L/min for maximum air volume of 25 L (TWA) or 2 L (peak). The samples may be shipped or stored under refrigeration until analysis can be performed. The samples are desorbed with toluene, and analysis is by gas chromatography (flame photometric detection). The detection limit is 0.02 mg per sample (42). The NIOSH Analytical method #1600 more completely describes this assessment method (30b).

21.4 Toxic Effects

Carbon disulfide is absorbed by the lungs; some dermal absorption can also occur. Inhaled carbon disulfide reaches equilibrium in 1 to 2 h and 40 to 50% is retained. Once absorbed, more carbon disulfide is carried in the red blood cells than in the plasma. About 70 to 90% is metabolized. Carbon dioxide and carbon monoxide are metabolites.

21.4.1 Experimental Studies

Rats and rabbits were exposed to 20 ppm or 62.3 mg/m^3 (recommended occupational exposure limit) and 40 ppm or 124.6 mg/m^3 of carbon disulfide during the entire length of the pregnancy period and also 34 weeks before breeding to simulate occupational exposure (164). Hardin et al. (164) observed no effects on fetal development in rats or rabbits following inhalation exposure to 62.3 or 124.6 mg/m^3, which corresponds to estimated equivalent oral dosages of 5 and 10 mg/kg for rats and 11 and 22 mg/kg for rabbits. In an oral study in which rabbits received 25 mg/kg(day), fetal resorption occurred (165, 166). Fetotoxicity and fetal malformations in this study were not observed in rats at the lowest level 100 mg/(kg)(day) of carbon disulfide exposure. The data from this study also suggest that the rabbit fetus is more sensitive than the rat fetus to carbon disulfide-induced toxicity. Johnson et al. (166) reported an epidemiological study that employed a wide range of exposure to carbon disulfide, such as 0.04 to 5 ppm (mean 1.2 ppm, low exposure), 0.04 to 33.9 ppm (mean 5.1 ppm, medium exposure), and 0.04 to 216 ppm (mean 12.6 ppm, high exposure). In this study, the entire population was subjected to a combined exposure of 7.3 ppm over a period of 12 or more years. Of the several clinical findings, the exposed population showed significant alterations in sensory conduction velocity and peroneal motor conduction velocity. However, in the opinion of the authors, the data indicated that minimal neurotoxicity was evident because the reduction in nerve conduction velocity was still within a range of clinically normal values and thus not associated with specific health consequences. Additionally, the exposed population had blood lead levels < 40 mg/dL and the exposed air alone contained H_2S, H_2SO_4, and tin oxide. Therefore 7.3 ppm CS_2 can be considered a NOAEL for neurotoxicity. Bulgarian investigators (167) reported significant fetal malformations in rats exposed to a low carbon disulfide dose of 0.3 mg/m^3 during three generations. However, the details of the procedures used in the study were not sufficiently presented to allow one to fully validate the findings.

21.4.2 Human Experience

21.4.2.1 General Information. Although the total sulfur in the urine of exposed persons is considerably elevated, many other factors cause that same result. Carbon disulfide in the blood or urine, however, is indicative of exposure; and its concentration can give some indication of the severity of the exposure. Only about 1% of inhaled carbon disulfide is eliminated in urine. Other urinary metabolites are thiocarbamide (thiourea) and mercaptothioaxolinone (168, 169).

Peak plasma concentration occurs about 2 h after inhalation. The plasma half-life is approximately 1 h. Carbon disulfide may be stored in large amounts in the fat and liver, and it binds to microsomal enzymes, reducing their activity (119).

Carbon disulfide affects the central nervous system, cardiovascular system, eyes, kidneys, liver, and skin. It may be absorbed through the skin as a vapor or liquid, inhaled, or ingested. The probable oral lethal dose for a human is between 0.5 and 5 g/kg or between 1 oz and 1 pt (or 1 lb) for a 70-kg person (150). The lowest lethal dose for humans has been reported at 14 mg/kg.

Most industrial poisonings from carbon disulfide occur in the viscose rayon manufacturing industry. The neuropsychiatric signs include mania, hallucinations, memory loss, increased suicide rates, tremors, and optical and peripheral neuropathy, as well as accelerated atherogenic changes. Underlying changes in mineral balance may cause some of these changes (119). The effect of carbon disulfide exposure on the cardiovascular system was investigated in 162 workers with a minimum of 1 year of work in a viscose rayon factory in central Taiwan. The exposed group included 118 workers (113 men and 5 women) in the manufacturing areas such as viscose manufacturing, cellophane processing, ripening, and filament spinning. Nine percent of all subjects, exposed and control, were treated for existing cardiovascular conditions. In the exposure group, 8.5% of the subjects had hypertension and 5.8% had a previous diagnosis of renal disease. The equivalent figures for the control group were 11.4% and 5.9%, respectively. Workers in the cellophane processing area during heating were exposed to 54.60 parts per million (ppm) CS_2 compared with 167 parts per billion (ppb) in the laboratory testing area for the reference group. Persons in the filament spinning section were exposed to 28.21 ppm CS_2. The exposed group had a significantly higher incidence of abnormal sodium levels (12.8%) compared with the reference group (2.3%). The exposed group exhibited statistically significant abnormalities in both generalized ST-T segment changes (13.2%) and generally conducted disturbances (16.7%) compared with 3% and none for the reference group, respectively. Exposed workers had a 4.18 times greater risk of exhibiting ECG abnormalities (170).

Table 44.6 lists representative levels of effect upon humans, and corresponding ranges in concentration of inhaled carbon disulfide. The predominant effect of high concentrations of carbon disulfide is narcosis, and death may result from respiratory failure. Less severe exposures may result in headache, giddiness, respiratory disturbances, precordial distress, and gastrointestinal disturbances. The possibility of injury to the CNS from a single severe acute exposure has been reported.

21.4.2.3.2 Chronic and Subchronic Toxicity. Repeated brief exposures to high concentrations or prolonged exposures to low concentrations at CS_2 are of much greater industrial importance than single acute exposures. Among the subjective complaints that characterize chronic carbon disulfide poisoning are fatigue, loss of memory, insomnia, listlessness, headache, excessive irritability, melancholia, vertigo, weakness, loss of appetite, gastrointestinal disturbances, and impairment of sexual functions. Visual

Table 44.6. Effects of Various Concentrations of Carbon Disulfide on Humans

Effects	Concentration mg/L	ppm
Slight or no effect	0.5–0.7	160–230
Slight symptoms after a few h	1.0–1.2	320–390
Symptoms after $\frac{1}{2}$ h	1.5–1.6	420–510
Serious symptoms after $\frac{1}{2}$ h	3.6	1150
Dangerous to life after $\frac{1}{2}$ h	10.0–12.0	3210–3850
Fatal in $\frac{1}{2}$ h	15.0	4815

disturbances, loss of reflexes, hallucinations, mania, or chronic dementia may occur. Lung irritation has been reported. Degenerative changes in the blood and blood-forming organs reportedly occur, sometimes after poisoning has progressed. Dermatitis, and even blistering, may result from contact with vapor or liquid on the skin or mucous surfaces. One hundred workers were exposed for 4 years to average concentrations of 1.0 to 5.5 ppm hydrogen sulfide and 1.9 to 26.4 ppm carbon disulfide, or a combined sulfide gas and vapor concentration of 2.9 to 31.9 ppm. There was no indication of intoxication. Improvement or complete recovery can be expected if the exposure is discontinued before severe damage results. Barthelemy (171), reporting on 10 years of experience in the manufacture of viscose rayon, cites three cases of poisoning due to excessive exposure to carbon disulfide: one of mental derangement and two with impaired motor nerves that adversely affected the leg muscles. All three workmen recovered completely within a few months after exposure terminated. He further states that when the carbon disulfide in the air was kept below 30 ppm and the hydrogen sulfide below 20 ppm, no trouble whatsoever was experienced.

21.4.2.3.4 Reproductive and Development. Recent epidemiological studies of female workers in China suggested that there were increased teratogenic effects. Congenital heart defects in offspring were most frequently observed. When the workers were divided on the basis of exposure above or below 10 mg/m^3, there was no difference between the groups, suggesting that the threshold for these effects was below the exposures encountered or was unrelated to dose (172).

21.5 Standards, Regulations, or Guidelines of Exposure

The OSHA PEL-TWA is 20 ppm with a ceiling of ACGIH TLV for carbon disulfide is 10 ppm (31 mg/m^3). It was lowered from 20 ppm to 10 ppm on the basis of cardiovascular effects. The 20 ppm limit was based on CNS effects (26). NIOSH recommends a TWA of 1 ppm (3 mg/m^3) and a STEL of 10 ppm (24). NIOSH indicates that 500 ppm is immediately dangerous to life or health (29).

The current PEL has little margin of safety for developmental effects. Alcoholics and those suffering from neuropsychic trouble are at special risk.

22.0 Sulfur Monochloride

22.0.1 CAS Number: [10025-67-9]

22.0.2 Synonyms:
Sulfur chloride; chloride of sulfur; sulfur subchloride; disulfur dichloride; sulfur chloride (mono); sulfur monochloride, 97%

22.0.4 Molecular Weight: 135.05

22.0.5 Molecular Formula: S_2Cl_2

22.0.6 Molecular Structure: Cl–S–S–Cl

22.1 Chemical and Physical Properties
See Table 44.4.

22.1.1 Odor and Warning Properties
Pungent nauseating irritating odor.

22.2 Production and Use
Sulfur monochloride (S_2Cl_2) is used in vulcanizing and in curing rubber. Sulfur chloride has been used in oven "curing" atmospheres, in the manufacturing of rubber-coated fabrics and in some such operations, has been poured into open containers and placed on steam coils on the floor of the curing oven with little or no ventilation. The leakage into the room in such instances caused pronounced irritation to the eyes and nose of anyone working there. The distribution and collection of open containers involve brief exposures to relatively high concentrations. Men who do this work sometimes unadvisedly rely upon holding their breath during the period of exposure to high vapor concentration, rather than wearing respirators. Sulfur monochloride is also used in the manufacture of organic chemicals, printers' inks, varnishes, and cements, in hardening soft woods, and as an agricultural insecticide (1).

22.3 Exposure Assessment

22.3.1 Air
Sulfur monochloride vapor in the air may be determined by scrubbing through scrubbers containing a measured quantity of 0.1 N silver nitrate acidified with nitric acid. When sampling has been completed, add 0.1 N sodium chloride solution equivalent to the silver nitrate used; and titrate the excess chloride with additional 0.1 N silver nitrate. The reaction is as follows:

$$2\,S_2Cl_2 + 2\,H_2O + 4\,AgNO_3 \rightarrow 4\,AgCl + 3\,S + SO_2 + 4\,HNO_3$$

22.4 Toxic Effects

Sulfur monochloride has a suffocating odor and is strongly irritating to the eyes, nose, and throat. A concentration of 150 ppm was fatal to mice after an exposure of 1 min, but the degree of toxicity has not yet been well established. The irritant effects are due to the sulfur dioxide and hydrochloric acid liberated by hydrolysis. Because this occurs rather readily, most of the irritant action likely is expended upon the upper respiratory tract. However, if the hydrolysis is not completed in the upper respiratory tract, injury to the bronchioles and alveoli would result.

22.5 Standards, Regulations, or Guidelines of Exposure

The ACGIH ceiling concentration for sulfur monochloride is 1 ppm (6.0 mg/m^3). OSHA has established a TWA PEL of 1 ppm and NIOSH recommends this limit as a ceiling value (24). NIOSH indicates that 5 ppm is immediately dangerous to life and health (29).

23.0 Thionyl Chloride

23.0.1 CAS Number: [7719-09-7]

23.0.2 Synonyms: Sulfurous oxychloride; sulfur chloride oxide; thionyl dichloride; sulfur oxychloride; sufinyl chloride; sulfurous dichloride; sulfurous chloride

23.0.3 Trade Names: NA

23.0.4 Molecular Weight: 118.97

23.0.5 Molecular Formula: Cl$_2$OS

23.0.6 Molecular Structure: $\text{Cl}-\underset{\|}{\overset{O}{S}}-\text{Cl}$

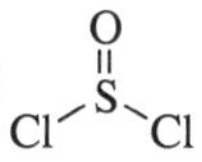

23.1 Chemical and Physical Properties

See Table 44.4.

23.2 Production and Uses

Thionyl chloride (SOCl$_2$) is used as a chlorinating agent in manufacturing organic compounds. It is also used as a solvent in high-energy lithium batteries (1).

23.3 Exposure Assessment

23.3.1 Air

Thionyl chloride may be determined by the same method as sulfur monochloride.

23.4 Toxic Effects

Thionyl chloride is an irritant to the eyes, mucous membrane, and skin. The information on this compound in the open literature is very limited. Hooker Chemical Company recommended 1.0 ppm as an industrial standard (26). A 20-min exposure to 17.5 ppm is said to have proved fatal to cats (149).

23.5 Standards, Regulations, or Guidelines of Exposure

The ACGIH ceiling value for thionyl chloride ($SOCl_2$) is 1 ppm (4.9 mg/m^3). OSHA does not have a standard, however, NIOSH also recommends a ceiling limit of 1 ppm. The production of 3 ppm of chlorine from 1 ppm of thionyl chloride is given as the rationale for the ACGIH ceiling. Thionyl chloride also produces sulfur dioxide, another irritant. Thus it is assumed that the 1-ppm ceiling would prevent the irritant effects of thionyl chloride's reaction products (26).

24.0 Sulfuric Acid

24.0.1 CAS Number: [7664-93-9]

24.0.2 Synonyms: Oil of vitriol; Dipping Acid; Vitriol Brown Oil; sulfuric; Acid Mist; hydrogen sulfate; sulfur acid

24.0.4 Molecular Weight: 98.07

24.0.5 Molecular Formula: H_2SO_4

24.0.6 Molecular Structure:
$$HO-\underset{\underset{O}{\|}}{\overset{\overset{O}{\|}}{S}}-OH$$

24.1 Chemical and Physical Properties

See Table 44.4.

24.1.2 Odor and Warning Properties

Sharp, acrid odor.

24.2 Production and Use

Sulfuric acid is the most widely used industrial chemical employed in fertilizer manufacture, petroleum refining, electroplating, acid cleaning, and storage batteries. (173) It is a starting material for a wide variety of chemicals.

Sulfuric acid has been an important item of commerce since the early to middle 1700s. It has been known and used since the Middle Ages. In the eighteenth and nineteenth centuries, it was produced almost entirely by the chamber process, in which oxides of nitrogen (as nitrosyl compounds) are used as homogeneous catalysts for the oxidation of sulfur dioxide. The product made by this process is of rather low concentration (typically 60° Baumé, or 77–78 wt% H_2SO_4). This is not high enough for many of the commercial uses of the 1990s. The chamber process is therefore considered obsolete for primary sulfuric acid production. However, more recently, modifications to the chamber process have been used to produce sulfuric acid from metallurgical off-gases in several European plants.

During the first part of the twentieth century, the chamber process was gradually replaced by the contact process. The primary impetus for development of the contact process came from a need for high strength acid and oleum to make synthetic dyes and organic chemicals. Two contact process employing platinum catalysts began to be used on a large scale late in the nineteenth century. The pace of its development was accelerated during World War I to provide concentrated mixtures of sulfuric and nitric acid for explosives production.

In 1875, a paper by Winkler awakened interest in the contact process, first patented in 1831. Winkler claimed that successful conversion of SO_2 to SO_3 could be achieved only with stoichiometric, undiluted ratios of SO_2 and O_2. Although erroneous, this belief was widely accepted for more than 20 years and the process was employed by a number of firms. Meanwhile, other German firms expanded a tremendous amount of time and money on research. This culminated in 1901 with Knietsch's lecture before the German Chemical Society revealing some of the investigations carried out by the Badische Anilin-und-Soda-Fabrik. This revealed the abandonment of Winkler's theory and further described principles necessary for successful application of the contact process.

In 1915, an effective vanadium catalyst for the contact process was developed and used by Badische in Germany. This type of catalyst was employed in the United States starting in 1926 and gradually replaced platinum catalysts over the next few decades. Vanadium catalysts have the advantages of exhibiting superior resistance to poisoning and being relatively abundant and inexpensive, compared to platinum. After World War II, the typical size of individual contact plants increased dramatically in the United States and around the world to supply the rapidly increasing demands of the phosphate fertilizer industry. The largest sulfur burning plants as of the mid-1990s produce approximately 3300 metric tons of acid per day. Plants using sulfur in other forms, especially SO_2 from smelting operations (metallurgical plants), have also increased in size. One metallurgical plant has been built to produce 3500 metric tons of acid per day.

Another significant change in the contact process occurred in 1963, when Bayer AG announced the first large-scale use of the double-contact (double-absorption) process. In this process, SO_2 gas that has been partially converted to SO_3 by catalysis is cooled, passed through sulfuric acid to remove SO_3, reheated and then passed through another one or two catalyst beds. Through these means, overall conversions can be increased from about 98% to > 99.7%, thereby reducing emissions of unconverted SO_2 to the atmosphere. Because of worldwide pressures to reduce SO_2 emissions, most plants as of the mid-1990s utilize double-absorption. An early U.S. patent disclosed the general concept of this process, but apparently it was not reduced to practice at that time.

Sulfuric acid may be produced by the contact process from a wide range of sulfur-bearing raw materials by several different process variants, depending largely on the raw material used. In some cases sulfuric acid is made as a by-product of other operations, primarily as an economical or convenient means of minimizing air pollution (qv) or disposing of unwanted by-products.

The contact process remained virtually unchanged from its introduction in the 1800s until the 1960s, when the double absorption process was introduced to reduce atmospheric SO_2 emissions. Double absorption did not, however, substantially change the nature of the process or the process equipment. In the 1970s and 1980s the increased value of energy and production of sulfuric acid from a variety of waste products, including off-gases and spent sulfuric acid, led to a number of process and equipment modifications.

The principal direct raw materials used to make sulfuric acid are elemental sulfur, spent (contaminated and diluted) sulfuric acid, and hydrogen sulfide. Elemental sulfur is by far the most widely used. In the past, iron pyrites or related compounds were often used; but as of the mid-1990s this type of raw material is not common except in southern Africa, China, Kazakhstan, Spain, Russia, and Ukraine. A large amount of sulfuric acid is also produced as a by-product of nonferrous metal smelting, roasting sulfide ores of copper, lead, molybdenum, nickel, zinc, or others.

Historically, consumption of sulfuric acid has been a good measure of a country's degree of industrialization and also a good indicator of general business conditions. This is far less valid in the 1990s, because of the heavy sulfuric acid usage by the phosphate fertilizer industry. Of total U.S. sulfuric acid consumption in 1994 of 42.5×10^6 metric tons, more than 70% went into phosphate fertilizers compared to 45% in 1970 and 64% in 1980. Uses other than fertilizer have grown only slowly or declined. This trend is expected to continue.

Other uses of sulfuric acid are in producing textile fibers, explosives, alkylate for gasoline production, pulp and paper, detergents, inorganic pigments, other chemicals, as a leaching agent for ores, a pickling agent for iron and steel, and as a component of lead storage batteries. Sulfur trioxide, either as liquid or from oleum, finds significant use as a sulfonating agent for surfactants.

World production of sulfuric acid in 1993 was 135.3 million tons. United States consumption of sulfuric acid in 1993 was 38.2 million tons.

24.3 Exposure Assessment

Workplace assessment of exposures may use NIOSH analytical method # 7903 (30b). A number of ASTM methods can also be used for determining H_2SO_4 concentrations.

24.4 Toxic Effects

24.4.1 Experimental Studies

Guinea pigs are more sensitive to sulfuric acid than rats, mice, or rabbits (174). The LC values for an 8-hour exposure to sulfuric acid range from 20 to 60 mg/m^3, depending on the age of the animals and the particle size of the sulfuric acid mist (175, 176). Concentrations of sulfuric acid lower than 5 mg/m^3 produce some bronchoconstriction in guinea pigs exposed for 1 hour (177, 178) Small particles of sulfuric acid mist produced the greater effect.

Amdur (177) exposed guinea pigs to concentrations of 2 to 200 mg of sulfuric acid mist/m^3 in the atmosphere for 1-h periods. Using particle sizes of 0.8, 2.5, and 7 mm,

Amdur (177) found that the smallest size was the most effective at the lowest level (2 mg/m^3). The 2.5-mm droplets caused the greatest response in the higher concentrations and at 200 mg/m^3 caused the deaths of all four animals exposed to this concentration within 1 h. The 7-mm particles in concentrations up to 30 mg/m^3 produced only a slight response because they did not penetrate beyond the upper respiratory tract.

24.4.1.2 Chronic and Subchronic Toxicity. Alarie et al. (179) allowed monkeys to inhale continuously several concentrations of sulfuric acid mists for 2 years. A concentration of 0.38 mg/m^3 (2.15 µm, mass median diameter [MMD] produced minimal histopathology: 0.48 mg/m^3 (0.65 µm, MMD) resulted in a slight alteration in ventilation; 2.43 mg/m^3 (3.6 µm, MMD) produced moderate changes and a slight decrease in arterial oxygen. Concentrations of 4.79 mg/m^3 (0.73 µm, MMD) produced moderate to severe histopathology and moderate changes in the function.

24.4.1.7 Other: Neurological, Pulmonary, Skin Sensitization, etc. Cutaneous application of a 10% aqueous solution of sulfuric acid to intact or abraded skin of rabbits or guinea pigs produced negligible irritation (180).

24.4.2 Human Experience

24.4.2.2.1 Acute Toxicity. Numerous case reports (173, 181, 182) and reviews of the literature (183, 184) document the toxicology of ingested sulfuric acid, in particular esophageal and gastric necrosis (185). Amdur et al. (186) reported the results of the inhalation exposure of human subjects to sulfuric acid mist at, concentrations ranging from 0.35 to 5 mg/m^3 for 5 to 15 minutes. Concentrations below 1 mg/m^3 could be detected by odor, taste, or irritation, For two subjects, the threshold was 1 mg/m^3, a concentration of 3 mg/m^3 was noticed by all, and 5 mg/m^3 was considered very objectionable to some but less so to others. Inhalation at the last concentration usually produced coughing.

24.4.2.3.5 Carcinogenesis. The International Agency for Research on Cancer (IARC) determined that there is sufficient evidence that occupational exposure to strong inorganic acid mists containing sulfuric acid is carcinogenic and, accordingly, has categorized such acids as Group 1, carcinogenic to humans (187).

24.5 Standards, Regulations, or Guidelines of Exposure

The ACGIH TLV-TWA is 1 mg/m^3 with a STEL of 3 mg/m^3. The OSHA PEL-TWA and the NIOSH REL is 1 mg/m^3.

BIBLIOGRAPHY

1. *Kirk-Othmer Concise Encyclopedia of Chemical Technology*, 4th ed., Vol. 18, Wiley-Interscience, New York, 1996.
2. D. E. Rushing, *Am. Ind. Hyg. Assoc. J.* **23**, 383 (1962).

3. M. J. Prager and W. R. Seitz, *Anal. Chem.* **47**, 148 (1975).
4. N. A. Tavilie, E. Perez, and D. P. Illustre, *Anal. Chem.* **34**, 866 (1962).
5. J. M. Cameron and R. S. Patrick, *Med. Sci. Law* **8**, 209 (1966).
6. J. M. Cameron and R. S. Patrick, *Med. Sci. Law* **6**, 209–214 (1966).
7. T. W. Clarkson, In: W. J. Hayes, Jr. and E. R. Laws, Jr. eds., *Handbook of Pesticide Toxicology*, Vol. 2, Academic Press, New York, 1991, pp. 552–555.
8. R. S. Diaz-Rivera et al., *Medicine (Baltimore)* **29**, 269 (1950).
9. F. A. Simon and L. K. Pickering, *J. Am. Med. Assoc.* **235**, 1343 (1976).
10. M. M. McCarron, G. P. Gaddus, and A. T. Trotter, *Clin. Toxicol* **18**, 693 (1981).
11. J. P. Hughes et al., *J. Ind. Med.* **19**, 83 (1962).
12. H. Heimann, *J. Ind. Hyg. Toxicol.* **28**, 142–150 (1946).
13. T. Nomura, *J. Sci. Labour* **32**, 1099–116 (1956).
14. S. Blumenthal and A. Lesser, *Am. J. Dis. Child.* **55**, 1280–1287 (1938).
15. Tennesse Valley Authority, Unpublished Report No. 11954, 1947, pp. 18, 22–24.
16. L. G. Welt and W. B. Blythe, In: L. S. Goodman and A. Galman, eds. *The Pharmacological Basis of Therapeutics*, 4th ed., The Macmillan, London, 1970 pp. 820–821.
17. C. O. Adams and B. G. Sarnat, *Arch. Pathol.* **30**, 1192 (1940).
18. L. L. Ashburn, A. J. McQueeney, and R. R. Faulkner, *Proc. Soc. Exp. Biol. Med.* **67**, 351 (1948).
19. T. Dalhamm and B. Holma, *Arch. Pathol.* **20**, 429 (1050).
20. T. J. Orcult and B. A. Pruitt, *Major Probl. Clin. Surg.* **19**, 84 (1976).
21. T. E. Bowen, T. J. Whelan, Jr. and T. G. Nelson, *Ann. Surg.* **174**, 779–784 (1971).
22. V. S. Aizenshtadt, S. M. Neruabai, and I. I. Voronin, *Gig. Tr. Prof. Zabol.* **15**, 48 (1971).
23. S. A. White and C. C. Armstrong, *Chemical Warfare Service*, Proj. A5.2-1, Edgewood Arsenal, MD, 1935.
24. American Conference of Governmental Industrial Hygienists (ACGIH), *1999 TLVs and BELs Threshold Limit Values for Chemical Substances and Physical Agents*, ACGIH, Cinicinnati, OH, 1999.
25. C. B. Dutton et al., *J. Occup. Med.* **35**,(10), 1028–1033 (1993).
26. American Conference of Governmental Industrial Hygienists (ACGIH), *Documentation of the Threshold Limit Values and Biological Exposure Indices*, 5th ed., ACGIH, Cincinnati, OH, 1986.
27. R. T. Sterner, *J. Toxicol. Environmen. Health* **39**,(2), 287–308 (1993).
28. R. T. Sterner, *J. Toxicol. Environmen. Health* **39**,(4), 497–515 1993.
29. National Institute for Occupational Safety and Health (NIOSH), *Pocket Guide to Chemical Hazards*, DHHS (NIOSH) Publ. No. 97–140, U.S. Department of Health and Human Services, Washington, DC, 1997.
30. J. F. Burgess, *Can. Med. Assoc. J.* **65**, 567 (1951).
30a. *NIOSH Manual of Analytical Methods*, 2nd ed., Vol. 5.
30b. *NIOSH Manual of Analytical Methods*, 4th ed.
31. R. Dechant, G. Sanders, and R. Graul, *Am. Ind. Hyg. Assoc. J.* **27**, 75 (1966).
31a. OSHA *Analytical Methods Manual*, 1990, 1993.
32. B. Berk, W. E. Westlake, and F. A. Gunther, *J. Agric. Food Chem.* **18**, 143 (1970).

33. D. L. Morgan, et al., *Inhalation Toxicol.* **7**,(2), 225–238 (1995).
34. A. D. Kligerman, et al., *Environmen. Mol. Mutagen.* **24**,(4), 301–306 (1994).
35. A. Glass, *J. R. Nav. Med. Serv.* **42**, 184 (1956).
36. R. N. Harger and L. W. Spoylar, *Arch. Ind. Health* **18**, 497 (1958).
37. S. Singh et al., *Br. Med. J.* **290**, 1111 (1985).
38. R. Wilson et al., *J. Am. Med. Assoc.* **244**, 148 (1980).
39. J. O. Nriagu, *Environment* **32**, 7 (1990).
40. P. Pentel, D. Fletcher, and J. Jentzen, *J. Forensic Sci.* **30**, 556 (1985).
41. U.S. Bureau of Mines, *Mineral Commodity Summaries*, U.S. Bureau of Mines, Washington, DC, 1986.
42. National Institute for Occupational Safety and Health (NIOSH), *Manual Of Analytical Methods*, 3rd ed., NIOSH, Cincinnati, OH, 1984.
42a. *NIOSH Manual of Analytical Methods*, 4th ed., Vol. 3.
43. Agency for Toxic Substances and Disease Registry (ATSDR), *Toxicological Profile for Selenium*, ATSDR, Washington, DC, 1989.
44. Z. Gregus and O. Klaassen, *Toxicol. Appl. Pharmacol.* **85**,(1), 24–38 (1986).
45. L. Friberg et al., *Handbook of the Toxicology of Metals*, 2nd ed., Vols. I and II, Elsevier, Amsterdam, 1986, Vol 2, p. 482.
46. J. R. Harr, in F. W. Oehme, ed., *Toxicity of Heavy Metals in the Environment*, F. W. Oehme, Ed., Dekker, New York, 1978, p. 393.
47. M. J. Cukierski et al., *Fundam. Appl. Toxicol.* **13**, 26 (1989).
48. S. M. Barlow and F. M. Sullivan, *Reproductive Hazards of Industrial Chemicals: An Evaluation of Animal and Human Data*, Academic Press, London, (1982), pp. 483–497.
49. W. C. Hawkes et al., *Fed. Proc., Fed. Am. Soc. Exp. Biol.* **5**,(4), A714 (1991).
50. C. C. Willhite, V. H. Ferm, and L. Zeise, *Teratology* **42**, 359–371 (1990).
51. V. H. Ferm et al., *Reprod. Toxicol.* **4**, 183–190 (1990).
52. D. S. F. Robertson, *Lancet*, **I**, 518 (1970).
53. A. F. Tarantal et al., *Fundam. Appl. Toxicol.* **16**, 147 (1991).
54. International Agency for Research on Cancer (IARC), *Monographs on the Evaluation of Carcinogenic Risk of Chemicals to Man: Some Aziridine, N-, S-, and O-Mustards and Selenium*, Vol. 9, IARC, Lyon, France, 1975.
55. National Toxicology Program (NTP), *Bioassay of Selenium Sulfide (Dermal Study) for Possible Carcinogenicity*, NCI Tech. Rep. Ser. No. 197, NTP No. 8018, National Cancer Institute, National Institutes of Health, Bethesda, MD, 1980.
56. National Toxicology Program (NTP), *Bioassay of Selsun for Possible Carcinogenicity*, NCI Tech. Rep. Ser. No. 194, NTP No. 8019, National Cancer Institute, National Institutes of Health, Bethesda, MD, 1980.
57. National Toxicology Program (NTP), *Bioassay of Selenium Sulfide (Gavage) for Possible Carcinogenicity*, NCI Tech. Rep. Ser. No. 194, NTP No. 8017, National Cancer Institute, National Institutes of Health, Bethesda, MD, 1980.
58. M. F. Robinson et al., *J. Br. Nutr.* **39**, 589–600 (1978).
59. B. H. Patterson et al., *Am. J. Physiol.* **257**, (Reg. Int. Comp. Physiol. 26), RE556–RE567 (1989).
60. R. P. Spencer and M. Blau, *Science* **136**, 155–156 (1962).

61. J. A. Butler, et al., *J. Nutr.* **120**, 751–759.
62. N. M. Griffiths, R. H. D. Stewart, and M. F. Robinson, *Br. J. Nutr.* **35**, 373–382 (1976).
63. J. R. Glover, *Ind. Med. Surg.* **39**, 50 (1970).
64. S. E. Koppel et al., *Clin. Toxicol.* **24**, 21 (1986).
65. A. K. Srivastava et al., *Vet. Hum. Toxicol.* **37**,(5), 468–469 (1995).
66. A. Hamilton, *Industrial Toxicology*, Harper, New York, 1934.
67. J. H. Ray and L. C. Altenburg, *Mutat. Res.* **78**, 261 (1980).
68. A. J. Natal, M. Brown, and P. Dery, *Vet. Hum. Toxicol.* **27**, 531 (1985).
69. E. Holstein, *Arbeitsmed.* **1**, 102 (1951).
70. I. S. Palmer, R. C. Arnold, and C. W. Carlson, *Poult. Sci.* **52**, 1984 (1973).
71. V. H. Ferm, *Adv. Teratol.* **6**, 51 (1972).
72. R. J. Shamberger, *Lancet* **2**, 1316 (1971).
73. G. W. R. Walker and A. M. Bradley, *Can. J. Genet. Cytol.* **11**, 677 (1969).
74. P. Sentein, *Chromosoma* **23**, 95 (1967).
75. H. A. Schroeder and M. Mitchner, *Arch. Environ. Health* **23**, 102 (1971).
76. R. J. Shamberger and C. E. Willis, *Crit. Rev. Clin. Lab. Sci.* **2**, 211–221 (1971).
77. R. J. Shamberger, et al., *J. Natl. Cancer Inst.* **50**, 863–870 (1973).
78. W. L. Broghamer, K. P. McConnell, and A. L. Blotcky, *Cancer (Philadelphia)* **37**, 1384–1388 (1976).
79. Epidemiological Aspects and Implications for Clinical Trials. *J. Am. Coll. Toxicol.* **5**, 29–36 (1986).
80. U. Reinhold, et al., *Acta Derm.-Venereol.* **69**, 132–136 (1989).
81. J. Ringstad, et al., *J. Clin. Pathol.* **41**, 454–457 (1988).
82. R. J. Shamberger and D. V. Frost, *Can. Med. Assoc. J.* **100**, 682 (1969).
83. National Toxicology Program (NTP), *Sixth Annual Report on Carcinogens*, Summary 1991, NTP, Research Triangle Park, NC, 1991, pp. 355–357.
84. A. Schecter, W. Shanske, and A. Stenzier, *Chest*, **77**, 554 (1980).
85. R. F. Buchan, *Occup. Med.* **3**, 439 (1947).
86. I. Rosenfeld and O. A. Beath, *Selenium-Geobotany, Toxicity, and Nutrition*, Academic Press, New York, 1964.
87. R. Rohmer, E. Carrot, and J. Gouffault, *Bull. Soc. Chim. Fr.* p. 275 (1950).
88. M. Windholz, ed., *The Merck Index*, 10th rev. ed., Merck & Co., Rahway, NJ, 1983, p. 1213.
89. C. G. Wilbur, *Clin. Toxicol.* **17**, 171–218 (1980).
90. G. Kimmerle, *Arch. Toxikol.* **18**, 140–144 (1960).
91. H. C. Dudley, *Public Health Rep.* **53**, 94 (1938).
92. L. M. Cummins, and E. T. Kimuna, *Toxicol. Appl. Pharmacol.* **20**, 89–96 (1971).
93. P. A. McAdam, and O. A. Levander, *Nutr. Res.* **7**, 601–610 (1987).
94. R. R. Herigstad, C. K. Whitehair, O. E. Olson, *Am. J. Vet. Res.* **34**, 1227–1238 (1973).
95. M Usami, and Y. Ohmo, *Teratog., Carcinog., Mutagen.* **16**, 27–36, 1996.
96. S. Biswas, G. Talukder, and A. Sharma: *Mutat. Res.* **390**, (3), 201–205 (1997).
97. National Fire Protection Association (NFPA), *Fire Protection Guide on Hazardous Materials*, 7th ed. NFPA, Quincy, MA, 1978.

98. R. A. Johnson and F. P. Kwan, *Anal. Chem.*, **25**, 11017 (1953).
99. R. A. Johnson, F. P. Kwan, and D. Westlake, *Anal. Chem.* **25**, 1017 (1953).
100. H. H. Steimberg et al., *J. Ind. Hyg. Toxicol.* **24**, 183 (1942).
101. R. H. DeMeio and F. C. Henriques, Jr., *J. Biol. Chem.* **169**, 609 (1947).
102. B. Venugopal and T. D. Luckey, *Metal Toxicity in Mammals*, Vol. 2, Plenum, New York, 1978, p. 246.
103. C. Thienes and T. J. Haley, *Clinical Toxicology*, 5th ed., Lea & Febiger, Philadelphia, PA, 1972, p. 199.
104. L. Friberg, G. R. Nordberg, and V. B. Vouk, *Handbook on the Toxicology of Metals*, Elsevier-North Holland, New York, 1979, p. 591.
105. M. I. Amdur, *Arch. Ind. Health*, **17**, 665 (1958).
106. P. Lampert, F. Garro, and A. Pentschew, *Acta Neuropathol.* **15**, 308 (1970).
107. A. D. Toews et al., *J. Neurosci. Res.* **26**, 501 (1990).
108. T. W. Bouldin et al., *Neuro toxicology*, **10**, 79 (1989).
109. S. Duckett, *Experientia* **26**, 1239 (1970).
110. K. F. Jackson et al., *Acta Neuropathol.* **78**, 301 (1989).
111. F. Dijkstra, *Water Sci. Technol.* **20**, 83 (1988).
112. International Labour Office (ILO) *Encyclopedia of Occupational Health and Safety*, Vols. I and II, McGraw-Hill, New York, 1971, p. 1391.
112a. *NIOSH Manual of Analytical Methods*, 4th ed., Vol. 3.
113. M. Sittig, *Handbook of Toxic and Hazardous Chemicals and Carcinogens*, 3rd ed., Vol.2, Noyes Publications, Park Ridge, NJ, 1991, pp. 1511–1512.
114. G. Kimmerele, *Arch. F. Toxikol.* **18**, 40 (1960).
115. G. J. Hathaway, et al., In G. J. Hathaway et al., eds., *Proctor and Hughes' Chemical Hazards of the Workplace*, 3rd. ed., Van Nostrand-Reinhold, New York, 1991, p. 531.
116. M. E. A. Cerwenka, Jr. and W. C. Cooper, *Arch. Environ. Health* **3**, 189–200 (1961).
117. E. S. Blackadder and W. G. Manderson, *Br. J. Ind. Med.* **32**, 59 (1975).
118. T. Sollmann, *A Mannual of Pharmacology*, Saunders, Philadelphia, PA, 1957, p. 150.
119. M. J. Ellenhorn and D. G. Barceloux, *Medical Toxicology*, Elsevier, Amsterdam, 1988.
120. National Research Council of Canada (NRCC), *Sulphur and its Inorganic Derivatives in the Canadian Environment*, NRCC, 1977, p. 56.
121. Occupational Safety and Health Administration (OSHA), *Analytical Methods Manual*, U.S. Department of Labor, OSHA Analytical Laboratory, Salt Lake City, UT, 1985.
122. World Health Organization, *WHO Tech. Rep. Ser.* **707**, 1115 (1984).
123. L. J. Roger et al., *J. Appl. Physiol.* **59**, 784 (1985).
124. T. J. Witek et al., *Int. Arch. Occup. Environ. Health*, **55**, 179 (1985).
125. R. S. Chapman, D. C. Calafiorne, and V. Hasselblad, *Am. Rev. Respir. Dis.* **132**, 261 (1985).
126. A. M. Lee and J. F. Fraumeni, *J. Natl. Cancer Inst.* **42**, 1045 (1969).
127. S. Laskin et al., *Combined Carcinogen-irritant Animal Inhalation Studies,* presented at OHOLO Biological Conference, New Zion, Israel, 1975.
128. R. A. Kehoe et al., *J. Ind. Hyg.* 14, 159 (1932).
129. I. O. Skaple, *Br. J. Ind. Med.* **21**, 69 (1964).
130. B. G. Ferris, S. Puleo, and H. Y. Chen, *Br. J. Ind. Med.* **36**, 127 (1979).

131. T. J. Smith et al., *Am. Rev. Respir. Dis.* **37**, 149 (1987).
132. U.S. Environmental Protection Agency (USEPA), *Pesticide Index*, USEPA, Washington, DC, 1985.
133. R. O. Beauchamp et al., *CRC Crit. Rev. Toxicol.* **13**, 25 (1984).
134. J. Kangas, P. Jappinen, and H. Savolainen, *Am. Ind. Hyg. Assoc. J.* **45**, 787 (1984).
135. Savolainen, *Biol. Monit.* **1**, 27 (1991).
136. M. Campanya, P. Sanz, and R. Reig, *Med. Lav.* **80**, 251 (1989).
137. W. Burnett et al., *Am. J. Ind. Med.* **117**, 1277 (1977).
138. M. Thoman, *Clin. Toxicol.* **2**, 383 (1969).
139. M. Arnold et al., *J. Occup. Med.* **27**, 373 (1985).
140. T. H. Milby, *J. Occup. Med.* **4**, 431 (1962).
141. R. Frank, in J. A. Merchant, ed., *Occupational Respiratory Diseases*, NIOSH, DHHS (NIOSH) Publ. No. 86-102, Division of Respiratory Disease Studies, Appalachian Laboratory for Occupational Safety and Health, 1986, pp. 571-605.
142. Chemical Industry Institute of Toxicology, *90-Day Vapor Inhalation Toxicity Study of Hydrogen Sulfide in Fischer-344 Rats*, Fiche No. 0000255-0, Doc. No. FYI-OTS-0883-0255, U.S. Environmental Protection Agency, Office of Toxic Substances Public Files, 1983.
143. Chemical Industry Institute of Toxicology, *90-Day Vapor Inhalation Toxicity Study of Hydrogen Sulfide in Sprague-Dawley Rats*, Fiche Number 0000255-0, Doc. No. FYI-OTS-0883-0255, U.S. Environmental Protection Agency, Office of Toxic Substances Public Files, 1983.
144. Chemical Industry Institute of Toxicology, *90-Day Vapor Inhalation Toxicity Study of Hydrogen Sulfide in B6C3F1 Mice*, Fiche No. 0000255-0, Doc. No. FYI-OTS-0883-0255, U.S. Environmental Protection Agency, Office of Toxic Substances Public Files, 1983.
145. A. Lopez et al., *Fundam. Appl. Toxicol.* **9**, 753 (1987).
146. A. Lopez et al., *Vet. Pathol.* **25**, 376 (1988).
147. A. Lopez et al., *Am. J. Vet. Res.* **49**, 1107 (1988).
148. A. P. Saillenfait, P. Bonnet, and J. deCeaurriz, *Toxicol. Lett.* **48**, 57 (1989).
149. N. I. Sax, Dangerous *Properties of Industrial Materials*, 5th ed., Van Nostrand, New York, 1979.
150. R. E. Gosselin, *Clinical Toxicology of Commercial Products*, Williams & Wilkins, Baltimore, MD, 1984.
151. L. Adelson and I. Sunshine, *Arch. Pathol.* **81**, 375 (1966).
152. I. H. Ohya, H. Komoriya, and Y. Bunai, *Res. Pract. Forensic Med.* **28**, 119 (1985).
153. L. W. Spolyar, *Ind. Health Mon.* **11**, 116 (1951).
154. J. M. McDonald and A. P. McIntosh, *Arch. Ind. Hyg. Occup. Med.* **3**, 445 (1951).
155. J. Deng and S. Chang, *Am. J. Ind. Med.* **11**, 447 (1987).
156. National Institute for Occupational Safety and Health (NIOSH), *Criteria for a Recommended Standard, Occupational Exposure to Hydrogen Sulfide*, NIOSH, Cincinnati, OH, 1977.
157. H. W. Haggard, *J. Ind. Hyg. Toxicol.* **7**, 113 (1925).
158. S. Kosmider, E. Rogala, and A. Pacholek, *Arch. Immunol. Ther. Exp.* **15**, 731 (1967).
159. G. Ahlborg, *Arch. Ind. Hyg. Occup. Med.* **3**, 247 (1951).
160. M. S. Hua and C. C. Huang, *J. Clin. Exp. Neuropsychol.* **10**, 328 (1988).

161. H. H. Wasch et al., *Arch. Neurol. (Chicago)* **46**, 902 (1989).
162. U.S. Environmental Protection Agency (USEPA), *Health Assessment Document for Hydrogen Sulfide*. External Review Draft, EPA/600/8-86/026A, prepared by the Office of Health and Environmental Assessment, Environmental Criteria and Assessment Office, Research Triangle Park, NC, 1990.
163. G. A. Poda, *Arch. Environ. Health* **12**, 795 (1966).
164. B. D. Hardin et al, *Scand. J. Work Environ. Health* **7**, (Suppl. 4), 66 (1981).
165. C. Jones-Price et al., *Teratologic Evaluation of Carbon Disulfide (CAS No. 75-15-0) Administered to New Zealand White Rabbits on Gestational Days 6 through 15* Gov. Rep. Announce. Index, Issue 15, NTIS PB 84-192350, National Center for Toxicological Research, Jefferson, AR, 1984.
166. B. L. Johnson et al., *Neuro. Toxicology* **4**, 53 (1983).
167. S. Tabacova, B. Nikiforov, and L. Balabaeva, *J. Appl. Toxicol.* **3**, 223 (1983).
168. M. Pergal et al., *Arch. Environ. Health* **25**, 38 (1972).
169. M. Pergal, N. Vukojevic, and D. Djuric, *Arch. Environ. Health* **25**, 42 (1972).
170. H.-W. Kuo et al., *Int. Archi. Occup. Environ. Health* **10**(1), 61–66 (1997).
171. H. L. Barthelmey, *J. Ind. Hyg. Toxicol.* **21**, 141 (1939).
172. Y. S. Bao et al., *Teratology* **43**, 451 (1991).
173. D. P. Degenhardt, and R. G. Henderson, *Lancet* **2**, 425 (1942).
173a. Kirk Othmer Concise Encyclopedia of Technology, 4th ed., Vol. 23, Wiley-Interscience, New York, 1997.
174. J. F. Treon et al., *Arch. Hyg. Occup. Med.* **2**, 716–734 (1950).
175. M. O. Amdur, R. Z. Schultz, and P. Drinker, *Arch. Hyg. Occup. Med.* **5**, 318–328 (1952).
176. R. E. Pattle, F. Burgess, and H. Cullumbine, *J. Pathol. Bacteriol.* **72**, 219–232 (1956).
177. M. O. Amdur, *Arch. Ind. Health* **18**, 407–414 (1958).
178. M. O. Amdur, M. Dubriel, and D. A. Creasia, *Environ. Res.* **15**, 418–423 (1978).
179. Y. Alarie et al., *Arch. Environ. Health* **27**, 16–24 (1973).
180. G. A. Nixon, C. A. Tyson, and W. C. Wertz, *Toxicol. Appl. Pharmacol.* **31**, 481–490 (1975).
181. E. G. Bovill, F. A. Bulawa, and R. G. Ollivetti, *Gastroenterology* **17**, 436–441 (1951).
182. R. Stewart-Harrison, *Br. J. Radio.* **7**, 48 (1934).
183. C. J. Polson, and R. N. Tattersall, *Clinical Toxicology*, Lippincott, Philadelphia PA 1959.
184. R. E. Gosselin, R. P. Smoith, and H. C. Hodge, *Clinical Toxicology of Commercial Products*, 5th ed., Sec. III, Therapeutics Index, Williams & Wilkins, Baltimore, MD, 1984, pp. 8–12.
185. D. B. Hawkins, M. J. Demeter, and T. E. Barnett, *Layrngoscope* **90**, 98–109 (1980).
186. M. O. Amdur, L. Silverman, and P. Drinker, *Arch. Ind. Hyg. Occup. Med.* **6**, 305–313 (1952).
187. International Agency for Research on Cancer (IARC), *Monographs on the Evaluation of Carcinogenic Risks to Humans*, Vol.54, IARC, Lyon, France, 1992, pp. 41–130.

CHAPTER FORTY-FIVE

Boron

B. Dwight Culver, MD, Philip L. Strong, Ph.D., and Jay F. Murray, Ph.D.

A INTRODUCTION

Boron is the fifth element of the periodic table and the only electron-deficient nonmetallic element. Consequently, boron has a high affinity for electronegative atoms such as oxygen and forms strong covalent boron–oxygen bonds in compounds known as *borates*. The boron atoms can bond to oxygen to form either planar trigonal BO_3 or negatively charged tetrahedral BO_4^- units. Essentially all boron-containing minerals are inorganic borate salts containing a mixture of trigonal and tetrahedral boron atoms. These naturally occurring borates are converted to a large variety of commercial boron-containing products, 98% of which are also borates.

Boron in elemental form is an expensive refractory material that is produced by the costly reduction of borates, and used in relatively small quantities. In this chapter this material is referred to as *elemental boron*.

The term *boron* alone is used in a generic sense here to mean boron as it is included in any chemical combination in boron-containing materials. The inorganic boron-containing materials share many chemical and biological characteristics thought to be due to unique properties of the boron element. For this reason, this chapter deals rather extensively with these shared characteristics first. Following this, individual inorganic boron-containing compounds that have a significant use in industry are considered and to the extent that they have unique attributes, are discussed here.

Borates are ubiquitous in nature in the form of a variety of borate minerals. Yet, in relatively few locations have they been found in quantities sufficiently concentrated to

---

Patty's Toxicology, Fifth Edition, Volume 3, Edited by Eula Bingham, Barbara Cohrssen, and Charles H. Powell.
ISBN 0-471-31934-1   © 2001 John Wiley & Sons, Inc.

mine commercially. Borate minerals are typically found in nature in low concentrations as alkali-metal and alkaline-earth borate and borosilicate minerals, and, very rarely, as boric acid. The element boron has been known to be an essential nutrient in vascular plants since the mid-1920s (1) and is commonly used in fertilizer applications. It is also generally accepted to be essential for diatoms and marine algal flagellates (2). More recently, several reports have been published on the essentiality of boron in animals. It has been reported to be essential in the frog, *Xenopus laevis*, for reproduction and development (3–5). Embryos from frogs cultured in low boron environments showed increased necrosis and poor viability. The ovaries and testes of adult frogs maintained on low boron for 120 days were distinctly atrophied. In fish (trout and zebra fish) development and retinal health was reported to be adversely affected (6–8). Embryonic trout growth is stimulated in a dose-dependent manner. Most zebra fish embryos did not survive the early blastula development stage in a low boron environment. Those that did survive were reported to be photophobic and were shown to have retinal abnormalities. Studies are currently under way in mice and rats (9, 10). Similar to the frog and fish studies, mice embryos from the low boron exposure were affected at a very early development stage as shown by reduced blastula cell counts. Boron stimulates growth in vitamin D–deficient chicks (11), including modulation of cartilage calcification and positive affects on the growth plate of one-day-old chicks (12, 13).

Numerous reports from the USDA Grand Forks Human Nutrition Research Center in North Dakota have indicated a nutritional role for boron in humans and animals (14–24). Positive benefits from nutritional boron have been reported for bone health (15, 17, 22), cell membrane function (14, 18), psychomotor skills and cognitive processes of attention and memory (16, 21), response to estrogen therapy (14, 18), control of inflammatory disease (17, 20, 24), enzyme regulation (18, 20), energy metabolism (17), and macroscale mineral metabolism (14, 18, 19, 22). A World Health Organization (WHO) expert committee has concluded that boron is "probably essential" (25). Since boron is a required nutrient in plants, it occurs naturally in all foods. Fruits, vegetables, and nuts contain the highest levels. The mean dietary boron consumption in the United States is reported to be 1.17 mg/day for adult males and 0.96 mg/day for adult females (26).

Boron occurs naturally as a mixture of the stable isotopes, boron-10 and boron-11, in a ratio of approximately 4:1 and an atomic weight of approximately 10.81 (27, 28). Slight variations in this ratio are observed depending on the source of the original boron mineral, but these variations are too small to affect normal analytical determinations.

Some of the more important physical and chemical properties of boron compounds of commercial significance and widespread exposure are listed in Table 45.1. These compounds are listed in decreasing order of exposure, but the order is approximate, since in some cases the available information on production and usage is incomplete. Although inorganic borates represent by far the most important class of commercial compounds, the total list includes a number of examples with diverse chemical, physical, and toxicological properties.

Much of the chemistry of boron is governed by the electron-deficient character of the boron atom and its resulting tendency to form highly stable bonds with electronegative atoms, particularly oxygen. Therefore, the most common boron-containing minerals and commercial products are inorganic boron–oxygen compounds or borates. This property

Table 45.1. Physical Properties of Commercial Boron-Containing Compounds

Compound	Empirical Formula	Formula Weight	Wt% B_2O_3	Physical Properties	Specific Gravity[a]	MP (°C)	BP (°C)	Solubility
Group 1 Used Worldwide (In many thousands of tons/year)							
Disodium tetraborate pentahydrate	$Na_2B_4O_7 \cdot 5H_2O$	291.4	48.70	White, trigonal, crystalline solid	1.815	<200 closed space	—	In H_2O, 1.52% at 0°C, 3.2% at 20°C, 51.2% at 100°C; soluble in glycerol, ethylene glycol
Disodium tetraborate decahydrate (borax)	$Na_2B_4O_7 \cdot 10H_2O$	381.4	36.51	Colorless, monoclinic crystalline solid	1.73	62 in closed space	—	In H_2O, 1.18% at 0°C, 2.58% at 20°C, 9.55% at 50°C; soluble in EtOH, MeOH, glycerol; very soluble in ethylene glycol
Ulexite	$NaCaB_5O_9 \cdot 8H_2O$	405.23	42.95	Colorless triclinic crystalline solid	1.955	—	—	0.5% as $NaCaB_5O_9$ at 25°C in H_2O
Colemanite	$Ca_2B_6O_{11} \cdot 5H_2O$	411.08	50.81	Colorless monoclinic crystalline solid	2.42	—	Decrepitates at 480	In H_2O, 0.1% at 25°C, 0.38% at 100°C
Sodium perborate tetrahydrate	$NaBO_3 \cdot 4H_2O$	153.86	22.6	White crystalline powder	—	63[b]	250[b]	In H_2O, 2.3% at 20°C, 3.7% at 30°C
Sodium perborate monohydrate	$NaBO_3 \cdot H_2O$	99.81	34.9	White crystalline powder	—	—	—	In H_2O 1.5% at 20°C, 2.4% at 30°
Boric acid (orthoboric acid)	H_3BO_3	61.83	56.30	White waxy triclinic solid plates	1.435	170.9 in closed space	—	In H_2O, 2.52% at 0°C, 4.72% at 20°C, 27.5% at 100°C, soluble in MeOH, EtOH, slightly soluble in acetone, dimethyl ether
Disodium tetraborate	$Na_2B_4O_7$	201.27	69.2	Light gray glass	2.367 glass	741 crystalline	1575[b]	16.7% in MeOH at 25°C, 30% in ethylene glycol at 25°C
Fertilizer borate granular®	Mixture	—	46.04	Colorless solid	1.75	—	—	—
Disodium octaborate tetrahydrate [Solubor® (Polybor®, Timbor®)]	$NaB_8O_3 \cdot 4H_2O$	412.52	67.5	White crystalline granutes	—	815	—	In H_2O, 2.4% at 0°C, 9.7% at 20°C, 45.3% at 94°C

Table 45.1. (*Continued*)

Compound	Empirical Formula	Formula Weight	Wt% B_2O_3	Physical Properties	Specific Gravity[a]	MP (°C)	BP (°C)	Solubility
\multicolumn{9}{c}{*Group 2: Used Worldwide (In hundreds of tons to a few thousand tons)*}
Sodium borohydride (most sold as solution in 40% NaOH)	$NaBH_4$	37.83	28.6	White cubic crystalline solid	1.074	400[b] (dec)	—	550 g/L in H_2O at 25°C, 40 g/L in EtOH, and 164 g/L in MeOH at 25°C
Methyl borate	$B(OCH_3)_3$	103.92	10.40	Colorless liquid	0.915	−29	68.7	Soluble with decomposition in H_2O, soluble in EtOH, Et_2O
Boric oxide	B_2O_3	69.64	100	A White crystalline granules or powder	2.46 (crystals) 1.85 (powder)	450	1500	slightly soluble in cold H_2O, soluble in hot H_2O; 4.0% at 20°C H_2O
Zinc borate hydrate (2335)	$2ZnO·3B_2O·3.5H_2O$	434.66	48.05	A White powder	2.7			<0.28% in H_2O at 25°C
Barium metaborate	$BaB_2O_4·4H_2O$	295.04	23.60	Colorless monoclinic crystalline solid	3.25–3.35	900–1050		~0.3% in H_2O at 21°C
Sodium metaborate dihydrate	$NaBO_2·2H_2O$	101.83	34.19	White crystalline granules crystalline solid	1.91	90–95	>1230	In H_2O, 22.4% at 0°C, 31.0% at 20°C, 59.3% at 60°C, 81.1% at 100°C slightly soluble in MeOH, EtOH
Sodium metaborate tetrahydrate	$NaBO_2·4H_2O$	137.86	25.25	White crystalline granules	1.91	90–95	>1230	In H_2O, 14.5% at 0°C, 31% at 20°, 81.0% at 100°C
Boron trifluoride	BF_3	67.82	15.94%B	Colorless gas	2.99	−126.7	−99.9	3322/1002 at 0°C, H_2O: very soluble in cold H_2O, decomposes in hot H_2O, EtOH; soluble in concentrated H_2SO_4
Potassium fluoborate	KBF_4	125.9	8.59%B	Colorless rhombic or cubic crystalline solid	2.498	530[b]		4.4 g/L in H_2O at 20°C, 62.7 g/L in H_2O at 100°C, slightly soluble in EtOH, Et_2O
\multicolumn{9}{c}{*Group 3: Used Worldwide (In pounds to a few hundred tons per year)*}
Ferroboron	Fe_xB_y	Varies	15–20%B	Black amorphous alloy				Insoluble in H_2O, organic solvents
Biobor® JF	Mixed dioxaborinanes	23.20		Colorless liquid	1.05[68]			

Name	Formula	%B	Appearance	Density	MP (°C)	BP (°C)	Solubility
Boron trichloride	BCl_3	9.22%B	Colorless gas	0.738	−107	12.4	Reacts violently with H_2O
Diammonium tetraborate tetrahydrate	$(NH_4)_2B_4O_7 \cdot 4H_2O$	52.87	Colorless solid	1.58			In H_2O, 3.75% at 0°C, 7.63% at 30°C, 21.2% at 50°C
Ammonium pentaborate tetrahydrate	$NH_4B_5O_8 \cdot 4H_2O$	63.95	Colorless solid	1.57			In H_2O, 4% at 0°C, 7.07% at 20°C, 14.4% at 50°C
Trimethyl amine borane	$C_3H_{12}BN$	14.82%B	Colorless solid		94	172	Very soluble in aqueous acetone, soluble in $CHCl_3$, Et_2O
Sodium tetraphenyl boron	$NaB(C_6H_5)_4$	4.15%B	Colorless solid		>300		Soluble in H_2O, Soluble in EtOH, Et_2O
Triethylborane	$C_6H_{15}B$	11.03%B	Colorless liquid	0.865	−92.9	$0^{12.5\ mm}$	Insoluble in H_2O, EtOH, soluble in fused alkali
Boron carbide	B_4C	78.2%B	Refractory black rhombohedral crystalline solid	2.52	2350	>3500	
Boron nitride	BN		White hexagonal crystalline solid				
Dipotassium tetraborate tetrahydrate	$K_2B_4O_7 \cdot 4H_2O$	45.58	Colorless orthorhombic crystalline solid	1.919			In H_2O, 12.1% at 20°C, 24.0% at 50°C
Potassium pentaborate	$KB_5O_8 \cdot 4H_2O$	59.36	Colorless orthorhombic prisms	1.74			In H_2O, 1.56% at 0°C, 2.82% at 20°C, 22.3% at 100°C, soluble in hot acids
Elemental boron	B	10.81	Brown amorphous or black crystalline solid	2.34–2.37 (amorphous)	2300	2550 sublimes	Insoluble H_2O, slightly soluble in HNO_3
Boron tribromide	BBr_3	4.31%B	Colorless fuming liquid	2.698^{18}	−46	91.3	Decomposes in H_2O, soluble in EtOH, CCl_4, $SiCl_4$
Diborane	B_2H_6	78.1%B	Colorless gas	0.447^{-11} liquid 0.21 at 15°C	−165	−92.5	Decomposes H_2O, soluble in NH_4OH
Decaborane	$B_{10}H_{14}$	88.4%B	Colorless crystalline solid	0.94	99.5	213	Slightly soluble in cold H_2O; decomposes in hot H_2O; soluble in EtOH, Et_2O, C_6H_6; very soluble in CS_2

[a] Unless otherwise given in superscript, values are at 20°C.
[b] Decomposition point.

also manifests itself in the tendency for boron to expand from trigonal covalent bonding by filling its fourth valence orbital to form tetrahedral structures. This is apparent in naturally occurring borate minerals, which in every case include both trigonal and tetrahedral boron atoms.

In Table 45.2 three alternative formulas for commercially significant borate minerals are included, whereas borates in Table 45.1 are identified only by their empirical formulas, the ones most commonly used in commerce. These formulas are misleading because the borates are represented as salts of fictitious ions. For example, borax is written as $Na_2B_4O_7 \cdot 10H_2O$. The oxide formula is easier to remember but also gives a distorted view of the borate crystal structure. Using this designation borax is $Na_2O \cdot 2B_2O_3 \cdot 10H_2O$. The structural formula for borax, $Na_2[B_4O_5(OH)_4] \cdot 8H_2O$, correctly differentiates hydroxyl groups from waters of hydration and gives information as to the size and nature of the boron–oxygen rings present in the crystal structure. The structure of the cyclic boron–oxygen dianion in crystalline borax as determined by X-ray diffraction is shown in Fig. 45.1. The crystal structures of all inorganic borates consist of monomeric and polymeric structures containing boron–oxygen rings of this type.

These general structural characteristics, which are so common in the crystal structures of inorganic borates, have also been found in aqueous solutions of the commercially important water-soluble compounds, in borate glasses, and more recently in alkali borate vapors. The structures of the soluble borates in aqueous solution depend on their concentrations and pH. For example, the structure of crystalline borax, sodium tetraborate decahydrate, has been shown by X-ray diffraction studies to be the disodium salt of the cyclic dianion illustrated in Fig. 45.1. When this material is dissolved in water, a series of soluble structures result, depending on the concentration and pH. At a high pH and low concentration the predominant species is $B(OH)_4^-$. At a low pH and low concentration the trigonal boric acid molecule, $B(OH)_3$, is formed. Between these extremes of pH, and increasingly at higher concentrations, a series of monomeric and polymeric borate anions have been identified as illustrated by the examples shown in Fig. 45.2. (For a more detailed discussion of this phenomenon, see Ref. 29.)

Reduced boron compounds such as halides, hydrides, alkyls, and aryls tend to oxidize and hydrolyze readily except in those cases where stearic hindrance restricts the approach of the reactive atoms or molecules. This general reactivity and the expense of reducing the high energy boron–oxygen bonds in the readily available borates in many cases have hindered the commercial development of reduced boron-containing materials, and general exposure to these compounds is minimal compared to inorganic borates.

1 Production and Use

Although boron-containing minerals are distributed widely throughout the world, deposits that are sufficiently concentrated to mine profitably are relatively rare as noted above. Worldwide production of boron minerals in 1990 is summarized in Table 45.3 taken from the 1990 U.S. Bureau of Mines summary (30). The relatively few minerals mined are all inorganic borate salts of alkali- and alkaline-earth metals as shown in Table 45.2.

The major production facilities for the conversion of these minerals to commercial products are located in Turkey and the United States. The ores are dissolved in water, and

Table 45.2. Commercially Significant Borate Minerals

Mineral	Empirical Formula	Oxide Formula	Structural Formula	Molecular Weight Oxide Formula	Wt% B_2O_3
Ascharite (Szaibelyiite)	$Mg_2B_2O_5 \cdot H_2O$	$2MgO \cdot B_2O_3 \cdot H_2O$	$Mg_2(OH)[B_2O_4(OH)]$	168.26	41.4
Colemanite	$Ca_2B_6O_{11} \cdot 5H_2O$	$2CaO \cdot 3B_2O_3 \cdot 5H_2O$	$Ca[B_3O_4(OH)_3] \cdot H_2O$	411.08	50.8
Datolite	$Ca_2Si_2B_2O_9 \cdot H_2O$	$2CaO \cdot B_2O_3 \cdot 2SiO_2 \cdot H_2O$	$Ca_4[B_4(SiO_4)_4(OH)_4]$	319.95	21.8
Howlite	$Ca_4Si_2B_{10}O_{23} \cdot 5H_2O$	$4CaO \cdot 5B_2O_3 \cdot 2SiO_2 \cdot 5H_2O$	$Ca_2[B_3O_4(OH)_2SiB_2O_5(OH)_3]$	782.66	44.4
Hydroboracite	$CaMgB_6O_{11} \cdot 6H_2O$	$CaO \cdot MgO \cdot 3B_2O_3 \cdot 6H_2O$	$CaMg[B_3O_4(OH)_3]_2 \cdot 3H_2O$	413.35	50.5
Inderite	$Mg_2B_6O_{11} \cdot 15H_2O$	$2MgO \cdot 3B_2O_3 \cdot 15H_2O$	$Mg[B_3O_3(OH)_5] \cdot 5H_2O$	559.72	37.3
Kernite	$Na_2B_4O_7 \cdot 4H_2O$	$Na_2O \cdot 2B_2O_3 \cdot 4H_2O$	$Na_2[B_4O_6(OH)_2] \cdot 3H_2O$	273.27	50.9
Meyerhofferite	$Ca_2B_6O_{11} \cdot 7H_2O$	$2CaO \cdot 3B_2O_3 \cdot 7H_2O$	$Ca[B_3O_3(OH)_5] \cdot H_2O$	447.13	46.7
Tincal (borax)	$Na_2B_4O_7 \cdot 10H_2O$	$Na_2O \cdot 2B_2O_3 \cdot 10H_2O$	$Na_2[B_4O_5(OH)_4] \cdot 8H_2O$	381.36	36.5
Ulexite	$NaCaB_5O_9 \cdot 8H_2O$	$Na_2O \cdot 2CaO \cdot 5B_2O_3 \cdot 16H_2O$	$NaCa[B_5O_6(OH)_6] \cdot 5H_2O$	810.50	43.0

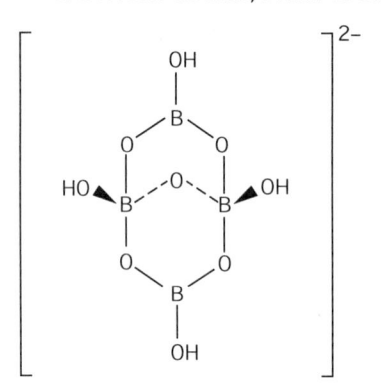

Figure 45.1. Tetraborate anion structure.

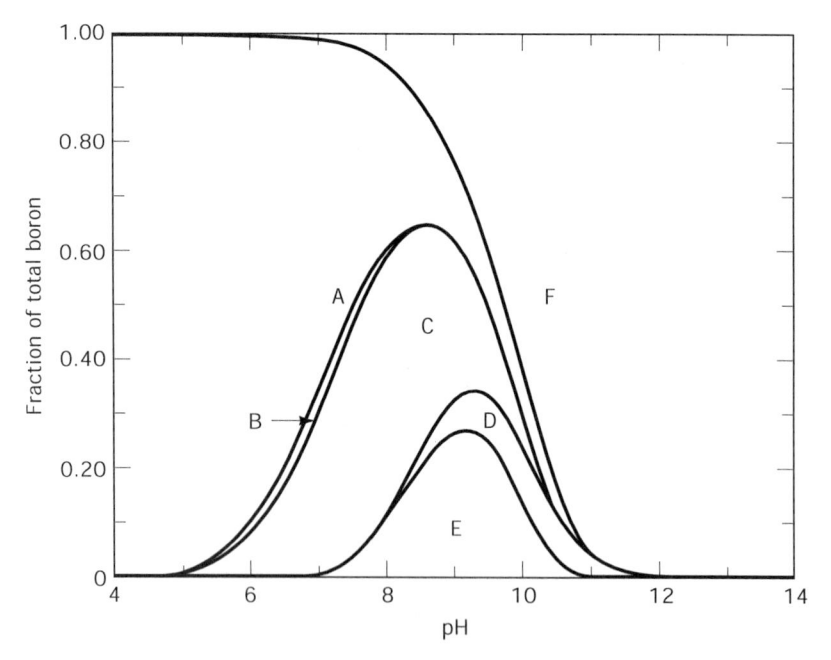

Figure 45.2. Distribution of boron (**A**) B(OH)$_3$; (**B**) B$_5$O$_6$(OH)$_4^-$; (**C**) B$_3$O$_3$ (OH)$_4^-$; (**D**) B$_3$O$_3$-(OH)$_5^{2-}$ (**E**) B$_3$O$_3$(OH)$_4^{2-}$; (**F**) B(OH)$_4^-$; where total B$_2$O$_3$ concentration is 13.93 g/L. At a given pH, the fraction of the total boron in a given ion is represented by the portion of a vertical line falling within the corresponding range.

the borates are recovered by crystallization. One producer in the United States extracts borates along with other salable inorganic salts from complex lake brines. Together, Turkey and the United States account for approximately 80% of the world's borate production.

Workers and consumers can be exposed to a variety of boron minerals and chemicals. It is difficult to quantify exactly the total exposure because of the proprietary and

Table 45.3. Worldwide Production of Boron Minerals in 1990

Country	Volume (thousands of metric tons)	Minerals
Turkey	1200	Tincal, colemanite, ulexite
United States	1094	Tincal, kernite, brines
Argentina	260	Tincal, ulexite, colemanite, howlite
Russia	175	Ascharite, datolitle, inderite, hydroboracite
Chile	132	Ulexite
China	27	Ascharite, tincal, meyerhofferite, kernite
Peru	18	Ulexite
Total tons	2906	

fragmentary information sources available in most countries. In addition, a number of minerals are handled in large quantities and converted to large-volume chemicals, which, in turn, are converted to other commercial products. For example, tincal and kernite ores are processed to sodium tetraborate pentahydrate in very large quantities, and this material, in turn, is converted in large tonnages to sodium perborates for use in household bleaches. Table 45.1 lists the major boron-containing minerals and products involved in human exposure. Specific amounts are not included because of the inaccuracy of consumption data available, but the compounds are listed in general exposure categories and in an estimated decreasing order of exposure. Group 1, compounds consumed worldwide in many thousands of tons per year, includes well over 90% of the total tonnage involved. The general class of inorganic borates represents over 98% of the total. Group 2 compounds are used in quantities from a few hundred tons to several thousand tons per year, and group 3 includes low volume compounds that are consumed at levels of pounds/year to a few hundred tons/year. If a number of group 3 compounds of the same structural type are used in relatively small quantities commercially, Table 45.1 includes only one example of each class; examples are trialkylboranes (triethylborane), zinc borates (zinc borate 2335), and amine boranes (trimethylamine borane).

As noted above, worldwide usage of boron compounds and minerals is difficult to quantify. The Bureau of Mines summary for 1993 and 1994 (30) includes the consumption figures for the United States listed in Table 45.4. International consumption would generally follow the same trends with one major exception. The use of perborate bleaches rather than chlorine bleaches in home laundry detergents in western Europe would lead to a large increase in the "soaps and detergents" category. This increase would raise worldwide consumption in this category to the second position behind glass and fiberglass. This glass category, including insulation and textile fiberglass, continues to be the largest worldwide market for borate minerals and chemicals. Following soaps and detergents are fire retardants, enamels and agriculture; in decreasing order. A major difficulty in constructing an accurate borate use picture is the large quantity of borates sold through independent distributors for unknown applications. In the United States this category amounted to approximately 17% of borate consumption in 1994.

Table 45.4. U.S. Consumption of Boron Minerals and Compounds (metric tons of B_2O_3, including imports)

End Use	1993	1994
Insulation fiberglass	106,000	97,000
Textile fiberglass	21,300	26,000
Borosilicate glasses	64,500	27,400
Soaps and detergents	13,800	14,000
Fire retardants		
Cellulosic insulation	9,670	15,800
Other	286	1,360
Enamels, frits, and glazes	10,300	15,400
Agriculture	14,800	21,200
Metallurgy	1,870	1,950
Nuclear applications	8,870	373
Sold to distributors, use unknown	37,200	51,300
Miscellaneous	32,600	23,800
Total	321,000	296,000

2 Borates in the Environment

Howe (31) has reviewed boron in the environment. Borates are found in ocean water, sedimentary rocks, coal, shale, and soils. They are released into the atmosphere from oceans and the weathering of sedimentary rocks. Boron is also released to the environment by anthropogenic sources and as noted in Table 45.4, is present in many products. Because it is present in glass, caution must be exercised in selecting containers for sampling and analysis. Concentrations of boron in air range from < 0.5 to 80 ng/m^3 with a mean of 20 ng/m^3 and in soils from 10 to 300 mg/kg with an average of 30 mg/kg. Seawater contains approximately 5 ppm boron. Concentrations of boron in domestic water supplies are generally < 1 ppm, but in borate-rich regions of Chile they have been measured as high as 15 ppm (32) and in Turkey as high as 29 ppm (33).

2.1 Methods for Measurement of Borates and Boron in the Workplace

Air samples for borates are collected and measured as total dust, most frequently using a 37-mm mixed cellulose ester filter, NIOSH method 0500 (92). However, with the change to size-selective sampling and OELs for borates that are different from nonreactive dust (PNOC), the industrial hygienist should collect air samples using a personal air sampler designed for inhalable particulate mass with a preweighed PVC filter and analyze for boron using inductively coupled plasma atomic emission spectroscopy (ICP). The inhalable particle mass fraction is selected since borates are soluble in the liquid layer overlying the respiratory tract mucosa and thus become available for systemic absorption no matter where they are deposited. The sodium borates do not appear to cause pulmonary injury, so that the respirable mass fraction will not estimate risk.

Since chemical methods do not distinguish among the separate species of borates but provide a measure of boron present, quantitation of the specific substance depends on knowledge of the source and calculation based on percent boron in the parent substance. It should be noted that in the case of the sodium salts of boron, the hydration state is dependent on temperature and humidity, so the species that is found on an air-sampling filter may be different from the material for which a sample is being taken (34). For this reason recommendations have been made to the ACGIH TLV committee to establish TWA based on boron.

2.2 Methods for Chemical Analysis of Inorganic Boron in Biological Samples

Techniques available for quantitative analysis of boron compounds in biological matrices are seldom compound specific. Most solution methods rely on analysis related to a common boron species obtained by hydrolysis or decomposition using acid or base. The common analyte species in dilute solution is an equilibrium mixture consisting primarily of boric acid, $B(OH)_3$, in acid solution, and borate ion, $B(OH)_4^-$, in basic solution. At the physiological pH (7.4) of plasma, boric acid ($pK_a = 9.2$) is the predominant species, existing $>98\%$ as undissociated $B(OH)_3$. In the case of emission spectra and neutron activation techniques, the analytical species is usually the element boron. Recently speciation of sugar alcohol borate complexes from plant extracts has been reported using size-exclusion high pressure liquid chromatography/inductively coupled plasma/mass spectrometry (SE-HPLC/ICP-MS) techniques (35), and by matrix-assisted laser desorption/ionization Fourier transform mass spectrometry (MALDI/FTMS) (36). Boron-bound plasma proteins were separated by gel chromatography and analyzed by thermal neutron activation and mass spectrometric assay (37).

Results are normally reported in percent (or milligrams per liter, micrograms per gram, *etc.*) relative to boron content or may be calculated in terms of boric acid or the specific boron compound of interest. There is some confusion in the literature regarding which boron species is being reported. When reporting analytical data for boron content, the specific boron species should be clearly stated to avoid confusion as to whether results refer to boron content, boric acid content, or whichever species is used in the calculations. In addition, there is sometimes confusion as to whether "boron" refers to elemental boron or to the boron-equivalent content of a specific sample.

Few published techniques for boron analysis have adequate sensitivity for the low boron levels found in biological samples (from ~30 µg/kg to several hundred micrograms per gram). The most popular current analytical procedure for analysis of boron in biological systems involves sample digestion in hot acid or base, possibly involving an oxidizing agent, followed by analysis of the resulting solution using ICP atomic emission spectroscopy (AES), (3, 38–42). These strongly acidic or basic digestion procedures are applicable to all boron compounds in biological systems, both organic and inorganic. Closed microwave digestion systems are preferred (43, 44). An increase in analytical sensitivity for boron has been reported using mannitol in the analyte solution (45, 46) or by adding methyl borate ester (47).

Acidic solutions of boric acid may lose some boron content to volatilization when heated. Sample procedures for samples containing a few milligrams of boron equivalent or less should include analysis of spiked samples or appropriate reference materials to be sure that significant quantities of boron are not being lost. In addition, nearly all glass contains leachable boron. Therefore, biological samples for boron analysis should not be stored in glass nor come in contact with glass during analysis. Polyethylene, and polypropylene vessels and tubing are satisfactory. Teflon can adsorb boric acid, causing losses during chemical analysis, and the adsorbed boron can be a source of contamination as well. Reviews of the analytical difficulties encountered during analysis of low concentrations of boron found in biological samples have appeared in the literature (48, 49). Reports of analytical procedures often confuse "detection limits" with "quantitation limits." "Detection limits" are the lowest levels of the particular component of concern, boron in this case, that can be detected at a significant level above background levels. It is not generally practical to analyze for a component at its "detection limit" because the variability is too great. The real issue of concern in toxicology studies is the "quantitation limit." This is the lowest level for which the analytical variation is acceptable and therefore the analytical results are meaningful. The analytical precision at the "quantitation limit" should be reported; however, most references quote detection limits only. Sometimes values reported as detection limits are actually quantitation limits, and one has to read carefully to determine which is actually meant. It is also important in interpreting analytical results to understand that precision varies with the magnitude of the value being reported. Precision, which is usually reported as percent variation, both plus and minus, from the reported analytical number, will be smaller for larger quantities and greater for smaller quantities. Samples in which percent levels of analyte are reported may vary only ±2%; values reported as a few micrograms per gram may show variation of ±100%, which seems large, but may represent only ±1 or 2 µg/g. The latter may not be significant even though the reported variation appears large.

As mentioned above, ICP methods are usually preferred for boron analysis in biological matrices. Although instrumentation is expensive and not generally available, inductively coupled plasma–mass spectroscopy (ICP/MS) gives about an order of magnitude improvement in sensitivity over ICP alone (50, 51). This technique also provides isotopic analyses. One useful application is to use the ^{10}B isotope as a tracer element since there are no stable radioisotopes of boron (52, 53). ICP/MS procedures developed specifically for boron in human serum are reported (54–56).

Another important analytical technique for boron analysis in biological systems is neutron activation analysis (NAA), which includes prompt gamma spectroscopy) (57–62). This technique does not require digestion of the sample and can be used on living subjects. The detection limit is about 1 mg ^{10}B/g of tissue and may be a limitation for applications that require greater sensitivity. Requirements for a neutron source and the ^{10}B isotope also restrict its use. For those who have access to nuclear reactor methods, prompt neutron activation gives accurate results and serves as an alternate check procedure. A study comparing ICP/AES, ICP/MS, and neutron capture radiography indicates that results are more consistent from the two ICP methods because of calibration difficulties with neutron capture radiography (63). Neutron activation combined with a mass spectrometer gives significant improvement in both sensitivity and accuracy and is the procedure used in the United States by the National Institute of Standards Technology, along with others, for determining boron levels in biological matrices (64–66).

Analytical methods for boron determination have been reviewed (67). A round-robin analytical study was carried out involving ten experienced and widely distributed research laboratories (68). Results showed good agreement among the laboratories with boron at the mg/kg level, but only three laboratories participated at the µg/kg level, and agreement was poor. This round-robin study pointed out the lack of certified standard reference materials at the µg/kg level (68).

Colorimetric analytical procedures do not require the expensive ICP instrumentation (69–74). However, colorimetric procedures are more time-consuming and require significantly more laboratory care and technical skill. Each different matrix requires individual developmental research to eliminate interferences and standardize results. Carefully developed colorimetric procedures can give analytical results with precision and accuracy equal to those from the more recent instrumental techniques and with sensitivity equal to that of ICP. Many reported results by colorimetric procedures are believed to be high in comparison to results from ICP and NAA procedures. Current colorimetric procedures should be subjected to reference analysis by ICP or NAA to be sure that reported results are in agreement. Colorimetric analytical procedures for boron are very sensitive to changes. Careful checking with known reference standards must be carried out before applying previously developed colorimetric techniques to new sample matrices.

Determination of boron in biological samples utilizing electrothermal atomic absorption has not gained wide support, partially due to formation of thermally stable boron carbides with the graphite tubing, the tendency for memory effects and the rather labor-intensive techniques required. Results utilizing this technique for analysis of boron in two standard reference materials (75), in tumor cell suspensions (76), and in urine samples (77) have been reported.

The α-track Etching (78) imaging procedure detects the ^{10}B distribution in histological test samples using a neutron-capture radiography procedure (79, 80). Use of a computer-controlled microscope stage resulted in resolution in the range of 2–4 µm and a boron lower detection limit of 0.03 ppm. A quantitative assessment of the boron distribution in whole-body animal sections using this procedure has been reported (81, 82).

Secondary-ion mass spectrometry (SIMS) electron probe microanalysis (83–85), is used for small tissue samples that are placed in the path of a high energy primary-ion beam. Surface atoms are removed by sputtering and the resulting secondary ions are detected in a mass spectrometer. The small sample size of this method is a major advantage; ICP requires about 1 mL of sample whereas SIMS requires only a few picoliters. Intracellular component analyses are being reported. However, a special probe is required for analysis of low atomic weight elements, including boron.

2.3 Use of Blood and Urine for Biomonitoring

Blood and urine boron levels can provide evidence of exposure but do not provide accurate information about the severity of exposure (86). Because boron is excreted almost entirely by the kidney and since boron has a biological half-life of about 20 h, a 24-h urine will estimate daily absorption in individuals in equilibrium with a steady-state environment. According to the U.S. 1994 Annual Reports, the total daily contribution of boron in the average diet is >0.5–3 mg.

Blood and urine samples should be collected in plastic containers and analyzed according to the methods described under analytical methods for biological samples, as described above. Estimates of whole-blood boron of unexposed adults range from a median and (range) of 0.057 (0.008–0.170) µg/mL (87) to a mean and (standard deviation) of 0.031 (0.006) µg/mL (88). Older studies with uncertain methods of sample preparation reported values 2–10 times higher. Boron concentration in urine of unexposed adults vary from a median and (range) of 0.70 (0.04–6.6) µg/mL (89) to 0.075 (0.15-2.98) µg/mL (87). Urine samples taken in the work environment are easily contaminated. Culver et al. (90) compared blood and urine boron concentration with measures of borate exposures on the job in borate workers and calculated regression equations for both: Blood B(µg) = 0.07 + 0.008 mg B(inhaled/workshift); for urine, urine B(µg/mg creatinine) = 0.6 + 0.46 mg B(inhaled/workshift). There was no progressive accumulation of boron in soft tissues across the workweek as measured by blood or urine boron.

3 Toxic Effects of Boron and Borates

In this section, boric acid and many inorganic borates will be considered together. In dilute aqueous solution at physiological pH levels, inorganic borates are present as molecular boric acid (see section on chemical and physical properties, above). Further support for joint consideration comes from the similarity of effects of boric acid and certain inorganic borates in biological systems; toxic doses are similar when measured in terms of boron content. Much of the discussion of health effects will be in terms of "boron" used as a generic designation rather than a designation of the elemental substance. However, not all boron compounds exhibit toxicity similar to that of boric acid. Where boron compounds have unique toxic effects (e.g., boron trihalides, metaborates, perborates), they will be considered separately.

3.1 Acute Toxicity

Boric acid and the borates are generally of low acute oral toxicity in mammalian species (91). For most borates, acute oral LD_{50} values range from 2000 to 5000 mg/kg. Symptoms of acute oral toxicity typically include depression, ataxia, convulsions, and death.

The acute oral toxicity of boric acid and several specific borates is summarized in Table 45.5 (93–107). Boric acid had acute oral LD_{50} values of 2660, 3450, and 2000 in rats, mice, and dogs, respectively (92). In dogs, doses in the range of 200–2000 mg/kg induced emesis, so morphine was given to prevent vomiting before boric acid could be given orally. In another study, the acute oral LD_{50} of boric acid in rats was 3000–4000 mg/kg (93). Symptoms of toxicity were CNS depression, ataxia, convulsion, and death.

For borax, the acute oral LD_{50} was 4500–6000 in rats, equivalent to boric acid when expressed in terms of mg B/kg bw (93). Signs of toxicity were the same for boric acid and borax.

The acute oral LD_{50} of disodium tetraborate pentahydrate in rats was 2400–4200 mg/kg (94). Symptoms of toxicity were comparable to those described for boric acid and borax.

Table 45.5. Acute Oral Toxicity of Boric Acid and Inorganic Borates

Compound	Species	LD$_{50}$ (mg/kg)	Ref.
Boric acid	Rat	2660	92
	Rat	3000–4000	93
	Mouse	3450	92
	Dog	2000	92
Disodium tetraborate decahydrate (borax)	Rat	4500–6000	93
Disodium tetraborate pentahydrate	Rat	2400–4200	94
Anhydrous disodium tetraborate	Rat	>2000	95
Disodium octaborate tetrahydrate	Rabbits	2550	96
Potassium tetraborate tetrahydrate	Rat	3700	97
Zinc borate	Rat	>5000	98,99
Sodium metaborate dihydrate	Rat	>2000	101
Sodium metaborate tetrahydrate	Rat	2300	102
Disodium perborate tetrahydrate	Rat	2250	103
	Rat	2100	104
	Mouse	3425	105
Disodium perborate monohydrate	Rat	1600–2100	106
	Rat	1120	107

The acute dermal toxicity of boric acid, disodium tetraborate pentahydrate, boric oxide, disodium octaborate, anhydrous sodium tetraborate, and zinc borate was evaluated in rabbits (108–112). The acute dermal LD$_{50}$ was greater than 2000 mg/kg for all of these compounds. The acute dermal LD$_{50}$ of borax in rabbits was greater than 1000 mg/kg (113).

Low acute inhalation toxicity was observed for boric acid and inorganic borates. The 4-h LC$_{50}$ in rats was >2 mg/m^3 for boric acid, borax, disodium tetraborate pentahydrate, and disodium octaborate tetrahydrate (114–117). The 4-h LC$_{50}$ of zinc borate in rats was >4.9 mg/m^3 (118).

Boric acid and inorganic borates generally are not skin irritants or are mild skin irritants (119–126). Boric acid and borax are used at concentrations of ≤5% in cosmetics in the United States and in talc in Europe, ≤3% in other cosmetics in Europe, and ≤0.5% in oral hygiene products in Europe and elsewhere (127, 128).

Certain inorganic borates produce eye irritation in rabbits, probably as a result of physicochemical properties. Severe eye irritation was observed in rabbits with borax and disodium tetraborate pentahydrate (105, 129–131). In contrast, boric acid, disodium octaborate tetrahydrate, and zinc borate did not induce eye irritation in rabbit studies (127, 132–136).

3.2 Chronic and Subchronic Toxicity

In a 13-week study, mice were exposed to boric acid in the diet at concentrations sufficient to produce doses of approximately 0, 200, 400, 800, 1600, or 3200 mg/kg per day (137).

Increased mortality was noted at the two highest doses. At 800 mg/kg of boric acid per day, testicular effects (degeneration and atrophy of the seminiferous tubules) were observed. At all doses, extramedullary hematopoiesis of the spleen of minimal to mild degree was seen.

In a 90-day study, boric acid or borax were administered to rats in the diet at equivalent daily doses on the basis of mg B/kg (93). At a daily dose of 1500 mg/kg of boric acid, all animals died in 3–6 weeks. A daily dose of 500 mg/kg of boric acid produced decreased body weight and a pronounced decrease in testicular weight, as well as decreases in other organ weights. Testicular effects are discussed in greater detail in a later section.

Boric acid or borax was fed to beagle dogs for 90 days or for 2 years (93). These studies demonstrated qualitative evidence of testicular atrophy in dogs. However, confidence is low in the quantitative results of these studies because of limitations in experimental design.

In a 2-year study, rats were administered doses of boric acid or borax in the diet at doses equivalent to approximately 0, 33, 100, or 333 mg/kg of boric acid per day (93). Animals receiving the high dose exhibited coarse hair, scaly tails, hunched posture, swollen and desquamated pads of the paws, abnormally long toenails, shrunken scrotum, inflamed eyelids, and bloody eye discharge. Hematocrit and hemoglobin levels were significantly lower than in controls. Testicular weight (absolute and relative) was significantly reduced at the high dose, and the relative weights of the brain and thyroid gland were significantly increased compared to controls. No significant effects were observed at the lower dose levels.

In a 2-year study in mice, boric acid in the diet provided doses of 0, 275, and 550 mg/kg of boric acid per day (93). The high dose was associated with a significant decrease in body weight in both males and females. In males, increased mortality and testicular histolopathologic effects were observed at the high dose.

Inhalation studies of inorganic borates are limited. Mice exposed to amorphous elemental boron at 72 mg/m^3 for 7 h/day, 5 days/week for 6 weeks showed no evidence of toxicity (138). A subchronic inhalation study of boric acid was inadequately reported (139). Boron oxide was evaluated in rats and dogs exposed by inhalation for 6 h/day, 5 days/week (140). Rats were exposed to 77, 125, or 470 mg/m^3 of boric oxide for 24, 12, or 10 weeks, respectively; dogs were exposed to 57 mg/m^3 of boric oxide for 23 weeks. Mass median diameters were measured by cascade impactor and ranged from 1.9 to 2.5 μm. Air sampling was done hourly. No signs of toxicity were observed. Chemical analysis of dog blood and rat blood and urine showed no changes from control values, except for increased urinary volume, boron content, and creatinine excretion and lower urinary pH. In the rat, there were no tissue changes, no change in bone fragility or radiological appearance, and no changes in organ weights. In the dog, hematology and sulfobromophthalein retention were within normal limits.

3.3 *Pharmacokinetics, Metabolism, and Mechanisms*

The pharmacokinetics of boric acid and inorganic borates have been reviewed extensively (141–144).

3.3.1 Absorption.
In animals, boric acid has been shown to be readily absorbed from the GI tract. Among the species studied were rats (145), rabbits (146), sheep (147), and cattle (148, 149).

Dermal absorption of boric acid in animals is minimal. Boric acid (5%) was applied topically to 10–15% of the body surface of rabbits with an occlusive dressing for 1.5 h per day for 4 consecutive days (146). Minimal amounts of boric acid were absorbed across intact skin and slightly abraded skin of rabbits as measured by excretion of B in urine. Absorption was greater in rabbits with more seriously damaged skin. In rats given boric acid in an ointment, urinary excretion accounted for only 1% of the administered dose (150). However, boric acid applied to the skin of rats in an aqueous jelly was absorbed, with 23% of the administered dose appearing in the urine.

3.3.2 Distribution.
Boric acid is rapidly distributed throughout body water. After administration of boric acid, B levels in soft tissues are equivalent to those found in plasma, whereas bone B levels appear to be higher than those found in plasma or soft tissues.

In rats, Ku et al. (145) studied the tissue distribution of B in reproductive, accessory sex organs, and other selected tissues in adult males given boric acid in the diet, providing about 600 mg/kg per day for ≤ 7 days. Bone achieved the highest concentration of B in rats, reaching levels 2–3 times those observed in plasma, and bone B levels continued to increase throughout the 7 days of exposure. In contrast, adipose tissue concentration of B was only 20% of the plasma value, a finding consistent with the lack of accumulation of B. Except for bone and adipose tissue, all tissues examined exhibited B levels comparable to the levels found in plasma. Accumulation of B in bone has also been reported in older studies in rats (151).

In rats, tissue levels of B reached steady state within 3–4 days when boric acid was given in the diet or drinking water for 28 days (152) or 9 weeks (124, 153). Thus, B does not accumulate with time in rats.

Levels of B were also determined in mice. The highest concentration of B in untreated mice was found in bone, 26 µg/g dry weight (154). In another study in mice (155), investigators observed similar concentrations of B in brain, heart, liver, muscle, spleen, and renal cortex. The highest concentration was reported in papillary region of the kidney. Bone concentration was not determined in this study.

3.3.3 Metabolism.
Boric acid is not metabolized in animals. The metabolism of boric acid by biological systems is not possible because of the high energy requirements (523 kJ/mol) needed to break the boron–oxygen bond (156).

Other inorganic borates convert to boric acid at physiological pH in the aqeous layer overlying mucosal surfaces prior to absorption. Additional support for this derives from studies in which more then 90% of administered doses of inorganic borates are excreted in the urine as boric acid.

3.3.4 Excretion.
Boric acid is excreted unchanged in the urine regardless of the route of administration. The major determinant of boric acid excretion is expected to be renal clearance. In animals, elimination half-lives have not been state explicitly in the scientific

literature, but they may be estimated from data in the literature. Using the data of Ku et al. (145) and assuming first-order kinetics for elimination, the half-life in rats may be estimated to be less than 14–19 h. Similarly, the pharmacokinetic data of Farr and Konikowski (157) may be used to estimate a half-life of boric acid of approximately one hour in mice.

3.4 Reproductive and Developmental Toxicity

Male reproductive effects have been observed in studies of boric acid and borax at high doses. Developmental toxicity was reported in studies of boric at high doses in rats, mice, and rabbits. Reproductive and developmental toxicity in animal studies has been used as the critical endpoint in many recent risk assessments (142–144, 158, 159). Generally, these risk assessments indicate a minimal risk of reproductive or developmental toxicity in humans.

3.4.1 Reproductive Toxicity.
Relatively high doses of boric acid and borax have been shown to adversely affect the testis and male fertility in animal studies in rats, mice, and dogs. A three-generation reproductive study was conducted to evaluate the effects of boric acid and borax in rats (93). Dietary administration of boric acid or borax at a daily dose equivalent to 330 mg/kg of boric acid (58 mg B/kg per day) had a clear effect on male fertility; microscopic evaluation of the atrophied testes of all males in this group revealed no viable sperm. No effect on reproduction was reported in rats receiving boric acid or borax in the diet at a daily dose equivalent to 100 mg/kg of boric acid (17.5 mg B/kg per day).

Male rats were administered diets containing borax for 30 or 60 days (160). Daily doses of borax equivalent to 570 or 1140 mg/kg of boric acid (100 or 200 mg B/kg per day) produced decreased testicular weight after 60 days. In addition, reduced spermatogenesis, decreases in selected testicular enzymes, and impaired fertility were also observed at these dose levels. Fertility was unaffected at a lower dose of borax (equivalent to 285 mg/kg of boric acid per day). In general, although recovery was seen at the lower doses, there was no recovery from testicular atrophy at higher doses, once the germ cells were depleted.

Male rats exposed to lower doses of boric acid for shorter time periods exhibit reversible inhibition of spermiation (sperm release in the tubules) (153). These early effects have been observed in rats at a daily dose of 217 mg/kg of boric acid for 14 days. At a lower dose of 149 mg/kg of boric acid per day, similar reversible effects on spermiation were noted after 28 days of exposure. Higher doses of boric acid in rats have been associated with testicular atrophy, reduced spermatogenesis, and infertility. A single oral dose of 500 mg/kg of boric acid per day was reported to affect spermiation in rats (161). Inhibition of spermiation was the earliest reproductive effect produced by boric acid, appearing by day 7 in rats receiving the substance in the diet at a dose level of approximately 350 mg/kg of boric acid per day (152).

Similar testicular findings were seen in mice. In a continuous breeding study of boric acid in mice, testicular effects (testicular atrophy, decreased sperm count, reduced motility, and abnormal sperm morphology) were observed in a dose-related manner (162). Fertility was partially and completely impaired in mice at doses of approximately 630 and

1260 mg/kg of boric acid per day, respectively. The NOAEL (no-observed-adverse-effect level) was considered to be 154 mg/kg of boric acid per day; this dose had no effect on fertility, but appeared to cause a slight decrease in motility of epididymal sperm.

Studies in dogs also demonstrate the potential for boric acid and borax to affect the testis (93). These dog studies are of limited value due to (*1*) the small number of dogs used, (*2*) different times of sacrifice such that the group size was further reduced, and (*3*) the presence of testicular atrophy in 3 of 4 dogs in the concurrent control for the high dose group. Despite these limitations, the dog studies provide additional evidence of male reproductive toxicity.

3.4.2 Developmental Toxicity. In the late 1980s, the National Toxicology Program (NTP) initiated a series of developmental toxicology studies of boric acid in mice, rats, and rabbits. Developmental toxicity was observed in all three species, and the rat was the most sensitive species. The developmental toxicity of boric acid was evaluated in rats fed estimated daily doses of 0, 78, 165, and 330 mg/kg, of boric acid in the diet on gestation days 0–20 (163). To limit prenatal mortality, an additional group of rats was fed boric acid at an estimated daily dose of 536 mg/kg of boric acid on gestation days 6–15 only. Maternal toxicity was observed at 165, 330, and 536 mg/kg of boric acid per day as demonstrated by increased relative liver and kidney weights; altered water and/or food consumption and decreased maternal weight gain were noted at 330 and 536 mg/kg per day. The NOAEL for maternal toxicity was 78 mg/kg per day.

Prenatal mortality was increased among rats at 330 and 536 mg/kg of boric acid per day (163). Average fetal body weight was reduced at all dose levels, ranging from a 6% decrease at the lowest dose to about a 50% decrease at the highest dose. The incidence of fetal malformations was increased at doses of ≥ 165 mg/kg per day; the most frequently observed malformations were enlarged ventricles of the brain and agenesis or shortening of the 13th rib. An increase in enlarged ventricles of the brain was observed among fetal rats of dams administered 330 and 536 mg/kg of boric acid per day. Skeletal variations were noted to occur with a lower incidence than controls at 78 and 165 mg/kg of boric acid per day, due primarily to a reduction in the incidence of full or rudimentary first lumbar rib(s); skeletal variations are common in rats, and they are not considered to be malformations. In contrast, the incidence of variations was increased above control levels at 330 mg/kg per day primarily because of an increase of another skeletal variation: wavy rib(s). The lowest dose tested, 78 mg/kg of boric acid per day, was considered by the authors to be the LOAEL rather than the NOAEL for developmental toxicity in rats because of the 6% decrease in fetal body weight.

More recently, the same investigators responsible for the original rat developmental study were involved in conducting a follow-up developmental toxicity study of boric acid in rats in order to (*1*) determine a NOAEL and (2) evaluate the potential for postnatal recovery from the effects of prenatal exposure to boric acid (164). As in the earlier study, Sprague–Dawley rats were exposed to boric acid in the diet on gestational days (GDs) 0–20. In phase I (teratology evaluation), approximately 30 dams/group were terminated, and examinations were performed on GD 20. In phase II (postnatal evaluation), additional dams (approximately 30/group) were allowed to deliver and rear their litters until postnatal day 21.

No evidence of developmental toxicity was observed in this study (phases I and II) in the offspring of rats fed diets delivering a dose of 55 mg/kg per day of boric acid or less from GDs 0—20 (164). At approximately 75 mg/kg of boric acid per day, reduced fetal body weight and morphological changes in the fetal skeleton (specifically an increased incidence of short rib XIII and wavy rib) were observed on GD 20 (phase I). None of these effects were observed at a comparable dose in the postnatal period (phase II) of the study; pup body weight recovered by postnatal day 0. At the highest daily dose ($\sim$145 mg/kg of boric acid), the same effects observed at 75 mg/kg per day were more pronounced (phase I); in addition, a slight (but not statistically significant) decrease in the incidence of extra rib on lumbar 1 was observed. None of these effects was discerned postnatally at the highest daily dose ($\sim$145 mg/kg) with the exception of an increase in short rib XIII. There was little evidence of maternal toxicity at any of the doses tested. Thus, the NOAELs for developmental toxicity in the prenatal and postnatal phases of this study (phases I and II) were 55 and 75 mg of boric acid per day, respectively.

A developmental toxicity study of boric acid in mice produced results similar to those seen in rats. Mice were given 0, 245, 450, or 1000 mg/kg of boric acid per day on GDs 0–20 in the diet (163). Maternal toxicity was observed at all dose levels. Signs of developmental toxicity included decreased fetal body weight at the middle and increased resorptions, malformations, and skeletal variations, as well as decreased fetal weight, at the high dose. The NOAEL for developmental effects in mice was 245 mg/kg of boric acid per day.

In rabbits administered daily doses 0, 63, 125, or 250 mg/kg of boric acid by gavage on GDs 6–19, developmental effects were seen at the highest dose only, including an increase incidence of resorptions and malformations (163). No evidence of developmental toxicity was observed at lower doses. The NOAEL for maternal and developmental toxicity was 125 mg/kg of boric acid per day.

In summary, boric acid was shown to be a developmental toxicant in all three mammalian species tested. The rat was the most sensitive species, with a NOAEL of 55 mg/kg of boric acid per day for developmental toxicity; the NOAEL for developmental toxicity in rabbits and mice was 125 and 245 mg/kg of boric acid per day, respectively.

3.5 Carcinogenesis

There was no evidence of carcinogenicity in a 2-year NTP carcinogenesis bioassay in male and female B6C3F$_1$ mice on diets of boric acid resulting in estimated doses of approximately 0, 430 and 1145 mg/kg/day of boric acid (137). An earlier 2-year rat study of boric acid produced no evidence of a carcinogenic effect (93). EPA has classified boric acid as a "group E" carcinogen, indicating that it shows "evidence of non-carcinogenicity" for humans (166).

3.6 Genetic and Related Cellular Effects Studies

Boric acid and borax were not mutagenic in *Salmonella* with or without metabolic activation (137, 167–169). Boric acid did not induce chromosome aberrations or sister chromatid exchanges in Chinese hamster ovary cells (137). In addition, boric acid was not mutagenic in the mouse lymphoma forward mutation assay (170), in the unscheduled

DNA synthesis assay in primary rat hepatocytes (171, 172), or in the *in vivo* micronucleus assay in mice (173).

3.7 Human Experience

3.7.1 General Information. Boron in the form of boric acid and borax has a long history of use in clinical medicine as an antiseptic agent, a sedative for epilepsy and colic, and a buffering agent in talcum powder and cosmetics. These applications and the misuse of boric acid and borax have been reviewed extensively (127, 174). Following the report of Goldbloom and Goldbloom in 1953 (175) boric acid was largely removed from hospital use and the incidence of tragic, often fatal, accidents declined.

3.8 Clinical Cases and Experimental Studies

3.8.1 Acute Toxicity. The medical literature reporting 109 cases of acute toxicity caused by boric acid was reviewed in 1953 by Goldbloom and Goldbloom (175). In this series 35% were children less than 1 year of age; their mortality rate was 70%. The mortality rate for the entire series was 55%. Of the 80 cases for which information was available, 73% had GI disturbances, 67% had CNS effects, and 76% had cutaneous lesions. Ingestion, inappropriate gastric lavage and topical application to damaged skin accounted for 76% of these cases. More recent clinical experience with boric acid has presented an entirely different picture, one in which only mild effects were seen. This comes from a retrospective review of 782 cases of boric acid ingestion reported to the National Capital Poison Center and the Maryland Poison Center between 1981 and 1985 (86). All except two cases were acute ingestions; 88.3% were asymptomatic. Among the remaining 11.7%, frequent symptoms included vomiting, abdominal pain, and diarrhea; less frequent findings included lethargy, headache, lightheadedness, and rash. Among the children less than 6 years of age, 21 ingested more than the estimated lethal dose of 2 g and among the adults, 8 ingested more than the estimated lethal dose of 15 g, all without severe manifestations of toxicity or life-threatening symptoms. In this series only minimal toxicity was seen at serum levels of ≤ 640 mg/mL (86).

3.8.2 Chronic and Subchronic Toxicity. There are a few reports in the literature of chronic boron poisoning in the human. Kliegel (176) assembled reports of cases of boron poisoning, some of which were the results of use of boric acid or borax medication. Daily treatment doses for epilepsy (in boron equivalents) ranged from 2.5 to 24.8 mg B/kg for periods up to many years. Signs and symptoms most frequently seen were dermatitis, alopecia, anorexia, and indigestion. These occurred it patients receiving doses of ≥ 5 mg B/kg per day (2 g of boric acid). In one case, when treatment was reduced to 2.5 mg B/kg (1 g of boric acid) per day, all symptoms and signs subsided. In all cases of drug effect reported by Kliegel (176), withdrawal was followed by recovery, and no sequelae were observed.

3.8.3 Pharmacokinetics, Metabolism, and Mechanisms. The pharmacokinetics and metabolism of boric acid have been studied by both the parenteral (177) and oral (178) routes. In the one study IV doses of 562–611 mg boric acid were given as a single injection over 25 min; in the other, doses of 740–1476 mg boric acid in an ointment base were ingested over a short time period. In the IV study, plasma boron concentration curves were followed for 3 days and urine was collected with a 120-h urinary excretion of 98.7 ± 9.1% of the dose (177). Clearance was 54.6 ± 8.0 mL min^{-1} 1.73 m^{-3} with a mean biological half-life of 21.0 ± 4.9 h. In the ingestion study where boric acid was administered in both a aqueous solution and a water emulsifying ointment, absorption was essentially complete as measured by urine content collected over a 96-h period, 93.9 and 92.4%, respectively (178).

3.8.4 Skin Absorption. Several earlier studies attempted to measure boron skin absorption but gave conflicting results primarily because of the lack of analytical sensitivity in biological matricies. Wester et al. (179), in a carefully conducted human *in vivo* study, found that only 0.23, 0.21, and 0.12% of a saturated dose of boric acid, borax, and disodium octaborate tetrahydrate, respectively, were absorbed. The borates used were made of the stable ^{10}B isotope and its recovery in urine following 24-h application to the skin was measured by ICP/MS as a change in isotope ratio. This method was necessary because of the large amount of naturally occurring boron present in food and the environment. The investigators found no evidence of skin irritation with any of the three borates tested.

3.8.5 Skin Effects. Skin effects associated with systemic boron poisoning were reported frequently in the cases collected by Kliegel (176). These range from erythema to exfoliation. A case where 28 g (70 mg B/kg) of boric acid was mistakenly administered by subcutaneous infusion to an adult patient in postoperative recovery exhibited severe cutaneous manifestations but recovered with only the maintenance of fluid balance and promotion of diuresis (180)

3.9 Epidemiology Studies

3.9.1 Acute Toxicity

3.9.1.1 Measurement of Exposure Causing Irritation. Because of concerns expressed by the ACGIH TLV committee that borax and the other sodium tetraborates were irritants to skin and mucosa and should bear TLVs commensurate with such concern, Wegman et al. (181) undertook an epidemiological study of acute effects in workers employed in a California borate production plant. Exposures were to one or more of the three sodium tetraborates: the anhydrous ($Na_2B_4O_7$), the pentahydrate ($Na_2B_4O_7 \cdot 5H_2O$), and the decahydrate ($Na_2B_4O_7 \cdot 10H_2O$). A repeated measurement study design recording occupational exposure and irritant symptoms in the eye, nose, and throat was used. Symptoms were recorded hourly for 79 exposed and 27 unexposed subjects beginning prior to the start of the shift and continuing throughout the shift for 4 sequential days. Concurrent air sampling was done using a real-time, personal aerosol monitor (MINIRAM) backed up with a 37-mm cassette filter. The aerosol monitor had previously been calibrated for the rather large-size particles of the exposure environment (78). Full-shift and 15-min exposure profiles were used in the analysis of the symptom reports. Dose-related mild symptoms of nose, eye, and throat irritation were found using both exposure profiles. The 15-min intervals gave a somewhat steeper exposure response slope. Daily exposures ranged from means of 1.0 to above 14 mg/m^3 of dust. Multivariate logistic regression analysis of additional variables indicated that smokers were less sensitive to irritation than were nonsmokers. There was no indication that the anhydrous borate was more irritating than the other forms. Previous concern that heat of hydration would render the anhydrous and pentahydrate forms more irritating was founded on misinformation. Solution of this compound is an endothermic, not exothermic, reaction. The study team considered by these findings to be compatible with a TLV of 10 mg/m^3 TWA without a STEL (short-team exposure limit) for all the sodium borates.

3.9.2 Chronic and Subchronic Toxicity. In humans, the only epidemiological studies of chronic effects, other than for reproductive effects (see discussion below) have been for pulmonary changes caused by borate dust exposure. Garabrant et al. (182) performed chest X rays on 621 borate workers in a major California plant. These were read by three NIOSH-certified readers using the ILO International Classification for Pneumoconiosis; no association was found with borate exposure. A study of pulmonary function was performed on workers from the same borate and boric acid facility with the same exposures to boron-containing dusts as the respiratory irritation study (181). Because Garabrant (182) had done a careful pulmonary function study of the workers in this plant in the previous 1981 study, there was an opportunity to compare them with current (1988) pulmonary function test results and search for decrements occurring over the 7-year interval that could be related to borate exposure. Of the 631 workers who participated in 1981, 371 were still available for retesting in 1988. Workers not available for retesting had 1981 pulmonary function test results similar to those who were available for retesting. Retesting was completed by 336 who had acceptable tests from both years. FEV$_1$ differences between the two test periods appeared to be accounted for by the differences in smoking. On the average, subjects were 44 years old and had worked with borates for 19 years, and the group included 42% current smokers 28% ex-smokers, and 30% non-smokers. The average 1988 FEV$_1$ and FVC values were 96 and 104% of predicted, respectively. The average annual loss in both FEV$_1$ and FVC was close to 30 mL/year, the rate of loss expected on the basis of cross-sectional studies of nonsmoking normal subjects. Thus there was no indication of chronic pulmonary effect relatable to long-term chronic borate exposure.

3.9.3 Pharmacokinetics, Metabolism, and Mechanisms.

3.9.3.1 Biological Measures of Work Exposure. An understanding of the relationship between borate dust exposure and blood and urine boron concentrations comes from the study of 14 workers engaged in the packing and shipping of sodium borates at the plant site of the two previous investigations (90). Daily dietary boron intake and on-the-job inhaled boron were compared with end-of-shift blood and urine boron concentrations. Workers were sampled throughout full shifts for 5 consecutive days each. Airborne sodium borate concentrations ranged from arithmetic means of 3.3 to 18 mg/m^3, measured gravimetrically. End-of-shift mean blood boron concentrations ranged from 0.11 to 0.26 µg/g; end-of-shift mean urine concentrations ranged from 3.16 to 10.72 µg/mg creatinine. There was no progressive increase in end-of-shift blood or urine boron concentrations across the days of the week, and blood boron concentrations remained within the range (0.04–0.36 µg/g of blood) reported for the general working population (89). Urine testing done at the end of the work shift gave a somewhat better estimate of borate exposure than did blood testing and was more easily sampled and analytically less difficult to perform. Personal air samplers of two types were used: (*1*) the 37-mm closed-face two-piece cassette, to estimate "total dust" and (*2*) the Institute of Occupational Medicine (IOM) sampler, to estimate "inhalable particulate mass." Under the conditions of this study, the IOM air sampler more nearly estimated human exposure as measured by blood and urine boron levels than did the 37-mm cassette sampler, which measures "total dust."

The mean dietary intake of the workers was 1.37 mg B/day. Total estimated boron intake, diet plus occupational exposure, for the five workers with the highest borate dust exposure had a mean daily boron intake of 27.90 mg/day or, based on the body weights of the subjects, 0.38 mg B/kg per day. These subjects had a mean blood boron level of 0.26 µg B/g blood. The relationship between boron as measured during a workshift and blood boron level is expressed by the regression: For *blood*, B(µg) = 0.07 + 0.008 mg B(inhaled/shift); for *urine*, urine B(µ/mg creatinine) = 0.6 + 0.46 mg B (inhaled/shift). There was no progressive accumulation of boron in soft tissues across the workweek as measured by blood or urine boron. Earlier measures of the relationship between blood and urine boron are not reported here because of limitations of the analytical method used (183).

3.9.4 Reproductive and Developmental. Because of concern that boron could cause infertility and developmental effects in humans as seen in rats, mice, rabbits, and dogs as discussed above, there have been several attempts to find sufficiently highly exposed human populations. Tarasenko (139) reported that 28 boric acid workers had decreased sexual activity and 6 had semen studies with reduced sperm counts or motility. He gave no information about how the subjects were chosen. Analysis of the workplace air revealed boric acid concentrations of ≤10 mg/m^3 in 50% of the samples but could be as high as 20–83 mg/m^3 in the remaining samples. No information about other exposures of this population was given.

Another study attempted to evaluate the reproductive performance of workers exposed to borates in a borate ore processing plant for several years. Reproductive performance of workers was measured in terms of live births to the wives of workers subsequent to specific occupational exposures to sodium borates and boric acid (184). The results were calculated as a "standardized birth ratio" (SBR). Participating male workers, 542 (72% of the 753 eligible workers) responded to a questionnaire with information on marital status, marital history, wife's date of birth, and number, sex, and dates of birth of offspring from each marriage. Personal air sampler data were available for most jobs beginning in 1981, and it was assumed that exposures on jobs prior to that time were equal to or greater than those measured since. Categories of exposure to boron containing dust were arbitrarily assigned as "high" (>8 mg/m^3), "medium" (3–8 mg/m^3), and "low" (<3 mg/m^3). There was no relationship between exposure category and SBR. For the total cohort the SBR was 113 (529 observed births compared with 466.6 births expected), based on national fertility tables published for successive birth cohorts of women in the United

States. The high exposure group with ≥2 years of consecutive high exposures had mean exposures per year of 19.7 mg/m^3 (~2.9 mg B/m^3). During the time period they worked in high exposure jobs (adjusted for gestation time and allowing for one to three spermatogenic cycles), the SBRs ranged from 121 to 125. Thus, from this retrospective fertility study, it was concluded that work involving high sodium borate exposure did not decrease fertility. This lack of effect is not surprising when one considers what a low dose of boron 19.7 mg/m^3 of borate dust (2.9 mg B/m^3) represents. The maximum-likelihood absorbed dose allowing for total respiratory retention and absorption and a 10-m^3 inhaled volume per 8-h workshift for a 70-kg man would be only 0.4 mg B/kg per day.

In Turkey, a country with extensive borate ore deposits and the world's largest mining and borate production facilities, Şayli et al. (33) and Tüccar et al. (185) have done extensive studies of fertility rates in the villages in borate regions and among the employees of the Turkish borate mines and plants and have found the rates to be as high as in other parts of the country: >95%.

3.9.5 Carcinogenesis. The epidemiological literature contains no information about borate carcinogenesis. However, because *in vivo* and *in vitro* studies of borates have been uniformly negative, there has been no basis for generating epidemiological hypotheses about cancer.

3.9.6 Skin Sensitization. The role of borates as direct irritants or sensitizers was reviewed for the FDA (127) Human experience is summarized:

> In clinical studies, cosmetic formulations containing 1.0–3.2% Sodium Borate were nonirritating to moderately irritating and nonsensitizing when applied to human skin (620 subjects, total). Products containing 0.4% and 2.4% Boric Acid were moderately irritating and practically nonirritating, respectively (39 subjects, total). Results of photopatch-testing indicate that formulations containing 1.1% or 1.7% Sodium Borate are non-photosensitizing (446 subjects, total).

Boric acid is used in topically applied cosmetics in a concentration up to 5%. Borax is also present in several ophthalmic preparations (186).

3.9.7 Guidelines for Occupational Exposure. At present, there is no OSHA PEL for boric acid or any of the sodium tetraborates. The OSHA PEL for boron oxide, the anhydrous form of boric acid is 1 mg/m^3. In 1989 OSHA established a PEL TWA of 10 mg/m^3 total dust for the borates on the basis of the conclusion that this level would protect against significant irritation to mucous membranes and the respiratory tract. In a decision in 1992 the U.S. Court of Appeals of the Eleventh Circuit vacated the PELs of 1989. The NIOSH REL TWA for anhydrous and pentahydrate forms of sodium tetraborate of 1 and 5 mg/m^3 for the decahydrate are the same as the ACGIH values, based on the misconception that heat of hydration would cause irritation. The PEL TWA established in California's Occupational Safety and Health Administration (CAL OSHA) is 5 mg/m^3. There are no U.S. OELs or exposure recommendations for boric acid. Neither OSHA, NIOSH, nor ACGIH have given the borates a carcinogen designation. The ACGIH TLV

TWA for boron oxide is 10 mg/m³. For the borates, the ACGIH TLV TWAs are 1 mg/m³ for anhydrous tetra sodium salts; 5 mg/m³ for decahydrate and 1 mg/m³ for the pentahydrate.

3.9.8 Studies on Environmental Impact. Borates are present in the aqueous environment generally as $B(OH)_3$ or $B(OH)_4^-$ depending on the pH of the water (29). Thus, since, as far as is known, these two species do not have different biological effects, environmental studies report effects in terms of boron. The literature on effects of boron on microorganisms, invertebrates, aquatic vertebrates in freshwater streams and lakes has been reviewed by the IPCS (142), ECETOC (144) and Howe (31). This extensive body of literature is summarized briefly below. Environmental boron comes from several sources. The largest source is seawater evaporation (seawater boron concentration ~4.5 ppm) and is estimated to amount globally to 800,000–4,000,000 metric tons/year, whereas total anthropogenic release to the atmosphere is in the order of 180,000–650,000 metric tons/year (187). Boron levels as high as 1900 mg/kg have been reported in coal fly ash (188). Despite these releases, atmospheric concentration of boron is low: <0.5–80 ng/m³.

Several aquatic organisms require boron, including cyanobacteria and diatoms, as do higher plants. Deficiencies lie below 3.2 ppm, while toxicity is found above a soil solution of 6 ppm (189). Despite the narrow range of concentrations for optimum growth. In general, boron deficiency is much more frequent than toxicity. Phytotoxicity, when it occurs, is usually due to anthropogenic sources such as boron-contaminated irrigation water, excessive use of boron-rich fertilizers, sewage sludge, or fly ash. Most frequently boron-containing irrigation water will lead to soil buildup through evaporation.

Boron toxicity for fish ranges from 14.2 ppm for zebra fish to 979 ppm for mosquito fish. Early life forms of fish and amphibians were found to be the most sensitive to boron; rainbow trout were extremely sensitive. Successful trout populations have been found in several locations with boron concentrations around 1 ppm. Eckhart (6) has shown that boron in ultrapure water increases trout larva length in a dose-dependent manner up to and including 11 ppm.

A number of studies of ducks have been done in the Kesterson Reservoir in the Kesterson National Wildlife Reserve in the San Joaquin Valley, California because of concerns generated by high boron concentrations there. The source is boron from runoff from irrigated fields. In mallard ducks, duckling growth is reduced at concentrations of 30 mg/kg in the diet (190). A subsequent study that included one of the original authors found duckling growth inhibition at 900 mg/kg but none at 450 mg/kg. There was no interaction between boron and selenium (191).

Although boron accumulates in plant and animal tissue, there is no indication, of biomagnification in the food chain (31).

B ELEMENTAL BORON AND SELECTED INORGANIC COMPOUNDS CONTAINING BORON

1.0 Elemental Boron

1.0.1 CAS Number: *[7440-42-8]*

BORON

1.0.2 Synonyms: NA

1.0.3 Trade Name: NA

1.0.4 Molecular Weight: 10.81

1.0.5 Molecular Formula: B

1.1 Chemical and Physical Properties

1.1.1 General

Elemental boron can be present in one of two polymorphic forms: (*1*) α-rhombohedral, clear red crystals with a density of 2.46 and (*2*) β-rhombohedral, black, opaque crystals with a density of 2.35 or an amorphous form, of black or dark brown powder with a density of 2.30.

The melting point of boron is 2200°C; vapor pressure at 2140°C is 1.56×10^{-5} atm. Boron is insoluble in water but soluble in boiling nitric and sulfuric acids and most molten metals.

1.2 Production and Use

Until the late 1990s elemental boron had not found widespread use in industry, where cost of production was a major obstacle. Now, there is increasing use as new applications for the element are developed in material composites.

1.3 Exposure Assessment

1.3.1 Air

Use of an inhalable particulate-mass air sampler and a respirable dust sampler and analyze gravimetric analysis for both respirable mass and inhalable boron are recommended.

1.3.2 Background Levels: NA

1.3.3 Workplace Methods: NA

1.3.4 Community Methods: NA

1.3.5 Biomonitoring/Biomarkers

Blood levels can provide evidence of exposure but do not provide information about the severity of exposure (86). Since absorbed boron has a biological half-life of about 20 h and is excreted almost entirely by the kidney, a 24-h urine will estimate total absorption. However, there is no information about rate of absorption of elemental from the lung. Recognition must be given to the fact that diet contributes 0.5–3 mg of boron or more per day. Blood and urine samples should be collected in plastic containers and analyzed according to the description provided under analytic methods in the introduction to this chapter. In a pilot health study conducted in an elemental boron production plant, Keil and

Miller (192) found exposures to elemental boron dust ranging up to 7.34 mg B/m³ resulted in blood boron levels of 0.1–0.64 µg B/g. Blood boron levels increased with increasing exposure in the three workers studied. The values found are more than an order of magnitude above values reported for the general population when IPCS/MS is used but is similar to those found by older analytical methods.

1.4 Toxic Effects

There have been no reports of toxicity in humans. Since the element is very insoluble, the dust should be treated as such, with a respirable dust limit applied.

1.4.1 Experimental Studies

1.4.1.1 Acute Toxicity. The mouse oral LD$_{50}$ for presumably amorphous boron was reported to be 2000 mg/kg by Stokinger (193), who recommended that it should be controlled at the "nuisance" dust level of 10 mg/m³.

1.4.1.2 Chronic and Subchronic Toxicity. Elemental boron inhalation studies in the mouse were done by Stokinger (193). No effects were seen at air concentrations of 72.8 mg/m³ over a 30-day period of 7 h/day, 5 days/week. However, boron concentrations in the lung were 792 µg/g; in the kidney, 252 µg/g; in the liver, 106 µg/g; and in the GI tract, 73 µg/g. The elemental boron studies in humans by Keil and Miller (192) and the inhalation study in the mouse emphasize the ability of the respiratory tract to absorb highly insoluble substances; solubility of elemental boron is 0.72 µg/mL in water and 1.3 µg/mL in plasma.

1.4.1.3 Pharmacokinetics, Metabolism, and Mechanisms. It is assumed that once elemental boron goes into solution, the boron will be in the form B(OH)$_3$ and thus will be absorbed, distributed, and excreted as boric acid. See toxicity information about boric acid above.

2.0 Sodium Tetraborate Decahydrate

2.0.1 CAS Number: [1303-96-4]

2.0.2 Synonyms: Borax; barax, 10 mol; disodium tetraborate decahydrate, borax decahydrate; borates, tetrasodium salts, decahydrate; sodium diborate decahydrate; tetraborate, decahydrade; sodium borate decahydrate; fused borax; borax glass; fused sodium borate; borax decahydrate; Borascu; Borax (B$_4$Na$_2$O$_7$·10H$_2$O); Boricin; Gerstley borate; sodium pyroborate decahydrate; sodium tetraborate decahydrate (Na$_2$B$_4$O$_7$·10H$_2$O); solubor; sodium borates

2.0.3 Trade Names: NA

2.0.4 Molecular Weight: 381.4, 11.34% boron

2.0.5 Molecular Formula: N$_2$B$_4$O$_7$·10H$_2$O

2.0.6 Molecular Structure:

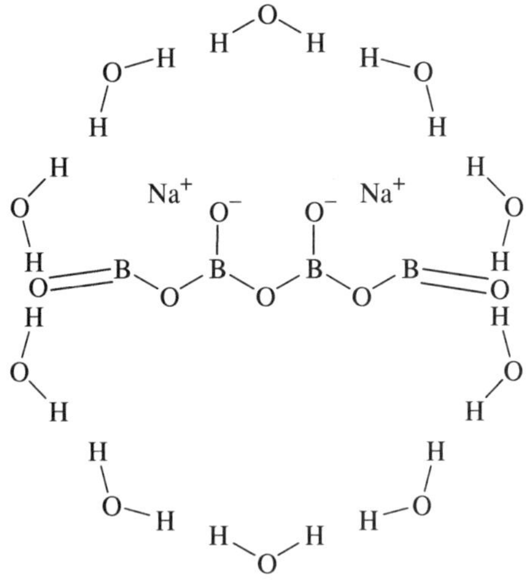

2.1 Chemical and Physical Properties

Appearance: White, odorless, monoclinic crystalline solid
Specific gravity: 1.73
Vapor pressure: Negligible at 20°C
Solubility in water: 5.92% at 25°C, 65.64% at 100°C
pH at 20°C: 9.3 (0.1% solution), 9.2 (1.0% solution)
Melting point: 62°C

2.1.1 General: NA

2.1.2 Odor and Warning Properties

The substance is odorless and has no warning properties.

2.2 Production and Use

See the introductory discussion (Section A) on borates.

2.3 Exposure Assessment

The assessment of borax (10 mol) is the same as for other borates. For those reports given in terms of boron, conversion to borax requires multiplication by a factor of 8.82. As mentioned in the section above on methods for measurement of borates and boron in the workplace, the hydration states of the sodium tetraborates are unstable and affected by temperature and humidity. This can lead to gravimetric and interpretation errors in reports of occupational hygiene field sampling. In dry air [0% relative humidity (RH)] and 100°F, particles of borax with a mean size of 7 μm dehydrated within an hour to about 3 mol and were fond to be composed of both crystalline borax 5 mol and an amorphous phase seen by powder pattern X-ray diffraction. In dry air at 70°F, dehydration is also rapid, attaining 5 mol in 1 h and continuing to drop to less than 4 mol in 4 h (34).

Human *in vivo* skin absorption studies by Wester et al. (179) showed a 0.20% uptake of a saturated solution applied to the skin surface and left for 24 h.

2.4 Toxic Effects

Toxic effects are those common to all inorganic borates, due to their conversion to $B(OH)_3$ in aqueous solutions. See toxicity discussion in the introduction to this chapter.

2.5 Standards, Regulations, or Guidelines of Exposure

OSHA PELs no longer exist as a result of the decision in 1992 by the U.S. Court of Appeals of the Eleventh Circuit, which vacated the PELs of 1989. However, Cal OSHA regulations set a PEL at 5 mg/m^3, which is the same as the NIOSH REL. The ACGIH TLV is 5 mg/m^3. There are no recommended short-term limits.

2.6 Studies on Environmental Impact

Literature search failed to identify any studies of environmental impact for borax. Boron in the environment has been studied, and a fraction of the anthropogenic boron in these studies will have come from the use of borax, predominantly in soap powders. A method has been developed by Vengosh (194) to distinguish between natural and anthropogenic sources of boron water contamination. The $^{11}B/^{10}B$ isotope ratios for seawater, sodium borates, and calcium-containing borates and the application of this method is being explored in California and Israel.

3.0 Sodium Tetraborate Pentahydrate

3.0.1 CAS Number: [12179-04-3]

3.0.2 Synonyms: Borax pentahydrate; tetraborate, pentahydrate; sodium tetraborate pentahydrate; boric acid ($H_2B_4O_7$), disodium salt, pentahydrate; boron sodium oxide ($B_4Na_2OO_7$), pentahydrate; Mule Team Borascu

3.0.3 Trade Names: Neobor®

3.0.4 Molecular Weight: 291.4

3.0.5 Molecular Formula: $Na_2B_4O_7 \cdot 5H_2O$

3.1 Chemical and Physical Properties

Appearance: White, odorless, crystalline solid
Specific gravity: 1.815
Vapor pressure: Negligible at 20°C
Solubility in water: 3.2% at 20°C, 51.2% at 100°C
pH at 20°C: 9.3 (3% solution)
Melting point: 200°C

3.1.1 *General:* NA

3.1.2 *Odor and Warning Properties*

The substance is odorless and has no warning properties.

3.2 Production and Use

See the introductory discussion (Section A) on borates.

3.3 Exposure Assessment

The assessment of borax (5 mol) is the same as for other borates. For those reports given in terms of boron, conversion to borax requires multiplication by a factor of 6.07. As mentioned in the Section A on methods for measurement of borates and boron in the workplace, the hydration states of the sodium tetraborates are unstable and affected by temperature and humidity. This can lead to gravimetric and interpretation errors in reports of occupational hygiene field sampling. However, borax is the most stable of the disodium tetraborates. It is quite stable to dehydration in dry air (0% RH) at both 70 and 100°F. On the other hand, when exposed to 80% RH at 100°F, there is a rapid increase in weight to 9.4 mol after 7 h, approaching pure borax 10 mol (34).

3.4 Toxic Effects

Toxic effects are those common to all inorganic borates, due to their conversion to $B(OH)_3$ in aqueous solutions at biologic pHs. See toxicity discussion on borates in general in the introduction to this chapter.

3.5 Standards, Regulations, or Guidelines of Exposure

OSHA PELs no longer exist as a result of the decision in 1992 by the U.S. Court of Appeals of the Eleventh Circuit, which vacated the PELs of 1989. However, Cal OSHA regulations set a PEL at 5 mg/m^3. The NIOSH REL is 1 mg/m^3. The ACGIH TLV is 1 mg/m^3. There are no recommended short-term limits.

3.6 Studies on Environmental Impact

Literature search failed to identify any studies of environmental impact for sodium tetraborate pentahydrate.

4.0 Sodium Tetraborate

4.0.1 CAS Number: [1330-43-4]

4.0.2 Synonyms: Anhydrous borax, borax glass, disodium tetraborate; sodium borate; sodium tetraborate; tetraborate; boron sodium oxide ($B_4Na_2O_7$); borax, fused; sodium tetraborate ($Na_2B_4O_7$); disodium tetraborate

4.0.3 Trade Names: Dehybor®

4.0.4 Molecular Weight: 201.27

4.0.5 Molecular Formula: $Na_2B_4O_7$

4.0.6 Molecular Structure:

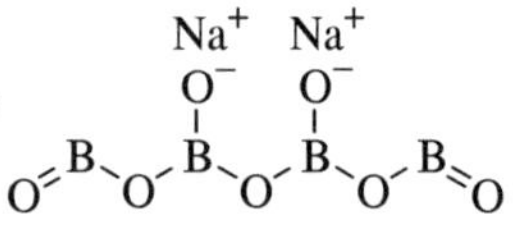

$$\begin{array}{c} Na^+ \quad Na^+ \\ O^- \quad O^- \\ | \quad\quad | \\ O=B-O-B-O-B-O-B=O \end{array}$$

4.1 Chemical and Physical Properties

Appearance: Light gray, odorless, vitreous granules
Specific gravity: 2.367
Vapor pressure: Negligible at 20°C
Solubility in water: 3.1% at 25°C, 2.5% at 20°C
pH at 20°C: 9.3 (3% solution)
Heat of solution: 2.13×10^5 J/kg (absorbed)
Melting point: 741°C
Decomposition point: 1575°C

4.2 Production and Use

Anhydrous borax is produced from borax through high temperature fusion. On cooling, the clear, glass-like material is ground into fine white granules. Because of its higher bulk density, it is preferred where storage space is limited. It is used principally in the manufacture of glass, ceramics, and enamel (29).

4.3 Exposure Assessment

Measurement of exposure is the same as for other borates and is described above. Anhydrous borax contains 21.49% boron. To obtain a measure of anhydrous borax from an analytical measure reported as boron, multiply by 4.65.

4.4 Toxic Effects

Toxic effects are those common to all sodium borates, since on going into solution at biological pH, they all exist as B(OH)$_3$. The concept that the anhydride on going into solution would give off heat is wrong. Solution of anhydrous borax is an endothermic reaction. Indeed, Wegman et al. (181) found that the anhydride did not cause an increase in respiratory irritation over that of other borates.

4.5 Standards, Regulations, or Guidelines of Exposure

OSHA PELs no longer exist as a result of the decision in 1992 by the U.S. Court of Appeals of the Eleventh Circuit, which vacated the PELs of 1989. However, Cal OSHA regulations set a PEL at 5 mg/m^3. The NIOSH REL is 1 mg/m^3. The ACGIH TLV is 1 mg/m^3. There are no recommended short-term limits.

4.6 Studies on Environmental Impact

There are no studies of environmental impact specifically relating to anhydrous borax. Any effect of the environment would be the same as for any other borate.

5.0 Boric Acid

5.0.1 CAS Number: [10043-35-3]

5.0.2 Synonyms: Boracic acid, Borofax; boron trihydroxide; Kjel-sorb; hydrogen orthoborate; three elephant; Kill-off; Orthoboric acid (H$_3$BO$_3$); trihydroxyborane

5.0.3 Trade Names: NA

5.0.4 Molecular Weight: 61.83

5.0.5 Molecular Formula: H$_3$BO$_3$

5.0.6 Molecular Structure: HO-B(OH)-OH

5.1 Chemical and Physical Properties

　　Appearance: White, odorless, crystalline granules or powder
　　Specific gravity: 1.435
　　Vapor pressure: Negligible at 20°C
　　Solubility in water: 4.72% at 20°C, 27.5% at 100°C
　　pH at 20°C: 6.1 (0.1% solution); 5.1 (1.0% solution); 3.7 (4.5% solution)
　　Melting point: 170.9°C

5.1.1 General: NA

5.1.2 Odor and Warning Properties

None. Granules or powder create slippery surfaces.

5.2 Production and Use

Boric acid is produced from borax or sodium or calcium containing ore by reaction with sulfuric acid in a hot aqueous solution. The resultant liquor is concentrated, crystalized, and dried. Principal use is in the manufacture of low sodium glass, enamels, and glazes. The high purity material is used for the control and emergency shutdown of nuclear reactors (29).

Boric acid and inorganic borates generally are not skin irritants or are mild skin irritants (119–126). Boric acid and borax are used at concentrations of $\leq 5\%$ in cosmetics in the United States and in talc in Europe, $\leq 3\%$ in other cosmetics in Europe, and $\leq 0.5\%$ in oral hygiene products in Europe and elsewhere (127, 128). Boric acid, together with borax, is used as a buffer in eye lotions.

5.3 Exposure Assessment

Measurement of exposure is the same as for the borates and is described above. Boric acid contains 17.48% boron. To obtain a measure of boric acid from an analytical measure reported as boron, multiply by 5.72.

Human *in vivo* skin absorption studies of Wester et al. (179) showed a 0.23% uptake of a saturated solution applied to the skin surface and left for 24 h.

5.4 Toxic Effects

Toxic effects are those common to the sodium borates, since on going into solution at biological pH, both boric acid and the borates exist as the undissociated $B(OH)_3$.

5.5 Standards, Regulations, or Guidelines of Exposure

There is no OSHA PEL, NIOSH REL, or ACGIH TLV for boric acid. However, since boric acid is the hydrated form of the easily hydrated boron oxide, it is reasonable to apply boron oxide exposure limits to boric acid. The U.S. OSHA PEL is 15 mg/m^3; the ACGIH TLV is 10 mg/m^3, and the NIOSH REL is 10 mg/m^3. The Cal OSHA PEL is 5 mg/m^3.

5.6 Studies on Environmental Impact

Specific studies were not encountered. Environmental impact is a function of B released to the environment and appears to be the same for all borates.

6.0 Boric Oxide

6.0.1 CAS Number: [1303-86-2]

BORON

6.0.2 Synonyms: Anhydrous boric acid, boron trioxide, diborane trioxide; diboron trioxide; boron oxide; boron trioxide; B–O; Boria; boric acid, anhydride; Boron oxide (B$_2$O$_3$); boron sesquioxide; boron oxide, 99.999%

6.0.3 Trade Names: NA

6.0.4 Molecular Weight: 69.64

6.0.5 Molecular Formula: B$_2$O$_3$

6.0.6 Molecular structure:

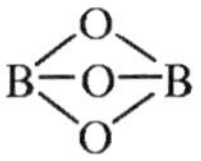

6.1 Chemical and Physical Properties

Appearance: White, odorless, crystalline granules or powder
Specific gravity: 2.46 (crystals), 1.85 (powder)
Vapor pressure: Negligible at 20°C
Solubility in water: 4.0% at 20°C
Heat of hydration: 75.94 kJ/mol (469 Btu/lb) (evolved)
pH at 20°C: 5.1 (1.0% solution)
Melting point: 450°C (crystals)
Boiling point: 1500°C (crystals)

6.1.1 General: NA

6.1.2 Odor and Warning Properties

No odor, but slightly irritating on inhalation at high concentration.

6.2 Production and Use

Boric oxide is produced by thermal fusion of boric acid, forming a clear transparent glass-like solid that is subsequently ground to white vitreous granules. It is used principally in the manufacture of glass and vitreous products (29).

6.3 Exposure Assessment

Boron oxide is measured in the environment and in biological matrices in the same ways as the borates, by analyzing for boron and calculating boron oxide using the relationship of boron:boron oxide weight ratio. Since 31% of boron oxide is boron, one has a measure of boron in a sample of boron oxide, multiplying that measure by factor of 3.22 will give the amount of boron oxide present.

6.4 Toxic Effects

Systemic effects of boron oxide are the same as those of the borates. Inhalation effects were studied during the 1950s, when major efforts were under way to find increasingly powerful rocket propellents. Because boron oxide was the combustion produce of one of these it was evaluated in rats and dogs exposed by inhalation for 6 h/day, 5 days/week by Wilding et al. (140). Rats were exposed to 77, 125, or 470 mg/m^3 of boric oxide for 24, 12, or 10 weeks, respectively; dogs were exposed to exposed to 57 mg/m^3 of boric oxide for 23 weeks. Mass median diameters were measured by cascade impactor and ranged from 1.9 to 2.5 μm. Air sampling was done hourly. No signs of toxicity were observed. Chemical analysis of dog blood and rat blood and urine showed no changes from control values, except for increased urinary volume, boron content and creatinine excretion, and lower urinary pH. In the rat, there were no tissue changes, no change in bone fragility or radiological appearance, and no changes in organ weights. In the dog, hematology and sulfobromophthalein retention were within normal limits.

6.5 Standards, Regulations, or Guidelines of Exposure

The U.S. OSHA PEL is 15 mg/m^3; the ACGIH TLV is 10 mg/m^3 and the NIOSH REL is 10 mg/m^3. The Cal OSHA PEL is 5 mg/m^3.

6.6 Studies on Environmental Impact

Because boron oxide is used mainly in the manufacture of specialty glass, where increased thermal stability, mechanical strength, and enhanced chemical and aqueous durability are required, it has little opportunity to enter the general environment (29). The small amount that may enter the environment will contribute to the boron pool generated largely by the inorganic borates.

7.0 Disodium Octaborate Tetrahydrate

7.0.1 CAS Number: [12280-03-4]

7.0.2 Synonyms: Disodium octaborate; Tim-Bor; Bora-Care

7.0.3 Trade Names: Octabor, Polybor, Solubor

7.0.4 Molecular Weight: 412.52

7.0.5 Molecular Formula: $Na_2B_8O_{13} \cdot 4H_2O$

7.1 Chemical and Physical Properties

Appearance: White, odorless, crystalline granules
Bulk density: 320–480 kg/m$_3$
Vapor pressure: Negligible at 20°C
Solubility in water: 9.7% at 20°C, 34.3% at 50°C
Heat of hydration: 75.94 kJ/mol (469 Btu/lb) (evolved)
pH at 20°C: 8.3 (3.0% solution)
Melting point: 815°C

BORON

7.2 Production and Use

The material is produced by spray-drying mixtures of boric acid and borax. Its greater water solubility gives it advantages in some applications over the tetraborates. Its uses include applications as pesticides, as flame retardants, and as an agricultural micronutrient (29).

7.3 Exposure Assessment

Measurement of exposure is the same as for other borates and is described above. Disodium octaborate tetrahydrate contains 20.96% boron. To obtain a measure of disodium octaborate tetrahydrate from an analytical measure reported as boron, multiply by a factor of 4.77.

Human *in vivo* skin absorption studies of Wester et al. (179) showed a 0.12% uptake of a saturated solution applied to the skin surface and left for 24 h.

7.4 Toxic Effects

Toxic effects are those common to all sodium borates since on going into solution at biological pH, they all exist as $B(OH)_3$.

7.5 Standards, Regulations, or Guidelines of Exposure

There are no occupational standards or regulations for disodium octaborate tetrahydrate. General industrial hygiene prudence will lead to control at 10 mg/m^3 despite the fact that its water solubility disqualifies it for use as a dust standard.

7.6 Studies on Environmental Impact

There are no studies of environmental impact specifically relating to disodium octaborate tetrahydrate. Any effect on the environment would be the same as for any other borate.

8.0 Sodium Metaborate Dihydrate

8.0.1 CAS Number: [16800-11-6]

8.0.2 Synonyms: Sodium metaborate, sodium metaborate tetrahydrate; Kodak balanced alkali

8.0.3 Trade Names: NA

8.0.4 Molecular Weight: 101.83

8.0.5 Molecular Formula: $NaBO_2 \cdot 2H_2O$

8.0.6 Molecular Structure:

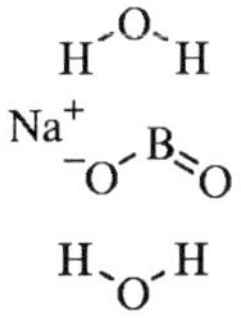

8.1 Chemical and Physical Properties

Appearance: White, odorless, crystalline granules
Specific gravity: 1.91
Vapor pressure: Negligible at 20°C
Solubility in water: 31.0% at 20°C, 81.1% at 100°C
pH at 20°C: 10.6 (0.1% solution), 11.1 (1.0% solution), 11.5 (4.0% solution)
Melting point: 90–95°C

8.2 Production and Use

Sodium metaborate dihydrate may be made by heating a slurry of the tetrahydrate above 54°C, by crystallizing metaborate solutions at 54–80°C or by dehydrating the tetrahydrate in vacuum. It is used in textile finishing as a components of sizing, scouring compositions, adhesives, and detergents. It is also used in photography and agriculture (29).

8.3 Exposure Assessment

Exposure assessment should use methods common to the other borates. The content of boron in sodium metaborate dihydrate is 10.6%.

8.4 Toxic Effects

Metaborates (*e.g.*, sodium metaborate tetrahydrate, sodium metaborate dihydrate) are highly alkaline in solution. A 0.1 molar solution has a pH of 10.5, and a 15-molar solution has a pH of 11.9. Because of their high pH, the metaborates may cause skin irritation and eye damage.

8.5 Standards, Regulations, or Guidelines of Exposure

There are no occupational exposure standards or regulations governing exposure to the metaborates. Until there are data from animal or human experience, exposures should be limited by good housekeeping, exhaust ventilation practices, and personal protection.

8.6 Studies on Environmental Impact

There are no substance specific studies. Environmental impact probably will be a function of boron concentrations.

9.0 Sodium Metaborate Tetrahydrate

9.0.1 CAS Number: *[10555-76-7]*

9.0.2 Synonyms: Sodium metaborate 8 mol; Sodium metaborate tetrahydrate; Kodak balanced alkali

9.0.3 Trade Names: NA

9.0.4 Molecular Weight: 137.86

9.0.5 Molecular Formula: $NaBO_2 \cdot 4H_2O$

9.1 Chemical and Physical Properties

Appearance: White, odorless, crystalline granules
Specific gravity: 1.74
Vapor pressure: Negligible at 20°C
Solubility in water: 41.9% at 20°C, 109.8% at 100°C
pH at 20°C: 10.5 (0.1% solution), 11.0 (1.0% solution), 11.4 (4.0% solution)
Melting point: 53.5°C

9.2 Production and Use

Sodium metaborate tetrahydrate may be made by cooling a solution containing borax and an amount of sodium hydroxide just in excess or the theoretical value. It is used in textile finishing as a component of sizing, scouring compositions, adhesives, and detergents. It is also used in photography and agriculture (29).

9.3 Exposure Assessment

Exposure assessment should use methods common to the other borates. The content of boron in sodium metaborate dihydrate is 7.8%.

9.4 Toxic Effects

Metaborates (*e.g.*, sodium metaborate tetrahydrate, sodium metaborate dihydrate) are highly alkaline in solution. A 0.1 molar solution has a pH of 10.5, and a 15 molar solution has a pH of 11.9. Because of their high pH, the metaborates may cause skin irritation and eye damage.

9.5 Standards, Regulations, or Guidelines of Exposure

There are no occupational exposure standards or regulations governing exposure to the metaborates. Until there are data from animal or human experience, exposures should be limited by good housekeeping and exhaust ventilation practices.

9.6 Studies on Environmental Impact

There are no substance-specific studies. Environmental impact, either negative or positive, probably will be a function of boron concentrations in the environment.

10.0 Sodium Perborate Tetrahydrate

10.0.1 CAS Number: *[10486-00-7]*

10.0.2 Synonyms: PBS4, PBST; Perboric acid, sodium salt, tetrahydrate; metaborate; perboric acid (HBO$_3$), sodium salt, tetrahydate; perboric acid [HBO(O$_2$)], sodium salt, tetrahydrate; sodium perborate (NaBO$_3$) tetrahydrate

10.0.3 Trade Names: NA

10.0.4 Molecular Weight: 153.86

10.0.5 Molecular Formula: Na$_2$BO$_3$·4H$_2$O

10.0.6 Molecular structure:

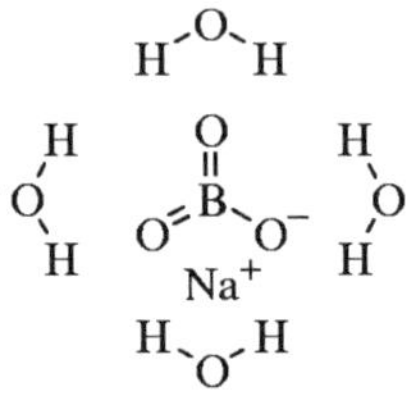

10.1 Chemical and Physical Properties

Appearance: White, odorless, crystalline powder
Bulk density: 700–900 kg/m^3
Solubility in water: 23 g/L at 20°C, 37 g/L at 30°C
pH at 20°C: 10 (1.5% solution)
Melting point: Decomposes

10.2 Production and Use

Sodium perborate tetrahydrate is prepared by reacting disodium tetraborate pentahydrate with hydrogen peroxide, with subsequent crystallization (29). The tetrahydrate is used primarily in European detergent powders as a bleaching agent in place of chlorine-containing bleaches and as a bleach for fabrics. However, perborate use is increasing in the United States as it is being introduced in washing powders as a color-safe bleaching agent This change is being encouraged by environmental concerns for chlorine (29).

10.3 Exposure Assessment

Air sampling and analytic methods used for other borates should be directly transferable to the perborates. Perborates do not exist in measurable quantities in the general environment since they are manufactured products and since they are unstable in the presence of moisture.

10.4 Toxic Effects

10.4.1 Experimental Studies

10.4.1.1 Acute Toxicity. The acute oral toxicity of perborates may be different from that of boric acid and other inorganic borates because of the generation of hydrogen peroxide as a breakdown product. Primary skin irritation studies of perborates have been performed in rabbits and guinea pigs. Sodium perborate tetrahydrate was classified as nonirritant when 500 mg was applied for 4 h to rabbit skin using the Draize method (195, 196). A 4-h administration of sodium perborate tetrahydrate was nonirritating to the skin of guinea pigs (120). Sodium perborate tetrahydrate caused severe eye irritation in rabbits when 100 mg of the granular material was applied to the conjuctival sac (195). Rinsing of the eye within 30 s prevented eye irritation in rabbits (105). Instilling a 1% solution of sodium perborate tetrahydrate to the rabbit eye produced no evidence of irritation (197). The LD_{50} in rats ranged from 2100 mg/kg (104) to approximately 2250 mg/kg (103). The acute inhalation toxicity of sodium perborate tetrahydrate was assessed in rats (120). The minimum concentration required to cause pulmonary irritation was 39 mg/m^3. The LC_{50} was 74 mg/m^3.

10.4.1.2 Chronic and Subchronic Toxicity. In a 28-day study, rats were given an oral dose of 1000 mg/kg of sodium perborate tetrahydrate per day (104). Clinical symptoms included increased salivation. The organ weights of brain, heart, kidneys, and testes were slightly reduced in males. Histological changes in the stomach were observed that are probably attributable to the formation of hydrogen peroxide in the stomach. No effect on the histology of the testes was found.

10.4.2 Human Experience

There have been no reports of systemic health effects, which is to be expected because of perborates' low toxicity and because they are not generally available to the public in concentrated form.

10.5 Standards, Regulations, or Guidelines of Exposure

There are no specific OELs for the perborates, and since they are soluble dusts, the PNOR or PNOC exposure limits do not apply. However, there is one acute rat inhalation study of the tetrahydrate that at 39 mg/m^3 caused minimum pulmonary irritation (120). Until there are data from animal or human experience, exposures should be limited through good housekeeping and exhaust ventilation practices, and workers should be observed for signs and symptoms of respiratory effects.

In animals, skin irritation and sensitization studies of perborates generally have been negative.

10.6 Studies on Environmental Impact

There are no substance-specific studies. Environmental impact, either negative or positive, probably will be a function of boron concentrations in the environment. Studies have been made of boron in the environment, and a fraction of the anthropogenic boron in these studies will have come from the use of borax, predominantly in soap powders. A method has been developed by Vengosh (194) to distinguish between natural and anthropogenic sources of boron water contamination. The $^{11}B/^{10}B$ isotope ratios for seawater, sodium borates, and calcium-containing borates and the application of this method are being explored in California and Israel. For information on the effects of boron in the environment, see the section on studies on environmental impact in the introductory sections dealing generically with boron.

11.0 Sodium Perborate Monohydrate

11.0.1 CAS Number: [10332-33-9]

11.0.2 Synonyms: PBSI, PBSM; Perboric acid (HBO), sodium borate, monohydate

11.0.3 Trade Names: NA

11.0.4 Molecular Weight: 99.8

11.0.5 Molecular Formula: $NaBO_3 \cdot H_2O$

11.1 Chemical and Physical Properties

White, crystalline, odorless powder
Decomposes on heating
Solubility in water 1.5% at 20°C, 2.4% at 30°C
pH 10 in a 1.5% solution

11.2 Production and Use

Sodium perborate monohydrate is produced by thermal dehydration of the tetrahydrate or by reaction of sodium metaborate with hydrogen peroxide in a fluidized-bed dryer. Uses are similar to those of the tetrahydrate (29).

11.4 Toxic Effects

11.4.1 Experimental Studies

11.4.1.1 Acute Toxicity. The acute oral toxicity of perborates may be different from that of boric acid and other inorganic borates due to the generation of hydrogen peroxide as a breakdown product. Primary skin irritation studies of perborates have been performed in rabbits and guinea pigs. Sodium perborate monohydrate was classified as nonirritant when 500 mg was applied for 4 h to rabbit skin using the Draize method (195, 196). An OECD test for irritation/corrosion on the skin of rabbits revealed that sodium perborate monohydrate was slightly irritating (106). Sodium perborate monohydrate caused severe eye irritation in rabbits when 100 mg of the granular material was applied to the conjunctival sac (195). Rinsing of the eye within 30 s prevented eye irritation in rabbits (105).

The acute oral toxicity of sodium perborate monohydrate in rats is in the range of 1120 mg/kg (107) and 1600–2100 mg/kg (106). Symptoms of toxicity included depression, bloated abdomen, and diarrhea. In rabbits, decomposition of sodium perborate monohydrate caused bloating of the stomach, which prevented determination of the oral LD_{50} (198). In rats, sodium perborate monohydrate in a 4% aqueous solution produced an oral LD_{50} of 3600 and 3250 mg/kg in males and females, respectively (105). Symptoms of toxicity included diarrhea and GI tract irritation. An oral dose of 50 mg/kg of sodium perborate monohydrate in dogs elicited vomiting, which was attributed to the generation of hydrogen peroxide in the stomach (103). The acute dermal LD$_{50}$ of sodium perborate monohydrate in rabbits was >2000 mg/kg (106). One rabbit died with signs of systemic toxicity, including diarrhea and congestion of internal organs.

11.5 Standards, Regulations, or Guidelines of Exposure

There are no specific OELs for the perborates, and since they are soluble dusts, the PNOR or PNOC exposure limits would not apply. However, there is one acute rat inhalation study of the tetrahydrate, which at 39 mg/m^3 caused minimum pulmonary irritation (120). Until there are data from animal or human experience, exposures should be limited through good housekeeping and exhaust ventilation practices, and workers should be observed for signs and symptoms of respiratory effects.

In animals, skin irritation and sensitization studies of perborates generally have been negative. However, one OECD test of sodium perborate monohydrate for irritation/corrosion was reported to be slightly irritating (106). For this reason protection of workers from skin irritation should be given when either the tetrahydrate or monohydrate is being handled.

11.6 Studies on Environmental Impact

There are no substance-specific studies. Environmental impact, either negative or positive, probably will be a function of boron concentrations in the environment. In studies of boron in the environment, a fraction of the anthropogenic boron will have come from the use of borax, predominantly in soap powders. Vengosh (194) has developed a method to distinguish between natural and anthropogenic sources of boron water contamination. The ^{11}B/^{10}B isotope ratios for seawater, sodium borates, and calcium-containing borates and the application of this method is being explored in California and Israel. For information on the effects of boron in the environment, see the section on studies on environmental impact in the introductory sections dealing generically with boron.

12.0 Zinc Borate Hydrate (2335)

12.0.1 CAS Number: [138265-88-0]

12.0.2 Synonyms: Dodecaboron, tetrazinc docosaoxide, heptahydrate

12.0.3 Trade Names: Firebrake® ZB, Borogard® ZB

12.0.4 Molecular Weight: 434.66

12.0.5 Molecular Formula: 2ZnO·3B$_2$O$_3$·3.5H$_2$O

12.1 Chemical and Physical Properties

    Appearance: A white, odorless powder
    Specific gravity: 2.7
    Vapor pressure: Negligible at 20°C

Solubility in water: < 0.28% at 25°C
pH at 20°C: 6.8–7.5 in aqueous solution
Thermal stability: Retains water of hydration to 290°C

12.2 Production and Use

Production is through the reaction of zinc oxide with boric acid at 90–100°C. Thermal stability makes it attractive as a fire-retardant additive for plastics and rubbers that require high processing temperatures. It is also used as a corrosion inhibitor in coatings (29).

12.3 Exposure Assessment

Air sampling can be done as inhalable dust on a mixed cellulose ester filter measured gravimetrically or chemically by flame atomic absorption spectrophotometry.

12.4 Toxic Effects

The acute oral LD_{50} toxicity of zinc borate was evaluated in rats. The acute dermal toxicity was evaluated in rabbits (108–112). The acute dermal LD_{50} was > 2000 mg/kg for all of these compounds. Zinc borate did not induce eye irritation in rabbit studies (127, 132–136). The 4-h LC_{50} of zinc borate in rats was greater than 4.9 mg/m^3 (118).

12.5 Standards, Regulations, or Guidelines of Exposure

There are no occupational exposure limits directly applicable zinc borate. However, since it has low acute toxicity and is an insoluble particulate, the OSHA PEL PNOR 15 mg/m^3 total dust, 5 mg/m^3 respirable dust or the ACGIH TLV PNOC 10 mg/m^3 total dust, 3 mg/m^3 respirable dust exposure limits should be protective.

12.6 Studies on Environmental Impact

There are no substance-specific studies.

13.0 Boron Tribromide

13.0.1 CAS Number: [10294-33-4]

13.0.2 Synonyms: Boron bromide; Tribromoborane; Boron tribromide, 1M solution in methylene chloride

13.0.3 Trade Names: NA

13.0.4 Molecular Weight: 250.57

13.0.5 Molecular Formula: BBr$_3$

13.0.6 Molecular Structure:

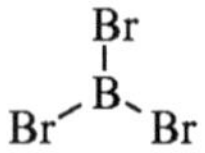

13.1 Chemical and Physical Properties

Appearance: Colorless fuming liquid
Specific gravity: 2.698
Vapor pressure: 40 torr at 14°C
Boiling point: 91.3°C
Solubility in water: Decomposes
Melting point: −46°C

13.1.1 General: NA

13.1.2 Odor and Warning Properties

Boron tribromide has a sharp, irritating odor. By analogy with hydrogen bromide, which is probably formed by hydrolysis with moisture in air, odor should be acrid and detectable at 2 ppm (199).

13.2 Production and Use

Boron tribromide is used in the production of diborane and high purity elemental boron.

13.3 Exposure Assessment

According to the ACGIH documentation of boron tribromide, its decomposition should produce 3 mol of hydrogen bromide (200).

13.4 Toxic Effects

The acute local effects of boron tribromide are considered to be high for irritation, ingestion, and inhalation. There are no toxicity data for boron tribromide listed in the *Registry of Toxic Effects of Chemical Substances*. Toxic effects are probably those of hydrogen bromide (200).

13.5 Standards, Regulations, or Guidelines of Exposure

There is no OSHA PEL. The ACGIH TLV ceiling is 1 ppm (10 mg/m^3), and NIOSH makes the same recommendation.

13.6 Studies on Environmental Impact

There are no substance-specific studies.

14.0 Boron Trifluoride

14.0.1 CAS Number: [7637-07-2]

14.0.2 Synonyms: Trifluoroborane, boron fluoride, trifluoro boron

14.0.3 Trade Names: NA

14.0.4 Molecular Weight: 67.82

14.0.5 Molecular Formula: BF_3

14.0.6 Molecular Structure:

$$\begin{array}{c} F \\ | \\ F{-}B{-}F \end{array}$$

14.1 Chemical and Physical Properties

Appearance: Colorless, nonflammable gas that fumes in moist air
Specific gravity: 2.99
Vapor pressure: > 1.0 torr at $20°C$
Solubility in water: 332 g/100 g at $0°C$
Boiling point: $-99.9°C$ (TLV)
Melting point: $-126.7°C$ (TLV)

14.1.1 General: NA

14.1.2 Odor and Warning Properties

It has a strong pungent, suffocating odor.

14.2 Production and Use

Its largest use is as a catalyst. It is also used to protect molten magnesium and its alloys from oxidation, as a flux for soldering magnesium, as a fumigant, and as a weak neutron detector in ionization chambers (201).

14.3 Exposure Assessment

OSHA recommends a method developed at its Salt Lake City Laboratory that requires sampling in a midget fritted glass bubbler containing 10 mL of 0.1 N NH_4F (202).

14.4 Toxic Effects

In inhalation studies (203), repeated exposure of rats, rabbits, and guinea pigs to boron trifluoride resulted in respiratory irritation. Boron trifluoride elicited such a degree of respiratory irritation as to cause the death of guinea pigs from respiratory failure, which occurred after the 19th exposure day at the nominal concentration of 12.8 ppm as boron trifluoride. Deaths still occurred in guinea pigs, but not in rats, exposed at a concentration of 3–4 ppm, but all three species exposed at a concentration of 1.5 ppm were only minimally affected; average body weights of the guinea pigs were only 85% of that of controls and occasional pneumonitis. Exposed rabbits did not differ histologically from controls. Fluorosis of rat teeth was observed at the highest dose.

BORON

Because of difficulty with earlier studies including limited use of controls, Rusch et al. (204) conducted additional inhalation studies. Because boron trifluoride reacts with moisture in the air of animal exposure chambers to form the stable dihydrate, these studies were conducted with the dihydrate. With this material using male and female Fischer 344 rats, the acute toxicity was a 4-h LC_{50} of 1.21 mg/L. Rats were exposed 6 h/day for 13 weeks to a respirable, liquid aerosol of BF_3 at 0, 2.0, 6.0, and 17 mg/m^3. One rat died at the high dose. At this dose there was necrosis of the proximal tubular epithelium of the kidneys, rales, and excessive lacrimation and increases in bone fluoride. At the 6 mg/m^3 (2.14 ppm) concentration there were no signs of toxicity, but there were increases in fluoride levels.

14.5 Standards, Regulations, or Guidelines of Exposure

A ceiling designation of 1 ppm for boron trifluoride was set by ACGIH in 1963, and that recommendation is still in effect. The NIOSH REL is the same.

14.6 Studies on Environmental Impact

There are no substance-specific studies.

15.0 Diborane

15.0.1 CAS Number: [19287-45-7]

15.0.2 Synonyms: Boroethane, boron hydride, Diboron hexahydride, Diborane (6), diborane hexanhydride, boron hydride

15.0.3 Trade Names: NA

15.0.4 Molecular Weight: 27.69

15.0.5 Molecular Formula: B_2H_6

15.0.6 Molecular Structure:
$$\begin{array}{c} H \quad H \quad H \\ \diagdown \diagup \diagdown \diagup \\ B \quad\quad B \\ \diagup \diagdown \diagup \diagdown \\ H \quad H \quad H \end{array}$$

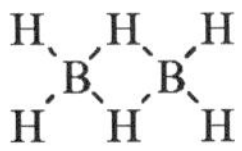

15.1 Chemical and Physical Properties

Appearance: Colorless, combustible gas
Specific gravity: 0.21 at 15°C
Melting point: -165°C
Boiling point: -92.5°C
Solubility in water: Hydrolyzes to H_3BO_3 and H_2
Vapor pressure: >1 atm

15.1.1 General: NA

15.1.2 Odor and Warning Properties

It has a repulsive, sickly sweet odor that is not protective at the TLV level. The median detectable concentration by odor for humans is 2–4 mg/m^3 (2.2–4.4 ppm) (205).

15.2 Production and Use

Diborane is used as a catalyst for olefin polymerization, as a rubber vulcanizer, as a reducing agent, as a flame-speed accelerator, in an intermediate in preparation of boron hydrides, and as a doping gas (206).

15.3 Exposure Assessment

Stokinger (193) recommended that air samples be collected by absorption in a glass bubbler containing bromine in glacial acetic acid saturated with potassium bromide. The determination of diborane is made by spectrophotometric measurement at 270 micrometers. OnoOgasawara et al. (207) studied methods of collecting diborane using several absorbents and analytical methods. They concluded that collection on silica gel impregnated with potassium permanganate, desorption into hydroxylamine hydrochloride solution, and ultimate determination by the chromotropic acid HPLC method was superior. The NIOSH method 6006 is recommended for determining workplace exposures (92).

15.4 Toxic Effects

15.4.1 Experimental Studies

The 4-h rat inhalation LC$_{50}$ is about 50 ppm; the corresponding LC$_{50}$ for the mouse is 30 ppm, and the guinea pig is somewhat less sensitive. Chronically, diborane showed the same order of toxicity among the animal species. Daily, 6-h exposures of dogs at about 7 ppm resulted in death after 10–25 exposures. Rats at this concentration survived somewhat longer, up to 113 exposures. Some dogs survived exposures of 1–2 ppm for a period of 6 months, whereas guinea pigs survived 95 exposures without showing any effects. The primary effect of diborane poisoning was concluded to be the production of pulmonary edema, resulting from the local irritation of the compound set off by the exothermic reaction of hydrolysis (208).

15.4.2 Human Studies

Observations made between 1956 and 1960 include 26 cases of acute diborane toxicity with 33 cases of subacute exposures. These cases showed predominately a bronchopulmonary involvement. Chronic respiratory distress was present in two patients from recurrent diborane exposure. Because these appeared clinically as an asthmatic bronchitis, the authors concluded that the chronic disability was on a hypersensitivity basis (209).

15.5 Standards, Regulations, or Guidelines of Exposure

The OSHA PEL, the NIOSH REL, and the ACGIH TLV are all 0.1 ppm (0.1 mg/m^3). The ACGIH value was set in 1956 and has remained at that level.

BORON

15.6 Studies on Environmental Impact

There are no substance-specific studies.

16.0 Pentaborane

16.0.1 CAS Number: [19624-22-7]

16.0.2 Synonyms: Pentaborane, pentaboron nonahydride, stable pentaborane, pentaborane (9)

16.0.3 Trade Names: NA

16.0.4 Molecular Weight: 63.12

16.0.5 Molecular Formula: B_5H_9

16.0.6 Molecular Structure:

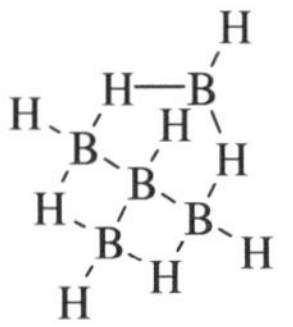

16.1 Chemical and Physical Properties

Appearance: Colorless, volitile liquid
Specific gravity: 0.61 at 0°C
Melting point: -46.6°C
Boiling point: -60°C
Vapor pressure: 171 torr at 20°C, 66 torr at 0°C
Flash point: 30°C, closed cup
Solubility in water: Decomposes; hydrolyzes in water after long heating

16.1.1 General: NA

16.1.2 Odor and Warning Properties

It has a pungent odor detectable at 0.96 ppm; thus odor is not protective at the TLV level. The median detectable concentration by odor for humans is 2.5 mg/m^3 (1 ppm) (205).

16.2 Production and Use

There appears to be no commercial market for pentaborane. In the 1950s it was explored as a potential rocket fuel.

16.3 Exposure Assessment

Stokinger (193) was unable to cite any sampling methods for pentaborane, and, insofar as can be determined, none have been published since then.

16.4 Toxic Effects

16.4.1 Experimental Studies

The highly acute toxic nature of B_5H_9 prompted exposure of animals for unusually brief periods of 0.5, 2, 5, and 15 min. The calculated LC_{50} for mice, monkeys, and dogs after 2-min inhalation exposures were respectively 342, 640, and 734 mg/m^3, indicating the mouse to be the most sensitive species on an acute basis. Studies indicate that acute effects are limited to CNS. In chronic exposures, a 6-month, daily inhalation at 0.2 ppm by monkeys, dogs, rabbits, rats, and hamsters showed toxic signs and symptoms conspicuously different from those seen in acute studies. All species showed weight loss, apathy and insensitivity to pain, loss of limb mobility, muscle tremors, and impaired coordination (210). A clue to the mechanism of action of pentaborane in the CNS was found in the inhibition of glycolysis (211).

16.4.2 Human Experience

Despite pentaborane's extreme, acute toxicity, Yarbrough et al. (212) cite the fact that until 1985, only 32 severe cases of pentaborane poisoning were described in the literature. In these cases, myoclonic spasms, convulsions, opisthotonos, and coma were reported. In only one case did severe symptoms begin as soon as 60 min following exposure. In the others, serious symptoms were delayed up to 40 h and followed an intervening period characterized by lethargy, anxiety, weakness, personality changes, dizziness, headaches, and inability to concentrate. The three new cases reported by Yarbrough et al. (212) resulted from a common exposure that produced rapid, severe toxicity resulting in one death and one permanent CNS damage. All patients showed impressive elevation in transaminases, both ALT, and AST, indicating liver damage. Rhabdomyolysis was also present. Medical personnel treating these cases also sustained mild to moderate toxicity (212).

16.5 Standards, Regulations, or Guidelines of Exposure

Pentaborane has the lowest TLV of any of the boranes, with a TLV TWA of 0.005 ppm proposed in 1961. In 1976, a further restraint on exposures resulted from the adoption of a STEL of 0.015 ppm, and both TLVs continue to be listed in the ACGIH 1999 list of TLVs.

OSHA PELs are the same as the TLVs, although the STEL was removed by the March 1992 ruling, which reversed those changes made by OSHA in 1989. NIOSH recommended the same values and in addition established a STEL/ceiling value of 3 ppm for pentaborane.

Pentaborane, B_5H_9 a highly toxic volatile liquid, is spontaneously flammable in air and highly explosive, forming shock-sensitive mixtures with chlorinated hydrocarbons and carbonyl-containing compounds. Minor spills can result in acute toxic effects in nearby workers. The median detectable concentration by odor for humans is reported to be in the range of 2.5 mg/m^3 is not protective. It is recommended that respiratory protection be provided by air-supplied respirators and skin exposure be avoided (205).

16.6 Studies on Environmental Impact

There are no substance-specific studies.

BORON

17.0 Decaborane

17.0.1 CAS Number: [17702-41-9]

17.0.2 Synonyms:
Decaboron tetradecahydride, boron hydride, nido-decaborane, decaborane (14), tetradecaborane

17.0.3 Trade Names: NA

17.0.4 Molecular Weight: 122.31

17.0.5 Molecular Formula: $B_{10}H_{14}$

17.0.6 Molecular Structure:

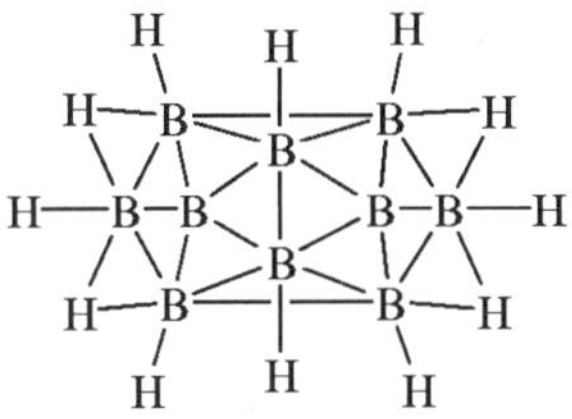

17.1 Chemical and Physical Properties

Appearance: Colorless, crystalline solid
Specific gravity: 0.94 at 25°C
Boiling point: 213°C
Vapor pressure: 0.05 torr at 25°C
Flash point: 80°C, closed cup
Solubility in water: Slightly soluble in cold water, decomposes in hot water
Reactivity: Explosive on contact with flame, heat, or oxygenated or chlorinated solvents

17.1.1 General

17.1.2 Odor and Warning Properties

It has a pungent odor detectable at 0.06 ppm; thus odor threshold is at the TLV and cannot be relied on to protect at the TLV level (205).

17.2 Production and Use

Decaborane is used as a catalyst in the polymerization of olefins.

17.3 Exposure Assessment

Stokinger (193) described a number of colorimetric, microanalytic methods developed in relation to the use of decaborane as a high energy fuel. One of these methods required collection in glass scrubbers containing a triethanolamine solution, with the subsequent development of a red color resulting from the formation of a complex with quinoline. No more recent methods are known to have been published.

17.4 Toxic Effects

17.4.1 Experimental Studies

The 4-h LC$_{50}$ for decaborane in mice was found to be 25.7 ppm and in rats was well above 95 ppm (213). Histopathological examination of the dog following intraperitoneal injection gave evidence of liver and kidney damage. Chronic, 5–6-h daily inhalation exposures for as long as 6 months at 4.5 ppm were fatal to rabbits after a few exposures, to the dog and monkey after 4–15 exposures, to mice after 10–100 exposures, and to rats after 135 exposures, demonstrating a wide difference in susceptibility among species. Pathological changes in the liver were prominent in the areas of active cellular metabolism, the central zones of the liver, and the tubules of the kidney. Valerino et al. (214) show that the boranes interact with hepatic microsomal enzymes but found that pyridoxal phosphate, a coenzyme that alters the inhibitory actions of decaborane on decarboxylases and glutamic oxaloacetic transaminase, had no effect on the interaction of decaborane with hepatic microsomal enzymes. They also attribute CNS effects to depression of norepinephrine, dopamine, and serotonin in the brain.

17.4.2 Human Experience

Cases have demonstrated dizziness, drowsiness, and headache with a decrease in cardiac output and persistent bradycardia (215). As a result of studies of the performance of macaque monkeys injected with small doses (2 or 4 mg/kg), inferences can be made about humans (216). Workers exposed to such levels of decaborane may be expected to show a performance decrement or clinical symptoms during the first 50 h. Usually, the first indication is a performance decrement on a task requiring continuous motor behavior, or a series of discriminations within the first 30 h. Tasks of a discrete nature may, at lower levels, reflect no decrement or clinical symptoms. When, however, performance decrements do occur, return to baseline performance levels may be expected within 3–10 days.

17.5 Standards, Regulations, or Guidelines of Exposure

An ACGIH TLV of 0.05 ppm was established in 1957, the notation "skin" was added in 1961, and a STEL of 0.15 ppm was added in 1976. These same values are included in the 1999 TLV list (217). OSHA adopted these values, and NIOSH did likewise.

Australia adopted the same levels in 1990, while the Federal Republic of Germany selected 0.05 ppm as a TWA, but chose a short-term level of 0.1 ppm for 5-min periods, 8 times per workshift, with skin notation, in 1990 (105).

Because of the high toxicity of decaborane, its ability to be absorbed through the skin, as well as its physical properties, industrial hygiene and safety measures are extremely important, and all possible measures must be taken to prevent exposures.

17.6 Studies on Environmental Impact

There are no substance-specific studies.

BIBLIOGRAPHY

1. K. Warington, Effect of boric acid and borax on the broad bean and certain other plants. *Ann. Bot. (London)* **37**, 629–672 (1923).
2. W. D. Loomis and R. W. Durst, Chemistry and biology of boron. *BioFactors* **3**, 229–239 (1992).
3. D. J. Fort et al., Adverse reproductive and developmental effects in Xenopus from insufficient boron. *Biol. Trace Elem. Res.* **66**(1–3), 237–259 (1998).
4. D. J. Fort et al., Adverse effects from low dietary and environmental boron exposure on reproduction, development, and maturation in *Xenopus laevis*. *J. Trace Elem. Exp. Med.* **12**(3), 175–186 (1999).
5. D. J. Fort et al., The effect of boron deprivation on reproductive parameters in *Xenopus laevis*. *J. Trace Elem. Exp. Med.* **12**(3), 187–204 (1999).
6. C. D. Eckhart, Boron stimulates embryonic trout growth. *J. Nutr.* **128**, 2488–2493 (1998).
7. R. I. Rowe et al., The response of trout and zebrafish embryos to low and high boron concentrations is U-shaped. *Biol. Trace Elem. Res.* **66**(1–3), 261–270 (1998).
8. C. D. Eckhert and R. I. Rowe, Embryonic dysplasia and adult retinal dystrophy in boron deficient zebrafish. *J. Trace Elem. Exp. Med.* **12**(3), 213–220 (1999).
9. L. Lanoue et al., Assessing the effects of low boron diets on embryonic and fetal delvelopment in rodents using *in vitro* and *in vivo* model systems. *Biol. Trace Elem. Res.* **66**(1–3), 271–298 (1998).
10. L. Lanoue, P. L. Strong, and C. L. Keen, Adverse effects of a low boron environment on the preimplantation development of mouse embryos *in vitro*. *J. Trace Elem. Exp. Med.* **12**(3), 235–250 (1999).
11. C. D. Hunt and F. H. Nielsen, Interaction between boron and cholecalciferol in the chick. In J. Gawthorne and C. White, eds., *Trace Element Metabolism in Man and Animals*, 4th ed., Australian Academy of Science, Canberra, 1981, pp. 597–600.
12. C. D. Hunt, Dietary boron modified the effects of magnesium and molybdenum on mineral metabolism in the cholecalciferol deficient chick. *Biol. Trace Elem. Res.* **22**, 201–220 (1989).
13. C. D. Hunt, J. L. Herbel, and J. P. Idso, Dietary boron modifies the effects of vitamin D3 nutriture on indices of energy substrate utilization and mineral metabolism in the chick. *J. Bone Miner. Res.* **9**, 171–181 (1994).
14. F. H. Nielsen, Biochemical and physiologic consequences of boron deprivation in humans. *Environ. Health Perspect.* **102**(Suppl. 7), 59–63 (1994).
15. C. D. Hunt, The biochemical effects of physiologic amounts of dietary boron in animal nutrition models. *Environ. Health Perspect.* **102**(Suppl. 7), 35–43 (1994).
16. J. G. Penland, Dietary boron, brain function, and cognitive performance. *Environ. Health Perspect.* **102**(Suppl. 7), 65–72 (1994).
17. C. D. Hunt, Biochemical effects of physiological amounts of dietary boron. *J. Trace Elem. Exp. Med.* **9**, 185–213 (1996).
18. F. H. Nielsen, Evidence for the nutritional essentiality of boron. *J. Trace Elem. Exp. Med.* **9**, 215–229 (1996).

19. C. D. Hunt, J. L. Herbel, and F. H. Nielsen, Metabolic response of postmenopausal women to supplemental dietary boron and aluminum during usual and low magnesium intake: Boron, calcium and magnesium absorption and retention and blood mineral concentrations. *Am. J. Clin. Nutr.* **65**, 803–813 (1997).
20. C. D. Hunt, Regulation of enzymatic activity: One possible role of dietary boron in higher animals and humans. *Biol. Trace Elem. Res.* **66**(1–3), 205–225 (1998).
21. J. G. Penland, The importance of boron nutrition for brain and psychological function. *Biol. Trace Elem. Res.* **66**(1–3), 299–317 (1998).
22. F. H. Nielsen, The justification for providing dietary guidance for the nutritional intake of boron. *Biol. Trace Elem. Res.* **66**(1–3), 319–330 (1998).
23. F. H. Nielsen and J. G. Penland, Boron supplementation of peri-menopausal woment affects boron metabolism and indices associated with macromineral metabolism, hormonal status and immune function. *J. Trace Elem. Res.* **12**(3), 251–262 (1999).
24. C. D. Hunt and J. P. Idso, Dietary boron as a physiological regulator of the normal inflammatory response: A review and current research progress. *J. Trace Elem. Res.* **12**(3), 221–234 (1999).
25. World Health Organization, *Trace Elements in Human Nutrition and Health*, WHO, Geneva, 1996.
26. C. J. Rainey et al., Daily boron intake from the American diet. *J. Am. Diet. Assoc.* **99**(3), 335–340 (1999).
27. *CRC Handbook of Chemistry and Physics*, 68th ed. CRC Press, inc. Boca Raton, FL, 1987–1988, p. B222.
28. E. J. Catanzaro, Boric acid. Isotopic and assay standard and reference materials. *NBS Spec. Publ. (U.S.)* **260-17**, (1970).
29. R. A. Smith and R. B. McBroom, Boron compounds. *In Kirk-Othmer's Encyclopedia of Chemical Technology*, 4th ed., Vol. 4, Wiley, New York, 1992.
30. P. A. Lyday, Boron. *Mineral Industry Surveys*, 1998 Annual Report, 1994 Annual Report U.S. Department of Interior, Bureau of Mines, Washington, DC, 1998.
31. P. D. Howe, A review of boron effects in the environment. *Biol. Trace Elem. Res.* **66**(1–3), 153–156 (1998).
32. R. D. Barr et al., Regulation of lithium and boron levels in normal human blood; environmental and genetic considerations. *J. Lab. Clin. Med.* **12**(4), 614–619 (1993).
33. B. S. Şayli, E. Tüccar, and A. H. Elhan, An assessment of fertility in boron-exposed Turkish subpopulations. *Reprod. Toxicol.* **12**, 297–304 (1998).
34. R. A. Smith and F. M. Ascherl, Issues concerning the measurement of borate in occupational environments. *Am. Ind. Hyg. Assoc. J.* **60**(5) (1999).
35. T. Matsunaga, T. Ishii, and H. Watanabe, Speciation of water soluble boron compounds in radish roots by size exclusion HPLC/ICP-MS. *Anal. Sci.* **12**, 673–675 (1996).
36. S. G. Penn et al., Direct analysis of sugar alcohol borate complexes in plant extracts by matrix-assisted laser desorption/ionization Fourier transform mass spectrometry. *Anal. Chem.* **69**, 2471–2477 (1997).
37. W. B. Clarke et al., Binding of lithium and boron to human plasma proteins. *Biol. Trace Elem. Res.* **65**, 237–249 (1998).
38. R. F. Barth et al., Determination of boron in tissues and cells using direct-current plasma atomic emission spectroscopy. *Anal. Chem.* **63**, 890–893 (1991).

39. C. D. Hunt and T. R. Schuler, Concentration of boron and other elements in human foods and personal-care products. *J. Micronutr. Anal.* **6**, 161–174 (1989).
40. S. R. Tamat and D. E. Moore, Determination of boron in biological tissues by inductively coupled plasma atomic emission spectrometry. *Anal. Chem.* **59**, 2161–2164 (1987).
41. D. Pollmann et al., Determination of boron in biological tissues by inductively coupled plasma optical emission spectrometry (ICP-OES). *Fresenius' J. Anal. Chem.* **346**, 441–445 (1993).
42. S. Evans and U. Krahenbuhl, Boron analysis in biological material: Microwave digestion procedure and determination by different methods. *Fresenius' J. Anal. Chem.* **349**, 454–459 (1994).
43. A. M. S. Nyomora et al., Boron determination in biological materials by inductive coupled plasma atomic emission and mass spectrometry: Effects of sample dissolution. *Fresenius' J. Anal. Chem.* **357**, 1185–1191 (1997).
44. D. Sun, J. K. Waters, and T. P. Mawhinney, Microwave digestion for determination of aluminum, boron, and 13 other elements in plants by inductively coupled plasma atomic emission spectrometry. *J. Assoc. Off. Anal. Chem. Int.* **80**(3), 647–650 (1997).
45. D. Sun, J. K. Waters, and T. P. Mawhinney, Microwave digestion and ultrasonic nebulization for determination of boron in animal tissues by inductively coupled plasma atomic emission spectrometry with internal standardization and addition of mannitol. *J. Anal. At. Spectrom.* **12**, 675–679 (1997).
46. D. Sun, J. K. Waters, and T. P. Mawwhinney, Determination of boron in plants and plant-derived foods by ultrasonic nebulization-inductively coupled plasma atomic emission spectrometry with addition of mannitol. *J. Assoc. Off. Anal. Chem. Int.* **80**(1), 20–24 (1997).
47. D. A. Johnson, D. D. Siemer, and W. F. Bauer, Determination of nanogram levels of boron in milligram-sized tissue samples by inductively coupled plasma-atomic emission spectroscopy. *Anal. Chim. Acta.* **270**, 223–230 (1992).
48. R. G. Downing et al., Considerations in the determination of boron at low concentrations. *Biol. Trace Elem. Res.* **66**(1–3), 3–21 (1998).
49. R. G. Downing and P. L. Strong, Analytical challenges of low level boron analysis in biological matrices. *J. Trace Elem. Exp. Med.* **12**(3), 205–212 (1999).
50. F. G. Smith et al., Measurement of boron concentration and isotope ratios in biological samples by inductively coupled plasma mass spectrometry with direct injection nebulization. *Anal. Chim. Acta* **248**, 229–234 (1991).
51. S. Evans and U. Krahenbuhl, Improved boron determination in biological material by inductively coupled plasma mass spectrometry. *J. Anal. At. Spectrom.* **9**, 1249–1253 (1994).
52. R. N. Sah and P. H. Brown, Isotope ratio determination in boron-analysis. *Biol. Trace Elem. Res.* **66**(1–3), 39–53 (1998).
53. E. Marentes, R. A. Vanderpool, and B. J. Shelp, Boron-isotope fractionation in plants. *Can. J. Plant Sci.* **77**, 627–629 (1997).
54. H. Vanhoe et al., Determination of boron in human serum by inductively coupled plasma mass spectrometry after a simple dilution of the sample. *Anal. Chim Acta* **281**, 401–411 (1993).
55. H. Vanhoe et al., Role of inductively coupled plasma mass spectrometry (ICP-MS) in the assessment of reference values for ultra-trace elements in human serum. *Trace Elem. Electrolytes* **12**(2), 81–88 (1995).
56. K. Usuda, K. Kono, and Y. Yoshida, Serum boron concentration from inhabitants of an urban area in Japan. *Biol. Trace Elem. Res.* **56**(2), 167–178 (1997).

57. N. I. Ward, The determination of boron in biological materials by neutron irradiation and prompt gamma-ray spectrometry. *J. Radioanal. Nucl. Chem., Articl.* **110**(2), 633–639 (1987).
58. T. Kobayashi and K. Kanda, Microanalysis system of PPM-order 10 B concentrations in tissue for neutron capture therapy by prompt gamma-ray spectrometry. *Nucl. Instrum. Methods* **204**, 525–531 (1983).
59. Y. S. Khrbish and N. M. Spyrou, Prompt gamma-ray neutron activation analysis by the absolute method. *J. Radioanal. Nucl. Chem., Artic.* **151**(1), 55–61 (1991).
60. A. E. Pillay and M. Peisach, Application of prompt gamma-ray spectrometry to helium-3-induced activation of boron. *J. Radioanal. Nucl. Chem., Artic.* **151**(2), 379–386 (1991).
61. M. Rossbach and N. T. Hiep, Prompt gamma cold neutron activation analysis applied to biological materials. *Fresenius' J. Anal. Chem.* **344**, 59–62 (1992).
62. L. G. L. Meneses et al., Neutron-capture-radiography study of boron absorption by leaves of coffee plants. *J. Trace Microprobe Tech.* **15**, 335–340 (1997).
63. T. U. Probst et al., Comparison of inductively coupled plasma atomic emission spectrometry and inductively coupled plasma mass spectrometry with quantitative neutron capture radiography for the determination of boron in biological samples from cancer therapy. *J. Anal. At. Spectrom.* **12**, 1115–1122 (1997).
64. W. B. Clarke et al., Analysis of ultratrace lithium and boron by neutron activation and mass-spectrometric measurement of 3 He and 4 He. *Appl. Radiat. Isot.* **38**, 735–743 (1987).
65. W. B. Clarke et al., Lithium and boron in human blood. *J. Lab. Clin. Med.* **109**(2), 155–158 (1987).
66. G. V. Iyegar, W. B. Clarke, and R. G. Downing, Determination of boron and lithium in diverse biological matrices using neutron activation-mass spectrometry (NA-MS). *Fresenius' J. Anal. Chem.* **338**, 562–566 (1990).
67. R. N. Sah and P. H. Brown, Boron determination: A review of analytical methods. *Microchem. J.* **56**, 285–304 (1997).
68. R. G. Downing and P. L. Strong, A round-robin determination of boron in botanical and biological samples. *Biol. Trace Elem. Res.* **66**(1–3), 23–37 (1998).
69. K. Yoshino et al., Spectrophotometric determination of trace boron in biological materials after alkali fusion decomposition. *Anal. Chem.* **56**, 839–842 (1984).
70. A. Kaczmarczyk, J. R. Messer, and C. E. Peirce, Rapid method for determining boron in biological materials. *Anal. Chem.* **43**(2), 271–272 (1971).
71. S. Ogawa et al., Colorimetric determination of boric acid in prawns, shrimp, and salted jelly fish by chelate extraction with 2-Ethyl-1,3-Hexanediol. *Anal. Chem.* **62**(3), 610–614 (1979).
72. J. Ciba and A. Chrusciel, Spectrophotometric determination of boron in human hair with Azomethine H. *Fresenius' J. Anal. Chem.* **342**, 147–149 (1992).
73. N. Chimpalee, D. Chimpalee, and B. Boonyanitchayakul, Flow-injection spectrofluorimetric determination of boron using alizarin red S in aqueous solution. *Anal. Chim. Acta* **282**, 643–646 (1993).
74. D. L. Callicoat and J. D. Wolszon, Carminic acid procedure for determination of boron. *Anal. Chem.* **31**(8), 1434–1437 (1959).
75. G. M. A. Botelho, A. J. Curtius, and R. C. Campos, Determination of boron by electrothermal atomic absorption spectrometry: Testing different modifiers, atomization surfaces and potential interferences. *J. Anal. At. Spectrom.* **9**, 1263–1267 (1994).

76. M. Papaspyrou and L. E. Feinendegen, Determination of boron in cell suspensions using electrothermal atomic absorption spectrometry. *J. Anal. At. Spectrom.* **9**, 791–795 (1994).
77. R. Kobayashi et al., Systematic study of boron determination by graphite furnace atomic absorption spectrometry. *Anal. Sci.* **13**(Suppl.), 31–34 (1997).
78. R. G. Zamenhof et al., Monte Carlo treatment planning and high resolution alpha-track autoradiography for neutron capture therapy. *Strahlenther. Onkol.* **165**, 188–192 (1989).
79. J. P. Pignol et al., Neutron capture radiography applied to the investigation of boron-10 biodistribution in animals: Improvements in techniques of imaging and quantitative analysis. *Nucl. Instrum. Methods Phys. Res., Sect. B* **94**, 516–522 (1994).
80. M. Takagaki and Y. Mishima, Boron-10 quantitative analysis of neutron capture therapy on malignant melanoma by spectrophotometric α-track reading. *Nucl. Tracks Radiat. Meas.* **17**(4), 531–535 (1990).
81. H. F. Arlinghaus and M. T. Spaar, Imaging of boron in tissue at the cellular level for boron neutron capture therapy. *Anal. Chem.* **69**, 3169–3176 (1997).
82. M. Laurent-Peterson, B. Delpech, and M. Thellier, The mapping of natural boron in histological sections of mouse tissues by the use of neutron-capture radiography. *Histochem. J.* **24**, 939–950 (1992).
83. W. A. Ausserer et al., Quantitative imaging of boron, calcium, magnesium, potassium, and sodium distributions in cultured cells with ion microscopy. *Anal. Chem.* **61**, 2690–2695 (1989).
84. G. H. Morrison and G. Slodzian, Ion microscopy. *Anal. Chem.* **47**(11), 932A–943A (1975).
85. M. Bendayan et al., Electron spectroscopic imaging for high-resolution immunocytochemistry: Use of boronated protein A. *J. Histochem. Cytochem.* **37**(5), 573–580 (1989).
86. T. L. Litovitz et al., Clinical manifestations of toxicity in a series of 784 boric acid ingestions. *Am. J. Emerg. Med.* **6**(3), 209–213 (1988).
87. F. R. Abou-Shakra, J. J. Havercroft, and N. I. Ward, Lithium and borox in biological tissues and fluids. *Trace Elem. Med.* **6**, 142 (1989).
88. W. B. Clarke, C. E. Webber, and M. Koecebakker, Lithium and boron in human blood. *J. Lab. Clin. Med.* **109**, 155 (1987).
89. H. R. Imbus et al., Boron, cadmium, chromium, and nickel in blood and urine. *Arch. Environ. Health* **6**, 286 (1963).
90. B. D. Culver et al., The relationship of blood and urine-boron to boron exposure in borax workers and the usefulness of urine-boron as an exposure marker. *Environ. Health Perspect.* **102**(Suppl. 7), 133–137 (1994).
91. S. A. Hubbard, Comparative toxicology of borates. *Biol. Trace Elem. Res.* **66**(1–3), 343–357 (1998).
92. NIOSH, *Manual of Analytical Methods*, 4th ed., 1994.
93. C. C. Pfeiffer, L. F. Hallman, and I. Gersh, Boric acid ointment. A study of possible intoxication in the treatment of burns. *J. Am. Med. Assoc.* **128**(4), 266–273 (1945).
94. J. Weir, Jr. and R. S. Fisher, Toxicological studies on borax and boric acid. *Toxicol. Appl. Pharmacol.* **23**, 351–364 (1974).
95. E. L. Reagan and P. L. Becci, *Acute Oral LD$_{50}$ Study of 20 Mule Team Lot No. USB 12-84 Sodium Tetraborate Pentahydrate in New Zealand White Rabbits*, Unpublished report to U.S. Borax, Food and Drug Research Laboratories, Waverly, NY, 1985.

96. S. M. Denton, *Dehybor Anhydrous Borax. Acute Oral Toxicity Study in the Rat*, Study No. 1341/9-1032 and 1341/3-1032, Unpublished report to Borax Europe Limited, Corning Hazleton (Europe), Harrogate, North Yorkshire, UK, 1996.

97. R. L. Doyle, *Acute Oral Toxicity in Rabbits of Disodium Octaborate Tetrahydrate*, Study No. 88-3197-21, Unpublished report to U.S. Borax, Hill Top Biolabs, Miamiville, OH, 1989.

98. J. Shipp, C. F. Folk, and J. A. Young, *Acute Oral Toxicity of TS-7504-14-1 Potassium Tetraborate Granular*, Study No. 75-550-21. Unpublished report to U.S. Borax, International Bio-Research, Miamiville, OH, 1975.

99. C. Daniels, M. J. Thompson, and R. H. Teske, *Acute Toxicity and Irritation Studies on Zinc Borate 2335*, Rep. T-298, Unpublished report to U.S. Borax, Hill Top Research, Miamiville, OH, 1969.

100. D. R. Cerven, *Zinc Borate 4.1.1. Single Dose Oral Toxicity in Rats/LD$_{50}$*, Study No. MB 92-1203, Unpublished report to U.S. Borax, AMB Research Laboratories, Spinnerstown, PA, 1992.

101. S. M. Denton, *Sodium Metaborate 4-Mol. Acute Oral Toxicity Study in the Rat*, Study No. 1341/5-1032 and 1341/8-1032, Unpublished report to Borax Europe Limited, Corning Hazleton (Europe), Harrogate, North Yorkshire, U.K., 1996.

102. R. L. Doyle, *Acute Oral Administration of 8-Mol Sodium Mtaborate to Rats*, Study No. N-118, Unpublished report to U.S. Borax, Hill Top Biolabs, Miamiville, OH, 1963.

103. J. J. Dufour, C. Rogg, and A. Cerioli, *Rapport Final d'une Etude des Effets Toxiques et Vomitifs du Perborate de Sodium*, Institut Batelle, Centre de Recherche, Carouge/Geneva, 1971.

104. B. Chater, *Sodium Perborate. Acute Oral Toxicity with Histology and Skin and Eye Irritation*, Unpublished Report No. CTL/T/1150, ICI Central Toxicology Laboratory, 1978.

105. J. Momma et al., Acute oral toxicity and ocular irritation of chemicals in bleaching agents. *J. Food Hyg. Soc. Jpn.* **27**, 553–560 (1986).

106. O. M. Moreno et al., *Sodium Perborate Monohydrate - LD$_{50}$ in Rats*, Rep. MB 87-8676 A to Solvay/Interox, MB Research Laboratories, Spinnerstown, PA, 1987.

107. S. M. Glaza, *Acute Oral Toxicity Study of Sodium Perborate Monohydrate (Grade A) in Rats*, Rep. HLA 71002522 to Solvay/Interox, Hazleton Laboratories America, Madison, WI, 1988.

108. A. S. Weiner, D. L. Conine, and R. L. Doyle, *Acute Dermal Toxicity Screen in Rabbits: Primary Skin Irritation Study in Rabbits of Boric Acid*, Rep. No. TX-82-10, Unpublished report to U.S. Borax, Hill Top Research, Cincinnati, OH, 1982.

109. E. L. Reagan and P. L. Becci, *Acute Dermal Toxicity Study of 20 Mule Team Lot No. USB 12-84 Sodium Tetraborate Pentahydrate in New Zealand White Rabbits*, Unpublished report to U.S. Borax, Food and Drug Research Laboratories, Waverly, NY, 1985.

110. R. L. Doyle, *Acute Dermal Toxicity in Rabbits of Disodium Octaborate Tetrahydrate*, Unpublished report to U.S. Borax, Hill Top Biolabs, Miamiville, OH, 1989.

111. J. J. Kreuzmann, *Acute Dermal Toxicity in Rabbits - Limit Test XPI-187 Zinc Borate*, Study No. 90-4186-21, Unpublished report to U.S. Borax, Hill Top Biolabs, Miamiville, OH, 1990.

112. D. L. Cerven, *Acute Dermal Toxicity in Rabbits/LD$_{50}$ in Rabbits. Zinc Borate 4.1.1*, Proj. No. MB 92-1301 B, Unpublished report to U.S. Borax, MB Research Laboratories, Spinnerstown, PA, 1992.

113. E. L. Reagan and P. L. Becci, *Acute Dermal Toxicity Study of 20 Mule Team Lot No. USB 11-84 Sodium Tetraborate Decahydrate in New Zealand White Rabbits*, Unpublished report to U.S. Borax, Food and Drug Research Laboratories, Waverly, NY, 1985.
114. G. Wnorowski, *Acute Inhalation Toxicity Limit Test on Boric Acid*, Study No. 3311, Unpublished report to U.S. Borax, Product Safety Labs, East Brunswick, NJ, 1994.
115. G. Wnorowski, *Acute Inhalation Toxicity Limit Test on Disodium Tetraborate Decahydrate*, Study No. 3309, Unpublished report to U.S. Borax, Product Safety Labs, East Brunswick, NJ, 1994.
116. G. Wnorowski, *Acute Inhalation Toxicity Limit Test on Disodium Octaborate Tetrahydrate*, Unpublished report to U.S. Borax, Study No. 3313, Product Safety Labs, East Brunswick, NJ, 1994.
117. G. Wnorowski, *Acute Inhalation Toxicity Limit Test on Disodium Tetraborate Pentahydrate*, Study No. 3307, Unpublished report to U.S. Borax, Product Safety Labs, East Brunswick, NJ, 1994.
118. S. Blagden, *Firebrake 415: Acute Inhalation Toxicity (Nose Only) Study in the Rat*, Study No. 801/011, Unpublished report to Borax Europe Ltd., SafePharm Laboratories, UK, 1996.
119. L. Roudabush et al., Comparative acute effects of some chemicals on the skin of rabbits and guinea pigs. *Toxicol. Appl. Pharmacol.* **7**, 559–565 (1965).
120. A. A. Silaev, Experimental determination of the maximum permissible concentration of sodium perborate in workplace air. *Gig. Tr. Prof. Zabol.* **6**, 44 (1984).
121. E. L. Reagan and P. L. Becci, *Primary Dermal Irritancy Study of 20 Mule Team Lot No. USB 11-84 Sodium Tetraborate Decahydrate in New Zealand White Rabbits*, Unpublished report to U.S. Borax, Food and Drug Research Laboratories, Waverly, NY, 1985.
122. E. L. Reagan and P. L. Becci, *Primary Dermal Irritancy Study of 20 Mule Team Lot No. USB 12-84 Sodium Tetraborate Pentahydrate in New Zealand White Rabbits*, Unpublished report to U.S. Borax, Food and Drug Research Laboratories, Waverly, NY, 1985.
123. R. L. Doyle, *Primary Skin Irritation in Rabbits of Disodium Octaborate Tetrahydrate*, Unpublished report to U.S. Borax, Hill Top Biolabs, Miamiville, OH, 1989.
124. J. A. Young and R. L. Doyle, *Corrosivity Study on a Series of Ten Materials*, 73-630-21, Unpublished report to U.S. Borax, Hill Top Biolabs, Miamiville, OH, 1973.
125. D. L. Cerven, *Primary Dermal Irritation in Albino Rabbits Zinc Borate 4.1.1*, Proj. No. MB 92-1301 C, Unpublished report to U.S. Borax, MB Research Laboratories, Spinnerstown, PA, 1992.
126. J. J. Kreuzmann, *Primary Skin Irritation Study in Rabbits of: XPI-187 Zinc Borate*, Study No. 90-4186-21, Unpublished report to U.S. Borax, Hill Top Biolabs, Miamiville, OH, 1990.
127. K. H. Beyer et al., FDA cosmetic ingredient review expert panel, final report on the safety assessment of sodium borate and boric acid. *J. Am. Coll. Toxicol.* **2**, 87–125 (1983).
128. EEC Council, *Directive 93/35/EEC Amending for the Sixth Time Directive 76/768/EEC on the Approximation of the Laws of the Member States Relating to Cosmetic Products*, OJ L151, 23.06.93, p. 32, EEC Council, 1993.
129. E. L. Reagan and P. L. Becci, *Primary Eye Irritation Study of 20 Mule Team Lot No. USB 11-84 Sodium Tetraborate Decahydrate in New Zealand White Rabbits*, Unpublished report to U.S. Borax, Food and Drug Research Laboratories, Waverly, NY, 1985.

130. E. L. Reagan and P. L. Becci, *Primary Eye Irritation Study of 20 Mule Team Lot No. USB 12-84 Sodium Tetraborate Pentahydrate in New Zealand White Rabbits*, Unpublished report to U.S. Borax, Food and Drug Research Laboratories, Waverly, NY, 1985.

131. R. L. Doyle, *Primary Eye Irritation in Rabbits of 10 Borax Mol*, Unpublished report to U.S. Borax, Hill Top Biolabs, Miamiville, OH, 1989.

132. R. L. Doyle, *Primary Eye Irritation in Rabbits of Boric Acid*, Unpublished report to U.S. Borax, Hill Top Biolabs, Miamiville, OH, 1989.

133. R. L. Doyle, *Primary Eye Irritation in Rabbits of Disodium Octaborate Tetrahydrate*, Unpublished report to U.S. Borax, Hill Top Biolabs, Miamiville, OH, 1989.

134. R. L. Doyle, *Eye Irritation Study without Rinsing in Rabbits of Disodium Octaborate Tetrahydrate*, Unpublished report to U.S. Borax, Hill Top Biolabs, Miamiville, OH, 1989.

135. D. L. Cerven, *Primary Eye Irritation and/or Corrosion in Rabbits. Zinc Borate 4.1.1*, Proj. No. MB 92-1301 D, Unpublished report to U.S. Borax, MB Research Laboratories, Spinnerstown, PA, 1992.

136. J. J. Kreuzmann, *Primary Eye Irritation Study in Rabbits of: XPI-187 Zinc Borate*, Study No. 90-4186-21, Unpublished report to U.S. Borax, Hill Top Biolabs, Miamiville, OH, 1990.

137. National Toxicology Program (NTP), *Toxicology and Carcinogenesis Studies of Boric Acid (CAS no. 10043-35-3) in B6C3F$_1$ Mice* (Food Studies), NTP Tech. Rep. Ser. No. 324, U.S. Department of Health and Human Service, National Institutes of Health, Research Triangle Park, NC, 1987.

138. H. E. Stockinger and C. J. Spiegel, Special materials. Part A. Inhalation-toxicity studies of boron halides and certain fluorinated hydrocarbons. In C. Voegtlin and H. C. Hodge, eds., *Pharmacology and Toxicology of Uranium Compounds: Chronic Inhalation and Other Studies*, McGraw-Hill, New York, 1953, pp. 2291–2321.

139. N. Y. Tarasenko, A. A. Kasparov, and O. M. Strongina, Effect of boric acid on the generative function in males. *Gig. Tr. Prof. Zabol.* **11**, 13–16 (1972).

140. J. L. Wilding et al., The Toxicity of Boron Oxid. *Am. Ind. Hyg. Assoc. J.* **20**, 284–289 (1959).

141. F. J. Murray, A comparative review of the pharmacokinetics of boric acid in rodents and humans. *Biol. Trace Elem. Res.* **66**(1–3), 331–342 (1998).

142. *Environmental Health Criteria 204: Boron*, International Programme on Chemical Safety (IPCS), World Health Organization, Geneva, 1998.

143. World Health Organization (WHO), *Guidelines for Drinking-water Quality*, Addendum to Vol. 2, WHO, Geneva, 1998, pp. 15–30.

144. European Centre for Exotoxicology and Toxicology of Chemicals, *Toxicology and Reproductive Toxicity of Some Inorganic Borates and Risk Assessment for Man*, Tech. Rep. No. 63, ECETOC, Brussels, Belgium, 1995.

145. W. W. Ku et al., Tissue disposition of boron in male fischer rats. *Toxicol. Appl. Pharmacol.* **111**, 145–151 (1991).

146. J. H. Draize and E. A. Kelley, The urinary excretion of boric acid preparations following oral administration and topical applications to intact and damaged skin of rabbits. *Toxicol. Appl. Pharmacol.* **1**, 267–276 (1959).

147. T. F. Brown et al., Effects of dietary boron on mineral balance in sheep. *Nutr. Res.* **9**, 503–512 (1989).

148. E. C. Owen, The excretion of borate by the dairy cow. *J. Dairy Res.* **13**, 243–248 (1944).

149. H. J. Weeth, C. F. Speth, and D. R. Hanks, Boron content of plasma and urine as indicators of boron intake in cattle. *Am. J. Vet. Res.* **42**, 474–477 (1981).
150. G. H. Nielsen, Percutaneous absorption of boric acid from boron-containing preparations in rats. *Acta Pharmacol. Toxicol.* **28**, 413–424 (1970).
151. R. M. Forbes and H. H. Mitchell, Accumulation of dietary boron and strontium in young and adult albino rats. *Arch. Ind. Health* **16**, 489–492 (1957).
152. K. A. Treinen and R. E. Chapin, Development of testicular lesions in F344 rats after treatment with boric acid. *Toxicol. Appl. Pharmacol.* **107**, 325–335 (1991).
153. W. W. Ku et al., Testicular toxicity of boric acid (BA): Relationship of dose to lesion development and recovery in the F344 rat. *Reprod. Toxicol.* **7**, 305–319 (1993).
154. H. R. Massie et al., Calcium, iron, copper, boron, collagen and density changes in bone with aging in C57BL/65 mice. *Exp. Gerontol.* **25**, 469–481 (1990).
155. M. Laurent-Pettersson, B. Delpech, and M. Hellier, The mapping of natural boron in histological sections of mouse tissues by the use of neutron-capture radiography. *Histochem. J.* **24**, 939–950 (1992).
156. J. Emsley, *The Elements*, Clarendon Press, Oxford, UK, 1989 p. 32.
157. L. E. Farr and T. Konikowski, The renal clearance of sodium pentaborate in mice and men. *Clin. Chem.* **9**, Winston-Salem, NC, 717–726 (1963).
158. J. A. Moore, and an Expert Scientific Committee, Assessment of boric acid and borax using the IEHR evaluative process for assessing human development and reproductive toxicity of agents. *Reprod. Toxicol.* **11**, 123–160 (1997).
159. F. J. Murray, A human health risk assessment of boron (boric acid and borax) in drinking water. *Regul. Toxicol. Pharmacol.* **22**, 221–230 (1995).
160. I. P. Lee, R. J. Sherins, and R. L. Dixon, Evidence for induction of germinal aplasia in male rats by environmental exposure to boron. *Toxicol. Appl. Pharmacol.* **45**, 577–590 (1978).
161. R. E. Linder, L. F. Strader, and G. L. Rehnberg, Effect of acute exposure to boric acid on the male reproductive system of the rat. *J. Toxicol. Environ. Health* **31**, 133–146 (1990).
162. P. A. Fail et al., Reproductive toxicity of boric acid in Swiss (CD-1) mice: assessment using the continuous breeding protocol. *Fundam. Appl. Toxicol.* **17**(2), 225–239 (1991).
163. J. J. Heindel et al., Developmental toxicity of boric acid in mice and rats. *Fundam. Appl. Toxicol.* **18**, 266–277 (1992).
164. C. J. Price et al., Developmental toxicity NOAEL and postnatal recovery in rats fed boric acid during gestation. *Fundam. Appl. Toxicol.* **32**, 179–193 (1996).
165. C. J. Price et al., The developmental toxicity of boric acid in rabbits. *Fundam. Appl. Toxicol.* **34**, 176–187 (1996).
166. U.S. Environmental Protection Agency (USEPA), *Reregistration Eligibility Decision (RED): Boric Acid and its Sodium Salts*, Rep. EPA 738-R-93-017, Office of Prevention, Pesticides and Toxic Substances, Washington, DC, 1993.
167. S. Haworth et al., Salmonella mutagenicity test results for 250 chemicals. *Environ. Mutagen.* Suppl. **1**, 1–142 (1983).
168. W. H. Benson, W. J. Birge, and H. W. Dorough, Absence of mutagenic activity of sodium borate (borax) and boric acid in the salmonella preincubation test. *Environ. Toxicol. Chem.* **3**, 209–214 (1984).
169. K. R. Stewart, Salmonella/*Microsome Plate Incorporation Assay of Boric Acid*, Study No. 2389-A200-91, SRI International, Menlo Park, CA, 1991.

170. C. J. Rudd, *Mouse Lymphoma Cell Mutagenesis Assay of Boric Acid*, Study No. 2389-G300-91, SRI International, Menlo Park, CA, 1991.

171. J. P. Bakke, *Evaluation of the Potential of Boric Acid to Induce Unscheduled DNA Synthesis in the in vitro Hepatocyte DNA Repair Assay Using the Male F-344 Rat*, Study No. 2389-V500-91, SRI International, Menlo Park, CA, 1991.

172. J. P. Bakke, *Evaluation of the Potential of Boric Acid to Induce Unscheduled DNA Synthesis in the in vitro Hepatocyte DNA Repair Assay Using the Male F-344 Rat*, Study No. 2389-V500-91, Amendment 1 to the Original Report, SRI International, Menlo Park, CA, 1992.

173. K. G. O'Loughlin, *Bone Marrow Erythrocyte Micronucleus Assay of Boric Acid in Swiss-Webster Mice*, Study No. 2389-C400-91, SRI International, Menlo Park, CA, 1991.

174. C. C. Pfeiffer and E. H. Jenney, The pharmacology of boric acid and boron compounds. *Bull. Nat. Formul. Comm.* **18**, 57–809 (1950).

175. R. B. Goldbloom and A. Goldbloom, Boric acid poisoning. *J. Pediatr.* **43**, 361 (1953).

176. W. Kliegel, *Bor in Biologie, Medizin, und Pharmazie*, Springer-Verlag, Berlin and New York, 1980.

177. J. A. Jansen, J. Andersen, and J. S. Schou, Boric acid single dose pharmacokinetics after intravenous administration to man. *Arch. Toxicol.* **55**, 64–67 (1984).

178. J. A. Jansen, J. S. Schou, and B. Aggerbeck, Gastro-intestinal absorption and *in vitro* release of boric acid from water-emulsifying ointments. *Food Chem. Toxicol.* **22**, 49–53 (1984).

179. R. C. Wester et al., *In vivo* percutaneous absorption of boric acid, borax, and disodium octaborate tetrahydrate in humans compared to *in vitro* absorption in human skin from infinite and finite doses. *Toxicol. Sci.* **45**, 42–51 (1998).

180. H. A. Payton and D. Green, Boric acid poisoning: Case report. *South. Med. J.* **34**, 1286–1288 (1941).

181. D. H. Wegman et al., Acute and chronic respiratory effects of sodium borate particulate exposures. *Environ. Health Perspect.* **102**((Suppl. 7), 119–128 (1994).

182. D. H. Garabrant et al., Respiratory effects of borax dust. *Br. J. Ind. Med.* **42**, 831–837 (1985).

183. C. Job, Resorption und Ausscheidung von Peroral Zugefuhrtem Bor. *Z. Angew. Baeder-Klimaheilkd.* **20**, 137–142 (1973).

184. M. D. Whorton et al., Reproductive effects of sodium borates on male employees; birth rate assessment. *Occup. Environ. Med.* **51**, 761–767 (1984).

185. E. Tüccar, A. H. Elhan, Y. Yavuz, and B. S. Sayli, Comparison of infertility rates in communities from boron-rich and boron-poor territories. *Biol. Trace Elem. Res.* **66**, 401–407 (1998).

186. *Physicians' Desk Reference*, Medical Economics Company, Montvale, NJ, 1997.

187. D. L. Anderson et al., Sources of atmospheric distribution of particulate and gas phase boron. *Atmos Environ.* **28**, 1401–1410 (1994).

188. J. A. Cox et al., Leaching of boron from coal ash. *Environ. Sci. Technol.* **12**, 722–723 (1978).

189. C. C. Gupta et al., Boron toxicity and deficiency: A review. *Can. J. Soil Sci.* **65**, 381–409 (1985).

190. G. J. Smith and V. P. Anders, Toxic effects of boron on mallard reproduction. *Environ. Toxicol. Chem.* **8**, 943–954 (1989).

191. T. R. Stanley et al., Effects of boron and selenium on mallard reproduction and duckling growth and survival. *Environ. Toxicol. Chem.* **15**, 1124–1132 (1996).

192. C. B. Keil and A. K. Miller, A pilot study of the health effects of elemental boron on production workers. *AIHCE Conf.*, Boston, MA, 1992, Abstr. Pap. No. 249.
193. H. E. Stokinger, In G. Clayton, and F. Clayton, eds., *Patty's Industrial Hygiene and Toxicology*, 3rd Rev. Ed., Vol. 2 B, Wiley, New York, 1981, pp. 2978–3005.
194. A. Vengosh, The isotopic composition of anthropogenic boron and its potential impact of the environment. *Biol. Trace Elem. Res.* **66**, 145–151 (1998).
195. J. Southwood, *Sodium Perborate Tetrahydrate. Skin Irritation and Eye Irritation Studies*, Rep. No. CTL/T/2427, ICI CTL, 1986.
196. J. Southwood, *Sodium Perborate Monohydrate. Skin Irritation and Eye Irritation Studies*, Rep. No. CTL/T/2423, ICI CTL, 1986.
197. A. G. Degussa, *Sodium Perborate Tetrahydrate. 4-Week Oral Toxicity Study after Repeated Administration in Rats*. Study No. 867666, Degussa AG, Hanau, Germany, 1989.
198. M. G. Mullinos, G. K. Higgins, and G. J. Christakis, On the toxicity of sodium perborate. *J. Soc. Cosmet. Chem.* **3**, 297–302 (1952).
199. J. E. Moore and E. Hautala, Odor as an aid to chemical safety: Odor thresholds compared with threshold limit values and Volatilities for 214 industrial chemicals in air and water dilution. *J. Appl. Toxicol.* **3**, 272–290 (1983).
200. American Conference of Governmental and Industrial Hygienists (ACGIH), *Documentation of the Threshold Limit Values and Biological Exposure Indices*, 6th ed., Vol. I, ACGIH, Cincinnati, OH, 1991.
201. M. Windholz, ed., *The Merck Index*, 10th ed., Merck & Co., Rahway, NJ, 1983, p. 187.
202. *Boron Trifluoride*, OSHA Instruction CPL 2-2,43A, U.S. Department of labor, Washington, DC, 1991, pp. II-51–11-52.
203. T. R. Torkelson, S. E. Sadek, and V. K. Rowe, The toxicity of boron trifluoride when inhaled by laboratory animals. *Am. Ind. Hyg. Assoc. J.* **22**, 263–270 (1961).
204. G. M. Rusch et al., Inhalation toxicity studies with boron trifluoride. *Toxicol. Appl. Pharmacol.* **83**, 69–78 (1986).
205. B. A. Krackow, Toxicity and health hazards of boron hydrides. *AMA Arch. Ind. Hyg. Occup. Med.* **8**, 335–339 (1953).
206. American Conference of Governmental and Industrial Hygienists (ACGIH), *Documentation of the Threshold Limit Values and Biological Exposure Indices*, 6th ed., Vol. I, ACGIH, Cincinnati, OH, 1991, pp. 393–394.
207. M. Ono-Ogasawara et al., Determination of diborane by adsorption sampling using modified gel and the chromotropic acid-hplc method. *Ind. Health* **30**, 35–45 (1992).
208. C. C. Comstock et al., Res. Rep. No. 258, U.S. Army Chemical Corps Medical Laboratories, 1954.
209. E. M. Cordasco et al., Pulmonary aspects of some toxic experimental space fuels. *Dis. Chest* **41**, 68–74 (1982).
210. G. J. Levinskas, M. R. Paslian, and W. R. Beckman, Chronic toxicity of pentaborane vapor. *Am. Ind. Hyg. Assoc. J.* **19**, 46–53 (1958).
211. A. A. Tomas, *State of the Art Report—Health Hazards of Borane Fuels and Their Control*, Aerosp. Med. Res. Lab. Wright-Patterson AFB, Dayton, OH, 1958.
212. B. E. Yarbrough et al., Severe central nervous system damage and profound acidosis in persons exposed to pentaborane. *J. Toxicol. Clin. Toxicol.* **23**, 519–536 (1985–1986).

213. J. L. Svirbely, Toxicity tests of decaborane for laboratory animals. II. Effect of repeated doses. *Arch. Ind. Health* **11**, 132–137 (1955).
214. D. M. Valerino et al., Studies on the interaction of several boron hydrides with liver microsomal enzymes. *Toxicol. Appl. Pharmacol.* **29**, 358–366 (1974).
215. A. S. Tadepalli and J. P. Buckley, Cardiac and peripheral vascular effects of decaborane. *Toxicol. Appl. Pharmacol.* **29**, 210–222 (1974).
216. H. H Reynolds et al., AMRL-Tdr-64-74, Wright-Patterson AFB, Dayton, OH, 1964.
217. American Conference of Governmental Industrial Hygienists (ACGIH), *Threshold Limit Values and Other Occupational Exposure Values*, ACGIH, Cincinnati, OH, 1999.

CHAPTER FORTY-SIX

Alkaline Materials: Sodium, Potassium, Cesium, Rubidium, Francium, and Lithium

Stevan W. Pierce, CIH, SCP, PE(CA)

A INTRODUCTION

The primary health hazards arising out of undue exposures to the alkaline chemicals are mainly those of irritation and corrosion of sensitive body tissues that come into direct contact with the agent. The most commonly encountered alkaline materials include the alkaline salts of ammonia, calcium, potassium, and sodium. The tissues most susceptible to biological damage are the eyes, which, if not treated rapidly after exposure, can be severely damaged. Most exposures come about as a result of accidental spills by way of splashes of solid and liquid materials through mishandling. The alkaline metals readily form hydroxides from moisture/humidity and are all strong bases. Employee training, process control, and the use of appropriate personal protective equipment are the most effective ways to reduce or minimize exposures.

In addition to appropriate training, process control, and proper use and care of personal protective equipment, all facilities using or handling alkaline materials should have emergency eye washes available in readily located areas for immediate and prolonged washing of the eyes with water in the case of exposure. It is extremely important that these emergency washing devices be periodically checked for proper working order; many instances of serious injuries resulting from accidental splashes have been reported that could have been avoided had the emergency wash devices been in working order. Direct contact of the alkaline materials with the skin or respiratory tract may result in irritation and corrosion, because these chemical agents react directly with tissue proteins to form aluminates. Long-term respiratory exposure, even for low airborne concentrations, can result in permanent damage to the respiratory system.

Cesium, rubidium, and lithium are indicated as clinical treatments for psychiatric disorders (1).

B ALKALINE MATERIALS AND THEIR COMPOUNDS

1.0 Potassium

1.0.1 CAS Number: [7440-09-7]

1.0.2 Synonyms: Kalium

1.0.3 Trade Names: Various, especially for commercial products

1.0.4 Molecular Weight: 39.09

1.0.5 Molecular Formula: K

1.1 Chemical and Physical Properties

Melting point: 63°C
Boiling point: 770°C
Vapor density: 1.4
Solubility: Decomposes in H_2O

1.1.1 General

Potassium is a soft, shiny solid metallic material. When the metal is freshly cut, this ductile metal is soft and waxy with a metallic sheen that is soon lost by reaction with atmospheric oxygen, carbon dioxide, or moisture. Metallic potassium is a dangerous fire hazard. (See also Table 46.1.)

Table 46.1. Physical and Chemical Properties of Potassium and Potassium Compounds

Compound	Molecular Formula	Molecular Weight	Boiling Point (°C)	Melting Point (°C)	Specific Gravity at 20°C	Solubility in Water at (at 68°F)	Vapor Pressure (mm Hg)
Potassium	K	39.09	770	63	0.86	Reacts	8 mm at 432°C
Potassium hydroxide	KOH	56.10	1320–1324	360	2.044	Very soluble	1 mm at 719°C

Potassium reacts readily with most gases and liquids. It does not react with the noble gases such as helium or argon or with hydrocarbons such as hexane or high alkanes. It reacts violently with water to release hydrogen, which may ignite from the heat of the reaction. The metal can form potassium peroxide and superoxide, which can explode violently when handled, even when stored under mineral oil.

1.2 Production and Use

Uses of potassium include fertilizers, laboratory reagent, heat-transfer alloys in nuclear reactors, fireworks, oxygen breathing apparatus as a source of oxygen, liquid detergents, and low sodium salt substitutes.

Potassium compounds are widely found in the earth's crust. Potassium is an essential element for both plant and human growth.

1.3 Exposure Assessment

Potassium exposure and toxicity is almost always that of the anion species. Given the violent nature of this element, exposure to metallic potassium is seldom encountered in industry and research or academia.

1.3.1 *Air*

Exposure to metalic potassium is rare. Since elemental potassium reacts sometimes violently with moisture in the air, airborne exposures are seldom encountered.

1.3.2 *Background Levels:* NA

1.3.3 *Workplace Methods:* NA

1.3.4 *Community Methods:* NA

1.3.5 *Biomonitoring/Biomarkers*

1.3.5.1 Blood, Blood/SST, SRT, or LGT, PST. Normal serum potassium values are 3.3–5.1 mmol/L (1a). Method is ISE (ion-selective electrode) procedure.

1.3.5.2 Urine. Urine levels are 24-h catch. Normal urine potassium values are 25–125 mmol/L (1a). Method is ISE procedure.

1.4 Toxic Effects

Potassium's high reactivity makes it strongly caustic and corrosive in contact with tissues. The products of potassium combustion include oxides. It is an essential element, commonly found in most foods as a salt, which regulates osmotic pressure within cells, maintains the acid–base balance, and is necessary for many enzymatic reactions, especially those involving energy transfer.

1.5 Standards, Regulations, or Guidelines of Exposure

Clinical blood and urine studies (1a) indicate the serum and urine levels of potassium per paragraphs 1.3.5.1 and 1.3.5.2.

There are no threshold limit values (ACGIH TLVs), permissible exposure limits (OSHA PEL), and/or recommended exposure limit (NIOSH REL) established for elemental potassium.

1.6 Studies on Environmental Impact

Potassium is the seventh most abundant element found in nature and comprises an estimated 1.5%, by weight, of the earth's crust. No significant environmental impact references were found.

2.0 Potassium Hydroxide

2.0.1 CAS Number: [1310-58-3]

2.0.2 Synonyms: Lye, potash, caustic, potassa, potassium hydrate, caustic potash, potassium hydroxide [K(OH)]

2.0.3 Trade Names: Numerous commercial products

2.0.4 Molecular Weight: 56.10

2.0.5 Molecular Formula: HKO

2.0.6 Molecular Structure: K^+ OH^-

2.1 Chemical and Physical Properties

2.1.1 General

A white deliquescent solid, potassium hydroxide has a molecular weight of 56.10 and a specific gravity of 2.044. It may be in the form of pellets, sticks, lumps, or flakes. The melting point is about 360°C. Potassium hydroxide is soluble in water, alcohol, and glycerin and slightly soluble in ether, and its boiling point is 1320–1324°C. Potassium is a strong alkali and an aqueous solution may have a pH $\geq$ 13.

Potassium hydroxide absorbs water and carbon dioxide directly from the air; hence it should be stored away from the open air. It produces a strong exothermic reaction when dissolved in water. Potassium hydroxide is incompatible with a number of materials, including organic halogen compounds, metals, and acids (2).

2.2 Production and Use

Potassium hydroxide, KOH (caustic potash), is produced by electrolysis of potassium chloride solution. Its principal use is in the manufacture of soft and liquid soaps. It is also used to make high-purity potassium carbonate, K_2CO_3, for use in the manufacture of glass.

2.3 Exposure Assessment

2.3.1 Air

For air sampling in the workplace for alkaline dust, NIOSH method 7401 is recommended (2a).

2.3.2 Background Levels: NA

2.3.3 Workplace Methods: NA

2.3.4 Community Methods: NA

2.3.5 Biomonitoring/Biomarkers

2.3.5.1 Blood. See Section 1.0 (on potassium).

2.3.5.2 Urine. See Section 1.0.

2.4 Toxic Effects

Potassium hydroxide, when inhaled in any form, is strongly irritating to the upper respiratory tract. Severe injury is usually avoided by the self-limiting sneezing, coughing, and discomfort. Contact with eyes or other tissues can produce serious injury, as described at the beginning of this chapter. Rubber gloves should be worn when handling potassium hydroxide to prevent irritation, burns, or contact dermatitis. When exposed to air, potassium hydroxide forms the bicarbonate and carbonate. Very little is known of their biological effects. Because they are less alkaline in aqueous solutions they may be expected to be less irritating or corrosive to skin and eyes. Bailey and Morgareidge (3), in a study for the U.S. Food and Drug Administration, found potassium carbonate to be nonteratogenic in mice when they were given daily oral intubations of ≤ 290 mg/kg on days 6–15 of gestation. Accidental ingestion of a solution of potassium hydroxide may be expected to produce rapid corrosion and perforation of the esophagus and stomach (4). Frequent applications of aqueous solutions (3–6%) of potassium hydroxide to the skin of mice for 46 weeks produced tumors identical to those from coal tar (5). Acute exposures involving the inhalation of dust or mist may cause symptoms in the respiratory tract, including severe coughing and pain. Additionally, lesions may develop along with burning of the mucous membranes. Pulmonary edema can develop within a latency period of 5–72 h.

Chronic exposures may cause inflammatory and ulcerative changes in the mouth and possibly bronchial and gastrointestinal (GI) disorders.

2.5 Standards, Regulations, or Guidelines of Exposure

See Sections 1.3.5.1 and 1.3.5.2 for potassium blood and urine normal levels.

2.5.1 NIOSH REL

2.0 mg/m^3.

2.5.2 OSHA PEL

None established (PEL of 2.0 mg/m^3 was vacated).

2.5.3 ACGIH TLV

2.0 mg/m^3 expressed as a ceiling value.

2.6 Studies on Environmental Impact

Potassium hydroxide is listed in the U.S. Environmental Protection Agency's Toxic Substances Control Act (TSCA) chemical inventory.

3.0 Sodium

3.0.1 CAS Number: *[7440-23-5]*

3.0.2 Synonyms: Natrium

3.0.3 Trade Names: Various commercial products

3.0.4 Molecular Weight: 22.98977

3.0.5 Molecular Formula: Na

3.1 Chemical and Physical Properties

Specific gravity: 0.9
Melting point: 97.8°C
Boiling point: 892°C
Flash point: 4°C
Solubility: Decomposed in H$_2$O

3.1.1 General

Metallic sodium is a light, ductile material having a high electrical conductivity (See also Table 46.2).

Sodium reacts violently with carbon dioxide, water, and most oxygenated and halogenated organic compounds. It may ignite spontaneously in air at temperatures above 115°C, producing a sodium peroxide fume (Section 8), which is strongly alkaline and is thus a serious hazard for inhalation or skin and eye contact. Procedures for handling sodium and similar materials are described in the *Atomic Energy Commission Liquid Metals Handbook* (6). Firefighting and waste-disposal methods are included. The behavior of the aerosol from a sodium fire has been described by Clough and Garland (7).

Table 46.2. Physical and Chemical Properties of Sodium and Sodium Compounds

Compound	Molecular Formula	Molecular Weight	Boiling Point (°C)	Melting Point (°C)	Specific Gravity	Solubility in Water (at 68 °F)	Vapor Pressure (mm Hg)
Sodium	Na	22.98977	892	97.80	0.9 20°C	Reacts	1 mm Hg at 432°C
Sodium carbonate	CNa_2O_3	105.99	1600	851	2.532 20°C	Soluble	
Sodium hydroxide	NaOH	40.01	1390	318.4	2.13 25°C	Very soluble	1 mm Hg 739°C
Sodium peroxide	Na_2O_2	77.9783	657	460	2.805 20°C	Reacts	
Trisodium phosphate	Na_3O_4P	163.941	Decomp.	73	1.62 20°C	Soluble	
Sodium metasilicate	Na_2O_3Si	122.06		1089	2.614	Soluble	

3.2 Production and Use

Sodium (Na) is manufactured by electrolysis of a molten mixture of sodium and calcium chlorides. It is a soft, waxy material having a silvery sheen on freshly cut surfaces. These surfaces quickly change to a coating of the white peroxide, Na_2O_2, by reaction with oxygen in the air. One-pound bricks and smaller amounts may be encountered in chemistry laboratories, where it is usually protected from the air by immersion in an aliphatic oil. Larger amounts are shipped in drums or tank cars. The manufacture of organometallic compounds, such as tetraethyllead, consumes the bulk of the production. Significant amounts are also used as a heat-exchange medium, frequently as the alloy with potassium known as NaK. Metal descaling baths also use a sodium–sodium hydride mixture.

3.3 Exposure Assessment

3.3.1 Air

Airborne exposure to elemental sodium is rare. Since elemental sodium reacts violently with moisture in the air, exposure assessment, if undertaken, should be for sodium hydroxide.

3.3.2 Background Levels: NA

3.3.3 Workplace Methods: NA

3.3.4 Community Methods: NA

3.3.5 Biomonitoring/Biomarkers

3.3.5.1 Blood. Blood/SST or PST. Normal sodium values are 133–135 mmol/L. Method is ISE (1).

3.3.5.2 Urine. Urine levels are 24-h catch. Normal values are 40–220 mmol/L. Method is ISE (1).

3.4 Toxic Effects

Vapors and fumes arising from sodium are strongly alkaline and are highly irritating and corrosive to the respiratory tract, eyes, and skin. Physiologically, sodium is an essential element encountered as a salt in most foodstuffs. Its ion is the principal electrolyte in extracellular fluid, which is excreted in the urine. Prolonged dietary excess may lead to renal hypertension. Sodium metal may react with moisture to form sodium hydroxide. Acute exposures may cause irritation and burning of the respiratory tract with severe coughing and pain. Pulmonary edema may develop within 72 h following exposure. Severe cases may be fatal. Chronic exposures may cause inflammation, ulcerate changes in the mouth, and bronchial and gastrointestinal disturbances.

ALKALINE MATERIALS

3.5 Standards, Regulations, or Guidelines of Exposure

Clinical blood and urine studies (1) indicate the serum and urine levels of sodium per Sections 3.3.5.1 and 3.3.5.2.

There are no TLVs (ACGIH), PELs (OSHA), and/or RELs (NIOSH) established for elemental sodium.

4.0 Sodium Carbonate

4.0.1 CAS Number: [497-19-8]

4.0.2 Synonyms: Sal soda, washing soda, soda monohydride, crystal carbonate, soda ash, disodium carbonate, soda, calcined soda, ASH, carbonic acid disodium salt

4.0.3 Trade Names: Various commercial products

4.0.4 Molecular Weight: 105.989

4.0.5 Molecular Formula: CNa_2O_3

4.0.6 Molecular Structure: $^-O-C(=O)-O^-$ Na^+ Na^+

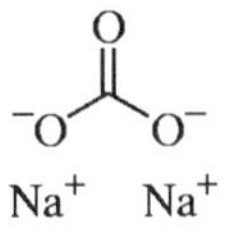

4.1 Chemical and Physical Properties

Specific gravity: 2.532
Melting point: 851°C
Boiling point: 1600°C

4.1.1 General

This compound, Na_2CO_3 (soda ash), is usually encountered as the decahydrate, $Na_2CO_3 \cdot 10H_2O$, commonly called *sal soda* or *washing soda.*

This hygroscopic, white powder is strongly caustic. Alkalinity pH is 11.5 for a 1% aqueous solution.

4.2 Production and Use

Sodium carbonate occurs naturally in large deposits in Africa and the United States as either the carbonate or trona, a mixed ore of equal molar amounts of the carbonate and bicarbonate. Soda ash is manufactured primarily by the Solvay process, whereby ammonia is added to a solution of sodium chloride and carbon dioxide and is then bubbled through to precipitate the bicarbonate, $NaHCO_3$. Calcination of the bicarbonate produces sodium carbonate. It may also be produced by injecting carbon dioxide into the cathode compartment, containing sodium hydroxide, of the diaphragm electrolysis of sodium chloride.

The glass industry consumes about one-third of the total production of sodium carbonate. About one-fourth is used to make sodium hydroxide by the double-decomposition reaction with slaked lime, $Ca(OH)_2$. Large amounts are also used in soaps and strong cleansing agents, water softeners, pulp-and-paper manufacture, textile treatments, and various chemical processes.

4.3 Exposure Assessment

4.3.1 Air: NA

4.3.2 Background Levels: NA

4.3.3 Workplace Methods: NA

4.3.4 Community Methods: NA

4.3.5 Biomonitoring/Biomarkers

4.3.5.1 Blood. See Section 3.0 (on sodium).

4.3.5.2 Urine. See Section 3.0.

4.4 Toxic Effects

Male rats were given an aerosol exposure of 2 h/day, 5 days/week, for $3\frac{1}{2}$ months. The particle size of the aerosol was less than 5 mm in diameter. A concentration of 10–20 mg/m^3 did not cause any pronounced effect. In observations from exposure at 70 ± 2.9 mg/m^3, the weight gain of the exposed group was 24% less than that of controls. There were no differences in hematological parameters. Histological examination of the lungs showed thickening of the intraalveolar walls, hyperemia, lymphoid infiltration, and desquamation (8).

An aqueous solution, 50% (w/v) of sodium carbonate, was applied to the intact and abraded skins of rabbits, guinea pigs, and human volunteers. The sites were examined at 4, 24, and 48 h and scored for erythema, edema, and corrosion. The solution produced no erythema and edema. The rabbit and human skins showed tissue destruction at the abraded sites (9).

Pregnant mice were dosed daily by oral intubation with aqueous solutions of sodium carbonate at levels of 3.4–340 mg/kg on days 6–15 of gestation. There were no effects on nidation or survival of the dams of fetuses. The number of abnormalities in soft and skeletal tissues in the experimental group did not differ from those of sham-treated controls. Positive controls gave the expected results (10). Similar studies at doses of ≤ 245 mg/kg in rats and ≤ 179 mg/kg in rabbits produced similar negative results.

Sodium bicarbonate, $NaHCO_3$, was evaluated for teratological effects by the same procedures as for sodium carbonate. Maximum dose levels were as follows: mice, 580 mg/kg; rats, 340 mg/kg; and rabbits, 330 mg/kg. No effects were found in any of these species (11).

Twenty-seven U.S. Army inductees assigned to dishwashing immersed their bare hands for 4–8 h in hot water containing a strong detergent blend of sodium carbonate, sodium metasilicate, and sodium tripolyphosphate. All developed irritation of the exposed

ALKALINE MATERIALS 593

surfaces. Six developed vesicles and giant bullae within 10–12 h after exposure. Three also had subungual purpura. Secondary infections were noted in several individuals (12). Acute exposures of dusts or vapors of sodium carbonate may cause irritation of mucous membranes with subsequent coughing and shortness of breath.

Chronic exposures may lead to perforation of the nasal septum; skin exposure may cause irritation and redness with concentrated solutions causing erythema. Chronic skin exposures may cause dermatitis and ulceration.

4.5 Standards, Regulations, or Guidelines of Exposure

Clinical blood and urine studies (1) indicate the serum and urine levels of sodium per Sections 3.3.5.1 and 3.3.5.2.

No TLVs (ACGIH), PELs (OSHA), and/or RELs (NIOSH) have been established for sodium carbonate. The level of 5 mg/m^3 has been tentatively recommended in the (former) Soviet Union for sodium carbonate (9).

5.0 Sodium Hydroxide

5.0.1 CAS Number: [1310-73-2]

5.0.2 Synonyms: Caustic soda; caustic flake; lye; liquid caustic; sodium hydrate; white caustic; soda lye; lye; white caustic; lye, caustic; augus hot rod

5.0.3 Trade Names: Numerous commercial products

5.0.4 Molecular Weight: 40.01

5.0.5 Molecular Formula: NaOH

5.0.6 Molecular Structure: Na$^+$ OH$^-$

5.1 Chemical and Physical Properties

Specific gravity: 2.13 at 25°C
Melting point: 318.4°C
Boiling point: 1390°C

5.1.1 General

Solubility 42–347 g in 100 mL H$_2$O at 0°C. Soluble in aliphatic alcohols. The refractive index is 1.3576. Alkalinity of a 1% aqueous solution has a pH of $\sim$13.

5.2 Production and Use

The primary source of sodium hydroxide, NaOH (caustic soda, caustic flake, lye, liquid caustic), is the electrolysis of sodium chloride solutions, which also yields chlorine. In this process the anode may be surrounded by an asbestos diaphragm to isolate the chlorine. The caustic soda so produced may contain a significant amount of asbestos fibers. As noted above, sodium hydroxide may also be produced from sodium carbonate.

The millions of tons of sodium hydroxide produced annually in the United States are used in the manufacture of chemicals, rayon, soap, and other cleansers; pulp and paper; petroleum products; textiles; and explosives. Caustic soda is also used in metal descaling and processing and in batteries.

As indicated in the synonyms, sodium hydroxide may be encountered as solids in various forms (pellets, flakes, sticks, cakes) and as solutions, usually 45–75% in water. Mists are frequently formed when dissolving sodium hydroxide in water, which is an exothermic process.

5.3 Exposure Assessment: NA

5.3.1 Air: NA

5.3.2 Background Levels: NA

5.3.3 Workplace Methods: NA

5.3.4 Community Methods: NA

5.3.5 Biomonitoring/Biomarkers

5.3.5.1 Blood. See Section 3.0.

5.3.5.2 Urine. See Section 3.0.

5.4 Toxic Effects

This strong alkali is irritating to all tissues and requires extensive washing to remove it. Eye splashes are especially serious hazards. Protective equipment is essential and treatment must be prompt. A 5% aqueous solution of sodium hydroxide produced severe necrosis when applied to the skin of rabbits for 4 h (13). Rats were exposed to an aerosol of 40% aqueous sodium hydroxide whose particles were less than 1 mm in diameter. Exposures were for 30 min, twice a week. The experiment was terminated after 3 weeks when two of the 10 rats died. Histopathological examination showed mostly normal lung tissue with foci of enlarged alveolar septa, emphysema, bronchial ulceration, and enlarged lymph adenoidal tissues (14). Nagao and co-workers (15) examined skin biopsies from volunteers having 1-N sodium hydroxide applied to their arms for 15–180 min. There were progressive changes beginning with dissolution of the cells in the horny layer and progressing through edema to total destruction of the epidermis in 60 min.

Sodium hydroxide concentrations of 250 mg/m^3 are considered immediately dangerous to life or health.

Acute exposures involving inhalation of dusts or mist may cause mucous membrane irritation with subsequent cough and dyspnea. Intense exposure may result in pulmonary edema and shock may result.

Prolonged or chronic exposures to high concentrations of sodium hydroxide may lead to ulceration of the nasal passages. All skin contact with this corrosive materials should be avoided. Acute skin exposures may cause cutaneous burns and skin fissures. Chronic skin exposures can lead to dermatitis.

5.5 Standards, Regulations, or Guidelines of Exposure

5.5.1 NIOSH Recommended Exposure Limit

2.0 mg/m^3 as a 15-min TWA ceiling.

5.5.2 OSHA Permissible Exposure Limit

None established (PEL of 2.0 mg/m^3 ceiling was vacated).

5.5.3 ACGIH Threshold Limit Value

2.0 mg/m^3 expressed as a ceiling value.

6.0 Sodium Peroxide

6.0.1 CAS Number: [1313-60-6]

6.0.2 Synonyms: Sodium dioxide, sodium superoxide, sodium binoxide

6.0.3 Trade Names: Various commercial names

6.0.4 Molecular Weight: 77.9783

6.0.5 Molecular Formula: Na_2O_2

6.0.6 Molecular Structure: Na–O–O–Na

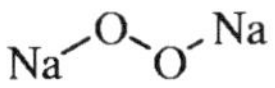

6.1 Chemical and Physical Properties

Specific gravity: 2.805
Melting point: 460°C
Boiling point: 657°C

6.1.1 General

Solubility is 42–347 g in 100 mL H_2O at 0°C. Soluble in aliphatic alcohols. Refractive index is 1.3576. Alkalinity of a 1% aqueous solution has a pH of ~13.

This white powder is a very strong oxidizing agent. A vigorous exothermic reaction takes place with water to form sodium hydroxide and oxygen.

6.2 Production and Use

Metallic sodium reacts in dry air to form sodium monoxide and sodium peroxide, Na_2O_2 (sodium dioxide, sodium superoxide). It may be encountered as an oxidant in chemical processes or as a bleaching agent, for example, of textiles. Its reactivity with carbon dioxide finds utility in self-contained breathing apparatus. It may also be encountered in the aerosol from sodium fires.

6.3 Exposure Assessment

6.3.1 Air: NA

6.3.2 Background Levels: NA

6.3.3 Workplace Methods: NA

6.3.4 Community Methods: NA

6.3.5 Biomonitoring/Biomarkers

6.3.5.1 Blood. See Section 3.0.

6.3.5.2 Urine. See Section 3.0.

6.4 Toxic Effects

In order to simulate the products of an alkali metal fire, sodium vapor was used to form a fresh aerosol composed mainly of sodium peroxide with some sodium monoxide. It was passed into an aging chamber to reach equilibrium with atmospheric carbon dioxide and water and to stabilize its particle size. The resulting aerosol was thought to represent the product of an accidental sodium metal fire in a nuclear reactor. Juvenile and adult rats were exposed for ≤ 2 h to various dilutions of the aerosol. The final particle size was ≤ 2.5 mm and was predominantly sodium carbonate with some sodium hydroxide. At necropsy following sacrifice, the only lesion observed was necrosis of the surface of the larynx; the area of affected epithelium and the depth of penetration were related to increased concentrations of aerosol. The ED_{50}, based on the number of animals affected, was about 510 mg/L for adults and 489 mg/L for juveniles. The severity of the injury was significantly greater in the juvenile rats. Animals sacrificed 4–7 days postexposure had no exposure-related lesions, suggesting a healing process (16). The characteristics of the experimental aerosol are in accord with those projected by Clough and Garland (7). Hughes and Anderson (17) measured the decreased visibility encountered in a sodium fire. They collected data on volunteers exposed for a short time to the fumes. They concluded that "Short term exposures of unprotected workers up to 40 mg/m^3 NaOH in air is unlikely to result in any serious discomfort. A concentration of 100 mg/m^3 produced serious discomfort promptly and precluded continuing work." Acute exposures to sodium peroxide via inhalation may cause respiratory irritation and difficulty in breathing.

Chronic airborne exposures result in severe irritation of the eye, mucous membranes, and skin. Skin contact, either acute or chronic, may result in severe irritation, redness, and pain. High concentrations can burn the skin.

6.5 Standards, Regulations, or Guidelines of Exposure

Clinical blood and urine studies (1) indicate the serum and urine levels of potassium per Sections 3.3.5.1 and 3.3.5.2.

There are no TLVs (ACGIH), PELs (OSHA), and/or RELs (NIOSH) established.

ALKALINE MATERIALS

7.0 Trisodium Phosphate

7.0.1 CAS Number: [7601-54-9]

7.0.2 Synonyms:
Sodium orthophosphate; sodium phosphate, tribasic; sodium phosphate; trisodium orthophosphate; phosphoric acid; trisodium salt; sodium phosphate, ACS, 98.0–102.0% (assay); TSP

7.0.3 Trade Names: Numerous Commercial Names

7.0.4 Molecular Weight: 163.941

7.0.5 Molecular Formula: Na_3O_4P

7.0.6 Molecular Structure:

$$O=P(O^-)(O^-)(O^-) \cdot 3\,Na^+$$

7.1 Chemical and Physical Properties

7.1.1 General

This strongly basic, tertiary salt is usually seen as colorless crystals. Solubility is 25.8 g/100 g water at 20°C and 157 g/100 g water at 70°C. Alkalinity of a 1% aqueous solution has a pH of 11.6.

7.2 Production and Use

Trisodium phosphate, $Na_3PO_4 \cdot 12H_2O$ (TSP, sodium orthophosphate) is produced by neutralization of disodium phosphate with sodium hydroxide. The disodium salt is produced from phosphoric acid and sodium carbonate. Trisodium phosphate is an important ingredient in soap powders, detergents, and cleaning agents. It is also used as a water softener to remove polyvalent metals and in the manufacture of paper and leather. Products for removing or preventing boiler scale often contain trisodium phosphate, as do those for removing insecticide residues from fruit and inhibiting mold.

7.3 Exposure Assessment

7.3.1 Air: NA

7.3.2 Background Levels: NA

7.3.3 Workplace Methods: NA

7.3.4 Community Methods: NA

7.3.5 Biomonitoring/Biomarkers

7.3.5.1 Blood. See Section 3.0.

7.3.5.2 Urine. See Section 3.0.

7.4 Toxic Effects

The toxicity of trisodium phosphate has not been investigated, but it may theoretically be related only to its alkalinity since its ions are normal constituents of all living matter. Its alkalinity is close to that of sodium carbonate.

Acute exposures to trisodium phosphate may cause irritation of the respiratory system with subsequent coughing and pain. In severe exposures pulmonary edema may develop.

7.5 Standards, Regulations, or Guidelines of Exposure

Clinical blood and urine studies (1) indicate the serum and urine levels of potassium per Sections 3.3.5.1 and 3.3.5.2.

There are no TLVs (ACGIH), PELs (OSHA), and/or RELs (NIOSH) established for elemental potassium.

8.0 Sodium Metasilicate

8.0.1 CAS Number: [6834-92-0]

8.0.2 Synonyms: Disodium metasilicate, disodium monosilicate; orthosil; Metso Beads, Drymet; silicic acid (H_2SiO_3), disodium salt; Silicic acid disodium salt; water glass

8.0.3 Trade Names: Various

8.0.4 Molecular Weight: 122.06 (anhydrous, 122.07; pentahydrate, 212.15; nonhydrous, 282.21).

8.0.5 Molecular Formula: Na_2O_3Si

8.0.6 Molecular Structure:

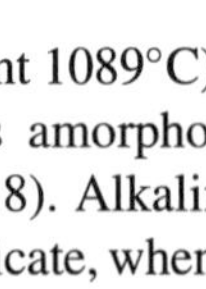

$$\text{}^{-}O-\underset{\underset{Na^+}{O^-}}{\overset{\overset{Na^+}{O}}{Si}}$$

8.1 Chemical and Physical Properties

8.1.1 General

The metasilicates (melting point 1089°C) are highly water-soluble. The anhydrous and the pentahydrate are produced as amorphous beads, whereas the nonahydrate appears as effluorescent sticky crystals (18). Alkalinity of a 1% aqueous solution has a pH of ~13.

Solutions of sodium metasilicate, when heated or acidified, are hydrolyzed to free sodium ions and silicic acid. The latter polymerizes through oxygen bridges to form amorphous silica. At high initial concentrations a gel is formed, whereas silica sols arise from dilute solutions. When dried, these colloidal forms of silica have great absorbing power.

8.2 Production and Use

The various hydrates of sodium metasilicate, $Na_2SiO_3 \cdot nH_2O$, range from the anhydrous to the nonahydrate; the anhydrous and the penta- and nonahydrate forms are the most common. The sodium metasilicates are differentiated from other sodium salts of silicic acid by the molar ratio of the Na_2O and SiO_2 components. In the metasilicates this ratio is 1 : 1. A continuum of sodium silicates of other ratios may also be commonly encountered. Their chemical and biological properties are essentially similar to those of the metasilicates. Fusing silica (sand) with sodium carbonate at 1400 °C produces sodium metasilicate. A major use is as a builder in soaps and detergents. It is also used extensively as an anticorrosion agent in boiler feedwater. The metasilicates should not be confused with other, less alkaline silicates used as adhesives of corrugated paper and as an additive to alfalfa cattle feed. Annual U.S. production of the metasilicates exceeds 400 million pounds.

8.3 Exposure Assessment

8.3.1 Air: NA

8.3.2 Background Levels: NA

8.3.3 Workplace Methods: NA

8.3.4 Community Methods: NA

8.3.5 Biomonitoring/Biomarkers

8.3.5.1 Blood. See Section 3.0.

8.3.5.2 Urine. See Section 3.0.

8.4 Toxic Effects

The acute oral toxicity (LD_{50}) of sodium metasilicate to rats is 1280 mg/kg as a 10% aqueous solution, and for mice the LD_{50} is 2400 mg/kg (18). The intraperitoneal injection in rats of the nonahydrate solution in amounts of 300 mg on gestation day 1 and 200 mg on days 2 and 3 produced lesions in the spleen and lymph nodes and caused mitotic changes in the nuclei of cells resembling those from ionizing radiation or hypoxia (19). Weaver and Griffith (20), in their studies of detergentemesis in 11 dogs by gastric intubation, found that 8 mg/kg as a 10.5% aqueous solution of a sodium silicate ($SiO_2 : Na_2O$, 3.2 : 1) produced emesis in 6 min, which continued for ≤ 33 min. Dogs were fed sodium silicate in their diet at a dose of 2.4 g/kg per day for 4 weeks (21). Polydipsia and polyuria were observed in some animals. Damage to renal tubules was observed in 15/16 dogs.

Radiolabeled (^{31}Si) sodium metasilicate, partially neutralized, when given orally, was rapidly absorbed and excreted in the urine, but a significant amount was retained in the tissues (21). These findings are consistent with the recognition that silicon is an essential trace element for bone formation in animals (22).

As might be expected, detergents containing sodium metasilicate and other alkaline materials are strong irritants to the skin, eyes, and respiratory tract (23–25). Seabaugh (23) has also shown that, of 134 detergents containing alkaline silicates, 81% were irritant or corrosive. Automatic dishwashing detergents were the most frequently corrosive.

"Soluble silica" in the drinking water of rats at 1200 ppm as silicon dioxide from time of weaning through reproduction reduced the numbers of offspring by 80% and decreased the number of pups surviving to weaning by 24% (26).

Acute exposures involving the inhalation of dusts of sodium metasilicate may result in irritation of the respiratory tract, and corrosive damage may result from contact with mucous membranes.

Prolonged exposures can lead to inflammatory changes and ulcerative problems in the mouth. Possible bronchial and gastrointestinal problems can exist, depending on concentration and duration of exposure.

Sodium metasilicates have not been evaluated in humans. Experience has shown that skin contact with solutions of strong detergents containing this builder produces severe skin irritation (12). However, other components of these detergents undoubtedly contribute to the irritancy.

Inhalation of dusts from soluble silicate powders is irritating to the upper respiratory tract (27). Exposure to such dusts is not related to the development of silicosis because their solubility permits them to be readily eliminated. Confirmation of this is found in the work of Svinkina (28), who was unable to produce immunologic sensitization in rabbits with sodium silicate bound to protein.

8.5 Standards, Regulations, or Guidelines of Exposure

Clinical blood and urine studies (1) indicate the serum and urine levels of potassium per Sections 3.3.5.1 and 3.3.5.2.

No TLVs (ACGIH), PELs (OSHA), and/or RELs (NIOSH) have been established for sodium metasilicate.

9.0 Cesium

9.0.1 CAS Number: [7440-46-2]

9.0.2 Synonyms: NA

9.0.3 Trade Names: NA

9.0.4 Molecular Weight: 132.9054

9.0.5 Molecular Formula: Cs

9.1 Chemical and Physical Properties

Specific Gravity: 1.873
Melting point: 28.5°C
Boiling point: 705°C

9.1.1 General

See Table 46.3.

ALKALINE MATERIALS

Table 46.3. Physical and Chemical Properties of Cesium and Cesium Compounds

Compound	CAS #	Molecular Formula	Molecular Weight	Boiling Point (°C)	Melting Point (°C)	Specific Gravity	Solubility in Water (at 68°F)
Cesium	[7440-46-2]	Cs	132.9054	705	25.5	1.873	Reacts violently
Cesium hydroxide	[21351-79-1]	CsOH	149.912		272.3		Very soluble
Cesium chloride	[7647-17-8]	CsCl	168.358	1280	646		Soluble
Cesium iodide	[7789-17-5]	CsI	259.81	1280	621		Soluble
Cesium nitrate	[7789-18-6]	CsNO$_3$	194.92	Decomp.	414		Soluble

9.2 Production and Use

Cesium is used in photoelectric cells, vacuum tubes, and atomic clocks. The isotype ^{137}Cs serves as a fission product in nuclear reactors.

9.3 Exposure Assessment

Although cesium has a low toxicity in animal experiments, it can act as a potentially harmful potassium analog.

9.4 Toxic Effects

The hydroxide which is one of the strongest known bases, can be formed from moisture with resulting alkaline toxic effects. Cesium has been studied as indicated medically in depressive disorders (1).

9.5 Standards, Regulations, or Guidelines of Exposure

No TLVs (ACGIH), PELs (OSHA), or RELs (NIOSH) have been established for cesium.

10.0 Rubidium

10.0.1 CAS Number: [7440-17-7]

10.0.2 Synonyms: Rubidium, powder, 99.8%

10.0.3 Trade Names: NA

10.0.4 Molecular Weight: 85.4678

10.0.5 Molecular Formula: Rb

Table 46.4. Physical and Chemical Properties of Rubidium and Rubidium Compounds

Compound	CAS #	Molecular Formula	Molecular Weight	Boiling Point (°C)	Melting Point (°C)	Specific Gravity	Solubility in Water (at 68°F)
Rubidium	[7440-17-7]	Rb	85.4678	700	39	1.532	Decomposes
Rubidium chloride	[7791-11-9]	RbCl	120.92	1390	715		
Rubidium carbonate	[584-09-8]	Rb_2CO_3	230.94		837		
Rubidium iodide	[7790-29-6]	RbI	212.37		642		Soluble
Rubidium hydroxide	[1310-82-3]	RbOH	102.48	300		1.74	Soluble

10.1 Chemical and Physical Properties

10.1.1 General

Rubidium is one of the most electropositive elements. It reacts spontaneously and violently with water liberating hydrogen. See Table 46.4.

10.2 Production and Use

Rubidium is used in photoelectric cells and as a reagent in zeolite catalysts.

10.3 Exposure Assessment

Rubidium can act as a potentially harmful potassium analog.

10.4 Toxic Effects

Rubidium has been indicated as exerting antidepressant affects and increasing a feeling of well being. Rubidium exhibits neurochemical and behavior effects (1).

10.5 Standards, Regulations, or Guidelines of Exposure

No TLVs (ACGIH), PELs (OSHA), or RELs (NIOSH), have been established for rubidium.

11.0 Francium

11.0.1 *CAS Number:* [7440-73-5]

11.0.2 *Synonyms:* EKa-cesium

11.0.3 *Trade Names:* NA

ALKALINE MATERIALS

11.0.4 Molecular Weight: 223.0

11.0.5 Molecular Formula: Fr

11.1 Chemical and Physical Properties

11.1.1 General

Francium has not been prepared or isolated in weighable quantities.

11.2 Production and Use

Francium is found in uranium minerals and can also be obtained by proton bombardment of thorium.

11.3 Exposure Assessment: NA

11.4 Toxic Effects: NA

11.5 Standards, Regulations, or Guidelines of Exposure

No TLVs (ACGIH), PELs (OSHA), or RELs (NIOSH) have been established for francium.

12.0 Lithium

12.0.1 CAS Number: [7439-93-2]

12.0.2 Synonyms: Cibalith; Eskalith; Lithane; Lithobid; Lithonate; Lithotabs; Carbolith; Cibalith-S; Duralith; Lithizine; Lithicarb

12.0.3 Trade Names: NA

12.0.4 Molecular Weight: 6.94

12.0.5 Molecular Formula: Li

12.1 Chemical and Physical Properties

12.1.1 General

    Boiling point: 1336°C
    Melting point: 179°C
    Specific gravity: 0.534
    Solubility in water (68°F): Decomp.

12.2 Production and Use

Lithium is used in production of aluminum-based alloys in the aircraft and aerospace industries, in polymerization catalysts in the polyolefin plastics industry, as an anode in electrochemical cells and batteries, and as a sedative in the pharmaceutical industry (lithium carbonate).

12.3 Exposure Assessment: NA

12.4 Toxic Effects

Lithium has significant neurological and psychiatric manifestations and is used in the treatment of unipolar, bipolar, and schizophrenic disorders (1).

12.5 Standards, Regulations, or Guidelines of Exposure

No TLVs (ACGIH), PELs (OSHA), or RELs (NIOSH), have been established for lithium.

BIBLIOGRAPHY

1. C. Y. Yung, *Pharmacology, Biochemistry, and Behavior*, **21**(Suppl. 1), 71 (1984).
1a. *Clinical Pathology Laboratory Manual*, Veterans Affairs North Texas Health Care System, 1999.
2. E. Legna, *The Sigma-Aldrich Library of Chemical Safety Data*, 1st ed., Sigma-Aldrich Corp., Milwaukee, WI, 1985, p. 1535, No. C.
2a. *NIOSH Manual of Analytical Methods*, 4th ed., 1994.
3. D. E. Bailey and K. Morgareidge, *Teratologic Examination of FED 73-76 (K_2CO_3) in Mice*, NTIS PB-245522, National Technical Information Service, Springfield, VA, 1975.
4. Food and Drug Administration, *Evaluation of the health effects of sodium hydroxide and potassium hydroxide as food ingredients*, NTIS, PB-265507, National Technical Information Service, Springfield, VA, 1976.
5. J. K. Narat, *J. Cancer Res.* **9**, 135 (1925).
6. C. B. Jackson, *Liquid Metals Handbook*, Sodium, NaK Suppl., Atomic Energy Commission, Washington, DC, 1955.
7. W. S. Clough and J. A. Garland, *J. Nucl. Energy* **25**, 425 (1971).
8. A. L. Reshetyuk and L. S. Shevchenko, *Hyg. Sanit.* **33** (1–3), 129 (1968).
9. G. A. Nixon, C. A. Tyson, and W. C. Wertz, *Toxicol. Appl. Pharmacol.* **31**, 481 (1975).
10. K. Morgareidge, *Teratologic Evaluation of Sodium Carbonate in Mice, Rats, and Rabbits*, NTIS PB-234868, National Technical Information Service, Springfield, VA, 1974.
11. K. Morgareidge, *Teratologic Evaluation of Sodium Bicarbonate in Mice, Rats and Rabbits*, NTIS PB-234871, National Technical Information Service, Springfield, VA, 1974.
12. N. Goldstein, *J. Occup. Med.* **10**, 423 (1968).
13. E. Horton, Jr. and R. R. Rawl, *Toxicological and Skin Corrosion Testing of Selected Hazardous Materials*, NTIS PB-264975, National Technical Information Service, Springfield, VA, 1976.
14. M. Dluhos, B. Sklensky, and J. Vyskocil, *Vnitr. Lek.* **15**(1), 38 (1969).
15. S. Nagao et al., *Acta Derm. Venereol.* **52**, 11 (1972).
16. G. M. Zwicker, M. D. Allen, and D. L. Stevens, *J. Environ. Pathol. Toxicol.* **2**, 1139 (1979).
17. G. W. Hughes and N. R. Anderson, *Int. At. Energy Agency Int. Working Group Fast Reactor Meet., 1971*, IAE-NPR-12.
18. A. Weissler, *Monograph on Sodium Metasilicate, NTIS* PB287766, National Technical Information Service, Springfield, VA, 1978.

19. L. Nanetti, *Zacchia* **9**(1), 96 (1973).
20. J. E. Weaver and J. F. Griffith, *Toxicol. Appl. Pharmacol.* **14**, 214 (1969).
21. F. Sauer, D. H. Laughland, and W. M. Davidson, *J. Biochem. Physiol.* **37** 1173 (1959).
22. E. M. Carlisle, Silicon: An essential element for the clack. *Science* **178**, 619 (1972).
23. V. M. Seabaugh, *Detergent Survey Toxicity Testing*, NTIS PB-264698hAS, National Technical Information Service, Springfield, VA, 1977.
24. G. E. Morris, *Arch. Ind. Hyg. Occup. Med.* **7**, 411 (1953).
25. L. G. Scharpf, Jr., I. D. Hill, and R. E. Kelly, *Food Cosmet. Toxicol.* **10**, 829 (1972).
26. G. S. Smith et al., *J. Anim. Sci.* **36**, 271 (1973).
27. Philadelphia Quartz Company, *Soluble Silicates Bulletin T-17-65*, Philadelphia Quartz Company, Valley Forge, PA, 1965.
28. N. V. Svinkina, *Lab. Hyg. Occup. Dis. (USSR)* **10**, 20 (1966).

CHAPTER FORTY-SEVEN

Inorganic Compounds of Carbon, Nitrogen, and Oxygen

George D. Leikauf, Ph.D. and Daniel R. Prows, Ph.D.

This chapter reviews the toxicology of some of the most commonly encountered chemicals in environmental and occupational settings. Although these chemicals are often generated by industrial processes such as combustion, several are generated by natural processes including endogenous production within the body. These substances are basic to the biological process and therefore life itself. Nonetheless, excessive exposures can be life-threatening and must be controlled. This chapter is modeled after the excellent, previous chapter in this edition written by Michael J. Lipsett, Dennis J. Shusterman, and Rodney R. Beard.

1.0 Carbon Monoxide

1.0.1 CAS Number: [630-08-0]

1.0.2 Synonyms: Carbonic oxide; monoxide; carbon oxide; carbon monoxide, various grades

1.0.3 Trade Names: NA

1.0.4 Molecular Weight: 28.010

1.0.5 Molecular Formula: CO

1.0.6 Molecular Structure: $^+C{\equiv}O^-$

Patty's Toxicology, Fifth Edition, Volume 3, Edited by Eula Bingham, Barbara Cohrssen, and Charles H. Powell.
ISBN 0-471-31934-1 © 2001 John Wiley & Sons, Inc.

1.1 Chemical and Physical Properties

Physical state: Colorless, odorless gas
Molecular weight: 28.010
Specific density: 0.967 (air = 1.0)
Melting point: $-207°C$
Boiling point (condensation point): $-190°C$
Solubility (in water): 3.5 mL/100 mL at 0°C, 2.3 mL/100 mL at 20°C, 1.5 mL/100 mL at 60°C
Flammability limits: 12–75% (in air)

Carbon monoxide (CO) is a colorless, odorless, and nonirritating gas. Because it can disrupt oxygen transport and delivery throughout the body by interfering with oxygen binding, it is classified as a chemical asphyxiant. A product of incomplete combustion, carbon monoxide can be encountered in many occupations and environments. The combination of a lack of warning properties and widespread exposure makes recognition and prevention of CO intoxication a common problem in industrial hygiene.

1.2 Production and Use

Carbon monoxide is formed during combustion of carbonaceous materials in oxygen (when carbon is in excess), or it can be formed (with oxygen) by thermal decomposition of carbon dioxide (> 2000°C). It can be generated by improperly vented cooking and heating appliances including coal stoves, furnaces, and gas appliances when the oxygen supply is insufficient. Other sources include exhaust of internal-combustion engines, structural fires, and tobacco products. Carbon monoxide can also be formed endogenously by normal heme turnover or during the metabolism of selected hydrocarbons, like methylene chloride. Not surprisingly, CO is one of the most common agents of inadvertent human intoxication in both occupational and non-occupational environments (2–4).

Carbon monoxide is one of the oldest known poisons. Accounts of CO poisonings date back to Aristotle's observation that "Coal fumes lead to heavy head and death" in third century B.C. Greece. Another early account of CO's toxicity involves the suffocation of the inhabitants of Nuceria, as noted in Hannibal's Carthage, 247–183 B.C. In Rome, coal fumes were commonly used for suicide and execution as recorded by Cicero (106–43 B.C.) and in the suicide of the Roman author, Seneca (65 A.D.).

Later, along with his formal description of carbonated water and identification of several other gases, Joseph Priestly (1733–1804) is credited with the initial description and purification of CO in 1772. He described CO as a combustible gas with a characteristic ability to convert the color of a burning flame to a bright blue. These and other observations led Henry Cavendish (1731–1810) to prove that water consists only of hydrogen and oxygen (after he had witnessed Priestly explode the two gases in a "random experiment to entertain a few philosophical friends").

Carbon monoxide was a major component of coal gas, and its usage led to one of the earliest clinical descriptions of coal gas poisoning made in 1775 by Harmant in

France. The introduction of coal gas illumination in the 1790s marshaled one of the early environmental laws to protect Prussian citizens against its hazards. In France, LeBlanc subsequently identified CO as the toxic constituent of coal gas in 1842. Shortly thereafter, in 1857 Claude Bernard demonstrated that CO reversibly combines with hemoglobin (5).

Since the midnineteenth century, combustion technology has improved considerably, but CO remains a persistent threat to health. Occupational CO exposures can occur in mines (particularly after blasting or fires), petroleum refineries (near catalytic cracking units), pulp mills (near lime kilns and kraft recovery furnaces), and boiler rooms, or wherever internal-combustion engines are used or repaired (6). In metallurgy, CO is used to reduce the oxygen content of iron and other metals (1) and in the steel industry, producer gas and blast-furnace effluent can contain 25–30% CO (7). In the chemical industry, CO is used as a feedstock for the synthesis of acrylates, aldehydes, ethylene, isocyanates, methanol, and phosgene. During periods of petroleum shortage, synthetic gases consisting of $\leq 90\%$ CO have been used as a combustion fuel, particularly in chemical manufacturing (8). As noted above, historically coal gas and water gas contained CO (8–30%) and at one time posed hazards in both occupational and home settings (9, 10).

Carbon monoxide exposure remains a particular concern for firefighters, who often enter enclosed (and therefore poorly ventilated) spaces in structural fires. Lethal CO concentrations can be encountered during the initial "knockdown" (when materials are actively burning) and subsequent "overhaul" (searching for "hot spots" among smoldering materials) phases of firefighting (11, 12). Unfortunately, the use of respiratory protective gear is often limited to the early phase of firefighting. Exposure assessments during wildlands (outdoor) firefighters, in contrast to urban fires, often report lower CO concentrations (13,14). Among smoke inhalation victims (both fatalities and survivors), CO poisoning is the rule rather than the exception (15).

Vehicular CO exposure can involve virtually any work with or near an internal-combustion engine. Occupations involving exposure include passenger car (taxi), ambulance, bus, and truck drivers. Mechanics, toll takers, garage attendants, and police officers encounter CO exposure (10, 16, 17). The placement of exhaust pipes near the front bumpers of aircraft refueling trucks apparently produced at least one CO-related death in an airport employee (18). The indoor use of propane-powered forklifts is another common source of CO exposure in industry, giving rise to the term "warehouse-worker's headache" (19, 20).

Among the public, vehicular emissions are a significant cause of CO poisoning. Malfunctioning exhaust systems, rusted or damaged automobile bodies, and the use of pickup truck campers and camper shells as passanger compartments have all been linked with serious CO intoxication, particularly among children (21–24). The indoor use of gasoline- or propane-powered ice resurfacing (Zamboni) machines has caused CO-related symptoms among both athletes and spectators (25, 26). The passenger compartments of some ambulances have exposed both patients and emergency medical personnel to excessive CO levels (27).

Although automotive emission controls have reduced CO emissions substantially since 1968, vehicular exhaust continues to be the principal contributor to atmospheric CO pollution nationwide (28). Conditions favoring incomplete combustion of gasoline (with consequent raised CO emissions) include lack of engine repair or tuning, high-altitude and cold weather operation, and excessive idling or stop-and-go driving. Reformulated gasoline with methanol, ethanol, and other oxygenated compounds (e.g., methyl *tert*-butyl ether) has been initiated in many areas to reduce CO emissions (29).

Within residential and commercial buildings, combustion appliances are the principal source of CO exposure. Hundreds to thousands of fatal and nonfatal human CO poisonings occur yearly throughout the United States because of improperly functioning (or inadequately vented) water heaters, furnaces, and kerosene space heaters (30–34). Poisonings with influenza-like symptoms (headaches, nausea, and lightheadedness) are often linked to the use of gas stoves and ovens as space heaters during the wintertime (35, 36). Another source of indoor CO exposure is entrainment of vehicular exhaust through improper placement of building air intakes (37).

Formerly of considerable concern, the use of CO-containing domestic fuels has been phased out in both the United States and western Europe, nearly eliminating both accidental and suicidal poisonings due to cooking gas exposure (38). In the developing world, by contrast, the use of charcoal as heating and cooking fuel continues the problem of indoor CO intoxication (39). Similar problems in the United States have been reported when charcoal is mistakenly used indoors (40).

The largest source of CO exposure in the United States is tobacco smoke. Although mainstream smoke contains about 5% CO by volume (41), sidestream smoke is the predominant source of environmental exposures and typically produces 70–90% CO yield per cigarette. This is because most of each cigarette is not smoked and the lower smoldering temperature of sidestream smoke produces 2–11 times the amount of CO than does mainstream smoke (42). Indoor CO levels in smoking areas may exceed 11 ppm, compared to < 2 ppm in most nonsmoking areas (43). Cigarette smokers normally exhibit significant elevations in the percentage of their hemoglobin combined with CO (carboxyhemoglobin, or COHb) when compared to nonsmokers. In a large, population-based study, for example, 95% of nonsmokers had COHb levels of < 2%, whereas the 95th percentile for COHb among smokers was nearly 9% (44).

An additional source of both occupational and personal CO exposure occurs among individuals using methylene chloride–containing products, particularly paint strippers (45). Methylene chloride is converted to CO by hepatic enzymes (46). Exposure to 500 ppm methylene chloride produces COHb levels in excess of 12%, which is more than 2–3 times higher than the COHb level produced by exposure to 25 ppm CO alone [the current recommended American Conference of Governmental Industrial Hygienists' (ACGIH) threshold level value (TLV)] (47).

An unavoidable source of CO exposure is that produced within the body. In erthyrocytes, normal heme turnover yields porphyrin degradation that releases 0.5–1.0 mL of CO per hour in adults. This endogenous production elevates baseline COHb levels to 0.3–0.7%, even absent any source of external exposure. Conditions that increase red blood cell breakdown, including hemolytic anemia, polycythemia, and blood transfusions, may add to circulating COHb levels (48–50). These conditions can be initiated by other chemicals that can produce additive risk from occupational exposures, like that of CO and methylene chloride.

1.3 Exposure Assessment

Carbon monoxide is absorbed from inspired air through the gas-exchange portion of the pulmonary system including the respiratory bronchioles, alveolar ducts, and alveoli. Carbon monoxide diffuses readily across normal alveolar tissue. It is used in trace amounts to clinically assess pulmonary gas diffusion [i.e., the diffusion lung capacity—carbon monoxide (DLCO)]. Once absorbed into the pulmonary capillary circulation, CO binds mainly to hemoglobin (Hb) molecules within red blood cells (RBCs) with the remaining small fraction of absorbed CO molecules in solution in the blood plasma or extracellular fluid (51). Diffusion of CO is concentration-driven; CO elimination via expired air represents the major mechanism of clearance of both inspired and endogenously produced CO. A very small fraction of the body's CO burden is oxidized to CO_2 (52).

Once bound to Hb molecules within RBCs, CO gains access to essentially all tissues of the body. About 10–15% of the body burden resides in the extravascular space, chiefly bound to other hemoproteins (myoglobin and cytochromes) (51). As noted in Section 1.5.3 on therapy, considerable controversy remains regarding CO's mechanism (or mechanisms) of toxicity; nevertheless, most investigators still believe that alterations in hemoglobin–oxygen binding explain the majority of CO-related health effects.

The understanding of CO's interaction with the Hb molecule has developed since the nineteenth century. As noted above, Claude Bernard first described CO's reversible interaction with Hb in the midnineteenth century. In 1895, John Haldane and associates quantified the competitive relationship between CO and oxygen (O_2) binding to Hb (53, 54). In a simplified situation of equilibrium conditions, Hb was found to bind CO with more than 200 times the avidity with which it binds O_2 as summarized in the following equation:

$$\frac{[COHb]}{[O_2Hb]} = M \times \frac{P_{A_{CO}}}{P_{A_{O_2}}}$$

where [COHb] = CO concentration in blood (vol/vol)
  [O_2Hb] = O_2 concentration in blood (vol/vol)
  $P_{A_{CO}}$ = alveolar partial pressure of CO (torr)
  $P_{A_{O_2}}$ = alveolar partial pressure of O_2 (torr)
  M: Haldane constant of CO's affinity relative to O_2 for Hb

Experimentally, observed values for M (the Haldane constant) vary between approximately 210 and 290. Thus, small concentrations of inspired CO can displace a much larger fraction of O_2 molecules from hemoglobin. Of equal importance toxicologically, however, is that when CO binds to a heterodimeric hemoglobin molecule, it alters the cooperative properties of the protein. The remaining O_2 molecules become more tightly bound and are less readily released to tissues. Conditions of impeded O_2 delivery to tissues, because of either reduced O_2 loading or release from Hb, are termed *hypoxias*, and are accompanied by metabolic and functional impairment of the organism.

The preceding simple formula defining the Haldane constant describes the static situation in which CO and O_2 compete for binding sites on Hb molecules under equilibrium conditions. This situation might be approximated by an animal breathing a fixed-composition atmosphere for a long period of time or, more likely, by a static *in vitro* situation like Hb in a test tube. However, under dynamic *in vivo* conditions other factors must be taken into account including endogenous CO production, ventilation rate, variations in alveolar O_2 pressure related to altitude, and individual differences in Hb concentration and total blood volume. The principal equation taking all of these variables into account is the Coburn–Forster–Kane (CKF) equation (55):

$$\frac{A[COHb]_t - B\dot{V}_{CO} - P_{I_{CO}}}{A[COHb]_0 - B\dot{V}_{CO} - P_{I_{CO}}} = e^{-tA/V_bB}$$

where

$$A = \frac{\bar{P}_{C_{O_2}}}{M[O_2Hb]}$$

$$B = \frac{1}{D_{L_{CO}}} + \frac{P_L}{\dot{V}_A}$$

and

M = relative affinity of Hb for CO versus O_2
$[O_2Hb]$ = O_2 concentration in blood (mL/mL)
$[COHb]_t$ = CO concentration in blood (mL/mL) at time t
$[COHb]_0$ = CO concentration in blood at beginning of exposure period
$\bar{P}_{C_{O_2}}$ = average partial pressure of O_2 in lung capillaries, torr
$\dot{V}_{CO}$ = rate of endogenous CO production, mL/min
$D_{L_{CO}}$ = diffusivity of the lung for CO, mL/min·torr
P_L = barometric pressure − vapor pressure of H_2O at body temperature, torr
V_b = blood volume, mL
$P_{I_{CO}}$ = partial pressure of CO in inspired air, torr
$\dot{V}_A$ = alveolar ventilation rate, mL/min
t = exposure duration, minutes

This equation is valid over a wide range of conditions, including large differences in CO levels, durations of exposure, and exercise conditions. For example, Peterson and Stewart conducted chamber studies to empirically test the predictions of the equation using exposures of 50–500 ppm for 15 to nearly 2000 min (56–58) (Fig. 47.1). This equation has several important features to environmental toxicology. One is that CO uptake is more rapid among workers who are actively exercising (have higher ventilation rates) than among sedentary individuals. Since the introduction of the CFK equation in 1965, its various assumptions have been examined critically, and a number of competing mathematical models for CO uptake and elimination have been proposed (59–63).

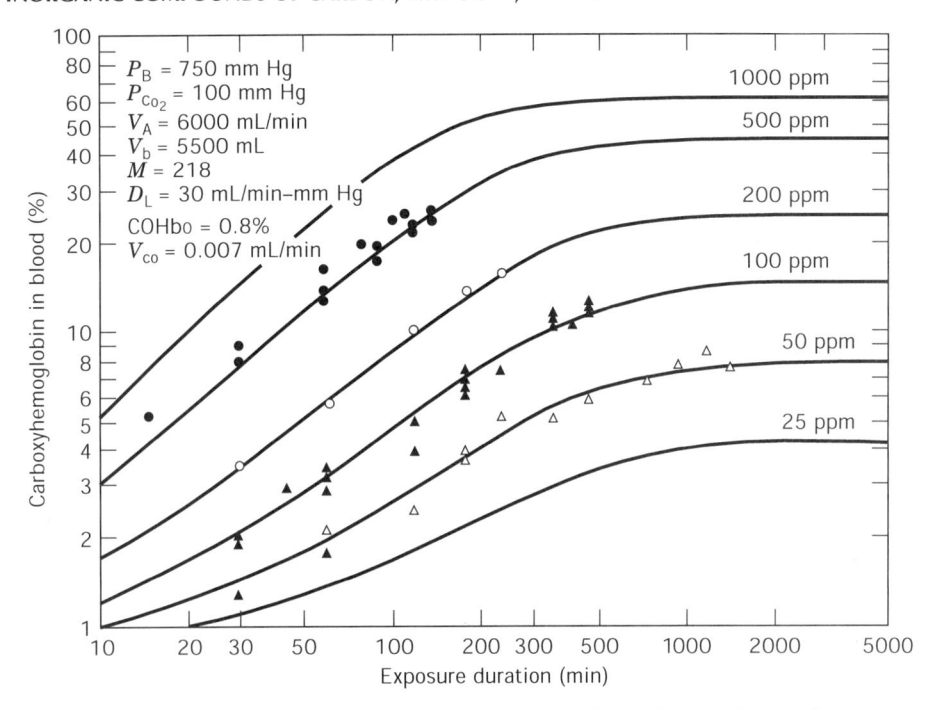

INORGANIC COMPOUNDS OF CARBON, NITROGEN, AND OXYGEN

Figure 47.1. Predicted and observed CO uptake (COHb formation) under varying exposure conditions. (From R. D. Stewart, "The Effect of Carbon Monoxide on Humans," *Ann. Rev. Pharmacol.*, **15**, 409–423 (1975). Used with permission.)

Although experimental validation of the CFK equation has concentrated on CO uptake, it can also be used to model CO elimination. Variations in CO elimination rates (e.g., with differing inspired O_2 concentrations) are pertinent to treatment of CO intoxication. Winter and Miller (64) modeled CO elimination from an initial COHb concentration of 50% under varying conditions of oxygenation, including "room air" (21% O_2), 100% O_2 at 1 atm, and 100% O_2 at 2.5 atm pressure. The ability to accelerate CO elimination with progressively increasing O_2 partial pressures is evidenced by a reduction of half-lives from about 5, to 2, and to $\frac{1}{2}$ hr for each of the three conditions, respectively (Fig. 47.2). The effect of CO on Hb's role in O_2 delivery is illustrated in Figure 47.3. Normally Hb provides the blood with the ability to carry about 19.4 mL O_2 per 100 mL of blood at O_2 tension of the blood traveling through the pulmonary capillary bed ($p_{O_2} = 104$ mm Hg). Leaving the lung and moving to the tissue, the venous blood oxygen tension falls to a $p_{O_2} = 40$ mm Hg. The normal Hb–O_2 dissociation curve in the absence of CO permits the delivery of ~ 5 mL O_2/100 mL blood. Under conditions of heavy exercise the venous blood oxygen tension falls even further to $p_{O_2} = 15$ mm Hg and the amount of O_2 delivered increases to 15 mL O_2 per 100 mL of blood.

This situation can be altered when a 50% loss in O_2-carrying capacity occurs as a result of anemia. Under these condiditions, the blood can carry only 9.7 mL O_2, per 100 mL, and will deliver only ~ 3 mL O_2 per 100 mL of blood. A lowering of tissue venous blood

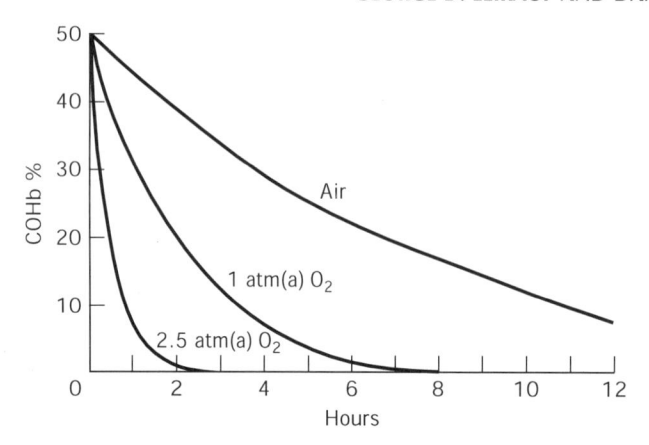

Figure 47.2. Carbon monoxide elimination (COHb clearance) under varying conditions of oxygenation. (From P. M. Winter and J. N. Miller, "Carbon Monoxide Poisoning," *J. Am. Med. Assoc.*, **126**, 1502–1504 (1976). Copyright 1976, American Medical Association. Used with permission).

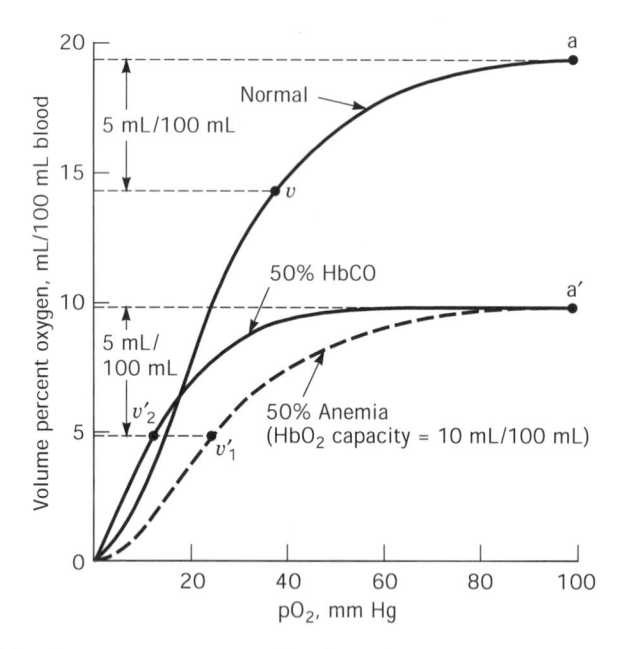

Figure 47.3. Blood oxygen content as function of oxygen tension in (1) normal state (Hb = 14.4 g/dL), (2) 50% anemia (Hb = 7.2 g/dL), and (3) normal Hb content but 50% carboxyhemoglobinemia. a and v refer to the arterial and venous points in normal state; a' is the arterial point for both anemia and carboxyhemoglobinemia; v'_1 is the venous point for anemia and v'_2 the venous point for carboxyhemoglobinemia. Assuming constant tissue oxygen extraction (5 mL/100 mL blood), note progressively lower venous oxygen tension from normal state of anemia to carboxyhemoglobinemia. (From Bartlett, "Effect of Carbon Monoxide on Human Physiological Processes," in *Proceedings of the Conference on Health Effects of Air Pollutants*, Serial No. 93–15, U.S. Government Printing Office, 1973.)

oxygen tension to only 25 mm Hg yields an adequate O_2 delivery. However, when 50% of Hb is bound to CO, the cooperative interactions of Hb subunits are altered and the dissociation changes from a sigmoidal to a hyperbolic function. The curve for 50% carboxyhemoglobinemia lies to the left of the curve for 50% anemia. Under this condition, a venous O_2 drop to 15 mm Hg is required to yield 5 mL O_2 per 100 mL blood. This is equivalent to the tension produced during heavy exercise in normal individuals. Thus the O_2 delivering capacity of the blood is lower and the tissues become more hypoxic in cases of CO poisoning than in anemia, even though the decrease in O_2-carrying molecules (Hb) is the same for both.

The mechanism the CO-induced left shift involves an alteration of the cooperativity in O_2–Hb binding. Hemoglobin consists of a pair of heterodimers that contains one α-globin unit and one β unit. Each unit has one iron(Fe)-containing porphyrin ring (heme) that has a binding site for either O_2 or CO. Each Hb molecule can therefore bind four O_2 molecules. Because the tertiary structure of Hb changes with O_2 binding as successive sites fill (the two α units pivot relative to the two β units), the energy of binding (and hence reluctance of Hb to release O_2) also varies. Starting with all four binding sites occupied by O_2 molecules, the release of O_2 (desaturation) of each site eases the release of the next O_2 molecule. This cooperativity of O_2–Hb binding produces the sigmoid shape of the O_2–Hb dissociation curve. A ligand bound to the O_2 site with greater affinity than O_2 (e.g., CO) interferes with the cooperative binding characteristics of the Hb molecule, and hence alters the shape of the O_2–Hb dissociation curve. Functionally, this shift means tissue O_2 tensions must fall to a lower level before the CO-poisoned Hb tetramer will yield its carried O_2 molecule.

Interestingly, an isolated heme ring (Fe–protoporphyrin IX) in solution without the carrier protein (globin) binds CO about 25,000 times more avidly than O_2. If Hb shared this characteristic, then M in the Haldane equation would equal 25,000 instead of approximately 250. Under this condition even the small amount of endogenously CO production would be life-threatening. However, Stryer (65) noted that Hb's protein structure prevents this from happening. An amino acid residue of the globin chain, histidine E7, lies immediately opposite the Fe atom's position in the center of the heme ring. This tertiary structure alters the shape of the binding site, but does not impede binding of O_2 molecules, which preferentially bind to the Fe at an angle. However, the CO molecule prefers linear binding with the Fe atom. The resulting steric hindrance produced by the histidine across from the binding site diminishes the binding affinity of Hb with CO. This steric hindrance reduces Hb–CO affinity by approximately 100-fold compared to heme–CO binding this remarkably subtle difference in protein structure is thus absolutely essential for life.

Finally, some investigators have suggested that all the toxicity of CO is not due solely to its ability to disrupt O_2 transport by Hb (66). In studies with dogs, transfusion of blood with highly saturated carboxyhemoglobin but minimal free CO did not reproduce the effects that CO exposure did. This suggests that even the small fraction of free CO dissolved in the plasma may have an important role in its toxicity. Thus, an alternative mechanism of toxicity may result from the cytochrome-mediated events on dissolved CO in the plasma (67, 68). Significant CO binding to cytochromes occurs during extreme anoxic conditions (as might occur with shock and severe asphyxia), and that with subsequent reperfusion the cytochrome–CO complex is relatively slow to dissociate, creating a potential for further tissue damage (69). Similarly, Thom measured lipid peroxidation in rat brain tissue following severe CO intoxication, and suggested that damage may be prevented (paradoxically) by hyperbaric O_2 (70).

1.4 Toxic Effects

1.4.1 Experimental Studies

1.4.1.1 General Information. The disruption of O_2 delivery by CO can have numerous pathophysiological consequences, many of which have been well studied in experimental animals (71). Initially, the CO challenged organism senses fatique and limits its activity, thus reducing O_2 demand by the tissue. This strategy can reduce O_2 demand only by voluntary (skeletal) muscle, but cannot affect resting functions in vital organs as the heart or brain. In the heart and brain (and some other organs), blood flow increases through vasodilation to increase O_2 delivery to almost normal rates (72). However, vasodilation has the unfortunate side effect of lowering systemic blood pressure, potentially compromising circulation. This leads to a compensatory increase in heart rate, initially restoring O_2 extraction by other organs by providing them with a greater quantity of (O_2-poor) blood per unit time (73). However, this aggravates the O_2 delivery situation for the heart, which must work harder in the face of a dwindling O_2 supply. Finally, the lack of O_2 increases anaerobic metabolism throughout the body, leading to an accumulation of lactic and other metabolic acids (74). Ultimately, insufficient O_2 delivery impairs central nervous system (CNS) function (with prolonged sleep, confusion, coma, and convulsions), which can impact survival or ability to escape exposure. This can be accompanied by loss of cardiac function with hypotension and cardiac dysrhythmias leading to death.

1.4.1.2 Acute Toxicity. The exposure conditions leading to lethality in experimental animals vary between the species, the rapidity with which a given COHb level is reached, and the activity state of the animal. Although an LD_{50} (lethal dose for 50% of animals studied) varies between studies, COHb levels of at least 60% typically produce a large percentage of death, with all animals succumbing at COHb concentrations of 65–75%. Besides lethality and acute physiological alterations, toxicological studies have examined a number of chronic health effects. Foremost among these have been cardiovascular dysfunction and the question of CO-related effects on the offspring of pregnant animals.

The role of CO in the acute aggravation of cardiac ischemia and cardiac rhythm disorders is well known in humans, and animal data support these findings (75, 76).

1.4.1.3 Chronic and Subchronic Toxicity. The possible role of chronic CO exposures in the development of hypertension and atherosclerosis has been reviewed by Penny (75, 76) and Smith (77). Effects of CO on the atherogenic process have principally been observed only in models of atherogenesis involving highly elevated serum cholesterol levels (e.g., 20 times normal). Findings in models of hypertension are also variable, even among species predisposed to the development of an elevated blood pressure. In general, CO is likely to produce hypotension by the mechanism described above to increase blood flow during hypoxia. Together there remains a lack of agreement as to whether CO is a causative agent for hypertension or atherosclerosis. The literature on this topic is also confused by studies with tobacco smoke in which exposures are mixed. Tobacco smoke clearly is associated with these outcomes but contains several opposing agents on the cardiovascular system (e.g., vasoconstrictors such as nicotine, which may contribute to hypertension).

1.4.1.4 Reproductive and Developmental. Animal studies of the reproductive and developmental effects of maternal CO exposure have been reviewed by Fechter (78–80). Carbon monoxide exposure during pregnancy produces offspring with decreased birth weight and increased neonatal mortality (81–84). Surviving offspring can develop persistent neurobehavioral deficits after intrauterine exposures (85). These effects occurred mainly at maternal COHb levels exceeding 15%, which is within a factor of 2 of COHb levels sometimes observed among smokers and permitted under the OSHA PEL (see Section 1.5.4.1). Equilibrium COHb levels are heightened and clearance of CO delayed in the fetus (86, 87). The apparent fetal susceptibility to CO may derive from a limited cardiovascular reserve and because both venous and arterial fetal O_2 tensions are shifted by CO loading (88). However, it should be noted that pregnant females have higher O_2-carrying capacity (to 25 mL O_2 per 100 mL blood) and fetal blood has even a higher Hb level than maternal blood.

1.4.2 Human Experience

1.4.2.1 General Information

1.4.2.1.1 Acute Toxicity. The effects of acute human exposures to CO involve mainly those tissues that are highly oxygenated (Table 47.1). Initial symptoms of CO intoxication include headache, fatigue, and lightheadedness through compromised CNS function. At higher COHb levels, signs of CO intoxication include flushing, tachycardia, and lowered systemic blood pressure resulting from vasodilation and the compensatory response to decreased O_2-carrying capacity. Dilated vessels are also responsible for a characteristic symptom of a throbbing headache, and cerebral vasodilation is associated with cerebral edema. With progressively increasing COHb levels, CNS effects increase with decrements in vigilance, visual perception, manual dexterity, and performance of complex sensorimotor tasks. Higher concentrations lead to severe nausea, vomiting, disabled coordination, syncope (fainting), coma, convulsions, pulmonary edema, and finally death.

There truly are no pathognomonic signs or symptoms to CO poisoning as some suppose. Victims can exhibit a "cherry red" discoloration of the skin and mucous membranes (lips), or retinal hemorrhages due to severe vasodilation, but these effects are rare. It is also important to note that severity of signs and symptoms do not always correlate with carboxyhemoglobin levels in a substantial number of cases. The length of exposure for periods longer than one hour increased morbidity over shorter exposures to equivalent concentration–time relationships (89).

Table 47.1. Acute Health Effects of Carbon Monoxide Exposure[a]

Blood Saturation COHb (%)	Response of Healthy Adults[b]	Response of Patients with Severe Coronary Artery Disease
0.3–0.7	Normal range due to endogenous CO production; no known detrimental effect	
2–5		Less exertion required to induce chest pain
5–10	Compensatory increase in CNS and coronary blood flow	Greater frequency and complexity of ventricular ectopic beats during exercise
	Possible neurobehavioral impairment (see text)	
10–20	Slight headache, fatigue, lightheadedness	Exertion may precipitate myocardial infarction
20–30	Moderate headache, nausea, fine manual dexterity impaired, visual evoked response abnormal, flushing and tachycardia	
30–40	Severe headache, nausea and vomiting, blood pressure changes (hypotension), ataxia (gross incoordination)	
40–50	Syncope	
50–65	Coma and convulsions	
>65–70	Lethal if not treated	

[a]Modified from Reference 58.
[b]Exposure to CO at high concentrations (> 50,000 ppm) can result in a fatal cardiac arrhythmia and death before the COHb is significantly elevated.

For individuals with severe coronary artery disease, relatively low COHb levels (2.9–4.5%) are accompanied by decreased exercise tolerance due to a shortening of the time to onset of angina and abnormal cardiac rhythms (see section 1.4.2.1.1.2). Even healthy individuals have decreased aerobic work capacity with COHb levels exceeding 4–5% (90).

Low-level CO poisonings are often associated with a flu-like syndrome, with headache, nausea, and lightheadedness. The consequence in many cases is a misdiagnosis, particularly during wintertime flu epidemics. Because of the nonspecificity of these symptoms, continued attention must be given to occult CO intoxication (91–93). In such situations, accurate diagnosis depends on careful detection by the physician and environmental hygienist with attention to symptoms among co-workers (or family members) and sources of CO exposure. It also should be remembered that warning signs of CO poisoning like cherry-red fingers may be infrequent and should not be relied on as a screening criterion for CO intoxication.

1.4.2.1.1.1 CNS EFFECTS IN HUMANS. As noted above, CNS symptoms of headache and lightheadedness are prominent as the earliest manifestations of CO intoxication. In addition, changes in psychomotor performance and electrophysiological response to external stimuli may possibly accompany relatively low COHb levels. At higher COHb levels, loss of coordination, syncope, coma, and convulsions occur. Persistent sequelae of severe CO intoxication involve the CNS, and include an impressive array of complications ranging from subtle personality changes to blindness, deafness, dementia, psychosis, and Parkinsonism. Recovery from these delayed neuropsychiatric syndromes occurs in about 50–75% of affected individuals within one year (39).

Various neurobehavioral and electrophysiological variables can change in humans after experimental CO exposure (94, 95). For the visual system, absolute visual threshold, visual acuity, depth perception, color vision, night vision, glare recovery, flicker fusion threshold, and visually evoked responses have been altered by exposure. Although transient alterations in visual evoked response accompanied by COHb levels of 20% in human volunteers (96), most findings at lower levels of exposure are inconsistent across studies. Acute (and reversible) impairment of auditory time discrimination occurred at an estimated COHb threshold of 3–5% (97–99), although others have been unable to replicate these findings with COHb levels of 10–20% (100). Other perceptual, motor, and cognitive variables have also been investigated.

Like many toxic substances with marked acute effects, it is often difficult to distinguish between chronic and persistent effects from repetitive acute exposures. This is somewhat a matter of semantics. For example, Grut (101) and others (102) have referred to chronic CO poisoning to describe chronic fatigue, sleep disturbances, and depression (neurasthenia) following repeated symptomatic CO exposures. Many of these cases, however, were either recurrent acute CO poisoning or patients who had experienced loss of consciousness on at least one occasion. Lindgren (103) disagreed with a diagnosis of chronic CO poisoning after surveying more than 600 Swedish workers. Although headache was reported more frequently among exposed than control workers, no other subjective neurological or psychometric abnormalities were linked with exposure. Acute and reversible CO headaches may develop an apparently persistent course if muscle tension headaches supervene. This can be assessed best when the location alters and quality of pain differs during unexposed periods (104). In contrast to the controversy surrounding patients who report symptoms from repeated, low level CO exposures, patients who have experienced CO-related syncope or coma are likely to be at risk for a variety of adverse neurological outcomes.

About 10–30% of all victims who lose consciousness during CO intoxication develop a delayed neuropsychiatric syndrome over the next 3–240 days. This effect has known since the midtwentieth century (105). In Korea, Choi noted that 5% of all patients evaluated for CO intoxication (or 24% of those hospitalized) experienced either prolonged coma or neuropsychiatric sequelae of immediate or delayed onset (39). In Great Britain, Smith and Brandon found that 11% of hospitalized CO cases developed gross neuropsychiatric sequelae and an additional 20–30% developed less profound personality changes or memory loss (106).

Neurological and psychological findings after significant CO intoxication includes personality and memory disturbances and Parkinsonism (including tremor, rigidity, and abnormal postural reflexes) (107, 108). Severe psychiatric disturbances were reported among survivors of CO poisoning, including obsessive–compulsive disorder, psychic akinesia, and Tourette's syndrome (109–111). In some individuals, periods when the patient is lucid and asymptomatic can separate periods when the patient has related symptoms and manifestation of delayed neurological or neuropsychiatric deficits (112–114). Pathological examination in fatalities may show necrosis within the basal ganglia, globus pallidus, and lesions of the periventricular white matter. Less commonly, diffuse white and gray-matter destruction may be found. Lesions in the basal ganglia demonstrated by computed tomography, molecular resonance imaging, or single-proton-emission computed tomography frequently are accompanied by a poor clinical outcome (115–118).

Peripheral neuropathy can develop with altered sensory or mixed sensorimotor function (119, 120). Severe CO intoxication has led to various ophthalmological sequelae, including retrobulbar neuritis, cortical blindness (121–123), retinal hemorrhages (124, 125), and ischemic retinopathy (126). Deafness has also been reported after severe CO intoxication, and a symmetrical "U-shaped" audiogram with sensorineural defect developed in one case of severe CO poisoning with otherwise complete neurological recovery (127).

1.4.2.1.1.2 CARDIOVASCULAR EFFECTS IN HUMANS. Nearly a million deaths from cardiovascular diseases occur annually in the United States; two-thirds of them involve atherosclerotic disease, and more than half result from myocardial infarctions (heart attacks) or lethal arrhythmias ("sudden death") (128). A considerable fraction of the workforce at any given time has established coronary artery disease (CAD). Not surprisingly, then, the cardiovascular effects of CO are of substantial regulatory interest and (129), in light of the findings reviewed below (129), provide a major rationale for exposure standards in both the workplace and the general environment.

Severe CO poisoning can induce myocardial ischemia through the diminished O_2 delivery capacity of Hb to the heart. As noted above, this often occurs during periods of increased cardiac output induced by systemic hypoxia. This situation can culminate into a complete myocardial infarction (MI) in individuals with normal coronary arteries (130, 131). Further, persons with established CAD can develop MIs following moderate CO concentrations (132). Low level CO exposures can lower the exercise threshold for angina in CAD patients. When exposures were calculated on an individual basis to produce either 2 or 4% COHb during the exercise period, exercise tolerance (time to angina) decreases at both target COHb levels (133–135). These levels of COHb also lower threshold for claudication in patients with peripheral vascular disease (136) and dyspnea in patients with chronic obstructive pulmonary disease (137, 138). In patients with CAD who experience arrhythmias, levels of COHb $\geq 6\%$ can increase the number and complexity of premature ventricular contractions during exercise (139–141). Frequent and multifocal premature ventricular contractions are significant in that they often precede fatal arrhythmias in CAD patients.

1.4.2.2 Epidemiology Studies. Apart from CO-related exacerbation of existing heart disease, it is unclear whether repetitive CO exposure influences the genesis of heart disease. Epidemiological studies suggesting a relationship between chronic CO exposures and either hypertension or atherosclerosis typically suffer from some degree of confounding. For example, in studies of cigarette smoking, CO inhalation is accompanied by exposure to nicotine and a variety of other toxicants. Similarly, in few occupations studied workers were exposed to CO alone. Potential confounders in occupational studies with automotive exhaust include concomitant-exposure to nitrosoamines and other aromatic compounds or stress (heat and noise). Thus, the role of CO alone in cardiovascular disease is difficult to assess from studies of bridge and tunnel workers, bus mechanics and drivers, or foundry workers (142–144). Historical associations include tatami mat makers in Japan in which a high CAD prevalence is associated with indoor use of CO-emitting charcoal heaters (145), but among New Guinea highlanders, atherosclerotic heart disease is rare despite repetitive indoor CO exposure (146). In summary, the epidemiological literature and animal experimental data are somewhat inconclusive with respect to the role of chronic CO exposures in the genesis of cardiovascular disease. Nonetheless, clinical studies associate acute low CO levels with exacerbation of pre-existing diseases of sufficient severity.

1.4.2.3 Reproductive and Developmental. The relationship between CO and adverse effects during fetal development are uncertain. Most studies involve clinical case reports of CO poisoning during pregnancy or epidemiological studies of cigarette-smoking mothers who produce offspring with decreased birth weight and increased perinatal mortality, consistent with animal studies of CO exposure (81, 147). These studies, however, cannot rule out the simultaneous action of other toxicants found in cigarette smoke (e.g., nicotine). Among clinical accounts of CO poisoning, stillbirths have been, reported, at times with fetal COHb levels as low as 23–24% (148–150). Perinatal asphyxia and congenital malformations have also been reported (151). Normal birth outcomes also commonly have been reported (152).

1.4.2.4 Other. Other health effects possibly associated with CO exposure include cutaneous bullae (blisters), particularly among comatose patients. This could be a specific effect of CO or a nonspecific effect of altered blood pressure in an immobile body (153–155). Myonecrosis (skeletal muscle destruction) has been reported following CO-related coma and in an elderly patient with subacute poisoning but no loss of consciousness (156, 157). Skeletal muscle damage typically results in some degree of myoglobinuria that can precipitate acute renal failure (158).

1.5 Standards, Regulations, or Guidelines of Exposure

1.5.1 Effects of Mixtures

Concomitant exposure of CO with other chemicals are important in numerous situations. The first and most common concomitant exposure is that produced by cigarette smoking. Adequate control of occupational CO levels requires assessment of the smoking habits of an employee.

The second involves exposure to other chemical asphyxiants that interfere with O_2 transport or utilization. For example, the combined exposure of CO and hydrogen cyanide (hydrocyanic acid, HCN), can uncouple oxidative phosphorylation. Concomitant exposure often occurs during fires caused by nitrogen (N_2)-containing plastic fuels that produce

HCN as a pyrolysis or combustion product. Animal testing suggested that the compounds produce death either as an additive or a synergistic function of CO and HCN concentration (159). In addition, blood levels of COHb and CN$^-$ do not correlate with the onset of incapacitation; thus postmortem blood COHb and CN$^-$ levels must be evaluated with care in fire victims (160).

Another example of dangerous combined exposures occurs when CO exposure is combined with a simple asphyxiant. Again this frequently happens during fires where O_2 concentration is diminished by combustion. An example is CO with carbon dioxide (CO_2). Simple asphyxiant, such as CO_2 and other inert compounds (e.g., N_2), can displace O_2 from inspired air, thereby augmenting the effects of CO-induced hypoxia. Carbon dioxide is mentioned here because it is a product of complete combustion of carbonaceous materials and therefore is common to fires that also produce CO. Another occupational situation where O_2 concentration can be displaced by simple asphyxiants and CO is generated by combustion is in confined spaces. This includes welding of tanks and pipes where sometimes even nontoxic volatile chemicals can be dangerous in the presence of CO.

An occupational situation where the hazard of CO exposure can be enhanced is when another, unsuspected source of CO is present. For example, metal carbonyls such as nickel carbonyl [$Ni(CO)_4$] can degrade in industrial processes to produce CO (161). Thus, exposure from both sources must be controlled.

Finally, CO exposure may occur as a result of the metabolism of absorbed chemicals such as methylene chloride (CH_2Cl_2), which is widely used in paint remover. Because CH_2Cl_2 is a hydrocarbon solvent and a CNS depressant, CNS effects can be additive with those produced by CO. In addition, CO is a product of hepatic metabolism of CH_2Cl_2 (162). To control COHb levels below target values, exposures limits for CO should be adjusted downward during periods of combined exposure to CH_2Cl_2 (see Section 1.5.4.1).

1.5.2 Measurement Techniques

1.5.2.1 Measurement in Air. In situations where rapid screening is needed, CO concentrations in air can be approximated utilizing indicator tubes. Other detection methods include several direct-reading, portable instruments that detect CO via an electrochemical reaction (163, 164). The standard method for performing ambient air-quality measurements is via nondispersive infrared spectrophotometry (absorption at 4.6 µm), and an alternative method utilizes dual column/dual thermal conductivity detector chromatography (165).

1.5.2.2 Measurement in Biologic Samples

1.5.2.2.1 Carbon Monoxide in Blood. Determinations of CO in blood typically measures the fraction of total Hb combined with CO (COHb) level. Methods for the determination of COHb in whole blood include (*1*) manometric or volumetric methods, (*2*) reduction by sodium dithionite, (*3*) microdiffusion with reduction of palladium chloride, (*4*) infrared spectrophotometry, (*5*) gas chromatography, and (*6*) differential spectrophotometry (166–169). In clinical situations, detection methods include differential spectrophotometry (CO oximetry) used when COHb exceeds 2.5% and gas chromatography when more accurate measurements are needed (170, 171).

The principle of CO oximetry depends on the simultaneous determination of the absorbance of a hemolyzed blood specimen at two or more absorbance wavelengths (172). Newer instruments utilize at least four test wavelengths to permit the simultaneous measurement of oxyhemoglobin, deoxyhemoglobin, carboxyhemoglobin, and methemoglobin. Interferences with this method may occur with hyperlipemia and may produce a falsely elevated COHb level (173). Conventional arterial blood gases can detect decreased O_2 saturation and possible metabolic acidosis in CO poisoning, but do not detect O_2 tension (5). Pulse oximetry cannot distinguish between COHb and O_2Hb and is not recommended in the diagnosis of CO poisoning (174).

1.5.2.2.2 Carbon Monoxide in Exhaled Breath. Electrochemical, gas chromatographic, or infrared absorption methods exist to measure CO in expired breath (175). These instruments employ an empirical relationship of exhaled breath CO and whole-blood COHb (176). False positives can occur with sample containing hydrogen (H_2) gas, which increases in exhaled breath during lactose intolerance and other types of intestinal malabsorption (177).

1.5.3 Therapy

The first line of treatment is removal and limitation of exposure, the second is increasing adequate O_2 supply, and the third is hyperbaric O_2 therapy. The first two approaches can be used whenever CO intoxication is suspected or confirmed. Protective breathing apparatus should be equipped to supply O_2 and remove CO. On removal from an area where CO is present, victims should be supplied with high flow O_2 immediately, because O_2 shortens the elimination half-life of carboxyhemoglobin by competing with the binding sites in Hb and improves tissue oxygenation (178). Oxygen should be provided until COHb levels become normal.

The third approach, hyperbaric O_2, remains controversial, and is recommended for persons who have lost consciousness, when COHb exceeds 40% in most people, 20% in persons with ischemic heart diseases, or 15% during pregnancy (5, 67, 179–184). Hyperbaric O_2 can help resolve symptoms, but it remains unclear whether it influences the rate of delay sequalea or mortality in non-life-threatening cases (185, 186). Raphael et al. (186) studied 629 nonoccupational CO poisonings and found no benefit from HBO among patients who had not lost consciousness, regardless of their presenting COHb level. Among patients with only brief initial loss of consciousness, there was no difference in outcome between those receiving one versus two HBO sessions. Among patients with initial coma, there was a small (but not statistically significant) difference in outcome favoring the treatment group (186). Significantly, this study did not compare hyperbaric with normobaric O_2 in patients who had initial loss of consciousness.

The potential for adverse effects of hyperbaric O_2 treatment must be considered. Some chambers are single-place cylinders that do not permit continuous hands-on care. Seizure can occur and lead to disruption of an intravenous line or endotracheal tube or cardiac arrest may be difficult to address because several minutes are required to depressurize a chamber from 2.5 atm back to sea level. Most of these concerns have been corrected by multiplace chambers, which can accommodate nursing or medical staff along with the patient. In addition, this treatment has resulted in tension pneumothorax and tympanic membrane rupture, and in conscious patients, ear and sinus pain (187, 188). Over 300 single-occupant chambers are available in the United States, and information can be obtained from Duke University's Divers Alert Network (telephone 919-684-8111).

1.5.4 Exposure Standards

1.5.4.1 Occupational Exposure Standards. A number of occupational exposure standards have been adopted (189–193). The current Occupational Safety and Health Administration (OSHA) PEL Time weighted average (TWA) 8 h/day, 40 h/week) is 35 ppm with a STEL (15 min) of 200 ppm. The American Conference of Governmental Industrial Hygienists' (ACGIH) threshold limit value (TLV) is 25 ppm (29 mg/m^3), and biological exposure indices (BEIs) are 3.5% COHb in blood, and 20 ppm CO in expired breath (end of shift). This standard represents a reduction in the target maximum COHb level from 8 to 3.5%, based on protection from cardiovascular, neurobehavioral, and reproductive endpoints (194). In addition, NIOSH's immediately-dangerous-to-life-or-health (IDLH) value is 1200 ppm, which is used in setting respiratory protection requirements (189). Although OSHA PELs and NIOSH IDLH values have the force of law, individual states with OSHA-designated programs may enforce stricter standards.

1.5.4.2 Environmental Exposure Standards. The U.S. Environmental Protection Agency (EPA) maintains a primary (health-based) ambient-air-quality standard for CO of 35 ppm (1-h average exposure) and 9 ppm (8-h average exposure). Ambient-air concentrations improved between 1986 and 1995; the national average CO concentrations decreased by 37% while CO emissions decreased 16%. This decrease occurred despite a 31% increase in vehicle miles traveled in the United States during that period. In 1995, transportation sources (includes highway and off highway vehicles) accounted for 81% of national total CO emissions. Corresponding standards promulgated by the State of California are 20 ppm (1-h average) and 9 ppm (8-h average). These standards seek to maintain COHb levels among the nonsmoking public below approximately 2.0%, and target those individuals with coronary artery disease as the most sensitive subpopulation of the public.

1.5.5 Prevention

Awareness of the sources and public education are essential to decreasing CO-induced morbidity and mortality. In addition, commercially available CO detectors can be installed. Operated by household current or batteries, CO detectors have two types of sensors. Detectors using household current typically employ a solid-state sensor that purges itself and resamples for CO periodically. This cycling of the sensor is the source of its increased power demands. Detectors powered by batteries typically use a passive sensor that reacts to the prolonged CO exposure. Minimum sensitivity and alarm characteristics standards have been recommended by Underwriters Laboratory (UL 2034). This revision specified additional requirements regarding identification of detector type, low level (nuisance) alarm sensitivity and alarm silencing. The Consumer Product Safety Commission recommends a detector on each floor of a residence. At a minimum, a single detector should be placed on each sleeping floor with an additional detector in the area of any major gas-burning appliances such as a furnace or water heater. In general, carbon monoxide detectors should be placed high (near the ceiling) for most effective use. Detectors should also not be placed within 5 ft of gas-fueled appliances or near cooking or bathing areas. Although CO detectors are inexpensive and readily available, there are no standard recommendations for their usage in occupational settings.

In addition, prevention of CO intoxication requires continuous maintenance of combustion appliances and motor vehicle exhaust systems, prompt investigation of indoor-air-quality complaints, and the maintenance of a high index of suspicion in clinical settings. These approaches should not be diminished or replaced by a reliance on CO detectors. When indoor vehicles (e.g., forklifts) are used, area ventilation must be increased for adequate protection, although substitution of electric vehicles is preferable. In repair garages, CO exposures should be minimized by connecting the tailpipe of a running vehicle to an active ventilation system (a negative-pressure exhaust hose).

In emergencies, exposures above the OSHA PEL (and ≤ 350 ppm) require use of a supplied air respirator or a self-contained breathing apparatus. Exposures of ≤ 875 ppm require a supplied air respirator or a self-contained breathing apparatus operated in continuous-flow mode. Exposures of ≤ 1500 ppm (or where levels are undetermined) require a supplied air respirator or a self-contained breathing apparatus with a full-face mask operated in continuous-flow mode with an auxiliary self-contained breathing apparatus (189).

2.0 Carbon Dioxide

2.0.1 CAS Number: [124-38-9]

2.0.2 Synonyms: Dry ice; Makr carbon dioxide; carbonic anhydride; carbon dioxide, 99.99%

2.0.3 Trade Names: NA

2.0.4 Molecular Weight: 44.010

2.0.5 Molecular Formula: CO_2

2.0.6 Molecular Structure: O=C=O

2.1 Chemical and Physical Properties

Physical state: Colorless gas
Molecular weight: 44.01
Boiling point: $-78.4°C$ (sublimes)
Melting point: Sublimes at $-78.33°C$ at 76 torr

Density: 1.997 at 0°C, or 1.527 (air = 1)
Solubility in water: 171 mg/100 mL at 0°C and 36 mL/100 mL at 60°C

2.2 Production and Use

Carbon dioxide (carbonic acid gas, Dry Ice) is normally present in the atmosphere at concentrations of 0.03%(v/v) above the ocean and from 0.0325–0.06% in urban areas. These concentrations are low in comparison to the 3.8% of exhaled human breath, which can be as high as 5.6% CO_2 (195).

Indoor CO_2 concentrations commonly are used as an indicator of the adequacy of ventilation, because CO_2 can be measured quickly and easily. Because low concentrations of CO_2 can induce mild discomfort, the American Society of Heating, Refrigerating and Air-Conditioning Engineers (ASHRAE) has set a standard recommendation of 0.1% (1000 ppm) CO_2 as a criterion of adequate ventilation (196). This standard relates the ventilation requirement primarily to the population density in the enclosed space. Such estimates are useful for moisture and odor removal and thermal comfort (heating and cooling), but may not provide adequate ventilation when toxic chemicals are present.

Carbon dioxide was recognized as a unique gas early in the seventeenth century by a Belgian chemist, Jan Baptist van Helmont, who noted that it was a product of both fermentation and combustion. Today, CO_2 is generated commercially by combustion of carbon-containing materials, fermentation, lime, or cement kilns, from flue gases and natural-gas wells. Once purified, it can be dehydrated and liquefied by compression [75 kg/cm^2 (−15°C)]. When liquid CO_2 is allowed to expand at atmospheric pressure, it cools and partially freezes to a solid "snow" that can be compacted into Dry Ice. Dry ice will sublime (passing directly into vapor without melting) at −78.5°C (1.0 atm). Over a wide range of temperatures, CO_2 is essentially inert, but above 1700°C, it partially decomposes into CO and O_2. (Addition of hydrogen or carbon can augment the conversion of CO_2 to CO at high temperatures.)

Carbon dioxide is used in carbonated beverages and is slightly soluble in water (1.79 v/v) at 0°C (1.0 atm), forming a weakly acidic solution of carbonic acid (H_2CO_3). Ammonia reacts with CO_2 under pressure to form ammonium carbamate, then urea, used in fertilizers and plastics.

Carbon dioxide is used in blasting coal, as a refrigerant, to promote growth of plants in greenhouses, in fire extinguishers, and in inflating life rafts and life jackets. Dry ice is used for preserving foods and chemicals, especially during transporation. Incidental uses include chilling aluminum rivets and shrinking cylinder liners or bearing inserts.

2.3 Exposure

Industrial exposures occur in the manufacture and use of dry ice or enclosed spaces where fermentation processes may have depleted the O_2 with formation of CO_2. This includes mines, tunnels, wells, the holds of ships, tanks, or vats. Because CO_2 is heavier than air, it collects in wells or tanks and will persist unless ventilated. Exposure to CO_2 can occur when fire extinguishers are operating in confined areas.

2.4 Toxic Effects

2.4.1 Experimental Studies and Human Experience

2.4.1.1 General Information. Endogenously generated (by respiratory metabolism) or inhaled CO_2 is carried in the bloodstream mainly in red blood cells. Under normal resting conditions, 4 mL CO_2 is transported by every 100 mL blood from the tissue to the lung. Most of CO_2 in the blood is carried in the form of bicarbonate ion (70%; 2.8 mL/100 mL blood). This ion is formed by the reaction of CO_2 with water to form carbonic acid, which is catalyzed by carbonic anhydrase, an enzyme inside each red blood cell. In addition to reacting with water, a portion of CO_2 (23%; 0.9 mL/100 mL blood) combines with hemoglobin (forming carbaminohemoglobin) and plasma proteins. Finally, only a small portion of the CO_2 (7%; 0.3 mL/100 mL blood) is carried in the form of dissolved state by the blood.

Increases in CO_2 in blood cause O_2 to dissociate and are important in O_2 unloading in the tissues. The loading of O_2 causes the unloading of CO_2 in the lung. This is known as the *Haldane effect*. The levels of CO_2 in the blood are tightly regulated by the nervous system through regulation of ventilation. The nervous system regulates ventilatory volume and rate by sensing increases in P_{CO_2} and hydrogen ion (decreases in pH) and increases are driven by decreases in P_{O_2}.

2.4.1.2 Acute Toxicity. Concentrations of $\geq 2\%$ CO_2 (20,000 ppm) deepen breathing (increasing tidal volume) and $\geq 4\%$ markedly increases in respiratory rate. At 4.5–5.0%, breathing becomes labored and distressing to some individuals, although some persons are capable of tolerating 10% for an hour (197–203). Exposure to $\geq 10\%$ can cause visual disturbances, tremors, perspiration, increased blood pressure, and loss of consciousness, and $\geq 25\%$ results in depression, convulsions, coma, and death (204, 205). Levy and Wegman (206) noted that visual disturbances can be associated with retinal degeneration. Although thought by some to be a simple asphyxiant, these concentrations are not sufficient to produce these effects simply by displacing O_2. Even 25% CO_2 would lower P_{O_2} to only 120 mm/Hg, equal to an altitude of 2500 m (about 8200 ft) (see Table 47.2). Instead, the toxicity of CO_2 is a result of its role in acid-base balance, which is tightly regulated in the body. Therefore, it is incorrect to consider CO_2 to be a physiologically inert gas.

Inhalation of CO_2 initially stimulates respiration and blood flow. This stimulatory phase is marked by vasodilation with flushing of the skin and increases in cerebral blood flow. More severe and prolonged exposures can lead to nausea or diarrhea. The secondary phase of intoxication leads to depression culminating in respiratory and cardiac failure. In the stimulatory phase, removal from the exposure and protection from other injuries are usually sufficient to prevent irreversible effects. Individuals with renal, hepatic, cardiovascular, or pulmonary disease may need additional attention. In the depressive phase, the main concern is hypoxia, treatable with O_2, and acidosis.

During pregnancy, maternal breathing is increased, reducing the partial pressure of CO_2 (P_{CO_2}) in the blood. The resulting mild alkalosis (increased pH) is balanced by a reduction in the concentration of bicarbonate and retention of hydrogen ions by the kidneys.

Table 47.2. Air and Oxygen Pressures, Ambient and Bronchial, by Altitude[a]

Altitude Meters	Altitude Feet	Ambient Pressure (millibars)	Ambient Pressure (mm Hg)	Ambient P_{O_2} (mm Hg)	Bronchial P_{Air}[a] (mm Hg)	Bronchial P_{O_2}[b] (mm Hg)
0	0	1013	760	160	713	150
500	1640	956	717	151	670	141
1000	3281	901	676	142	629	132
1500	4921	849	637	134	590	124
2000	6562	800	600	126	553	116
2500	8202	755	566	119	519	109
3000	9842	712	534	112	487	102
3500	11483	671	503	106	456	96
4000	13123	632	474	100	427	90
4500	14763	596	447	94	400	84
5000	16404	563	422	89	375	79
5500	18045	531	398	84	351	74
6000	19685	499	375	79	328	69
6500	21325	470	353	74	306	64
7000	22966	444	333	70	286	60
7500	24606	419	314	66	267	56
8000	26247	395	296	62	249	52
8500	27887	372	279	59	232	49
9000	29528	351	263	55	216	45

[a] Total air pressure in terminal bronchioles, taking into account pressure of water vapor, 47 mm Hg.
[b] Oxygen partial pressure in terminal bronchioles, taking into account pressure of water vapor, 47 mm Hg. Calculated from Zuntz equation,

$$\log b = \log B - \frac{h}{72(256.4 + t)}$$

where h = altitude in meters, from B to b,
   B = barometric pressure in mm Hg at lower level,
   b = barometric pressure in mm Hg at higher level, and
   t = temperature in degrees Celsius (15°C average assumed).

Inhalation of 2–4% CO_2 causes the P_{CO_2} and pH to assume values approximating those of nonpregnant women, and causes no apparent injury to the near-term fetus. Because the fetus responds by increasing the chest wall movements normally present in near-term pregnancy, CO_2 administration can be used to test fetal health. In the latter weeks of pregnancy, fetal P_{CO_2} is higher than maternal P_{CO_2}, as measured in umbilical cord blood (207). The fetus may be sensitive early in pregnancy to P_{CO_2} injury by an increased rate of blood flow in the brain, so it would be prudent to avoid exposures that exceed 0.5% (5000 ppm) (208).

Two natural disasters in Cameroon, on the western coast of Africa, have been associated with high level CO_2 exposures. Each involved the massive release of gas from volcanic crater lakes. The largest release occurred in August 1986, when Lake Nyos erupted and emitted a cloud of gas that blanketed a 300-km² area. A total of 1746 persons and over

3000 cattle died, many with cutaneous erythema and bullae. Over 800 survivors developed signs and symptoms compatible with asphyxiation. Of these, 161 patients also had cutaneous lesions that were initially believed to be from acidic gases. Further investigation suggested that they were associated with coma, possibly caused by CO_2 exposure. Although much of the toxicity has been ascribed to 1.2 km^3 CO_2 released, other toxic compounds were also released (including SO_2 and H_2S) (209). However, survivors recovered without sequelae of the eyes or respiratory system, which would be expected with exposure to volcanic gases (210). Geologic examination of the lake revealed that the gas was mainly CO_2 postulated to be derived from the underlying rocks. These investigations found that CO_2 accumulated to near saturation in the deepest part of the lake. At the time of the eruption, a portion of this high density water rose and rapidly released its gas in a massive bubble burst just beneath the surface (211). The disaster at Lake Nyos was similar to an eruption at Lake Monoun 2 years earlier. About 100 km southeast of Nyos, the explosion at Lake Monoun killed 37 people (212).

2.4.1.3 Chronic and Subchronic Toxicity. Reports in the medical literature of chronic CO_2 exposure are limited. In 1951, Schaefer (213) reported that exposures in World War II submarines often consisted of 3% CO_2 with reduced (15–17%) O_2. These exposures were prolonged over days and weeks and led initially to brief periods of excitation, followed by progressive depression. Cutaneous blood flow increased, core body temperature fell, blood pressure fell, and the rate of blood flow (measured from arm vein to tongue) increased. Other indices of circulatory function were depressed, the rate of breathing was slowed, and mental functions were impaired, in some instances to a disabling extent.

Speculative mechanisms of the neurotoxicity of CO_2 have been postulated by Max (214). He notes that the reaction of specific amino acids, such as alanine, with bicarbonate can produce toxic intermediates, such as β-(*N*-methylamino) alanine. Reactions with other amino acids may also form neurotoxic carbamates.

2.5 Standards, Regulations, or Guidelines of Exposure

The ACGIH TLV (TWA) is 5000 ppm (9000 mg/m^3) with a STEL of 30,000 ppm (54,000 mg/m^3) (201). However, the documentation of the threshold limit values and biological exposure indices stated that "medically fit" persons in special circumstances may tolerate daily exposures to 1.5% (15,000 ppm) (194). The definition of "medically fit" excludes all persons over the age of 65 and persons with current endocrine disorders.

The National Institute for Occupational Safety and Health has recommended a TWA of 5000 ppm as a limit, with short-term excursions up to 3% (30,000 ppm) for 10 min (198).

3.0 Phosgene

3.0.1 CAS Number: *[75-44-5]*

3.0.2 Synonyms:
Carbonic dichloride; carbon oxychloride; chloroformyl chloride; carbonyl chloride; CG; carbonyl dichloride; cytosine–guanine; Phosgene-13C (∼1 M solution in benzene)

3.0.3 Trade Names: NA

3.0.4 Molecular Weight: 98.916

3.0.5 Molecular Formula: CCl$_2$O

3.0.6 Molecular Structure:

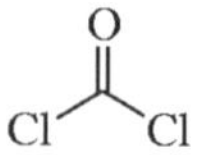

$$\underset{Cl \quad\quad Cl}{\overset{O}{\|}}$$

3.1 Chemical and Physical Properties

Physical state: Colorless gas at standard temperature and pressure; light yellow liquid refrigerated or compressed
Odor description: Similar to "new-mown hay"
Molecular weight: 98.916
Boiling point: 8.27°C
Melting point: −118°C
Vapor density: 3.48 (air = 1.0)
Solubility: Decomposes in water (to HCl and CO$_2$); soluble in organic solvents (e.g., benzene, toluene)

See also Ref. 1.

Phosgene is a colorless gas and a severe pulmonary irritant at high concentrations, but at low concentrations, it faintly smells like green corn or newly mowed hay and is only weakly irritating initially. These furtive properties provide only a scant warning of its potential toxicity, and led, regrettably, to phosgene exploitation as a chemical warfare agent during World War I. Currently, phosgene is used mainly as an intermediate in chemical synthesis.

3.2 Production and Use

Phosgene is produced commercially by the reaction of carbon monoxide and chlorine gas catalyzed by activated carbon. Although a gas at atmospheric temperature and pressure, phosgene is often supplied to industry in liquid form in pressurized steel cylinders or in limited quantities as a solid triphosgene (215). It is used in the manufacture of a variety of organic chemicals, including dyestuffs, isocyanates, carbonic acid esters (polycarbonates), acid chlorides, insecticides, and pharmaceuticals (216). In metallurgy, it is used to refine ores by chlorination of metal oxides.

Phosgene can be generated by thermal decomposition of chlorinated hydrocarbons and photooxidation of chloroethylenes in the ambient air. Occupational exposures have resulted from heating paint removers, degreasers, and welding on freshly degreased parts (217–219). Chlorinated hydrocarbons, such as chloroform, also can degrade spontaneously. One example involved laboratory personnel who became ill when working with 3-year-old chloroform. Subsequent analysis found 15,000 ppm phosgene in the headspace of the bottle and a 1.1% phosgene concentration in the bulk solution (220, 221). Decomposition of chlorinated hydrocarbons can produce other toxic chemicals, including hydrogen chloride, chlorine, and dichloroacetyl chloride (222).

3.3 Exposure Assessment

The principal route of phosgene exposure is inhalation. Secondary routes of exposure are dermal and ocular from splashes of the compound when in liquid form. Phosgene is only sparingly soluble in water, decomposing to HCl and CO_2 on prolonged contact with aqueous media. Because the respiratory passage is covered by a thin aqueous lining fluid that can remove inhaled gases, phosgene's low water solubility enables deep penetration through the respiratory tract. Nonetheless, little unreacted phosgene will enter the body. Another practical consequence of phosgene's physicochemical properties is that it has poor immediate (odor or upper respiratory tract irritation) warning characteristics. Thus, unsuspecting victims have remained in a phosgene-contaminated atmosphere for sufficient time to injure the lower respiratory tract, with massive exposures (≥ 200 ppm) producing severe pulmonary edema (223).

3.4 Toxic Effects

3.4.1 Experimental Studies

The pulmonary toxicology of phosgene has been studied in a number of animal species, including mice, rats, rabbits, dogs, and sheep. Dose–response relationships from these studies have been expressed in terms of the concentration–time product (usually in ppm × min). Lethal concentration–time products for one-half of experimental animals ($LC \times T_{50}$) are approximately 250–3300 ppm × min (224, 225). At lower concentration–time products, substantial lung damage occurs. Exposures of 60–120 ppm × min (0.5 ppm for 2–4 h), for example, increased edema (lung wet:dry weight ratios) and increased protein and cellular content of bronchoalveolar lavage (BAL) fluid (226).

Because of phosgene's low water solubility and high chemical reactivity, acute exposure produces lesions throughout the respiratory tract of experimental animals. In the conducting airway, vacuolization occurs in ciliated cells and Clara cells, and epithelial desquamation develops. In respiratory bronchioles, interstitial edema and bronchiolar constriction can occur. In the alveolar region, focal distention or atelectasis can occur, and at high doses, severe pulmonary edema (sometimes hemorrhagic) is noted (223). Physiologically, animals exposed to less than 1 ppm may not show any of the acute responses characteristic of exposure to respiratory tract irritants (i.e., changes in respiratory rate or depth of respiration). Nevertheless, exposed animals may exhibit severe pulmonary pathology after a "latent" period. As noted below, the phenomenon of delayed pulmonary edema has triggered considerable interest regarding the underlying mechanism of phosgene-induced lung injury.

The mechanism of phosgene's toxicity was once thought to involve phosgene hydrolysis in the respiratory tract lining fluid to yield HCl. However, this process is slow compared to direct acylation of macromolecules in the airways (223). The latter is now thought to initiate epithelial cell and macrophage injury and activation. This damage is, in turn, amplified by subsequent release of chemotactic factors. This produces inflammation with infiltrating polymorphonuclear leukocytes, which subsequently release proteolytic enzymes important in tissue remodeling (227). Edema is also aggravated by decreased sodium–potassium ATPase activity, with impaired water and electrolyte transport across cell membranes (228). Phosgene impairs cellular redox status, depletes of glutathione and free sulfhydryls, and diminishes the level of other antioxidants (229). It also reacts with lung proteins such as pulmonary surfactant, and alters surfactant synthesis and release (230). The loss of surfactant function combined with impaired natural-killer (lymphocyte) cell function, in turn, predisposes the exposed animal to infection (231). Enhanced susceptibility to influenza virus infection has been demonstrated among rats exposed to phosgene (1.0 ppm × 4 h, or 240 ppm × min) (232).

3.4.2 Human Experience

The odor of phosgene (sometimes described as that of newly mowed hay) can be detected at concentrations between 0.5 and 1.5 ppm (233, 234). Concentrations ≥ 3 ppm produce acute upper respiratory tract irritation affecting the eyes, nose, throat, and upper portion of the tracheobronchial tree (223). Phosgene exposures in higher concentrations produce bronchitis and pulmonary edema. Human phosgene exposures frequently have resulted from heated chlorinated solvents, more so than after work with preformed phosgene (219, 235, 236).

As with experimental animals, significant lower respiratory tract pathology can occur at concentrations that do not initially cause severe irritation. This effect may be predicted better by the concentration–time product than by the peak exposure concentration. Acute phosgene exposures (< 25 ppm × min) in humans seldom produce lower respiratory tract problems. Exposures of 25–50 ppm × min, however, have been associated with bronchitis, moderate to severe pulmonary edema, and death. Severe pulmonary edema is noted following exposures 50–150 ppm × min (226). However, these relationships assume that time and concentration are interchangeable risk factors, which is invalid at low exposure concentrations.

Pulmonary responses to phosgene exposure can be separated into four phases. The first or immediate phase (0–4 h) is dominated by direct reaction of phosgene with macromolecules present in cells constitutively present in the airways and alveolar regions (epithelial cells and macrophage). This response is initially marked by upper respiratory symptoms (nasal burning, increased mucus secretion, and eye irritation) and possibly disturbances in breathing. This can be followed by a second, latency phase, ranging from 4 to 24 h. During this period symptoms may subside, but inflammatory mediators, including neutrophil chemotactic factors, are generated by injured cells. The third phase (24–72 h) is an early response phase marked by moderate to severe dyspnea and hypoxemia associated with noncardiogenic pulmonary edema (5). The lungs are likely to be highly inflamed by activated neutrophils. Often patients display a delayed development of a mixed obstructive and restrictive ventilatory defect. This may signal the subsequent development of bronchiolitis obliterans (a condition more commonly associated with nitrogen dioxide exposure, as in "silo filler's disease"), although this diagnosis has not been confirmed pathologically (235). The fourth or late phase (> 72 h) consists of monocytic infiltrates,

INORGANIC COMPOUNDS OF CARBON, NITROGEN, AND OXYGEN

mucus cell hyperplasia, and bronchitis. Persistent bronchial hyperreactivity, consistent with the diagnosis of "reactive airways dysfunction syndrome" (RADS — also known as "irritant-induced asthma"), has also been reported after phosgene exposure (237).

3.5 Standards, Regulations, or Guidelines of Exposure

3.5.1 Effects of Mixtures

Phosgene can exhibit additive toxicity with the acute effects of respiratory tract irritants. Mixed exposures can occur as a result of the pyrolysis of chlorinated solvents that can produce HCl, Cl_2, and dichloroacetyl chloride.

3.5.2 Measurement Techniques

Airborne concentrations of phosgene can be measured by semiquantitative methods including test paper or indicator tubes (238). Air samples can be collected with sampling tubes containing XAD-2 adsorbent coated with 2-(hydroxymethyl) piperidine. Following desorption with toluene, the sample can be analyzed by gas chromatography using a nitrogen selective detector. Spectrophotometric measurements include a visible method involving reaction with 4-(4′-nitrobenzyl) pyridine (224, 239) and an ultraviolet method involving reaction with aniline (238). In addition, gas chromatographic and laser photoacoustic methods have been developed (240–242).

3.5.3 Therapy

Therapy for phosgene inhalation is largely supportive. Individuals with mild initial symptoms should remain under observation for at least 24 h from exposure to detect potential delayed pulmonary edema, which may or may not be detectable by chest radiography (243). Acute lung injury can be treated by oxygen, diuretics, glucocorticoids (anti-inflammatory), and exogenous surfactant, and assisted ventilation with positive end-expiratory pressure (PEEP) should be considered, depending on the severity of the clinical course (244, 245). Numerous specific therapies (e.g., hexamethylenetetramine) have been attempted in animal experiments, but, at present, none has been shown to produce consistent and unequivocal clinical benefit (246, 247). *In vitro*, isolated rabbit lungs were protected by pretreatment with aminophylline, terbutaline, dibutyryl cAMP, or post-exposure treatment with aminophylline or terbutaline (247).

3.5.4 Occupational Exposure Standards (192, 193)

The recommended occupational standards (OSHA PEL TWA, NIOSH REL TWA, and ACGIH TLV TWA) are in agreement for phosgene exposure and are set at 0.1 ppm (0.4 mg/m^3). The NIOSH REL STEL (ceiling, 15 min) is 0.2 ppm (0.8 mg/m^3).

3.5.5 Prevention

To prevent phosgene toxicity, avoidance or minimization of exposure is essential. Standard industrial hygiene practice includes substitution of a less hazardous product, engineering controls, administrative controls, and finally and least desirable, personal protective equipment. For example, in degreasing operations involving heated solvents or the subsequent heating of degreased parts, the substitution of nonhalogenated solvents is recommended. Alternatively, thermal shutdown controls on heated degreasers and strictly enforced work practices pertaining to drying time of parts can be instituted. Operations involving preformed phosgene should be totally enclosed, with leakage detectors, local exhaust ventilation, and respiratory escape gear provided.

For situations in which phosgene exposures potentially exceed the 0.1-ppm PEL, respirator use may be authorized on a selected basis through the OSHA variance process. For exposures below 1 ppm, NIOSH-approved respirators include any supplied air or self-contained breathing apparatus. Above 1 ppm and below 2 ppm, any supplied air or self-contained breathing apparatus with full faceplate is recommended. For emergency or planned entry into IDLH or unknown concentration, the use of either a self-contained breathing apparatus or a supplied air respirator is recommended, with each having a full faceplate and operating in positive-pressure mode.

The spontaneous generation of dangerous quantities of phosgene from chloroform can occur even when it is alkene-preserved and stored properly in the absence of heat and light (in a brown glass container). Alkene-preserved chloroform bottles can be tested for phosgene using filter paper strips. This can be generated by wetting with 5% diphenylamine and 5% dimethylaminobenzaldehyde, and then drying; this strip will then turn yellow in phosgene vapor. Alternatively, the use of ethanol-preserved chloroform is desirable because ethanol, unlike alkenes, reacts quickly with phosgene.

4.0 Nitrogen

4.0.1 CAS Number: *[7727-37-9]*

4.0.2 Synonyms: LN2

4.0.3 Trade Names: NA

4.0.4 Molecular Weight: 28.0134

4.0.5 Molecular Formula: N_2

4.0.6 Molecular Structure: N≡N

4.1 Chemical and Physical Properties

Physical state: Colorless, odorless gas
Atomic weight: 14.0067
Boiling point: $-196°C$
Melting point: $-210°C$
Density: 1.2506 g/L at 0°C, or 0.967 (air = 1)
Solubility in water: 0.0294 g/kg at 0°C; 0.0146 g/kg at 37°C
Specific gravity: 0.965 (gas)

4.2 Production and Use

Nitrogen is the main component of the atmosphere, 78.1% (v/v). Industrially, nitrogen is used for the displacement of O_2 or explosive gases from enclosed spaces such as large (fuel tanks) and small (communications cables and material packaging) vessels. Compressed nitrogen is mixed with oxygen, helium, or other gases for deep-sea diving. Liquefied nitrogen is used in cryogenic metallurgy (to alter the physical characteristics of metals) and in biomedical research (to quick-freeze and store tissues and microorganisms). Skin contact with liquid nitrogen can cause a serious burn. At high temperatures, nitrogen will combine with O_2 to form nitrogen oxides, with hydrogen to form ammonia, or with carbon (in the presence of bases or barium oxide) to form cyanide. Nitrogen also can form nitrides in the presence of lithium, barium, silicon, calcium, or strontium. Nitrogen can oxidize explosively with ozone.

4.3 Exposure Assessment: NA

4.4 Toxic Effects

4.4.1 Experimental Studies and Human Experience

A simple asphyxiant, nitrogen's main toxicity arises from its ability to displace O_2 and generate an atmosphere that does not support the chemical reactions needed for maintenance of life. The displacement of O_2 can be complete or incomplete, leading to varying degrees of hypoxia.

In addition, under certain circumstances nitrogen also has a direct toxic action of its own, affecting brain functions and inducing a stupor or euphoria. Nitrogen narcosis ("rapture of the deep" or "the martini effect") results from a direct toxic effect of high nitrogen pressure on nerve conduction and produces effects similar to alcohol intoxication. Complex reasoning, decision-making ability, motor function, and manual dexterity decrease. Individuals vary in this response widely, but it typically can be noticed among divers at depths exceeding 100 ft (30 m). For example, certain individuals experience no effect at depths of ≤ 130 ft, whereas others feel some effect at around 80 ft. Nonetheless, the narcotic effect increases with increasing depth so that each additional 50 ft incrementally produces the effect of "another martini."

Early in the 1910s and 1920s, nitrogen poisoning was recognized in divers. In 1919, Professor Elihu Thompson proposed that helium, as a codiluent with nitrogen, could produce a safe O_2 breathing mixture that would reduce narcosis. Unfortunately, helium cost over $ 2500/ft^3, so this was viewed as economically unfeasible at the time. Shortly thereafter, however, C. J. Cooke applied to patent the use of helium as a breathing gas, and a series of experimental dives were actually conducted using a "heliox" mixture to about ∼45 m. Heliox became commercially available following the discovery of helium in natural-gas wells. Helium is less soluble in tissues than nitrogen, and therefore is less likely to impair behavior (divers using helium, however, still have to decompress to prevent decompression sickness).

Later, in his 1956 film *The Silent World*, the renowned Jacques-Yves Cousteau described one of his earliest experiences with this effect:

I am personally quite receptive to nitrogen rapture. I like it and fear it like doom. It destroys the instinct of life. Tough individuals do not overcome as soon as neurasthenic persons like me, but they have difficulty extricating themselves. Intellectuals get drunk early and suffer acute attacks on all the senses, which demand hard fighting to overcome. When they have beaten the foe, they recover quickly. The agreeable glow of depth rapture resembles the giggle-party jags of the nineteen-twenties when flappers and sheiks convened to sniff nitrogen protoxide. "*L'ivresse des grandes profoundeurs*" [the intoxication of the grand profounder] has one salient advantage over alcohol — no hangover. If one is able to escape from its zone, the brain clears instantly and there are no horrors in the morning.

The U.S. Navy and other military naval organizations have compiled an in-depth analysis of this problem, which is addressed in the *U.S. Navy Diving Manual* (248). This effect, more fully investigated, is not unique to nitrogen but due to a slowing of nerve impulses that can result form several inert gases (but not helium) under high pressure. The effect is similar to what patients experience inhaling an anesthetic such as nitrous oxide (N_2O). With increasing N_2O concentration, symptoms progress from an initial feeling of euphoria to drunkenness and finally to unconsciousness. Other sources of information include Piantadosi (249), Bennett and Elliot (250), and the comprehensive collection of reports published by the Undersea Medical Association (1984) (251).

As was noted by Jacques-Yves Cousteau, prompt reduction of the nitrogen partial pressure gives relief. The main problem is to prevent divers from doing something foolish before they can be returned to normal pressure. For shallow dives, the problem is avoided by increasing the proportion of O_2 in the breathing mixture, but this introduces the problem of O_2 toxicity. For deep dives, carefully balanced mixtures of nitrogen, oxygen, and helium are used. The effects of nitrogen narcosis are compounded by simultaneous exposure to carbon dioxide, alcohol, or hypoxia, but do not seem to be affected by repeated exposure (252). However, professional divers must be taught to recognize the early symptoms and to take precautions against their progression (253).

On resurfacing, decompression sickness can arise from the subsequent release of nitrogen from body tissues. Also known colloquially as "the bends" because of the contorted postures of the agonized sufferers, decompression sickness is attributed to the formation of gas bubbles, mainly nitrogen, in tissues with the rapid release of pressure. The decompressed gases coming out of solution have been compared to the formation of carbon dioxide bubbles in a freshly opened carbonated beverage. This can be avoided by slower resurfacing, preliminary breathing of O_2 (to wash the dissolved nitrogen out of the body), the use of breathing mixtures that contain the least soluble gases (helium, neon), or decompression in a specially designed chamber. In very deep-sea work (deeper than 100 m) decompression may take several days. This prolonged time for decompression has led to the use of "saturation diving," in which people stay at the working depth continuously for days, resting, eating, and sleeping in a ventilated hyperbaric chamber. Additional information about the hyperbaric chambers is discussed above under carbon monoxide therapy.

4.5 Standards, Regulations, or Guidelines of Exposure

As a simple asphyxiant, the concentration of nitrogen should be controlled so that a minimal O_2 content of 18% be maintained ($P_{O_2} = 135$ torr). Atmospheres deficient in O_2

INORGANIC COMPOUNDS OF CARBON, NITROGEN, AND OXYGEN

do not provide adequate warning, and nitrogen is odorless. Several professional and private organizations have developed elaborate schedules for the control of diving and work under increased atmospheric pressure. The *Diving Manual of the U.S. Navy* is a good example (248).

5.0a Nitric Oxide

5.0.1a CAS Number: [10102-43-9]

5.0.2a Synonyms: Nitrogen monoxide; nitrogen oxide; nitrogen oxide (NO), NO; mononitrogen monoxide

5.0.3a Trade Names: NA

5.0.4a Molecular Weight: 30.0061

5.0.5a Molecular Formula: NO

5.0.6a Molecular Structure: ·N=O

5.0b Nitrogen Dioxide

5.0.1b CAS Number: [10102-44-0]

5.0.2b Synonyms: Nitrogen peroxide; nitrogen oxide; nitrogen oxide (NO_2); nitrogen dioxide (+Cyl.), 99.5%

5.0.3b Trade Names: NA

5.0.4b Molecular Weight: 46.0055

5.0.5b Molecular Formula: NO_2

5.0.6b Molecular Structure: O–N=O

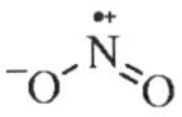

5.1 Physical and Chemical Properties

	Nitric Oxide	Nitrogen Dioxide
Physical state	Colorless gas	Reddish-brown gas, Colorless liquid
Molecular weight	30.01	46.01
Density (air = 1)	1.04	1.58
Melting point	−164°C	−9.3°C
Boiling point	−152°C	21.15°C
Solubility (mL/100 mL H_2O, STP)	4.6 mL/100 mL at 20°C	Slight[a]
Specific gravity at 20°C	1.448	1.448

[a]Reacts with water to form nitrous and nitric acids.
See also Refs. 254 and 255.

5.2 Production and Use

Nitric oxide (nitrogen monoxide, mononitrogen monoxide) and nitrogen dioxide [nitrogen peroxide, nitrogen tetroxide (NTO)] are often found in dynamic equilibrium, Historically, these compounds sometimes have been erroneously described as "nitrous fumes." In air, NO is readily oxidized to NO_2, and liquefied NO_2 (existing principally as its dimer nitrogen tetroxide, N_2O_4) releases NO_2 at room temperature (256). Thus, these compounds are often grouped as nitrogen oxides (NO_x). Other nitrogen oxides include nitrogen trioxide (NO_3), dinitrogen trioxide (N_2O_3), and dinitrogen pentaoxide (N_2O_5). Of all the oxides of nitrogen, NO_2 is the most actuely toxic and has been most extensively studied. Accordingly, much of this section focuses on the toxicity of this compound.

Nitric oxide and NO_2 occur naturally by bacterial degradation of nitrogenous compounds and to a lesser extent from fires, volcanic action, and fixation by lightning. NO has been the subject of intense and extensive research in a vast array of fields including chemistry, molecular biology, pharmaceuticals, and gene therapy. Formed endogenously, NO has a physiological role in blood-flow regulation, thrombosis, and neurotransmission, and a pathophysiological role in inflammation, oxidative stress, and host defense. NO is derived from the amino acid L-arginine by five-electron oxidation catalyzed by NO synthetase (requiring reduced pyridine nucleotides, reduced biopteridines, and calmodulin). The by-product, citrulline, is recycled back to L-arginine. In the bloodstream, NO binds primarily hemoglobin, is converted to NO_3, and is eliminated in the urine with a half-life of 5–8 h.

The main anthropogenic source of NO_x emissions in ambient air is the high temperature combustion of fossil fuels in motor vehicles and industry (especially power plants). Nitrogen in the fuel (e.g., coal or heavy oil) or the supply air of fuel combustors can be oxidized to NO and NO_2 under oxygen-rich combustion conditions. Under fuel-rich combustion conditions, molecular nitrogen tends to be formed preferentially from nitrogen in the fuel. The efficiency of thermal NO_x formation is highly temperature-sensitive and depends on the air:fuel ratio, residence time of gases in the combustion chamber, and other factors (257).

Atmospheric reactions that lead to ozone formation are initiated by the photolysis of NO_2, whereas NO acts as an ozone scavenger. (See Section 8.0, "Ozone.") The diurnal NO_x pattern in cities consists of a low background concentration with peaks in the morning and late afternoon corresponding to periods of rush-hour traffic (258). High local concentrations of NO_x can result from motor vehicle exhaust near busy streets and intersections. Historically, $\frac{1}{2}$ and 24-h averages as high as 0.45 and 0.21 ppm, respectively, have been reported (257). In Los Angeles, the urban area in the United States that had regularly violated the federal ambient air-quality standard for NO_2, 1–h maximum concentrations have been as high as 0.54 ppm (259). Average NO_2 concentrations decreased in the 1990s; for instance, the 1995 level was 14% lower than the average concentrations recorded in 1986. The two primary sources of NO_x emissions in 1995 were fuel combustion (46%) and transportation (49%). Between 1986 and 1995, emissions from fuel combustion decreased by 6% and from highway vehicles, 2%. Today, annual average concentrations of NO_2 in urban areas typically range between 0.01 and 0.05 ppm. Since 1991 all monitoring locations across the nation, including Los Angeles, have met the

INORGANIC COMPOUNDS OF CARBON, NITROGEN, AND OXYGEN

Federal NO_2 air-quality standard (0.053 ppm measured as an annual arithmetic mean concentration).

Nitric oxide is manufactured by passing air through an electric arc or by oxidation of ammonia over platinum gauze. Nitric oxide is used in nitric acid production, rayon bleaching, and organic chemical synthesis (as an oxidant). It is used to stabilize (against free-radical decomposition) methyl ether, propylene, and other substances. Nitrogen dioxide is manufactured by oxidation of nitric oxide (254).

Nitrogen oxide exposures have occurred in numerous occupational environments, mainly where fossil fuels are combusted. Combustion of fossil fuels in enclosed or poorly ventilated spaces (e.g., mines or skating rinks) may also produce toxic quantities of NO_2 (260–263). It is a common contaminant of indoor air whenever natural-gas stoves and furnaces are employed (264). In addition, production, transportation, and use of nitric acid can lead to NO_x exposure. Nitrogen dioxide is formed when nitric acid reacts with metals (as in bright dipping, pickling, and etching), or with organic materials (as in the nitration of cotton or other cellulose). It is also a by-product of the manufacture of many chemicals, including explosives, dyes, lacquers, and celluloid. Arc welding (gas or electric) causes NO_x formation via oxidation of atmospheric nitrogen (see Table 47.3).

In agriculture, NO_x is a well-known hazard and produces a condition known as "silo-filler's disease." Microbial degradation of stored silage (such as alfalfa or corn) for use as feed for livestock can produce NO_2 from nitrate in a silo or pit. Nitrate concentrations in plants are augmented by drought, nitrogen-containing fertilizers, and sunlight exposure. These nitrates ferment to form nitrites, which, in turn, react with organic acids to form nitrous acid. Decomposition of the nitrous acid yields NO that when oxidized produces NO_2, sometimes in extremely high (> 1000 ppm) concentrations (1, 265, 266). The practice of burning propane and kerosene to heat and add CO_2 to greenhouses also produces NO_x in potentially toxic concentrations, although much lower than those found in silos (1, 258).

In nonoccupational settings, NO_x exposure is usually greater indoors than outside (267, 268). Natural-gas appliances, including unvented stoves, furnaces, and water heaters, are the principal sources. In the past, historical peak NO_2 concentrations in kitchens where gas stoves are used have exceeded 1 ppm (258, 269), but this problem has been markedly reduced with the advent of the microwave oven. Kerosene space heaters, tobacco smoke, and motor vehicle exhaust penetrating from attached garages can also be contributing factors (270–272).

5.3 Exposure Assessment

Inhalation is by far the most common route of NO_x exposure, although occasional accidents (e.g., missile fuel spills) may involve contact of the skin and mucous membranes with nitrogen tetroxide (273). Overt dermal and mucous membrane toxicity from contact with the liquid takes the form of chemical burns.

Deposition of NO_2 in the respiratory tract is substantial. Healthy volunteers exposed to 0.29–7.2 ppm NO_2 retained 81–92% in each breath (274). In another study of persons with asthma expose to 0.30 ppm NO_2, retention was 72% at rest and 87% with exercise (275). The low water solubility of NO_2 allows it to reach the distal respiratory passages,

Table 47.3. Occupational and Nonoccupational Occurrence and Exposure to Nitrogen Oxides

Occupational

Combustion of fossil fuels (e.g., automobile garages, ice resurfacing machines in skating rinks, other internal combustion engines, boilers)
Nitric acid production, transportation, and use (including acid dipping of metals)
Electric arc fixation of nitrogen
Decomposition of aqueous nitrous acid
High-temperature oxidation of ammonia
Manufacture of lacquers and dyes
Other chemical manufacturing uses (nitrating agent, oxidizing agent, catalyst, inhibitor of acrylate polymerization)
Manufacture or use of explosives
Missile fuel oxidizer
Agriculture (silo filling)
Mining (diesel exhaust, shot-firing at coal seams)
Arc welding
Firefighting (including exposure to smoke from burning plastics, shoe polish, nitrocellulose film, or fossil fuels)

Nonoccupational

Gas and oil-fired household appliances
Kerosene heaters
Motor vehicle exhaust
Cigarette smoke
Ice-skating rinks
Industrial boilers

Source: Adapted from Reference 25

which is a principal site of toxicity. Brief exposures (7–9 min) of rhesus monkeys to radiolabeled NO_2 revealed that ^{13}N activity (measured over the thorax) remained virtually constant during the immediate postexposure period, whereas the activity of ^{125}Xe (an inert control gas) declined rapidly. These findings suggest that most of the reactive NO_2 is bound to respiratory macromolecules, whereas the inert Xe is exhaled (276). In isolated rat lung, uptake of NO_2 from the pulmonary airspace is saturable and temperature-dependent, again confirming that chemical reaction with lung surface constituents is the mechanism of NO_2 retention (277). Metabolic elimination (a process that is also likely to be saturable) and tissue reactions of NO_2 have not been extensively investigated, however, It has been proposed that inhaled NO_2 reacts with intrapulmonary water to form nitric and nitrous acids (276). In perfused rat lung *in vitro*, most NO_2 was reported to have converted to nitrite ion (NO_2^-), probably in dissociative equilibrium with HNO_2 (278).

In the presence of red blood cells, nitrite is rapidly oxidized to nitrate (NO_3^-) (279). Nitrogen dioxide possesses an unpaired electron that can be donated to receptor moieties such as unsaturated carbon–carbons like those in lipids. This can result in oxidation of polyunsaturated fatty acids (PUFAs), particularly in cell membranes and in lung surfactant. In addition, NO_2 can abstract an allylic hydrogen from other biological molecules. Both processes initiate one kinetic chain, but hydrogen abstraction results in formation of HNO_2, which can nitrosate organic amines to form nitrosamines (280).

With low water solubility, NO also penetrates the upper airways and deposits deeply in the respiratory passages. Like NO_2, most (85–93%) of the inhaled NO is retained (274, 281, 282). The affinity of NO for hemoglobin is about 1000 times greater than that of carbon monoxide. When absorbed into the blood it rapidly forms nitrosylhemoglobin, which is then converted into methemoglobin. Hemoglobin can then be regenerated in the presence of methemoglobin reductase and oxygen, both of which are found in abundance in red blood cells, resulting in the rapid formation of nitrate (NO_3^-) and nitrite (NO_2^-). Some inhaled NO reacts with O_2 to form NO_2, which is also metabolized to nitrate and nitrite (283).

Within tissues, mainly in activated inflammatory cells, NO works in consort with other reactive O_2 and nitrogen species, including superoxide anion (O_2^-), an O_2 free radical. Reactive O_2 species have long been known to be important in host defense. When cells are presented with bacteria, a foreign body, or other chemical injury, metabolism shifts by increasing the activity of the pentose phosphate pathway. This results in the increased production of reduced nicotinamide adenine dinucleotide phosphate (NADPH), and a dramatic increase in O_2 consumption that is referred to as the "respiratory burst." The NADPH generated participates in a transmembrane electron transport chain, NADPH:O_2-oxidoreductase (NADPH-oxidase), where cytosolic NADPH is the donor and O_2 is the acceptor of a single electron. The product is a free radical, superoxide anion (O_2^-). Through a process catalyzed by superoxide dismutase (SOD), O_2^- can be converted to hydrogen peroxide. This molecule is useful in the generation of bactericide compounds such as hypochlorous acid. Under ideal conditions O_2^- is released totally into the phagolysosome and thereby held inside the cell. Other reactive O_2 species produced during the respiratory burst include singlet O_2 and hydroxyl radicals. All of these intermediates are bactericidal.

Additional information on the role of NO in antimicrobial defense and other pathophysiological responses has been described. As stated above, the oxidation of L-arginine produces NO endogenously by a reaction catalyzed by nitric oxide synthetase. (The enzyme exists in constitutive and inducible isoforms.) In aqueous solution, freshly generated NO persists for several minutes in micromolar concentrations before it reacts with O_2 to form much stronger oxidants such as NO_2. Nonetheless, in the body the half-life of NO is only seconds because NO diffuses ($>100\,\mu m$) from the tissue and enters red blood cells and reacts with oxyhemoglobin. The direct toxicity of endogenous NO is modest but is greatly enhanced by reacting with superoxide to form peroxynitrite ($ONOO^-$). Interestingly, NO is produced in sufficiently high concentrations to outcompete SOD for superoxide. Compared to other reactive O_2 and nitrogen species, peroxynitrite reacts relatively slowly with most biological molecules, making peroxynitrite a selective oxidant. Peroxynitrite mainly modifies tyrosine in proteins to create nitrotyrosines, leaving a biological signature. Nitration of structural proteins, including neurofilaments and actin, can disrupt filament assembly with major pathological consequences. Nitrotyrosines have been detected in human atherosclerosis, myocardial ischemia, acute lung injury, inflammatory bowel disease, and amyotrophic lateral sclerosis

(284–286). Most inhaled NO is excreted as nitrate in the urine within 48 h. Other metabolites include nitrogen gas, ammonia, and urea (281, 283).

5.4 Toxic Effects

5.4.1 Experimental Studies and Human Experience

5.4.1.1 Acute Toxicity

5.4.1.1.1 Nitrogen Dioxide. As stated above, the major site of NO_2 toxicity is the respiratory tract. The level of NO_2 detectable by odor [described as a "sweetish, acrid" (234), or "bleach" smell (287)] varies greatly between individuals and ranges from 0.04 to 5 ppm. Upper respiratory tract irritation is perceptible at higher concentration (1–13 ppm) (234, 288–290). Individuals unfamiliar with the visual (red-brown color) or olfactory characteristics of NO_2, or in whom olfactory fatigue develops, may unwittingly inhale toxic doses, putting them at risk for delayed-onset pulmonary edema (291). The severity of the clinical effects following acute exposures may depend more on the concentration inhaled than on the duration of exposure (292).

Low concentrations may induce transient cough, dyspnea, headache, nausea, vertigo, fatigue, and somnolence, which typically dissipate over the following hours to days. These effects may persist for up to 2 weeks without significant decrements in pulmonary function.

Exposures to high NO_2 concentrations have caused sudden death from severe constriction of the airways or larynx (293). Survivors of high level exposures may develop a multiphasic response. Symptoms initially can be marked and then wane rapidly after exposure. This may be followed by a symptom-free period that can lead to a secondary response, marked by pulmonary edema. Because the pulmonary edema is delayed, it can evade detection, which may even lead to fatalities due to a lack of medical support. Initial response during or shortly after exposure can be mild to moderate, and include cough, dyspnea, wheeze, chest pain, tachycardia, weakness, sweating, nausea, vomiting, headache, and eye irritation. Certain individuals may describe a "choking" or "smothering" sensation, while others may be mildly irritated or symptom-free (294). Removal to fresh air often brings alleviation of symptoms, which may induce the exposed individual to forego medical care, or even reenter an area of exposure. (This behavior may be reinforced when others have had greater exposure or greater initial responses.) After an interval of a few hours (usually 4–12 h), however, a severe pulmonary edema may develop, accompanied by rapid breathing, dyspnea, cough, hemoptysis, cyanosis, chest pain, rales, fine crackles, and tachycardia (290, 291, 293–300). The interval between exposure and the development of pulmonary edema also varies between individuals and some may even be symptom-free. Chest X-rays taken soon after NO_2 exposure may be normal and cannot rule out the subsequent development of pulmonary edema.

During the pulmonary edema stage, arterial O_2 partial pressure will decrease, due to impaired diffusion capacity, ventilation–perfusion defects, and in some cases, methemoglobinemia (293). Methemoglobin levels, normally less than 1%, may increase, as a result of nitrite reacting with hemoglobin, to 2–3% (found in arc welding) or as high as 44% (found in exposure to silo gas containing NO and NO_2) (301). Metabolic acidosis may occur because of increased lactic acid production in response to hypoxemia. An early appearance of findings consistent with pulmonary edema indicates severe exposure and a worse prognosis. Except in massive exposures, most individuals receiving appropriate, timely therapy survive NO_2-induced acute lung injury. However, the overall fatality rate has historically been reported to be substantial (29%) (293).

Bronchiolitis obliterans can develop as a sequela following acute illness and can involve a rapidly progressive inflammatory occlusion of the small airways. Typically developing 10–30 days after the exposure, bronchiolitis obliterans may occur in the absence of a prior episode of pulmonary edema (265, 295, 300). This development can be heralded by flu-like symptoms (fever and chills) that rapidly deteriorate. Other presenting signs and symptoms may include fatigue, progressive shortness of breath and dyspnea, rapid breathing, cough, hemoptysis, chest tightness, and cyanosis (290, 293, 297, 300). Auscultation of the chest often reveals wheezing and other abnormal breath sounds, although occasionally minimal or no unusual findings are observed on physical examination (265, 290). A precipitous fall in forced expiratory volume is consistent with a diagnosis of asthma; however, the diminution of lung function in bronchiolitis obliterans is not intermittent or reversible as in asthma. Laboratory findings include an increased white blood cell count and an elevated sedimentation rate (265). Chest X-rays and high resolution computer tomography may show a patchy consolidation with linear opacities, a ground-glass pattern, or focal pseudotumoral, nodular infiltrates similar to those seen in tuberculosis, with confluence of nodules in severe cases (265, 291, 299).

Acute NO_2 exposures have often been associated with ice-resurfacing machines (Zambonis) used in inadequately ventilated arenas (260–262, 302, 303). Exposures can be episodic and have led to incidences where 50 to > 100 ice-hockey players, cheerleaders, and spectators experienced various respiratory symptoms (cough, chest pain, hemoptysis, shortness of breath) during or shortly after hockey games (260, 304). Persons on the ice can have more severe symptoms than spectators seated in the bleachers, presumably because of increased breathing during exercise, and proximity to the source near the ice. Air monitoring following these incidents has recorded NO_2 concentrations of 1.0–3.5 ppm, often combined with carbon monoxide. Resurfacers produce varying amounts of NO_2 depending on how they are powered; propane, gasoline, or diesel power produces increasing NO_2 concentrations, respectively. Electric resurface machines produce little or no NO_2.

Effects of acute exposure to low levels of NO_2 have been studied extensively with human volunteers. Results from several inhalation chamber studies have shown that healthy subjects may experience changes in lung function for short exposures at concentrations exceeding the range 1.5–2.0 ppm, although there are noticeable interindividual differences in susceptibility (129, 305–307). Persons with chronic obstructive pulmonary disease may be more sensitive and respond to exposure concentrations as low as 0.30 ppm (4 h) (307–310). Brief exposure to 5.0 ppm has also been reported to result in a decreased level of O_2 in arterial blood (306).

In addition to changes in lung function measurements, acute exposure to 1–2 ppm for ≤ 4 h can alter inflammatory mediators in the lung in healthy human volunteers (311–313). In one study (314), exercising healthy volunteers exposed to 2.0 ppm NO_2 for 4 h developed increases in polymorphonuclear neutrophils (PMNs), interleukin 6 (IL-6), IL-8, α_1-antitrypsin, and tissue plasminogen activator in their bronchoalveolar lavage (BAL) fluid the following morning. These changes are consistent with epithelial injury, inflammatory mediator release, and recruitment of leukocytes into the lung. Other control exposures at slightly higher concentrations (2.25–5 ppm) have also resulted in increased inflammatory cells (311–313, 315), but these reactions are often less than those found after low concentrations of ozone (316, 317). In addition, exposures to air followed by 2 ppm NO_2 for 4 h on 4 consecutive days resulted in a small but persistent increase in neutrophils and myeloperoxidase (MPO), indicative of migration and activation of neutrophils in the airways (318).

At these levels (2 ppm for 4 h), NO_2 can react with and transiently diminish the protective antioxidant pool present in respiratory tract lining fluid (319–321). Concentration of antioxidants (uric acid, ascorbic acid, and reduced glutathione) and a marker of lipid peroxidation (malondialdehyde) have been measured in BAL fluid. Uric acid and ascorbic acid decrease rapidly (1.5–6 h) and then rebound to levels above or equal to control (6–24 h). In contrast, glutathione levels increased at 1.5 and 6 h after exposure to NO_2 and returned to control levels by 24 h. These findings suggest that NO_2 reacts with antioxidants in the lung lining fluid, which modulate the acute effects of NO_2 on the lung. Supportive results include *ex vivo* exposure of lung fluid to NO_2 (322). However, changes in antioxidants and decrements in lung function may be attenuated on repeated exposures (323).

At slightly higher concentrations (3.0 or 4.0 ppm for 3 h), the activity of α_1-proteinase inhibitor ($\alpha_1 PI$) in lavage fluid has been found to be inhibited by as much as 45% in one study of healthy volunteers (324). This acute–phase protein is generated in the liver, travels to the lung, and is important in preventing degradation of the extracellular matrix, like that in emphysema. Persons genetically deficient in $\alpha_1 PI$ (a homozygous recessive trait) are at added risk from early-onset emphysema when the activity of this protein falls below 20%. Because this predisposition is relatively common, the frequency of heterozygotes (with resting levels of enzymatic activity of 50%) is also likely to be high in the general population. Thus, this group may represent a susceptible subpopulation. In another study, lower concentrations (1.5 ppm × 3 h or 0.05 ppm × 3 h with three 15-min, 2-ppm peaks) failed to demonstrate this effect (325).

At concentrations of 2–5 ppm, acute exposure may produce extrapulmonary changes. For example, the 2-ppm exposure that has been associated with hematological alterations, includes changes in hematocrit, hemoglobin, red cell acetylcholinesterase activity, and circulating aldehydes (consistent with lipid peroxidation) (326).

At lower concentrations (0.10–0.50 ppm), acute NO_2 exposures may transiently enhance bronchial hyperreactivity in persons with asthma (327–333). Bronchial hyperreactivity (or hyperresponsiveness) is a tendency of the airways to constrict to subthreshold concentrations of physiological mediators (acetylcholine or histamine) and may be immunospecific (e.g., antigen-induced) or nonspecific (e.g., exercise-induced). A pathognomonic feature of asthma, bronchial hyperreactivity, can occur reversibly in healthy individuals during and after respiratory tract infection or after exposure to irritant gases. However, in asthma it is persistent even during symptom-free periods. Although immediate symptoms or changes in lung function may be small in persons with asthma, these persons may be at greater risk for exacerbation of their condition with exposures to other respiratory irritants.

Findings of NO_2-induced airway reactivity may vary across asthma populations, possibly because of differences in disease severity of the study subjects and in study protocols (331–335). In one investigation (336), the effects of NO_2 were investigated examining the response to allergen following repeated exposures. For 4 subsequent days, 16 subjects with mild asthma and allergy to birch or grass pollen were exposed at rest to either purified air or 0.27 ppm (500 µg/m^3) NO_2 for 30 min. Four hours later, an individually determined nonsymptomatic allergen dose was inhaled, and lung function was decreased after repeated exposure to NO_2 and allergen compared to air. The effects were small, however. Nonetheless, the results indicate that a repeated short exposure to NO_2 enhances the airway response to an allergen among immunosensitive individuals. This mechanism has been supported by experiments with laboratory animals, although at higher (5 ppm) NO_2 concentration (337). In addition, high level exposures ($>$ 80 ppm for 1 h) may augment initial processes that lead to sensitization in rats (338).

High ($\geq$ 1.5 ppm) NO_2 concentrations also induce increased airway reactivity in healthy subjects (339, 340) and laboratory animals (341). The mechanisms for increased reactivity attributable to NO_2 exposure are unknown, but could involve antioxidants, because it can be partially inhibited by oral vitamin C (ascorbic acid) (342).

In laboratory animals, acute exposure to 0.3–10 ppm NO_2 has been associated with alterations in host defenses against infection. Mucociliary clearance can be impaired with decreases in number of cilia, changes in ciliary morphology, and depressed ciliary motility and beat frequency. Alveolar macrophage clearance can also be inhibited, leading to decrements in their engulfability to kill inhaled microorganisms, superoxide production, removing other foreign particles, and participating in various immune functions (343–353).

Studies with human alveolar macrophages have produced similar but less consistent results, even at high levels. For example, Pinkston et al. (354) exposed macrophages isolated from BAL fluid obtained from 15 healthy adult volunteers to 5, 10, and 15 ppm NO_2 for 3 h and found no effect on selected functional assays (354). Similiarly, Frampton et al. (355) found that 0.6 ppm NO_2 (for 3 h) decreased the capacity of macrophages to inactivate influenza virus, but this effect varied between subjects (affecting 4 of 9 subjects) and failed to reach statistical significance ($p < .07$). This was a lower exposure, but no such effects were observed with an exposure protocol involving three 15-min spikes of 2.0 ppm NO_2 superimposed on a continuous (3-hr) 0.05 ppm exposure. In contrast, Devlin et al. (314) reported decreases in macrophage antimicrobial function and superoxide production following 2.0 ppm exposures (with exercise).

When laboratory animals are exposed to NO_2 (usually at concentrations exceeding 1 ppm), susceptibility to respiratory challenge with bacteria, viruses, or mycoplasma can increase (343–345, 356). Superimposition of short (1.0-h) NO_2 high concentrations on a lower continuous level can enhance this effect in mice (357). However, repeated exposures to 0.5 ppm have not resulted in increased mortality from bacterial infection in rodents (345, 358). In addition, in human volunteers infections with attenuated influenza virus was marginally effected by 1.0–3.0-ppm exposures (359).

Epidemiological studies have suggested that exposure from gas stoves (which produce NO_2 and other substances) are associated with increases in respiratory illness among preschool children, decreases in lung function in school children, and exacerbation of respiratory symptoms in asthmatic adults (360–362). Several epidemiological studies suggest that NO_2 may also increase human susceptibility to respiratory illnesses, although others have not found such an effect (360, 363–365). This issue has been difficult to resolve because exposure sources are often hard to quantify and exposure–effect relationships are often unclear (366). However, epidemiological studies suggest that chronic exposure is associated with increases in respiratory illness and symptoms in children (325, 367, 368). This topic has been reviewed extensively (369–371).

5.4.1.1.2 Nitric Oxide. Compared with the extensive scientific literature on the effects of NO_2, much less has been published regarding the inhalation toxicity of NO. However, the few published reports indicate that acute NO inhalation is much less toxic (at least 30-fold) compared with NO_2 (372–374). For example, NO exposures of rates to ≤ 1500 ppm (15 min) or ≤ 1000 ppm (30 min) produced no major pulmonary histopathology. In contrast, inflammation has been noted following 30-min exposure to 25 ppm NO_2 (372). In addition, animals exposed to 1000 ppm for 30 min became cyanotic and died shortly thereafter, presumably because of methemoglobin formation. The basis of the current TLV is the ability of NO to form methemoglobin (194). In a study of extended 14-day exposures to 5.0 ppm NO, the principal histological findings in rabbits included alveolar membrane thickening and arterial cuffing in the lungs, without an accompanying inflammatory reaction (375). However, when NO or NO_2 exposures were extended to 9 weeks at 0.5 ppm (with twice daily, 1-h spikes to 1.5 ppm), the effects of NO were greater than NO_2 in rats. Formation of alveolar fenestrae (with focal degeneration of interstitial cells, interstitial matrix, and connective tissue fibers) resulted from low level NO exposure. The production of these defects in the interstitial spaces of alveolar septa, although limited in number and size, is thought to be the initial step in an emphysema-like destruction of alveolar septa.

As stated above, NO can react with superoxide to form peroxynitrite, and endogenous NO formation has a role in normal lung physiology (376–383). Peroxynitrite is involved in the killing of microbes by activated phagocytosing macrophages. In severe inflammation, peroxynitrite may be responsible for damaging proteins, lipids, and DNA.

Peroxynitrite added to surfactant *in vitro* is capable of decreasing the surface activity, inducing lipid peroxidation, decreasing the function of surfactant proteins, SP-A and SP-B, and inducing protein-associated nitrotyrosine. In animals, exposures (2–3 days) to 80–120 ppm NO resulted in decreases in surface activity, caused by binding of surfactant to iron proteins that are modified by NO (particularly methemoglobin), or by peroxynitrite-induced damage of surfactant (384, 385). In addition, exposure of isolated surfactant complex to NO during surface cycling decreases the inactivation of surfactant, preventing the conversion of surfactant to small vesicles that are no longer surface-active, and preventing lipid peroxidation. This finding is consistent with a mechanism whereby NO acts as a lipid-soluble chain-breaking antioxidant.

When healthy volunteers were exposed to NO for 15 min, they developed altered arterial O_2 content and lung function at concentrations ≥ 15 ppm (386). Although some earlier investigations using an exposure concentration of 1 ppm reported minor, isolated effects, these may have been of questionable biological significance (387). More recently, lower doses of inhaled nitric oxide (≤ 5 ppm) have been used therapeutically, and can improve the oxygenation and (to a lesser extent) ventilation during respiratory failure

(388–390). Higher doses have been used up to 100 ppm, but these concentrations produce unpredictable responses and necessitate individualized dosing of inhaled nitric oxide, starting at concentrations of ≤ 1 ppm. Inhaled nitric oxide at ≤ 20 ppm may alter bronchial smooth-muscle tone. The benefits of this therapy are somewhat controversial, however, and the long-term effects of prolonged inhaled nitric oxide remain unknown (391–393).

5.4.1.2 Chronic and Subchronic Toxicity. Episodic high level NO_2 exposures can lead to persistent respiratory impairment consistent with a diagnosis of chronic bronchitis or bronchiolitis obliterans (273, 293, 294, 394, 395). Persistent nonrespiratory effects from an acute overexposure to NO_x are less common (396). Survivors of episodic exposures, sometimes reported headache and other subjective neuropsychiatric complaints for up to a year after the accident (273).

The pulmonary pathology following exposure to sublethal (2–17-ppm) NO_2 concentrations has been examined in laboratory animals (397). In the alveolus, the type I epithelial cell sustains the greatest damage and necrotic cells are replaced through the proliferation of type II cells. The type II cells, which normally produce surfactant, have a lower surface:volume ratio and are more resistant to the damaging effects of NO_2. In addition, epithelial cells at the bronchiole–alveolar junction are effected. With continued NO_2 exposure, focal infiltrates of leukocytes and interstitial thickening develop. These histological changes appear to be somewhat reversible following acute and subacute exposures if the animals are allowed to recover in clean air (397).

However, continuing exposures of > 1.0 ppm NO_2 can induce nonreversible focal emphysema-like lesions in experimental animals (398–403). These studies were also marked by an initial transient inflammation of the small airways and alveoli, but subsequently patchy areas of alveolar enlargement and mild interstitial fibrosis developed (398, 400). Subchronic exposure to 2 ppm NO_2 in hamsters (8 h/day, 5 days/week for 8 weeks) also can exacerbate experimentally induced emphysema (404). Low concentrations, 0.6 ppm NO_2 and 0.25 ppm NO for 16 h/day for 68 months, also produced irreversible emphysematous changes in beagles (405). These authors initially attributed this effect solely to the NO_2 present in the mixture; however, a more recent study suggests that NO may have contributed substantially to this pathology (402). In several laboratory species, subchronic and chronic exposure to concentrations ≥ 0.34 ppm have resulted in findings of varying degrees of tissue damage, inflammation, and repair (406–412).

Whether long-term NO_2 exposure affects the progression of chronic lung disease in humans has not been adequately studied. In many populations, NO_2 is often present in complex mixtures. For example, the occupational literature has suggested that NO_2 exposure among coal miners may contribute to the development of chronic obstructive pulmonary disease (COPD) (413, 414). However, NO_2 is clearly not the only respiratory toxicant present in a coal mine. Similar difficulties in exposure assessment affect long-term studies of the general population exposed to NO_2 in the ambient air or in the indoor environment (366).

Numerous epidemiological studies of ambient-air pollution have investigated the acute and chronic effects of NO_2 exposure. For example, Euler et al. studied the relationship of 10-year exposures to several ambient-air pollutants to the prevalence of self-reported symptoms of COPD in California Seventh Day Adventists (415). These investigators found statistically significant relationships between COPD symptom prevalence and other pollutants (airborne particles and oxidants, primarily ozone), but not NO_2. The UCLA chronic obstructive respiratory disease study involved cross-city comparisons of lung function and respiratory symptoms in relation to mean ambient-air pollution levels in several communities in the greater metropolitan Los Angeles area. In an initial 5-year prospective study of lung function in two cities, residents (particularly children) of an area with higher levels of NO_2, sulfur oxides, and particles had a greater deterioration of lung function (416). In this study, no single pollutant could be assigned an etiological role in the deterioration of lung function, but neither could an effect of NO_2 be excluded. Subsequently, public school children were studied in 12 demographically similar communities with historic monitoring information on the extremes of exposure to one or more pollutants. Wheeze prevalence was positively associated with levels of both ambient particulate acid and NO_2 in boys (417). In another study, the effects of low levels of air pollution and weather conditions on the number of patients admitted to hospitals for exacerbation of chronic bronchitis or emphysema was studied in Helsinki during a 3-year period. Increases in the daily number of admissions via the emergency room was associated with prevailing levels of SO_2 (24 h mean = 0.0067 ppm) and nitrogen dioxide (24 h mean = 0.021 ppm). The effect of NO_2 was strongest after a 6-day lag and was significant only among those over 64 years old (418).

A case report also suggests that occupational NO_2 exposure may result in pulmonary alveolar proteinosis, a rare condition of uncertain origin characterized by an accumulation of lipid–protein material in the alveoli (419). Pulmonary alveolar proteinosis is recognized to be associated with silica exposure, and can involve chronic overproduction of pulmonary surfactant or inappropriate surfactant recycling. This case is of interest in light of the likelihood of surfactant damage that has been noted following exposure of animals.

Chronic NO_2 exposures of laboratory animals have produced contradictory results with respect to effects involving the blood or other organs (420, 421). Although significance of these results for human health is difficult to assess, these findings suggest that NO_2 or its reaction products enter and affect hematological functions including antibody formation and lymphocyte subpopulations (420–428). These studies suggest that an additional subpopulation susceptible to NO_2 exposure may be patients undergoing immunosuppression, as in cancer chemotherapy or individuals with acquired immunodeficiency syndrome (AIDS).

Although NO_2 can oxidize a variety of biological molecules (429), the evidence that NO_2 can directly generate free-radical reactions (lipid peroxidation) is currently uncertain (430–432). Reported increases of aldehyde degradation products may be due to an indirect mechanism involving inflammatory cells. Some have suggested that NO_2 might be genotoxic *in vitro* through formation of HNO_2, which, in turn, may form nitrosamines; however, this remains uncertain (433–437).

5.5 Standards, Regulations, or Guidelines of Exposure

5.5.1 Exposure Limits

For occupational NO_2 exposure, NIOSH recommends 20 ppm (94 mg/m^3) as immediately dangerous to life and health (defined as the maximal concentration one could escape within 30 min without a respirator), OSHA (and ACGIH) recommends 5 ppm (9.4 mg/m^3) as a ceiling permissible exposure limit (defined as the instantaneous maximum not to be exceeded at any time), and ACGIH recommends 3 ppm (5.6 mg/m^3) as a time-weighted average (TWA) (defined as the TWA concentration for a 8-h day, 40-h/week period). ACGIH — rates NO$_2$ as A4 — not classified as a human carcinogen. The level set by the USEPA for ambient exposure is 0.053 ppm (0.1 mg/m^3) as the national ambient-air quality standard to be averaged annually.

For occupational NO exposure, NIOSH recommends 100 ppm (123 mg/m^3) as immediately dangerous to life and health (defined as the maximal concentration one could escape within 30 minutes without a respirator). OSHA (and ACGIH) recommends 25 ppm (31 mg/m^3) as a TWA (defined as the TWA concentration for a 8-h/day, 40-h/week period.

5.5.2 Prevention

Because delayed responses can occur following acute exposures, even when initial responses are minimal, workers should be taught to seek medical attention even if a given exposure did not seem excessive. Medical attention should be protracted to prevent delayed onset of severe lower respiratory injury. Occupations at greater risk (e.g., welding) should be provided adequate ventilation. Entry into a recently filled silo (or crop storage pit) should be avoided (especially if the silo has been closed for 1–2 weeks) (438). In all cases, the silo should be vented prior to entry. Appropriate respiratory protective equipment should be available. Farm workers should be educated about the hazards of entering a recently filled silo, including appropriate safety practices, as well as visual and olfactory cues to the recognition of NO$_2$ and symptoms due to silo gas exposure. Warning signs should be prominently placed, and children should never be permitted to play in or near a silo.

Nitrogen dioxide (and CO) exposures can be prevented in ice arenas by the use of electrical resurfacing machines. If other types of machines are used, they must be maintained regularly and the heating system inspected constantly. The arena needs to be adequately ventilated, and an air-monitoring system is recommended (262).

NIOSH recommends several classes of respirators for protection against NO$_2$ inhalation toxicity (189). In addition, protective clothing and goggles must be work to protect against skin or eye contact with NO$_2$ (or more importantly, N$_2$O$_4$). Eyewash and quick drench equipment should be available for immediate use. Contaminated clothing should be removed immediately. NIOSH indicates that, except for respiratory protection at elevated concentrations, other protective clothing is not required for exposure to NO.

5.5.3 Treatment

Management of NO$_2$ inhalation is supportive and directed at improving gas exchange and preventing complications (infections) as is common for therapy directed at acute lung injury (acute respiratory distress syndrome) (298). The diagnosis may not be obvious in the initial acute phase, and will depend on eliciting a history suggestive of NO$_x$ exposure. Carbon monoxide toxicity often accompanies exposure. In addition, farm personnel may also present with flu-like symptoms from an accompanying hypersensitivity to thermophilic actinomycetes (438). Supportive care includes supplemental oxygen, broncho- and/or vasodilators, and ventilatory support. If methemoglobinemia is present, it may exacerbate hypoxemia and should be treated with methylene blue (2 mg/kg) (301).

The mainstays of therapy of pulmonary edema and bronchiolitis obliterans involve treatment with O_2 (as dictated by the patient's blood gas profile) and corticosteroids. The use of steroids may require aggressive vigilance for infection (bronchoscopic pneumonia screening). No controlled clinical trials have tested the efficacy of steroid administration in NO_2 inhalation. However, case reports indicate that such treatment can result in improvement (265, 295, 299, 300, 394, 439). Some authors recommend continuation of oral steroids for 6–8 weeks after a severe initial episode to retard or abort the development of bronchiolitis obliterans (290, 293).

5.5.4 Measurement

Numerous methods are available for measurement of NO and NO_2 in occupational and environmental settings. Many existing methods for analysis of NO_2 can also be used to measure NO. The OSHA recommended method for NO_2 involves drawing a known volume of air through a sampling tube containing a triethanolamine-impregnated molecular sieve. The sample is desorbed with an aqueous solution and the resultant nitrite is analyzed by ion chromatography (440). Measurement requires initial trapping of coexisting NO_2 to prevent overestimation of NO. This is accomplished with a three-tube sampling train: the first (which traps NO_2) and third tubes contain a triethanolamine-impregnated molecular sieve, separated by the middle tube containing chromate to oxidize NO to NO_2 (441).

Direct-reading instruments and reliable passive sampling devices [Palmes passive sampler (442–445)] have been developed. The latter device consists of a stainless-steel tube with three triethanolamine-impregnated sampling grids secured at one end with a plastic cap. The other end is sealed with a similar removable cap except during the sampling period, during which the tube is clipped to a worker's shirt pocket or lapel with the open end facing down. At the end of the sampling period, the tube is recapped until analysis. Triethanolamine forms a stable complex with NO_2 that can be analyzed following a reaction with sulfanilamide and N- (1-naphthylamido) diamine dihydrochloride using a spectrophotometer (540 nm) or by an ion chromatographic method. To measure NO, a chromic acid-impregnated disk to convert NO to NO_2 is added to a separate tube. Analysis of the contents of the second tube yields the NO_x concentration (NO + NO_2), and NO is obtained by subtraction followed by a diffusional correction (446). In addition, a badge-type personal sampler is available consisting of hydrophobic filters through which NO_2 diffuses before reacting with a triethanolamine-coated cellulose filter. The filter is extracted and analyzed colorimetrically as described above. By replacing three of the hydrophobic filters with glass fiber filters containing a strong oxidant (chromium trioxide), this sampler can also be used to measure NO (447).

Direct-reading instruments for ambient-air sampling measure NO_x use chemical luminescence, absorption spectroscopy, or laser-induced fluorescence (448, 453). The most common method for ambient-air monitoring is gas-phase chemiluminescence, which has been designated as the standard reference method by USEPA (450, 451). This method measures NO; thus NO_2 must be converted to NO prior to analysis by a heated metal catalyst, surfaces coated with ferrous sulfate or carbon, or photolysis. Conversion of NO_2 to NO by methods other than photolysis may generate significant interferences from other nitrogen-containing compounds in the air (448). Two wet chemical techniques (the sodium arsenite and the triethanolamine guaiacol sulfite methods) have been designated by the USEPA as equivalent monitoring methods for ambient NO_2 (451). Additional manual methods (the Griess–Saltzman and Jacobs–Hochheiser methods) are no longer recommended (450).

6.0 Nitrous Oxide

6.0.1 CAS Number: [10024-97-2]

6.0.2 Synonyms: Dinitrogen monoxide, laughing gas, hyponitrous acid anhydride, nitrogen oxide, dinitrogen oxide

6.0.3 Trade Names: NA

6.0.4 Molecular Weight: 44.0128

6.0.5 Molecular Formula: N_2O

6.0.6 Molecular Structure: $^-O-N^+\equiv N$

6.1 Chemical and Physical Properties

Physical state: Colorless gas, slightly sweet odor and taste
Molecular weight: 44.02
Melting point: $-90.81°C$
Boiling point: $-88.46°C$
Density: 1.53 g/mL
Solubility in water at 25°C: 1.30 gas/L H_2O
Flammability: Nonflammable, but supports combustion

See also Ref. 454.

6.2 Production and Use

A colorless gas with a slightly sweet odor and taste, nitrous oxide (nitrogen protoxide, nitrogen oxide, dinitrogen monoxide, hyponitrous acid anhydride, factitious air, laughing gas) is used primarily as an analgesic and anesthetic in surgery and dental procedures. Nitrous oxide was discovered by Joseph Priestley with several other gases in 1772. By 1799, its psychoactive properties were described by Sir Humphrey Davy in his treatise, *Researches, Chemical and Philosophical, Chiefly Concerning Nitrous Oxide, or Dephlogisticated Nitrous Air, and Its Respiration*. Throughout the nineteenth century use became widespread, even among the illustrious. Devotees including William James

(Harvard physiologist, psychologist, and philosopher), William Ramsey (1904 Nobel laureate and the discoverer of the inert gases), Peter M. Roget (physician and author of the *Thesaurus*), and Joniah Wedgwood (potter) and numerous poets, including John Addington Symonds and Samuel Taylor Coleridge and Robert Southey. It was Southey who mused: "It made me laugh and tingle in every toe and finger tip. It makes one strong, and so happy! So gloriously happy! Oh excellent gas bag!... I am sure the air in heaven must be this wonder working gas of delight" (453). Phineas T. Barnum and others organized public demonstrations of the effects of N_2O administration. Although N_2O is relatively nontoxic in comparison with the other nitrogen oxides, excessive occupational exposure has been associated with toxicity, and abuse and improper usage have occasionally resulted in death by asphyxiation.

Denitrification in soil by microorganisms is the primary source of background concentrations of N_2O, which range from 0.3 to 0.5 ppm in the troposphere (455, 456). Although it is present only in low concentrations, N_2O is a major greenhouse gas. Contributing about 5% to total atmospheric absorption of heat radiation, it is about 200 times more effective in heat absorbency than CO_2 and persists for $\sim$150 years in the stratosphere (456).

Nitrous oxide is commercially prepared by heating (245–270°C) aqueous (83%) solutions or solid ammonium nitrate ($NH_4NO_3 = N_2O + 2\,H_2O$). This is an exothermic decomposition reaction. Caution must be exercised (particularly with the solid) because NH_4NO_3 explodes at 290°C and this has led to over 5000 injures or deaths (455). Other, less common, manufacturing processes include reacting nitric and sulfamic acids or catalytically oxidizing ammonia. N_2O can be formed in the production of various chemicals (such as adipic acid) when nitric acid is used to oxidize organic compounds (455). N_2O is normally stored and shipped as a compressed liquid.

Its primary use, as noted above, is as an analgesic and as an adjunct in general anesthesia in human and veterinary surgery and in dentistry. Occupational exposure exceeds 200,000 persons (mainly dental, operating room, and veterinary personnel) (457). N_2O is used as a propellant in whipped cream, an oxidant for organic compounds, a nitrating agent for alkali metals, and a component of some rocket fuels. It is used to enhance ignition in high performance automobile racing, where it is usually mixed with hydrogen sulfide (454, 458).

Prior to the 1980s, N_2O was used indiscriminately, often without adequate ventilation or waste-gas scavenging. Peak breathing zone concentrations under these conditions exceeded 2.0% (24,000 ppm) and average concentrations ranged from 130 to nearly 7000 ppm (459, 460). More recent exposure assessments indicate that current levels are substantially lower. Occasionally exposures exceed the NIOSH-recommended standard due to the lack of scavenging equipment, inadequate ventilation, use of open masks (rather than endotracheal intubation), and leaks from the exposure system (461–465).

6.3 Exposure Assessment

N_2O is highly lipid soluble and rapidly absorbed and distributed throughout the body, particularly the vessel-rich regions, including the brain, heart, kidney, splanchnic circulation, and endocrine glands. The rate of N_2O uptake during the first 1 or 2 min is about 1.0 L/min (at an inspired concentration of 80%), with later uptake inversely proportional to the square root of time (466). N_2O is relatively nonreactive and poorly soluble in blood. This provides the desired properties of rapid onset of anesthesia and rapid clearance mostly through exhalation. Little hepatic or renal metabolism is detectable in experimental animals, although intestinal bacteria can reduce small quantities of inhaled N_2O to nitrogen gas (467). Small amounts of inhaled N_2O are also eliminated through the skin and urine (466, 468).

6.4 Toxic Effects

The major toxicological effect associated with N_2O is the depletion of vitamin B_{12} (cyanocobalamin), which is an essential cofactor in mammals for methionine synthetase and methyl malonyl CoA mutase (a mitochondrial enzyme that converts methylmalonic acid to succinic acid). Inactivation of methionine synthetase affects metabolism of folate, which, in turn, interferes with DNA synthesis and produces effects similar to those associated with vitamin B_{12} deficiency (particularly neuropathy and myelotoxicity, including megaloblastosis) (469). It is also possible that this mechanism may have effects during pregnancy (470–473). Even routine anesthesia causes significant oxidation of the vitamin B_{12} component of methionine synthetase, and particular concern arises in repetitive or extended (> 3-h) exposures or in individuals with vitamin B_{12} deficiencies (474). Occupational N_2O exposure of medical and dental personnel may produce adverse effects in B_{12}-deficient individuals or in those routinely exposed to high N_2O levels (475). Administration of folinic acid or methionine can protect against megaloblastosis and neurotoxicity occurring after N_2O administration.

In humans, chronic abuse has led to myeloneuropathy with symmetrical abnormal signals in the posterior columns of the cervical cord (476). This observation is supported by studies in rats with prolonged exposure to 1000 ppm N_2O, in which methionine synthetase is inactivated. However, at lower concentrations (150–450 ppm) blood methionine levels were unchanged in operating room personnel exposed routinely (469). These results must be reviewed with caution because blood methionine levels may not adequately reflect methionine synthetase activity, because dietary intake or methylation of homocysteine could offset decreased methionine production (477).

Repeated high level exposure to N_2O in occupational settings (dentistry) or in relation to N_2O abuse may lead to neurological damage similar to that seen in vitamin B_{12} deficiency, with progressive symptoms including numbness or tingling in the hands, legs, and trunk; poor balance; loss of finger dexterity; unsteady gait, Lhermitte's sign (a feeling like an electrical shock going up or down the back and legs caused by flexing the neck); leg weakness; impotence, spastic paralysis of bladder and bowel; personality change, depression; and memory loss (478–481). Nitrous oxide may also act as an asphyxiant if administered without supplemental oxygen, which has led to abuse-related fatalities in a variety of workplace settings where this substance is readily available, including healthcare facilities, laboratories, and restaurants (458).

Epidemiological investigations suggest that repeated exposures to anesthetic gases (including N_2O) may result in adverse reproductive outcomes including spontaneous abortion and possibly congenital anomalies among medical and dental personnel (459,

482, 483). Although most studies involved mixed exposures to several anesthetic agents, one large study examined several outcomes in dentists and dental assistants with exposure to N_2O alone (481, 482). In this investigation, N_2O exposure was associated with an increased rate of spontaneous abortion and congenital anomalies among assistants. In several animal studies N_2O exposure has produced fetal resorptions (similar to spontaneous abortion) and congenital anomalies (459, 484, 485).

Chronic exposure may also produce hepatic and renal effects (481, 482). Uncertainty remains whether N_2O is mutagenic or carcinogenic to humans or animals due to inadequate data (485, 486).

6.5 Standards, Regulations, or Guidelines of Exposure

The Occupational Safety and Health Administration (OSHA) lists N_2O as a simple asphyxiant. (The limiting factor is the available oxygen, which shall be at least 18% and be within the requirements addressing explosion in subpart B of part 1915.) NIOSH recommended exposure limit (REL) for N_2O is 25 ppm as a time-weighted average (TWA) during the period of anesthetic administration (487, 488). This REL is intended to prevent decreases in mental performance, audiovisual ability, and manual dexterity during exposures to N_2O. An REL to prevent adverse reproductive effects has not been established (awaiting additional data).

The American Conference of Governmental Industrial Hygienists (ACGIH) threshold limit value (TLV) for N_2O is 50 ppm as an 8-h TWA (194). The Documentation of Threshold Limit Values and Biological Exposure Indices states that "control to this level should prevent embryofetal toxicity in humans and significant decrements in human psychomotor and cognitive functions or other adverse health effects in exposed personnel" (489). There is no recommended short-term (peak) exposure standard (459). Urinary N_2O has also been suggested as a biologic index of exposure in operating room personnel (468).

6.5.1 Measurement (Exposure Monitoring)

NIOSH recommended guidelines to minimize worker exposures includes monitoring for N_2O when the anesthetic equipment is installed and every 3 months thereafter. These tests should include leak testing of equipment, monitoring of air in the worker's personal breathing zone, and environmental (room air) monitoring. A written monitoring and maintenance plan for each facility that uses N_2O should be developed by knowledgeable persons who consider the equipment manufacturers' recommendations, frequency of use, and other circumstances that might affect the equipment.

Older sampling methods include monitoring by grab sampling (with gas bag or syringe) for subsequent analysis by infrared spectroscopy or gas chromatography (490) or direct-reading instruments (491). During real-time sampling of personal exposure data, the sampling train inlet should be attached to the lapel of the worker on the side closest to the patient. N_2O concentrations in this location are most representative of those in the worker's breathing zone. Diffusive samplers (referred to as "passive dosimeters") are commercially available and may be useful as initial indicators of exposure (492). Testing of active cartridges and passive monitors over a wide range of concentrations, OSHA has deemed to have sufficient accuracy, precision, and convenience for use in the field (493).

6.5.2 Prevention

Prevention of N_2O exposure of personnel during biomedical applications includes use of gastight equipment (this excludes older equipment that cannot be rendered gastight) and testing for leakage before each use. All equipment should have a record of routine, periodic maintenance. Perhaps most important is a scavenging system that collects directly from the anesthetic breathing circuitry and vents N_2O and waste gases (457, 459, 484). Exhausts can be connected to outlets using a one-way, pressure-relief valve, to ventilators, and to nasal masks. Waste gas can be disposed by connection with the exhaust of nonrecirculating air-conditioning systems or by direct, dedicated exhaust lines connecting the breathing circuitry with the atmosphere, either passively or assisted by a vacuum system (458). The efficacy of control measures should be assessed frequently with appropriate air monitoring (457, 459, 465, 484). To reduce the hazards of N_2O unrecorded loss or abuse, inventory control and limited access should be instituted by employers. Cylinders containing N_2O should bear warning labels indicating the potential hazards of suffocation and neurological damage (458).

7.0 Oxygen

7.0.1 CAS Number: *[7782-44-7]*

7.0.2 Synonyms: *GOX, LOX, dioxygen*

7.0.3 Trade Names: *NA*

7.0.4 Molecular Weight: *31.9988*

7.0.5 Molecular Formula: *O_2*

7.0.6 Molecular Structure: *O=O*

7.1 Chemical and Physical Properties

Atomic weight: 32.00
Physical state: Colorless, odorless, tasteless gas
Boiling point: $-183.17°C$
Melting point: $-218.4°C$
Density: 1.429 g/L at STP, or 1.105 relative to air
Solubility in water: 0.0694 g/kg at 0°C
                           0.0325 g/kg at 37°C

7.2 Production and Use

Oxygen is the most prevalent element in the earth's crust, making up 49.2% by weight (494). It accounts for 20.95% by volume of the earth's atmosphere, and accounts for approximately 65% by weight of the human body. Industrially, O_2 has many uses, but its greatest importance is the dependence of most life forms on O_2 as a source of cellular energy. Because of this absolute requirement, a major use of O_2 is in the clinical treatment of disorders arising from, or resulting in, a lack of O_2 delivery to the cells.

7.3 Exposure Assessment

Oxygen enters the body primarily through the lungs, but may also be taken up by mucous membranes of the GI tract, the middle ear, and the paranasal sinuses. It diffuses from the alveoli into the pulmonary capillaries, dissolves in the blood plasma, enters the red blood cells, and binds to hemoglobin. The red cells transport bound O_2 to tissues throughout the body via the circulatory system. In tissues where the partial pressure of O_2 is lower than that of the blood, the O_2 diffuses out of the red cells, through the capillaries and plasma, and into the cells. As the O_2 plasma concentration diminishes, it is replaced by that contained in the red cells. The red blood cells are then circulated back to the lungs in a continuous recycling process.

On entry into the cells, the mitochondria use the diffused O_2 in chemical reactions that ultimately supply the energy required for cellular functions. Most O_2 combines with carbon and hydrogen atoms from glucose molecules to form cellular energy, known as adenosine triphosphate or ATP, along with carbon dioxide (CO_2) and water. The remaining O_2 is combined with various compounds to synthesize cellular structures or elimination products. The CO_2 generated in the cells then diffuses back to the red blood cells and returns to the lungs, where it is exhaled. The metabolic water combines with ingested water and the excess is eliminated by excretion through the kidneys or by evaporation from the lungs and skin.

For each liter of O_2 used, approximately 5 kcal (~ 20 J) of energy is liberated (511). However, with increasing exertion (walking to competitive skiing or running), this figure progressively increases from 4.8 to 5.6 kcal/L (511). Similarly, Grandjean reported an average of 4.8 kcal/L, but a wide variation among individuals has been reported (512).

The amount of O_2 consumed in various activities ranges from about 0.1 L/min for resting in bed, fasting, to as much as 5 L/min for brief periods of maximal exertion. Light work consumes less than 1 L/min, whereas manual labor (as in farming, mining, or brisk walking) consumes about 1.5 L/min. Heavy manual labor, such as climbing stairs or running 5.5 mph (mi/h), burns about 2 L/min. Very heavy work, like running 7 mph or vigorous swimming, uses almost 3 L/min, and running 8 mph consumes about 3.5 L/min. For a given activity level, larger, heavier persons will use more O_2, whereas physically fit persons will use less O_2 and be capable of greater maximal exertion.

The maximum capability for O_2 use, and hence the maximum rate of physical work, increases up to an age of about 25 years, then slowly declines. By age 65, about 25% of the maximal work capacity has been lost. Women on average have about the same work capacity as men in proportion to their muscle mass. However, because women have a greater proportion of body fat, they have lower capacities with respect to their size and weight. European studies reported before 1972 indicated that the maximum for women was 65–75% that for men, but the applicability of these estimates to other times and places is unknown. For additional information on the physiology of work, publications by Grandjean (512), Rodahl (511), and Åstrand and Rodahl (513) are suggested.

7.3.1 Oxygen Deficiency

Tissue oxygen deficiency (hypoxia) can arise in numerous environmental and clinical situations. Potentially dangerous areas include any poorly ventilated space where the air may be diluted or displaced by gases or vapors of volatile materials, such as tanks, vats, holds of ships, silos, and mines. Similarly, areas where O_2 may be consumed by chemical or biologic reactions can be dangerous. It is recommended that a worker entering any such place wear a lifeline that is continually attended by another person who has no other duty. In addition, the affects of decreased O_2 pressure are a concern in high altitude conditions, including aviation, mountain climbing, or inhabitants of high altitude areas. Clinically, a variety of acute and chronic medical disorders can result in a lack of delivered O_2 to the tissues.

7.3.1.1 Effects of Hypoxia. The normal 21% of O_2 in the air provides an O_2 partial pressure (P_{O_2}) of ~160 mm Hg at sea level. If the P_{O_2} is reduced below 120 mm Hg, equivalent to a sea-level atmosphere with 16% O_2, impairment of mental performance is soon evident. *Hypoxia* may be defined as a decrease in O_2 delivery to the cells resulting in an energy production that is below the cellular requirements. The effects of progressive hypoxia are summarized in Table 47.4.

Hypoxic effects are usually undetectable when P_{O_2} is greater than 120 mm Hg. Interference with the adaptation of the eye to darkness, however, can occur at this pressure. Several textbooks suggest that a minimum safe level is 17–18% O_2 (495–497), yet the OSHA standard is 19.5% (498).

In the automatic control of breathing, the need for ventilation is measured by chemical sensors called *chemoreceptors*, which influence the respiratory control center in the brain. The control center regulates the respiratory muscles for breathing. The chemoreceptors are located in the carotid bodies (closely attached to the carotid arteries in the neck) and aortic bodies (located near the aortic arch) and send impulses to the brain to modulate the rate of breathing. These sensors respond weakly to changes of P_{O_2} greater than 200 mm Hg. When the arterial P_{O_2} falls below 100 mm Hg, the rate of chemoreceptor signaling increases rapidly and, at P_{O_2} 30–60 mm Hg, the ventilation rate increases three- to sixfold. The peripheral chemoreceptors are responsible for all the increase in ventilation that occurs as a result of arterial hypoxemia. The sensors are influenced mainly by the partial pressures of

Table 47.4. Response of Humans to the Inhalation of Atmospheres Deficient in Oxygen

Stage	Oxygen Vol. (%)	Symptoms or Phenomena
1	12–16	Breathing and pulse rate increased, muscular coordination slightly disturbed
2	10–14	Consciousness continues, emotional upsets, abnormal fatigue upon exertion, disturbed respiration
3	6–10	Nausea and vomiting, inability to move freely, loss of consciousness may occur; may collapse and although aware of circumstances be unable to move or cry out
4	Below 6	Convulsive movements, gasping respiration; respiration stops and a few minutes later heart action ceases

CO_2 and O_2 in the arterial blood and its acidity (pH). These factors are affected by the level of physical activity, individual physical fitness, nutritional state, emotional state, ambient temperature, and air supply, and many other factors. Also, these factors affect each other: blood pH is strongly influenced by P_{CO_2}, for example, and the flow of blood through the sensors is affected by peripheral vasoconstriction associated with hypoxia.

The physiological response to hypoxia is complex. Hypoxia acts to increase the respiratory rate, heart rate, and cardiac output to maintain arterial P_{O_2} and ensure adequate tissue oxygenation. Constriction of peripheral arteries increases the supply of blood to the brain and other central organs. This is counterbalanced by constriction of brain arteries due to P_{CO_2} reduction that follows increased ventilation. An increase in P_{CO_2} enhances the sensitivity of the P_{O_2} receptors synergistically, so if P_{CO_2} is increased to 48.7 mm Hg, a P_{O_2} near 100 mm Hg increase ventilation. A strong element of physiological adaptation or acclimatization also affects these relationships, and compensatory adjustments to low P_{O_2} conditions occurs within hours or a few days in many cases (499–501).

A ventilatory response to hypoxia usually occurs under extreme conditions, as in work or exercise at high altitudes or in diseases associated with compromised respiratory or cardiac function. The uptake of O_2 through the lung decreases linearly with the decrease in ambient P_{O_2}. If the deficiency of O_2 is sudden and severe, consciousness is lost within 5–10 s and permanent brain injury is likely within 2 min (500, 501). If the onset of hypoxia is gradual, as in an ascending aircraft, talkativeness and a false sense of well-being often precede loss of mental coordination and physical signs (502).

According to the pioneer physiologist Paul Bert in 1878, a dulling of perception and slowing of intellectual activity were among the earliest consequences of hypoxia observed in mountain climbers and balloonists (503). The brain is the organ most susceptible to the effects of hypoxia because it has a high metabolic rate, has no O_2 storage reserves, and is incapable of anaerobic metabolism (501, 504). Air travel, space exploration, military aviation, and physicians for high altitude sports have focused attention on the effects of hypoxia on intellectual performance. The time of useful consciousness associated with exposure to various altitudes has been estimated by several investigators (505, 506). A wide variation in responses among individuals has been described, and wide diurnal variations probably also exist. Generally, some impairment will gradually develop, even in healthy people, at altitudes of $\geq 10,000$ ft ($\sim$3050 m). At higher altitudes, with greater O_2 deprivation, there is less delay and the effects are more severe. Table 47.5 summarizes some estimates of the effects of sudden exposures to high altitudes.

An imperceptible effect of hypoxia is impairment of visual sensitivity, which has been demonstrated with an ambient P_{O_2} of 142 mm Hg (equivalent to 1000 m or about 3300 ft altitude). This suggests that other CNS functions may be impaired, but extensive testing of various psychomotor activities has shown that few healthy young men have detectable changes in performance below an altitude of 10,000 ft (3050 m), where the ambient P_{O_2} is 112 mm Hg (502). A review of the uses of high altitude to study the effects of hypoxia has recently been reported (507).

Hypoxia has also been shown to affect the action of drugs and their metabolism. Jones et al. reviewed important effects of hypoxia on the actions of various medications, which may be either inhibitory or synergistic in complex ways (508). The effects of hypoxia on drug metabolism and drug resistance have also been reviewed (509, 510).

Table 47.5. Altitude, Hypoxia, and "Time of Useful Consciousness" (TUC)

Altitude (ft)	Altitude (m)	Barometric Pressure (mm Hg)	O_2 Partial Pressure (mm Hg)	TUC (sec)
50,000	15,240	126	26	15
40,000	12,192	181	38	20
35,000	10,668	216	45	26
30,000	9,144	259	54	65
25,000	7,620	310	65	132

Source: Modified from a graphic presentation by U.S. Air Force, 1968. Other sources indicate similar or slightly longer times.

7.3.1.2 Mountain Sickness. Although the percentage of O_2 in the inspired air remains constant at different altitudes, the lower atmospheric pressure at higher altitudes decreases the P_{O_2}, and thus the driving force for gas exchange in the lungs. Rapid ascent by humans to an altitude of ≥ 2500 m (~ 8200 ft) causes an illness known as *acute mountain sickness* [acute benign mountain sickness (AMS), puna, soroche] in many people (514). The etiology of AMS is poorly understood, but the likely cause is the hypoxia brought about by exposure to a $\leq P_{O_2}$ of ≤ 119 mm Hg, for periods of time greater than about 4 h. AMS can be avoided by ascending gradually over several days to give your body time to acclimate (acclimatize) to the higher altitude conditions. AMS can also be avoided by ascending slowly with frequent or continuous exercise, or less effectively, by minimizing physical exertion during the first 2 or 3 days after reaching diminished O_2 exposure. At altitudes above 3000 m, it is recommended that climbers ascend no more than 300 m per day with a rest every third day.

AMS is dispelled by a dual process of acclimatization (physical) and accommodation (psychosocial); people do achieve a partial physical adjustment, and they learn their limits and observe them. The prescription drug Diamox® (acetazolamide) has been shown to accelerate the acclimatization process and can be taken shortly before and during the ascent. Acetazolamide aids in the retention of CO_2, which helps to avoid the alkalosis that results from increased breathing. It thereby promotes the release of O_2 from hemoglobin and acts to correct other physiological disturbances (515). One study determined that low dose dexamethasone with Diamox® relieves symptoms of AMS better than does Diamox® alone (516), although use of dexamethasone in AMS is still controversial.

During ascent to higher altitudes, an increased rate and depth of breathing, along with shortness of breath on mild exertion, and an increased pulse rate may be noticed. These normal, healthy manifestations of accommodation to hypoxia can be relieved promptly by breathing an atmosphere with enhanced O_2 content, which is at least 160 mm Hg (514).

After an interval of about 6–12 h to several days of sustained hypoxia, insomnia, headache, loss of appetite, impaired mental concentration, nausea, vomiting, muscular weakness and fatigue, cough, chest pain, and heart palpitation may ensue. A wide variation among individuals in the severity and duration of illness has been reported. The condition is most severe 2 or 3 days after ascent; and most people recover completely by the fourth or fifth day. Some persons, however, are troubled for weeks or even months after a severe hypoxic episode.

Among apparently healthy people, no one has yet found reliable clues to identify susceptible individuals. Strength and physical fitness offer no protection; in fact, athletes may attempt to "work off" the early symptoms and thereby make things worse for themselves. Almost anyone may be susceptible if the ascent to an extreme altitude is rapid. Women and men appear to be affected equally.

The normal responses of acclimatization to high altitude include, in addition to increased respiration and increased heart output, an increase in the number and concentration of red blood cells and hemoglobin, with a consequent improvement in the delivery of O_2 to the brain and other tissues. Additionally, several complex adjustments of blood and tissue acid–base balance and water and electrolyte distribution occur. An increase in urine production is also observed. For a detailed discussion of AMS and physiologic responses to high altitudes, see Ward et al. (514), Hultgren (517), or Houston (515).

A life-threatening acute illness that may be encountered at altitudes over about 3000 m (or 10,000 ft) is *high altitude pulmonary edema* (HAPE), which may or may not be preceded by symptoms of AMS. HAPE is associated with vigorous exertion and, fortunately, is uncommon. Persons responsible for the health of workers and others in high altitudes should be aware of its risks. HAPE is characterized by cough, labored breathing, and collapse. If left untreated, HAPE can progress rapidly and be fatal. Oxygen administration, competent medical treatment, and removal to a lower altitude are urgently required (514, 515, 517, 518).

When immediate descent is impossible and supplemental O_2 is not available, treatment with portable hyperbaric chambers may be used (519), and initiation of calcium channel blockers (nifedipine) or vasodilators (hydralazine or phentolamine) is recommended until descent is possible (520, 521).

High altitude cerebral edema (HACE) is another uncommon manifestation of high altitude illness, appearing initially as an exaggeration of AMS, but progressing to severe headache, muscular incoordination (ataxia), irrationality, drowsiness, confusion, stupor, and coma. Other signs and symptoms that may occur include nausea, vomiting, hallucinations, cranial nerve palsies, seizures, and sometimes visual symptoms. Progression of symptoms also creates the potential for fatal accidents, such as falls. Neurological symptoms of altitude illness can progress from those of AMS to coma within 12–72 h. Like HAPE, HACE calls for O_2 administration, emergency medical care, and prompt removal to a lower altitude (514). Additionally, administration of dexamethasone, 10 mg intravenously, followed by 4 mg every 6 h appears beneficial if initiated early in the course of illness. For concise clinical reviews of altitude-related illnesses, see Klocke et al. (522) and Peacock (523).

Chronic mountain sickness (Monge's disease) is a disorder that affects high mountain natives and other long-term residents of high altitude regions (>4000 m). The primary characteristic is inadequate breathing (hypoventilation), which lowers the already inadequate oxygenation of the blood and leads to excessive production of red blood cells (polycythemia), with consequent thickening of the blood. Damage to the lungs and heart may follow, including right ventricular hypertrophy and the corresponding electrocardiogram alterations. Removal to a lower altitude can stop the process. If descent is not possible, venesection has been shown to be beneficial, improving exercise performance, pulmonary gas exchange, and many of the neuropsychological symptoms (524). Persons with arteriosclerosis and high blood pressure may be more susceptible to the disease than others. Monge's disease is described in a comprehensive monograph by Winslow and Monge (525).

Atmospheric pressures, O_2 partial pressures, and pressures of inspired air in the distal airways of the lung are indicated in Table 47.2. Atmospheric pressures change exponentially with altitude and are influenced not only by the depth of the column of air above the place of measurement but also by the temperature of the air and other factors. Although the standard atmosphere tables prepared by U.S. and international agencies are commonly used for prediction of pressures at various altitudes, Pugh has shown that the computation described by Zuntz et al. in 1906 has been more accurate for ground-based measurements (526). The differences are small, but are significant at altitudes greater than 6000 m.

7.4 Toxic Effects

7.4.1 General Information

Oxygen, essential for mammalian life, is paradoxically harmful. If O_2 is given at high enough concentrations for long enough times, the body's protective mechanisms are overwhelmed, leading to cellular injury and, with continued exposure, even death. In the course of O_2 metabolism, several toxic substances are generated, including superoxide anion (O^{2-}), hydrogen peroxide (H_2O_2), hydroxyl radical (OH^-), lipid peroxides, and others. Without the availability of several enzymes that destroy these toxic intermediary compounds, cell death quickly occurs. The protective enzymes include superoxide dismutases (SODs), catalase (CAT), and glutathione peroxidase (GP). Glutathione reductase (GR) participates by re-forming glutathione, which is preferentially oxidized, thereby sparing sulfhydryl-bearing proteins and cell wall constituents. Other contributors to the control of oxidant toxicity include vitamin C (ascorbic acid), vitamin E (α-tocopherol), vitamin A, and selenium, a cofactor for GP.

Normally, a balance exists between the production of toxic oxidants and their destruction by antioxidant mechanisms. Some individuals may lack the ability to produce sufficient antioxidants and suffer a slow progressive tissue deterioration as a result. Although the benefits are equivocal, antioxidant supplies can be enhanced by the administration of vitamins A, C, and E or by specific antioxidant enzymes such as SOD.

Lodato and Floyd have provided detailed discussions of the chemistry involved (531, 535). Other helpful reviews include those of Balentine (504), Clark (536), Fridovich (537), Jamieson (530), Klein (532), Lambertson (538), Olson and Kobayashi (539), and Witschi (533).

Industrial exposures to high O_2 pressure are uncommon. Sea diving and aeronautics are probably the most frequent. For prolonged deep-sea operations, the techniques of using special gas mixtures of helium (He), O_2, and nitrogen (N_2) have become general practice and are highly developed. In some instances, astronauts are exposed to pure O_2 at a pressure of ~0.2 ATA, simulating sea-level P_{O_2}. Caisson workers and tunnel makers may also be exposed to pressures that are high enough to cause lung damage. For those people concerned with these operations, it is imperative to know the associated risks of O_2 toxicity. These are discussed in Chapter 9, "Physiological Effects of Abnormal Atmospheric Pressures," in Volume 1 of the third edition of this series.

It appears that inhalation of pure O_2 for ≤ 12 h will do no serious harm. Its use in some emergencies is justified. However, to use O_2 automatically in certain circumstances may not be prudent. The risks of handling cylinders of compressed gas and the possibility of fire must always be remembered. It is also important that appropriate precautions not be overlooked in the haste to commence O_2 therapy.

The increase of O_2 partial pressure in the blood associated with hyperoxia leads to slower, shallower breathing. This, in turn, leads to accumulation of CO_2 in the tissues and the blood, with the further effect of increasing acidity. The decreased respiration rate in combination with the increased P_{O_2} can also cause some alveoli to collapse completely, a condition called *absorption atelectasis*. With this, the adjustment of the flow of blood through the lung to match the air supply is upset, producing a ventilation/perfusion mismatch. The pulmonary arterioles dilate, the heart rate is slowed, and the rate of blood flow from the heart decreases. This is only one example of the complex array of interactions attributed to hyperoxia, as reviewed by Balentine (504).

Jamieson points out that the foregoing constellation of oxidant injuries is not unique to hyperoxia, but is encountered also with reperfusion injury following an ischemic episode, cancer, the action of some anticancer drugs, inflammation, surgical shock, and in humans, Parkinson's disease (530). If the O_2 exposure has not been lethal, the epithelial cells and the capillaries regenerate, fibroblasts multiply, and healing follows, but with a price — the interstitial portions of the lung are thickened with fibrous scar tissue (fibrosis), and the elasticity of the lungs is impaired (531–534).

7.4.1.1 Injury Mechanisms. Endogenous oxidants, generated within the lung as part of normal metabolic processes, damage cells by modifying essential materials such as deoxyribonucleic acid (DNA), proteins, and lipids, and by adversely affecting cellular organelles including the mitochondria and nucleus. These oxidants can also leak from damaged cells and attack adjacent membrane lipids. The amount of hydrocarbons (ethane, pentane) in one's expired air has been suggested as an indicator of oxidative damage (i.e., lipid peroxidation) (541).

The most prominent adverse respiratory consequences of O_2 exposures with partial pressures of < 3 ATA include epithelial swelling and breakdown, blood vessel congestion, interstitial edema and alveolar exudate. These changes, known as the *Lorain Smith effect*, were first described in mice in 1899 (542). In contrast, the CNS-induced convulsive reactions associated with higher O_2 pressures (hyperbaric O_2) were described earlier by Paul Bert in 1878.

Microscopic injury to the epithelial cells lining the trachea and bronchi occurs early in the sequence of O_2 damage and response. Oxygen exposure depresses mucus secretion and the "upstream" movement of mucus (mucociliary transport) to the throat; such decreases have been observed after only 3 h of exposure to 90–95% O_2 (531, 532, 534). Other early manifestations of O_2-induced lung injury include leakage of proteins from the blood into the air spaces and increased permeability of the lung epithelium. These have been demonstrated by studies on bronchoalveolar lavage (BAL) fluid retrieved from humans. Davis et al. exposed healthy subjects to >95% O_2 for 17 h; they detected increases in the blood derivatives albumin and transferrin, although the recovered lung cells appeared normal (543). Griffith et al. reported dose-related increases in albumin in BAL fluids and increased epithelial permeability after 45-h exposures to 30–50% O_2 (544).

Protein accumulation in the BAL fluid results from damage to the pavement-like cells that line the alveolar sacs, known as type I cells, which cover ~95% of the alveolar surface. Type I cells are generally incapable of dividing, but when damaged, can be replaced by the type II alveolar cells interspersed among them. Type II cells are less susceptible to toxic injury, can proliferate rapidly, and can be transformed into type I cells. Toxic injuries that affect only type I cells can be repaired by this proliferative process. To the extent that type II cells are also injured, the effects are more severe and may lead to permanent changes. Other types of cells in the lung are also affected, especially the capillary endothelial cells, leading to leakage of blood plasma into the interstitial tissue between the alveoli, and ultimately into the alveoli. Blood cells in the capillaries may also form a clot or may leak into alveolar spaces (hemorrhage). Other cells in the interstitium, such as fibroblasts, are damaged. An inflammatory response, with infiltration of white blood cells, proliferation of fibroblasts, and subsequent fibrosis may follow (504, 531, 532, 534).

Alveolar epithelial and capillary endothelial cells are the main metabolic contributors of free radicals in response to hyperoxia. Other cell types are also involved in free-radical generation, including interstitial macrophages, lymphocytes, and polymorphonuclear neutrophils, which accumulate in response to irritation. Synergistic interactions were reported between hyperoxia and neutrophils in producing lung injury (545). However, Kubo et al. showed that the presence of neutrophils is not required for the progression of hyperoxic injury in sheep (546). On the other hand, the edema induced by hyperoxia could be reduced by depleting rabbits of granulocytes (547). Therefore, the role of the neutrophil in hyperoxia-induced injury is still unresolved.

The activity of intracellular antioxidant enzymes is quickly increased in response to hyperoxia exposure. Jornot and Junod observed that these increases in enzymatic activity are associated with increases in the messenger ribonucleic acids (mRNAs), leading to the production of SOD, CAT, and GP proteins (548). This was shown by *in vitro* cultures of human endothelial cells exposed to 95 percent O_2 for 3 or 5 days. The mRNA changes were accompanied by complex changes in enzyme protein concentrations and activities, generally in the direction of protective reactions to hyperoxia. The mechanism of stimulating RNA activity was not explained (548).

7.4.2 Experimental Studies

Experimental studies of O_2 toxicity have been performed to a limited extent in humans, but in much greater detail in other mammalian species, including hamsters, rabbits, cats, dogs, sheep, monkeys, and, most commonly, mice and rats. Because the outcome of hyperoxia can be devastating, animal models are often used to determine pathologies and mechanisms, so that parallels in humans can be tested.

The biochemical and metabolic basis for O_2 poisoning of animal lungs was reviewed by Mustafa and Tierney (540). In this report, they also discussed the similarities of O_2, ozone, and NO_2 effects on the lung, and the likely mechanisms of these oxidants to induce pulmonary injuries.

Oxygen poisoning was described by Paul Bert in 1878 (503). He observed convulsions in animals exposed to pure O_2 at 3–4 ATA (atmosphere absolute) pressure, as well as other effects at lower pressures. He concluded that O_2, at pressures greater than 1 ATA, is a poison. Several excellent reviews of O_2 toxicity have been published (527–529).

7.4.2.1 Acute toxicity. Porte et al. performed detailed histological studies of lungs from adult rats exposed to 100% O_2, and found bronchiolar and vascular constriction after only 24 h, confirming several earlier reports. They suggested a nerve-mediated reflex action to explain this prompt response and further suggested that even a 24-h exposure to 100% O_2 is unsafe (561).

Crapo et al. reported that rats, at the time of death from 100% O_2 exposure, showed destruction of 44% of the capillary endothelial cells. No loss of alveolar epithelial cells was mentioned in this report, but animals exposed to 85% O_2 showed proliferation and hypertrophy of type II alveolar epithelial cells, along with 41% destruction of capillary endothelial cells (562).

De los Santos reported that in baboons, alveolar type I epithelium was almost completely destroyed after 4 days in 100% O_2, and after 7 days, type II cells had almost completely replaced the alveolar lining (563).

Relatively late in the course of normobaric hyperoxic exposures, the amount of surfactant in the lungs is reduced (532). This leads to an increased surface tension, making the alveoli resistant to expansion and more likely to collapse, thereby reducing the surface available for gas diffusion. Matalon et al. exposed rabbits to 100% O_2 and showed pathological changes after 24 h. A progressive increase in pulmonary capillary permeability was seen by 48 h, and after 64 h, decreased phospholipids (surfactant), decreased lung capacity, pulmonary edema, and high surface tension in lung fluids were noted (555). Instillation of calf lung surfactant into the rabbit lungs during the O_2 exposure diminished the injury (555). Matalon et al. subsequently reported that surfactant has scavenging powers for reactive oxidants in addition to its primary surface-tension reducing function (556).

Horowitz et al. observed an increase of mRNA encoding for surfactant protein in the lungs of newborn rabbits exposed to 100% O_2 for 96 h (557), and Veness-Meehan et al. reported an increase of mRNA encoding for surfactant apoprotein in O_2-exposed type II alveolar cells (558). In rats, a 12-h exposure of 95% O_2 was enough to cause a sharp decline in surfactant mRNA and protein expression (559).

Fracica et al. reported comparative studies of pre- and postexposure lung biopsies from rats and baboons, using 40–60% O_2 for periods of ≤ 80 h. Changes were hard to detect in the animals exposed to 40% O_2, but injuries became evident when the animals were subsequently exposed to a different form of lung stress. It was concluded that the basic reaction pattern was the same for both species, but the primate response was slower. They also observed that rats earlier exposed to 40 or 60% O_2 died sooner than did control animals when they were later exposed to 100% O_2 (560).

7.4.2.2 Strain and Species Differences. Marked differences in susceptibility to O_2 poisoning have been reported between mice and rats; little is known about comparisons among other species. Larger animals, such as dogs and monkeys, are less susceptible to hyperoxic injury than smaller ones, especially rodents. Older rats are more sensitive than were younger rats, and heavier rats are more sensitive than lighter ones. Rats reduced in weight by dietary restrictions are also less sensitive than heavier rats of the same age. Northway et al. (564) and Bonikos et al. (565) observed that newborn mice are notably less susceptible to the lethal effect of undiluted O_2 than were adult animals.

Interstrain variations in hyperoxic injury have been reported in mice and rats. Among common laboratory strains of inbred mice, differential responses have been shown for hyperbaric and normobaric oxygen exposures. Hill et al. found significant differences in survival time for 20 inbred mouse strains in hyperbaric oxygen conditions (566). Hudak et al. reported a significant difference between common inbred mouse strains in their susceptibility to an inflammatory response to hyperoxic injury (567). Protein in BAL was 10-fold higher in the most susceptible mouse strain compared to the most resistant. In rats, several investigators have reported on variation in response to hyperoxia (568, 569).

As stated previously, prevention of cell injury and death by oxidant products of normal metabolism rests on the actions of several antioxidant enzymes, SOD, CAT, GP, and others. However, the results of various individual experiments with different species are not consistent. Frank et al. (570) exposed newborn and adult animals of five species to $> 95\%$ O_2 for 7 days. Measurements of antioxidant enzymatic activity in the lungs (SOD, CAT, GP) showed no significant responses in adult animals to a 24-h exposure, nor were these enzymes increased in newborn guinea pigs or hamsters. Newborn mice, rabbits, and rats showed significant increases in all the enzymes ranging from 12 to 36% (except that mice showed only a 5% increase in CAT). The exposure time to cause 50% mortality was much greater for the newborn mice, rats, and rabbits than for the adults, but the newborn guinea pigs and hamsters only lived as long as the adults of the same species.

Significant variation in lung antioxidant activity has been reported for GP, SOD, CAT, and glutathione *S*-transferase (GST) in rat, hamster, baboon, and human lung (571). SOD activity was similar for all four groups, and GP activity was higher in rat lung than baboon or hamster lung. CAT activity was variable; the highest activity was present in the baboon, which revealed a lung CAT activity 10 times higher than activity present in the rat. Lung GST activities were higher in hamster, baboon, and human lung than in rat lung. From these data it was concluded that the hamster was the best model of the animals studied for mimicking human lung antioxidant enzymatic activity and that rat lung antioxidant enzymatic activity was markedly different from that in any of the other species examined. In a separate report, the oxidative capacity of alveolar macrophage from rats and hamsters differed significantly, whereas the antioxidant capacity was similar for the two species (572).

Arieli concluded that O_2 toxicity is not related to body size, and that small mammals can serve as good models for O_2 toxicity in humans (573). However, Witschi reported substantial differences in the patterns of posthyperoxic lung cell proliferation among mice, rats, hamsters, and marmosets, with the reactions in mice and marmosets resembling the alveolar epithelial responses seen in humans (533). Rats predominantly showed a capillary endothelial response. Jackson cited similar reports of species differences (534). This suggests that extrapolation of observations in rats may give misleading estimates of O_2 toxicity in human lungs. This correlates with findings of Bryan and Jenkinson described above for the differences in antioxidant activities in rats compared to other species (574). Nevertheless, observations in rats have been widely cited as predictors of human responses.

7.4.2.3 Mouse Models of Oxidative Injury. A number of studies have utilized the biotechnological breakthroughs in the 1990s to eliminate (knockout) or boost (by overexpressing a gene) important antioxidant function in laboratory animals. Examples of this technology, as it relates to the role of antioxidants in O_2 toxicity, are studies of transgenic and knockout mouse lines for some of the SOD enzymes. Transgenic mice that overexpress copper/zinc-SOD (549) or manganese-SOD (550) are partially protected from hyperoxic injury. However, Ho did not demonstrate protection against O_2 toxicity in transgenic mice overexpressing manganese-SOD, although he used a different promoter to drive expression (551). When the overexpression of extracellular copper/zinc-SOD was specifically targeted to alveolar type II cells and nonciliated bronchial, epithelial cells, mice showed reduced inflammation and lung toxicity after hyperoxia (552). Correlating with a protective role for this enzyme, knockout mice lacking extracellular copper/zinc SOD were more sensitive to hyperoxia (553). In total, these studies highlight the great potential of the molecular technologies for gaining insights into the complex lung response to O_2 insult. For a review of the role of SODs in O_2 toxicity, see Tsan (554).

7.4.2.4 Interactions. Witschi (533) urges research aimed at understanding the biological mechanisms leading to the pathological consequences of exposures to two or more agents, simultaneously or in close association. For example, in rats, hyperoxia enhanced the damage of paraquat to type II cells, while capillary endothelial cells were relatively unaffected; if paraquat had enhanced O_2 toxicity, the capillary cells should have been the more affected, because in rats these cells are the ones most affected by hyperoxia alone. Also, some anticancer drugs, for example, bleomycin and cyclophosphamide, show increased acute lung damage in animals in combination with hyperoxia. Whether this is also true in humans is uncertain. It appears that direct interaction between O_2 and the anticancer drugs occurs, rather than interference by O_2 with the repair of damage done by the drugs. Witschi also gives other examples (533).

Hyperoxia and nitric oxide act synergistically in their cytotoxicity to A549 cells (human alveolar epithelial cells) (586). This interaction is of human relevance because nitric oxide is being tested clinically for infants and children with pulmonary hypertension, and these patients are also frequently receiving simultaneous O_2 therapy (587).

Coalson et al. exposed carefully maintained, anesthetized, intubated, and mechanically ventilated baboons for 11 days to 40% O_2, 80% O_2, 100% O_2, 100% O_2 preceded by oleic acid inhalation, or 80% O_2 followed by inoculation with *Pseudomonas aeruginosa* bacteria. The groups exposed to 100% O_2 or 80% O_2 *plus* bacterial inoculation showed mixed exudative–reparative diffuse alveolar lesions, altered morphology of type II cells, increased numbers of type II cells and interstitial cells, and decreased numbers of type I and endothelial cells. The animals exposed to 40% O_2 or 80% O_2 showed increased numbers of alveolar macrophages and focal widening of alveolar walls. The most striking finding was that 80% O_2 plus infection caused responses as severe as 100% O_2. It was suggested that the combination with infection reflects the evolution of adult respiratory

Table 47.6. Agents That Increase Oxygen Toxicity

Name	Refs.
Catecholamines	
Epinephrine	90
Norepinephrine	
Corticosteroids	
Dexamethasone	91, 92
Methylprednisolone	93, 94
Hormones	
Testosterone	95
Thyroxine	96
Chemotherapeutics	
Bleomycin	97
Cyclophosphamide	98
1,3,-bis(2-chloroethyl)-1-nitrosourea	99
Antibiotics	
Nitrofurantoin	100, 101

Source: Modified from *Clin. Chest Med.* **11**, 73–86 (1990).

distress syndrome that occurs in some human patients in intensive medical care units (and other situations involving lung injury) (588).

Jackson (534) states that many drugs affect O_2 tolerance, most of them unfavorably. These include drugs and chemicals that increase tissue O_2 consumption, agents that themselves produce free radicals, or those that interfere with the activity of antioxidant enzymes. These compounds are summarized in Table 47.6.

Species differences are also important in the investigation of interactions of agents with hyperoxia. Whereas bleomycin potentiated lung disease in the rat (589), it did not alter toxicity of O_2 in rabbits (590). A difference in drug administration (subcutaneous versus intravenous injection) could explain all or part of this differential response.

7.4.2.5 Prevention of Oxidative Injury. Bacterial lipopolysaccharide (LPS), or endotoxin, a constituent of the cell wall of Gram-negative pathogenic bacteria, has been one of the most effective agents against hyperoxic lung injury in experimental animals (591). Even when injected 36 h after the beginning of exposure to 95% O_2, it has provided 95% protection against death in rats, which otherwise showed only 76% survival after 72 h. Frank summarized these observations and similar previous results reported by others (592). Endotoxin itself, however, is highly toxic; Frank has succeeded in producing a modified endotoxoid that retains the protective action against hyperoxic injury, while the general toxicity is reduced to $\frac{1}{250}$ th of the original. Endotoxin has also been found to extend the survival time of adult mice in hyperoxia, although the number of mice used in this study was small (593).

Smith (594) reported that diphosphoryl lipid A extracted from the endotoxin produced by *Salmonella minnesota* protected rats from hyperoxic toxicity as effectively as did the parent endotoxin, and was less acutely toxic. Nontoxic lipid fractions from the endotoxin potentiated the hyperoxic injury (594). In a subsequent study, the protection from O_2 effects by the diphosphoryl lipid A component of endotoxin was confirmed, and the monophosphoryl lipid A component was shown not to have O_2 protective properties (595).

Kobayashi et al. (596) reported marked prolongation of survival time and partial prevention of increased lung vascular permeability in sheep given endotoxin just prior to 100% O_2 exposure. However, impairment of gas exchange, loss of hypoxic pulmonary vasoconstriction (normally seen with low O_2 concentrations), and ultimate respiratory failure were not prevented.

The effects of repeated exposures to high concentrations of O_2 with intervals of air breathing have been equivocal. Several studies have shown that a sublethal exposure to 80–85% O_2 for 5–7 days could increase the tolerance for a subsequent exposure to $>95\%$ O_2 (562, 597, 598). However, Fracica et al. reported that rats previously exposed to 40 or 60% O_2 died sooner when they were subjected to 100% O_2 than did rats not previously exposed to O_2 (560). Paradoxically, Frank also observed that preexposure of rats to *hypoxia* (10% O_2) for 3–5 days stimulated increases in lung SOD, CAT, GP, and glucose-6-phosphate dehydrogenase activities and was strongly protective against lung damage by hyperoxia (599). Mice and hamsters similarly treated, however, showed no increases in enzyme activities, nor were they protected from injury. Frank cites evidence for the production of superoxide anion and peroxide as a response to hypoxia; this would be a stimulus for the production of antioxidant enzymes.

After a review of the available literature, Frank et al. proposed a protocol for experiments utilizing intermittent room air breathing during continuous O_2 exposures. The preliminary exposure should be 48 h in $>95\%$ O_2, followed by a "rest period" of breathing 21% O_2, before continuous exposure to $>95\%$ O_2 again. With rest intervals of 6 or 12 h, 100% survival of rats was achieved for 72-h exposures to $>95\%$, whereas the survival for control animals without preexposure was 33%. Rest periods in which the rats breathed 50% or 70–75% O_2 for 24 h were equally effective (600).

Winter reported a twofold increase in survival time of rabbits from hyperoxic exposure following the intravenous injection of the lung irritant oleic acid (601). In this report, Winter suggested that a variety of lung injuries can increase production of antioxidants. Efforts have been made to enhance the supply of antioxidant materials in the lungs and other tissues by feeding or treating with antioxidants or their precursors. This has been done both to demonstrate the role of the antioxidants in O_2 poisoning and to search for protective agents.

Specific antioxidant enzymes such as SOD and CAT have been used experimentally in rats, by intratracheal instillation or insufflation or liposome-encapsulated particles. They produced significant protection, as shown by Hoidal et al. (602) and by Thibeault et al. (603). Hoidal et al. also reported effective protection by the intratracheal instillation in rats of washed human or rat erythrocytes. They characterized the erythrocytes as "biologic packets of encapsulated antioxidants" (602).

Acetylcysteine is known to have an antioxidant activity. Wagner et al. gave *N*-acetylcysteine intravenously to anesthetized, mechanically ventilated dogs, just before and during 100% O_2 exposure for 54 h. Using both functional and structural criteria, the treated dogs showed significant protection compared to controls. *N*-acetylcysteine functions as a surfactant when administered by inhalation, but it is not clear whether such action affected this experiment (604). N-acetylcysteine also protected guinea pigs from hyperoxia-induced lung injury (605) but did not protect rats from lung injury induced by hyperoxia (606).

Protein replacement studies have also determined the importance of a variety of genes in hyperoxia-induced injury. Surfactant proteins have long been thought to play a protective role in lung injury. A *surfactant* is an agent that lowers surface tension of fluids in the lung, thereby allowing the lung to expand freely. It is produced by type II cells and is lost during exposure to hyperoxia. Matalon et al. found that intratracheal instillation of a calf lung surfactant extract to rabbits during exposure to 100% O_2 lessened the progression of hyperoxic injury (555). Ghio et al. reported that a synthetic surfactant (Exosurf), and its non-surface-active components tyloxapol and cetyl alcohol, can function as antioxidants, and their *in vivo* instillation is associated with decreased hyperoxic injury in rats (607).

Loewen et al. (608) gave rabbits calf lung surfactant extract at 24-h intervals while they were exposed to 100% O_2. Eight controls were dead after 72 h, whereas six of eight treated animals survived. The mean survival time of treated animals was 120 ± 4 h, that of saline-solution-instilled controls was 102 ± 4 h, and that of untreated controls was 95 ± 6 h. After sacrifice at 72 h, the treated animals had lower surface tension in their lungs and showed less damage than the controls (608).

King et al. studied specimens obtained by washing out material from the small air passages (BAL) from baboons exposed to hyperoxia and found a progressive reduction of the surfactant concentration (609).

Survival of a hyperoxic exposure improves tolerance of rabbits to a subsequent exposure, associated with increased lung surfactant. Baker et al. reported that rabbits exposed to 100% O_2 for 64 h, then returned to room air for 8 days, showed twice the control level of alveolar phospholipids, a measure of pulmonary surfactant. They survived twice as long as controls when again exposed to 100% O_2, although their lungs showed no difference in SOD, CAT, or GP, and no differences in alveolar type II cell numbers (610).

Other protein replacement studies have also been reported. Pretreatment of rats with recombinant tumor necrosis factor plus interleukin 1 (IL-1) decreased mortality to hyperoxia, suggesting a possible treatment strategy (611). In addition, mice treated intravenously, and rats treated intratracheally (612) with recombinant human keratinocyte growth factors had improved lung protection against O_2 injury and decreased mortality. Aerosol administration of manganese-SOD to the lungs of primates decreased hyperoxia-induced lung injury (targeted gene overexpression studies have been performed in cultured cells and animals) (613).

Methods to overexpress and eliminate protein or gene function include manipulating cells in culture or altering expression in laboratory animals. These methods have become powerful strategies to determine important proteins and genes involved in O_2-induced injury. A summary of some of the studies that used SOD transgenic and knockout mice was mentioned earlier (Section 7.4.1.7.2) and results demonstrated the importance of these antioxidant enzymes in O_2-induced lung injury.

Respiratory cells engineered to express high levels of the antioxidant heme oxygenase-1 (614) or heat-shock protein 70 (615) had increased protection from O_2-induced injury. Similarly, transgenic mice with targeted overexpression of interleukin-11 in the lung showed markedly diminished hyperoxic lung injury (616).

Studies designed to test disruption of normal gene/protein function have also been performed in laboratory animals. Knockout mice with no heme oxygenase-2 expression were sensitized to hyperoxia-induced lung injury and had shorter survival times, suggesting heme oxygenase-2 has a protective role in O_2-induced lung injury (617). Similarly, mice lacking Clara cell secretory protein showed increase sensitivity to hyperoxia (618). Mice with a mutated plasminogen activating inhibitor-1 gene were also more susceptible to O_2 injury (619).

A new area of lung research has focused on gene therapy; methods that attempt to increase gene function in cells previously containing a dysfunctional gene. Two studies have reported gene transfer of antioxidants using an adenovirus vector. Danel et al. transferred SOD and CAT activity into lung cells of rats and showed an increased survival time (620). Otterbein et al. administered heme oxygenase-1 to rats and demonstrated protection from O_2 injury (621).

Thurnham discussed nutritional factors in hyperoxidant injury. Deficiencies of micronutrients, notably zinc, vitamins A, B complex, and C, contribute to such injury by diminishing tissue stores of antioxidants and their precursors. Thurnham cited evidence that such deficiencies may contribute to cancer, rheumatoid arthritis, drug toxicity, atherosclerosis, and the process of aging (622–626). He suggested that oxidant injury is likely to be involved in a variety of pathological changes in undernourished populations. His emphasis was on an adequate intake of micronutrients (627). Additional comments on free radicals and aging can be found in works by Pacifici and Davies (628) and Sohal and Allen (629).

Burns et al. tested the hypothesis that a diet rich in unsaturated fatty acids would protect mice from O_2 toxicity and lipid peroxidation, but found no differences in a 3-week comparison with a diet high in saturated fatty acids, as measured by tissue lipid peroxides and hyperoxic lung injuries (630).

Sosenko et al. found that diets rich in polyunsaturated fatty acids fed to female rats before and throughout pregnancy and lactation gave their newborn offspring a measure of protection from the effects of continuous exposure to 95% O_2. Survival was improved by 12–42%. Use of a diet low in polyunsaturated fatty acids increased susceptibility to hyperoxia. This could be countered, in part, by having the infants nursed by foster mothers who were receiving the enriched diet (631).

Because the sympathetic nervous system had been implicated in O_2 toxicity, Akers and Crittenden gave guinea pigs the catecholamine inhibitors reserpine or phenoxybenzamine. They exposed the animals to O_2 at 500 mm Hg partial pressure (equivalent to 66% O_2) for 6 days. Scanning and transmission electron microscopy showed extensive changes in the control animals, including alveoli clogged by proliferation of alveolar type II epithelial cells and macrophages, and thickening of the air–blood barrier. Animals pretreated with reserpine showed little change from normal after 6 days, supporting the hypothesis of neural participation in hyperoxic toxicity. Phenoxybenzamine was less effective (632).

7.4.3 Human Experience

The concerns for O_2 toxicity have become apparent with the increased therapeutic uses for O_2 in medical practice. Any patient with impaired O_2 passage through the lungs to the

INORGANIC COMPOUNDS OF CARBON, NITROGEN, AND OXYGEN

blood, or in whom the general circulation of the blood is impaired, is a candidate for O_2 administration. In many cases, supplemental O_2 has been life-saving. Conversely, it has also been used under the illusion that it could do no harm; when there was only a hope that it might be helpful and no better alternative was available.

7.4.3.1 Hyperoxia. In healthy men, the earliest indicators of O_2 toxicity are substernal pain and cough after a ≤ 4 h exposure to $> 95\%$ O_2. Inflammatory changes in the respiratory mucous membranes appear within 6 h and, after a 12 h exposure, the forced vital capacity (FVC), the amount of air that can be forcibly exhaled after taking a deep breath, is reduced (633).

At an early stage, water begins to accumulate in the interstices of lung tissue and within the pulmonary cells. This accumulated water causes a thickening of the alveolar–capillary membrane, impairment of gaseous diffusion, and interference with the elasticity of the organ. The combined effect is to diminish the FVC (504). The foregoing observations were the topic of several reports. Not all the results, however, were concordant. Two studies found no effects on pulmonary function after 6–12 h (634) or after 17 h (635) of breathing 100% O_2; subjects in the second group, however, did complain of chest pain. Most of these studies were done in healthy persons at rest; physical exertion would increase the O_2 exposure because of increased depth and rate of breathing.

Kapanci et al. (636) reported detailed examinations of the lungs of six persons who died after 14 h to 13 days of mechanical ventilation with 60–100% O_2. After 14 h, swelling of type I alveolar epithelial cells was most obvious, accompanied by bronchial hyperemia. After 3 days, endothelial and type I cell damage, interstitial edema, thickened and ruptured septa, perivascular hemorrhage, and alveolar hemorrhage were present. After 6 days, capillary destruction and interstitial hemorrhage were noted and after 13 days marked septal fibrosis, proliferation of type II cells, distorted alveoli, and arteriolar clots were present (636).

Balentine (504) prepared a monograph on the pathology of O_2 toxicity, including a summary of effects in humans. He emphasized that, although the respiratory tract is most severely injured, the heart, skeletal muscle, kidney, and most other organ systems have been affected in experimental exposures. Lodato (531) commented that the few reports of pathological observations in humans were consistent with animal studies, but in doing so, implied that the animal studies were consistent among themselves. This implication is only partially true (531).

In patients with chronic congestive heart failure, increased inspired O_2 (30 or 50%) had a beneficial effect on exercise performance (637). A significant increase was determined for arterial O_2 saturation and significant decreases in minute volume, cardiac output, and subjective scores for fatigue and breathlessness.

Because of the severe lung pathology induced by O_2, many studies use human cells in culture to gain insights to *in vivo* responses. Martin and Kachel (638) have contributed to the search for a threshold level of O_2 required to cause injury to human lungs by exposing cultures of human pulmonary artery endothelial cells to O_2. Impaired cell replication was evident after ≥ 8 h at 95% O_2 or after 48 h at 60%. No reduction of antioxidant activity in the cultures was evident, and glutathione production was increased. Recovery occurred after returning to a normal O_2 environment. The relevance of these observations to human health is questionable because the authors began with the premise that the "pulmonary capillary endothelium appears to be the earliest site of significant O_2 injury in the lung," which seems to be true in rats, but not necessarily in humans.

To test the vulnerability of airway epithelial cells to O_2-induced injury, antioxidant mRNA levels (SODs and CAT) were measured in bronchial cells retrieved from normal human volunteers at baseline and after nearly 15 h of 100% O_2 exposure (639). CAT activity did not change after hyperoxia, and total SOD increase could not protect the epithelium from damage. In a follow-up study, human bronchial epithelial cells were tested in culture for their ability to up-regulate CAT activity after hyperoxia (640). Constitutive expression of CAT in these cells was low, and hyperoxia did not induce increased expression. The authors concluded that this inability to up-regulate antioxidants may play a role in the sensitivity of the human airway epithelium to hyperoxia, and could be a strategy for protecting these cells from such injury.

7.4.3.2 Hyperbaric Oxygen Exposures. "Hyperbaric" is an atmospheric pressure greater than that at sea level. This increase in pressure, often associated with diving, is opposite of the decrease in atmospheric pressure associated with aviation and mountain climbing (hypobaric). Partial pressures of O_2 are higher in hyperbaric, and lower in hypobaric, atmospheres. Increased O_2 pressures result in hyperoxygenation of the blood (increased dissolved O_2) and allows for improved O_2 delivery to tissues. Hyperbaric O_2 has a number of important uses in medicine. However, the toxicity of increased O_2 pressures are also evident in human and animal studies.

Although it appears that all living cells can be damaged by O_2, the respiratory system and the central nervous system (CNS) are most affected by its toxic actions. Whereas the respiratory system is more susceptible to injury, the effects on the CNS are more dramatic. Animals or humans exposed to a $\geq$4-ATA pressure of O_2 are likely to show muscular twitching or general convulsions (like those of epilepsy) within 1 h. Pure O_2, at atmospheric pressure or less, can cause pulmonary irritation and edema after 24-h exposures; higher pressures cause damage in a shorter time. Half an atmosphere of pure O_2 can probably be tolerated indefinitely without lung damage; 3 ATA is probably safe for healthy adults for 1 h. Intermittent breathing of air or another gas mixture containing O_2 at the accustomed 0.2 ATA pressure lessens the risk of lung injury.

7.4.3.2.1 CNS Effects. Functionally, the CNS shows a high tolerance for increased P_{O_2}. In healthy men at rest, CNS effects usually appear only after exposures to ≥ 3 ATA of pressure. The effects of O_2 inhaled under higher-than-atmospheric pressures on birds and dogs was described by Paul Bert in 1878, using 5–10 times normal ATA containing 73–81% O_2. The most prominent effects were convulsive seizures (503). This CNS response has been called the *Paul Bert effect* (in contrast to the Lorain Smith effect mentioned above). CNS effects are uncommon even after several hours of exposure with $P_{O_2} < 3$ ATA, and susceptibility is species-dependent. Bert also made observations on himself and other humans.

In humans, the seizures closely resemble those of *grand mal* epilepsy, and show similar electroencephalogram (EEG) patterns. If the excess P_{O_2} exposure is not quickly ended, severe convulsions can lead to death. As little as 20 min O_2 exposure at < 4 ATA can

INORGANIC COMPOUNDS OF CARBON, NITROGEN, AND OXYGEN

cause convulsions, and the time required is less for higher pressures. The convulsions are preceded by uncontrollable twitching and various other psychomotor signs and symptoms that are not immediately relieved by a return to a normal atmosphere. Based on the recoveries of early investigators who experiemented on themselves (Paul Bert and J. B. S. Haldane), no apparent lasting damage occurs, even after convulsions, if normal P_{O_2} is quickly reestablished (504). Some evidence exists that food and/or water deprivation before exposure to hyperbaric O_2 may postpone the hyperoxia-induced seizures (575). In addition, use of a cytochrome P450 2E1 enzyme inhibitor (diethyldithiocarbamate), 21-aminosteroid, and propranolol have been shown to produce significant delays in the onset of seizures by CNS O_2 toxicity (576).

Extensive structural damage to the CNS has been described in rats and cats after high pressure O_2 exposures, with lesions in both the spinal cord and the brain (504). Microscopic evidence suggested that CNS lesions can be induced by tissue P_{O_2} levels similar to those affecting lung cells. It was also reported that the CNS is more susceptible than was previously recognized (504). This is supported in a review by Kontos (577).

The CNS effects of O_2 were reviewed by Wood (578), Edwards et al. (579), and Lambertson (580). As shown in Figure 47.4. These effects appear more quickly than the pulmonary symptoms, but only under higher O_2 pressure.

Although the CNS of animals is resistant at birth to the Paul Bert effect and to neurological degeneration caused by hyperbaric O_2, it soon becomes susceptible. Exposure to even 1 ATA of pure O_2 causes destruction of brain cells. The human infant may also be more susceptible to hyperbaric O_2 than the adult (504).

7.4.3.2.2 Respiratory Effects. Clark reported the effects of O_2 exposures at pressures ranging from 3.0 ATA for 3.4 h to 1.5 ATA for 17.7 h continuously in healthy, resting men. He observed that flow in peripheral airways was impeded more than diffusion was

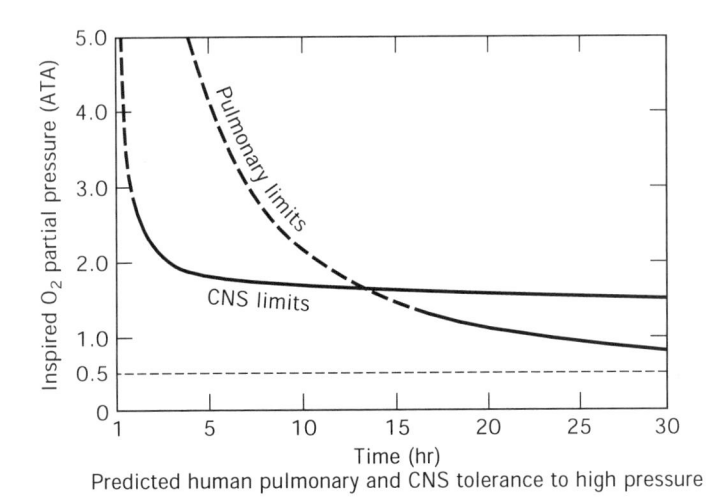

Figure 47.4. Time to onset of pulmonary or central nervous symptoms with increasing oxygen pressure. From Ref. 580.

impaired. No single measure of function was satisfactory for comparing response rates. There was marked variation among individual subjects. Most showed complete recovery of mechanical functions within 24 h, but some took longer (536, 581).

Webb et al. exposed 21 men and women, aged 19–37 years, to 100% O_2 at an atmospheric pressure of 9.5 lb/in^2 ("equivalent to 11,500 ft, 491 mm Hg or 65.7 kPa"), 8 h/day for 5 consecutive days, while they performed moderate exercise. No decompression sickness, symptoms of O_2 toxicity, changes in pulmonary function, blood oxygenation, or other manifestations of injury were observed (582).

The Undersea Medical Society collected and published five volumes of reports on the effects of increased atmospheric pressure on animals and humans (251).

7.4.3.2.3 Hyperbaric Oxygen Therapy. For many of us, memories of a Jacques Cousteau special expedition or an episode of *Sea Hunt* come to mind when thinking of hyperbaric chambers used for divers surfacing too quickly and developing "the bends" (decompression sickness). Although hyperbaric O_2 therapy is still used by divers in distress, such treatment now encompasses a large number of diverse conditions.

Hyperbaric O_2 therapy is a medical treatment in which the entire patient is enclosed in a pressure chamber while breathing 100% O_2 at elevated ambient pressures. Hyperbaric O_2 therapy has been effective in treating decompression sickness and air or gas embolism by mechanically decreasing the size of air bubbles, and increasing dissolved O_2 levels in the blood. It is also a treatment of choice in carbon monoxide poisoning and smoke inhalation. Other approved indications for adjunctive hyperbaric O_2 therapy include acute traumatic ischemia (crush injuries depriving tissues of O_2), large blood loss (anemia), intracranial abscess, osteomyelitis, enhancement of healing in selected problem wounds, gas gangrene, radiation tissue damage, and thermal burns. This therapy has also been used in preparation and preservation of skin grafts or flaps in compromised tissue. Grim et al. has provided a complete review on hyperbaric O_2 therapy and is recommended as a starting point for additional information (583). Also, a variety of Internet websites can be investigated for more information on hyperbaric therapy, including those in Refs. 584 and 585.

Portable hyperbaric chambers, including the Gamow® bag, the Certec® bag, or the PAC® portable altitude chamber, can be used to treat hypoxia in high altitude illnesses (see section on hypoxia). Advantages, disadvantages, indications, and contraindications of this equipment can be found at http://www.gorge.net/hamg/hyperbaric.html. These items provide an on-site atmosphere of high O_2 pressure to aid in the delivery of much needed O_2 to the tissues of individuals unable to move to lower grounds.

7.4.3.3 Prevention of Oxidative Injury. Since the 1980s, a significant amount of research has focused on the prevention of O_2-induced injury. A considerable number of pharmaceuticals and biological agents have been tested for their ability to eliminate, reduce, or control the oxidative damage associated with hyperoxia and O_2 therapy. Most of these studies have been performed in cell culture or animal models of O_2-induced injury, with the hopes that insights toward the prevention of the human pathology can be achieved.

A number of drugs have been tested for their ability to protect the lung against hyperoxia-induced injury. For example, cimetidine, cyclosporin A, dopamine, isoproterenol, lisofylline, pentoxifylline, and the lazaroid U-74389G each show protective actions against hyperoxia-induced injury (641–647). In addition, low-level carbon monoxide may protect against oxygen-induced lung injury (648).

7.4.3.4 Carcinogenesis. Exposure to ionizing radiation is recognized as a cause of cancer, and the production of free radicals in the exposed cells is a part of the process. If hyperoxia causes free-radical formation, it may contribute to the occurrence of cancer. Although a direct relation of hyperoxic injury and cancer has not been proven, numerous instances of association in experimental animals exist. Reactive oxidative intermediates have been shown to cause chromosome breaks and damage to DNA that can initiate carcinogenesis (533, 535).

Experimental work done with mouse skin tumors (as models of human tumors) has been both revealing and confusing. A substance may act as a cancer initiator or as a promoter, or sometimes both, depending on intensity and duration of exposure, and the presence of other carcinogenic materials. The same substance can also inhibit cancer growth. Hyperoxia has clearly been involved in modifying the course of tumor development, but the effects have differed under varying circumstances. Witschi has reviewed this rapidly changing field (533, 649).

7.4.3.5 Other

7.4.3.5.1 Eye Injuries. Immature, growing tissues, capillary blood vessels in particular, are susceptible to hyperoxic damage. Premature infants have organs that are not fully developed at birth, including the lungs and eyes, where growth of new capillaries is still very active. These infants may have difficulty in getting sufficient O_2 into their blood, so they are placed in enclosed incubators or on respirators with an increased P_{O_2}. In the early history of this life-saving procedure, high O_2 concentrations were used. The immediate benefit of tissue oxygenation was followed by a high incidence of pulmonary damage. It was soon observed that these babies were blind, the result of hyperoxic damage to the retinal capillaries. This injury was followed by proliferation of fibrous tissue. Careful adjustment of the O_2 pressure to meet the needs of the infant has been effective in controlling the damage to the eye (504).

Following the identification of O_2 as the principal cause, the incidences of retrolental fibroplasia and other eye injuries have been greatly reduced; however, the exact pathological mechanisms of O_2-induced eye injuries are still unknown. Numerous animal models for retinopathies have been developed, including mice, rats, and dogs, which will help in the research to determine the pathogenesis of O_2-induced eye injuries in humans (650–652). Moreover, these animal models are now being used to determine preventive and treatment strategies to be tested later in human clinical trials (653–655).

Transient impairment of visual fields and visual acuity have been reported in adult humans in association with hyperbaric O_2. Cultures of animal eye tissues, especially the lens and the retina, are susceptible to hyperoxic injury at normal pressure. NIOSH has recommended that the minimum O_2 concentration for working spaces be 19.5% (498). Paradoxically, hyperoxia has also been used to treat specific eye injuries with some degree of success (656).

8.0 Ozone

8.0.1 CAS Number: *[10028-15-6]*

8.0.2 Synonyms: Triatomic oxygen; trioxygen

8.0.3 Trade Names: NA

8.0.4 Molecular Weight: 47.9982

8.0.5 Molecular Formula: O_3

8.0.6 Molecular Structure: $O=\overset{+}{O}-O^-$

8.1 Chemical and Physical Properties

Physical state: Bluish gas or liquid
Molecular weight: 47.9982
Boiling point: $-111.9°C$
Melting point: $-193°C$
Density: 1.66 (air = 1) at 25°C
Solubility in water: 0.49 mL/100 mL at 0°C
: 30 μg/100 mL at 20°C

See also Ref. 255.

8.2 Production and Use

Ozone (triatomic oxygen) is a light blue gas with a characteristic odor (reminiscent to some individuals of an electrical discharge such as lightening) (1, 657). Ozone was first described in 1840 by Christian Friedrich Schonbein (1799–1868), who produced it from phosphorus and electrolysis of water. Schonbein also developed a colorimetric assay involving starch and potassium iodide–impregnated paper that was widely used to measure atmospheric ozone concentrations. Interestingly, Schonbein's studies were interrupted when he discovered the acute toxicity of ozone in 1851 and noted that ozone caused "a really painful affection of the chest, a sort of asthma, connected with a violent cough" (658). Concern about ozone's toxicity dates back since the midtwentieth century, when it was recognized as a major air pollutant in urban areas, with additional concern in the 1980s and 1990s about its depletion in the stratosphere.

Ozone can be found naturally in the troposphere during electrical storms and in the stratosphere. Background levels of ozone in nonurban areas average about 10–20 ppb and are due mainly to intrusion of stratospheric ozone into the lower atmosphere (659).

8.3 Exposure Assessment

Occupational exposures can occur when ozone is used in chemical manufacturing (e.g., lipid ozonolysis, bleaching processes, peroxide production) and the disinfection and deodorization of water. It is a unwanted by-product of photocopying machines, electric-arc welding, high voltage electrical equipment, X-ray generators, mercury vapor lamps, linear accelerators, and indoor ultraviolet sources. Because ozone mistakenly has been thought to be therapeutic, ozone generators periodically are marketed as air fresheners for use in commercial operations (restaurants, bars, childcare centers, and bowling lanes) and in the home.

Ozone is a problem in aviation because it is present at high attitudes, with maximal concentrations occurring at about 75,000 ft (~23 km). It is detectable at all altitudes below ~300,000 ft (~90 km), above which stratospheric ozone can reach up to 10 ppm. Here ozone regulates the flux of high energy ultraviolet solar radiation (virtually all UV-C and a portion of UV-B) through the atmosphere to the ground. This regulation of stratospheric ultraviolet is critical to numerous tropospheric biological processes, including photosynthesis, and may be involved in the incidence of skin cancer. Nitrogen oxides and halogenated organic gases used as propellants and refrigerants, such as chlorofluorcarbons (CFCs) can catalyze the degradation of stratospheric ozone (which has been noted particularly over Antarctica).

Ozone control in the troposphere involves the regulation of nitrogen oxides and hydrocarbons, primarily from mobile sources. Little or no ozone is released from mobile (vehicles) or industrial sources, but ozone is formed in a complex series of photochemical reactions. These reactions involve nitrogen oxides and reactive organic compounds that are derived mainly from anthropogenic (olefinic hydrocarbons, formaldehyde, and xylene), natural, and agricultural (methane) sources. These reactions are initiated by the photodissociation of nitrogen dioxide (NO_2) into nitric oxide (NO) molecules and oxygen atoms (O). Atomic oxygen then reacts with molecular oxygen (O_2) to form ozone. Nitric oxide, however, rapidly scavenges ozone, producing NO_2 and O_2. These reactions can be summarized as follows:

$$NO_2 + h\nu\,(295 - 430\,nm) \rightarrow NO + O$$
$$O + O_2 \rightarrow O_3$$
$$NO + O_3 \rightarrow NO_2 + O_2$$

The sum of this reaction sequence does not yield ozone. In addition, several other ozone scavengers, such as ethylene, are present in the atmosphere. Nonetheless, other reactions scavenge NO or facilitate conversion of NO to NO_2; thus ozone production exceeds ozone-scavenging reactions. Nitric oxide can react with hydroperoxyl radicals:

$$HO_2 + NO \rightarrow HO + NO_2$$

This and other reactions reduce O_3 scavenging by NO, allowing steady-state ozone levels to rise (659).

Because these reactions depend on solar radiation, ozone formation tends to be greatest during hot summer days. Because ozone is reactive, levels fall during the evening. This produces a diurnal pattern that peaks from the late morning until the late afternoon, and then peaks again in early evening following increases in automobile traffic.

Ozone and its precursors can be transported great distances from the precursor source areas, so elevated ozone concentrations may extend over thousands of square kilometers, including remote rural sites. The major formation factors include the baseline levels of ozone in the air aloft, the rate of photochemical production, and the rate of ozone and nitric oxide scavenger formation. Local concentrations are influenced by wind speed and direction, levels of precursors in air-mass trajectory, and the presence of thermal inversions. Together these factors affect the time of the peak concentration, which may occur anytime from noon until midnight (660).

Of all the current air pollutants in which a national ambient-air-quality standard has been set, ozone is the most often exceeded. A large percentage (over 40%) of the U.S. population lives in areas that regularly exceed the current standard for ozone. Millions of people are intermittently exposed to ozone concentrations that would exceed the occupational standard if such exposures were to occur in the workplace.

Ozone can penetrate buildings, but indoor levels tend to be lower because ozone is so reactive with many indoor surfaces (661). However, indoor:outdoor ratios often range from 0.4 to 0.8 (662). Modern buildings with low ventilation rates or those equipped with activated-carbon air-filtration systems can have ratios of < 0.1–0.25 (662). Reaction of ozone with furnishing and carpeting can increase indoor concentrations of other respiratory irritants, such as formaldehyde and acetaldehyde (663). Potential nonindustrial sources of ozone include equipment using high voltage or ultraviolet light, including electrostatic air cleaners, laser printers, and photocopy machines (664). The latter has led to levels mostly around 50 ppb, but in some instances as high as 100–200 ppb in breathing zones of photocopy machine operators (664, 665).

Ozone is commercially manufactured by ozone generators that use oxygen and ultraviolet irradiation. Occupational exposures to ozone can occur in electric-arc welding, in industries using ozone as an oxidizing agent, portland cement plants with kilns, and in aircraft cabins (666–671). In aircraft, ozone exposures have been reduced by ozone converters in ventilation systems and by modification of flight plans. Persons in outdoor occupations, letter carriers, farmers, and construction workers may also be at risk when ambient levels exceed occupational standards.

8.3.1 Air

Ozone is adsorbed by the lining fluid of anatomic sites of the air–liquid interface (e.g., the mucous membranes of the respiratory tract and eye). Ozone is a strong irritant, and its relatively low aqueous solubility facilitates penetration of the upper respiratory tract. The highest rate of transfer to the lining fluid occurs in the distal region, where the small diameter airways communicate with the alveoli (bronchioalveolar junction) (672, 673). Little systemic absorption or metabolism occurs because ozone rapidly reacts with the free sulfhydryls, cysteinic sulfur–sulfur crosslinks, and the carbon–carbon double bonds of biological macromolecules. Animal studies of high concentrations also suggest that a small portion may be absorbed into the blood, resulting in increased red blood cell fragility and alterations in blood chemistry (674).

On inhalation, approximately 40–50% of inhaled ozone is deposited in the nasal passages. About an equivalent level of deposition occurs in the mouth and throat during oral breathing at rest. However, during exercise the transit time of ozone through mouth and throat decrease and much higher levels penetrate to the lower respiratory tract. About 90% of ozone reaching the lungs is deposited in a single breath, and thus the overall deposition efficiency exceeds 95% (675–677). In the lungs, ozone removal efficiency appears to be related directly to concentration and inversely to the breathing rate (675). Direct readings of ozone concentrations in the throat have suggested that shallow breathing, a common response to ozone exposure, reduces in lower respiratory tract deposition (678).

Individuals vary greatly in their susceptibility to ozone induced pulmonary responses (679, 680). Acute symptoms and diminution of lung function are determined by the concentration, ventilation rate, and duration of exposure delivered to the lung (659, 681–684). Of these variables, ozone concentration appears to be most influential, followed by ventilation and duration of exposure (685–689). At concentrations ≤120 ppb, duration of exposure is nonetheless important for the induction of bronchoconstriction. At this concentration, for example, little effect was noted over the initial 1–2 h; however, progressive decreases in pulmonary function occurred when exposures were extended past 4 h and on to 6.6 h (690–692).

8.3.1.1 Reaction of Ozone with Biological Macromolecules. At concentrations < 200 ppb, most, if not all, ozone is likely to react with the biological macromolecules in the respiratory lining fluid (693–700). Ozone is a powerful oxidant and will react with amino acids (particularly cysteine, tryptophan, methionine, phenylalanine, and tyrosine) and with lipids (particularly the unsaturated fatty acids contained in membrane phospholipids). The former can yield disulfides and methionine sulfoxide; the latter can yield hydrogen peroxide, aldehydes, and hydroxyhydroperoxides (694, 701–704). Antioxidants in mucus and other fluids lining the respiratory tract, as well as those in the tissues themselves, may be protective (701,705–709). In the past, ozone has been purported to act as a free radical. However, although clearly a strong oxidant, ozone is not a free radical. In addition, supportive evidence for this mechanism at best is only indirect and comes from studies showing that vitamin E, a free-radical scavenger, retards or prevents ozone's effects on polyunsaturated fatty acids *in vitro* (706). In addition, vitamin E deficiency in experimental animals enhances ozone's toxicity. It is not known, however, whether supplemental vitamin E in the diet can protect humans against ozone's effects (707).

8.4 Toxic Effects

8.4.1 Experimental Studies and Human Experience

8.4.1.1 General Information. Ozone attacks macromolecules and induces cellular damage throughout the respiratory tract. Focal deposition and therefore regional responses depend on breathing pattern and exposure concentration and duration. At high concentrations (> 1200 ppb), ozone may cause acute lung injury with epithelial sloughing in the airways, perivascular cuffing, and hemorrhagic pulmonary edema (810, 811). At sublethal concentrations (< 1200 ppb), airway surface epithelial cells are damaged with loss of cilia. A principal site for injury in the airways is the junction of the bronchioles and the alveoli, or the centriacinar region. In the alveolus, type I epithelial cells are particularly susceptible to ozone toxicity, with damage evident within 4 h of exposure (397). Inflammation initially consists of polymorphonuclear (neutrophils) leukocytes, and then monocytes and/or macrophages are evident at the junction of the conducting airways and the gas-exchange zone; these findings have been reported consistently in studies of rodents, dogs, and nonhuman primates (659, 812, 813). The initial increase in polymorphonuclear leukocytes may contribute to the repair process by enhancing the proliferation of injured airway epithelial cells (814). Short-term exposure (several days) results in replacement of type I by type II cells as well as hypertrophy and hyperplasia of nonciliated cuboidal cells on the inner surface of the small airways. These changes can reverse when animals are allowed to recover from acute and subacute exposures in clean air (815, 816).

8.4.1.2 Acute Toxicity. Effects can develop rapidly on acute exposure of 1–2 h. Pulmonary symptoms caused by low level (60–200 ppb) ozone exposure include substernal pain and difficulty in deep inspiration, cough and dry throat, wheeze, and dyspnea (710). Other symptoms reported include headache, nausea, and malaise. At slightly higher concentrations (200–1000 ppb), ozone exposure has led to somnolence and extreme fatigue, dizziness, insomnia, decreased ability to concentrate, cyanosis, pulmonary edema, acrid taste and smell, and eye irritation (657, 711–714). Higher concentrations of ozone (3.2–12 ppm) can be fatal and can produce pulmonary edema and hemorrhage in experimental animals (711, 715, 716).

Multihour exposures involving moderate or intense exercise provokes ozone-related symptoms at concentrations as low as 80 ppb (316, 317, 690–692, 717, 718). In vigorously exercising young adults, scratchy throat, substernal discomfort, and other symptoms increased as a result of exposures to 120 ppb ozone lasting $<\frac{1}{2}$ h (719). Ozone can impair athletic performance and the ability to perform sustained exercise (719–722). Discomfort on inspiration is thought to contribute to the diminution of exercise performance (723, 724).

Acute exposure to ozone can decrease lung function and increase airway reactivity. The latter is observed when bronchoconstriction is induced at lower concentrations of an irritant stimulus or selected pharmacologic or physical agents. Ozone exposure alters tidal volume, ventilatory rates, specific airway resistance, and several measures of inspiratory and forced expiratory flow, including forced vital capacity and forced expiratory volume in 1.0 s (659, 690, 717, 725). Some of these responses are mediated by the irritant sensory fibers and the autonomic nervous system (725–728). Ozone also increases inflammatory mediators, including eicosanoids and cytokines (317, 729). The latter include chemokines such as interleukin 8 and macrophage inflammatory protein-2, which are important in directing neutrophil migration in humans and experimental animals (317, 730–735). In the past, it was thought that inflammatory responses mediated the effects of ozone on respiratory mechanics because they often occurred in responsive individuals; however, with additional study, this situation is now uncertain.

As noted above, considerable interindividual variability exists in ozone-induced decrements of lung function. In the past, investigators tried to explain this difference by several factors. These studies associated individual susceptibility with a history of allergies or asthma, greater baseline airway reaction to bronchoconstrictive agents, and self-reported sensitivity to air pollution, appear to be interrelated (736). Ozone-related susceptibility also was reported to decline with age (680), and may be influenced by gender, as females show a greater susceptibility than males for a given exposure (737).

However, more recent studies suggest that host (genetic) factors may determine responses. In one study, a portion of subjects experienced a decrease of >25% in their lung function whereas others had only a small (<10%) change in lung function (659). On a second test up to 14 months later, these intrasubject changes in lung function were reproducible, suggesting the existence of intrinsic factors that control responsiveness to ozone (738, 739). This hypothesis has been supported by animal studies that have identified at least two genetic loci with linkage to ozone-induced acute lung injury and inflammation (740, 741).

Adolescents and children can experience decreases in pulmonary function comparable or greater in magnitude to those observed in adults, but may not report symptoms to the same extent (742–744). In addition to the transient decrements in lung function (745–747), the effects in children may persist a few days after the day of peak exposure. Inflammation is also known to become maximal 14–18 h after exposure (748, 749). In one study of an air-pollution episode lasting several days, during which maximum daily ozone concentrations ranged between 0.12 and 0.185 ppm, peak flow decrements in some children lasted up to a week after the episode ended (750).

In human studies, acute exposure to as little as 80 ppb ozone can induce neutrophilic inflammation, peaking in bronchoalveolar lavage fluid or biopsies of the bronchial mucosa 12–18 h after a single exposure (316, 317,718, 817). These data indicate that the acute inflammatory damage of ozone repeatedly demonstrated in animals is likely to be duplicated in the human lung at concentrations lower than those used in animal studies. In addition, chronic pathological effects have been noted in lung specimens from accident victims in southern California. More than 25% of the tissues examined had severe and extensive injury to the centriacinar region (with monocytic infiltrates) (818–820).

8.4.1.3 Tolerance. Individuals also can vary in the attenuation of response that occurs with repeated daily exposures. In controlled chamber studies to moderate level exposure (200–500 ppb), maximal responses are typically observed the second day of exposure, but on subsequent days, ozone-related effects diminish (751, 752). Interestingly, when ozone levels are reduced to 120 ppb, the day of the peak response varies more (occurring on the first to the third day) and depends on the lung function measurement (753).

Under laboratory conditions, this tolerance to ozone toxicity can persist for 1–3 weeks following cessation of exposure (754–756). The effect of tolerance in ambient exposures is less clear. For example, in studies of children at summer camp attenuation was not found (747, 748). In contrast, Los Angeles adults present seasonal variability (greater response each spring), which suggests longer-term adaptation (757). Tolerance has also been noted in laboratory animals. In an early study, Stokinger (715) noted that pretreating animals to levels <1000 ppb, protected them from subsequent exposures to lethal concentrations (>3000 ppb). However, more recent studies suggested that tolerance to the irritant effects and thus the diminution of lung function may not be beneficial. Rats exposed to ozone at 350–1000 ppb for several days developed tolerance to changes in lung function, but developed progressive inflammation and tissue damage in the distal airways (758, 759).

8.4.1.4 Effects in Asthmatics. Persons with asthma may be added risk from ozone exposure. Measures of the possible threshold of decreased lung function do not seem to change at low concentrations among persons with asthma as compared to health control subjects (760, 761), as has been noted with sulfur dioxide (762). A few controlled exposure studies suggest that persons with chronic obstructive pulmonary disease also do not have enhanced sensitivity to ozone effects on lung function (763, 764). Similarly, subjects with a history of allergic rhinitis did not differ markedly from nonallergic subjects in ozone-related symptoms or pulmonary function tests, except for specific airway resistance (765). However, following relatively high exposure (500 ppb ozone for 4 h) subjects with allergic rhinitis had greater nasal symptoms (rhinorrhea and congestion) and inflammation of the lower respiratory tract (766). Subjects with asthma also can develop heightened inflammatory response in the upper respiratory tract following ozone exposure (767–769).

Likewise, at effective doses asthmatic subjects develop reduced lung function suggestive of bronchoconstriction and immediate symptomatic responses that are similar in intensity to control subjects (770). Nonetheless, numerous epidemiological studies find association of ozone with increased symptoms and hospital admission for asthma (768, 771–778). This may be partially explained by inhalation studies. In these studies, brief exposure to low levels of ozone augmented susceptibility of persons with asthma to the effects of other irritants (SO$_2$) or inhaled allergens (779–782). These observations may be relevant to the mixtures present in occupational environments where workers are exposed to other allergens. For example, repetitive low level exposure to a mixture of 40 ppb ozone and copper oxide has been associated with a severe exacerbation of asthma in a worker with a history of childhood asthma (783).

Ozone may also contribute in a complex mixture to the eye irritation during air-pollution episodes. Although much of the effect has been ascribed to peroxyacetylnitrate and aldehydes, like acrolein (659, 784, 785), in industrial settings, ozone alone can induce eye and nasal irritation (786). Ozone concentrations > 2000 ppb can produce eye irritation within minutes (787).

8.4.1.5 Chronic and Subchronic Toxicity. One of the principal uncertainties about ozone toxicity is the relationship between repeated exposures and chronic lung disease. Guinea pigs and rats exposed to high ozone concentrations ($\geq$1000 ppb) for over 8 months developed chronic bronchiolitis, with bronchiolar fibrosis, pneumonitis, "mild to moderate" emphysema, and occasional lesions in the trachea and major bronchi (821). Exposure of rats to lower ozone concentrations (120–250 ppb) resulted in less severe alteration of the terminal bronchioles and alveolar septa, and a distribution of inflammation similar to that observed in short-term exposures (411, 822). The lungs of monkeys following chronic exposures manifested bronchiolitis, altered epithelial-cell proliferation, nasal secretory hyperplasia, and other effects, including focal lung lesions, which persisted after the cessation of exposure (823–826). Thus, unlike the case of acute ozone exposure, effects of chronic exposure become irreversible. After 3 months the degree of neutrophilic inflammation was less than that observed after the first week, suggesting that this is a transient response when concentrations are lowered. Nonetheless, monocytic inflammation persists during long-term exposures (823).

A major finding from animal experiments is that chronic exposure to ozone concentrations found in urban air can result in persistent inflammation and small-airway structural changes. Other lines of evidence support the concept that repeated ozone exposure may result in chronic lung disease. Ozone inactivates lysozyme, an antimicrobial protein secreted by airway cells, and human α_1-antitrypsin, a protease inhibitor that protects the lung from emphysema. It also increases the synthesis, deposition, and degradation of collagen in rat lung (827, 828).

8.4.1.6 Epidemiology Studies. Epidemiological studies suggest the existence of significant associations of photochemical oxidant exposure with an exacerbation of lung disease, accelerated decline in lung function, inflammation, and symptoms of chronic respiratory disease in nonsmokers (415, 829–834). Additional evidence suggests that persistent effects may also include increased risk of developing adult-onset asthma, which has been associated with increased outdoor concentrations of ozone over a 20-year period (835). Furthermore, lifetime exposures in areas with higher levels of ozone have been associated with diminished expiratory flows and increases in respiratory symptoms of chronic phlegm and wheeze apart from colds (836, 837). As with all epidemiological studies concerns exists about the confounding effects of other air pollutants. Nonetheless, these studies are consistent with observations from chronic animal studies.

Epidemiological studies have also reported an association between ozone (often with several other pollutants) and acute (daily) mortality in several cities throughout the world (802–804). For example, ozone increases of an interquartile range of 20 ppb were associated with increases in mortality of ~2% (805–808). Interestingly, these associations are strengthened when the averaging time is extended from 2 to >5 h, which is in agreement with decrements in lung function noted in chamber studies. However, the biological mechanisms for this association are uncertain. It seems likely that these effects involve susceptible subpopulations, such as the elderly with cardiovascular disease. In rats, ozone exposure (0.25 ppm for 5 days) increases oxidative reaction products and antioxidant enzymes in the heart and brain (809).

Mice exposed briefly (2–3 h) to ozone concentrations at or below 120 ppb develop decreased resistance to bacterial, but not viral, respiratory infections (788–794). Ozone also impairs the functional capabilities of alveolar macrophages involved in defense against infection (795–797). Epidemiological evidence supporting an association of ozone with respiratory infections in humans, however, is limited (798–801).

Ozone may cause chronic respiratory effects through complex interactions with other inhaled materials. For example, retention of asbestos is greater among rats continuously exposed to 60 ppb ozone (with diurnal peaking concentrations that gradually increase to and decrease from 250 ppb) than among air-exposed controls. This suggests that ozone exposure at concentrations found in urban areas may retard clearance of asbestos or other carcinogens (838).

8.4.1.7 Carcinogenesis. Ozone has been positive as a genotoxic substance in certain assay systems, but the results are inconsistent (659, 839, 840). For example *in vitro* assays have noted that ozone can induce bacterial mutations, plasmid DNA strand breakage, chromatid and chromosome aberations in lymphocytes, and a doubling the frequency of preneoplastic variants compared with control cultures. However, *in vivo* assays of similar endpoints produced mixed results (841, 842). For example, alveolar macrophages from rats exposed for 270–800 ppb ozone developed chromatid damage, but no chromosomal changes (843). In human subjects exposed to 500 ppb ozone (6–10 h), a slight increase in sister chromatid exchange persisted for ≤6 weeks (844). In contrast, no significant changes in chromosome or chromatid breaks were observed in lymphocytes of subjects exposed to 400 ppb (4 h) (845). Cultured human epidermal cells exposed to 500 ppb ozone for 10 min showed no evidence of DNA strand breakage (846).

Short (5-min) exposure of hamster embryo fibroblasts to 5000 ppb ozone can enhance cell proliferation (neoplastic transformation), and ozone enhanced cell proliferation induced by irradiation with ionizing and nonionizing radiation (847, 848). Repeated ozone exposures (700 ppb × 40 min) of cultured rat tracheal cells biweekly for $4\frac{1}{2}$ weeks resulted in similar proliferative (preneoplastic) changes, which, however, were not observed in rats exposed to ozone (0.14, 0.6, or 1.2 ppm) 6 h/day for ≤4 weeks (849).

Other investigators have suggested that chronic ozone exposure may facilitate the development of benign pulmonary tumors (adenomas) in mice and other hyperplastic nodules in the lungs of nonhuman primates (824, 850). As is true of hyperoxia, ozone exposure may enhance or retard lung tumorigenesis by other agents in rodents, depending on the exposure protocol (849, 851). In mice, ozone has been associated with increased lung tumors in two strains (A and B6C3F1 mice). Inhalation exposure for 24 or 30 months to 1000 ppb resulted in an increased incidence of alveolar/bronchiolar adenoma or carcinoma in female mice. Male and female mice exposed to 500 or 1000 ppb ozone for 2 years also developed mild, site-specific, nonneoplastic lesions in the nasal cavity and centriacinar lung that persisted with continued exposure to 30 months (852, 853). Tumors from B6C3F1 mice had two unique mutations in *K-ras*, one (in codon 61) consistent with a direct genotoxic event and a second (in codon 12) consistent with an indirect genotoxic effect (854). However, 2-year and lifetime exposures to 1000 ppb ozone did not cause an increased incidence of lung neoplasms in Fischer 344/N rats.

Other investigators have suggested that *in vitro* assays indicate that ozone may exert indirect genotoxic effects. Ozone has been purported to affect the integrity of immune system defenses against tumor development and progression (855, 856). In addition, arylamines found in tobacco smoke (e.g., naphthylamine and toluidine isomers) can be chemically altered by brief exposures (1 h) to 100–400 ppb ozone. The unidentified stable products of this reaction cause single-strand DNA breaks in cultured human lung cells equivalent to that produced by 100 rads of irradiation (857). However, an *in vivo* cocarcinogenicity study failed to find similar effects. For example, ozone exposure did not have an additive carcinogenic effect with the subcutaneous injection of 4-(*N*-methyl-*N*-nitrosamino)-1-(3-pyridyl)-1-butanone in male rats (858, 859) and others have also failed to observe cocarcinogenic effects (860, 861).

8.5 Standards, Regulation, or Guidelines of Exposure

For occupational O_3 exposure, the current OSHA permissible exposure limit (PEL) is 0.1 ppm (0.2 mg/m^3) as a time-weighed averaged (TWA). [The 1989 OSHA STEL is 0.3 ppm (0.6 mg/m^3)]. NIOSH recommends 0.1 ppm (0.2 mg/m^3) as a recommended exposure limit (REL), ceiling value. The NIOSH immediately dangerous to life and health (defined as the maximal concentration that one could escape within 30 min without a respirator) is 5 ppm. This is based on human health effects including pulmonary edema developed in welders who had a severe acute exposure to an estimated 9 ppm ozone plus other air pollutants (862, 863). It is also based on the theory that 15–20 ppm is lethal to small animals within 2 h (864). Other existing short-term exposure guidelines include the recommendation of the National Research Council Emergency Exposure Guidance Levels (EEGLs) for 1 h of 1 ppm and for 24 h of 0.1 ppm (865).

The ACGIH standards take into account workload and include *heavy work* = 0.05 ppm (0.1 mg/m^3), *moderate work* = 0.08 ppm (0.16 mg/m^3), and *light work* = 0.10 ppm (0.2 mg/m^3) as a time weight average (defined as the TWA concentration for a 8-h/day, 40-h/week period). There is an additional ≤ 2 h ACGIH recommendation for light, moderate, or heavy workloads of 0.20 ppm (0.39 mg/m^3). The basis for this TLV is the critical effects of pulmonary function (lung function), irritation, and headache. The ACGIH rates O_3 as A4 — not classified as a human carcinogen (stating that inadequate data exist to classify ozone in terms of its carcinogenicity in humans and/or animals).

In July 1997, USEPA revised the national ambient-air-quality standards (NAAQS) to phase out and replace the 1-h ozone standard with an 8-h standard to protect against longer exposure periods. The previous, 1-h standard of 0.12 ppm will be revoked when an area has achieved 3 consecutive years of air-quality data meeting the 1-h standard. The new, primary and secondary 8-h standard is 0.08 ppm, measured as maximum daily 8-h average concentrations.

To attain the ozone NAAQS, the 3-year average of the annual fourth highest daily maximum 8-h ozone concentration must be less than or equal to 0.08 ppm. An area meets the ozone NAAQS when the highest hourly value exceeds the threshold for no more than one day per year. (If monitoring did not take place every day because of equipment malfunction or other operational problems, actual measurements are prorated for the missing days. The estimated total number of above-threshold days must be ≤ 1.0.)

To be in attainment, an area must meet the ozone NAAQS for three consecutive years. On November 6, 1991, most areas of the country were designated nonattainment or unclassifiable/attainment. These terms are defined follows:

> *Nonattainment*: Any area that does not meet (or that contributes to ambient-air quality in a nearby area that does not meet) the national primary or secondary ambient-air quality standard for the pollutant.
>
> *Attainment*: Any area that meets the national primary or secondary ambient-air-quality standard for the pollutant.
>
> *Unclassifiable*: Any area that cannot be designated on the basis of available information as meeting or meeting the national primary or secondary ambient-air-quality standard for the pollutant.

Those areas designated nonattainment were also classified, based on their design value. The design value was typically the fourth highest monitored value with 3 complete years of data, because the standard allows one exceedance for each year. Generally the 1987–1989 period was used.

> *Extreme*: The area has a design value of ≥ 0.280 ppm.
>
> *Severe 17*: The area has a design value of 0.190–0.280 ppm and has 17 years to attain this level.

Severe 15: The area has a design value of 0.180–0.190 ppm and has 15 years to attain this value.

Serious: The area has a design value of 0.160–0.180 ppm.

Moderate: The area has a design value of 0.138–0.160 ppm.

Marginal: The area has a design value of 0.121–0.138 ppm.

Transitional: An area designated as an ozone nonattainment area as of the date of enactment of the Clean Air Act Amendments that did not violate NAAQS between January 1, 1987 and December 31, 1989.

Incomplete (or no) data: An area designated as an ozone nonattainment area as of the date of enactment of the Clean Air Act Amendments of 1990 that did not have sufficient data to determine if it was or was not meeting the NAAQS.

Although USEPA promulgated a new 8-h standard for ozone, the 0.12-ppm 1-h standard will not be revoked in a given area until that area has achieved 3 consecutive years of air-quality data meeting the 1-h standard.

8.5.1 Personal Protection

In most industrial applications, containment and avoidance are the best management strategies. It also is important to consider management of workload. Corrective measures should include adequate engineering controls (exposure isolation by entirely enclosing a process or through instillation of local exhaust ventilation), thorough worker education about appropriate work practices (such as the use of personal protective equipment when adequate ventilation is impractical), and recognition of ozone-related symptoms [e.g., eye irritation (189, 787)], and strict adherence to health and safety practices. A variety of supplied air and cartridge respirator classes are considered acceptable by NIOSH, with the stringency of protection geared to the likely intensity of exposure (189).

In areas where current ambient exposures can exceed occupational standards, individuals should be advised to avoid aerobic exercise during peak ozone hours (typically late morning until early evening in many urban areas) and to conform to health advisories accompanying the declaration of alerts. However, it should be noted that signs and symptoms of ozone toxicity have been repeatedly demonstrated to occur in exercising adults at ozone concentrations lower than the current recommended stage 1 smog alert level (690, 691).

Alert levels are based on the air-pollution levels that are currently reported as a *pollution standard index* (PSI). This index is an indicator of air quality developed by USEPA to provide information about the daily levels of air pollution to the public and is a percentage of the federal health standard (Table 47.7). A PSI of > 100 is considered unhealthy, and the higher the number, the more unhealthy the air. The American Lung Association recommends that at ozone levels above 100, children, persons with asthma, and other sensitive groups should limit strenuous exercise. Even otherwise healthy people should consider limiting vigorous exercise when ozone levels are at or above the standard. If the PSI is above 200, the pollution level is judged unhealthy for everyone. At this level, air pollution is a serious health concern. Everyone should avoid strenuous outdoor activity, as respiratory tract irritation can occur.

Table 47.7. Pollution Standard Index for Ozone, Including Description and Possible Health Effects

Pollution Standard Index (Ozone)	Index Description	General Health Effects
≤ 50 (60 ppb: 1 h) (40 ppb: 8 h)	Good	None for general population
50–100 (60–120 ppb: 1 h) (40–80 ppb: 8 h)	Moderate	Few/none for general population
100–200 (120–240 ppb: 1 h) (80–160 ppb: 8 h)	Unhealthful	Mild aggravation of symptoms for susceptible population
		Irritation symptoms for healthy population
200–300 (240–360 ppb: 1 h) (160–240 ppb: 8 h)	Very unhealthful	Significant aggravation of symptoms and decreased exercise tolerance for populations with heart or lung disease; widespread symptoms for healthy population;
> 300 (> 360 ppb: 1 h) (> 240 ppb: 8 h)	Hazardous	Early onset of certain diseases; significant aggravation of symptoms and decreased exercise tolerance for healthy population

8.5.2 Treatment

Diagnosis of ozone-related toxicity is based on a history of probable exposure and recognition of symptoms compatible with exposure. Ozone-related symptoms may mimic several cardiorespiratory illnesses, including influenza, the common cold, sinusitus, asthma, bronchopneumonia, pulmonary embolism, and myocardial infarction (713). Persons with asthma may have exacerbations, which should be monitored and treated. Severe industrial overexposure should be managed as with acute lung injury. Ozone-related symptoms can be self-limited; recovery in milder cases generally occurs within hours after termination of exposure. However, because ozone has many similarities (including the capacity to irritate the distal lung) to nitrogen dioxide, vigilance for delayed pulmonary edema is recommended. Industrial ozone intoxication has led to a prolonged convalescence, in which resolution of symptoms required up to 2 weeks (713). In mild to moderate cases, symptomatic treatment should include analgesics for headache and chest pain, and cough suppressants if appropriate.

8.5.3 Measurement

From its discovery by Schonbein in the nineteenth century, until the mid-1970s, ozone was measured by collection in a potassium iodide solution ($O_3 + 3KI + H_2O \rightarrow O_2 +$

$KI_3 + 2KOH$). The iodine produced by this reaction was measured by either colorimetric or electrochemical means. Mainly because of interference with other pollutants (nitrogen dioxide and sulfur dioxide), this method been replaced with chemiluminescence and ultraviolet photometry for routine monitoring.

The USEPA reference method, gas-phase chemiluminescence, involves mixing ambient air with ethylene, which reacts with ozone to produce an electronically excited species emitting visible light (451). A photomultiplier detects the light and produces a photocurrent directly proportional to the ozone concentration with a detection limit of 0.004 ppm (866).

Ozone analyzers require periodic calibration with reference standards containing known concentrations of ozone. Some instruments are available with internal calibration systems. Because standard samples of ozone cannot be stored or distributed, calibration systems require an ozone generator, which is typically a mercury vapor lamp irradiating clean air passing through a quartz tube. The output of the ozone generator itself must be calibrated by a primary reference method before it can be used to calibrate other ozone analyzers for field use. The primary reference methods in current use for calibration of ozone generators include ultraviolet (UV) photometry and gas-phase titration with nitric oxide, although a variety of wet chemical methods using potassium iodide have been used (659).

Equivalent USEPA methods include UV photometry and gas–solid chemiluminescence; both methods are highly specificity with limited interferences with other common pollutants (659). Ultraviolet photometry measures the amount of light absorbed (254 nm). Commercial UV photometers alternate from a zero-air sample (created by diverting the airstream through an internal ozone scrubber before detection) and the environmental sample (in which the ozone is to be measured). The analyzer calculates the ozone concentration from the difference and has a detection limit of 0.001 ppm (659, 867). Gas–solid chemiluminescence also has a high specificity and a low detection limit. In these analyzers, chemiluminescence is produced by ozone's reaction with Rhodamine B adsorbed on activated silica. The intensity of the chemiluminescence is proportional to the concentration of ozone (659). Finally, passive sampling devices for ozone have been developed in order to evaluate personal exposures, but have been used only for research purposes (868–870).

BIBLIOGRAPHY

1. N. I. Sax and R. Lewis, *Hawley's Condensed Chemical Dictionary*, 11th ed., Van Nostrand-Reinhold, New York, 1987.
2. Centers for Disease Control, Carbon monoxide intoxication — A preventable environmental health hazard. *Morbid. Mortal. Wkly. Rep.* **31**, 529–531 (1982).
3. N. Cobb and R. A. Etzel, Unintentional carbon monoxide-related deaths in the United States, 1979 through 1988. *J. Am. Med. Assoc.* **266**, 659–663 (1991).
4. A. Woolf et al., Serious poisonings among other adults: A study of hospitalization and mortality rates in Massachusetts 1983-85. *Am. J. Public Health* **80**, 867–869 (1990).
5. K. K. Jain, *Carbon Monoxide Poisoning*, Warren H. Green, St. Louis, MO, 1990.

6. National Institute for Occupational Safety and Health (NIOSH), *Occupational Diseases: A Guide to Their Recognition*, rev. ed., DHEW (NIOSH) 77–181, U.S. Department of Health, Education, and Welfare, Washington, DC, 1977.
7. D. Hunter, *The Diseases of Occupations*, 6th ed., Hodder & Stoughton, London, 1978.
8. R. L. Pruett, Synthesis gas: A raw material for Industrial chemicals. *Science* **211**, 11–16 (1981).
9. N. H. Black and J. B. Conant, *Practical Chemistry*, Macmillan, New York, 1931.
10. R. R. Sayers and S. J. Davenport, *U.S. Public Health Service, Review of Carbon Monoxide Poisoning: 1936*, Public Health Bull. No. 195, U.S. Government Printing Office, Washington, DC, 1937.
11. P. W. Brandt-Rauf et al., Health hazards of fire fighters: Exposure assessment. *Br. J. Ind. Med.* **45**, 606–612 (1988).
12. J. Jankovic et al., Environmental study of firefighters. *Ann. Occup. Hyg.* **35**, 581–601 (1991).
13. J. R. Brotherhood et al., Fire fighters' exposure to carbon monoxide during Australian bushfires. *Am. Ind. Hyg. Assoc. J.* **51**, 234–240 (1990).
14. B. L. Materna et al., Occupational exposures in California wildland fire fighting. *Am. Ind. Hyg. Assoc. J.* **53**, 69–76 (1992).
15. R. A. Anderson, A. A. Watson, and W. A. Harland, Fire deaths in the Glasgow area: II. The role of carbon monoxide. *Med. Sci. Law* **21**, 288–294 (1981).
16. S. M. Ayers et al., Health effects of exposure to high concentrations of automotive emissions: Studies in bridge and tunnel workers in New York City. *Arch. Environ. Health* **27**, 168–178 (1973).
17. F. B. Stern, R. A. Lemen, and R. A. Curtis, Exposure of motor vehicle examiners to carbon monoxide: A historical prospective mortality study. *Arch. Environ. Health* **36**, 59–66 (1981).
18. Centers for Disease Control, Carbon monoxide exposure in aircraft fuelers — New York City. *Morbid. Mortal. Wkly. Rep.* **28**, 254–255 (1979).
19. Centers for Disease Control, Carbon monoxide poisoning in a garment-manufacturing plant — North Carolina. *Morbid. Mortal. Wkly. Rep.* **36**, 543–545 (1987).
20. T. A. Fawcett et al., Warehouse workers' headache: Carbon monoxide poisoning from propane-fueled forklifts. *J. Occup. Med.* **34**, 12–15 (1992).
21. Centers for Disease Control, Fatal carbon monoxide poisoning in a camper-truck — Georgia. *Morbid. Mortal. Wkly. Rep* **40**, 154–155 (1991).
22. N. B. Hampson and D. M. Norkool, Carbon monoxide poisoning in children riding in the back of pickup trucks. *J. Am. Med. Assoc.* **267**, 538–540 (1992).
23. J. P. Piatt et al., Occult carbon monoxide poisoning in an infant. *Pediatr. Emerg. Care* **6**, 21–23 (1990).
24. H. Venning, D. Roberton, and A. D. Milner, Carbon monoxide poisoning in an infant. *Br. Med. J.* **284**, 651 (1982).
25. Centers for Disease Control, Carbon monoxide intoxication associated with use of a gasoline-powered resurfacing machine in an ice-skating rink — Pennsylvania. *Morbid. Mortal. Wkly. Rep.* **33**, 49–50 (1984).
26. L. J. Paulozzi, F. Satink, and R. F. Spengler, A carbon monoxide mass poisoning in an ice arena in Vermont. *Am. J. Public Health* **81**, 222 (1991).
27. R. Iglewicz et al., Elevated levels of carbon monoxide in the patient compartment of ambulances. *Am. J. Public Health* **74**, 511–512 (1984).

28. U.S. Environmental Protection Agency (USEPA), *National Air Pollutant Emission Estimates 1940-1989*, EPA-450/4-91-004, USEPA, Washington, DC, 1991.
29. F. D. Stump, K. T. Knapp, and W. D. Ray, Seasonal impact of blending oxygenated organics with gasoline on motor vehicle tailpipe and evaporative emissions. *J. Air Waste Manage. Assoc.* **40**, 872–880 (1990).
30. R. E. Burney, S. C. Wu, and M. J. Nemiroff, Mass carbon monoxide poisoning: Clinical effects and results of treatment in 184 victims. *Ann. Emerg. Med.* **11**, 394–399 (1982).
31. Y. H. Caplan et al., Accidental poisonings involving carbon monoxide, heating systems, and confined spaces. *J. Forensic Sci.* **31**, 117–121 (1986).
32. Centers for Disease Control, Carbon monoxide inhalation—Florida. *Morbid. Mortal. Wkly. Rep.* **29**, 574 (1980).
33. B. P. O'Sullivan, Carbon monoxide poisoning in an infant exposed to a kerosene heater. *J. Pediatr.* **103**, 249–215 (1983).
34. M. Wharton et al., Fatal carbon monoxide poisoning at a motel. *J. Am. Med. Assoc.* **261**, 1177–1178 (1989).
35. P. S. Heckerling et al., Predictors of occult carbon monoxide poisoning in patients with headache and dizziness. *Ann. Intern. Med.* **107**, 174–176 (1987).
36. P. S. Heckerling, J. B. Leikin, and A. Maturen, Occult carbon monoxide poisoning: Validation of a prediction model. *Am. J. Med.* **84**, 251–256 (1988).
37. L. A. Wallace, Carbon monoxide in air and breath of employees in an underground office. *J. Air Pollut. Control Assoc.* **33**, 678–682 (1983).
38. D. Lester, The effects of detoxification of domestic gas on suicide in the United States. *Am. J. Public Health* **80**, 80–81 (1990).
39. I. S. Choi, Delayed neurologic sequelae in carbon monoxide intoxication. *Arch. Neurol. (Chicago)* **40**, 433–435 (1983).
40. J. D. Gasman, J. Varon, and J. P. Gardner, Revenge of the barbecue grill-carbon monoxide poisoning. *West. J. Med.* **153**, 656–657 (1990).
41. L. H. Hawkins, P. V. Cole, and J. R. W. Harris, Smoking habits and blood carbon monoxide levels. *Environ. Res.* **11**, 310–318 (1976).
42. W. S. Rickert, J. C. Robinson, and N. Collishaw, Yields of tar, nicotine, and carbon monoxide in the sidestream smoke from 15 brands of Canadian cigarettes. *Am. J. Public Health* **74**, 228–231 (1984).
43. T. D. Sterling, C. W. Collett, and J. A. Ross, Levels of environmental tobacco smoking under different conditions of ventilation and smoking regulation. In J. P. Harper, ed., *Combustion Processes and the Quality of the Indoor Environment*, Air & Waste Management Association, Pittsburgh, PA, 1989, pp. 223–235.
44. E. P. Radford and T. A. Drizd, *Blood Carbon Monoxide Levels in Persons 3–74 Years of Age: United States, 1976-80*, Advance data, Vol. 76, National Center for Health Statistics, U.S. Department of Health and Human Services, Washington, DC, 1982, pp. 1–24.
45. R. D. Stewart and C. L. Hake, Paint-remover hazard. *J. Am. Med. Assoc.* **235**, 398–401 (1976).
46. V. L. Kubic and M. W. Anders, Metabolism of dihalomethanes to carbon monoxide: II. In vitro studies. *Drug Metab. Dispos.* **3**, 104–112 (1975).
47. G. D. DiVincenzo and C. J. Kaplan, Uptake, metabolism, and elimination of methylene chloride vapor by humans. *Toxicol. Appl. Pharmacol.* **59**, 141–148 (1981).

48. T. Sjöstrand, Endogenous formation of carbon monoxide in man under normal and pathological conditions. *Scand. J. Clin. Lab. Invest.* **1**, 201–214 (1949).
49. M. Delivoria-Papadapoulos, R. F. Coburn, and R. E. Forster, Cyclic variation of rate of carbon monoxide production in normal women. *J. Appl. Physiol.* **36**, 49–51 (1974).
50. R. F. Coburn, Endogenous carbon monoxide production. *N. Engl. J. Med.* **282**, 207–209 (1980).
51. R. F. Coburn, The carbon monoxide body stores. *Ann. N. Y. Acad. Sci.* **174**, 1–22 (1970).
52. K. Luomanmaki and R. F. Coburn, Effects of metabolism and distribution of carbon monoxide on blood and body stores. *Am. J. Physiol.* **217**, 354–363 (1969).
53. J. Haldane, The relation of the action of carbonic oxide to oxygen tension. *J. Physiol. (London)* **18**, 201–217 (1895).
54. C. G. Douglas, J. S. Haldane, and J. B. S. Haldane, The laws of combination of haemoglobin with carbon monoxide and oxygen. *J. Physiol. (London)* **44**, 275–304 (1912).
55. R. F. Coburn, R. E. Forster, and P. B. Kane, Considerations of the physiological variables that determine the blood carboxyhemoglobin concentration in man. *J. Clin. Invest.* **44**, 1899–1910 (1965).
56. J. E. Peterson and R. D. Stewart, Absorption and elimination of carbon monoxide by inactive young men. *Arch. Environ. Health* **21**, 165–171 (1970).
57. J. E. Peterson and R. D. Stewart, Predicting the carboxyhemoglobin levels resulting from carbon monoxide exposures. *J. Appl. Physiol.* **39**, 633–638 (1975).
58. R. D. Stewart, The effect of carbon monoxide on humans. *Annu. Rev. Pharmacol.* **15**, 409–423 (1975).
59. A. H. Marcus, Mathematical models for carboxyhemoglobin. *Atmos. Environ.* **14**, 841–844 (1980).
60. R. Joumard et al., Mathematical models of the uptake of carbon monoxide on hemoglobin at low carbon monoxide levels. *Environ. Health Perspect.* **41**, 277–289 (1981).
61. H. Hauck and M. Neuberger, Carbon monoxide uptake and the resulting carboxyhemoglobin in man. *Eur. J. Appl. Physiol.* **53**, 186–190 (1984).
62. P. Tikuisis et al., A critical analysis of the use of the CFK equation of predicting COHb formation. *Am. Ind. Hyg. Assoc. J.* **48**, 208–213 (1987).
63. M. L. McCartney, Sensitivity analysis applied to Coburn-Forster-Kane models of carboxyhemoglobin formation. *Am. Ind. Hyg. Assoc. J.* **51**, 169–177 (1990).
64. P. M. Winter and J. N. Miller, Carbon monoxide poisoning. *J. Am. Med. Assoc.* **236**, 1502–1504 (1976).
65. L. Stryer, *Molecular Design of Life*, Freeman, New York, 1989.
66. L. R. Goldbaum, R. G. Ramirez, and B. A. Karel, What is the mechanism of carbon monoxide toxicity? *Aviat. Space Environ. Med.* **46**, 1289–1291 (1975).
67. K. R. Olson and C. E. Becker, Hyperbaric oxygen for carbon monoxide poisoning. *J. Am. Med. Assoc.* **248**, 172–173 (1982).
68. R. P. Geyer, The design of artificial blood substitutes. In E. J. Ariens, ed., *Drug Design*, Academic Press, New York, 1976, pp. 1–58.
69. C. A. Piantadosi, Carbon monoxide, oxygen transport, and oxygen metabolism. *J. Hyperbaric Med.* **2**, 27–44 (1987).
70. S. R. Thom, Antagonism of carbon monoxide-mediated brain lipid peroxidation by hyperbaric oxygen. *Toxicol. Appl. Pharmacol.* **105**, 340–344 (1990).

71. D. G. Penney, Acute carbon monoxide poisoning: Animal models: A review. *Toxicology* **62**, 123–160 (1990).
72. B. R. Pitt et al., Interaction of carbon monoxide and cyanide on cerebral circulation and metabolism. *Arch. Environ. Health* **34**, 354–359 (1979).
73. D. G. Penney, Hemodynamic response to carbon monoxide. *Environ. Health Perspect.* **77**, 121–130 (1988).
74. J. A. Sokal and E. Kralkowska, The relationship between exposure duration, carboxyhemoglobin, blood glucose, puruvate and lactate and the severity of intoxication in 39 cases of acute carbon monoxide poisoning in man. *Arch. Toxicol.* **57**, 196–199 (1985).
75. D. G. Penney, The cardiac toxicity of carbon monoxide. In S. I. Baskin ed., *Principles of Cardiac Toxicology*, CRC Press, Boca Raton, FL, 1991, pp. 573–605.
76. D. G. Penney and J. W. Howley, Is there a connection between carbon monoxide exposure and Hypertension? *Environ. Health Perspect.* **95**, 191–198 (1991).
77. C. J. Smith and T. J. Steichen, The atherogenic potential of carbon monoxide. *Atherosclerosis* **99**, 137–149 (1993).
78. L. D. Fechter, Toxicity of carbon monoxide exposure in early development. In S. Kacew and M. J. Reasor, eds., *Toxicology and the Newborn*, Elsevier, Amsterdam, 1984, pp. 122–140.
79. C. F. Mactutus, Developmental neurotoxicity of nicotine, carbon monoxide, and other tobacco smoke constituents. *Ann. N. Y. Acad. Sci.* **562**, 105–122 (1989).
80. C. F. Mactutus and L. D. Fechter, Prenatal exposure to carbon monoxide: Learning and memory deficits. *Science* **223**, 409–411 (1984).
81. P. Astrup et al., Effect of moderate carbon-monoxide exposure on fetal development. *Lancet* **2**, 1221–1222 (1972).
82. J. Singh and L. H. Scott, Threshold for carbon monoxide induced fetotoxicity. *Teratology* **30**, 253–257 (1984).
83. L. D. Fechter and Z. Annau, Toxicity of mild prenatal carbon monoxide exposure. *Science* **197**, 680–682 (1977).
84. L. D. Fechter and Z. Annau, Prenatal carbon monoxide exposure alters behavioral development. *Neurobehav. Toxicol.* **2**, 7–11 (1980).
85. C. F. Mactutus and L. D. Fechter, Moderate prenatal carbon monoxide exposure produces persistent, and apparently permanent, memory deficits in rats. *Teratology* **31**, 1–12 (1985).
86. L. D. Longo, Carbon monoxide: Effects on oxygenation of the fetus in utero. *Science* **194**, 523–525 (1976).
87. E. P. Hill et al., Carbon monoxide exchanges between the human fetus and mother: A mathematical model. *Am. J. Physiol.* **232**, H311–H323 (1977).
88. L. D. Longo, The biological effects of carbon monoxide on the pregnant woman, fetus, and newborn infant. *Am. J. Obstet. Gynecol.* **129**, 69–103 (1977).
89. M. Bogusz et al., A comparison of two types of acute carbon monoxide poisoning. *Arch. Toxicol.* **33**, 141–149 (1975).
90. S. M. Horvath et al., Maximal aerobic capacity at different levels of carboxyhemoglobin. *J. Appl. Physiol.* **38**, 300–303 (1975).
91. T. W. Grace and F. W. Platt, Subacute carbon monoxide poisoning: Another great imitator. *J. Am. Med. Assoc.* **246**, 1698–1700 (1981).
92. M. C. Dolan et al., Carboxyhemoglobin levels in patients with flu-like symptoms. *Ann. Emerg. Med.* **16**, 782–786 (1987).

93. J. N. Kirkpatrick, Occult carbon monoxide poisoning. *West. J. Med.* **146**, 52–56 (1987).
94. V. G. Laties and W. H. Merigan, Behavioral effects of carbon monoxide on animals and man. *Annu. Rev. Pharmacol. Toxicol.* **19**, 357–392 (1979).
95. V. A. Benignus, K. E. Muller, and C. M. Malott, Dose-effects functions for carboxyhemoglobin and behavior. *Neurotoxicol. Teratol.* **12**, 111–118 (1990).
96. M. J. Hosko, The effect of carbon monoxide on the visual evoked response in man. *Arch. Environ. Health* **21**, 174–180 (1979).
97. R. R. Beard and G. A. Wertheim, Behavioral impairment associated with small doses of carbon monoxide. *Am. J. Public Health* **57**, 2012–2022 (1967).
98. G. R. Wright and R. J. Shephard, Carbon monoxide exposure and auditory duration discrimination. *Arch. Environ. Health* **33**, 226–235 (1978).
99. V. R. Putz, B. L. Johnson, and J. V. Setzer, A comparative study of the effects of carbon monoxide and methylene chloride on human performance. *J. Environ. Pathol. Toxicol.* **2**, 97–112 (1979).
100. R. D. Stewart et al., Effect of carbon monoxide on time perception. *Arch. Environ. Health* **27**, 155–160 (1973).
101. A. Grut, *Chronic Carbon Monoxide Poisoning*, Munksgaard, Copenhagen, 1949.
102. G. J. Gilbert and G. H. Glaser, Neurologic manifestations of chronic carbon monoxide poisoning. *N. Engl. J. Med.* **261**, 1217–1220 (1959).
103. S. A. Lindgren, A study of the effect of protracted occupational exposure to carbon monoxide, with special reference to the occurrence of so-called chronic carbon monoxide poisoning. *Acta Med. Scand.* **356**(Suppl.), 1–83 (1961).
104. D. Shusterman, Problem-solving techniques in occupational medicine. *J. Fam. Pract.* **21**, 195–199 (1985).
105. F. H. Schillito, C. K. Drinker, and T. J. Shaughnessy, The problem of nervous and mental sequelae in carbon monoxide poisoning. *J. Am. Med. Assoc.* **106**, 669–674 (1936).
106. J. S. Smith and S. Brandon, Morbidity from acute carbon monoxide poisoning at three-year follow-up. *Br. Med. J.* **1**, 318–321 (1973).
107. H. L. Klawans et al., A pure Parkinsonian syndrome following acute carbon monoxide intoxication. *Arch. Neurol. (Chicago)* **39**, 302–304 (1982).
108. R. S. Jaeckle and H. A. Nasrallah, Major depression and carbon monoxide-induced Parkinsonism: Diagnosis, computerized axial tomography, and response to L-dopa. *J. Nerv. Ment. Dis.* **173**, 503–508 (1985).
109. D. Laplane et al., Obsessive-compulsive and other behavioral changes with bilateral basal ganglia lesions. A neuropsychological, magnetic resonance imaging and positron tomography study. *Brain* **112**, 699–725 (1989).
110. A. Lugaresi et al., 'Psychic akinesia' following carbon monoxide poisoning. *Eur. Neurol.* **30**, 167–169 (1990).
111. S. M. Pulst, T. M. Walshe, and J. A. Romero, Carbon monoxide poisoning with features of Gilles de la Tourette's syndrome. *Arch. Neurol. (Chicago)* **40**, 443–444 (1983).
112. C. R. Norris, J. M. Trench, and R. Hook, Delayed carbon monoxide encephalopathy: Clinical and research implications. *J. Clin. Psychiatry* **43**, 294–295 (1982).
113. G. M. Sawa et al., Delayed encephalopathy following carbon monoxide intoxication. *Can. J. Neurol. Sci.* **8**, 77–79 (1981).

114. B. Werner et al., Two cases of acute carbon monoxide poisoning with delayed neurological sequelae after a 'free' interval. *J. Toxicol. Clin. Toxicol.* **23**, 249–265 (1985).
115. K. S. Kim et al., Acute carbon monoxide poisoning: Computed tomography of the brain. *Am. J. Neuroradiol.* **1**, 399–402 (1980).
116. L. R. Nardizzi, Computerized tomographic correlate of carbon monoxide poisoning. *Arch. Neurol. (Chicago)* **36**, 38–39 (1979).
117. Y. Sawada et al., Computerized tomography as an indication of long-term outcome after acute carbon monoxide poisoning. *Lancet* **1**, 783–784 (1980).
118. A. S. Zagami, A. K. Lethlean, and R. Mellick, Delayed neurological deterioration following carbon monoxide exposure: MRI findings. *J. Neurol.* **240** 113–116 (1993).
119. G. Wilson and N. W. Winkleman, Multiple neuritis following carbon monoxide-poisoning. *J. Am. Med. Assoc.* **82**, 1407–1410 (1924).
120. I. S. Choi, Peripheral neuropathy following acute carbon monoxide poisoning. *Muscle Nerve* **9**, 265–266 (1986).
121. Y. Katafuchi et al., Cortical blindness in acute carbon monoxide poisoning. *Brain Dev.* **7**, 516–519 (1985).
122. G. Quattrocolo et al., A case of cortical blindness due to carbon monoxide poisoning. *Ital. J. Neurol. Sci.* **8**, 57–58 (1987).
123. N. C. Reynolds and I. Shapiro, Retrobulbar neuritis with neuroretinal edema as a delayed manifestation of carbon monoxide poisoning: Case report. *Mil. Med.* **144**, 472–473 (1979).
124. L. C. Dempsey, J. J. O'Donnell, and J. T. Hoff, Carbon monoxide retinopathy. *Am. J. Ophthalmol.* **82**, 692–693 (1976).
125. J. S. Kelley and G. J. Sophocleus, Retinal hemorrhages in subacute carbon monoxide poisoning. *J. Am. Med. Assoc.* **239**, 1515–1517 (1978).
126. R. C. Bilchik, H. A. Muller-Bergh, and M. E. Freshman, Ischemic retinopathy due to carbon monoxide poisoning. *Arch. Ophthalmol. (Chicago)* **86**, 142–144 (1971).
127. S. R. Baker and D. J. Lilly, Hearing loss from acute carbon monoxide intoxication. *Ann. Otol. Rhinol. Laryngol.* **86**, 323–328 (1977).
128. American Heart Association, *Textbook of Advanced Cardiac Life Support*, 2nd ed., American Heart Association, Dallas, TX, 1990, pp. 2–3.
129. W. S. Linn et al., Effects of exposure to 4 ppm nitrogen dioxide in healthy and asthmatic volunteers. *Arch. Environ. Health* **40**, 234–239 (1985).
130. S. Ebisuno et al., Myocardial infarction after acute carbon monoxide poisoning: Case report. *Angiology* **37**, 621–624 (1986).
131. A. L. Marius-Nunez, Myocardial infarction with normal coronary arteries after acute exposure to carbon monoxide. *Chest* **97**, 491–494 (1990).
132. E. H. Atkins and E. L. Baker, Exacerbation of coronary artery disease by occupational carbon monoxide exposure: A report to two fatalities and a review of the literature. *Am. J. Ind. Med.* **7**, 73–79 (1985).
133. E. N. Allred et al., Short-term effects of carbon monoxide exposure on the exercise performance of subjects with coronary artery disease. *N. Engl. J. Med.* **321**, 1426–1432 (1989).
134. W. S. Aronow, Aggravation of angina pectoris by two percent carboxyhemoglobin. *Am. Heart J.* **101**, 154–157 (1981).
135. M. T. Kleinman et al., Effects of short-term exposure to carbon monoxide in subjects with coronary artery disease. *Arch. Environ. Health* **44**, 361–369 (1989).

136. W. S. Aronow, E. A. Stemmer, and M. W. Isbell, Effect of carbon monoxide exposure on intermittent claudication. *Circulation* **49**, 415–417 (1974).

137. W. S. Aronow, J. Ferlinz, and F. Glauser, Effect of carbon monoxide on exercise performance in chronic obstructive lung disease. *Am. J. Med.* **63**, 904–908 (1977).

138. P. M. A. Calverley, R. J. E. Leggett, and Flenley, Carbon monoxide and exercise tolerance in chronic bronchitis and emphysema. *Br. Med. J.* **283**, 878–880 (1981).

139. D. S. Sheps et al., Production of arrhythmias by elevated carboxyhemoglobin in patients with coronary artery disease. *Ann. Intern. Med.* **113**, 343–351 (1990).

140. N. L. Benowitz, Cardiotoxicity in the workplace. *Occup. Med. State Art Rev.* **7**, 465–478 (1992).

141. T. S. Kristensen, Cardiovascular diseases and the work environment. A critical review of the epidemiologic literature on chemical factors. *Scand. J. Work Environ. Health* **15**, 245–264 (1989).

142. S. Hernberg et al., Angina pectoris, ECG findings and blood pressure of foundary workers in relation to carbon monoxide exposure. *Scand. J. Work Environ. Health* **2**, 54–63 (1976).

143. F. B. Stern et al., Heart disease mortality among bridge and tunnel officers exposed to carbon monoxide. *Am. J. Epidemiol.* **128**, 1276–1288 (1988).

144. C. Edling and O. Axelson, Risk factors of coronary heart disease among personnel in a bus company. *Int. Arch. Occup. Environ. Health* **54**, 181–183 (1984).

145. J. R. Goldsmith, Carbon monoxide research — recent and remote. *Arch. Environ. Health* **21**, 118–120 (1970).

146. K. Master, Carbon monoxide, atherosclerosis, and natives of the highlands of New Guinea. *N. Engl. J. Med.* **294**, 556–557 (1976).

147. E. J. Wouters et al., Smoking and low birth weight: Absence of influence by carbon monoxide? *Eur. J. Obstet. Gynecol. Reprod. Biol.* **25**, 35–41 (1987).

148. G. L. Muller and S. Graham, Intrauterine death of the fetus due to accidental carbon monoxide poisoning. *N. Engl. J. Med.* **252**, 1075–1078 (1955).

149. D. P. Goldstein, Carbon monoxide poisoning in pregnancy. *Am. J. Obstet. Gynecol.* **92**, 526–528 (1965).

150. C. R. Cramer, Fetal death due to accidental maternal carbon monoxide poisoning. *J. Toxicol. Clin. Toxicol.* **19**, 297–301 (1982).

151. C. A. Norman and D. M. Halton, Is carbon monoxide a workplace teratogen? A review and evaluation of the literature. *Ann. Occup. Hyg.* **34**, 335–347 (1990).

152. J. L. Margulies, Acute carbon monoxide poisoning during pregnancy. *Am. J. Emerg. Med.* **4**, 516–519 (1986).

153. R. A. Myers, S. K. Snyder, and T. Majerus, Cutaneous blisters and carbon monoxide poisoning. *Ann. Emerg. Med.* **14**, 119–606 (1985).

154. U. W. Leavell, C. H. Farley, and J. S. McIntyre, Cutaneous changes in a patient with carbon monoxide poisoning. *Arch. Dermatol.* **99**, 429–433 (1969).

155. R. Nagy, K. E. Greer, and L. E. German, Cutaneous manifestations of acute carbon monoxide poisoning. *Cutis* **24**, 381–383 (1979).

156. J. Finely, A. VanBeek, and J. L. Glover, Myonecrosis complicating carbon monoxide poisoning. *J. Trauma* **17**, 536–540 (1977).

157. G. D. Herman, Myonecrosis in carbon monoxide poisoning. *Vet. Hum. Toxicol.* **30**, 28–30 (1988).

158. R. Bessoudo and J. Gray, Carbon monoxide poisoning and nonoliguric acute renal failure. *Can. Med. Assoc. J.* **119**, 41–44 (1978).
159. K. Yamamoto and C. Kuwahara, A study on the combined action of CO and HCN in terms of concentration-time products. *Z. Rechtsmed.* **86**, 287–294 (1981).
160. A. K. Chaturedi et al., Exposures to carbon monoxide, hydrogen cyanide, and their mixture: Interrelationship between gas exposure concentration, time to incapacitation, carboxyhemoglobin and blood cyanide in rats. *J. Appl. Toxicol.* **15**, 357–363 (1995).
161. D. G. Barceloux, Nickel. *J. Toxicol. Clin. Toxicol.* **37**, 239–258 (1999).
162. G. Winneke, The neurotoxicity of dichloromethane. *Neurobehav. Toxicol. Teratol.* **3**, 391–395 (1981).
163. National Institute for Occupational Safety and Health (NIOSH), *Manual of Analytical Methods*, 3rd ed., Method No. S340, U.S. Department of Health and Human Services, NIOSH, Cincinnati, OH, 1984.
164. J. E. Zatek, Direct-reading gas and vapor monitors. In B. A. Plog, ed., *Fundamentals of Industrial Hygiene*, 3rd ed., National Safety Council, Chicago, IL, 1988, pp. 435–453.
165. J. P. Lodge, ed., *Methods of Air Sampling and Analysis*, 3rd ed., Lewis Publishers, Chelsea, MI, 1989, pp. 296–306.
166. R. F. Coburn et al., Carbon monoxide in blood: Analytical method and sources of error. *J. Appl. Physiol.* **19**, 510–515 (1964).
167. E. E. Dahms and S. M. Horvath, Rapid, accurate technique for determination of carbon monoxide in blood. *Clin. Chem. (Winston-Salem, N.C.)* **20**, 533–537 (1974).
168. E. J. van Kampen and W. G. Zijlstra, Determination of hemoglobin and its derivatives. *Adv. Clin. Chem.* **8**, 141–187 (1965).
169. A. H. J. Maas, M. L. Hamelink, and R. J. M. De Leeuw, An evaluation of the spectrophotometric determination of HbO_2, HbCO, and Hb in blood with the CO-oximeter IL 182. *Clin. Chim. Acta* **29**, 303–309 (1970).
170. J. J. Mahoney et al., Measurement of carboxyhemoglobin and total hemoglobin by five specialized spectrophotometers (CO-oximeters) in comparison with reference methods. *Clin. Chem. (Winston-Salem, N.C.)* **39**, 1693–1700 (1993).
171. P. J. Mathews, Jr., Co-oximetry. *Respir. Care Clin. North Am.* **1**, 47–68 (1995).
172. J. A. Brunelle et al., Simultaneous measurement of total hemoglobin and its derivatives in blood using CO-oximeters: Analytical principles; their application in selecting analytical wavelengths and reference methods; a comparison of the results of the choices made. *Scand. J. Clin. Lab. Invest., Suppl.* **224**, 47–69 (1996).
173. J. E. Hodgkin and D. M. Chan, Diabetic ketoacidosis appearing as carbon monoxide poisoning. *J. Am. Med. Assoc.* **231**, 1164–1165 (1975).
174. M. Vegfors and C. Lennmarken, Carboxyhaemglobinaemia and pulse oximetry. *Br. J. Anaesth.* **66**, 625–626 (1991).
175. R. D. Stewart et al., Rapid estimation of carboxyhemoglobin level in fire fighters. *J. Am. Med. Assoc.* **235**, 390–391 (1976).
176. J. E. Peterson Postexposure relationship of carbon monoxide in blood and expired air. *Arch. Environ. Health* **21**, 172–173 (1970).
177. A. D. McNeill et al., Abstinence from smoking and expired-air carbon monoxide levels: Lactose intolerance as a possible source of error. *Am. J. Public Health* **80**, 1114–1115 (1990).

178. A. L. Ilano and T. A. Raffin, Management of carbon monoxide poisoning. *Chest* **97**, 165–169 (1990).
179. R. A. Myers et al., Value of hyperbaric oxygen in suspected carbon monoxide poisoning. *J. Am. Med. Assoc.* **246**, 2478–2480 (1981).
180. D. M. Norkool and J. N. Kirkpatrick, Treatment of acute carbon monoxide poisoning with hyperbaric oxygen: A review of 115 cases. *Ann. Emerg. Med.* **14**, 1168–1171 (1985).
181. K. B. Van Hoesen et al., Should hyperbaric oxygen be used to treat the pregnant patient for acute carbon monoxide poisoning? A case report and literature review. *J. Am. Med. Assoc.* **261**, 1039–1043 (1989).
182. A. T. Proudfoot, Carbon monoxide poisoning—recent advances. *Acta Clin. Belg., Suppl.* **13**, 61–68 (1990).
183. National Heart Lung and Blood Institute, NHLBI Workshop Summary. Hyperbaric oxygenation therapy. *Am. Rev. Respir. Dis.* **144**, 1414–1421 (1991).
184. P. M. Tibbles and J. S. Edelsberg, Hyperbaric oxygen therapy. *N. Engl. J. Med.* **334**, 1642–1648 (1996).
185. R. A. Myers, Carbon monoxide poisoning. *J. Emerg, Med.* **1**, 245–248 (1984).
186. J. C. Raphael et al., Trial of normobaric and hyperbaric oxygen for acute carbon monoxide intoxication. *Lancet* **2**, 414–419 (1989).
187. E. P. Sloan et al., Complications and protocol considerations in carbon monoxide-poisoned patients who require hyperbaric oxygen therapy: Report from a ten-year experience. *Ann. Emerg. Med.* **18**, 629–634 (1989).
188. D. G. Murphy et al., Tension pneumothorax associated with hyperbaric oxygen therapy. *Am. J. Emerg. Med.* **9**, 176–179 (1991).
189. National Institute for Occupational Safety and Health (NIOSH), *Pocket Guide to Chemical Hazards*, DHHS (NIOSH) 90–117, U.S. Government Printing Office, Washington, DC, 1990.
190. National Institute for Occupational Safety and Health (NIOSH), *Criteria for a Recommended Standard: Occupational Exposure to Carbon Monoxide*, HSM 73-11000, U.S. Department of Health, Education, and Welfare, Washington, DC, 1972.
191. Occupational Safety and Health Administration (OSHA), Air contaminants; Proposed rule (29 CFR 1910). *Fed. Regis.* **53**, 21171 (1988).
192. Occupational Safety and Health Administration (OSHA), *Air Contaminants—Permissible Exposure Limits* (Title 29 Code of Federal Regulations, Part 1910.1000), OSHA 3112, U.S. Department of Labor, Washington, DC, 1989.
193. American Conference of Governmental Industrial Hygienists (ACGIH), *1991-1992 Threshold Limit Values for Chemical Substances and Physical Agents and Biological Exposure Indices*, ACGIH, Cincinnati, OH, 1991.
194. American Conference of Governmental Industrial Hygienists (ACGIH), *Documentation of the Threshold Limit Values and Biological Exposure Indices*, 6th ed., ACGIH, Cincinnati, OH, 1991.
195. J. E. Cone and M. J. Hodgson, eds., Problem buildings. *State Art Rev. Occup. Med.* **4**(4), (1989).
196. American Society of Heating, Refrigerating, Airconditioning Engineers (ASHRAE), *Ventilation for Acceptable Indoor Air Quality*, ASHRAE Stand. 62-1989, ASHRAE, Atlanta, GA, 1989.
197. T. Sollman, *A Manual of Pharmacology*, 6th ed., Saunders, Philadelphia, PA, 1944.

198. National Institute for Occupational Safety and Health (NIOSH), *Criteria for a Recommended Standard for Occupational Exposure to Carbon Dioxide* (NIOSH), 76-194, U.S. Government Printing Office, Washington, DC, 1976.
199. D. J. Cullen and E. L. Eger, Cardiovascular effects of carbon dioxide in man. *Anesthesiology* **41**, 345–349 (1974).
200. American Industrial Hygiene Association, Hygienic Guide Series: Carbon dioxide. *Am. Ind. Hyg. Assoc. J.* **25**, 519 (1964).
201. American Conference of Governmental Industrial Hygienists (ACGIH), *1990-1991 Threshold Limit Values for Chemical Substances and Physical Agents and Biological Exposure Indices*, ACGIH, Cincinnati, OH, 1990.
202. S. C. Thomas and C. W. Shilling, *Carbon Dioxide Effect in Mammalian Tissue: An Annotated Bibliography*, Undersea Medical Society, Bethesda, MD, 1980.
203. M. A. Frey et al., The effects of moderately elevated ambient carbon dioxide levels on human physiology and performance: A joint NASA-ESA-DARA study-overview. *Aviat. Space Environ. Med.* **69**, 282–284 (1998).
204. J. M. Arena and R. H. Drew, eds., *Poisoning; Toxicology, Symptoms, Treatments*, 5th ed., Thomas, Springfield, IL, 1986.
205. A. B. Raj, Behaviour of pigs exposed to mixtures of gases and the time required to stun and kill them: Welfare implications. *Vet. Rec.* **144**, 165–168 (1999).
206. B. S. Levy and D. H. Wegman, eds., *Occupational Health*, 2nd ed., Little, Brown, Boston, MA, 1988.
207. R. K. Creasy and R. Resnik, eds., *Maternal-Fetal Medicine: Principles and Practice*, 2nd ed., Saunders, Philadelphia, PA, 1989.
208. N. Yamashita, K. Kamiya, and H. Nagai, CO_2 reactivity and autoregulation in fetal brain. *Child's Nerv. Syst.* **7**(6), 327–331 (1991).
209. G. W. Kling et al., The 1986 Lake Nyos Gas Disaster in Cameroon, West Africa. *Science* **236**, 169–175 (1987).
210. Z. E. Afane et al., Respiratory symptoms and peak expiratory flow in survivors of the Nyos disaster. *Chest* **110**, 1278–1281 (1996).
211. P. J. Baxter, M. Kapila, and D. Mfonfu, Lake Nyos Disaster, Cameroon, 1986: The medical effects of large scale emission of carbon dioxide? *Br. Med. J.* **298**, 1341–1347 (1989).
212. S. J. Freeth et al., Conclusions from Lake Nyos Disaster. *Nature (London)* **348**, 201 (1990).
213. K. E. Schaefer, Studies of carbon dioxide toxicity. (1). Chronic CO_2 toxicity in submarine medicine. *U.S., Nav. Med. Res. Lab.*, pp. 156–176 (1951).
214. B. Max, This and that: The neurotoxicity of carbon dioxide. *Trends Pharmacol. Sci.* **12**(11), 408–411 (1991).
215. M. D. Hollingsworth, Triphosgene warning. *Chem. Eng. News* **70**(28), 4 (1992).
216. Occupational Safety and Health Administration (OSHA), *Industrial Exposure and Control Technologies for OSHA Regulated Hazardous Substances*, U.S. Department of Labor, Washington, DC, 1989, pp. 1579–1582.
217. M. H. Noweir, E. A. Pfitzer, and T. F. Hatch, Decomposition of chlorinated hydrocarbons: A review. *Am. Ind. Hyg. Assoc. J.* **33**, 454–460 (1972).
218. L. C. Rinzema and L. G. Silverstein, Hazards from chlorinated hydrocarbon decomposition during welding. *Am. Ind. Hyg. Assoc. J.* **33**, 35–40 (1972).

219. R. W. Snyder, H. S. Mishel, and G. C. Christensen Pulmonary toxicity following exposure to methylene chloride and its combustion product, phosgene. *Chest* **101**, 860–861 (1992).
220. J. Jaeger, *Phosgene Generated from Chloroform*, Old Knowledge Re-learned, Environ. Health Saf. Newsl. Article 53, University of Maryland, College Park, 1998. Available at http://www.ehs.umaryland.edu/News.Articles/pr53.html
221. U.S. Department of Commerce, Danger of phosgene generation from unstabilized chloroform. Available at http://www.nwsfe.noa.gov/protocols/chloroform.html
222. J. A. Dahlberg and L. M. Myrin, The formation of dichloroacetyl chloride and phosgene from trichloroethylene in the atmosphere of welding shops. *Ann. Occup. Hyg.* **14**, 269–274 (1974).
223. W. F. Diller, Pathogenesis of Phosgene poisoning. *Toxicol Ind. Health* **1**, 7–15 (1985).
224. National Institute for Occupational Safety and Health (NIOSH), *Criteria for a Recommended Standard: Occupational Exposure to Phosgene*, U.S. Department of Health, Education and Welfare, Washington, DC, 1976, p. 52.
225. J. R. Keeler et al., Estimation of the LCt50 of phosgene in sheep. *Drug Chem. Toxicol.* **13**, 229–239 (1990).
226. W. D. Currier, G. E. Hatch, and M. F. Frosolono, Pulmonary alterations in rats due to acute phosgene inhalation. *Fundam. Appl. Toxicol.* **8**, 107–114 (1987).
227. A. J. Ghio et al., Reduction of neutrophil influx diminishes lung injury and mortality following phosgene inhalation. *J. Appl. Physiol.* **71**, 657–665 (1991).
228. W. D. Currie, P. C. Pratt, and M. F. Frosolono, Response of pulmonary energy metabolism to phosgene. *Toxicol. Ind. Health* **1**, 17–27 (1985).
229. R. H. Jaskot et al., Effects of inhaled phosgene on rat lung antioxidant systems. *Fundam. Appl. Toxicol.* **17**, 666–674 (1991).
230. M. F. Frosolono and W. D. Currie, Response of the pulmonary surfactant system to phosgene. *Toxicol. Ind. Health* **1**, 29–35 (1985).
231. G. R. Burleson and L. L. Keyes, Natural killer activity in Fischer-344 rat lungs as a method to assess pulmonary immunocompetence: Immunosuppression by phosgene inhalation. *Immunopharmacol. Immunotoxicol.* **11**, 421–443 (1989).
232. J. P. Ehrlich and G. R. Burleson, Enhanced and prolonged pulmonary influenza virus infection following phosgene inhalation. *J. Toxicol. Environ. Health* **34**, 259–273 (1991).
233. J. E. Amoore and E. Hautala, Odor as an aid to chemical safety: Odor thresholds compared with threshold limit values and volatilities for 214 industrial chemicals in air and water dilution. *J. Appl. Toxicol.* **3**, 272–290 (1983).
234. J. H. Ruth, Odor thresholds and irritation levels of several chemical substances: A review. *Am. Ind. Hyg. Assoc. J.* **47**, A142–A151 (1986).
234a. M. Sun, EPA bars use of Nazi data. *Science* **240**, 21 (1988).
235. B. Sjögren et al., Pulmonary reactions caused by welding-induced decomposed trichloroethylene. *Chest* **99**, 237–238 (1991).
236. A. Selden and L. Sundell, Chlorinated solvents, welding and pulmonary edema. *Chest* **99**, 263 (1991).
237. W. F. Diller, Late sequelae after phosgene poisoning: A literature review. *Toxicol. Ind. Health* **1**, 129–136 (1985).
238. American Industrial Hygiene Association, (AIHA), Hygienic Guide Series: Phosgene. *Am. Ind. Hyg. Assoc. J.* **29**, 308–311 (1968).

239. J. P. Lodge, ed., *Methods of Air Sampling and Analysis*, 3rd ed., Lewis Publishers, Chelsea, MI, 1989, pp. 606–607.

240. K. Bachmann and J. Polzer, Determination of tropospheric phosgene and other halocarbons by capillary gas chromatography. *J. Chromatogr.* **481**, 373–379 (1989).

241. J. P. Hendershott, The simultaneous determination of chloroformates and phosgene at low concentrations in air using a solid sorbent sampling-gas chromatographic procedure. *Am. Ind. Hyg. Assoc. J.* **47**, 742–746 (1986).

242. X. Luo, F. Y Shi, and J. X. Lin, CO-laser photoacoustic detection of phosgene ($COCl_2$). *Int. J. Infrared Millimeter Waves* **12**, 141–147 (1991).

243. W. F. Diller, Early diagnosis of phosgene overexposure. *Toxicol. Ind. Health* **1**, 73–80 (1985).

244. W. F. Diller, Therapeutic strategy in phosgene poisoning. *Toxicol. Ind. Health* **11**, 93–99 (1985).

245. W. F. Diller and R. Zante, A literature Review: Therapy for phosgene poisoning. *Toxicol. Ind. Health* **1**, 117–128 (1985).

246. M. F. Frosolono, Prophylactic and antidotal effects of hexamethylenetetramine against phosgene poisoning in rabbits. *Toxicol. Ind. Health* **1**, 101–116 (1985).

247. T. P. Kennedy et al., Dibutyryl cAMP, Aminophylline, and beta-adrenergic agonists protect against pulmonary edema caused by phosgene. *J. Appl. Physiol.* **67**, 2542–2552 (1989).

248. U.S. Naval Sea Systems Command, *U.S. Navy Diving Manual*, Revision 3, U.S. Government Printing Office, Washington DC, 1991.

249. C. A. Piantadosi, Physiological effects of altered biometric pressure. In G. D. Clayton and F. E. Clayton, eds., *Patty's Industrial Hygiene and Toxicology*, 4th ed., Vol. 1, Wiley, New York, 1991.

250. P. B. Bennett and D. H. Elliott, eds., *The Physiology and Medicine of Diving and Compressed Air Work*, 2nd ed., Baillière Tindall, London, 1975.

251. *Key Documents of the Biomedical Aspects of Deep-sea Diving; Selected from the World's Literature 1608–1982*, Undersea Medical Society, Bethesda, MD, 1983.

252. W. H. Rogers and G. Moeller, Effect of brief, repeated hyperbaric exposure on susceptibility to nitrogen narcosis. *Undersea Biomed. Res.* **16** (3), 227–232 (1989).

253. B. Fowler, E. Pang, and I. Mitchell, On controlling inert gas narcosis. *Hum. Factors* **34**(1), 115–120 (1992).

254. S. Budavari, ed., *The Merck Index: An Encyclopedia of Chemicals, Drugs, and Biologicals*, 11th ed., Merck & Co., Rahway, NJ, 1989.

255. R. C. Weast, *Handbook of Chemistry and Physics*, 69th ed., CRC Press, Cleveland, OH, 1988.

256. U. Bauer et al., Nickle RA: Acute effects of nitrogen dioxide after accidental release. *Public Health Rep.* **113**, 62–70 (1998).

257. J. H. Seinfeld, *Atmospheric Chemistry and Physics of Air Pollution*, Wiley, New York, 1986, pp. 79–86.

258. World Health Organization, *Air Quality Guidelines for Europe*, Eur. Ser. No. 23, WHO Regional Publications, Copenhagen, 1987.

259. South Coast Air Quality Management District, *1988 Air Quality Data*, South Coast Air Quality Management District, El Monte, CA, 1989.

260. K. Hedberg et al., An outbreak of nitrogen-dioxide-induced respiratory illness among ice-hockey players. *JAMA, J. Am. Med. Assoc.* **262**, 3014–3017 (1989).

261. Anonymous, Ice hockey lung: NO_2 poisoning. *Lancet* **335**, 1191 (1990).

262. W. Smith et al., Nitrogen dioxide and carbon monoxide intoxication in an indoor ice arena — Wisconsin, 1992. *Morbid. Mortal. Wkly. Rep.* **41**, 383–385 (1992).
263. J. I. Levy et al., Determinants of nitrogen dioxide concentrations in indoor ice skating rinks. *Am. J. Public Health* **88**, 1781–1786 (1998).
264. M. H. Garrett et al., Respiratory symptoms in children and indoor exposure to nitrogen dioxide and gas stoves. *Am. J. Respir. Crit. Care Med.* **158**, 891–895 (1998).
265. T. Lowry and L. M. Schuman, Silo-Filler's Disease — A syndrome caused by nitrogen dioxide. *J. Am. Med. Assoc.* **162**, 153–160 (1956).
266. B. T. Cummins, F. J. Raveney, and M. W. Jesson, Toxic gases in tower silos. *Ann. Occup. Hyg.* **14**, 275–283 (1971).
267. J. D. Spengler et al., Nitrogen dioxide inside and outside 137 homes and implications for ambient air quality standards and health effects research. *Environ. Sci. Technol.* **17**, 164–168 (1983).
268. M. C. Marbury et al., Indoor residential NO_2 concentrations in Albuquerque, New Mexico. *J. Air Pollut. Control Assoc.* **38**, 392–398 (1988).
269. I. F. Goldstein and L. R. Andrews, Peak exposures to nitrogen dioxide and study design to detect their acute health effects. *Environ. Int.* **13**, 285–291 (1987).
270. T. Lindvall, Health effects of nitrogen dioxide and oxidants. *Scand. J. Work Environ. Health* **11**(Suppl. 13), 10–28 (1985).
271. B. P. Leaderer, Air pollutant emissions from kerosene space heaters. *Science* **218**, 1113–1115 (1982).
272. C. Borland and T. Higenbottam, Nitric oxide yields of contemporary U.K., U.S. and French cigarettes. *Int. J. Epidemiol.* **16**, 31–34 (1987).
273. C. C. Yockey, B. M. Eden, and R. B. Byrd, The McConnell missile accident: Clinical spectrum of nitrogen dioxide exposure. *J. Am. Med. Assoc.* **244**, 1221–1223 (1980).
274. H.-M. Wagner, Absorption of NO and NO_2 in MIK and MAK concentrations during inhalation. *Staub — Reinhalt. Luft* **30**, 25–26 (1970).
275. M. A. Bauer et al., Inhalation of 0.30 ppm nitrogen dioxide potentiates exercise-induced bronchospasm in asthmatics. *Am. Rev. Respir. Dis.* **134**, 1203–1208 (1986).
276. E. Goldstein et al., Absorption and transport of nitrogen oxides. In S. D. Lee, ed., *Nitrogen Oxides and Their Effects on Health*, Ann Arbor Sci. Publ., Ann Arbor, MI, 1983, pp. 143–160.
277. E. M. Postlethwaite and A. Bidani, Reactive uptake governs the pulmonary air space removal of inhaled nitrogen dioxide. *J. Appl. Physiol.* **68**, 594–603 (1990).
278. E. M. Postlethwaite and A. Bidani, Pulmonary disposition of inhaled NO_2-nitrogen in isolated rat lungs. *Toxicol. Appl. Pharmacol.* **98**, 303–312 (1989).
279. E. M. Postlethwaite and M. G. Mustafa, Fate of inhaled nitrogen dioxide in isolated perfused rat lung. *J. Toxicol. Environ. Health* **7**, 861–872 (1981).
280. W. A. Pryor, Mechanism and detection of pathology caused by free radicals, tobacco smoke, nitrogen dioxide, and ozone. In J. D. McKinney, ed., *Environmental Health Chemistry. The Chemistry of Environmental Agents as Potential Human Hazards*, Ann Arbor Sci. Publ., Ann Arbor, MI, 1981, pp. 445–466.
281. K. Yoshida and K. Kasama, Biotransformation of nitric oxide. *Environ. Health Perspect.* **73**, 201–206 (1987).

282. M. Meyer and J. Piiper, Nitric Oxide (NO), a new test gas for study of alveolar-capillary diffusion. *Eur. Respir. J.* **2**, 494–496 (1989).
283. H. Kosaka, M. Uozumi, and I. Tyuma, The interaction between nitrogen oxides and hemoglobin and endothelium-derived relaxing factor. *Free Radical Biol. Med.* **7**, 653–658 (1989).
284. J. S. Beckman et al., Apparent hydroxyl radical production by peroxynitrite: Implications for endothelial injury from nitric oxide and superoxide. *Proc. Natl. Acad. Sci. U.S.A.* **87**, 1620–1624 (1990).
285. B. A. Freeman, H. Gutierrez, and H. Rubbo, Nitric oxide: A central regulatory species in pulmonary oxidant reactions. *Am. J. Physiol.* **268**, 697–698 (1995).
286. G. L. Squadrito and W. A. Pryor, Oxidative chemistry of nitric oxide: The roles of superoxide, peroxynitrite, and carbon dioxide. *Free Radical Biol. Med.* **25**, 392–403 (1998).
287. American Industrial Hygiene Association (AIHA), *Odor Thresholds for Chemicals with Established Occupational Health Standards*, AIHA, Fairfax, VA, 1989.
288. G. Bylin et al., Effects of short-term exposure to ambient nitrogen dioxide concentrations on human bronchial reactivity and lung function. *Eur. J. Respir. Dis.* **66**, 205–217 (1985).
289. J. E. Amore and E. Hautala, Odor as an aid to chemical safety: Odor thresholds compared with threshold limit values and volatilities for 214 industrial chemicals in air and water dilution. *J. Appl. Toxicol.* **3**, 272–290 (1983).
290. R. L. Tse and A. A. Bockman, Nitrogen dioxide toxicity. Report of four cases in firemen. *J. Am. Med. Assoc.* **212**, 1341–1344 (1970).
291. T. Lindquist, Nitrous gas poisoning among welders using acetylene flame. *Acta Med. Scand.* **119**, 210–243 (1944).
292. B. E. Lehnert et al., Lehnert, LR Gurley, DM Stavert, Lung injury following exposure of rats to relatively high mass concentrations of nitrogen dioxide. *Toxicology* **89**, 239–277 (1994).
293. E. P. Horvath et al., Nitrogen dioxide-induced pulmonary disease. Five new cases and a review of the literature. *J. Occup. Med.* **20**, 103–110 (1978).
294. G. M. P. Leib et al., Chronic pulmonary insufficiency secondary to Silo-Filler's disease. *Am. J. Med.* **24**, 471–474 (1958).
295. G. R. Jones, A. T. Proudfoot, and J. I. Hall, Pulmonary effects of acute exposure to nitrous fumes. *Thorax* **28**, 61–65 (1973).
296. T. Hirotani et al., Adult respiratory distress syndrome caused by inhalation of oxides of nitrogen. *Keio J. Med.* **36**, 315–320 (1987).
297. J. E. H. Milne, Nitrogen dioxide inhalation and bronchiolitis obliterans. *J. Occup. Med.* **11**, 538–547 (1969).
298. T. L. Guidotti, The higher oxides of nitrogen: Inhalation toxicology. *Environ. Res.* **15**, 443–472 (1978).
299. A. Seaton and W. K. C. Morgan, Toxic gases and fumes. In W. K. C. Morgan and A. Seaton, eds., *Occupational Lung Diseases*, 2nd ed., Saunders, Philadelphia, PA, 1984, pp. 609–642.
300. R. L. Moskowitz, H. A. Lyons, and H. R. Cottle, Silo filler's disease. *Am. J. Med.* **36**, 457–462 (1964).
301. J. A. Fleetham, B. W. Tunicliffe, and P. W. Munt, Methemoglobinemia and the oxides of nitrogen. *N. Engl. J. Med.* **298**, 1150 (1978).
302. C. Karlson-Stiber et al., Nitrogen dioxide pneumonitis in ice hockey players. *J. Intern. Med.* **239**, 451–456 (1996).

303. M. Brauer et al., Nitrogen dioxide in indoor ice skating facilities: An international survey. *J. Air Waste Manage. Assoc.* **47**, 1095–1102 (1997).

304. M. Rosenlund and G. Bluhm, Health effects resulting from nitrogen dioxide exposure in an indoor ice arena. *Arch. Environ. Health* **54**, 52–57 (1999).

305. G. von Nieding et al., Effect of experimental and occupational exposure to NO_2 in sensitive and normal subjects. In S. D. Lee, ed., *Nitrogen Oxides and Their Effects on Health*, Ann Arbor Sci. Publ., Ann Arbor, MI, 1980, pp. 315–331.

306. G. von Nieding and H. M. Wagner, Experimental studies on the short-term effect of air pollutants on pulmonary function in man: Two-hour exposures to NO_2, O_3, and SO_2 alone and in combination. *4th Int. Clean Air Congr.*, Tokyo, Japan. *1977*, pp. 5–8.

307. P. E. Morrow et al., Pulmonary performance of elderly normal subjects and subjects with chronic obstructive pulmonary disease exposed to 0.3 ppm nitrogen dioxide. *Am. Rev. Respir. Dis.* **145**, 291–300 (1992).

308. W. S. Linn et al., Controlled exposure of volunteers with chronic obstructive pulmonary disease to nitrogen dioxide. *Arch. Environ. Health* **40**, 313–317 (1985).

309. J. D. Hackney et al., Exposures of older adults with chronic respiratory illness to nitrogen dioxide. *Am. Rev. Respir. Dis.* **146**, 1480–1486 (1992).

310. H. D. Kerr et al., Effects of nitrogen dioxide on pulmonary function in human subjects: An environmental chamber study. *Environ. Res.* **19**, 392–404 (1979).

311. T. Sandström et al., Bronchoalveolar mastocytosis and lymphocytosis after nitrogen dioxide exposure in men: A time-kinetic study. *Eur. Respir. J.* **3**, 138–143 (1990).

312. T. Sandström et al., Inflammatory cell response in bronchoalveolar lavage fluid after nitrogen dioxide exposure of healthy subjects: A dose-response study. *Eur. Respir. J.* **3**, 332–339 (1991).

313. M. W. Frampton et al., Effects of nitrogen dioxide exposure on bronchoalveolar lavage proteins in humans. *Am. J. Respir. Cell. Mol. Biol.* **1**, 499–505 (1989).

314. R. B. Devlin et al., Inflammatory response in humans exposed to 2.0 ppm nitrogen dioxide. *Inhalation Toxicol.* **11**, 89–109 (1999).

315. I. Rubinstein, et al., Effects of 0.60 nitrogen dioxide on circulating and bronchoalveolar lavage lymphocyte phenotypes in healthy subjects. *Environ. Res.* **55**, 18–30 (1991).

316. H. S. Koren et al., Ozone-induced inflammation in the lower airways of human subjects. *Am. Rev. Respir. Dis.* **139**, 407–415 (1989).

317. J. Seltzer et al., O_3-Induced change in bronchial reactivity to methacholine and airway inflammation in humans. *J. Appl. Physiol.* **60**, 1321–1326 (1986).

318. A. Blomberg et al., Persistent airway inflammation but accommodated antioxidant and lung function responses after repeated daily exposure to nitrogen dioxide. *Am. J. Respir. Crit. Care Med.* **159**, 536–543 (1999).

319. E. M. Postlethwait et al., NO_2 reactive absorption substrates in rat pulmonary surface lining fluids. *Free Radical Biol. Med.* **19**, 553–563 (1995).

320. L. W. Velsor and E. M. Postlethwait, NO_2-induced generation of extracellular reactive oxygen is mediated by epithelial lining layer antioxidants. *Am. J. Physiol.* **273**, L1265–L1275 (1997).

321. F. J. Kelley et al., Antioxidant kinetics in lung lavage fluid following exposure of humans to nitrogen dioxide. *Am. J. Respir. Crit. Care Med.* **154**, 1700–1705 (1996).

322. F. J. Kelly and T. D. Tetley, Nitrogen dioxide depletes uric acid and ascorbic acid but not glutathione from lung lining fluid. *Biochem. J.* **325**, 95–99 (1997).

323. A. Blomberg et al., Persistent airway inflammation but accommodated antioxidant and lung function responses after repeated daily exposure to nitrogen dioxide. *Am. J. Respir. Crit. Care Med.* **159**, 536–543 (1999).

324. S. Idell and A. B. Cohen, The pathogenesis of emphysema. In J. E. Hodgkin and T. L. Petty, eds., *Chronic Obstructive Pulmonary Disease Current Concepts*, Saunders, Philadelphia, PA, 1987, pp. 7–21.

325. V. Hasselblad, D. M. Eddy, and D. J. Kotchmar, Synthesis of environmental evidence: Nitrogen dioxide epidemiology studies. *J. Air Waste Manage. Assoc.* **42**, 662–671 (1992).

326. C. Posin et al., Nitrogen dioxide inhalation and human blood biochemistry. *Arch. Environ. Health* **33**, 318–324 (1978).

327. J. Orehek et al., Effect of short-term, low-level nitrogen dioxide exposure on bronchial sensitivity of asthmatic patients. *J. Clin. Invest.* **57**, 301–307 (1976).

328. L. J. Roger et al., Pulmonary function, airway responsiveness and respiratory symptoms in asthmatics following exercise in NO_2. *Toxicol. Ind. Health* **6**, 155–171 (1990).

329. M. T. Kleinman et al., Effects of 0.2 ppm nitrogen dioxide on pulmonary function and response to bronchoprovocation in asthmatics. *J. Toxicol. Environ. Health* **12**, 815–826 (1983).

330. V. Mohsenin, Airway responses to nitrogen dioxide in asthmatic subjects. *J. Toxicol. Environ. Health* **22**, 371–380 (1987).

331. E. L. Avol et al., Laboratory study of asthmatic volunteers exposed to nitrogen dioxide and to ambient air pollution. *Am. Ind. Hyg. Assoc. J.* **49**, 143–149 (1988).

332. I. Rubinstein et al., Short-term exposure to 0.3 ppm nitrogen dioxide does not potentiate airway responsiveness to sulfur dioxide in asthmatic subjects. *Am. Rev. Respir. Dis.* **141**, 381–385 (1990).

333. M. J. Hazucha et al., Effects of 0.1 ppm nitrogen dioxide on airways of normal and asthmatic subjects. *J. Appl. Physiol.* **54**, 730–739 (1983).

334. M. J. Utell, Asthma and nitrogen dioxide: A review of the evidence. In M. J. Utell and R. Frank, eds., *Susceptibility to Inhaled Pollutants*, ATSM STP 1024, Am. Soc. Test. Mater., Philadelphia, PA, 1989, pp. 218–223.

335. B. Vagaggini et al., Effect of short-term NO_2 exposure on induced sputum in normal, asthmatic and COPD subjects. *Eur. Respir J.* **9**, 1852–1857 (1996).

336. V. Strand et al., Repeated exposure to an ambient level of NO_2 enhances asthmatic response to a nonsymptomatic allergen dose. *Eur. Respir. J.* **12**, 6–12 (1998).

337. M. I. Gilmour, P. Park, and M. J. Selgrade, Increased immune and inflammatory responses to dust mite antigen in rats exposed to >5 ppm NO_2. *Fundam. Appl. Toxicol.* **31**, 65–70 (1996).

338. P. D. Siegel et al., Adjuvant effect of respiratory irritation on pulmonary allergic sensitization: Time and site dependency. *Toxicol. Appl. Pharmacol.* **144**, 356–362 (1997).

339. M. W. Frampton et al., Effect of nitrogen dioxide exposure on pulmonary function and airway reactivity in normal humans. *Am. Rev. Respir. Dis.* **143**, 522–527 (1991).

340. V. Mohsenin, Airway responses to 2.0 ppm nitrogen dioxide in normal subjects. *Arch. Environ. Health* **43**, 242–246 (1988).

341. A. Papi et al., Bronchopulmonary inflammation and airway smooth muscle hyperresponsiveness induced by nitrogen dioxide in guinea pigs. *Eur. J. Pharmacol.* **374**, 241–247 (1999).

342. V. Mohsenin, Effect of vitamin C on NO_2-induced airway hyperresponsiveness in normal subjects. *Am. Rev. Respir. Dis.* **136**, 1408–1411 (1987).

343. R. M. Rose et al., The pathophysiology of enhanced susceptibility to murine cytomegalovirus respiratory infection during short-term exposure to 5 ppm nitrogen dioxide. *Am. Rev. Respir. Dis.* **137**, 912–917 (1988).

344. R. F. Parker et al., Short-term exposure to nitrogen dioxide enhances susceptibility to murine respiratory mycoplasmosis and decreases intrapulmonary killing of *Mycoplasma pulmonis*. *Am. Rev. Respir. Dis.* **140**, 502–512 (1989).

345. J. E. Pennington, Effects of automotive emissions on susceptibility to respiratory infections. In S. Y. Watson, R. R. Bates, and D. Kennedy, eds., *Air Pollution, the Automobile, and Public Health*, National Academy Press, Washington, DC, 1988, pp. 499–518.

346. R. B. Schlesinger, Comparative toxicity of ambient air pollutants: Some aspects related to lung defense. *Environ. Health Perspect.* **81**, 123–128 (1989).

347. T. Suzuki et al., Decreased phagocytosis and superoxide anion production in alveolar macrophages of rats exposed to nitrogen dioxide. *Arch. Environ. Contam. Toxicol.* **15**, 733–739 (1986).

348. G. J. Jakab, Modulation of pulmonary defense mechanisms by acute exposures to nitrogen dioxide. *Environ. Res.* **42**, 215–228 (1987).

349. S. V. Dawson and M. B. Schenker, Health effects of inhalation of ambient concentrations of nitrogen dioxide. *Am. Rev. Respir. Dis.* **120**, 281–292 (1979).

350. R. F. Heller and R. E. Gordon, Chronic effects of nitrogen dioxide on cilia in hamster bronchioles. *Exp. Lung Res.* **10**, 137–152 (1986).

351. M. A. Amoruso, G. Witz, and B. D. Goldstein, Decreased superoxide anion radical production by rat alveolar macrophages following inhalation of ozone or nitrogen dioxide. *Life Sci.* **28**, 2215–2221 (1981).

352. C. Aranyi et al., Scanning electron microscopy of alveolar macrophages after exposure to oxygen, nitrogen dioxide and ozone. *Environ. Health Perspect.* **16**, 180 (1976).

353. T. W. Robison et al., Depression of stimulated arachidonate metabolism and superoxide production in rat alveolar macrophages following in vivo exposure to 0.5 ppm NO_2. *J. Toxicol. Environ. Health* **38**, 273–292 (1993).

354. P. Pinkston et al., Effects of in vitro exposure to nitrogen dioxide on human alveolar macrophage release of neutrophil chemotactic factor and interleukin-1. *Environ. Res.* **47**, 48–58 (1989).

355. M. W. Frampton et al., Nitrogen dioxide exposure in vivo and human alveolar macrophage inactivation of influenza virus in vitro. *Environ. Res.* **48**, 179–192 (1989).

356. R. Ehrlich, J. C. Findlay, and D. E. Gardner, Health effects of short-term inhalation of nitrogen dioxide and ozone mixtures. *Environ. Res.* **14**, 223–231 (1977).

357. J. A. Graham et al., Influence of exposure patterns of nitrogen dioxide and modifications by ozone on susceptibility to bacterial infectious disease in mice. *J. Toxicol. Environ. Health* **21**, 113–125 (1987).

358. R. Ehrlich and M. C. Henry, Chronic toxicity of nitrogen dioxide: I. Effect on resistance to bacterial pneumonia. *Arch. Environ. Health* **17**, 860–865 (1968).

359. S. A. J. Goings et al., Effect of nitrogen dioxide on susceptibility to influenza. A virus infection in healthy adults. *Am. Rev. Respir. Dis.* **139**, 1075–1081 (1989).

360. F. E. Speizer et al., Respiratory disease rates and pulmonary function in children associated with NO_2 exposure. *Am. Rev. Respir. Dis.* **121**, 3–10 (1980).

361. B. D. Ostro et al., Asthmatic responses to airborne acid aerosols. *Am. J. Public Health* **81**, 694–702 (1991).

362. C. Infante-Rivard, Nitrogen dioxide and allergic asthma. *Lancet* **345**, 931 (1995).
363. L. C. Koo et al., Personal exposure to nitrogen dioxide and its association with respiratory illness in Hong Kong. *Am. Rev. Respir. Dis.* **141**, 1119–1126 (1990).
364. M. Jacobsen et al., Respiratory infections in coal miners exposed to nitrogen oxides. *Health Eff. Inst. Res. Rep.* **18**, 1–64 (1988).
365. J. M. Samet, M. C. Marbury, and J. D. Spengler, Health effects and sources of indoor air pollution. Part I. *Am. Rev. Respir. Dis.* **136**, 1486–1508 (1987).
366. J. D. Spengler et al., Nitrogen dioxide and respiratory illness in children. Part IV: Effects of housing and meteorologic factors on indoor nitrogen dioxide concentrations. *Res. Rep. Health Eff. Inst.* **58**, 1–36 (1996).
367. J. M. Samet et al., A study of respiratory illnesses in infants and nitrogen dioxide exposure. *Arch. Environ. Health* **47**, 57–63 (1992).
368. L. S. Pilotto et al., Respiratory effects associated with indoor nitrogen dioxide exposure in children. *Int. J. Epidemiol.* **26**, 788–796 (1997).
369. L. S. Pilotto and R. M. Douglas, Indoor nitrogen dioxide and childhood respiratory illness. *Aust. J. Public Health* **16**, 245–250 (1992).
370. L. S. Pilotto, R. M. Douglas, and J. M. Samet, Nitrogen dioxide, gas heating and respiratory illness. *Med. J. Aust.* **167**, 295–296 (1997).
371. A. J. Chauhan et al., Exposure to nitrogen dioxide (NO_2) and respiratory disease risk. *Rev. Environ. Health* **13**, 73–90 (1998).
372. D. M. Stavert and B. E. Lehnert, Nitric oxide and nitrogen dioxide as inducers of acute pulmonary injury when inhaled at relatively high concentrations for brief periods. *Inhalation Toxicol.* **2**, 53–67 (1990).
373. D. Hess, L. Bigatello, and W. E. Hurford, Toxicity and complications of inhaled nitric oxide. *Respir. Care Clin. North Am.* **3**, 487–503 (1997).
374. E. Troncy, M. Francoeur, and G. Blaise, Inhaled nitric oxide: Clinical applications, indications, and toxicology. *Can. J. Anaesth.* **44**, 973–988 (1997).
375. C. Hugod, Ultrastructural changes of the rabbit lung after a 5 ppm nitric oxide exposure. *Arch. Environ. Health* **34**, 12–17 (1979).
376. S. Moncada, R. M. J. Palmer, and E. A. Higgs, Nitric oxide: Physiology, pathophysiology, and pharmacology. *Pharmacol. Rev.* **43**, 109–142 (1991).
377. E. Anggard, Nitric oxide: Mediator, murderer, and medicine. *Lancet* **343**, 1119–1206 (1994).
378. R. Radi et al., Peroxynitrite oxidation of sulfhydryls. *J. Biol. Chem.* **266**, 4244–4250 (1991).
379. R. Radi et al., Peroxynitrite-induced membrane lipid peroxidation: The cytotoxic potential of superoxide and nitric oxide. *Arch. Biochem. Biophys.* **288**, 481–487 (1991).
380. R. D. Curran et al. Nitric oxide and nitric oxide-generating compounds inhibit hepatocyte protein synthesis. *FASEB J.* **5**, 2085–2092 (1991).
381. V. L. Dawson et al., Nitric oxide mediates glutamate neurotoxicity in primary cortical cultures. *Proc. Natl. Acad. Sci. U.S.A.* **88**, 6368–6371 (1991).
382. J. S. Beckman et al., Apparent hydroxyl radical production by peroxynitrite: Implications for endothelial injury from nitric oxide and superoxide. *Proc. Natl. Acad. Sci. U.S.A.* **87**, 1620–1624 (1990).
383. J. Pepke-Zaba et al., Inhaled nitric oxide as a cause of selective pulmonary vasodilatation in pulmonary hypertension. *Lancet* **338**, 1173–1174 (1991).

384. M. Hallman, K. Bry, and U. J. Lappalainen, A mechanism of nitric oxide-induced surfactant dysfunction. *J. Appl. Physiol.* **80**, 2035–2043 (1996).

385. S. Matalon et al., Inhaled nitric oxide injures the pulmonary surfactant system of lambs in vivo. *Am. J. Physiol.* **270**, L273–L280 (1996).

386. G. von Nieding, H. M. Wagner, and H. Krekeler, Investigation of the acute effects of nitrogen monoxide on lung function in man. *Proc. 3rd Int. Clean Air Conf.*, Dusseldorf, Germany, *1973*, pp. A14–A16.

387. J. Kagawa, Respiratory effects of 2-hr exposures to 1.0 ppm nitric oxide in normal subjects. *Environ. Res.* **27**, 485–490 (1982).

388. R. W. Day et al., Inhaled nitric oxide in children with severe lung disease: Results of acute and prolonged therapy with two concentrations. *Crit. Care Med.* **24**, 215–221 (1996).

389. R. P. Dellinger et al., Effects of inhaled nitric oxide in patients with acute respiratory distress syndrome: Results of a randomized phase II trial. Inhaled nitric oxide in ARDS study group. *Crit. Care Med.* **26**, 15–23 (1998).

390. C. M. Hart, Nitric oxide in adult lung disease. *Chest* **115**, 1407–1417 (1999).

391. M. A. Matthay, J. F. Pittet, and C. Jayr, Just say NO to inhaled nitric oxide for the acute respiratory distress syndrome. *Crit. Care Med.* **26**, 1–2 (1998).

392. C. Jones, Inhaled nitric oxide: Are the safety issues being addressed? *Intensive Crit. Care Nurs.* **14**, 271–275 (1998).

393. R. S. Ream et al., Low-dose inhaled nitric oxide improves the oxygenation and ventilation of infants and children with acute, hypoxemic respiratory failure. *Crit. Care Med.* **27**, 989–996 (1999).

394. M. R. Becklake et al., The long-term effects of exposure to nitrous fumes. *Am. Rev. Tuberc.* **76**, 398–409 (1957).

395. B. Muller, Nitrogen dioxide intoxication after a mining accident. *Respiration* **26**, 249–261 (1969).

396. U. Bauer et al., Acute effects of nitrogen dioxide after accidental release. *Public Health Rep.* **113**, 62–70 (1998).

397. M. J. Evans, Oxidant gases. *Environ. Health Perspect.* **55**, 85–95 (1984).

398. G. Freeman et al., Pathogenesis of the nitrogen dioxide-induced lesion in the rat lung: A review and presentation of new observations. *Am. Rev. Respir. Dis.* **98**, 429–443 (1968).

399. G. B. Haydon, G. Freeman, and N. J. Furiosi, Covert pathogenesis of NO_2-induced emphysema in the rat. *Arch. Environ. Health* **11**, 776–783 (1965).

400. J. E. Glasgow et al., Neutrophil recruitment and degranulation during induction of emphysema in the rat by nitrogen dioxide. *Am. Rev. Respir. Dis.* **135**, 1129–1136 (1987).

401. P. J. Barth et al., Diffuse alveolar damage in the rat lung after short and long term exposure to nitrogen dioxide. *Pathol. Res. Pract.* **190**, 33–41 (1994).

402. R. R. Mercer, D. L. Costa, and J. D. Crapo, Effects of prolonged exposure to low doses of nitric oxide or nitrogen dioxide on the alveolar septa of the adult rat lung. *Lab. Invest.* **73**, 20–28 (1995).

403. P. J. Barth et al., Quantitiative analysis of parenchymal and vascular alterations in NO_2-induced lung injury in rats. *Eur. Respir. J.* **8**, 1115–1121 (1995).

404. C. Lafuma et al., Effect of low-level NO_2 chronic exposure on elastase-induced emphysema. *Environ. Res.* **43**, 75–84 (1987).

405. D. Hyde et al., Morphometric and morphologic evaluation of pulmonary lesions in beagle dogs chronically exposed to high ambient levels of air pollutants. *Lab. Invest.* **38**, 455–469 (1978).

406. R. P. Sherwin and V. Richters, Hyperplasia of type 2 pneumocytes following 0.34 ppm nitrogen dioxide exposure: Quantitation by image analysis. *Arch. Environ. Health* **37**, 306–315 (1982).

407. K. Kubota et al., Effects of long-term nitrogen dioxide exposure on rat lung: Morphological observations. *Environ. Health Perspect.* **73**, 157–169 (1987).

408. Y. Hayashi, T. Kohno, and H. Ohwada, Morphological effects of nitrogen dioxide on the rat lung. *Environ. Health Perspect.* **73**, 135–145 (1987).

409. W. H. Blair, M. C. Henry, and R. Ehrlich, Chronic toxicity of nitrogen dioxide. II. Effect on histopathology of lung tissue. *Arch. Environ. Health* **18**, 186–192 (1969).

410. P. J. A. Rombout et al., Influence of exposure regimen on nitrogen dioxide induced morphological changes in the rat lung. *Environ. Res.* **41**, 466–480 (1986).

411. J. D. Crapo et al., Alterations in lung structure caused by inhalation of oxidants. *J. Toxicol. Environ. Health*, **13**, 301–321 (1984).

412. P. E. Morrow, Toxicological data on NOx: An overview. In F. J. Miller and D. B. Menzel, eds., *Fundamentals of Extrapolation Modeling of Inhaled Toxicants: Ozone and Nitrogen Dioxide*, Hemisphere Publishing, Washington, DC, 1984, pp. 205–227.

413. M. C. S. Kennedy, Nitrous fumes and coal-miners with emphysema. *Ann. Occup. Hyg.* **15**, 285–300 (1972).

414. A. Robertson et al., Exposure to oxides of nitrogen: Respiratory symptoms and lung function in British coalminers. *Br. J. Ind. Med.* **41**, 214–219 (1984).

415. G. L. Euler et al., Chronic obstructive pulmonary disease symptom effects of long-term cumulative exposure to ambient levels of total oxidants and nitrogen dioxide in California Seventh-Day Adventist residents. *Arch. Environ. Health* **43**, 279–285 (1988).

416. R. Detels et al., The UCLA population studies of CORD: X. A cohort study of changes in respiratory function associated with chronic exposure to SOx, NOx, and hydrocarbons. *Am. J. Public Health* **81**, 350–359 (1991).

417. J. M. Peters et al., A study of twelve southern California communities with differing levels and types of air pollution. I. Prevalence of respiratory morbidity. *Am. J. Respir. Crit. Care Med.* **159**, 760–767 (1999).

418. A. Ponka and M. Virtanen, Chronic bronchitis, emphysema, and low-level air pollution in Helsinki, 1987–1989. *Environ. Res.* **65**, 207–217 (1994).

419. S. A. Dawkins, H. Gerhard, and M. Nevin, Pulmonary alveolar proteinosis: A possible sequel of NO_2 exposure. *J. Occup. Med.* **33**, 638-641 (1991).

420. National Research Council (NRC), *Nitrogen Oxides*, National Academy Press, Washington, DC, 1977.

421. U.S. Environmental Protection Agency (USEPA), *Air Quality Criteria for Oxides of Nitrogen*, 3 vols., EPA/600/8–91/049aA, USEPA, Washington, DC, 1994.

422. H. Fujimaki, F. Shimizu, and K. Kubota, Effect of subacute exposure to NO_2 on lymphocytes required for antibody responses. *Environ. Res.* **29**, 280–286 (1982).

423. E. Azoulay-Dupuis et al., Evidence for humoral immunodepression in NO_2-exposed mice: Influence of food restriction and stress. *Environ. Res.* **42**, 446–454 (1987).

424. A. Richters and K. S. Damji, Changes in T-lymphocyte subpopulations and natural killer cells following exposure to ambient levels of nitrogen dioxide. *J. Toxicol. Environ. Health* **25**, 247–256 (1988).

425. K. V. Kuraitis and A. Richters, spleen cellularity shifts from the inhalation of 0.250.35 ppm nitrogen dioxide. *J. Environ. Pathol. Toxicol. Oncol.* **9**, 1–11 (1989).

426. A. Richters, Effects of nitrogen dioxide and ozone on blood-borne cancer cell colonization of the lungs. *J. Toxicol. Environ. Health* **25**, 383–390 (1988).

427. A. Richters and V. Richters, Nitrogen dioxide (NO_2) inhalation, formation of microthrombi in lungs and cancer metastasis. *J. Environ. Pathol. Toxicol. Oncol.* **9**, 45–51 (1989).

428. A. Richters, V. Richters, and W. P. Alley, The mortality rate from lung metastases in animals inhaling nitrogen dioxide (NO_2). *J. Surg. Oncol.* **28**, 63–66 (1985).

429. S. F. P. Man and W. C. Hulbert, Airway repair and adaptation to injury. In J. Loke, ed., *Pathophysiology and Treatment of Inhalation Injuries*, Dekker, New York, 1988, pp. 1–47.

430. M. Sagai and T. Ichinose, Lipid peroxidation and antioxidative protection mechanism in rat lungs upon acute and chronic exposure to nitrogen dioxide. *Environ. Health Perspect.* **73**, 179–189 (1987).

431. M. G. Mustafa and D. F. Tierney, Biochemical and metabolic changes in the lung with oxygen, ozone, and nitrogen dioxide toxicity. *Am. Rev. Respir. Dis.* **118**, 1061–1090 (1990).

432. E. J. Calabrese and H. M. Horton, The effects of vitamin E on ozone and nitrogen dioxide toxicity. *World Rev. Nutr. Diet.* **46**, 124–147 (1985).

433. S. Gorsdorf et al., Nitrogen dioxide induces DNA single-strand breaks in cultured Chinese hamster cells. *Carcinogenesis (London)* **11**, 37–41 (1990).

434. K. Victorin et al., Genotoxic activity of 1, 3-butadiene and nitrogen dioxide and their photochemical reaction products in drosophila and in the mouse bone marrow micronucleus assay. *Mutat. Res.* **228**, 203–209 (1990).

435. K. Victorin and M. Stahlberg, Mutagenic activity of ultraviolet-irradiated mixtures of nitrogen dioxide and propene or butadiene. *Environ. Res.* **49**, 271–282 (1989).

436. T. Ichinose, K. Fujii, and M. Sagai, Experimental studies on tumor promotion by nitrogen dioxide. *Toxicology* **67**, 211–225 (1991).

437. H. P. Witschi, Ozone, nitrogen dioxide and lung cancer: A review of some recent issues and problems. *Toxicology* **48**, 1–20 (1988).

438. W. W. Douglas, N. G. Hepper, and T. V. Colby, Silo-Filler's disease. *Mayo Clin. Proc.* **64**, 291–304 (1989).

439. J. Gailitis, L. E. Burns, and J. B. Nally, Silo-Fillers disease. Report of a case. *N. Engl. J. Med.* **258**, 543–544 (1958).

440. Occupational Safety and Health Administration (OSHA), *Nitrogen Dioxide in Workplace Atmospheres (Ion Chromatography)*, Method ID-182, revised, OSHA SLC Analytical Laboratory, Salt Lake City, UT, 1991.

441. Occupational Safety and Health Administration (OSHA), *Nitric Oxide in Workplace Atmospheres (Ion Chromatography)*, Method ID-190, revised, OSHA SLC Analytical Laboratory, Salt Lake City, UT, 1991.

442. J. Namiesnik, T. Gorecki, and E. Kozlowski, Passive dosimeters — An approach to atmospheric pollutants analysis. *Sci. Total Environ.* **38**, 225–258 (1984).

443. E. D. Palmes et al., Personal sampler for nitrogen dioxide. *Am. Ind. Hyg. Assoc. J.* **37**, 570–577 (1976).

444. D. P. Miller, Ion chromatographic analysis of palmes tubes for nitrite. *Atmos. Environ.* **18**, 891–892 (1984).

445. D. P. Miller, Low-level determination of nitrogen dioxide in ambient air using the Palmes tube. *Atmos. Environ.* **22**, 945–947 (1988).

446. E. D. Palmes and C. Tomczyk, Personal sampler for NOx. *Am. Ind. Hyg. Assoc. J.* **40**, 588–591 (1979).

447. Y. Yanagisawa and H. Nishimura, A badge-type personal sampler for measurement of personal exposure to NO_2 and NO in ambient air. *Environ. Int.* **8**, 235–242 (1982).

448. F. C. Fehsenfeld et al., Intercomparison of NO_2 measurement techniques. *J. Geophys. Res.* **95**, 3579–3597 (1990).

449. J. M. Hoell, Jr. et al., An intercomparison of nitric oxide measurement techniques. *J. Geophys. Res.* **90**, 12843–12851 (1985).

450. L. J. Purdue and T. R. Hauser, Review of U.S. Environmental Protection Agency NO_2 Monitoring Methodology Requirements. In S. D. Lee, ed., *Nitrogen Oxides and Their Effects on Health*, Ann Arbor Sci. Publ., Ann Arbor, MI, 1980, pp. 51–76.

451. 40 Code of Federal Regulations Part 50, Appendix D, Measurement principles and calibration procedures for the measurement of ozone in the atmosphere 1991.

452. T. J. Kelly, C. W. Spicer, and G. F. Ward, An assessment of the luminol chemiluminescence technique for measurement of NO_2 in ambient air. *Atmos. Environ.* **24A**, 2397–2403 (1990).

453. E. A. M. Frost, A history of nitrous oxide. In E. I. Eger, II, ed., *Nitrous Oxide N_2O*, Elsevier, New York, 1985, pp. 1–22.

454. S. Budavari, ed., *The Merck Index: An Encyclopedia of Chemicals, Drugs, and Biologicals*, 11th ed., Merck & Co., Rahway, NJ, 1989.

455. J. M. Wynne, Physics, chemistry, and manufacture of nitrous oxide. In E. I. Eger, II, ed., *Nitrous Oxide N_2O*, Elsevier, New York, 1985, pp. 23–40.

456. H. Rodhe, A comparison of the contribution of various gases to the greenhouse effect. *Science* **248**, 1217–1219 (1990).

457. C. Whitcher, Controlling occupational exposure to nitrous oxide. In E. I. Eger, II, ed., *Nitrous Oxide N_2O*, Elsevier, New York, 1985, pp. 313–337.

458. A. J. Suruda and J. D. McGlothlin, Fatal abuse of nitrous oxide in the workplace. *J. Occup. Med.* **32**, 682–684 (1990).

459. National Institute for Occupational Safety and Health (NIOSH), *Criteria for a Recommended Standard... Occupational Exposure to Waste Anesthetic Gases and Vapors*, U.S. Department of Health, Education, and Welfare, Washington, DC, 1977.

460. R. L. Campbell et al., Exposure to anesthetic waste gas in oral surgery. *J. Oral Surg.* **35**, 625–630 (1977).

461. J. A. Ship, A survey of nitrous oxide levels in dental offices. *Arch. Environ. Health* **42**, 310–314 (1987).

462. R. J. Gardner, J. Hampton, and J. S. Causton, Inhalation anaesthetics — exposure and control during veterinary surgery. *Ann. Occup. Hyg.* **35**, 377–388 (1991).

463. R. J. Gardner, Inhalation anaesthetics — exposure and control: A statistical comparison of personal exposures in operating theatres with and without anaesthetic gas scavenging. *Ann. Occup. Hyg.* **33**, 159–173 (1989).

464. A. M. Sass-Kortsak et al., Exposure of hospital operating room personnel to potentially harmful environmental agents. *Am. Ind. Hyg. Assoc. J.* **53**, 203–209 (1992).

465. W. M. Gray, Occupational exposure to nitrous oxide in four hospitals. *Anesthesia* **44**, 511–514 (1989).
466. E. I. Eger, II. Pharmacokinetics. In E. I. Eger, II, ed., *Nitrous Oxide N_2O*, Elsevier, New York, 1985, p. 107.
467. J. R. Trudell, Metabolism of nitrous oxide. In E. I. Eger, II, ed., *Nitrous Oxide N_2O*, Elsevier, New York, 1985, pp. 203–210.
468. M. Imbriani et al., Nitrous oxide (N_2O) in urine as biological index of exposure in operating room personnel. *Appl. Ind. Hyg.* **33**, 223–226 (1988).
469. J. F. Nunn and I. Chanarin, Nitrous oxide inactivates methionine synthetase. In E. I. Eger, II, ed., *Nitrous Oxide N_2O*, Elsevier, New York, 1985, pp. 211–233.
470. D. L. Hansen and R. E. Billings, Effects of nitrous oxide on maternal and embryonic folate metabolism in rats. *Dev. Pharmacol. Ther.* **8**, 43–54 (1985).
471. M. J. Landon et al., Influence of vitamin B_{12} status on the inactivation of methionine synthase by nitrous oxide. *Br. J. Anaesth.* **69**, 81–86 (1992).
472. D. S. Ostreicher, Vitamin B_{12} supplements as protection against nitrous oxide inhalation. *N. Y. State Dent. J.* **60**, 47–49 (1994).
473. S. P. Rothenberg, Increasing the dietary intake of folate: Pros and cons. *Semin. Hematol.* **36**, 65–74 (1999).
474. T. S. Flippo and W. D. Holder, Jr., Neurologic degeneration associated with nitrous oxide anesthesia in patients with vitamin B_{12} deficiency. *Arch. Surg. Chicago* **128**, 1391–1395 (1993).
475. R. T. Louis-Ferdinand, Myelotoxic, neurotoxic and reproductive adverse effects of nitrous oxide. *Adverse Drug React. Toxicol. Rev.* **13**, 193–206 (1994).
476. P. J. Pema, H. A. Horak, and R. H. Wyatt, Myelopathy caused by nitrous oxide toxicity. *Am. J. Neuroradiol.* **19**, 894–896 (1998).
477. J. F. Nunn et al., Serum methionine and hepatic enzyme activity in anaesthetists exposed to nitrous oxide. *Br. J. Anaesth.* **54**, 593–597 (1982).
478. R. B. Layzer, Myeloneuropathy after prolonged exposure to nitrous oxide. *Lancet* **2**, 1227–1230 (1978).
479. R. B. Layzer, Nitrous oxide abuse. In E. I. Enger, II, ed., *Nitrous Oxide N_2O*, Elsevier, New York, 1985, pp. 249–257.
480. E. J. Heyer et al., Nitrous oxide: Clinical and electrophysiologic investigation of neurologic complications. *Neurology* **36**, 1618–1622 (1986).
481. E. N. Cohen et al., Occupational disease in dentistry and chronic exposure to trace anesthetic gases. *J. Am. Dent. Assoc.* **101**, 21–31 (1980).
482. J. B. Brodsky, Toxicity of nitrous oxide. In E. I. Eger, II, ed., *Nitrous Oxide N_2O*, Elsevier, New York, 1985, pp. 259–279.
483. T. N. Tannenbaum and R. J. Goldberg, Exposure to anesthetic gases and reproductive outcome. *J. Occup. Med.* **27**, 659–668 (1985).
484. Canadian Centre for Occupational Safety and Health, CCOSH, *Anaesthetic Gases and Vapours*, Hamilton, Ontario, 1986.
485. J. M. Baden, Mutagenicity, carcinogenicity, and teratogenicity of nitrous oxide. In E. I. Eger, II, ed., *Nitrous Oxide N_2O*, Elsevier, New York, 1985, pp. 235–247.

486. W. B. Coate, B. M. Ulland, and T. R. Lewis, Chronic exposure to low concentrations of halothane-nitrous oxide: Lack of carcinogenic effect in the rat. *Anesthesiology* **50**, 306–309 (1979).
487. National Institute for Occupational Safety and Health (NIOSH), *Control of Occupational Exposure to N_2O in the Dental Operatory*, DHEW (NIOSH) Publ. No. 77B171, U.S. Department of Health, Education, and Welfare, Public Health Service, Center for Disease Control, Cincinnati, OH, 1977.
488. National Institute for Occupational Safety and Health (NIOSH), *Controlling Exposures to Nitrous Oxide During Anesthetic Administration*, NIOSH ALERT, Publ. No. 94–100, U.S. Department of Health and Human Services, Public Health Service, Centers for Disease Control, Cincinnati, OH, 1994.
489. American Conference of Governmental Industrial Hygienists (ACGIH), *Documentation of the Threshold Limit Values and Biological Exposure Indices: Nitrous Oxide Revision*, 6th ed., pp. 1134B–1138B. ACGIH, Cincinnati, OH, 1992.
490. National Institute for Occupational Safety and Health (NIOSH), Nitrous oxide. In P.M. Eller, ed., *Manual of Analytical Methods*, 3rd ed., Method 6600, DHHS (NIOSH) Publ. No. 84B100, U.S. Department of Health and Human Services, Public Health Service, Centers for Disease Control, Cincinnati, OH, 1984.
491. J. D. McGlothlin et al., *In-depth Survey Report: Control of Anesthetic Gases in Dental Operatories at Children's Hospital Medical Center Dental Facility*, Rep. No. ECTB 166B11b, U.S. Department of Health and Human Services, Public Health Service, Centers for Disease Control, National Institute for Occupational Safety and Health, Cincinnati, OH, 1989.
492. E. C. Bishop and M. A. Hossain, Field comparison between two nitrous oxide (N_2O) passive monitors and conventional sampling methods. *Am. Ind. Hyg Assoc. J.* **45** (12), 812–816 (1984).
493. Occupational Safety and Health Administration (OSHA), *Evaluation of Landauer Nitrous Oxide Monitors* OSHA, Salt Lake Technical Center, Salt Lake City, UT, 1986.
494. D. R. Lide, *CRC Handbook of Chemistry and Physics*, CRC, Boca Raton, FL, 1993.
495. M. J. Ellenhorn and D. G. Barceloux, *Medical Toxicology*, Elsevier, New York, 1988.
496. C. Zenz, ed., *Occupational Medicine*, Year Book, Chicago, IL, 1988.
497. J. B. Sullivan and G. R. Krieger, eds., *Hazardous Materials Toxicology: Clinical Principles of Environmental Health*, Williams & Wilkins, Baltimore, MD, 1992.
498. T. Pettit and H. Linn, *A Guide to Safety in Confined Spaces*, National Institute for Occupational Safety and Health, Cincinnati, OH, 1987.
499. J. B. West, *Respiratory Physiology: The Essentials*, 4th ed., Williams & Wilkins, Baltimore, MD, 1990.
500. A. C. Guyton, *Textbook of Medical Physiology*, Saunders, Philadelphia, PA, 1991.
501. J. B. West, ed., *Best and Taylor's Physiological Basis of Medicine*, Williams & Wilkins, Baltimore, MD, 1991.
502. R. L. De Hart, ed., *Fundamentals of Aerospace Medicine*, Lea & Febiger, Philadelphia, PA, 1985.
503. M. A. Hitchcock and F. A. Hitchcock, *Barometric Pressure. Researches in Experimental Physiology by Paul Bert*, College Book Co., Columbus, OH, 1943, and Undersea Medical Society, Bethesda, MD, 1978.

504. J. D. Balentine, *Pathology of Oxygen Toxicity*, Academic Press, New York, 1982.

505. S. Izraeli et al., Determination of the "time of useful consciousness" (TUC) in repeated exposures to simulated altitude of 25,000 ft (7,620 m). *Aviat. Space Environ. Med.* **59**, 1103–1105 (1988).

506. J. Sen Gupta, L. Mathew, and P. M. Gopinath, Effect of physical training at moderate altitude (1850 m) on hypoxic tolerance. *Aviat. Space Environ. Med.* **50**, 714–716 (1979).

507. J. W. Severinghaus, Uses of high altitude for studies of effects of hypoxia. *Adv. Exp. Med. Biol.* **454**, 17–28 (1998).

508. D. P. Jones, T. Y. Aw, and X. Q. Shan, Drug metabolism and toxicity during hypoxia. *Drug Metab. Rev.* **20**, 247–260 (1989).

509. P. W. Angus, D. J. Morgan, R. A. Smallwood, Review article: Hypoxia and hepatic drug metabolism—clinical implications. *Aliment. Pharmacol. Ther.* **4**, 213–225 (1990).

510. B. A. Teicher, Hypoxia and drug resistance. *Cancer Metastasis Rev.* **13**, 139–168 (1994).

511. K. K. Rodahl, *The Physiology of Work*, Taylor & Francis, London, 1989.

512. E. Grandjean, *Fitting the Task to the Man*, Taylor & Francis, London, 1981.

513. P.-O. Åstrand and K. Rodahl, *Textbook of Work Physiology*, McGraw-Hill, New York, 1986.

514. M. P. Ward, J. S. Milledge, and J. B. West, *High Altitude Medicine and Physiology*, University of Pennsylvania, Philadelphia, 1989.

515. C. S. Houston, Mountain sickness. *Sci. Am.* **267**(4), 58–66 (1992).

516. W. N. Bernhard et al., Acetazolamide plus low-dose dexamthasone is better than acetazolamide alone to ameliorate symptoms of acute mountain sickness. *Aviat. Space Environ. Med.* **69**, 883–886 (1998).

517. H. Hultgren, High altitude medical problems. In E. Rubenstein and D. Federman, eds., *Scientific American Medicine*, Scientific American, New York, 1991.

518. P. Bärtsch, High altitude pulmonary edema. *Med. Sci. Sports Exerc.* **31**(Suppl. 1), S23–S27 (1999).

519. R. L. Taber, Protocols for the use of portable hyperbaric chamber for the treatment of high altitude disorders. *J. Wilderness Med.* **1**, 181–192 (1990).

520. O. Oelz et al., Nifedipine for high altitude pulmonary oedema. *Lancet* **2**, 1241–1244 (1989).

521. P. H. Hackett et al., The effect of vasodilators on pulmonary hemodynamics in high altitude pulmonary edema: A comparison. *Int. J. Sports Med.* **13**, S68–S71 (1992).

522. D. L. Klocke, W. D. Wyatt, and J. Stepanek, Altitude-related illnesses. *Mayo Clin. Proc.* **73**, 988–993 (1998).

523. A. J. Peacock, ABC of oxygen: Oxygen at high altitude. *Br. Med. J.* **317**, 1063–1066 (1998).

524. J. B. West, High altitude. In R. G. Crystal et al., eds., *The Lung: Scientific Foundations*, 2nd ed., Lippincott-Raven, Philadelphia, PA, 1997, pp. 2653–2666.

525. R. M. Winslow and C. Monge, *Hypoxia, Polycythemia, and Chronic Mountain Sickness*, Johns Hopkins, University Press, Baltimore, MD, 1987.

526. L. G. C. E. Pugh, Resting ventilation and alveolar air on Mount Everest: With remarks on the relation of barometric pressure to altitude in mountains. *J. Physiol. (London)*, **135**, 590–610, (1957).

527. K. J. Davies, Oxidative stress: The paradox of aerobic life. *Biochem. Soc. Symp.* **61**, 1–31 (1995).

528. C. K. Sen, Oxygen toxicity and antioxidants, State of the art. *Indian J. Physiol. Pharmacol.* **39**, 177–196 (1995).

529. M. S. Carraway and C. A. Piantadosi, Oxygen toxicity. *Respir. Care Clin. North Am.* **5**, 265–295 (1999).

530. D. Jamieson, Oxygen toxicity and reactive oxygen metabolites in mammals. *Free Radical Biol.Med.* **7**(1), 87–108 (1989).

531. R. F. Lodato, Oxygen toxicity. *Crit. Care Clin.* **6**(3), 749–765 (1990).

532. J. Klein, Normobaric oxygen toxicity. *Anesth. Analg.* **70**, 195–207 (1990).

533. H. Witschi, Responses of the lung to toxic injury. *Environ. Health Perspect.* **85**, 5–13 (1990).

534. R. M. Jackson, Molecular, pharmacologic, and clinical aspects of oxygen-related lung injury. *Clin. Chest Med.* **11**(1), 73–86 (1990).

535. R. A. Floyd, Role of free oxygen radicals in carcinogenesis and brain ischemia. *FASEB J.* **4**, 2587–2597 (1990).

536. J. M. Clark, Pulmonary limits of oxygen tolerance in man. *Exp. Lung Res.* **14**, 897–910 (1988).

537. I. Fridovich, Oxygen toxicity: A radical explanation. *J. Exp. Biol.* **201**, 1203–1209 (1998).

538. C. J. Lambertson, Extension of oxygen tolerance in man: Philosophy and significance. *Exp. Lung Res.* **14**, 1035–1058 (1988).

539. J. H. Olson and S. Kobayashi, Antioxidants in health and disease. *Proc. Soc. Exp. Biol. Med.* **2001**, 245–247 (1992).

540. M. S. Mustafa and D. F. Tierney, Biochemical and metabolic changes in the lung with oxygen, ozone, and nitrogen dioxide toxicity. *Am. Rev. Respir. Dis.* **118**, 1061–1090 (1978).

541. C. A. Reily, G. Cohen, and M. Lieberman, Ethane evolution; A new index of lipid peroxidation. *Science* **183**, 208–210 (1974).

542. J. L. Smith, The pathological effects due to increase of oxygen tension in the air breathed. *J. Physiol. (London)* **24**, 19–35 (1899), through Ref. 51.

543. W. Davis, S. Rennard, and P. Bitterman, Early reversible changes in human alveolar structures induced by hyperoxia. *N. Engl. J. Med.* **309**, 878–883 (1983).

544. D. E. Griffith et al., Effects of common therapeutic concentrations of oxygen on lung clearance of 99 mTc DTPA and bronchoalveolar lavage albumin concentration. *Am. Rev. Respir. Dis.* **144**, 233–237 (1986).

545. B. P. Krieger et al., Mechanisms of interaction between oxygen and granulocytes in hyperoxic lung injury. *J. Appl. Physiol.* **58**, 1326–1330 (1985).

546. K. Kubo et al., Differing effects of nitrogen mustard and hydroxyurea on lung O_2 toxicity in adult sheep. *Am. Rev. Respir. Dis.* **145**(1), 13–18 (1992).

547. D. M. Shasby et al., Reduction of the edema of acute hyperoxic lung injury by granulocyte depletion. *J. Appl. Physiol.* **52**, 1237–1244 (1982).

548. L. Jornot and A. F. Junod, Response of human endothelial cell antioxidant enzymes to hyperoxia. *Am. J. Respir. Cell Mol. Biol.* **6**, 107–115 (1992).

549. C. W. White et al., Transgenic mice with expression of elevated levels of copper/zinc-superoxide dismutase in the lungs are resistant to pulmonary oxygen toxicity. *J. Clin. Invest.* **87**, 2162–2168 (1991).

550. J. R. Wispe et al., Human Mn-superoxide dismutase in pulmonary epithelial cells of transgenic mice confers protection from oxygen injury. *J. Biol. Chem.* **267**, 23937–23941 (1992).

551. Y. S. HO, Transgenic models for the study of lung biology and disease. *Am. J. Physiol.* **266**, L319–L353 (1994).

552. R. J. Folz, A. M. Abushamaa, and H. B. Suliman, Extracellular superoxide dismutase in the airways of transgenic mice reduces inflammation and attenuates lung toxicity following hyperoxia. *J. Clin. Invest.* **103**, 1055–1066 (1999).

553. L. M. Carlsson et al., Mice lacking extracellular superoxide dismutase are more sensitive to hyperoxia. *Proc. Natl. Acad. Sci. U.S.A.* **92**, 6264–6268 (1995).

554. M.-F. Tsan, Superoxide dismutase and pulmonary oxygen toxicity. *Proc. Soc. Exp. Biol. Med.* **214**, 107–113 (1997).

555. S. Matalon et al., Sublethal hyperoxic injury to the alveolar epithelium and the pulmonary surfactant system. *Exp. Lung Res.* **14**, 1021–1033 (1988).

556. S. Matalon et al., Characterization of antioxidant activities of pulmonary surfactant mixtures. *Biochim. Biophys. Acta* **1035**(2), 121–127 (1990).

557. S. Horowitz et al., Changes in gene expression in hyperoxia-induced neonatal lung injury. *Am. J. Physiol.* **258** (Lung Cell. Mol. Physiol., 2), L107–L111 (1990).

558. K. A. Veness-Meehan et al., Cell specific alterations in expression of hyperoxia-induced m-RNAs of lung. *Am. J. Respir. Cell Mol. Biol.* **5**(6), 516–521 (1991).

559. T. F. Allred et al., Brief 95% O_2 exposure effects on surfactant protein and mRNA in rat alveolar and bronchiolar epithelium. *Am. J. Physiol.* **276**, L999-L1009 (1999).

560. P. J. Fracica, M. J. Knapp, and J. D. Crapo, Patterns of progression and markers of lung injury in rodents and subhuman primates exposed to hyperoxia. *Exp. Lung Res.* **14**, 868–895 (1988).

561. A. Porte et al., Early bronchopulmonary lesions in rat lung after normobaric 100% oxygen exposure and their evolution. A light and electron microscopic study. *Virchows Arch. A: Pathol. Anat. Histopathol.* **414**(2), 135–145 (1989).

562. J. D. Crapo et al., Structural and biochemical changes in rat lungs occurring during exposures to lethal and adaptive doses of oxygen. *Am. Rev. Respir. Dis.* **122**, 123–143 (1978).

563. R. De los Santos, 100% oxygen lung injury in adult baboons. *Am. Rev. Respir. Dis.* **136**, 657–661 (1987).

564. W. H. Northway, Jr. et al., Oxygen toxicity in the newborn lung; reversal of inhibition of DNA synthesis in the mouse. *Pediatrics.* **57**, 41–46 (1976).

565. D. S. Bonikos et al., Oxygen toxicity in the newborn; the effect of 100% O_2 exposure on the lung of newborn mice. *Lab. Invest.* **32**(5), 619–635 (1975).

566. G. B. Hill, S. Osterhout, and W. M. O'Fallon, Variation in response to hyperbaric oxygen among inbred strains of mice. *Proc. Soc. Exp. Biol. Med.* **129**, 687–689 (1968).

567. B. B. Hudak, L.-Y. Zhang, and S. R. Kleeberger, Inter-strain variation in susceptibility to hyperoxic injury of murine airways. *Pharmacogenetics* **3**, 135–143 (1993).

568. J. D. Stenzel et al., Hyperoxic lung injury in Fischer-344 and Sprague-Dawley rats *in vivo*. *Free Radical Biol. Med.* **14**, 531–539 (1993).

569. L. S. He et al., Lung injury in Fischer but not Sprague-Dawley rats after short-term hyperoxia. *Am. J. Physiol.* **259**, L451–L458 (1990).

570. L. Frank, J. R. Bucher, and R. J. Roberts, Oxygen toxicity in neonatal and adult animals of various species. *J. Appl. Physiol.* **45**(5), 699–704 (1978).

571. C. L. Bryan and S. G. Jenkinson, Species variation in lung antioxidant enzyme activities. *J. Appl. Physiol.* **63**, 597–602 (1987).

572. M. Dörger et al., Interspecies comparison of rat and hamster alveolar macrophage antioxidative and oxidative capacity. *Environ. Health Perspect.* **105S**(Suppl. 5), 1309–1312 (1997).

573. R. Arieli, Oxygen toxicity is not related to mammalian body size. *Comp. Biochem. Physiol.* **91**(2), 221–223 (1988).
574. C. L. Bryan and S. G. Jenkinson, Species variation in lung antioxidant enzyme activities. *J. Appl. Physiol.* **63**, 597–602 (1987).
575. N. Bitterman, E. Skapa, and A. Gutterman, Starvation and dehydration attenuate CNS oxygen toxicity in rats. *Brain Res.* **761**, 146–150 (1997).
576. H. T. Whelan et al., Use of cytochrome-P450 mono-oxygenase 2 E1 isozyme inhibitors to delay seizures caused by central nervous system oxygen toxicity. *Aviat. Space Environ. Med.* **69**, 480–485 (1998).
577. H. A. Kontos, Oxygen radicals in CNS damage. *Chem.-Biol. Interact.* **72**(3), 229–255 (1989).
578. J. D. Wood, Oxygen toxicity. In P. B. Bennett and D. H. Elliott, eds., *The Physiology and Medicine of Diving and Compressed Air Work*, 2nd ed., Baillière Tindall, London, 1975.
579. C. Edwards, C. Lowry, and J. Pennefather, *Diving and Subaquatic Medicine*, Diving Medical Centre, Sydney, Australia, 1976.
580. C. J. Lambertson, Effects of oxygen at high partial pressure. In W. O. Fenn and H. Rahn, eds., *Handbook of Physiology*, Vol II, Sect. 3, American Physiological Society, Washington, DC, 1965, pp. 1027–1046.
581. J. M. Clark et al., Pulmonary function in men after oxygen breathing at 3.0 ATA for 3.5 h. *J. Appl. Physiol.* **71**(3), 878–885 (1991).
582. J. T. Webb et al., Human tolerance to 100% oxygen at 9.5 psia, during five daily simulated 8-hr EVA exposures. *Aviat. Space Environ. Med.* **60**(5), 415–421 (1989).
583. P. S. Grim et al., Hyperbaric oxygen therapy. *J. Am. Med. Assoc.* **263**, 2216–2220 (1990).
584. Undersea and Hyperbolic Medical Society. Available at: http://www.uhms.org
585. Healthcare Professionals reference, Wright-Patterson Air Force Base. Available at: http://wpmcl.wpafb.af.mil/pages/hbo/hcpflyer.htm
586. P. Narula et al., Synergistic cytotoxicity fron nitric oxide and hyperoxia in cultured lung cells. *Am. J. Physiol.* **274**, L411–L416 (1998).
587. C. L. Dent and J. J. Perez Fontan, Long-term therapy for pulmonary hypertension in children. *Curr. Opin. Pediatr.* **11**, 218–222 (1999).
588. J. J. Coalson et al., O_2- and pneumonia-induced lung injury. I. Pathological and morphometric studies. *J. Appl. Physiol.* **67**(1), 346–356 (1989).
589. D. B. Coursin and H. P. Cihla, Non-lethal hyperoxic potentiation of lung disease in the rat following subcutaneous administration of bleomycin. *Toxicology* **16**, 45–55 (1988).
590. S. Matalon et al., Intravenous bleomycin dose not alter the toxic effects of hyperoxia in rabbits. *Anesthesiology* **64**, 614–619 (1986).
591. L. Frank, J. Yam, and R. J. Roberts, The role of endotoxin in protection of adult rats from oxygen-induced lung toxicity. *J. Clin. Invest.* **61**, 269–275 (1978).
592. L. Frank, Extension of oxygen tolerance by treatment with endotoxin: Means to improve its potential therapeutic safety in man. *Exp. Lung Res.* **14**(Suppl.), 987–1003 (1998).
593. J. T. Berg, R. C. Allison, and A. E. Taylor, Endotoxin extends survival of adult mice in hyperoxia. *Proc. Soc. Exp. Biol. Med.* **193**, 167–170 (1990).
594. R. M. Smith, Pulmonary oxygen toxicity in rats; prevention by pyrogenic diphosphoryl lipid A and potentiation by nontoxic monophosphoryl lipid A and lipid X. *Res. Commun. Chem. Pathol. Pharmacol.* **62**(2), 221–234 (1988).

595. C. L. Byran et al., Diphosphoryl lipid A protects rats from lethal hyperoxia. *J. Lab. Clin. Med.* **120**, 444–452 (1992).
596. T. Kobayashi et al., Simultaneous exposure of sheep to endotoxin and 100% oxygen. *Am. Rev. Respir. Dis.* **144**(3, Pt. 1), 600–605 (1991).
597. F. J. C. Smith et al., Morphologic changes in the lungs of rats under compressed air conditions. *J. Exp. Med.* **56**, 79–92 (1932), through Ref. 55.
598. J. D. Crapo and D. F. Tierney, Superoxide dismutase and pulmonary oxygen toxicity. *Am. J. Physiol.* **226**, 1401–1407 (1974).
599. L. Frank, Protection from pulmonary oxygen toxicity by preexposure of rats to hypoxia: Role of the lung antioxidant enzyme systems. *J. Appl. Physiol.* **53**, 475–482 (1982).
600. L. Frank et al. New rest period' protocol for inducing tolerance to high O_2 exposure in adult rats. *Am. J. Physiol.* **257** (lung Cell. Mol. Physiol., 1), L226-L231 (1989).
601. P. M. Winter, G. Smith, and R. F. Wheelis, The effect of prior pulmonary injury on the rate of development of fatal oxygen toxicity. *Chest* **66**(1, Suppl., pt. 2), 18–45 (1974).
602. J. R. Hoidal et al., Therapy with red blood cells decreases hyperoxic pulmonary injury. *Exp. Lung Res.* **14**, 977–985 (1988).
603. D. W. Thibeault et al., Prevention of chronic pulmonary toxicity in young rats with liposome-encapsulated catalase administered intratracheally. *Pediatr. Pulmonol.* **11**(4), 318–327 (1991).
604. P. D. Wagner et al., Protection against pulmonary O_2 toxicity by N-acetylcysteine. *Eur. Respir. J.* **2**(2), 116–126 (1989).
605. S. C. Langley and F. J. Kelly, *N*-acetylcysteine ameliorates hyperoxic lung injury in the preterm guinea pig. *Biochem. Pharmacol.* **45**, 841–846 (1993).
606. R. J. van Klaveren et al., *N*-acetylcysteine does not protect against type II cell injury after prolonged exposure to hyperoxia in rats. *Am. J. Physiol.* **273**, L548–L555 (1997).
607. A. J. Ghio et al., Synthetic surfactant scavenges oxidants and protects against hyperoxic lung injury. *J. Appl. Physiol.* **77**, 1217–1223 (1994).
608. G. M. Loewen et al., Injury in rabbits receiving exogenous surfactant. *J. Appl. Physiol.* **66**(3), 1087–1092 (1989).
609. R. J. King et al., O_2- and pneumonia-induced lung injury. II. Properties of pulmonary surfactant. *J. Appl. Physiol.* **67**(1), 357–365 (1989).
610. R. R. Baker et al., Development of O_2 tolerance in rabbits with no increase in antioxidant enzymes. *J. Appl. Physiol.* **66**(4), 1679–1684 (1989).
611. C. W. White et al., Recombinant tumor necrosis factor/cachectin and interleukin-1 pretreatment decreases lung oxidized glutathione accumulation, lung injury, and mortality in rats exposed to hyperoxia. *J. Clin. Invest.* **79**, 1868–1873 (1987).
612. C. Barazzone et al., Keratinocyte growth factor protects alveolar epithelium and endothelium from oxygen-induced injury in mice. *Am. J. Pathol.* **154**, 1479–1487 (1999).
613. K. E. Welty-Wolf et al., Aerosolized manganese SOD decreases hyperoxic pulmorary injury in primates. II. Morphometric analysis. *J. Appl. Physiol.* **83**, 559–568 (1997).
614. D. M. Suttner et al., Protective effects of transient HO-1 overexpression on susceptibility to oxygen toxicity in lung cells. *Am. J. Physiol.* **276**, L443–L451 (1999).
615. H. R. Wong et al., Increased expression of heat shock protein-70 protects A549 cells against hyperoxia. *Am. J. Physiol.* **275**, L836–L841 (1998).

616. A. B. Waxman et al., Targeted lung expression of interleukin-11 enhances murine tolerance of 100% oxygen and diminishes hyperoxia-induced DNA fragmentation. *J. Clin. Invest.* **101**, 1970–1982 (1998).

617. P. A. Dennery et al., Oxygen toxicity and iron accumulation in the lungs of mice lacking heme oxygenase-2. *J. Clin. Invest.* **101**, 1001–1111 (1998).

618. C. J. Johnston et al., Altered pulmonary response to hyperoxia in Clara cell secretory protein deficient mice. *Am. J. Respir. Cell. Mol. Biol.* **17**, 147–155 (1997).

619. C. Barazzone et al., Plasminogen activator inhibitor-1 in acute hyperoxic mouse lung injury. *J. Clin. Invest.* **98**, 2666–2673 (1996).

620. C. Danel et al., Gene therapy for oxidant injury-related diseases. Adenovirus-mediated transfer of superoxide dismutase and catalase cDNAs protects against hyperoxia but not against ischemia-reperfusion lung injury. *Hum. Gene Ther.* **9**, 1487–1496 (1998).

621. L. E. Otterbein et al., Exogenous administration of heme oxygenase-1 by gene transfer provides protection against hyperoxia-induced lung injury. *J. Clin. Invest.* **103**, 1047–1054 (1999).

622. T. F. Slater, Free radical mechanisms in tissue injury. *Biochem. J.* **222**, 1–15 (1984).

623. D. R. Blake et al., Hypoxic-reperfusion injury in the inflamed human joint. *Lancet* **1**, 289–293 (1984).

624. S. Sherlock, The spectrum of hepatotoxicity due to drugs. *Lancet* **2**, 440–444 (1986).

625. K. F. Gey, On the antioxidant hypothesis with regard to arteriosclerosis. *Bibl. Nutr. Diet.* **37**, 53–91 (1986).

626. D. Harman, Free radical theory of Aging; the free radical' diseases. *Age*, **7**, 111–131 (1984).

627. D. C. Thurnham, Antioxidants and prooxidants in malnourished populations. *Proc. Nutr. Soc.* **49**, 274–359 (1990).

628. R. E. Pacifici and K. F. Davies, Protein, lipid and DNA repair systems in oxidative stress: The free-radical theory of aging revisited. *Gerontology* **37**(1-3), 166–180 (1991).

629. R. S. Sohal and R. G. Allen, Oxidative stress as a causal factor in differentiation and aging: A unifying hypothesis. *Exp. Gerontol.* **25**(6), 499–522 (1990).

630. A. Burns et al., The effect of a fish oil enriched diet on oxygen toxicity and lipid peroxidation in mice. *Biochem. Pharmacol.* **42**(7), 1353–1360 (1991).

631. I. R. Sosenko, S. M. Innis, and L. Frank, Polyunsaturated fatty acids and protection of newborn rats from oxygen toxicity. *J. Pediatr.* **112**(4), 630–637 (1988).

632. I. K. Akers and D. J. Crittenden, Into deepest lung with drug and camera. *Neurosci. Biobehav. Rev.* **12**(3/4), 315–320 (1988).

633. W. S. Beckett and N. D. Wong, Effect of normorbaric hyperoxia on airways of normal subjects. *J. Appl. Physiol.* **64**(4), 1683–1687 (1988).

634. J. Van de Water et al., Response of the lung to six to twelve hours of 100% oxygen inhalation in normal men. *N. Engl. J. Med.* **2832**, 621–626 (1970).

635. A. B. Montgomery, J. M. Luce, and J. F. Murray, Retrosternal pain is an early indicator of oxygen toxicity. *Am. Rev. Respir. Dis.* **139**(6), 1548–1550 (1989).

636. Y. Kapanci et al., Oxygen pneumonitis in Man. *Chest* **62**(2), 162–169 (1972).

637. D. P. Moore et al., Effects of increased inspired oxygen concentrations on exercise performance in chronic heart failure. *Lancet* **339**, 850–853 (1992).

638. W. J. Martin, 2nd and D. L. Kachel, Oxygen-mediated impairment of human pulmonary epithelial cell growth: Evidence for a specific threshold of toxicity. *J. Lab. Clin. Med.* **113**(4), 413–421 (1989).

639. S. C. Erzurm et al., In vivo antioxidant gene expression in human airway epithelium of normal individuals exposed to 100% O_2. *J. Appl. Physiol.* **75**, 1256–1262 (1993).

640. J. H. Yoo et al., Vulnerability of the human airway epithelium to hyperoxia. Constitutive expression of the catalase gene in human bronchial epithelial cells despite oxidant stress. *J. Clin. Invest.* **93**, 297–302 (1994).

641. T. A. Hazinski et al., Cimetidine reduces hyperoxic lung injury in lambs. *J. Appl. Physiol.* **67**, 2586–2592 (1989).

642. E. Matthew et al., Cyclosporin A protects lung function from hyperoxic damage. *Am. J. Physiol.* **276**, L786–L795 (1999).

643. F. J. Saldias et al., Dopamine restores lung ability to clear edema in rats exposed to hyperoxia. *Am. J. Respir. Crit. Care Med.* **159**, 626–633 (1999).

644. F. J. Saldias et al., Isoproterenol improves ability of lung to clear edema in rats exposed to hyperoxia. *J. Appl. Physiol.* **87**, 30–35 (1999).

645. C. L. George et al., Effects of lisofylline on hyperoxia-induced lung injury. *Am. J. Physiol.* **276**, L776–L785 (1999).

646. H. J. Lindsey et al., Pentoxifylline attenuates oxygen-induced lung injury. *J. Surg. Res.* **56**, 543–548 (1994).

647. R. J. van Klaveren et al., Protective effects of the lazaroid U-74389G against hyperoxia in rat type II pneumocytes. *Pulm. Pharmacol. Ther.* **11**, 23–30 (1998).

648. L. E. Otterbein, L. L. Mantell, and A. M. Choi, Carbon monoxide provides protection against hyperoxic lung injury. *Am. J. Physiol.* **276**, L688–L694 (1999).

649. H. Witschi and H. M. Schuller, Diffuse and continuous cell proliferation enhances radiation-induced carcinogenesis in hamster lungs. *Cancer Lett.* **60**, 193–197 (1991).

650. L. E. Smith et al., Oxygen-induced retinopathy in the mouse. *Invest. Ophthalmol. Visual. Sci.* **35**, 101–111 (1994).

651. J. S. Penn et al., Oxygen-induced retinopathy in the rat: Hemorrhages and dysplasias may lead to retinal detachment. *Curr. Eye Res.* **11**, 939–953 (1992).

652. D. S. McLeod, S. A. D'Anna, and G. A. Lutty, Clinical and histopathologic features of canine oxygen-induced proliferative retinopathy. *Invest. Ophthalmol. Visual Sci.* **39**, 1918–1932 (1998).

653. M. R. Niesmann, K. A. Johnson, and J. S. Penn, Therapeutic effect of liposomal superoxide dismutase in an animal model of retinopathy of prematurity. *Neurochem. Res.* **22**, 597–605 (1997).

654. R. D. Higgins et al., Diltiazem reduces retinal neovascularization in a mouse model of oxygen induced retinopathy. *Curr. Eye Res.* **18**, 20–27 (1999).

655. B. N. Nandgaonkar et al., Indomethacin improves oxygen-induced retinopathy in the mouse. *Pediatr. Res.* **46**, 184–188 (1999).

656. J. Kiryu, and Y. Ogura, Hyperbaric oxygen treatment for macular edema in retinal vein occlusion: Relation to severity of retinal leakage. *Ophthalmologica* **210**, 168–170 (1996).

657. L. S. Jaffe, The biological effects of ozone on man and animals. *Am. Ind. Hyg. Assoc. J.* **28**, 267–277 (1967).

658. D. V. Bates, Ozone — myth and reality. *Environ. Res.*, **50**, 230–237 (1989).

659. U.S. Environmental Protection Agency (USEPA), *Air Quality Criteria for Ozone and Other Photochemical Oxidants*, 5 vols., EPA Rep. No. 600/8-84/020 a-eF, USEPA, Research Triangle Park, NC, 1986.
660. P. J. Lioy and R. V. Dyba, Tropospheric ozone: The dynamics of human exposure. *Toxicol. Ind. Health* **5**, 493–504 (1989).
661. T. E. Graedel, Ambient levels of anthropogenic emissions and their atmospheric transformation products. In *Air Pollution, the Automobile and Public Health*, Health Effects Institute, National Academy Press, Washington, DC, 1988, pp. 133–160.
662. C. J. Wechsler, H. C. Shields, and D. V. Nalk, Indoor ozone exposures. *J. Air Pollut. Control Assoc.* **39**, 1562–1568 (1989).
663. C. J. Wechsler, A. T. Hodgson, and J. D. Wooley, Indoor chemistry: Ozone, volatile organic compounds, and carpets. *Environ. Sci. Technol.* **26**, 2371–2377 (1992).
664. R. J. Allen, R. A. Wadden, and E. D. Ross, Characterization of potential indoor sources of ozone, *Am. Ind. Hyg. Assoc. J.* **39**, 466–471 (1978).
665. T. B. Hansen and B. Andersen, Ozone and other air pollutants from photocopying machines. *Am. Ind. Hyg. Assoc. J.* **47**, 659–665 (1986).
666. W. K. C. Morgan and A. Seaton, *Occupational Lung Diseases*, 2nd eds., Saunders, Philadelphia, PA, 1984, pp. 625–626.
667. D. A. Schwartz, Acute Inhalational injury. In L. Rosenstock, ed., *State of the Art Reviews: Occupational Pulmonary Disease*, Hanley & Belfus, Philadelphia, PA, 1987, pp. 297–318.
668. D. Reed, S. Glaser, and J. Kaldor, Ozone toxicity symptoms among flight attendants. *Am. J. Ind. Med.* **1**, 43–54 (1980).
669. B. Sjögren and U. Ulfvarson, Respiratory symptoms and pulmonary function among welders working with aluminum, stainless steel and railroad tracks. *Scand. J. Work Environ. Health.* **11**, 27–32 (1985).
670. T. Kauppinen et al., International data base of exposure measurements in the pulp, paper and paper product industries. *Int. Arch. Occup. Environ. Health.* **70**, 119–127 (1997).
671. W. T. Sanderson, D. Almaguer, L. H. Kirk, 3rd Ozone-induced respiratory illness during the repair of a portland cement kiln. *Scand. J. Work Environ. Health* **25**, 227–232 (1999).
672. F. J. Miller et al., A model of the regional uptake of gaseous pollutants in the lung. 1. The sensitivity of the uptake of ozone in the human lung to lower respiratory tract secretions and exercise. *Toxicol. Appl. Pharmacol.* **79**, 11–27 (1985).
673. J. H. Overton, R. C. Grahm, and F. J. Miller, A model of the regional uptake of gaseous pollutants in the lung. II. The sensitivity of ozone uptake in laboratory animal lungs to anatomical and ventilatory parameters. *Toxicol. Appl. Pharmacol.* **88**, 418–432 (1987).
674. R. D. Buckley et al., Ozone and human blood. *Arch. Environ. Health* **30**, 40–43 (1975).
675. T. R. Gerrity et al., Extrathoracic and Intrathoracic removal of O_3 in tidal-breathing humans. *J. Appl. Physiol.* **65**, 393–400 (1988).
676. S. C. Hu, A. Ben-Jebria, and J. S. Ultman, Longitudinal distribution of ozone absorption in the lung: Quiet respiration in healthy subjects. *J. Appl. Physiol.* **75**, 1655–1661 (1992).
677. W. C. Adams, E. S. Schelegle, and J. D. Shaffrath, Oral and oronasal breathing during continuous exercise produce similar responses to ozone inhalation. *Arch. Environ. Health* **44**, 311–316 (1989).
678. T. R. Gerrity and W. F. McDonnell, Do functional changes in humans correlate with the airway removal efficiency of ozone? In T. Schneider et al., eds. *Atmospheric Ozone Research*

and Its Policy Implications, Proc. 3rd U.S.-Dutch Int. Symp. Nijmegen, The Netherlands, 1988, Elsevier, Amsterdam, 1989, pp. 293–300.

679. W. F. McDonnell, Intersubject variability in human acute ozone responsiveness. *Pharmacogenetics.* **1**, 110–113 (1991).

680. W. F. McDonnell et al., Predictors of individual differences in acute response to ozone exposure. *Am. Rev. Respir. Dis.* **147**, 818–825 (1993).

681. F. Silverman et al., Pulmonary function changes in ozone-interaction of concentration and ventilation. *J. Appl. Physiol.* **41**, 859–864 (1976).

682. W. C. Adams, W. M. Savin, and A. E. Christo, Detection of ozone toxicity during continuous exercise via the effective dose concept. *J. Appl. Phyisol.* **51**, 415–422 (1981).

683. T. R. Gerrity, W. F. McDonnell, and D. E. House, The relationship between delivered ozone dose and functional responses in humans. *Toxicol. Appl. Pharmacol.* **124**, 275–283 (1994).

684. W. F. McDonnell et al., Prediction of ozone-induced FEV1 changes. Effects of concentration, duration, and ventilation. *Am. J. Respir. Crit. Care Med.* **156**, 715–722 (1997).

685. M. Hazucha, Relationship between ozone exposure and pulmonary function changes. *J. Appl. Physiol.* **62**, 1671–1680 (1987).

686. D. M. Dreschler-Parks, S. M. Horvath, and. J. F. Bedi, 'The effective dose' concept in older adults exposed to ozone. *Exp. Gerontol.* **25**, 107–115 (1990).

687. D. Kriebel and T. J. Smith, A Nonlinear pharmacologic model of the acute effects of ozone on the human lungs. *Environ. Res.* **51**, 120–146 (1990).

688. H. R. Kehrl et al., Ozone exposure increases respiratory epithelial permeability in humans. *Am. Rev. Respir. Dis.* **135**, 1124–1128 (1987).

689. P. D. Miller et al., Effect of ozone and histamine on airway permeability to horseradish peroxidase in guinea pigs. *J. Toxicol. Environ. Health.* **18**, 121–132 (1986).

690. L. J. Folinsbee, W. F. McDonnell, and D. H. Horstman, Pulmonary function and symptom responses after 6.6 hour exposure to 0.12 ppm ozone with moderate exercise. *J. Air. Pollut. Control Assoc.* **38**, 28–35 (1988).

691. D. Horstman et al., Ozone concentration and pulmonary response relationships for 6.6-hour exposures with five hours of moderate exercise to 0.08, 0.10 and 0.12 ppm. *Am. Rev. Respir. Dis.* **142**, 1158–1163 (1990).

692. W. F. McDonnell et al., Respiratory response of humans exposed to low levels of ozone for hours. *Arch. Environ. Health* **46**, 145–150 (1991).

693. J. B. Mudd et al., Reaction of ozone with amino acids and proteins. *Atmos. Environ.* **3**, 669–682 (1969).

694. B. Teige, T. T. McManus, and J. B. Mudd, Reaction of ozone with phosphatidylcholine liposomes and the lytic effect of products on red blood cells. *Chem. Phys. Lipids.* **12**, 153–171 (1974).

695. B. A. Freeman and J. B. Mudd, Reaction of ozone with sulfhydryls of human erythrocytes. *Arch. Biochem. Biophys.* **208**, 212–220 (1981).

696. W. A. Pryor, How far does ozone penetrate into the pulmonary air/tissue boundary before it reacts. *Free Radical Biol. Med.* **12**, 83–88 (1992).

697. W. A. Pryor and R. M. Uppu, A kinetic model for the competitive reactions of ozone with amino acid residues in proteins in reverse micelles. *J. Biol. Chem.* **268**, 3120–3126 (1993).

698. R. M. Uppu et al., What does ozone react with at the air/lung interface? Model studies using human red blood cell membranes. *Arch. Biochem. Biophys.* **319**, 257–266 (1995).

699. J. B. Mudd, P. J. Dawson, and J. Santrock, Ozone does not react with human erythrocyte membrane lipids. *Arch. Biochem. Biophys.* **341**, 251–258 (1997).

700. E. M. Postlethwait et al., O_3-induced formation of bioactive lipids: Estimated surface concentrations and lining layer effects. *Am. J. Physiol.* **274**, L1006–L1016 (1998).

701. W. A. Pryor, B. Das, and D. F. Church, The ozonation of unsaturated fatty acids: Aldehydes and hydrogen peroxide as products and possible mediators of ozone toxicity. *Chem. Res. Toxicol.* **4**, 341–348 (1991).

702. J. F. Kanofsky and P. Sima, Singlet oxygen production from the reactions of ozone with biological molecules. *J. Biol. Chem.* **266**, 9039–9042 (1991).

703. J. Santrock, R. A. Gorski, and J. F. O'Gara, Products and mechanism of the reaction of ozone with phospholipids in unilamellar phospholipid vesicles. *Chem. Res. Toxicol.* **5**, 134–141 (1992).

704. G. D. Leikauf et al., Ozonolysis products of membrane fatty acids activate eicosanoid metabolism in human airway epithelial cells. *Am. J. Respir. Cell Mol.Biol.* **9**, 594–602 (1993).

705. C. E. Gross et al., Oxidative damage to plasma constituents by ozone. *FEBS Lett.* **298**, 269–272 (1992).

706. M. G. Mustafa, Biochemical basis of ozone toxicity. *Free Radical Biol. Med.* **9**, 245–265 (1990).

707. W. A. Pryor, Can vitamin E protect humans against the pathological effects of ozone in smog? *Am. J. Clin. Nutr.* **53**, 702–722 (1991).

708. D. B. Menzel, Ozone: An overview of its toxicity in man and animals. *J. Toxicol. Environ. Health* **13**, 183–204 (1984).

709. S. F. P. Man and W. C. Hulbert, Airway repair and adaptation to inhalation injury. In J. Loke, ed. *Pathophysiology and Treatment of Inhalation Injuries*, Dekker, New York, 1988, pp. 1–47.

710. M. Lippmann, Health effects of ozone: A critical review. *J. Air Pollut. Control Assoc.* **39**, 672–695 (1989).

711. H. E. Stokinger, ozone toxicology: A review of research and industrial experience: 1954–1964. *Arch. Environ. Health* **10**, 719–731 (1965).

712. S. S. Griswold, L. A. Chambers, and H. L. Motley, Report of a case of exposure to high ozone concentrations for two hours. *Arch. Ind. Health* **15**, 108–110 (1957).

713. A. N. Nasr, Ozone poisoning in man: Clinical manifestations and differential diagnosis. A review. *Clin. Toxicol.* **4**, 461–466 (1971).

714. M. Kleinfeld and C. P. Giel, Clinical manifestations of ozone poisoning: Report of a new source of exposure. *Am. J. Med. Sci.* **231**, 638–643 (1956).

715. H. E. Stokinger, Evaluation of the hazards of ozone and oxides of nitrogen. *Arch. Ind. Health*, **15**, 181–190 (1957).

716. D. R. Prows et al., Ozone-induced acute lung injury: Genetic analysis of F(2) mice generated from A/J and C57BL/6J strains. *Am. J. Physiol.* **277**, L372–L380 (1999).

717. W. F. McDonnel et al., Pulmonary effects of ozone exposure during exercise: Dose-response characteristics. *J. Appl. Physiol.* **54**, 1345–1352 (1983).

718. R. B. Devlin et al., Exposure of humans to ambient levels of ozone for 6.6 hours causes cellular and biochemical changes in the lung. *Am. J. Respir. Cell Mol. Biol.* **4**, 72–81 (1991).

719. J. Linder et al., Die Wirkung Von Ozon auf die Korperliche Leistungsfahigkeit (The Effect of Ozone on Physical Activity). *Schweiz. Z. Sportmed.* **36**, 5–10 (1988).

720. E. S. Schelegle and W. C. Adams, Reduced exercise time in competitive simulations consequent to low level ozone exposure. *Med. Sci. Sports Exercise* **18**, 408–414 (1986).

721. W. S. Wayne, P. F. Wehrle, and R. E. Carroll, Oxidant air pollution and athletic performance. *J. Am. Med. Assoc.* **199**, 151–154 (1967).

722. H. Gong, Jr. et al., Impaired exercise performance and pulmonary function in elite cyclists during low-level ozone exposure in a hot environment. *Am. Rev. Respir. Dis.* **134**, 726–733 (1986).

723. W. C. Adams, Effects of ozone exposure at ambient air pollution episode levels on exercise performance. *Sports Med.* **4**, 395–424 (1987).

724. L. J. Folinsbee, F. Silverman, and R. J. Shephard, Decrease of maximum work performance following ozone exposure. *J. Appl. Physiol.* **42**, 531–536 (1977).

725. J. A. Golden, J. A. Nadel, and H. A. Boushey, Bronchial hyperirritability in healthy subjects after exposure to ozone. *Am. Rev. Respir. Dis.* **118**, 287–294 (1978).

726. W. S. Beckett et al., Role of the parasympathetic nervous system in acute lung response to ozone. *J. Appl. Physiol.* **59**, 1879–1885 (1985).

727. M. J. Hazucha, D. V. Bates, and P. A. Bromberg, Mechanism of Ozone on the human lung. *J. Appl. Physiol.* **67**, 1535–1541 (1989).

728. A. N. Passannante et al., Nociceptive mechanisms modulate ozone-induced human lung function decrements. *J. Appl. Physiol.* **85**, 1863–1870 (1998).

729. G. D. Leikauf, K. E. Driscoll, and H. E. Wey, Ozone-induced augmentation of eicosanoid metabolism in cultured epithelial cells from bovine trachea. *Am. Rev. Respir. Dis.* **137**, 435–442 (1988).

730. W. M. Hulbert, T. McLean, and J. C. Hogg, The effect of acute airway inflammation on bronchial reactivity in guinea pigs. *Am. Rev. Respir. Dis.* **132**, 7–11 (1985).

731. M. J. Holtzman et al., Importance of airway inflammation for hyperresponsiveness induced by ozone. *Am. Rev. Respir. Dis.* **127**, 686–690 (1983).

732. K. E. Driscoll et al., Ozone inhalation stimulates expression of the neutrophil chemotactic peptide, MIP-2. *Toxicol. Appl. Pharmacol.* **119**, 306–309 (1993).

733. I. Jaspers, E. Flescher, and L. C. Chen. Ozone-induced IL-8 expression and transcription factor binding in respiratory epithelial cells. *Am. J. Physiol.* **272**, L504–L511 (1997).

734. I. Jaspers, L. C. Chen, and E. Flescher, Induction of interleukin-8 by ozone is mediated by tyrosine kinase and protein kinase A, but not by protein kinase C. *J. Cell. Physiol.* **177**, 313–323 (1998).

735. M. M. Chang et al., IL-8 is one of the major chemokines produced by monkey airway epithelium after ozone-induced injury. *Am. J. Physiol.* **275**, L524–L532 (1998).

736. J. D. Hackney et al., Responses of selected reactive and nonreactive volunteers to ozone in high and low pollution seasons. In T. Schneider et al., eds., *Atmospheric Ozone Research and Its Policy Implications*, Proc. 3rd U.S.-Dutch Int. Symp., Nijmegen, The Netherlands, 1988, Elsevier, Amsterdam, 1989, pp. 311–318.

737. T. D. Messineo and W. C. Adams, Ozone inhalation effects in females varying widely in lung size: Comparison with males. *J. Appl. Physiol.* **69**, 96–103 (1990).

738. W. F. McDonnell et al., Reproducibility of individual responses to ozone exposure. *Am. Rev. Respir. Dis.* **131**, 36–40 (1985).

739. J. A. Gliner, S. M. Horvath, and L. J. Folinsbee, Pre-exposure to low ozone concentrations does not diminish the pulmonary function response on exposure to higher ozone concentrations. *Am. Rev. Respir. Dis.* **127**, 51–55 (1983).

740. D. R. Prows et al., Genetic analysis of ozone-induced acute lung injury in sensitive and resistant strains. *Nat. Genet.* **17**, 471–474 (1997).

741. S. R. Kleeberger et al., Linkage analysis of susceptibility to ozone-induced lung inflammation in inbred mice. *Nat. Genet.* **17**(4), 475–478 (1997).

742. E. L. Avol et al., Respiratory effects of photochemical oxidant air pollution in exercising adolescents. *Am. Rev. Respir. Dis.* **132**, 619–622 (1985).

743. E. L. Avol et al., Short-term respiratory effects of photochemical oxidant exposure in exercising children. *J. Am. Med. Assoc.* **37**, 158–162 (1987).

744. W. F. McDonnell et al., Respiratory responses of vigorously exercising children to 0.12 ppm ozone exposure. *Am. Rev. Respir. Dis.* **132**, 875–879 (1985).

745. P. L. Kinney et al., Short-term pulmonary function change in association with ozone levels. *Am. Rev. Respir. Dis.* **139**, 56–61 (1989).

746. D. M. Spektor et al., Effects of ambient ozone on respiratory function in active normal children. *Am. Rev. Respir. Dis.* **137**, 313–320 (1988).

747. I. T. T. Higgins et al., Effect of exposures to ambient ozone on ventilatory lung function in children. *Am. Rev. Respir. Dis.* **141**, 1136–1146 (1990).

748. D. M. Spektor et al., Effects of single- and multiday ozone exposure on respiratory function in active normal children. *Environ. Res.* **55**, 107–122 (1991).

749. M. Castillejos et al., Effects of ambient ozone on respiratory function and symptoms in Mexico City schoolchildren. *Am. Rev. Respir. Dis.* **145**, 276–282 (1992).

750. P. J. Lioy, T. A. Vollmuth, and M. Lippmann, Persistence of peak flow decrement in children following ozone exposures exceeding the national ambient air quality standard. *J. Air Pollut. Control Assoc.* **35**, 1068–1071 (1985).

751. B. P. Farrell et al., Adaptation in human subjects to the effects of inhaled ozone after repeated exposure. *Am. Rev. Respir. Dis.* **119**, 725–730 (1979).

752. L. J. Folinsbee, J. F. Bedi and S. M. Horvath, Respiratory responses in humans repeatedly exposed to low concentrations of ozone. *Am. Rev. Respir. Dis.* **121**, 431–439 (1980).

753. L. J. Folinsbee et al., Respiratory responses to repeated prolonged exposure to 0.12 ppm ozone. *Am. J. Respir. Crit. Care Med.* **149**, 98–105 (1994).

754. W. S. Linn et al., Persistence of adaptation to ozone in volunteers exposed repeatedly for six weeks. *Am. Rev. Resp. Dis.* **125**, 491–495 (1982).

755. S. M. Horvath, J. A. Gliner, and L. J. Folinsbee, Adaptation to ozone: Duration of effect. *Am. Rev. Respir. Dis.* **123**, 496–499 (1981).

756. T. J. Kulle et al., Duration of pulmonary function adaptation to ozone in humans. *Am. Ind. Hyg. Assoc. J.* **43**, 832–837 (1982).

757. W. S. Linn et al., Repeated laboratory ozone exposure of volunteer Los Angeles residents: An apparent seasonal variation of response. *Toxicol. Ind. Health* **4**, 505–520 (1988).

758. J. S. Tepper et al., Unattenuated structural and biochemical alterations in the rat lung during functional adaptation to ozone. *Am. Rev. Respir. Dis.* **140**, 493–501 (1989).

759. W. A. van der Wal et al., Attenuation of acute lung injury by ozone inhalation-the effect of low level pre-exposure. *Toxicol. Lett.* **72**, 291–298 (1994).

760. J. Q. Koenig et al. Acute effects of 0.12 ppm ozone or 0.12 ppm nitrogen dioxide on pulmonary function in healthy and asthmatic adolescents. *Am. Rev. Respir. Dis.* **132**, 648–651 (1985).

761. W. S. Linn et al., Health effects of ozone exposure in asthmatics. *Am. Rev. Respir. Dis.* **117**, 835–843 (1978).

762. D. Sheppard et al., Lower threshold and greater bronchomotor responsiveness of asthmatic subjects to sulfur dioxide. *Am. Rev. Respir. Dis.* **122**, 873–878 (1980).

763. H. R. Kehrl et al., Responses of subjects with chronic obstructive pulmonary disease after exposure to 0.3 ppm ozone. *Am. Rev. Respir. Dis.* **131**, 719–724 (1985).

764. J. J. Solic, M. J. Hazucha, and P. A. Bromberg, The acute effects of 0.2 ppm ozone in patients with chronic obstructive pulmonary disease. *Am. Rev. Respir. Dis.* **125**, 664–669 (1982).

765. W. F. McDonnell et al., The respiratory responses of subjects with allergic rhinitis to ozone exposure and their relationship to nonspecific airway reactivity. *Toxicol. Ind. Health.* **3**, 507–517 (1987).

766. R. Bascom et al., Effect of ozone inhalation on the response to nasal challenge with antigen of allergic subjects. *Am. Rev. Respir. Dis.* **142**, 594–601 (1990).

767. D. E. McBride et al., Inflammatory effects of ozone in the upper airways of subjects with asthma. *Am. J. Respir. Crit. Care Med.* **149**, 1192–1197 (1994).

768. H. S. Koren, Associations between criteria air pollutants and asthma. *Environ. Health Perspect.* **103**(Suppl. 6), 235–242 (1995).

769. H. S. Koren, Environmental risk factors in atopic asthma. *Int. Arch. Allergy Immunol.* **113**, 65–68 (1997).

770. J. W. Kreit et al., Ozone-induced changes in pulmonary function and bronchial responsiveness in asthmatics. *J. Appl. Physiol.* **66**, 217–222 (1989).

771. A. W. Whittemore and E. L. Korn, Asthma and air pollution in the Los Angeles area. *Am. J. Public Health* **70**, 687–696 (1980).

772. A. H. Holguin et al., The effects of ozone on asthmatics in the Houston area. In S. D. Lee, ed., *Evaluation of the Scientific Basis for Ozone/Oxidants Standards*, APCA Int. Spec. Conf. Trans., TR-4, Air Pollution Control Association, Pittsburgh, PA, 1985, pp. 262–280.

773. C. E. Schoettlin and E. Landau, Air pollution and asthmatic attacks in the Los Angeles area. *Public Health Rep.* **76**, 545–548 (1961).

774. R. P. Cody et al., The effect of ozone associated with summertime photochemical smog on the frequency of asthma visits to hospital emergency departments. *Environ. Res.* **58**, 184–194 (1992).

775. J. R. Stedman et al., Emergency hospital admissions for respiratory disorders attributable to summer time ozone episodes in Great Britain. *Thorax* **52**, 958–963 (1997).

776. A. Ponka and M. Virtanen, Asthma and ambient air pollution in Helsinki. *J. Epidemiol. Commun. Health.* **50**(Suppl. 10), s59–s62 (1996).

777. C. P. Weisel, R. P. Cody, and P. J. Lioy, Relationship between summertime ambient ozone levels and emergency department visits for asthma in central New Jersey. *Environ. Health Perspect.* **103**(Suppl. 2), 97–102 (1995).

778. D. V. Bates, and R. Sizto, Air pollution and hospital admissions in Southern Ontario: The acid summer haze effect. *Environ. Res.* **43**, 317–331 (1987).

779. J. Q. Koenig et al., Prior exposure to ozone potentiates subsequent response to sulfur dioxide in adolescent asthmatic subjects. *Am. Rev. Respir. Dis.* **141**, 377–380 (1990).

780. N. A. Molfino et al., effect of low concentrations of ozone on inhaled allergen responses in asthmatic subjects. *Lancet* **338**, 199–203 (1991).

781. R. Jorres, D. Nowak, and H. Magnussen, The effect of ozone exposure on allergen responsiveness in subjects with asthma or rhinitis. *Am. J. Respir. Crit. Care Med.* **153**, 56–64 (1996).

782. H. S. Jenkins et al., The effect of exposure to ozone and nitrogen dioxide on the airway response of atopic asthmatics to inhaled allergen: Dose- and time-dependent effects. *Am. J. Respir. Crit. Care Med.* **160**, 33–39 (1999).

783. H. S. Lee, Y. T. Wang, and K. T. Tom, Occupational asthma due to ozone. *Singapore Med. J.* **30**, 485–487 (1989).

784. K. W. Wilson, *Survey of Eye Irritation and Lachrymation in Relation to Air Pollution*, Copley International Corporation, La Jolla, CA (prepared for the Coordinating Research Council, Inc., New York), 1974.

785. D. L. Hammer et al., Los Angeles student nurse study: Daily symptom reporting and photochemical oxidants. *Arch. Environ. Health* **28**, 255–260 (1974).

786. P. J. R. Challen, D. E. Hickish, and J. Bedford, An investigation of some health hazards in an inert-gas tungsten-arc welding shop. *Br. J. Ind. Med.* **15**, 276–282 (1957).

787. W. M. Grant, *Toxicology of the Eye*, 3rd ed., Thomas, Springfield, IL, 1986, pp. 693–694.

788. M. J. K. Selgrade et al., Evaluation of effects of ozone exposure on influenza infection in mice using several indicators of susceptibility. *Fundam. Appl. Toxicol.* **11**, 169–180 (1988).

789. J. A. Wolcott, Y. C. Zee, and J. W. Osebold, Exposure to ozone reduces influenza disease severity and alters distribution of influenza viral antigens in murine lungs. *Appl. Environ. Microbiol.* **44**, 723–731 (1982).

790. D. L. Coffin and E. J. Blommer, Alteration of the pathogenic role of streptococci group C in mice conferred by previous exposure to ozone. In I. H. Silver, ed., *Aerobiology*, Proc. 3rd Int. Symp. University of Sussex, England, Academic Press London, 1970, pp. 54–61.

791. R. Ehrlich, J. C. Findlay, and D. E. Gardner, Effects of repeated exposures to peak concentrations of nitrogen dioxide and ozone on resistance to streptococcal pneumonia. *J. Toxicol. Environ. Health* **5**, 631–642 (1979).

792. D. E. Gardner, Oxidant-induced enhanced sensitivity to infection in animal models and their extrapolations to man. *J. Toxicol. Environ. Health* **13**, 423–439 (1984).

793. F. J. Miller, J. W. Illing, and D. E. Gardner, Effect of urban ozone levels on laboratory-induced respiratory infections. *Toxicol. Lett.* **2**, 163–169 (1978).

794. M. I. Gilmour et al., Factors that influence the suppression of pulmonary antibacterial defenses in mice exposed to ozone. *Exp. Lung Res.* **19**, 299–314 (1993).

795. K. E. Driscoll, T. A. Vollmuth, and R. B. Schlesinger, Acute and subchronic ozone inhalation in the rabbit: Response of alveolar macrophages. *J. Toxicol. Environ. Health.* **21**, 27–43 (1987).

796. J. Soukup, H. S. Koren, and S. Becker, Ozone effect on respiratory syncytial virus infectivity and cytokine production by human alveolar macrophages. *Environ. Res.* **60**, 178–186 (1993).

797. S. Becker et al., Modulation of human alveolar macrophage properties by ozone exposure in vitro. *Toxicol. Appl. Pharmacol.* **110**, 403–415 (1991).

798. M. E. Pearlman et al., Chronic oxidant exposure and epidemic influenza. *Environ. Res.* **4**, 129–140 (1971).

799. W. S. Wayne and P. F. Wehrle, Oxidant air pollution and school absenteeism. *Arch. Environ. Health* **19**, 315–322 (1969).

800. F. W. Henderson, D. M. Elliott, and G. S. Orlando, The immune response to rhinovirus infection in human volunteers exposed to ozone. In S. D. Lee, M. G. Mustafa, and M. A. Mehlman, eds., *International Symposium on the Biomedical Effects of Ozone and Related Photochemical Oxidants*, Princeton Scientific Publishers, Princeton, NJ, 1983, pp. 253–254.

801. F. W. Henderson et al., Experimental rhinovirus infection in human volunteers exposed to ozone. *Am. Rev. Respir. Dis.* **137**, 1124–1128 (1988).

802. P. L. Kinney and H. Ozkaynak, Associations of daily mortality and air pollution in Los Angeles county. *Environ. Res.* **54**, 99–121 (1991).

803. G. Hoek et al., Effects of ambient particulate matter and ozone on daily mortality in Rotterdam, The Netherlands. *Arch. Environ. Health* **52**, 455–463 (1997).

804. R. W. Simpson et al., Associations between outdoor air pollution and daily mortality in Brisbane, Australia. *Arch. Environ. Health* **52**, 442–454 (1997).

805. J. E. Kelsall et al., Air pollution and mortality in Philadelphia, 1974–1988. *Am. J. Epidemiol.* **146**, 750–762 (1997).

806. G. Touloumi et al., Short-term effects of ambient oxidant exposure on mortality: A combined analysis within the APHEA project. Air pollution and health: A European approach. *Am. J. Epidemiol.* **146**, 177–185 (1997).

807. L. A. Baxter et al., Comparing estimates of the effects of air pollution on human mortality obtained using different regression methodologies. *Risk Anal.* **17**, 273–278 (1997).

808. S. H. Moolgavkar et al., Air pollution and daily mortality in Philadelphia. *Epidemiology* **6**, 476–484 (1995).

809. I. U. Ruhman, G. D. Massaro, and D. Massaro, Exposure of rats to ozone: Evidence of damage to heart and brain. *Free Radical. Biol. Med.* **12**, 323–326 (1992).

810. D. L. Coffin and H. E. Stokinger, Biological effects of air pollutants. In A. C. Stern, ed., *Air Pollution*, 3rd ed., Vol. 2, Academic Press, London, 1977, pp. 231–360.

811. D. R. Prows et al., Ozone-induced acute lung injury: Genetic analysis of F(2) mice generated from A/J and C57BL/6J strains. *Am. J. Physiol.* **277**, L372–L380 (1999).

812. G. D. Leikauf et al., Airway epithelial cell responses to ozone injury. *Environ. Health Perspect.*, **103**, 91–95 (1995).

813. Q. Zhao et al., Chemokine regulation of ozone-induced neutrophil and monocyte inflammation. *Am. J. Physiol.* **274**, L39–L46 (1998).

814. K. R. Vesely et al., Breathing pattern response and epithelial labeling in ozone-induced airway injury in neutrophil-depleted rats. *Am. J. Respir. Cell Mol. Biol.* **20**, 699–709 (1999).

815. C. G. Plopper et al., Effects of low levels of ozone on rat lungs. II Morphological responses during recovery and reexposure. *Exp. Mol. Pathol.* **29**, 400–411 (1978).

816. C. K. Chow et al., Effect of low levels of ozone on rat lungs. I. Biochemical responses during recovery and reexposure. *Exp. Mol. Pathol.* **25**, 182–188 (1976).

817. R. M. Aris et al., Ozone-induced airway inflammation in human subjects as determined by airway lavage and biospy. *Am. Rev. Respir. Dis.* **148**, 1363–1372 (1993).

818. R. P. Sherwin and V. Richters, Centriacinar region (CAR) disease in the lungs of young adults, a preliminary report. In R. L. Berglund, D. R. Lawson, and D. J. McKee, eds., *Tropospheric Ozone and the Environment*, TR-19, Air & Waste Management Association, Pittsburgh, PA, 1991, pp. 178–196.

819. R. P. Sherwin, Air pollution: The pathobiologic issues. *J. Toxicol. Clin. Toxicol.* **29**, 385–400 (1991).

820. M. Lippmann, Use of human lung tissue for studies of structural changes associated with chronic ozone exposure: Opportunities and critical issues. *Environ. Health Perspect.* **101**(Suppl. 4), 209–212 (1993).

821. H. E. Stokinger, W. D. Wagner, and O. J. Dobrogorski, Ozone toxicity studies. III. Chronic injury to lungs of animals following exposure at a low level. *Arch. Ind. Health* **16**, 514–522 (1957).

822. B. E. Barry, F. J. Miller, and J. D. Crapo, Effects of inhalation of 0.12 and 0.25 parts per million ozone on the proximal alveolar region of juvenile and adult rats. *Lab. Invest.* **53**, 692–704 (1985).

823. S. L. Eustis et al., Chronic bronchiolitis in nonhuman primates after prolonged ozone exposure. *Am. J. Pathol.* **105**, 121–137 (1981).

824. L. E. Fujinaka et al., Respiratory Bronchiolitis following long-term ozone exposure in bonnet monkeys: A morphometric study. *Exp. Lung Res.* **8**, 167–190 (1985).

825. J. R. Harkema et al., Effects of an ambient level of ozone on primate nasal epithelial mucosubstances. *Am. J. Pathol.* **127**, 90–96 (1987).

826. W. S. Tyler et al., Comparison of daily and seasonal exposures of young monkeys to ozone. *Toxicology* **50**, 131–144 (1988).

827. D. A. Johnson, Ozone inactivation of human al-antiproteinase inhibitor. *Am. Rev. Respir. Dis.* **121**, 1031–1038 (1980).

828. K. M. Rieser et al., Long-term consequences of exposure to ozone. II. Structural alterations in lung collagen of monkeys. *Toxicol. Appl. Pharmacol.* **89**, 314–322 (1987).

829. R. Detels et al., The UCLA population studies of chronic obstructive lung disease. 9. Lung function changes associated with chronic exposure to photochemical oxidants; a cohort study among never-smokers. *Chest* **92**, 594–603 (1987).

830. M. C. White et al., Exacerbations of childhood asthma and ozone pollution in Atlanta. *Environ. Res.* **65**, 56–68 (1994).

831. I. Romieu et al., Effects of air pollution on the respiratory health of asthmatic children living in Mexico City. *Am. J. Respir. Crit. Care Med.* **154**, 300–307 (1996).

832. P. L. Kinney et al., Biomarkers of lung inflammation in recreational joggers exposed to ozone. *Am. J. Respir. Crit. Care Med.* **154**, 1430–1435 (1996).

833. G. D. Thurston et al., Summertime haze air pollution and children with asthma. *Am. J. Respir. Crit. Care Med.* **155**, 654–660 (1997).

834. M. Lippmann and D. M. Spektor, Peak flow rate changes in O_3 exposed children: Spirometry vs miniWright flow meters. *J. Exp. Anal. Environ. Epidemiol.* **8**, 101–107 (1998).

835. J. R. Greer, D. E. Abbey, and R. J. Burchette, Asthma related to occupational and ambient air pollutants in nonsmokers. *J. Occup. Med.* **35**, 909–915 (1993).

836. N. Kunzli et al., Association between lifetime ambient ozone exposure and pulmonary function in college freshmen — results of a pilot study. *Environ. Res.* **72**, 8–23 (1997).

837. A. Galizia and P. L. Kinney, Long-term residence in areas of high ozone: Associations with respiratory health in a nationwide sample of nonsmoking young adults. *Environ. Health Perspect.* **107**, 675–679 (1999).

838. K. E. Pinkerton et al., Exposure to low levels of ozone results in enhanced pulmonary retention of inhaled asbestos fibers. *Am. Rev. Respir. Dis.* **140**, 1075–1081 (1989).

839. J. J. Steinberg, J. L. Gleeson, and D. Gil, The Pathobiology of ozone-induced damage. *Arch. Environ. Health* **45**, 80–87 (1990).
840. K. Victorin, Genotoxicity and carcinogenicity of ozone. *Scand. J. Work Environ. Health* **22**(Suppl. 3), 42–51 (1996).
841. D. G. Thomassen et al., The role of ozone in tracheal cell transformation. *Res. Rep. Health Eff. Inst.* **50**, 1–32 (1992).
842. J. T. Haney, Jr. T. H. Connor, and L. Li, Detection of ozone-induced DNA single strand breaks in murine bronchoalveolar lavage cells acutely exposed in vivo. *Inhalation Toxicol.* **11**, 331–341 (1999).
843. K. Rithidech et al., Chromosome damage in rat pulmonary alveolar macrophages following ozone inhalation. *Mutat. Res.* **241**, 67–73 (1990).
844. T. Merz et al., Observations of aberrations in chromosomes of lymphyocytes from human subjects exposed to ozone at a concentration of 0.5 ppm for 6 and 10 hours. *Mutat. Res.* **31**, 299–302 (1975).
845. W. H. McKenzie et al., Cytogenetic effects of inhaled ozone in man. *Mutat. Res.* **48**, 95–102 (1977).
846. C. Borek, A. Ong, and J. E. Cleaver, DNA damage from ozone and radiation in human epithelial cells. *Toxicol. Ind. Health* **4**, 547–553 (1988).
847. C. Borek et al., Ozone acts alone and synergistically with ionizing radiation to induce in vitro neoplastic transformation. *Carcinogenesis (London)* **7**, 1611–1613 (1986).
848. C. Borek, A. Ong, and H. Mason, Ozone and ultraviolet light act as additive cocarcinogens to induce in vitro neoplastic transformation. *Teratog., Carcinog., Mutagen.* **9**, 71–74 (1989).
849. D. G. Thomassen et al., Preneoplastic transformation of rat tracheal epithelial cells by ozone. *Toxicol. Appl. Pharmacol.* **109**, 137–148 (1991).
850. C. Hassett et al., Murine lung carcinogenesis following exposure to ambient ozone concentrations. *JNCI, J. Natl. Cancer Inst.* **75**, 771–777 (1985).
851. H. Witschi, Effects of oxygen and ozone on mouse lung tumorigenesis. *Exp. Lung Res.* **17**, 473–483 (1991).
852. National Toxicology Program, (NTP), *Technical Report on the Toxicology and Carcinogenesis Studies of Ozone and Ozone/NNK in F344/N Rats and B6C3F1 Mice*, NTP TR 440, NIH Publ. No. 93–3371, U.S. Department of Health and Human Services, Washington, DC, 1993.
853. R. A. Herbert et al., Two-year and lifetime toxicity and carcinogenicity studies of ozone in B6C3F1 mice. *Toxicol. Pathol.* **24**, 539–548 (1996).
854. R. C. Sills et al., Increased frequency of K-ras mutations in lung neoplasms from female B6C3F1 mice exposed to ozone for 24 or 30 months. *Carcinogenesis (London)* **16**, 1623–1628 (1995).
855. J. T. Zelikoff et al., Immunomodulating effects of ozone on macrophage functions important for tumor surveillance and host defense. *J. Toxicol. Environ. Health* **34**, 449–467 (1991).
856. A. F.-Y. Li and A. Richters, Ambient level ozone effects on subpopulations of thymocytes and spleen T lymphocytes. *Arch. Environ. Health* **46**, 57–63 (1991).
857. W. J. Kozumbo and S. Agarwal, Induction of DNA damage in cultured human lung cells by tobacco smoke arylamines exposed to ambient levels of ozone. *Am. J. Respir. Cell Mol. Biol.* **3**, 611–618 (1990).
858. G. A. Boorman et al., Long-term toxicity studies of ozone in F344/N rats and B6C3F1 mice. *Toxicol. Lett.* **82–83**, 301–306 (1995).

859. G. A. Boorman et al., Toxicology and carcinogenesis studies of ozone and ozone 4- (N-nitrosomethylamino) -1- (3-pyridyl) -1-butanone in Fischer-344/N rats. *Toxicol. Pathol.* **22**, 545–554 (1994).
860. H. Witschi, M.A. Breider, and H. M. Schuller, Failure of ozone and nitrogen dioxide to enhance lung tumor development in hamsters. *Res. Rep. Health Eff. Inst.* **60**, 1–38 (1993).
861. H. Witschi, D. W. Wilson, and C. G. Plopper, Modulation of N-nitrosodiethylamine-induced hamster lung tumors by ozone. *Toxicology* **77**, 193–202 (1993).
862. M. Kleinfeld, C. Giel, and I.R. Tabershaw, Health hazards associated with inert-gas-shielded metal arc welding. *AMA Arch. Ind. Health* **15**, 27–31 (1957).
863. Ozone. In *Toxicity of Drugs and Chemicals*, Academic Press, New York, 1969, pp. 446–448.
864. W. N. Witheridge and C. P. Yaglou, Ozone in ventilation: Its possibilities and limitations. *Trans. Am. Soc. Heat Vent. Eng.* **5**, 509–522 (1937).
865. National Research Council: (NRC) *Emergency and Continuous Exposure Limits for Selected Airborne Contaminants.*, Vol. 1, National Academy Press, Committee on Toxicology, Board on Toxicology and Environmental Health Hazards, Commission on Life Sciences, Washington, DC, 1984.
866. Intersociety Committee, Determination of ozone in the atmosphere by gas-phase chemiluminescence instruments. In J. P. Lodge, ed., *Methods of Air Sampling and Analysis*, 3rd ed., Lewis Publishers, Chelsea, MI, 1989, pp. 407–411.
867. Intersociety Committee, Contiuous monitoring of ozone in the atmosphere by ultraviolet photometric instruments. In J. P. Lodge, ed., *Methods of Air Sampling and Analysis*, 3rd ed., Lewis Publishers, Chelsea, MI, 1989, pp. 422–426.
868. P. Koutrakis et al., Measurement of ambient ozone using a nitrite-coated filter. *Anal. Chem.* **65**, 209–214 (1993).
869. C. Monn and M. Hangartner, Passive sampling for ozone. *J. Air Waste Manage. Assoc.* **40**, 357–361 (1990).
870. D. Grosjean and M. W. M. Hisham, A passive sampler for atmospheric ozone. *J. Air Waste Manage. Assoc.* **42**, 169–173 (1992).

CHAPTER FORTY-EIGHT

The Halogens

Daniel Thau Teitelbaum, MD

The halogens are those elements in group XVII of the periodic table, and include fluorine, chlorine, bromine, iodine, and astatine, the latter of which is a radioactive element of no industrial importance. The physical properties of the halogens are shown in Table 48.1, which indicates an almost perfect doubling of atomic weights progressing from fluorine to bromine, paralleled by increases in specific gravity and melting and boiling points and by decreases in water solubility.

Chemically, fluorine is the most powerful oxidizing agent known. Elements become progressively less electronegative and have less oxidizing potential as atomic weight increases. Each halogen forms an acid in water and combines with metals to form salts; the reactivity of these compounds shows the same relationship as the elemental halogens.

Because each halogen has seven electrons in its outer shell, the normal valence is minus 1 in binary compounds (HX-1). With the exception of fluorine, halogens may also have valences of $+1$, $+3$, $+5$, and $+7$, such as in industrially important halogen compounds containing oxygen; examples of these include XO_n and KXO to KXO_4. Interhalogen compounds are of some industrial interest and include those of the type XX_n, where n may be 1, 3, 5, or 7 (for e.g., ClF_3). Interhalogen compounds, as well as some of the halogen compounds containing oxygen, demonstrate even higher reactivities than the elemental halogens.

Chlorine and fluorine exist in the earth's crust in almost equal proportions (770 ppm for fluorine, 550 ppm for chlorine). The relative abundance of bromine and iodine are only about 2 and 0.04%, respectively, of that for chlorine. Seawater contains almost 19,000 ppm chlorine, compared to 65 ppm for bromine and less than 2 ppm for fluorine. Iodine exists only in trace quantities in seawater (about 0.05 ppm).

Patty's Toxicology, Fifth Edition, Volume 3, Edited by Eula Bingham, Barbara Cohrssen, and Charles H. Powell.
ISBN 0-471-31934-1 © 2001 John Wiley & Sons, Inc.

Table 48.1. Selected Physical Properties of Elemental Halogens[a]

Property	Fluorine	Chlorine	Bromine	Iodine
Color	Pale yellow	Greenish-yellow	Reddish-brown	Violet
Atomic number	9	17	35	53
Atomic weight	18.9984	35.453	79.904	126.905
Boiling point (°C)	−187.9	−34.6	58.78	184.35
Freezing point (°C)	−217.9	−100.98	−7.2	113.5
Specific gravity of liquid (water = 1)	1.11 at BP	1.56 at BP	3.12(20°C)	4.93 (solid, 20°C)
Critical temperature (°C)	−129	143.9	315	546
Critical pressure (atm)	51.4	78.7	101.6	—
Vapor pressure at 20 °C (atm)	> 10	> 5	> 0.5	< 0.002
Solubility in water (mol/L)	Decomposes	0.092 (25°C)	0.214 (20°C)	0.0013 (25°C)
Ionic radius (Å, Valence of −1)	1.33	1.81	1.96	2.2
Valence(s)	1	1, 3, 5, 7	1, 3, 5, 7	1, 3, 5, 7

[a]From Ref. 1–3.

A FLUORINE AND ITS COMPOUNDS

1.0 Fluorine

1.0.1 CAS Number: [7782-41-4]

1.0.2 Synonyms: Fluorine-19

1.0.3 Trade Names: NA

1.0.4 Molecular Weight: 37.99680

1.0.5 Molecular Formula: F_2

1.0.6 Molecular Structure: F—F

1.1 Chemical and Physical Properties

Because of the small atomic radius (Table 48.1), the effective surface charge density of the fluorine atom is greater than that of any other element, making fluorine the most electronegative and the most reactive of the elements (1–3). Under appropriate conditions, it forms compounds with all other elements except argon, helium, and neon (4). Fluorine is univalent, and no compounds are known in which fluorine has a valence of > 1. Asbestos and finely dispersed water, glass, ceramics, carbon, and metals all burn in fluorine. In contact with massive mild steel, copper, nickel, or Monel metal, fluorine forms a film that prevents further attack, which means that these materials can be used to handle fluorine at ordinary temperatures (3). Fluorine reacts with water to form hydrogen fluoride and the highly toxic oxygen difluoride, OF_2. Fluorine in combination exists in either the ionic form or the covalent tetrahedral form. Compounds resulting from the interaction of fluorine and metals are usually ionic and have high melting and boiling points. Many fluorides, for example, the fluorides of lithium, aluminum, strontium, barium, magnesium, and manganese, are sparingly soluble or insoluble in water. Nonmetallic elements react with fluorine to yield covalent compounds, for examples are silicon tetrafluoride, sulfur hexafluoride, and complex anionic forms. These covalent compounds are characterized by low melting points and high volatility. The inorganic chemistry of fluorine is discussed in detail by Glemser (5) and by the World Health Organization in its Environmental Health Criteria 36 (6).

1.2 Production and Use

Fluorine is the most reactive of all the elements. Free fluorine is rarely, if ever, found in nature. Elemental fluorine is produced on a commercial scale by the electrolysis of anhydrous hydrogen fluoride in a molten solution of potassium fluoride (essentially the same process as that used by Moissan in 1886 to isolate fluorine for the first time). Fluorine gas is formed at the anode and hydrogen at the cathode. The principal impurity is hydrogen fluoride, most of which is removed by passing the gas stream through a condensation trap; the remaining traces are converted to sodium bifluoride ($NaHF_2$) on exposure to sodium fluoride pellets.

Fluorine is used in the nuclear energy industry to produce gaseous uranium hexafluoride by the direct fluorination of solid uranium tetrafluoride. The desired ^{235}U is then separated as the hexafluoride by a gaseous diffusion process. As the most potent chemical oxidizer known, fluorine has been used as an oxidizer of liquid rocket fuels. Direct fluorination procedures with elemental fluorine are used to produce useful hydrocarbon derivatives in which some or all of the hydrogen is replaced by fluorine. These are commonly referred to as *fluorocarbons*, although the term properly applies only to those compounds in which hydrogen has been totally replaced by fluorine. Most commercially available fluorine containing alkanes are also brominated or chlorinated or contain all three halogens. Extensive use of fluorocarbons of all types as cleaning and degreasing agents in the metalworking industries, the semiconductor industry, and the electronics assembly industry has led to widespread exposure of workers to these volatile compounds. Environmental contamination and groundwater pollution with degreasing compounds is widespread. The elimination of halogenated compounds from degreasing operations, and their substitution by nonvolatile, nonhalogenated detergents has led to a significant decline in worker exposure and environmental contamination with fluorine-containing degreasing agents.

Common operations where fluorine exposure occurs include the manufacture of fluorochemicals and plastics, rocket propellants, and fluorinated intermediates, metal production such as aluminum potroom work, the fluorination of pharmaceuticals and consumer products such as dentifrices, and the fluoridation of public drinking-water supplies. Environmental contamination and air pollution with fluorine and fluorides may occur as a result of emissions from aluminum production facilities, glass and ceramic manufacture, fertilizer manufacture, and the processing of fluorspar.

An estimated 12,854 employees were exposed to fluorine in 1983 (7).

1.3 Exposure Assessment

1.3.1 Air

NIOSH has not published a method for measuring airborne gaseous fluorine. However, a method for the measurement of fluorides as aerosol and gas by ion chromatography was published as NIOSH method 7906 in 1994 (10). Method 7902 ("fluorides by IC") offers an alternative for measurement of fluorides using ion-selective electrode techniques. NIOSH method 7903 ("Acids, inorganic") is an alternative method for HF. Lorberau noted problems that were inherent in NIOSH method 7902 and published a method for determination of gaseous and particulate fluorides by ion chromatography in 1993 (11, 12). NIOSH notes the need for special precautions in the laboratory during the use of these methods because of the extreme corrosivity of fluorides to skin, eyes, and mucous membranes.

In their studies of the effects of fluorine on laboratory animals, Keplinger and Suissa (13) determined fluorine concentrations in the exposure chamber by drawing a known volume of air through an alkaline solution of potassium iodide and measuring the iodine displaced by the fluorine. AOAC method 961.16 allows for the microchemical determination of fluorine by a titrimetric method.

THE HALOGENS

1.4 Toxic Effects

Because of the reactivity of fluorine, exposures of humans and animals, and environmental contamination problems are almost always the result of fluorides, rather than fluorine gas. Very little reliable contemporary study of elemental fluorine is found in the literature. Most studies in which fluorine itself has been studied are highly artificial and of academic rather than practical interest. Functionally, acute exposures to fluorine gas must be regarded as severe and potentially lethal corrosive exposures with the added nuances of disturbances in calcium metabolism because of the reaction between calcium ion and fluoride ion.

1.4.1 Experimental Studies

1.4.1.1 Acute Toxicity: NA

1.4.1.2 Chronic and Subchronic Toxicity: NA

1.4.1.3 Pharmacokinetics, Metabolism, and Mechanisms: NA

1.4.1.4 Reproductive and Developmental: NA

1.4.1.5 Carcinogenesis: NA

1.4.1.6 Genetic and Related Cellular Effects Studies: NA

1.4.1.7 Other: Neurological, Pulmonary, Skin Sensitization

1.4.1.7.1 Skin. A comparison of the burns produced on the shaved backs of anesthetized rabbits by fluorine under pressure, by an oxyacetylene flame, and by aqueous hydrofluoric acid demonstrated that fluorine produces a thermal burn, not a chemical burn. New, mildly hyperemic skin was evident approximately 13–14 days later. The hydrofluoric acid burn required an additional 2 weeks to heal (14).

1.4.1.7.2 Respiratory System. Investigations into the short-term toxicity of inhaled fluorine have been conducted by Keplinger and Suissa (13). The LC_{50} values determined in these studies are listed in Table 48.2. As would be expected, the LC_{50} decreased with increasing exposure time for each of the four species. However, the LC_{50}s at each exposure time were not statistically significantly different among the species, with one exception. After 5 min of exposure, the LC_{50} for the mouse (600 ppm) was significantly different from that for the rabbit (820 ppm). Signs of intoxication in these experimental animals included conjunctivitis, increased nasal secretions, sneezing, dyspnea, loss of body weight, and general weakness. Except in the LC_{90} to LC_{100} range, deaths occurred about 12–18 h after exposure. Animals surviving for the first 48 h after exposure generally survived the 14-day observation period.

Gross pathological examination of animals exposed to fluorine concentrations near the mean lethal concentration revealed congestion, hemorrhage, and atelectasis in the lungs and some congestion and/or swelling in the liver. Histological examination of the lungs revealed massive hemorrhage into the alveolar spaces, coagulation necrosis of the alveoli, and peribronchial lymphocytic proliferation. Seven days after exposure, proliferation of septal cells, macrophages, and lymphocytes were present. Histological evidence of kidney

Table 48.2. LC$_{50}$ Values for Animals Exposed to Fluorine[a]

Species	Exposure (min)	LC$_{50}$ mg/m^3	ppm
Rat	5	1088	700
Mouse	5	932	600
Rabbit	5	1274	820
Rat	15	606	390
Mouse	15	583	375
Guinea pig	15	614	395
Rat	30	420	270
Mouse	30	350	225
Rabbit	30	420	270
Rat	60	287	185
Mouse	60	233	150
Guinea pig	60	264	170

[a]From Ref. 13.

and liver damage was not apparent until 7–14 days postexposure. In the kidney, focal areas of coagulation necrosis were seen in the cortex, and focal areas of lymphocytic infiltration were apparent throughout the cortex and medulla. In the liver, coagulation necrosis, periportal hemorrhage, and diffuse cloudy swelling were evident.

Complete blood counts in rats exposed to fluorine for 15 min at 107–329 ppm and for 60 min at 98–142 ppm, and in dogs exposed to fluorine for 15 min at 39–93 ppm and for 60 min at 38–109 ppm, did not show any significant changes.

Keplinger and Suissa's (13) data show no apparent effects in animals exposed to fluorine concentrations at or below 100 ppm for 5 min, 70 ppm for 15 min, 55 ppm for 30 min, or 45 ppm for 60 min. The prime importance of fluorine's pulmonary effects and the lack of fluorine-induced hematological changes agree with the earlier observations of Stokinger (14).

Keplinger (15) has presented some evidence suggesting that repeated exposure to fluorine may lead to the development of a tolerance or a degree of protection. The LC$_{50}$s of previously exposed animals were higher, and the degree of lung injury was less, than in control animals. As Keplinger states, however, this evidence is "not very remarkable" (15).

1.4.2 Human Experience

There is little information on human exposure to fluorine gas in the literature. Lyon (16) has reported his experience in a gaseous diffusion plant over a 9-year period. In this study "fluorine was considered to be present if during sampling (1) the characteristic odor of fluorine was detected or (2) the characteristic odor of HF was not detected in circumstances where the contaminant was likely to be fluorine." Sixty-one employees were considered to have been intermittently exposed under these conditions to average yearly fluorine concentrations of 0.3–1.4 ppm, with peak concentrations of 3.8–24.7 ppm. Lyon

(16) found no evidence of impaired health in these workers. However, these studies were done many years ago, and utilized diagnostic techniques and standards which must be considered obsolescent. The negative findings may give a false sense of comfort in the light of today's knowledge of the effects of chronic absorption of fluorides on the lung, the skeletal system, the kidney, and the liver. Moreover, the occurrence of asthma in aluminum potroom workers, which may be due in part to the fluoride in their environment, would suggest that pulmonary surveillance would be appropriate. Benke et al. (17) have reviewed the issue of exposures in the alumina and aluminum industry and have addressed the occurrence of potroom asthma and its relationship to fluoride levels in the workplace. In the event that exposure to fluorine gas at levels that do not produce intolerable acute working conditions is suspected, medical surveillance is indicated. NIOSH suggests that medical surveillance is appropriate for all fluorine workers. Such surveillance should be focused on the skin, eyes, and respiratory system, as well as the kidneys, the liver, and the skeletal system. Pulmonary function studies are indicated, as are laboratory assessment of biomarkers of fluoride accumulation.

Keplinger and Suissa (13) recorded the subjective impressions of five human volunteers exposed briefly to known concentrations of fluorine. Eye irritation proved to be the most sensitive index. At a concentration of 10 ppm for intervals of 3–15 min, fluorine was not particularly irritating to the eye or nose, although the odor was prominent. A concentration of 23 ppm was slightly irritating to the eyes but could be inhaled without respiratory difficulty; at 50 ppm for 3 min, fluorine irritated the eyes and was slightly irritating to the nose; at a concentration of 67–100 ppm, fluorine was very irritating to the eye and nose and this sensation became uncomfortable after a few seconds. The test subjects recommended immediate evacuation of areas when the concentration reached 100 ppm.

According to Rickey (18), brief exposure to 40 mg/m^3 (25 ppm) was intolerable and breathing was impossible at 75 mg/m^3 (43 ppm). Belles et al. (19, cited by Ref. 20) reported that exposure to a concentration between 25 and 40 mg/m^3 (16–25 ppm) caused some eye and nasal irritation after two or three breaths and that repeated short-term exposures to concentrations of ≤ 15 mg/m^3 (10 ppm) were tolerated without discomfort. Dermal irritation occurred at concentrations in the range of 150–300 mg/m^3 (97–194 ppm). Odor thresholds reported by these volunteers ranged between 0.15 and 0.30 mg/m^3 (0.1–0.2 ppm).

Fatalities have been reported following relatively small acute respiratory exposures to inorganic fluorine compounds, including hydrofluoric acid and most of the inorganic fluorinated gases. Acute dermal absorption and ingestion has also led to severe illness and to lethal outcomes. The gases and liquids cause severe irritation, corrosive injury, frostbite, and metabolic changes that are the result of interference with calcium metabolism and the avid binding of calcium by the fluoride ion. Exposures to inorganic fluorine-containing compounds by any route should be considered an urgent medical problem, and assistance of the nearest poison control center, and emergency medical care should be undertaken at once. Because the signs and symptoms of severe and even lethal fluoride poisoning may be delayed for several hours, first responders should not be tempted to delay vigorous emergency decontamination and prompt referral for definitive medical care. A relatively asymptomatic presentation may be followed by rapid and irreversible decline in the clinical status of those exposed, if medical intervention is delayed.

Table 48.3. Thresholds and Standards for Exposure to Fluorine[a]

Standard/Response	Concentration ppm	mg/m^3
NIOSH REL-TWA	0.1	0.2
OSHA PEL-TWA		
ACGIH TLV-TWA	1.0	1.6
ACGIH TLV-STEL	2.0	3.1
Odor threshold; odor pungent, irritating	0.1–0.2	0.16–0.30
Short-term; repeated exposures are tolerated	≥ 10.0	≥ 16.0
Not particularly irritating to eyes for up to 15 min	10.0	16.0
Some nasal, eye irritation	16–25	25–40
No respiratory discomfort; slight eye irritation, 5 min exposure; brief exposures intolerable; NIOSH IDLH level	25.0	40.0
Breathing impossible	48.0	75.0
Eye irritation, slight nasal irritation	50.0	79.0
Skin irritation	95–100	150–300
Very slight skin irritation; marked irritation of eye, nose; 1 min exposure extremely uncomfortable	100.0	158.0

[a]From Refs. 1, 21 and 22.

1.5 Standards, Regulations, or Guidelines of Exposure

Standards for occupational exposure to fluorine, as well as thresholds for several biological responses in human subjects, are listed in Table 48.3 (21, 22). The current OSHA standard is 0.1 ppm (0.2 mg/m^3.) (8). The ACGIH-8 h TLV-TWA is 1 ppm, 1.6 mg/m^3; the ACGIH STEL is 2 ppm, 3.1 mg/m^3 (8). NIOSH and OSHA (1) consider fluorine to have poor warning properties. The odor and irritation thresholds do not fall at or below the permissible exposure limit (PEL). Discrepancies are evident in the responses reported by subjects at 25 ppm (40 mg/m^3) these ranged from no discomfort to intolerable. The recommendation of NIOSH (21) is that 25 ppm (40 mg/m^3) be considered a concentration that is immediately dangerous to life or health (IDLH).

Releases of fluorine must be reported under CERCLA reporting requirements section 103, in 40 CFR 302. Fluorine is considered an "extremely hazardous substance (EHS)" subject to reporting requirements when stored in amounts in excess of its "threshold planning quantity (TPQ)" of 500 lbs.

Numerous NPL sites contaminated with fluorides have been identified by EPA. In its toxicological profile for fluorides, hydrogen fluoride, and fluorine, ATSDR notes that water and soil concentrations of fluoride are heavily influenced by background of fluorides in nature as well as by point emission sources. Fluoride levels in surface water generally range within 0.01–0.3 mg/L. Groundwater levels may be higher and may reach 1.5 ppm, in part because of the contribution of mineral derived fluoride from rocks through which they flow. Highly mineralized water such as the Great Salt Lake has greatly elevated concentrations of fluoride, and may reach 14 ppm. Seawater levels may be between 1.4 and 1.5 ppm (8).

THE HALOGENS

Fluoride in ambient air according to ATSDR is largely dependent on the amount emitted from nearby point sources. Generally elevated air fluoride levels are concentrated in proximity to industrial sources such as those that utilize or produce fluorides (9).

General control methods are listed in the *National Institute for Occupational Safety and Health/Occupational Safety and Health Administration (NIOSH/OSHA) guideline for fluorine* (1). Because most inorganic fluorine compounds are corrosive per se, or form corrosive hydrofluoric acid when they are wet or exposed to humid environments, precautions against corrosive, skin, eye, and respiratory exposures are necessary when handling these compounds. Encapsulating and vapor protective clothing and respiratory protection using SCBA or airline respirators will usually be required when there are releases of fluorine and fluorine-containing inorganics as dust, gases, or liquids.

2.0 Hydrogen Fluoride

2.0.1 CAS Number: [7664-39-3]

2.0.2 Synonyms: Etching acid; HF A; fluorohydric acid; fluoric acid; hydrofluoric acid; Antisal 2B; deuteriumfluoride

2.0.3 Trade Names: NA

2.0.4 Molecular Weight: 20.0

2.0.5 Molecular Formula: HF

2.0.6 Molecular Structure: H—F

2.1 Chemical and Physical Properties

Hydrogen bonding between HF molecules results in polymerization of the molecules; the extent to which this occurs depends on the partial pressure of the HF and the temperature. Under workplace conditions where the concentration of HF approximates the PEL or threshold limit value (TLV), the partial pressure of the gas is low enough to ensure that polymerization is negligible; under these conditions, the HF is assumed to exist as a monomer of molecular weight 20.

Selected physical properties of HF are listed in Table 48.4. According to Banks and Goldwhite (3), anhydrous HF is highly associated in the liquid state, as indicated by its high boiling point. The association persists in the vapor phase as well; at 1 atm of pressure, the mean degree of association corresponds to (HF) 3.45. As stated earlier, at 80°C and 1 atm, the vapor is probably monomeric. The dielectric constant of liquid HF is very near that of water and, like water, HF is an excellent solvent. It is an extremely strong acid. However, in dilute aqueous solution it is weakly acidic; 0.1 M solutions are ionized only to the extent of ~10%.

2.2 Production and Use

Hydrogen fluoride is manufactured by treating fluorspar (fluorite, CaF_2) with concentrated sulfuric acid in heated kilns. The gaseous HF evolved is purified by distillation, condensed

Table 48.4. Selected Physical Properties of Hydrogen Fluoride

Property	Anhydrous HF Liquid	Anhydrous HF Gas	Aqueous HF
Color	Colorless	Colorless	Colorless
Molecular weight (monomer)	20	20	20
Boiling point (1 atm)	19.5°C	—	Varies with concentration
Melting point	−83.7°C	—	Varies with concentration
Specific gravity (1 atm)	1.0 (4°C) (air = 1)	1.27 (34°C)	Varies with concentration
Constant boiling mixture (35.35%)	—	—	120°C
Dielectric constant (water = 87.9 at 0°C)	83.6 (°C)	—	—
Solubility in water		Miscible in all proportions	

as liquid anhydrous HF, and stored in steel tanks and cylinders (4). In addition to anhydrous HF 99.9% pure, 38, 47, 53, and 70% HF solutions are available as commodity chemicals. The 38% (wt/wt HF) solution is a binary azeotrope.

The principal uses of HF are in the manufacture of artificial cryolite, for use in the production of aluminum, fluorocarbons, and uranium hexafluoride as a catalyst in alkylation processes in petroleum refining, in the manufacture of fluoride salts, and in stainless-steel pickling operations. It is also used in the stimulation of oil production in sandstone formations. Hydrogen fluoride also finds use as an etching agent in the glass and ceramic industries. Major uses of HF have developed in high technology operations such as the manufacture of semiconductors, integrated circuits, circuit boards, and other electronic components. In these industries, etching of silicon wafers and other substrates with liquid and gaseous HF is a common practice. In addition, many inorganic fluorides that are functionally equivalent to HF because of their behavior in aqueous environments are also used in high technology industries. Many opportunities for dermal and inhalation exposure to HF exist in the high technology industries. Numerous reports of injuries and some fatalities due to HF have occurred (23).

Exposure to HF releases, as well as to particulate fluorides, can occur in a variety of industries. Among these are the manufacture of phosphate fertilizers, phosphate-containing animal feed supplements, phosphoric acid, elemental phosphorus, uranium tetrafluoride, brick, tile, pottery, cement, and glass, and in arc welding, metal casting, and brazing. The use of HF as a metal brightening agent in industrial metal cleaning and finishing operations, and as a rust remover in laundry detergents in place of oxalic acid, remains a significant source of HF exposure. NIOSH (24) lists 57 occupations in which workers have potential exposure to HF, and another NIOSH publication (1) details common operations where exposure to HF can occur and lists general control methods used to reduce or eliminate such exposures. NIOSH estimated that 189,051 workers were exposed to HF in 1983 (7).

2.3 Exposure Assessment

2.3.1 Air

Standard procedures for the collection and analysis of airborne samples of fluoride gas and aerosols have been described by NIOSH in methods 7902 and 7906 (25). The sampler consists of a prefilter cellulose ester membrane followed by a Na_2CO_3-treated cellulose pad. Particulate fluoride is retained on the prefilter membrane, and gaseous fluoride is absorbed on the alkaline cellulose pad. The membrane filter is placed in 20% sodium hydroxide solution and then dried; the residue is fused, and the soluble fluoride is taken up in total ionic strength activity buffer (TISAB). Gaseous fluoride retained on the alkaline cellulose pad is removed by soaking in TISAB. Fluoride is then measured in the two TISAB solutions using a fluoride ion–specific electrode. The more recent NIOSH method 7906 offers an ion chromatography method for fluoride aerosol and gas (10).

2.3.2 Biomonitoring/Biomarkers

2.3.2.1 Blood: NA

2.3.2.2 Urine. An analytical method for the determination of fluoride in urine has been described by NIOSH (24).

2.4 Toxic Effects

2.4.1 Experimental Studies

2.4.1.1 Acute Toxicity: NA

2.4.1.2 Chronic and Subchronic Toxicity: NA

2.4.1.3 Pharmacokinetics, Metabolism, and Mechanisms: NA

2.4.1.4 Reproductive and Developmental: NA

2.4.1.5 Carcinogenesis: NA

2.4.1.6 Genetic and Related Cellular Effects Studies: NA

2.4.1.7 Other: Neurological, Pulmonary, Skin Sensitization

2.4.1.7.1 Skin. The nature of the chemical burn produced by the action of HF solution on animal skin has been described by Alhassan and Zink (30). Application of three to four drops of 40% HF or 35% H_2SiF_6 on the intact skin of rats, guinea pigs, or miniature swine for 1 s to 1 h generally caused no effect. When the skin had been scarified by several small incisions prior to application of the acid, however, a penetrating necrosis developed, evident histologically an hour after the application. Macroscopic reaction was not yet apparent at this time. The principal histological changes were a hypocellular necrosis and edema penetrating to the subcutaneous connective tissue.

Kono et al. (31) have shown with convincing clarity that sufficient fluoride can be absorbed through rat skin to produce the characteristic picture of acute fluoride toxicity.

After a 0.5-mL application of 50% HF solution for 5 min to a shaved area (5.8 cm^2) approximating 1.7% of the total body surface, spontaneous movement and ventilation decreased, and loss of coordination and of the righting reflex became evident. The injured skin area gradually became necrotic, but blisters did not form. In the serum, an increase in ionized fluoride and decreases in total and ionized calcium were evident within 30 min. A decrease in sodium and increases in potassium, phosphate, and magnesium appeared over the next 24 h. Parathyroid hormone concentrations increased but began to decline again after about 6 h, when calcium concentrations began to increase again. A slowed heartbeat (bradycardia) was evident on the electrocardiogram within the first 30 min. Mortality was approximately 80% within 24 h.

2.4.1.7.2 Respiratory System. Machle et al. (32) demonstrated HF to be a severe irritant to guinea pigs and rabbits. On exposure, the animals' eyes were kept closed, paroxysms of coughing and sneezing were frequent, respiration was slowed, and there were copious discharges from the eyes and nose. Pulmonary damage included massive hemorrhage, edema, congestion, and emphysema. Concentrations of 1500 mg/m^3 (1835 ppm) were lethal to both species within 5 min. Exposure to 1000 mg/m^3 (1200 ppm) for 30 min was not fatal, nor was exposure to concentrations below 100 mg/m^3 (122 ppm) for 5 h. Concentrations not exceeding 50 mg/m^3 (61 ppm) were mildly irritating.

Several investigators have made LC$_{50}$ measurements; these data are listed in Table 48.5. The findings are not strictly comparable among authors, inasmuch as the experimental protocols differed in several important aspects. For example, Rosenholtz et al. (33) observed young male rats during a 14-day postexposure interval; Higgins et al. (34) observed older rats of unspecified sex for a 7-day period; Wohlslagel et al. (35) also

Table 48.5. LC$_{50}$ Values for Animals Exposed to Hydrogen Fluoride Vapors

Exposure (min)	Rat	Mouse	Guinea Pig	Reference
5	4970[b]			33
	18200[c]			34
		6247[e]		34
15	2689[b]		4327[g]	33
30	2042[b]			33
60	1307[b]			33
	1395[d]			35
		342[f]		35

LC$_{50}$ (ppm)[a]

[a] 1.223 ppm is equivalent to 1 mg/m^3.
[b] 100–200-g males, Wistar-derived; 14 days postexposure.
[c] 250–275-g Wistar-derived, sex not specified; 7 days postexposure.
[d] 250–325-g males, Sprague-Dawley derived; 14 days postexposure.
[e] 30–35-g ICR-derived, sex not specified; 7 days postexposure.
[f] 25–32-g females, ICR-derived; 14 days postexposure.
[g] 340–360-g males, Hartley-derived; 14 days postexposure.

observed older male rats for 14 days. The latter investigators studied slightly younger mice than did Higgins et al. and observed them for 14 days, whereas Higgins et al. observed the animals for 7 days; both teams of investigators studied mice of the same strain, but Higgins et al. did not specify the sex of the animals in their study. Nevertheless, one can conclude that the toxicity of airborne HF increases for the rat as exposure time increases. Also, at 60 min of exposure, age did not significantly affect lethality in the rats. The greater sensitivity of mice exposed for 60 min compared with those exposed for 5 min may be due, at least in part, to a difference in sex, or it may reflect the greater number of deaths that occurred over the longer exposure period. The latter explanation seems unlikely in view of the fact that in the Higgins et al. study most deaths occurred about 24 h after exposure. The limited data in Table 48.5 suggest that the guinea pig may be the most resistant of the three species investigated. Hydrogen fluoride appears to be less toxic than fluorine to laboratory rodents. Signs of toxicity in rats, rabbits, guinea pigs, and dogs exposed to high concentrations of HF include irritation of the conjunctiva, the nasal tissues, and the respiratory system, as evidenced by reddened conjunctivas, pawing at the nose, marked lacrimation, nasal secretion, coughing, and sneezing. Within an hour after a lethal exposure, the noses and nasal vestibules appeared black. The severity of these signs decreased with decreasing airborne concentrations, becoming very mild and transient at 307 ppm (251 mg/m^3) and 103 ppm (84 mg/m^3) after 15 and 60 min of exposure, respectively (33). Histopathological changes in rats exposed to lethal concentrations included mucosal and submucosal necrosis in the nose and nasal vestibule, associated with acute inflammation, renal tubular necrosis, presence of cytoplasmic globules in the liver, dermal collagen changes, and possibly, myeloid hyperplasia of the bone marrow. Changes in the lung were not described, although the lungs of these animals were examined (33). Pulmonary edema and hemorrhage were also reported by Higgins et al. (34) and by Wohlslagel et al. (35).

In the experience of Stavert et al. (36), tissue injury in the respiratory tract of rats inhaling a 1300-ppm concentration of HF for 30 min was confined to the anterior regions of the nasal passages and consisted of epithelial and submucosal necrosis, moderate to severe fibronecrotic rhinitis, and hemorrhage. No histopathology was seen in the lungs of these animals. During exposure there was a marked reduction in breathing frequency, leading to an ~ 27% reduction in minute ventilation (mL air/min). No deaths occurred during exposure, but 25% of the rats died in the next 24 h.

Stokinger (14) and colleagues exposed rats, mice, guinea pigs, rabbits, and dogs to 7 and to 25 mg/m^3 HF 6 h/day for 30 days. The higher concentration was lethal to the rats and mice but not to the other species. Moderate hemorrhage and edema of the lungs were evident in the dogs, rabbits, and rats at autopsy. Renal cortical degeneration and necrosis were seen in the rat. At a concentration of 7 mg/m^3, localized hemorrhages were seen in the lungs of one of five dogs, and no changes were evident in the rabbit or rat.

The absence of injury to the rat lung observed by Rosenholtz et al. (33), Morris and Smith (37), and Stavert et al. (36) is attributed to the extraordinary efficiency of the upper respiratory tract in absorbing inhaled HF. As reported by Morris and Smith (37), in rats exposed for 6 h to 36, 96, or 176 mg F (measured as HF/m^3), virtually all of the inhaled HF is deposited in the upper respiratory tract. It may then be absorbed and translocated to other sites. The rat is an obligate nose-breather; this allows the protective scrubbing effect of the upper respiratory airways and nasal turbinates to come into full play. When Stavert et al. (36) forced their rats to become pseudo-mouth-breathers by inserting tracheal tubes connected to mouthpieces, the protective effect of the upper airways was lost, and peripheral lung damage was evidenced by increased wet weight of the lungs and by histopathological damage, primarily in the larger conducting airways. The fact that Higgins et al. (34) and Wohlslagel et al. (35) observed pulmonary edema and hemorrhage may be attributed to different exposure conditions, different strains of rats, and/or other unknown factors. The increased sensitivity of the mouse may reflect less efficient scrubbing in this species.

Dalbey and colleagues (38, 39) conducted a series of acute inhalation exposure studies with HF in rats to assist in the derivation of a short-term exposure limit in humans. They exposed four groups of animals to concentrations of HF that ranged from 135 ppm to 8621 ppm for periods of 2–10 min. The experiments were performed with a mouth breathing cannula apparatus to assure delivery of the HF to the lower airways and to obviate certain of the problems inherent in the study of respiratory exposures in nose breathing species such as rats and mice. Extensive postexposure studies were performed. Some animals were followed for 14 weeks, by which time complete histopathological recovery of the airways was apparent. A concentration curve for nonlethal exposure to HF in these female rats was developed. On the basis of this study, Dalbey et al. concluded that most people could be exposed to 130 ppm HF for 10 min without permanent injury. Irritation would occur, but on the basis of histological studies they believe that the effects on the respiratory tract of short-term exposure at this level would not be "serious" and would be expected to be reversible.

2.4.1.7.3 Kidney and the Renal Excretion of Fluoride. As the major excretory organ for fluoride, the kidney can be the site of locally high fluoride concentrations. These concentrations may result in the production of a vasopressin-resistant polyuria that resembles that of diabetes insipidus. Investigations into the mechanism underlying this effect have shown that fluoride inhibits the resorption of water in the nephron of the kidney, resulting in diuresis. In the acidotic state, the lower pH of the renal tubular fluid favors the resorption of more undissociated HF from the fluid and less fluoride is excreted in the urine. In alkalosis, tubular resorption of fluoride is decreased and more fluoride remains in the tubular fluid to be excreted in the urine (37, 40–42).

Usuda et al. (43) studied the biomarkers of fluoride exposure, which were expressed in Wistar rats given single oral doses of sodium fluoride. Urinary fluoride ion, α-glutathione-S-transferase (α-GST) and N-acetyl-β-D-glucosaminidase (NAG) and creatinine (CR) concentrations were assessed. α-GST was quite useful in the determination of S3 proximal tubule injury, which is found in fluoride intoxication in rats. Most prominent injury in the animals occurred in the S3 segment of the proximal tubule rather than in the S1 or S2 segments or the glomeruli. α-GST proved useful as a measure of acute injury and as a long-term follow-up to the initial exposure. The authors suggested that measurement of α-GST may be useful in the monitoring of workers exposed over long periods of time to fluorides.

Because of widespread fluoridation of public water supplies for the purpose of dental health, background levels of fluoride in serum and in urine may be elevated, and the assessment of the workplace contribution to the total measured fluoride in blood or in urine may be difficult. Torra et al. (44, 45) determined serum and urine fluoride concentrations in a nonfluoridated population on an age, sex, and renal function basis. The mean serum fluoride in this population was 17.5 (± 9.5) µg/L. Concentrations that were 3–4 times higher were noted in patients with identified renal disorders. Urine fluoride concentrations in normal persons were 671 (± 373) µg/24 h.

2.4.1.7.4 Bone. Skeletal storage of absorbed fluoride is a characteristic response to fluoride exposure. Stokinger et al. (14) found approximately 10-fold increases in fluoride concentrations in the femurs of dogs exposed for 95 hr to 25 mg HF/m^3 or to 7.2 mg HF/m^3 for 166 h.

Swarup et al. (46) reported skeletal changes in bovines due to industrial point source fluoride exposure. Affected animals demonstrated skeletal and dental fluorosis. Urinary fluoride concentrations were elevated. The source of the exposure was believed to be pasture grass contaminated by a nearby aluminum smelter.

2.4.2 Human Experience

2.4.2.1 General Information. In the occupational setting, the principal concern for route of entry for both gaseous and particulate fluorides is the respiratory tract. In practice, however, many more cutaneous exposures occur than inhalation exposures. Deaths due to cutaneous exposure are reported in patients whose initial presentation after decontamination was quite unremarkable except for the cutaneous burns, but who deteriorated over a few hours after arrival at the hospital (23). Often, the area of initial skin contact appears only minimally affected when the original examination is performed. However, as fluoride is absorbed and penetrates to nerve and bone, the pain becomes intense and severe tissue damage may occur. While damage due to anhydrous HF is immediately apparent in most cases, damage due to more dilute solutions may be delayed, although in the end its impact may be quite severe.

Splashes of HF into the eyes and onto the skin cause frequent problems in the industries where engineering controls and closed processes are not common such as the semiconductor industry, circuit-board manufacture, laundries, and metal preparation. Although inhalation of HF is a critical route of exposure, and its toxicity potentially massive, the lack of awareness of the severity and lethality of dermal HF exposure may lead to inattention to these exposures and substantial morbidity, and some mortality may result. Corrosive injury to the respiratory tree, including the upper airways and the lungs may be immediate, or delayed. Deep lung injury and metabolic disturbance may be manifested at once, or may develop over the first 12–24 h following the exposure. Burns that accompany inhalation exposure are important. Burns over more than 10–15% of body surface area may lead to absorption of such large quantities of fluoride ion through the damaged surfaces that severe disturbances of calcium and potassium metabolism occur. If these physiologic alterations are not immediately treated, lethal outcome is likely. Fatal systemic effects have occurred as a consequence of burns to 2.5–10% of the body surface area, when the contactant was 70% HF.

Although concern for respiratory inflammation and pulmonary edema may overshadow concern about the systemic toxicity of the absorbed fluoride, inattention to the metabolic consequences may lead to death even if adequate decontamination and emergency respiratory intervention is made. HF is a direct irritant of all tissues, and is a highly corrosive tissue toxin. It causes rapid tissue dehydration and coagulation necrosis of all contact areas. This necrosis probably depends on hydrogen ion release. Later, necrosis from deeper penetration occurs in the skin or in other tissues. As fluoride complexes with calcium a later phase of necrosis will occur. At this time, depletion of local calcium stores may lead to potassium release in affected nerve endings, which will produce very severe pain. If large areas of skin are contacted, or the inhalation leads to absorption of a lethal dose of HF, systemic complexing of calcium and depletion of calcium stores may lead to cardiac arrhythmia and arrest. Ellenhorn and Barceloux suggest that the following general guidelines be used to assess the seriousness of HF exposure. When these criteria have been met, serious risks of hypocalcemia exist: exposure of one hand to $\geq$ 50% concentration of HF, any dermal exposure of > 5% of the body surface, inhalation of vapors from $\geq$ 60% concentration HF solution, and any ingestion of HF solution (26). All the mechanisms of HF toxicity that have been demonstrated in animals have been observed in humans acutely exposed to HF or to substances that are functionally equivalent to HF. When there is documentation or serious suspicion of HF inhalation or ingestion via dermal, buccal, or ocular routes, an acute emergency response with decontamination and immediate intervention to assess and control metabolic abnormalities, burns, and respiratory distress is mandatory. Comprehensive reviews of the management of local and systemic injury due to hydrofluoric acid have been published by Upfal and Doyle (27), Kirkpatrick et al. (28), and Sheridan et al. (29). These reviews provide definitive management protocols for HF poisoning by all routes of exposure.

2.4.2.2 Clinical Cases

2.4.2.2.1 Acute Toxicity

SKIN. Injuries from hydrofluoric acid may range from burns of the fingers from pinhole leaks in protective gloves to burns caused by splashes on the torso, arms, neck, or face. Initially, because HF is a weak acid in dilute solutions, there is a corrosive burn similar to but less severe than the burns produced by other mineral acids. There may be little or no initial pain, and the burn may not be discovered for some time. Eventually a characteristically severe throbbing pain sets in, which may not be adequately controlled by parenteral narcotics. The undissociated HF penetrates into the deep subdermal layers, causing severe destruction and liquefaction necrosis. The pain is thought to result from the release of cellular potassium and the intense stimulation of nerve endings (47). According to Bertolini (47), HF burns have been categorized according to the concentration in the offending solution: concentrations exceeding 50% cause immediate pain and tissue destruction; concentrations of 20–50% cause burns becoming apparent within several hours; for concentrations < 20%, pain and erythema may be delayed for up to 24 h.

Instances of fatal systemic poisoning following the absorption of lethal amounts of fluoride from HF-burned surfaces have been reviewed by Bertolini (47), by Mayer and Goss (48), and by Chan et al. (49). These burns involved areas from as little as 2.5% of the body surface to as much as 22% of the body surface. According to Himes (50), persons with burns less than 50 cm² in area may be treated on an outpatient basis if vital areas are not involved. Persons with burns involving a 50–100 cm² area should be hospitalized. If the burn exceeds an area of 100 cm², the patient should have access to the facilities of an intensive care unit. However, the decision to treat any HF burn without a period of hospital observation must be made only after complete identification of the product and the concentration of the HF is known with certainty. If material safety data sheets are not immediately available, and the stated contents of the inhalant or contactant is not known, observation or inpatient treatment until the possibility of metabolic derangement and cardiac dysrhythmia and arrest is past.

Gallerani et al. (51) have reported a suicidal subcutaneous self administration of 5 cm³ of a domestic rust removal solution which contained 7% HF. The patient was decontaminated in the emergency department and was given both cutaneous and subcutaneous treatment with magnesium chloride, which has been recommended for management of HF burns with some limited success. He was treated with 10% calcium gluconate and 2% xylocaine locally while the burn site was continuously irrigated with calcium gluconate solution, as is recommended. The patient developed severe hypocalcemia, hypokalemia, hyponatremia, and hypochloremia. He was then treated with IV calcium gluconate and with local surgery to the arm injury, and he survived. Although this is an extreme case, the severity of the complications from a small dose of a 7% solution of HF is notable (51).

To quote Modly and Burnett (52), "The dermatologic manifestations of fluoride/fluorine exposure have not been established definitely. There is no evidence of allergic dermatologic reactions to fluoride." However, Apted has recently reported delayed dermatitis from contact with HF, which may represent a contact allergy (53).

The eye is especially vulnerable to damage by HF, due less to the acidity of the agent than to the rapid and deep penetration of the undissociated HF. Conjunctival inflammation and edema, corneal opacification, and decreased visual acuity may result. One report cited by NIOSH (54) indicated recovery and repair that was complete at 35 days postexposure of the corneas to HF used at an alkylation plant. Other reports are less favorable. Beiran et al. reported on a study of animals treated with traditional saline eye irrigation following instillation of 2% HF in rabbit eyes compared to irrigation with 1% calcium gluconate. No advantage to the use of 1% calcium gluconate as an eye irrigating solution following dilute HF injury could be demonstrated over the more traditional use of saline. Injection of 1% CG subconjunctivally offered no therapeutic advantage in this study (55).

RESPIRATORY SYSTEM. Hydrogen fluoride has a sharp, irritating odor. Russian investigators have reported an odor threshold of 0.03–0.11 mg/m³ (0.04–0.14 ppm) for different individuals and a threshold of 0.02–0.04 mg/m³ (0.02–0.05 ppm) for sensitive subjects (56, 57). According to Lindberg (57), the "maximum imperceptible concentration" is 0.02 mg/m³ (0.02 ppm).

2.4.2.2.2 Chronic and Subchronic Toxicity. Experience with human subjects exposed to HF in laboratory trials (32, 58) showed that repeated daily exposures to 1.2 mg/m³ (1.5 ppm) were without noticeable adverse effect; 2.1–3.9 mg/m³ (2.6–4.9 ppm) caused slight irritation of the skin, eyes, and nose; 26 mg/m³ (32 ppm) was tolerable for several minutes; at 50 mg/m^3 (61 ppm), irritation of the eyes, nasal passages, and middle respiratory tract was marked; and 100 mg/m^3 (122 ppm) was the highest concentration that could be tolerated for > 1 min.

Reports involving workers exposed only to HF, rather than to mixtures with particulate inorganic fluorides such as cryolite, fluorspar, or rock phosphate, are uncommon. However, one such study has been described by NIOSH (24). Airborne concentrations of HF ranged within 0.05–8.2 mg/m^3 (0.07–10 ppm) and averaged 0.85 mg/m^3 (1.03 ppm). Pulmonary function tests on 305 chemical workers, including 11 who were exposed to HF, and on 88 control subjects, were all within normal limits; there were no significant differences between the chemical workers and the control subjects.

Two instances have been described in which HF, inadvertently released from industrial sources, drifted across nearby populated areas. In the first instance, 36 persons were treated for chemical exposure at area hospitals (50). It was estimated that airborne concentrations of HF could have reached 20 ppm. In the second incident, approximately 53,000 lb of anhydrous HF drifted across a community of 41,000 residents (59). Persons within a half-mile of the release were evacuated within 20 min, and eventually 3000 individuals were evacuated from a 5-mi^2 area. Airborne concentrations downwind 1 h later were 10 ppm; traces were still evident 2 h after the release. The next day damage to vegetation, car windows, and car paint was evident. Two nearby hospitals received 939 persons, 10% of whom were hospitalized. The most commonly reported symptoms were eye irritation (41.5%), burning throat (21%), headache (20.6%), and shortness of breath (19.4%). Decreased pulmonary function, measured by forced expiratory volume and by the ratio of forced expiratory volume to forced vital capacity, was noted in both nonhospitalized and hospitalized patients, although this finding was more common in the latter group.

Sjstrand et al. (60) tested a chamber designed to perform experimental studies on HF inhalation in humans. Acceptable control of concentrations of HF at or close to 1.5 mg/m^3 were maintained. Lund et al. (61) utilized this apparatus in a study of 20 healthy male volunteers who were exposed to HF concentrations of 0.2–5.2 mg/m^3, concentrations of HF in air which duplicated airborne HF levels in aluminum potroom work. They found that the total symptom score was significantly increased in all subjects at concentrations below the Norwegian standard for total fluorides, of 0.6 mg/m^3. No changes were detected in the FEV 15 in this group, but the FVC was found to be significantly decreased in the group exposed below the Norwegian standard. They measured plasma fluoride concentrations in the exposed workers, and concluded that a strong relationship existed between inhaled HF and concentrations of fluoride in plasma. Upper airway and eye symptoms occurred after one hour of exposure even at concentrations that were below the Norwegian hygienic standard.

Numerous reports of asthma in aluminum potroom workers exposed to fluorides have appeared in the literature. It is not entirely clear whether elevation of fluoride levels in the workplace is a cause of potroom asthma, an aggravating exposure, or merely a marker for other causes of asthma in that environment. Sordrager et al. (62) reported atopy to be a predisposing factor to the development of potroom asthma, but other authors do not regard this condition as a prerequisite or precondition for the development of occupational asthma in potroom workers. In a retrospective, nested case-control study, Sordrager et al. (63) reviewed the preemployment eosinophil counts of workers with and without potroom asthma. They found that although the total eosinophil counts of both groups were within normal limits in the preemployment data, the group with potroom asthma had higher total counts. Moreover, of the group with preemployment counts of 220 cells/cm^3, 39 eventually developed potroom asthma. Multiple regression analyses to control for other confounding factors showed that only the preemployment eosinophil count was related to the eventual development of potroom asthma. Potroom asthma in aluminum production workers is discussed in further detail below.

KIDNEY AND THE RENAL EXCRETION OF FLUORIDE. Excretion through the kidney is the major pathway for the removal of absorbed fluoride, and the determination of fluoride concentration in the urine is used as an indicator of fluoride exposure or lack thereof in the work environment. Much of the basis for this relationship has come from investigations in the aluminum industry, where exposures have been to fluoride-containing dusts or to mixtures of gaseous and particulate fluorides.

There are far fewer studies of occupational exposure to HF only. Zober and Welte (64) found a high correlation between atmospheric fluoride concentration and renal excretion in workers exposed to HF in the glass industry and emphasized that exposure in this industry, that is, to gaseous HF, differed from exposure in the aluminum industry, which involves fluoride-containing dusts as well as gaseous fluorides.

Steinegger and Schlatter (65) have reviewed the state of the art in evaluation of fluoride exposure in the aluminum smelter industry. They considered urinary fluoride measurement to be highly dependent on urinary excretion rates and recommended that all values be corrected to reflect the 24-h creatinine excretion. Sharma et al. (66) studied urinary fluorides in workers engaged in the manufacture or inorganic fluoride chemicals. Urinary fluorides were highly dependent on the job assignment of the workers. Preworkshift urinary fluoride ranged from 0.5–4.54 ppm. Postworkshift fluorides ranged from 0.5 to 13 ppm, with a mean of 4.19 ppm. No evidence of skeletal fluorosis was found in this population.

Kono et al. (67) examined fluoride excretion in workers using HF to clean television tubes and to etch semiconductors. The geometric mean urinary fluoride concentration for 82 unexposed workers was 0.59 ppm (standard deviation(SD) 0.37–0.81). For 19 workers exposed to 0.6–1.6 ppm HF in the workplace atmosphere, mean preworkshift and postworkshift urinary concentrations were 1.68 (SD 0.81–3.48) and 2.34 (SD 0.94–5.84) ppm, respectively. For 141 workers exposed to 0.3–5.0 ppm HF, mean urinary concentrations ranged from 0.91 (SD 0.65–1.27) to 6.50 (SD 4.42–9.35) ppm. All urine concentrations were adjusted to a standard urine specific gravity of 1.024. From a plot of the data, it was estimated that an airborne concentration of 3 ppm (2.46 mg/m^3), the maximal allowable concentration recommended by the Japanese Association of Industrial Health, corresponded to a urinary concentration of 4 ppm (4 mg/L). Hogstedt (68) found a good correlation between the concentration of fluoride in postworkshift urine samples and airborne concentrations of fluoride. The airborne fluoride varied between 0.01 and 1.42 mg/m^3 and never contained more than 0.05 mg/m^3 of particulate fluoride. The airborne concentrations also correlated with the total 24-h excretion of fluoride and with the concentration in the 24-h urine sample, but there was no significant correlation with the total amount excreted during the workshift. The data indicate that an airborne concentration of 0.5 mg/m^3 HF and F corresponds roughly to 4 mg F/L (4 ppm) in the urine. Ehrnebo and Ekstrand (69) investigated fluoride excretion in aluminum workers exposed to atmospheres in which 34% (0.3 mg/m^3) of the total fluoride was gaseous. The total fluoride excreted during the 8-h workshift plus the following 8 h (total: 16 h) correlated better with the gaseous exposure than did the mean urinary fluoride concentration and did not correlate well with the concentration of airborne particulate fluoride. It was also noted that wearing a respirator during the entire shift reduced the inhalation of fluoride to 30–40% of that inhaled in the absence of the mask.

Because of the central role of the kidney in excreting fluoride, it is also important to consider the effect of fluoride on this organ. Polyuria in some patients anesthetized with methoxyflurane (Penthrane) [76-38-0] has been attributed to the inorganic fluoride released by the *in vivo* metabolism of the anesthetic agent (70). Biopsy and autopsy specimens from patients with permanent renal failure following methoxyflurane anesthesia showed proximal tubular dilation, interstitial fibrosis, and deposits of oxalate as a metabolic product. These effects are related to the peak concentration of blood F and the duration of exposure of the kidney to high concentrations. Cousins and Mazza (70) concluded, for patients anesthetized with methoxyflurane, that peak serum fluoride concentrations below 0.76 mg/mL are not associated with nephrotoxicity; concentrations of 0.95–1.5 mg/mL produce subclinical nephrotoxicity; concentrations of 1.7–2.3 mg/mL cause mild clinical toxicity, and concentrations of 1.5–33 mg/mL produce clinical toxicity.

Burke et al. (71) described a severe polyuria persisting for at least 3 days in a laboratory technician who was severely poisoned following a burn by 100% anhydrous HF. Serum fluoride levels were not reported.

Patients with chronic renal failure have been shown to excrete fluoride less efficiently than do subjects with normal renal function. This is reflected in higher serum fluoride concentrations in these patients and decreased amounts of fluoride excreted per 24 h in the urine, although urinary concentration of fluoride may be normal (72, 73). Because bone fluoride concentrations are known to increase with increasing blood fluoride concentrations (74), the risk of osteofluorosis is greater in patients with decreased renal function. Kono et al. (73) suggest that, because urinary fluoride concentrations may nevertheless be normal in these subjects, screening of a workforce should include the measurement of serum fluoride concentrations in order to detect those workers at greater risk. Ekstrand and Ehrnebo (75) and Ehrnebo and Ekstrand (69) also have advocated including plasma fluoride determinations in screening programs.

BONE. Regardless of the source or route of entry, it is fluoride from the circulating blood that is deposited in bone. Because the bulk of our information regarding occupational skeletal fluorosis has come from investigations involving exposure to mixtures of gaseous and particulate fluorides, or to particulate fluorides only, osteofluorosis is considered in later sections.

2.4.2.2.3 Pharmacokinetics, Metabolism, and Mechanisms: NA

METABOLISM OF FLUORIDE. In occupational exposures to fluorides, the major portal of entry of gaseous or particulate fluoride is via inhalation. Hydrogen fluoride is soluble in water and is readily absorbed in the upper respiratory tract. Particulate fluorides are deposited in the nasopharynx, the tracheobronchial tree, and the lower airways. The finer particles penetrate farthest into the respiratory system. Depending on the solubility of the particles, some absorption occurs at the sites of deposition. Some of the particles on the ciliated portion of the respiratory tract are moved up to the pharynx, and these, as well as the larger particles unable to penetrate deeply, are swallowed and absorbed from the gastrointestinal (GI) tract hours to days later. Absorption is a pH-dependent event. The acid medium of the stomach, for example, favors the existence of nondissociated HF, which is absorbed across the gastric mucosa far more readily than is the fluoride ion.

About 75% of the fluoride absorbed into the blood is carried in the plasma. Fluoride is not bound to the soft tissues and is present in organs in proportion to their vascularity and the concentration in the blood. Accumulation in the kidney occurs as fluoride is concentrated in the urine for elimination. Of the body burden of fluoride, 99% is found in the skeleton, where fluoride substitutes easily for the hydroxyl group of hydroxyapatite, the principal mineral component of bone. Deposition occurs in three stages. Initially fluoride in the extracellular fluid bathing the bone crystal surface enters the hydration shell of the crystal. The fluoride then exchanges with hydroxyl groups on the crystal surface, forming fluorapatite. Finally, fluoride can migrate more deeply into the crystal as it continues to exchange with hydroxyl ions during recrystallization and mineralization. The movement of fluoride into and out of the hydration shell is relatively rapid. Movement of the more deeply buried fluoride is slower, and thus bone fluoride continues to increase with time. The fluoridated bone is less soluble and more dense than is otherwise the case. The formation of new bone is also stimulated by fluoride. When fluoride exposure is reduced and concentrations in the blood are consequently decreased, fluoride is mobilized from the bone, although the process is slow (years).

Renal excretion of fluoride, discussed earlier, is prompt and represents the major route for the removal of fluoride from the body. Renal excretion and skeletal storage of fluoride are the principal defense mechanisms by which absorbed fluoride is detoxified. The latter process is the more rapid.

Fecal excretion of fluoride accounts for 5–10% of the total fluoride entering the body. Loss of fluoride in perspiration can be considerable; potroom workers in the aluminum industry may lose an average of 6 L of perspiration per day in the summer.

For more detailed descriptions of the physiology of fluoride, the reader is referred to Dinman et al. (76), Farley et al. (77), Grynpas (78), Hogstedt (68), Whitford (79), and the ATSDR Toxicological Profile for Fluorides, Hydrogen Fluoride and Fluorine (9).

2.5 Standards, Regulations, or Guidelines of Exposure

Standards for occupational exposure to hydrogen fluoride, as well as the airborne concentrations associated with respiratory and skin effects, are listed in Table 48.6.

The standards for airborne concentrations are intended to prevent irritation of the skin, eyes, and respiratory tract, and to prevent the deleterious effects of skeletal fluorosis defined as increased bone density or osteosclerosis due to retention of fluoride. In cases where the average of pre-shift urinary fluoride concentrations for the preceding 6 years exceeds 4 mg F/L, the physician may wish to consider pelvic X rays.

Table 48.6. Thresholds and Standards for Exposure to Hydrogen Fluoride

Standard/Response	Concentration ppm	mg/m^3
ACGIH TLV-TWA	3.0	2.5
NIOSH REL-TWA	3.0	2.5
OSHA PEL	3.0	2.5
NIOSH ceiling	6.0	5.0
Maximal imperceptible concentration	0.02	0.02
Odor threshold; odor is strong, irritating	0.04–0.014	0.03–0.11
Without noticeable adverse effect during repeated exposures	1.5	1.2
Slight irritation of skin, eyes, nose	2.6–4.9	2.1–3.9
Tolerable for several minutes	32.0	26.0
Marked irritation of eyes, nasal passages; mild respiratory distress	61.0	50.0
Highest concentration tolerable for more than 1 min	122.0	100.0

Urine Standards[a]
(Concentrations corrected to specific gravity of 1.024)

Preshift sample taken at start of work shift at least 48 hrs after last occupational exposure	4 mg/L
Postshift sample taken at end of shift after 4 or more consecutive days of exposure	7 mg/L

[a]From Ref. 24.

3.0 Sulfur Fluorides

3.0a Sulfur Hexafluoride

3.0.1a CAS Number: *[2551-62-4]*

3.0.2a Synonyms: Sulfur fluoride, (OC-6-11)-

3.0.3a Trade Names: NA

3.0.4a Molecular Weight: 146.05

3.0.5a Molecular Formula: SF$_6$

3.0.6a Molecular Structure:
$$\begin{array}{c} F \\ F\diagdown | \diagup F \\ S \\ F \diagup | \diagdown F \\ F \end{array}$$

3.0b Sulfur Tetrafluoride

3.0.1b CAS Number: *[7783-60-0]*

3.0.2b Synonyms: Sulfur fluoride; sulfur fluoride, (T-4)-; sulfur fluoride (SF4), (T-4)-

3.0.3b Trade Names: NA

3.0.4b Molecular Weight: 108.05

3.0.5b Molecular Formula: SF₄S

3.0.6b Molecular Structure:

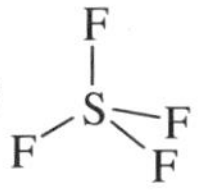

3.1 Chemical and Physical Properties

Sulfur tetrafluoride melts at $-121°C$ and boils at $-40.4°C$. The corresponding physical constants for disulfur decafluoride are $-92°C$ and $29°C$, respectively.

3.2 Production and Use

Compounds of interest in this group include sulfur hexafluoride (SF_6) used as an electrical insulating material in circuit breakers, cables, capacitors, and transformers, and its degradation products, which are produced when electrical arcing occurs. The specific compounds produced depend on the arcing conditions. In anaerobic and anhydrous circumstances, sulfur tetrafluoride (SF_4) is produced; if moisture is present, the tetrafluoride may hydrolyze to form thionyl fluoride (SOF_2) and HF. Sulfuryl fluoride (SO_2F_2), disulfur decafluoride (S_2F_{10}), sulfur pentafluoride (SF_5), and sulfur dioxide are also formed from sulfur hexafluoride (80). Some of these compounds are also produced as contaminants in the commercial production of sulfur hexafluoride by burning sulfur in fluorine gas, for example, sulfur tetrafluoride and disulfur decafluoride, as well as sulfur monofluoride (S_2F_2) (81). Griffin et al. (82) question whether disulfur decafluoride does in fact exist in significant amounts in "electrically stressed" sulfur hexafluoride atmospheres. Sulfur tetrafluoride has been identified among the degradation products of sulfur hexafluoride (80), and sulfuryl fluoride was identified in the atmosphere of a sulfur hexafluoride storage tower in which two men had been overcome after entering the previously vented tower (83). According to Kraut and Lilis (80), more than half of all U.S. workers exposed to sulfur hexafluoride are electrical and electronic equipment repair workers. NIOSH estimated that, in 1983, 9282 workers were exposed to sulfur hexafluoride and 5136 were exposed to sulfuryl fluoride (7).

3.3 Exposure Assessment

Analytical procedures for sulfur hexafluoride has been published by NIOSH, method #6602 (25). Truhaut et al. (84) have described gas chromatographic methods for measuring thionyl fluoride and sulfuryl fluoride in animal inhalation studies, and Pilling and Jones (83) used a Miran 104 infrared analyzer to measure airborne concentrations of sulfur hexafluoride and sulfuryl fluoride.

3.4 Toxic Effects

3.4.1 Experimental Studies

Exposure of rats to atmospheres containing as much as 80% sulfur hexafluoride plus 20% oxygen for up to 24 h caused no perceptible indication of irritation or other toxicological

Table 48.7. LC_{50} Values for Animals Exposed to Disulfur Decafluoride[a]

Species	Exposure (min)	LC_{50} (ppm)
Mouse	1	133–203
	10	9.5–19
Rat	1	218
	10	19–28
Guinea pig	10	38–57
Rabbit	10	38–57
Rhesus monkey	10	86

[a]From Ref. 82.

effects (85). These authors therefore concluded that the gas is physiologically inert and that, at higher concentrations, it behaves as an asphyxiant simply by excluding oxygen.

Disulfur decafluoride presents a far different toxicity profile. In rats exposed for 16–18 h, death occurred at a concentration of 1 ppm, severe pulmonary lesions developed at 0.5 ppm, and irritation was present at 0.1 ppm; no discernible effects were seen at 0.01 ppm (86). The LC_{50}s reported by Griffin et al. (82) for several animal species exposed for 1 or 10 min are shown in Table 48.7. Both Griffin et al. (82) and Greenberg and Lester (86) emphasize the insidious nature of the action of S_2F_{10}. The skin and eyes are not irritated, and there is no warning of toxic exposure. Pure S_2F_{10} is claimed by some to be odorless and by others to have an odor that resembles that of sulfur dioxide (82). Disulfur decafluoride exerted a profound effect on the circulatory and respiratory systems of dogs, with death resulting from a fulminant pulmonary edema that was often hemorrhagic (81). The major pathological lesion was diffuse endothelial damage to the cardiovascular system, which was responsible for the pulmonary edema. All animals showed renal changes resembling those of acute glomerulonephritis. In another study, Saunders et al. (81) administered the S_2F_{10} IV as a lecithin emulsion. Truhaut et al. (84) suggest the following toxicological classification for the sulfur fluorides: simple asphyxiants such as sulfur hexafluoride; pulmonary irritants such as sulfur tetrafluoride, disulfur decafluoride, and thionyl fluoride; and convulsants such as sulfuryl fluoride.

3.4.2 Human Experience

Pilling and Jones (83) describe what they believe to be the first reported instance of "degraded SF_6" toxicity in humans. In this instance two maintenance workers entered a sulfur hexafluoride storage tower that had been incompletely vented. Both collapsed within minutes and were subsequently removed by rescuers equipped with safety breathing apparatus. Simultaneously, the bottom of the tower was opened, permitting gas to escape and fresh air to enter the tower. One victim was rescued approximately 19 min after collapsing, the other after about 30 min. At the hospital 90 min later, the latter individual developed florid pulmonary edema, which was successfully treated with oxygen. Analyses of the tower atmosphere after the workers had been rescued, and after the bottom of the tower had been opened, showed 21 and 18.6% oxygen at the top and at the base of the tower, respectively. The SF_6 concentration was 1500 ppm, and the SO_2F_2 level was 50 ppm. Clinical observations of both workers suggested a direct effect on the alveolar–capillary membrane, leading to the pulmonary edema, hypoxemia, and mild tissue acidosis. The victim who developed pulmonary edema was discharged after 3 days; the other, after 2 days. No significant long-term effects were seen in either person. Pilling and Jones (83) believe that, because of possible delayed effects, persons asphyxiated with industrial SF_6 should be hospitalized and observed for the next 12–24 h to permit early detection and prompt treatment of pulmonary edema.

Kraut and Lilis (80) describe an incident in which a six-member team was working in an enclosed space on the repair of cable after the burnout of a transformer. Sulfur tetrafluoride caused by the thermal decomposition of SF_6 was believed to be present in the pipe carrying the cable, and exposure occurred when the pipe was cut. About 1 h after the team began to work, a "burning battery"–like odor was evident and the workers experienced eye irritation with tearing, dry and burning throat, and tightness of the chest. Exposure was intermittent for a total of about 6 h over a 12-h period as the crew attempted to complete the work. Efforts were stopped after five of the workers again experienced chest tightness, shortness of breath, headache, fatigue, nosebleed, nausea, and vomiting. The safety officer, who remained outside the confined space, experienced only eye irritation. Sulfur tetrafluoride was identified in air samples taken at the partly opened pipe, but quantitative measurements were not made. Four workers remained symptomatic for 1 week to 1 month after this incident. Intermittent nosebleed was the most persistent symptom. Chest X-rays showed several areas of transitory plate-like atelectasis in one worker and a slight diffuse infiltrate in the lower left lobe of another. One individual showed transient obstructive changes in pulmonary function tests. Follow-up after 1 year showed no permanent abnormalities.

Ostlund et al. (87) investigated the comparative narcotic potency of sulfur hexafluoride and nitrous oxide and nitrogen gas in humans. The purpose of their experiment was to elucidate the relationship between lipid solubility of inert gases when compared to anesthetic gases and their narcotic potency. The order of narcotic potency when compared to nitrogen and nitrous oxide was consistent with the lipid solubility of the three gases.

3.5 Standards, Regulations, or Guidelines of Exposure

Standards for exposure to sulfur fluorides in the workplace are shown in Table 48.8.

4.0 Particulate Fluorides Alone or in Combination with Gaseous Fluoride

In the occupational setting, exposures to airborne fluorides are likely to occur in the form of exposure to mixtures of particulate and gaseous fluorides. In the production of aluminum, for example, particulate matter may constitute as much as 80% of the total airborne fluoride; the remainder is gaseous in nature, or represents gaseous fluoride absorbed on aerosols (76, 88–96). In the processing of rock phosphate, 60–80% of the airborne fluoride may be present as dust. The biological effects of chief concern caused by long-term exposure are those produced in the skeleton and the respiratory system. They are attributable to the inhaled and ingested fluoride, rather than to the unique features of any

Table 48.8. Standards for Exposure to Sulfur Fluorides

Compound	ACGIH TLV ppm	ACGIH TLV mg/m^3	NIOSH REL ppm	NIOSH REL mg/m^3	OSHA PEL ppm	OSHA PEL mg/m^3	NIOSH IDLH ppm	NIOSH IDLH mg/m^3
Sulfur hexafluoride	1000 (8-h TWA)	5970 (8-h TWA)			1000	6000		
Sulfur pentafluoride (disulfur decafluoride, sulfur decafluoride)	C 0.01	C 0.10	C 0.01	C 0.10	0.025 (8-h TWA)	0.25 (8-h TWA)	1	10
Sulfur tetrafluoride	C 0.10	C 0.40						
Sulfuryl fluoride (sulfur difluoride dioxide)	5 (8-h TWA) 10[b]	20 (8-h TWA) 42[b]	5 (10-h TWA) 10[b]	20 (10-h TWA) 40[b]	5 (8-h TWA)	20 (8-h TWA)	200	800

[a] From Refs. 21 and 22.
[b] Short-term exposure limit.

THE HALOGENS

given industrial operation. Contaminant fluorides encountered include cryolite, fluorapatite, fluorspar, sodium fluoride, ammonium fluoride, hydrofluorosilicic acid, and silicon tetrafluoride. Toxicity data relating to the industrial uses of these fluorides are minimal or nonexistent. These fluorides are discussed in the following sections as a group, to avoid redundancy. More detailed descriptions of the uses of inorganic fluorides are to be found in Ref. 97. NIOSH (1) has listed operations, as well as suitable general control methods, for exposures to fluoride dusts.

4.0a Cryolite

4.0.1a CAS Number: [15096-52-3]

4.0.2a Synonyms: Kryocide sodium fluoaluminate; sodium aluminofluoride; sodium aluminum fluoride; trisodium hexafluoroaluminate(3-), (OC-6-11)-

4.0.3a Trade Names: NA

4.0.4a Molecular Weight: NA

4.0.5a Molecular Formula: Na_3AlF_6

4.0b Calcium Fluoride

4.0.1b CAS Number: [7789-75-5]

4.0.2b Synonyms: Fluorite; Acid-Spar; Calcium Difluoride; Fluorspan; Met-Spar

4.0.3b Trade Names: NA

4.0.4b Molecular Weight: 78.08

4.0.5b Molecular Formula: CaF_2

4.0.6b Molecular Structure: $\quad Ca^{2+}$
$\qquad\qquad\qquad\qquad\qquad\quad F^- \quad F^-$

4.1 Chemical and Physical Properties

Selected physical properties of cryolite and calcium fluoride are listed in Table 48.9.

4.2 Production and Use

Cryolite, sodium aluminum fluoride (Na_3AlF_6; 54.2% F) is essential in the production of aluminum by the Hall–Heroult process, in which alumina is electrolytically reduced to aluminum using carbon electrodes; cryolite, to which calcium fluoride usually is added, is the flux in which the alumina is dissolved. Until they were severely depleted, naturally occurring deposits in Greenland were the principal source of cryolite; this substance is now synthesized by one of several processes. For example, sodium aluminate ($NaAlO_2$) can be treated with sodium hydroxide and hydrogen fluoride to yield sodium aluminum fluoride. Sodium carbonate can be substituted for the sodium hydroxide, and aluminum

Table 48.9. Selected Physical Properties of Cryolite and Calcium Fluoride

Property	Cryolite	Calcium Fluoride
Melting point	1009°C	1402°C
Boiling point	—	2513°C
Density		
Solid at 25°C	—	3.181 g/cm^3
Solid at MP	2.52 g/cm^3	—
Liquid at MP	2.087 g/cm^3	2.52 g/cm^3
Solubility in water (°C)		
At 25°	0.042 g/100 g	0.0017 g/100 g
At 100°	0.135 g/100 g	—
At 175°	—	0.0018 g/100 g

sulfate can be substituted for the sodium aluminate. In another process, the fluorosilicic acid wastes from the manufacture of phosphorus and phosphate fertilizers are treated with ammonia to remove the silica. The ammonium fluoride formed is then added to a solution of sodium sulfate and ammonium sulfate to yield sodium aluminum fluoride. Old carbon pot liners, which serve as both the cathode and the container for the melt, are another source of cryolite. When the cells are shut down for maintenance, the liners may contain as much as 30% cryolite by weight.

A variety of emissions are released from aluminum electrolysis cells. These include carbon monoxide, carbon dioxide, sulfur dioxide, silicon tetrafluoride, hydrogen fluoride, carbon sulfide, carbon disulfide, hydrocarbons, alumina, carbon, cryolite, aluminum trifluoride, sodium aluminum tetrafluoride, calcium fluoride, chiolite (Na$_5$Al$_3$F$_{14}$), and ferric oxide. Most of these are removed by filters, centrifugal collectors, electrostatic precipitators, and various wet scrubbing systems.

Fluorspar (CaF$_2$; 48.5% F) is used as a flux in the production of steel and magnesium and is added to the cryolite flux in the production of aluminum. Another major use is in the production of hydrogen fluoride by treatment of the mineral with sulfuric acid. Fluorspar is encountered in the production of elemental phosphorus, steel, cement, enamel coatings, and glass.

Fluorapatite [Ca$_{10}$F$_2$(PO$_4$)$_6$; 3.8% F] is a constituent of rock phosphate, which is, in turn, an important source of phosphate fertilizers, phosphoric acid, and elemental phosphorus.

Several miscellaneous inorganic fluoride compounds are used as coatings for sand molds in metal casting operations, for example, ammonium fluoride, ammonium bifluoride, ammonium fluoroborate, potassium fluoroborate, and ammonium fluorosilicate. On contact with the hot metal, the coatings vaporize and partially decompose to compounds such as hydrogen fluoride, silicon tetrafluoride, and boron trifluoride. Electrodes coated with calcium fluoride or other inorganic fluorides are commonly used in arc welding of ferrous and nonferrous metals. Depending on the type of electrode, current intensity, and the potential, gaseous as well as particulate fluorides are released. These may include fluorine, hydrogen fluoride, fluorosilicates, and the fluorides of titanium, calcium,

THE HALOGENS 759

and potassium. Sodium fluoride is the most important of the alkali-metal fluorides. It is prepared by neutralizing aqueous hydrogen fluoride with sodium bicarbonate, and is used as a flux, for example, in the manufacture of steel, for scrubbing hydrogen fluoride from fluorine, as a pesticide, and for the fluoridation of water supplies in dental health programs.

Common operations that may involve exposure to fluoride dusts, together with generally applicable control methods, are listed by NIOSH and OSHA (1). The estimated numbers of workers exposed to selected fluoride compounds are as follows: sodium fluoride, 84,776; calcium fluoride, 344,071; ammonium fluoride, 17,582; and potassium fluoride, 22,283 (1).

4.3 Exposure Assessment

4.3.1 Air

A standard procedure for the collection and analysis of airborne fluorides has been described by NIOSH (25) for the collection of aerosol and gaseous fluorides (method 7902). In this method, air is collected on a cellulose ester membrane prefilter and Na_2CO_3-treated cellulose pad. The filters are extracted with a buffer solution and fluoride ion content is quantified using an ion-specific electrode. This method has an estimated limit of detection of 3 mg per sample; the overall precision (coefficient of variation) of the method has not been evaluated. An ion chromatography method for determination of fluoride aerosol and gases was published by NIOSH as method 7906, which represents a technological advance over the earlier method. The collection methods are essentially the same as that for 7902, but working ranges for HF and water insoluble fluorides are improved. Some analytical error may be introduced if aerosols other than the fluoride of interest are present as a result of absorption onto or reaction with these particles, which may lead to an overestimation of the particulate/gaseous fluoride ratio. The method is applicable to mining samples, aluminum reduction, ceramic and glass work, electroplating, and the semiconductor and fluorochemical industries.

4.4 Toxic Effects

4.4.1 Experimental Studies: NA

4.4.2 Human Experience

4.4.2.1 General Information. Inasmuch as the pertinent information derives chiefly from experience with human subjects, effects in animals are not considered separately.

4.4.2.1.1 Signs and Symptoms. The certainly lethal oral dosage of fluoride for humans has been estimated to be 32–65 mg F/kg of body weight. Following ingestion of soluble fluoride, the victim may experience a salty or soapy taste, vomiting, abdominal pain, diarrhea, shortness of breath, difficulty in speaking, thirst, weakened pulse, disturbed color vision, muscular weakness, convulsion, loss of consciousness, and death.

Reported signs of acute systemic poisoning may be grouped in several categories, for example, general malaise, weakness, and pallor; cardiopulmonary tachycardia, hypotension, prolonged QT interval on the electrocardiogram, ventricular fibrillation, and pulmonary edema; neurological/neuromuscular and CNS depression, respiratory depression and paralysis, seizures, carpopedal spasm, tetany; metabolic hypocalcemia, hyperkalemia, and hypomagnesemia. All of these signs and symptoms are not necessarily seen in every case of poisoning.

4.4.2.1.2 Skin. Effects on the skin associated with occupational exposure to fluorides have been reported, although, as indicated by Modly and Burnett (52), fluoride-induced dermatologic manifestations have not been established definitively.

Roholm (98) reported a 12% prevalence rate for skin rashes among cryolite workers. Friis et al. (99) reported a high prevalence of skin eruptions or eczema among cryolite production workers, as well as irritation of the eye and mucous membranes, for example, mouth and pharynx. Jahr et al. (100) also reported inflammation of the pharynx and pain in the throat in aluminum smelter workers. Mean airborne fluoride concentrations were less than 2.5 mg/m^3, with a few short-term excursions above 5 mg/m^3 during specific operations. Kaltreider et al. (96) found no difference in the incidence of dermatitis in control subjects and potroom workers in the prebake process for the reduction of alumina. Theriault et al. (101, 102) described skin telangiectasia in potroom workers exposed during the Soderberg process; the incidence increased with increasing years of exposure. Theriault et al. suggested that the causative agent might be a gas containing both hydrocarbon and fluoride components emitted from the electrolytic cells.

4.4.2.1.3 Respiratory System. The lack of significant differences between fluoride workers and control subjects in the incidence of various pulmonary diseases such as bronchitis, bronchiectasis, emphysema, pneumonia, and carcinoma has been noted by Kaltreider et al. (96), Zober and Welte (64), and Friis et al. (99).

Effects of airborne fluorides on the respiratory system have seldom been investigated at the cellular level. Concern for the carcinogenicity of welding fumes because of their chromium content led to a concomitant interest in the effects of the fluoride component. Baker et al. (103) demonstrated that both the water-soluble and water-insoluble fractions of fumes generated by arc welding using a variety of electrodes induced sister chromatid exchanges (SCEs) in Chinese hamster lung cells in proportion to the chromium content, whereas the contribution from chromium(III), fluoride, nickel, manganese, and other constituents was minor.

Suspensions of fluoride mine dust containing 33.5% calcium fluoride and 47.5% silica produced only a typical foreign-body reaction when injected into the trachea of rabbits through a slit in the neck (104). Fluoride did not stimulate pulmonary alveolar macrophages to release fibrogenic factors *in vitro*. Eklund et al. (95) also noted an absence of fibroblast activation by alveolar macrophages. In this instance the macrophages were obtained by bronchoalveolar lavage of potroom workers whose workplace atmosphere was known to contain 0.10–0.50 mg total fluoride/m^3 (mean, 0.31 mg/m^3), well below the Swedish threshold limit value.

Potroom asthma has been reported among fluoride-exposed workers, although, as observed by Kongerud et al. (105), the reported incidence appears to be lower in North American studies than in European investigations. These authors noted that the incidence of work-related asthmatic symptoms increased with increasing years of employment. Air flow limitation, as reflected in FEV_1 (forced expiratory volume in 1 s), also progressed with longer exposure. In a later study, Soyseth and Kongerud (106) demonstrated a greater prevalence of respiratory symptoms and abnormal lung function (FEV_1) in workers exposed to total atmospheric fluorides at concentrations of ≥ 0.5 mg/m^3 than in workers exposed to levels below 0.5 mg/m^3. These concentrations are below the Norwegian standard of 2.5 mg/m^3. Bronchial hyperresponsiveness was not associated with exposure to fluoride. These authors note, however, that the prevalence of respiratory symptoms in the exposed workers was not higher than the incidence reported for the general population. Saric et al. (94) concluded that the potroom workers' asthma seen in some workers is most probably caused by bronchial hyperreactivity rather than an allergic mechanism. Exposures of the workers tested by Saric et al. were 0.17–1.72 mg HF/m^3 (mean, 0.56 mg/m^3) and 0.01–0.25 mg/m^3 (mean, 0.15 mg/m^3) of particulate fluorides. Bronchial asthma and bronchial hyperreactivity were seen in aluminum welders using potassium aluminum tetrafluoride as a flux. There is no specific standard for this compound, and these authors suggest that such a standard should be set at well below 1 mg/m^3. Simonsson et al. (107) noted hyperreactivity in workers who developed nocturnal wheezing and dyspnea, with reversible airway obstruction, as a result of exposure to aluminum trifluoride [7784-18-1] or aluminum sulfate. Dyspnea in potroom workers has been reported by a number of investigators (91).

Chan-Yeung et al. (92) have reported extensively on the effects of potroom exposure on the respiratory system. Two groups of potroom workers were studied: those who spent more than 50% of their working time in the potroom (high exposure) and those who worked less than 50% of their time in the potroom (medium exposure). A control group had no significant exposure to contaminants. Total, gaseous, and particulate fluorides in the potroom atmosphere were 0.48 0.35, 0.20 0.17, and 0.28 0.03 mg/m^3, respectively. Potroom asthma was not seen in any group of workers included in this study. Workers in the high (longer) exposure group showed an increased frequency of cough and wheeze, and a lowered FEV_1 and maximal midexpiratory flow rate (FEF 25–75%) relative to these indexes in the control groups. These parameters were shown to decline during the course of an 8-h work shift, but the changes were not significantly different from those seen before and after the shift in the controls. These changes also did not correlate with pre- and postshift (pre- and postworkshift) urinary fluoride concentrations. The medium exposure workers showed an insignificantly greater prevalence of respiratory systems and decreased lung function compared with the controls. Jahr et al. (100) previously had noted a decline in FEV_1 during the course of a shift and the lack of correlation with urinary fluoride. Kaltreider et al. (96) found little difference in the incidence of chronic obstructive pulmonary disease in potroom workers and control subjects. Discher and Breitenstein (108) compared the prevalence of respiratory symptoms and of changes in FEV_1 and forced vital capacity (FVC) in potroom workers and control subjects and found no differences. These authors concluded that potroom workers are not subject to a major risk of chronic pulmonary disease. More recently, Friis et al. (99) reported a lack of significant correlation between length of worker exposure in refining cryolite operations and FEV_1 and FVC. In addition, these authors found no relation between exposure and dyspnea and no evidence of pneumoconiosis in the chest X-rays of these workers.

Soyseth et al. (109) compared the bronchial responsiveness of workers removed from potroom work, and those who remained in the exposed areas. The subjects were regularly examined for 2 years after relocation from potroom work. Statistically significant improvement occurred in the removed worker's bronchial reactivity when compared to the workers who remained in the potroom environment. Again, although the precise inciting cause of the increased bronchial reactivity remained unresolved, the clear association between asthma and potroom work was demonstrated by a classic dechallenge procedure, in this study.

Akbar-Khanzadeh (110) reported on 213 Iranian male potroom workers in an aluminum smelter. Questionnaires, physical and laboratory assessment, and ventilatory functions were evaluated. Personal air samples were collected and analyzed by the ion-specific method. Of the total particulate air samples, 35% exceeded 10 mg/m^3; 75% exceeded 5 mg/m^3. Fluoride levels were below 2.5 mg/m^3, the ACGIH TLV, but exceeded the European standards of 1 mg/m^3. Workers with $\geq$ 7 or more years of longevity reported increased chest symptoms, and had significant reductions in ventilatory function, which were particularly striking in nonsmokers. The authors noted the elevated fluoride levels in the plant, and observed that the fact that the plant was in a third world country may account for some of its technological failings. They did not ascribe the asthma to the fluoride exposure, in particular.

A review of the state-of-the-art evidence on asthma and respiratory diseases in the aluminum industry by O'Donnell (111) concluded that on the basis of the evidence available in 1995, asthma in potroom workers was irritant-induced and due to the inhalation of gaseous or particulate fluoride compounds. He noted that medical surveillance, respiratory protection, engineering improvements, and transfer of asthmatic workers has resulted in the improvement of signs and symptoms of those workers who have been affected. He also noted that chronic bronchitis and chronic obstructive pulmonary disease is increased in workers in the electrode bake areas, and that tobacco smoking has a significant additive adverse effect. He noted, as well, that the occurrence of fibrotic disease was "trivial" in the potroom worker studies.

4.4.2.1.4 Kidney and the Renal Excretion of Fluoride. Several large-scale studies convincingly demonstrate the absence of fluoride-induced renal effects in potroom workers. Tourangeau (112) identified no differences in the incidence of renal disease between control subjects and potroom workers, and Kaltreider et al. (96) found no difference between the incidence of albuminuria in potroom workers and control subjects. The incidences of renal calculi and urinary tract infections also were comparable in the two groups. Dinman et al. (89) observed no correlation between urinary fluoride concentrations and the presence of albuminuria in more than 16,000 urinary tests for protein. More recently, Chan-Yeung et al. (92) found no evidence of impaired renal function in aluminum smelter workers, as reflected in normal findings for blood urea nitrogen and creatinine.

As discussed above, removal into the urine is one means by which the body is protected against adverse effects of absorbed fluoride. The process is rapid, and elevated urinary fluoride concentrations are seen within minutes of absorption. Fluoride concentration in the urine has been used for many years as a biological index of fluoride exposure. For example, Jahr et al. (100) described a linear correlation between fluoride concentrations in post-shift urine samples and airborne fluoride concentrations in the workforce in three Swedish aluminum smelters. Dinman et al. (76) also demonstrated a good correlation between these two measures in workers in eight aluminum smelters, and Kono et al. (39) found a similar relationship for hydrogen fluoride workers.

Both Dinman (76, 88) and Hodge and Smith (91) have pointed out that the results obtained by analyzing a single-spot urine sample should not be used to evaluate working conditions because the fluoride concentration in any one sample can be influenced by a variety of factors, including normal biological variations among individuals, differences in tasks performed by different persons and differences in the associated exposures, and timing of the sample collection in relation to these fluctuations. Accordingly, mean urinary concentrations are used to characterize general working conditions. Preshift samples taken no later than 48 h after the last occupational exposure are considered to be a measure of the worker's skeletal burden of fluoride, whereas postshift samples taken after four or more consecutive exposures can be used to determine the adequacy of engineering controls and industrial hygiene practices.

4.4.2.1.5 Bone. The second mechanism that protects exposed individuals against the effects of absorbed fluoride is deposition of the fluoride in the skeletal system; indeed, skeletal deposition is the principal health hazard associated with chronic fluoride exposure. Continued occupational exposure leads to skeletal fluorosis, a condition that is characterized by a progressive increase in fluoride concentrations in the bone accompanied by histological changes, increased density on X-rays, morphometric changes, and bony outgrowths on long bones and vertebrae (90, 92–94, 113–117). On the basis of an extensive review of the literature prior to 1977, Hodge and Smith (91) concluded that the incidence of osteosclerosis was often high among potroom workers when airborne fluoride concentrations exceeded 2.5 mg/m^3 and urinary fluoride equaled or exceeded 9 mg/L. At the higher concentrations and with prolonged employment, osteosclerosis develops more rapidly, and the extent, degree, and incidence of the condition all increase. Not all signs and symptoms are seen in all patients; some individuals remain asymptomatic, whereas others, especially older workers exposed for 10–20 years in poorly controlled work environments, become crippled and increasingly immobile.

The early stages of osteofluorosis are difficult to detect radiologically. Boillat et al. (114) stated that, in the absence of a bone biopsy, the diagnosis of fluorosis requires that at least three criteria be met, that is, polyarthralgia, elevated urinary fluoride, and ossification of ligamentous insertion on two different regions of the skeleton, for example, heel and knee. A bone fluoride concentration greater than ~ 4000 mg F/g in a biopsy sample (iliac crest) and the accompanying histological changes would confirm a diagnosis of fluorosis (113). Increased density of the bone on X-ray is associated with a bone fluoride concentration of 4000–5000 mg/g (91). Obtaining a bone biopsy sample is an extreme measure. In the absence of a biopsy sample, a diagnosis of fluorosis can be made if urinary fluoride concentrations approach 8–10 mg/L (90, 117). On cessation of occupational fluoride exposure, fluoride in the skeleton is released slowly over time, decreasing to about half the original concentration over a 8–20 year period (113, 118). Evidence for so-called musculoskeletal fluorosis in potroom workers, based on complaints of back and neck pain, could not be found by Chan-Yeung et al. (93).

In recent years, the incidence of fluorosis has declined markedly. Conditions in the aluminum industry have improved to the point where Schlatter and Steinegger (119) and Jost et al. (120) suggest that industrial fluorosis can be considered an occupational disease of the past.

4.4.2.1.6 Metabolism. The discussion of fluoride metabolism presented in Section 2.4.2 is also appropriate here. In addition, Dinman et al. (76) have developed a model describing the biological disposition of inhaled particulate fluoride generated during smelter operations.

Hodge and Smith (121) estimated that exposure to an atmosphere containing 2.5 mg F/m^3 during an 8-h workshift would result in a daily intake of 5–6 mg of fluoride, assuming retention of 25% of the inhaled particles.

4.5 Standards, Regulations, or Guidelines of Exposure

Standards for airborne fluoride-containing dusts, and for fluoride in urine, are listed in Table 48.10. According to the NIOSH recommendation, if an individual's postshift urinary fluoride level exceeds 7.0 mg/L, preshift samples should be taken within 2 weeks; samples should be taken at the start of a shift and at least 48 h after a previous occupational exposure, and a repeat postshift sample should be taken at the conclusion of the shift at the end of the week in which the preshift sample was taken. If fluoride in the second sample exceeds either the preshift limit of 4.0 mg/L or the postshift limit of 7.0 mg/L, then dietary sources, personal hygiene, work practices, and environmental controls should be evaluated. If the median postshift urinary fluoride level for the entire group of shift workers exceeds 7.0 mg F/L, an industrial hygiene survey is essential.

The OSHA, and American Conference of Governmental Industrial Hygienists (ACGIH) limits are intended to prevent the deleterious effects of skeletal fluorosis (osteosclerosis or increased bone density due to excessive absorption and retention of fluorides) (122).

Table 48.10. Standards for Exposure to Fluoride Dusts

Standard	Source	
Airborne dust		
2.5 mg/m^3	ACGIH TLV-TWA	
	OSHA PEL	
Urine, sp. gr. 1.024[a]		
Preshift sample taken at start of workshift at least 48 h after last occupational exposure	4 ppm	4 mg/L
	3 mg/g creatinine[b]	
Postshift sample taken at end of shift after 4 or more consecutive days of exposure	7 ppm	7 mg/L
	10 mg/g creatinine[b]	

[a]From Ref. 122.
[b]From Ref. 22.

THE HALOGENS

Some investigators have suggested that the airborne standard might be altered in specific circumstances. Hjortsberg et al. (123) noted that potassium aluminum tetrafluoride causes bronchial asthma at concentrations of about 1 mg/m^3 and suggested that this concentration would be a more suitable standard for this compound. Both Rees et al. (124) and Rama and Yousefi (125) observed that exposure to airborne fluorspar at concentrations many times greater than the standard rarely or never resulted in urinary fluoride concentrations exceeding safe upper limits. These authors suggested that the airborne standard might be raised for the less soluble compounds.

Levi et al. (126) point out that industries using fluorides are well aware of the need for monitoring and suggest that other industries also should carry out occasional urinary surveys for fluoride to detect unexpected or unknown exposures.

Ehrnebo and Ekstrand (69) and Ekstrand and Ehrnebo (75) suggest that, because blood fluoride concentrations correlate well with bone fluoride and with amounts of fluoride excreted (better than with urinary concentration), blood fluoride measurements could be useful in determining the fluoride status of individuals. Kono et al. (127) noted that fluoride concentrations in the hair of hydrogen fluoride workers were elevated relative to those of control subjects and correlated well with the postshift urinary fluoride concentrations of these workers.

Grandjean and colleagues (128) in Denmark observed significantly elevated rates of cancer in cryolite workers. A total of 119 incident cases of cancer occurred among male cryolite workers, with 103.6 expected. Excesses of cancer of the lung, larynx, and urinary bladder were noted. Maximum morbidity occurred after 10–19 years of exposure. Two cases of bladder cancer occurred in female workers. Workers in this environment were considered by the authors to have had no exposure to suspected carcinogens other than fluoride dust. They hypothesized that heavy exposure to fluoride dust increased the cancer risk of the cohort.

5.0 Miscellaneous Fluorides

There are a number of fluoride compounds for which an OSHA standard has not been established. However, recommended threshold limits have been developed for some of these by the ACGIH, using the available data. These limits are listed in Table 48.11 (129, 130).

B CHLORINE AND ITS COMPOUNDS

6.0 Chlorine

6.0.1 CAS Number: *[7782-50-5]*

6.0.2 Synonyms: Dichlorine, molecular chlorine, chlorinated water

6.0.3 Trade Names: NA

6.0.4 Molecular Weight: 70.906

6.0.5 Molecular Formula: Cl$_2$

6.0.6 Molecular Structure: Cl—Cl

Table 48.11. Exposure Limits for Miscellaneous Fluorides[a]

Compound/Limits	Major Uses	Principal Effects
Bromine trifluoride ACGIH TLV (ceiling) 0.1 ppm (0.72 mg/m^3) NIOSH REL (ceiling) 0.1 ppm (0.72 mg/m^3)	Fluorinating agent; oxidizer in rocket propellant systems	In human subjects, painful, deep-seated, long-lasting burns; short-term high concentrations: serious lung injury; lower concentrations: watering of eyes, difficulty in breathing after few minutes; no specific concentrations given
Carbonyl fluoride ACGIH TLV 2 ppm (5.4 mg/m^3) ACGIH TLV-STEL 5 ppm (13 mg/m^3) NIOSH REL 2 ppm (5.4 mg/m^3) NIOSH STEL 5 ppm (13 mg/m^3)	Intermediate in organic synthesis; encountered as thermal decomposition product of fluoropolymers	Effects resemble those of exposure to HF, because HF is produced by hydrolysis in the moist respiratory tract. 1-hr LC$_{50}$s for 8-week- and 23-week- old rats were 360 and 460 ppm, respectively; concentration × time constant for 1-4 h
Chlorine trifluoride ACGIH TLV (ceiling) 0.1 ppm (0.38 mg/m^3) NIOSH REL (ceiling) 0.1 ppm (0.38 mg/m^3) OSHA PEL (ceiling) 0.1 ppm (0.38 mg/m^3) IDLH 20 ppm	Fluorinating agent; igniter and propellant for rocket systems; processing nuclear reactor fuels; pyrolysis inhibitor for fluorocarbon polymers	In rats, inflammation of mucosal surfaces; burning of skin; corneal ulceration; lung damage; vapor decomposes to yield Cl$_2$, HF, ClO$_2$, and other products. Sweet, suffocating odor
Nitrogen trifluoride ACGIH TLV 10 ppm (29 mg/m^3) NIOSH REL 10 ppm (29 mg/m^3) OSHA PEL 10 ppm (29 mg/m^3) IDLH 1000 ppm	Intermediate in chemical synthesis; oxidizer for high-energy fuels	Produces methemoglobin and cyanosis; no odor warning properties at dangerous concentrations; moldy odor
Oxygen difluoride ACGIH TLV (ceiling) 0.05 ppm (0.1 mg/m^3) NIOSH REL (ceiling) 0.05 ppm (0.1 mg/m^3) OSHA PEL 0.05 ppm (0.1 mg/m^3) IDLH 0.5 ppm	Proposed oxidizer for rocket propellants	Extremely potent irritant to respiratory tract, kidney, internal genitalia; 1-h LC$_{50}$s are 1.5, 2.6, 16, and 26 ppm for mouse, rat, monkey, and dog, respectively; effects resemble those of ozone in animals, humans; intractable headache; moldy odor detectable at about 0.1 ppm

Perchloryl fluoride
  ACGIH TLV   3 ppm (13 mg/m^3)
  ACGIH STEL 6 ppm (25 mg/m^3)
  NIOSH REL   3 ppm (14 mg/m^3)
  NIOSH STEL 6 ppm (25 mg/m^3)
  OSHA PEL   3 ppm (14 mg/m^3)
  IDLH   385 ppm

Selenium hexafluoride
  ACGIH TLV   0.05 ppm (0.4 mg/m^3) (measured as Se)
  NIOSH REL   0.05 ppm (0.4 mg/m^3) (measured as Se)
  OSHA PEL   0.05 ppm (0.4 mg/m^3) (measured as Se)
  IDLH   5 ppm

Tellurium hexafluoride
  ACGIH TLV   0.02 ppm (0.1 mg/m^3) (measured as Te)
  NIOSH REL   0.02 ppm (0.1 mg/m^3) (measured as Te)
  OSHA PEL   0.02 ppm (0.2 mg/m^3) (measured as Te)
  IDLH   1 ppm

Fluorinating agent; oxidant in rocket fuels — Produces methemoglobin, cyanosis, alveolar edema, bronchopneumonia; characteristic sweet odor

Gaseous electric insulator — Concentrations of 1.00–10 ppm for 4 h lethal for rabbit, guinea pig, rat, mouse; pulmonary edema at 5 ppm but animals survived; no effects at 1 ppm; about half as toxic as ozone for laboratory rodents

Concentrations of 5–100 ppm for 4 h lethal for rabbit, guinea pig, rat, mouse; pulmonary edema at 1 ppm; two human subjects exposed to leakage from 10 g TeF$_6$ experienced metallic taste, anorexia, lassitude, sleepiness, skin rash and blue-black patches, garlic-like odor in breath, sweat, urine; no significant pulmonary effects; repulsive odor

[a]From Refs. 21, 22, 129, and 130.

6.1 Chemical and Physical Properties

Chlorine is associated with the largest array of industrially useful compounds of all the halogens; it is the ninth highest volume chemical produced in the United States (131). Chemically, chlorine is more reactive than either bromine or iodine; it displaces bromine and iodine from their salts and enters into substitution and addition reactions with both inorganic and organic substances. Moist, but not dry, chlorine unites directly with most elements. The physical properties of chlorine appear in Table 48.1. Recognition of widespread environmental contamination problems associated with persistent chlorinated organics has led to pressure to reduce the utilization of chlorinated compounds in industrial chemicals. The outcome of this concern is unclear at this time, but it is likely to lead to substitution of other compounds for those containing environmentally persistent and toxic chlorinated materials. Although most of the identified compounds that present environmental contamination problems are higher molecular weight organics, the use of chlorinated solvents in cleaning and degreasing operations has already been reduced and will be further reduced because of their impact on the ozone layer. Higher molecular weight organics containing chlorine have been identified as possible xenoestrogens. The concern of the scientific community is high, and much research is under way to clarify the role of halogenated xenoestrogens in biological systems. It is likely that if the halogenated organics are confirmed to be biologically detrimental, there will be a further reduction in the production and use of these compounds. The introduction of chemically reactive inorganic halogens, principally chlorinated compounds, into the environment in groundwater and air may lead to active halogenation of natural products to xenoestrogens. If this pathway is confirmed, this, too, will lead to a reduction in the use of chlorinated compounds.

Chlorine is noncombustible in air but will support the combustion of other materials. It reacts explosively or forms explosive mixtures with many common materials, including acetylene, turpentine, ammonia gas, fuel gas, hydrocarbons, hydrogen, and finely divided metals. Chlorine may also combine with water or steam to produce hydrogen chloride fume (132).

6.2 Production and Use

The reactivity of chlorine is so great that none is found free in nature. Instead, it is found in landlocked lakes as sodium chloride (NaCl) [7647-14-5] (Great Salt Lake, Utah), in underground rock salt deposits, in brines in southern Michigan, central New York, Louisiana, and Mississippi and along the Gulf Coast of Texas, in Stassfurt, Germany, and elsewhere in an almost inexhaustible supply. In addition to NaCl, chlorine is found naturally in deposits of carnallite ($KMgCl_3$) and sylvite (KCl) [7447-40-7].

Chlorine is principally produced by electrolysis of NaCl or KCl brine in either diaphragm, mercury, or membrane cathode cells. In these processes, gaseous chlorine is released at the anode and caustic is a by-product. Chlorine may also be produced by electrolysis of hydrochloric acid (HCl) [7647-01-0], by oxidation of HCl in the presence of nitrogen oxide as a catalyst (Kel-Chlor process), or as a coproduct from metal production (131, 133). Chlorine is currently produced at 49 facilities operated by 29 companies in the

United States. Of these, 17 facilities use diaphragm cells, 15 use mercury cells, 6 use membrane cells, and 2 use a combination of diaphragm and membrane cells. Six other facilities produce chlorine as a by-product of metals manufacture, and one by electrolysis of HCl. Two facilities produce chlorine together with bromine (134).

The production of chlorine by the electrolytic process has proven to be a serious industrial hazard both to those employed in the industry and to the environment within which the plants are located. The corrosivity of the liquid and gaseous products leads to rapid decay of plant equipment and releases of chlorine gas and hydrochloric acid as well as hypochlorites and other compounds. The releases of mercury to the environment from leaking mercury cells or from contaminated solutions can be massive. In a case tried in the U.S. District Court for the Southern District of Georgia, sitting in Brunswick (GA, USA), the owners and operators of a chloralkali plant were convicted of several criminal violations of the CERCLA act for releases of mercury from their poorly maintained and operated plant, to estuarine waters with major resulting environmental damage (*United States of America v. Christian A. Hansen, et al.*, CR 298-23). Chloralkali workers are potentially heavily exposed to both acid gas vapor and aerosol and may be seriously poisoned by mercury on an acute and chronic basis (135–138).

The largest use for chlorine is for the manufacture of a variety of nonagricultural chemicals. Among these are chlorinated hydrocarbons such as carbon tetrachloride, vinyl chloride, ethylene and propylene oxides, trichloroethylene, perchloroethylene, chloroform, vinylidene chloride (for plastics production), ethylene dichloride, metal chlorides, chlorinated benzenes, phosgene, chlorinated paraffins, HCl, chlorine dioxide for pulp bleaching, and many others (139).

Another important use is in the pulp-and-paper industry, where elemental chlorine or chlorine dioxide is used as an oxidizing bleaching agent to make white paper. Oxygenated chlorine compounds such as chlorates and hypochlorites are used as commercial and household bleaching agents for textiles. A significant reduction in the use of chlorine bleaching processes in all industries is under way, because of the persistence of chlorinated residues in the environment. Ozonation has also been advocated as a replacement for chlorination in the disinfection of public water supplies, and as the cost of this technological advance is reduced, it may be expected to replace many of the currently used procedures for the treatment of water.

Chlorine is used as a biocide in water-purification and waste-treatment systems, in cooling systems, and in the dairy industry. A limited amount is used in processing meats, fish, and fresh produce (140).

Total U.S. production of chlorine in 1995 was 25.09 billion lb. The 1999 projected chlorine demand for the United States was 13.2 million tons, slightly more than the 1995 demand of 12.6 million tons (141). While demand has remained fairly stable in the short term, pressure to reduce persistent chlorinated compound contamination of the environment in the United States and in the European Community is likely to lead to reduction of demand for chlorine in these areas. Of all chlorine produced in 1988, 28% was used for vinyl chloride and ethylene dichloride production, 13% for the production of chloromethanes and chloroethanes, 20% for the production of other organic chemicals, 18% for the pulp-and-paper industry, 8% for the production of inorganic chemicals, 6% for water treatment, and 7% for miscellaneous uses (142). Miscellaneous uses of chlorine include the manufacture of pharmaceuticals, cosmetics, lubricants, flame proofers, adhesives, food additives, and hydraulic fluids. NIOSH estimated that there were 4768 workers occupationally exposed to chlorine in 1983 (7).

6.3 Exposure Asessment

Employee exposures to chlorine gas can be measured using NIOSH method 6011 (25). This method involves collecting 2–90 L of air through a silver membrane filter mounted in a carbon-filled polypropylene cassette and analyzing the sample using ion chromatography to quantify the chloride ion. Samples collected on the silver membrane filter are stable for 30 days at 25°C or 60 days at 5°C, although the filter must be protected from light. This method has a working range of 0.007–0.5 ppm for a 90-L air sample. The estimated limit of detection for this method is 0.6 mg chloride ion per sample, and the overall precision (coefficient of variation) is 0.075. Hydrogen sulfide gives a negative interference and HCl gives a positive interference up to a maximum of 15 mg per sample.

The OSHA method for sampling chlorine uses 10 mL of 0.1% sulfamic acid solution in a glass bubbler as the collection medium (143). Samples are treated with acidic potassium iodide, which is oxidized by chlorine to form iodine; chlorine ion content is quantified using an ion-specific electrode. Samples collected in sulfamic acid are stable for up to 30 days at room temperature. This method has a reliable quantification limit of 6 mg per sample, and an overall precision (coefficient of variation) of 0.03. Strong oxidizing agents including iodate, bromine, cupric ion, and manganese dioxide may interfere with the analysis.

6.4 Toxic Effects

6.4.1 Experimental Studies

The results of acute exposure studies conducted on experimental animals are comparable to those obtained from acutely exposed humans. The LC_{50}s for rats and mice are 293 and 137 ppm, respectively (114). Barrow and Smith (145) and Barrow et al. (146) reported that chlorine exposure causes altered pulmonary function in rabbits and a reduced respiratory rate in mice following 10-min exposures. The concentration of chlorine required to decrease the respiratory rate in mice by 50% (RD_{50}) was about 10 ppm. Exposure of rats to 3 or 9 ppm chlorine for 6 h/day, 5 days/week for 6 weeks was associated with decreased body weight, increased hematocrit and white cell count (in females exposed to 9 ppm only), and increases in clinical chemistry measures suggestive of altered renal function (alkaline phosphatase, blood urea nitrogen, GGTP, and SGPT) (147). There was gross evidence of an inflammatory reaction in the upper and lower respiratory tract among animals exposed to a 9-ppm concentration of chlorine; animals exposed to 3 ppm exhibited these effects to a lesser degree.

Demnati et al. (148) studied the time course of functional and pathological changes in rats. They evaluated lung resistance, and responsiveness to methacholine in the acute period, and followed bronchoalveolar lavage and airway epithelial biopsy over 3 months following a 5-min exposure to 1500 ppm of chlorine. They found that maximal changes were present 1–3 days after exposure, and that the changes slowly resolved in most animals. Some however had persistent changes at the end of the study. They postulated that the acute changes observed may have been related to those that provoke RADS in humans.

Rats exposed to 1 ppm chlorine have been found to exhibit tolerance to chlorine-induced sensory irritation (149).

6.4.2 Human Experience

6.4.2.1 Clinical Cases

6.4.2.1.1 Acute Toxicity. Chlorine is a severe irritant of the eyes, nose, throat, and lining of the respiratory tract. Stokinger (20) reported that the action of chlorine in the respiratory tract is due to its strong oxidizing capability, in which chlorine splits hydrogen from water in moist tissue, causing the release of nascent oxygen and hydrogen chloride. The oxygen produces the major tissue damage, which is enhanced by the hydrogen chloride. Another mode of action involves the conversion of chlorine to hypochlorous acid, which easily penetrates the cell wall and reacts with cytoplasmic proteins to form N-chloro derivatives that destroy cell structure.

Ross and McDonald (161) reviewed the results of the SWORD project (surveillance of work-related and occupational respiratory disease). In 1996 they reported that chlorine gas was the most frequently reported exposure that led to occupational asthma. Both typical asthma and RADS were found to be sequelae of chlorine exposure. Furthermore, once asthma developed, symptoms were exacerbated by exposure to many known irritants, infections, and nonspecific agents. The authors concluded that accidental inhalation of irritants can initiate occupational asthma. Reilly and Rosenman (162) reviewed Michigan hospital discharge data and concluded that one-third of the hospital discharges for chemically related respiratory disease were due to occupational exposures. The most common work exposures were to chlorine and sulfur dioxide, and to industrial cleaning agents. Follow-ups were then carried out at the workplaces in which disease had occurred and the authors found that 61/261 (61 out of 261) fellow workers of those hospitalized had new onset occupational asthma or were bothered by asthma like symptoms.

The response of humans to lowlevel acute exposure to chlorine has been well studied. Acute exposures to chlorine concentrations above 5 ppm are severely irritating and cannot be tolerated for more than a few minutes (163). Anglen (164) examined the physiological response of 29 students exposed to 0, 0.5, 1.0, or 2.0 ppm chlorine for 4–8 h. Sensations of itching or burning of the nose and eyes and general discomfort were reported, primarily during exposures to 1.0 or 2.0 ppm. Few subjects reported experiencing feelings of nausea, headache, dizziness, or fatigue at these concentrations, although several reported experiencing shortness of breath for several hours after the 8-h exposure to 1 ppm. No subjects reported any sensory perception related to exposure the morning after the exposure. Statistically significant reductions in pulmonary function were found following the 8-h exposure to 1 ppm, but not following exposure to 0.5 ppm. Another later study reported findings consistent with these among eight healthy unacclimatized individuals (165); that is, exposure to 1.0 ppm but not to 0.5 ppm chlorine was associated with statistically significant changes in FVC, FEV_1, peak expiratory flow rate, forced expiratory flow rate, and airway resistance. More recent studies have addressed the problem of hyperreactive persons who may be exposed to concentrations of irritants such as chlorine, which would not cause problems in most individuals. D'Alessandro et al. (166) exposed volunteer subjects both with and without known nonspecific airway hyperresponsiveness to low levels of chlorine gas and monitored their lung response. Airflow, lung volumes, diffusing capacity, airway resistance, and methacholine responsiveness were studied. In both normal and hyperresponsive patients who were exposed to 1 or 0.4 ppm chlorine for 60 min (the existing PEL for chlorine is 1 ppm), forced expiratory volume fell significantly, and specific airway resistance increased. Two of those with known hyperresponsiveness developed significant respiratory symptoms. At 24 h, both groups had recovered. The authors concluded that persons with preexisting airway hyperresponsiveness would develop exaggerated responses to low levels of chlorine exposure. Such studies cast doubt on the protective nature of the existing PEL of 1 ppm.

The threshold for odor perception of chlorine is about 0.3 ppm, with considerable variation among subjects (167). However, among chlorine workers, odor does not provide adequate warning of exposure. In a group of 13 workers exposed to chlorine for a 2–5-year period, all had some degree of olfactory deficiency, which was severe in 11 workers. Of 4 workers exposed for 1 year or less, 1 had a severe deficiency, one had a moderate degree of deficiency, and 2 had slight olfactory deficiency (155). Furthermore, Meggs et al. (168) have identified a pathological basis and clinical symptomatology of a variant of RADS that is expressed in the upper airways, termed *reactive upper airways dysfunction syndrome* (RUDS). This disorder is characterized by chronic rhinitis with intermittent exacerbation, mucosal congestion, and sudden sneezing and watering of the eyes after irritant inhalation exposures. Clearly, chlorine and volatile or aerosolized chlorine compounds may precipitate attacks of RUDS and RADS.

In addition to irritation, other symptoms are associated with acute exposure to chlorine; these include nausea, headache, dizziness, muscle weakness, pulmonary edema, and respiratory distress. The extent of injury is dependent on the concentration and duration of exposure as well as the water content of the tissue exposed. Ellenhorn and Barceloux estimate that at 1–3 ppm mild mucous membrane irritation will occur. Exposures of 5–15 ppm will cause moderate irritation of the upper respiratory tract; 30 ppm will produce immediate chest pain, vomiting, dyspnea, and cough (150). A 30-min exposure to ≥ 430 ppm of chlorine can be fatal to humans (151), as are longer exposures (1–1.5 h) to concentrations in the range of 34–51 ppm (152, 153). Exposure to concentrations in excess of 14 ppm for ≥ 30 min results in severe pulmonary damage. Exposure to 1000 ppm usually results in death in a few minutes.

Symptoms of pulmonary congestion and edema may develop after a latency period of several hours following severe acute exposure to chlorine. In a study of 838 World War I soldiers exposed to chlorine gas, 28 died, and of these deaths four were attributed to the delayed effects of chlorine exposure; these effects included bronchopneumonia, lobar pneumonia, purulent pleurisy, and tubercular meningitis. A total of 48 others of this group had exposure-related disabilities at time of discharge; effects included a case involving a flare-up of pulmonary tuberculosis, bronchitis, pleurisy, tachycardia, dyspnea, and nephritis (154). However, all of these cases occurred in an era that antedated modern antibiotics, steroids, and intensive respiratory care, so that their relevance to contemporary management and outcome is likely small.

Severe exposures to chlorine resulting from industrial accidents have also been associated with delayed-onset pulmonary symptoms. While clinical presentation in pulmonary distress, with edema and imminent respiratory failure is common, some patients are asymptomatic on first evaluation. These patient may deteriorate over several hours and become critically ill. No person exposed to significant concentrations of chlorine gas should be released from medical care before a definitive evaluation at 24 h has been completed. Pulmonary edema is manageable in a hospital setting, but lethal when untreated.

It was long believed that acute exposures to chlorine gas that did not end fatally did not produce longlasting pulmonary impairment. However, recent studies in both animals (148) and humans have proved conclusively that acute exposures to chlorine gas produce sequelae that persist for 3 months to several years, in some patients. Although acute effects often subside within 72 h, exceptions among workers with preexisting cardiac conditions were noted (155). Ploysongsang and colleagues (156) studied the effects of an accidental chlorine exposure lasting 2–5 min in four healthy adults and reported finding reduced pulmonary function that cleared within 1 month without residual effect. Symptoms of respiratory distress appeared in these individuals within 14–16 h after exposure. All exhibited impaired diffusing capability and evidence of small-airway obstruction. In contrast to these reports, Kowitz et al. (157) reported finding decreased lung capacity among a group of workers 3 years following an accidental exposure to chlorine.

6.4.2.2 Epidemiology Studies

6.4.2.2.1 Acute Toxicity. Leroyer et al. (158) studied changes in airway function and bronchial responsiveness after acute occupational exposures to chlorine that were severe enough to require treatment in a first aid unit, but were not life threatening. They followed a cohort of 278 workers who had baseline spirometry and methacholine challenges. Following acute exposure to chlorine in the course of employment, 13 workers from this cohort were assessed with the same tests. Three workers had significant drops in forced expiratory volume, (FEV_1) during methacholine challenge. Three months later, these patients appeared to have returned to baseline levels or pulmonary function. On the other hand, Schonhofer et al. (159) noted persistent long-term sequelae following acute, sublethal exposure to chlorine in 3 patients with no past history of respiratory disease. These patients complained of shortness of breath on exposure to irritants and with exertion 2.5 years after exposure to chlorine gas. Moderate to severe nonspecific bronchial hyperreactivity was assessed at 4, 20, and 30 months after exposure. Bronchial lavage in these patients demonstrated inflammatory cell reaction acutely, but was nearly normal at 16 months. The patients were considered to have typical reactive airways dysfunction syndrome (RADS), a variant of occupational asthma, as a result of a single, high dose chlorine gas exposure. Rivoire and colleagues (160) also studied patients who were acutely exposed to single massive inhalation exposures of chlorine. They noted that symptoms evolved over 5–84 months after the initial accident. Although only mild airflow obstruction was noted, all subjects were found to be hyperresponsive to methacholine. They noted that secondary aggravation by other irritants and progression was the rule even in the absence of further exposure to the inciting agent. They recommended evaluation and surveillance of the workers and removal to another job to minimize the effects over time.

6.4.2.2.2 Chronic and Subchronic Toxicity. At exposure levels below those associated with severe irritant effects, long-term exposure to chlorine has generally not been associated with the development of adverse health effects. An in-depth study of 25 chloralkali plants by Patil et al. was reported in 1970 (169). This was a study involving 332 male workers engaged in chlorine production and 382 unexposed age-matched control workers. Information on chlorine exposure levels was available from bimonthly air sampling conducted at representative plant locations; 8-h time-weighted average (TWA) exposures were determined for the exposed workers from these data.

TWA exposures to chlorine ranged from 0.006 to 1.42 ppm, with a mean of 0.15 ppm. All except 6 of the 332 workers had TWA exposures below 1 ppm, and 21 had TWA exposures above 0.52 ppm; the average number of years of exposure to chlorine for all workers was 10.9 years. Medical histories showed no dose-related increase in the prevalence of colds, dyspnea, palpitation, or chest pain among exposed workers. There was also no evidence of permanent lung damage among exposed workers, as indicated by chest X-rays or measurements of ventilatory capacity, maximal ventilatory volume, or FEV_3. However, 9.4% of the 329 electrocardiograms taken on the 332 exposed workers were abnormal, compared with 8.5% among controls. The incidence of fatigue was greater among those exposed to concentrations above but not below, 0.5 ppm, and anxiety and dizziness showed a modest correlation ($p=.02$) with exposure levels. Leukocytosis ($p < .05$) and lower hematocrit ($p < .017$) showed some relation to chlorine exposure. No neoplasia or serious pulmonary disease was reported among exposed workers.

A study of 52 Italian electrolytic cell workers with about the same average duration of exposure (10 years) but exposed at twice the average chlorine concentration (0.298 ppm, SD 0.181) found no clinically significant pulmonary effect, even temporarily (170). Using five tests of respiratory function, only carbon monoxide diffusing capacity (DL_{CO}) showed a significant difference between exposed and control workers. This value was significantly lower in chlorine-exposed smokers compared with nonexposed smokers ($p < .02$), lower in exposed smokers than in exposed nonsmokers ($p < .04$), and lower in exposed smokers than in nonexposed nonsmokers ($p < .003$), indicating that chlorine exposure was associated with a slightly significant effect on respiratory function over and above that from smoking alone.

In contrast to these findings, Drobnic et al. (171) studied the impact of low level chlorine exposure on swimmers during training. They demonstrated the presence of asthma-like symptoms and airway hyperresponsiveness at concentrations of chlorine that were below the TLV. However, because of the level of activity in the swimmers in training, the gradual buildup of chlorine over the pool during the day, and low air turnover, they postulated that chlorine in the pool area was responsible for the swimmers' respiratory problems. Fjellbirkeland et al. (172) found similar asthmatic responses in swimmers whom they studied. They noted that more swimmers were symptomatic when the pool area was warm and there was a strong smell of chlorine. Two athletes who were troubled by the problem of asthma when they swam indoors reported no problems when they swam in outdoor pools.

The findings from the epidemiological studies described above are supported by a 1-year inhalation study conducted on rhesus monkeys (173). Exposure to 2.3 ppm chlorine was associated with eye irritation in the monkeys during the exposure period, and only

THE HALOGENS

mild histological changes in the respiratory epithelium of the nasal passages and trachea. Monkeys were found to be less sensitive than rats to the toxic effects of chlorine.

6.5 Standards, Regulations, or Guidelines of Exposure

The OSHA limit for exposure to chlorine is 1 ppm as a ceiling limit (174). The ACGIH established TLVs for chlorine of 0.5 ppm as an 8-h TWA and 1 ppm as a 15-min short term exposure limit (STEL) (22); NIOSH has published the same limits as ACGIH recommended exposure limits (RELs) (175). NIOSH has also established an IDLH value of 10 ppm IDLH for chlorine (21).

7.0 Hydrogen Chloride

7.0.1 CAS Number: [7647-01-0]

7.0.2 Synonyms: Muriatic acid, chlorohydric acid, hydrochloride, spirits of salts, hydrogen chloride (acid), hydrochloric acid, hydrogen chloride gas only

7.0.3 Trade Names: NA

7.0.4 Molecular Weight: 36.461

7.0.5 Molecular Formula: HCl

7.0.6 Molecular Structure: H—Cl

7.1 Chemical and Physical Properties

Hydrogen chloride is a colorless, corrosive, nonflammable gas that fumes in air. It has a characteristic suffocating, pungent odor. Hydrogen chloride has a melting point of −114.8°C, and a boiling point of −84.9°C and is highly soluble in water to form hydrochloric acid.

Increases in temperature and pressure can result when HCl is mixed with many common materials, including acetic anhydride, ammonium or sodium hydroxide, 2-aminoethanol, chlorosulfonic acid, ethylenediamine, sulfonic acid, and vinyl acetate (132). Hydrogen chloride in the presence of aldehydes and epoxides causes violent polymerization reactions. Hydrogen gas can be released when HCl comes into contact with metals or is heated to high temperatures.

7.2 Production and Use

Commercially significant processes for producing HCl include decomposition of sodium and potassium chloride salts by sulfuric acid, direct synthesis from hydrogen and chlorine, by-product formation from a number of organic chemical processes, and production from waste gases (133). Commercial hydrochloric acid is produced as aqueous solutions containing 38% HCl (muriatic acid) or 20% HCl (constant-boiling acid).

Anhydrous HCl is used predominantly as a chemical feedstock in hydrochlorination, polymerization, alkylation, and isomerization reactions. Uses of hydrochloric acid include metal cleaning (steel pickling), metallurgical processes, chlorine and chlorine compound production, food processing, leather tanning, electroplating, and as a laboratory reagent (131, 133). NIOSH has estimated that 1,238,572 workers are exposed to HCl or hydrochloric acid (7).

7.3 Exposure Assessment

7.3.1 Air

NIOSH method 7903 for measuring exposures to inorganic acids can be used to sample for HCl gas (25). Air samples ranging in volume from 3 to 100 L are collected on a dual-section washed silica gel tube (400 mg/200 mg), desorbed with a mixture of sodium bicarbonate and sodium carbonate buffer, and analyzed by ion chromatography. Samples are stable at ambient conditions and require no special handling or shipping. The method has been evaluated over the range of 0.5–200 mg per sample, and has an overall precision (coefficient of variation) of 0.059. The estimated limit of detection for this method is 2 mg per sample. Compounds that may interfere with this method include particulate salts of HCl, chlorine, and hypochlorite.

Another method described by OSHA (176) involves collecting air samples in a fritted-glass bubbler containing 0.1 N NaOH. Analysis is performed with an ion-specific electrode. The standard analytic error for this method is reported to be 0.12.

A variety of colorimetric detector tubes are available for quantifying airborne concentrations of HCl; depending on the specific make of tube, these devices can detect HCl concentrations in the range of from less than 1 to 5000 ppm.

7.4 Toxic Effects

7.4.1 Experimental Studies

7.4.1.1 Acute Toxicity. The 30-min LC_{50}s for HCl gas in rats and mice are 4701 ppm and 2644 ppm, respectively (177); pathological findings in these animals revealed emphysema, atelectasia, and pulmonary edema. A concentration of 304 ppm administered for 6 h/day over 3 days was fatal to a group of Swiss Webster mice; this concentration is near the RC50 level for these animals (178). No persistent alteration in pulmonary function was observed in adult male baboons exposed for 15 min to HCl concentrations as high as 10,000 ppm.

In a series of experiments designed to evaluate the extent to which exposure to HCl impairs escape, Kaplan et al. (179) exposed baboons for 5 min to concentrations ranging from 190 to 17,290 ppm, and rats for 5 min to concentrations ranging from 11,800 to 87,660 ppm. Irritant effects were evident in all animals except baboons exposed to the lowest concentration. With the exception of rats exposed to the highest concentration, all animals were able to perform a learned signal-avoidance task. At the highest exposure levels, animals experienced persistent respiratory effects and died after exposure. Thus even at concentrations of HCl that caused severe respiratory damage, animals were able to perform the escape task. According to Kaplan et al., these results were surprising because adverse effects have been reported in humans exposed at much lower levels. In an evaluation of the Kaplan et al. study, the National Research Council (180) noted that the escape task performed by the animals was a relatively simple task, and that the animals were severely affected by exposure to HCl at most of the concentrations tested. A later study by Kaplan (181) showed that severe hypoxia developed rapidly in baboons exposed to HCl at concentrations of 3500–4000 ppm, a finding that confirms the earlier view of Henderson and Haggard (182) that exposure to HCl at concentrations of 1000–2000 ppm is dangerous even for a short time.

Pavlova (183) exposed female rats for 1 h to 300 ppm HCl 12 days before or 9 days after pregnancy. Survival of the progeny was not affected in rats exposed before pregnancy, but the offspring showed reduced weight gain. The mortality rate among the progeny of the rats exposed during pregnancy was more than fivefold higher than among controls. The progeny of both exposed groups showed disturbances in kidney function, as measured by diuresis and proteinuria.

7.4.1.2 Chronic and Subchronic Toxicity. In a lifetime exposure study, a group of 100 male Sprague–Dawley rats were exposed to 10 ppm HCl for 6 h/day, 5 days/week, for life (maximum 128 weeks). There was no significant increase in mortality or tumor response among exposed animals compared with either colony control or air-exposed control groups. The exposed group of animals did exhibit an increased incidence of hyperplasia of the larynx and trachea (22/99 and 26/99, respectively) (184).

7.4.2 Human Experience

7.4.2.1 General Information: NA

7.4.2.2 Clinical Cases

7.4.2.2.1 Acute Toxicity. Male volunteers exposed to HCl gas found concentrations of 50–100 ppm barely tolerable for 1 h. Exposure to 35 ppm caused throat irritation on brief exposure, and 10 ppm was considered the maximal acceptable level for prolonged exposure (182). Exposure to a 5-ppm concentration of HCl is immediately irritating to the nose and throat but is not associated with any longlasting effects (185). According to Heyroth (186), most people can detect the presence of HCl at concentrations between 1 and 5 ppm; Kamrin (187) estimated that the no-effect level for respiratory effects in humans ranges from 0.2 to 10 ppm. Because of its high water solubility, the major effects of acute exposure are confined to the upper respiratory tract and are severe enough to prompt voluntary withdrawal before more extensive tissue damage results. On contact with moisture in the mucous membranes, HCl dissociates almost completely; the hydrogen ions combine with water to form hydronium ions (H_3O^+) that can cleave organic molecules and cause cell death (180).

Inhalation of HCl at irritating concentrations causes coughing, pain, inflammation, and edema of the upper respiratory tract. At high concentrations, the gas causes necrosis of the bronchial epithelium, constriction of the larynx and bronchi, and closure of the glottis. Concentrations of 1000–2000 ppm and higher are immediately dangerous (182). One fatal case of overexposure has been reported; postmortem examination showed severe pulmonary hemorrhage, edema, and pneumonitis (188).

Exposure of the skin to gaseous HCl or strong solutions of hydrochloric acid causes burns and ulcerations, and leads to scarring in some instances. Repeated dermal contact with HCl solutions may cause dermatitis.

7.4.2.2.2 Chronic and Subchronic Toxicity. Prolonged exposure to HCl gas has frequently been associated with dental decay and erosion, bleeding of nose and gums, and ulceration of the nasal and oral mucosa (189). Dental erosion was observed in 90% of picklers in a zinc galvanizing plant who spent 27% of their time in areas where HCl concentrations exceeded 5 ppm (190).

Several cohort and case-control studies have been conducted to examine the mortality experience of workers exposed to HCl; in many of these studies, workers were exposed to mixtures of acids or other chemicals that confounded the study results. Collins et al. (191) reported an excess of lung cancer at one of four chemical plants in which workers were exposed to acrylamide. The excess was attributed to an increased number of lung cancer deaths occurring among workers using muriatic acid. Beaumont et al. (192) detected an increased incidence of lung cancer among U.S. steel pickling workers exposed primarily to hydrochloric acid for at least 6 months between 1940 and 1964 [standard mortality rate = 2.24, 95% confidence interval (CI) 1.02–4.25]. An excess of laryngeal cancer was observed among members of this same cohort in another study (relative risk 2.6, CI 1.2–5.0, nine cases) (193). However, many of these cases involved exposure to a mixture of acids.

Case-control studies that have been conducted on HCl-exposed workers report no association between exposure to HCl and increased incidences of primary intracranial neoplasms (194), renal cancer (195), or lung cancer (196, 197). In a population-based case-control study, Siemiatycki (198) reported odds ratios for oat cell carcinoma of the lung of 1.6 among HCl-exposed workers and 2.1 among the most highly exposed workers; neither of these findings was statistically significant. Among a French-Canadian subset of the study population, the odds ratio for non-Hodgkin's lymphoma was 1.6 (90% CI 1.0–2.5, 18 cases), and the odds ratio for rectal cancer was 1.9 (90% CI 1.1–3.4, 18 cases).

7.5 Standards, Regulations, or Guidelines of Exposure

From the animal and human evidence available for HCl, the International Agency for Research on Cancer (IARC) has concluded that hydrochloric acid is not classifiable with respect to its carcinogenicity (group 3) (199).

OSHA's PEL for HCl and hydrochloric acid is 5 ppm as a ceiling limit (174). The ACGIH has established a 5-ppm ceiling limit as a TLV (22). NIOSH also recommends that exposures not exceed 5 ppm as a ceiling limit (175). NIOSH has assigned an IDLH value of 50 ppm for HCl (21).

8.0 Phosgene

8.0.1 CAS Number: *[75-44-5]*

8.0.2 Synonyms: Carbonic dichloride, carbon oxychloride, chloroformyl chloride, carbonyl chloride, CG, carbonyl dichloride, cytosine-guanine, phosgene-13C ($\sim$1 M solution in benzene)

THE HALOGENS

8.0.3   Trade Names: NA

8.0.4   Molecular Weight: 98.916

8.0.5   Molecular Formula: CCl$_2$O

8.0.6   Molecular Structure: $\text{Cl}-\underset{}{\overset{O}{C}}-\text{Cl}$

8.1   Chemical and Physical Properties

Phosgene is a colorless gas or a fuming liquid at temperatures below 8.3°C. It has a suffocating odor said to be reminiscent of moldy hay or decaying fruit. The boiling point is 8.2°C at 760 mm Hg. The melting point is −118°C. Molecular weight is 98.92. Specific gravity is 1.381 at 20°C. Phosgene is slightly soluble in water and freely soluble in benzene, toluene, and glacial acetic acid and in most liquid hydrocarbons. Its vapor density is 3.4 relative to air, and its vapor pressure is 1420 mm Hg at 25°C. Phosgene is not corrosive when in an anhydrous condition, but in the presence of moisture, corrosive conditions develop rapidly. Under these conditions, specialty metal containers or glass lined vessels may be required.

8.2   Production and Use

Phosgene is an important commodity chemical used as an intermediate in the manufacture of dyestuffs, isocyanates and their derivatives, and many other organic chemicals. It was formerly used as a war gas, and much of the clinical information on phosgene poisoning has been developed in the context of its military applications (200). It is also used in the pharmaceutical industries, in the manufacture of some pesticides, and in metallurgy. Of the total phosgene produced, 62% is used to manufacture toluene diisocyanate, and varying amounts are used to manufacture related chemicals. Phosgene is encountered as a pyrolysis product when chlorinated organics burn or are heated to decomposition. It is also encountered as an environmental contaminant of toxicologic importance in welding (201, 202) in incineration fly ash (203), and in the refining of petroleum (204). Numerous deaths and injuries due to phosgene have been reported in industrial settings (205). Recent reports from China and Korea indicate that phosgene remains an industrial poison of importance (206, 207).

Phosgene is manufactured in many facilities in the United States and in many other countries. Typical production processes involve the reaction of carbon monoxide with nitrosyl chloride, or the reaction of carbon tetrachloride with oleum. The commercial grades are 99–100% pure as a compressed liquified gas.

Although NIOSH estimated that 5700 workers were exposed to phosgene in industry in the 1974 NOHS study, and that 2500 workers were exposed in industry in the 1981 NOES study in the United States, these figures must represent a gross underestimate of the persons exposed to phosgene as a decomposition product, pyrolysis product, or contamination product in industrial processes not identified as phosgene using facilities. For example, welding in the presence of 2–30 ppm of perchloroethylene used as a degreasing agent is reported to produce 0.20–1.7 ppm phosgene in the welder's breathing zone. Early studies by Hamilton and Hardy showed that when chlorohydrocarbon vapors are cracked in open flames or arcs associated with furnaces, boilers, or welding equipment, "sufficient phosgene may be produced to create a hazard" (208).

8.3 Exposure Assessment

8.3.1 Air

Sampling of phosgene in air is generally performed using a midget impinge technique. No official NIOSH analytical method is available, although the midget impinger method using nitrobenzylpyridine is referenced by NIOSH (209). The midget impinger techniques are quite cumbersome to use. Work by Nakano et al. has demonstrated that environmental monitoring for phosgene using a porous cellulose tape impregnated with nitrobenzylpyridine and benzaniline is highly reliable and sensitive as a detector of phosgene. This method is not official, but it seems more convenient and portable than the method that is generally in use (210). OSHA does, however, have a recommended method, OSHA #61 (210a).

It is imperative that a monitoring program be in place wherever phosgene may be encountered because occult exposure at concentrations readily encountered in the industrial environment may lead to serious or fatal acute pulmonary toxicity.

8.4 Toxic Effects

8.4.1 Experimental Studies

8.4.1.1 Acute Toxicity: NA

8.4.1.2 Chronic and Subchronic Toxicity. Kodavanti et al. studied very low level phosgene exposures and effects in Fisher 344 rats (211). They demonstrated that subchronic exposure to concentrations of phosgene that were as low as 0.1 ppm for 6 h/day for 5 days a week for 4 weeks resulted in significant increases in lung weight due to edema and fibrotic changes. At exposures of 0.2 ppm the findings included terminal bronchiolar thickening and inflammation with minimal alveolar changes. At 1.0 ppm, all the above findings were present and were more intensely expressed, and alveolar wall thickening was also present. Since these concentrations of phosgene are likely to be encountered in welding and in other operations where chlorohydrocarbons and heat coexist, it may well be that the current TLV/PEL are not protective of worker pulmonary integrity (212).

Studies noted by WHO in the recently issued Environmental Health Criterion Document for Phosgene (213) demonstrated alterations in the lung immune status following acute and chronic exposure to phosgene. Some alterations in mortality in rodents due to influenza virus and melanoma of the lung was also noted by WHO following single exposures to phosgene in rats. Similar findings were made by Yang et al. in their studies of pulmonary hosts defenses in rats and resistance to infection (214). No developmental or reproductive effects were found in animals or humans. Skin, eye, and lung targets which have long been recognized in humans were acknowledged to be correct in the WHO document. Although some epidemiologic studies of workers chronically exposed to phosgene at undefined levels were found by WHO's review panel, the studies were not sufficiently informative either as to exposure or outcome to provide meaningful data.

8.4.2 Human Experience

8.4.2.1 General Information: NA

8.4.2.2 Clinical Cases

8.4.2.2.1 Acute Toxicity. Phosgene is highly irritating and corrosive and corrosive to all body tissues. It is toxic by all routes of exposure: inhalation, ingestion, and dermal absorption. The NFPA health classification is 4. A health classification of 4 implies that a few whiffs of the gas or vapor can be fatal. Under these circumstances, phosgene exposed individuals may not achieve adequate protection with ordinary protective clothing. SCBA and impervious clothing will be needed to make safe entry into phosgene contaminated areas for first responders and firefighters if the IDLH is exceeded.

Phosgene is a severe pulmonary, eye, and skin irritant. Because of its relative insolubility in water, penetration to deep lung tissues may occur, and delayed occurrence of pulmonary edema of massive proportions occurs. Patients who have been exposed to phosgene should be evaluated promptly and expectantly. Admission to hospital and observation for the development of pulmonary edema, acute respiratory distress syndrome, and rapid decline in clinical status is required. In patients who have been exposed to low concentrations, ($\leq$ 3 ppm) fatal outcomes have occurred after an asymptomatic "golden" period. Patients exposed to higher concentrations of phosgene, close to 50 ppm for even brief periods, may present in pulmonary edema and die very rapidly. Even in patients exposed to low airborne concentrations of phosgene, who do not develop pulmonary edema or other acute respiratory disorders, later occurrence of bronchiolitis obliterans has been reported (215).

Phosgene-exposed patients should be managed expectantly following decontamination in a facility with comprehensive respiratory care services. Early institution of maximal modern pulmonary management with ventilation, PEEP (positive end-expiratory pressure), and drugs may lead to a favorable outcome, although the prognosis for the pulmonary edema of phosgene poisoning is grim. Some research on management of phosgene poisoning has been done by military medicine specialists. Use of exogenous pulmonary surfactant (216), glutathione, *n*-acetylcysteine (217), and aminophylline (218) has been helpful in the prevention of mortality in phosgene poisoned animals, but there is no evidence that these materials help in the management of phosgene-poisoned human patients. Glutathione, acetylcysteine, and aminophylline are well understood medications of relatively low toxicity. They are readily available, and in a critical case, the judgment of the intensivist on scene to use these medications to save a life, could not be faulted.

8.4.2.2.2 Chronic and Subchronic Toxicity. Little useful information of human populations exposed chronically to phosgene is available. Frank reviewed the chronic effects of exposure to inhaled toxic agents in the industrial setting (219). His review of animal studies, human reports, and basic science investigations confirmed that phosgene shares many of the aspects of the toxicity and outcomes of exposure to irritant gases. In general, the reported effects of phosgene are those of gases sparingly soluble in aqueous solution. Analogous substances such as nitrous oxide [100-249-72] produce mild to moderate upper airway irritation, but intense perturbation of the peripheral airways. Typical bronchiolitis and alveolitis results with damage to type I and type II pneumocytes, severe acute edema and later occurrence of fibrosis. As noted above, Boswell has reported bronchiolitis obliterans, a progressive fibrotic disorder of the transmitting airways with a potentially lethal outcome in persons exposed to flight contaminated with phosgene produced in the incineration of chlorine contaminated wastes (215).

Concern about the potential for community exposure to phosgene is significant. The storage of large quantities of halocarbons in proximity to urban centers has led to concern that in the event of fire that might involve these storage facilities significant human exposure to phosgene as a pyrolysis product might occur. Williams et al. studied the potential impact of range fires at the Aberdeen Proving Grounds that might involve the release of phosgene. They concluded that the risk of human health effects from such fires, was low (220). On the other hand, Faucher et al. studied the risk to general populations in Montreal, Canada from fires of high quantities of 1,1,1-trichloroethane in the urban zone. They concluded from their modeling exercises that "results showed that if such a fire should arise, the surrounding population could be exposed to levels of chlorine, hydrogen chloride and phosgene associated with mucous sensorial irritation, pulmonary inflammations, edema, and even death. These effects could reach populations up to many kilometers from the fire and the death rate could be high for those in the toxic cloud's axis (221)."

8.5 Standards, Regulations, Guidelines of Exposure

The current OSHA PEL NIOSH REL for phosgene is 0.1 ppm (0.4 mg/L) on an 8-h TWA basis. A 15-min ceiling value of 0.2 ppm is recommended by NIOSH. The 8-h ACGIH TLV is 0.1 ppm. In view of some of the recently completed animal studies, these values may be insufficiently protective of the pulmonary integrity of phosgene-exposed workers. The NIOSH IDLH for phosgene is 2 ppm. Phosgene is listed as an air pollutant generally considered to cause serious health problems under the Clean Air Act.

9.0 Binary Salts of Alkali and Alkaline-Earth Metals

9.0a Calcium Chloride

9.0.1a CAS Number: [10043-52-4]

9.0.2a Synonyms: Calcosan, calcium dichloride

9.0.3a Trade Names: NA

9.0.4a Molecular Weight: 111.0

9.0.5a Molecular Formula: $CaCl_2$

9.0.6a Molecular Structure: Ca^{2+} Cl^- Cl^-

THE HALOGENS

9.0b Ammonium Chloride

9.0.1b CAS Number: *[12125-02-9]*

9.0.2b Synonyms: Sal ammoniac, ammonium muriate, Amchlor, Darammon, Salammonite

9.0.3b Trade Names: NA

9.0.4b Molecular Weight: 53.491

9.0.5b Molecular Formula: ClH$_4$N

9.0.6b Molecular Structure:

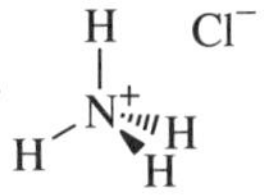

$$H-\overset{H}{\underset{H}{N^+}}-H \quad Cl^-$$

9.0c Sodium Chloride

9.0.1c CAS Number: *[7647-14-5]*

9.0.2c Synonyms: Salt; Nacl; table salt; saline solution; stat trak plus; common salt; sea salt; Rock salt; halite; saline; dendritis; extrafine 200 salt; extrafine 325 salt; h.g. blending; sterling; top flake; white crystal; sodium chloride, 99.999%

9.0.3c Trade Name: Purex

9.0.4c Molecular Weight: 58.443

9.0.5c Molecular Formula: NaCl

9.0.6c Molecular Structure: Na$^+$ Cl$^-$

9.1 Chemical and Physical Properties

These properties are summarized in Table 48.12.

9.2 Production and Use

The compounds in this class of industrial importance include sodium chloride (NaCl), calcium chloride (CaCl), and ammonium chloride (NH$_4$Cl); the physical properties of these materials appear in Table 48.12 (222). Sodium chloride is one of the more significant compounds used in the chemical industry and is the source of nearly all industrial materials containing chlorine or sodium. Sodium chloride, or rock salt, is obtained from underground room and pillar mining or solution mining (in which water is pumped into a rock salt deposit, brought back to the surface, and evaporated). In addition to being an essential nutrient and finding use in the preservation and flavoring of meat and fish, NaCl is an important chemical feedstock and is used in ceramic glazes, metallurgy, soap manufacture, water softeners, highway deicing, regeneration of ion-exchange columns, photography, and herbicides (131, 133).

Table 48.12. Properties of Chloride Salts[a]

Compound	CAS Number	Mol. Wt.	Specific Gravity	MP (°C)	BP (°C)	Solubility
Ammonium chloride	[12125-02-9]	53.5	1.527^{25}	Sublimes without melting	—	283 g/L H$_2$O, 25°C, 396 g/L H$_2$O, 80°C; soluble in methanol, ethanol; insoluble in acetone, ether
Potassium chloride	[7447-40-7]	74.55	1.98	773	—	357 g/L H$_2$O, 25°C, 555 g/L H$_2$O, 100°C; 0.4% in alcohol; insoluble in acetone, ether
Sodium chloride	[7647-14-5]	58.45	2.17^{25}	804–1600	—	357 g/L H$_2$O, 25°C, 385 g/L H$_2$O, 100°C; slightly soluble in alcohol; insoluble in HCl

[a]From Ref. 222.

THE HALOGENS

Calcium chloride is derived principally from underground brines. It is used as a deicer and in road stabilization, dust control, and production of concrete products.

Ammonium chloride, produced from double decomposition reactions of NaCl, is considerably less stable than the other alkaline-earth metal salts of chlorine. At elevated temperatures, ammonium chloride sublimes into a vapor consisting of ammonia and hydrogen chloride. Over prolonged storage, solid ammonium chloride tends to lose ammonia and become more acidic. The chief uses of ammonium chloride include manufacture of ammonia compounds and dry batteries; as a mordant in dyeing and printing; as a soldering flux; in electroplating; in ureaformaldehyde resins, adhesives, and medicines; as a fertilizer; and as a pickling agent in zinc coating and tinning (131). Large amounts of ammonium chloride fume are typically released during galvanizing operations. NIOSH has estimated that more than one million workers were exposed to alkali-metal chloride salts in the United States in 1985 (7).

9.3 Exposure Assessment: NA

9.4 Toxic Effects

The toxicity of NaCl is so low that only mild nasal irritation is experienced by drillers in salt mines even when dust levels exceed the nuisance dust ACGIH TLV of 10 mg/m^3 (20). The main systemic effect of excess NaCl intake is on blood pressure elevation. The lowest toxic dose (TD$_{LO}$) for an adult man with normal blood pressure is 8.2 g/kg (157); in the rat, the oral LD$_{50}$ is 3 g/kg. The corresponding LD$_{50}$s for CaCl$_2$ and NH$_4$Cl are 1.0 and 1.65 g/kg, respectively (157).

9.5 Standards, Regulations, or Guidelines of Exposure

For chloride salts of alkali and alkaline-earth metals, the nuisance dust limits apply. The OSHA PELs for nuisance dust are 15 mg/m^3 as an 8-h TWA for the total dust fraction and 5 mg/m^3 TWA for the respirable fraction (174). The ACGIH has established a 10-mg/m^3 TWA as the TLV for nuisance dusts (22). The TLV for ammonium chloride fume is 10 mg/m^3 TWA and 20 mg/m^3 as a 15-min STEL (22).

10.0 Chlorine Oxides and Fluorides

10.0a Chlorine Trifluoride

10.0.1a CAS Number: [7790-91-2]

10.0.2a Synonyms: Chlorotrifluoride, chlorine fluoride

10.0.3a Trade Names: NA

10.0.4a Molecular Weight: 92.448

10.0.5a Molecular Formula: ClF$_3$

10.0.6a Molecular Structure:
$$\underset{F}{\overset{F}{\underset{|}{Cl}}}{\diagdown} F$$

10.0b Chlorine Pentafluoride

10.0.1b CAS Number: [13637-63-3]

10.0.2b Synonyms: NA

10.0.3b Trade Names: NA

10.0.4b Molecular Weight: 130.445

10.0.5b Molecular Formula: ClF$_5$

10.0.6b Molecular Structure:

$$\begin{array}{c} F \\ | \\ F-Cl^{\cdots F} \\ | \searrow F \\ F \end{array}$$

10.0c Chlorine Dioxide

10.0.1c CAS Number: [10049-04-4]

10.0.2c Synonyms: Chlorine peroxide, chloroperoxyl, Doxcide 50, chlorine oxide

10.0.3c Trade Names: NA

10.0.4c Molecular Weight: 67.452

10.0.5c Molecular Formula: ClO$_2$

10.0.6c Molecular Structure: O=Cl−O

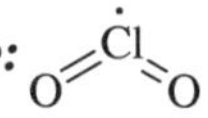

10.1 Chemical and Physical Properties

Of this group, the industrially important compounds include chlorine dioxide (ClO$_2$) and chlorine trifluoride (ClF$_3$); chlorine pentafluoride (ClF$_5$) has no significant industrial use. These are all extremely reactive compounds. Chlorine dioxide exists primarily as the monomeric free radical and is explosive at temperatures above $-40°C$. In aqueous solution, it is quite stable if protected from heat and light (133). Chlorine dioxide reacts violently with organic matter and will cause explosions when the mixture is subjected to sparks or shock. In the presence of sulfur or phosphorus, spontaneous reactions can cause ignition and/or explosions (223). Chlorine trifluoride and pentafluoride are both nonflammable gases that are dangerously reactive. They explode on contact with organic materials, are violently hydrolyzed by water, and react vigorously with metals. Explosions result from reactions with ammonia, carbon monoxide, hydrogen sulfide, sulfur dioxide, or hydrogen gas (132). In the vapor phase, these compounds decompose by hydrolysis to form chlorine, chlorine monofluoride ClOF, ClO$_2$F, ClO$_2$, and hydrofluoric acid.

The physical properties of ClO$_2$, ClF$_3$, and ClF$_5$ are shown in Table 48.13.

10.2 Production and Use

Chlorine dioxide is manufactured from the oxidation of chlorite or the reduction of chlorate. The latter method is used for large-volume production and is carried out in strongly acidic solution using reducing agents such as NaCl, HCl, sulfur dioxide, and

THE HALOGENS

Table 48.13. Properties of Chlorine Dioxide and Fluorides[a]

Compound	Description	Mol. Wt.	MP (°C)	BP (°C)	Sp. Gr. (Liquid)	Solubility
Chlorine dioxide	Yellow to reddish-yellow gas; unpleasant odor similar to that of chlorine	67.46	−59.00	11	1.642	Soluble in water, alkalies, H_2SO_4
Chlorine trifluoride	Corrosive, odorless gas; sweet, suffocating odor	92.46	−76.34	11.75	1.77^{13}	Violently hydrolyzed by water
Chlorine pentafluoride	Colorless gas with a suffocating odor	130.45	—	—	—	Reacts explosively with water

[a]From Refs. 222 and 223.

methanol (133). Chlorine dioxide is used primarily as a bleaching agent for pulp and paper, textiles, leather, fats, and oils. It is also used as a water purifier, as an antiseptic, and as a chemical feedstock for chlorite salts.

The fluorides ClF_3 and ClF_5 are manufactured from reactions of chlorine and fluorine. The trifluoride, and to a much lesser extent the pentafluoride, is used as fluorinating and oxidizing reagents. Chlorine trifluoride is also used in nuclear reactor fuel processing, as an igniter and propellant for rockets, and as a pyrolysis inhibitor for fluorocarbon polymers (222, 223).

In 1983, NIOSH estimated that there 7892 and 357 workers were exposed to ClO_2 and ClF_3, respectively (7).

10.3 Exposure Assessment

According to OSHA (176), high concentrations of ClO_2 in the percent range can be measured using a dual-pH iodometric technique. This method is not recommended for use in personal monitoring. Low concentrations of ClO_2 can be measured using one of several makes of colorimetric detector tubes, which can quantify concentrations in the range of 0.05–15 ppm. OSHA recommends the use of OSHA method #10202 for determining workplace exposures (210a).

10.4 Toxic Effects

CHLORINE DIOXIDE. Chlorine dioxide is a severe respiratory irritant. Because it is increasingly used as a water disinfectant in place of more traditional chlorination techniques, some interest in research on this compound has yielded new information on its toxicologic importance. Elkins found that a concentration of 5 ppm is definitely irritating and that 19 ppm of the gas was sufficient to cause the death of a worker inside a bleach tank (185). Bronchitis and emphysema developed in a chemist repeatedly exposed for several years to ClO_2; symptoms of increasing dyspnea and asthmatic bronchitis were evident even without further exposure (224).

Clinical investigations have been reported on a small group of men who showed respiratory symptoms after having worked in a sulfite–cellulose plant in Sweden in the mid-1950s. Spot air samples indicated that concentrations of chlorine and ClO_2 were normally below 0.1 ppm, with low exposures on rare occasions to sulfur dioxide. Higher concentrations of ClO_2 and chlorine did occur from occasional leaks; workers exposed during these incidents were examined for possible sequelae (225).

Bronchoscopy and biopsy revealed slight chronic bronchitis in 7 of 12 workers. Only two workers who had been exposed just prior to examination showed physical signs of respiratory effects. In one case, an earlier-observed bronchitis disappeared, indicating that the effect is reversible in the absence of continued exposure. Complaints of irritation of the eyes and respiratory tract, and in some cases the GI tract, indicated that ClO_2 is a more severe irritant than other irritating gases. No signs or symptoms of CNS effects were noted. The irritant effects reported were believed to be associated with brief exposures to ClO_2 and chlorine in excess of 0.1 ppm.

Ferris et al. (226) found no significant differences in respiratory symptoms or pulmonary function between a group of 147 pulp-mill workers exposed to ClO_2, chlorine, and sulfur dioxide and a group of 124 paper-mill workers who were exposed for only one-sixth as long as the pulp-mill workers. Average concentrations of ClO_2 on different days ranged from the limit of detection to 0.25 ppm; peak exposures in the pulp mill ranged from nondetectable to 2 ppm.

Salisbury et al. reported on a cohort of workers in a pulpmill who were exposed to gassing incidents in which chlorine and chlorine dioxide were involved (221). There was a significant decrease in the FEV_1/FVC ratio in age-matched, smoking-matched, gassed workers over those not so exposed. Between 1981 and 1988 there was a significantly greater decline in FEV_2/FVC ratios and MMF in the gassed group than in the nonexposed group. This study suggests that previous incomplete findings on long-term respiratory impact of exposure to these compounds may not have been sufficiently focused to detect the changes that were taking place over years of employment exposure to chlorine and chlorine dioxide fume (221).

Kanitz et al. studied somatic parameters at birth and associations with water disinfectants to discern whether there were any adverse effects of chlorine dioxide used for water disinfection purposes (227). A number of parameters were noted to be marginally abnormal in the group exposed to disinfected water. The children born of mothers who drank disinfected water treated with either chlorine dioxide or hypochlorite were noted to have an increased incidence of small cranial circumference, small body length, and neonatal jaundice. All of these findings were statistically significant. They suggest that chlorine dioxide and hypochlorite and their disinfection by-products such as chlorites, chlorates, and trihalomethanes had adverse impacts on pregnancy that had not been fully foreseen (227).

Meggs et al. reported that workers acutely exposed to chlorine dioxide developed both reactive airways dysfunction syndrome (RADS), a form of occupational asthma, and an upper airways reactive disorder that they called *reactive upper airways dysfunction syndrome* (RUDS). Pathological examination of biopsy specimens from the workers who developed RUDS demonstrated changes in the nasal epithelium that the investigators believed provided a basis for the ongoing syndrome of upper airway distress, rhinitis, sinusitis, and conjunctivitis that was observed in these patients. There were increases in the number of nerve fibers and increased nasal epithelial desquamation was noted. Permeability of epithelial junctions also appeared increased. An overlap between RADS and RUDS appeared to be present in the opinion of the investigators (168).

IARC currently considers chlorine dioxide to fall into category D, not classifiable as to human carcinogenicity. There is an absence of human and animal data on this compound.

Concentrations of 0.1 ppm chlorine dioxide in air produced no respiratory or other abnormalities in rats exposed for 5 h/day for 10 weeks (228). Rats exposed to 10 ppm died within 14 days.

Smith and Willhite (229) reviewed several studies demonstrating an association between exposure of experimental animals to ClO_2 in drinking water and effects on developmental function. Relatively high concentrations (ranging from 100 to 500 mg/L) caused decreased fetal weight, increased incorporation of tritiated thymidine into cell nuclei, and a significant decrease in serum thyroxine and triiodothyronine in male and female pups. These same authors reported that ClO_2 exhibits mutagenic potential, evidenced by induction of micronuclei in mice and positive results in an Ames test. Chlorine dioxide was negative on a chromosome test.

Astrakianakis et al. reported on industrial hygiene aspects of sampling surveys at a bleached kraft pulp mill (230). In the course of a study designed to evaluate associations between exposure and cancer risk, they performed many personal measurements for 46 job titles throughout the facility. Although chlorine dioxide was the only bleaching agent used in the facility, 77 area samples indicated that only chlorine and not chlorine dioxide was present in all areas apart from the chemical preparation area. Electrochemical determinations performed in the chemical preparation area demonstrated that most samples exceeded the STEL of 0.3 ppm at least once in each data logging period. The authors suggest that data logging equipment be used to monitor chlorine dioxide exposures because the toxicity of short-term exposures are well known, and excessive exposures appear to occur even in controlled conditions (230).

CHLORINE TRIFLUORIDE. Chlorine trifluoride is also a severe respiratory irritant. Despite its instability in moist air and its tendency to form a variety of decomposition products, the respiratory damage is considered to result mainly from ClF_3 itself, with chlorine, ClO_2, and hydrogen fluoride contributing to the damage (231). At high concentrations detectable by odor, exposure can cause gasping, swelling of the eyes and eyelids, cloudiness of the cornea, lacrimation, respiratory distress, and convulsions within a few minutes (232). Direct contact with the skin causes ulcers and burns that heal with difficulty. The pungent odor and irritant effects of ClF_3 are detectable at such a low concentration that exposed individuals would generally be able to escape before experiencing severe effects from overexposure. In a report of one human exposure, a worker was exposed to ClF_3 for 1–2 min while eating lunch downwind from the effluent of a chlorine trifluoride–charcoal reactor. Symptoms experienced by the individual included headache, abdominal pain, and dyspnea that lasted about 2 h. Except for some fatigue, there were no apparent aftereffects (233).

Exposure of rats to 800 ppm for 15 min was uniformly lethal, as was exposure to 400 ppm for 35 min (231). Chlorine trifluoride given intraperitoneally at doses higher than those administered by inhalation was not lethal, indicating that respiratory damage is the principal cause of death. Pulmonary release of 14CO$_2$ from injected 14C-bicarbonate decreased after inhalation of ClF$_3$, as did blood pH, indicating pulmonary impairment. Chlorine trifluoride lethality was considered by the authors to be comparable to that of hydrogen fluoride on a fluorine-equivalent basis.

Rats exposed to 480 ppm ClF$_3$ exhibited increased activity immediately and showed signs of respiratory difficulty within a few minutes (234). All were in acute respiratory distress in 20 min, and all died within 70 min. At a concentration of 96 ppm, the same signs developed but at a slower rate; 70% of the animals died within 4.5 h. Pathological examination of rats exposed to either concentration revealed emphysema, pulmonary edema, vascular congestion, and fusion of the cells lining the bronchi. In this same series of experiments, rats and dogs were exposed to 21 ppm ClF$_3$ for 6 h on the first day and 4.5 h on the second. During the first day, both rats and dogs showed signs of respiratory irritation and conjunctivitis. Except for eye inflammation in the dogs, animals appeared normal the morning after the first day of exposure. Signs appearing during exposure on the second day were essentially the same as on the first day, except for the development of severe bilateral corneal ulcers in one dog. All animals appeared normal within a few days after the second exposure. In a subchronic study conducted by the same investigators, two dogs and 20 rats were exposed to an average concentration of 5.15 ppm for 6 h/day, 5 days/week for 6 weeks. The same signs and symptoms developed as in the acute study, but were less severe. Respiratory distress developed in the animals at the midpoint of the study; one dog died after 17 days and the other after 26 days, apparently from pneumonia. Animals killed at the end of the experiment had lung hyperemia, hemorrhage, pulmonary edema, and vascular congestion of the liver and kidney.

In the only chronic exposure study conducted in animals, two dogs and 20 rats were exposed to an average CHF$_3$ concentration of 1.17 ppm 6 h/day, 5 days/week for 6 months (235). Dogs exhibited early signs of irritation, but rats were not affected until the ninth day, when they started preening and appeared depressed. Six rats and one dog died during the study. Pathological findings were similar to those seen in the subchronic study.

CHLORINE PENTAFLUORIDE. Administration of microliter amounts of ClF$_5$ to four species of experimental animals caused traumatic explosions and death in most animals (236). Among those that survived, there was inhibition or absence of SGOT (serum glutamic–oxalacetic transaminase) activity and alterations of protein (macroglobulin) structure near the site of application. Dermal application of 10 mL ClF$_5$ caused severe irritation and massive destruction of the skin. Inhalation of ClF$_5$ for brief periods produced a graded response; no rats survived longer than 10 min at 400 ppm, 30% exposed to 200 ppm survived for 24 h, and almost all survived exposure to 100 ppm for 15 min daily for 5 days but had lost weight by the end of that period. No pharmacological action was noted at exposure levels that were tolerated by the animals; where effects were noted, they were attributed to the rapid evolution of energy caused by the reaction of ClF$_5$ with body tissues and fluids.

THE HALOGENS

11.5 Standards, Regulations, or Guidelines of Exposure

The Occupational Safety and Health Administration (OSHA) has established PELs for ClO_2 and ClF_3 of 0.1 ppm as an 8-h TWA and 0.1 ppm as a ceiling limit for ClF_3 (174). The American Conference of Governmental Industrial Hygienists (ACGIH) established TLVs for ClO_2 of 0.1 ppm as an 8-h TWA and 0.3 ppm as a 15-min STEL. The ACGIH TLV for ClF_3 is 0.1 ppm as a ceiling limit (22). The NIOSH RELs for ClO_2 and ClF_3 are the same as the ACGIH TLVs (175). The IDLH values established by NIOSH for ClO_2 and ClF_3 are 5 and 20 ppm, respectively (21).

12.0 Ternary Salts of Alkali Metals Containing Oxygen

12.0a Chloric Acid

12.0.1a **CAS Number:** [7790-93-4]

12.0.2a **Synonyms:** Chloric-acid-, chloric acid solution

12.0.3a **Trade Names:** NA

12.0.4a **Molecular Weight:** 84.459

12.0.5a **Molecular Formula:** $HClO_3$

12.0.6a **Molecular Structure:**
$$\text{HO}-\overset{\overset{\displaystyle O}{\|}}{\text{Cl}}=\text{O}$$

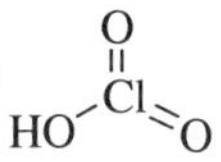

12.0b Hypochlorous Acid

12.0.1 **CAS Number:** [7790-92-3]

12.0.2 **Synonyms:** NA

12.0.3 **Trade Names:** NA

12.0.4 **Molecular Weight:** 52.460

12.0.5 **Molecular Formula:** HOCl

12.0.6 **Molecular Structure:** HO−Cl

12.0c Sodium Chlorate

12.0.1c **CAS Number:** [7775-09-9]

12.0.2c **Synonyms:** Soda chlorate; chlorate of soda; chloric acid; sodium salt; Drop-Leaf; Fall; Harvest-Aid; Tumbleaf; Altacide; Chlorax; shed-A-Leaf 'L'

12.0.3c **Trade Names:** NA

12.0.4c **Molecular Weight:** 106.44

12.0.5c Molecular Formula: NaClO₃

12.0.6c Molecular Structure: $\overset{O}{\underset{O}{Cl}}-O^-\ Na^+$

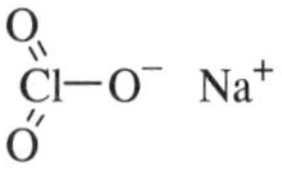

12.0d Potassium Chlorate

12.0.1d CAS Number: [3811-04-9]

12.0.2d Synonyms: Chloric acid; potassium salt; Berthollet salt; chlorate of potash; potash chlorate

12.0.3d Trade Names: NA

12.0.4d Molecular Weight: 122.54

12.0.5d Molecular Formula: KClO₃

12.0.6d Molecular Structure: $\overset{O}{\underset{O}{Cl}}-O^-\ K^+$

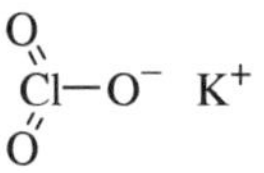

12.0e Sodium Hypochlorite

12.0.1e CAS Number: [7681-52-9]

12.0.2e Synonyms: Clorox; bleach; liquid bleach; sodium oxychloride; Javex; antiformin; showchlon; chlorox; B-K; Carrel–dakin solution; Chloros; Dakin's solution; hychlorite; Javelle water; Mera Industries 2MOM3B; Milton; modified dakin's solution; Piochlor; sodium hypochlorite; 13% active chlorine

12.0.3e Trade Names: NA

12.0.4e Molecular Weight: 74.442

12.0.5e Molecular Formula: NaOCl

12.0.6e Molecular Structure: Na⁺ ⁻O–Cl

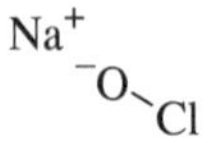

12.0f Calcium Hypochlorite

12.0.1f CAS Number: [7778-54-3]

12.0.2f Synonyms: Losantin; calcium hypochloride; hypochlorous acid; calcium salt; BK powder; Hy-Chlor; Lo-Bax; chlorinated lime; lime chloride; choride of lime; calcium oxychloride; HTH; mildew remover X-14; perchloron; pittchlor

12.0.3f Trade Names: NA

12.0.4f Molecular Weight: 142.98

THE HALOGENS

12.0.5f **Molecular Formula:** CaCl$_2$O$_2$

12.0.6f **Molecular Structure:** Cl–O$^-$ Ca^{2+} $^-$O–Cl

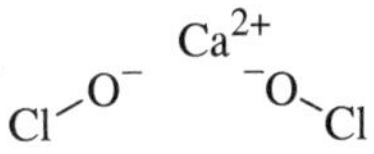

12.0g Sodium Chlorite

12.0.1g **CAS Number:** [7758-19-2]

12.0.2g **Synonyms:** Sodium chlorite, unstabilized

12.0.3g **Trade Names:** NA

12.0.4g **Molecular Weight:** 90.442

12.0.5g **Molecular Formula:** NaClO$_2$

12.0.6g **Molecular Structure:** Na$^+$ $^-$O–Cl=O

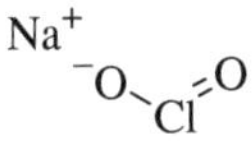

12.0h Potassium Perchlorate

12.0.1h **CAS Number:** [7778-74-7]

12.0.2h **Synonyms:** Perchloric acid, potassium salt

12.0.3h **Trade Names:** NA

12.0.4h **Molecular Weight:** 138.54

12.0.5h **Molecular Formula:** KClO$_4$

12.0.6h **Molecular Structure:** O=Cl(=O)–O$^-$ K$^+$

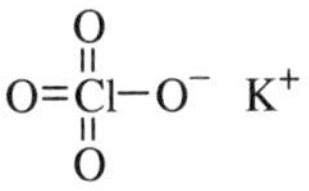

12.0i Ammonium Perchlorate

12.0.1i **CAS Number:** [7790-98-9]

12.0.2i **Synonyms:** Perchloric acid, ammonium salt

12.0.3i **Trade Names:** NA

12.0.4i **Molecular Weight:** 117.49

12.0.5i **Molecular Formula:** NH$_4$ClO$_4$

12.0.6i **Molecular Structure:** O=Cl(=O)–O$^-$ NH$_4^+$

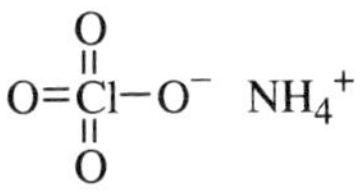

12.1 Chemical and Physical Properties

This group of chlorinated compounds includes the alkali-metal salts of chloric acid ($HClO_3$), chlorous acid ($HClO_2$), hypochlorous acid (HOCl), and perchlorous acid ($HClO_4$). The chief items of industrial importance include sodium, calcium, and potassium chlorate, sodium and calcium hypochlorite, sodium chlorite, sodium potassium, and ammonium perchlorate. These materials are strong oxidizers and chlorinating agents. None are flammable, but all present a fire and explosion risk when in contact with organic materials, including oil, grease, wood, cloth, and paper (132). Mixtures of hypochlorite with a variety of materials can result in the formation of explosive reaction products, for example, chloramines from primary amines, nitrogen trichloride from urea or ammonium salts, or N-chloro compounds from ethylenimine (237). Perchlorates can produce shock-sensitive explosive mixtures when in the presence of reducing materials. All the oxygenated chlorine compounds release ClO_2 in the presence of heat, thus increasing the risk of explosion. Table 48.14 shows some of the physical properties of these compounds.

12.2 Production and Use

Chlorates are produced by electrolysis of sodium or potassium chloride, or by the chlorination of hot lime slurry to manufacture calcium chlorate (131, 133). Perchlorate and chlorite are produced from the electrooxidation or reduction, respectively, of chlorate. Sodium and calcium hypochlorite are manufactured by the chlorination of sodium hydroxide or lime (131).

Sodium and calcium hypochlorite are used primarily as oxidizing and bleaching agents; the most common use for dilute sodium hypochlorite solutions (5%) is as a household bleaching agent. In addition, hypochlorite compounds are used as disinfectants for glass, ceramics, and water systems; as an algicide for cooling water in power stations; as a disinfectant in dairy systems; and as a sanitizer in swimming pools. Sodium chlorate is also used in oil refineries as a sweetening agent.

Because of their strong oxidizing properties, perchlorate and chlorate compounds are used as oxidizers in rocket propellants and jet fuels, and in pyrotechnics, matches, and explosives. Sodium chlorite is used in water purification systems and as a bleaching agent for textiles, pulp, and fabrics (131, 221).

The discovery of widespread water pollution with perchlorate ion, believed to be derived principally from the manufacture of ammonium perchlorate in Nevada and Utah, and its utilization in the production of rocket fuels and other munitions in many areas of the United States has led to increased concern about the toxicology of this compound. In addition to ammonium perchlorate, potassium perchlorate, and lithium perchlorate have been used in these operations. The perchlorate ion has been used in pharmaceutical preparations for the suppression of thyroid function, for many years, both in the United States and in other countries. It has been recognized that perchlorates produce alterations in thyroid function, aplastic anemia, renal dysfunction and other disorders, at pharmacologically common dosages. Therefore, considerable concern exists about its widespread presence in potable-water supplies.

Table 48.14. Properties of Chlorine Salts Containing Oxygen[a]

Compound	Formula	Description	Mol. Wt.	Sp. Gr.	MP (°C)	Solubility
Calcium chlorate	Ca(ClO$_3$)$_2$	Hygroscopic	206.99	2.71	100	Soluble in water, alcohol
Potassium chlorate	KClO$_3$	Colorless, lustrous crystals	122.55	2.32	368	60.6 g/L H$_2$O, 25 °C; 555 g/L H$_2$O, 100 °C; almost insol. in alcohol
Sodium chlorate	NaClO$_3$	Colorless crystals	106.45	2.5	248	Soluble in 1 part water, 0.5 parts boiling water; slightly soluble in alcohol
Sodium hypochlorite	NaOCl	Crystals; anhydrous form is explosive	74.44	—	18	293 g/L H$_2$O, 0°C
Sodium chlorite	NaClO$_2$	Slightly hygroscopic crystals or flakes	90.45	—	Dec. at 180–200°C	390 g/L H$_2$O, 30°C
Ammonium perchlorate	NH$_4$ClO$_4$	Orthorhombic crystals	117.49	1.95	Dec. on heating	Soluble in water, methanol; slightly soluble in ethanol, acetone; insoluble in ether
Potassium perchlorate	KClO$_4$	Colorless crystals or white powder	138.55	2.52	Dec. at 400°C	Soluble in 65 parts water; practically insoluble in alcohol
Sodium perchlorate	NaClO$_4$	White, deliquescent crystals	122.44	2.02	Dec. at 130°C	Very soluble in water

[a] From Ref. 222.

A series of cross-sectional studies performed in the final days of operation of a large ammonium perchlorate manufacturing facility near Las Vegas, Nevada (USA) did not demonstrate significant changes in thyroid function, hematological parameters, or liver function in the perchlorate exposed workers then employed at the facilities. Assumptions of absorption of ammonium perchlorate from airborne dust were based on analogies to beagle dog studies with cesium chloride aerosols, because of the author's beliefs that similar solubilities implied similar absorption. Whether these and other assumptions are valid cannot currently be evaluated (238).

A second investigation performed on the same group of workers by Lamm et al, studied the absorption of perchlorate from doses of 0.004–167 mg total airborne particulate perchlorate per day. These exposures resulted in mean absorbed perchlorate doses of 1–34 mg perchlorate per day. These absorbed doses are well below the administered pharmacological doses of perchlorate currently in use. The study found no changes in thyroid function across the ranges of dose absorbed, nor did it find any hematological alterations. Renal function was not evaluated in detail. Chronic impact was not assessed in this study except as reflected in current laboratory values. Carcinogenicity and adverse reproductive effects were not evaluated. The outcome that was noted on the basis of industrial airborne exposure in a working population may not be relevant to prolonged, low level oral ingestion from waterborne sources of perchlorate ion in a general population of varying health, age, gender, and ethnicity (239).

Perchlorate ion in water is now quantitated by ion chromatography at levels well below those previously considered quantifiable. Jackson et al. reported on an improved quantitation method that detects perchlorates at ppb levels (240).

The numbers of workers estimated by NIOSH to be exposed to these compounds in 1983 are shown in Table 48.15 (7).

12.3 Exposure Assessment: NA

12.4 Toxic Effects

The toxicity of $NaClO_3$ and $NaClO_4$ stems from their strong oxidizing reaction on body tissues, particularly their destructive action on red blood cells. The lowest oral toxic dose

Table 48.15. Estimated Number of Workers Exposed to Oxygen-Containing Chlorine Salts in 1983[a]

Compound	Number of Workers
Calcium chlorate	19,597
Potassium chlorate	8,679
Sodium chlorate	28,583
Calcium hypochlorite	39,878
Sodium hypochlorite	562,423
Sodium chlorite	18,585
Ammonium perchlorate	1,445
Potassium perchlorate	2,640
Sodium perchlorate	1,452

[a]From Ref. 7.

Table 48.16. Acute Toxicity Values for Oxygen-Containing Chlorine Salts[a]

Compound	Species	Oral LD$_{50}$ (mg/kg)
Calcium chlorate	Rat	4500
Potassium chlorate	Rat	1870
Sodium chlorate	Rat	1200
	Mouse	8350
	Rabbit	7200
Sodium chlorite	Rat	165
	Mouse	350
	Guinea pig	300
Sodium hypochlorite	Rat	8910
Sodium perchlorate	Rat	2100

[a]From Ref. 151.

(TD$_{LO}$) reported for this action in humans is 800 mg/kg. Massive hemolysis arising from the accidental introduction of sodium hypochlorite into the blood has been reported as a result of hypochlorite disinfection of dialysis machines (241). In experimental animals, the chlorates, perchlorates, and hypochlorites exhibit low to moderate acute toxicity; the toxicity of sodium chlorite is at least 10-fold higher. Table 48.16 provides reported LD$_{50}$ values for various species treated with these compounds.

Since publication of the previous edition of this volume, much new information regarding the toxicities of sodium chlorite and sodium hypochlorite has become available, primarily as a result of safety concerns about chlorinated drinking water. In general, human population studies and animal bioassays have not found an association between exposure to these compounds and an increased risk of cancer, reproductive, or teratogenic effects.

SODIUM CHLORITE. Michael et al. (242) examined 197 individuals for blood changes after a new ClO$_2$ water disinfectant system was installed in their village. Blood cell and chemistry values were compared with those for 112 unexposed individuals. Chlorite levels in the water ranged from 3.19 to 6.96 mg/L, ClO$_2$ levels from 0.33 to 1.11 mg/L, and chlorate levels from 0.34 to 1.82 mg/L. No significant difference was observed between the two groups in hematocrit, hemoglobin levels, mean corpuscular volume, red and white blood cell counts, methemoglobin levels, blood urea nitrogen, bilirubin, or serum creatinine levels 70 days after the ClO$_2$ system was installed.

In other studies, no physiologically important findings were noted among male volunteers who drank two 500-mL portions of water containing up to 2.4 mg/L chlorite and chlorate and 24 mg/L ClO$_2$ (243). There were also no physiologically significant findings among individuals who drank 500 mL water containing 5 mg/L chlorite, chlorate, or ClO$_2$ each day for 12 weeks (244).

Tuthill et al. (245) compared the neonatal morbidity and mortality experiences of two communities, one of which utilized ClO$_2$ for water disinfection and the other, chlorination.

Monthly average chlorite levels in the water from the ClO_2 system were 0.32 mg/L; this was later reduced to 0.16 mg/L. In the community with the ClO_2 system, the authors found a higher number of infants judged to be premature or to have greater weight loss after birth compared with these values in the community using chlorination. Rates for jaundice, birth defects, and neonatal mortality did not differ between the two communities.

Rats treated with much higher concentrations of chlorite than are found in water systems (100–500 mg chlorite per liter) for 30 and 60 days developed decreased hemoglobin concentrations, loss of packed blood cell volume, and hemolysis (246). After 90 days of treatment, hemoglobin concentrations and red blood counts returned to normal; however, blood cell glutathione concentrations remained significantly depressed. Administration of chlorite in the drinking water did not induce methemoglobinemia, in contrast to a study in which methemoglobinemia was produced in rats given 10 mg/kg chlorite by intraperitoneal injection (247). Drinking-water studies conducted on cats, mice, and monkeys resulted in hematologic findings similar to those seen in rats (246, 248).

There was no statistically significant increase in hyperplastic nodules in the liver or hepatocellular carcinomas in groups of 50 male and 50 female B6C3F1 mice given 0.025 or 0.05% sodium chlorite in drinking water for 80 weeks (249). There was a statistically significant increase in lung adenomas among male mice (0/35 control, 2/47 low-dose, and 5/43 high dose, $p < .05$), but not in female mice. Groups of 50 male and 50 female Fischer 344 rats given 0.03 or 0.06% sodium chlorite in drinking water for 85 weeks did not exhibit an increase in tumor response. However, the study was terminated early owing to a Sendai virus infection among the animals.

The International Agency for Research on Cancer (248) has concluded that there is inadequate evidence for the carcinogenicity of sodium chlorite in animals, and that sodium chlorite is not classifiable as to its carcinogenicity in humans (group 3).

In a study to investigate the reproductive toxicity of sodium chlorite, female A/J mice were given sodium chlorite (100 mg/L) in drinking water throughout gestation and through 28 days of lactation (250). A smaller proportion of treated mice became pregnant compared with controls. There was no exposure-related effect on litter size or neonatal survival, but pups in the treated group exhibited reduced weight gain. In another reproductive study, male and female Long–Evans rats were given sodium hypochlorite in the drinking water ($\leq$ 100 mg/L for females and $\leq$ 500 mg/L for males). There was no effect on fertility, litter size, or neonatal survival, or on testis or epididymis weight. No treatment-related histological abnormalities were seen in either the female or male reproductive tracts. There was a small increase in the number of abnormal sperm and a decrease in progressive sperm motility among males administered 100 or 500 mg/L chlorite (251).

Couri et al. (252) reported finding no evidence of a teratogenic effect among Sprague–Dawley rats treated on gestation days 8–15 with $\leq$ 2% sodium chlorite in drinking water, $\leq$ 50 mg/kg chlorite by intraperitoneal injection, or $\leq$ 200 mg/kg by gavage. Maternal toxicity, including excess mortality, was observed in groups receiving the higher doses by injection or gavage. In another study, female Sprague–Dawley rats were pretreated with $\leq$ 10 mg/L sodium chlorite in drinking water for 2.5 months prior to being mated with untreated males. Treatment with chlorite continued up to the 20th day of gestation. There was no evidence of maternal toxicity and no treatment-related effect on litter size or weight. However, there was a dose-related increase in the number of fetuses with variation of the sternum and with increased crown–rump length.

SODIUM HYPOCHLORITE. Small but significant increases in body weights were observed in male Wistar rats given sodium hypochlorite ($\leq$ 80 mg/L free chlorine) in drinking water for $\leq$ 6 weeks (253). In female rats treated by gavage with 200 mg/kg, there was an increase in kidney weights. Fidler (254) reported finding an absence of phagocytic activity in macrophages recovered from the lungs of mice given 25–30 mg/L sodium hypochlorite in drinking water for 3 weeks. Subsequent experiments showed a decrease in phagocytic activity of macrophages recovered from the spleen and liver (255). These findings were not confirmed in rats given $\leq$ 30 mg/L hypochlorite in drinking water (256); however, rats treated at the high dose were found to have reduced spleen weights and delayed hypersensitivity reactions.

Among rats treated with sodium hypochlorite in drinking water, no effect was observed on survival, serum chemistry, or organ weights until concentrations reached 4000 mg/L (257–259).

There was no observed increase in tumor incidence among male and female B6C3F1 mice given 500 or 1000 mg/L sodium hypochlorite in drinking water for 103 weeks (259). Groups of 50 male and 50 female Fischer 344 rats were given 0, 500, or 1000 mg/L (males) or 0, 1000, or 2000 mg/L (females) sodium hypochlorite in drinking water for 104 weeks. No increase in tumor incidence was seen in treated animals compared with controls (258). Repeated dermal application of sodium hypochlorite in mice has not been found to increase tumor incidence (260, 261). However, application of sodium hypochlorite following initiating doses of 4-nitroquinoline 1-oxide caused tumors in 9 of 32 mice, compared to none in animals given sodium hypochlorite or the initiator alone (260). Sodium hypochlorite administered after the initiator 7,12-dimethylbenz[*a*]anthracene caused a squamous-cell carcinoma in only 1 of 20 mice (261).

The International Agency for Research on Cancer has concluded that there is inadequate evidence for the carcinogenicity of sodium hypochlorite in animals, and that sodium hypochlorite is not classifiable as to its carcinogenicity in humans (group 3) (262).

12.5 Standards, Regulations or Guidelines

The current OSHA PEL for perchlorates as nuisance dust is 15 mg/m^3. The California groundwater standard is 18 µg/L.

C BROMINE AND ITS COMPOUNDS

13.0 Bromine

13.0.1 CAS Number: [7726-95-6]

13.0.2 Synonyms: NA

13.0.3 Trade Names: NA

13.0.4 Molecular Weight: 159.81

13.0.5 *Molecular Formula:* Br$_2$

13.0.6 *Molecular Structure:* Br—Br

13.1 Chemical and Physical Properties

Bromine is the only nonmetallic element that is liquid at normal temperatures and pressures. It is a heavy, mobile, reddish-brown liquid that volatilizes readily to a red vapor. The most stable valence states of bromine are -1 and $+5$, although valence states of $+1$, $+3$, and $+5$ are known. The diatomic nature of bromine persists in the solid, liquid, or gaseous state. The physical properties of bromine are given in Table 48.1.

The reactivity of bromine lies between that of chlorine and iodine. Bromine will cause ignition of organic materials, including wood, cotton, and straw. It reacts violently on contact with natural rubber and reacts explosively with a number of common substances, including aldehydes, ketones, carboxylic acids, acetylene, acrylonitrile, ammonia, ethyl phosphine, hydrogen, nickel carbonyl, ozone, oxygen difluoride, phosphorus, potassium, sodium, and sodium carbide (132, 237).

Facilities where bromine is manufactured or used should be designed to dispose rapidly of liquid bromine spills. Solutions or slurries of 10–50% potassium carbonate, 10–13% sodium carbonate, and 5–10% sodium bicarbonate or saturated "hypo' solution (prepared by dissolving 4 kg technical-grade sodium thiosulfate in 9.5 L water and adding 113 g of soda ash) are preferred neutralizing agents for liquid bromine (20). A 5% lime slurry or a 5% sodium hydroxide solution may be used, but the heats of reaction are higher for these reagents.

13.2 Production and Use

Although it is estimated that 1015–1016 tons of bromine are contained in the earth's crust, the element is widely distributed and found only in low concentrations in the form of bromide salts. The most readily recoverable form of bromine occurs as soluble salts in salt lakes, inland seas, brine wells such as those in Michigan and Arkansas, and seawater. Today, little bromine is extracted from seawater, which contains bromide salts only in a concentration of 65 ppm.

Recovery of bromine from brines involves the oxidation of the bromide ions to free elemental bromine, which is then vaporized from the solution by either air or steam. Oxidation of bromide is accomplished using chlorine.

From sources that contain low concentrations of bromide salts, bromine must first be concentrated before the steam vaporization process becomes practical. Consequently, air is used to vaporize the bromine from the chlorinated brine; this is followed by injection of sulfur dioxide into the dilute bromine-laden air. Subsequent absorption of the hydrogen bromide in a controlled amount of water produces a much more concentrated bromide solution, according to the equation

$$SO_2 + Br_2 + 2H_2O \rightarrow 2HBr + H_2SO_4$$

Bromine must then be released again by chlorine and steamed from the solution. The byproducts hydrochloric and sulfuric acids are recycled to acidify the incoming brine to the pH necessary for efficient chlorination.

THE HALOGENS 801

The air-blowing process can also be used when an alkali bromide is the desired product. Absorption of the bromine from the air stream by sodium carbonate, for example, produces sodium bromide and sodium bromate as shown:

$$3\,Br_2 + 3\,Na_2CO_3 \rightarrow 5\,NaBr + NaBrO_3 + 3\,CO_2$$

The sodium bromate may then be crystallized from solution or reduced with iron, if NaBr is the only desired product.

Typical, freshly prepared bromine from a modern plant is likely to be at least 99.9% pure. Probable impurities are chlorine, moisture, and organic material at levels of 50 ppm each. Specifications of the *U.S. Pharmacopoeia* and of the American Chemical Society for reagent-grade bromine allow up to 0.3% chlorine, 0.05% iodine, 0.002% sulfur, and 0.015% nonvolatile matter.

U.S. production of bromine has fallen over the past two decades, with 350 million lb produced in 1985, compared with 400 million lb in 1973 (263). More recent figures are not available. Large quantities of bromine are imported to fulfill U.S. industrial needs.

Today, the largest single use of bromine is for the production of fire retardants. About 30% of bromine production was used for the manufacture of flame retardants in 1985 (263). Another 28% was used to manufacture oil- and gas-well completion drilling fluids such as calcium bromide.

Production of ethylene dibromide, a gasoline antiknock compound for leaded fuels, from bromine has fallen substantially since the late 1970s as a result of the phaseout of leaded gasoline. About 20% of bromine production was consumed in the manufacture of ethylene dibromide in 1985, compared to 66% in 1974. Ethylene dibromide was banned for use as a fumigant by the USEPA and was withdrawn from domestic U.S. use in 1984.

Agricultural chemical production consumed about 10% of bromine production, primarily as methyl bromide. However, recent restrictions on the use of brominated pesticides, such as ethylene dibromide and DBCP and proposed limitations on such fumigants as methyl bromide are expected to further reduce the use of these brominated chemicals in the future. Sanitation preparations, together with miscellaneous uses, accounted for an estimated 8% of bromine demand in 1985. A large part of this usage was elemental bromine added directly to water in swimming pools, cooling towers, and the like for control of algae, bacteria, and odors. Among the products containing bromine or requiring it as an intermediate in their manufacture are photographic films and papers, dyes, inks, sedatives, anesthetics, hydraulic fluids, refrigerating and dehumidifying agents, hair-waving preparations, and laboratory reagents. The remaining 4% of bromine production in 1985 was used for exports.

13.3 Exposure Assessment

Employee exposures to bromine vapors can be measured using NIOSH method 6011, revision 2, issued in 1994 (25). This method involves collecting 2–90 L of air through a silver membrane filter mounted in a carbon-filled polypropylene cassette and analyzing the sample using ion chromatography to quantify the bromide ion. Samples collected on the silver membrane filter are stable for 30 days at 25°C or 60 days at 5°C, although the filter must be protected from light. This method has a working range of 0.008–0.4 ppm for a 90-L air sample. The estimated limit of detection for this method is 1.6 mg bromide ion per sample, and the overall precision (coefficient of variance) is 0.069. Hydrogen sulfide gives a negative interference, and hydrogen chloride gives a positive interference at a maximum 15 µg per sample. Hydrogen bromide gives a positive interference as it is sampled continuously. The revised method has sufficient sensitivity for STEL determinations.

The OSHA method for sampling bromine uses 10 mL of a standard eluent carbonate-bicarbonate solution (0.003 M NaHCO$_3$/0.0024 M Na$_2$CO$_3$) in a midget fritted-glass bubbler as the collection medium (143). In basic solution, bromine disproportionates to bromide and bromate ions, which are quantified using an ion-specific electrode. Samples collected in the solution are stable for up to 30 days at room temperature. This method has a reliable quantification limit of 0.9 mg per sample, and an overall precision (coefficient of variance) of 0.06. Potential interfering substances include nitrate (if present in greater concentration than bromide) and anions such as Cl$^-$, ClO$_3^-$, and IO$_3^-$.

13.4 Toxic Effects

Information on the toxicity of bromine in humans is limited to the effects caused by acute exposures. Henderson and Haggard (182) stated that exposure to 1000 ppm is rapidly fatal; 40–60 ppm is dangerous for brief exposures; 4 ppm is the maximal allowable concentration for exposures not exceeding 1 h; and 0.1–0.15 ppm is the maximal allowable for prolonged exposure. Flury and Zernik (264) quoted Lehmann, who reported that exposure to a 0.75-ppm concentration of bromine in a workroom caused no symptoms in 6 h. However, Elkins (185) reported that exposure to 1 ppm in a plant handling liquid bromine was considered to be excessively irritating.

According to Stokinger (20), the signs and symptoms associated with inhalation of bromine vapor include coughing, nosebleed, a feeling of oppression, dizziness, and headache, followed several hours later by abdominal pain and diarrhea, and sometimes by a measles-like eruption on the trunk and extremities. Workers exposed regularly to bromine concentrations of approximately 0.3–0.6 ppm for 1 year experienced headache, pain in the joints, stomach, and chest, irritability, and loss of appetite (265).

Oppenheim (266) mentioned the frequency with which discharging pustules and furuncles appear in exposed areas of the skin of those who handle bromine. Brief contact of the liquid with the skin leads to the formation of vesicles and pustules. If not removed at once, bromine induces deep, painful ulcers.

Necropsy of guinea pigs and rabbits following a 3-h exposure to 300 ppm bromine revealed edema of the lungs, a pseudomembranous deposit on the trachea and bronchi, and hemorrhages of the gastric mucosa. Foci of bronchopneumonia were found in animals that died several days after exposure, and there was evidence of functional disturbances (267). Oral LD$_{50}$s in four species of animal (rats, mice, rabbits, and guinea pigs) range from 2.6 to 5.5 mg/kg, rats are the most sensitive species (151).

Bitron and Aharonson (268) exposed male albino mice once to 750 ppm bromine for 15–30 min or 240 ppm for 15–270 min. The duration of exposure associated with a 50% mortality rate (LT$_{50}$) was determined to be 9 min at 750 ppm and 100 min at 240 ppm. A delayed mortality effect was observed at these concentrations; the mean times of death estimated for exposure durations at the LT$_{50}$ were 11 days at 750 ppm and 6 days at

240 ppm. For each of the two concentrations tested, the mean time of death increased with decreasing exposure duration.

Exposure of rats, mice, and rabbits to bromine concentrations of 0.2 ppm for 4 months resulted in respiratory, nervous, and endocrine system dysfunction; exposure to 0.02 ppm for the same time period produced no adverse response (269). Changes in conditioned responses and in several blood indices were observed in rats given 0.01 mg/kg bromine orally for 6 months.

13.5 Standards, Regulations, or Guidelines of Exposure Limit

The current limit established by OSHA is 0.1 ppm as an 8-hr TWA (174). The TLV Committee of the ACGIH established an 8-h TWA TLV of 0.1 ppm and a 15-min STEL of 0.3 ppm (22) for this substance. The recommended exposure limits of NIOSH are the same as the TLVs (175). NIOSH has also established an IDLH level of 3 ppm for bromine (21).

Monte Carlo uncertainty analysis was used to assess worker potential for exposure to halogen gases during dosing of brominators with 500-lb sacks of a bromine-based biocide. Indoor and outdoor dosing scenarios were studied. The diffusion model employed described a concentration gradient of halogen as a function of distance from the operation and time of exposure. The model demonstrated that this operation was unlikely to result in exceedence of the working occupational exposure limit for halogens when the facilities were out of doors. However, in indoor settings, exceedences may occur. Field trials and measurements generally confirmed the predictions of the model (270).

Eldan et al. monitored the exposure of 848 bromide production workers to brominated compounds between July 1990 and March 1991 through the determination of serum bromide levels using X-ray fluorescence methodology (271). The method used measured total bromine content of blood. Content of bromine was determined using a silicon/lithium fluorescence detector. A detection limit of 0.6 ppm bromine was obtained by this method. Its accuracy and precision was cross checked using HPLC on 99 duplicate samples. Because the serum half-life of bromine is only 12–14 days, the method used is considered reliable by the authors for quantitation only of recent bromine exposures. No relationships were found between duration (longevity) in job and serum bromine, although there were significant differences among job titles for the short-term studies. Such differences reflected the exposures at work in recent weeks. Although the concentrations of bromine found in this working population were greater than those found in the general population, no levels believed to be in the toxic range were detected. The average concentration of bromine in blood in this group was 22.9 ppm (8.3–428.6 ppm) and the toxicologically relevant benchmark is about 700 ppm (271).

14.0 Inorganic Compounds

The inorganic bromine compounds of industrial interest are listed in Table 48.17 along with their physical properties and the estimated numbers of workers exposed in 1983.

14.0a Hydrogen Bromide

14.0.1a CAS Number: [10035-10-6]

Table 48.17. Chemical Properties of Bromine Compounds and Estimated Number of Workers Exposed[a]

Compound	Formula	Description	Mol. Wt.	Sp. Gr.	MP (°C)	BP (°C)	Solubility	Number of Workers Exposed in 1983
Hydrogen bromide	HBr	Colorless gas	80.92	3.5 g/l	−86.9	−66.8	Saturated solution at 66% HBr	20,571
Ammonium bromide	NH_4Br	White, slightly hygroscopic crystals	97.96	2.43^{25}	Sublimes	—	Soluble in water, alcohol, acetone; slightly soluble in ether	—
Potassium bromide	KBr	Colorless crystals or white powder	119.01	2.75	730	—	Soluble in water; slightly soluble in alcohol	50,375
Sodium bromide	NaBr	White crystals	102.91	3.21	755	—	Soluble in water, alcohol	124,815
Potassium bromate	$KBrO_3$	White crystals	167.01	3.27	350 (dec. 370)	—	Soluble in water; slightly soluble in alcohol	26,562
Sodium bromate	$NaBrO_3$	Colorless crystals	150.91	3.34	381 (dec.)	—	Soluble in water	27,921
Bromine trifluoride	BF_3	Colorless or pale yellow liquid	136.92	2.80^{25}	8.77	125.75	Very reactive	357
Bromine pentafluoride	BF_5	Fuming liquid	174.92	2.46^{25}	−60.5	40.76	Explodes on contact with water	357

[a]From Refs. 7 and 222.

THE HALOGENS

14.0.2a   Synonyms: HBr, hydrobromic acid

14.0.3a   Trade Names: NA

14.0.4a   Molecular Weight: 80.92

14.0.5a   Molecular Formula: HBr

14.1a   Chemical and Physical Properties

Gaseous HBr and aqueous hydrobromic acid, usually as the constant-boiling-point solution (48% HBr), are the major intermediates in the production of bromine compounds. The principal means of producing HBr is by reacting H_2 and Br_2 at high temperatures in the absence of air, although HBr is also formed as a by-product of an organic substitution reaction. Liquefied HBr is available commercially in cylinders, and various strengths of aqueous hydrobromic acid are available in drums or tankcars.

Information on human responses to HBr vapors appears to be limited to one report (272). The subjective responses of six volunteers exposed to HBr concentrations ranging from 2 to 6 ppm for several minutes showed that exposure to 5 or 6 ppm caused nasal irritation in all individuals and throat irritation in one. None of the subjects experienced eye irritation at any of the concentrations tested. At a 3-ppm concentration, only one individual experienced nasal or throat irritation. The odor of HBr was detectable at all concentrations administered.

Orlando et al. described the occurrence of both chronic upper and chronic lower airway disorders including anosmia, and aphonia together with nonspecific bronchial hyperreactivity in a patient who was exposed to low, weakly irritant concentrations of hydrobromic acid. This worker remained in the exposure area after he became symptomatic with cough and shortness of breath, and the authors postulate that the severity of his disease was related to the insidious nature of the toxin, which was not sufficiently obnoxious to force the worker to leave the exposure zone. This worker subsequently developed progressive bronchiolitis obliterans, and the author's were considering him as a candidate for lung transplant (273).

One-hour rat and mouse LC_{50} values for HBr vapor are 2858 and 814 ppm, respectively (151).

The PEL established by OSHA is 3 ppm as an 8-hr TWA (143). The ACGIH TLV for HBr is 3 ppm as a ceiling limit (22). The National Institute for Occupational Safety and Health has also recommended that exposures to HBr not exceed 3 ppm as a ceiling (175). The IDLH value for HBr is 30 ppm (21).

14.0b   Metal Salts, Bromides

The industrially important bromides include those of sodium, potassium, and ammonium. They are produced by the reaction of the metal hydroxide or carbonate with HBr. The bromides have a number of applications, including fireproofing, photography, lithography, analytical chemistry, and corrosion inhibition (131, 222).

The physiological effects of the bromides are similar, and all are attributed to the bromide ion. Most of the available data come from their medical use as oral sedatives, diuretics, and antiepileptics. An oral dose of 3 g (30–60 mg/kg for an adult), which causes the blood bromide level to rise to 50 mg%, was considered a 'no ill effect' level (274), and typical signs of bromine intoxication may be absent even if blood bromide levels reach 200 mg% or more (275). Acute overdoses may cause vomiting or profound stupor, and chronic, prolonged use often leads to depression, ataxia, and psychosis, especially with low sodium intake.

The lethal oral dose of NaBr is high: the LD_{50}s in rats and mice are 3500 and 7500 mg/kg, respectively (151).

Metabolically, bromide has a biologic half-life of about 12 days, is not incorporated into fat or blood proteins, and none is extractable from plasma or hemolyzed blood cells by ether (276, 277). Bromide ion does not interfere with thyroid activity even if large daily doses are given for an extended period of time (278).

The dermal toxicity of hydrobromic acid and other bromides is quite similar to that of hydrochloric acid. Primary irritant contact dermatitis is the rule from skin contamination of strong solutions of this acid, although allergic contact dermatitis has also been reported from HBr and from other bromine-containing compounds. Eliaz et al. studied the nature of bromine chemical burns on full thickness human skin *in vitro* (279). They found that the progression of the *in vitro* contact burns tracked the course of burns that had been clinically observed. There was pronounced destruction in the epidermis with disruption of the stratum corneum, and vacuolation of the keratinocytes, and damage to the dermis, as well. Coagulation of the collagen occurred which was dependent on the concentration of the bromine and the duration of exposure. On the basis of their studies, the authors recommend early medical intervention and prompt treatment for the successful management of both severe and mild cases of bromine cutaneous injuries (279).

Use of bromides as water disinfectants in place of chlorine and hypochlorite has led to the development of dermal and respiratory toxicity which had not previously been observed. Fitzgerald et al. reported spa/pool contact dermatitis of an allergic nature from bromides in swimming-pool water (280). Burns et al. reported acute pneumonitis and subsequent reactive airways dysfunction syndrome (RADS) in two patients who were exposed to a bromine containing water disinfectant in a hot tub. Bromine and hydrobromic acid produced from the disinfectant were the putative agents for the toxic outcome (281).

14.0c   Metal Salts, Bromates

14.0c(i)   Sodium Bromate

14.0c(i).1   CAS Number: [7789-38-0]

14.0c(i).2   Synonyms: NA

14.0c(i).3   Trade Names: NA

14.0c(i).4   Molecular Weight: 150.892

14.0c(i).5   Molecular Formula: NaBrO$_3$

14.0c(i).6   Molecular Structure: $\overset{O}{\underset{O}{\overset{\|}{\underset{\|}{Br}}}}-O^-\ Na^+$

14.1c Chemical and Physical Properties

See Table 48.17.

14.2c Production and Use

The bromates of chief industrial interest are those of sodium and potassium. Sodium bromate is used in mixtures with sodium bromide to extract gold from ore. It is also used as an analytical reagent, as a cleaning agent for boilers, and as a component in hair-waving formulations (133). Potassium bromate is used primarily as a conditioner for flour and dough; some of its nonfood uses include use as an oxidizing agent for analytic chemistry and as a brominating agent. The bromates of calcium and barium have limited use as oxidizers, maturing agents in flour, and analytic reagents.

Combinations of finely divided bromate with combustible material or aluminum, arsenic, carbon, copper, phosphorus, or sulfur [7704-34-9] may explode on contact with heat or shock (132). Bromates may also ignite combustible materials.

14.3c Exposure Assessment: NA

14.4c Toxic Effects

Bromates exhibit higher acute toxicity in animals than do bromides. The oral $LD_{50}s$ in mice and rats are reported to be 289 and 321 mg/kg, respectively. Bromate produces moderate eye irritation and slight corneal injury when instilled into rabbit eyes and causes superficial burns on prolonged contact with intact rabbit skin (20). Both effects are transient and heal within a few days.

Potassium bromate dust is irritating to the mucous membranes of the upper respiratory tract, but there was no evidence of any cumulative effect on the lung of workers after 18 years of exposure to levels as high as 15 mg/m^3 (20). Where fatal poisonings have occurred, death was reportedly caused by renal failure (282, 283). DeVriese et al. reported a case of severe acute renal failure due to bromate intoxication and outlined principles and guidelines for clinical management based on a review of the literature (284).

Acute animal studies confirm the kidney as the target organ for potassium bromate; guinea pigs received subcutaneous injections of 100 mg/kg potassium bromate manifested cortical damage within 24 h. Cloudy swelling and necrosis of epithelial cells in renal tubules were seen in animals killed after treatment (285, 286). Kidney and stomach damage was also seen in guinea pigs and dogs treated with sodium bromate (287).

Peripheral neuropathy and nerve deafness as a result of potassium bromate poisoning has been reported. Tono et al. noted deafness following $KBrO_4$ intoxication (288). Wang and Tsai reported polyneuropathy as a result of bromate intoxication (289).

Potassium bromate has been investigated for carcinogenic activity in animal studies. There are no human studies available from which to evaluate the carcinogenic potential of potassium bromate.

Male and female mice (Theiller's original strain) and male and female Wistar rats were fed diets containing 79% breadcrumbs made from flour that contained 0, 50, or 75 mg/kg potassium bromate for 80 weeks (mice) or 104 weeks (rats). Survival of the animals through the end of the study ranged from 50 to 65% mortality was not treatment-related. There was no increase in tumor response in either mice or rats given the low or highdose diets (290, 291). The IARC Working Group that evaluated these studies noted that bromates are substantially degraded during the baking of bread (286).

Groups of male and female Fischer 344 rats were administered potassium bromate at a dose of 0, 250, or 500 mg/L in the drinking water for 110 weeks (292). Males administered the high dose experienced a decreased survival rate, but survival rates of all other groups were comparable. Treatment-related increases in several types of tumors were observed in both sexes, including renal adenocarcinoma, renal adenoma, mesothelioma of the peritoneum (males only), and benign and malignant thyroid tumors (females only). Increases in the incidences of these tumor types were all dose-related and statistically significant. Potassium bromate administered in drinking water after administration of N-nitrosoethyl-N-hydroxyethylamine (NE-HEA) increased the incidence of renal cell tumors and dysplastic loci of the renal tubules in rats compared with results where NEHEA was administered alone (293).

Potassium bromate is produced in the course of ozonation of water containing bromide compounds. Carcinogenicity of potassium bromate contaminated drinking water was studied in male $B6C3F_1$ mice and F344/N rats by DeAngelo et al. (294). They reported that $KbrO_3$ is a renal carcinogen in rodents at water concentrations as low as 0.02g/L, a dose equivalent to 20 ppm or 1.5 mg/kg per day. They propose the use of these data for estimation of human health risk associated with the shift from chlorination to ozonation of drinking water (294).

Umemura et al. demonstrated oxidative DNA damage and cell proliferation in the kidneys of male and female rats during 13 weeks of exposure to potassium bromate. They found significant differences in the liability to oxidative stress in the kidneys of male and female rats. Female rats seemed more resistant to the impact of potassium bromate than did male rats (295). Chipman suggested that the apparent oxidation of DNA by potassium bromate might be linked to lipid peroxidation. Studies which they performed did not support a direct pathway for DNA oxidation in the kidney on the basis of their observation that kidney DNA oxidation was not inhibited by glutathione and required sustained exposure at near toxic levels characteristic of lipid peroxidation (296).

In 1987 the International Agency for Research on Cancer concluded that there is sufficient evidence for the carcinogenicity of potassium bromide in animals and has placed it in group 2B (possibly carcinogenic to humans) (286).

14.0d Bromine Halides

The commercially available bromine halides include $BrCl$, BrF_3, and BrF_5. Their physical properties, given in Table 48.17, indicate high chemical reactivity. The fluorinated bromine compounds react explosively with water, strong acids, ammonium salts, and carbon monoxide (at high temperatures or concentrations) (132). Bromine and fluorine fumes are released on contact with acids or acid fumes. Fire or explosion is likely from reactions with acetic acid, ammonia, benzene, cellulose in paper, ethanol, or organic matter.

The tri- and pentafluorides are strong fluorinating agents, useful in organic synthesis and in forming uranium fluorides. With alkali fluorides they form double salts; $MBrF_4$, $MBrF_6$,

THE HALOGENS

and various quaternary ammonium compounds are used pharmaceutically as bromides because of their suitable solubility, crystalline character, and ease of preparation (133).

Skin or eye contact with vapor or liquid bromine pentafluoride causes painful, deep-seated, long-lasting burns. The acute effect of this substance on the lung is similar to that of phosgene (129, 130). The American Conference of Governmental Industrial Hygienists has established a TLV of 0.1 ppm as an 8-h TWA for BrF_5 (22). This limit is based on toxicological analogy with chlorine trifluoride. The National Institute for Occupational Safety and Health has recommended the same exposure limit as the ACGIH TLV (175).

D IODINE AND ITS COMPOUNDS

15.0 Iodine

15.0.1 CAS Number: [7553-56-2]

15.0.2 Synonyms: NA

15.0.3 Trade Names: NA

15.0.4 Molecular Weight: 253.8090

15.0.5 Molecular Formula: I_2

15.0.6 Molecular Structure: I—I

15.1 Chemical and Physical Properties

Iodine is the heaviest of the halogens that are of industrial interest. Under ordinary conditions, iodine takes the form of gray-black plates or granules that have a metallic, crystalline luster. It volatilizes at room temperature to yield a sublimed, violet vapor. Iodine's physical properties are shown in Table 48.1.

Although iodine resembles other members of the halogen group, it is the least electronegative; it is thus the least chemically reactive of the halogens and forms the weakest bonds with more electropositive elements. Like other halogens, iodine unites with all elements except sulfur, selenium, and the noble gases. It reacts directly with most elements (except carbon, nitrogen, oxygen, and some of the more unreactive metals), and reacts with numerous organic compounds that are of pharmaceutical interest. Iodine also forms compounds with other halogens, such as iodine bromide and iodine monochloride; these interhalogen compounds are used in organic synthesis (131). Iodine occurs in valence states from 1 to 7 (223). The most stable of the positive iodine compounds are the iodates, in which the valence of iodine is +5, and the periodates, in which it is +7. Thirty isotopes of iodine have been identified, although only one, 127I, occurs in nature (2).

15.2 Production and Use

Iodine is the 47th most abundant element in the earth's crust (133). The name iodine derives from the Greek word for violet-colored, *ioeides*, which was used to describe the purple vapor generated by heating iodine (133).

The major U.S. sources of iodine are natural and oil-field brines, such as those near Shreveport, Louisiana (oil field), in the Los Angeles basin of California, and in the natural brines of Midland, Michigan and Woodward, Oklahoma. Iodine occurs in trace quantities in seawater and igneous rocks (223).

Large iodine resources exist in foreign countries. Japan's natural gas-well brines are credited with as much as four-fifths of the world's iodine reserve. An estimate of the indicated reserve of Chilean nitrate minerals, from which iodine is obtained as a by-product, is about 1 billion short tons ($\sim 0.04\%$ I_2). Unmeasured quantities of iodine are contained in brines in Indonesia, Germany, France, Italy, the United Kingdom, Norway, Ireland, and the former Soviet Union (297). In the United States, the principal method used to recover iodine from oil brines involves the oxidation of iodide by chlorine, followed by removal of the volatile iodine from solution with an airstream. The iodine is reabsorbed in solution and reduced to hidrotic acid with sulfur dioxide. The solution is then chlorinated to precipitate free iodine, which is further purified by treatment with concentrated sulfuric acid. The same process is used to recover iodine from natural brines. In the recovery of iodine from Chilean nitrate deposits, solutions containing the iodates are reduced with sodium bisulfite to precipitate the iodine, which is then purified by sublimation.

In 1984, an estimated 500 trillion lb of crude or resublimed iodine products were produced in the United States, and more than 2 billion lb were imported in 1985 (263). Approximately 22% of this total was used in organic synthesis, and another 20% found use in pharmaceutical applications. Other uses, and their percentages of total consumption, were as follows in 1984: animal feed supplements (18%), sanitary and industrial disinfectants (12%), stabilizers (11%), inks and colorants (6%), photographic chemicals (5%), and miscellaneous uses (6%) (263).

Iodine is used both in animal and human medicine, where its disinfectant and antiseptic properties are valued. The lack of iodine causes goiter (compensatory hypertrophy of the thyroid gland), and iodine is used both to treat iodine deficiency and hyperthyroidism (131). Table 48.18 shows the principal iodine compounds and their industrial uses. NIOSH has estimated that in 1983, 204,902 workers were exposed to iodine (7).

16.0 Inorganic Compounds

The inorganic iodine compounds of commercial interest, and their physical properties, are shown in Table 48.19. The iodides, an important class of inorganic iodine compounds, have less tendency to form complexes than the other halides. Chlorine and bromine freely displace iodine from the iodides (133). Iodine forms industrially useful and important compounds with hydrogen, metals, the other halogens, and oxygen. The ones presented below are typical.

16.0a Hydrogen Iodide

Hydrogen iodide [HI (hidrotic acid)] gas dissolves in water at 10°C and 1 atm pressure to the extent of 70 wt% to form hidrotic acid. The technical grade, 47% HI, is a highly corrosive liquid that fumes in moist air. Its solutions, like others containing the iodide ion, can dissolve in large quantities of iodine (e.g., tincture of iodine, $KI\acute{u}I_2$). Hidrotic acid is

THE HALOGENS

Table 48.18. Principal Iodine Compounds and Their Uses[a]

Compound	Principal Uses
Sodium iodide	Photography, organic chemical cloud seeding, iodized salt, medication, wet extraction of silver, feed additive, laboratory reagent
Potassium iodide	Photographic emulsions, treatment of radiation accidents, laboratory analysis, animal and poultry feed
Hidrotic acid	Reducing agent, manufacture of organic iodides, laboratory analysis, disinfectant
Potassium iodate	Oxidizing agent in chemical analysis, topical antiseptic, nutrient iodine source
Iodine chloride	Organic synthesis, laboratory analysis
Iodine monobromide	Organic synthesis
Iodine trichloride	Chlorinating agent, oxidizing agent, topical antiseptic
Iodine pentafluoride	Fluorinating agent, component of explosives
Hydrogen iodide	Manufacture of hidrotic acid and organic iodo compounds

[a]From Refs. 2, 131, 222.

more stable than the gas and is one of the strongest acids; it dissolves metals, oxides, carbonates, and salts of the other weak, monoxidizing acids, causing the formation of iodides. The major uses of hidrotic acid are shown in Table 48.18.

16.0b Iodides

Of the commercially available iodides, potassium iodide (KI) is the most important. All iodides are highly water-soluble, stable, and high melting solids (Table 48.19). The tetraiodides of titanium and zirconium, however, decompose to their elements at elevated temperatures, yielding very high purity zirconium metal. According to NIOSH (7), there were 243,989 workers exposed to potassium iodide in 1983.

16.0c Iodates and Periodates

The iodates and periodates are among the most stable and well known of the iodine compounds. Except for the salts of the alkali metals, most iodates are sparingly soluble in water. Both iodates and periodates are powerful oxidizers in acid solution and are thus used as disinfectants. Other uses are as feed additives and in medicine. Approximately 193,692 workers were exposed to potassium iodate in 1983; the corresponding figure for sodium iodate is 8972 (7).

16.0c Iodine Halides

The four iodine halides of interest, iodine monobromide, iodine chloride, iodine trichloride, and iodine pentafluoride, are shown in Table 48.19.

16.1 Chemical and Physical Properties

Table 48.19 show their physical properties. All are high density substances.

Table 48.19. Properties of Industrially Useful Inorganic Iodine Compounds

Compound	CAS Number	Formula	Mol. Wt.	Physical Properties	Sp. Gr. (°C)	MP (°C)	BP (°C)	Solubility
Hydrogen iodide (hidrotic acid)	[10034-85-2]	HI	127.91	Colorless gas	Gas 5.66^0	−50.8	−35.88	425 cm^3/L H$_2$O, 0°C; soluble alcohol
Ammonium iodide	[12027-06-4]	NH$_4$I	144.94	Colorless, cubic, hygr.	2.514^{25}	Subl. 551	220 vac.	1.542 kg/L H$_2$O, 0°C; 2.503 kg/L, H$_2$O, 100°C; very soluble alcohol, acetone, NH$_3$
Potassium iodide	[7681-11-0]	KI	166.01	Colorless or white, cubic	3.13	681	1330	1.275 kg/L H$_2$O, 0°C; 2.08 kg/L, H$_2$O, 100°C; 18.8 g/L, alcohol, 25°C; slightly soluble ether
Sodium iodide	[7681-82-5]	NaI	149.89	Colorless cubic	3.667^{25}	661	1304	1.84 kg/L H$_2$O, 25°C; 3.02 kg/L H$_2$O, 100°C; 425 g/L, 25°C; sol. alcohol
Iodic acid	[7782-68-5]	HIO$_3$	175.91	Colorless or pale yellow cryst.	4.629^0	Dec. 110	—	2.8 kg/L H$_2$O, 0°C; 4.73 kg/L H$_2$O, 80°C; very soluble alcohol 87%
Potassium iodate	[7758-05-6]	KIO$_3$	214.0	Colorless, monoclinic	3.93^{32}	560	Dec.->100	47.4 g/L H$_2$O, 0°C; 323 g/L H$_2$O, 100°C; soluble KI; insol. alcohol
Sodium iodate	[7681-55-2]	NaIO$_3$	197.89	White, rhombic	4.277$^{17.5}$	Dec.	—	90 g/L H$_2$O, 20°C; 340 g/L H$_2$O, 100°C; insoluble. alcohol; soluble acetic acid
Periodic acid	[10450-60-9]	H$_5$IO$_6$	191.91	Colorless	—	Subl. 110 (42) subl. 50	Dec. 138	Very soluble (decomposes) cold H$_2$O
Iodine bromide	[7789-33-5]	IBr	206.81	Dark gray cryst.	4.4157^0		Dec. 116	Soluble (dec.) cold H$_2$O; soluble alcohol, ether, chloroform, CS$_2$
Iodine chloride	[7790-99-0]	ICl	162.36	Dark red needles	3.1822^0	27.2	97.4	Decomposes to HIO$_3$ cold H$_2$O; soluble alcohol, ether, CS$_2$, HCl
Iodine tribromide	—	IBr$_3$	366.63	Brown liquid	—	—	—	Soluble cold H$_2$O; soluble alcohol
Iodine pentafluoride	[7783-66-6]	IF$_5$	221.90	Colorless liquid	3.75	9.6	98	Decomposes cold H$_2$O; dec. acid, alkali

16.2 Production and Use

The chief use of these halides is in organic synthesis and as halogenation catalysts. Iodine pentafluoride is a fluorinating and incendiary agent. The reaction of iodine pentafluoride with organic substances must be carefully controlled because explosions may occur.

16.3 Exposure Assessment

Occupational exposure to airborne iodine can be measured using a charcoal tube impregnated with alkali-metal hydroxide. The iodine is desorbed using sodium nitrate, and analysis is conducted by ion chromatography (143). This method has a limit of detection of 0.01 ppm for a 7.5-L air sample. The overall precision of the method (coefficient of variation) is 0.11. At iodine concentrations of 0.5 times the OSHA PEL of 0.1 ppm, the overall error of the method is 27%, which is slightly above the maximum acceptable error of 25%. At iodine concentrations of 1 and 2 times the limit, overall error is acceptable.

The NIOSH method for sampling iodine (method 6005) is similar to that described above, except that sodium carbonate is used as the desorption agent (25). The estimated limit of detection for this method is 1 mg per sample, and the overall precision is slightly better (coefficient of variation 0.085) than is obtained from the use of sodium nitrate as the desorption agent.

16.4 Toxic Effects

Because iodine is an essential nutrient that is required for the healthy development and functioning of the thyroid gland, it has been the subject of numerous metabolic studies in human volunteers and experimental animals. However, information on the toxic effects of iodine and iodine compounds in the occupational setting is scarce.

In general, the vapor of iodine is reported to be more irritating than that of chlorine or bromine (129, 130), although case reports are few because iodine does not find wide industrial use. In one of the few available animal studies, dogs exposed to iodine vapor (concentration not specified) developed signs of pulmonary edema (129, 130). In humans, exposure to the vapor causes irritation of the eyes and eyelids, upper respiratory tract, lungs, and skin (129, 130).

Four volunteers exposed to a 0.57-ppm concentration of iodine vapor for 5 min reported no ill effects; at a concentration of 1.63 ppm, however, all volunteers reported experiencing eye irritation after 2 min (298). Exposures to the vapor have also been associated with headaches, lacrimation, chest tightness, and sore throat (129, 130). In a study of workers exposed to vapors emanating from a tank containing an iodine solution, concentrations of 1.0 ppm were reported to be 'highly' irritating (129,130), and early studies report that working in an atmosphere containing 0.15–0.2 ppm of the vapor was difficult and that work had to be discontinued when the concentration reached 0.3 ppm (129, 130).

Topical applications of iodine solutions have caused redness and inflammation; strong solutions may cause thermal burns (129, 130). There are reports of fatalities caused by skin absorption of iodine tinctures applied to wounds (298), and a fatality has been attributed to the use of an iodine solution to provide continuous postoperative irrigation of a hip wound (33). Two Dutch eel fishermen developed contact dermatitis attributed to the iodine in Japanese Sargasso weed in the local lake; both cases were confirmed by patch test (299).

Matt studied the effects of halogen gases on humans and animals. In a series of open-room exposure experiments he confirmed the known cutaneous, mucosal, and respiratory toxicities of the halogen gases, at relatively low concentrations. He estimated that workers could tolerate iodine exposures of 0.001%, but that concentrations of 0.005% could not be tolerated. He observed preferential irritation of the ocular membranes by iodine in this study (300).

Chronic absorption of iodine can lead to "iodism," a condition characterized by sleeplessness, tremor, rapid heart rate, diarrhea, weight loss, conjunctivitis, rhinitis, and bronchitis (298). However, this syndrome is usually associated with long-term ingestion of iodine-containing medications (expectorants, diuretics) rather than with occupational exposure (129, 130).

There is little information on the acute toxicity of iodine and its compounds. Table 48.20 shows the available data, as well as the signs and symptoms associated with exposure to each substance.

Table 48.20 Acute Toxicity and Exposure-Related Signs and Symptoms for Iodine and Some of Its Compounds[a]

Substance	Acute Toxicity Data (Route, Species)	Signs and Symptoms of Exposure
Iodine	Oral LD_{50}, rat: 14 g/kg Inhalation LC_{LO}, rat: 800 mg/m^3 for 1 hr	Tearing, nasal secretions, sore throat, chest tightness, headache, skin burns, rash, allergic dermatitis
Iodine monochloride	Oral LD_{LO}, rat: 50 mg/kg Dermal LD_{LO}, rat: 500 mg/kg	Severe eye, skin, and respiratory tract irritation; pulmonary edema; burns of the eye and skin
Potassium iodate	Oral LD_{LO}, guinea pig: 400 mg/kg	Dust causes eye, skin irritaion
Potassium iodide	Oral LD_{LO}, mouse: 1862 mg/kg	Eye, skin irritation; in animals, teratogenic effects
Sodium iodide	Oral LD_{LO}, rat: 4340 mg/kg	Eye, skin irritation; cough
Hydrogen iodide	—	Severe eye, nose, throat irritation; laryngeal and broanchial spasms (sometimes fatal); pulmonary edema; necrosis of skin
Iodine pentafluoride	—	Severe eye, nose, and throat irritation, bronchial and laryngeal edema
Hidrotic acid	—	Lacrimation, sore throat, difficult breathing, cough, edema; burns, blisters, and necrosis of the skin

[a]From Refs. 129, 130, 151, 223, and 298.

THE HALOGENS

The most significant concerns about iodine and iodine exposure have arisen from air contamination and food and waterborne .radioiodine (radioactive iodine) from nuclear releases. These releases from atmospheric nuclear testing, and from accidents such as that at Three Mile Island (Pennsylvania, USA) and Chernobyl (Russia) have profound implications for the occurrence of thyroid cancer in the surrounding populations. Drift of radioiodine and contamination of potable-water and food supplies at great distances from the source of the material is also a major concern. The kinetics of these exposures, and their impact is beyond the scope of this chapter, but in any discussion of iodine as an industrial poison, the radioiodine forms produced and released by all of the uses of nuclear energy should be considered.

16.5 Standards, Regulations, or Guidelines of Exposure

OSHA (174), NIOSH (175), and the ACGIH (22) all have established a ceiling limit of 0.1 part per million (ppm) for iodine. NIOSH has also calculated an IDLH value for this element of 2 ppm (21). No limits have been established for the principal compounds of iodine.

E ASTATINE

17.0 Astatine

17.0.1 CAS Number: [7440-68-8]

17.0.2 Synonyms: NA

17.0.3 Trade Names: NA

17.0.4 Molecular Weight: 210

17.0.5 Molecular Formula: At

17.1 Chemical and Physical Properties

The fifth halogen — astatine — is little known and has to date found few uses. The physical and chemical properties of astatine are as follows:

Atomic number: 85
Atomic weight: 210
Melting point: 302°C
Boiling point: 337°C (est.)

The name *astatine* derives from the Greek word for unstable, *astatos*. Astatine is believed to have four valence states: 1, 3, 5, and 7. Twenty-four isotopes have so far been identified, and the longest-lived of these, (301), At, has a half-life of only 8.1 h. The total amount of astatine in the earth's crust is believed to be less than 1 ounce (2).

Astatine behaves chemically in a manner much like iodine; however, it is believed to be more metallic than iodine. It is likely to accumulate in the thyroid gland, although this has not been demonstrated (2).

BIBLIOGRAPHY

1. National Institute for Occupational Safety and Health/Ocupational Safety and Health Administration (NIOSH/OSHA), *Occupational Health Guidelines for Chemical Hazards*, DHHS (NIOSH) Pub. No. 81-123, U.S. Department of Health and Human Services, U.S. Department of Labor, Washington, DC, 1981.
2. D. R. Lide, ed., *CRC Handbook of Chemistry and Physics*, 73rd ed., CRC Press, Boca Raton, FL, 1992-1993.
3. R. E. Banks and H. Goldwhite, in F. A. Smith, ed., *Handbook of Experimental Pharmacology*, Vol. 20, Part 1, Springer, New York, 1966.
4. R. E. Banks, *J. Fluorine Chem.* **33**, 3 (1986).
5. O. Glemser, *J. Fluorine Chem.* **33**, 45 (1986).
6. International Programme on Chemical Safety (IPCS), *Environmental Health Criteria 36: Fluorine and Fluorides*, World Health Organization, Geneva, 1984.
7. National Institute for Occupational Safety and Health (NIOSH), *National Occupational Exposure Survey*, U.S. Department of Health and Human Services, Washington, DC, 1983.
8. National Library of Medicine, *TOXNET: Hazardous Substances Data Bank*, Record No. 541: Fluorine, April, NLM, 1999 (on-line).
9. Agency for Toxic Substances and Disease Registry (ATSDR), *Toxicological Profile: Fluorides, Hydrogen Fluoride, and Fluorine*, ATSDR, Washington, DC, 1993.
10. National Institute for Occupational Safety and Health (NIOSH), *Manual of Analytical Methods*, 4th ed., Method 7906, Issue 1, NIOSH, Washington, DC, 1994.
11. C. Lorberau, *Appl. Occup. Environ. Hyg.* **8**(9), 775 (1993).
12. C. Lorberau and K. J. Mulligan, *Appl. Ind. Hyg.* **3**(11), 302 (1988).
13. M. L. Keplinger and L. W. Suissa, *Am. Ind. Hyg. Assoc. J.* **29**, 10 (1968).
14. H. E. Stokinger, in C. Voegtlin and H. C. Hodge, eds., *Pharmacology and Toxicology of Uranium Compounds, with a Section on the Pharmacology and Toxicology of Fluorine and Hydrogen Fluoride*, Vol. 2, Natl. Nucl. Energy Ser., Manhattan Proj. Tech. Sect., Div. VI, Vol. 1, McGraw-Hill, New York, 1949.
15. M. L. Keplinger, *Toxicol. Appl. Pharmacol.* **14**, 192 (1969).
16. J. S. Lyon, *J. Occup. Med.* **4**, 199 (1962).
17. G. Benke, M. Abrahamson, and M. Sim, *Ann. Occup. Hyg.* **42**(3), 173 (1998).
18. R. P. Rickey, *USAF Flight Training Command Tech. Rep.* **59-21** (1959), cited in Ref. 6.
19. F. Belles et al., eds., *Fluoride Handbook*, NASA Lewis Research Center, Cleveland, OH 1965.
20. H. E. Stokinger, in G. D. Clayton and F. E. Clayton, eds., *Patty's Industrial Hygiene and Toxicology*, 3rd ed., Vol. 2B, Wiley, New York, 1981.
21. National Institute for Occupational Safety and Health (NIOSH), *Pocket Guide to Chemical Hazards*, DHHS (NIOSH) Publ. No. 97-140, U.S. Department of Health and Human Services, Public Health Service, Centers for Disease Control, Washington, DC, 1997.

92. M. Chan-Yeung et al., *Am. Rev. Respir. Dis.* **127**, 465 (1983).
93. M. Chan-Yeung et al., *Arch. Environ. Health* **38**, 35 (1983).
94. M. Saric et al., *Am. J. Ind. Med.* **9**, 239 (1986).
95. A. Eklund et al., *Br. J. Ind. Med.* **46**, 782 (1989).
96. N. L. Kaltreider et al., *J. Occup. Med.* 14, 531 (1972).
97. National Academy of Sciences (NAS), *Fluorides, National Academy of Sciences Committee on Biologic Effects of Atmospheric Pollutants*, National Research Council, Division of Medical Sciences, Washington, DC, 1971.
98. K. Roholm, *Fluorine Intoxication. A Clinical-Hygienic Study with a Review of the Literature and Some Experimental Investigations*, H. K. Lewis, London, 1937.
99. H. Friis, J. Clausen, and F. Gyntelberg, *Occup. Med.* **39**, 133 (1989).
100. J. Jahr et al., in *Light Metals*, Proc. 103rd AIME Annu. Meet., 1974, Metallurgical Society of AIME, New York, 1974.
101. G. Thériault, S. Gingras, and S. Provencher, *Br. J. Ind. Med.* **41**, 367 (1984).
102. G. Thériault, S. Cordier, and S. Harvey, *N. Engl. J. Med.* **303**, 1278 (1980).
103. R. S. U. Baker et al., *J. Appl. Toxicol.* **6**, 357 (1986).
104. X. Ying-Han et al., *Environ. Res.* **1**, 350 (1987).
105. J. Kongerud, J. K. Gronnesby, and P. Magnus, *Scand. J. Work Environ. Health* **16**, 270 (1990).
106. V. Soyseth and J. Kongerud, *Br. J. Ind. Med.* **49**, 125 (1992).
107. B. G. Simonsson et al., *Eur. J. Respir. Dis.* **66**, 105 (1985).
108. D. P. Discher and B. D. Breitenstein, *J. Occup. Med.* **18**, 379 (1976).
109. V. Soyseth et al., *Int. Arch. Occup. Environ. Health* **67**(1), 53 (1995).
110. F. Akbar-Khanzadeh, *Am. Ind. Hyg. Assoc. J.* **56**(10), 1008 (1995).
111. T. V. O'Donnell, *Sci. Total Environ.* **163**, 137 (1995).
112. F. J. Tourangeau, *Lav. Med.* **9**, 548 (1944).
113. C.-A. Baud et al., *Virchow's Arch. A: Pathol. Anat. Histol.* **380**, 283 (1978).
114. M. A. Boillat et al., *Schweiz. Med. Wochenschr.* **109**, (Suppl. 8), 5 (1979).
115. M. A. Boillat, J. Garcia, and L. Velebit, *Skeletal Radiol.* **5**, 161 (1980).
116. E. Czerwinski et al., *Arch. Environ. Health* **43**, 340 (1988).
117. L. Nemeth and E. Zsgn, *Baillieres Clin. Rheumatol.* **3**, 81 (1989).
118. P. Grandjean and G. Thomsen, *Br. J. Ind. Med.* **40**, 456 (1983).
119. C. Schlatter and A. Steinegger, *Soz.- Preventivmed.* **33**, 112 (1988).
120. J. Jost, G. Rudaz, and B. Liecht, in *Soz.- Preventivmed.* **33**, 112 (1988).
121. H. C. Hodge and F. A. Smith, *J. Air Pollut. Control Assoc.* **20**, 226 (1970).
122. National Institute for Occupational Safety and Health (NIOSH), *Criteria for a Recommended Standard: Occupational Exposure to Inorganic Fluorides*, HEW Publ. No. (NIOSH) 76-103, U.S. Department of Health, Education and Welfare, Public Health Service, Centers for Disease Control, Cincinnati, OH, 1975.
123. U. Hjortsberg et al., *Scand. J. Work Environ. Health* **12**, 223 (1986).
124. D. Rees, D. B. K. Rama, and V. Yousefi, *Am. J. Ind. Med.* **17**, 311 (1990).
125. D. B. K. Rama and V. Yousefi, *S. Afr. Med. J.* **77**, 112 (1990).
126. S. Levi et al., *Am. J. Ind. Med.* **9**, 153 (1986).

127. K. Kono et al., *Int. Arch. Occup. Environ. Health* **62**, 85 (1990).
128. P. Grandjean, J. H. Olsen, and K. Juel, *Scand. J. Work Environ. Health* **19** (Suppl. 1), 108 (1993).
129. American Conference of Governmental and Industrial Health (ACGIH), *Documentation of the Threshold Limit Values and Biological Exposure Indices*, 5th ed., ACGIH, Cincinnati, OH, 1986.
130. American Conference of Governmental and Industrial Health (ACGIH), *Documentation of the Threshold Limit Values and Biological Exposure*, 6th ed., ACGIH, Cincinnati, OH, 1991.
131. R. J. Lewis, *Hawley's Condensed Chemical Dictionary*, 12th ed., Van Nostrand-Reinhold, New York, 1993.
132. National Fire Protection Association, (NFPA), *Fire Protection Guide on Hazardous Materials*, 9th ed., NFPA, Quincy, MA, 1986.
133. *Kirk-Othmer's Concise Encyclopedia of Chemical Technology*, 3rd ed., Wiley, New York, 1985.
134. *Directory of Chemical Producers*, SRI International, Menlo Park, CA, 1992.
135. R. E. Bluhm and R. A. Branch, *Int. Arch. Occup. Environ. Health* **68**, 421 (1996).
136. R. E. Bluhm et al., *Hum. Exp. Toxicol.* **11**, 201 (1992).
137. R. Matiussi, G. Armeli, and V. Bareggi, *Am. J. Ind. Med.* **3**, 335 (1982).
138. G. Sallsten, L. Barregard, and B. Jarvholm, *Ann. Occup. Hyg.* **34**, 205 (1990).
139. U.S. Environmental Protection Agency (USEPA), *Ambient Water Quality Criteria Document: Chlorine*, EPA No. 440/3-78-005, USEPA, Cincinnati, OH, 1981.
140. A. W. Kotula et al., *J. Toxicol. Environ. Health* **20**, 401 (1987).
141. National Library of Medicine, *TOXNET: Hazardous Substances Data Bank:* Chlorine, August, NLM, 1999 (on-line).
142. A. R. Kavaler, *Chem. Mark. Rep.* **235**(24), 50 (1989).
143. Occupational Safety and Health Administration (OSHA), *Analytical Methods Manual*, U.S. Department of Labor, OSHA Analytical Laboratory, Salt Lake City, UT, 1985.
144. K. C. Back, A. A. Thomas, and J. D. MacEwen, *Reclassification of Materials Listed as Transportation Health Hazards*, Rep. No. TSA-20-72-3, Aerospace Medical Research Laboratory, Wright-Patterson AFB, OH, 1972.
145. R. E. Barrow and R. G. Smith, *Am. Ind. Hyg. Assoc. J.* **36**, 398 (1975).
146. C. S. Barrow et al., *Arch. Environ. Health* **32**, 68 (1977).
147. C. S. Barrow, R. J. Kociba, and L. W. Rampy, *Toxicol. Appl. Pharmacol.* **45**, 290 (1978).
148. R. Demnati et al., *Eur. Respir. J.* **11**, 922 (1998).
149. C. S. Barrow and W. N. Steinhagen, *Toxicol. Appl. Pharmacol.* **65**, 383 (1982).
150. M. J. Ellenhorn and D. G. Barceloux, *Medical Toxicology: Diagnosis and Treatment of Human Poisoning*, Elsevier, New York, 1988. p. 878.
151. National Institute for Occupational Safety and Health (NIOSH), *Registry of Toxic Effects of Chemical Substances for 1992*, on-line data system available from the National Library of Medicine, Bethesda, MD, 1992.
152. Freitag, *Z. Gesamte Schiess-Spregstoffwes.* (1941), through National Research Council (153).
153. National Research Council (NRC), *Emergency and Continuous Exposure Limits for Selected Airborne Contaminants*, Vol. 2: Chlorine, National Academy Press, Washington, DC, 1984.
154. H. L. Gilchrist and P. B. Matz, *Med. Bull. Vet. Admin.* **9**, 229 (1933).

155. National Institute for Occupational Safety and Health (NIOSH), *Criteria for a Recommended Standard: Occupational Exposure to Chlorine*, HEW Publ. No. (NIOSH) 76-170, U.S. Department of Health, Education and Welfare, Public Health Service, Centers for Disease Control, Washington, DC, 1976.
156. Y. Ploysongsang, B. C. Beach, and R. E. DiLiso, *South. Med. J.* **75**, 23 (1982).
157. T. A. Kowitz et al., *Arch. Environ. Health* **14**, 545 (1967).
158. C. Leroyer et al., *Occup. Environ. Med.* **55**, 356 (1998).
159. B. Schonhofer, T. Voshaar, and D. Kohler, *Respiration* **63**, 155 (1996).
160. B. Rivoire et al., *Rev. Mal. Respir.* **12**, 471 (1995).
161. D. J. Ross and J. C. McDonald, *Ann. Occup. Hyg.* **40**, 645 (1996).
162. M. J. Reilly and K. D. Rosenman, *Arch. Environ. Health* **50**, 26 (1995).
163. R. L. Zielhuis, *Ann. Occup. Hyg.* **13**, 171 (1970).
164. D. M. Anglen, Doctoral Dissertation, University of Michigan, Ann Arbor, 1981, through ACGIH (129, 130).
165. H. H. Rotman et al., *J. Appl. Physiol.* **54**, 1120 (1983).
166. A. D'Alessandro et al., *Chest* **109**, 331 (1996).
167. J. E. Amoore and E. Hautala, *J. Appl. Toxicol.* **3**, 272 (1983).
168. W. J. Meggs et al., *J. Toxicol. Clin. Toxicol.* **34**, 383 (1996).
169. L. R. S. Patil et al., *Am. Ind. Hyg. Assoc. J.* **31**, 678 (1970).
170. E. Capodaglio et al., *Med. Lav.* **60**, 192 (1970).
171. F. Drobnic et al., *Med. Sci. Sports Exercise.* **28**, 271 (1996).
172. L. Fjellbirkeland, A. Gulsvik, and A. Walloe, *Tidsskr. Nor. Laegeforen.* **115**, 2051 (1995).
173. D. R. Klonne et al., *Fundam. Appl. Toxicol.* **9**, 557 (1987).
174. *Code of Federal Regulations*, Title 29 (Department of Labor), Part 1910.1000, U.S. Government Printing Office, Office of the Federal Register, Washington, DC.
175. National Institute for Occupational Safety and Health (NIOSH) *Recommendations for Occupational Safety and Health: Compendium of Policy Documents and Statements*, DHHS (NIOSH) Publ. No. 92-100, U.S. Department of Health and Human Services, Cincinnati, OH, 1992.
176. Occupational Safety and Health Administration (OSHA) *Chemical Information Manual*, Instruction CPL 2-2.43, U.S. Department of Labor, Washington, DC, 1987.
177. K. I. Darmer, E. R. Kinkead, and L. C. DiPasquale, *Am. Ind. Hyg. Assoc. J.* **35**, 623 (1974).
178. L. A. Buckley et al., *Toxicol. Appl. Pharmacol.* **74**, 417 (1984).
179. H. L. Kaplan et al., *J. Fire Sci.* **3**, 228 (1985).
180. National Research Council, in *Emergency and Continuous Exposure Guidance Levels for Selected Airborne Contaminants*, Vol. 7: *Hydrogen Chloride*, National Academy Press, Washington, DC, 1987.
181. H. L. Kaplan, *Toxicologist* **6**, 52 (1986).
182. Y. Henderson and H. W. Haggard, *Noxious Gases*, Reinhold, New York, 1943, p. 126.
183. T. E. Pavlova, *Bull. Exp. Biol. Med. (Engl. Transl.)* **82**, 1078 (1976).
184. A. R. Sellakumar et al., *Toxicol. Appl. Pharmacol.* **81**, 401 (1985).
185. H. B. Elkins, *The Chemistry of Industrial Toxicology*, Wiley, New York, 1959, p. 79.

186. F. F. Heyroth, in D. W. Fassett and D. E. Irish, eds., *Patty's Industrial Hygiene and Toxicology*, 2nd ed., Vol. 2, Wiley-Interscience, New York, 1963.
187. M. A. Kamrin, *Regul. Toxicol. Pharmacol.* **15**, 73 (1992).
188. R. F. Dyer and V. H. Esch, *J. Am. Med. Assoc.* **235**, 393 (1976).
189. W. Ludewig, *Arch. Gewerbepathol. Gewerbehyg.* **11**, 296 (1942).
190. B. Remijn et al., *Ann. Occup. Hyg.* **25**, 299 (1982).
191. J. J. Collins et al., *J. Occup. Med.* **31**, 614 (1989).
192. J. J. Beaumont et al., *J. Natl. Cancer Inst.* **79**, 911 (1987).
193. K. Steenland et al., *Br. J. Ind. Med.* **45**, 766 (1988).
194. G. G. Bond et al., *J. Occup. Med.* **25**, 377 (1983).
195. G. G. Bond et al., *Am. J. Ind. Med.* **7**, 123 (1985).
196. G. G. Bond et al., *Am. J. Epidemiol.* **124**, 53 (1986).
197. G. G. Bond et al., *J. Occup. Med.* **33**, 958 (1991).
198. J. Siemiatycki, ed., *Risk Factors for Cancer in the Workplace*, CRC Press, Boca Raton, FL, 1991.
199. International Agency for Research on Cancer (IARC), *IARC Monogr. Eval. Carcinog. Risk Chem. Hum.* **54**, 189 (1992).
200. G. H. Gurtner et al., *Development and Testing of Treatments for Battlefield Phosgene Poisoning, New York Medical College, Valhalla*, Contract DAMD17-94-C-4043, 7, NTIS/AD-A299 739/3, Government Reports Announcements and Index, Washington, DC, 1996.
201. W. S. Beckett, in P. Harber, M. B. Schenker, and J. R. Balmes, eds., *Occupational and Environmental Respiratory Disease*, Mosby, St. Louis, MO, 1996, pp. 704-717.
202. H. S. Lee et al., *Singapore Med. J.* **31**, 506 (1990).
203. M. Gochfed, *Mt. Sinai J. Med.* **62**, 365 (1995).
204. P. J. Nigro and W. B. Bunn, in P. Harber, M. B. Schenker and J. R. Balmes, eds., *Occupational and Environmental Respiratory Disease*, Mosby-Year Book, St. Louis, MO, 1996.
205. J. P. Wyatt and C. A. Allister *J. Accid. Emerg. Med.* **12**, 212 (1995).
206. S. C. Lim et al., *Korean J. Intern. Med.* 11, 87 (1996).
207. L. Wang, X. Ding, and H. Zhang, *Chung-hua Yu Fang I Hsueh Tsa Chih* **32**, 235 (1998).
208. National Library of Medicine, *TOXNET: Hazardous Substances Data Bank*, Phosgene, August, NLM, 1999 (on-line).
209. M. Sittig, *Handbook of Toxic and Hazardous Chemicals and Carcinogens*, 2nd ed., Noyes Publications, Park Ridge, NJ, p. 714.
210. N. Nakano et al., *Talanta* **42**, 641 (1995).
210a. *OSHA Analytical Methods Manual*, U.S. Government Printing Office, Washington, DC, 1990, 1993.
211. U. P. Kodavanti et al., *Fundam. Appl. Toxicol.* **37**, 54 (1997).
212. International Programme on Chemical Safety (IPCS), *Environmental Health Criteria 193: Phosgene*, World Health Organization, Geneva, 1997.
213. Y. G. Yang et al., *Inhalation Toxicol.* **7**, 393 (1995).
214. G. P. Williams et al., *Potential Health Impacts from Range Fires at Aberdeen Proving Ground, Maryland*, ANL/EAD/TM-79, Contract W-31109-ENG-38, Gov. Rep. Announce. Index, 09,

1999, NTIS/DE98004910, Department of Defense, Department of Energy, Washington DC, 1999.
215. R. T. Boswell and R. J. McCunney, *J. Occup. Environ. Med.* **37**, 850 (1995).
216. W. D. Currie, *Attenuation of Phosgene Toxicity,* Duke University Medical Center, Durham, NC., *18*, 1996 NTIS/AD-A306 936/6, Government Reports Announcements and Index, Washington, DC, 1996.
217. A. M. Sciuto et al., *Am. J. Respir. Crit. Care Med.* **151**, 768 (1995).
218. A. M. Sciuto et al., *Exp. Lung Res.* **23**, 317 (1997).
219. R. Frank, in J. A. Merchant, ed., *Occupational Respiratory Diseases*, DHHS (NIOSH) Publ. No. 86-102, Division of Respiratory Disease Studies, Appalachian Laboratory for Occupational Safety and Health, NIOSH, U. S. Department of Health and Human Services, Washington, DC, 1986, pp. 571–605.
220. F. Faucher, G. Carrier, and J. C. Panisset, *J. Hazard. Mater.* **45**, 141 (1996).
221. D. A. Salisbury et al., *Am. J. Ind. Med.* **20**, 71 (1991).
222. M. Windholtz, ed., *The Merck Index: An Encyclopedia of Chemicals, Drugs, and Biologicals*, 10th ed., Merck & Co., Rahway, NJ, 1983.
223. P. Patnaik, *A Comprehensive Guide to the Hazardous Properties of Chemical Substances*, Van Nostrand-Reinhold, New York, 1992.
224. H. Petry, *Arch. Gewerbepathol. Gewerbehyg.* **13**, 363 (1954).
225. J. Gloemme and K. D. Lundgren, *Arch. Ind. Health* **16**, 169 (1957).
226. B. G. Ferris et al., *Br. J. Ind. Med.* **24**, 26 (1964).
227. S. Kanitz et al., *Environ. Health Perspect.* **104**, 516 (1996).
228. T. Dalhamm, *Arch. Ind. Health* **15**, 101 (1957).
229. R. P. Smith and C. C. Willhite, *Regul. Toxicol. Pharmacol.* **11**, 42 (1990).
230. G. Astrakianakis et al., *Am. Ind. Hyg. Assoc. J.* **59**, 694 (1998).
231. F. N. Dost et al., *Toxicol. Appl. Pharmacol.* **27**, 527 (1974).
232. D. R. Cloyd and W. J. Murphy, *Handling of Hazardous Materials: Technology Survey*, National Aeronautics and Space Administration, Washington, DC, 1965.
233. National Research Council, in *Emergency and Continuous Exposure Limits for Selected Airborne Contaminants*, Vol. 2: *Chlorine Trifluoride*, National Academy Press, Washington, DC, 1984.
234. H. J. Horn and R. J. Weir, *AMA Arch. Ind. Health* **12**, 515 (1955).
235. H. J. Horn and R. J. Weir, *AMA Arch. Ind. Health* **13**, 340 (1956).
236. M. S. Weinberg and R. E. Goldhamer, *Pharmacology, Toxicology, and Metabolism of ClF5*, AMRL-TR-66-238, Aerospace Medical Research Laboratory, Wright-Patterson AFB, OH, 1966.
237. L. Bretherick, *Handbook of Reactive Chemical Hazards*, 3rd ed., Butterworth, London, 1985.
238. J. P. Gibbs et al., *J. Occup. Environ. Med.* **40**, 1072 (1998).
239. S. H. Lamm et al., *J. Occup. Environ. Med.* **41**, 248 (1999).
240. P. E. Jackson, M. Laikhtman, and J. S. Rohrer, *Abstr. Pap., 216th Nat. Meet. Am. Chem. Soc.*, Boston, MA, Anyl. 103, 1998.
241. R. H. Hoy, *Am. J. Hosp. Pharm.* **38**, 1512 (1981).
242. G. E. Michael et al., *Arch. Environ. Health* **36**, 20 (1981).

243. J. R. Lubbers and J. R. Bianchine, *J. Environ. Pathol. Toxicol. Oncol.* **5**, 215 (1984).
244. J. R. Lubbers et al., *J. Environ. Pathol. Toxicol. Oncol.* **5**, 229 (1984).
245. R. W. Tuthill et al., *Environ. Health Perspect.* **46**, 39 (1982).
246. W. P. Heffernan, C. Guion, and R. J. Bull, *J. Environ. Pathol. Toxicol.* **2**, 1487 (1979).
247. J. Musil et al., *Technol. Water* **8**, 327 (1964).
248. International Agency for Research on Cancer (IARC), in *IARC Monographs on the Evaluation of Carcinogenic Risks to Humans*, Vol. 52, World Health Organization, Geneva, 1991, p. 145.
249. Y. Yokose et al., *Environ. Health Perspect.* **76**, 205 (1987).
250. G. S. Moore and E. J. Calabrese, *J. Environ. Pathol. Toxicol.* **4**, 513 (1980).
251. B. D. Carlton et al., *Environ. Res.* **42**, 238 (1987).
252. D. Couri et al., *Environ. Health Perspect.* **46**, 25 (1982).
253. H. M. Cunningham, *Am. J. Vet. Res.* **41**, 295 (1980).
254. I. J. Fidler, Nature. (London) **270**, 735 (1977).
255. I. J. Fidler et al., *Cancer Res.* **42**, 496 (1982).
256. J. H. Exon et al., *Toxicology* **44**, 257 (1987).
257. F. Furukawa et al., Berll. *Natl. Inst. Hyg. Sci.* **98**, 62 (1980).
258. R. Hasegawa et al., *Food Chem. Toxicol.* **24**, 1295 (1986).
259. Y. Kurokawa et al., *Environ. Health Perspect.* **69**, 211 (1986).
260. H. Hayatsu, H. Hoshino, and Y. Kawazoe, *Nature* **233**, 495 (1971).
261. Y. Kurokawa et al., *Cancer Lett.* **24**, 299 (1984).
262. International Agency for Research on Cancer (IARC), in *IARC Monographs on the Evaluation of Carcinogenic Risks to Humans*, Vol. 52, World Health Organization, Geneva, 1991, p. 159.
263. *Hazardous Substances Data Bank*, on-line data system available from the National Library of Medicine, National Institutes for Health, Bethesda, MD, 1992.
264. F. Flury and F. Zernik, *Schilddliche Gase*, Springer, Berlin, 1931, p. 538, through ACGIH, (130,131).
265. C. Parmeggiani, ed., *Encyclopedia of Occupational Health and Safety*, 3rd ed., Vol. 1, International Labour Organization, Geneva, 1983, p. 327.
266. M. Oppenheim, *Wien. Klin. Wochenschr.* **28**, 1273 (1915), through Stokinger (20).
267. K. B. Lehmann and R. Hess, *Arch. Hyg.* **7**, 335 (1887).
268. M. D. Bitron and E. F. Aharonson, *Am. Ind. Hyg. Assoc. J.* **39**, 129 (1978).
269. National Research Council, *Prudent Practices for Handling Hazardous Chemicals in Laboratories*, National Academy Press, Washington, DC, 1981.
270. W. D. Shade and M. A. Jaycock, *Am. Ind. Hyg. Assoc. J.* **58**, 418 (1997).
271. M. Eldan et al., *J. Occup. Environ. Med.* **38**, 1026 (1996).
272. Connecticut State Department of Health, Hartford, unpublished data, 1955, through H. E. Stokinger (20).
273. J. P. Orlando et al., *Rev. Pneumol. Clin.* **53**, 339 (1997).
274. F. B. Flinn, *J. Lab. Clin. Med.* **26**, 1325 (1941).
275. W. Deichmann and H. Gerarde, *Toxicity of Drugs and Chemicals*, Academic Press, New York, 1969, through Stokinger (20).

276. R. Sremark, *Acta Physiol. Scand.* **50**, 119 (1960).
277. R. Sremark, *Acta Physiol. Scand.* **50**, 306 (1960).
278. R. R. Grayson, *Am. J. Med.* **28**, 397 (1960).
279. R. Eliaz et al., *J. Burn Care Rehabil.* **19**, 18 (1998).
280. D. A. Fitzgerald et al., *Contact Dermatitis* **33**, 53 (1995).
281. M. J. Burns and C. H. Linden, *Chest* **111**, 816 (1997).
282. I. Dunsky, *Am. J. Dis. Child.* **74**, 730 (1947).
283. N. Ohashi et al., *Jpn. J. Urol.* **62**, 639 (1971).
284. A. De Vriese, R. Vanholder, and N. Lameire, *Nephrol. Dial. Transplant.* **12**, 204 (1997).
285. I. Matsumoto, *Otol. Fukuoka* **19**, 220 (1973), through International Agency for Research on Cancer (286).
286. International Agency for Research on Cancer (IARC), in *IARC Monographs on the Evaluation of Carcinogenic Risks of Chemicals to Humans*, Vol. 40, World Health Organization, Geneva, 1986.
287. C. G. Santesson and G. Wickberg, *Skand. Arch. Physiol.* **30**, 337 (1913).
288. T. Tona et al., *Adv. Oto-Rhino-Laryngol.* **52**, 315 (1997).
289. V. Wang et al., *J. Neurol., Neurosurg. Psychiatry* **58**, 516 (1995).
290. N. Fisher et al., *Food Cosmet. Toxicol.* **17**, 33 (1979).
291. A. V. Ginocchio et al., *Food Cosmet. Toxicol.* **17**, 41 (1979).
292. Y. Kurokawa et al., *J. Natl. Cancer Inst.* **71**, 965 (1983).
293. Y. Kurokawa et al., *Gann* **74**, 607 (1983).
294. A. B. DeAngelo et al., *Toxicol. Pathol.* **26**, 587 (1998).
295. T. Umemura et al., *Arch. Toxicol.* **72**, 264 (1998).
296. J. K. Chipman et al., *Toxicology* **126**, 93 (1998).
297. *Mineral Facts and Problems*, Bur. Mines Bull. 667, U.S. Department of the Interior, Washington, DC, 1975.
298. G. J. Hathaway et al., *Proctor and Hughes' Chemical Hazards of the Workplace*, 3rd ed., Van Nostrand-Reinhold, New York, 1991.
299. A. H. van der Willigen et al., *Contact Dermatitis* **18**, 250 (1988).
300. L. Matt, *Inaugural Dissertation for Achieving a Degree in Medicine, Surgery and Obstetrics*, Hygiene Institute, Wurzburg University, Germany, Julius Waldkirch Publishing, Ludwigshafen am Rhine, 1989.
301. J. D'Auria, S. Lipson, and J. M. Garfield, *J. Trauma* **30**, 353 (1990).

Subject Index

ACGIH. *See* American Conference of Governmental Industrial Hygienists
Acid mist. *See* Sulfuric acid
AIHA. *See* American Industrial Hygiene Association
Alcohol of sulfur. *See* Carbon disulfide
Alkaline materials, 583–604
　earth metals, 782–785
　toxic effects, and safety, 583–584
Alnico. *See* Nickel
Altacide. *See* Sodium chlorate
Amchlor. *See* Ammonium chloride
American Conference of Governmental Industrial Hygienists (ACGIH), exposure standards
　boric acid, 552
　boric oxide, 554
　boron, 543–544
　boron tribromide, 563
　boron trifluoride, 565
　bromine trifluoride, 766t
　calcium chloride, 785
　carbon dioxide, 629
　carbon disulfide, 505
　carbon monoxide, 624
　carbonyl fluoride, 766t
　carbonyl trifluoride, 766t
　chlorine trifluoride, 791
　chromium compounds, 92t
　cobalt, 190
　cryolite, 764, 764t
　decaborane, 570
　diborane, 566
　flourine, 738, 738t
　hydrogen chloride, 778

　hydrogen fluoride, 752t
　hydrogen selenide, 481
　hydrogen sulfide, 501
　iodine, 815
　manganese, 151, 152t, 153
　nickel, 212t
　nickel carbonate, 223t
　nickel carbonyl, 239t
　nickel chloride, 245t
　nickel hydroxide, 247t
　nickel oxide, 220t
　nickel subsulfide, 229t
　nickel sulfate, 253t
　nitric oxide, 649
　nitrogen trifluoride, 766t
　nitrous oxide, 654
　osmium tetroxide, 289, 290t
　oxygen difluoride, 766t
　ozone, 685
　pentaborane, 568
　perchloryl fluoride, 767t
　phosgene, 633, 782
　phosphine, 473
　phosphoric acid, 466
　phosphorus oxychloride, 468
　phosphorus pentasulfide, 469
　phosphorus trichloride, 470
　platinum, 306, 308t
　potassium hydroxide, 588
　radioactive ruthenium, 264, 264t
　rhodium, 269t–270t
　selenium, 480
　selenium hexafluoride, 482, 767t

SUBJECT INDEX

ACGIH (*Continued*)
  sodium hydroxide, 595
  sodium tetraborate, 551
  sodium tetraborate pentahydrate, 549
  sulfur dioxide, 494
  sulfur hexafluoride, 756t
  sulfuric acid, 511
  sulfur monochloride, 507
  sulfur tetrafluoride, 756t
  tantalum, 61t
  tellurium, 487
  tellurium hexafluoride, 488, 767t
  tungsten, 121
  uranium, 406–408, 407t
  vanadium, 36t
American Industrial Hygiene Association (AIHA), exposure standards, phosphorus pentoxide, 465
Ammonium bromide, chemical and physical properties, 804t
Ammonium chloride *[12125-02-9]*
  chemical and physical properties, 783, 784t
  exposure standards, 785
  production and use, 783–785
  toxic effects, 785
Ammonium hexafluorovanadate *[13815-31-1]*, chemical and physical properties, 6t
Ammonium iodide *[12027-06-4]*, chemical and physical properties, 812t
Ammonium metavanadate *[7803-55-6]*
  chemical and physical properties, 6t
  toxicity, 23t, 25, 26
Ammonium molybdate, chemical and physical properties, 94t
Ammonium muriate. *See* Ammonium chloride
Ammonium pentaborate tetrahydrate, chemical and physical properties, 523t
Ammonium perchlorate *[7790-98-9]*
  chemical and physical properties, 793, 794, 795t
  exposure standards, 799
  production and use, 794–796, 796t
  toxic effects, 796–797, 797t
Ammonium tetrachloropalladate(II) *[13820-40-1]*, 271
  biomonitoring/biomarkers, 273
  carcinogenesis, 279
  chemical and physical properties, 271–272, 272t
  exposure assessment, 273
  exposure standards, 281
  genetic and cellular effects, 279
  neurological, pulmonary, skin sensitization effects, 279–280, 280t
  pharmacokinetics, metabolism, and mechanisms, 274–279
  production and use, 271–272, 272t
  reproductive and developmental effects, 279
  toxic effects
    acute and chronic, 273–274, 274t, 275t, 276t–277t
    experimental studies, 273–280
    human experience, 279–280, 280t
Ammonium vanadate, toxicity, 22, 23, 24, 26
Antibiotics, oxygen interaction, 667t
Antiformin. *See* Sodium hypochlorite
Antisal 2B. *See* Hydrogen fluoride
Aquacat. *See* Cobalt
ASH. *See* Sodium carbonate
Astatine *[7440-68-8]*, chemical and physical properties, 815–816
ATSDR. *See* Agency fot Toxic Substances and Disease Registry
Augus hot rod. *See* Sodium hydroxide
Australia, exposure standards, decaborane, 570

Barax, 10 mol. *See* Sodium tetraborate decahydrate
Barium metaborate, chemical and physical properties, 522t
Binary salts of alkali, 782–785
Biobor JF, chemical and physical properties, 522t
Bipotassium tetrachloroplatinate. *See* Potassium tetrachloroplatinate(II)
1,3,-Bis(2-chloroethyl)-1-nitrosourea, oxygen interaction, 667t
Bismuth orthovanadate *[53801-77-7]*, toxicity, 11, 15t
Bis($N^5$2,4-cyclopentadien-l-yl)nicke1. *See* Nickelocene
B-K. *See* Sodium hypochlorite
BK powder. *See* Calcium hypochlorite
Bleach. *See* Sodium hypochlorite
Bleomycin, oxygen interaction, 666, 667t
Bora-Care. *See* Disodium octaborate tetrahydrate
Borax. *See* Sodium tetraborate decahydrate anhydrous (*See* Sodium tetraborate)
Borax pentahydrate. *See* Sodium tetraborate pentahydrate
Boric acid *[10043-35-3]*
  carcinogenesis, 538, 543
  chemical and physical properties, 521t, 551–552
  environmental impact studies, 544, 552
  exposure assessment, 552
  exposure guidelines, 543–544
  exposure standards, 552
  genetic and cellular effects, 538–539
  neurological, pulmonary, skin sensitization effects, 539–540, 543
  odor and warning properties, 552

SUBJECT INDEX

pharmacokinetics, metabolism, and mechanisms, 534–536, 539–540, 541–542
production and use, 552
reproductive and developmental effects, 536–538, 542–543
toxic effects, 532–533, 533t, 552
   acute and chronic, 533–534, 539, 540–541
   epidemiology studies, 540–544
   human experience, 539–544

Boricin. See Sodium tetraborate decahydrate
Boric oxide [1303-86-2]
   chemical and physical properties, 522t, 552–553
   environmental impact studies, 554
   exposure assessment, 553
   exposure standards, 554
   odor and warning properties, 553
   production and use, 553
   toxic effects, 533, 553–554
Boroethane. See Diborane
Borofax. See Boric acid
Borogard ZB. See Zinc borate hydrate
Boron, elemental [7440-42-8]
   biomonitoring/biomarkers, 545–546
   chemical and physical properties, 523t, 544–545
   exposure assessment, 545–546
   pharmacokinetics, metabolism, and mechanisms, 546
   production and use, 545
   toxic effects, 546

Boron and borates
   biomonitoring/biomarkers, 531–532
   carcinogenesis, 538, 543
   chemical and physical properties, 519–524, 521t–523t, 526f
   environmental impact studies, 544
   exposure assessment, 528–532
   exposure guidelines, 543–544
   genetic and cellular effects, 538–539
   neurological, pulmonary, skin sensitization effects, 539–540, 543
   pharmacokinetics, metabolism, and mechanisms, 534–536, 539–540, 541–542
   production and use, 519, 524–527, 525t, 527t, 528t
   reproductive and developmental effects, 536–538, 542–543
   toxic effects
      acute and chronic, 532–533, 533–534, 533t, 539, 540–541
      epidemiology studies, 540–544
      human experience, 539–544
Boron carbine, chemical and physical properties, 523t
Boron nitride, chemical and physical properties, 523t
Boron sodium oxide. See Sodium tetraborate

Boron tribromide [10294-33-4]
   chemical and physical properties, 523t, 562–563
   environmental impact studies, 563
   exposure assessment, 563
   exposure standards, 563
   odor and warning properties, 563
   production and use, 563
   toxic effects, 563
Boron trifluoride [7637-07-2]
   chemical and physical properties, 523t, 563–564
   environmental impact studies, 565
   exposure assessment, 564
   exposure standards, 565
   odor and warning properties, 564
   production and use, 564
   toxic effects, 564–565
Boron trihydroxide. See Boric acid
Britain. See United Kingdom
Bromine [7726-95-6]
   chemical and physical properties, 731, 732t, 799–800
   exposure assessment, 801–802
   exposure standards, 803
   halides, 804t, 808–809
   inorganic compounds, 803–809, 804t
   production and use, 800–801
   toxic effects, 802–803
Bromine pentafluoride, chemical and physical properties, 804t
Bromine trifluoride
   chemical and physical properties, 804t
   exposure standards, 766t
Bunsenite. See Nickel oxide

CACP. See cis-Diamminedichloroplatinum(II)
Cadmium, chemical and physical properties, 424t
Calcinated soda. See Sodium carbonate
Calcium chloride [10043-52-4]
   chemical and physical properties, 782
   exposure standards, 785
   production and use, 783–785
   toxic effects, 785
Calcium chlorite [7758-19-2]
   chemical and physical properties, 793, 794, 795t
   exposure standards, 799
   production and use, 794–796, 796t
   toxic effects, 796–799, 797t
Calcium dichloride. See Calcium chloride
Calcium fluoride [7789-75-5]
   carcinogenesis, 765
   chemical and physical properties, 757, 758t
   exposure assessment, 759
   exposure standards, 764–765, 764t

Calcium fluoride *[7789-75-5]* (*Continued*)
  neurological, pulmonary, skin sensitization effects, 760–762
  pharmacokinetics, metabolism, and mechanisms, 764
  production and use, 757–759
  toxic effects, 759–764
Calcium hypochlorite *[7778-54-3]*
  chemical and physical properties, 792–793, 794, 795t
  exposure standards, 799
  production and use, 794–796, 796t
  toxic effects, 796–797, 797t
Calcium molybdate, chemical and physical properties, 94t
Calcium oxychloride. *See* Calcium hypochlorite
Calcium salt. *See* Calcium hypochlorite
Calcosan. *See* Calcium chloride
Carbolith. *See* Lithium
Carbon dioxide *[124-35-9]*
  chemical and physical properties, 625–626
  exposure standards, 629
  production and use, 626
  toxic effects
    acute and chronic, 627–629
    experimental studies, 627–629
Carbon disulfide *[75-15-0]*
  chemical and physical properties, 490t, 502
  exposure assessment, 502–503
  exposure standards, 505
  odor and warning properties, 502
  production and use, 502
  reproductive and developmental effects, 505
  toxic effects
    acute and chronic, 504–505, 505t
    experimental studies, 503
    human experience, 503–505, 505t
Carbonic acid. *See* Nickel carbonate
Carbonic acid disodium salt. *See* Sodium carbonate
Carbonic dichloride. *See* Phosgene
Carbonic oxide. *See* Carbon monoxide
Carbon monoxide *[630-08-0]*
  chemical and physical properties, 607–608
  exposure assessment, 611–616, 613f, 614f
    measurement techniques, 622–623
    mixtures and, 621–622
  exposure standards, 624
  exposure treatment
    prevention, 624–625
    therapy, 623–624
  production and use, 608–610
  reproductive and developmental effects, 617, 621
  toxic effects
    acute and chronic, 616–617
    epidemiology studies, 620–621
    experimental studies, 616–617
    human experience, 617–621, 618t
Carbon oxychloride. *See* Phosgene
Carbonyl [di]chloride. *See* Phosgene
Carbonyl fluoride, exposure standards, 766t
Carbonyls
  chemical and physical properties, 171
  of cobalt, preparation of, 182
Carboplatin. *See cis*-Diammine(1,1-cyclobutanedicarboxylato)platinum(II)
Carrel-Dakin solution. *See* Sodium hypochlorite
Catecholamines, oxygen interaction, 667t
Caustic soda. *See* Sodium hydroxide
CDDP. *See cis*-Diamminedichloroplatinum(II)
Cerium *[7440-45-1]*
  carcinogenesis, 441–442
  chemical and physical properties, 424t, 425, 428–430, 429t
  environmental impact studies, 453–454
  exposure assessment, 434
  exposure standards, 453
  genetic and cellular effects, 442–443
  neurological, pulmonary, skin sensitization effects, 443–449, 447t
  pharmacokinetics, metabolism, and mechanisms, 437–440, 451–452
  production and use, 430–434, 431t
  reproductive and developmental effects, 440–441
  toxic effects
    acute and chronic, 435–437, 451
    epidemiology studies, 452–453
    experimental studies, 434–449
    human experience, 450–453
Cesium *[7440-46-2]*
  chemical and physical properties, 600, 601t
  exposure assessment, 601
  exposure standards, 601
  production and use, 601
  toxic effects, 601
Cesium chloride *[7647-17-8]*, chemical and physical properties, 601t
Cesium hydroxide *[21351-79-1]*, chemical and physical properties, 601t
Cesium iodine *[7789-17-5]*, chemical and physical properties, 601t
Cesium nitrate *[7789-18-6]*, chemical and physical properties, 601t
CG. *See* Phosgene
Chemotherapeutic drugs, oxygen interaction, 666, 667t
Chlorate of soda. *See* Sodium chlorate

SUBJECT INDEX

Chlorax. *See* Sodium chlorate
Chloric acid *[7790-93-4]*
   chemical and physical properties, 791, 794, 795t
   exposure standards, 799
   production and use, 794–796, 796t
   toxic effects, 796–797, 797t
Chloride of lime. *See* Calcium hypochlorite
Chloride of phosphorus. *See* Phosphorus trichloride
Chloride of sulphur. *See* Sulfur monochloride
Chlorine *[7782-50-5]*
   chemical and physical properties, 731, 732t, 765, 768
   exposure assessment, 770
   production and use, 768–770
   toxic effects
      acute and chronic, 771–773, 773–775
      epidemiology studies, 773–775
      experimental studies, 770–771
      human experience, 771–775
Chlorine dioxide *[10049-04-4]*
   chemical and physical properties, 786, 787t
   exposure assessment, 787
   exposure standards, 766t, 791
   production and use, 786–787
   toxic effects, 787–789
Chlorine pentafluoride *[13637-63-3]*
   chemical and physical properties, 786, 787t
   exposure assessment, 787
   exposure standards, 766t, 791
   production and use, 786–787
   toxic effects, 790
Chlorine trifluoride *[7790-91-2]*
   chemical and physical properties, 785, 786, 787t
   exposure assessment, 787
   exposure standards, 766t, 791
   production and use, 786–787
   toxic effects, 789–790
Chloroformyl chloride. *See* Phosgene
Chlorohydric acid. *See* Hydrogen chloride
Chloroplatinic acid. *See* Potassium tetrachloroplatinate(II)
Chromic acid, chemical and physical properties, 76t
Chromic chloride, chemical and physical properties, 76t
Chromic oxide, chemical and physical properties, 76t
Chromic trioxide, chemical and physical properties, 76t
Chromium *[7440-47-3]*
   carcinogenesis, 82–83, 86–87, 90–91
   chemical and physical properties, 75–76, 76t
   environmental impact studies, 93
   exposure assessment, 78–79
   exposure levels, by industrial process, 77, 78t
   exposure standards, 92, 92t
   genetic and cellular effects, 83, 87
   neurological, pulmonary, skin sensitization effects, 83, 91
   odor and warning properties, 76
   pharmacokinetics, metabolism, and mechanisms, 79–82, 85–86, 90
   production and use, 76–78
   reproductive and developmental effects, 82, 86
   toxic effects
      acute and chronic, 79
      epidemiology studies, 87–91
      experimental studies, 79–83
      human experience, 83–91
Chromium sesquioxide, chemical and physical properties, 76t
Chromous chloride, chemical and physical properties, 76t
Chromyl chloride
   chemical and physical properties, 76t
   exposure standards, 92
C.I. 77320. *See* Cobalt
C.I. 77775. *See* Nickel
C.I. 77777. *See* Nickel oxide
Cibalith. *See* Lithium
Cisplatin. *See* cis-Diamminedichloroplatinum(II)
Cisplatyl. *See* cis-Diamminedichloroplatinum(II)
Clorox. *See* Sodium hypochlorite
CMT. *See* Cyclopentadienyl manganese tribcarbonyl
Cobalt *[7440-48-4]*
   carcinogenesis, 190
   chemical and physical properties, 179–180, 179t
   exposure assessment, 182–183
   exposure standards, 190
   neurological, pulmonary, skin sensitization effects, 185, 187–188, 190
   pharmacokinetics, metabolism, and mechanisms, 183–185, 186–187
   production and use, 180–182
   toxic effects
      acute and chronic, 183, 185–186, 188–190
      epidemiology studies, 188–190
      experimental studies, 183–185
      human experience, 185–190
Cobalt carbonyl
   chemical and physical properties, 179t
   exposure standards, 190
Cobalt hydrocarbonyl
   chemical and physical properties, 179t
   exposure standards, 190
Cobaltic-cobaltous oxide, chemical and physical properties, 179t

Cobaltous oxide, chemical and physical properties, 179t
Colemanite, chemical and physical properties, 521t
Columbium. *See* Niobium
Corticosteroids, oxygen interaction, 667t
CPDC. *See cis*-Diamminedichloroplatinum(II)
CPDD. *See cis*-Diamminedichloroplatinum(II)
Cryolite *[15096-52-3]*
   carcinogenesis, 765
   chemical and physical properties, 757, 758t
   exposure assessment, 759
   exposure standards, 764–765, 764t
   neurological, pulmonary, skin sensitization effects, 760–762
   pharmacokinetics, metabolism, and mechanisms, 764
   production and use, 757–759
   toxic effects, 759–764
Crystal carbonate. *See* Sodium carbonate
Cyclopentadienyl manganese tribcarbonyl (CMT)
   exposure limits, 152t
   toxic effects, 135–136
Cyclophosphamide, oxygen interaction, 666, 667t
Cytosine-guanine. *See* Phosgene

Dakin's Solution. *See* Sodium hypochlorite
Darammon. *See* Ammonium chloride
dCDP. *See cis*-Diamminedichloroplatinum(II)
DDP. *See cis*-Diamminedichloroplatinum(II)
cis-DDP. *See cis*-Diamminedichloroplatinum(II)
DDPt. *See cis*-Diamminedichloroplatinum(II)
Decaborane *[17702-41-9]*
   chemical and physical properties, 523t, 569
   exposure assessment, 569–570
   exposure standards, 570
   odor and warning properties, 569
   production and use, 569
   toxic effects, 570
2′-Deoxycytidine diphosphate. *See cis*-Diamminedichloroplatinum(II)
Department of Defense (DoD), uranium used by, 384, 384t
Detia gas EX-B. *See* Phosphine
Deuteriumfluoride. *See* Hydrogen fluoride
Dexamethasone, oxygen interaction, 667t
DHHS, exposure standards, uranium, 406–408, 407t
Diamineedichloroplatinum. *See cis*-Diamminedichloroplatinum(II)
(SP-4-2)-diaminodichloroplatinum. *See cis*-Diamminedichloroplatinum(II)
cis-Diamminedichloroplatinum(II) *[15663-27-1]*
   carcinogenesis, 326–327, 328t–330t, 336
   chemical and physical properties, 322–323
   environmental impact studies, 338
   exposure assessment, 323
   exposure standards, 308t, 338
   genetic and cellular effects, 327–331, 336–337
   neurological, pulmonary, skin sensitization effects, 331–333
   pharmacokinetics, metabolism, and mechanisms, 325, 335
   production and use, 323
   reproductive and developmental effects, 325–326, 335–336
   toxic effects
      acute and chronic, 323–325, 324t, 335
      epidemiology studies, 337–338
      experimental studies, 323–333
      human experience, 333–338
cis-Diammine(1,1-cyclobutanedicarboxylato)platinum(II) *[41575-94-4]*
   carcinogenesis, 317, 321
   chemical and physical properties, 315
   environmental impact studies, 322
   epidemiology studies, 322
   exposure assessment, 315–316
   exposure standards, 308t, 322
   genetic and cellular effects, 317–318, 321–322
   neurological, pulmonary, skin sensitization effects, 318–320
   pharmacokinetics, metabolism, and mechanisms, 316–317, 321
   production and use, 315
   reproductive and developmental effects, 317, 321
   toxic effects
      acute and chronic, 316
      experimental studies, 316–320
      human experience, 320–322
Diammonium tetraborate tetrahydrate, chemical and physical properties, 523t
Diborane *[19287-45-7]*
   chemical and physical properties, 523t, 565–566
   environmental impact studies, 567
   exposure assessment, 566
   exposure standards, 566
   odor and warning properties, 566
   production and use, 566
   toxic effects, 566
Dicyclopentadienylnickel. *See* Nickelocene
Dierbium trioxide. *See* Erbium trioxide
Dieuropium trioxide. *See* Europium oxide
Digadolinium trioxide. *See* Gadolinium oxide
Dilanthanum oxide. *See* Lanthanum oxide
Dineodymium trioxide. *See* Neodymium oxide
Dinitrogen monoxide. *See* Nitrous oxide

SUBJECT INDEX

Di-*p*-cyclopentadienyinickel. *See* Nickelocene
Diphosphorus pentaoxide. *See* Phosphorus pentoxide
Diphosphorus pentasulfide. *See* Phosphorus pentasulfide
Dipotassium salt. *See* Potassium tetrachloroplatinate(II)
Dipotassium tetraborate tetrahydrate, chemical and physical properties, 523t
Dipping acid. *See* Sulfuric acid
Disamarium trioxide. *See* Samarium oxide
Disodium carbonate. *See* Sodium carbonate
Disodium metasilicate. *See* Sodium metasilicate
Disodium octaborate, toxic effects, 533
Disodium octaborate tetrahydrate [12280-03-4]
   chemical and physical properties, 521t, 522t, 554
   environmental impact studies, 555
   experimental studies, 555
   exposure standards, 555
   production and use, 555
   toxic effects, 540, 555
Disodium perborate tetrahydrate, toxic effects, 533t
Disodium salt. *See* Sodium selenite
   pentahydrate (*See* Sodium tetraborate pentahydrate)
Disodium selenite. *See* Sodium selenite
Disodium tetraborate. *See* Sodium tetraborate
Disodium tetraborate decahydrate. *See* Sodium tetraborate decahydrate
Disodium tetraborate pentahydrate. *See* Sodium tetraborate pentahydrate
Disulfur dichloride. *See* Sulfur monochloride
Dithiocarbonic anhydride. *See* Carbon disulfide
DoD. *See* Department of Defense
Dodecaboron. *See* Zinc borate hydrate
Drop-Leaf. *See* Sodium chlorate
Drumet. *See* Sodium metasilicate
Dysprosia. *See* Dysprosium oxide
Dysprosium, chemical and physical properties, 424t
Dysprosium oxide [1308-87-8]
   carcinogenesis, 441–442
   chemical and physical properties, 427, 428–430, 429t
   environmental impact studies, 453–454
   exposure assessment, 434
   exposure standards, 453
   genetic and cellular effects, 442–443
   neurological, pulmonary, skin sensitization effects, 443–449, 447t
   pharmacokinetics, metabolism, and mechanisms, 437–440, 451–452
   production and use, 430–434, 431t
   reproductive and developmental effects, 440–441
   toxic effects
      acute and chronic, 435–437, 451
      epidemiology studies, 452–453
      experimental studies, 434–449
      human experience, 450–453

Environmental Protection Agency (EPA)
   exposure standards
      carbon monoxide, 624
      niobium, 49
      ozone, 685–686
      uranium, 408
      vanadium, 36t
   and flourine, 738
EPA. *See* Environmental Protection Agency
Epinephrine, oxygen interaction, 667t
Erbia. *See* Erbium trioxide
Erbium, chemical and physical properties, 424t
Erbium sesquioxide. *See* Erbium trioxide
Erbium trioxide [12061-16-4]
   carcinogenesis, 441–442
   chemical and physical properties, 427, 428–430, 429t
   environmental impact studies, 453–454
   exposure assessment, 434
   exposure standards, 453
   genetic and cellular effects, 442–443
   neurological, pulmonary, skin sensitization effects, 443–449, 447t
   pharmacokinetics, metabolism, and mechanisms, 437–440, 451–452
   production and use, 430–434, 431t
   reproductive and developmental effects, 440–441
   toxic effects
      acute and chronic, 435–437, 451
      epidemiology studies, 452–453
      experimental studies, 434–449
      human experience, 450–453
Eskalith. *See* Lithium
Etching acid. *See* Hydrogen fluoride
Europium, chemical and physical properties, 424t
Europium oxide [1308-96-9]
   carcinogenesis, 441–442
   chemical and physical properties, 426, 428–430, 429t
   environmental impact studies, 453–454
   exposure assessment, 434
   exposure standards, 453
   genetic and cellular effects, 442–443
   neurological, pulmonary, skin sensitization effects, 443–449, 447t

Europium oxide *[1308-96-9]* (*Continued*)
  pharmacokinetics, metabolism, and mechanisms, 437–440, 451–452
  production and use, 430–434, 431t
  reproductive and developmental effects, 440–441
  toxic effects
    acute and chronic, 435–437, 451
    epidemiology studies, 452–453
    experimental studies, 434–449
    human experience, 450–453

Ferric chloride, chemical and physical properties, 170t
Ferric sulfate, chemical and physical properties, 170t
Ferroboron, chemical and physical properties, 522t
Ferrocene, chemical and physical properties, 170t
Ferrochromium, toxic effects, 89–90
Ferrous oxide, black, chemical and physical properties, 170t
Ferrous sulfate, chemical and physical properties, 170t
Ferrovanadium *[12604-58-9]*
  chemical and physical properties, 6t
  toxicity, 11, 12, 13t, 15t, 32
Fertilizer borate granular, chemical and physical properties, 521t
Firebrake ZB. *See* Zinc borate hydrate
Fluorine *[7782-41-4]*
  chemical and physical properties, 731, 732t, 733
  control methods, 739
  exposure assessment, 734
  exposure standards, 738–739, 738t
  production and use, 733–734
  toxic effects
    experimental studies, 735–736, 736t
    human experience, 736–737
Fluorohydric acid. *See* Hydrogen fluoride
Francium *[7440-73-5]*
  exposure standards, 602–603, 603
  production and use, 603

Gadolinia. *See* Gadolinium oxide
Gadolinium *[7440-54-2]*
  carcinogenesis, 441–442
  chemical and physical properties, 424t, 426, 428–430, 429t
  environmental impact studies, 453–454
  exposure assessment, 434
  exposure standards, 453
  genetic and cellular effects, 442–443
  neurological, pulmonary, skin sensitization effects, 443–449, 447t
  pharmacokinetics, metabolism, and mechanisms, 437–440, 451–452
  production and use, 430–434, 431t
  reproductive and developmental effects, 440–441
  toxic effects
    acute and chronic, 435–437, 451
    epidemiology studies, 452–453
    experimental studies, 434–449
    human experience, 450–453
Gadolinium oxide *[12064-62-9]*
  carcinogenesis, 441–442
  chemical and physical properties, 427, 428–430, 429t
  environmental impact studies, 453–454
  exposure assessment, 434
  exposure standards, 453
  genetic and cellular effects, 442–443
  neurological, pulmonary, skin sensitization effects, 443–449, 447t
  pharmacokinetics, metabolism, and mechanisms, 437–440, 451–452
  production and use, 430–434, 431t
  reproductive and developmental effects, 440–441
  toxic effects
    acute and chronic, 435–437, 451
    epidemiology studies, 452–453
    experimental studies, 434–449
    human experience, 450–453
Gadolinium sesquioxide. *See* Gadolinium oxide
Gadolinium trioxide. *See* Gadolinium oxide
Gas EX-B. *See* Phosphine
Germany, exposure standards
  cobalt, 190
  decaborane, 570
  manganese, 152t
  nickel, 212t
  nickel carbonate, 223t
  nickel carbonyl, 239t
  nickel chloride, 245t
  nickel oxide, 220t
  nickel sulfate, 253t
  osmium tetroxide, 289, 290t
  platinum, 307, 308t
  tantalum, 61t
  vanadium, 36t
Gerstley borate. *See* Sodium tetraborate decahydrate

Halogens, 731–816, 732t
Harvest Aid. *See* Sodium chlorate
Heazlewoodite. *See* Nickel subsulfide
Hepatic acid. *See* Hydrogen sulfide
Hexachloroplatinic(IV) acid, environmental impact, 309

SUBJECT INDEX

HF A. *See* Hydrogen fluoride
Hidrotic acid. *See also* Hydrogen iodide
  toxic effects, 814t
  uses, 811t
Holmium
  carcinogenesis, 441–442
  chemical and physical properties, 424t
  environmental impact studies, 453–454
  exposure assessment, 434
  exposure standards, 453
  genetic and cellular effects, 442–443
  neurological, pulmonary, skin sensitization effects, 443–449, 447t
  pharmacokinetics, metabolism, and mechanisms, 437–440, 451–452
  production and use, 430–434, 431t
  reproductive and developmental effects, 440–441
  toxic effects
    acute and chronic, 435–437, 451
    epidemiology studies, 452–453
    experimental studies, 439–449
    human experience, 450–453
Hormones, oxygen interaction, 667t
HTH. *See* Calcium hypochlorite
Hy-Chlor. *See* Calcium hypochlorite
Hydrofluoric acid. *See* Hydrogen fluoride
Hydrogen bromide *[10035-10-6]*, chemical and physical properties, 803–805, 804t
Hydrogen chloride *[7647-01-0]*
  carcinogenesis, 778
  chemical and physical properties, 775
  exposure assessment, 776
  exposure standards, 778
  production and use, 775–776
  toxic effects
    acute and chronic, 776–777, 777–778
    experimental studies, 776–777
    human experience, 777–778
Hydrogen fluoride *[7664-39-3]*
  biomonitoring/biomarkers, 741
  chemical and physical properties, 739, 740t
  exposure assessment, 741
  exposure standards, 751, 752t
  neurological, pulmonary, skin sensitization effects, 741–745, 742t
  odor and warning properties, 747
  pharmacokinetics, metabolism, and mechanisms, 750–751
  production and use, 739–740
  toxic effects
    acute and chronic, 746–750
    experimental studies, 741–745, 742t

  human experience, 745–751
Hydrogen iodide *[10034-85-2]*
  chemical and physical properties, 810–811, 812t
  toxic effects, 814t
  uses, 811t
Hydrogen orthoborate. *See* Boric acid
Hydrogen phosphide. *See* Phosphine
Hydrogen selenide *[7783-07-5]*
  chemical and physical properties, 475t, 480
  exposure standards, 481
  odor and warning properties, 480
  production and use, 480
  toxic effects, 481
Hydrogen sulfate. *See* Sulfuric acid
Hydrogen sulfide *[7783-06-4]*
  chemical and physical properties, 490t, 495
  exposure assessment, 496
  exposure standards, 501–502
  odor and warning properties, 495
  pharmacokinetics, metabolism, and mechanisms, 496–498, 497t
  production and use, 495–496
  reproductive and developmental effects, 499–500
  toxic effects, 496–501
    acute and chronic, 497–498, 497t, 498–499, 500–501
    experimental studies, 498–500
    human experience, 500–501
Hydrogen telluride *[7783-09-7]*
  chemical and physical properties, 488, 488t
  exposure standards, 489
  odor and warning properties, 489
  production and use, 489
  toxic effects, 489
Hydroselenic acid. *See* Hydrogen selenide
Hydrosulfuric acid. *See* Hydrogen sulfide
Hypochlorous acid. *See* Calcium hypochlorite
Hypochlorous acid *[7790-92-3]*
  chemical and physical properties, 791, 794, 795t
  exposure standards, 799
  production and use, 794–796, 796t
  toxic effects, 796–797, 797t

IARC. *See* International Agency for Research on Cancer
ICRP. *See* International Commission on Radiological Protection
IDHL, exposure standards
  nitrogen trifluoride, 766t
  oxygen difluoride, 766t
  perchloryl fluoride, 767t
  selenium hexafluoride, 767t
  tellurium hexafluoride, 767t

Institute for Occupational Safety and Health
(NIOSH), exposure standards, manganese, 152t
International Agency for Research on Cancer (IARC),
exposure standards
hydrogen chloride, 778
iron, 178
selenium, 480
International Commission on Radiological Protection
(ICRP), exposure standards, uranium, 406–408,
407t
Iodates, chemical and physical properties, 811, 812t
Iodic acid [7782-68-5], chemical and physical
properties, 812t
Iodides, chemical and physical properties, 811, 812t
Iodine [7553-56-2]
  chemical and physical properties, 731, 732t, 809
  exposure assessment, 813
  exposure standards, 815
  halides
    chemical and physical properties, 811, 812t
    production and use, 813
  inorganic compounds, 810–815, 812t
  production and use, 809–810, 811t
  toxic effects, 813–815, 814t
Iodine bromide [7789-33-5], chemical and physical
properties, 812t
Iodine chloride [7790-99-0]
  chemical and physical properties, 812t
  uses, 811t
Iodine monobromide, uses, 811t
Iodine monochloride, toxic effects, 814t
Iodine pentafluoride [7783-66-6]
  chemical and physical properties, 812t
  toxic effects, 814t
  uses, 811t
Iodine tribromide, chemical and physical properties,
812t
Iodine trichloride, uses, 811t
Ionizing radiation, exposure standards, 264, 264t
IPCS, exposure standards, platinum, 306
Iron [7439-89-6]
  carcinogenesis, 177
  chemical and physical properties, 169–171, 170t
  exposure assessment, 173–174
  exposure standards, 178–179
  pharmacokinetics, metabolism, and mechanisms,
  176–177
  production and use, 172–173
  reproductive and developmental effects, 175
  toxic effects
    acute and chronic, 175–176
    epidemiology studies, 177–178
    experimental studies, 174–175
    human experience, 175–177
Iron oxide, magnetite, red, chemical and physical
properties, 170t
Iron pentacarbonyl, chemical and physical properties,
170t, 171

Javell water. See Sodium hypochlorite
Javex. See Sodium hypochlorite

Kalium. See Potassium
Khislevudite. See Nickel subsulfide
Kill-off. See Boric acid
Kjel-sorb. See Boric acid
Kodak balanced alkali. See Sodium metaborate
dihydrate

Lanthana. See Lanthanum oxide
Lanthanides, 423–454, 424t
  carcinogenesis, 441–442
  chemical and physical properties, 423–424, 424–
  425, 424t, 428–430, 429t
  environmental impact studies, 453–454
  exposure assessment, 434
  exposure standards, 453
  exposure standards
    epidemiology studies, 452–453
    genetic and cellular effects, 442–443
  neurological, pulmonary, skin sensitization effects,
  443–449, 447t
  pharmacokinetics, metabolism, and mechanisms,
  437–440, 451–452
  production and use, 430–434
  reproductive and developmental effects, 440–441
  toxic effects
    acute and chronic, 435–437, 451
    experimental studies, 434–449
    human experience, 450–453
Lanthanum, chemical and physical properties, 424,
424t
Lanthanum nitrate [10099-59-9]
  carcinogenesis, 441–442
  chemical and physical properties, 425, 428–430,
  429t
  environmental impact studies, 453–454
  exposure assessment, 434
  exposure standards, 453
  genetic and cellular effects, 442–443
  neurological, pulmonary, skin sensitization effects,
  443–449, 447t
  pharmacokinetics, metabolism, and mechanisms,
  437–440, 451–452
  production and use, 430–434, 431t
  reproductive and developmental effects, 440–441

SUBJECT INDEX

toxic effects
acute and chronic, 435–437, 451
epidemiology studies, 452–453
experimental studies, 434–449
human experience, 450–453
Lanthanum oxide [1312-81-8]
carcinogenesis, 441–442
chemical and physical properties, 425, 428–430, 429t
environmental impact studies, 453–454
exposure assessment, 434
exposure standards, 453
genetic and cellular effects, 442–443
neurological, pulmonary, skin sensitization effects, 443–449, 447t
pharmacokinetics, metabolism, and mechanisms, 437–440, 451–452
production and use, 430–434, 431t
reproductive and developmental effects, 440–441
toxic effects
acute and chronic, 435–437, 451
epidemiology studies, 452–453
experimental studies, 434–449
human experience, 450–453
Lead, chemical and physical properties, 424t
Lead chromate, exposure standards, 92t
Lime chloride. *See* Calcium hypochlorite
Lithane. *See* Lithium
Lithium [7439-93-2]
chemical and physical properties, 603
exposure standards, 604
production and use, 603
toxic effects, 604
Lithizine. *See* Lithium
Lithobid. *See* Lithium
LN2. *See* Nitrogen
Lo-Bax. *See* Calcium hypochlorite
Losantin. *See* Calcium hypochlorite
Lutetium, chemical and physical properties, 424t
Lye. *See* Potassium hydroxide; Sodium hydroxide

Manganese [7439-96-5]
carcinogenesis, 139
chemical and physical properties, 129, 130, 131t
exposure assessment, 132
exposure standards, 151–153, 152t
neurological, pulmonary, skin sensitization effects, 137–138
odor and warning properties, 130
pharmacokinetics, metabolism, and mechanisms, 139–142, 150–151
production and use, 132
reproductive and developmental effects, 138–139

toxic effects, 129–130, 133–151
acute and chronic, 134–151
animal studies, 149–151
epidemiology studies, 144–149, 150t
neurotoxicity, 142–151
Manganese acetate [638-38-0], chemical and physical properties, 131t
Manganese carbonate [592-62-9], chemical and physical properties, 131t
Manganese chloride [7773-01-5], chemical and physical properties, 131t
Manganese dioxide [1313-13-9]
chemical and physical properties, 131t
production and use, 132
Manganese oxide, production and use, 132
Manganese sulfate [7785-87-7], chemical and physical properties, 131t
Manganese sulfate, production and use, 132
Manganese tetroxide [1317-35-7]
chemical and physical properties, 131t
exposure limits, 125t
Metaborate. *See* Sodium perborate tetrahydrate
Metavanadate, toxicity, 20–21
Methyl borate, chemical and physical properties, 522t
Methylcyclopentadienyl manganese tricarbonyl (MMT) [12108-13-3]
chemical and physical properties, 131t
exposure limits, 152t
production and use, 132
toxic effects, 134–135
Methylprednisolone, oxygen interaction, 667t
Metso Beads. *See* Sodium metasilicate
MMT. *See* Methylcyclopentadienyl manganese tricarbonyl
Molybdenum [7439-98-7]
biomonitoring/biomarkers, 97–98
carcinogenesis, 101–102
chemical and physical properties, 93–95, 94t
environmental impact studies, 106
exposure assessment, 97–98
exposure standards, 105–106
genetic and cellular effects, 102
neurological, pulmonary, skin sensitization effects, 102
odor and warning properties, 95
pharmacokinetics, metabolism, and mechanisms, 100–101, 104–105
production and use, 95–96
reproductive and developmental effects, 101
toxic effects
acute and chronic, 98–100
experimental studies, 98–102
human experience, 102–105

Molybdenum disulfide, chemical and physical properties, 94t
Molybdic oxide, chemical and physical properties, 94t
Mononickel oxide. *See* Nickel oxide
Muriatic acid. *See* Hydrogen chloride

National Institute for Occupational Safety and Health (NIOSH), 36t
  exposure standards
    boric acid, 552
    boric oxide, 554
    boron, 543
    boron tribromide, 563
    boron trifluoride, 565
    bromine trifluoride, 766t
    carbon dioxide, 629
    carbon disulfide, 505
    carbon monoxide, 624
    carbonyl fluoride, 766t
    carbonyl trifluoride, 766t
    chlorine trifluoride, 791
    chromium compounds, 92t
    cobalt, 190
    decaborane, 570
    diborane, 566
    fluorine, 738, 738t
    hydrogen chloride, 778
    hydrogen fluoride, 752t
    hydrogen selenide, 481
    hydrogen sulfide, 501–502
    iodine, 815
    manganese, 151–153, 152t
    nickel, 212t
    nickel carbonate, 223t
    nickel carbonyl, 239t
    nickel chloride, 245t
    nickel hydroxide, 247t
    nickel oxide, 220t
    nickel subsulfide, 229t
    nickel sulfate, 253t
    nitric oxide, 649
    nitrogen trifluoride, 766t
    nitrous oxide, 654
    osmium tetroxide, 289, 290t
    oxygen difluoride, 766t
    ozone, 684–685
    pentaborane, 568
    perchloryl fluoride, 767t
    phosgene, 633, 782
    phosphine, 473
    phosphorus oxychloride, 468
    phosphorus pentasulfide, 469
    phosphorus trichloride, 470
    platinum, 308t
    potassium hydroxide, 588
    rhodium, 269t–270t
    selenium hexafluoride, 482, 767t
    sodium hydroxide, 595
    sodium tetraborate, 551
    sodium tetraborate pentahydrate, 549
    sulfur dioxide, 494
    sulfur hexafluoride, 756t
    sulfur monochloride, 507
    sulfur tetrafluoride, 756t
    tantalum, 61t
    tellurium, 487
    tellurium hexafluoride, 488, 767t
    tungsten, 121
    uranium, 406–408, 407t
National Toxicology Program (NTP), exposure standards, iron, 178
Natrium. *See* Sodium
Neobor. *See* Sodium tetraborate pentahydrate
Neodymia. *See* Neodymium oxide
Neodymium *[7440-00-8]*
  carcinogenesis, 441–442
  chemical and physical properties, 424t, 425, 428–430, 429t
  environmental impact studies, 453–454
  exposure assessment, 434
  exposure standards, 453
  genetic and cellular effects, 442–443
  neurological, pulmonary, skin sensitization effects, 443–449, 447t
  pharmacokinetics, metabolism, and mechanisms, 437–440, 451–452
  production and use, 430–434, 431t
  reproductive and developmental effects, 440–441
  toxic effects
    acute and chronic, 435–437, 451
    epidemiology studies, 452–453
    experimental studies, 434–449
    human experience, 450–453
Neodymium oxide *[1313-97-9]*
  carcinogenesis, 441–442
  chemical and physical properties, 426, 428–430, 429t
  environmental impact studies, 453–454
  exposure assessment, 434
  exposure standards, 453
  genetic and cellular effects, 442–443
  neurological, pulmonary, skin sensitization effects, 443–449, 447t
  pharmacokinetics, metabolism, and mechanisms, 437–440, 451–452

SUBJECT INDEX 839

production and use, 430–434, 431t
reproductive and developmental effects, 440–441
toxic effects
acute and chronic, 435–437, 451
epidemiology studies, 452–453
experimental studies, 434–449
human experience, 450–453
Neodymium sesquioxide. *See* Neodymium oxide
Neodymium trioxide. *See* Neodymium oxide
Neoplatin. *See cis*-Diamminedichloroplatinum(II)
Nickel *[7440-02-0]*
carcinogenesis, 203–204, 207
chemical and physical properties, 195–196, 197t, 198t
environmental impact studies, 211–213
exposure assessment, 198–199
exposure standards, 212t
genetic and cellular effects, 204, 210–211
isotopes of, 198t
pharmacokinetics, metabolism, and mechanisms, 200–203, 205–207
production and use, 196–197
reproductive and developmental effects, 203
toxic effects
acute and chronic, 199–200, 200t, 204–205
epidemiology studies, 207–210
experimental studies, 199–204
human experience, 204–211
Nickel carbonate *[3333-67-3]*
carcinogenesis, 221
chemical and physical properties, 197t, 219–221
exposure assessment, 221
exposure standards, 223t
genetic and cellular effects, 221–222
production and use, 221
reproductive and developmental effects, 221
toxic effects
acute and chronic, 221
experimental studies, 221–222
human experience, 222
Nickel carbonyl *[13463-39-3]*
carcinogenesis, 236, 238
chemical and physical properties, 197t, 231–233
exposure assessment, 233–234
exposure standards, 239t
genetic and cellular effects, 236, 238
odor and warning properties, 233
pharmacokinetics, metabolism, and mechanisms, 235–236, 238
production and use, 233
reproductive and developmental effects, 236, 238
toxic effects
acute and chronic, 234–235, 237–238

epidemiology studies, 237–238
experimental studies, 234–236, 234t
human experience, 236–238
Nickel chloride *[7718-54-9]*
carcinogenesis, 242, 243
chemical and physical properties, 197t, 238–239, 238–240
exposure assessment, 240
exposure standards, 245t
genetic and cellular effects, 242, 243
pharmacokinetics, metabolism, and mechanisms, 241
production and use, 240
reproductive and developmental effects, 241–242, 243
toxic effects
acute and chronic, 240–241, 243
experimental studies, 240–242
human experience, 242–243
Nickel hydroxide *[12054-48-7]*
carcinogenesis, 246
chemical and physical properties, 197t, 244
exposure assessment, 244
exposure standards, 247t
genetic and cellular effects, 246
pharmacokinetics, metabolism, and mechanisms, 244
production and use, 244
reproductive and developmental effects, 246
toxic effects
acute and chronic, 244
experimental studies, 244–246
human experience, 246
Nickelocene *[1271-28-9]*
carcinogenesis, 230
chemical and physical properties, 197t, 228
exposure assessment, 230
exposure standards, 232t
genetic and cellular effects, 231
pharmacokinetics, metabolism, and mechanisms, 230
production and use, 230
reproductive and developmental effects, 230
toxic effects
acute and chronic, 230
experimental studies, 230–231
human experience, 231
Nickel oxide *[1313-99-1]*
carcinogenesis, 215, 216t–217t, 218–219
chemical and physical properties, 197t, 213
exposure assessment, 214
exposure standards, 219, 220t
genetic and cellular effects, 215, 219

840 SUBJECT INDEX

Nickel oxide *[1313-99-1]* (Continued)
  pharmacokinetics, metabolism, and mechanisms, 215
  production and use, 213–214
  reproductive and developmental effects, 215, 218
  toxic effects
    acute and chronic, 214–215, 214t, 218
    experimental studies, 214–215
    human experience, 215–219
Nickel subsulfide *[12035-72-2]*
  carcinogenesis, 224–225, 226t–227t, 228
  chemical and physical properties, 197t, 222
  exposure assessment, 224
  exposure standards, 229t
  genetic and cellular effects, 225, 228
  production and use, 222–224
  reproductive and developmental effects, 224–225, 225
  toxic effects
    acute and chronic, 224, 225
    experimental studies, 224–225
    human experience, 225–228
    pharmacokinetics, metabolism, and mechanisms, 224
Nickel sulfate *[7786-81-4]*
  carcinogenesis, 250, 252
  chemical and physical properties, 197t, 246–248
  exposure assessment, 248
  exposure standards, 253t
  genetic and cellular effects, 250, 252
  pharmacokinetics, metabolism, and mechanisms, 249
  production and use, 248
  reproductive and developmental effects, 249–250, 252
  toxic effects
    acute and chronic, 248–249, 248t, 250–252
    experimental studies, 248–250
    human experience, 250–252
Nickel tetracarbonyl. *See* Nickel carbonyl
(T-4)-nickel tetracarbonyl. *See* Nickel carbonyl
Niobium *[7440-03-1]*
  biomonitoring/biomarkers, 41–42
  carcinogenesis, 46
  chemical and physical properties, 1–4, 2t, 38–40, 39t
  environmental impact studies, 49
  exposure assessment, 40–42
  exposure standards, 48–49
  genetic and cellular effects, 46
  neurological, pulmonary, skin sensitization effects, 46–47

  occupational exposure limits, 4
  odor and warning properties, 40
  pharmacokinetics, metabolism, and mechanisms, 43–46, 47–48
  production and use, 40
  reproductive and developmental effects, 46
  toxic effects, 2–3, 4
    acute and chronic, 42–43
    epidemiology studies, 47–48
    experimental studies, 42–47
    human experience, 47–48
Niobium carbide *[12069-94-2]*, chemical and physical properties, 39t
Niobium dioxide *[12034-59-2]*, chemical and physical properties, 39t
Niobium hydride *[12655-93-5]*, chemical and physical properties, 39t
Niobium pentachloride *[10026-12-7]*, chemical and physical properties, 39t
Niobium pentafluoride *[7783-68-8]*, chemical and physical properties, 39t
Niobium pent[a]oxide *[1313-96-8]*, chemical and physical properties, 39t
Niobium potassium fluoride *[17523-77-2]*, chemical and physical properties, 39t
NIOSH. *See* National Institute for Occupational Safety and Health
Nitric acid. *See* Terbium trinitrate
Nitric oxide *[10102-43-9]*, 642–647
  chemical and physical properties, 637
  exposure assessment, 639–642
    measurement, 650–651
  exposure prevention, 649
  exposure standards, 648–649
  exposure treatment, 649–650
  oxygen interaction, 666
  production and use, 638–639, 640t
  toxic effects, 646–648
Nitrofurantoin, oxygen interaction, 667t
Nitrogen *[7727-37-9]*
  chemical and physical properties, 634
  exposure standards, 636–637
  production and use, 635
  toxic effects, 635–636
Nitrogen dioxide *[10102-44-0]*, 642–647
  chemical and physical properties, 637
  exposure assessment, 639–642
    measurement, 650–651
  exposure prevention, 649
  exposure standards, 648–649
  exposure treatment, 649–650
  production and use, 638–639, 640t
  toxic effects, 642–646

SUBJECT INDEX

Nitrogen trifluoride, exposure standards, 766t
Nitrous oxide *[10024-97-2]*
   chemical and physical properties, 651
   exposure assessment, 652-653
      measurement, 654-655
   exposure prevention, 655
   exposure standards, 654
   production and use, 651-652
   toxic effects, 653-654
Norepinephrine, oxygen interaction, 667t
NRC. *See* Nuclear Regulatory Commission
NSC 119875. *See cis*-Diamminedichloroplatinum(II)
NTP. *See* National Toxicology Program
   exposure standards, selenium, 480
Nuclear Regulatory Commission (NRC), exposure standards, uranium, 406-408, 407t

Occupational Safety and Health Administration (OSHA), exposure standards
   boric acid, 552
   boric oxide, 554
   boron, 543-544
   calcium chloride, 785
   carbon disulfide, 505
   carbon monoxide, 624
   carbonyl trifluoride, 766t
   chloric acid, 799
   chlorine trifluoride, 791
   chromium compounds, 92t
   cobalt, 190
   cryolite, 764, 764t
   decaborane, 570
   diborane, 566
   fluorine, 738, 738t
   hydrogen chloride, 778
   hydrogen fluoride, 752t
   hydrogen selenide, 481
   hydrogen sulfide, 502
   iodine, 815
   iron, 178-179
   manganese, 151, 152t
   nickel, 212t
   nickel carbonate, 223t
   nickel carbonyl, 239t
   nickel chloride, 245t
   nickel hydroxide, 247t
   nickel oxide, 220t
   nickel subsulfide, 229t
   nickel sulfate, 253t
   nitric oxide, 649
   nitrogen trifluoride, 766t
   nitrous oxide, 654
   osmium tetroxide, 289, 290t

   oxygen difluoride, 766t
   ozone, 684
   pentaborane, 568
   perchloryl fluoride, 767t
   phosgene, 633, 782
   phosphine, 473
   phosphorus oxychloride, 468
   phosphorus pentasulfide, 469
   phosphorus trichloride, 470
   platinum, 308t
   potassium hydroxide, 588
   rhodium, 269t-270t
   selenium, 480
   selenium hexafluoride, 482, 767t
   sodium hydroxide, 595
   sodium tetraborate, 551
   sodium tetraborate pentahydrate, 549
   sulfur dioxide, 494
   sulfur hexafluoride, 756t
   sulfuric acid, 511
   sulfur monochloride, 507
   sulfur tetrafluoride, 756t
   tantalum, 61t
   tellurium, 487
   tellurium hexafluoride, 488, 767t
   uranium, 406-408, 407t
   vanadium, 36t
Octabor. *See* Disodium octaborate tetrahydrate
Oil of vitriol. *See* Sulfuric acid
Oleic acid, oxygen interaction, 666
Orthoboric acid. *See* Boric acid
Orthophosphoric acid. *See* Phosphoric acid
Orthosil. *See* Sodium metasilicate
Orthovanadate, toxicity, 24
OSHA. *See* Occupational Safety and Health Administration
Osmic acid. *See* Osmium tetroxide
Osmium *[7440-04-2]*
   carcinogenesis, 283
   chemical and physical properties, 281-282
   environmental impact studies, 283-284
   epidemiology studies, 283
   exposure assessment, 282
   exposure standards, 283
   genetic and cellular effects, 283
   odor and warning properties, 282
   pharmacokinetics, metabolism, and mechanisms, 283
   production and use, 282
   reproductive and developmental effects, 283
   toxic effects
      experimental studies, 282-283
      human experience, 283

Osmium tetroxide [20816-12-0]
  carcinogenesis, 287, 289
  chemical and physical properties, 284
  exposure assessment, 285
  exposure standards, 289, 290t
  genetic and cellular effects, 287, 289
  odor and warning properties, 284
  pharmacokinetics, metabolism, and mechanisms, 286, 288
  production and use, 284–285
  reproductive and developmental effects, 286–287, 288
  toxic effects
    acute and chronic, 285–286, 287–288
    epidemiology studies, 289
    experimental studies, 285–287
    human experience, 287–289
Oxygen [7782-44-7]
  carcinogenesis, 675
  chemical and physical properties, 655
  hypoxia, 657–661, 657t, 659t
  mountain sickness, 659–661
  pharmacokinetics, metabolism, and mechanisms, 656–661, 662–663
  production and use, 655–656
  toxic effects (hyperoxia)
    experimental studies, 663–670
    human experience, 670–675
    hyperbaric exposures, 672–674, 673t
    mechanisms, 661–663
    mixtures and, 666–667, 667t
    prevention, 667–670, 674–675
Oxygen difluoride, exposure standards, 766t
Ozone [10028-15-6]
  carcinogenesis, 683–684
  chemical and physical properties, 676
  exposure assessment, 676–679
    measurement, 687–688
  exposure standards, 684–686
  production and use, 676
  toxic effects, 679
    acute, 680–681
    in asthmatics, 681–682
    chronic and subchronic, 682–683
    epidemiology studies, 683
    prevention, 687t, 886
    reaction with biological macromolecules, 679–680
    tolerance, 681
    treatment, 687

Palladex 600. See Palladium
Palladium [7440-05-3], 271

biomonitoring/biomarkers, 273
carcinogenesis, 279
chemical and physical properties, 271–272, 272t
exposure assessment, 273
exposure standards, 281
genetic and cellular effects, 279
neurological, pulmonary, skin sensitization effects, 279–280, 280t
pharmacokinetics, metabolism, and mechanisms, 274–279
production and use, 271–272, 272t
reproductive and developmental effects, 279
toxic effects
  acute and chronic, 273–274, 274t, 275t, 276t–277t
  experimental studies, 273–280
  human experience, 279–280, 280t
Palladium(II) chloride [7647-10-1], 271
  biomonitoring/biomarkers, 273
  carcinogenesis, 279
  chemical and physical properties, 271–272, 272t
  exposure assessment, 273
  exposure standards, 281
  genetic and cellular effects, 279
  neurological, pulmonary, skin sensitization effects, 279–280, 280t
  pharmacokinetics, metabolism, and mechanisms, 274–279
  production and use, 271–272, 272t
  reproductive and developmental effects, 279
  toxic effects
    acute and chronic, 273–274, 274t, 275t, 276t–277t
    experimental studies, 273–280
    human experience, 279–280, 280t
Paraquat, oxygen interaction, 666
PBS4. See Sodium perborate tetrahydrate
PBSI. See Sodium perborate monohydrate
PBSM. See Sodium perborate monohydrate
PBST. See Sodium perborate tetrahydrate
Pentaborane [19624-22-7]
  chemical and physical properties, 567
  environmental impact studies, 568
  exposure assessment, 567
  exposure standards, 568
  odor and warning properties, 567
  production and use, 567
  toxic effects, 568
Pentachlorophosphorane. See Phosphorus pentachloride
Pentasulfide, undistilled. See Phosphorus pentasulfide
Perchloron. See Calcium hypochlorite
Perchloryl fluoride, exposure standards, 767t

SUBJECT INDEX

Periodates, chemical and physical properties, 811, 812t
Periodic acid [10450-60-9], chemical and physical properties, 812t
Perosmic acid anhydride. See Osmium tetroxide
Peyrone's salt/chloride. See cis-Diamminedichloroplatinum(II)
Phosgene [75-44-5]
　chemical and physical properties, 629–630, 778–779
　exposure assessment, 631, 780
　　measurement techniques, 633
　　mixtures and, 633
　exposure standards, 633, 782
　exposure treatment
　　prevention, 633–634
　　therapy, 633
　production and use, 630–631, 779–780
　toxic effects
　　acute and chronic, 780, 781–782
　　experimental studies, 631–632, 780–781
　　human experience, 632–633, 781–782
Phosphine [7803-51-2]
　chemical and physical properties, 460t, 470–471
　exposure assessment, 471
　exposure standards, 473
　odor and warning properties, 471
　production and use, 471
　toxic effects
　　experimental studies, 471–472
　　human experience, 472–473
Phosphorated hydrogen. See Phosphine
Phosphoric. See Phosphorus pentoxide
Phosphoric acid. See Trisodium phosphate
Phosphoric acid [7664-38-2]
　chemical and physical properties, 460t, 465
　exposure assessment, 466
　exposure standards, 466
　odor and warning properties, 465
　production and use, 465–466
　toxic effects, 466
Phosphoric anhydride. See Phosphorus pentoxide
Phosphoric perchloride. See Phosphorus pentachloride
Phosphoric sulfide. See Phosphorus pentasulfide
Phosphorus [7723-14-0]
　chemical and physical properties, 459–461, 460t
　exposure assessment, 462
　exposure standards, 464
　neurological, pulmonary, skin sensitization effects, 463–464
　odor and warning properties, 461

　pharmacokinetics, metabolism, and mechanisms, 462
　production and use, 461
　toxic effects
　　acute and chronic, 462–463
　　experimental studies, 462–464
Phosphorus hydride. See Phosphine
Phosphorus oxychloride [10025-87-3]
　chemical and physical properties, 460t, 468
　exposure standards, 468
　odor and warning properties, 468
　production and use, 468
　toxic effects, 468
Phosphorus pentachloride [10026-13-8]
　chemical and physical properties, 460t, 467
　exposure assessment, 467
　exposure standards, 468
　production and use, 467
　toxic effects, 468
Phosphorus pentasulfide [1314-80-3]
　chemical and physical properties, 460t, 469
　exposure standards, 469
　production and use, 469
　toxic effects, 469
Phosphorus pentoxide [1314-56-3]
　chemical and physical properties, 460t, 464
　exposure standards, 465
　production and use, 464
　toxic effects, 464–465
Phosphorus sesquisulfide. See Tetraphosphorus trisulfide
Phosphorus trichloride [7719-12-2]
　chemical and physical properties, 460t, 469–470
　exposure assessment, 470
　exposure standards, 470
　odor and warning properties, 470
　production and use, 470
　toxic effects, 470
Phosphorus trihydride. See Phosphine
Phosphoryl chloride. See Phosphorus oxychloride
PICI. See Phosphorus trichloride
Piochlor. See Sodium hypochlorite
Pittchlor. See Calcium hypochlorite
Platiblastin. See cis-Diamminedichloroplatinum(II)
Platinate(2-), tetrachloro-, dipotassium, (SP-4-1)-. See Potassium tetrachloroplatinate(II)
Platinex. See cis-Diamminedichloroplatinum(II)
Platinol. See cis-Diamminedichloroplatinum(II)
Platinous potassium chloride. See Potassium tetrachloroplatinate(II)
Platinum [7440-06-4]
　biomonitoring/biomarkers, 293–294
　carcinogenesis, 299–300, 305

SUBJECT INDEX

Platinum *[7440-06-4]* (*Continued*)
  chemical and physical properties, 289–291, 292t
  environmental impact studies, 307–309
  exposure assessment, 293–294
  exposure standards, 306–307, 308t
  genetic and cellular effects, 300, 305
  neurological, pulmonary, skin sensitization effects, 300–301
  pharmacokinetics, metabolism, and mechanisms, 295–299, 297t, 303–304
  production and use, 291–293
  reproductive and developmental effects, 299, 304
  toxic effects
    acute and chronic, 294–295, 294t, 296t, 302–303
    epidemiology studies, 305–306
    experimental studies, 294–301
    human experience, 301–306
Platinum(II) chloride *[10025-65-7]*
  chemical and physical properties, 292t
  toxic effects, 294t, 295, 296t
Platinum(IV) chloride *[13454-96-1]*
  carcinogenesis, 310
  chemical and physical properties, 292t, 309
  exposure assessment, 309
  exposure standards, 308t, 311
  genetic and cellular effects, 300, 310–311
  neurological, pulmonary, skin sensitization effects, 300, 311
  pharmacokinetics, metabolism, and mechanisms, 298
  production and use, 309
  reproductive and developmental effects, 310
  toxic effects, 294t, 295–298, 296t
    acute and chronic, 294t, 296t, 310
    experimental studies, 310–311
    human experience, 311
Platinum diamminochloride cisplatyl. *See cis*-Diamminedichloroplatinum(II)
Platinum(II) oxide, chemical and physical properties, 292t
Platinum(IV) oxide *[1314-15-4]*
  chemical and physical properties, 292t, 311–312
  environmental impact studies, 313
  exposure assessment, 312
  exposure standards, 313
  pharmacokinetics, metabolism, and mechanisms, 297t, 312
  production and use, 312
  toxic effects, 292t, 296t
    acute and chronic, 294t, 296t, 312
    experimental studies, 312
    human experience, 312–313

Platinum(IV) sulfate tetrahydrate *[69102-79-0]*
  carcinogenesis, 314
  chemical and physical properties, 292t, 313
  exposure assessment, 313
  exposure standards, 308t
  genetic and cellular effects, 300, 314
  neurological, pulmonary, skin sensitization effects, 301, 314
  pharmacokinetics, metabolism, and mechanisms, 297t, 298, 314
  production and use, 313
  reproductive and developmental effects, 314
  toxic effects, 292t, 295, 296t
    acute and chronic, 294t, 296t, 313–314
    experimental studies, 313–314
    human experience, 314–315
Poland, exposure standards
  manganese, 152t
  nickel, 212t
  nickel carbonate, 223t
  nickel carbonyl, 239t
  nickel chloride, 245t
  nickel hydroxide, 247t
  nickelocene, 232t
  nickel oxide, 220t
  nickel subsulfide, 229t
  nickel sulfate, 253t
  osmium tetroxide, 289, 290t
  tantalum, 61t
  vanadium, 36t
Polybor. *See* Disodium octaborate tetrahydrate
Potash. *See* Potassium hydroxide
Potassium *[7440-09-7]*, 584
  biomonitoring/biomarkers, 585
  chemical and physical properties, 584–585, 584t
  environmental impact studies, 586
  exposure assessment, 585
  exposure standards, 586
  production and use, 585
  toxic effects, 585
Potassium bromate, chemical and physical properties, 804t
Potassium bromide, chemical and physical properties, 804t
Potassium chlorate *[3811-04-9]*
  chemical and physical properties, 792, 794, 795t
  exposure standards, 799
  production and use, 794–796, 796t
  toxic effects, 796–797, 797t
Potassium chloroplatinite. *See* Potassium tetrachloroplatinate(II)
Potassium dichromate, chemical and physical properties, 76t

SUBJECT INDEX

Potassium fluoborate, chemical and physical properties, 522t
Potassium hexachloroplatinate(IV) [16921-30-5]
  carcinogenesis, 339
  chemical and physical properties, 292t, 338–339
  environmental impact studies, 341
  exposure assessment, 339
  exposure standards, 308t, 341
  genetic and cellular effects, 339–340
  pharmacokinetics, metabolism, and mechanisms, 301, 339
  production and use, 339
  reproductive and developmental effects, 339
  toxic effects
    acute and chronic, 339
    epidemiology studies, 340
    experimental studies, 339–340
    human experience, 340
Potassium hydroxide [1310-58-3]
  biomonitoring/biomarkers, 587
  chemical and physical properties, 584t, 586
  environmental impact studies, 588
  exposure assessment, 587
  exposure standards, 587–588
  production and use, 586
  toxic effects, 587
Potassium iodate [7758-05-6]
  chemical and physical properties, 812t
  toxic effects, 814t
  uses, 811t
Potassium iodide [7681-11-0]
  chemical and physical properties, 812t
  toxic effects, 814t
Potassium iodine, uses, 811t
Potassium niobate [12030-85-2], chemical and physical properties, 39t
Potassium pentaborate, chemical and physical properties, 523t
Potassium perchlorate [7778-74-7]
  chemical and physical properties, 793, 794, 795t
  exposure standards, 799
  production and use, 794–796, 796t
  toxic effects, 796–797, 797t
Potassium permangante [7722-64-7], chemical and physical properties, 131t
Potassium perrhenate, toxic effects, 156
Potassium platinochloride. See Potassium tetrachloroplatinate(II)
Potassium tantalum fluoride [16924-00-8], toxic effects, 55t
Potassium tetraborate tetrahydrate, toxic effects, 533t
Potassium tetrachloropalladate(II) [10025-98-6], 271
  biomonitoring/biomarkers, 273
  carcinogenesis, 279
  chemical and physical properties, 271–272, 272t
  exposure assessment, 273
  exposure standards, 281
  genetic and cellular effects, 279
  neurological, pulmonary, skin sensitization effects, 279–280, 280t
  pharmacokinetics, metabolism, and mechanisms, 274–279
  production and use, 271–272, 272t
  reproductive and developmental effects, 279
  toxic effects
    acute and chronic, 273–274, 274t, 275t, 276t–277t
    experimental studies, 273–280
    human experience, 279–280, 280t
Potassium tetrachloroplatinate(II) [10025-99-7]
  carcinogenesis, 342
  chemical and physical properties, 292t, 341
  environmental impact studies, 343
  exposure assessment, 341
  exposure standards, 308t, 343
  genetic and cellular effects, 300, 342
  neurological, pulmonary, skin sensitization effects, 300–301, 342
  pharmacokinetics, metabolism, and mechanisms, 342, 343
  production and use, 341
  reproductive and developmental effects, 342
  toxic effects, 294t, 296t
    acute and chronic, 341–342, 343
    epidemiology studies, 343
    experimental studies, 341–342
    human experience, 342–343
Praseodymium, chemical and physical properties, 424t
Promethium
  carcinogenesis, 441–442
  chemical and physical properties, 424t
  environmental impact studies, 453–454
  exposure assessment, 434
  exposure standards, 453
  genetic and cellular effects, 442–443
  neurological, pulmonary, skin sensitization effects, 443–449, 447t
  pharmacokinetics, metabolism, and mechanisms, 437–440, 451–452
  production and use, 430–434, 431t
  reproductive and developmental effects, 440–441
  toxic effects
    acute and chronic, 435–437, 451
    epidemiology studies, 452–453
    experimental studies, 439–449
    human experience, 450–453

cis-PT(II). See cis-Diamminedichloroplatinum(II)
PT-01. See cis-Diamminedichloroplatinum(II)

Radiation, ionizing, exposure standards, 264, 264t
Radioactive ruthenium (Ruthenium 106 *[13967-48-1]*; Ruthenium 103 *[13968-53-1]*)
  carcinogenesis, 262, 263
  chemical and physical properties, 261
  environmental impact studies, 264–265
  exposure assessment, 262
  exposure standards, 264, 264t
  genetic and cellular effects, 262, 263
  pharmacokinetics, metabolism, and mechanisms, 262, 263
  production and use, 261–262
  reproductive and developmental effects, 262, 263
  toxic effects
    acute and chronic, 262, 263
    experimental studies, 262
    human experience, 262–263
Radon, 402–403, 405–406
Rare earth metals, 423–454, 423t
  carcinogenesis, 441–442
  chemical and physical properties, 423–424, 424–425, 424t, 428–430, 429t
  environmental impact studies, 453–454
  exposure assessment, 434
  exposure standards, 453
  exposure standards
    epidemiology studies, 452–453
  genetic and cellular effects, 442–443
  neurological, pulmonary, skin sensitization effects, 443–449, 447t
  pharmacokinetics, metabolism, and mechanisms, 437–440, 451–452
  production and use, 430–434
  reproductive and developmental effects, 440–441
  toxic effects
    acute and chronic, 435–437, 451
    experimental studies, 434–449
    human experience, 450–453
Rhenium *[7440-15-5]*
  chemical and physical properties, 129, 130, 153
  environmental impact studies, 158
  exposure assessment, 154
  exposure standards, 157
  pharmacokinetics, metabolism, and mechanisms, 157
  production and use, 153–154
  toxic effects, 130
    experimental studies, 154–157
    human experience, 157
Rhenium trichloride, toxic effects, 156

Rhodium *[7440-16-6]*
  biomonitoring/biomarkers, 266
  carcinogenesis, 268
  chemical and physical properties, 265–266, 265t
  exposure assessment, 266
  exposure standards, 269t–270t
  genetic and cellular effects, 268
  neurological, pulmonary, skin sensitization effects, 268
  pharmacokinetics, metabolism, and mechanisms, 267–268
  production and use, 266
  reproductive and developmental effects, 268
  toxic effects
    acute and chronic, 266–267, 267t
    experimental studies, 266–268, 267t
    human experience, 268
Rhodium chloride, chemical and physical properties, 265t
Royer@R. See Ruthenium
Rubidium *[7440-17-7]*
  chemical and physical properties, 601–602, 602t
  exposure assessment, 602
  exposure standards, 602
  production and use, 602
  toxic effects, 602
Rubidium carbonate *[584-09-8]*, chemical and physical properties, 602t
Rubidium chloride *[7791-11-9]*, chemical and physical properties, 602t
Rubidium hydroxide *[1310-82-3]*, chemical and physical properties, 602t
Rubidium iodine *[7790-29-6]*, chemical and physical properties, 602t
Ruthenium *[7440-18-8]*. See also Radioactive ruthenium
  biomonitoring/biomarkers, 256–257
  carcinogenesis, 259
  chemical and physical properties, 254, 255t
  exposure assessment, 256–257
  exposure standards, 261
  genetic and cellular effects, 259
  neurological, pulmonary, skin sensitization effects, 259–260
  pharmacokinetics, metabolism, and mechanisms, 258–259, 261
  production and use, 254–256
  reproductive and developmental effects, 259
  toxic effects
    acute and chronic, 257–258, 258t
    epidemiology studies, 260–261
    experimental studies, 257–260
    human experience, 260–261

SUBJECT INDEX

Ruthenium(III) chloride, chemical and physical properties, 255t
Ruthenium chloride hydroxide, chemical and physical properties, 255t
Ruthenium hydroxychloride, exposure standards, 261
Ruthenium(IV) oxide, chemical and physical properties, 255t
Ruthenium(VIII) oxide, chemical and physical properties, 255t
Ruthenium oxychloride, ammoniated (Ruthenium Red), chemical and physical properties, 255t

Sal ammoniac. See Ammonium chloride
Salammonite. See Ammonium chloride
Sal soda. See Sodium carbonate
Samaria. See Samarium oxide
Samarium, chemical and physical properties, 424t
Samarium oxide [12060-58-1]
   carcinogenesis, 441–442
   chemical and physical properties, 426, 428–430, 429t
   environmental impact studies, 453–454
   exposure assessment, 434
   exposure standards, 453
   genetic and cellular effects, 442–443
   neurological, pulmonary, skin sensitization effects, 443–449, 447t
   pharmacokinetics, metabolism, and mechanisms, 437–440, 451–452
   production and use, 430–434, 431t
   reproductive and developmental effects, 440–441
   toxic effects
      acute and chronic, 435–437, 451
      epidemiology studies, 452–453
      experimental studies, 434–449
      human experience, 450–453
Samarium sesquioxide. See Samarium oxide
Samarium trioxide. See Samarium oxide
Scandium chloride [10361-84-9]
   carcinogenesis, 441–442
   chemical and physical properties, 423–424, 424–425, 424t, 428–430, 429t
   environmental impact studies, 453–454
   exposure assessment, 434
   exposure standards, 453
   exposure standards
      epidemiology studies, 452–453
   genetic and cellular effects, 442–443
   neurological, pulmonary, skin sensitization effects, 443–449, 447t
   pharmacokinetics, metabolism, and mechanisms, 437–440, 451–452
   production and use, 430–434

reproductive and developmental effects, 440–441
toxic effects
   acute and chronic, 435–437, 451
   experimental studies, 434–449
   human experience, 450–453
Seld-A-Leaf. See Sodium chlorate
Seleninyl chloride [oxide]. See Selenium oxychloride
Selenious acid. See Sodium selenite
Selenium [7782-49-2]
   carcinogenesis, 477, 479
   chemical and physical properties, 473–474, 475t
   exposure assessment, 476
   exposure standards, 480
   genetic and cellular effects, 479
   neurological, pulmonary, skin sensitization effects, 479–480
   odor and warning properties, 474
   production and use, 474
   reproductive and developmental effects, 477
   toxic effects
      acute and chronic, 476–477, 479
      epidemiology studies, 478–480
      experimental studies, 476–477
      human experience, 477–480
Selenium dioxide [7446-08-4], chemical and physical properties, 475t
Selenium hexafluoride [7783-79-1]
   chemical and physical properties, 475t, 481
   exposure standards, 482, 767t
   production and use, 481
   toxic effects, 481–482
Selenium hydride. See Hydrogen selenide
Selenium oxychloride [7791-23-3]
   chemical and physical properties, 475t, 482
   exposure standards, 483
   production and use, 482
   toxic effects, 482–483
Selenium trioxide [13768-86-0], chemical and physical properties, 475t
Showchlon. See Sodium hypochlorite
Silicic acid. See Sodium metasilicate
Smoking
   and platinum allergy, 302
   and uranium exposure, 401–405
Soda ash. See Sodium carbonate
Soda lye. See Sodium hydroxide
Soda monohydride. See Sodium carbonate
Sodium [7440-23-5], 588
   biomonitoring/biomarkers, 590
   chemical and physical properties, 588–590, 589t
   exposure assessment, 590
   exposure standards, 590

Sodium *[7440-23-5], (Continued)*
  production and use, 590
  toxic effects, 590
Sodium borate. *See* Sodium tetraborate
Sodium borohydride, chemical and physical
  properties, 522t
Sodium bromate *[7789-38-0]*
  chemical and physical properties, 804t, 806–807
  production and use, 807
  toxic effects, 807–808
Sodium carbonate *[497-19-8]*
  biomonitoring/biomarkers, 592
  chemical and physical properties, 589t, 591
  exposure standards, 593
  production and use, 591–592
  toxic effects, 592–593
Sodium chlorate *[7775-09-9]*
  chemical and physical properties, 791–792, 794, 795t
  exposure standards, 799
  production and use, 794–796, 796t
  toxic effects, 796–797, 797t
Sodium chloride *[7647-14-5]*
  chemical and physical properties, 783, 784t
  exposure standards, 785
  production and use, 783–785
  toxic effects, 785
Sodium dioxide. *See* Sodium peroxide
Sodium(IV) hexachloroplatinate
  neurological, pulmonary, skin sensitization effects, 300–301
  pharmacokinetics, metabolism, and mechanisms, 299
Sodium hydroxide *[1310-73-2]*
  biomonitoring/biomarkers, 594
  chemical and physical properties, 589t, 593
  exposure standards, 595
  production and use, 593–594
  toxic effects, 594
Sodium hypochlorite *[7681-52-9]*
  chemical and physical properties, 792, 794, 795t
  exposure standards, 799
  production and use, 794–796, 796t
  toxic effects, 796–797, 797t, 799
Sodium iodate *[7681-55-2]*, chemical and physical properties, 812t
Sodium iodide *[7681-82-5]*
  chemical and physical properties, 812t
  toxic effects, 814t
  uses, 811t
Sodium metaborate dihydrate *[16800-11-6]*
  chemical and physical properties, 555–556
  environmental impact studies, 556
  exposure assessment, 556
  exposure standards, 556
  production and use, 556
  toxic effects, 533t, 556
Sodium metaborate tetrahydrate *[10555-76-7]*
  chemical and physical properties, 522t, 556–557
  environmental impact studies, 557
  exposure assessment, 557
  exposure standards, 557
  production and use, 557
  toxic effects, 533t, 557
Sodium metasilicate *[6834-92-0]*
  biomonitoring/biomarkers, 599
  chemical and physical properties, 589t, 598
  exposure standards, 600
  production and use, 598–599
  toxic effects, 599–600
Sodium metavanadate *[13718-26-8]*
  chemical and physical properties, 6t
  toxicity, 11, 12, 13t, 15t, 20, 21, 26
Sodium molybdate, chemical and physical properties, 94t
Sodium molybdenate, toxic effects, 156
Sodium orthophosphate. *See* Trisodium phosphate
Sodium orthovanadate *[13721-39-6]*
  chemical and physical properties, 6t
  toxicity, 24–25
Sodium perborate monohydrate *[10332-33-9]*
  chemical and physical properties, 521t, 560
  environmental impact studies, 561
  exposure standards, 561
  production and use, 560
  toxic effects, 560–561
Sodium perborate tetrahydrate *[10486-00-7]*
  chemical and physical properties, 521t, 557–558
  environmental impact studies, 559–560
  exposure assessment, 558
  exposure standards, 559
  production and use, 558
  toxic effects, 558–559
Sodium peroxide *[1313-60-6]*
  biomonitoring/biomarkers, 596
  chemical and physical properties, 589t, 595
  exposure standards, 596
  production and use, 595
  toxic effects, 596
Sodium perrhenate, toxic effects, 155–156
Sodium phosphate. *See* Trisodium phosphate
Sodium pyroborate decohydrate. *See* Sodium tetraborate decahydrate
Sodium salt. *See* Sodium perborate tetrahydrate
Sodium selenite *[10102-18-8]*
  chemical and physical properties, 475t, 483

SUBJECT INDEX

exposure standards, 484
genetic and cellular effects, 484
reproductive and developmental effects, 483-484
toxic effects
  acute and chronic, 483
  experimental studies, 483-484
Sodium tetraborate, anhydrous, toxic effects, 533, 533t
Sodium tetraborate *[1330-43-4]*
  chemical and physical properties, 550
  environmental impact studies, 551
  exposure assessment, 550
  exposure standards, 551
  production and use, 550
  toxic effects, 551
Sodium tetraborate decahydrate *[1303-96-4]*
  carcinogenesis, 543
  chemical and physical properties, 521t, 546-547
  environmental impact studies, 544, 548
  exposure assessment, 547-548
  exposure guidelines, 543-544
  exposure standards, 548
  genetic and cellular effects, 538-539
  neurological, pulmonary, skin sensitization effects, 539-540, 543
  odor and warning properties, 547
  pharmacokinetics, metabolism, and mechanisms, 539-540, 541-542
  production and use, 547
  reproductive and developmental effects, 542-543
  toxic effects, 533, 533t, 534, 536-537, 548
    acute and chronic, 539, 540-541
    epidemiology studies, 540-544
    human experience, 539-544
Sodium tetraborate pentahydrate *[12179-04-3]*
  chemical and physical properties, 521t, 548-549
  environmental impact studies, 550
  exposure assessment, 549
  exposure standards, 549
  odor and warning properties, 549
  production and use, 549
  toxic effects, 532-533, 533t, 549
Sodium tetraphenyl boron, chemical and physical properties, 523t
Sodium tetravanadate *[12058-74-1]*, chemical and physical properties, 6t
Solubor. *See* Disodium octaborate tetrahydrate; Sodium tetraborate decahydrate
Sonac. *See* Phosphoric acid
Soviet Union, exposure standards
  niobium, 48-49
  ruthenium, 261
Steel. *See* Iron

Stink Damp. *See* Hydrogen sulfide
Strontium chromate, exposure standards, 92t
Sufinyl chloride. *See* Thionyl chloride
Sulfocarbonic anhyride. *See* Carbon disulfide
Sulfur *[63705-05-5]*
  chemical and physical properties, 489, 490t
  exposure assessment, 491
  exposure standards, 491
  production and use, 489
  toxic effects, 491
Sulfur chloride oxide. *See* Thionyl chloride
Sulfur dioxide *[7446-09-5]*
  carcinogenesis, 494
  chemical and physical properties, 490t, 491-492
  exposure assessment, 493
  exposure standards, 494
  odor and warning properties, 492
  production and use, 492
  toxic effects, 493-494
    acute and chronic, 494
Sulfuretted hydrogen. *See* Hydrogen sulfide
Sulfur hexafluoride *[2551-62-4]*
  chemical and physical properties, 752-753
  exposure assessment, 753
  exposure standards, 755, 756t
  production and use, 753
  toxic effects
    experimental studies, 753-755, 754t
    gasous *vs.* particulate states, 755-757
    human experience, 753-755, 754t
Sulfur hydride. *See* Hydrogen sulfide
Sulfuric acid *[7664-93-9]*
  acute and chronic, 511
  carcinogenesis, 511
  chemical and physical properties, 490t, 508
  exposure assessment, 510
  exposure standards, 511
  neurological, pulmonary, skin sensitization effects, 511
  odor and warning properties, 508
  production and use, 508-510
  toxic effects
    acute and chronic, 511
    experimental studies, 510-511
    human experience, 511
Sulfuric acid nickel salt. *See* Nickel sulfate
Sulfur monochloride *[10025-67-9]*
  chemical and physical properties, 490t, 506
  exposure assessment, 506
  exposure standards, 507
  odor and warning properties, 506
  toxic effects, 506-507

SUBJECT INDEX

Sulfurous oxychloride. *See* Thionyl chloride
Sulfur phosphide. *See* Phosphorus pentasulfide
Sulfur tetrafluoride *[7783-60-0]*
  chemical and physical properties, 752–753
  exposure assessment, 753
  exposure standards, 755, 756t
  production and use, 753
  toxic effects
    experimental studies, 753–755, 754t
    gasous *vs.* particulate states, 755–757
    human experience, 753–755, 754t
Sulfur trioxide *[7446-11-9]*, chemical and physical properties, 490t
Sweden, exposure standards
  manganese, 152t
  nickel, 212t
  nickel carbonate, 223t
  nickel carbonyl, 239t
  nickel chloride, 245t
  nickel oxide, 220t
  nickel subsulfide, 229t
  nickel sulfate, 253t
  osmium tetroxide, 289, 290t
  vanadium, 36t
Switzerland, exposure standards, niobium, 48

Tantalum *[7440-25-7]*
  biomonitoring/biomarkers, 53–54
  carcinogenesis, 58, 60
  chemical and physical properties, 1–4, 2t, 49–51, 50t
  environmental impact studies, 61–62
  exposure assessment, 52–54
  exposure standards, 61, 61t
  genetic and cellular effects, 58, 60
  molecular structure, 49
  neurological, pulmonary, skin sensitization effects, 58, 60
  occupational exposure limits, 4
  odor and warning properties, 51
  pharmacokinetics, metabolism, and mechanisms, 56–58, 59–60
  production and use, 51–52
  reproductive and developmental effects, 60
  toxic effects, 2–3
    acute and chronic, 54–56
    epidemiology studies, 60–61
    experimental studies, 54–58, 55t
    human experience, 58–61
Tantalum carbide [12070-06-3], chemical and physical properties, 50t
Tantalum chloride *[7721-01-9]*
  chemical and physical properties, 50t
  toxic effects, 55t
Tantalum fluoride *[7783-71-3]*
  chemical and physical properties, 50t
  toxic effects, 55t
Tantalum oxide *[1314-61-0]*
  chemical and physical properties, 50t
  toxic effects, 55t
Tellurium *[13494-80-9]*
  chemical and physical properties, 484, 488t
  exposure assessment, 485
  exposure standards, 487
  production and use, 485
  reproductive and developmental effects, 486
  toxic effects, 485–486
    acute and chronic, 486
Tellurium hexafluoride *[7783-80-4]*
  chemical and physical properties, 487, 488t
  exposure assessment, 487
  exposure standards, 488, 767t
  odor and warning properties, 487
  production and use, 487
  toxic effects, 488
Terbium, chemical and physical properties, 424t
Terbium (+3) salt. *See* Terbium trinitrate
Terbium trinitrate *[10043-27-3]*
  carcinogenesis, 441–442
  chemical and physical properties, 427, 428–430, 429t
  environmental impact studies, 453–454
  exposure assessment, 434
  exposure standards, 453
  genetic and cellular effects, 442–443
  neurological, pulmonary, skin sensitization effects, 443–449, 447t
  pharmacokinetics, metabolism, and mechanisms, 437–440, 451–452
  production and use, 430–434, 431t
  reproductive and developmental effects, 440–441
  toxic effects
    acute and chronic, 435–437, 451
    epidemiology studies, 452–453
    experimental studies, 434–449
    human experience, 450–453
Testosterone, oxygen interaction, 667t
Tetraborate. *See* Sodium tetraborate
  decohydrade (*See* Sodium tetraborate decahydrate)
  pentahydrate (*See* Sodium tetraborate pentahydrate)
Tetracarbonylnickel. *See* Nickel carbonyl
Tetrachloroplatinic(II) acid, environmental impact, 309
Tetrachloroplatinum. *See* Platinum(IV) chloride
Tetraphosphorus trisulfide *[1314-85-8]*

SUBJECT INDEX

chemical and physical properties, 460t, 466
exposure standards, 467
production and use, 467
toxic effects, 467
Tetrasodium salts, decahydrate. *See* Sodium tetraborate decahydrate
Thionyl chloride *[7719-09-7]*
  chemical and physical properties, 490t, 507
  exposure assessment, 507
  exposure standards, 508
  production and use, 507
  toxic effects, 508
Thorium *[7440-29-1]*
  biomonitoring/biomarkers, 413–414
  carcinogenesis, 415
  chemical and physical properties, 409, 410t
  exposure assessment, 412–414
  exposure standards, 417
  genetic and cellular effects, 415
  odor and warning properties, 409
  pharmacokinetics, metabolism, and mechanisms, 414–415
  production and use, 409–412
  toxic effects
    epidemiology studies, 415–417
    experimental studies, 414–415
    human experience, 415–417
Thorium dioxide, chemical and physical properties, 410t
Thorium fluoride, chemical and physical properties, 410t
Thorium hydroxide, chemical and physical properties, 410t
Thorium nitrate tetrahydrate, chemical and physical properties, 410t
Thorium oxalate, chemical and physical properties, 410t
Thorium sulfate, chemical and physical properties, 410t
Thorotrast, toxic effects, 416–417
Three Elephant. *See* Boric acid
Thulium
  carcinogenesis, 441–442
  chemical and physical properties, 424t
  environmental impact studies, 453–454
  exposure assessment, 434
  exposure standards, 453
  genetic and cellular effects, 442–443
  neurological, pulmonary, skin sensitization effects, 443–449, 447t
  pharmacokinetics, metabolism, and mechanisms, 437–440, 451–452
  production and use, 430–434, 431t

reproductive and developmental effects, 440–441
toxic effects
  acute and chronic, 435–437, 451
  epidemiology studies, 452–453
  experimental studies, 439–449
  human experience, 450–453
Thyroxine, oxygen interaction, 667t
Tim-Bor. *See* Disodium octaborate tetrahydrate
Triatomic oxygen. *See* Ozone
Tribromoborane. *See* Boron tribromide
Triethylborane, chemical and physical properties, 523t
Trifluoroborane. *See* Boron trifluoride
Trihydroxyborane. *See* Boric acid
Trimethyl amine borane, chemical and physical properties, 523t
Trinickel disulfide. *See* Nickel subsulfide
Trisodium phosphate *[7601-54-9]*
  biomonitoring/biomarkers, 597
  chemical and physical properties, 589t, 597
  exposure standards, 598
  production and use, 597
  toxic effects, 598
Trisodium salt. *See* Trisodium phosphate
Trisulfurated phosphorus. *See* Tetraphosphorus trisulfide
Triuranium octoxide, chemical and physical properties, 382t
Tumbleaf. *See* Sodium chlorate
Tungsten *[7440-33-7]*
  biomonitoring/biomarkers, 110–111
  carcinogenesis, 113, 118
  chemical and physical properties, 106–108, 107t
  exposure assessment, 109–111
  exposure standards, 121
  genetic and cellular effects, 113, 119
  neurological, pulmonary, skin sensitization effects, 113–115, 119–120
  odor and warning properties, 108
  pharmacokinetics, metabolism, and mechanisms, 112–113, 118
  production and use, 108–109
  reproductive and developmental effects, 113, 118
  toxic effects
    acute and chronic, 111–112, 117–118
    epidemiology studies, 120–121
    experimental studies, 111–115
    human experience, 115–121
Tungsten acid, chemical and physical properties, 107t
Tungsten carbide, chemical and physical properties, 107t
Tungsten disulfide, chemical and physical properties, 107t

SUBJECT INDEX

Tungsten hexachloride, chemical and physical properties, 107t
Tungsten oxytetrachloride, chemical and physical properties, 107t
Tungsten trioxide, chemical and physical properties, 107t

Ulexite, chemical and physical properties, 521t
United Kingdom, exposure standards
  manganese, 152t
  nickel, 212t
  nickel carbonate, 223t
  nickel carbonyl, 239t
  nickel chloride, 245t
  nickel hydroxide, 247t
  nickel oxide, 220t
  nickel subsulfide, 229t
  nickel sulfate, 253t
  osmium tetroxide, 289, 290t
  platinum, 308t
  tantalum, 61t
  vanadium, 36t
Uranium *[7440-61-1]*
  biomonitoring/biomarkers, 390–392
  carcinogenesis, 398, 401–405
  chemical and physical properties, 381–384, 382t
  enriched, 384, 384t
  exposure assessment, 387–392, 389t
  exposure standards, 406–408, 407t
  genetic and cellular effects, 399, 405–406
  pharmacokinetics, metabolism, and mechanisms, 395–398, 396t
  production and use, 287f, 384–387
  reproductive and developmental effects, 398
  toxic effects
    acute and chronic, 392–395, 399–401
    epidemiology studies, 401–406
    experimental studies, 392–399
    human experience, 399–406
Uranium dioxide, chemical and physical properties, 382t
Uranium hexafluoride, chemical and physical properties, 382t
Uranium oxyfluoride. *See* Uranyl fluoride
Uranium tetrachloride, chemical and physical properties, 382t
Uranium tetrafluoride, chemical and physical properties, 382t
Uranyl acetate, dihydrate, chemical and physical properties, 382t
Uranyl chloride, chemical and physical properties, 382t
Uranyl fluoride, chemical and physical properties, 382t
Uranyl nitrate, chemical and physical properties, 382t
Uranyl oxide, chemical and physical properties, 382t

Vanadate, toxicity, 20, 22
Vanadium *[7440-62-2]*
  biomonitoring/biomarkers, 10
  carcinogenesis, 21–22, 31, 34–35
  chemical and physical properties, 1–4, 2t, 4–5, 6t
  environmental impact studies, 35–38
  exposure assessment, 7–10
  exposure standards, 35, 36t
  genetic and cellular effects, 22–24, 23t, 31, 35
  mitogenic potential of, 3
  neurological, pulmonary, skin sensitization effects, 24–26, 35
  occupational exposure limits, 4
  odor and warning properties, 5
  pharmacokinetics, metabolism, and mechanisms of, 12–19, 19t, 30, 34
  production and use, 5–7
  reproductive and developmental effects, 19–21, 31, 34
  toxic effects, 2–3
    acute and chronic, 3–4, 10–12
    epidemiology studies, 31–35
    experimental studies, 10–26
    human experience, 26–35
Vanadium dichloride *[10580-52-6]*, chemical and physical properties, 6t
Vanadium gluconate, 25
Vanadium oxydichloride *[10213-09-9]*
  chemical and physical properties, 6t
  toxicity, 23t
Vanadium oxytrichloride *[7727-18-6]*, chemical and physical properties, 6t
Vanadium pentoxide *[1314-62-1]*
  chemical and physical properties, 6t
  toxicity, 10–12, 13t–15t, 20, 21, 23t, 24, 26, 28–30, 32–33
Vanadium sulfate, toxicity, 24
Vanadium tetrachloride *[7632-51-1]*
  chemical and physical properties, 6t
  odor and warning properties, 5
Vanadium tribromide *[13470-26-3]*, chemical and physical properties, 6t
Vanadium trichloride *[7718-98-1]*
  chemical and physical properties, 6t
  odor and warning properties, 5
  toxicity, 26

SUBJECT INDEX

Vanadium trioxide [1314-34-7]
　chemical and physical properties, 6t
　toxicity, 25, 26
Vanadium trioxybromide [13520-90-6], chemical and physical properties, 6t
Vanadyl sulfate [27774-13-6]
　chemical and physical properties, 6t
　toxicity, 14t, 20
Vanadyl sulfate pentahydrate [12439-96-2], chemical and physical properties, 6t
VCl₃. See Vanadium trichoride
VCl₄. See Vanadium tetrachloride
Vitriol brown oil. See Sulfuric acid

Weeviltox. See Carbon disulfide
Wolfram. See Tungsten
WP. See Phosphorus

X-14. See Calcium hypochlorite

Ytterbium [7440-64-4]
　carcinogenesis, 441–442
　chemical and physical properties, 424t, 427, 428–430, 429t
　environmental impact studies, 453–454
　exposure assessment, 434
　exposure standards, 453
　genetic and cellular effects, 442–443
　neurological, pulmonary, skin sensitization effects, 443–449, 447t
　pharmacokinetics, metabolism, and mechanisms, 437–440, 451–452
　production and use, 430–434, 431t
　reproductive and developmental effects, 440–441
　toxic effects
　　acute and chronic, 435–437, 451
　　epidemiology studies, 452–453
　　experimental studies, 434–449
　　human experience, 450–453
Yttrium [7440-65-5]
　carcinogenesis, 441–442
　chemical and physical properties, 423–424, 424t, 427, 428–430, 429t
　environmental impact studies, 453–454
　exposure assessment, 434
　exposure standards, 453
　genetic and cellular effects, 442–443
　neurological, pulmonary, skin sensitization effects, 443–449, 447t
　pharmacokinetics, metabolism, and mechanisms, 437–440, 451–452
　production and use, 430–434, 431t
　reproductive and developmental effects, 440–441
　toxic effects
　　acute and chronic, 435–437, 451
　　epidemiology studies, 452–453
　　experimental studies, 434–449
　　human experience, 450–453

Zinc borate, toxic effects, 533, 533t
Zinc borate hydrate [138265-88-0]
　chemical and physical properties, 522t, 561–562
　environmental impact studies, 562
　exposure assessment, 562
　exposure standards, 562
　production and use, 562
　toxic effects, 562
Zinc chromates, exposure standards, 92t

Chemical Index

Acid mist. *See* Sulfuric acid
Alcohol of sulfur. *See* Carbon disulfide
Alkaline-earth metals, 782–785
Alkaline materials, 583–584, 583–604
Alnico. *See* Nickel
Altacide. *See* Sodium chlorate
Amchlor. *See* Ammonium chloride
Ammonium bromide, 804t
Ammonium chloride *[12125-02-9]*, 784t
Ammonium hexafluorovanadate *[13815-31-1]*, 6t
Ammonium iodide *[12027-06-4]*, 812t
Ammonium metavanadate *[7803-55-6]*, 6t, 23t, 25, 26
Ammonium molybdate, 94t
Ammonium muriate. *See* Ammonium chloride
Ammonium pentaborate tetrahydrate, 523t
Ammonium perchlorate *[7790-98-9]*, 793–799, 795t, 796t, 797t
Ammonium tetrachloropalladate(II) *[13820-40-1]*, 271–281, 272t, 274t–277t, 280t
Ammonium vanadate, 22, 23, 24, 26
Antiformin. *See* Sodium hypochlorite
Antisal 2B. *See* Hydrogen fluoride
Aquacat. *See* Cobalt
ASH. *See* Sodium carbonate
Astatine *[7440-68-8]*, 815–816
Augus hot rod. *See* Sodium hydroxide

Barax, 10 mol. *See* Sodium tetraborate decahydrate
Barium metaborate, 522t
Binary salts of alkali, 782–785
Biobor JF, 522t
Bipotassium tetrachloroplatinate. *See* Potassium tetrachloroplatinate(II)
1,3,-Bis(2-chloroethyl)-1-nitrosourea, 667t
Bismuth orthovanadate *[53801-77-7]*, 11, 15t
Bis($N^5$2,4-cyclopentadien-l-yl)nickel. *See* Nickelocene
B-K. *See* Sodium hypochlorite
BK powder. *See* Calcium hypochlorite
Bleach. *See* Sodium hypochlorite
Bleomycin, 666, 667t
Bora-Care. *See* Disodium octaborate tetrahydrate
Borax. *See* Sodium tetraborate decahydrate
Borax, anhydrous. *See* Sodium tetraborate
Borax pentahydrate. *See* Sodium tetraborate pentahydrate
Boric acid *[10043-35-3]*, 521t, 532–544, 533t, 551–552
Boricin. *See* Sodium tetraborate decahydrate
Boric oxide *[1303-86-2]*, 522t, 552–554
Boroethane. *See* Diborane
Borofax. *See* Boric acid
Borogard ZB. *See* Zinc borate hydrate
Boron, and borates, 519–544, 521t–523t, 525t–528t, 533t
Boron carbine, 523t
Boron, elemental *[7440-42-8]*, 523t, 544–546
Boron nitride, 523t
Boron sodium oxide. *See* Sodium tetraborate
Boron tribromide *[10294-33-4]*, 523t, 562–563
Boron trifluoride *[7637-07-2]*, 523t, 563–565
Boron trihydroxide. *See* Boric acid
Bromine *[7726-95-6]*, 731, 732t, 799–809, 804t
Bromine pentafluoride, 804t

855

856   **CHEMICAL INDEX**

Bromine trifluoride, 766t, 804t
Bunsenite. *See* Nickel oxide

CACP. *See cis*-Diamminedichloroplatinum(II)
Cadmium, 424t
Calcinated soda. *See* Sodium carbonate
Calcium chloride *[10043-52-4]*, 782–785
Calcium chlorite *[7758-19-2]*, 793–799, 795t, 796t, 797t
Calcium dichloride. *See* Calcium chloride
Calcium fluoride *[7789-75-5]*, 757–765, 758t, 764t
Calcium hypochlorite *[7778-54-3]*, 792–799, 795t–797t
Calcium molybdate, 94t
Calcium oxychloride. *See* Calcium hypochlorite
Calcium salt. *See* Calcium hypochlorite
Calcosan. *See* Calcium chloride
Carbolith. *See* Lithium
Carbon dioxide *[124-35-9]*, 625–629
Carbon disulfide *[75-15-0]*, 490t, 502–505, 505t
Carbonic acid. *See* Nickel carbonate
Carbonic acid disodium salt. *See* Sodium carbonate
Carbonic dichloride. *See* Phosgene
Carbonic oxide. *See* Carbon monoxide
Carbon monoxide *[630-08-0]*, 607–625, 613f, 614f, 618t
Carbon oxychloride. *See* Phosgene
Carbonyl [di]chloride. *See* Phosgene
Carbonyl fluoride, 766t
Carbonyls, 171, 182
Carboplatin. *See cis*-Diammine(1,1-cyclobutanedicarboxylato)platinum(II)
Carrel-Dakin solution. *See* Sodium hypochlorite
Catecholamines, 667t
Caustic soda. *See* Sodium hydroxide
CDDP. *See cis*-Diamminedichloroplatinum(II)
Cerium *[7440-45-1]*, 424t, 425, 428–454, 429t, 431t, 447t
Cesium *[7440-46-2]*, 600, 601, 601t
Cesium chloride *[7647-17-8]*, 601t
Cesium hydroxide *[21351-79-1]*, 601t
Cesium iodine *[7789-17-5]*, 601t
Cesium nitrate *[7789-18-6]*, 601t
CG. *See* Phosgene
Chemotherapeutic drugs, 666, 667t
Chlorate of soda. *See* Sodium chlorate
Chlorax. *See* Sodium chlorate
Chloric acid *[7790-93-4]*, 791, 794–799, 795t, 796t, 797t
Chloride of lime. *See* Calcium hypochlorite
Chloride of phosphorus. *See* Phosphorus trichloride
Chloride of sulphur. *See* Sulfur monochloride
Chlorine *[7782-50-5]*, 731, 732t, 765, 768–775

Chlorine dioxide *[10049-04-4]*, 766t, 786–791, 787t
Chlorine pentafluoride *[13637-63-3]*, 766t, 786–791, 787t
Chlorine trifluoride *[7790-91-2]*, 766t, 785–791, 787t
Chloroformyl chloride. *See* Phosgene
Chlorohydric acid. *See* Hydrogen chloride
Chloroplatinic acid. *See* Potassium tetrachloroplatinate(II)
Chromic acid, 76t
Chromic chloride, 76t
Chromic oxide, 76t
Chromic trioxide, 76t
Chromium *[7440-47-3]*, 75–93, 76t, 78t, 92t
Chromium sesquioxide, 76t
Chromous chloride, 76t
Chromyl chloride, 76t, 92
C.I. 77320. *See* Cobalt
C.I. 77775. *See* Nickel
C.I. 77777. *See* Nickel oxide
Cibalith. *See* Lithium
Cisplatin. *See cis*-Diamminedichloroplatinum(II)
Cisplatyl. *See cis*-Diamminedichloroplatinum(II)
Clorox. *See* Sodium hypochlorite
CMT. *See* Cyclopentadienyl manganese tribcarbonyl
Cobalt *[7440-48-4]*, 179–190, 179t
Cobalt carbonyl, 179t, 190
Cobalt hydrocarbonyl, 179t, 190
Cobaltic-cobaltous oxide, 179t
Cobaltous oxide, 179t
Colemanite, 521t
Columbium. *See* Niobium
CPDC. *See cis*-Diamminedichloroplatinum(II)
CPDD. *See cis*-Diamminedichloroplatinum(II)
Cryolite *[15096-52-3]*, 757–765, 758t, 764t
Crystal carbonate. *See* Sodium carbonate
Cyclopentadienyl manganese tribcarbonyl (CMT), 135–136, 152t
Cyclophosphamide, 666, 667t
Cytosine-guanine. *See* Phosgene

Dakin's Solution. *See* Sodium hypochlorite
Darammon. *See* Ammonium chloride
dCDP. *See cis*-Diamminedichloroplatinum(II)
DDP. *See cis*-Diamminedichloroplatinum(II)
cis-DDP. *See cis*-Diamminedichloroplatinum(II)
DDPt. *See cis*-Diamminedichloroplatinum(II)
Decaborane *[17702-41-9]*, 523t, 569–570
2′-Deoxycytidine diphosphate. *See cis*-Diamminedichloroplatinum(II)
Detia gas EX-B. *See* Phosphine
Deuteriumfluoride. *See* Hydrogen fluoride

CHEMICAL INDEX

Dexamethasone, 667t
Diamineedichloroplatinum. See cis-
　Diamminedichloroplatinum(II)
(SP-4-2)-diaminodichloroplatinum. See cis-
　Diamminedichloroplatinum(II)
cis-Diamminedichloroplatinum(II) [15663-27-1],
　308t, 322–338, 324t, 328t–330t
cis-Diammine(1,1-
　cyclobutanedicarboxylato)platinum(II)
　[41575-94-4], 308t, 315–322
Diammonium tetraborate tetrahydrate, 523t
Diborane [19287-45-7], 523t, 565–567
Dicyclopentadienylnickel. See Nickelocene
Dierbium trioxide. See Erbium trioxide
Dieuropium trioxide. See Europium oxide
Digadolinium trioxide. See Gadolinium oxide
Dilanthanum oxide. See Lanthanum oxide
Dineodymium trioxide. See Neodymium oxide
Dinitrogen monoxide. See Nitrous oxide
Di-p-cyclopentadienyinickel. See Nickelocene
Diphosphorus pentaoxide. See Phosphorus pentoxide
Di-phosphorus pentasulfide. See Phosphorus
　pentasulfide
Dipotassium salt. See Potassium
　tetrachloroplatinate(II)
Dipotassium tetraborate tetrahydrate, 523t
Dipping acid. See Sulfuric acid
Disamarium trioxide. See Samarium oxide
Disodium carbonate. See Sodium carbonate
Disodium metasilicate. See Sodium metasilicate
Disodium octaborate, 533
Disodium octaborate tetrahydrate [12280-03-4],
　521t, 522t, 540, 554, 555
Disodium perborate tetrahydrate, 533t
Disodium salt. See Sodium selenite
Disodium salt, pentahydrate. See Sodium tetraborate
　pentahydrate
Disodium selenite. See Sodium selenite
Disodium tetraborate. See Sodium tetraborate
Disodium tetraborate decahydrate. See Sodium
　tetraborate decahydrate
Disodium tetraborate pentahydrate. See Sodium
　tetraborate pentahydrate
Disulfur dichloride. See Sulfur monochloride
Dithiocarbonic anhydride. See Carbon disulfide
dNRC. See Nuclear Regulatory Commission
Dodecaboron. See Zinc borate hydrate
Drop-Leaf. See Sodium chlorate
Drumet. See Sodium metasilicate
Dysprosia. See Dysprosium oxide
Dysprosium, 424t
Dysprosium oxide [1308-87-8], 427–454, 429t, 431t,
　447t

Epinephrine, 667t
Erbia. See Erbium trioxide
Erbium, 424t
Erbium sesquioxide. See Erbium trioxide
Erbium trioxide [12061-16-4], 427–454, 429t, 431t,
　447t
Eskalith. See Lithium
Etching acid. See Hydrogen fluoride
Europium, 424t
Europium oxide [1308-96-9], 426–454, 429t, 431t,
　447t

Ferric chloride, 170t
Ferric sulfate, 170t
Ferroboron, 522t
Ferrocene, 170t
Ferrochromium, 89–90
Ferrous oxide, black, 170t
Ferrous sulfate, 170t
Ferrovanadium [12604-58-9], 6t, 11, 12, 13t, 15t, 32
Fertilizer borate granular, 521t
Firebrake ZB. See Zinc borate hydrate
Fluorine [7782-41-4], 731–739, 732t, 736t, 738t
Fluorohydric acid. See Hydrogen fluoride
Francium [7440-73-5], 602–603

Gadolinia. See Gadolinium oxide
Gadolinium [7440-54-2], 424t, 426–454, 429t, 431t,
　447t
Gadolinium oxide [12064-62-9], 427–454, 429t,
　431t, 447t
Gadolinium sesquioxide. See Gadolinium oxide
Gadolinium trioxide. See Gadolinium oxide
Gas EX-B. See Phosphine
Gerstley borate. See Sodium tetraborate decahydrate

Halogens, 731–816, 732t
Harvest Aid. See Sodium chlorate
Heazlewoodite. See Nickel subsulfide
Hepatic acid. See Hydrogen sulfide
Hexachloroplatinic(IV) acid, 309
HF A. See Hydrogen fluoride
Hidrotic acid, See Hydrogen iodide
Holmium, 424t, 430–454, 431t, 447t
HTH. See Calcium hypochlorite
Hy-Chlor. See Calcium hypochlorite
Hydrofluoric acid. See Hydrogen fluoride
Hydrogen bromide [10035-10-6], 803–805, 804t
Hydrogen chloride [7647-01-0], 775–778
Hydrogen fluoride [7664-39-3], 739–751, 740t, 742t,
　752t
Hydrogen iodide [10034-85-2], 810–811, 811t, 812t,
　814t

CHEMICAL INDEX

Hydrogen orthoborate. *See* Boric acid
Hydrogen phosphide. *See* Phosphine
Hydrogen selenide *[7783-07-5]*, 475t, 480, 481
Hydrogen sulfate. *See* Sulfuric acid
Hydrogen sulfide *[7783-06-4]*, 490t, 495–502, 497t
Hydrogen telluride *[7783-09-7]*, 488–489, 488t
Hydroselenic acid. *See* Hydrogen selenide
Hydrosulfuric acid. *See* Hydrogen sulfide
Hypochlorous acid *[7790-92-3]*, 791, 794–799, 795t, 796t, 797t

Iodates, 811, 812t
Iodic acid *[7782-68-5]*, 812t
Iodides, 811, 812t
Iodine *[7553-56-2]*, 731, 732t, 809–815, 811t, 812t, 814t
Iodine bromide *[7789-33-5]*, 812t
Iodine chloride *[7790-99-0]*, 811t, 812t
Iodine monobromide, 811t
Iodine monochloride, 814t
Iodine pentafluoride *[7783-66-6]*, 811t, 812t, 814t
Iodine tribromide, 812t
Iodine trichloride, 811t
Ionizing radiation, 264
Iron *[7439-89-6]*, 169–179, 170t
Iron oxide, magnetite, red, 170t
Iron pentacarbonyl, 170t, 171

Javell water. *See* Sodium hypochlorite
Javex. *See* Sodium hypochlorite

Kalium. *See* Potassium
Khislevudite. *See* Nickel subsulfide
Kill-off. *See* Boric acid
Kjel-sorb. *See* Boric acid
Kodak balanced alkali. *See* Sodium metaborate dihydrate

Lanthana. *See* Lanthanum oxide
Lanthanides, 423–454, 424t, 429t, 447t
Lanthanum, 424
Lanthanum nitrate *[10099-59-9]*, 425, 428–454, 429t, 431t, 447t
Lanthanum oxide *[1312-81-8]*, 425, 428–454, 429t, 431t, 447t
Lead, 424t
Lead chromate, 92t
Lime chloride. *See* Calcium hypochlorite
Lithane. *See* Lithium
Lithium *[7439-93-2]*, 603, 604
Lithizine. *See* Lithium
Lithobid. *See* Lithium
LN2. *See* Nitrogen

Lo-Bax. *See* Calcium hypochlorite
Losantin. *See* Calcium hypochlorite
Lutetium, 424t
Lye. *See* Potassium hydroxide; Sodium hydroxide

Maganese acetate *[638-38-0]*, 131t
Maganese carbonate *[592-62-9]*, 131t
Maganese chloride *[7773-01-5]*, 131t
Maganese dioxide *[1313-13-9]*, 131t, 132
Maganese sulfate *[7785-87-7]*, 131t
Maganese tetroxide *[1317-35-7]*, 125t, 131t
Manganese *[7439-96-5]*, 129–153, 131t, 150t, 152t
Manganese oxide, 132
Manganese sulfate, 132
Metaborate. *See* Sodium perborate tetrahydrate
Metavanadate, 20–21
Methyl borate, 522t
Methylcyclopentadienyl manganese tricarbonyl (MMT) *[12108-13-3]*, 131t, 132–135, 152t
Methylprednisolone, 667t
Metso Beads. *See* Sodium metasilicate
MMT. *See* Methylcyclopentadienyl manganese tricarbonyl
Molybdenum [7439-98-7], 93–106, 94t
Molybdenum disulfide, 94t
Molybdic oxide, 94t
Mononickel oxide. *See* Nickel oxide
Muriatic acid. *See* Hydrogen chloride

Natrium. *See* Sodium
Neobor. *See* Sodium tetraborate pentahydrate
Neodymia. *See* Neodymium oxide
Neodymium *[7440-00-8]*, 424t, 428–454, 429t, 431t, 447t
Neodymium oxide *[1313-97-9]*, 426, 428–454, 429t, 431t, 447t
Neodymium sesquioxide. *See* Neodymium oxide
Neodymium trioxide. *See* Neodymium oxide
Neoplatin. *See cis*-Diamminedichloroplatinum(II)
Nickel *[7440-02-0]*, 195–213, 197t, 198t, 200t, 212t
Nickel carbonate *[3333-67-3]*, 197t, 219–222, 223t
Nickel carbonyl *[13463-39-3]*, 197t, 231–238, 234t, 239t
Nickel chloride *[7718-54-9]*, 197t, 238–243, 245t
Nickel hydroxide *[12054-48-7]*, 197t, 244–246, 247t
Nickelocene *[1271-28-9]*, 197t, 228, 230–231, 232t
Nickel oxide *[1313-99-1]*, 197t, 213–219, 214t, 216t–217t, 220t
Nickel subsulfide *[12035-72-2]*, 197t, 222–228, 226t–227t, 229t
Nickel sulfate *[7786-81-4]*, 197t, 246–252, 248t, 253t
Nickel tetracarbonyl. *See* Nickel carbonyl

CHEMICAL INDEX

(T-4)-Nickel tetracarbonyl. *See* Nickel carbonyl
Niobium *[7440-03-1]*, 1–4, 2t, 38–49, 39t
Niobium carbide *[12069-94-2]*, 39t
Niobium dioxide *[12034-59-2]*, 39t
Niobium hydride *[12655-93-5]*, 39t
Niobium pentachloride *[10026-12-7]*, 39t
Niobium pentafluoride [7783-68-8], 39t
Niobium pent[a]oxide *[1313-96-8]*, 39t
Niobium potassium fluoride *[17523-77-2]*, 39t
Nitric acid. *See* Terbium trinitrate
Nitric oxide *[10102-43-9]*, 637–651, 640t, 666
Nitrofurantoin, 667t
Nitrogen *[7727-37-9]*, 634–637
Nitrogen dioxide *[10102-44-0]*, 637–651, 640t
Nitrogen trifluoride, 766t
Nitrous oxide *[10024-97-2]*, 651–655
Norepinephrine, 667t
NSC 119875. *See cis*-Diamminedichloroplatinum(II)

Octabor. *See* Disodium octaborate tetrahydrate
Oil of vitriol. *See* Sulfuric acid
Oleic acid, 666
Orthoboric acid. *See* Boric acid
Orthophosphoric acid. *See* Phosphoric acid
Orthosil. *See* Sodium metasilicate
Orthovanadate, 24
Osmic acid. *See* Osmium tetroxide
Osmium *[7440-04-2]*, 281–284
Osmium tetroxide *[20816-12-0]*, 284–289, 290t
Oxygen *[7782-44-7]*, 655–675, 657t, 659t, 667t, 673t
Oxygen difluoride, 766t
Ozone *[10028-15-6]*, 676–688, 687t

Palladex 600. *See* Palladium
Palladium *[7440-05-3]*, 271–281, 272t, 274t–277t, 280t
Palladium(II) chloride *[7647-10-1]*, 271–281, 272t, 274t–277t, 280t
Paraquat, 666
PBS4. *See* Sodium perborate tetrahydrate
PBSI. *See* Sodium perborate monohydrate
PBSM. *See* Sodium perborate monohydrate
PBST. *See* Sodium perborate tetrahydrate
Pentaborane *[19624-22-7]*, 567, 568
Pentachlorophosphorane. *See* Phosphorus pentachloride
Pentasulfide, undistilled. *See* Phosphorus pentasulfide
Perchloron. *See* Calcium hypochlorite
Perchloryl fluoride, 767t
Periodates, 811, 812t
Periodic acid *[10450-60-9]*, 812t
Perosmic acid anhydride. *See* Osmium tetroxide

Peyrone's salt/chloride. *See cis*-Diamminedichloroplatinum(II)
Phosgene *[75-44-5]*, 629–634, 778–782
Phosphine *[7803-51-2]*, 460t, 470–473
Phosphorated hydrogen. *See* Phosphine
Phosphoric. *See* Phosphorus pentoxide
Phosphoric acid *[7664-38-2]*, 460t, 465–466, 466
Phosphoric anhydride. *See* Phosphorus pentoxide
Phosphoric perchloride. *See* Phosphorus pentachloride
Phosphoric sulfide. *See* Phosphorus pentasulfide
Phosphorus *[7723-14-0]*, 459–464, 460t
Phosphorus hydride. *See* Phosphine
Phosphorus oxychloride *[10025-87-3]*, 460t, 468
Phosphorus pentachloride *[10026-13-8]*, 460t, 467, 468
Phosphorus pentasulfide *[1314-80-3]*, 460t, 469
Phosphorus pentoxide *[1314-56-3]*, 460t, 464–465
Phosphorus sesquisulfide. *See* Tetraphosphorus trisulfide
Phosphorus trichloride *[7719-12-2]*, 460t, 469–470
Phosphorus trihydride. *See* Phosphine
Phosphoryl chloride. *See* Phosphorus oxychloride
PICl. *See* Phosphorus trichloride
Piochlor. *See* Sodium hypochlorite
Pittchlor. *See* Calcium hypochlorite
Platiblastin. *See cis*-Diamminedichloroplatinum(II)
Platinate(2-), tetrachloro-, dipotassium, (SP-4-1)-. *See* Potassium tetrachloroplatinate(II)
Platinex. *See cis*-Diamminedichloroplatinum(II)
Platinol. *See cis*-Diamminedichloroplatinum(II)
Platinous potassium chloride. *See* Potassium tetrachloroplatinate(II)
Platinum *[7440-06-4]*, 289–309, 292t, 294t, 296t, 297t, 308t
Platinum(II) chloride *[10025-65-7]*, 292t, 294t, 295, 296t
Platinum(IV) chloride *[13454-96-1]*, 292t, 294t, 295–298, 296t, 300, 308t, 309–311
Platinum diamminochloride cisplatyl. *See cis*-Diamminedichloroplatinum(II)
Platinum(II) oxide, 292t
Platinum(IV) oxide *[1314-15-4]*, 292t, 294t, 296t, 297t, 311–313
Platinum(IV) sulfate tetrahydrate *[69102-79-0]*, 292t, 294t, 295, 296t, 297t, 298, 300, 301, 308t, 313–315
Polybor. *See* Disodium octaborate tetrahydrate
Potash. *See* Potassium hydroxide
Potassium *[7440-09-7]*, 584–586, 584t
Potassium bromate, 804t
Potassium bromide, 804t

CHEMICAL INDEX

Potassium chlorate *[3811-04-9]*, 792–799, 795t, 796t, 797t
Potassium chloroplatinite. *See* Potassium tetrachloroplatinate(II)
Potassium dichromate, 76t
Potassium fluoborate, 522t
Potassium hexachloroplatinate(IV) *[16921-30-5]*, 292t, 308t, 338–341
Potassium hydroxide *[1310-58-3]*, 584t, 586–588
Potassium iodate *[7758-05-6]*, 811t, 812t, 814t
Potassium iodide *[7681-11-0]*, 812t, 814t
Potassium iodine, 811t
Potassium niobate [12030-85-2], 39t
Potassium pentaborate, 523t
Potassium perchlorate *[7778-74-7]*, 793–799, 795t, 796t, 797t
Potassium permanganate *[7722-64-7]*, 131t
Potassium perrhenate, 156
Potassium platinochloride. *See* Potassium tetrachloroplatinate(II)
Potassium tantalum fluoride *[16924-00-8]*, 55t
Potassium tetraborate tetrahydrate, 533t
Potassium tetrachloropalladate(II) *[10025-98-6]*, 271–281, 272t, 274t–277t, 280t
Potassium tetrachloroplatinate(II) *[10025-99-7]*, 292t, 294t, 296t, 300–301, 308t, 341–343
Praseodymium, 424t
Promethium, 424t, 430–454, 431t, 447t
cis-PT(II). *See cis*-Diamminedichloroplatinum(II)
PT-01. *See cis*-Diamminedichloroplatinum(II)

Radioactive ruthenium (Ruthenium 106 *[13967-48-1]*; Ruthenium 103 *[13968-53-1]*), 261–265, 264t
Radon, 402–403, 405–406
Rare earth metals, 423–454, 423t, 424t, 429t, 447t
Rhenium *[7440-15-5]*, 129, 130, 153–158
Rhenium trichloride, 156
Rhodium *[7440-16-6]*, 265–268, 265t, 267t, 269t–270t
Rhodium chloride, 265t
Royer@R. *See* Ruthenium
Rubidium *[7440-17-7]*, 601–602, 602t
Rubidium carbonate *[584-09-8]*, 602t
Rubidium chloride *[7791-11-9]*, 602t
Rubidium hydroxide *[1310-82-3]*, 602t
Rubidium iodine *[7790-29-6]*, 602t
Ruthenium *[7440-18-8]*, 254–261, 255t, 258t. *See also* Radioactive ruthenium
Ruthenium(III) chloride, 255t
Ruthenium chloride hydroxide, 255t
Ruthenium hydroxychloride, 261
Ruthenium(IV) oxide, 255t

Ruthenium(VIII) oxide, 255t
Ruthenium oxychloride, ammoniated (Ruthenium Red), 255t

Sal ammoniac. *See* Ammonium chloride
Salammonite. *See* Ammonium chloride
Sal soda. *See* Sodium carbonate
Samaria. *See* Samarium oxide
Samarium, 424t
Samarium oxide *[12060-58-1]*, 426–454, 429t, 431t, 447t
Samarium sesquioxide. *See* Samarium oxide
Samarium trioxide. *See* Samarium oxide
Scandium chloride *[10361-84-9]*, 423–425, 424t, 428–454, 429t, 447t
Seld-A-Leaf. *See* Sodium chlorate
Seleninyl chloride [oxide]. *See* Selenium oxychloride
Selenious acid. *See* Sodium selenite
Selenium *[7782-49-2]*, 473–480, 475t
Selenium dioxide *[7446-08-4]*, 475t
Selenium hexafluoride *[7783-79-1]*, 475t, 481–482, 767t
Selenium hydride. *See* Hydrogen selenide
Selenium oxychloride *[7791-23-3]*, 475t, 482–483
Selenium trioxide *[13768-86-0]*, 475t
Showchlon. *See* Sodium hypochlorite
Silicic acid. *See* Sodium metasilicate
Soda ash. *See* Sodium carbonate
Soda lye. *See* Sodium hydroxide
Soda monohydride. *See* Sodium carbonate
Sodium *[7440-23-5]*, 588–590, 589t
Sodium borate. *See* Sodium tetraborate
Sodium borohydride, 522t
Sodium bromate *[7789-38-0]*, 804t, 806–808
Sodium carbonate *[497-19-8]*, 589t, 591–593
Sodium chlorate *[7775-09-9]*, 791–799, 795t–797t
Sodium chloride *[7647-14-5]*, 783–785, 784t
Sodium dioxide. *See* Sodium peroxide
Sodium(IV) hexachloroplatinate, 299, 300–301
Sodium hydroxide *[1310-73-2]*, 589t, 593–595
Sodium hypochlorite *[7681-52-9]*, 792–799, 795t–797t
Sodium iodate *[7681-55-2]*, 812t
Sodium iodide *[7681-82-5]*, 811t, 812t, 814t
Sodium metaborate dihydrate *[16800-11-6]*, 533t, 555–556
Sodium metaborate tetrahydrate *[10555-76-7]*, 522t, 533t, 556–557
Sodium metasilicate *[6834-92-0]*, 589t, 598–600
Sodium metavanadate *[13718-26-8]*, 6t, 11, 12, 13t, 15t, 20, 21, 26
Sodium molybdate, 94t
Sodium molybdenate, 156

CHEMICAL INDEX

Sodium orthophosphate. *See* Trisodium phosphate
Sodium orthovanadate *[13721-39-6]*, 6t, 24–25
Sodium perborate monohydrate *[10332-33-9]*, 521t, 560–561
Sodium perborate tetrahydrate *[10486-00-7]*, 521t, 557–560
Sodium peroxide *[1313-60-6]*, 589t, 595, 596
Sodium perrhenate, 155–156
Sodium phosphate. *See* Trisodium phosphate
Sodium pyroborate decohydrate. *See* Sodium tetraborate decahydrate
Sodium salt. *See* Sodium perborate tetrahydrate
Sodium selenite *[10102-18-8]*, 475t, 483–484
Sodium tetraborate *[1330-43-4]*, 550, 551
sodium tetraborate, anhydrous, 533, 533t
Sodium tetraborate decahydrate *[1303-96-4]*, 521t, 533–548, 533t
Sodium tetraborate pentahydrate *[12179-04-3]*, 521t, 532–533, 533t, 548–550
Sodium tetraphenyl boron, 523t
Sodium tetravanadate *[12058-74-1]*, 6t
Solubor. *See* Disodium octaborate tetrahydrate; Sodium tetraborate decahydrate
Sonac. *See* Phosphoric acid
Stink Damp. *See* Hydrogen sulfide
Strontium chromate, 92t
Sufinyl chloride. *See* Thionyl chloride
Sulfocarbonic anhyride. *See* Carbon disulfide
Sulfur *[63705-05-5]*, 489, 490t, 491
Sulfur chloride oxide. *See* Thionyl chloride
Sulfur dioxide *[7446-09-5]*, 490t, 491–494
Sulfuretted hydrogen. *See* Hydrogen sulfide
Sulfur hexafluoride *[2551-62-4]*, 752–757, 754t, 756t
Sulfur hydride. *See* Hydrogen sulfide
Sulfuric acid *[7664-93-9]*, 490t, 508–511
Sulfuric acid nickel salt. *See* Nickel sulfate
Sulfur monochloride *[10025-67-9]*, 490t, 506–507
Sulfurous oxychloride. *See* Thionyl chloride
Sulfur phosphide. *See* Phosphorus pentasulfide
Sulfur tetrafluoride *[7783-60-0]*, 752–757, 754t, 756t
Sulfur trioxide *[7446-11-9]*, 490t

Tantalum *[7440-25-7]*, 1–4, 2t, 49–62, 50t, 55t, 61t
Tantalum carbide [12070-06-3], 50t
Tantalum chloride *[7721-01-9]*, 50t, 55t
Tantalum fluoride *[7783-71-3]*, 50t, 55t
Tantalum oxide *[1314-61-0]*, 50t, 55t
Tellurium *[13494-80-9]*, 484–487, 488t
Tellurium hexafluoride *[7783-80-4]*, 487, 488, 488t, 767t
Terbium, 424t
Terbium (+3) salt. *See* Terbium trinitrate

Terbium trinitrate *[10043-27-3]*, 427–454, 429t, 431t, 447t
Testosterone, 667t
Tetraborate. *See* Sodium tetraborate
Tetraborate, decohydrade. *See* Sodium tetraborate decahydrate
Tetraborate, pentahydrate. *See* Sodium tetraborate pentahydrate
Tetracarbonylnickel. *See* Nickel carbonyl
Tetrachloroplatinic(II) acid, 309
Tetrachloroplatinum. *See* Platinum(IV) chloride
Tetraphosphorus trisulfide *[1314-85-8]*, 460t, 466, 467
Tetrasodium salts, decahydrate. *See* Sodium tetraborate decahydrate
Thionyl chloride *[7719-09-7]*, 490t, 507, 508
Thorium *[7440-29-1]*, 409–417, 410t
Thorium dioxide, 410t
Thorium fluoride, 410t
Thorium hydroxide, 410t
Thorium nitrate tetrahydrate, 410t
Thorium oxalate, 410t
Thorium sulfate, 410t
Thorotrast, 416–417
Three elephant. *See* Boric acid
Thulium, 424t, 430–454, 431t, 447t
Thyroxine, 667t
Tim-Bor. *See* Disodium octaborate tetrahydrate
Triatomic oxygen. *See* Ozone
Tribromoborane. *See* Boron tribromide
Triethylborane, 523t
Trifluoroborane. *See* Boron trifluoride
Trihydroxyborane. *See* Boric acid
Trimethyl amine borane, 523t
Trinickel disulfide. *See* Nickel subsulfide
Trisodium phosphate *[7601-54-9]*, 589t, 597, 598
Trisodium salt. *See* Trisodium phosphate
Trisulfurated phosphorus. *See* Tetraphosphorus trisulfide
Triuranium octoxide, 382t
Tumbleaf. *See* Sodium chlorate
Tungsten *[7440-33-7]*, 106–121, 107t
Tungsten acid, 107t
Tungsten carbide, 107t
Tungsten disulfide, 107t
Tungsten hexachloride, 107t
Tungsten oxytetrachloride, 107t
Tungsten trioxide, 107t

Ulexite, 521t
Uranium *[7440-61-1]*, 287f, 381–408, 382t, 389t, 396t, 407t
Uranium dioxide, 382t

CHEMICAL INDEX

Uranium hexafluoride, 382t
Uranium oxyfluoride. *See* Uranyl fluoride
Uranium tetrachloride, 382t
Uranium tetrafluoride, 382t
Uranyl acetate, dihydrate, 382t
Uranyl chloride, 382t
Uranyl fluoride, 382t
Uranyl nitrate, 382t
Uranyl oxide, 382t

Vanadate, 20, 22
Vanadium *[7440-62-2]*, 1–38, 2t, 6t, 19t, 23t, 36t
Vanadium dichloride *[10580-52-6]*, 6t
Vanadium gluconate, 25
Vanadium oxydichloride *[10213-09-9]*, 6t, 23t
Vanadium oxytrichloride *[7727-18-6]*, 6t
Vanadium pentoxide *[1314-62-1]*, 6t, 10–12, 13t–15t, 20, 21, 23t, 24, 26, 28, 29–30, 32–33
Vanadium sulfate, 24
Vanadium tetrachloride *[7632-51-1]*, 5, 6t
Vanadium tribromide *[13470-26-3]*, 6t
Vanadium trichloride *[7718-98-1]*, 5, 6t, 26

Vanadium trioxide *[1314-34-7]*, 6t, 25, 26
Vanadium trioxybromide *[13520-90-6]*, 6t
Vanadyl sulfate *[27774-13-6]*, 6t, 14t, 20
Vanadyl sulfate pentahydrate *[12439-96-2]*, 6t
VCl$_3$. *See* Vanadium trichoride
VCl$_4$. *See* Vanadium tetrachloride
Vitriol brown oil. *See* Sulfuric acid

Weevitox. *See* Carbon disulfide
WHO. *See* World Health Organization
Wolfram. *See* Tungsten
WP. *See* Phosphorus

X-14. *See* Calcium hypochlorite

Ytterbium *[7440-64-4]*, 424t, 427–454, 429t, 431t, 447t
Yttrium *[7440-65-5]*, 423–424, 424t, 427–454, 429t, 431t, 447t

Zinc borate, 533, 533t
Zinc borate hydrate *[138265-88-0]*, 522t, 561–562
Zinc chromates, 92t

Ref.
RA
1229
.P38

46602

SOUTH UNIVERSITY
709 MALL BLVD.
SAVANNAH, GA 31406

Ref.
RA
1229
.P38

-46602

SOUTH UNIVERSITY
709 MALL BLVD.
SAVANNAH, GA 31406